Revised Edition

Algebra and Trigonometry

Functions and Applications

Revised Edition

Algebra and Trigonometry

Functions and Applications

Paul A. Foerster

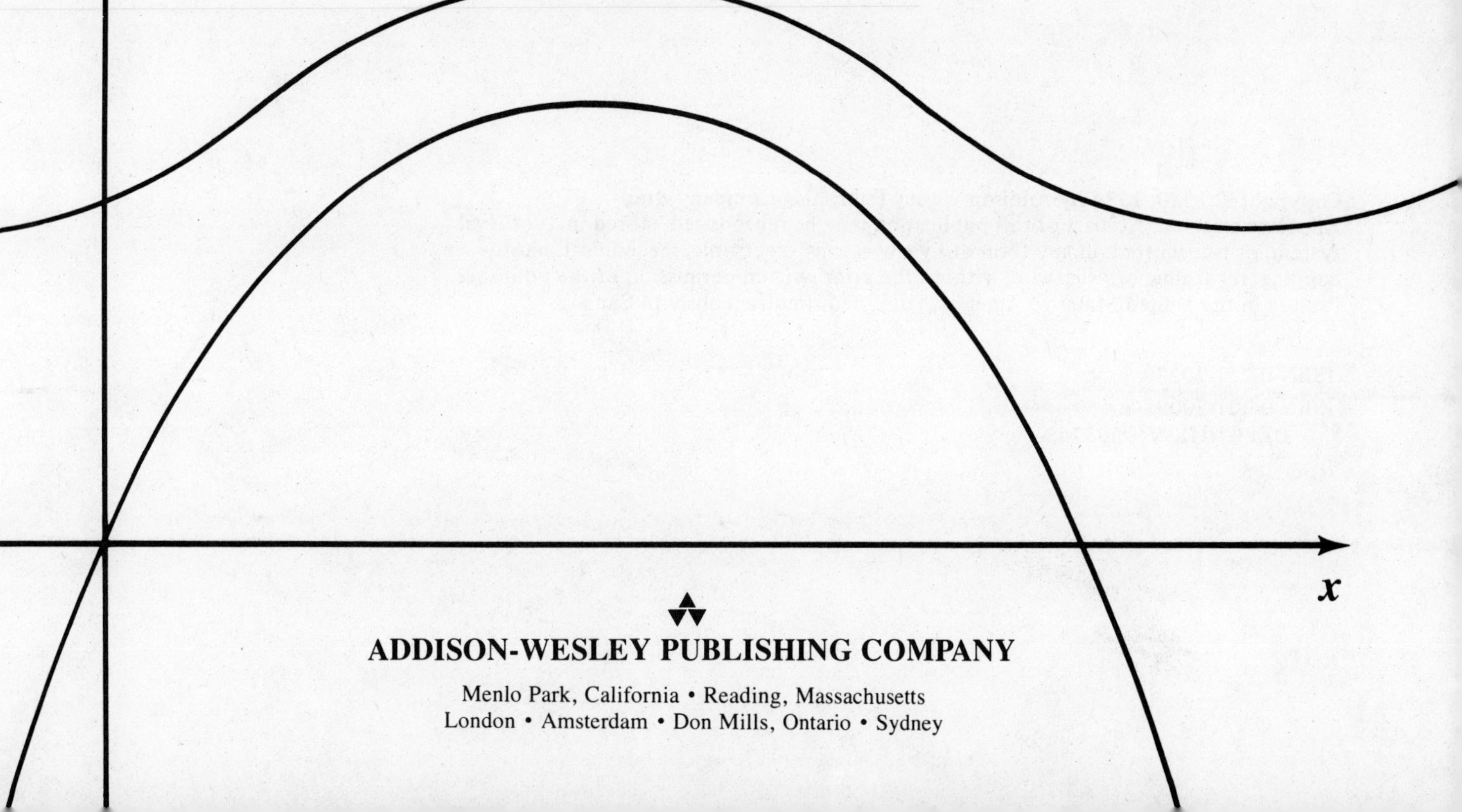

ADDISON-WESLEY PUBLISHING COMPANY

Menlo Park, California • Reading, Massachusetts
London • Amsterdam • Don Mills, Ontario • Sydney

This book is published by the ADDISON-WESLEY INNOVATIVE DIVISION.

Printed in the United States of America. Published simultaneously in Canada.

ISBN-0-201-20239-5
ISBN 0-201-20098-8
DEFGHIJK-VH-898765

To A. W. Foerster, who helped me to understand the real world,
to Admiral Rickover, who taught me how to write about it, and
to Jo Ann, who helps me to live in it.

Foreword

Algebra and Trigonometry: Functions and Applications is designed for a course in intermediate algebra, advanced algebra, and trigonometry. The book can be used in two different ways:

1. As an algebra and trigonometry book, with applications,
2. As an applications book, with supporting algebra and trig.

In either case, there are two possible sequences of presentation:

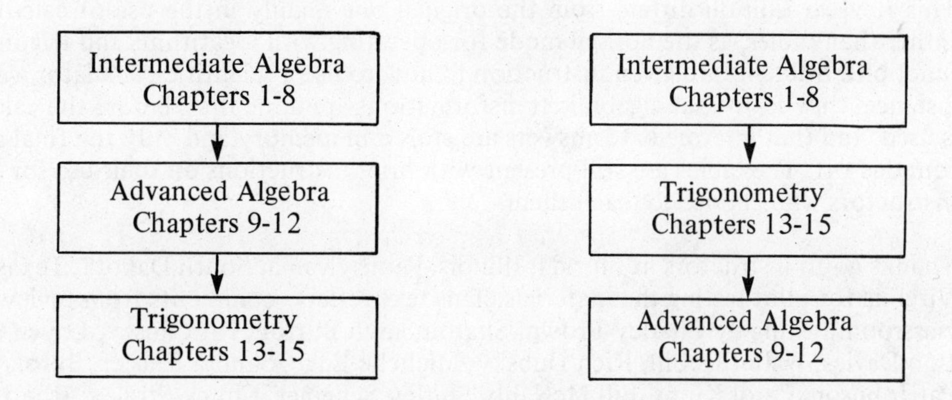

As a chemical engineer and mathematician, it has been my experience that a thorough understanding of theory is a prerequisite for the intelligent application of mathematics. As an instructor, I have evolved over the past 22 years an approach to algebra and trigonometry that both imparts the theory and handles applications in a realistic way. In the first chapter students learn the "basics," namely, the Field Axioms and the resulting properties necessary to operate algebraically. In the second chapter, students move right into the function concept. Thereafter, each new topic is fit into the framework of studying a particular class of function. Since students handle trigonometric functions the same way as algebraic ones, they can learn the essential unity of these two subjects.

Applications are handled by creating mathematical models of phenomena in the real world. Students must select a kind of function that fits a given situation, and derive an equation that suits the information in the problem. The equation is then used to predict values of y when x is given, or values of x when y is given. Sometimes students must use the results of their work to make interpretations about the real world, such as what "slope" means, or why there cannot be people as small as in Gulliver's Travels. The problems require the students to use *many* mathematical concepts in the *same* problem. This is in contrast to the traditional "word problems" of elementary algebra, in which the same one concept is used in many problems.

Henry Pollak of Bell Telephone Labs claims that there are just two kinds of numbers: "real" numbers such as encountered in everyday life, and "fake" numbers such as encountered in most mathematics classes! Since this book has many problems involving untidy decimals ("real" numbers), a calculator or computer are called for where appropriate. There are also problems that have small-integer answers ("fake" numbers) so that students may gain confidence in their work when they are just learning a new technique.

Since all educators share the responsibility of teaching students to read and write, there are discovery exercises so that students may wrestle with a new concept before it is reinforced by classroom discussion. The students are helped with this reading by the fact that much of the wording came from the mouths of my own students. Special thanks go to Susan Cook, Brad Foster, and Nancy Carnes, whose good class notes supplied input for certain sections. Students Lewis Donzis and David Fey wrote computer programs for some of the problems.

This Revised Edition differs from the original one mainly in the use of calculators, rather than tables, as the normal mode for operating with logarithms and trigonometric functions. Students are given instruction in how to use a scientific calculator *well.* For instance, they learn that algebraic transformations are done first, before the calculator is used, and that intermediate answers are stored in memory and only the final answer is rounded off. The tables are still present with brief instructions on their use for those instructors who choose to teach them.

Thanks go to instructors in Florida, Illinois, Pennsylvania, South Dakota, Texas, and Virginia for pilot testing the materials. This text reflects comments from review and classroom testing by Charley Brown, Sharon Sasch Button, Pat Causey, Loyce Collenback, Bob Davies, Walter DeBill, Rich Dubsky, Michelle Edge, Sandra Frasier, Byron Gill, Pat Johnson, Carol Kipps, Bill McNabb, Shirley Scheiner, Chuck Straley, Rhetta Tatsch, Susan Thomas, Kay Thompson, Zalman Usiskin, Jim Wieboldt, Marv Wielard, Mercille Wisakowsky, and Martha Zelinka. Calvin Butterball and Phoebe Small appear with the kind permission of their parents, Richard and Josephine Andree.

Paul A. Foerster

Contents

CHAPTER 1 PRELIMINARY INFORMATION

1-1 Sets of Numbers 1
1-2 The Field Axioms 3
1-3 Variables and Expressions 5
1-4 Polynomials 9
1-5 Equations 11
1-6 Inequalities 15
1-7 Properties Provable from the Axioms 17
1-8 Chapter Review and Test 24

CHAPTER 2 FUNCTIONS AND RELATIONS 28

2-1 Graphs of Equations with Two Variables 29
2-2 Mathematical Relations 30
2-3 Mathematical Functions 34
2-4 Chapter Review and Test 39

CHAPTER 3 LINEAR FUNCTIONS 42

3-1 Introduction to Linear Functions 43
3-2 Graphs of Linear Functions from their Equations 44
3-3 Other Forms of the Linear Function Equation 47
3-4 Equations of Linear Functions from their Graphs 49
3-5 Linear Functions as Mathematical Models 53
3-6 Chapter Review and Test 60

CHAPTER 4 SYSTEMS OF LINEAR EQUATIONS AND INEQUALITIES 64

4-1 Systems of Two Linear Equations with Two Variables 65
4-2 Solution of Systems by Linear Combination 66
4-3 Second-Order Determinants 69
4-4 f(x) Terminology, and Systems as Models 72
4-5 Linear Equations with Three or More Variables 76
4-6 Systems of Linear Equations with Three or More Variables 79
4-7 Higher Order Determinants 81
4-8 Matrix Solution of Linear Systems 83
4-9 Systems of Linear Inequalities 86
4-10 Linear Programming 88
4-11 Chapter Review and Test 95

CHAPTER 5 QUADRATIC FUNCTIONS 98
5-1 Introduction to Quadratic Functions 99
5-2 Quadratic Function Graphs 100
5-3 Completing the Square to Find the Vertex 102
5-4 Quadratic Equations with One Variable 105
5-5 The Quadratic Formula 108
5-6 Evaluating Quadratic Functions 111
5-7 Equations of Quadratic Functions from their Graphs 114
5-8 Quadratic Functions as Mathematical Models 117
5-9 Chapter Review and Test 125
5-10 Cumulative Review: Chapters 1 through 5 126

CHAPTER 6 EXPONENTIAL AND LOGARITHMIC FUNCTIONS 130
6-1 Introduction to Exponential Functions 131
6-2 The Operation of Exponentiation 132
6-3 Properties of Exponentiation 133
6-4 Negative and Zero Exponents 135
6-5 Radicals and Fractional Exponents 137
6-6 Powers by Calculator 141
6-7 Scientific Notation 144
6-8 Inverses of Functions 147
6-9 Logarithmic Functions 149
6-10 Properties of Logarithms 151
6-11 Special Properties of Base 10 Logarithms 154
6-12 Exponential Equations, and Logarithms with Other Bases 156
6-13 Evaluation of Exponential Functions 159
6-14 Exponential Functions as Mathematical Models 162
6-15 Chapter Review and Test 173

CHAPTER 7 RATIONAL ALGEBRAIC FUNCTIONS 176
7-1 Introduction to Rational Algebraic Functions 177
7-2 Graphs of Rational Functions 177
7-3 Special Products 178
7-4 Factoring Special Kinds of Polynomials 181
7-5 Division of Polynomials 187
7-6 Factoring Higher Degree Polynomials 188
7-7 Products and Quotients of Rational Expressions 193
7-8 Sums and Differences of Rational Expressions 197
7-9 Graphs of Rational Algebraic Functions, Again 200
7-10 Fractional Equations 204
7-11 Variation Functions 207
7-12 Chapter Review and Test 221

CHAPTER 8 IRRATIONAL ALGEBRAIC FUNCTIONS 224
8-1 Existence of Irrational Numbers 225
8-2 Simple Radical Form 226
8-3 Radical Equations 229
8-4 Irrational Variation Functions 233
8-5 Functions of More than One Independent Variable 238
8-6 Chapter Review and Test 243
8-7 Cumulative Review, Chapters 6, 7, and 8 245

CHAPTER 9 QUADRATIC RELATIONS AND SYSTEMS 250
9-1 Introduction to Quadratic Relations 251
9-2 Circles 252
9-3 Ellipses 255
9-4 Hyperbolas 259
9-5 Parabolas 263
9-6 Equations from Geometrical Definitions 264
9-7 Quadratic Relations – xy-Term 268
9-8 Systems of Quadratics 269
9-9 Chapter Review and Test 275

CHAPTER 10 HIGHER DEGREE FUNCTIONS AND COMPLEX NUMBERS 278
10-1 Introduction to Higher Degree Functions 279
10-2 Imaginary Numbers 280
10-3 Complex Numbers 283
10-4 Complex Solutions of Quadratic Equations 286
10-5 Graphs of Higher Degree Functions – Synthetic Substitution 289
10-6 Higher Degree Functions as Mathematical Models 294
10-7 Chapter Review and Test 297

CHAPTER 11 SEQUENCES AND SERIES 300
11-1 Introduction to Sequences 301
11-2 Arithmetic and Geometric Sequences 303
11-3 Arithmetic and Geometric Means 308
11-4 Introduction to Series 310
11-5 Arithmetic and Geometric Series 312
11-6 Convergent Geometric Series 315
11-7 Sequences and Series as Mathematical Models 319
11-8 Factorials 328
11-9 Introduction to Binomial Series 330
11-10 The Binomial Formula 331
11-11 Chapter Review and Test 335

CHAPTER 12 PROBABILITY, AND FUNCTIONS OF A RANDOM VARIABLE 338

12-1 Introduction to Probability 339
12-2 Words Associated with Probability 340
12-3 Two Counting Principles 341
12-4 Probabilities of Various Permutations 342
12-5 Probabilities of Various Combinations 350
12-6 Properties of Probability 355
12-7 Functions of a Random Variable 361
12-8 Mathematical Expectation 367
12-9 Chapter Review and Test 372
12-10 Cumulative Review, Chapters 9 through 12 374

CHAPTER 13 TRIGONOMETRIC AND CIRCULAR FUNCTIONS 378

13-1 Introduction to Periodic Functions 379
13-2 Measurement of Arcs and Rotation 380
13-3 Definitions of Trigonometric and Circular Functions 383
13-4 Approximate Values of Trigonometric and Circular Functions 390
13-5 Graphs of Trigonometric and Circular Functions 393
13-6 General Sinusoidal Graphs 398
13-7 Equations of Sinusoids from their Graphs 402
13-8 Sinusoidal Functions as Mathematical Models 405
13-9 Inverse Circular Functions 412
13-10 Evaluation of Inverse Relations 419
13-11 Inverse Circular Relations as Mathematical Models 422
13-12 Chapter Review and Test 426

CHAPTER 14 PROPERTIES OF TRIGONOMETRIC AND CIRCULAR FUNCTIONS 430

14-1 Three Properties of Trigonometric Functions 431
14-2 Trigonomeric Identies 434
14-3 Properties Involving Functions of More than One Argument 438
14-4 Multiple Argument Properties 444
14-5 Half-Argument Properties 447
14-6 Sum and Product Properties 450
14-7 Linear Combination of Cosine and Sine with Equal Arguments 452
14-8 Simplification of Trigonometric and Expressions 453
14-9 Trigonomeric Equations 456
14-10 Chapter Review and Test 460

CHAPTER 15 TRIANGLE PROBLEMS 464
15-1 Right Triangle Problems 465
15-2 Oblique Triangles – Law of Cosines 470
15-3 Area of a Triangle 474
15-4 Oblique Triangles – Law of Sines 476
15-5 The Ambiguous Case 479
15-6 General Solution of Triangles 482
15-7 Vectors 485
15-8 Vectors – Resolution into Components 489
15-9 Real-World Triangle Problems 494
15-10 Chapter Review and Test 499
15-11 Cumulative Review, Chapters 13, 14, and 15 501

FINAL EXAMINATION 503

APPENDIX A REVIEW OF COMPUTER PROGRAMMING 509
A-1 Reading Flow Charts 509
A-2 Writing Flow Charts 516
A-3 The BASIC Computer Language 519

APPENDIX B ADDITIONAL TOPICS 527
B-1 Operations with Matrices 527
B-2 Descartes' Rule of Signs and the Upper Bound Theorem 531
B-3 Mathematical Induction 533

TABLES 542
I Squares and Square Roots, Cubes and Cube Roots 542
II Four-Place Logarithms of Numbers 544
III Trigonometric Functions, and Degrees-to-Radians 546
IV Circular Functions, and Radians-to-Degrees 551

GLOSSARY 554

INDEX OF PROBLEM TITLES 556

GENERAL INDEX 558

ANSWERS TO SELECTED PROBLEMS 563

PHOTOGRAPH ACKNOWLEDGEMENTS 608

1 Preliminary Information

Contrary to former views of mathematics, numbers were ***invented*** *by people, rather than simply being discovered. In this book you will see how things invented mainly to form a complete mathematical system can be used to describe things that happen in the real world. First, however, you must be sure that you and your instructor are speaking the same language! The first chapter is designed with this purpose in mind.*

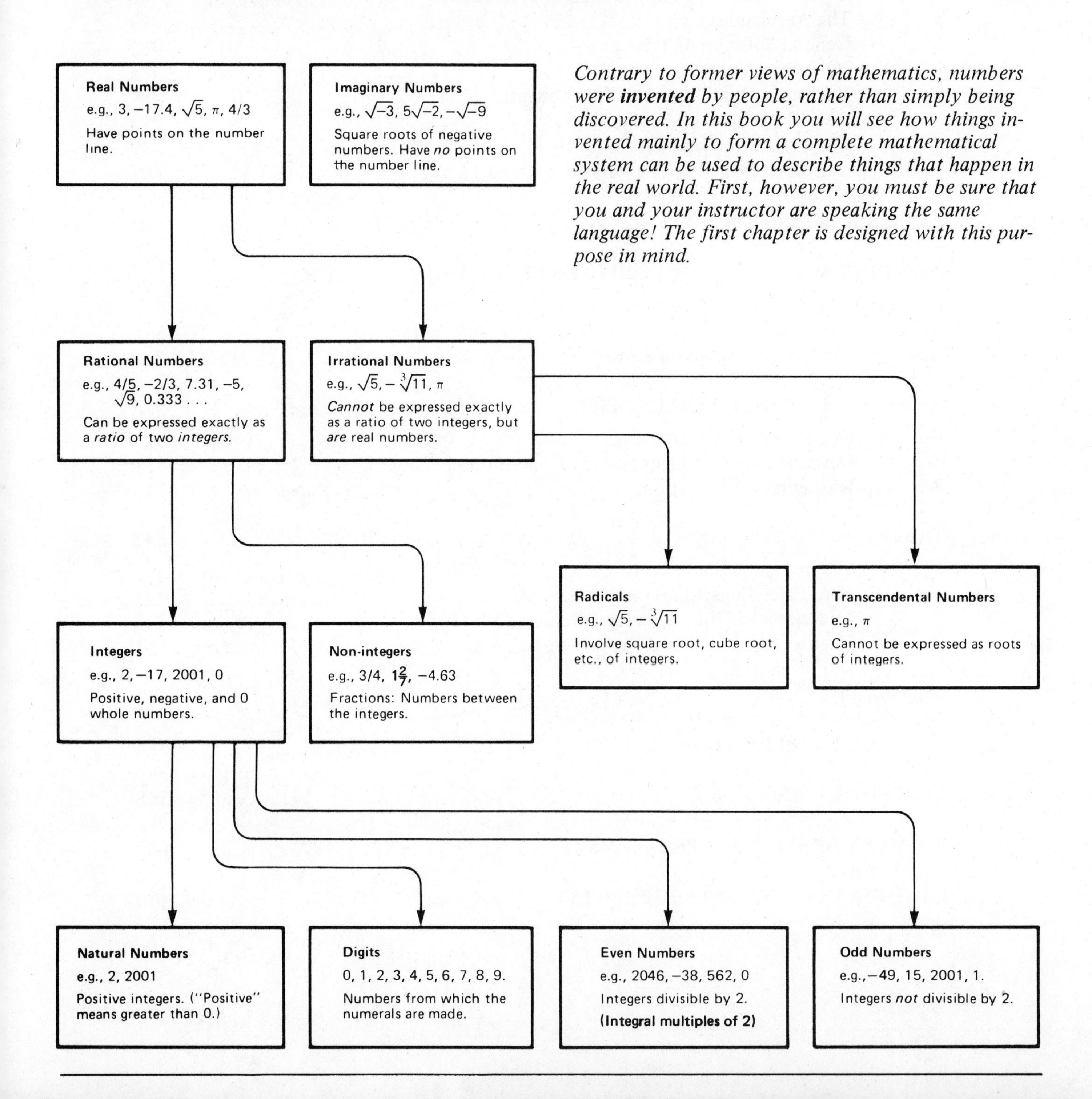

Real Numbers
e.g., 3, −17.4, √5, π, 4/3
Have points on the number line.
Imaginary Numbers
e.g., √−3, 5√−2, −√−9
Square roots of negative numbers. Have no points on the number line.
Rational Numbers
e.g., 4/5, −2/3, 7.31, −5, √9, 0.333 . . .
Can be expressed exactly as a ratio of two integers.
Irrational Numbers
e.g., √5, −∛11, π
Cannot be expressed exactly as a ratio of two integers, but are real numbers.
Radicals
e.g., √5, −∛11
Involve square root, cube root, etc., of integers.
Transcendental Numbers
e.g., π
Cannot be expressed as roots of integers.
Integers
e.g., 2, −17, 2001, 0
Positive, negative, and 0 whole numbers.
Non-integers
e.g., 3/4, 1 2/7, −4.63
Fractions: Numbers between the integers.
Natural Numbers
e.g., 2, 2001
Positive integers. ("Positive" means greater than 0.)
Digits
0, 1, 2, 3, 4, 5, 6, 7, 8, 9.
Numbers from which the numerals are made.
Even Numbers
e.g., 2046, −38, 562, 0
Integers divisible by 2.
(Integral multiples of 2)
Odd Numbers
e.g.,−49, 15, 2001, 1.
Integers not divisible by 2.

1-1 SETS OF NUMBERS

From previous work in mathematics you should recall the names of different kinds of numbers (positive, even, irrational, etc.). In this section you will refresh your memory so that you will know the exact meaning of these names.

Objective:

Given the name of a set of numbers, provide an example; or given a number, name the sets to which it belongs.

There are two major sets of numbers you will deal with in this course, the *real* numbers and the *imaginary* numbers. The real numbers are given this name because they are used for "real" things such as measuring and counting. The imaginary numbers are square roots of negative numbers. They are useful, too, but you must learn more mathematics to see why.

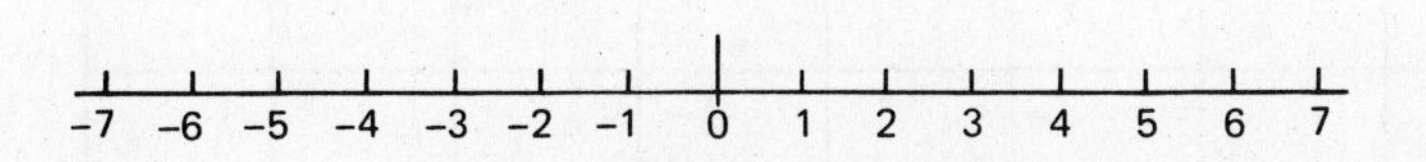

The real number line

Figure 1-1.

The real numbers are all numbers which you can plot on a number line (see Figure 1-1). They can be broken into subsets in several ways. For instance, there are positive and negative real numbers, integers and non-integers, rational and irrational real numbers, and so forth. The diagram facing this page shows some subsets of the set of real numbers.

The numbers in the diagram were invented in *reverse* order. The natural (or "counting") numbers came first because mathematics was first used for counting. The negative numbers (those less than zero) were invented so that there would always be answers to subtraction problems. The rational numbers were invented to provide answers to division problems, and the irrational ones came when it was shown that numbers such as $\sqrt{2}$ could not be expressed as a ratio of two integers.

Other operations you will invent, such as taking logarithms and cosines, lead to irrational numbers which go beyond even extracting roots. These are called "transcendental" numbers, meaning "going beyond." When all of these various kinds of numbers are put together, you get the set of *real* numbers. The imaginary numbers were invented because no real number squared equals a negative number. Later, you will see that the real and imaginary numbers are themselves simply subsets of a larger set, called the "complex numbers."

The following exercise is designed to help you accomplish the objectives of this section.

Exercise 1-1

1. Write a definition for each of the following sets of numbers. Try to do this *without* referring to the diagram opposite page 1. Then look to make sure you are correct.

a. {integers}
b. {digits}
c. {even numbers}
d. {positive numbers}
e. {negative numbers}
f. {rational numbers}
g. {irrational numbers}
h. {imaginary numbers}
i. {real numbers}
j. {natural numbers}
k. {counting numbers}
l. {transcendental nos.}

2. Write an example of each type of number mentioned in Problem 1.

3. Copy the following chart. Put a check mark in each box for which the number on the left of the chart belongs to the set across the top.

		Integers	Digits	Even Numbers	Positive Numbers	Negative Numbers	Rational Numbers	Irrational Numbers	Imaginary Numbers	Real Numbers	Natural Numbers	Counting Numbers	Transcendental Numbers
a.	5												
b.	2/3												
c.	−7												
d.	$\sqrt{3}$												
e.	$\sqrt{16}$												
f.	$\sqrt{-16}$												
g.	$\sqrt{-15}$												
h.	44												
i.	π												
j.	1.765												
k.	−10000												
l.	$-1\frac{1}{2}$												
m.	$-\sqrt{6}$												
n.	0												
o.	1												
p.	1/9												

4. Write another name for {natural numbers}.

5. Which of the sets of numbers in Problem 1 do you suppose was the *first* to be invented? Why?

6. One of the sets of numbers in Problem 1 contains all but one of the others as subsets.

a. Which one *contains* the others?
b. Which one is left out?

7. Do decimals such as 2.718 represent *rational* numbers or *irrational* numbers? Explain.

8. Do repeating decimals such as 2.3333 . . . represent *rational* numbers or *irrational* numbers? Explain.

9. What real number is neither positive nor negative?

1-2 THE FIELD AXIOMS

From previous mathematics courses you probably remember names such as, "Distributive Property," "Reflexive Property," and "Multiplication Property of Zero." Some of these properties, called *axioms*, are accepted without proof and are used as a starting point for working with numbers. From a small number of rather obvious axioms, you will derive all the other properties you will need. In this section you will concentrate on the axioms that apply to the *operations* with numbers such as + and ×. In Section 1-7 you will find the axioms that apply to the *relationships* between numbers, such as = and <.

Objective:

Given the name of an axiom that applies to + or ×, give an example that shows you understand the meaning of the axiom; and vice versa.

There are eleven axioms that apply to adding and multiplying real numbers. These are called the *Field Axioms*, and are listed in the following table. If you already feel familiar with these axioms, you may go right to the problems in Exercise 1-2. If not, then read on!

The Field Axioms

Closure: {real numbers} is *closed* under addition and under multiplication. That is, if x and y are real numbers, then

$x + y$ is a *unique, real* number,
xy is a *unique, real* number.

Commutativity: Addition and multiplication of real numbers are *commutative* operations. That is, if x and y are real numbers, then

$x + y$ and $y + x$ are *equal* to each other,
xy and yx are *equal* to each other.

Associativity: Addition and multiplication of real numbers are *associative* operations. That is, if x, y, and z are real numbers, then

$(x + y) + z$ and $x + (y + z)$ are *equal* to each other,
$(xy)z$ and $x(yz)$ are *equal* to each other.

Distributivity: Multiplication *distributes* over addition. That is, if x, y, and z are real numbers, then

$x(y + z)$ and $xy + xz$ are *equal* to each other.

Identity Elements: {real numbers} contains:

A *unique* identity element for *addition*, namely 0. (Because $x + 0 = x$ for any real number x.)

A *unique* identity element for *multiplication*, namely 1. (Because $x \cdot 1 = x$ for any real number x.)

Inverses: {*real numbers*} contains:

A *unique additive* inverse for every real number x. (Meaning that every real number x has a real number $-x$ such that $x + (-x) = 0$.)

A *unique multiplicative* inverse for every real number x except zero.
(Meaning that every non-zero number x has a real number $\frac{1}{x}$ such that $x \cdot \frac{1}{x} = 1$.)

Notes:

1. Any set that obeys all eleven of these axioms is a *field*.
2. The eleven Field Axioms come in 5 pairs, one of each pair being for addition and the other for multiplication.

The Distributive Axiom expresses a relationship between these two operations.
3. The properties $x + 0 = x$ and $x \cdot 1 = x$ are sometimes called the "Addition Property of 0" and the "Multiplication Property of 1," respectively, for obvious reasons.
4. The number $-x$ is called, "the *opposite* of x," "the *additive inverse* of x," or "negative x."
5. The number $1/x$ is called the "multiplicative inverse of x," or the "reciprocal of x."

Closure – By saying that a set is "closed" under an operation, you mean that you cannot get an answer that is *out* of the set by performing that operation on numbers *in* the set. For example, $\{0, 1\}$ is closed under multiplication because $0 \times 0 = 0$, $0 \times 1 = 0$, $1 \times 0 = 0$, and $1 \times 1 = 1$. All the answers are *unique*, and are *in* the given set. This set is *not* closed under addition because $1 + 1 = 2$, and 2 is *not* in the set. It is not closed under the operation "taking the square root" since there are *two* different square roots of 1: $+1$ and -1.

Commutativity – The word "commute" comes from the Latin word "commutare," which means "to exchange." People who travel back and forth between home and work are called "commuters" because they regularly exchange positions. The fact that addition and multiplication are commutative operations is somewhat unusual. Many operations such as subtraction and exponentiation (raising to powers) are *not* commutative. For example,

$2 - 5$ does *not* equal $5 - 2$,

and

2^3 does *not* equal 3^2.

Indeed, most operations in the real world are not commutative. Putting on your shoes and socks (in that order) produces a far different result from putting on your socks and shoes!

Associativity – You can remember what this axiom states by remembering that to "associate" means to "group." Addition and multiplication are associative, as shown by

$$(2 + 3) + 4 = 9 \quad \text{and} \quad 2 + (3 + 4) = 9.$$

But subtraction is *not* associative. For example,

$$(2 - 3) - 4 = -5 \quad \text{and} \quad 2 - (3 - 4) = 3.$$

Distributivity – Parentheses in an expression such as $2 \times (3 + 4)$ mean, "Do what is inside *first*." But you don't *have* to do $3 + 4$ first. You could "distribute" a 2 to each term inside the parentheses, getting $2 \times 3 + 2 \times 4$. The Distributive Axiom expresses the fact that you get the same answer either way. That is,

$$2\times(3 + 4) = 14 \quad \text{and} \quad 2\times 3 + 2\times 4 = 14.$$

Note that multiplication does *not* distribute over multiplication. For example,

$2\times(3 \times 4)$ does *not* equal $2\times 3 \times 2\times 4$,

as you can easily check by doing the arithmetic.

Identity Elements – The numbers 0 and 1 are called "identity elements" for adding and multiplying, respectively, since a number comes out "identical" if you add 0 or multiply by 1. For example,

$$5 + 0 = 5 \quad \text{and} \quad 5 \times 1 = 5.$$

Inverses – A number is said to be an *inverse* of another number for a certain operation if it "undoes" (or inverts) what the other number did. For example, $1/3$ is the multiplicative inverse of 3. If you start with 5 and multiply by 3 you get

$$5 \times 3 = 15.$$

Multiplying the answer, 15, by $1/3$ gives

$$15 \times \frac{1}{3} = 5,$$

which "undoes" or "inverts" the multiplication by 3. It is easy to tell if two numbers are *multiplicative inverses* of each other because their product is always equal to 1, the multiplicative identity element. For example,

$$3 \times \frac{1}{3} = 1.$$

Similarly, two numbers are *additive inverses* of each other if adding them to each other gives 0, the additive identity element. For example, 5/7 and −5/7 are additive inverses of each other because

$$5/7 + (-5/7) = 0.$$

The following exercise is designed to familiarize you with the names and meanings of the Field Axioms.

Exercise 1-2

1. Tell what is meant by

a. additive identity element,
b. multiplicative identity element.

2. What is

a. the *additive* inverse of 2/3?
b. the *multiplicative* inverse of 2/3?

3. Using variables (x, y, z, etc.) to stand for numbers, write an example of each of the eleven field axioms. Try to do this by writing all eleven examples first, then checking to be sure you are right. Correct any which you left out or got wrong.

4. Explain why 0 has *no* multiplicative inverse.

5. The Closure Axiom states that you get a *unique* answer when you add two real numbers. What is meant by a "unique" answer?

6. You get the same answer when you add a column of numbers "up" as you do when you add it "down." What axiom(s) show that this is true?

7. Calvin Butterball and Phoebe Small use the distributive property as follows:

Calvin: $3(x + 4)(x + 7) = (3x + 12)(x + 7)$.
Phoebe: $3(x + 4)(x + 7) = (3x + 12)(3x + 21)$.

Who is right? What mistake did the other one make?

8. Write an example which shows that:

a. Subtraction is *not* a commutative operation.
b. {negative numbers} is *not* closed under multiplication.
c. {digits} is *not* closed under addition.
d. {real numbers} is *not* closed under the $\sqrt{\ }$ operation (taking the square root).
e. Exponentiation ("raising to powers") is *not* an associative operation. (Try 4^{2^3}.)

9. For each of the following, tell which of the Field Axioms was used, and whether it was an axiom for *addition* or for *multiplication*. Assume that x, y, and z stand for real numbers.

a. $x + (y + z) = (x + y) + z$
b. $x \cdot (y + z)$ is a real number
c. $x \cdot (y + z) = x \cdot (z + y)$
d. $x \cdot (y + z) = (y + z) \cdot x$
e. $x \cdot (y + z) = xy + xz$
f. $x \cdot (y + z) = x \cdot (y + z) + 0$
g. $x \cdot (y + z) + (-[x \cdot (y + z)]) = 0$
h. $x \cdot (y + z) = x \cdot (y + z) \cdot 1$
i. $x \cdot (y + z) \cdot \dfrac{1}{x \cdot (y + z)} = 1$

10. Tell whether or not the following sets are *fields* under the operations + and ×. If the set is not a field, tell which one(s) of the Field Axioms do not apply.

a. {rational numbers} b. {integers}
c. {positive numbers} d. {non-negative numbers}

1-3 VARIABLES AND EXPRESSIONS

In previous mathematics courses you have seen *expressions*, such as

$$3x^2 + 5x - 7,$$

that stand for numbers. Just *what* number an expression stands for depends on what value you pick for the *variable* (x in this case). The name "variable" is picked because x can stand for various different numbers at different times. The numbers 3, 5, −7, 9/10, $\sqrt{11}$ etc., are called *constants* because they stand for the *same* number *all* the time.

In this section you will *evaluate* expressions by substituting values for the variable. In order to do this more easily, you can *simplify* the expression using the axioms of the previous section.

Objective:

Given an expression containing a variable,

a. *evaluate* it by substituting a given number for the variable, and finding the value of the expression,
b. *simplify* it by using the Field Axioms to transform it to an equivalent expression that is easier to evaluate.

DEFINITION

A ***variable*** is a letter which stands for an *unspecified* number from a *given* set.

For example, if the set you have in mind is {digits}, and x is the variable, then x could stand for any one of the numbers 0, 1, 2, 3, 4, 5, 6, 7, 8, or 9. In this case, {digits} is called the *domain* of x. The word comes from the Latin "domus," meaning "house." So the domain of a variable is "where it lives." Since the domain of most variables in this course will be {real numbers}, you make the following agreement:

Agreement: Unless otherwise specified, the domain of a variable will be assumed to be the set of all real numbers.

DEFINITION

An ***expression*** is a collection of variables and constants connected by operation signs (+, −, ×, ÷, etc.) which stands for a *number*.

To find out *what* number an expression stands for, you must substitute a value for each variable, then do the indicated operations.

Example 1: Evaluate $3x^2 + 5x - 7$ if $x = 4$.

$3x^2 + 5x - 7 = 3 \cdot 4^2 + 5 \cdot 4 - 7$
Substitute 4 for x.

$= 3 \cdot 16 + 5 \cdot 4 - 7$
Square the 4.

$= 48 + 20 - 7$
Do the multiplication.

$= \underline{\underline{61}}$
Do the adding and subtracting from *left* to *right*.

There are several things you should realize about the preceding calculations. First, you must substitute the *same* value of x *everywhere* it appears in the expression. Although a variable can take on different values at different times, it stands for the *same* number at any *one* time. This fact is expressed in the Reflexive Axiom, which states, "$x = x$."

The second thing you should realize is that this expression involves *subtraction* and *exponentiation* (raising to powers). These operations, as well as *division*, can be defined in terms of addition and multiplication.

DEFINITIONS

Subtraction: $x - y$ means $x + (-y)$.

Division: $x \div y$ means $x \cdot \frac{1}{y}$. (The symbols $\frac{x}{y}$ and x/y are also used for $x \div y$.)

Exponentiation: x^n means n, x's *multiplied* together. For example,

x^3 means $x \cdot x \cdot x$.

The third thing you should realize is that the answer you get depends on the *order* in which you do the operations. So that there will be no doubt about what an expression such as $3x^2 + 5x - 7$ means, you make the following agreement:

Agreement: *Sequence of Operations*

1. Do any operations inside parentheses *first*.
2. Do any exponentiating next.
3. Do multiplication and division in the order in which they occur, from left to right.
4. Do addition and subtraction last, in the order in which they occur, from left to right.

Example 2: Carry out the following operations:

a. $3 + 4 \times 5 = 3 + 20$ — Multiply *first*.
$= \underline{\underline{23}}$ — Add *last*.

b. $3 + 4 \times 5 \div 2$
$= 3 + 20 \div 2$ — Multiply and divide from left to right.
$= 3 + 10$ — Divide *before* adding.
$= \underline{\underline{13}}$ — Add *last*.

c. $3 - 4 \times 5 \div 2 + 9$
$= 3 - 20 \div 2 + 9$ — Multiply and divide from left to right.
$= 3 - 10 + 9$ — Divide *before* + and −.
$= -7 + 9$ — Add and subtract from left to right.
$= \underline{\underline{2}}$ — Add and subtract *last*.

An expression might contain the *absolute value* operation. The symbol $|x|$ means the *distance* between the number x and the origin of the number line. For example, $|-3|$ and $|3|$ are both equal to 3, since both 3 and −3 are located 3 units from the origin (Figure 1-3).

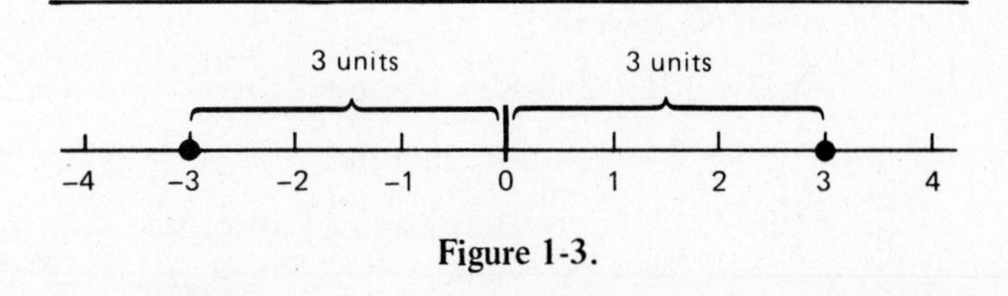

Figure 1-3.

Similarly,

$$|5| = 5$$
$$|-7| = 7$$
$$|0| = 0,$$

and so forth.

The absolute value of a *variable* presents a problem. If x is a *positive* number, then $|x|$ is equal to x. But if x is a *negative* number, then $|x|$ is equal to the *opposite* of x. For instance, if $x = -9$, then

$$|x| = |-9| = -(-9) = 9.$$

A precise definition of absolute value can be written as follows:

DEFINITION

$|x| = x$ if x is *positive* (or 0),
$|x| = -x$ if x is *negative*.

Example 3: Evaluate $|17 - 4x| - 2$ if

a. $x = 5$
b. $x = -3$.

a. $|17 - 4x| - 2$
$= |17 - 20| - 2$ — Substitute 5 for x.
$= |-3| - 2$ — Arithmetic.
$= 3 - 2$ — Definition of absolute value.
$= \underline{\underline{1}}$ — Arithmetic.

b. $|17 - 4x| - 2$

$= |17 + 12| - 2$	Substitute -3 for x.
$= |29| - 2$	Arithmetic.
$= 29 - 2$	Definition of absolute value.
$= \underline{\underline{27}}$	Arithmetic.

Two expressions are equivalent if they *equal* each other for *all* values of the variable. For example, $3x + 8x$ and $11x$ are equivalent expressions. *Simplifying* an expression means transforming it to an equivalent expression that is in some way simpler to work with. The expression $11x$ is considered to be simpler than $3x + 8x$ because it is easier to evaluate when you pick a value of x. Adding the $3x$ and $8x$ is called "collecting like terms." It is justified by using the Distributive Axiom *backwards*.

$3x + 8x = (3 + 8)x$	Distributivity.
$= 11x$	Arithmetic.

Example 4: Simplify $7x \cdot 2 \div x$.

Since the Field Axioms apply to multiplication rather than division, you would treat "$\div x$" as "$\cdot 1/x$", commute the multiplication, and get

$7x \cdot 2 \div x$

$= 7x \cdot 2 \cdot \dfrac{1}{x}$	Definition of division.
$= \left(7x \cdot \dfrac{1}{x}\right) \cdot 2$	Commutativity and associativity.
$= 7 \cdot 2$	Associativity, and multiplicative inverses.
$= \underline{\underline{14}}$	Arithmetic.

Example 5: Simplify $2 - 3[x - 2 - 5(x - 1)]$.

Here you must observe the agreed-upon sequence of operations. The first thing to do is *start inside the innermost parentheses* and work your way out (like a termite!).

$2 - 3[x - 2 - 5(x - 1)]$

$= 2 - 3[x - 2 - 5x + 5]$	Distributivity.
$= 2 - 3[-4x + 3]$	Collecting like terms.
$= 2 + 12x - 9$	Distributivity.
$= \underline{\underline{12x - 7}}$	Commutativity and associativity.

Notes:

1. You must remember some things from previous mathematics courses. For example, a negative number times a negative number is a *positive* number. This sort of thing can be proved using the Field Axioms, as you will see in Section 1-7.
2. There are several kinds of symbols of inclusion.

( ) Parentheses.
[ ] Brackets.
{ } Braces (also used for *set* symbols).
— Vinculum (an overhead line, used in fractions and elsewhere, such as in $\dfrac{x-3}{x+7}$).

To avoid so many different symbols, sometimes "nested" parentheses are used. For example, the expression

$$2-(3+4(5-6(7+x)))$$

would be simplified by starting with the *innermost* parentheses.

In the following exercise, you will get practice simplifying and evaluating expressions. If the going gets difficult, just tell yourself that no matter how complicated an expression looks, it just stands for a *number*. And people *invented* numbers!

Exercise 1-3

For Problems 1 through 10, carry out the indicated operations in the agreed-upon order.

1. $5 + 6 \times 7$
2. $3 + 8 \times 7$
3. $9 - 4 + 5$
4. $11 - 6 + 4$
5. $12 \div 3 \times 2$
6. $18 \div 9 \times 2$

7. $7 - 8 \div 2 + 4$

8. $24 - 12 \times 2 + 4$

9. $16 - 4 + 12 \div 6 \times 2$

10. $50 - 30 \times 2 + 8 \div 2$

For Problems 11 through 24, evaluate the given expression (a) for $x = 2$ and (b) for $x = -3$.

11. $4x - 1$

12. $3x - 5$

13. $|3x - 5|$

14. $|4x - 1|$

15. $5 - 7x - 8$

16. $8 - 5x - 2$

17. $|8 - 5x| - 2$

18. $|5 - 7x| - 8$

19. $x^2 - 4x + 6$

20. $x^2 + 6x - 9$

21. $4x^2 - 5x - 11$

22. $5x^2 - 7x + 1$

23. $5 - 2 \cdot x$

24. $3 + 4 \cdot x$

For Problems 25 through 40, simplify the given expression.

25. $6 - [5 - (3 - x)]$

26. $2x - [3x + (x - 2)]$

27. $7(x - 2(3 - x))$

28. $3(6x - 5(x - 1))$

29. $7 - 2[3 - 2(x + 4)]$

30. $8 + 4[5 - 6(x - 2)]$

31. $3x - [2x + (x - 5)]$

32. $4x - [3x - (2x - x)]$

33. $6 - 2[x - 3 - (x + 4) + 3(x - 2)]$

34. $7[2 - 3(x - 4) + 4(x - 6)]$

35. $6[x - \frac{1}{2}(x - 1)]$

36. $8[2x - \frac{1}{4}(6x + 5)]$

37. $x^2 + y^2 - [x(x + y) - y(y - x)]$

38. $4x^2 - 2x(x - 2y) + 2y(2y + x) - 2x^2$

39. $-(-(-(-x)))$

40. $x - [x - (x - \overline{x - y})]$

41. Calvin Butterball and Phoebe Small evaluate the expression $|x - 3|$ for $x = 7$, getting:

Calvin: $|x - 3| = |7 - 3| = 7 + 3 = \underline{\underline{10}}$

Phoebe: $|x - 3| = |7 - 3| = |4| = \underline{\underline{4}}$

Who is right? What mistake did the other one make?

42. Kay Oss evaluates the expression $|x + 2| - 5x$ by substituting 7 for the first x and 3 for the second x. What axiom did Kay violate?

1-4 POLYNOMIALS

Polynomials are algebraic expressions that involve only the operations of *addition, subtraction,* and *multiplication*. For example,

$$3x^2 + 5x - 7, \quad x + 2, \quad \text{and} \quad xy^3z^2$$

are polynomials. They involve no non-algebraic operations such as absolute value, and no operations under which the set of real numbers is not closed, such as division and square root. Thus, polynomials stand for *real* numbers no matter what real values you substitute for the variables.

Objectives:

1. Given an expression, tell whether or not it is a polynomial. If it is, then *name* it by "degree" and by number of terms.
2. Given two binomials, multiply them together.

Notes:

1. The expression $\dfrac{3}{x - 5}$ is *not* a polynomial since it involves *division* by a *variable*. If x were 5, the expression would have the form 3/0, which is *not* a real number.
2. The expression $\sqrt{x}$ is *not* a polynomial since it involves the *square root* of a *variable*. If x were less than 0, the expression would stand for an imaginary number rather than a real number.
3. The expression $|x - 7|$ is *not* a polynomial since it involves the *non-algebraic operation* "absolute value."
4. Expressions such as $\sqrt{3}x$ and $x/3$ (which equals $\frac{1}{3} \cdot x$) *are* considered to be polynomials since the operations $\div$ and $\sqrt{\ }$ are performed on *constants* rather than variables.
5. The operation exponentiation ("raising to powers") is *not* listed among the polynomial operations. If the exponent is an integer, such as in x^4, then exponentiation is just repeated multiplication. So expressions with

only *integer* exponents *are* polynomials. In Chapter 6 you will learn what happens when the exponent is not an integer.

"Terms" in an expression are parts of the expression that are *added* or *subtracted.* For example, the expression

$$3x^2 + 5x - 7$$

has three terms, namely, $3x^2$, $5x$, and 7. Special names are used for expressions that have 1, 2, or 3 terms.

No. of terms	Name	Example
1	monomial	$3x^2y^5$
2	binomial	$3x^2 + y^5$
3	trinomial	$3 - x^2 + y^5$
4 or more	(no special name)	$3x^5 - 2x^4 + 5x^3 - 6x^2 + 2x$

The word "polynomial" originally meant "many terms." However, it is possible to get a *monomial* by adding two polynomials. For example,

$$(3x^2 + 5x - 7) + (8x^2 - 5x + 7) = 11x^2,$$

a *monomial*. By calling monomials, binomials, and trinomials "polynomials," too, the set of polynomials has the desirable property of being *closed* under addition. It is also closed under multiplication.

"Factors" in an expression are parts of the expression that are *multiplied* together. For example, $5x^2$ has *three* factors, 5, x, and x. Special names are given to polynomials depending on how many *variables* are *multiplied* together.

For example, $3x^2y^5$ is *seventh* degree because seven variables are multiplied together ($x \cdot x \cdot y \cdot y \cdot y \cdot y \cdot y$). But $3x^2 + y^5$ is only *fifth* degree because at most five variables are multiplied together ($y \cdot y \cdot y \cdot y \cdot y$). An expression such as $17x$ that has only *one* variable is called *first* degree, and a constant such as 17 which has *no* variable is called *zero* degree.

DEFINITION

The ***degree*** of a polynomial is the maximum number of *variables* that appear as *factors* in any one term.

Various degrees are given special names, as follows:

Degree	Name	Example	Memory Aid
0	constant	13	Constants do not vary.
1st	linear	$5x$	A line has *one* dimension.
2nd	quadratic	$7x^2$	A square is a quadrangle.
3rd	cubic	$4x^3$	A cube has *three* dimensions.
4th	quartic	x^4	A *quart* is a *fourth* of a gallon.
5th	quintic	$9x^5$	Quintuplets are *five* children.
6th or more	(no special name)	$3x^{17}$	(Make up your *own* names, Hectic, Septic, etc.)

Notes:

1. Various parts of a monomial such as $3x^2$ have special names.

   3 is the *numerical coefficient.*
   x is the *base.*
   2 is the *exponent*.
   x^2 is a *power* (the second power of x).

2. "Zero" could have *any* degree, because 0 equals $0x^3$, $0x^{15}$, $0x^{1066}$, etc. To avoid this difficulty, 0 is usually called a polynomial with *no* degree.

Multiplying Binomials: Multiplying binomials requires a *double* use of the distributive property.

For example,

$(x - 3)(2x + 5)$

can be thought of as

number $\times$ $(2x + 5)$.

Distributing the "number", you get

number $\times$ $2x$ + number $\times$ 5.

Recalling that "number" is actually $(x - 3)$, you get

$(x - 3) \times 2x + (x - 3) \times 5.$

Distributing the $2x$ and the 5 gives

$2x^2 - 6x + 5x - 15,$

which can be simplified by collecting like terms to give

$2x^2 - x - 15.$

Once you understand the procedure, you can multiply two binomials *quickly*, in your head. Just multiply each term of one binomial by each term of the other, and write down the answer.

The following exercise is designed to give you practice identifying and naming polynomials, and multiplying binomials.

Exercise 1-4

For Problems 1 through 24, tell whether or not the given expression is a polynomial. If it *is*, then name it according to its *degree* and number of *terms* (e.g., "quadratic trinomial"). If it is *not* a polynomial, then tell *why* not.

1. $3x^4 - 2x + 51$
2. $3^2x^3 + 5y^4$
3. $9xy + 2z$
4. $9xyz + 2$
5. $3x^2y - \frac{57}{x}$
6. $3x^2y - \sqrt{57x}$
7. $\frac{x}{2} - 1$
8. $\sqrt{x} + 5$
9. $\frac{2}{x} - 1$
10. $x + \sqrt{5}$
11. -13
12. 19
13. $8x^3 + 5x^2 - 2x + 11$
14. $3x^2 + 5x - 7$
15. $x^3y^2 - \pi$
16. $xy^5 - \pi^7$
17. $\sqrt{x} + 11$
18. $\frac{x}{4} - 17$
19. $x + \sqrt{11}$
20. $\frac{4}{x} - 17$
21. $5^2x^3y^7 + 8z^9$
22. $6^3x^5y^2 - 3z^6$
23. 0
24. $0x^5$

For problems 25 through 34, multiply the two binomials.

25. $(x - 3)(x + 7)$
26. $(x - 6)(x + 5)$
27. $(x + 4)(2x - 1)$
28. $(3x + 1)(x - 2)$
29. $(3x - 8)(2x - 7)$
30. $(4x - 3)(7x - 5)$
31. $(2x - 5)(2x - 5)$
32. $(3x - 10)(3x - 10)$
33. $(2x - 5)^2$
34. $(3x - 10)^2$

1-5 EQUATIONS

Evaluating an expression may be thought of as finding out what number the expression equals when you know the value of the variable. You are ready to *reverse* the process, and find out the value of the *variable* when you know what number the *expression* equals. For example, if

$3x - 5 = 16,$

then x must equal 7, since $3 \times 7 - 5$ equals 16. The statement "$3x - 5 = 16$" is an equation, and the process of writing down what x must equal is called *solving* the equation. Since there may be more than one solution, it is customary to write the solutions in a *solution set*. For the above equation,

$S = \{7\}.$

Agreement: *Solving* an equation means writing its solution set.

Most of the equations you will encounter are too complicated to solve by inspection. So the procedure is to transform them to one or more equations of the form

x = a constant.

Then the solution set *can* be written by inspection.

Example 1: Solve the equation $3x - 5 = 16$ by first transforming it to an equivalent equation of the form x = a constant.

The equation would be transformed by first *adding* 5 to both members.

$3x - 5 = 16$	Given equation.
$3x = 21$	Add 5 to both members.

This step gets rid of the -5 in the left member. To get rid of the 3 in the left member, you could *multiply* both members of the equation by 1/3, getting

$x = 7$.

This equation is *equivalent* to the original one, meaning it has the *same* solution set. By inspection, you can write

$S = \{7\}$.

Notes:

1. Before writing the solution set, you should substitute the value(s) of x back into the original equation to be sure you have made no mistakes.
2. You should not stop at the step "$x = 7$." Since you have agreed that solving an equation means writing the solution set, the equation is not solved until you write "$S = \{7\}$."
3. Adding (or subtracting) and multiplying (or dividing) both members of an equation by the same number are justified by the *Addition Property of Equality* and the *Multiplication Property of Equality*, respectively. These properties can be proved from the axioms, as you will see in Section 1-7.

Example 2: Solve $x + 3 = x$.

This equation has no solutions at all, since no number comes out the same when 3 is added to it. The solution set is *empty*, and you would write

$S = \emptyset$ or $S = \{\ \}$.

Example 3: Solve $(3x - 4)(x + 5) = 0$.

This equation contains a *product* which equals *zero.* The only way a product of two real numbers can equal zero is for one of the factors to equal zero. This fact is expressed in the converse of the *Multiplication Property of Zero*, which can be proved from the axioms as in Section 1-7.

Because of this fact, you can transform the equation to

$3x - 4 = 0$ or $x + 5 = 0$.

Further transformations give

$3x = 4$ or $x = -5$	Adding 4 or subtracting 5.
$x = 4/3$ or $x = -5$	Dividing by 3.
$\therefore S = \{4/3, -5\}$.	Solving the transformed equation.

Note: In order for a number to be a solution, it must be in the *domain* of the variable. If the domain of x were

$x \in$ {positive numbers},

then $S = \{4/3\}$. If the domain is

$x \in$ {integers},

then $S = \{-5\}$. If the domain is

$x \in$ {natural numbers},

then $S = \emptyset$, since neither of the possible solutions is a positive integer.

Example 4: Solve $|x - 2| = 3$.

By inspection you might see that x could equal 5. But x could also equal -1. To make sure you do not overlook any solutions, you try to transform the equation to equivalent equations of the form x = a constant. You can do this by realizing that there are just *two* numbers whose absolute value is 3, namely 3 and -3. So the expression inside the absolute value sign, $x - 2$, must equal one of these two numbers. You would write

$$x - 2 = 3 \text{ or } x - 2 = -3.$$

Adding 2 to both members of each equation gives

$$x = 5 \text{ or } x = -1,$$

so that

$$S = \{5, -1\}.$$

Extraneous Solutions and Irreversible Steps – Two equations are *equivalent* if they have the *same* solution set. Unfortunately, there are transformations which you can perform on an equation which seem perfectly correct, but which *change* the solution set. Since it is the *original* equation you are trying to solve, you must be aware of the types of transformations which might *not* produce an equivalent equation.

1. *Multiply by an expression which can equal zero* – For example, the equation

$$x = 5$$

has $S = \{5\}$. But if you multiply both members by $x - 2$ you get

$$x(x - 2) = 5(x - 2).$$

This transformed equation is still true when $x = 5$, as the Multiplication Property of Equality tells you it must be. But it is also true when $x = 2$ because

$$2(2 - 2) = 5(2 - 2) \text{ or } 2 \times 0 = 5 \times 0$$

is also a true statement. So 2 is a solution of the transformed equation which does *not* satisfy the original equation. Such a solution is called an *extraneous* solution, "extra-" meaning "added," and "-neous" meaning "new."

DEFINITION

An ***extraneous*** solution is a number which satisfies a transformed equation, but not the original equation.

Multiplying both members of an equation by an expression which can equal zero is called an *irreversible step*. You cannot go backwards and divide both members by that expression since division by zero is undefined. Irreversible steps sometimes produce extraneous solutions.

2. *Divide by an expression that can equal zero* – For example, the equation

$$x^2 = 4x$$

has the solution set $S = \{4, 0\}$, as you can see by substituting these numbers. But if you divide both members by x you get

$$x = 4,$$

which has only $S = \{4\}$. So dividing by a variable can *lose* a valid solution.

The following exercise is designed to give you practice solving equations of the type you should recall from previous mathematics courses. You will also demonstrate that you understand about transformations that do *not* produce equivalent equations.

Exercise 1-5

For Problems 1 through 22, solve the equation in the indicated domain.

1. $3x + 7 = -8$ {real numbers}
2. $4x - 6 = 10$ {real numbers}
3. $2x + 3 = x - 1$ {positive numbers}
4. $5x - 1 = 2x + 5$ {negative numbers}

5. $5x + 3 = 2x + 3$ {real numbers}
6. $5x + 3 = 5x - 4$ {real numbers}
7. a. $4x + 3 = x - 8$ {rational numbers}
   b. $4x + 3 = x - 8$ {integers}
8. a. $7x - 4 = 2x - 19$ {rational numbers}
   b. $7x - 4 = 2x - 19$ {integers}
9. a. $x^2 = 16$ {negative numbers}
   b. $x^2 = 16$ {real numbers}
10. a. $x^2 = 36$ {positive numbers}
    b. $x^2 = 36$ {real numbers}
11. a. $x^2 = 1/9$ {rational numbers}
    b. $x^2 = 1/9$ {integers}
12. a. $x^2 = 7$ {rational numbers}
    b. $x^2 = 7$ {irrational numbers}
13. $(x + 3)(3x - 2) = 0$ {real numbers}
14. $(2x - 5)(x + 1) = 0$ {real numbers}
15. $(2x - 5)(x + 1) = 0$ {integers}
16. $(x + 3)(3x - 2) = 0$ {integers}
17. $(x + 3)(3x - 2) = 0$ {positive numbers}
18. $(2x - 5)(x + 1) = 0$ {positive numbers}
19. $(2x + 3)(x - 5)(x + 1) = 0$ {real numbers}
20. $(4x - 3)(x + 2)(x - 6) = 0$ {real numbers}
21. $x(2x - 1)(x + 4) = 0$ {real numbers}
22. $x(5x - 3)(x - 6) = 0$ {real numbers}

For Problems 23 through 32, solve the equation assuming that the domain is $x \in$ {real numbers}.

23. $|x| = 7$
24. $|x| = 5$
25. $|x| = -6$
26. $|x| = -9$
27. $|x + 3| = 5$
28. $|x - 2| = 14$
29. $|4x - 1| = 11$
30. $|5x + 3| = 7$
31. $|9x - 17| = 1$
32. $|7x - 2| = 4$

33. Although they do not look much alike, there *is* a relationship between the equations $(x + 7)(3x - 5) = 0$ and $|3x + 8| = 13$.
   a. Find the solution set of each equation.
   b. What word best describes the relationship between the two equations?

34. Given the equation $(x - 3)(x - 7) = 0$:
   a. Write the solution set.
   b. Divide both members of the equation by $(x - 3)$, and write the solution set of the transformed equation.
   c. Is the transformed equation *equivalent* to the original one? Explain.
   d. Multiply both members of the original equation by $(x - 2)$. What number is a solution of the transformed equation which was *not* a solution of the original one?
   e. What name is given to the solution in part d?
   f. What name is given to the operation of multiplying both members of an equation by an expression which can equal 0?

35. *Introduction to Inequalities* – The statement $5 < 7$ is an *inequality*. It is *true* because 5 *is* less than 7.
   a. Add the positive number 3 to both members. Is the resulting inequality true?
   b. Add the negative number −8 to both members of the original inequality. Is the resulting inequality true?
   c. Use variables to write a statement of the *Addition Property of Order*, which states that you may add the same number to both members of an inequality *without* changing the order (from $<$ to $>$, for example).

36. *More about Inequalities* – Given the true inequality $5 < 7$:
   a. Multiply both members of the inequality by the positive number 3. Is the resulting inequality true?

b. Multiply both members of the original inequality by the negative number −8. Is the resulting inequality true?
c. Multiply both members of the original inequality by 0. Is the resulting inequality true?
d. Write a statement of the *Multiplication Property of Order* which takes into account what you have observed in parts a, b, and c.
e. In your own words, tell what happens to the order in an inequality when you multiply both members by a *negative* number. Tell also what *you* must do to make a true statement out of the resulting inequality.

1-6 INEQUALITIES

If the "=" sign in an equation is replaced by one of the order signs, $<$, $>$, $\leq$, or $\geq$, the resulting sentence is called an *inequality*. For example,

$$3x - 5 < 16.$$

The solution set of an inequality contains all values of the variable that make the sentence true. Since the solution set usually contains an *infinite* number of solutions, it is customary to draw a graph rather than write the set.

Objective:

Given an inequality, transform it to a simpler, equivalent inequality so that you can draw a graph of its solution set.

If you cannot see the solution set by inspection, you can transform the inequality by

1. adding the same number to both members, or
2. multiplying both members by the same number. (This is tricky! See Example 2.)

Example 1: Graph the solution set of $3x - 5 < 16$.

Starting with

$$3x - 5 < 16,$$

you would add 5 to both members, getting

$$3x < 21.$$

Then you would multiply both members by 1/3, getting

$$x < 7.$$

The graph would be plotted on a number line, looking like this:

−8 −7 −6 −5 −4 −3 −2 −1 0 1 2 3 4 5 6 7 8 9

The open circle at 7 indicates that the endpoint of the ray is *not* included. A *closed* circle would be used for inequalities with $\leq$ or $\geq$, where the endpoint *is* included.

Example 2: Graph the solution set of $-2x \geq 18$.

To get rid of the "−2" on the left, you must multiply both members of the inequality by −1/2. But multiplying both members of an inequality by a *negative* number makes the order *reverse*. You can see why, by multiplying both members of an inequality like $5 < 7$ by a negative number such as −8. The result will be that −40 is *greater* than −56. So *you* must reverse the order sign whenever you multiply both members of an inequality by a negative number. This fact is summarized in the Multiplication Property of Order.

Multiplication Property of Order: If $x < y$, then

$xz < yz$, if z is *positive*,
$xz > yz$, if z is *negative*,
$xz = yz$, if z is *zero*.

A similar property applies to the relationship $>$.

The steps in solving the inequality would be:

$-2x \geq 18$	Given inequality.
$-\frac{1}{2}(-2x) \leq -\frac{1}{2}(18)$	Multiply by −1/2 and reverse the order sign.
$x \leq -9$	Associate and do the arithmetic.

The graph would be

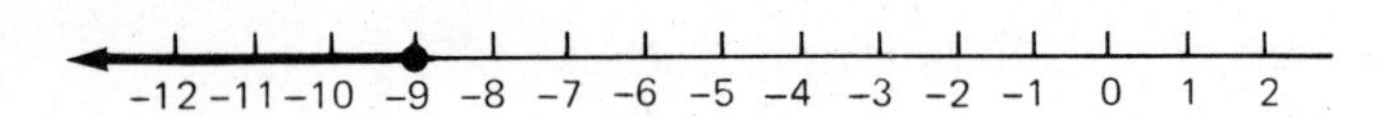

Example 3: Graph the solution set of $3 \leq 2x + 5 < 11$.

This inequality has *three* members. Starting with

$$3 \leq 2x + 5 < 11,$$

you would add -5 to *all three* members, getting

$$-2 \leq 2x < 6.$$

Multiplying all three members by 1/2 gives

$$-1 \leq x < 3.$$

The graph is all numbers between -1 and 3, including the -1 but not including the 3.

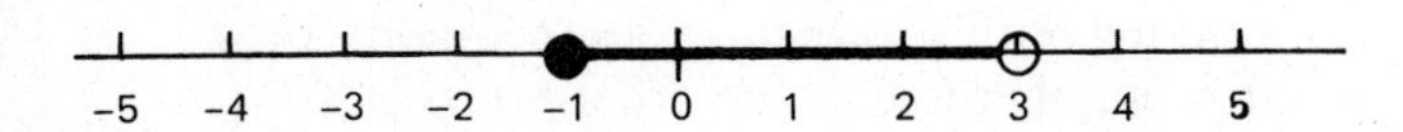

Example 4: Graph the solution set of $|x| > 29$.

$|x|$ means, "the distance between the origin and x." So the inequality really says, "x is a number that is *more than* 29 units from the origin." Therefore, one part of the graph will start at 29 and go "upward," and the other part will start at -29 and go "downward." The graph is

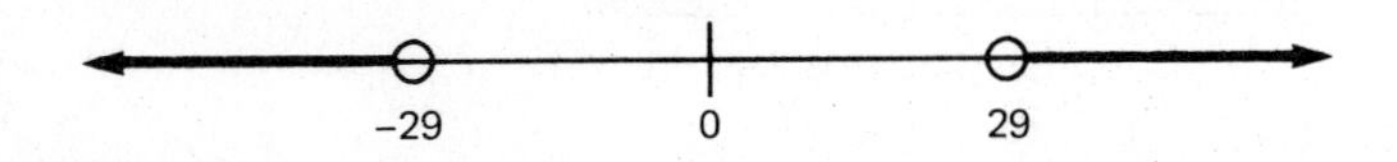

From the graph you should be able to tell that the inequality $|x| > 29$ is equivalent to the *combined* inequalities,

$$x > 29 \text{ or } x < -29.$$

Example 5: Graph the solution set of $|x| < 29$.

The solution set will contain all numbers that are *closer* to the origin than 29 units. So the graph will be all those points *between* -29 and 29. The graph is

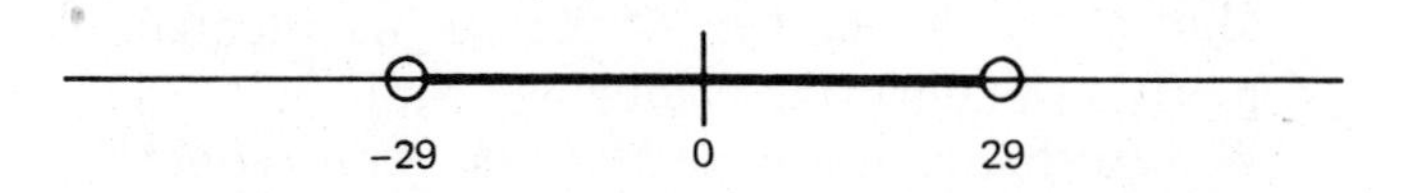

From this graph, you should be able to tell that $|x| < 29$ is equivalent to

$$x < 29 \text{ and } x > -29,$$

which is equivalent to

$$-29 < x < 29.$$

From Examples 4 and 5, you can figure out a way to transform inequalities to eliminate the absolute value sign.

Conclusion: If c is a non-negative constant, then $|\text{expression}| > c$ is equivalent to:

expression $> c$ *or* expression $< -c$.

$|\text{expression}| < c$ is equivalent to:

expression $< c$ *and* expression $> -c$.
(or to $-c <$ expression $< c$).

If you ever forget these transformations, you can think them up easily by recalling that the absolute value of a number is its distance from the origin.

Example 6: Graph the solution set of $|3x - 5| \leq 13$ if the domain of x is

a. {real numbers},
b. {positive numbers},
c. {integers}.

Starting with

$$|3x - 5| \leq 13,$$

you use the appropriate transformation, above, to write

$$-13 \leq 3x - 5 \leq 13.$$

Adding 5 to all three members,

$-8 \leq 3x \leq 18.$

Dividing all three members by 3 gives

$-8/3 \leq x \leq 6.$

The graphs are as follows:

a. {real numbers}

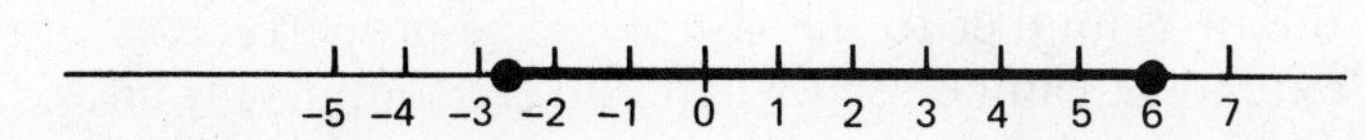

b. {positive numbers}

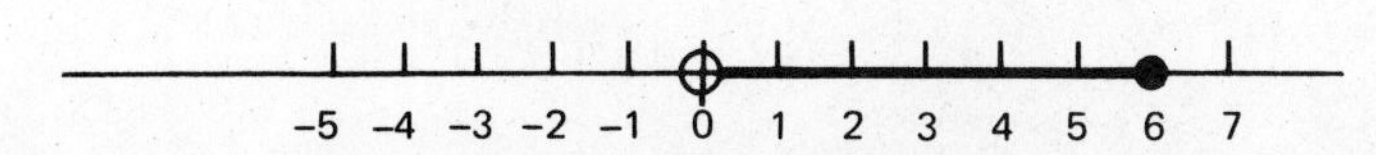

c. {integers}

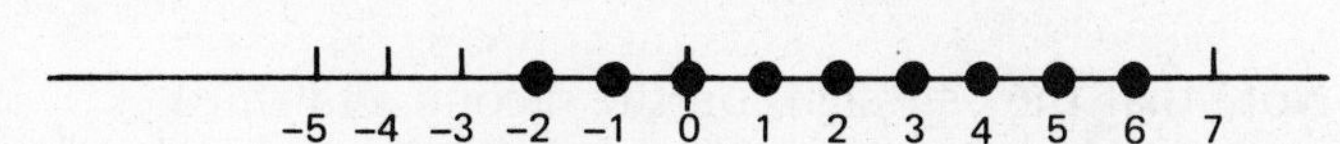

The following exercise is designed to give you practice in graphing the solution sets of inequalities.

Exercise 1-6

For Problems 1 through 14, graph the solution set of the inequality, observing the given domain.

1. $3x - 5 < 4$ {real numbers}
2. $4x - 7 > 1$ {real numbers}
3. $2 - 6x \leq 8$ {real numbers}
4. $5 - 3x \geq -4$ {real numbers}
5. $2x + 3 > 10$ {integers}
6. $3x - 5 < 2$ {integers}
7. $-6 < x - 5 \leq 2$ {real numbers}
8. $4 \leq x - 1 < 7$ {real numbers}
9. $x + 3 \geq 2$ or $x + 3 < -7$ {real numbers}
10. $x - 4 > -1$ or $x - 4 \leq -6$ {real numbers}
11. $x + 2 < 5$ or $x + 2 \geq 7$ {positive numbers}
12. $3 \leq x + 4 < 5$ {negative numbers}
13. $3 - 2x > 5$ {positive numbers}
14. $4 - 3x < -2$ {negative numbers}

For Problems 15 through 32, graph the solution set assuming that the domain is:

a. $x \in$ {real numbers}
b. $x \in$ {integers}
c. $x \in$ {positive numbers}

15. $|x| < 5$
16. $|x| \leq 7$
17. $|x| \geq 2$
18. $|x| > 3$
19. $|x + 2| \leq 3$
20. $|x - 1| < 4$
21. $|2x + 5| > 9$
22. $|3x + 7| \geq 7$
23. $|5x - 6| \leq 16$
24. $|4x - 3| < 7$
25. $|2x - 7| \geq 21$
26. $|5x - 3| > 22$
27. $|4x + 2| < 9$
28. $|2x + 4| \leq 9$
29. $|x - 1| < -3$
30. $|x + 3| > -6$
31. $3 \leq |x - 2| < 5$
32. $4 < |x - 3| \leq 7$

1-7 PROPERTIES PROVABLE FROM THE AXIOMS

The Field Axioms you studied in Section 1-2 express properties of addition and multiplication which should be obvious to you. However, there are some true properties which are *not* obvious, and some things which seem obvious, but are *not true*. Therefore, mathematicians seek ways of proving new properties from a small handful of axioms.

The properties you will be proving in this section form the basis for the mathematics you will study for the rest of the course. Some you already know. The others you can either prove *now*, or you can go on to Chapter 2 and return to these proofs as you need the properties.

Objectives:

1. Given the steps in the proof of a new property, name a property justifying each step.

2. Given the name or statement of a new property and some clues about how to prove it, prove the property giving reasons for each step.

You must first recall the axioms which apply to the relationships =, <, and >. These are as follows.

Reflexive Property (an axiom)

> If x is a real number, then $x = x$.

Equality is said to be a *reflexive* relationship. The name is picked because a number can "look into the '=' sign" and see its own "reflection" on the other side. It is this axiom which expresses the fact that a variable stands for the *same* number, wherever it appears in an expression. Note that order is *not* reflexive since statements like $5 < 5$ or $5 > 5$ are *false.*

Symmetry (an axiom)

> If $x = y$, then $y = x$.

Equality is said to be a *symmetric* relationship. The "=" sign looks the same when viewed from either direction. So it does not matter on which side of the "=" sign a number appears. The order relationships are *not* symmetrical. For example, the statement $4 < 5$ *cannot* have the numbers reversed. The statement $5 < 4$ is *false.*

Transitivity (an axiom)

> For Equality: If $x = y$ and $y = z$, then $x = z$.
> For Order: If $x < y$ and $y < z$, then $x < z$.
> If $x > y$ and $y > z$, then $x > z$.

The prefix "trans-" means, "across," "beyond," or "through." For example, a rapid *transit* system carries you *through* a city. The name is used here because the equality or order carries through from the first number to the last one. The property, extended, can be used when you are simplifying an expression to connect the original form to the final form. For example:

$$(x + 3)(x + 4) = x(x + 4) + 3(x + 4)$$
$$= x^2 + 4x + 3x + 12$$
$$= x^2 + 7x + 12.$$
$$\therefore (x + 3)(x + 4) = x^2 + 7x + 12. \leftarrow \text{Transitivity used } \textit{here}.$$

Note that the "=" signs on the second and third lines connect the expression to the one *just above* it, not to the original expression on the left. This is why the transitive step is needed at the end. See Appendix B for a proof of this "extended" transitive property.

Comparison Axiom (Trichotomy)

> If x and y are two real numbers,
> then *exactly one* of the following must be true:
> $y < x$
> $y > x$
> $y = x$.

This axiom tells you how any two given numbers must *compare* to each other. The words, "exactly one" mean two things. *One* of the statements *must* be true, and *only one* of the statements can be true. The word "trichotomy," meaning "cut in three," is used because placing a number x on the number line *cuts* it into *three* pieces.

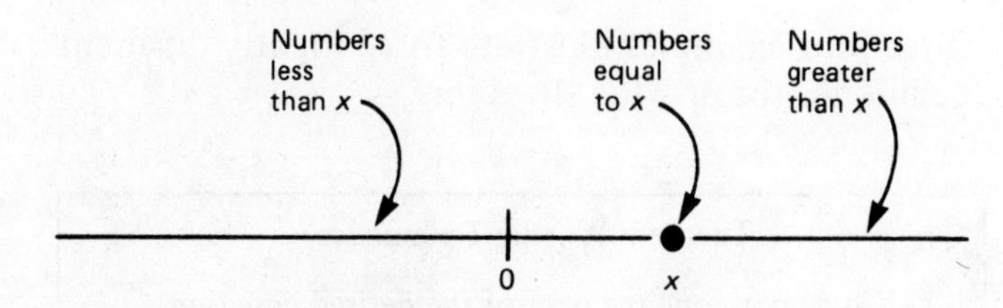

Any other number, y, must be on one of these three pieces.

These axioms help you with the mechanics of going from one step to the next in a proof. You are now armed with the tools you need to do some proving.

Example 1: *Substitution into Sums and Products* (*not* an axiom)

Prove that if $z = x$, then

$x + y$ and $z + y$ stand for *equal* numbers,
xy and zy stand for *equal* numbers.

(In plain English, this property says that equals can be substituted for equals in sums or products.)

Proof:

The Closure Axiom states that a sum such as $x + y$ or a product such as xy stands for a *unique* real number.For this reason, it does not matter what other symbol you use for the number x. Therefore, $x + y = z + y$, and $xy = zy$, Q.E.D.

This is called a *paragraph proof*. You have written a paragraph explaining why a certain property is true. The letters "Q.E.D." at the end stand for the Latin words, "quod erat demonstrandum," meaning, "which was to be proved."

Example 2: *Addition Property of Equality* (not an axiom)

Prove that if $x = y$, then $x + z = y + z$.

(In plain English, prove that you can add the same number to both members of an equation.)

Proof:

a. $x + z = x + z$	a. Reflexive Axiom.
b. $x = y$	b. Hypothesis (given).
c. $\therefore x + z = y + z$, Q.E.D.	c. Substitution into a sum.

Proofs such as this one are not easy to think up from "scratch." However, there is usually a "key step" that forms the heart of the proof. In this case, the thought process is, "I want an equation that has $x + z$ on the left side, so I'll *write* an equation like that and *transform* it to the one I want." The substitution property gives you a way to do the transformation.

The "if" part of a property is called the "hypothesis." The word comes from "hypo-" (as in hypodermic), meaning "under," and "thesis," (as in thesis sentence), meaning "main idea." The property you get by *reversing* the hypothesis and the conclusion is called the *converse* of the property. The converse of a true statement may or may not be true. For example,

"If you play varsity football, then you are male,"

is (probably!) a true statement. But its converse,

"If you are a male, then you play varsity football,"

is *not* true. In the next example is a proof of the converse of the Addition Property of Equality.

Example 3: *Converse of the Addition Property of Equality*

Prove that if $x + z = y + z$, then $x = y$.

Proof:

a. $x + z = y + z$	a. Given ("hypothesis").
b. $(x + z) + (-z) = (y + z) + (-z)$	b. Addition property of equality.
c. $x + (z + (-z)) = y + (z + (-z))$	c. Associativity for addition.

d. $x + 0 = y + 0$	d. Additive inverses.
e. $x = y$, Q.E.D.	e. Additive identity.

In this proof it was helpful to start with the hypothesis and work toward the conclusion. Note that you can *never* start with the conclusion, nor can you use the conclusion as a reason in the proof. Doing so is called "circular reasoning," for which your instructor will probably give you circular grades!

The heart of this proof is doing something to "cancel out" the unwanted z's. Adding $-z$ to both members in step b does this job. The remainder of the proof consists of tidying up the resulting equation. This property is sometimes called the *Cancellation Property of Equality for Addition.*

A theorem used to make the proof of a subsequent theorem easier is called a *lemma* for that theorem. A theorem which follows directly from a previous theorem is called a *corollary* of that theorem. Thus, the addition property of equality is used as a lemma for proving its converse.

If both the theorem and its converse are true, you may write both as a *single* statement. You use the words "if and only if":

$$x + z = y + z \text{ if and only if } x = y.$$

Example 4: *Adding Like Terms*

Prove, for example, that $2x + 3x = 5x$.

Proof:

a. $2x + 3x = (2 + 3)x$	Distributivity.
b. $\quad = 5x$	Arithmetic.
c. $\therefore 2x + 3x = 5x$, Q.E.D.	Transitivity.

In this proof you started with *one member* of the equation in the conclusion, and transformed it to the *other* member. The transitive step at the end connects the original expression with the final one, thus expressing the desired conclusion.

These four examples illustrate three slightly different techniques for proving theorems.

Theorem-Proving Techniques

1. Start with one member of the desired equation and transform it to the other member (Example 4).
2. Start with a given equation and transform it to the desired equation (Example 3).
3. Start somewhere else and use a clever series of transformations or a clever argument (Examples 1 and 2).

The following exercise contains most of the basic properties you learned in previous mathematics courses. Your main purpose in proving these properties is so that you will realize that they are all based on the *axioms*. You must also, of course, remember *what* they say so that you will be able to use them later on.

Exercise 1-7

1. Write an example of each of the following axioms:

a. Transitivity for equality.
b. Transitivity for order.
c. Symmetry for equality.
d. Reflexive axiom for equality.
e. Trichotomy.

2. Explain why the order relationship "<" is *not* symmetric and *not* reflexive.

3. What is the difference between an axiom and any other property?

4. Calvin Butterball sees the expression $x^2 + 3x - 5$. He decides to substitute 7 for the first x and 2 for the second x. What axiom has he violated?

5. What axiom tells you that a variable such as x can stand for only *one* number at a time?

6. State, without proof,

a. the Addition Property of Order, and
b. the Multiplication Property of Order.

For Problems 7 through 32, prove the property either from scratch or by supplying names of the properties justifying the given steps. As reasons for steps, you may use any axiom or definition, or any other property whose proof appears *before* the one you are doing.

7. Prove the *Multiplication Property of Equality* which states that you can multiply both members of an equation by the same number.

8. Prove the *converse* of the Multiplication Property of Equality.

9. If you proved the property in Problem 8, you proved a *false* theorem! For example, $3 \times 0 = 5 \times 0$, but 3 does *not* equal 5. Fix the hypothesis in Problem 8 so that the theorem *is* true. Then think up a good name for this property.

10. Can the Multiplication Property of Equality and its converse be written as a single statement using "if and only if"? Explain.

11. The symbol $-(-x)$ means the additive inverse of $-x$. Use this fact to prove that $-(-x) = x$.

12. The symbol $\frac{1}{\frac{1}{x}}$ means the multiplicative inverse (reciprocal) of $\frac{1}{x}$. Use this fact to prove that $\frac{1}{\frac{1}{x}} = x$.

For what value of x is the theorem *false*?

13. *Property of the Reciprocal of a Product.*

Prove that $\frac{1}{xy} = \frac{1}{x} \cdot \frac{1}{y}$.

Proof:

a. $\left(\frac{1}{xy}\right)(xy) = 1$

b. $\left[\left(\frac{1}{xy}\right)(xy)\right] \cdot \frac{1}{y} = 1 \cdot \frac{1}{y}$

c. $\left[\left(\frac{1}{xy}\right)(x)\right]\left(y \cdot \frac{1}{y}\right) = 1 \cdot \frac{1}{y}$

d. $\left[\left(\frac{1}{xy}\right)(x)\right] \cdot 1 = 1 \cdot \frac{1}{y}$

e. $\left(\frac{1}{xy}\right)(x) = \frac{1}{y}$

f. $\left[\left(\frac{1}{xy}\right)(x)\right] \cdot \frac{1}{x} = \frac{1}{y} \cdot \frac{1}{x}$

g. $\frac{1}{xy}\left(x \cdot \frac{1}{x}\right) = \frac{1}{y} \cdot \frac{1}{x}$

h. $\frac{1}{xy} \cdot 1 = \frac{1}{y} \cdot \frac{1}{x}$

i. $\frac{1}{xy} = \frac{1}{y} \cdot \frac{1}{x}$

j. $\therefore \frac{1}{xy} = \frac{1}{x} \cdot \frac{1}{y}$, Q.E.D.

14. *Multiplication Property of Zero*

Prove that for any real number x, $x \cdot 0 = 0$.

Proof:

a. $0 = 0$
b. $0 + 0 = 0$
c. $x(0 + 0) = x \cdot 0$
d. $x(0 + 0) = 0 + x \cdot 0$
e. $x \cdot 0 + x \cdot 0 = 0 + x \cdot 0$
f. $\therefore x \cdot 0 = 0$, Q.E.D.

15. *Converse of the Multiplication Property of Zero*

Prove that if $xy = 0$, then $x = 0$ or $y = 0$.

Proof:

a. Either $y = 0$ or $y \neq 0$

b. If $y = 0$, the conclusion is true.
(Only *one* clause of an "or" statement needs to be true.)

c. If $y \neq 0$, then $\frac{1}{y}$ is a real number.

d. $xy = 0$

e. $(xy) \cdot \frac{1}{y} = 0 \cdot \frac{1}{y}$

f. $(xy) \cdot \frac{1}{y} = 0$

g. $x\left(y \cdot \frac{1}{y}\right) = 0$

h. $x \cdot 1 = 0$

i. $x = 0$

j. $\therefore x = 0$ or $y = 0$, Q.E.D.
(From Steps b and i.)

16. The Multiplication Property of Zero can be stated, "If $x = 0$ or $y = 0$, then $xy = 0$." Write this property and its converse as a *single* statement using "if and only if" terminology.

17. *Lemma – Multiplication Property of Negative One*

Prove that $-1 \cdot x = -x$.

Proof:

a. $-1 \cdot x + x = -1 \cdot x + 1 \cdot x$
b. $\quad = (-1 + 1)(x)$*
c. $\quad = 0 \cdot x$
d. $\quad = 0$
e. $\quad = -x + x$
f. $\therefore -1 \cdot x + x = -x + x$
g. $\therefore -1 \cdot x = -x$, Q.E.D.

18. *Theorem – The Product of Two Negatives is Positive.*

Prove that $(-x)(-y) = xy$.

Proof:

a. $(-x)(-y) = (-1 \cdot x)(-1 \cdot y)$
b. $\quad = (-1)[x \cdot (-1)](y)$
c. $\quad = (-1)[-1 \cdot x](y)$
d. $\quad = [-1 \cdot (-1)](xy)$
e. $\quad = 1 \cdot xy$**
f. $\quad = xy$
g. $\therefore (-x)(-y) = xy$, Q.E.D.

19. *If Two Real Numbers are Equal, then their Additive Inverses are Equal.*

a. *State* this property using letters to stand for the numbers.
b. *Prove* this property. You may find that the Multiplication Property of -1 is useful as a lemma.

20. *If Two Real Numbers are Equal, then their Multiplicative Inverses are Equal.*

a. *State* this property using letters to stand for the numbers. Be sure that the hypothesis *excludes* the one number for which the property is false.
b. *Prove* the property.

21. Prove that *the square of a real number is never negative.* Do this by showing that whatever value you pick for x,

$x^2 \geq 0.$

---

*The "=" signs in Steps b, c, d, and e connect each expression with the one *just above* it, *not* with the original expression in Step a.

**Note that there *is* a reason why $-1 \cdot (-1) = 1$ in Step e which does not involve circular reasoning. You must treat each of the -1's *differently* to see how the reasoning works.
Note also that this property tells you the *real* reason why the product of two negatives is positive. If it came out any other way, it would contradict the Field Axioms. This is an example of a not-so-obvious property which turns out to be *true*.

The axiom of trichotomy and the Multiplication Property of Order should be helpful as lemmas.

22. *Multiplication Distributes over Subtraction.*

Prove that $x(y - z) = xy - xz$.

Proof:

a. $x(y - z) = x[y + (-z)]$
b. $= xy + x(-z)$
c. $= xy + x[(-1)(z)]$
d. $= xy + [(x)(-1)]z$
e. $= xy + [(-1)(x)]z$
f. $= xy + (-1)(xz)$
g. $= xy + (-xz)$
h. $= xy - xz$
i. $\therefore x(y - z) = xy - xz$, Q.E.D.

23. Prove that *division distributes over addition.* That is, prove that

$$\frac{x + y}{z} = \frac{x}{z} + \frac{y}{z}.$$

Do this by first transforming the division to multiplication using the Definition of Division. Then distribute the multiplication over the addition.

Note that this property when read backwards says that you can add two fractions when they have a *common* denominator. This explains why you must *find* a common denominator before you *can* add two fractions. Adding fractions any other way (such as adding the numerators and adding the denominators) would violate the field axioms.

24. *Multiplication Property of Fractions*

Prove that $\frac{xy}{ab} = \frac{x}{a} \cdot \frac{y}{b}$.

Proof:

a. $\frac{xy}{ab} = (xy) \cdot \frac{1}{ab}$

b. $= (xy)\left(\frac{1}{a} \cdot \frac{1}{b}\right)$

c. $= x\left(y \cdot \frac{1}{a}\right) \cdot \frac{1}{b}$

d. $= x\left(\frac{1}{a} \cdot y\right) \cdot \frac{1}{b}$

e. $= \left(x \cdot \frac{1}{a}\right)\left(y \cdot \frac{1}{b}\right)$

f. $= \frac{x}{a} \cdot \frac{y}{b}$

g. $\therefore \frac{xy}{ab} = \frac{x}{a} \cdot \frac{y}{b}$, Q.E.D.

Note that this property when read backwards says that the way to multiply two fractions is to multiply their denominators together and multiply their numerators together. Again, the reason you do not multiply fractions in other ways is because to do so would violate the field axioms.

25. Prove that *division distributes over subtraction.*

26. Prove that a *non-zero number divided by itself equals 1,* i.e., $\frac{n}{n} = 1$.

27. Prove that *1 is its own reciprocal.* That is, $\frac{1}{1} = 1$.

28. Prove that a *number divided by 1 is that number,* i.e., $\frac{n}{1} = n$.

29. Prove that a *negative number divided by a positive number is negative,* $\frac{-x}{y} = -\frac{x}{y}$.

30. Prove that a *positive number divided by a negative number is negative,* $\frac{x}{-y} = -\frac{x}{y}$.

31. Prove that *the negative of a sum equals the sum of the negatives.* That is, prove that $-(x + y) = -x + (-y)$.

32. Prove that $x - y$ *and* $y - x$ *are additive inverses of each other.* That is, prove that $-(x - y) = y - x$.

1-8 CHAPTER REVIEW AND TEST

The purpose of this chapter has been to refresh your memory about some of the words and techniques of mathematics so that you and your instructor will be speaking the same language. The objectives of the chapter can be summarized as follows:

1. *Name various kinds of numbers.*
2. *State and prove properties.*
3. a. *Recognize and name polynomials.*
   b. *Simplify expressions.*
   c. *Evaluate expressions.*
4. *Solve equations and inequalities.*

The Review Problems below give you a relatively straightforward test of these objectives. The problem numbers correspond to the objectives so that you will have no doubt about what is expected of you. The Concepts Test, on the other hand, has problems that may require you to use *several* of the objectives. In some cases, you will have the chance to *extend* your knowledge by applying what you know to *new* situations. For this reason, the Concepts Test may be longer and more difficult than a test your instructor might give you.

Review Problems

The following problems are numbered according to the four objectives listed above.

Review Problem 1a. Give an example of

i. a rational number that is not an integer,
ii. an irrational number that is not positive,
iii. an imaginary number,
iv. a transcendental number,
v. a negative even number,
vi. a positive integer that is not a digit,
vii. a natural number,
viii. a real number that is not a natural number,
ix. a real number that is neither positive nor negative,
x. an irrational number that is not a real number.

b. Name all the sets of numbers to which each of the following belongs.

i. 2
ii. −3
iii. $\sqrt{3}$
iv. $\sqrt{-3}$
v. 2.3

Review Problem 2a. What is meant by

i. an axiom?
ii. a lemma?
iii. a corollary?
iv. a hypothesis?

b. State each of the following, and tell whether or not it is an axiom. If so, is it a *Field* Axiom?

i. Definition of Subtraction.
ii. Multiplicative Inverses Property.
iii. Trichotomy.
iv. Closure of {real numbers} under addition.
v. Additive Identity Property.
vi. Multiplication Property of Order.

c. *Positive Divided by Negative* – Justify each step of the following proof.

Prove that $\frac{x}{-y} = -\frac{x}{y}$.

Proof:

i. $\frac{x}{-y} = x \cdot \frac{1}{-y}$

ii. $= x \cdot \frac{1}{-1 \cdot y}$

iii. $= x \cdot \left(\frac{1}{-1} \cdot \frac{1}{y}\right)$

iv. $= x \cdot (-1 \cdot \frac{1}{y})$

v. $= [x \cdot (-1)] \cdot \frac{1}{y}$

vi. $= [-1 \cdot x] \cdot \frac{1}{y}$

vii. $= -1 \cdot \left(x \cdot \frac{1}{y}\right)$

viii. $= -1 \cdot \frac{x}{y}$

ix. $= -\frac{x}{y}$

x. $\therefore \frac{x}{-y} = -\frac{x}{y}$, Q.E.D.

d. The Distributive Axiom states that multiplication distributes over a sum of *two* terms. That is, $a(b + c) = ab + ac$. Prove that multiplication also distributes over a sum of *three* terms. Starting with $a(b + c + d)$, you can use the Associative Property to write the expression inside the parentheses as *two* terms. Then you can use the Distributive Axiom *twice* to get the desired result.

Review Problem 3a. Name each polynomial by degree and by number of terms. If it is *not* a polynomial, tell *why* not.

i. $x^2y - 5$
ii. $x^2/y - 5$
iii. $x^2\sqrt{y - 5}$
iv. $x^2y^2 - \sqrt{5}$
v. $4^2r^2s^3$
vi. $6x^2 + 7x - 5$

b. Carry out the indicated operations and simplify.

i. $13 - 5 + 1$
ii. $40 \div 10 \times 2$
iii. $24 - 12 \div 3 + 1$
iv. $(3x + 7)(x - 8)$
v. $5 - 3[x - 7(2x - 6)]$

c. Evaluate the following expressions for $x = 5$ and for $x = -4$.

i. $3x - 8$
ii. $|2x - 10|$
iii. $3x^2 - 2x + 11$

Review Problem 4a. Write the solution set.

i. $2x + 7 = -5$, $x \in \{\text{integers}\}$
ii. $5x - 3 = 8$, $x \in \{\text{integers}\}$
iii. $(2x + 6)(3x - 2) = 0$, $x \in \{\text{rational numbers}\}$
iv. $|4x + 3| = 9$, $x \in \{\text{positive numbers}\}$
v. $x^2 = 81$, $x \in \{\text{real numbers}\}$

b. Graph the solution set.

i. $4x - 3 < 7$, $x \in \{\text{real numbers}\}$
ii. $2 - 5x \leq 17$, $x \in \{\text{real numbers}\}$
iii. $|x - 2| > 5$, $x \in \{\text{integers}\}$
iv. $|3 - 4x| \leq 9$, $x \in \{\text{integers}\}$

Concepts Test

Work each of the problems below. For each part of each problem, tell by number and letter which one or ones of the above objectives you used in that problem. Also, if the problem involves a new concept, write, "new concept."

Concepts Test 1. Give an example of

a. a cubic binomial with three variables,
b. a quintic monomial with two variables,
c. a quadratic trinomial,
d. an expression that is not a polynomial,
e. a rational number between -12 and -13,
f. a radical that represents a rational number,
g. the Multiplicative Identity Axiom,
h. the Additive Inverse Axiom.

Concepts Test 2. Write an example that shows why,

a. subtraction is *not* associative,
b. exponentiation is *not* commutative,
c. {real numbers} is *not* closed under the operation "square root,"
d. multiplication *does* distribute over subtraction.

Concepts Test 3. Given the expression $3x - 4[2x - (5x - 9)]$,

a. *evaluate* it by substituting 7 for x, and then doing the indicated operations,
b. *simplify* it, *without* substituting a value for x,

c. substitute 7 for x in the simplified expression of part b and show that you get the same value as in part a.

Concepts Test 4. Given the expression $(x + 7)(2x - 3)$,

a. carry out the indicated multiplication and simplify,
b. find the value(s) of x that make the expression equal to 0.

Concepts Test 5. Transform the following to equivalent inequalities that have *no* absolute value signs. Then graph the solution sets.

a. $|x - 3| < 7$, $x \in$ {integers}
b. $|7 - 2x| \leq 1$, $x \in$ {real numbers}

Concepts Test 6. What extraneous solution is created if you multiply both members of the equation $x = 17$ by the expression $(x - 23)$?

Concepts Test 7. There is a set of numbers that contains both the real numbers *and* the imaginary numbers. What is the name of this set?

Concepts Test 8. Write the Multiplication Property of Zero and its converse, as a *single* statement, using "if and only if."

Concepts Test 9a. If the converse of the Multiplication Property of Equality were true, what would it say?

b. Explain why this converse is *false*.

Concepts Test 10. Professor Snarff gives an algebra test on which students must add

$$\frac{5}{7} + \frac{6}{7}.$$

Calvin Butterball gets $\frac{11}{14}$ and Phoebe Small gets $\frac{11}{7}$.

a. Who is right?
b. Name the property that explains *why* he or she is right.

Concepts Test 11. You have learned what it means to say that equality is "reflexive," "symmetric," and "transitive." Answer the following questions about the relationship "$\neq$," which means, "is *not* equal to."

a. Is $\neq$ reflexive? Justify your answer.
b. Is $\neq$ symmetric? Justify your answer.
c. Write an example that shows why $\neq$ is *not* transitive.

Concepts Test 12. In Section 1-7 it was stated that the Transitive Property of Order is an *axiom*. By making precise definitions of $>$ and $<$, you can use other axioms to *prove* that order is transitive.

DEFINITION

Definition of $>$ and $<$:

$a > b$ if and only if $a - b$ is *positive*.
$a < b$ if and only if $b > a$.

The only other fact you need to know is that {positive numbers} is *closed* under + and ×. Supply reasons for the following proof.

Prove that if $x > y$ and $y > z$, then $x > z$.

Proof:

a. $x > y$ and $y > z$.
b. $x - y$ is positive and $y - z$ is positive.
c. $(x - y) + (y - z)$ is positive.
d. $x + (-y + y) - z$ is positive.
e. $x + 0 - z$ is positive.
f. $x - z$ is positive.
g. $x > z$, Q.E.D.

Concepts Test 13. Prove that if you multiply both members of an inequality by a *positive* number, the order does *not* change. That is, prove that if $x > y$ and z is positive, then $xz > yz$.

Concepts Test 14. Supply reasons that the order *reverses* in the steps in the following proof when you multiply both members of an inequality by a *negative* number.

Prove that if $x > y$ and z is negative, then $xz < yz$.

Proof:

a. z is negative.
b. $\therefore (-1)(z)$ is positive.
c. $\therefore -z$ is positive.
d. $x > y$.
e. $\therefore x - y$ is positive.
f. $\therefore (x - y)(-z)$ is positive.
g. $\therefore -xz + yz$ is positive.
h. $\therefore yz + (-xz)$ is positive.
i. $\therefore yz - xz$ is positive.
j. $\therefore yz > xz$.
k. $\therefore xz < yz$, Q.E.D.

Concepts Test 15. Use the result of Problem Concepts Test 14 as a lemma to prove that if $x < y$ and z is negative, then $xz > yz$.

Functions and Relations

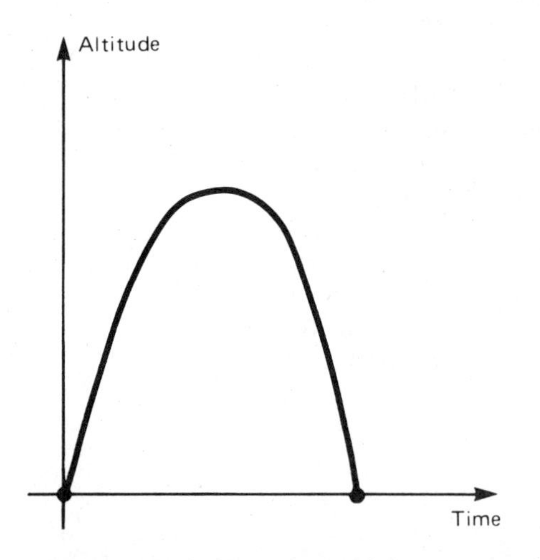

In Chapter 1 you refreshed your memory about properties of numbers, and how the properties are used to evaluate expressions and solve equations. Now you will concentrate on equations that have ***two*** *variables. These equations tell how one variable is* ***related*** *to the other. In Exercise 2-2 you will draw graphs showing the relationship between variables such as the altitude of a football and the time since it was kicked, or the temperature of the water and the time since the "hot" faucet was turned on.*

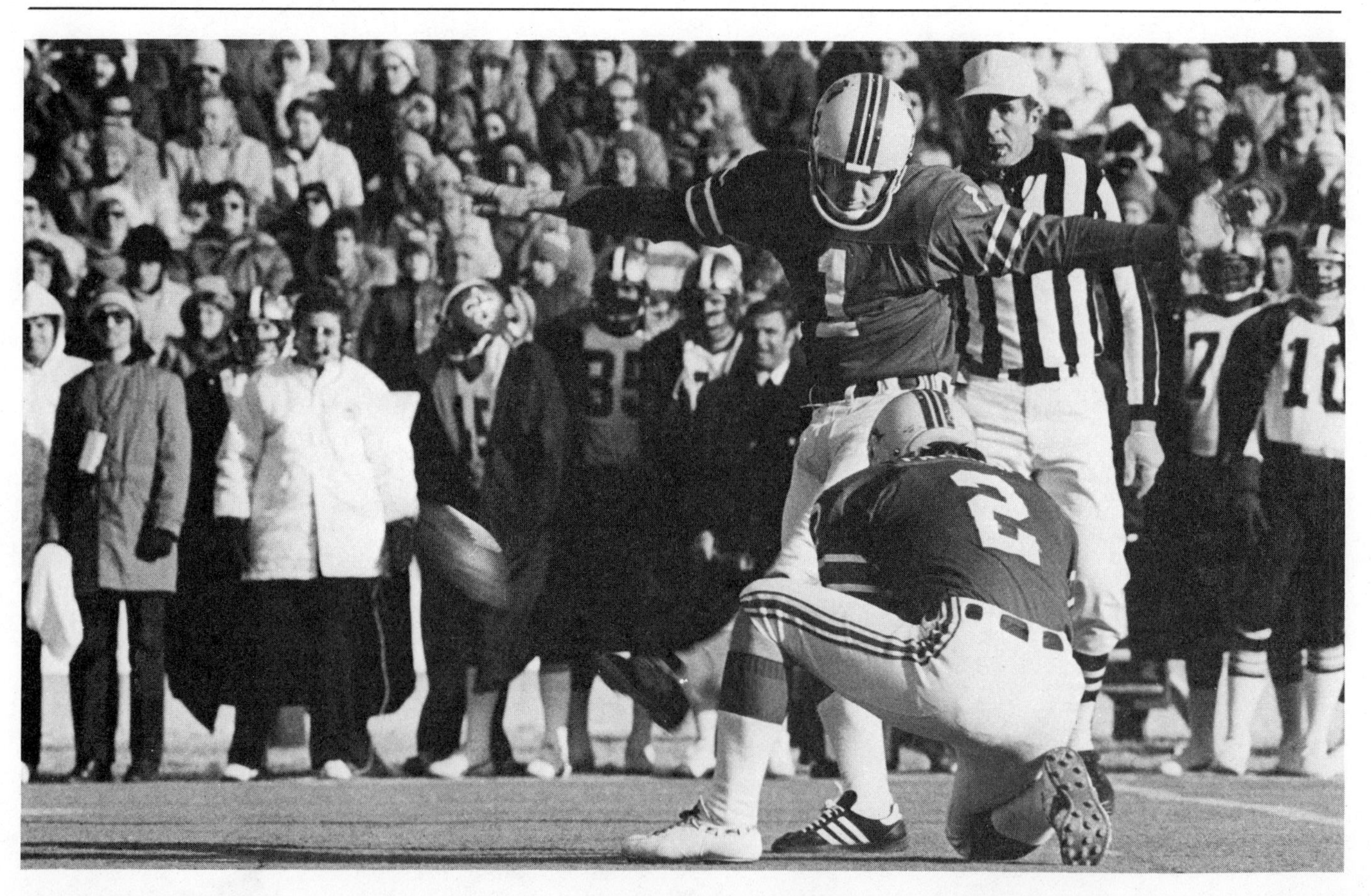

2-1 GRAPHS OF EQUATIONS WITH TWO VARIABLES

You have learned how to plot number-line graphs of equations with *one* variable. In this section you will plot a graph of an equation with *two* variables.

Objective:

Use what you recall from previous mathematics courses to draw the graph of an equation with two variables.

In order to satisfy the equation

$$2x - 3y = 13,$$

it is necessary to specify values for both x and y. For instance, if $x = 8$, then

$2 \cdot 8 - 3y = 13$	Substituting 8 for x.
$-3y = -3$	Subtracting 16.
$y = 1.$	Dividing by -3.

So $x = 8$ and $y = 1$ is a solution of $2x - 3y = 13$.

It is customary to write the values of the two variables as *ordered pairs,* such as (8, 1), where the first number stands for a value of x and the second for a value of y.

The solution set of an equation with two variables contains all the ordered pairs that make the equation true. Since there are *two* numbers in each ordered pair, the graph will consist of points in a *two*-dimensional plane, as in Figure 2-1, rather than a one-dimensional number line. The first coordinate ("abscissa") in the ordered pair is plotted horizontally and the second coordinate ("ordinate") is plotted vertically. You may recall from previous mathematics courses that this plane is called a Cartesian coordinate system, after the French mathematician Rene Descartes who lived from 1596 to 1650.

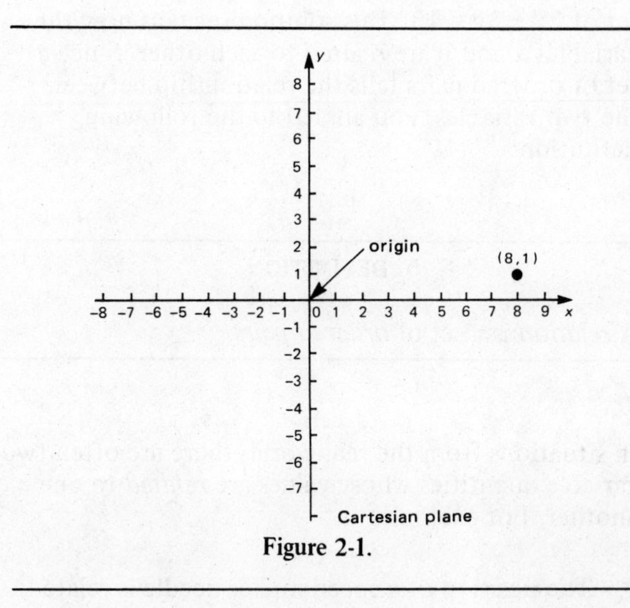

Figure 2-1.

In the following exercise you will plot more solutions of the equation $2x - 3y = 13$, and try to see what pattern the points follow.

Exercise 2-1

1. Show that the ordered pair (8, 1) satisfies the equation $2x - 3y = 13$. Do this by substituting 8 for x and 1 for y, and showing that you get a *true* statement.

2. Show that the ordered pair (1, 8) does *not* satisfy the equation $2x - 3y = 13$.

3. Substitute 5 for x in $2x - 3y = 13$, and solve for y. Then draw a Cartesian coordinate system as in Figure 2-1, and plot this point on it.

4. Repeat Problem 3 for $x = 2$, -1, and -4. Use the same Cartesian coordinate system.

5. Connect the points you have drawn on the Cartesian coordinate system of Problem 3. If they do not all lie on the same straight line, go back and check your work!

2-2 MATHEMATICAL RELATIONS

In Exercise 2-1, you plotted a graph of the solution set of $2x - 3y = 13$. This solution set tells how the variables x and y are *related* to each other. Since a set of ordered pairs tells the relationship between the two variables, you are led to the following definition:

DEFINITION

A ***relation*** is a set of *ordered pairs.*

In situations from the real world, there are often two variable quantities whose values are *related* to one another. For example:

1. The position of a speedometer needle is related to how fast the car is going.

2. The distance you travel is related to how long you have been traveling (and to how fast you go, also).

3. The weight of a person is related to his or her height.

4. How hard you hit your thumb with a hammer is related to how badly it hurts!

Very often, the value of one of the variables *depends* on the value of the other. For example, the speedometer needle position *depends* on how fast you are driving. The speed can be varied *independently,* and the needle follows along. But the speed does *not* depend on the needle position. Removing the glass and moving the needle would *not* change the speed of the car! Consequently, the needle position is called the *dependent* variable and the speed is called the *independent* variable.

Objective:

Given a situation from the real world similar to those above, sketch a reasonable graph showing how the dependent variable is related to the independent variable.

Example 1: The time it takes you to get home from the football game is related to how fast you drive. Sketch a reasonable graph showing how time and speed are related.

The first step is to decide which variable is dependent. You ask yourself which of the following is more reasonable:

"How long it takes depends on how fast I go."

"How fast I go depends on how long it takes."

Most people feel that the first sentence is more reasonable, and pick *time* as the dependent variable; speed is thus the independent variable.

The next step is deciding which axis to use for which variable. So that everybody will do the same thing, you simply make an agreement.

Agreement: The *independent* variable is plotted on the *horizontal* axis. The *dependent* variable is plotted on the *vertical* axis.

Note: By a previous agreement, you plot the *first* coordinate of an ordered pair (the "abscissa") on the *horizontal* axis, and the *second* coordinate (the "ordinate") on the *vertical* axis. Combining these two agreements, you must use the first coordinate of an ordered pair for the independent variable, and the second coordinate for the dependent variable.

Figure 2-2a shows the axes labeled with the correct variable names.

You decide what the graph looks like by thinking about what happens to the time as your speed varies. For example, at some moderate speed it will take you a moderate amount of time to get home. Therefore, you put a point somewhere in the first quadrant (Figure 2-2a).

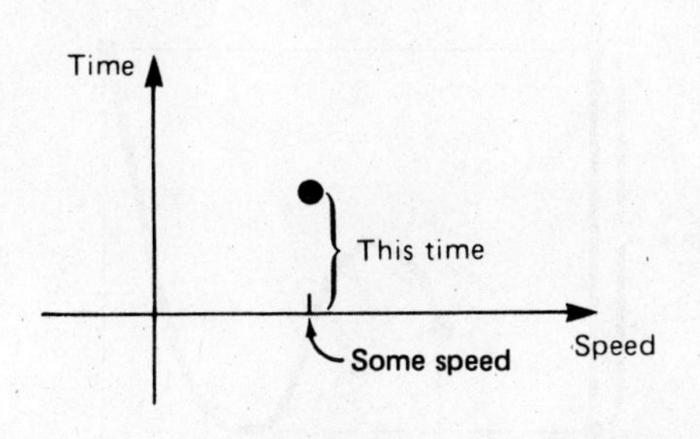

Figure 2-2a.

If you go *faster*, it will take you *less* time, so there will be another point to the *right* of and *below* the first one (Figure 2-2b).

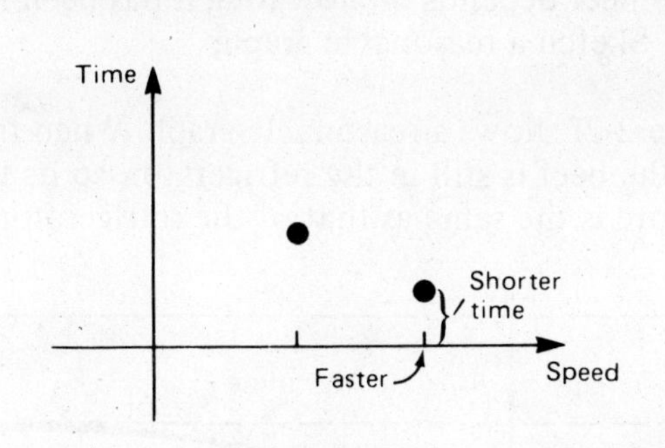

Figure 2-2b.

If you go *slower*, it will take *more* time, so there will be a point to the *left* of and *above* the first point (Figure 2-2c).

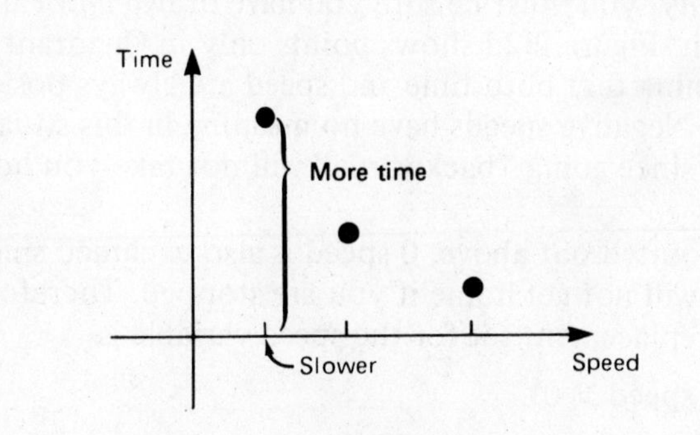

Figure 2-2c.

When you have enough points to tell what the graph looks like, you connect them with a line or curve. Figure 2-2d shows the completed graph. Since it always takes you *some* amount of time no matter *how* fast you drive, the graph never touches the horizontal axis or the vertical axis. Any point on the vertical axis has speed = 0, and you would *never* get home if you remained stopped!

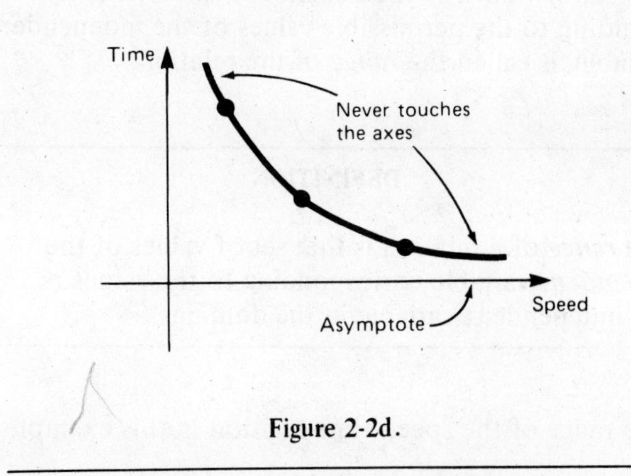

Figure 2-2d.

A line that a graph gets closer and closer to but never touches as x or y gets very large is called an *asymptote.* The time and speed axes in Figure 2-2d are both asymptotes.

Finally, you must be sure you have drawn *all* of the graph. Figure 2-2d shows points only in Quadrant I, meaning that both time and speed are always positive. Negative speeds have no meaning in this situation since going "backwards" will not take you home.

As pointed out above, 0 speed is also excluded since you will not get home if you are stopped. Therefore, the replacement set for the speed variable is

{speed > 0}.

The set of permissible values for the *independent* variable is called the *domain* of the relation.

DEFINITION

The ***domain*** of a relation is the set of values of the *independent* variable.

As you recall, the word "domain" comes from the Latin word "domus," meaning "home."

The set of values of the dependent variable, corresponding to the permissible values of the independent variable, is called the *range* of the relation.

DEFINITION

The ***range*** of a relation is that set of values of the *dependent* variable corresponding to the values of the independent variable in the domain.

The range of the speed-time relation in this example is

{time > 0}.

This range is reasonable since you could never get home *before* you started or at the *instant* you started.

Since the graph in Figure 2-2d uses all permissible values of speed in the domain, the graph is complete.

Figure 2-2e is intended to help you keep in mind the definitions of domain and range.

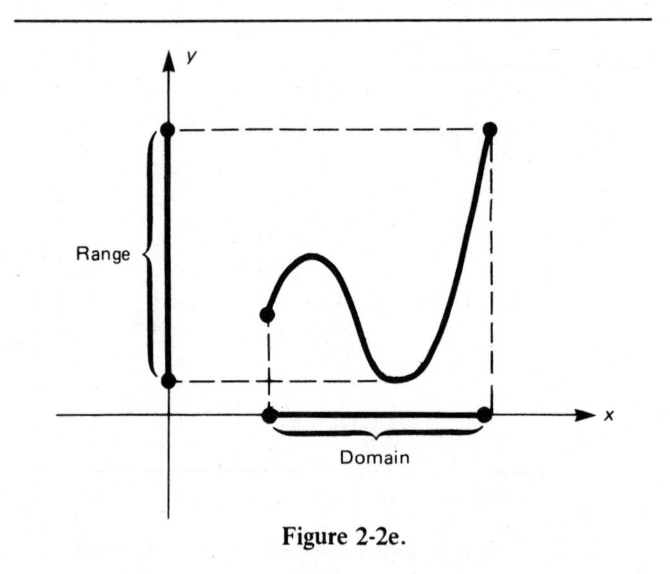

Figure 2-2e.

Example 2: You take a roast beef from the refrigerator and put it into a hot oven. The temperature of the beef depends on how long it has been in the oven. Sketch a reasonable graph.

Figure 2-2f shows a reasonable graph. When time < 0, the beef is still in the refrigerator, so its temperature is the same as that of the refrigerator. For

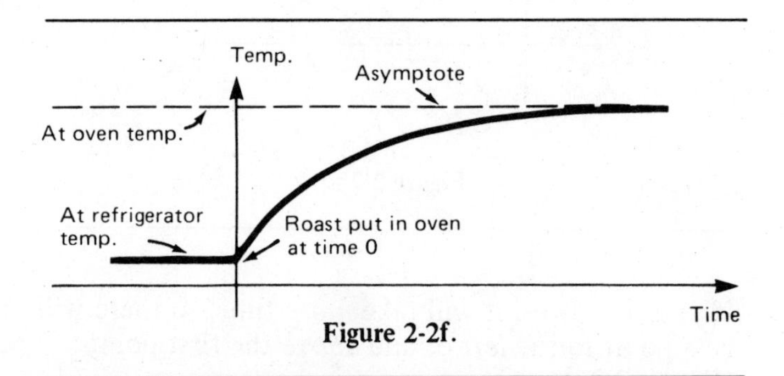

Figure 2-2f.

time > 0, the beef warms up rapidly at first, then more slowly, and finally approaches the oven temperature very gradually. It is debatable whether the beef ever actually *reaches* oven temperature, or just gets so close that nobody can tell the difference. Thus, the dotted line at oven temperature is an *asymptote.* The domain in this case includes both positive and negative values of time. The range is the set of temperatures between refrigerator temperature and oven temperature. The temperatures *could* be *negative* if the beef had been in the freezer. You might be able to think of ways to make the graph even more reasonable in that case!

In the exercise which follows, you will obtain practice sketching reasonable graphs of real-world situations in accordance with the objective of this section. Many of these real-world situations will appear in later chapters when you study about relations which have graphs like these.

Exercise 2-2

Sketch a reasonable graph.

1. a. The distance you have gone depends on how long you have been going (at a constant speed).
   b. The distance required to stop your car depends on how fast you are going when you apply the brakes.
   c. Your car is standing on a long, level highway. You start the motor and floorboard the gas pedal. The speed you are going depends on the number of seconds that have passed since you stepped on the gas pedal.
   d. The maximum speed your car will go depends on how steep a hill you are going up or down.
   e. The distance you are from the band and how loud it sounds to you are related.
   f. Your age and your height are related to one another.
   g. The price you pay for a carton of milk depends on how much milk the carton holds.
   h. The price you pay for a pizza depends on the diameter of the pizza.
   i. You climb to the top of the 190-meter tall Tower of the Americas and drop your algebra book off. The distance the book is above the ground depends on the number of seconds that have passed since you dropped it.
   j. The temperature of your cup of coffee is related to how long it has been cooling.
   k. The time of sunrise depends on the day of the year.

2. a. The number of used aluminum cans you collect and the number of dollars refunded to you are related.
   b. The mass of a person of average build depends on his or her height.
   c. The altitude of a punted football depends on the number of seconds since it was kicked.
   d. Dan Druff's age and the number of hairs he has growing on his head are related.
   e. You fill up your car's gas tank and start driving. The amount of gas you have left in the tank depends on how far you have driven.
   f. You pull the plug out of the bathtub. The amount of water remaining in the tub and the number of seconds since you pulled the plug are related to each other.
   g. Calvin Butterball desires to lose some weight, so he reduces his food intake from 8000 Calories per day to 2000 Calories per day. His weight depends on the number of days that have elapsed since he reduced his food intake.
   h. The distance you are from the reading lamp and the amount of light it shines on your book are related.
   i. As you blow up a balloon, its diameter and the number of breaths you have blown into it are related.
   j. You turn on the hot water faucet. As the water runs, its temperature depends on the number of seconds it has been since you turned on the faucet.

k. As you breathe, the volume of air in your lungs depends upon time.

3. a. You start running, and go as fast as you can for a long period of time. The number of minutes you have been running and the speed you are going are related to each other.
   b. You run the mile once each day. The length of time it takes you to run it depends on the number of times you have run it in practice.
   c. The rate at which you are breathing depends on how long it has been since you finished running a race.
   d. Milt Famey pitches his famous fast ball to Stan Dupp, who hits it for a home run. The number of seconds that have elapsed since Milt released the ball and its distance from the ground are related.
   e. The amount of water you have put on your lawn depends on how long the sprinkler has been running.
   f. A leading soft drink company comes out with a new product, Ms. Phizz. They figure that there is a relationship between how much of the stuff they sell and how many dollars they spend on advertising.
   g. The number of letters in the corner mailbox depends upon the time of day.
   h. The number of cents you pay for a long distance telephone call depends on how long you talk.
   i. You go from the sunlight into a dark room. The diameter of your pupils and the length of time you have been in the room are related.
   j. You pour some cold water from the refrigerator into a glass, but forget to drink it. As the water sits there, its temperature depends on the number of minutes that have passed since you poured it.
   k. You pour some popcorn into a popper and turn it on. The number of pops per second depends on how long the popper has been turned on.

4. a. As you play with a yo-yo, the number of seconds that have passed and the yo-yo's distance from the floor are related.
   b. When you dive off the 3-meter diving board, time, and your position in relation to the water's surface, are related to each other.
   c. Taryn Feathers catches the bus to work each morning. Busses depart every 10 minutes. The time she gets to work depends on the time she leaves home.
   d. The amount of postage you must put on a first-class letter depends on the weight of the letter.
   e. You plant an acorn. The height of the resulting oak tree depends on the number of years that have elapsed since the planting.
   f. Your car stalls, so you get out and push. The speed at which the car goes depends on how hard you push.
   g. The diameter of a plate and the amount of food you can put on it are related to each other.
   h. The grade you could make on a particular test depends upon how long you study for it.
   i. The grade you could make on a particular test depends on how much time elapses between the time you study for it and the time you take the test.
   j. Your efficiency at studying algebra depends on how late at night it is.
   k. Your feelings of affection for your date are related to the time of evening.

2-3 MATHEMATICAL FUNCTIONS

In the preceding section you drew graphs of mathematical relations. As it turns out, the graphs of all the relations in Exercise 2-2 have one thing in common. If you pick a value of the *independent* variable, you get *only one* value of the dependent variable. For example, a punted football is never at two different altitudes at the same time (Figure 2-3a). Relations that have this property are called *functions.*

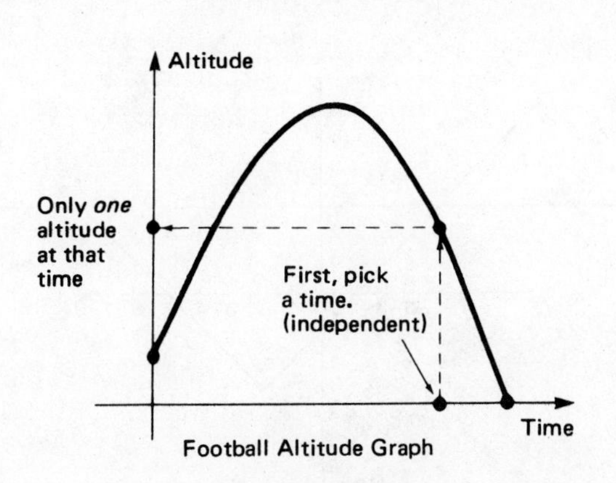

Figure 2-3a.

DEFINITION

A ***function*** is a relation in which there is *exactly one* value of the dependent variable for each value of the independent variable in the domain.

Figure 2-3b shows a relation that is *not* a funtion. If there is any place in the domain where you can pick a value of x and get *two* or more values of y, then the relation is *not* called a function.

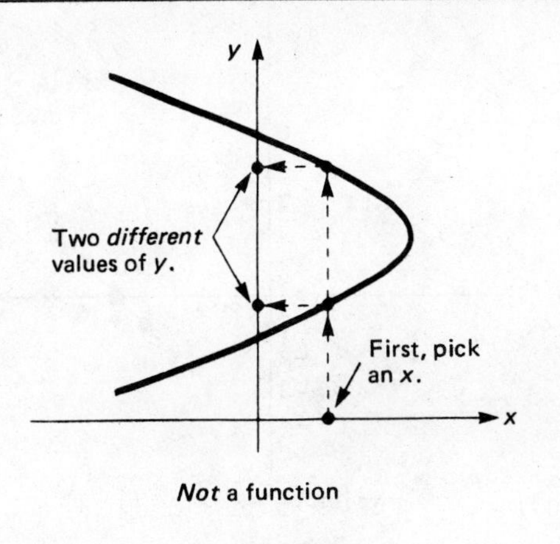

Figure 2-3b.

Objectives:

1. Given the graph of a relation, tell whether or not the relation is a function.

2. Given the equation of a relation, and the domain of the independent variable,

a. plot the graph of the relation,
b. tell the range of the relation,
c. tell whether or not the relation is a function.

Example 1: Tell whether or not the relations graphed in Figures 2-3c and 2-3d are functions.

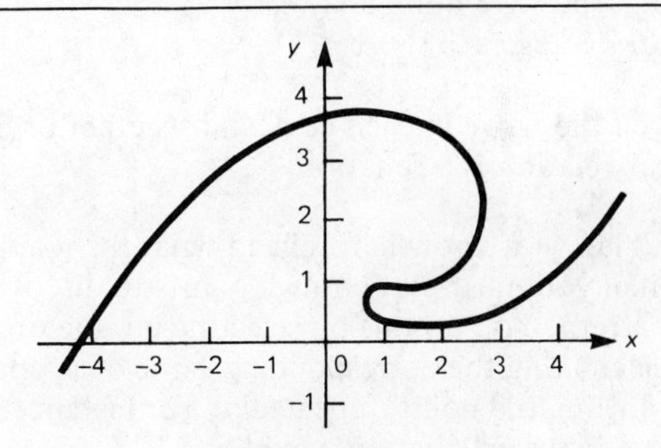

Figure 2-3c.

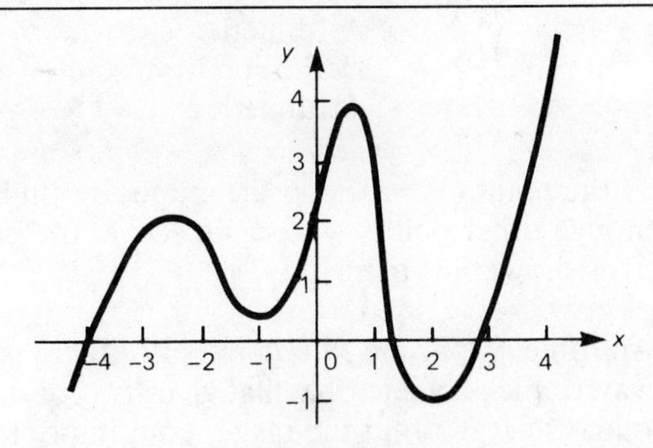

Figure 2-3d.

The relation in Figure 2-3c is *not* a function. For example, if you pick $x = 2$, you get *three* different values of y. The relation in Figure 2-3d *is* a function. There are no places in the domain where you can pick a value of x that has two or more different values of y. Note that even though Figure 2-3d has several values of x when y equals 2, this fact has *no* bearing on whether or not the relation is a function.

Example 2: a. Plot the graph of the relation whose equation is

$$y = |x - 2| - 3,$$

if the domain of x is

i. {real numbers},
ii. {positive numbers},
iii. {integers, $-1 \leq x \leq 7$}.

b. Tell the range in each case, and whether or not the relation is a function.

a. Unless you know beforehand what the graph looks like, you must plot enough points to discover a pattern. You do this by selecting a value of x, calculating the corresponding value of y, and plotting the point, repeatedly. For instance, if $x = 1$, the calculation would be:

$y = \|1 - 2\| - 3$	Substitute 1 or x.
$= \|-1\| - 3$	Arithmetic.
$= 1 - 3$	Definition of absolute value.
$= -2.$	Arithmetic.

So the point $(1, -2)$ is on the graph. By finding enough other points, you can discover the patterns shown in Figure 2-3e.

b. All three graphs are *functions,* since there is no way to pick a value of x that gives two or more values of y. To find the range, you simply look at the graph and see how high and how low it goes. For the first case, the graph goes as low as $y = -3$, and goes up forever. Therefore, the range is

$$\{y \geq -3\}$$

as shown in Figure 2-3f. The range is the same for the second case, for the same reasons. How-

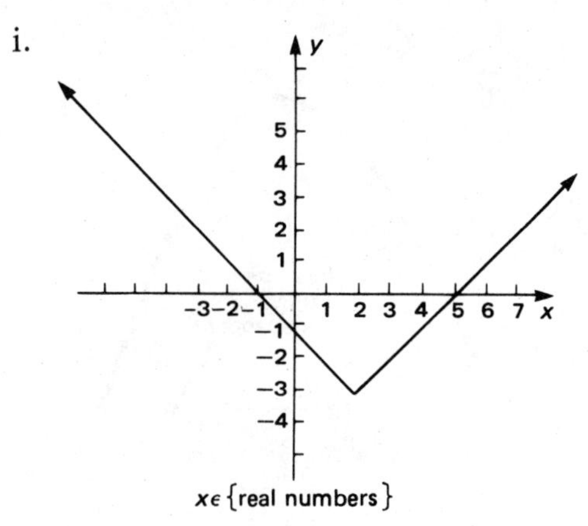

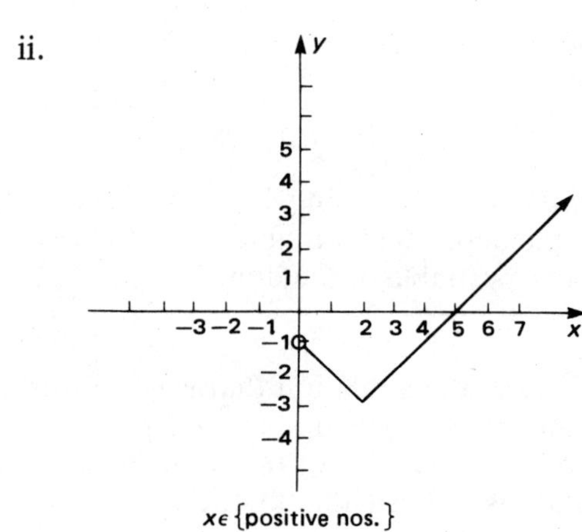

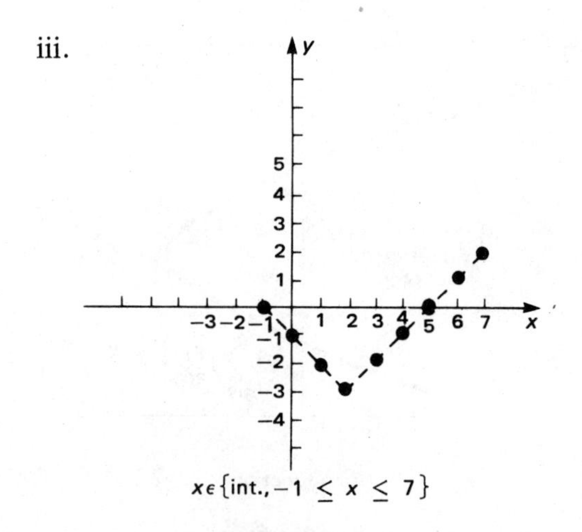

Figure 2-3e.

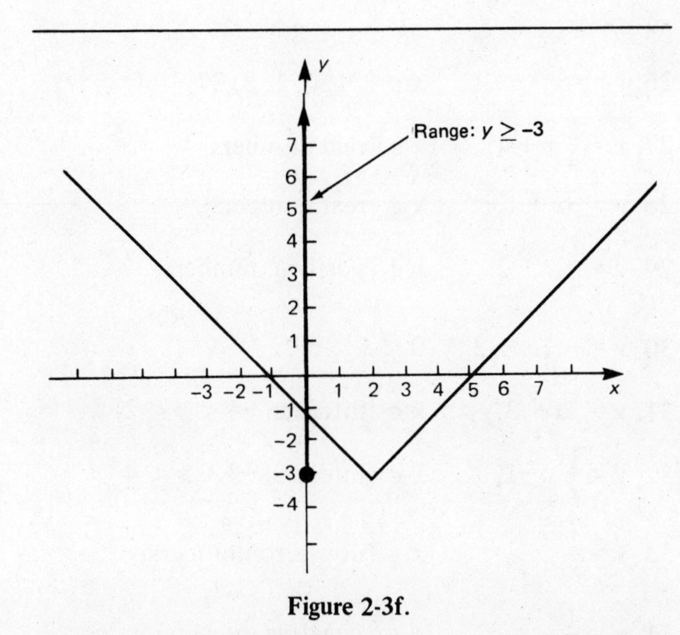

Figure 2-3f.

ever, for the last case, y takes on only the values −3, −2, −1, 0, 1, and 2. So for this relation, the range is

$$\{-3, -2, -1, 0, 1, 2\}.$$

The following exercise is designed to give you practice plotting graphs of relations, and identifying those that are functions.

Exercise 2-3

For Problems 1 through 16, tell whether or not the relation graphed is a function.

1.

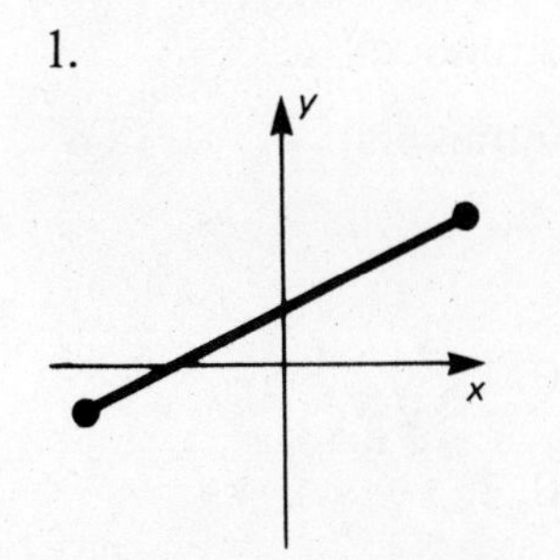

2.

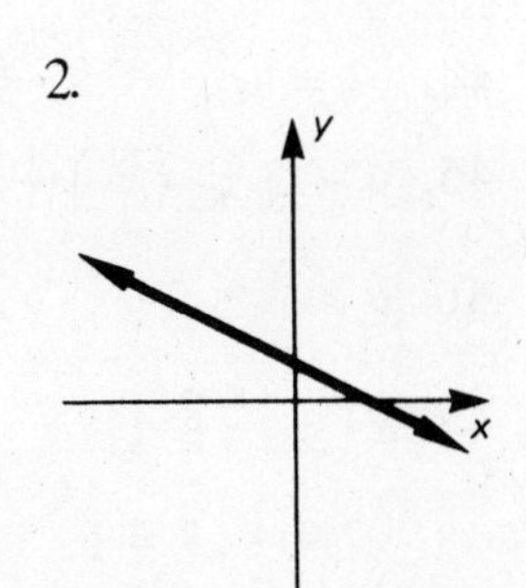

3.

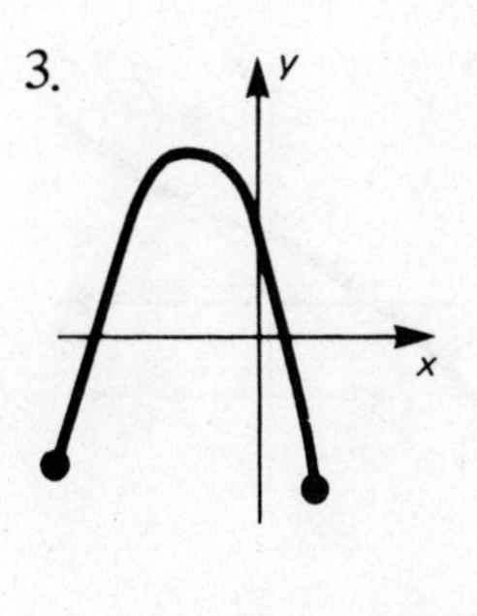

4.

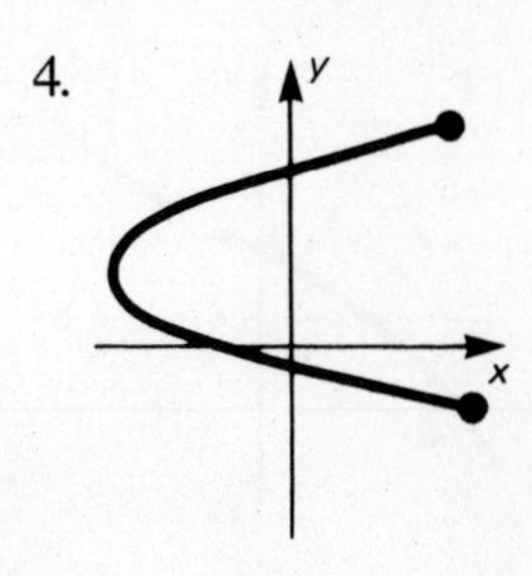

5.

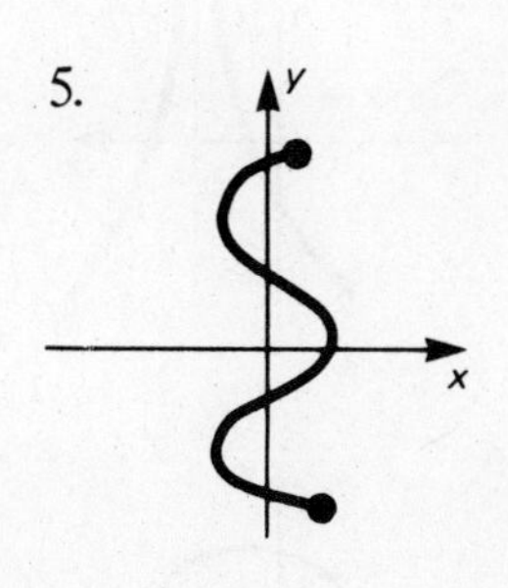

6.

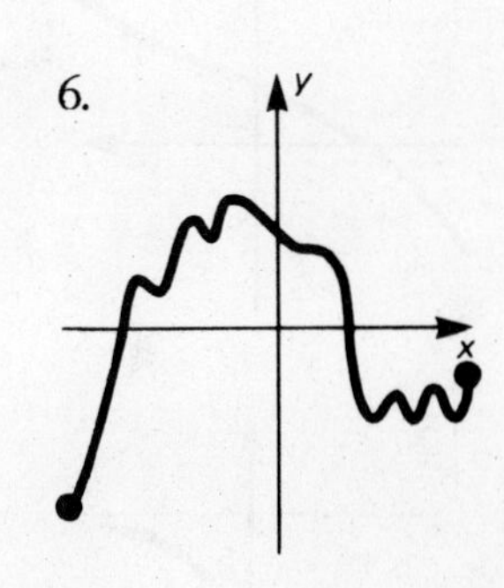

7.

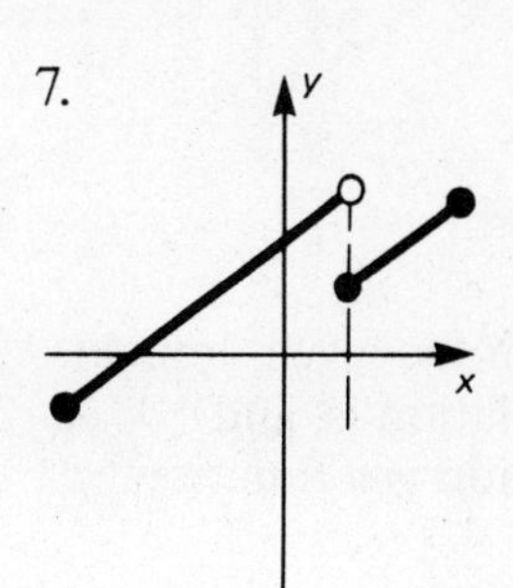

8.

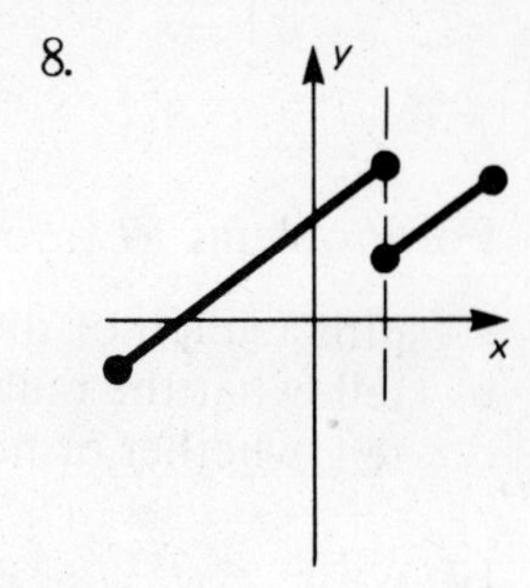

9.

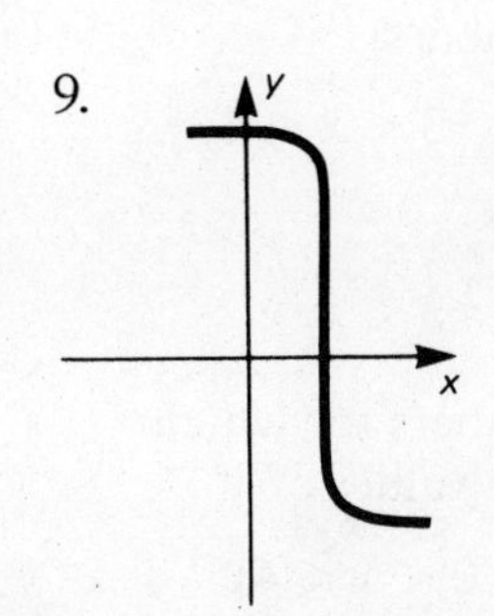

10.

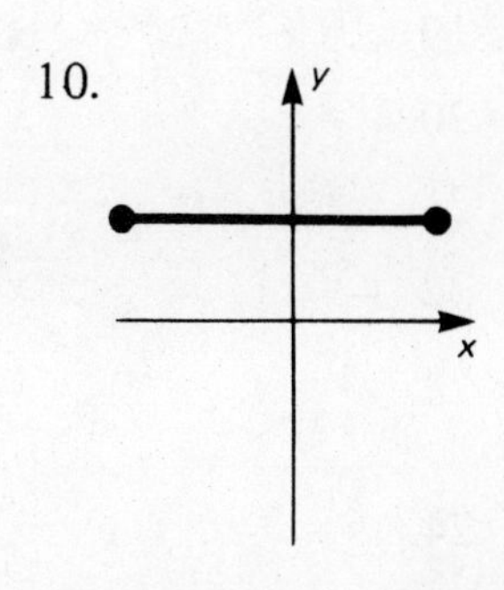

11.

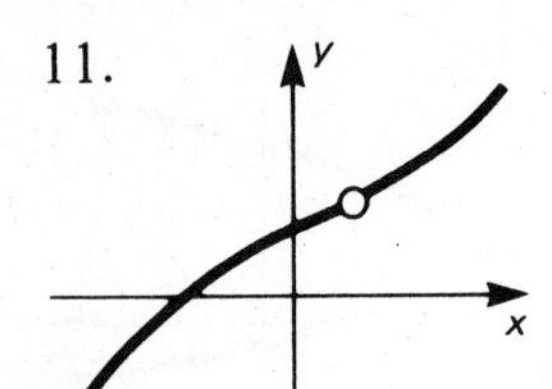

12.

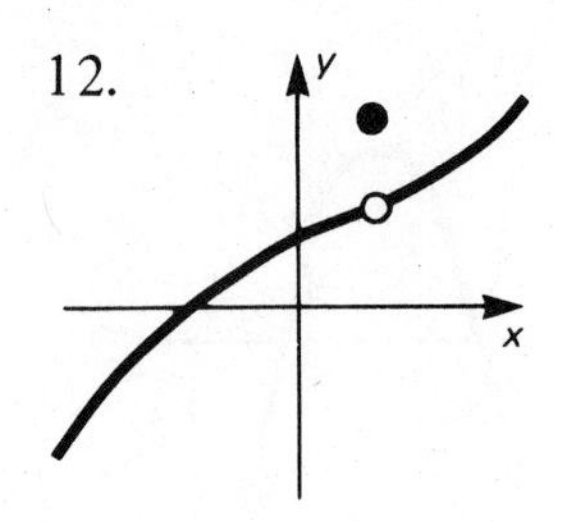

13.

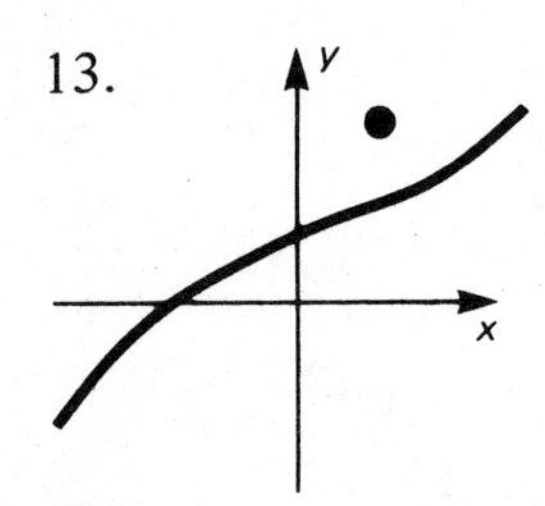

14.

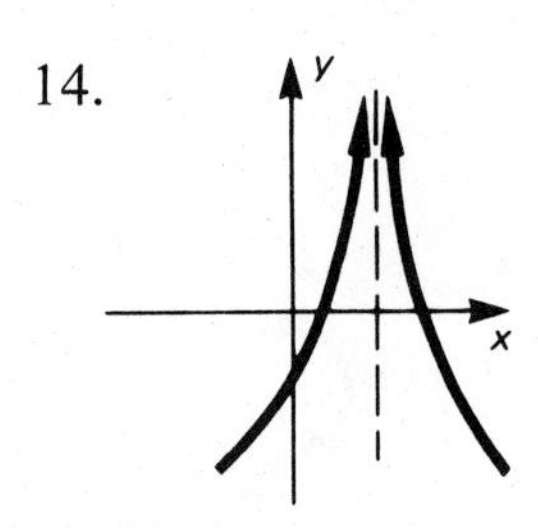

15.

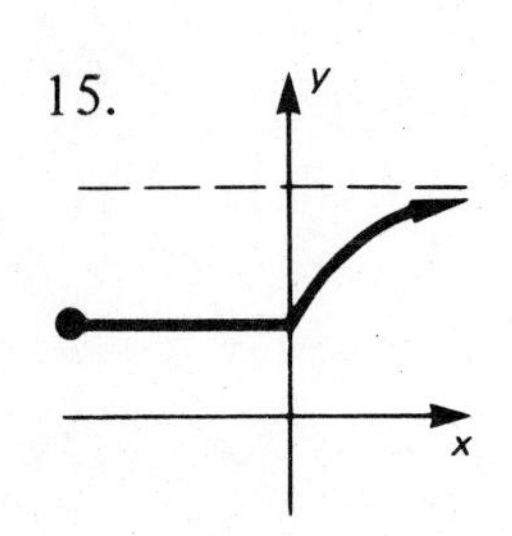

16.

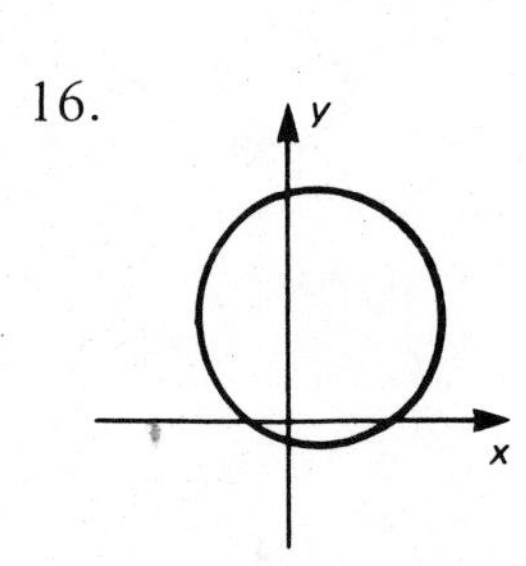

For Problems 17 through 48.

a. plot a graph of the relation in the given domain,
b. tell what the range of the relation is, and
c. tell whether or not the relation is a function.

17. $y = x$, $\quad x \in \{\text{real numbers}\}$

18. $y = x$, $\quad x \in \{\text{integers}\}$

19. $y = x^2$, $\quad x \in \{\text{real numbers}\}$

20. $y = x^2$, $\quad x \in \{\text{integers}, -3 \leq x \leq 4\}$

21. $y = x^3$, $\quad -2 \leq x \leq 2$

22. $y = x^3$, $\quad x \in \{\text{integers}, -3 \leq x \leq 2\}$

23. $y^2 = x$, $\quad x \in \{\text{real numbers for which there are values of } y\}$

24. $y^2 = x$, $\quad x \in \{\text{integers}, 0 \leq x \leq 4\}$

25. $y^3 = x$, $\quad -8 \leq x \leq 8$

26. $y^3 = x$, $\quad x \in \{-1, 0, 1, 8, 27\}$

27. $y = \frac{2}{3}x + 4$, $\quad x \in \{\text{real numbers}\}$

28. $y = -x + 1$, $\quad x \in \{\text{real numbers}\}$

29. $y = \frac{1}{4}x + 2$, $\quad x \in \{\text{positive numbers}\}$

30. $y = -\frac{1}{3}x + 2$, $\quad 0 \leq x \leq 6$

31. $y = 2x + 3$, $\quad x \in \{\text{integers}, -4 \leq x \leq 2\}$

32. $y = \frac{1}{2}x - 1$, $\quad x \in \{\text{integers}, -3 \leq x \leq 4\}$

33. $y = \frac{1}{x}$, $\quad x \in \{\text{non-zero numbers}\}$

34. $y = \frac{1}{x}$, $\quad x \in \{\text{negative integers}\}$

35. $y = |x| + x$, $\quad x \in \{\text{real numbers}\}$

36. $y = |x| - 1$, $\quad x \in \{\text{real numbers}\}$

37. $y = |x - 3| - 5$, $\quad x \in \{\text{positive integers}\}$

38. $y = |x + 2| - 1$, $\quad -5 < x < 3$

39. $|y| = x$, $\quad x \in \{\text{real numbers for which there are values of } y\}$

40. $|y| = |x|$, $\quad x \in \{\text{real numbers for which there are values of } y\}$

41. $y = |x^2 - 4|$, $\quad -3 \leq x \leq 3$

42. $y = x^2 - 4x$, $\quad x > 0$

43. $y = x - \left|\frac{1}{2}x\right|$, $\quad x \in \{\text{integers}, -4 \leq x \leq 4\}$

44. $y^2 = |x|$, $\quad x \in \{\text{real numbers}\}$

45. $y = \big|3 - |x - 1|\big|$, $\quad x \in \{\text{real numbers}\}$

46. $y = \frac{x}{|x|}$, $\quad x \neq 0$

47. $x^2 + y^2 = 1$, $\quad x \in \{-1, 0, 1\}$

48. $x^2 + \frac{1}{4}y^2 = 1$, $\quad x \in \{-1, 0, 1\}$

2-4 CHAPTER REVIEW AND TEST

In this chapter you have been introduced to the concept of a mathematical *function*. A function is a special kind of *relation*, or set of ordered pairs. A relation can be specified by an equation that tells how the two variables are related. The graph of such an equation can be plotted by calculating enough points to discover a pattern. Once you have plotted the graph, you can tell whether or not the relation is a function by seeing whether there are any values of x that have more than one value of y. Functions are important because they can be used to describe the relationship between two variable quantities in the real world.

The objectives of this chapter may be summarized as follows:

1. *Plot the graph of a given equation.*
2. *Tell whether or not a given graph is a function graph.*
3. *Draw reasonable graphs of real-world situations.*

Following are two sets of problems. The Review Problems are similar to those you have worked in this chapter. They are numbered according to the three objectives above. The Concepts Test requires you to put together two or more of these techniques (or to use concepts learned before), to work problems that are somewhat different. One of the most important things you should learn during your education is how to apply your knowledge to new situations!

Review Problems

The following problems are numbered according to the three objectives listed above.

Review Problem 1. Plot the graph of the given equation in the indicated domain. Tell the corresponding range and whether or not the relation is a function.

a. $x + y = 2,\quad -1 \leq x \leq 3$

b. $x^2 + y = 2,\quad x \in \{-2, -1, 0, 1, 2\}$

c. $x + y^2 = 2,\quad x \in \{-2, 1, 2\}$

d. $x + |y| = 2,\quad -2 \leq x \leq 2$

Review Problem 2. Tell whether or not the relation graphed is a function.

a.

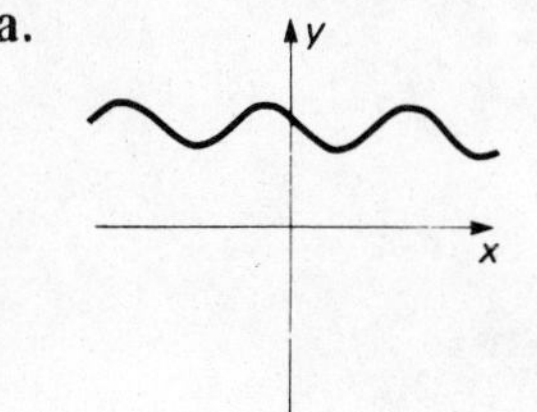

b.

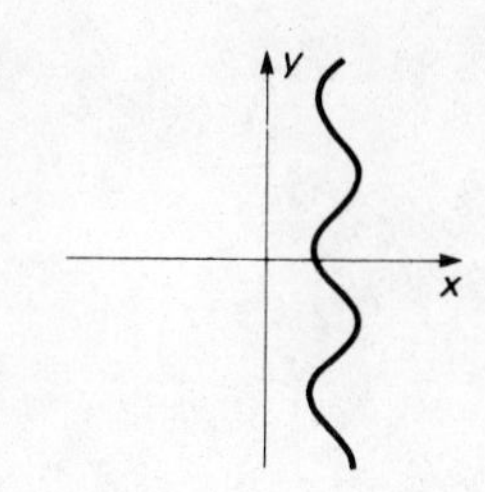

c.

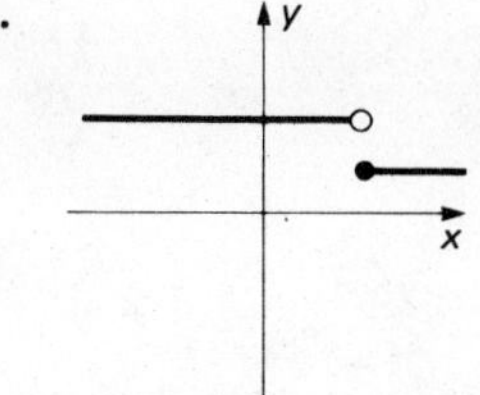

d.

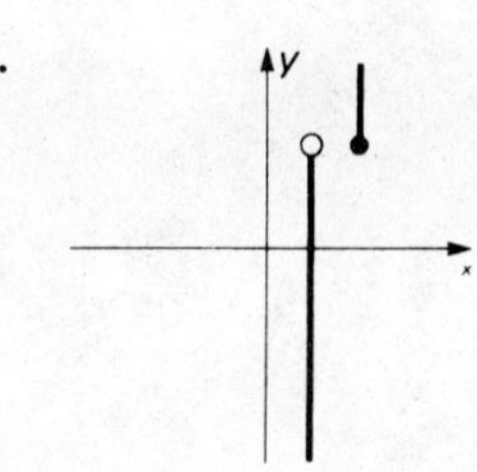

e.

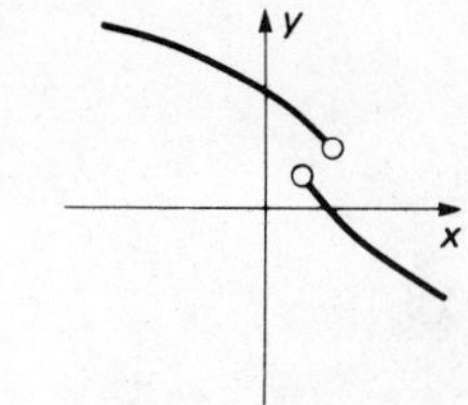

f.

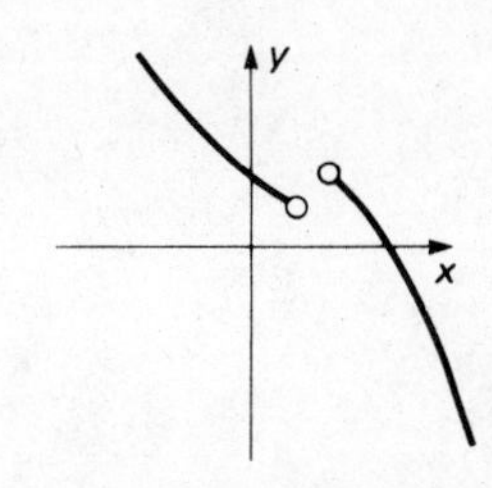

Review Problem 3. For each of the following, sketch a reasonable graph showing how the *dependent* variable is related to the *independent* variable.

a. Your car stalls, and you roll to a stop without putting on the brakes. The car's speed depends on how long it has been since it stalled.

b. The length of time an oven has been turned on and the oven's temperature are related.

c. The angle at which you have to look up to see the Sun depends on the time of day. Make the domain extend for *several* days.

d. The number of cars in the student parking lot and the time of day are related.

Concepts Test

The following problems combine all of the objectives of this chapter. For each part of each problem, do what is asked, then identify by number which one or ones of the objectives you used in working that part.

Concepts Test 1. Consider the relation whose equation is

$$y = 1 + 4x - x^2.$$

a. Plot the graph of this relation, assuming that the domain is the set of non-negative numbers, and that the range must also contain only non-negative numbers.

b. What is the range of this relation?

c. Is the relation a function? Justify your answer.

d. At what value of y does the graph touch the y-axis?

e. At approximately what value of x does the graph touch the x-axis?

f. Tell *two* different real-world situations having graphs looking like this one.

Concepts Test 2 – Introduction to Polynomial Functions. A function is called a "polynomial function" if it has an equation of the form

y = a polynomial involving x.

a. Which of the relations in Problems 16-47 of Exercise 2-3 are polynomial functions?

b. Explain how closure insures that the domain of a polynomial function can be the set of *all* real numbers.

c. Explain how closure insures that a polynomial function really *is* a function.

Linear Functions

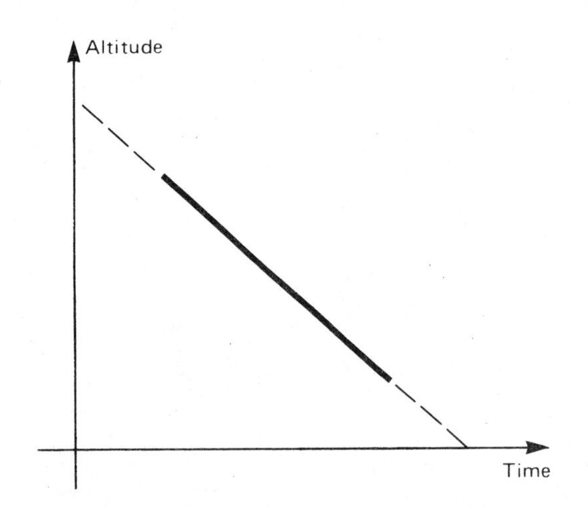

*In Chapter 2 you learned that a **function** relates two variables. In this chapter you will study a special kind of function, the **linear** function. Your ultimate objective is to be able to write the **equation** for a linear function from given information about its graph. In Exercise 3-5 you will use this kind of equation to predict such things as how high a sky diver is.*

3-1 INTRODUCTION TO LINEAR FUNCTIONS

A function is named according to the equation which specifies how the dependent variable is related to the independent one. For example, if y were related to x by the equation

$$y = 3x + 7,$$

then the function would be called a *linear* function. This name is picked because y equals a linear polynomial in the variable x.

DEFINITION

A ***linear function*** is a function specified by an equation of the form

$$\boxed{y = mx + b,}$$

where m and b stand for constants, and $m \neq 0$.

Thus, linear functions differ only by the values assigned to the two constants m and b. Note that if m were equal to zero, then the equation would degenerate to $y = b$. Since b is a *constant* (degree 0), the function would be called a "constant function."

Objective:

Discover what the graph of a linear function looks like, and what effects the values of m and b have.

The following exercise is designed to accomplish this objective.

Exercise 3-1

1. Plot graphs of the following functions by selecting values of x, calculating the corresponding values of y, and plotting the points.

a. $y = x + 5$
b. $y = 2x + 5$
c. $y = 3x + 5$
d. $y = -2x + 5$
e. $y = 2x - 5$

2. From your graphs in Problem 1, answer the following questions:

a. Why are first-degree functions called "linear" functions?
b. What is the effect on the graph of changing the coefficient of x?
c. What is the effect on the graph of changing the constant term?
d. From what you have learned in previous mathematics courses, tell the special names given to the constants m and b in the equation $y = mx + b$.

3. If m were 0, then the graph would be a horizontal straight line, yet the function would not be called "linear." Explain *why not*.

3-2 GRAPHS OF LINEAR FUNCTIONS FROM THEIR EQUATIONS

In Exercise 3-1 you found that the graphs of linear functions are straight lines. Once you know this fact, you can draw the graph quickly, without many calculations.

Objective:

Given the equation for a linear function, draw the graph *quickly*.

In order to accomplish this objective, you need to know some *properties* of linear functions. Figure 3-2a shows some of the graphs from Exercise 3-1.

From these graphs you should be able to observe the following properties:

1. The graph of $y = mx + b$ is a *straight line.*

2. The value of m tells you by how much the graph is *tilted.*

   If $m > 0$, then the graph slopes *up* as x increases.
   If $m < 0$, then the graph slopes *down* as x increases.

   If $m = 0$, then the graph is horizontal (constant function).

3. The value of b tells you where the graph crosses the y-axis.

Because of these properties, the constants m and b are given special names.

DEFINITION

The constant m in the equation $y = mx + b$ is defined to be the ***slope*** of the linear function.

Note that this definition takes the geometrical idea of slope and replaces it with a *number.* Consequently, a horizontal line *does* have a slope (equal to 0), even though it is not tilted. It turns out that a vertical line has *no* slope (even though it is as tilted as it can get!) simply because there is no number big enough for m to make the graph go straight up and down.

The geometrical idea of "crossing the axis" can also be replaced with the more algebraic idea of a number. All points on the y-axis have $x = 0$. Substituting 0 for x in $y = mx + b$ gives $y = b$. Therefore, b is the value of y when $x = 0$.

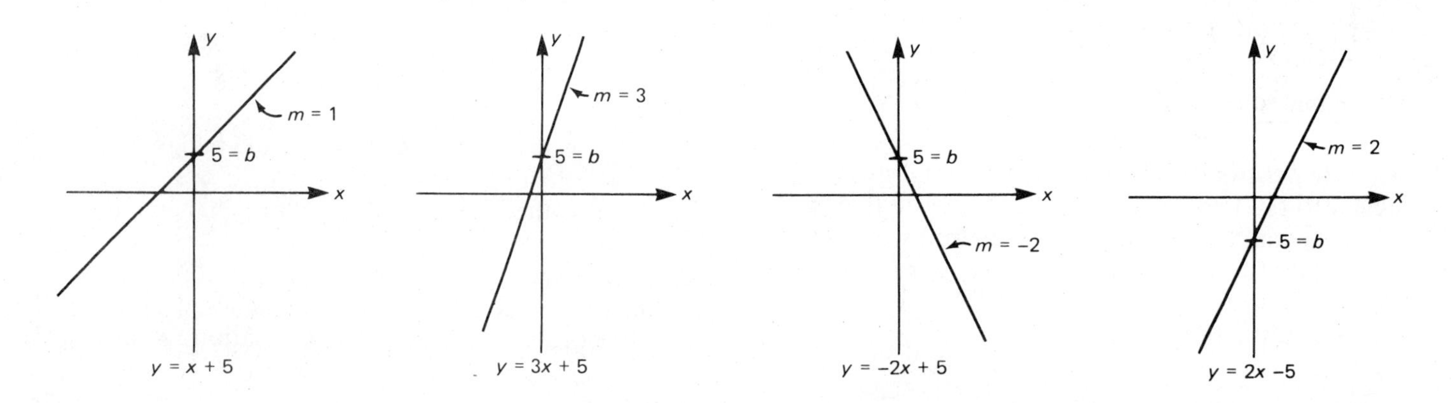

Figure 3-2a.

DEFINITION

The ***y-intercept*** of a function is the value of y when $x = 0$.

The places where the graph crosses the x-axis are defined in the same manner.

DEFINITION

The ***x-intercept*** of a function is a value of x when $y = 0$.

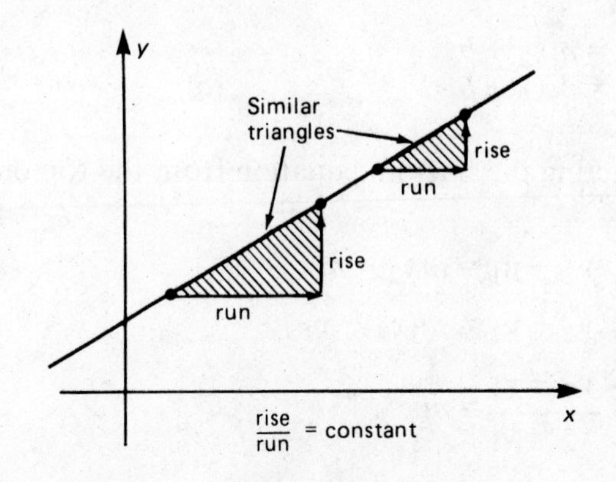

Figure 3-2b.

Another property of linear functions is illustrated in Figure 3-2b. If you start at any point on the graph and run along in the positive x direction, then rise up (or down) to another point on the graph, then the ratio

$$\frac{\text{rise}}{\text{run}}$$

will be *constant*, no matter what two points you pick. This property is a direct consequence of the properties of similar triangles which you learned in geometry.

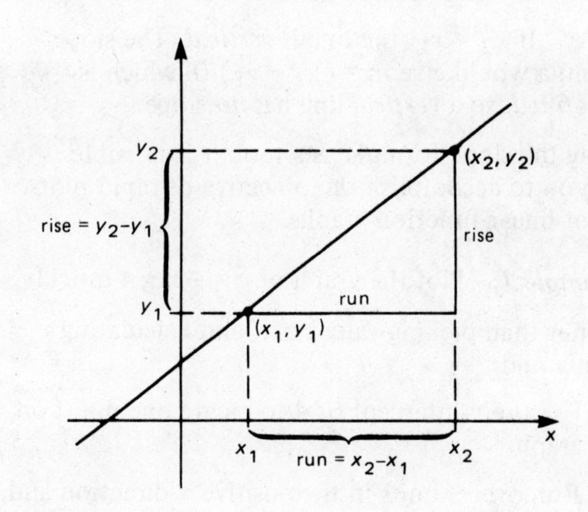

Figure 3-2c.

Although it is not obvious, the ratio rise/run also turns out to equal the slope, m, in the equation $y = mx + b$. To see why, you must first express rise and run as *numbers*.

Suppose that (x_1, y_1) and (x_2, y_2) are two points on the graph. As shown in Figure 3-2c, the *rise* is equal to the *difference* between the two y-values, $y_2 - y_1$. The run equals the difference between the two x-values, $x_2 - x_1$. The Greek letter Δ ("delta" for "difference") is sometimes used to emphasize this fact.

$$\text{rise} = y_2 - y_1 = \Delta y$$

$$\text{run} = x_2 - x_1 = \Delta x$$

The following is a proof that slope = rise/run. *You should determine the reason for each step.*

Theorem: The Slope Formula – If (x_1, y_1) and (x_2, y_2) are ordered pairs of two points on the graph of a linear function whose equation is $y = mx + b$, then

$$m = \frac{\text{rise}}{\text{run}} = \frac{y_2 - y_1}{x_2 - x_1} = \frac{\Delta y}{\Delta x}.$$

Proof:

$$y_2 = mx_2 + b$$
$$y_1 = mx_1 + b.$$

Subtracting the bottom equation from the top one gives:

$$y_2 - y_1 = mx_2 - mx_1$$
$$y_2 - y_1 = m(x_2 - x_1)$$
$$\frac{y_2 - y_1}{x_2 - x_1} = m$$
$$\therefore m = \frac{y_2 - y_1}{x_2 - x_1}, \text{ Q.E.D.}$$

Note: If $x_2 = x_1$, the line is *vertical.* The slope formula would give $m = (y_2 - y_1)/0$, which is *undefined,* so a *vertical* line has *no* slope.

Using the slope formula as a tool, it is possible for you to accomplish the objective of rapid plotting of linear function graphs.

Example 1: Plot the graph of $y = \frac{2}{3}x + 4$ quickly.

Rather than picking values of x and calculating y, you would:

1. Use the y-intercept of 4 to locate one point on the graph.
2. Run over 3 units in the positive x-direction and rise up 2 units in the y-direction to find a second point.
3. Draw a straight line through the points.

These three steps are illustrated in Figure 3-2d.

Example 2: Plot the graph of $5x + 7y = 14$ quickly.

Since the equation is not in the form $y = mx + b$, you can first transform it to that form.

$7y = -5x + 14$ Subtracting $5x$.

$y = -\frac{5}{7}x + 2$ Dividing by 7.

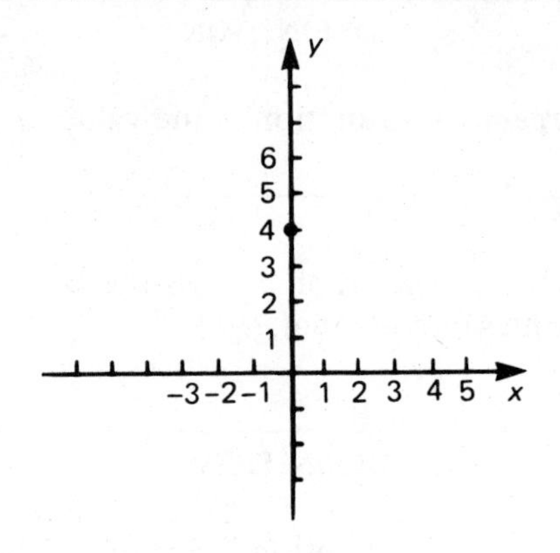

Find y-intercept

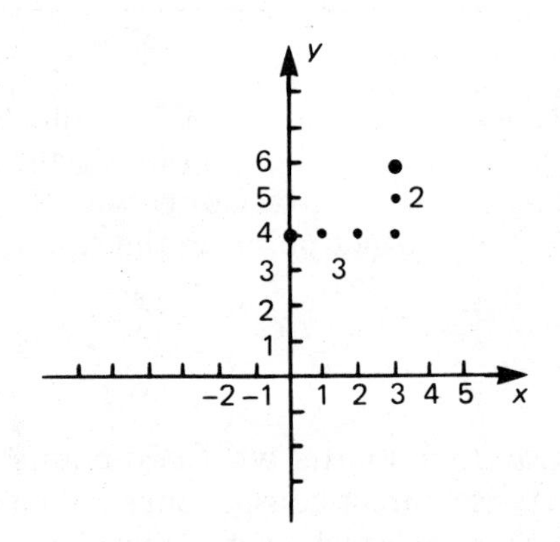

Find second point

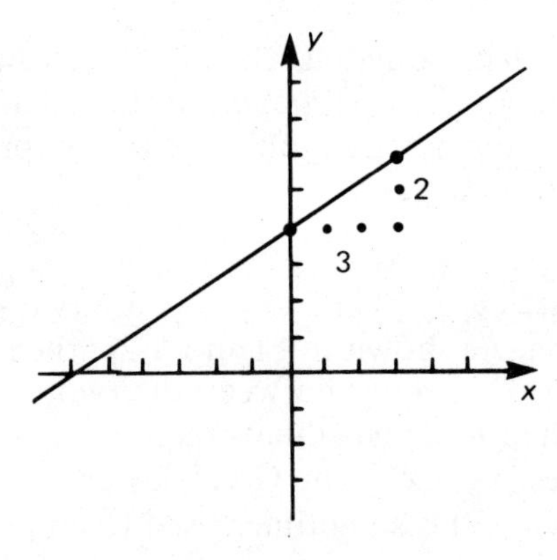

Draw the line

Figure 3-2d.

Since the slope is a *negative* number, either the rise or the run must be negative (and the other one must be positive). After you use the y-intercept to locate (0, 2) on the graph, you can either run 7 to the right and "rise" *downward* −5, or run *back* to the left −7 and rise up 5. The graph is shown in Figure 3-2e.

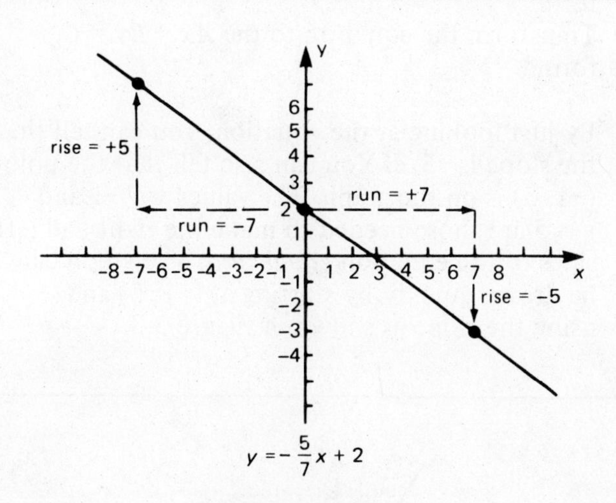

Figure 3-2e.

In the following exercise you will get practice plotting the graphs of linear functions quickly, using the slope and y-intercept.

Exercise 3-2

For Problems 1 through 14, plot the graph. If the equation is not in the form $y = mx + b$, first transform it to that form. Use graph paper, ruler, etc., and do a neat job, labeling axes and showing the scales you use. But do the plotting *quickly* with the aid of the y-intercept and the slope.

1. $y = \frac{3}{5}x + 3$
2. $y = \frac{5}{2}x - 1$
3. $y = -\frac{3}{2}x - 4$
4. $y = -\frac{1}{4}x + 3$
5. $y = 2x - 5$
6. $y = 3x - 2$
7. $y = -3x + 1$
8. $y = -2x + 6$
9. $7x + 2y = 10$
10. $3x + 5y = 10$
11. $x - 4y = 12$
12. $2x - 5y = 15$
13. $y = 3x$
14. $y = -2x$

For Problems 15 through 20, graph the relation specified.

15. $y = 3$
16. $y = -5$
17. $x = -4$
18. $x = 2$
19. $y = 0$
20. $x = 0$
21. Relations such as those in Problems 15 through 20 where x = constant or y = constant, have graphs that are straight lines. However, neither kind of relation is a linear function, though the reason is different for both. Explain *why* such relations are not called linear functions.
22. *Another Form of the Linear Function Equation*
    a. Show that the relation specified by the equation
       $y - 4 = 2(x - 3)$
       is a *linear function* by transforming the equation to the form $y = mx + b$.
    b. Plot a graph of this transformed equation.
    c. What does the number 2 in the original equation tell you about the graph?
    d. If you have drawn the graph accurately, you should be able to find a point on it whose coordinates appear in the equation in its *original* form. What point is it?

3-3 OTHER FORMS OF THE LINEAR FUNCTION EQUATION

Suppose that a relation has the equation

$$y - 4 = 2(x - 3).$$

If you substitute the ordered pair (3, 4) for (x, y), both sides of the equation become *zero*, so (3, 4) is a point on the graph. If you distribute the "2" on the right, then add 4 to both members, the equation becomes

$$y = 2x - 2.$$

Therefore, the relation is a *linear function* with slope 2. You may already have discovered this fact if you worked Problem 22 in Exercise 3-2.

In general, if a fixed point on the line has coordinates (x_1, y_1) and the line has slope m, then an equation of the line is

$$\boxed{y - y_1 = m(x - x_1).}$$

This equation is called the *point-slope form* of the linear function equation. The familiar form $y = mx + b$ is usually called the *slope-intercept form* of the linear function equation.

Sometimes it is convenient to write the equation for a linear function with both x and y on the left, and a constant on the right, such as

$$3x + 4y = 13.$$

This form of the linear function equation is called the "$Ax + By = C$" form. The A, B, and C stand for constants such as 3, 4, and 13 in the above equation.

The three forms of the linear function equation are listed below.

$y = mx + b$	Slope-intercept form.
$y - y_1 = m(x - x_1)$	Point-slope form.
$Ax + By = C$	"$Ax + By = C$" form.

Objective:

Given an equation in the point-slope form, be able to

a. plot the graph quickly, and
b. transform the equation to the slope-intercept form or to the $Ax + By = C$ form.

The following example shows you how to accomplish this objective.

Example: If $y - 5 = -\frac{3}{2}(x + 1)$, do the following things:

a. Plot the graph quickly.
b. Transform the equation to the slope-intercept form.
c. Transform the equation to the $Ax + By = C$ form.

a. By just looking at the equation, you can tell that the slope is $-3/2$. You can also tell that the point $(-1, 5)$ is on the graph. The values $x = -1$ and $y = 5$ are those needed to make the right and left sides of the equation *equal zero*. The graph can be drawn quickly by starting at $(-1, 5)$ and using the slope as shown in Figure 3-3.

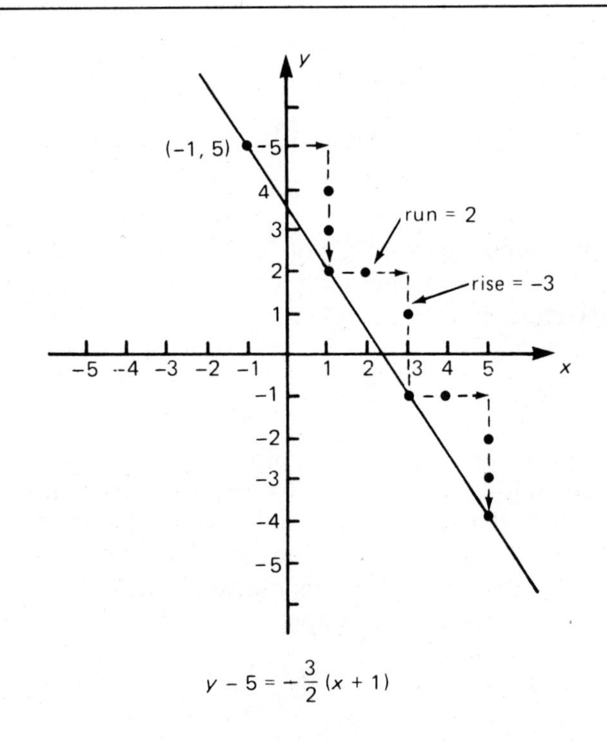

Figure 3-3.

b. The steps in transforming the equation to the slope-intercept form are shown below:

$y - 5 = -\frac{3}{2}(x + 1)$	Given equation.
$y - 5 = -\frac{3}{2}x - \frac{3}{2}$	Distribute the $-\frac{3}{2}$.
$\underline{\underline{y = -\frac{3}{2}x + \frac{7}{2}}}$	Add 5 to both members.

c. Transforming to the $Ax + By = C$ form can be done by continuing from the last step of part b.

$y = -\frac{3}{2}x + \frac{7}{2}$	Answer to part b.
$\frac{3}{2}x + y = \frac{7}{2}$	Add $\frac{3}{2}x$ to both members.
$\underline{\underline{3x + 2y = 7}}$	Multiply both members by 2 to make all coefficients *integers.*

The following exercise is designed to give you practice in plotting graphs from the point-slope form, and transforming to the other two forms.

Exercise 3-3

For Problems 1 through 10, do the following things:

a. Plot the graph quickly.
b. Transform the equation to slope-intercept form.
c. Transform the equation to $Ax + By = C$ form, where A, B, and C are all *integers.*

1. $y - 2 = \frac{3}{5}(x - 1)$
2. $y - 3 = \frac{2}{5}(x - 6)$
3. $y + 4 = \frac{7}{2}(x - 3)$
4. $y + 1 = \frac{7}{3}(x - 4)$
5. $y - 6 = -\frac{1}{4}(x + 2)$
6. $y - 2 = -\frac{1}{2}(x + 5)$
7. $y + 1 = -2(x + 4)$
8. $y + 6 = -3(x + 2)$
9. $y = \frac{1}{3}(x - 12)$
10. $y - 5 = \frac{2}{5}x$

If you understand the point-slope form, you can use it *backwards* to write the equation when you know the slope and a point. For Problems 11 through 14, write an equation in point-slope form for the linear function described.

11. Contains the point (5, 7), and has slope −3.
12. Contains the point (6, 3), and has slope 5.
13. Contains the point (−2, 5), and has slope 9/13.
14. Contains the point (7, −9), and has slope −22/7.

3-4 EQUATIONS OF LINEAR FUNCTIONS FROM THEIR GRAPHS

In the last two sections you started with the equation of a linear function, then drew its graph. If you understand the properties you learned, you can *reverse* the process. Starting with the graph, you can find the equation. It is this process that will allow you to apply linear functions to problems from the real world in the next section.

Objective:

Given the graph of a linear function, or some information about the graph, write the equation for that particular function.

Equations such as $y = mx + b$ or $y - y_1 = m(x - x_1)$ are called *general* equations. By using letters (m, b, x_1, y_1) to stand for the constants, these equations apply to *any* linear function. An equation such as

$$y = 3x - 5 \quad \text{or} \quad y - 4 = -2(x + 8)$$

is called a *particular* equation for a certain linear function since there is only *one* linear function that has these particular values of the constants.

Example 1: Find the particular equation of the linear function with slope of −3/2, containing the point (7, 5).

With the aid of the point-slope form, you can write the equation immediately. The ordered pair (x_1, y_1)

is (7, 5), and $m = -3/2$, so the particular equation is

$$y - 5 = -\frac{3}{2}(x - 7).$$

Example 2: Find the particular equation of the linear function with slope 6, containing the point (3, −8).

In this case, y_1 is *negative.* So you write

$$y - (-8) = 6(x - 3),$$

or more simply.

$$y + 8 = 6(x - 3).$$

Example 3: Find the particular equation of the linear function containing the points (−4, 5) and (6, 10).

This new problem can be turned into an *old* problem by first finding the slope, m. Using the slope formula,

$$m = \frac{10 - 5}{6 - (-4)} = \frac{5}{10} = \frac{1}{2}.$$

You can use *either* of the given points as (x_1, y_1), getting

$$y - 5 = \frac{1}{2}(x + 4) \quad \text{or} \quad y - 10 = \frac{1}{2}(x - 6).$$

These equations are *equivalent*, as you can see by transforming both of them to other forms. In both cases you get

$$y = \frac{1}{2}x - 7 \quad \text{or} \quad x - 2y = 14.$$

Example 4: Find the particular equation of the linear function containing (−2, 7) whose graph is parallel to the graph of $3x + 4y = 72$.

Here, again, you have a point but not the slope. But two parallel lines have the *same* slope. Transforming the given equation to $y = mx + b$, you get

$$y = -\frac{3}{4}x + 18.$$

So the slope must equal −3/4. The equation can now be written immediately.

$$y - 7 = -\frac{3}{4}(x + 2).$$

Note that the "18" in the given equation is *not* needed in this problem. It is simply the y-intercept of the *given* line, not the line you are seeking.

Example 5: Find the particular equation of the horizontal line containing the point (7, 8).

Since the rise is *zero* for any two points on a horizontal line, the slope will also be zero (Figure 3-4a). So you can write

$$y - 8 = 0(x - 7), \quad \text{or}$$

$$y = 8.$$

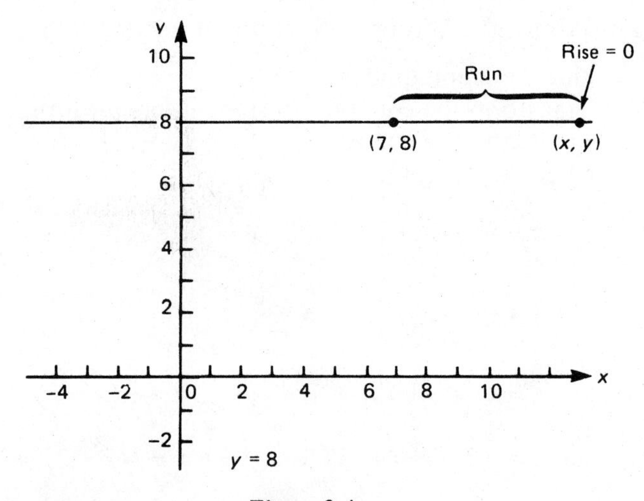

Figure 3-4a.

This equation says, "No matter what x equals, y is always 8." Once you see this fact, you can write the equation of a horizontal line in *one* step. The general equation for a horizontal line is

$$\boxed{y = \text{constant.}}$$

Example 6: Find the particular equation of the vertical line containing the point (7, 8).

For any two points on a vertical line, the *run* equals *zero* (Figure 3-4b). If you try to find the slope, you get

$$m = \frac{\text{rise}}{0},$$

which is *not* a real number. So vertical lines do *not* have a slope. The only way to find the particular equation is to be clever. From the graph you can see that every point has the *same* abscissa. Therefore, a vertical line must have the general equation

$$\boxed{x = \text{constant.}}$$

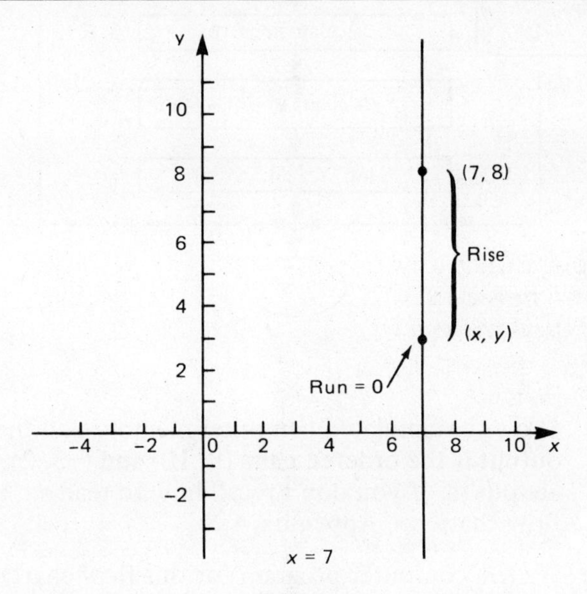

Figure 3-4b.

For this problem, the particular equation would be

$x = 7.$

This equation says, "No matter what y is, x always equals 7."

The following exercise gives you practice finding particular equations of linear functions from information about their graphs.

Exercise 3-4

For Problems 1 through 22,

a. find the particular equation of the line described,
b. transform the equation to the slope-intercept form,
c. transform the equation to the $Ax + By = C$ form, where A, B, and C are *integer* constants.

1. Has y-intercept of 21 and slope of -5.
2. Has y-intercept of -13 and slope of 7.
3. Contains (3, 7) and has slope of 11.
4. Contains (4, 9) and has slope of 5.73.
5. Contains (4, -5) and has slope of -6.
6. Contains (-3, 7) and has slope of $-8/5$.
7. Contains (1, 7) and (3, 10).
8. Contains (5, 2) and (8, 11).
9. Contains (2, -4) and (-5, -10).
10. Contains (-1, 4) and (-5, -4).
11. Contains (5, 8) and is parallel to the graph of $y = 7x - 6$.
12. Contains (7, 2) and is parallel to the graph of $y = -4x + 3$.
13. Contains (5, 8) and is parallel to the graph of $2x + 3y = 9$.

14. Contains (7, 2) and is parallel to the graph of $5x - 3y = 6$.

15. Has x-intercept of 5 and slope of $-2/3$.

16. Has x-intercept of 7 and y-intercept of 5.

17. Contains the origin and has slope of 0.315.

18. Contains the origin and has slope of 2.

19. Is horizontal and contains $(-8, 9)$.

20. Is horizontal and contains $(11, -13)$.

21. Is vertical and contains $(-8, 9)$.

22. Is vertical and contains $(11, -13)$.

In Problems 23 through 26 several ordered pairs are listed. For each problem, tell whether or not there is a linear function that contains *all* of the ordered pairs. If there *is* such a linear function, find its particular equation.

23. (6, 2), (5, 3), (1, 7).

24. (−3, 16), (1, 10), (9, −3).

25. (1, 4), (3, 7), (5, 10), (7, 13).

26. (4, 9), (20, 23), (13, 17), (29, 31).

27. *Intercept Form* – Another form of the linear function equation is

$$\frac{x}{a} + \frac{y}{b} = 1.$$

Do the following things:

a. Show that a and b are the x-intercept and y-intercept, respectively.

b. Transform the equation $\frac{x}{3} + \frac{y}{5} = 1$ to the following forms:

i. $Ax + By = C$, where A, B, and C are *integers.*
ii. $y = mx + b$.

c. Transform the equation $y = 4x - 12$ to the intercept form. Tell what the intercepts are.

28. *Computer Program for Linear Function Equations* – Following is a flow chart for a computer program to find the particular equation of a linear function. The input is two ordered pairs $(X1, Y1)$ and $(X2, Y2)$. The output is an equation of the form $y = mx + b$, $y =$ constant, or $x =$ constant.

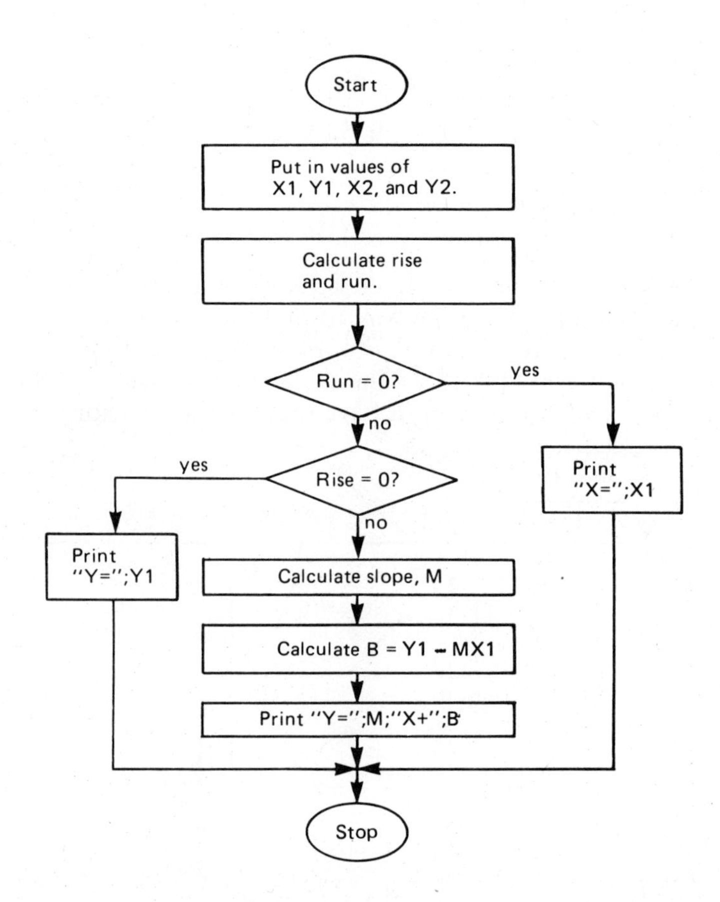

a. Show the simulated computer memory and the output if the ordered pairs (9, 10) and (−3, 2) are put in. If you don't recall how to read a flow chart, see Appendix A.

b. Write a computer program for this flow chart. Appendix A shows you how to do this in a computer language such as BASIC.

c. Test your program by putting in the following ordered pairs:

   i. (1, 7), (3, 10).
   ii. (−1, −4), (−5, 4).
   iii. (3, 8), (6, 8).
   iv. (−2, 13), (−2, 4).

d. What would happen if you tried running the program with the ordered pairs (0, 3) and (2, 7)? How would you modify your program to avoid this difficulty?

29. *Introduction to Linear Models* – Calvin Butterball drives from his home on the farm to the nearby town of Scorpion Gulch. As he drives, his distance from Scorpion Gulch depends on the number of minutes he has been driving. When he has been driving for 6 minutes, he is 17 kilometers away; when he has been driving for 15 minutes, he is 11 kilometers away.

Let y = number of kilometers Calvin is from Scorpion Gulch.
Let x = number of minutes he has been driving.

Do the following:

a. Write the information about distances and times as two ordered pairs.
b. Plot the two ordered pairs on a Cartesian coordinate system. Remember which axes are used for dependent and independent variables.
c. Assuming that the distance-time relation is a *linear function*, draw its graph on the Cartesian coordinate system of part b.
d. Write the particular equation of the linear function in part c. Use the point-slope form.
e. Transform the equation in part d to the slope-intercept form.
f. Find Calvin's distance from Scorpion Gulch when he has been driving for 24 minutes by substituting into your equation from part e.
g. Find the x- and y-intercepts for the function in part e.
h. How long does it take Calvin to get from the farm to Scorpion Gulch? Justify your answer.
i. How far does Calvin live from Scorpion Gulch? Justify your answer.

3-5 LINEAR FUNCTIONS AS MATHEMATICAL MODELS

In Section 2-2 you drew "reasonable" graphs relating two real-world variables. Some of these graphs turned out to be straight lines. With what you now know, you can find an *equation* relating two such variables. All that is needed is two known ordered pairs. Once you have found the equation, you can use it to calculate other values of the variables. Using mathematics to make such predictions is called "mathematical modeling." The linear function is called a *mathematical model* since it is built out of mathematics to represent something in the real world.

Objective:

Given a situation in which two real-world variables are related by a straight-line graph, be able to:

a. Find the particular equation from given points.
b. Use the equation to predict values of either variable.
c. Figure out what the slope and intercepts tell you about the real world.

For example, as you drive home from the football game, the number of kilometers you are away from home depends on how long you have been driving. Suppose that you are 11 kilometers from home when you have been driving for 10 minutes, and 8 kilometers from home when you have been driving for 15 minutes. If you assume that the graph is a *straight line*, you can figure out its equation. The first thing you must do is define the variables you would like to use.

Let t = number of minutes you have been driving.
Let d = number of kilometers you are away from home.

Since d depends on t, you can write the two times and distances as ordered pairs, (t, d).

$$(10, 11), \quad (15, 8)$$

Once you have assumed a linear function, defined the variables, and written the given ordered pairs, you have taken the problem out of the real world and put it into the mathematical world. The graph of the linear function is shown in Figure 3-5a.

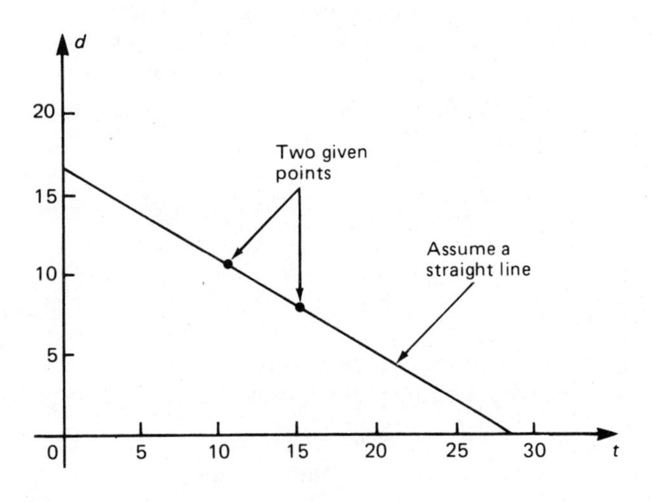

Figure 3-5a.

The general equation of a linear function with these two variables is

$$d - d_1 = m(t - t_1).$$

The known ordered pairs, (10, 11) and (15, 8), can be used to find the slope

$$m = \frac{d_2 - d_1}{t_2 - t_1} = \frac{8 - 11}{15 - 10} = -\frac{3}{5}.$$

So the particular equation is

$$d - 11 = -\frac{3}{5}(t - 10).$$

With this equation you can predict your distance at various times. For this purpose it is better to transform the equation to the "$d = mt + b$" form. That way, d will be expressed *explicitly* in terms of t.

$d - 11 = -\frac{3}{5}(t - 10)$ From above.

$d - 11 = -\frac{3}{5}t + 6$ Distribute the $-\frac{3}{5}$.

$\underline{\underline{d = -\frac{3}{5}t + 17}}$ Add 11 to both members.

The mathematical model is now ready to use. Suppose you want to predict your distances at times of 20, 25, and 30 minutes. The calculations would be as follows:

If $t = 20$, then $d = -\frac{3}{5}\cdot 20 + 17 = -12 + 17 = \underline{\underline{5}}$

If $t = 25$, then $d = -\frac{3}{5}\cdot 25 + 17 = -15 + 17 = \underline{\underline{2}}$

If $t = 30$, then $d = -\frac{3}{5}\cdot 30 + 17 = -18 + 17 = \underline{\underline{-1}}.$

The model can be used backwards to predict the time you would be a certain distance from home. For instance, if $d = 7$, then

$7 = -\frac{3}{5}t + 17$ Substitute 7 for d.

$\frac{3}{5}t = 10$ Add $\frac{3}{5}t$ and subtract 7.

$t = \frac{50}{3}$ Multiply by $\frac{5}{3}$.

So it would take you about 16-2/3 minutes to reach a point 7 kilometers from home.

The intercepts and slope have meaning in the real world. The d-intercept is the value of d when $t = 0$. Setting $t = 0$, gives $d = 17$. To figure out what this means, you can ask yourself, "What was going on in the real world when $t = 0$?" Since you were just leaving the game, you can conclude that there must be 17 kilometers between the stadium and your house.

The t-intercept is the value of t when $d = 0$. Setting $d = 0$, you get

$0 = -\frac{3}{5}t + 17$ Substitute 0 for d.

$\frac{3}{5}t = 17$ Add $\frac{3}{5}t$ to both members.

$t = 28\frac{1}{3}$ Multiply by $\frac{5}{3}$.

When $d = 0$, you are *at home.* So it must take you about 28-1/3 minutes to get all the way home.

The meaning of the slope can be figured out by checking its *units.*

$$\text{slope} = \frac{\text{rise}}{\text{run}} = \frac{\Delta d}{\Delta t} = \frac{\text{kilometers}}{\text{minutes}}.$$

The slope is thus a number of *kilometers per minute.* So it represents your *speed.* You are going 3/5 kilometers per minute (36 kilometers per hour). The "−" sign tells you that your distance is *decreasing* as time increases.

A mathematical model may give reasonable predictions for *some* values of the independent variable, but not for other values. For instance, before $t = 0$, you were still at the game. Your distance from home stayed 17 kilometers and did *not* vary. After you reach home your distance remains 0 (presumably!) rather than the −1 predicted by the model when $t = 30$.

A simple way around this difficulty is to say that the linear model works only in the *domain*

$$\left\{t\colon 0 \le t \le 28\frac{1}{3}\right\}.$$

The range corresponding to this domain is

$$\{d\colon 0 \le d \le 17\}.$$

Figure 3-5b shows the range and domain of the linear model, and suggests what the function would look like elsewhere.

You should realize that predictions you make with a mathematical model are no better than the assumptions you make in setting up the model. For this problem you assumed a *linear* function. Since the slope represents the speed, and the slope of a linear function is constant, you have assumed that you drive at a *constant speed.* If your speed actually varies, as shown in Figure 3-5c, you must either be

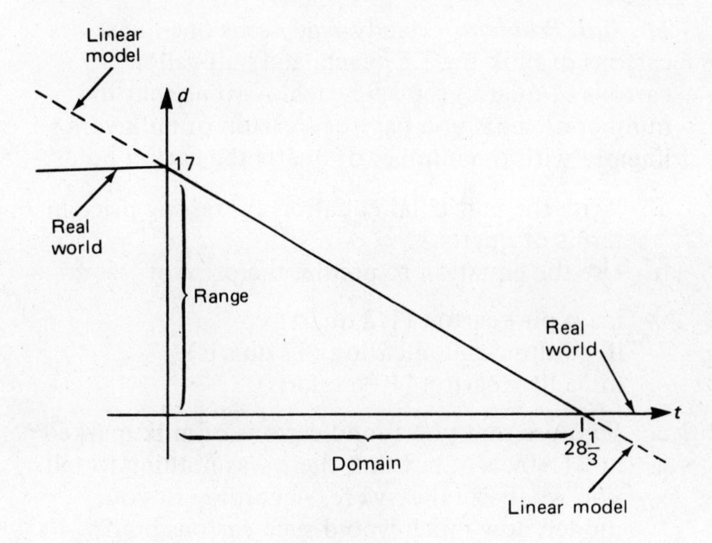

Figure 3-5b.

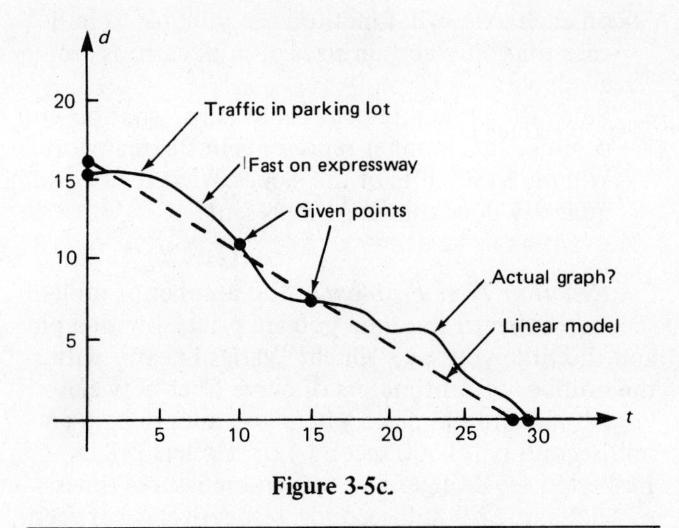

Figure 3-5c.

satisfied with *approximate* answers, or use a better mathematical model.

The following problems will give you experience in using linear functions as mathematical models.

Exercise 3-5

1. ***Milk Problem.*** Handy Andy sells one-quart cartons of milk for $.57 each, and half-gallon cartons (2 quarts) for $.95 each. Assume that the number of cents you pay for a carton of milk varies linearly with the number of quarts the carton holds.

a. Write the particular equation expressing price in terms of quarts.
b. Use the equation to predict the price of,
   i. a pint carton (1/2 quart),
   ii. a three-gallon carton (12 quarts),
   iii. a liter carton (1.06 quarts).
c. Suppose that you found cartons of milk marked at $1.90 each, but that there was nothing to tell you what size they were. According to your model, how much would such cartons hold?
d. Plot a graph of this function. Be sure to label the axes with the variable names and show the scales on each axis. What method can you use to indicate that only certain sizes of milk cartons are available?
e. Since the price-intercept is not zero, what do you suppose this number represents in the real world?
f. What are the units of the slope? What real-world quantity does this slope represent?

2. ***Reaction Time Problem.*** The number of milliseconds between the time you are pricked with a pin and the time you say, "Ouch!" varies linearly with the number of centimeters of nerve fiber between your brain and the place where you are pricked. (A millisecond is 1/1000 second.) Dr. Hollers pricks Leslie Morley's finger and toe, and measures times of 15.2 and 22.9 milliseconds, respectively, between the prick and the ouch. Leslie's finger is 100 centimeters from the brain, and toe is 170 centimeters from the brain.

a. Write the particular equation expressing time in terms of distance.
b. How long would it take Leslie to say, "Ouch!" if pricked in the neck, 10 centimeters from the brain?
c. What does the reaction time intercept equal, and what does it represent in the real world?
d. Plot the graph of this function.
e. The units of the slope are *milliseconds per centimeter.* How fast do impulses travel in Leslie's nerve fibers? Give your answer in centimeters per second.

3. ***Cricket Problem.*** Based on information in *Deep River Jim's Wilderness Trail Book,* the rate at which crickets chirp varies linearly with the temperature. At 59° F, they make 76 chirps per minute, and at 65° F, they make 100 chirps per minute.

a. Write the particular equation expressing chirping rate in terms of temperature.
b. Predict the chirping rate for 90° F and 100° F.
c. Plot the graph of this function.
d. Calculate the temperature-intercept. What significance does this number have in the real world?
e. Based on your answer to part d, what would be a suitable domain for this linear function? Make your graph in part c agree with this domain.
f. What is the real-world significance of the chirping rate intercept?
g. Transform the equation of part a so that temperature is expressed in terms of chirping rate.
h. What would you predict the temperature to be if you counted 120 chirps per minute? 300 chirps per minute?

4. ***Cost of Owning a Car Problem.*** The number of dollars it costs you a month to own a car depends on the number of kilometers you drive it that month. Based on figures in *Changing Times Magazine,* it costs

about \$158 a month if you drive 300 kilometers, and \$230 a month if you drive 1500 kilometers. Assume that a linear function is a reasonable mathematical model of the way cost varies with distance.

a. Write the particular equation expressing dollars per month in terms of kilometers per month.
b. Predict your monthly cost if you drive 500, 1000, and 2000 kilometers per month.
c. About how far could you drive in a month without exceeding a cost of \$250?
d. What does the slope of this function represent?
e. List all the reasons you can think of to explain why the dollars per month intercept is greater than zero.

5. ***Speed On a Hill Problem.*** Assume that the maximum speed your car will go varies linearly with the steepness of the hill it is climbing or descending. If the hill is 5° up, your car can go 77 kilometers per hour (kph). If the hill is 2° down (i.e., −2°), your car can go 154 kph.

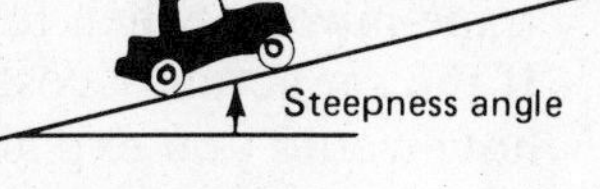

a. Write the particular equation expressing maximum speed in terms of the number of degrees.
b. How fast can you go on a 7° downhill?
c. If your top speed is 99 kph, how steep is the hill? Is it *up* or *down*? Justify your answer.
d. What does the speed-intercept equal, and what does it represent?
e. What does the steepness-intercept equal, and what does it represent?
f. Plot the graph of this linear function, using an appropriate domain.

6. ***Thermal Expansion Problem.*** Bridges on expressways often have "expansion joints" which are small gaps in the roadway between one bridge section and the next. The gaps are put there so that the bridge will have room to expand when the weather gets hot. (See sketch.)

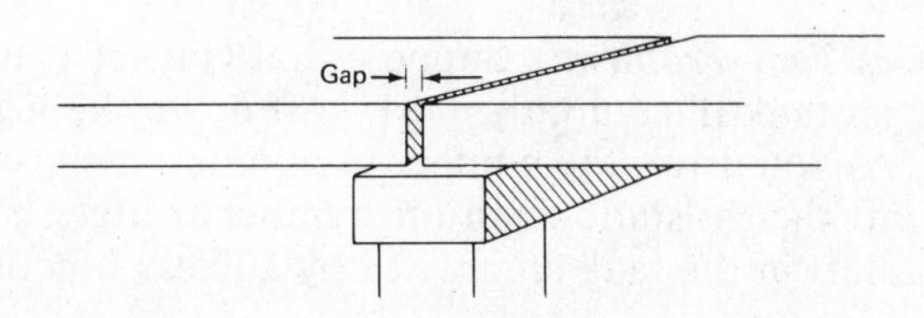

Suppose that a bridge has a gap of 1.3 centimeters when the temperature is 22° C, and that the gap narrows to 0.9 centimeter when the temperature warms to 30° C.

a. Assuming that the gap width varies linearly with temperature, write the particular equation for this function.
b. What would the gap width be at 35° C? −10° C?
c. At what temperature would the gap close completely? What mathematical name is given to this temperature?
d. Would the temperature ever be likely to get high enough to close the gap? Justify your answer.

7. ***Calorie Consumption Problem.*** H. E. Lansburg's *Weather and Health* reports data gathered during World War II which shows that people use about 30 more Calories per day for each 1° drop in the Celsius temperature. At a normal temperature of about 21°C, a working person uses about 3000 Calories per day.

a. Explain how you know that the Calorie use varies *linearly* with temperature. What is the slope? Write the particular equation.
b. How many Calories per day would a working person use
   i. in the Sahara Desert, when the temperature is 50° C?
   ii. in Antarctica in August, when the temperature is −50° C?
c. At what temperature does your mathematical model predict the working person would use no Calories at all? Is it ever likely to reach this temperature?
d. Plot the graph of this function from −50°C through 50° C.

8. ***Gas Tank Problem.*** Suppose that you get your car's gas tank filled up, then drive off down the highway. As you drive, the number of minutes, t, since you left the gas station and the number of liters, g, of gas left in the tank are related by a linear function.

a. Which variable should be dependent and which independent?
b. After 40 minutes you have 52 liters left, and after an hour you have 40 liters left. Find the slope of the function, and tell what it represents in the real world. Explain the significance of the sign of the slope.
c. Write the particular equation for this function.
d. From the equation, find the two intercepts and tell what each represents in the real world.
e. Plot the graph of this function.
f. What are the domain and range of the linear function?

9. ***Terminal Velocity Problem.*** If you jump out of an airplane at high altitude but do not open your parachute, you will soon fall at a constant velocity called your "terminal velocity." Suppose that at time $t = 0$ you jump. When $t = 15$ seconds, your wrist altimeter shows that your distance from the ground, d, is 3600 meters. When $t = 35$, you have dropped to $d = 2400$ meters. Assume that you are at your terminal velocity by the time $t = 15$.

a. Explain why d varies *linearly* with t after you have reached your terminal velocity.
b. Write the particular equation expressing d in terms of t.
c. If you neglect to open your parachute, when will you hit the ground?
d. According to your mathematical model, how high was the airplane when you jumped?
e. The plane was actually at 4200 meters when you jumped. How do you reconcile this fact with your answer to part d?
f. Plot the graph of d versus t. Show what the *actual* graph looks like for values of t less than 15. Also, show what the graph would look like for values of t *after* you open your parachute.
g. What is your terminal velocity in meters per second? In kilometers per hour?

10. ***Linear Depreciation Problem.*** Suppose that you own a car which is presently 40 months old. From an automobile dealer's "Blue Book" you find that its present trade-in value is $1650. From an old Blue Book, you find that it had a trade-in value of $2350 ten months ago. Assume that its value decreases linearly with time.

a. Write the particular equation expressing the trade-in value of your car as a function of its age in months.
b. You plan to get rid of the car before its trade-in value drops below $500. How much longer can you keep the car?
c. By how many dollars does the car "depreciate" (decrease in value) each month? What part of the mathematical model tells you this?
d. When do you predict your car will be worthless? What part of the mathematical model tells you this?
e. According to your model, what was the car's trade-in value when it was new?
f. If the car actually cost $5280 when it was new, how would you explain the difference between this number and your answer to part e?

11. ***Shoe Size Problem.*** The size of shoe a person needs varies linearly with the length of his or her foot. The smallest adult shoe is Size 5, and it fits a 9 inch long foot. An 11 inch foot requires a Size 11 shoe.

a. Write the particular equation which expresses shoe size in terms of foot length.
b. If your foot is a foot long, what size shoe do you need?
c. Bob Lanier of the Detroit Pistons wears a Size 22 shoe. How long is his foot?
d. Plot a graph of adult shoe size versus foot length. Use the given points and calculated points. Be sure that the domain is consistent with the information in this problem.

12. ***Hippopotamus Problem.*** In order to hunt hippopotami, a hunter must have a hippopotamus hunting license. Since the hunter can sell the hippopotami he catches, he can use the proceeds to pay for part or all of the cost of the license. If he catches only three hippos, he is still in debt by \$2050. If he catches seven hippos, he makes a profit of \$1550. The South African Game and Wildlife Commission allows a limit of ten hippos per hunter. Let h be the number of hippos he catches and d be the number of dollars of profit he makes. Assume that h and d are related by a linear function.

a. Which variable should be dependent and which should be independent?
b. Write a suitable domain for the independent variable.
c. Write the given pieces of information as ordered pairs.
d. Plot the graph of the function. You can use the two given ordered pairs and the fact that the function is *linear.* Be sure to observe the domain you specified in part b. Also, choose scales that make the entire graph fit on your graph paper. (You may use different scales for the two axes if this seems desirable.)
e. Find the particular equation for this function.
f. Use the equation to find the d-intercept and the h-intercept. Tell what these two numbers represent in the real world.
g. State what real-world quantity the slope represents.
h. What is the origin of the word "hippopotamus?"

13. ***Celsius-to-Fahrenheit Temperature Conversion.*** The Fahrenheit temperature, "F," and Celsius temperature, "C" of an object are related by a linear function. Water boils at 100°C, or 212°F, and freezes at 0°C, or 32°F.

a. Write an equation expressing F in terms of C.
b. Transform the equation so that C is in terms of F.
c. Lead boils at a temperature of 1620°C. What Fahrenheit temperature is this?
d. If you had a fever thermometer calibrated in Celsius degrees, what would "normal" (98.6°F) be on it?
e. If the weather forecaster says it will be 40°C today, is this hot, cold, or medium? Explain.
f. The coldest possible temperature is absolute zero, −273°C. What Fahrenheit temperature is this?
g. Plot the graph of this function, observing the domain implied by part f.
h. For what temperature is the number of Fahrenheit degrees equal to the number of Celsius degrees?

14. ***Charles's Gas Law.*** In 1787, the French scientist Jacques Charles observed that when he plotted a graph of the volume of a fixed amount of air versus the temperature of the air, the points lay along a straight line. Therefore, he decided that a linear function would be a reasonable mathematical model of the way in which the volume of a gas varies with temperature. Suppose that he found that at 27°C a certain amount of air occupied 500 cubic centimeters. Upon warming the air to 90°C he found that its volume had increased to 605 cubic centimeters.

a. Write the particular equation expressing volume in terms of temperature.
b. Predict the volume at 60°C.
c. The process of predicting a value *between* two given data points is called "interpolation." Determine the origin of this word.
d. Predict the volume at 300°C.
e. The process of predicting a value *beyond* any given data points is called "extrapolation." Determine the origin of this word.
f. Extrapolate your mathematical model back to the point where the volume is zero. That is, find the temperature-intercept.
g. Draw a graph of volume versus temperature.
h. Find out from your chemistry instructor or from someone taking chemistry, the special name given to this temperature-intercept.

15. ***Direct Variation, Pancake Problem.*** If the constant b in $y = mx + b$ equals zero, then y is said to vary *directly* with x. The amount of pancake batter you must mix up varies directly with the number of people who come to breakfast.

a. Suppose that it takes 7 cups of batter to serve 10 people. Write the particular equation expressing number of cups in terms of number of people.
b. How many cups must you prepare for 50 people?
c. About how many people can you serve with 12 cups of batter?
d. Plot the graph of this function. Through what special point does the graph of a direct variation function go?

3-6 CHAPTER REVIEW AND TEST

In this chapter you have studied one kind of function, the *linear* function. You *defined* linear function by the general equation $y = mx + b$. Then you found out that the *graph* is always a straight line. By learning *properties* such as the Slope Formula, you were able to plot the graphs quickly, and to get particular equations from given graphs. These techniques allowed you to use linear functions as *mathematical models* of some things in the real world where two variables were related by a straight-line graph.

The objectives of this chapter may be summarized as follows:

1. *Given the equation of a linear function, draw the graph.*

2. *Given information about the graph of a linear function, write the particular equation.*

3. *Use linear functions as mathematical models.*

Two kinds of exercise appear below. The Review Problems, numbered according to the three objectives above, test your skill with *familiar* problems. The Concepts Test requires you to combine several skills in working *one* rather involved problem. By working the Concepts Test, you will gain experience in applying your knowledge to new situations. It is this ability that is one of the most important goals of your education.

Review Problems

The following problems are numbered according to the objectives listed previously.

Review Problem 1. Plot the graph of:

a. $y = -\frac{5}{2}x + 7$
b. $y - 2 = \frac{1}{3}(x + 5)$
c. $4x + 7y = 21$
d. $y = -5$
e. $x = 11$

Review Problem 2. Write the equation of the particular linear function or relation described. Transform the equation to both the $y = mx + b$ form and to the $Ax + By = C$ form, where A, B, and C are integers.

a. The relation contains $(2, -7)$ and $(5, 3)$.
b. The relation contains $(-4, 1)$ and has a graph parallel to that of $2x - 9y = 47$.
c. The relation has x-intercept of 5 and y-intercept of -6.
d. The graph is vertical and contains $(-13, 8)$.
e. The graph is horizontal and contains $(22, \pi)$.
f. The relation has the x-axis as its graph.

Review Problem 3 – Computer Time Problem. If a computer program has a "loop" in it, the length of time it takes the computer to run the program varies linearly with the number of times it must go around the loop. Suppose that the computer takes 8 seconds to run a given program when it must go around the loop 100 times, and 62 seconds when it must loop 1000 times.

a. Write the particular equation for this function.
b. How long would it take the computer to run the program if it must loop 30 times? 10,000 times?
c. Suppose that the computer takes 23 seconds for a given run of the program. How many times did it go around the loop?
d. How long does it take the computer to run the rest of the program, excluding the loop? What part of the mathematical model tells you this?

e. How long does it take the computer each time it goes through the loop? What part of the mathematical model tells you this?
f. Plot the graph of this function. Choose suitable scales for the two axes.

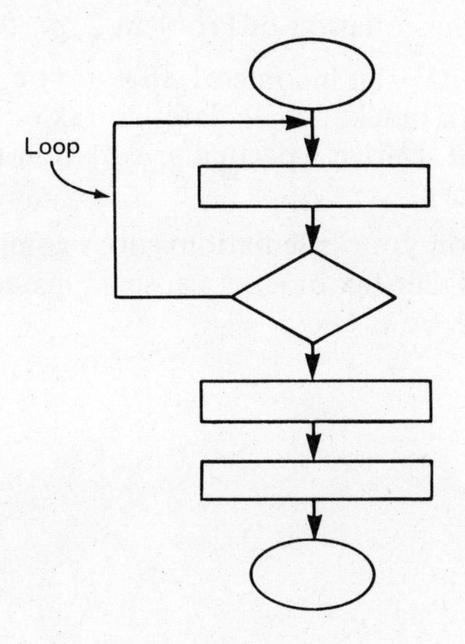

Concepts Test

The following problems concern income tax you must pay to the federal government on money that you earn. In working these problems, you will be called upon to use all of the concepts of this chapter. In some cases, you must put together these concepts in new ways to work problems that are somewhat different from any you have seen before. For this reason, the test is probably somewhat harder than the one your instructor will give you. For each part of each problem, tell by number which one or ones of the three objectives of this chapter you used in working that part. For problems different than you have seen before, write "new concept."

Concepts Test 1. The amount of income tax you must pay varies linearly with your taxable income. However, there are *different* linear functions that apply in different parts of the domain. The table below shows tax rates that became effective in 1979. One linear function applies for incomes from $2300 through $3400, another applies for incomes from $3400 through $4400, and so forth.

Tax Rates for Unmarried Individuals

If taxable income is:		The tax is:	
Over-	But not over-		of the amount over-
$0	$2300	No tax	
$2300	$3400	14%	$2300
$3400	$4400	$154 + 16%	$3400
$4400	$6500	$314 + 18%	$4400
$6500	$8500	$692 + 19%	$6500
$8500	$10800	$1072 + 21%	$8500

a. From the third and fourth lines of the table, you can see that the tax on $3400 would be $154, and the tax on $4400 would be $314. Write these pieces of information as ordered pairs. Then show that the tax rate in this interval really is 16% (as indicated by the table) by showing that the slope of the line connecting these points is 0.16.
b. Explain how you know that the function is *linear* between incomes of $3400 and $4400.
c. Write the particular equation for the linear function in this domain. Use the point-slope form, with the point (3400, 154). Then transform the equation so that tax is expressed in terms of income. It is better if you do *not* distribute the 0.16.
d. What amount of tax would you pay if your income were $4130?

Concepts Test 2a. What is the slope of the linear function for taxable incomes between $8500 and $10800? Use this slope and a suitable point to write the particular equation expressing tax in terms of income for this part of the domain.

b. What tax would you pay on a taxable income of $10800?

c. How much tax would you pay on an income of \$10800 if the equation of Problem 1, part c, applied in this part of the domain? By how many dollars does the actual tax exceed this amount?

Concepts Test 3. If you say, "I am in the 18% *tax bracket*," you mean that your taxable income is between \$4400 and \$6500.

a. Phil T. Rich paid a tax of \$927.60. What tax bracket was he in?
b. Derive the particular equation for this tax bracket.
c. What was Phil's taxable income?

Concepts Test 4. People sometimes worry that a *small* increase in income will make a *large* increase in tax if the increase puts them in a higher tax bracket.

a. If you have an income of \$8501, you are in the 21% tax bracket. Calculate the tax on this amount using the equation of Problem 2, part a.
b. If you have an income of \$8499, you are in the 19% tax bracket. Calculate the tax on this amount. Derive a particular equation first, if necessary.
c. Based on your calculations, does going to the next higher tax bracket cause an instantaneous jump in your tax?

d. True or false? "If you are in the 21% tax bracket, then your tax is 21% of your taxable income." Justify your answer.

Concepts Test 5a. Plot the graph of tax versus taxable income for incomes from $0 through $10800. Choose scales that make the graph fill most of a sheet of graph paper. There should be enough plotting data from the table and from your calculations.

b. What word could be used to describe the behavior of the graph as you go from one tax bracket to the next? See your conclusion from Problem 4, part c.

c. For incomes less than $2300, the graph is a straight line, but the function is *not* linear. Why not? What kind of function is it?

Systems of Linear Equations and Inequalities

Linear functions may be used as mathematical models of the real world. ***Two*** *or more linear functions with the* ***same*** *variables form what is called a* ***system*** *of equations. In this chapter you will learn how to find the intersection of the graphs, and how to tell what this intersection can represent. The techniques you develop can be used to find the "best" way to operate a business, such as raising race horses.*

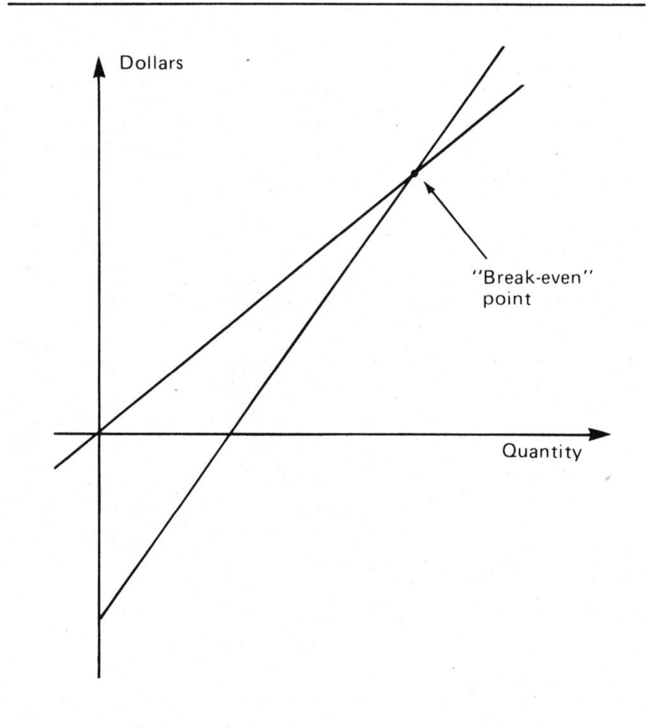

4-1 SYSTEMS OF TWO LINEAR EQUATIONS WITH TWO VARIABLES

A pair of *different* equations containing the same *variables*, such as

$$2x - y = 10$$
$$x + 3y = -9,$$

is called a *system* of equations. The word "system" is also used if there are *more* than two equations, if there are *more* than two variables, and if the open sentences are *inequalities* rather than equations.

DEFINITION

A ***system*** of equations or inequalities is a set of open sentences each of which contains the *same* variables.

In this section you will discover a way to "solve" such a system. That is, you will find the ordered pair where the graphs intersect.

DEFINITION

The ***solution set*** of a system is the set of *all* ordered pairs that satisfy *all* the open sentences in the system.

The following exercise is designed to lead you stepwise through transformations that will reduce the system above to one that can be solved by inspection.

Exercise 4-1

1. Plot *neat* graphs of the following equations, and find the intersection point:

$$2x - y = 10 \quad \text{——} \quad ①$$
$$x + 3y = -9 \quad \text{——} \quad ②.$$

2. Multiply both members of Equation ① by 3. Then *add* the transformed equation to Equation ②,–left member to left member and right member to right member. Simplify the resulting equation.

3. The equation you get from Problem 2 should be very simple! Graph it on the Cartesian coordinate system of Problem 1. Through what "interesting" point does the graph go?

4. Multiply both members of Equation ② by −2 and add the resulting equation to Equation ①. Simplify the resulting equation, then graph it on the Cartesian coordinate system of Problem 1.

5. Two equations are *equivalent* if they have the *same* solution set. Explain why *neither* equation you got in Problems 2 and 4 is equivalent to either Equation ① or Equation ②.

6. Two *systems* of equations are equivalent if they have the *same* solution set. Is the *system* formed by the equations you got in Problems 2 and 4 equivalent to the *system* of Equations ① and ② ? Explain.

4-2 SOLUTION OF SYSTEMS BY LINEAR COMBINATION

In Exercise 4-1 you solved the system

$2x - y = 10$ —— ①
$x + 3y = -9$ —— ②.

By a clever multiplication of one equation by a constant, followed by adding the two equations, you were able to *eliminate* one of the variables. The resulting equation had only *one* variable. By repeating the process, you were able to get an equation with only the *other* variable. Specifically, you got

$x = 3$ —— ③
$y = -4$ —— ④.

Equations ③ and ④ form another system. Neither Equation ③ nor Equation ④ is equivalent to ① or ② . But the *system* they form *is* equivalent to the system of ① and ② . As you can see from Figure 4-2a, both pairs of lines intersect at the point (3, −4). Obviously, the second system is simple enough to be solved by inspection!

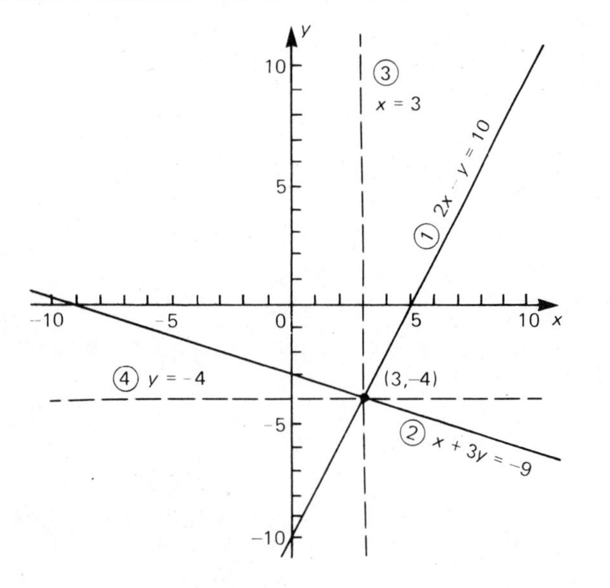

Figure 4-2a.

Objective:

Given a system of two linear equations with two variables, transform it to an equivalent system of the form

$x = \text{constant}$
$y = \text{constant},$

so that the solution set can be found by inspection.

Example: Solve the system

$3x + 4y = 2$ —— ①
$5x - 7y = 17$ —— ②.

The procedure is to multiply both members of one or both equations by constants. You choose constants that will make the coefficients of x or the coefficients of y be *opposites* of each other. If you multiply ① by 5 and ② by −3, then the x-coefficients will be 15 and −15. Adding the two equations will then *eliminate* x.

$$\begin{array}{r} 15x + 20y = 10 \\ -15x + 21y = -51 \\ \hline 41y = -41 \\ \therefore\ y = -1. \end{array}$$

Now that you know that $y = -1$, you can eliminate y by simply *substituting* −1 for y in one of the equations. Substituting −1 for y in ① gives

$3x - 4 = 2$
$\therefore\ x = 2$
$\therefore\ S = \{(2, -1)\}.$

The process of *multiplying* two variables by *constants* and then *adding* the results is called "linearly combining" the two variables. For example, $3r + 7s$ is a linear combination of r and s. Since the above process for solving systems of equations involves multiplying the two equations by constants and then adding the results, the process is often called "solving the system in *linear combination.*"

The exercise that follows will give you practice solving systems of linear equations with two variables. Some of the later problems introduce you to special cases in which the slopes of the two linear functions are equal. The last problem lets you prove the property that justifies the linear combination of equations process.

Exercise 4-2

For Problems 1 through 30, solve the system. Some of the later problems have additional instructions or suggestions.

1. $5x + 2y = 11$, $x + y = 4$
2. $x - y = -11$, $7x + 4y = -22$
3. $6x - 7y = 47$, $2x + 5y = -21$
4. $8x + 3y = 41$, $6x + 5y = 39$
5. $5x + 7y = -15$, $2x + 9y = -6$
6. $9x - 5y = 26$, $4x - 3y = 17$
7. $2x + 3y = 2$, $4x - 9y = -1$
8. $3x + 10y = -24$, $6x + 7y = -9$
9. $9x - 7y = 5$, $10x + 3y = -16$
10. $11x - 5y = -38$, $9x + 2y = -25$

For Problems 11 through 20, there is an *easy* multiplication that will work. In Problem 11, for example, x can be eliminated by multiplying the first equation by 3 and the second one by -2.

11. $14x + 31y = -6$, $21x + 17y = 50$
12. $12x - 7y = 59$, $8x + 11y = -39$
13. $13x + 10y = -7$, $17x - 15y = 47$
14. $13x + 20y = -1$, $19x - 15y = 87$
15. $22x - 19y = 28$, $55x - 29y = 107$
16. $37x - 36y = 180$, $41x - 24y = 120$
17. $-37x - 24y = 35$, $15x - 18y = 69$
18. $18x - 19y = 161$, $54x + 25y = 565$
19. $23x - 34y = -46$, $13x + 51y = -26$
20. $33x + 16y = -2$, $-29x + 20y = 138$

Problems 21 through 26 have "untidy" fractions for answers. You may want to use the linear combination to find *both* variables, rather than substituting a fraction back into one of the equations.

21. $4x - 3y = 11$, $5x - 6y = 9$
22. $3x + 4y = 18$, $9x + 6y = 17$
23. $3x + 4y = 8$, $2x - 2y = 7$
24. $5x - 3y = 22$, $6x - 7y = 41$
25. $-x + 5y = 22$, $7x - 2y = 19$
26. $8x + 5y = 23$, $3x - 2y = 37$

For Problems 27 through 30, you can write $2/x$ as $2(1/x)$, $-5/y$ as $-5(1/y)$, and so forth. Then you can solve for $1/x$ and $1/y$. To find x and y, recall that if two numbers are equal, then their reciprocals are equal.

27. $\frac{2}{x} - \frac{5}{y} = 5$, $\frac{3}{x} + \frac{10}{y} = 18$
28. $\frac{3}{x} + \frac{6}{y} = 1$, $\frac{3}{x} + \frac{7}{y} = 2$
29. $\frac{12}{x} + \frac{5}{y} = 25$, $\frac{2}{x} - \frac{15}{y} = -18$
30. $\frac{6}{x} - \frac{7}{y} = 8$, $\frac{15}{x} - \frac{14}{y} = 21$

31. *Inconsistent and Dependent Equations*—If the slopes of two linear functions are equal, their graphs will either be parallel or will coincide with each other, as shown in the first two sketches of Figure 4-2b.

In such cases, the two equations are said to be *inconsistent* or *dependent*, respectively. For the more normal case where the graphs intersect at a unique point, the equations are said to be *independent*. In this problem you will see what happens when you try to solve a system of inconsistent or dependent equations.

a. Try to solve the following systems by linear combination:

i. $3x + 2y = 7$, $6x + 4y = 8$.

ii. $3x + 2y = 7$, $6x + 4y = 14$.

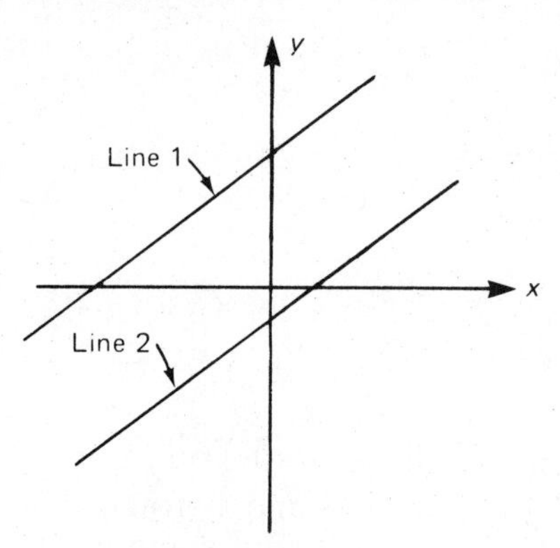

Inconsistent equations.
Parallel graphs.

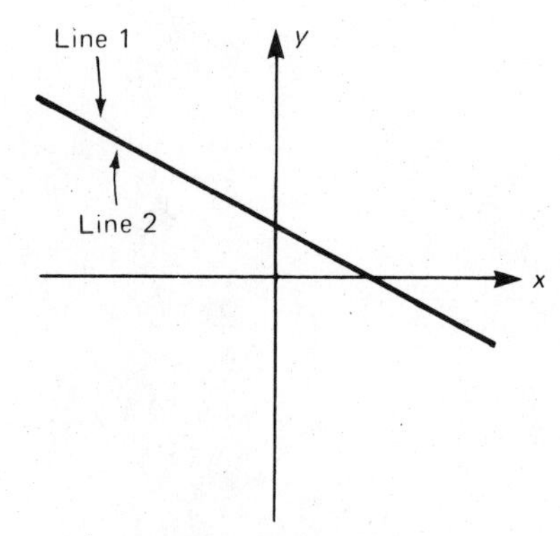

Dependent equations.
Coincident graphs.

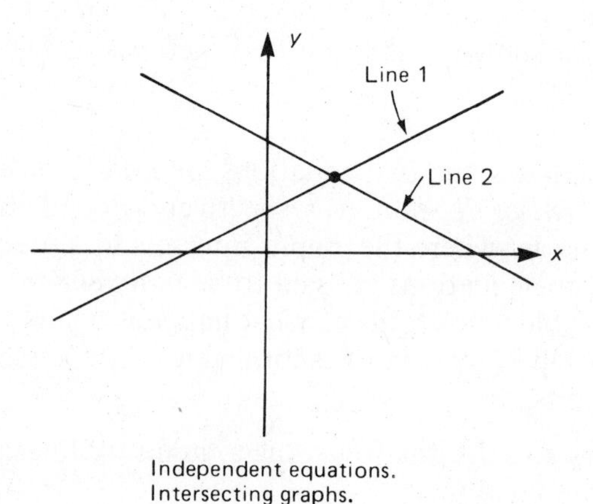

Independent equations.
Intersecting graphs.

Figure 4-2b.

b. On *separate* sets of axes, draw graphs of both systems. Tell which equations are *dependent*, and which are *inconsistent.*

c. When you eliminate a variable by linear combination, how can you tell when a pair of equations is inconsistent? Dependent?

d. For the following systems, eliminate a variable by linear combination. From the results of your calculations, tell whether the equations are *independent, dependent,* or *inconsistent.*

i. $15x + 12y = 8$, $10x + 8y = 13$ ii. $15x - 12y = 8$, $10x + 8y = 13$

iii. $15x - 12y = 18$, $10x - 8y = 12$ iv. $15x - 12y = 18$, $10x - 8y = 14$

32. If you draw three straight lines on a piece of paper, usually they will not all cross at the same point. Therefore, the solution set of a system of *three* linear equations with two variables is usually the *empty* set. For each of the following systems, either find an ordered pair which satisfies all three equations or demonstrate that the solution set is empty.

a. $5x + 3y = 17$, $2x - y = 9$, $4x - 3y = 19$ b. $3x - y = 5$, $2x + 3y = 7$, $x - 4y = 9$

33. The following system is difficult to solve because the numbers get so big. However, the answer is surprisingly pleasant because the values of x and y turn out to be small integers. Solve the system:

$$67x + 23y = 203$$
$$53x + 71y = 319.$$

34. Prove that if $p = q$ and $r = s$, then $p + r = q + s$. The addition property of equality, followed by substitution, should help. This is the property which allows you to "add" two equations in the linear combination process.

4-3 SECOND-ORDER DETERMINANTS

If you solved the system in Problem 33 of Exercise 4-2, you found that the arithmetic was quite tedious. If you could specify the sequence of steps to be taken in solving a system, you could let a *computer* do the tedious arithmetic. In this section you will derive general formulas for the values of x and y that satisfy a given system.

Objectives:

1. Given a *general* system of two linear equations with two variables, derive formulas for calculating the values of x and y that satisfy the system.

2. Discover a pattern whereby the formulas can be remembered.

3. Use the pattern to solve particular systems of linear equations.

The general system of linear equations has the form

$$\begin{aligned} ax + by &= c \quad \text{———} \quad ① \\ dx + ey &= f \quad \text{———} \quad ②, \end{aligned}$$

where $a, b, c, d, e,$ and f stand for constants. This general system can be solved by linear combination, the same way you have been solving particular systems. To eliminate y, you would multiply ① by e and ② by $-b$, then add, getting

$$\begin{array}{rcl} aex + bey &=& ce \\ -bdx - bey &=& -bf \\ \hline aex - bdx &=& ce - bf. \end{array}$$

Since x is a common factor of the two terms on the left, you can use the Distributive Property (backwards) to factor it out. You get

$$(ae - bd)x = ce - bf.$$

The algebraic form of this equation is really no worse than $3x = 7$. You simply divide both members of the equation by the coefficient of x, getting

$$x = \frac{ce - bf}{ae - bd}.$$

This is a formula which you could use to calculate the value of x just by substituting values for the six constants and doing the arithmetic.

Eliminating x and solving for y in the same manner gives

$$y = \frac{af - cd}{ae - bd}.$$

Three things appear when you compare these formulas with the original system of equations,

$$\begin{aligned} ax + by &= c \\ dx + ey &= f. \end{aligned}$$

They are:

1. Both denominators are the same, and contain only the *coefficients* of the *left* members of the equations.

2. The numerator for x does *not* contain the two x-coefficients, a and d.

3. The numerator for y does *not* contain the two y-coefficients, b and e.

There is a convenient way to remember how to get the denominators. If you write the four coefficients of x and y in the order they appear in the equations,

$$\begin{vmatrix} a & b \\ d & e \end{vmatrix},$$

you can get the denominator by multiplying the *top left* number by the *bottom right*, then subtracting the *top right* times the *bottom left*. The four-number symbol with the vertical bars is called a *second-order determinant.* The diagonal multiplication scheme is shown in the following diagram:

$$D = \begin{vmatrix} a & b \\ d & e \end{vmatrix} = ae - bd.$$

Looking back at the formulas for x and y, you can see that this is actually the correct denominator.

The two numerators may also be expressed as second-order determinants. Recalling that the x-numerator does not contain the x-coefficients, you simply replace the x-coefficients in the *left* members of the equations with the c and f from the *right* members.

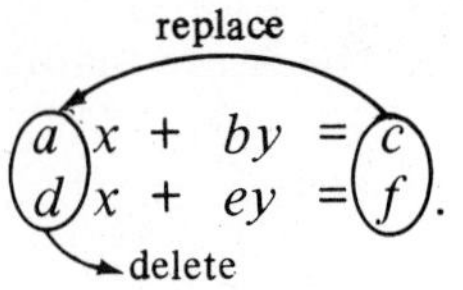
replace
delete

You get a new determinant, N_x (for "x-numerator"),

$$N_x = \begin{vmatrix} c & b \\ f & e \end{vmatrix}.$$

Expanding this determinant gives

$$N_x = \begin{vmatrix} c & b \\ f & e \end{vmatrix} = ce - bf,$$

which is the correct numerator for x.

The numerator determinant for y, N_y, is found by replacing the y-coefficients in the left members of the equations with the c and f from the right members.

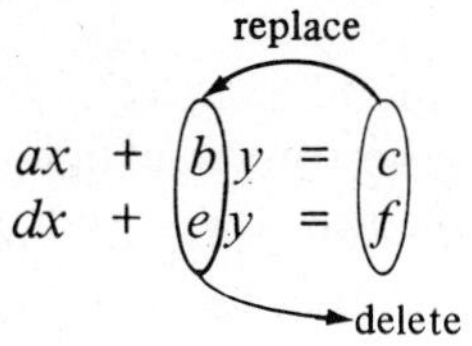
replace
delete

$$\therefore N_y = \begin{vmatrix} a & c \\ d & f \end{vmatrix} = af - \text{cd},$$

which is the correct numerator for y.

The solution of the system can now be written in *one* step.

$$x = \frac{N_x}{D} \quad \text{and} \quad y = \frac{N_y}{D}.$$

Example: Solve by determinants:

$$\begin{aligned} 3x + 4y &= 2 \\ 5x - 7y &= 17. \end{aligned}$$

The denominator determinant is found simply by writing the four coefficients in the left members of the equations in the order in which they appear.

$$D = \begin{vmatrix} 3 & 4 \\ 5 & -7 \end{vmatrix} = (3)(-7) - (5)(4) = -21 - 20 = -41.$$

The x-numerator is found by replacing the x-coefficients, 3 and 5, with the constants 2 and 17 from the right members of the equations.

$$N_x = \begin{vmatrix} 2 & 4 \\ 17 & -7 \end{vmatrix} = (2)(-7) - (17)(4) = -14 - 68 = -82.$$

The y-numerator is found by replacing the y-coefficients, 4 and -7, with the constants 2 and 17.

$$N_y = \begin{vmatrix} 3 & 2 \\ 5 & 17 \end{vmatrix} = (3)(17) - (5)(2) = 51 - 10 = 41.$$

Putting together these three numbers, you get the solution,

$$x = \frac{N_x}{D} = \frac{-82}{-41} = 2,$$

$$y = \frac{N_y}{D} = \frac{41}{-41} = -1,$$

$$\therefore S = \{(2, -1)\}.$$

Solving systems of linear equations by determinants is called using *Cramer's Rule.*

Once you understand the process, much of the work can be done in your head. For example, you might write simply

$$x = \frac{\begin{vmatrix} 2 & 4 \\ 17 & -7 \end{vmatrix}}{\begin{vmatrix} 3 & 4 \\ 5 & -7 \end{vmatrix}} = \frac{-14 - 68}{-21 - 20} = \frac{-82}{-41} = 2,$$

$$y = \frac{\begin{vmatrix} 3 & 2 \\ 5 & 17 \end{vmatrix}}{-41} = \frac{51 - 10}{-41} = \frac{41}{-41} = -1,$$

$$\therefore S = \{(2, -1)\}.$$

In the exercise which follows, you will use determinants to solve systems from Exercise 4-2. Then you will have a chance to write a computer program to do the arithmetic involved in expanding determinants.

Exercise 4-3

For Problems 1 through 16, solve the system using second-order determinants. Problems 1 through 10 are the same as Problems 1 through 10 in Exercise 4-2. Problems 11 through 16 are the same as Problems 21 through 26 in Exercise 4-2.

1. $5x + 2y = 11$
   $x + y = 4$
2. $x - y = -11$
   $7x + 4y = -22$
3. $6x - 7y = 47$
   $2x + 5y = -21$
4. $8x + 3y = 41$
   $6x + 5y = 39$
5. $5x + 7y = -15$
   $2x + 9y = -6$
6. $9x - 5y = 26$
   $4x - 3y = 17$
7. $2x + 3y = 2$
   $4x - 9y = -1$
8. $3x + 10y = -24$
   $6x + 7y = -9$
9. $9x - 7y = 5$
   $10x + 3y = -16$
10. $11x - 5y = -38$
    $9x + 2y = -25$
11. $4x - 3y = 11$
    $5x - 6y = 9$
12. $3x + 4y = 18$
    $9x + 6y = 17$
13. $3x + 4y = 8$
    $2x - 2y = 7$
14. $5x - 3y = 22$
    $6x - 7y = 41$
15. $-x + 5y = 22$
    $7x - 2y = 19$
16. $8x + 5y = 23$
    $3x - 2y = 37$

17\. a. Graph the following system of equations and write the ordered pair at which they seem to intersect.

$$5x - 2y = 4$$
$$3x + 7y = 26$$

b. Solve the above system using determinants and express the answers as mixed numbers.

c. Based on the answers to parts a and b, why do you suppose that it is safer to get answers by *calculation* than it is by graphing?

18\. a. Plot the graph of each of the following equations on the *same* Cartesian coordinate system:

$x + y = 5$ —— ①
$3x - 2y = 8$ —— ②
$x + 3y = 8$ —— ③

If you have done the work correctly, the graphs will all intersect at (or near) the *same* point.

b. Solve the systems formed by equation ① and ②, by ② and ③, and by ① and ③ using de determinants. Write the answers as mixed numbers and by comparing the three answers, tell whether or not all three *really* intersect at the same point.

19\. *Inconsistent Equations, Dependent Equations, and Determinants* – a. On *separate* sets of axes, plot the graphs of the following two systems:

i. $3x + 2y = 7$
   $6x + 4y = 8.$

ii. $3x + 2y = 7$
    $6x + 4y = 14.$

b. From your graphs, tell which pair of equations is *inconsistent* (graphs are parallel), and which pair of equations is *dependent* (graphs coincide).

c. Show that the denominator determinant for each system equals zero.

d. Find the numerator determinant, N_x, for each system. How can you tell from the value of this determinant whether the equations are inconsistent or dependent?

e. Solve the following systems by determinants. If the denominator determinant equals zero, tell whether the equations are inconsistent or dependent.

i. $15x + 12y = 8$
   $10x + 8y = 13$

ii. $15x - 12y = 8$
    $10x + 8y = 13$

iii. $15x - 12y = 18$
     $10x - 8y = 12$

iv. $15x - 12y = 18$
    $10x - 8y = 14$

20\. *Computer Solution of Linear Systems* – Refer to the following flow chart for a computer program to solve linear systems. If you have never read a flow chart before, or need a refresher, you might consult Appendix A for help with this problem.

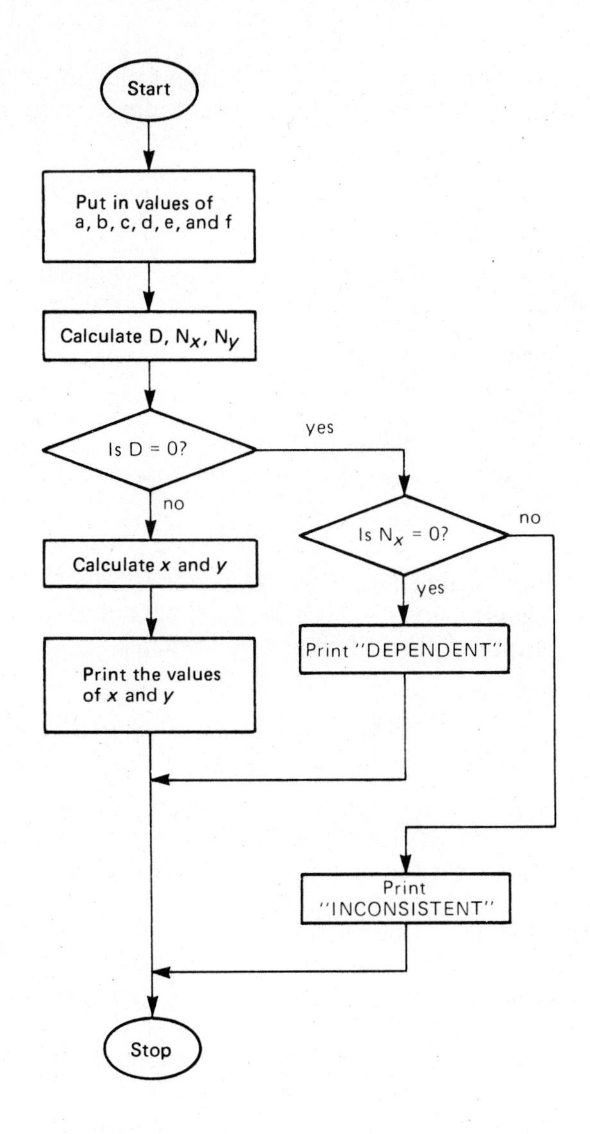

a. Show the simulated computer memory and the output if 3, 4, 2, 5, −7, and 17 are put in for *a, b, c, d, e,* and *f*, respectively.

b. Write a computer program for this flow chart. Appendix A shows you how to do this in BASIC.

c. Use your program to solve the four systems in Problem 19.e, above. Make sure that your program gives the correct answers. If it does not, you must go back and "debug" the program.

4-4 $f(x)$ TERMINOLOGY, AND SYSTEMS AS MODELS

In the preceding sections you have studied systems of equations such as

$$3x - 4y = -12$$
$$5x - 2y = 10.$$

For graphing purposes, each equation could be transformed to $y = mx + b$, giving

$$y = \frac{3}{4}x + 3$$

$$y = \frac{5}{2}x - 5.$$

In this form, the equations can be thought of as representing two *different* functions with the *same* independent variable, x. For example, a newspaper carrier's expenses and income both depend on the number of papers he or she delivers. The intersection point of the graphs could represent the number of papers for which the income equals the expenses.

Suppose someone asks, "What is y when $x = 4$?" You cannot answer this question, since there are two different functions. If you simply use "y," there is

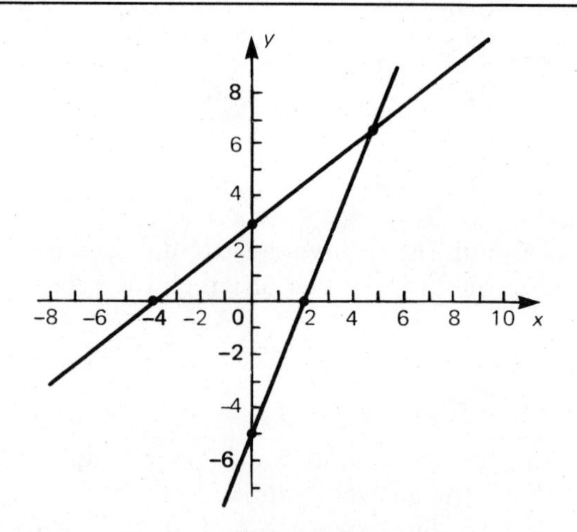

Figure 4-4a.

no way to tell which function is meant, and no way to tell what number was substituted for x. A better symbol is needed. Recalling that a function is a *set* of *ordered pairs*, you could let

$$f = \{(x, y): y = \frac{3}{4}x + 3\}$$

$$g = \{(x, y): y = \frac{5}{2}x - 5\}.$$

The letter "f" is picked because it is the first letter of "function." The letter "g" is picked because it comes next to "f" in the alphabet.

The symbol "$f(4)$" is used to mean, "The value of y in the function f when $x = 4$." The symbol $f(4)$ is pronounced "f of 4," or "f at 4." Similarly, $g(4)$ is the value of y in function g when $x = 4$. The symbols $f(x)$ and $g(x)$ are simply fancy names for the dependent variable y. Therefore, you can write the equations for the functions as

$$f(x) = \frac{3}{4}x + 3 \quad \text{and} \quad g(x) = \frac{5}{2}x - 5.$$

By substituting 4 for x in these equations, you get

$$f(4) = \frac{3}{4} \cdot 4 + 3 = 3 + 3 = 6,$$

$$g(4) = \frac{5}{2} \cdot 4 - 5 = 10 - 5 = 5.$$

Objectives:

1. Become "comfortable" using $f(x)$ terminology by using it to evaluate given functions for given values of x.

2. Use systems of linear functions as mathematical models.

For the first objective, all you need to remember is that something like "$f(7)$" simply means, "Substitute 7 everywhere you find x in the $f(x)$ equation."

Example 1: If $f(x) = 2x^2 + 4x - 1$, find $f(-3)$.

Substituting -3 for x, you get

$$\begin{aligned} f(-3) &= 2(-3)^2 + 4(-3) - 1 \\ &= 18 - 12 - 1 \\ &= \underline{\underline{5}}. \end{aligned}$$

For the second objective, you must keep firmly in mind that "$f(x)$" and "$g(x)$" are just symbols for *dependent variables*. They behave exactly like y in $y = mx + b$.

Example 2: Suppose that your family is going to purchase a new air conditioning unit. One brand costs \$900 to purchase, and \$30 per month to operate. A more expensive brand costs \$1300 to purchase. But it is more efficient, and costs only \$25 per month to operate.

Let x = number of months since you purchased the unit.

Let $f(x)$ = total number of dollars you will spend in x months if you purchase the cheaper unit.

Let $g(x)$ = total number of dollars you will spend in x months if you buy the more efficient unit.

a. Write particular equations expressing $f(x)$ and $g(x)$ in terms of x.
b. Find $f(100)$ and $g(100)$. What do these two numbers tell you about the relative costs of the two units after 100 months?
c. Plot graphs of functions f and g on the *same* Cartesian coordinate system.
d. Solve the system in part a to find the "break-even" point. That is, find the number of months at which the total cost of either unit would be the *same*.

a. Since it costs \$30 to run the cheaper unit for *one* month, it will cost $30x$ dollars to run it for x months. Adding the purchase price of \$900, the total cost is $30x + 900$. So

$$\underline{\underline{f(x) = 30x + 900.}}$$

Similarly, it costs $25x$ dollars to run the efficient unit for x months. So the total cost is

$g(x) = 25x + 1300.$

b. $f(100) = 30 \cdot 100 + 900 = 3000 + 900 = 3900.$

$g(100) = 25 \cdot 100 + 1300 = 2500 + 1300 = 3800.$

So it costs *less* to own the *expensive* (efficient) unit if you keep it for 100 months!

c. Since both functions are *linear*, the easiest way to plot the graphs is to plot the *known* points. Then you can connect the points with straight lines. The intercepts are 900 and 1300 for f and g, respectively. Using $f(100) = 3900$ and $g(100) = 3800$ from part b, you can plot a graph as shown in Figure 4-4b.

d. The easiest way to solve this system is to realize that where the graphs cross, $f(x) = g(x)$. From the equations in part a,

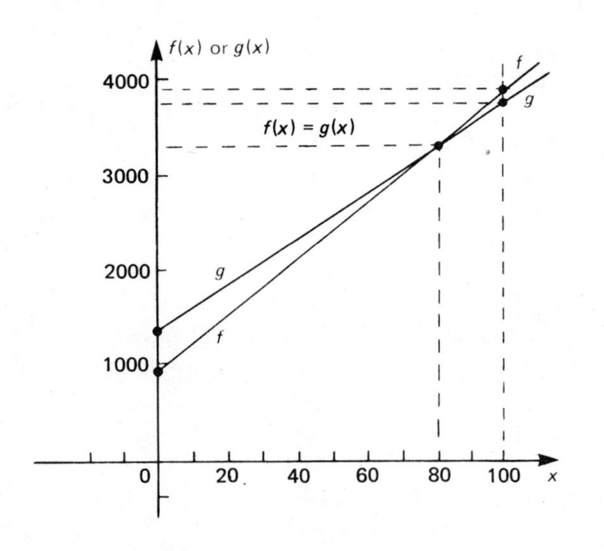

Figure 4-4b.

$30x + 900 = 25x + 1300$	Equate $f(x)$ and $g(x)$
$5x = 400$	Subtract $25x$ and subtract 900.
$x = 80$	Divide by 5.

So the "break-even" point comes at 80 months.

The following exercise gives you some practice using $f(x)$ terminology. After you gain confidence with this terminology, you will use it in real-world problems involving two functions.

Exercise 4-4

Problems 1 through 4 refer to the following functions:

$f = \{(x, y): y = 3x + 11\},$

$g = \{(x, y): y = x^2 + x + 1\}.$

1. Find the following function values:

a. $f(7)$ b. $f(-4)$ c. $f(0)$ d. $f(0.5)$
e. $g(5)$ f. $g(-3)$ g. $g(0)$ h. $g(1/3)$

2. Recalling that "$f(2)$" means, "Substitute 2 for x," find the following:

a. $f(r)$ b. $f(j)$ c. $g(n)$ d. $g(k)$
e. $f(s + t)$ f. $g(4 - a)$

3. The symbol "$f(5)$" does *not* mean, "f *times* 5," since f is a *set*, not a number. With this fact in mind, evaluate the following:

a. $\frac{f(5)}{g(5)}$ b. $\frac{g(2)}{f(2)}$ c. $\frac{f(6)}{g(3)}$ d. $\frac{g(1)}{g(0)}$

4. From Chapter 1 you should recall that parentheses mean, "Do what's inside *first*." Using this fact, evaluate the following:

a. $f(g(2))$ b. $g(f(-5))$ c. $f(f(-2))$ d. $g(g(0))$
e. $f(g(x))$ f. $g(f(x))$ g. $f(f(x))$ h. $g(g(x))$

5. ***Cops and Robbers Problem.*** Robin Banks robs a bank and drives off. A short time later he passes a truck stop at which police officer Willie Katchup is dining. Willie receives a call from his dispatcher, and takes off in pursuit of Robin.

Let t = number of minutes that have elapsed since Robin passed the truck stop.
Let $f(t)$ = number of kilometers Robin has gone past the truck stop.
Let $g(t)$ = number of kilometers Willie has gone past the truck stop.

a. Robin's equation is $f(t) = 3/4t$. Find $f(12)$, $f(4)$, and $f(-8)$.
b. Willie's equation is $g(t) = 2(t - 5)$. Find $g(7)$ and $g(15)$.
c. By calculation, find out where and when Willie Katchup will catch up with Robin Banks.
d. When did Willie leave the truck stop?
e. Plot the graphs of functions f and g on the same Cartesian coordinate system. You should choose scales that make the intersection point of the two graphs come on the graph paper.
f. The slope of a linear function is a rate of change. What are the units of the slopes of functions f and g?
g. How many kilometers per hour are Robin and Willie going?

6. ***Pedalboat Problem*** Eb and Flo go pedalboating on the San Antonio River. They check out a boat, head out along the river for awhile, then turn around and come back.

Let t = number of minutes that have elapsed since they left the boatdock.
Let $f(t)$ = number of meters they are from the boatdock on the way out.
Let $g(t)$ = number of meters they are from the boatdock on the way back in.

a. They find that the equation for function f is $f(t) = 32t$. Find $f(3)$, $f(7)$, and $f(10)$.
b. The equation for function g is $g(t) = -17t + 510$. Find $g(18)$ and $g(23)$.
c. On the same Cartesian coordinate system, plot graphs of functions f and g. You may use different scales for the two axes if this makes the graphs have better proportions. Make sure that the domain of each function agrees with the facts in this problem.
d. To the nearest minute and the nearest 10 meters, where do the graphs seem to intersect each other?
e. Calculate the ordered pair where the graphs intersect.
f. What real-world quantities do the ordinate and the abscissa of the intersection point represent?
g. When did they arrive back at the boatdock?
h. Assuming that they went the same speed *through the water* on both the trip out and the trip back, did they start out going *upstream* or *downstream*? Justify your answer.
i. Just for fun, see if you can calculate the speed of the current in the San Antonio River.

7. ***Diesel Car vs. Gasoline Car Problem.*** The July, 1974, issue of *Popular Science* magazine has an article comparing the cost of owning a car with a diesel engine (like big trucks use) with the cost of owning a car with a normal gasoline engine. According to this article, it costs about 2.5 cents per mile to operate a diesel car. Gasoline cars cost about 7 cents per mile to operate because they use more fuel, the fuel is more expensive, and they need more frequent tune-ups.

a. Let d be a variable equal to the number of miles you have driven your car. Let $f(d)$ be the number of cents you have spent driving a gasoline car for d miles. Write the particular equation expressing $f(d)$ in terms of d, recalling that f is a *linear* function, and that $f(0) = 0$.
b. Calculate $f(1000)$, $f(10{,}000)$, and $f(100{,}000)$.
c. How many *dollars* will you spend driving a gasoline car for 100,000 miles? Surprising??
d. Let $g(d)$ be the number of cents you have spent driving a *diesel* car for d miles. According to *Popular Science*, you must pay about \$2500 more to buy a diesel car. So in order to compare gasoline and diesel cars, you must add

this initial \$2500 to the diesel car's operating cost. That is, the $g(d)$-intercept must be 250,000. Thereafter, $g(d)$ varies *linearly* with d. Write the particular equation expressing $g(d)$ in terms of d.

e. Calculate $g(1000)$, $g(10{,}000)$, and $g(100{,}000)$.

f. Plot graphs of functions f and g on the *same* Cartesian coordinate system.

g. Which kind of car is more economical to own if you plan to drive

   i. many thousands of miles?
   ii. just a few thousand miles?

   Explain how you can tell.

h. Figure out how far you must drive in order to "break even." That is, find out where the cost of operating one kind of car just equals the cost of operating the other kind.

8. ***Hamburger Problem*** Sue Flay and Cassa Roll obtain a franchise to operate a hamburger stand for a well-known national hamburger chain. They pay \$20,000 for the franchise, and have additional expenses of \$250 per thousand hamburgers they sell. The price of a hamburger is \$.75, so they take in a revenue of \$750 per thousand burgers.

a. Let x be the number of thousands of burgers they sell, and let $r(x)$ be the number of dollars of revenue they receive. Write an equation expressing $r(x)$ in terms of x.

b. Find $r(20)$, $r(50)$, and $r(0)$.

c. Let $c(x)$ be the cost of running the hamburger stand, including the \$20,000 they paid at the beginning for the franchise and the additional expenses. Write an equation expressing $c(x)$ in terms of x.

d. Find $c(20)$, $c(50)$, and $c(0)$.

e. Plot graphs of functions r and c on the *same* Cartesian coordinate system. Use a scale small enough so that the point of intersection of the two graphs is on your graph paper.

f. Calculate the number of burgers Sue and Cassa must sell in order to "break even."

4-5 LINEAR EQUATIONS WITH THREE OR MORE VARIABLES

In Section 4-4 you studied situations in which there were *three* variables. Two dependent variables such as distance were related to one independent variable such as time. In this section you will study linear equations with three variables *without* being concerned with which are dependent and which are independent.

Objective:

Determine what the graph of a linear equation with three variables looks like, and be able to sketch the graph.

Since you are not concerned with whether the variables are dependent or independent, you usually write equations with the variables on one side and the constant on the other. For example,

$$2x + 3y + 4z = 12.$$

A solution to such an equation must contain *three* numbers, one for each of the three variables. It is customary to write the solutions as ordered *triples;* for example (4, 0, 1). The order in which the numbers appear tells which variable they stand for. If the equation contained four variables, the solutions would be called ordered *quadruples;* for five variables, ordered *quintuples;* for n variables, ordered n*-tuples.*

Agreement: Unless otherwise specified, variables in ordered n-tuples will come in alphabetical order.

Equations with three variables can be graphed on a three-dimensional Cartesian coordinate system. In addition to the normal x- and y-axes, there is a z-axis perpendicular to the xy-plane, passing through the origin. The positive portions of the three axes are shown in Figure 4-5a.

Plotting an equation such as

$$2x + 3y + 4z = 12,$$

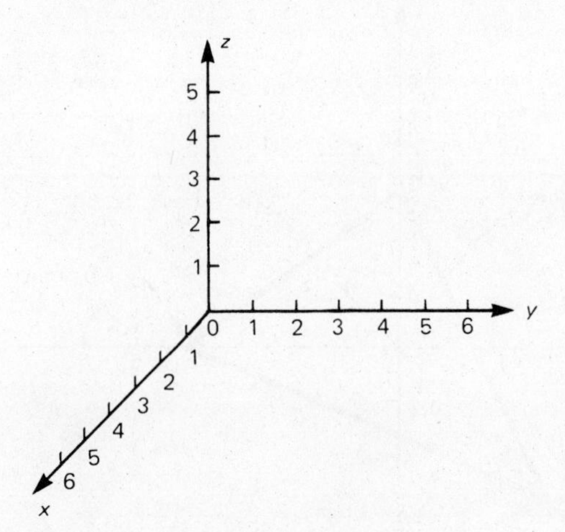

Figure 4-5a.

can be accomplished by picking values of *one* variable, and seeing what you get for the others.

If $z = 0$, then $2x + 3y = 12$ —— ①.

If $z = 1$, then $2x + 3y + 4 = 12$
$2x + 3y = 8$ —— ②.

If $z = 2$, then $2x + 3y + 8 = 12$
$2x + 3y = 4$ —— ③.

Graphs of Equations ①, ②, and ③ are shown in Figure 4-5b.

The three lines in Figure 4-5b are the parts of the graph at $z = 0$, at $z = 1$, and at $z = 2$. By stacking these three planes on top of each other (Figure 4-5c), you can see that the whole graph is a *plane* in *space*, containing these three lines.

Conclusion: The graph of a linear equation with three variables is a *plane* in *space*.

Once you realize what the graph looks like, you can draw it more quickly. The line where the graph cuts the xy-plane is called the graph of the *xy-trace*. It is obtained by setting $z = 0$. Similarly, the yz-trace

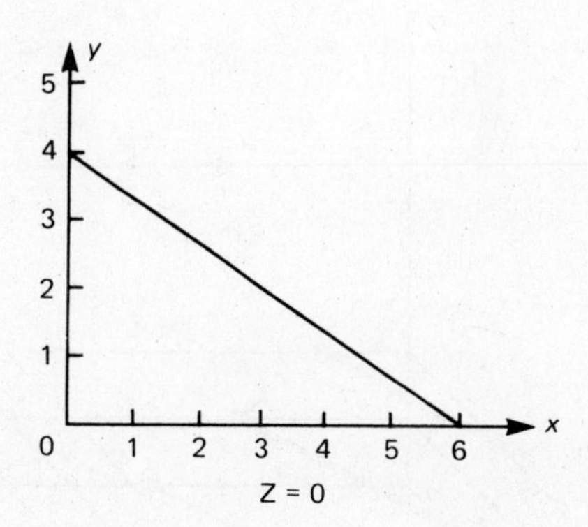

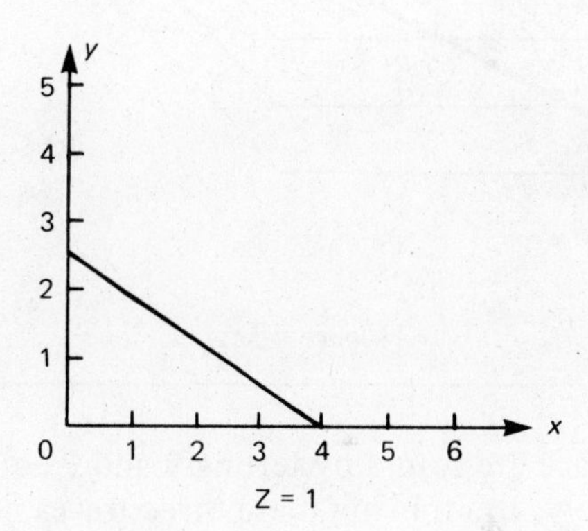

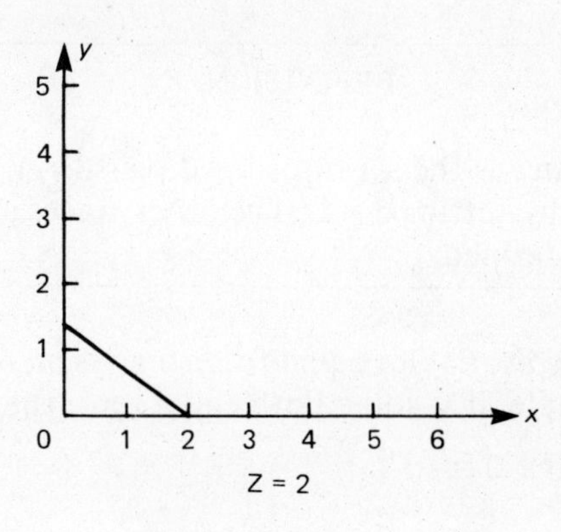

Figure 4-5b.

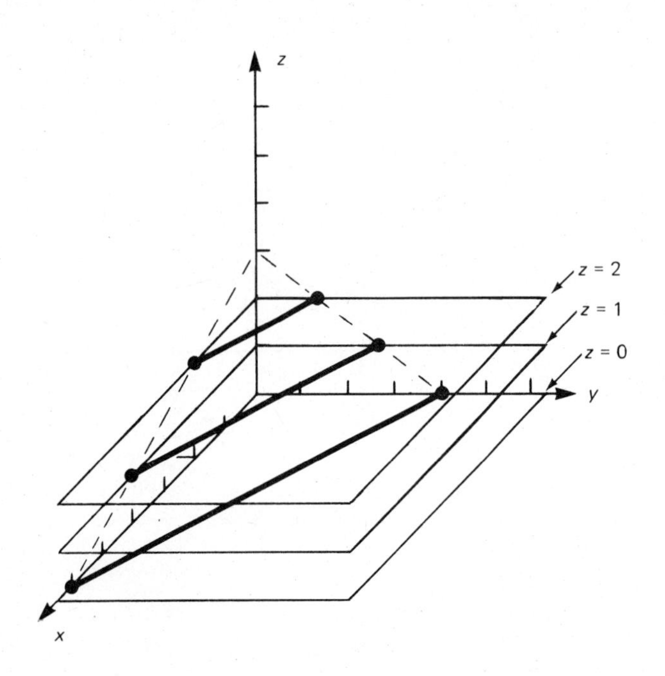

Figure 4-5c.

and xz-trace are found by letting x and y equal zero, respectively. By drawing these three traces, you get a reasonable picture of the plane (Figure 4-5d).

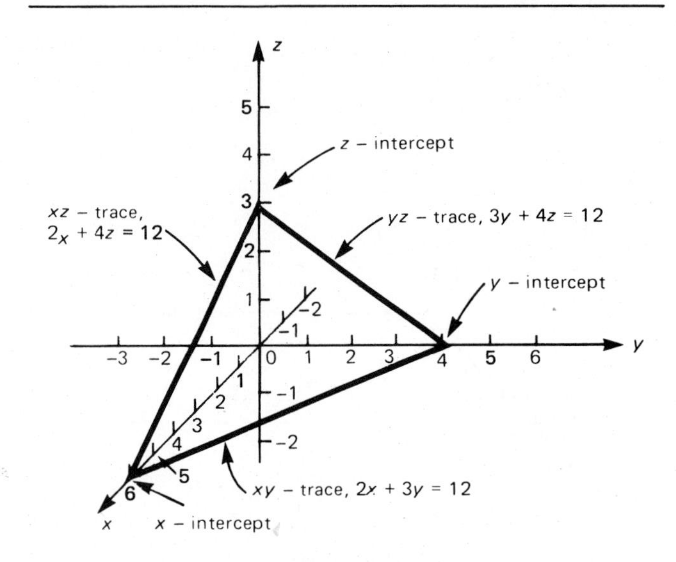

Figure 4-5d

DEFINITION

The ***xy-trace*** is the set of ordered pairs (x, y) obtained by setting $z = 0$. The other traces are similarly defined.

Setting *two* variables equal to zero gives an *intercept*. For example, if y and z both equal zero, then

$$2x + 0 + 0 = 12,$$
$$x = 6.$$

So the x-intercept equals 6. Similarly, the y- and z-intercepts are 4 and 3, respectively as shown in Figure 4-5d.

DEFINITION

The ***x-intercept*** is the value of x when y and z are both zero. The other intercepts are similarly defined.

The idea of traces and intercepts can be extended to equations with more than three variables. Since you have used an *algebraic* definition of these quantities, the fact that there can be no four or five dimensional graphs becomes insignificant. A trace is obtained simply by letting *one* variable equal zero, while an intercept is obtained by setting *all but one* variable equal to zero.

Exercise 4-5

1. Sketch a graph of each of the following equations by drawing its three traces as in Figure 4–5d.

a. $6x + 4y + 3z = 24$
b. $2x - 3y + z = 12$

c. $3x + 5y - 3z = 15$
d. $4x - 2y - z = 8$
e. $x + y + z = -7$

2. There are some interesting special cases in which the graphs turn out to be parallel, perpendicular, or coincident with the coordinate planes or axes. Sketch a graph of each of the following equations by drawing their traces as in Problem 1. Then tell what the graph is parallel to, or coincident with.

a. $x + y = 7$.
b. $y + z = 4$.
c. $x + y = 0$.
d. $y - z = 0$.
e. $x = 5$.
f. $y = -6$.

3. For the equation $7w + 3x - 4y + 6z = 42$,

a. find the four intercepts, and
b. find the equation of the xyz-trace.

4. Explain why a trace is the same as an intercept for an equation with two variables.

5. Obtain three pieces of cardboard; playing cards or index cards will do. These will represent graphs of equations with three variables.

a. Hold two of the cardboards together so that they meet along an edge. What do the points of intersection represent with regard to a system of two equations in three variables? How many ordered triples normally satisfy a system of two equations in three variables?
b. Hold the third cardboard so that its corner touches the line of intersection of the other two. Why do you suppose that a system of *three* linear equations in three variables has a *unique* solution, whereas a system with only two does *not*?
c. Hold the three cardboards in such a way that they intersect at an *infinite* number of points. There are at least two ways to do this. If this happens, the three equations are said to be "dependent."
d. Hold the three cardboards in such a way that there are *no* points common to all three. There are at least three ways of doing this besides the obvious way of three parallel planes. If this happens, the three equations are said to be "inconsistent."
e. From what you observed above, can you conclude that a system of three linear equations in three variables *always* has a unique solution? Explain.

4-6 SYSTEMS OF LINEAR EQUATIONS WITH THREE OR MORE VARIABLES

The graph of a linear equation with three variables is a plane in space. If you hold three index cards as shown in Figure 4-6, you can see that *three* such planes usually intersect at a *single* point. So a system of three linear equations with three variables will usually have a *single* ordered triple in its solution set.

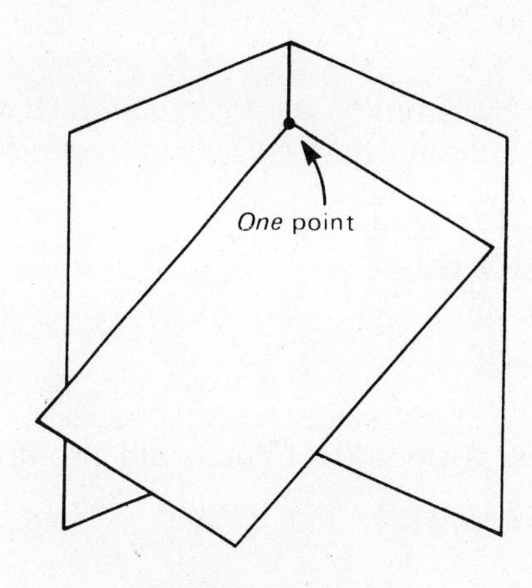

Three intersecting planes

Figure 4-6.

Objective: Be able to find the single ordered triple that satisfies a system of three linear equations with three variables.

Suppose that you must solve the system

$$\begin{aligned} 2x + 3y - z &= -1 \quad \text{———} \quad ① \\ -x + 5y + 3z &= -10 \quad \text{———} \quad ② \\ 3x - y - 6z &= 5 \quad \text{———} \quad ③. \end{aligned}$$

If you can *eliminate* a variable, you can reduce the problem to a system with *two* variables. Thus, you will change a new problem into an old problem. The technique of linear combination that you learned in Section 4-2 can be used to eliminate a variable. The following steps show how.

Multiply both members of Equation ② by 2 and add the resulting equation to Equation ①.

$$\begin{aligned} -2x + 10y + 6z &= -20 \\ 2x + 3y - z &= -1 \quad \text{———} \quad ① \\ \hline 13y + 5z &= -21 \quad \text{———} \quad ④. \end{aligned}$$

Multiply ② by 3 and add it to ③.

$$\begin{aligned} -3x + 15y + 9z &= -30 \\ 3x - y - 6z &= 5 \quad \text{———} \quad ③ \\ \hline 14y + 3z &= -25 \quad \text{———} \quad ⑤. \end{aligned}$$

Multiply ④ by 3 and ⑤ by −5, and *add*. (If you're brave, you could multiply ⑤ by +5, then *subtract*.)

$$\begin{aligned} 39y + 15z &= -63 \\ -70y - 15z &= 125 \\ \hline -31y \quad &= 62 \\ \therefore\ y &= -2. \end{aligned}$$

Substitute −2 for y in ④. (You could also use ⑤.)

$$\begin{aligned} -26 + 5z &= -21 \\ 5z &= 5 \\ \therefore\ z &= 1. \end{aligned}$$

Substitute −2 for y and 1 for z in ①. (You could use ② or ③.)

$$\begin{aligned} 2x - 6 - 1 &= -1 \\ 2x &= 6 \\ \therefore\ x &= 3 \end{aligned}$$

$\therefore S = \{(3, -2, 1)\}$.

Note that there are quite a few tedious computations to do. It is helpful, therefore, if you write yourself messages as you go along, such as

"Multiply ② by 3 and add ③."

These messages will help you retrace your steps if you make a mistake.

The following exercise will give you practice solving systems with three variables. You will also apply what you have learned to solving systems with *more* than three variables.

Exercise 4-6

Solve the following systems.

1. $\begin{aligned} x - 2y + 3z &= 3 \\ 2x + y + 5z &= 8 \\ 3x - y - 3z &= -22 \end{aligned}$

2. $\begin{aligned} 2x - y - z &= 7 \\ 3x + 5y + z &= -10 \\ 4x - 3y + 2z &= 4 \end{aligned}$

3. $\begin{aligned} 3x + 2y - z &= 10 \\ x + 4y + 2z &= 3 \\ 2x + 3y - 5z &= 23 \end{aligned}$

4. $\begin{aligned} 3x + 4y + 2z &= 6 \\ x + 3y - 5z &= -7 \\ 5x + 7y - 3z &= 3 \end{aligned}$

5. $\begin{aligned} 5x - 4y + 3z &= 15 \\ 6x + 2y + 9z &= 13 \\ 7x + 6y - 6z &= 6 \end{aligned}$

6. $\begin{aligned} 3x - 2y + 5z &= -17 \\ 2x + 4y - 3z &= 29 \\ 5x - 6y - 7z &= 7 \end{aligned}$

7. $\begin{aligned} 5x - 4y - 6z &= 21 \\ -2x + 3y + 4z &= -15 \\ 3x - 7y - 5z &= 15 \end{aligned}$

8. $\begin{aligned} 2x + 2y + 3z &= -1 \\ 3x - 5y - 2z &= 21 \\ 7x + 3y + 5z &= 10 \end{aligned}$

Problems 9 and 10 have some variables "missing." This makes the systems *easier*, if you are clever enough to figure out *why*.

9. $\begin{aligned} 3x + 4y \qquad &= 19 \\ 2y + 3z &= 8 \\ 4x \qquad - 5z &= 7 \end{aligned}$

10. $\begin{aligned} 2x - 3y \qquad &= 5 \\ 4y + 2z &= -6 \\ 5x \qquad + 7z &= -15 \end{aligned}$

The equations in Problems 11 and 12 are either *inconsistent* (*no* common solution), or *dependent* (an *infinite* number of common solutions). By solving the two systems, tell which is which.

11. $6x + 9y - 12z = 14$
$2x + 3y - 4z = -11$
$x + y + z = 1$

12. $3x + 2y - z = 4$
$5x - 3y + 2z = 1$
$9x - 13y + 8z = -5$

Problems 13 and 14 are systems with *four* variables.

13. $4w + x + 2y - 3z = -16$
$-3w + 3x - y + 4z = 20$
$-w + 2x + 5y + z = -4$
$5w + 4x + 3y - z = -10$

14. $w - 5x + 2y - z = -18$
$3w + x - 3y + 2z = 17$
$4w - 2x + y - z = -1$
$-2w + 3x - y + 4z = 11$

Problems 15 and 16 involve fractions. With *constants* in the denominators, as in Problem 15, you may simply *multiply* both members by some number that will get rid of the fractions. With *variables* in the denominators, as in Problem 16, you may first solve for $1/x$, $1/y$, and $1/z$, recognizing that $4/x$ is the same as $4(1/x)$, etc.

15. $\frac{x}{2} + \frac{y}{4} + \frac{z}{3} = 24$
$\frac{x}{4} + \frac{y}{3} + \frac{z}{2} = 29$
$\frac{x}{3} + \frac{y}{2} + \frac{z}{4} = 25$

16. $\frac{4}{x} - \frac{2}{y} + \frac{6}{z} = -5$
$-\frac{3}{x} - \frac{5}{y} + \frac{4}{z} = -3$
$\frac{2}{x} + \frac{7}{y} - \frac{10}{z} = 2$

Problems 17 and 18 require you to *find* a system of three equations in three variables. Then you must *solve* the system so that you can answer the problem.

17. A Watusi, a Ubangi, and a Pigmy compare the speeds at which each can run. The sum of the speeds of the natives is 30 miles per hour (mph). The Pigmy's speed plus one third of the Watusi's speed is 22 miles per hour more than the Ubangi's speed. Four times the Watusi's speed plus three times the Ubangi's speed minus twice the Pigmy's speed is 12 miles per hour. Find out how fast each native can run. Then tell what is unfortunate about the Ubangi.

18. The road from Tedium to Ennui is uphill for 5 miles, level for 4 miles, then downhill for 6 miles. John Garfinkle walks from Ennui to Tedium in 4 hours; later he walks halfway from Tedium to Ennui and back again in 3 hours and 55 minutes. Still later he walks from Tedium all the way to Ennui in 3 hours and 52 minutes. What are his rates of walking uphill, downhill, and on level ground, if these rates remain constant?

4-7 HIGHER ORDER DETERMINANTS

In Section 4-3 you learned how to use determinants to solve a system of two linear equations in two variables. In this section you will extend the technique to systems with three or more variables.

Objective: Be able to use determinants to solve a system of three (or more) linear equations with three (or more) variables.

You recall that the system

$3x + 4y = -7$
$2x - 5y = 8$,

has x and y values as follows:

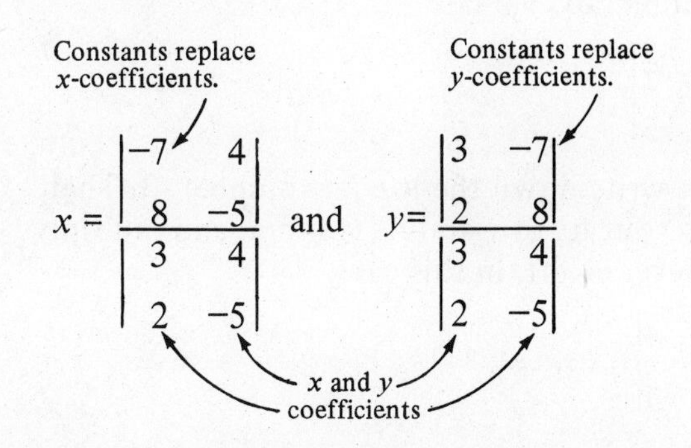

The denominator determinant contains the x- and y-coefficients from the equations. The x-numerator is obtained by replacing the x-coefficients with the constants -7 and 8 from the right members of the equations. The y-numerator is obtained by replacing the y-coefficients with these constants.

The solutions of a system of three or more linear equations in three or more variables can be written the same way. For instance, the system in the example of Section 4-6 has the following value for y:

System:

$$\begin{aligned} 2x + 3y - z &= -1 \\ -x + 5y + 3z &= -10 \\ 3x - y - 6z &= 5 \end{aligned}$$

y-value:

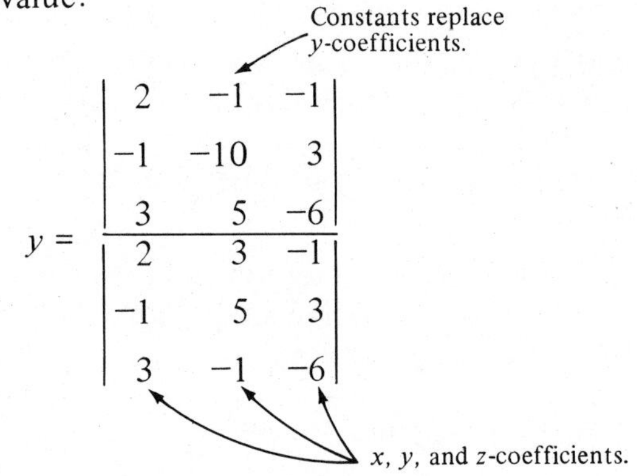

There is another way to evaluate a second-order determinant that carries over to higher-order ones. For example, to evaluate

$$\begin{vmatrix} 3 & 4 \\ 2 & -5 \end{vmatrix}$$

you first write down the *top left* number, 3. Then you cross out its row and its column, and multiply by whatever is left, in this case, -5.

$$\begin{vmatrix} 3 & 4 \\ 2 & -5 \end{vmatrix} = 3(-5)\ldots.$$

Next, you write the 4 from the *top right* position, cross out its row and its column, and multiply by what is left, 2. This is *subtracted* from the first term.

$$\begin{vmatrix} 3 & 4 \\ 2 & -5 \end{vmatrix} = 3(-5) - 4(2).$$

Completing the arithmetic, the determinant equals -23.

Third-order determinants are evaluated exactly the same way. First, write down the *top left* number, cross out its row and its column, and multiply by what is left. For example,

$$\begin{vmatrix} 2 & 3 & -1 \\ -1 & 5 & 3 \\ 3 & -1 & -6 \end{vmatrix} = 2\begin{vmatrix} 5 & 3 \\ -1 & -6 \end{vmatrix}\ldots.$$

Note that "what is left" is simply a *second*-order determinant. This is called a *minor* determinant. Proceeding across the top row, the next number is 3. Crossing out its row and column and multiplying by the minor determinant remaining,

$$\begin{vmatrix} 2 & 3 & -1 \\ -1 & 5 & 3 \\ 3 & -1 & -6 \end{vmatrix} = 2\begin{vmatrix} 5 & 3 \\ -1 & -6 \end{vmatrix} - 3\begin{vmatrix} -1 & 3 \\ 3 & -6 \end{vmatrix}\ldots.$$

Note that this minor determinant is *subtracted* from the first one. Finally the -1 at the end of the top row is multiplied by its minor determinant, found by crossing out its row and column, and the result is *added* to the other terms. The final answer is:

$$\begin{vmatrix} 2 & 3 & -1 \\ -1 & 5 & 3 \\ 3 & -1 & -6 \end{vmatrix} = 2\begin{vmatrix} 5 & 3 \\ -1 & -6 \end{vmatrix} - 3\begin{vmatrix} -1 & 3 \\ 3 & -6 \end{vmatrix} + (-1)\begin{vmatrix} -1 & 5 \\ 3 & -1 \end{vmatrix}.$$

Expanding the minor determinants gives

$$\begin{aligned}&2(-30-(-3))-3(6-9)+(-1)(1-15)\\&=-54+9+14\\&=-31.\end{aligned}$$

The denominators for x, y, and z are all the *same*. In this case, they equal -31. The numerator for y, shown above, equals 62. The numerators for x and z are found by replacing the x- and z-coefficients, respectively, with the constants on the right of the equations. Expanding these determinants by minors gives

$$x=\frac{-93}{-31}=3,\quad y=\frac{62}{-31}=-2,\quad \text{and} \quad z=\frac{-31}{-31}=1.$$

From these, the solution set is $S=\{(3,-2,1)\}$.

The procedure is easier than linear combination from the standpoint that you immediately know the answers. A lot of arithmetic must be done to simplify the answer, though. The "payoff" comes when you must solve equations that have "untidy" fractions for answers. Determinants work just as easily on these systems as on systems that have integer answers!

Systems with four, five, and more variables can be solved by determinants. A fourth order determinant can be expanded into four third-order minors, which, in turn, give 12 second-order minors. A tenth-order determinant would have almost two *million* second-order minors!

In the following exercise you will again solve the systems in Exercise 4-6. This time you will use determinants (sometimes called using Cramer's Rule).

Exercise 4-7

Solve the following systems by determinants (Cramer's Rule). Problems 1 through 14 are the same as those in Exercise 4-6.

1. $\begin{aligned}x-2y+3z&=3\\2x+y+5z&=8\\3x-y-3z&=-22\end{aligned}$

2. $\begin{aligned}2x-y-z&=7\\3x+5y+z&=-10\\4x-3y+2z&=4\end{aligned}$

3. $\begin{aligned}3x+2y-z&=10\\x+4y+2z&=3\\2x+3y-5z&=23\end{aligned}$

4. $\begin{aligned}3x+4y+2z&=6\\x+3y-5z&=-7\\5x+7y-3z&=3\end{aligned}$

5. $\begin{aligned}5x-4y+3z&=15\\6x+2y+9z&=13\\7x+6y-6z&=6\end{aligned}$

6. $\begin{aligned}3x-2y+5z&=-17\\2x+4y-3z&=29\\5x-6y-7z&=7\end{aligned}$

7. $\begin{aligned}5x-4y-6z&=21\\-2x+3y+4z&=-15\\3x-7y-5z&=15\end{aligned}$

8. $\begin{aligned}2x+2y+3z&=-1\\3x-5y-2z&=21\\7x+3y+5z&=10\end{aligned}$

9. $\begin{aligned}3x+4y\quad&=19\\2y+3z&=8\\4x\quad-5z&=7\end{aligned}$

10. $\begin{aligned}2x-3y\quad&=5\\4y+2z&=-6\\5x\quad+7z&=-15\end{aligned}$

11. $\begin{aligned}6x+9y-12z&=14\\2x+3y-4z&=-11\\x+y+z&=1\end{aligned}$

12. $\begin{aligned}3x+2y-z&=4\\5x-3y+2z&=1\\9x-13y+8z&=-5\end{aligned}$

13. $\begin{aligned}4w+x+2y-3z&=-16\\-3w+3x-y+4z&=20\\-w+2x+5y+z&=-4\\5w+4x+3y-z&=-10\end{aligned}$

14. $\begin{aligned}w-5x+2y-z&=-18\\3w+x-3y+2z&=17\\4w-2x+y-z&=-1\\-2w+3x-y+4z&=11\end{aligned}$

Problems 15 and 16 have solutions with untidy fractions. Use determinants (Cramer's Rule) to solve:

15. $\begin{aligned}2x+4y-3z&=7\\7x-3y+2z&=8\\5x-5y+7z&=-1.\end{aligned}$

16. $\begin{aligned}3x-7y+2z&=11\\8x+2y-5z&=-3\\5x-3y-3z&=4.\end{aligned}$

4-8 MATRIX SOLUTION OF LINEAR SYSTEMS

Higher-order determinants can be used to solve systems with more than two variables. However, the amount of work increases enormously with the number of variables. For example, solving a system with 5 variables is equivalent to evaluating 360 second-order determinants, and solving one with 10 variables is equivalent to evaluating 19,958,400 second-order determinants! Since real-world problems can lead to systems with thousands of variables, a more efficient procedure is needed.

The example solved in Section 4-6 was

$$\begin{aligned} 2x + 3y - z &= -1 \\ -x + 5y + 3z &= -10 \\ 3x - y - 6z &= 5. \end{aligned}$$

The important thing about the system is the *coefficients* rather than the variable names. So you strip away the variables and write the system as

$$\left[\begin{array}{ccc|c} 2 & 3 & -1 & -1 \\ -1 & 5 & 3 & -10 \\ 3 & -1 & -6 & 5 \end{array}\right].$$

This array of numbers is called a *matrix*. Each *row* represents one of the *equations*. Each *column* contains the coefficients of one *variable*. The last column is set off by a vertical line since it contains the righthand members of the equations rather than coefficients of variables. The large brackets, [ ], which hold it all together are symbols used to indicate a matrix.

Objective:

Be able to solve a linear system by performing operations on the corresponding matrix.

Since a matrix represents a linear system, you can perform any operation on it that you can perform on the system. Specifically, you can

1. multiply each element of a row by the same number, and

2. add the elements of one row to the corresponding elements of another row.

The result of such operations will be a matrix of an *equivalent* system.

Eliminating variables by linear combination is equivalent to getting *zeros* at various places in the matrix. In this case, you would write down the first row, then multiply the second row by 2, getting

$$\left[\begin{array}{ccc|c} 2 & 3 & -1 & -1 \\ -2 & 10 & 6 & -20 \\ 3 & -1 & -6 & 5 \end{array}\right] \begin{array}{l} \\ \leftarrow 2 \times \text{old row } ② . \\ \\ \end{array}$$

Now adding the first row to the second will produce "0" in the second row, giving

$$\left[\begin{array}{ccc|c} 2 & 3 & -1 & -1 \\ 0 & 13 & 5 & -21 \\ 3 & -1 & -6 & 5 \end{array}\right] \begin{array}{l} \\ \leftarrow \text{new row } ② + \text{old row } ① . \\ \\ \end{array}$$

To get a "0" in the third row, first column, you multiply the third row by −2, the *opposite* of the number in row 1, column 1, getting

$$\left[\begin{array}{ccc|c} 2 & 3 & -1 & -1 \\ 0 & 13 & 5 & -21 \\ -6 & 2 & 12 & -10 \end{array}\right] \begin{array}{l} \\ \\ \leftarrow -2 \times \text{old row } ③ . \end{array}$$

If you are confident enough, you can now multiply the numbers in row ① by 3 (in your head), and add the results to row ③ to get

$$\left[\begin{array}{ccc|c} 2 & 3 & -1 & -1 \\ 0 & 13 & 5 & -21 \\ 0 & 11 & 9 & -13 \end{array}\right] \begin{array}{l} \\ \\ \leftarrow 3 \times \text{row } ① + \text{new row } ③ . \end{array}$$

The last two rows now correspond to a system of *two* equations with *two* variables. You now try to make the number in row ③, column ② equal zero.

First you multiply row ③ by −13, the *opposite* of the number in row ②, column ②, getting

$$\left[\begin{array}{ccc|c} 2 & 3 & -1 & -1 \\ 0 & 13 & 5 & -21 \\ 0 & -142 & -117 & 169 \end{array}\right] \longleftarrow -13 \times \text{row } ③.$$

Multiplying the numbers in row ② by 11 (the number in old row ③, column ②) and adding the results to row ③ gives

$$\left[\begin{array}{ccc|c} 2 & 3 & -1 & -1 \\ 0 & 13 & 5 & -21 \\ 0 & 0 & -62 & -62 \end{array}\right] \longleftarrow 11 \times \text{row } ② + \text{row } ③.$$

The matrix is now in "triangular form" since the "0's" form a triangle in the lower left corner. The last row corresponds to the equation

$$-62z = -62,$$

from which you can see immediately that $z = 1$. The second row corresponds to the equation

$$13y + 5z = -21.$$

Substituting $z = 1$ gives

$$13y + 5 = -21$$
$$13y = -26$$
$$y = -2.$$

Substituting $y = -2$ and $z = 1$ into $2x + 3y - z = -1$ (the first row) gives

$$2x - 6 - 1 = -1$$
$$2x = 6$$
$$x = 3.$$
$$\therefore S = \{(3, -2, 1)\}$$

The matrix scheme is not really much simpler than straightforward linear combination if you are using just paper and pencil. However, it is worth learning because

1. it is adaptable to programming on the computer, and

2. it introduces you to the concept of a matrix, which has very broad applications later on in mathematics.

In the following exercise you will solve the systems of Exercise 4-5 using matrices (the plural of "matrix"). If you are familiar enough with computing, you can try writing a program for solving systems by matrices. See Appendix B-1 for more about operations with matrices.

Exercise 4-8

1.–14. Solve the systems in Problems 1 through 14 of Exercise 4-6 by writing the system as a matrix, then transforming the matrix to triangular form.

15. *Computer Program for Transforming Matrices*—A matrix can be represented as a doubly-subscripted variable, such as $M(R, C)$. The first subscript, R, stands for the row number, and the second subscript, C, stands for the column number. For example, $M(5, 3)$ stands for the number in the *fifth* row, *third* column of matrix M.

Write a computer program to transform a matrix for a system of linear equations into triangular form. The program should contain the following major blocks of instructions:

a. Read the coefficients of the equations into matrix M.
b. Transform the matrix into triangular form.
c. Print out the transformed matrix.

The heart of the transformation is one major step which is repeated many times, "Multiply $M(I, C)$ and $M(R, C)$ by the appropriate numbers, add the results, and store the answer as the new value of $M(I, C)$." Here, I is the number of the row you are *changing*, R is the number of the row you are combining with row I, and C is the column number. The rest of the transformation involves nesting this step inside loops that change the subscripts to the appropriate values. If you are highly skillful at programming, you can write the program yourself. Otherwise, your instructor may wish to show you the program and have you use it for the next problem.

16. *Using the Computer Program*—Use the computer program written in Problem 15 to solve the following systems:

a. Exercise 4-6, Problem 1.
b. Exercise 4-6, Problem 7.
c. Exercise 4-6, Problem 11.
d. Exercise 4-6, Problem 13.
e.

$$\begin{aligned} 3v - 5w + 2x + 4y + z &= 35 \\ 2v + 4w - x - 3y + 6z &= -16 \\ 4v - 2w - 3x + y + 2z &= 18 \\ -5v + w + 4x - y - 3z &= -18 \\ -2v + 5w + 6x - 2y + z &= -19. \end{aligned}$$

4-9 SYSTEMS OF LINEAR INEQUALITIES WITH TWO VARIABLES

You have extended the concept of a system to more than two equations and to more than two variables. Now you are ready to extend the concept to linear *inequalities*. Since it is difficult to write the solution set of a system of inequalities without just restating the inequalities in the solution set, it is customary to draw graphs of the solution sets. For this purpose, it is convenient to consider a piece of graph paper to be a series of horizontal and vertical x and y number lines, each located at the appropriate value of y or x, respectively.

For example, suppose that you must graph the inequality

$$y > 2x - 5.$$

As you do with any unfamiliar graph, you start by picking a value of x, and calculating y. Substituting 2 for x gives the inequality

$$y > -1.$$

This portion of the graph can be plotted on the y-number line which is at $x = 2$ (Figure 4-9a).

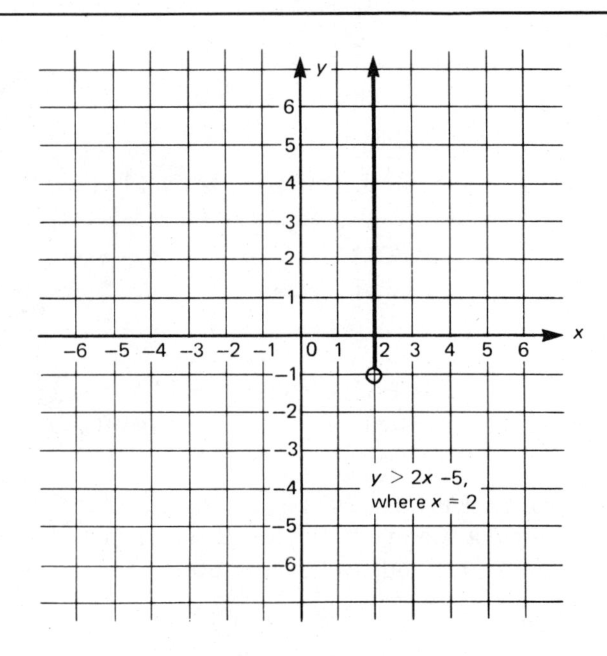

Figure 4-9a.

Selecting other values of x and plotting the corresponding y-inequalities gives a graph which looks like Figure 4-9b. From this graph you can see what the completed graph will look like. The *endpoints* of the y-inequalities will be on the *line* $y = 2x - 5$. The graph itself will be the entire region *above* this boundary line.

Once you know what the graph of a linear inequality looks like, you can draw the graph *quickly*.

Objective:

Be able to draw the graph of a system of linear inequalities with two variables *quickly*.

Example 1: Graph $y > 2x - 5$.

First you draw the boundary line, $y = 2x - 5$. It will be dotted because y cannot *equal* $2x - 5$. Since $y >$

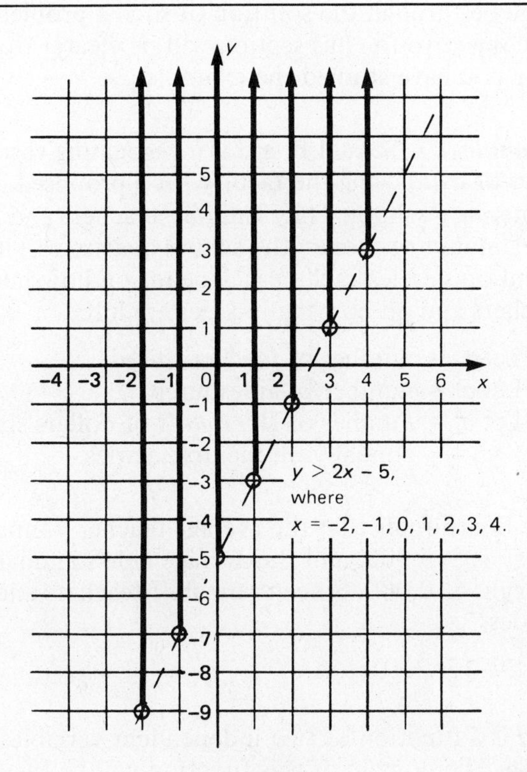

Figure 4-9b.

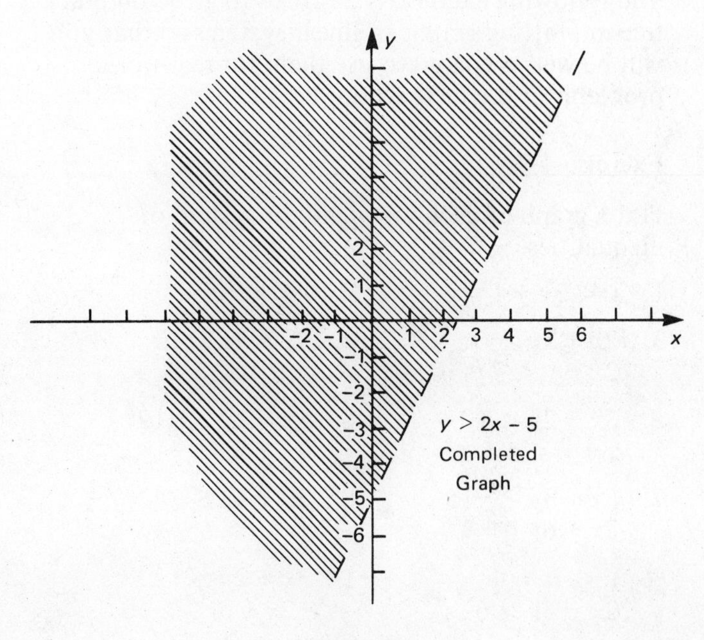

Figure 4-9c.

something, the desired region is *above* the boundary line. So you shade in the region. The graph is shown in Figure 4-9c.

Example 2: Graph the system

$$y > 2x - 5$$

$$y \leq -\frac{1}{3}x + 3.$$

For a system of inequalities, you first plot the boundary line of each inequality. Rather than shading in each inequality as you go, it is tidier to place small arrows on the boundary lines to show which side the region will be on. Then you shade the *intersection* of the regions, since solutions of a system must satisfy *all* of the inequalities. The completed graph is shown in Figure 4-9d.

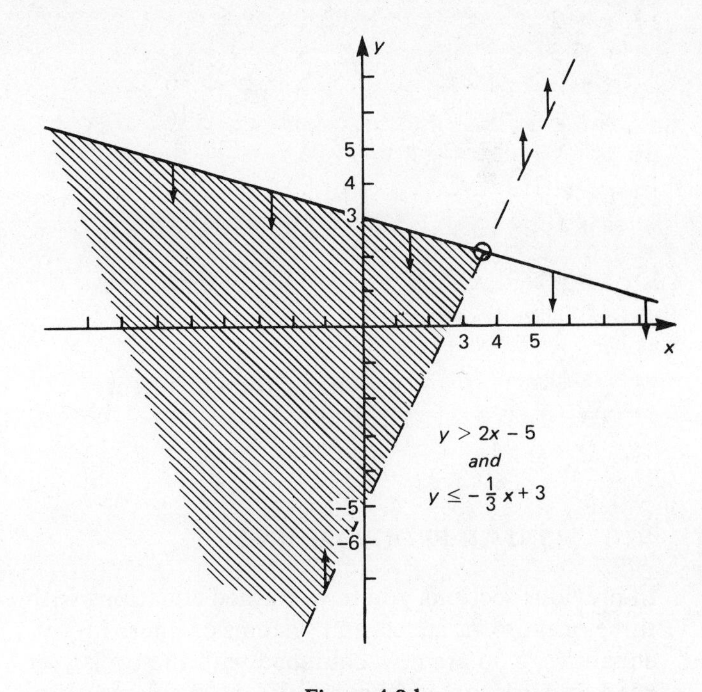

Figure 4-9d.

The following exercise is designed to give you practice in plotting graphs of linear systems so that you will be well prepared to use them for real-world problems in the next section.

Exercise 4-9

Plot a graph of each inequality or system of inequalities.

1. $y < 3x + 1$
2. $y \leq -x + 4$
3. $2x + 3y \geq 6$
   $2x - y < 7$
4. $3x + 4y > 12$
   $x - 4y \geq -8$
5. $5x - 2y < 10$
   $x + y < 4$
6. $5x + 2y > -12$
   $7x + 2y < -18$
7. $5x + 3y \geq -15$
   $2x + 6y < -9$
8. $3x + 4y \leq 24$
   $x - y > 5$
9. $y < 3x - 4$
   $2x - 3y > 6$
   $y - 2 \leq -\frac{3}{5}(x - 4)$
10. $3x + 2y \leq 12$
    $-2x + 5y > -10$
    $x - y \geq -1$
11. $x > 2$
    $y \leq 3$
    $x + y \geq -4$
    $x + y < -1$
12. $2 \leq x \leq 6$
    $y > -1$
    $x - 2y \geq -6$
    $y - 5 \leq -\frac{1}{2}(x - 4)$
13. $y \leq 3x$
    $x + 4y > 12$
14. $y > x$
    $2x - 5y \geq 10$
15. $x + y \leq 4$
    $2x - y > 6$
    $x - 2y = 8$
    (Watch out for a surprise!)
16. $x + y \geq 3$
    $3x - 2y > 6$
    $y \geq 4x + 1$
    (Watch out for a different surprise!)

4-10 LINEAR PROGRAMMING

In previous sections you have studied equations with three or more variables, and systems of linear inequalities. You are now equipped with the tools you need to work some rather significant problems from the real world. The following example leads you stepwise through the solution of such a problem. The objective for this section will be clearer to you after you have studied the example.

Example: A School Board is investigating various ways of composing the faculty for a proposed new elementary school. They can hire teachers and aides. The amount of money the school district will have to spend on salaries each year depends on how many teachers and on how many aides are hired.

Let t = number of teachers hired.
Let a = number of aides hired.
Let d = number of *thousands* of dollars spent annually on faculty salaries.

The Board finds that the average teacher's annual salary is \$15,000, and the average aide's annual salary is \$10,000. Since there are t teachers and a aides,

$$d = 15t + 10a\,.$$

So d is a function of *two* independent variables, t, and a. The domain of this function will be the set of all permissible *ordered pairs,* (t, a).

Suppose that the Board finds the following requirements concerning the permissible numbers of teachers and aides:

i. The building can accommodate no more than 50 faculty members, total. Since the total number of faculty members is $t + a$, this means that

$$t + a \leq 50.$$

ii. A minimum of 20 faculty members is needed to staff the school. This means that

$$t + a \geq 20.$$

iii. The school cannot be run entirely by aides, so there must be at least 12 teachers. Since "at least" means "greater than or equal to," you can write

$$t \geq 12.$$

iv. For a proper teacher-to-aide ratio, the number of teachers must be at least half the number of aides. Therefore,

$$t \geq \frac{1}{2}a.$$

v. It is impossible (obviously!) to hire a *negative* number of teachers or aides. Therefore,

$$t \geq 0 \quad \text{and} \quad a \geq 0.$$

The six inequalities, above, form a *system* that can be plotted using the techniques of Section 4-9. The graph is shown in Figure 4-10a. Ordered pairs (t, a) in this region are possible, or "feasible." So this region is called the *feasible* region. You may now use the equation

$$d = 15t + 10a$$

to find out *which* of the feasible ordered pairs gives the *minimum* cost.

Since d is a third variable, the d-axis should come up vertically out of the page! However, there is a clever way to reduce this three-dimensional problem to a two-dimensional one. Suppose that you wish to find the portion of the feasible region in which the cost is less than or equal to some fixed amount, say \$600,000. That is, you want to find out where $d \leq 600$. Since $15t + 10a$ is equal to d, you can write

$$15t + 10a \leq 600.$$

Solving for a in terms of t,

$$10a \leq -15t + 600.$$

$$a \leq -\frac{3}{2}t + 60.$$

This inequality can be graphed on the feasible region, as shown in Figure 4-10b. The darkly-shaded region represents the values of t and a that make the feasible cost less than or equal to \$600,000 per year.

Picking another fixed value of the cost, say \$400,000, you can find the values of t and a that lead to a cost no more than this:

$$15t + 10a \leq 400.$$

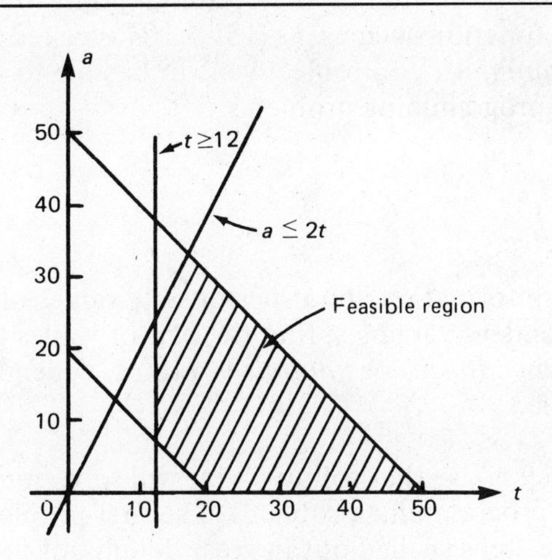

Figure 4-10a.

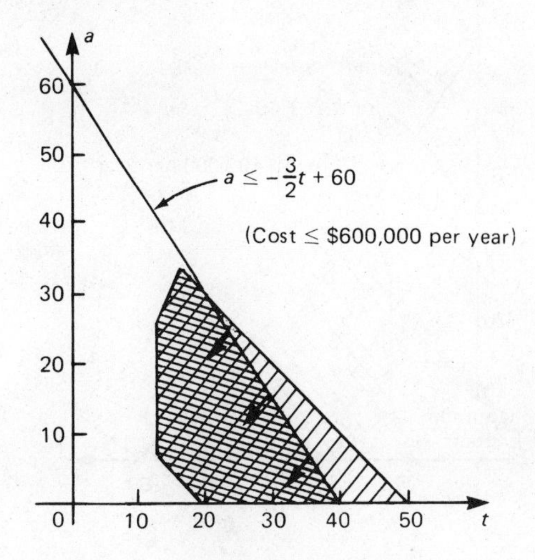

Figure 4-10b.

This can be transformed to

$$10a \leq -15t + 400$$

$$a \leq -\frac{3}{2}t + 40.$$

The portion of the feasible region in which the cost is less than or equal to $400,000 per year is shown in Figure 4-10c.

To find the *minimum* feasible cost, observe that the two "cost boundary lines,"

$$a = -\frac{3}{2}t + 60 \quad \text{and} \quad a = -\frac{3}{2}t + 40,$$

are *parallel* to each other. Any such cost line will have a slope of −3/2. As the cost gets smaller and smaller, the cost line moves closer and closer to the origin, always keeping a slope of −3/2. The minimum cost, therefore, occurs at the *last* point to disappear from the feasible region as the cost line moves toward the origin. From the graph, you can tell that this is the point where the boundary lines

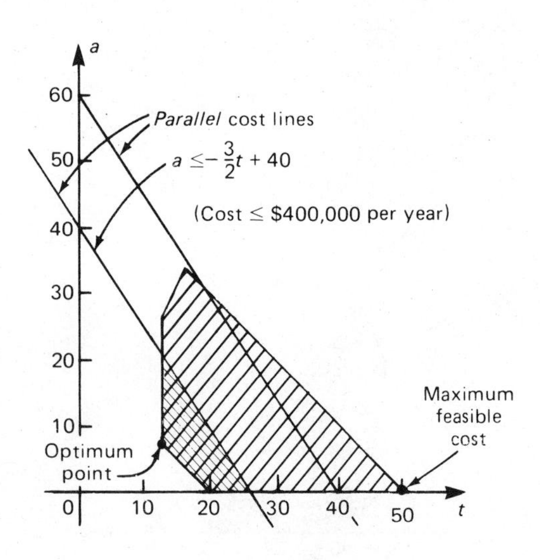

Figure 4-10c.

$$t = 12 \quad \text{and} \quad t + a = 20$$

intersect each other. If you cannot read the intersection point from the graph, you can solve this *system* of equations. Either way, the point is

$$(t, a) = (12, 8).$$

The cost can be found by substituting (12, 8) into the cost equation, $d = 15t + 10a$, getting

$$d = 15(12) + 10(8)$$

$$= 260.$$

So, the minimum feasible cost is $260,000 per year.

The maximum feasible cost could be found by sliding the cost line in the *opposite* direction, always keeping a slope of −3/2. From Figure 4-10c you can tell that the point would be (50, 0), for which d would equal 750.

The point (12, 8), at which the minimum cost occurs is called the *optimum* point, from the Latin word for "best." The lowest cost is the "best" from the point of view of the School Board. [The point (50, 0) might be considered to be "optimum" by the teachers since more money would go to teachers!] The process of finding the optimum point for a linear function such as $d = 15t + 10a$ is called *linear programming.* Your objective is to be able to work linear programming problems.

Objective:

Given information about permissible values of two independent variables, find the pair of values that either *maximizes* or *minimizes* a third, dependent variable.

In the exercise that follows, you will solve some linear programming problems. The first problem has the steps spelled out in great detail, but in later problems you will be expected to know what to do with progressively less instruction.

Exercise 4-10

1. ***Park Clean-Up Problem.*** Your mathematics club has arranged to earn some extra money by cleaning up Carr Park. The City Recreation Department agrees to pay each old member \$10 and each new member \$8 for his or her services. (The club did the same thing last year, so the old members are experienced.)

a. Define variables, then write an equation expressing the dollars the club earns in terms of the numbers of old and new members who work.
b. The equation in part a defines a function with three variables. Which is dependent, and which are independent?

You recall that the *domain* of a function is the set of values of the *independent* variable. Since there are *two* independent variables in this function, the domain will be a set of *ordered pairs.* The domain of a function is determined by things in the real world. In this case, the following things establish the domain:

i. The number of old members is non-negative, and so is the number of new members.
ii. The club has at most 9 old members and at most 8 new members who can work.
iii. The Department will hire at least 6 students, but no more than 15.
iv. There must be at least 3 new members.
v. The number of new members must be at least 1/2 the number of old members, but less than 3 times the number of old members.

c. Write inequalities for each of the above requirements.
d. Draw a graph of the solution set of this system of inequalities. Remember that club members come only in *integer* quantities!
e. The region you drew in part *d* is the domain of a function with two independent variables. Such a region is usually called the "feasible region." Why do you suppose that this name was picked?
f. Based on your graph, is it feasible to do without any old members at all? Explain.
g. Shade the portion of the feasible region in which the club would make *at least* \$100. Your answer to part *a* will help you figure out what the inequality should be.
h. Draw a line on the graph showing the number of old and new members needed to make \$160. Again, the answer to part a will help. Is it feasible to make \$160? Explain.
i. What numbers of old and new members would earn the *maximum* feasible amount? What would this amount be?
j. What is the *minimum* feasible amount the club could earn?
k. Suppose tradition was broken and the new members were paid more than the old members. What would be the maximum feasible earnings if new members get \$12 and old members get \$10?

2. ***Vitamin Problem.*** Suppose that you are Chief Mathematician for the Government's million dollar Vitamin Research Project. Your scientists have been studying the combined effects of vitamins A and B on the human system, and have turned to *you* for analysis of their findings:

i. The body can tolerate no more than 600 units per day of vitamin A, and no more than 500 units per day of vitamin B.
ii. The total number of units per day of the two vitamins must be between 400 and 1000, inclusive.
iii. Due to the combined effects of the two vitamins, the number of units per day of vitamin B must be more than 1/2 the number of units per day of vitamin A, but less than or equal to three times the number of units per day of vitamin A.

Let x and y be the numbers of units per day of vitamins A and B, respectively. Answer the following questions:

a. Let c be the number of cents per day it costs you for vitamins. Write an equation expressing c in terms of x and y if vitamin A costs 0.06

cents per unit (*not* \$0.06!), and vitamin B costs 0.05 cents per unit.

b. Write inequalities to represent each of the above requirements, and plot a graph of the solution set of the system.
c. What does a point lying in this solution set represent in the real world?
d. Plot lines on your graph showing where the cost per day is

   i. 60 cents,
   ii. 30 cents,
   iii. 15 cents.

   You might use a different colored pencil to make these lines show up more distinctly.
e. Is it feasible to spend 60 cents per day on vitamins? 30 cents per day? 15 cents per day? Explain why or why not.
f. What word describes the point at which the feasible cost is a *minimum*? Write the coordinates of this point, and find the minimum cost.

3. ***Aircraft Problem.*** Calvin Butterball is Chief mathematician for Fly-By-Night Aircraft Corp. He is responsible for mathematical analysis of the manufacturing of the Company's two models of planes, the Sopwith Camel and the larger Sopwith Hippopotamus. Each Department at Fly-By-Night has certain restrictions concerning the number of planes which can be manufactured per day.

   i. Production: No more than 7 Hippopotami and no more than 11 Camels can be manufactured per day.
   ii. Shipping: No more than 12 planes, total, can be manufactured per day.
   iii. Sales: the number of Hippopotami manufactured per day must be no more than twice the number of Camels.
   iv. Labor: You must use more than 1000 man-hours of labor per day. (It takes 100 man-hours to manufacture each Camel and 200 man-hours to manufacture each Hippo.)

a. Select variables to represent the numbers of Camels and Hippopotami manufactured per day, write inequalities expressing each of the above restrictions, and draw a graph of the feasible region (the solution set of the system).
b. If Fly-by-Night makes a profit of \$300 per camel and \$200 per Hippo, show the region on your graph in which the daily profit would be at least \$3000.
c. How many Hippos and how many Camels should be produced per day to give the *greatest* feasible profit? What would this profit be?

4. ***CB Radio Manufacturing Problem.*** Suppose that you work for a small company that makes high-quality CB radios. They wish to optimize the numbers of the two models they produce, "Breakers" and "Good Buddies." Since you are an expert at linear programming, your Boss assigns you the job. You find the following information:

   i. The assembly lines can produce no more than 25 Breakers and 16 Good Buddies per day.
   ii. No more than 32 radios, total, can be produced per day.
   iii. They can spend a total of no more than 189 man-hours a day assembling radios. It takes 7 man-hours to assemble a Breaker and 3 man-hours to assemble a Good Buddy.
   iv. They can spend at most 68 man-hours per day, total, testing radios. It takes 1 man-hour to test a Breaker and 4 man-hours to test a Good Buddy.
   v. The number of Breakers produced per day must be more than half the number of Good Buddies.

Using this information, answer the following questions:

a. Define variables, write inequalities for each of the above requirements, and plot the graph of the feasible region. (It may be easier if you plot Breakers on the vertical axis.)
b. The company makes a profit of \$40 on each Breaker and \$20 on each Good Buddy. Write an equation expressing dollars profit earned per

day in terms of the numbers of Breakers and Good Buddies produced each day.

c. Shade the portion of the feasible region in which the daily profit is at least \$960.

d. Find the optimum point at which the daily profit is a maximum, and the number of dollars they would earn per day by operating at this point.

5. ***Feedlot Problem.*** Butch Err owns a feedlot on which he fattens up cattle for market by feeding them a mixture of corn and pellets. He wants to find what mixture of corn and pellets will give the minimum feasible cost.

The pellets and corn contain the following ingredi-ants per 100 pound (lb) sack:

	Pounds per 100 lb. sack			
	Protein	Minerals	Starch	Bulk
Pellets*	25	12	20	12
Corn	10	10	32	48

*Pellets contain a "secret ingredient," not shown here.

For the number of cattle Mr. Err owns, the daily needs are

i. total protein: at least 250 lb/day,
ii. total minerals: at least 200 lb/day,
iii. total starch: at least 512 lb/day,
iv. total bulk: at least 480 lb/day,
v. total food: no more than 3000 lb/day.

a. Let x and y be the numbers of 100 lb sacks of pellets and corn, respectively, used per day. Since each sack of pellets has 25 lb of protein and each sack of corn has 10 lb of protein, the total amount of protein is $25x + 10y$. Write a system of inequalities expressing the five requirements above, and draw the graph of the feasible region.

b. Is it feasible to feed only corn? Only pellets? Explain.

c. Pellets cost \$16 per 100 lb sack and corn costs \$8 per 100 lb sack. Write an equation expressing the number of dollars, d, spent per day in terms of x and y.

d. Show the portion of the feasible region in which the daily cost is

i. no more than \$240,
ii. no more than \$200.

e. Show that there are four optimum points with integer coordinates, each of which gives the *same* minimum feasible cost.

f. How much would Mr. Err save by feeding at one of the optimum points of part e rather than by feeding the minimum feasible "all-corn" diet?

g. Find the exact point of intersection of the two boundary lines that meet at the *actual* (non-integer) optimum point. How much could Mr. Err save per day by feeding at this optimum point rather than one of the integer points of part e?

6. ***Thoroughbreds and Quarter Horses Problem.*** Suppose that you go into business raising Thoroughbreds and quarter horses. Having studied linear programming, you decide to maximize the feasible profit you can make. Let x be the number of Thoroughbreds, and let y be the number of quarter horses you raise each year.

a. Write inequalities expressing each of the following requirements:

i. Your supplier can get you at most 20 Thoroughbreds and at most 15 quarter horses to raise each year.
ii. You must raise at least 12 horses, total, each year to make the business worthwhile.
iii. A Thoroughbred eats 2 tons of food per year, but a quarter horse eats 6 tons per year. You can handle no more than 96 tons of food per year.
iv. A Thoroughbred requires 1000 hours of training per year, and a quarter horse only 250 hours per year. You have enough personnel to do at most 10,000 hours of training per year.

b. Draw a graph of the feasible region.

c. One of the inequalities has no effect on the feasible region. Which one? Tell what this means in the real world.
d. What is the minimum feasible number of quarter horses?
e. What is the maximum feasible number of Thoroughbreds?
f. Is it feasible to raise *no* Thoroughbreds? Explain.
g. You can make a profit of \$500 for each Thoroughbred and \$200 for each quarter horse. Shade the portion of the feasible region in which the profit would be at least \$5000 per year.
h. What is the maximum feasible profit you could make per year, and how would you operate in order to attain that profit?
i. How much more profit do you make per year by operating at the optimum point of part h rather than by operating at the *worst* feasible point?

7. ***Coal Problem.*** Skinflint Coal Company has a stock of 16 tons of Number Nine Coal. Ten tons of this is located at Warehouse No. 1 and the remaining 6 tons is located at Warehouse No. 2. Mr. Skinflint has orders from Weedies Cereal Co. for 5 tons, from Pikkitt Lock Co. for 7 tons, and from Treadwell Tire Co. for 4 tons. He wants to minimize the cost of shipping the coal.

Let x, y, and z represent the numbers of tons sent from Warehouse 1 to Pikkitt, Weedies, and Treadwell, respectively.

a. The remainder of each order must come from Warehouse 2. Write expressions in terms of x, y, and z representing how much must be shipped from Warehouse 2 to each customer.
b. Write an equation stating that all 10 tons of coal from Warehouse 1 must be shipped. Then transform the equation so that z is expressed in terms of x and y.
c. Substitute the value of z from part b into the appropriate expression from part (a) so that all three of these expressions are in terms of x and y alone.
d. Write inequalities representing each of the following conditions:
   i. x is non-negative.
   ii. y is non-negative.
   iii. Pikkitt gets no more than 7 tons from Warehouse 1.
   iv. Weedies gets no more than 5 tons from Warehouse 1.
   v. Treadwell gets no more than 4 tons from Warehouse 1.
   vi. The total shipped from Warehouse 1 to Weedies and Pikkitt is no more than 10 tons (the total contents of Warehouse 1).
e. Plot a graph of the system of inequalities in part d.
f. The distances from the warehouses to the customers are shown in Figure 4-10d. Let d be the number of dollars it costs to ship all the coal. Write an equation expressing d in terms of x and y if it costs \$3.00 to ship one ton one kilometer.
g. Mr. Skinflint desires to keep the total shipping cost below \$192. Write an inequality expressing

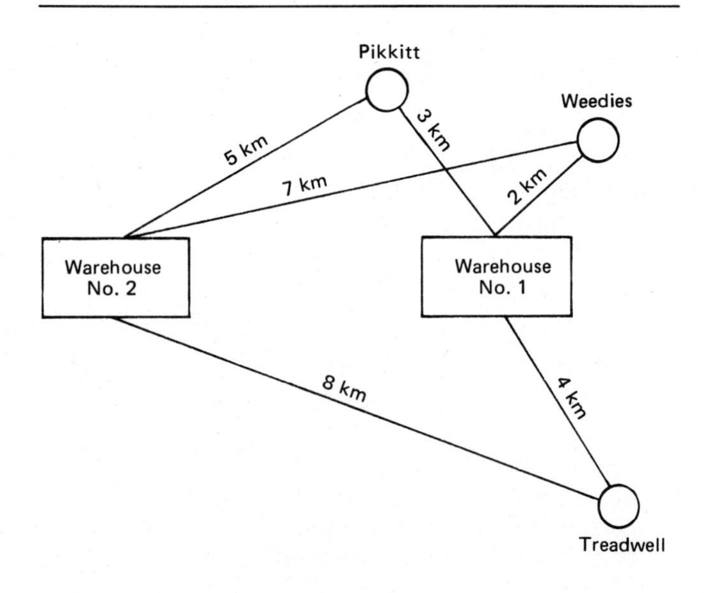

Figure 4-10d.

this requirement, plot it on your graph, and indicate its intersection with your present feasible region.

h. What values of x and y would result in a *minimum* shipping cost? A *maximum* shipping cost? What would these maximum and minimum costs be?

4-11 CHAPTER REVIEW AND TEST

In this chapter you have studied relations that involve more than one linear equation or inequality. The objectives for the chapter are summarized below.

1. *Solve systems of two linear equations with two variables.*
2. *Find intersections of real-world graphs, and use f(x) terminology.*
3. *Solve systems of three (or more) linear equations with three (or more) variables.*
4. *Graph systems of linear inequalities.*
5. *Use linear programming techniques to find optimum values of* ***two*** *independent variables.*

By working the Review Problems below, you will see how well you can accomplish these objectives one at a time. In the Concepts Test you will have to put together several concepts. Some parts of the problems allow you to extend your knowledge to concepts you have never seen before!

Review Problems

The following problems are numbered according to the five objectives listed above.

Review Problem 1a. Solve the system by linear combination:

$$\begin{aligned} 5x - 8y &= 47 \\ 2x + 7y &= -22. \end{aligned}$$

b. Solve the system by determinants:

$$\begin{aligned} 4x - 3y &= 5 \\ 7x + 2y &= -1. \end{aligned}$$

Review Problem 2. At Annie Moore's Coffee Shop, the waitress earns $16 and the cook earns $24 in a normal shift. In addition, the waitress gets 70% of the tip money received, and the cook gets 30%.

Let t = total dollars in tips received in a shift.
Let $w(t)$ = total dollars the waitress gets in a shift.
Let $c(t)$ = total dollars the cook gets in a shift.

a. Write equations expressing $w(t)$ and $c(t)$ in terms of t.
b. Calculate $w(5)$ and $c(5)$.
c. How much would have to be received in tips for the waitress and the cook to break even?
d. Plot graphs of functions w and c, and thus show that your answer to part c is correct.

Review Problem 3. Solve the system:

$$\begin{aligned} 3x - 2y + 5z &= -1 \\ 4x + 3y - 2z &= -13 \\ 2x + 5y - 4z &= -9. \end{aligned}$$

Review Problem 4. Graph the solution set of the system:

$$\begin{aligned} y &\geq 3x + 2 \\ 2x - 3y &> -15 \\ y + 2 &\geq -\tfrac{2}{5}(x + 3) \\ y &> -1. \end{aligned}$$

Review Problem 5 – Achievement Test. You are to take a mathematics achievement test. Two days before the test you receive the following instructions concerning the point values of the two kinds of questions, and how many of each you must answer.

i. There are 10 questions worth 7 points each, and 16 questions worth 5 points each.

ii. You can receive credit for a maximum of 20 questions. Any others you answer will not be scored.
iii. To receive any credit at all, you must answer at least 5 questions.
iv. The number of 5-point questions you answer must be no more than twice the number of 7-point questions.
v. The number of 5-point questions you answer must be more than $\frac{1}{2}$ the quantity (number of 7-point questions minus 5).

a. What are the optimum numbers of 5- and 7-point questions to answer in order to maximize your score?
b. What is the maximum feasible score?
c. What is the minimum feasible score, assumming that you satisfy all of the above requirements, and answer each question correctly?

Concepts Test

Answer the following two questions. For each part of each question, tell which one(s) of the five objectives you used. If the part of the problem involves a new concept, write "new concept."

Concepts Test 1 – Camping Trip Problem. Suppose that you are going on a prolonged camping trip. There are two types of condensed food you can take along, X-rations and Yummies. You wish to minimize the amount of money you spend on food for the trip while still meeting all your nutritional requirements.

a. Define variables for the numbers of pounds of X-rations and Yummies, and for the number of dollars spent on food.
b. Each pound of X-rations and Yummies contains the following:

	Vitamins	Calories	Protein	Carbohydrates
X-rations	400 units	800	1 ounce	4 ounces
Yummies	100 units	700	4 ounces	5 ounces

Your total needs for the trip are:

i. Vitamins, at least 4000 units.
ii. Calories, no less than 16,800.
iii. Protein, more than or equal to 28 ounces.
iv. Carbohydrates, 100 ounces or more.

Write inequalities for each of these requirements.

c. Write an inequality expressing the fact that the total weight of X-rations and Yummies must be less than 30 pounds.
d. Plot a graph of the feasible region formed by the inequalities of parts b and c.
e. If you took only X-rations, what is the smallest feasible number of pounds you could take?
f. If you took only Yummies, what is the smallest feasible number of pounds you could take?
g. X-rations cost $3 per pound and Yummies cost $2 per pound. Write an equation expressing the number of dollars spent for food in terms of the numbers of pounds of X-rations and Yummies.
h. Shade the portion of the feasible region in which your total cost would be at most $60.
i. Mark the optimum point on your graph.
j. The optimum point of part i does *not* have integer coordinates. Write *equations* for the two boundary lines that cross at the optimum point (replace the "$\geq$" signs with "=" signs). Then solve this system of equations to find the *exact* coordinates of the optimum point.
k. What is the minimum feasible cost?
l. What point with *integer* coordinates gives the minimum feasible cost? (Your graphs must be *very* carefully drawn because the answer is somewhat surprising!)

Concepts Test 2. So far you have studied *linear* functions and relations. In a *quadratic* function, y equals a *quadratic trinomial*. That is, the general equation of a quadratic function is

$$f(x) = ax^2 + bx + c,$$

where a, b, and c stand for *constants* (just like m and b stand for constants in a linear function).

a. Suppose that for a particular quadratic function, $f(3) = 7$. Explain why the following equation involving a, b, and c must be true:

$9a + 3b + c = 7.$

b. If this particular function also has $f(2) = 6$ and $f(1) = 3$, write two other equations involving a, b, and c, as in part a.

c. The three equations in parts a and b, above, form a system of three linear equations in the three "variables" a, b, and c. Solve this system to find the values of a, b, and c.

d. Write the particular equation for $f(x)$.

e. Find $f(6)$.

Quadratic Functions

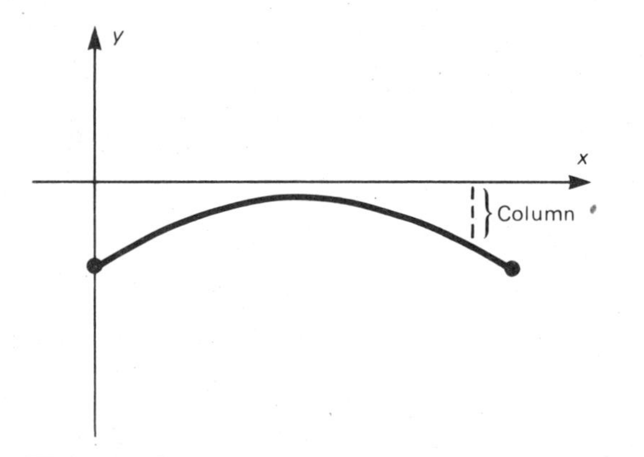

The linear functions in Chapters 3 and 4 had straight-line graphs. In this chapter you will encounter ***quadratic*** *functions whose graphs are* ***curved*** *lines. Again, your ultimate objective will be to find the particular equation of a quadratic function from information about its graph. In Exercise 5-8 you will use such equations in problems ranging from bridge design to predicting the price of pizza!*

5-1 INTRODUCTION TO QUADRATIC FUNCTIONS

In Chapter 3 you learned that a linear function has an equation like $y = 3x + 8$, where y equals a linear polynomial. A function with an equation such as

$$y = 3x^2 - 7x + 11,$$

is called a *quadratic* function because y equals a quadratic (*second* degree) polynomial.

DEFINITION

A ***quadratic function*** has a general equation of the form

$$\boxed{y = ax^2 + bx + c \quad ,}$$

where a, b, and c stand for constants, and $a \neq 0$.

After defining a new kind of function, your next task is to see what the graphs of some particular examples look like. Unless you know some shortcuts, the way to do this is to pick values of x, calculate the corresponding values of y, and plot the points. When you are reasonably sure of the pattern, you should connect the points with a smooth curve.

Objective:

Discover by pointwise plotting, what the graphs of several quadratic functions look like, and what different values of a, b, and c do to the graph.

The following exercise is designed to help you accomplish this objective.

Exercise 5-1

1. Plot the graphs of the following quadratic functions. Calculate enough points to establish a pattern, then connect them with a smooth curve. You should use the *same* scales for all six graphs, although the scale on the y-axis need not be the same as that on the x-axis.

a. $y = x^2$ b. $y = \frac{1}{2}x^2$

c. $y = 2x^2$ d. $y = -\frac{1}{2}x^2$

e. $y = x^2 - 5$ f. $y = x^2 + 6x - 5$

2. What is the effect on the graph of:

a. Changing the x^2-coefficient?
b. Adding an x-term?
c. Adding a constant term?

5-2 QUADRATIC FUNCTION GRAPHS

In Exercise 5-1 you plotted the graphs of several quadratic functions. Each graph was a "U-shaped" figure called a *parabola*. The prefix "para-" comes from the word parallel. Turn to section 9-6 to find out why!

From the graphs you should have been able to reach the following conclusions:

1. The constant a determines the *shape* of the parabola.

   a. If $a > 0$, the graph opens *upward*; e.g., $y = \frac{1}{2}x^2$.

   b. If $a < 0$, the graph opens *downward*; e.g., $y = -\frac{1}{2}x^2$.

   c. As a gets closer to zero, the graph becomes *flatter* (i.e., opens wider). This is not surprising since the function becomes linear when a *equals* zero.

2. The constant c equals the y-intercept since $y = c$ when $x = 0$. It determines the *vertical* position of the graph without affecting its shape.

3. The constant b determines the *horizontal* position of the graph without affecting its shape or y-intercept.

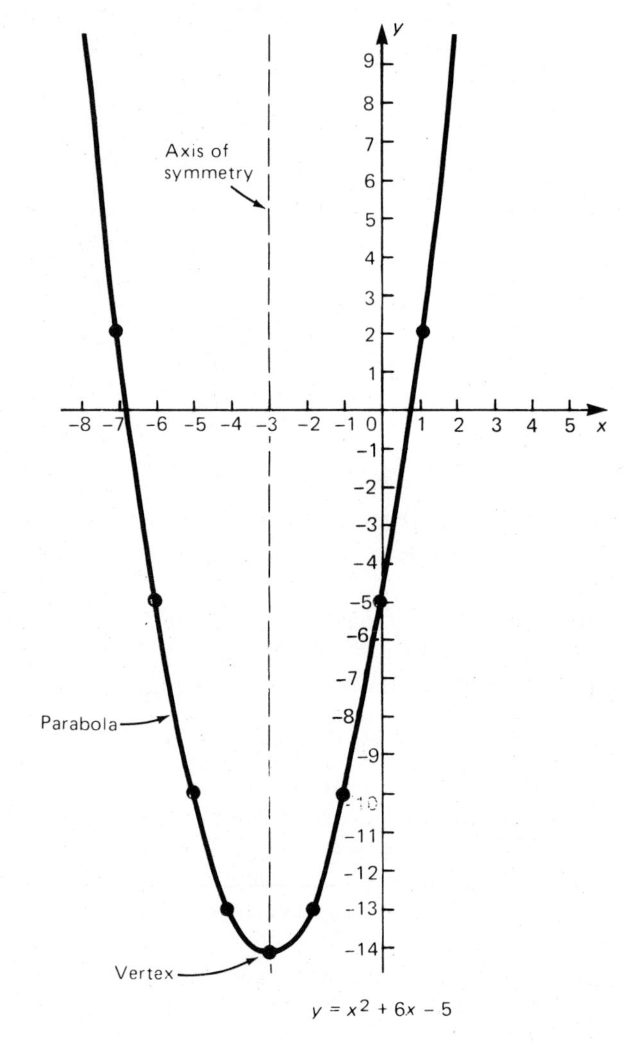

Figure 5-2a.

Figure 5-2a shows the graph of $y = x^2 + 6x - 5$, which you plotted in Problem 1.f. of Exercise 5-1. The lowest point is called the *vertex*. If the parabola opens downward instead of upward, then the vertex is the highest point. The vertical line through the vertex is called the *axis of symmetry*. If you were to fold the graph along this line, the two branches would fit right on top of each other.

Obviously, the vertex is the most important point on the graph of a quadratic function. If you know where the vertex is, you can sketch a reasonably good parabola with very little other information. For example, suppose that the vertex is at $(-3, 5)$, and the y-intercept is 2.

As shown in the left-hand part of Figure 5-2b, first you plot the vertex and the y-intercept. Then you draw the axis of symmetry through the vertex. Since this is an axis of *symmetry,* there will be a *third* point which is opposite the y-intercept, and the *same* distance from the dotted line. With these three points and the knowledge of what a parabola looks like, you can draw a good sketch, as shown in the right-hand part of Figure 5-2b.

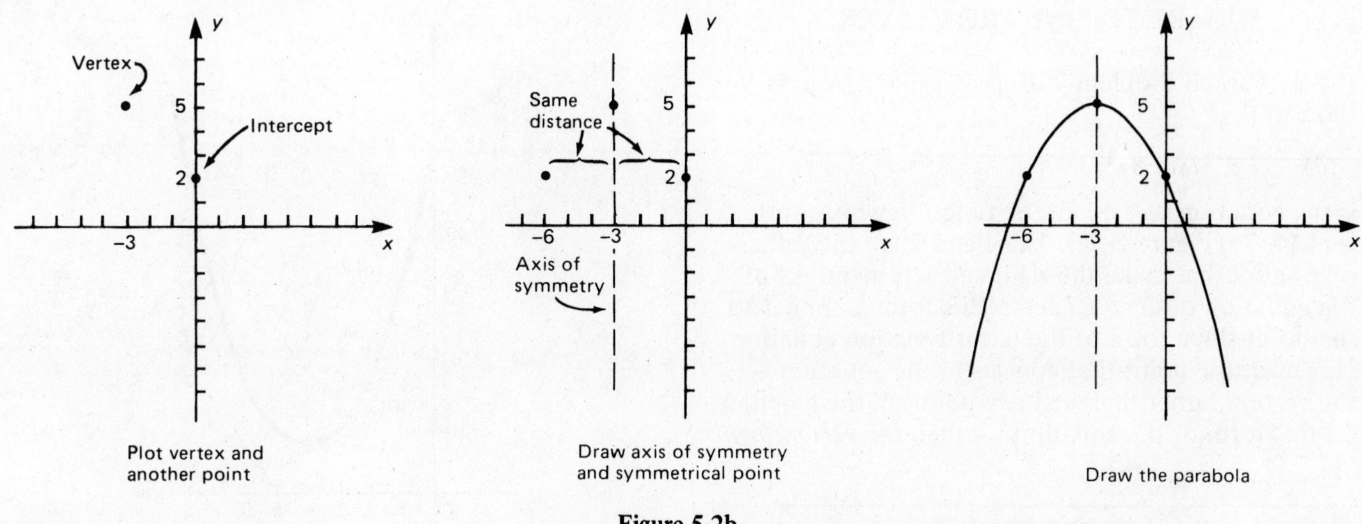

Figure 5-2b.

Objective:

Given the vertex and another point on the graph of a quadratic function, sketch the parabola *quickly*.

The following exercise has a few problems in which you will sketch parabolas quickly. The last problem introduces you to another form of the quadratic function equation from which you can find the vertex by inspection!

Exercise 5-2

For Problems 1 through 6, sketch the graph of the quadratic function with the given vertex and intercept.

1. Vertex: (2, 3), y-intercept: 7
2. Vertex: (−3, 2), y-intercept: −4
3. Vertex: (3, −1), y-intercept: −6
4. Vertex: (−3, −6), y-intercept: −4
5. Vertex: (−3, −4), x-intercept: 2
6. Vertex: (3, 4), x-intercept: 1

7. It is possible to find the vertex of a parabola quickly when the equation is written in a form such as

$$y - 7 = 3(x - 4)^2.$$

This form is similar to the point-slope form of the linear function equation, except that the $(x - 4)$ is *squared*.

a. Transform the above equation to $y = ax^2 + bx + c$. Remember that $(x - 4)^2$ equals $(x - 4)(x - 4)$.
b. Calculate values of y for $x = 0, 1, 2, 3, 4, 5,$ and 6. Use these points to plot the parabola on graph paper. You should probably compress the scale on the y-axis so that the graph will fit on the paper.
c. Mark the vertex on the graph, and write down the ordered pair.
d. Where do the coordinates of the vertex show up in the *equation*?
e. How could you find the values of y for $x = 7$ and $x = 8$ *without* substituting these numbers into the equation?
f. Find the vertex of $y - 2 = \frac{1}{2}(x - 6)^2$ *quickly*. Then sketch the graph.

5-3 COMPLETING THE SQUARE TO FIND THE VERTEX

If you worked Problem 7 of Exercise 5-2, you showed that

$$y - 7 = 3(x - 4)^2$$

is the equation of a quadratic function whose vertex is at (4, 7) (Figure 5-3a). The 4 and 7 are the values of x and y that make the right and left members of the equation both *equal zero.* This form is similar to the point-slope form of the linear function equation. However, the point that appears in the equation is the *vertex*, rather than just any point on the graph. So this form of the equation is called the *vertex form.*

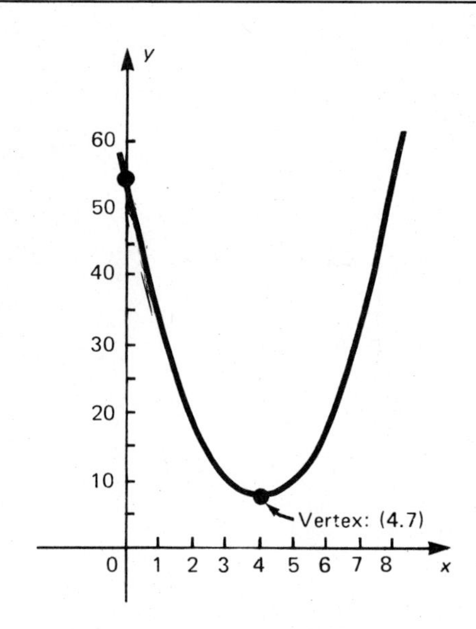

Figure 5-3a.

DEFINITION

A quadratic function equation is in ***vertex form*** if it has the form

$$\boxed{y - k = a(x - h)^2}$$

where a, h, and k stand for constants. The vertex is at the point (h, k), and $\boldsymbol{a}$ is the same constant as in $y = ax^2 + bx + c$.

If the equation is in vertex form, you can pick out the vertex quickly. Then you can use the technique in Section 5-2 to sketch the graph. In this section you will learn how to transform an equation to the vertex form.

Objective:

Given a particular equation in the form $y = ax^2 + bx + c$, be able to transform it to the vertex form, $y - k = a(x - h)^2$.

The transformation is somewhat tricky, and can best be understood by first going in the reverse direction, then retracing the steps. For example, if

$$y - 7 = 3(x - 4)^2, \qquad \text{Step (1)}$$

you would first *square* the binomial by multiplying $(x - 4)(x - 4)$ as you did in Chapter 1, getting

$$y - 7 = 3(x^2 - 8x + 16). \qquad \text{Step (2)}$$

Distributing the 3 gives

$$y - 7 = 3x^2 - 24x + 48, \qquad \text{Step (3)}$$

and adding 7 to both members produces the desired result,

$$y = 3x^2 - 24x + 55. \qquad \text{Step (4)}$$

The first difficulty you enounter in trying to reverse the steps is the fact that the 48 and the 7 have been added together to form 55 in the right member. Unless you *knew* that 7 was the number to subtract from both members, you would have difficulty getting from Step (4) back to Step (3). So the best thing to do is subtract 55 from both members in Step (4), getting

$$y - 55 = 3x^2 - 24x.$$

To get back to Step (2), first notice that the x^2 coefficient, 3, is factored out. So the next thing to do is factor out 3 from the right member also, getting

$$y - 55 = 3(x^2 - 8x).$$

The most crucial step is making the expression inside the parentheses a *perfect square*. From Step (2) you know that 16 must be added inside the parentheses. To see how to get this 16, it helps to take part of the *next* step *first*. You write empty parentheses with a "square" outside, getting

$$y - 55 \_\_\_ = 3(x^2 - 8x \_\_\_) \qquad \text{Leave spaces.}$$
$$\_\_\_\_ \ (\_\_\_)^2.$$

The first term inside the parentheses must be x, so you write

$$y - 55 \qquad 3(x^2 - 8x \qquad) \qquad \text{First term is } x.$$
$$(x \qquad)^2.$$

When you *squared* the $(x - 4)$ from Step (1) to Step (2), the middle term, $-8x$, was *twice* the *product* of the x and the -4. So, going in the other direction, you can reason that the -4 is *half* the *coefficient* of x; that is, half of -8. So you write this piece of information next.

$$y - 55 \qquad = 3(x^2 - 8x \qquad) \qquad -4 \text{ is } \textit{half} \text{ of } -8.$$
$$(x - 4)^2.$$

The number to add inside the original parentheses is now obvious. When you square $(x - 4)$, the constant term is $(-4)^2$, which equals 16. So you come back to the line above and add 16 inside the parentheses. This step is called "completing the square."

$$y - 55 \qquad = 3(x^2 - 8x + 16) \qquad 16 \text{ is } (-4)^2.$$
$$(x - 4)^2.$$

By adding 16 *inside* the parentheses, you have really added 3×16 to the right member of the equation. So you must add 3×16, or 48, to the *left* member, also, to obey the addition property of equality.

$$y - 55 + 48 = 3(x^2 - 8x + 16) \qquad \text{Add 48.}$$
$$(x - 4)^2.$$

You need only fill in the remainder of the last line, and the equation will be in vertex form.

$$y - 55 + 48 = 3(x^2 - 8x + 16) \qquad \text{Complete the last step.}$$
$$y - 7 = 3(x - 4)^2.$$

Although there is a lot of thinking to be done, the actual *writing* can be done in just *three* steps! In fact, you can find the 16 that is needed to complete the square by simply

1. taking *half* the coefficient of x, and
2. squaring it.

The three steps would look like this:

$y = 3x^2 - 24x + 55$	Given.
$y - 55 + 48 = 3(x^2 - 8x + 16)$	Subtract 55 from both members, factor 3 out of right member, complete square on right, add 3×16 to left.
$y - 7 = 3(x - 4)^2.$	Write right member as a square, tidy up the left number.

The exercise which follows is designed to give you practice in finding the vertex by completing the square, and using the vertex to help with rapid sketching of the graph. You will also have the chance to learn a "rise-and-run" technique for plotting *many* points *quickly* (see Problem 13).

Exercise 5-3

For Problems 1 through 12,

a. transform the equation to the vertex form,
b. write the coordinates of the vertex.
c. sketch the graph using the vertex, y-intercept, and the point across the axis of symmetry from the y-intercept.

1. $y = x^2 + 6x + 11$
2. $y = x^2 + 8x + 21$
3. $y = 3x^2 - 24x + 17$
4. $y = 5x^2 - 30x + 31$
5. $y = 4x^2 + 12x - 11$
6. $y = 8x^2 + 40x + 37$
7. $y = 2x^2 - 11x - 12$
8. $y = 2x^2 - 7x + 12$
9. $y = -5x^2 - 30x + 51$
10. $y = -3x^2 - 24x - 41$
11. $y = -x^2 + x + 1$
12. $y = -x^2 - x + 3$

13. *Rapid Parabola Plotting* – The sum of the first n consecutive, odd integers equals n^2. For example,

$$S_1 = 1 = 1^2,$$
$$S_2 = 1 + 3 = 4 = 2^2,$$
$$S_3 = 1 + 3 + 5 = 9 = 3^2.$$

This fact can be used to plot graphs of quadratic functions by a "rise-and-run" technique similar to that you used for linear functions.

a. Calculate S_4, S_5, and S_6, and show that they equal 4^2, 5^2, and 6^2, respectively.
b. The graph of $y = x^2$ can be plotted by starting at the vertex and going over 1, up 1; over 1, up 3; over 1, up 5; and so forth as shown in Figure 5-3b. Since any quadratic with equation $y = x^2 + bx + c$ has the same shaped graph as $y = x^2$, its graph can be plotted in the same way. Locate the vertex, then plot the graph of $y = x^2 - 6x + 2$ using this technique.
c. If the x^2 coefficient is not equal to 1, the graph can still be plotted in a similar way. For $y = ax^2 + bx + c$, you would start at the vertex, then go over 1, up a; over 1, up $3a$; over 1, up $5a$, and so on. Use this technique to plot graphs of the following functions:

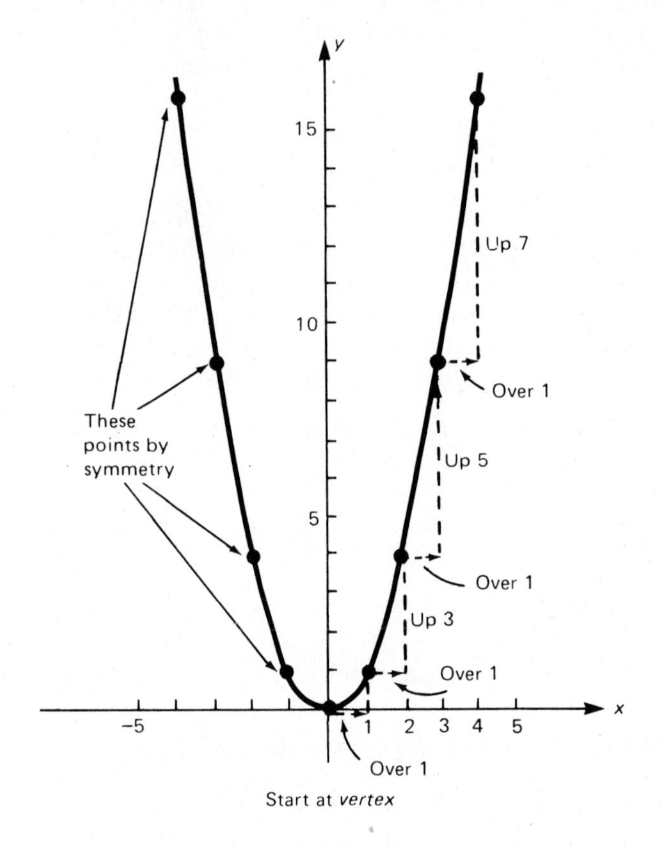

Figure 5-3b.

i. $y = 2x^2 - 8x + 7$.
ii. $y = -x^2 + 10x + 3$.
iii. $y = \frac{1}{2}x^2 + 3x - 1$.

d. If you have studied mathematical induction (Appendix B), *prove* that $1 + 3 + 7 + 9 + \ldots + (2n - 1) = n^2$.

14. *x-Intercepts of Quadratic Functions* – You recall that an x-intercept is a value of x for which $y = 0$. So an x-intercept tells you where the graph crosses the x-axis. From what you have observed in sketching graphs of quadratic functions in the previous problems, do the following:

a. Sketch a graph of a quadratic function having

   i. *two* x-intercepts,
   ii. *no* x-intercepts,
   iii. exactly *one* x-intercept.

b. Explain why no function can have more than one y-intercept.

c. Just for fun, see if you can *calculate* what the x-intercepts are for the function with the equation

$$y = x^2 - 5x - 6.$$

5-4 QUADRATIC EQUATIONS WITH ONE VARIABLE

Suppose that $y = x^2 + 10x + 33$, and you are asked to find x when $y = 21$. Substituting 21 for y gives

$$21 = x^2 + 10x + 33.$$

This is a quadratic equation with *one* variable. Using symmetry to write the equation with the variable terms on the left, you get

$$x^2 + 10x + 33 = 21.$$

Adding -21 to both members gives

$$x^2 + 10x + 12 = 0.$$

In previous mathematics courses you learned how to solve such equations by factoring the left member and setting each factor equal to zero. Unfortunately, $x^2 + 10x + 12$ will *not* factor. So you need another, more general method of solving quadratic equations with one variable.

Objective:

Be able to solve any equation of the form

$$\boxed{ax^2 + bx + c = 0}.$$

The procedure for accomplishing this objective is to transform the left member to a perfect square by completing the square. Then you take the square root of both members. If you have forgotten about square roots, Problems 1 through 4 in Exercise 5-4, following, will give you a refresher.

Example 1: Solve $x^2 + 3x - 5 = 0$.

First, you must make a space on the left in which to complete the square. Adding 5 to both members gives

$$x^2 + 3x \quad\quad = 5.$$

To complete the square, you take *half* the coefficient of x (1/2 of 3), *square* it, and *add* the result, 9/4, to both members.

$$x^2 + 3x + \frac{9}{4} = 5 + \frac{9}{4}.$$

Now, you can write the left member as a perfect square, and add the two terms on the right, getting

$$\left(x + \frac{3}{2}\right)^2 = \frac{29}{4}.$$

The symbol $\sqrt{n}$ means the *positive* square root of the number n. You may use as a lemma the fact that if two numbers are equal, then their positive square roots are equal. Therefore,

$$\sqrt{\left(x + \frac{3}{2}\right)^2} = \sqrt{\frac{29}{4}}.$$

Since $\sqrt{n^2} = |n|$ (as you can see by substituting numbers like 5 and -5 for n), and since $\sqrt{n/d} = \sqrt{n}/\sqrt{d}$, the equation becomes

$$\left|x + \frac{3}{2}\right| = \frac{\sqrt{29}}{2}.$$

You solved this type of equation in Section 1-5. You get

$$x + \frac{3}{2} = \frac{\sqrt{29}}{2} \quad \text{or} \quad x + \frac{3}{2} = -\frac{\sqrt{29}}{2},$$

or, more simply,

$$x + \frac{3}{2} = \pm\frac{\sqrt{29}}{2}.$$

The symbol "±" means "positive or negative." The last step is adding $-3/2$ to both members, getting

$$x = -\frac{3}{2} \pm \frac{\sqrt{29}}{2}$$

$$\therefore S = \left\{\frac{-3+\sqrt{29}}{2}, \frac{-3-\sqrt{29}}{2}\right\}.$$

If you need a decimal approximation for these values of x, you can find $\sqrt{29}$ by calculator, by Table I at the back of the book, or by "educated guesses." Using $\sqrt{29} \approx 5.4$, you get

$$x \approx \frac{-3+5.4}{2} \quad \text{or} \quad x \approx \frac{-3-5.4}{2}.$$

These reduce to

$$x \approx 1.2 \quad \text{or} \quad x \approx -4.2,$$

so that

$$S = \{1.2, -4.2\}.$$

Example 2: A sign change in the equation of Example 1 produces a surprising result. If you try to solve

$$x^2 + 3x + 5 = 0,$$

you get

$$x^2 + 3x = -5$$

$$x^2 + 3x + \frac{9}{4} = -5 + \frac{9}{4} = -\frac{11}{4}.$$

$$\sqrt{\left(x+\frac{3}{2}\right)^2} = \sqrt{-\frac{11}{4}}.$$

At this point, you realize that $\sqrt{-11/4}$ is an *imaginary* number. Since you have an agreement that variables will stand for *real* numbers (unless otherwise specified), the answer is

$$S = \emptyset.$$

In Chapter 10 you will learn how to deal with imaginary numbers and how to write solutions for such equations.

Example 3: Solve $x^2 - 2x + 7 = 4x + 6x^2 - 1$.

This equation is not in the form $ax^2 + bx + c = 0$. So you simply *transform* it to that form.

$x^2 - 2x + 7 = 4x + 6x^2 - 1$	Given.
$-5x^2 - 2x + 7 = 4x - 1$	Subtract $6x^2$.
$-5x^2 - 6x + 7 = -1$	Subtract $4x$.
$-5x^2 - 6x + 8 = 0$	Add 1.

Since the process for completing the square does not work if the x^2-coefficient does not equal 1, you must divide both members by -5.

$$x^2 + \frac{6}{5}x - \frac{8}{5} = 0.$$

From here on, the procedure is similar to Example 1.

$x^2 + \frac{6}{5}x = \frac{8}{5}$	Adding 8/5 to both members.
$x^2 + \frac{6}{5}x + \frac{9}{25} = \frac{8}{5} + \frac{9}{25}$	Completing the square.
$\left(x + \frac{3}{5}\right)^2 = \frac{49}{25}$	Writing left member as perfect square; combining terms on right.
$\left\lvert x + \frac{3}{5} \right\rvert = \frac{7}{5}$	Taking positive square root of both members.
$x + \frac{3}{5} = \pm\frac{7}{5}$	Definition of absolute value.
$x = -\frac{3}{5} \pm \frac{7}{5}$	Adding $-3/5$ to both members.
$x = \frac{4}{5}$ or -2	Arithmetic.
$\therefore S = \left\{\frac{4}{5}, -2\right\}$.	By inspection.

The following exercise is designed to give you practice in solving quadratic equations in one variable by completing the square. You will also use the technique to find x-intercepts of quadratic functions so that you may plot the graphs more accurately. The first four problems give you a brief review of square roots.

Exercise 5-4

Problems 1 through 4 concern the properties of radicals.

1. $\sqrt{x}$, read "square root of x," means "a number which when squared gives x for an answer."

a. Show that 3 is a square root of 9.
b. Show that -3 is also a square root of 9.
c. Explain why there can be *no* real number that is a square root of -9. (As you recall from Chapter 1, square roots of negative numbers are called "imaginary numbers." Their properties will be explored in Chapter 10.)
d. Since every positive number has *two* square roots, the symbol $\sqrt{x}$ is defined to be the *positive* square root of x. The symbol $-\sqrt{x}$ is used for the negative square root of x. In other words, $\sqrt{9} = 3$, and $-\sqrt{9} = -3$. Find:

i. $\sqrt{25}$. ii. $-\sqrt{36}$.
iii. $-\sqrt{81}$. iv. $\sqrt{100}$.
v. $-\sqrt{49}$. vi. $\sqrt{64}$.
vii. $\sqrt{-25}$.

2. The square root of a *product* of two non-negative numbers equals the product of the two square roots. That is, $\sqrt{xy} = \sqrt{x} \times \sqrt{y}$.

a. 36 is equal to 9×4. Show that $\sqrt{36} = \sqrt{9} \times \sqrt{4}$.
b. 100 is equal to 25×4. Show that $\sqrt{100} = \sqrt{25} \times \sqrt{4}$.
c. 75 is equal to 25×3. What should $\sqrt{75}$ equal?
d. Simplify the following, as in part c, above:

i. $\sqrt{12}$. ii. $\sqrt{18}$.
iii. $\sqrt{50}$. iv. $\sqrt{27}$.

3. The square root of a *sum* (or difference) of two numbers is *not* equal to the sum (or difference) of the square roots.

Show that:

a. $\sqrt{16+9} \neq \sqrt{16} + \sqrt{9}$.
b. $\sqrt{144+25} \neq \sqrt{144} + \sqrt{25}$.
c. $\sqrt{289-64} \neq \sqrt{289} - \sqrt{64}$.

4. The square root of a *perfect square* is surprising! By definition, $(\sqrt{x})^2 = x$. But $\sqrt{x^2}$, where the operations are done in reverse order, is *not* always equal to x. Recalling that $\sqrt{x^2}$ means the *positive* square root of x^2,

a. find $\sqrt{3^2}$,
b. find $\sqrt{(-3)^2}$, and
c. explain why $\sqrt{x^2} = |x|$.

For Problems 5 through 24, solve the quadratic equation by completing the square.

5. $x^2 - 3x + 2 = 0$
6. $x^2 - 5x + 6 = 0$
7. $x^2 + 7x + 12 = 0$
8. $x^2 + 8x + 15 = 0$
9. $x^2 - 2x - 8 = 0$
10. $x^2 - x - 12 = 0$
11. $x^2 + 4x - 3 = 0$
12. $x^2 - 6x + 4 = 0$
13. $2x^2 + 9x - 5 = 0$
14. $3x^2 + 7x + 2 = 0$
15. $3x^2 - 7x = 0$
16. $2x^2 - 15x = 0$
17. $4x^2 - 12x + 9 = 0$
18. $9x^2 + 30x + 25 = 0$
19. $x^2 + 2x + 13 = 0$
20. $x^2 - 10x + 26 = 0$
21. $2x^2 + 2x - 2 = x - x^2$
22. $4x^2 - 8x + 5 = x + 3$
23. $x^2 - 2x + 2 = 2x$
24. $2x^2 - 2x - 2 = x^2$

For Problems 25 through 38, locate the vertex and the x- and y-intercepts. Then use these points and any others obtainable by symmetry to sketch the graph. Find decimal approximations for any radicals you may encounter.

25. $y = x^2 - 6x + 8$
26. $y = x^2 + 4x + 3$
27. $y = x^2 - 2x - 15$
28. $y = x^2 + 2x - 8$
29. $y = -x^2 - 2x + 3$
30. $y = -x^2 + 4x + 5$
31. $y = 2x^2 + 7x + 3$
32. $y = 3x^2 - 7x + 2$

33. $y = -4x^2 + 4x - 1$ 34. $y = x^2 + 6x + 9$

35. $y = x^2 + 2x + 5$ 36. $y = -2x^2 + 4x - 3$

37. $y = x^2 + 2x - 5$ 38. $y = -x^2 + 4x - 1$

5-5 THE QUADRATIC FORMULA

The process of solving quadratic equations by completing the square is somewhat tedious. However, if you do it with the *general* quadratic equation,

$$ax^2 + bx + c = 0,$$

you can get a *formula* for x in terms of the three coefficients, a, b, and c. With this "Quadratic Formula" you (or a computer) need only do the arithmetic to get the solutions.

The steps below show how the formula is derived. A particular equation is solved on the right to help convince you that the derivation is really just an application of the technique of the preceding section.

Start with:

$$ax^2 + bx + c = 0 \qquad 3x^2 + 11x + 5 = 0.$$

Divide both members by a to make the x^2 coefficient equal 1:

$$x^2 + \frac{b}{a}x + \frac{c}{a} = 0 \qquad x^2 + \frac{11}{3}x + \frac{5}{3} = 0.$$

Add $-c/a$ to both members to make room to complete the square:

$$x^2 + \frac{b}{a}x \qquad = -\frac{c}{a} \qquad x^2 + \frac{11}{3}x \qquad = -\frac{5}{3}.$$

Complete the square by taking 1/2 the coefficient of x and squaring it:

$$x^2 + \frac{b}{a}x + \frac{b^2}{4a^2} = -\frac{c}{a} + \frac{b^2}{4a^2}$$

$$x^2 + \frac{11}{3}x + \frac{121}{36} = -\frac{5}{3} + \frac{121}{36}.$$

Write the left member as a perfect square, and simplify the right:

$$\left(x + \frac{b}{2a}\right)^2 = \frac{b^2 - 4ac}{4a^2} \qquad \left(x + \frac{11}{6}\right)^2 = \frac{61}{36}.$$

Take the positive square root of both members:

$$\left|x + \frac{b}{2a}\right| = \frac{\sqrt{b^2 - 4ac}}{2a} \qquad \left|x + \frac{11}{6}\right| = \frac{\sqrt{61}}{6}.$$

Remove the absolute value signs:

$$x + \frac{b}{2a} = \pm\frac{\sqrt{b^2 - 4ac}}{2a} \qquad x + \frac{11}{6} = \pm\frac{\sqrt{61}}{6}.$$

Subtract $b/2a$ from both members:

$$\boxed{x = \frac{-b \pm \sqrt{b^2 - 4ac}}{2a}} \qquad x = \frac{-11 \pm \sqrt{61}}{6}.$$

The equation above is called the *Quadratic Formula.*

Objective:

Be able to solve quadratic equations by substituting into the Quadratic Formula.

Example: Solve $3x^2 + 10x - 8 = 0$.

All you need to do is substitute 3 for a, 10 for b, and -8 for c, then do the arithmetic.

$$x = \frac{-10 \pm \sqrt{100 - 4(3)(-8)}}{2(3)}$$

$$= \frac{-10 \pm \sqrt{100 + 96}}{6}$$

$$= \frac{-10 \pm \sqrt{196}}{6}$$

$$= \frac{-10 \pm 14}{6}$$

$$= 2/3 \quad \text{or} \quad -4$$

$$\therefore S = \{2/3, -4\}.$$

In the exercise which follows, you will get practice using the Quadratic Formula. You will also discover that *part* of the formula can be used to tell what *kind* of numbers the solutions will be, *without* actually solving the equation. Next, you will discover a way to locate the vertex of a parabola *quickly* with the aid of the formula. Finally, you will write a computer program for solving quadratic equations.

Exercise 5-5

For Problems 1 through 20, solve the equation using the Quadratic Formula. Note that these are the same equations as in Exercise 5-4, Problems 5 through 24.

1. $x^2 - 3x + 2 = 0$
2. $x^2 - 5x + 6 = 0$
3. $x^2 + 7x + 12 = 0$
4. $x^2 + 8x + 15 = 0$
5. $x^2 - 2x - 8 = 0$
6. $x^2 - x - 12 = 0$
7. $x^2 + 4x - 3 = 0$
8. $x^2 - 6x + 4 = 0$
9. $2x^2 + 9x - 5 = 0$
10. $3x^2 + 7x + 2 = 0$
11. $3x^2 - 7x = 0$
12. $2x^2 - 15x = 0$
13. $4x^2 - 12x + 9 = 0$
14. $9x^2 + 30x + 25 = 0$
15. $x^2 + 2x + 13 = 0$
16. $x^2 - 10x + 26 = 0$
17. $2x^2 + 2x - 2 = x - x^2$
18. $4x^2 - 8x + 5 = x + 3$
19. $x^2 - 2x + 2 = 2x$
20. $2x^2 - 2x - 2 = x^2$

21. *The Discriminant* – The expression $b^2 - 4ac$ which appears in the Quadratic Formula is called the *discriminant*. For example, if

$$3x^2 + 2x - 5 = 0,$$

then the discriminant equals

$$\begin{aligned} b^2 - 4ac &= 2^2 - 4(3)(-5) \\ &= 4 + 60 \\ &= 64. \end{aligned}$$

The discriminant can be used to "discriminate" among the possible kinds of numbers the solutions of the equation can be, such as imaginary, irrational, or rational. In the example above, 64 is a *perfect square*, so its square root, 8, is a *rational* number. Thus the solutions of $3x^2 + 2x - 5 = 0$ will be rational numbers. Answer the following questions:

a. Find the discriminant of each of the following equations:

   i. $x^2 - x - 12 = 0.$
   ii. $4x^2 - 12x + 9 = 0.$
   iii. $x^2 - 2x + 13 = 0.$
   iv. $x^2 - 4x - 3 = 0.$

b. Based on the answers you get for the discriminant, above, tell for each equation whether the solutions will be *rational* numbers, *irrational* numbers, or *imaginary* numbers. (Clue: The discriminant appears under a *square root sign* in the Quadratic Formula.)

c. The following diagram shows possible kinds of numbers the solutions of a quadratic equation

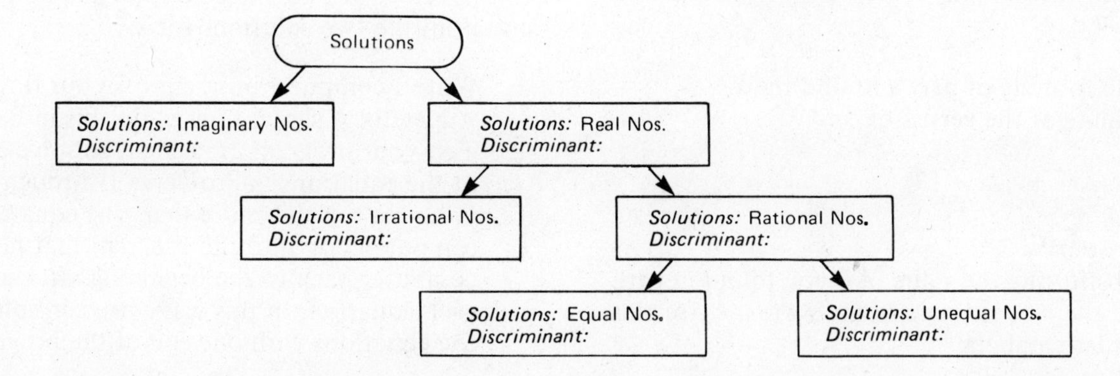

could be. Copy the diagram then write in each box what kind of *discriminant* leads to the indicated kind of *solutions.*

22. *Discriminant Practice* – For each of the following equations, calculate the discriminant (see Problem 21). Based on your answer, tell whether the solutions will be *real* numbers or *imaginary* numbers. If they are real numbers, tell whether they will be *rational* numbers or *irrational* numbers. If the solutions are rational numbers, tell whether or not the two solutions are *equal.*

a. $3x^2 - 5x + 6 = 0.$
b. $5x^2 + 7x - 3 = 0.$
c. $2x^2 - 13x + 15 = 0.$
d. $9x^2 + 6x + 1 = 0.$
e. $10x^2 + 19x + 7 = 0.$
f. $x^2 - 6x + 3 = 0.$
g. $-3x^2 + 5x - 2 = 0.$
h. $x^2 + x + 1 = 0.$
i. $-x^2 + 4x - 4 = 0.$
j. $x^2 + 6x + 10 = 0.$
k. $5x^2 - 8x + 2 = 0.$
l. $7x^2 - 10x + 3 = 0.$
m. $2x^2 - 3x + 5 = 0.$
n. $9x^2 + 12x + 13 = 0.$
o. $25x^2 + 10x + 1 = 0.$
p. $6x^2 - 13x + 6 = 0.$
q. $-3x^2 + x + 5 = 0.$
r. $x^2 + x - 1 = 0.$
s. $-x^2 - 15x - 56 = 0.$
t. $x^2 + 2x + 1 = 0.$

23. *Vertex, Quickly* – a. By transforming the *general* equation $y = ax^2 + bx + c$ to the vertex form, show that h, the x-coordinate of the vertex, is

$$h = -\frac{b}{2a}.$$

b. Use the formula of part a to find the x-coordinate of the vertex of

$$y = 3x^2 + 12x + 17,$$

in *one* step.

c. By substituting the value of h you found in part b, find the y-coordinate of the vertex, k, for this particular parabola.

d. Which seems to you to be the easier way of finding the vertex: Using the *formula* as in parts b and c, or completing the square as in Section 5-3?

24. *Quick Vertex Practice* – For each of the following functions, find the vertex, (h, k), by using the results of Problem 23, that if $y = ax^2 + bx + c$, then

$$\boxed{h = -\frac{b}{2a}.}$$

These functions are the same as in Exercise 5-3, Problems 1 through 12.

a. $y = x^2 + 6x + 11.$
b. $y = 3x^2 - 24x + 17.$
c. $y = 4x^2 + 12x - 11.$
d. $y = 2x^2 - 11x - 12.$
e. $y = -5x^2 - 30x + 51.$
f. $y = -x^2 + x + 1.$
g. $y = x^2 + 8x + 21.$
h. $y = 5x^2 - 30x + 31.$
i. $y = 8x^2 + 40x + 37.$
j. $y = 2x^2 - 7x + 12.$
k. $y = -3x^2 - 24x - 41.$
l. $y = -x^2 - x + 3.$

25. *Computer Solution of Quadratics* – a. Write a flow chart for a computer program to solve quadratic equations using the Quadratic Formula. The flow chart should begin with inputting values for the constants *a, b,* and *c.* The discriminant should then be checked to see whether or not it is negative. If it is negative the computer should print out a message to this effect. If not, the computer should calculate and print the two solutions for x.

b. Write a computer program for your flow chart. Appendix A shows how to do this in BASIC.
c. Test your program by using it to solve any or all of the equations in Problems 1 through 20. If you plan to solve more than one equation, you can put a loop into the program that takes the computer back to the beginning after solving each equation. In this way, you can solve *all* of the equations with *one* run of the program.

26. *Untidy Quadratic Equations* – The quadratic equations you have solved so far have had *integer* coefficients, usually fairly small integers. Equations arising from problems in the real world are seldom this tidy. However, solving such equations using the Quadratic Formula is straightforward as long as you have a calculator or a computer available. Solve the following equations.

a. $37.8x^2 + 42.4x - 17.9 = 0$.

b. $432.77x^2 - 1176.48x + 239.97 = 0$.

c. $0.235x^2 + 21.4x + 437.5 = 0$.

d. $43.672x^2 - 48.0392x - 152.852 = 0$.

e. $3.119x^2 - 3.547x + 1.023 = 0$.

f. $x^2 - 2001x + 25.41 = 0$.

g. $56.9x^2 - 77.4x - 33.2 = 0$.

h. $292.97x^2 + 732.88x - 147.22 = 0$.

i. $0.0764x^2 - 19.7x - 203.3 = 0$.

j. $37.922x^2 - 82.259x + 44.612 = 0$.

k. $10.56x^2 + 34.32x + 19.80 = 0$.

l. $2001x^2 - 1336x - 1001 = 0$.

5-6 EVALUATING QUADRATIC FUNCTIONS

When you use a function as a mathematical model, you must be able to calculate y for known values of x. It is also important to be able to calculate x for given values of y. In this section you will practice these skills in a purely mathematical context. Thus, you will be comfortable using these skills in later real-world problems.

Objective:

Given the equation for a quadratic function, be able to calculate the value of y for a known value of x, and the value of x for a known value of y.

From Section 4-4, you recall that "$f(x)$" terminology gives a convenient way to indicate what number was substituted for x. If $f(x) = 3x^2 + 2x - 11$, for example, then "$f(-4)$" means, "The value of y when $x = -4$." The following example shows how you can find values of y or x for this function.

Example 1: If $f(x) = 3x^2 + 2x - 11$, find

a. $f(-4)$,
b. x, when $f(x) = -6$, and
c. the x-intercepts.

a. Calculating y for a known value of x is easy. You simply substitute for x, then do the arithmetic. You must, of course, observe the order of operations agreed upon in algebra.

$f(-4) = 3(-4)^2 + 2(-4) - 11$	Substitution.
$= 3(16) + 2(-4) - 11$	*Squaring* comes *first.*
$= 48 - 8 - 11$	*Multiplying* is done *next.*
$= \underline{\underline{29}}$.	Adding and subtracting are done *last,* from left to right.

With practice, you should be able to do some of this work in your head and thus save writing.

b. Calculating x for a known value of y involves the quadratic equation techniques of the last two sections. To find x when $f(x) = -6$, you first substitute -6 for $f(x)$, getting

$$-6 = 3x^2 + 2x - 11.$$

You may be more comfortable with the variables on the *left* and the constant on the right, so you may use symmetry to get

$$3x^2 + 2x - 11 = -6.$$

This is a quadratic equation with *one* variable. But the right member does not equal zero, so you add 6 to both members, getting

$$3x^2 + 2x - 5 = 0.$$

From here on, this is an *old* problem. You may complete the square or use the Quadratic Formula.

$$x = \frac{-2 \pm \sqrt{2^2 - 4(3)(-5)}}{2(3)}$$ Quadratic Formula.

$$= \frac{-2 \pm \sqrt{64}}{6}$$ Arithmetic.

$$= \frac{-2 \pm 8}{6}$$ $\sqrt{64} = 8$.

$$= 1 \quad \text{or} \quad -\frac{5}{3}.$$ Arithmetic.

Since the question was, "Find the values of x ...," rather than, "Solve the equation ...," it is *not* necessary to write the answer as a solution set.

c. Finding the x-intercepts means finding the value(s) of x *when* $y = 0$. So you substitute 0 for $f(x)$ and solve the resulting equation.

$$0 = 3x^2 + 2x - 11$$ Substitution.

$$3x^2 + 2x - 11 = 0$$ Symmetry (optional step).

$$x = \frac{-2 \pm \sqrt{2^2 - 4(3)(-11)}}{2(3)}$$ Quadratic Formula.

$$= \frac{-2 \pm \sqrt{136}}{6}$$ Arithmetic.

The answers can either be left in this form, be simplified, or be approximated with decimals.

$$\sqrt{136} = \sqrt{4 \cdot 34} = \sqrt{4} \cdot \sqrt{34} = 2\sqrt{34}.$$

So you can write

$$x = \frac{-2 \pm 2\sqrt{34}}{6}$$ $\sqrt{136} = 2\sqrt{34}$.

$$= \frac{-1 \pm \sqrt{34}}{3}.$$ Multiply by 1 in the form ½/½.

An approximation for $\sqrt{34}$ can be obtained by calculator, Table I, or an educated guess. By calculator,

$$\sqrt{34} \approx 5.830951895.$$

So

$$x \approx 1.610317298 \quad \text{or} \quad -2.276983965.$$

Sometimes there are *no* values of x for a given value of y. As shown in Figure 5-6, this happens when a horizontal line drawn at the given value of y *misses* the parabola.

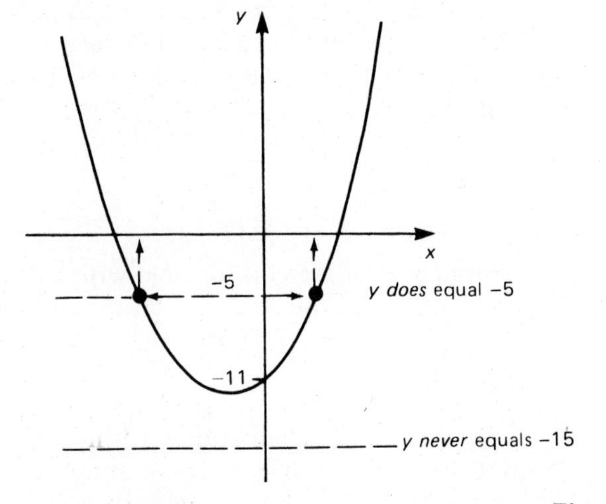

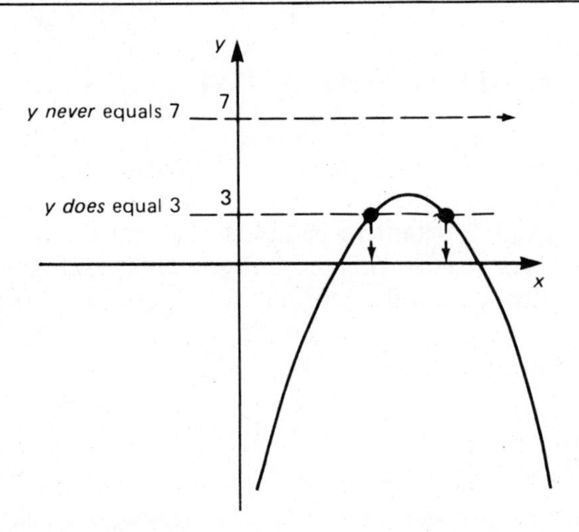

Figure 5-6.

In some real-world applications, the only thing you are concerned about is whether or not y ever *does* equal a certain given number. There is a quick way to tell this by finding the *discriminant*, which you learned about if you worked Problems 21 and 22 in Exercise 5-5.

Example 2: If $f(x) = 3x^2 + 2x - 11$, does y ever equal

a. -5?
b. -15?

a. The question, "Does y ever equal -5?" is equivalent to the question, "Are there *real* numbers, x, for which $f(x) = -5$?" Substituting -5 for $f(x)$, you get

$-5 = 3x^2 + 2x - 11.$	Substitution.
$3x^2 + 2x - 6 = 0.$	Add 5, then use symmetry.

This equation is now in the form $ax^2 + bx + c = 0$. So you find the discriminant,

$b^2 - 4ac = 2^2 - 4(3)(-6)$	Definition of discriminant.
$= 76.$	Arithmetic.

Since the discriminant is a *positive* number, its square root (which appears in the Quadratic Formula) will be a *real* number. Thus, there *are* real values of x for which $f(x) = -5$. So you can conclude that

$f(x)$ *does* equal -5 sometimes.

b. If $f(x) = -15$, the same line of reasoning leads to

$-15 = 3x^2 + 2x - 11$	Substitution.
$3x^2 + 2x + 4 = 0$	Add 15, then use symmetry.
$b^2 - 4ac = 2^2 - 4(3)(4)$	Definition of discriminant.
$= -44$	Arithmetic.
$\therefore$ there are *no* real solutions	$\sqrt{-44}$ is an *imaginary* number.
$\therefore f(x)$ *never* equals -15.	No real x when $f(x) = -15$.

In the exercise that follows, you will get practice finding y for known values of x, finding x for known values of y, and using the discriminant to tell quickly whether or not y ever equals certain given values.

Exercise 5-6

1. Suppose that $f(x) = 5x^2 + 8x - 7$.
   a. Find $f(-3)$.
   b. Find x when $f(x) = -3$.
   c. Find the x-intercepts.

2. Suppose that $g(x) = 2x^2 - 5x - 11$.
   a. Find $g(-4)$.
   b. Find x when $g(x) = -4$.
   c. Find the x-intercepts.

3. Suppose that $h(x) = -2x^2 + 3x - 10$.
   a. Find $h(-9)$.
   b. Find x when $h(x) = -9$.
   c. Find the x-intercepts.

4. Suppose that $f(x) = -4x^2 + 4x + 15$.
   a. Find $f(-3)$.
   b. Find x when $f(x) = 20$.
   c. Find the x-incercepts.

5. Suppose that $y = x^2 + 8x + 15$. Find the value(s) of x for which
   a. $y = 3$, b. $y = 2$, c. $y = 0$,
   d. $y = -1$, e. $y = -3$, f. $y = 15$.

6. Suppose that $y = -x^2 - 6x - 5$. Find the value(s) of x for which
   a. $y = 5$, b. $y = 4$, c. $y = 3$,
   d. $y = 2$, e. $y = 0$, f. $y = -5$.

For Problems 7 through 14, use the discriminant to tell whether or not the indicated function ever has the given values of y (for *real* values of x).

7. $y = 4x^2 - 7x + 2$; $y = 5$, $y = -3$.
8. $y = 3x^2 + 10x - 1$; $y = 6$, $y = -4$.
9. $y = 2x^2 + 3x + 6$; $y = 1$, $y = -5$.
10. $y = 5x^2 - 8x + 6$; $y = 3$, $y = -4$.
11. $y = -3x^2 + 5x + 1$; $y = 4$, $y = -3$.
12. $y = -2x^2 + 6x - 7$; $y = 10$, $y = -10$.
13. $y = -x^2 + 10x - 8$; $y = 7$, $y = 0$.
14. $y = -x^2 - 6x - 9$; $y = 1$, $y = 0$.

5-7 EQUATIONS OF QUADRATIC FUNCTIONS FROM THEIR GRAPHS

You now know enough properties of quadratic functions to be able to evaluate and graph them when you know the particular equation. To use a quadratic function as a mathematical model, you must be able to *find* the particular equation from information about the real world.

From Section 3-4, you should recall how to find the particular equation of a linear function from two points on its graph. For quadratic functions, you must be given *three* points.

Objective:

Given three points on the graph of a quadratic function, or the vertex and one other point, find the particular equation of the function.

Example 1: Find the particular equation of the quadratic function containing $(-2, -11)$, $(4, 13)$, and $(6, 29)$.

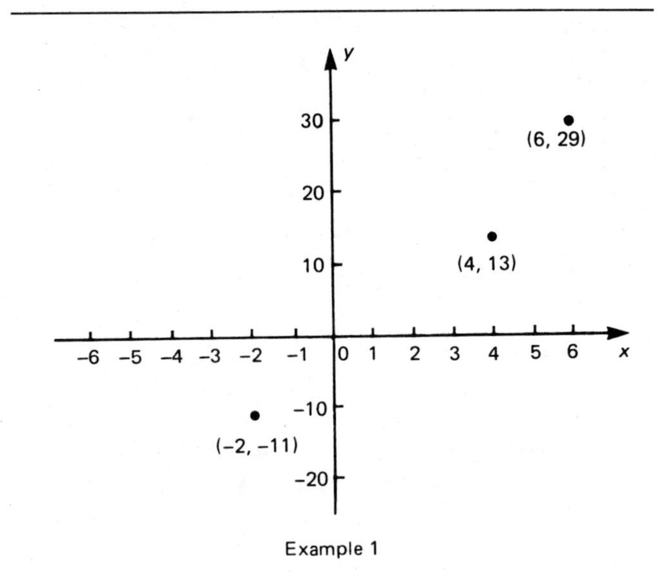

Figure 5-7a.

Since all quadratic functions have equations of the form $y = ax^2 + bx + c$, finding the particular equation is done by finding the appropriate values of a, b, and c.

Substituting the first ordered pair, $(-2, -11)$, for (x, y) in the equation $y = ax^2 + bx + c$ gives:

$$-11 = a(-2)^2 + b(-2) + c,$$

which can be transformed to

$$4a - 2b + c = -11 \quad \text{———} \quad ①.$$

This is simply a linear equation that relates the three "variables" a, b, and c. Substitution of the other two ordered pairs $(4, 13)$ and $(6, 29)$ gives two more linear equations in a, b, and c:

$$16a + 4b + c = 13 \quad \text{———} \quad ②$$

$$36a + 6b + c = 29 \quad \text{———} \quad ③.$$

Equations ①, ②, and ③ form a system of three linear equations in three variables, which you learned how to solve in Section 4-5. The constant c can be eliminated by multiplying ① by −1 and adding it to ② and ③, giving the two equations

$12a + 6b = 24$ Multiply ① by −1 and add ②.

$32a + 8b = 40.$ Multiply ① by −1 and add ③.

These can be simplified to

$2a + b = 4$ ——— ④

$4a + b = 5$ ——— ⑤.

Multiplying ④ by −1 and adding it to ⑤ eliminates b, giving

$2a = 1$

$a = \frac{1}{2}.$

Substituting 1/2 for a in ④ gives

$1 + b = 4$

$b = 3.$

Substituting 1/2 for a and 3 for b in ① gives

$2 - 6 + c = -11.$

$c = -7.$

So the desired equation is

$$y = \frac{1}{2}x^2 + 3x - 7.$$

Note that the answer is an *equation* rather than a solution set.

Example 2: Find the particular equation of the quadratic function containing (0, 5), (2, 13), and (3, 26).

The job of finding this equation is easier because you know the y-intercept. Substituting (0, 5) for (x, y) in $y = ax^2 + bx + c$ gives

$5 = a \cdot 0^2 + b \cdot 0 + c,$

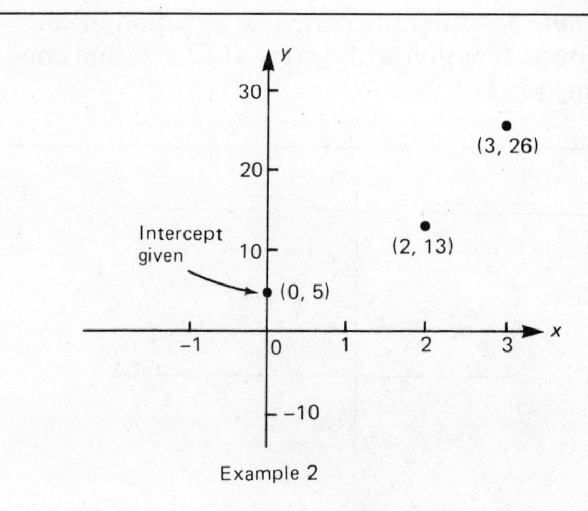

Figure 5-7b.

from which $c = 5$. Using this value of c when you substitute the other two ordered pairs gives *two* equations in a and b:

$13 = 4a + 2b + 5$ Substitute $(x, y) = (2, 13)$ and $c = 5$.

$26 = 9a + 3b + 5$ Substitute $(x, y) = (3, 26)$ and $c = 5$.

which can be transformed to

$2a + b = 4$ ——— ①

$3a + b = 7$ ——— ②.

Multiplying ① by −1 and adding it to ② gives

$a = 3.$

Substituting 3 for a in ① gives

$6 + b = 4,$

$b = -2.$

So the desired equation is

$$y = 3x^2 - 2x + 5.$$

Example 3: Find the particular equation of the quadratic function with vertex at (2, −5) and containing (3, 1).

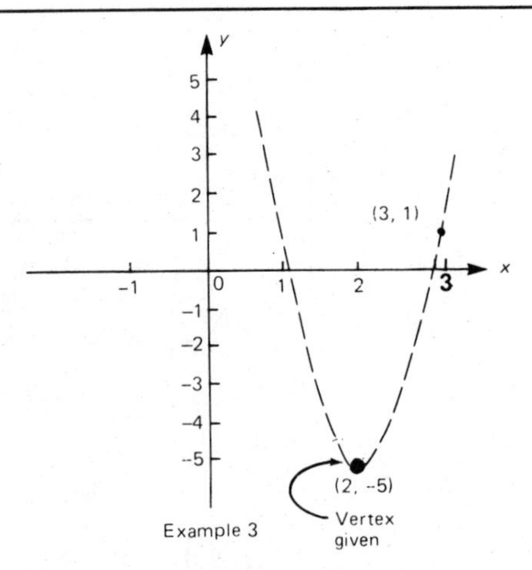

Figure 5-7c.

If you know that one of the given points is the *vertex*, then it would be easier to use the *vertex* form, $y - k = a(x - h)^2$. Substituting (2, −5) for the vertex (h, k) gives

$$y - (-5) = a(x - 2)^2.$$

The only constant left to be evaluated is a. This is why you need only *one* other ordered pair. Substituting (3, 1) for (x, y) gives

$$1 - (-5) = a(3 - 2)^2$$

$$6 = a.$$

Therefore, the equation is

$$y + 5 = 6(x - 2)^2.$$

If desired, this equation can be transformed to

$$y = 6x^2 - 24x + 19.$$

The exercise which follows is designed to give you practice in finding the equation of a quadratic function from information about its graph. This is the fundamental technique which you will use in Section 5-8 for making quadratic mathematical models of situations in the real world.

Exercise 5-7

For Problems 1 through 14, find the particular equation of the quadratic function containing the given ordered pairs. Write the equation in the form $y = ax^2 + bx + c$.

1. (1, 6), (3, 26), (−2, 21).
2. (1, 2), (−2, 23), (3, 8).
3. (−2, −41), (−3, −72), (5, −48).
4. (−3, 18), (6, −9), (12, −57).
5. (4, 7.3), (6, 12.7), (−3, 1.0).
6. (10, 1), (20, 22), (−30, −3).
7. (10, 40), (−20, 160), (−5, 10).
8. (2, −2.8), (−3, −6.3), (5, −17.5).
9. (−4, −37), (2, 11), (0, −1).
10. (0, 5), (4, 1), (−3, −13).
11. (0, 0), (−1, 7), (6, 42).
12. (0, 0), (−1, 4), (3, −48).
13. Vertex at (−4, 3), also containing (−6, 11).
14. Vertex at (−2, 3), also containing (4, 12).

15. Show that there is *no* quadratic function which contains the points (5, 2), (6, −4), and (5, −7). Explain what it is about these three points that prevents there being a quadratic function containing all of them.

16. Show that there is *no* quadratic function which contains the points (−2, −1), (1, 8), and (3, 14). Explain what it is about these three points that prevents there being a quadratic function containing all of them.

5-8 QUADRATIC FUNCTIONS AS MATHEMATICAL MODELS

In Exercise 2-2 of Chapter 2 you sketched reasonable graphs for a number of real-world situations showing how two variables were related to each other. Some of these graphs looked like parabolas. Therefore, it would seem reasonable to use quadratic functions as mathematical models for these real-world situations.

As was the case for linear functions in Chapter 3, the most important step in such problems is deriving an equation which shows how the dependent variable is related to the independent one. The technique for finding the equation of a particular quadratic function from given points is presented in Section 5-7. Once you have the equation, you can use it to make predictions and interpretations about the real world.

Objective:

Be able to use a quadratic function as a mathematical model for a real-world situation in which the graph is a *curved* line.

Example: An old menu from the Pizza Inn lists the following prices for plain cheese pizzas:

Small (8″ diameter) – \$ .85.
Medium (10″ diameter) – \$1.15.
Large (13″ diameter) – \$1.75.

Use these prices to predict the prices of *other* sizes of pizza.

If you let

d = number of inches in the diameter, and
p = number of cents a pizza costs,

and assume that the price depends on the diameter, then the ordered pairs (8, 85), (10, 115), and (13, 175) are in the price-diameter function. The graph of these ordered pairs in Figure 5-8a shows

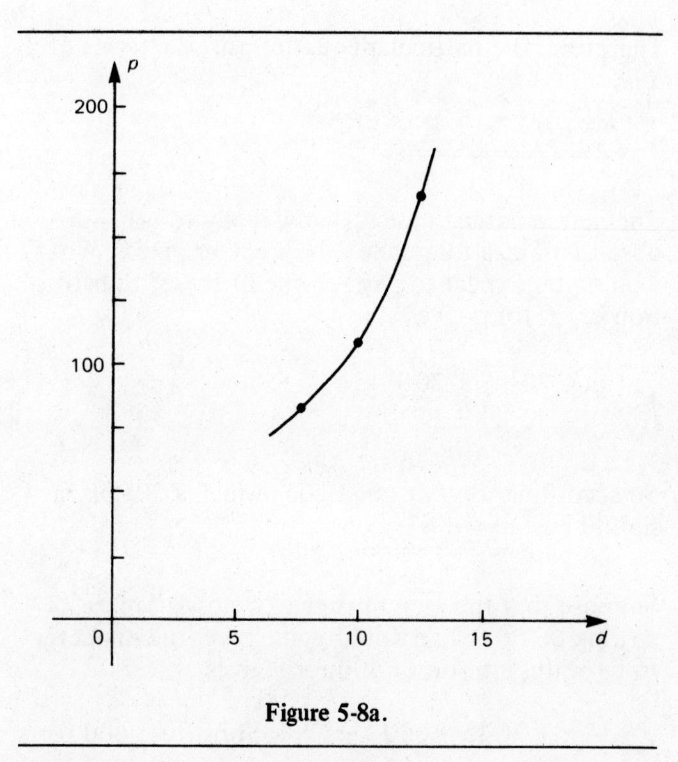

Figure 5-8a.

that the points do *not* lie on a straight line. Therefore, the function is not *linear.*

Since the graphs of quadratic functions are curved, you could assume that p varies *quadratically* with d. With this assumption, you can immediately write a general equation expressing p in terms of d,

$$p = ad^2 + bd + c$$

where a, b, and c stand for constants. Substitution of the three ordered pairs, above, for (d, p) yields the system of three linear equations in a, b, and c:

$$64a + 8b + c = 85 \quad \text{①}$$

$$100a + 10b + c = 115 \quad \text{②}$$

$$169a + 13b + c = 175 \quad \text{③}$$

This system may be solved as in Section 5-7 to give

$$a = 1, \quad b = -3, \quad \text{and} \quad c = 45.$$

Therefore, the particular equation for p in terms of d is

$$p = d^2 - 3d + 45.$$

The mathematical model is now ready to use. Suppose that Pizza Inn made 20″ diameter pizzas. What would you expect to pay for one of these? Substitution of 20 for d gives

$$p = 20^2 - 3 \cdot 20 + 45$$
$$\therefore p = 385.$$

So according to your quadratic model, a 20″ pizza should cost about $3.85.

Suppose that the menu listed a "Colossal" pizza costing $6.00. What would you expect the diameter to be? Substitution of 600 for p gives

$d^2 - 3d + 45 = 600$	Substitute 600 for p.
$d^2 - 3d - 555 = 0$	Subtract 600.
$d = \dfrac{3 \pm \sqrt{3^2 - 4(1)(-555)}}{2(1)}$	Quadratic Formula.
$= \dfrac{3 \pm \sqrt{2229}}{2}$	Arithmetic.
$\approx \dfrac{3 \pm 47.2}{2}$	$\sqrt{2229} \approx 47.2$.
$= 25.1$ or -22.1	Arithmetic.

Only the *positive* solution is meaningful this time, so the diameter would be about 25 inches.

If you substitute 0 for d, you find that $p = 45$. Why should you have to pay 45 cents if you buy *no* pizza? The interpretation is that there are *fixed* costs which do not depend on how big the pizza is. For example, it costs Pizza Inn just as much to take your order, cook the pizza, serve it, clean off the table, and wash the dishes when you order a small pizza as it does when you order a big one. Your model tells you that Pizza Inn charges 45 cents per pizza to pay for these fixed costs.

As a check on the reasonability of your model, there should be *no* diameter for which the price is zero. There are two ways to check your model for this feature. The first way is to find the vertex by completing the square as in Section 5-3, or by using $h = -b/2a$ as in Problems 23 and 24 of Exercise 5-5. Since

$$p = d^2 - 3d + 45,$$

the d-coordinate of the vertex is $-b/2a = -(-3)/2 = 3/2$. Substituting 3/2 for d gives

$$p = (3/2)^2 - 3(3/2) + 45 = 42\tfrac{3}{4}.$$

Since the p-coordinate of the vertex is 42-3/4, which is greater than zero, and since the vertex is the lowest point on this parabola, the price will never be zero.

A second, more direct way is to try to find the d-intercepts, the values of d when $p = 0$. Substituting 0 for p gives

$$d^2 - 3d + 45 = 0.$$

All you really need to know is whether or not there *are* any real solutions to this equation. You can check the discriminant and find out:

$$b^2 - 4ac = (-3)^2 - (4)(1)(45)$$
$$= -171.$$

Since the discriminant is *negative*, the equation has *no* real solutions, and thus there are *no* diameters for which the price is zero. Your model is thus "reasonable" in the sense that it does not predict that Pizza Inn would *give away* any size of pizza.

The completed graph in Figure 5-8b shows the given points and the predictions.

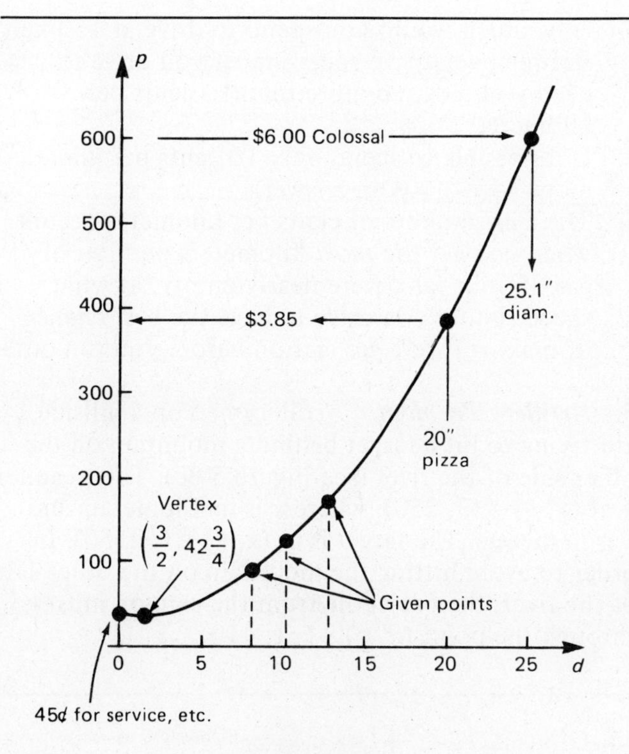

Figure 5-8b.

In the exercise which follows, you are to construct and use quadratic mathematical models of other real-world situations. Remember that the most important part of the problem is getting the equation correct. With an incorrect equation, other parts of the problem may turn out to be senseless.

Exercise 5-8

1. ***Phoebe Small's Rocket Problem.*** Phoebe Small is out Sunday driving in her spaceship. As she approaches Mars, she changes her mind, decides that she does not wish to visit that planet, and fires her retro-rocket. The spaceship slows down, and if all goes well, stops for an instant then starts pulling away. While the rocket motor is firing, Phoebe's distance, d, from the surface of Mars depends by a quadratic function on the number of minutes, t, since she started firing the rocket.

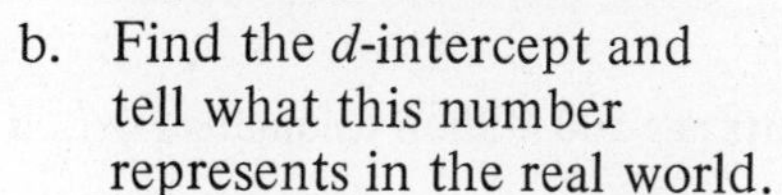

a. Phoebe finds that at times $t = 1$, 2, and 3 minutes, her distances are $d = 425$, 356, and 293 kilometers, respectively. Find the particular equation expressing d in terms of t.
b. Find the d-intercept and tell what this number represents in the real world.
c. According to the equation, where will Phoebe be when $t = 15$? When $t = 16$? Does this tell you she is pulling away from Mars when $t = 16$, or still approaching?

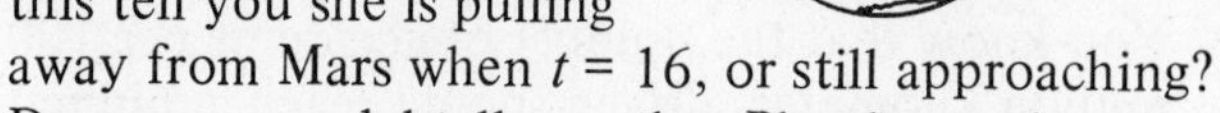

d. Does your model tell you that Phoebe crashes into the surface of Mars, just touches the surface, or pulls away before reaching the surface? Explain.
e. Draw the graph of this quadratic function. Show the vertex.
f. Based on your answers to the above questions, in what domain do you think this quadratic function will give reasonable values for d? Modify your graph in part e, if necessary, by using a *dotted* line for those parts of the graph that are out of this domain.

2. ***Bathtub Problem.*** Assume that the number of liters of water remaining in the bathtub varies quadratically with the number of minutes which have elapsed since you pulled the plug.

a. If the tub has 38.4, 21.6, and 9.6 liters remaining at 1, 2, and 3 minutes respectively, since you pulled the plug, write an equation expressing liters in terms of time.
b. How much water was in the tub when you pulled the plug?
c. When will the tub be empty?
d. In the real world, the number of liters would never be negative. What is the lowest number of liters the *model* predicts. Is this number reasonable?
e. Draw a graph of the function in the appropriate domain.
f. Why is a quadratic function more reasonable for this problem than a linear function would be?

3. ***Car Insurance Problem.*** Suppose that you are an actuary for F. Bender's Insurance Agency. Your company plans to offer a senior citizen's accident policy, and you must predict the likelihood of an accident as a function of the driver's age. From previous accident records, you find the following information:

Age (years)	Accidents per 100 Million Kilometers Driven
20	440
30	280
40	200

You know that the number of accidents per 100 million kilometers driven should reach a minimum then go up again for very old drivers. Therefore, you assume that a *quadratic* function is a reasonable model.

a. Write the particular equation expressing accidents per 100 million kilometers in terms of age.
b. How many accidents per 100 million kilometers would you expect for an 80-year-old driver?
c. Based on your model, who is safer; a 16-year-old driver or a 70-year-old driver?
d. What age driver appears to be the safest?
e. Your company decides to insure licensed drivers up to the age where the accident rate reaches 830 per 100 million kilometers. What, then, is the domain of this quadratic function?

4. ***Cost of Operating a Car Problem.*** The number of cents per kilometer it costs to drive a car depends on how fast you drive it. At low speeds the cost is high because the engine operates inefficiently, while at high speeds the cost is high because the engine must overcome high wind resistance. At moderate speeds the cost reaches a minimum. Assume, therefore, that the number of cents per kilometer varies *quadratically* with the number of kilometers per hour (kph).

a. Suppose that it costs 28, 21, and 16 cents per kilometer to drive at 10, 20, and 30 kph, respectively. Write the particular equation for this function.
b. How much would you spend to drive at 150 kph?
c. Between what two speeds must you drive to keep your cost no more than 13 cents per kilometer?
d. Is it possible to spend only 10 cents per kilometer? Justify your answer.
e. The *least* number of cents per kilometer occurs when you get the *most* kilometers per liter of gas. If your tank were nearly empty, at what speed should you drive to have the best chance of making it to a gas station before you run out?

5. ***Artillery Problem.*** Artillerymen on a hillside are trying to hit a target behind a mountain on the other side of the river (see Figure 5-8c). Their cannon is at $(x, y) = (3, 250)$, where x is in kilometers and y is in meters. The target is at $(x, y) = (-2, 50)$. In order to avoid hitting the mountain on the other side of the river, the projectile from the cannon must go through the point $(x, y) = (-1, 410)$.

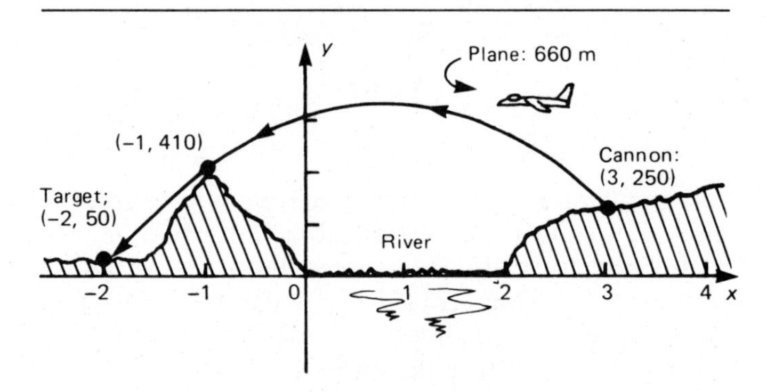

Figure 5-8c.

a. Write the particular equation of the parabolic path of the projectile.
b. How high above the river will the projectile be where it crosses
   i. the right riverbank, $x = 2$?
   ii. the left riverbank, $x = 0$?
c. Approximately where will the projectile be when $y = 130$?

d. A reconnaissance plane is flying at 660 meters above the river. Is it in danger of being hit by projectiles fired along this parabolic path? Justify your answer.

6. ***Football Problem.*** When a football is punted, it goes up into the air, reaches a maximum altitude, then comes back down. Assume, therefore, that a quadratic function is a reasonable mathematical model for this real-world situation.

Let t = number of seconds that have elapsed since the ball was punted.
Let d = number of feet the ball is above the ground.

a. When the ball was kicked it was 4 feet above the ground. One second later, it was 28 feet above the ground. Two seconds after it was kicked, it was 20 feet up. Write the particular equation expressing d in terms of t.
b. Find the d and t coordinates of the vertex, and tell what each represents in the real world.
c. Find the t-intercepts and tell what each represents in the real world. Use square root tables or an educated guess to find a decimal approximation for any square roots you may encounter.
d. Draw a graph of the function. Select scales which make the graph fill up most of the sheet of graph paper. You should have enough points with the vertex, intercepts, given points, and points obtainable easily from symmetry.
e. By looking at your graph and thinking about what it represents, figure out a *domain* for this function. Why would your model not give reasonable answers for d when the value of t is

 i. below the lower bound of the domain?
 ii. above the upper bound of the domain?

 Modify your graph, if necessary, to agree with this domain.

f. What influences in the real world might make your model slightly inaccurate *within* the domain?
g. From your graph and your calculations, tell what the *range* of this function is.

7. ***Color TV Problem.*** The following are prices of a popular brand of color TV as of June, 1976:

Size (in.)	Price
5″	\$450
9″	\$430
12″	\$400
15″	\$450
17″	\$510
19″	\$570
21″	\$700

a. Plot these ordered pairs (size, price) on a Cartesian coordinate system. Connect the dots with a smooth curve.
b. Assume that the price varies quadratically with screen size. Use the ordered pairs for 9″, 15″, and 19″ screens to derive the particular equation for this function.
c. If the manufacturer produced a 24″ TV set, how much would you expect to pay for one?
d. Use the equation to calculate the prices of 5″, 12″, 17″, and 21″ diagonal sets.
e. Plot the predicted points from part c on the Cartesian coordinate system of part a. Connect these points with a smooth curve. Use means such as a colored pencil to distinguish clearly between the two graphs.
f. Based on the two graphs, would you describe the quadratic model as *accurate, reasonable,* or *inaccurate*?
g. Why do you suppose the price goes *up* as the size gets very small?

8. ***Calvin Butterball's Gasoline Problem.*** Calvin Butterball is driving along the highway. He starts up a long, straight hill (see Figure 5-8d). 114 meters from the bottom of the hill Calvin's car runs out of gas. He doesn't put on the brakes, so the car keeps rolling for awhile, coasts to a stop, then starts rolling backwards. He finds that his distances from the bottom of the hill 4 and 6 seconds after he runs out of gas are 198 and 234 meters, respectively.

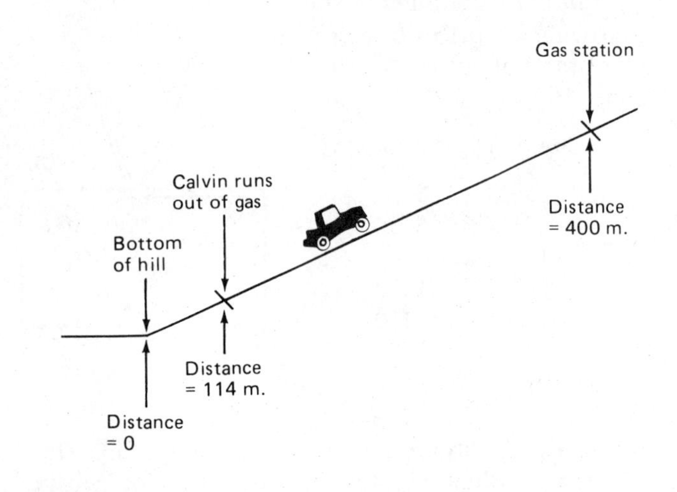

Figure 5-8d.

a. There are *three* ordered pairs of time and distance in the information given above. What are they?

b. Write an equation expressing Calvin's distance from the bottom of the hill in terms of the number of seconds which have elapsed since he ran out of gas. Assume that a quadratic function is a reasonable mathematical model.

c. Use the model to predict Calvin's distance from the bottom of the hill 10 and 30 seconds after he ran out of gas.

d. Draw a graph of this function. Use the given points, the calculated points from part c, and any other points you find useful.

e. Last Chance Texaco Station is located 400 meters from the bottom of the hill. Based on your model, will Calvin's car reach the station before it stops and starts rolling backwards? Justify your answer.

9. ***Spaceship Problem.*** When a spaceship is sent to the Moon, it is first put into orbit around the Earth. Then just at the right time, the rocket motor is fired to start it on its parabolic path to the Moon (see Figure 5-8e).

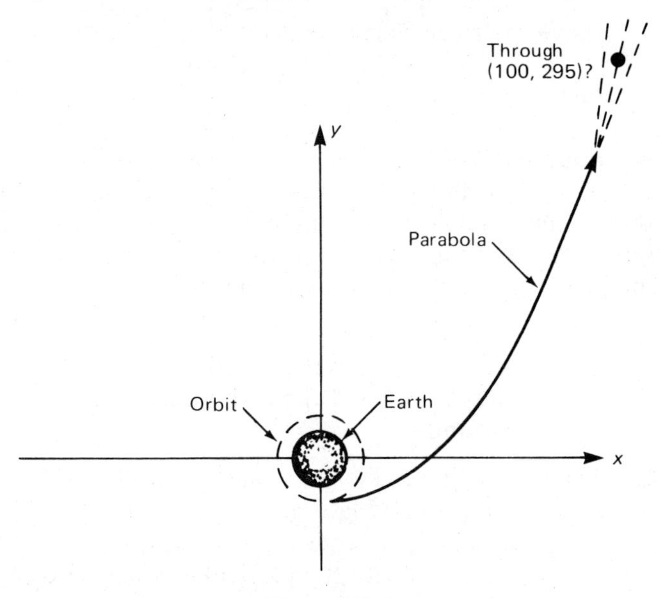

Figure 5-8e.

Let x and y be coordinates measuring the position of the spacecraft (in thousands of kilometers) as it goes Moonward. When the rocket is fired, $x = 0$ and $y = -7$. The tracking station measures the position at two later times, and finds that $y = -4$ when $x = 10$, and $y = 5$ when $x = 20$.

a. Find an equation of the parabolic path in Figure 5-8e.

b. In order to hit the Moon without a mid-course maneuver, the path of the spaceship must pass through the point (100, 295). Use your mathematical model to predict whether or not a mid-course maneuver will be necessary.

10. ***Gateway Arch Problem.*** On a trip to St. Louis you visit the Gateway Arch. Since you have plenty of time on your hands, you decide to estimate its altitude. You set up a Cartesian coordinate system with one end of the arch at the origin, as shown in Figure 5-8f. The other end of the arch is at $x = 162$ meters. To find a third point on the arch, you measure a value of $y = 4.55$ meters when $x = 1$ meter. You assume that the arch is parabolic.

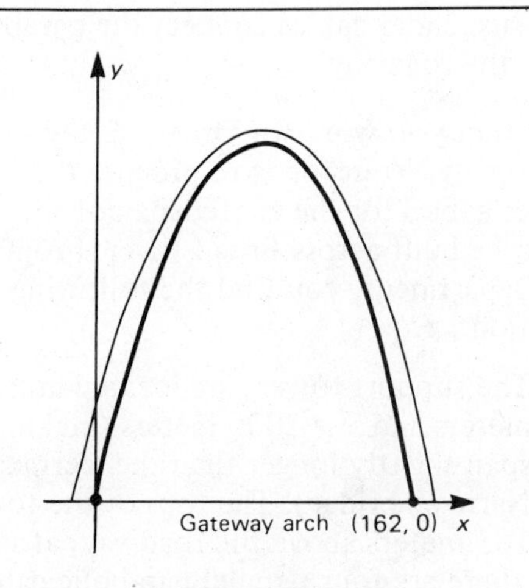

Figure 5-8f.

a. Find the particular equation of the underside of the arch.
b. What is the x-coordinate of the vertex? By substituting this number into the equation, predict the height of the arch.
c. An airplane with a wingspan of 40 meters tries to fly through the arch at an altitude of 170 meters. Could the plane possibly make it? Justify your answer.

11. ***Barley Problem.*** The number of bushels of barley an acre of land will yield depends on how many seeds per acre you plant. From previous planting statistics you find that if you plant 2 hundred thousand seeds per acre, you can harvest 22 bushels per acre, and if you plant 4 hundred thousand seeds per acre you can harvest 40 bushels per

acre. As you plant more seeds per acre, the harvest will reach a maximum, then decrease. This happens because the young plants crowd each other out and compete for food and sunlight. Assume, therefore, that the number of bushels per acre you can harvest varies *quadratically* with the number of millions of seeds per acre you plant.

a. Write *three* ordered pairs of (millions of seeds, bushels). The third ordered pair is not given above, but should be obvious.
b. Write the particular equation for this function.
c. How many bushels per acre would you expect to get if you plant 16 hundred thousand seeds per acre?
d. Based on your model, would it be possible to get a harvest of 70 bushels per acre? Justify your answer.
e. How much should you plant to get the *maximum* number of bushels per acre?
f. According to your model, is it possible to plant so many seeds that you harvest no barley at all?
g. Plot the graph of this quadratic function in a suitable domain.

12. ***Suspension Bridge Problem.*** In a suspension bridge such as the Golden Gate Bridge in California or the Verrazano-Narrows Bridge in New York, the roadway is supported by parabolic cables hanging from support towers, as shown in Figure 5-8g. Vertical suspender cables connect the parabolic cables to the roadway.

Suppose that you work for Ornery & Sly Construction Company. Your job is to procure the vertical suspender cables for the center span of the new bridge to be built across Broad River. From the Design Department, you find the following information:

i. The support towers are located at $x = 400$ meters and $x = 2000$ meters (making this span slightly longer than the Verrazano-Narrows bridge). The tops of the towers are 165 meters above the roadway, at $y = 165$.
ii. There are four parallel parabolic cables that go from tower top to tower top. They are 5 meters above the roadway at their lowest points, halfway between the two towers.
iii. The vertical suspender cables are spaced every 20 meters, starting at $x = 420$ and ending at $x = 1980$. Each of the four parabolic cables has *two* suspender cables at each value of x.

a. Demonstrate that you understand the above information by writing *three* ordered pairs for points on the parabolic cables.
b. Find the particular equation of the parabolic cables. You may use the vertex form or the $y = ax^2 + bx + c$ form, whichever seems more convenient.

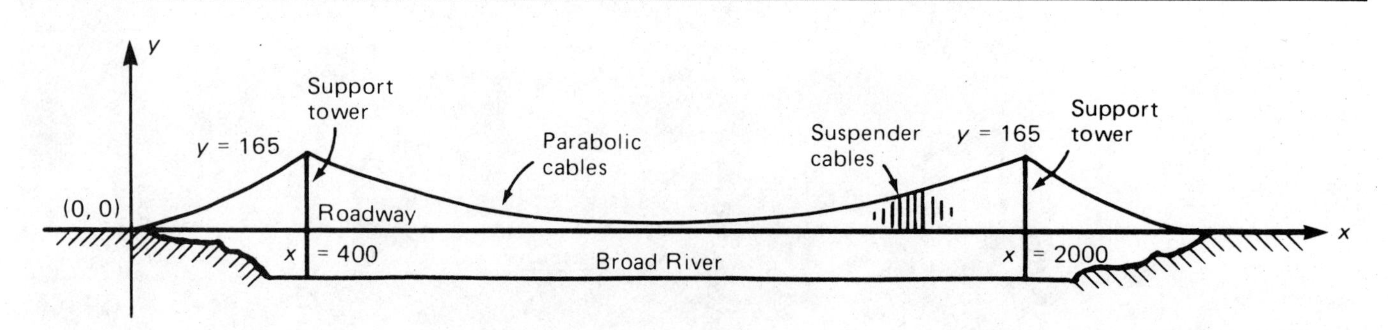

Figure 5-8g.

c. The success of the \$700,000,000 project (and whether or not you keep your job!) depends on the correctness of your calculations. Demonstrate that your equation gives correct values of y by substituting the three values of x from the given ordered pairs in part a.
d. As a further check on the correctness of your equation, substitute 0 for y, and show that the resulting values of x are consistent with the given information.
e. Write a computer program to calculate and print the length of each vertical suspender cable and its corresponding value of x. As each value is calculated, the computer should add it to the sum of the previous values of y. When all values are printed, the computer should print the total length of suspender cable that will be used for the center span so that the Purchasing Department may buy the correct amount of cable.
f. Remember, you will be *fired* if you present your boss with an incorrect answer. Perform a *quick* calculation to show that the total computed cable length is *reasonable*.

5-9 CHAPTER REVIEW AND TEST

In this chapter you have learned about a second type of function, the quadratic function. Unlike those of linear functions, quadratic function graphs are *curved*. They reach a maximum or minimum value at the vertex. The technique of completing the square allowed you to find the vertex. It also allowed you to solve quadratic equations with one variable so that you could calculate x for given values of y. Since this process is rather long and tedious, you did it in *general*, thus deriving the Quadratic Formula. With this formula you can write the solutions of a quadratic equation in *one* step. Using quadratic functions as mathematical models required you to find particular values for the three constants in the general equation $y = ax^2 + bx + c$. For this purpose, you called upon the linear systems techniques of Chapter 4.

The objectives of this chapter are summarized below:

1. For a given quadratic function,

*a. calculate **y** when you know **x**,*
*b. calculate **x** when you know **y**,*
c. find the vertex and draw the graph.

2. Find the particular equation for the quadratic function containing three given ordered pairs.

3. Use quadratic functions as mathematical models.

The review problems below are numbered according to these objectives. By working these problems you will see if you can accomplish the objective when you *know* which one is being tested. The **Concepts Test** combines *all* of these objectives in a *single* problem. The problem is somewhat different from those you have been working. By working it, you will get experience *choosing* which technique to use, as well as using the technique. In this way you will have applied your knowledge to a new situation.

Review Problems

Review Problem 1. Suppose that $y = -3x^2 + 8x - 5$.

a. Find y when $x = 0$, $x = 4$, and $x = -2$.
b. Find x when $y = 0$, $y = 1$, and $y = -2$.
c. Find the vertex, and use it and the above information to draw the graph.

Review Problem 2. Find the particular equation of the quadratic function containing (2, 1), (5, 31), and (−1, 7).

Review Problem 3 – Diving Board. Jack Potts dives off the high diving board. His distance from the surface of the water varies *quadratically* with the number of seconds that have passed since he left the board.

a. His distances at times of 1, 2 and 3 seconds since he left the board are 24, 18, and 2 meters above the water, respectively. Write the particular equation expressing distance in terms of time.

b. How high is the diving board? Justify your answer.
c. What is the highest Jack gets above the water?
d. When does he hit the water?
e. Draw the graph of the quadratic function using a suitable domain.

Concepts Test

The following problem involves all of the concepts of this chapter. You are also called upon to use information from previous chapters, and to think your way through new situations. For each part of the problem, tell by number which one or ones of the previous objectives of this chapter were used. Write "new concept" if a new concept is involved.

Concepts Test – Office Building Problem. Suppose that you work for a company that is planning to operate a new office building. You find that the monthly payments the company must make vary directly with the *square* of the number of stories in the building. That is, the general equation is payment = (constant)(stories)^2. The amount of money the company will take in each month from people renting office space will vary *linearly* with the number of stories. The *profit* the company will make each month equals the rent payments it takes in minus the monthly payments it pays out.

a. Write general equations expressing in terms of number of stories
   i. the monthly payments you pay out,
   ii. the rent payments you take in, and
   iii. the monthly profit.
b. Tell in words how the monthly profit varies with the number of stories.
c. You predict monthly profits of 4, 9, and 12 thousand dollars for buildings of 2, 3, and 4 stories, respectively. Find the particular equation expressing monthly profit as a function of the number of stories.
d. Will the profit ever equal $15 thousand per month? Justify your answer.
e. What number of stories gives the *maximum* monthly profit?
f. Between what two numbers of stories will your monthly profit be positive?
g. If the building is 7 stories high, how much would you *receive* per month in rent, and how much would you *pay* out per month in monthly payments?
h. Plot a graph of this function, using a suitable domain.

5-10 CUMULATIVE REVIEW: CHAPTERS 1 THROUGH 5

The following exercise may be considered to be a "final exam" covering the materials on linear and quadratic functions, and the properties leading up to these. If you are thoroughly familiar with the material in Chapters 1 through 5, you should be able to work all of these problems in about 2 to 3 hours.

Exercise 5-10

1. Calvin Butterball and Phoebe Small are studying for their algebra examination. In the table on page 127 they give the answers indicated to the questions on the left. Tell which, if either, of them is right. Use "Calvin," "Phoebe," "Both," or "Neither."

2. Given the statement, "If R is a relation, then R is a function:"

a. Draw a graph which shows clearly why the statement is *false*.
b. Write the *converse* of the statement.
c. Is the converse true or false? Explain.

3. You have learned that graphs of two linear functions are parallel if their slopes are equal. It is also true that the graphs of two linear functions are *perpendicular* if the slope of one is equal to the *negative* of the *reciprocal* of the other slope. Given the equations

	Problem	Calvin	Phoebe
a.	$17ax + 34a^2 = 17a(x + 2a)$	distributivity	associativity
b.	$(xy + z) + w = xy + (z + w)$	commutativity	associativity
c.	$2 + ab = 2 + ba$	commutativity	reflexive property
d.	Either $x > 0$, $x < 0$, or $x = 0$	comparison	trichotomy
e.	$xy \in \{\text{real numbers}\}$	closure	agreement
f.	$\frac{xy}{ab} = xy \cdot \frac{1}{ab}$	reciprocal of a product	definition of division
g.	$xy + (-xy) = 0$	additive inverses	multiplicative inverses
h.	A field axiom	reflexive	symmetric
i.	An axiom which is *not* a field axiom	transitive	closure
j.	Name for 4th degree	quadratic	quintic
k.	Degree of $3^5x^4y^3 + z^6$	6	12
l.	$\sqrt{x^2} =$	x	$\lvert x \rvert$
m.	Set containing $\sqrt{-7}$	{irrational numbers}	{imaginary numbers}
n.	If $-2x < 6$, then	$x > -3$	$x < -3$

$y = \frac{2}{3}x - 4$ (A), $3y - 2x = 6$ (C),

$2x + 3y = 12$ (B), $y - 5 = -\frac{3}{2}(x + 1)$ (D).

a. Find the slope of each graph.
b. From the slopes, tell which (if any) of the graphs are parallel, and which (if any) are perpendicular. Use symbols such as $A \parallel B$, or $A \perp B$.
c. Plot a graph of each equation on the *same* coordinate system.

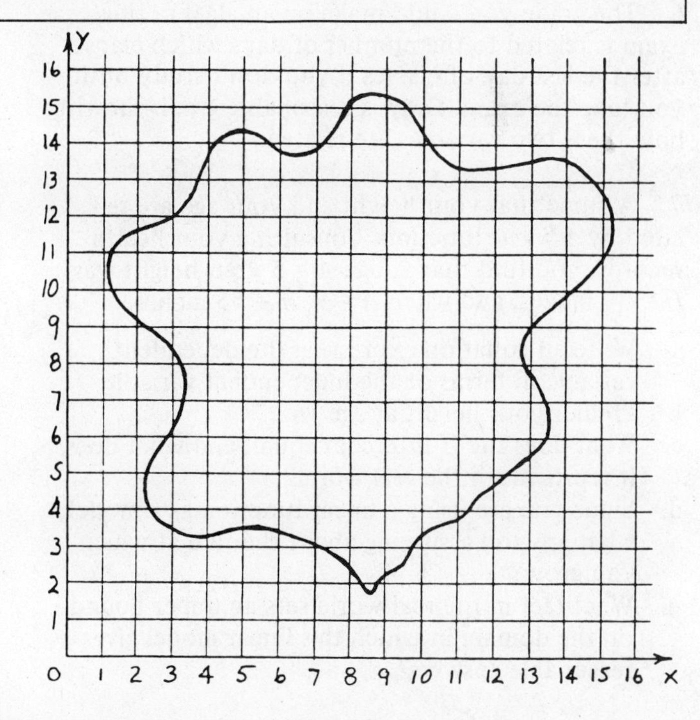

4. The adjacent sketch shows the feasible region for a problem involving the manufacture of two products. This is similar to a linear programming problem, but the boundaries of the region are not linear. If the profit for this process is given by

$$P = 30x - 50y + 500,$$

a. Trace or Xerox the graph. Then shade the portion of the feasible region in which the profit is more than 300.
b. Find the point with integer coordinates that gives the maximum profit, and tell what this profit equals.

5. The following is a proof that a number divided by itself equals 1. Supply a reason for each step. Tell whether or not the reason is an axiom, and if so, whether it is a field axiom.

Prove that

$$\frac{n}{n} = 1.$$

Proof:

$$\frac{n}{n} = n \cdot \frac{1}{n}$$

$$= 1$$

$$\therefore \frac{n}{n} = 1.$$

6. The grade you could make on an algebra final exam is related to the number of days which elapse after the last day of classes if you don't study until you take the exam. Draw a reasonable graph showing how these two variables are related.

7. Assume that your height and your age are related by a *linear* function. Consulting your health records, you find that at age $A = 5$ your height was $H = 39$ inches, and when $A = 9$, $H = 55$ inches.

a. Write an equation expressing the dependent variable in terms of the independent variable.
b. Predict your height at age 16.
c. What does the H-intercept equal, and what does it represent in the real world?
d. Since you are using a linear function as a model, what are you assuming about the rate at which you grow?
e. What fact in the real world sets an upper bound on the domain in which this linear model gives reasonable answers?

8. For the quadratic function $y = -2x^2 - 12x - 10$,

a. Transform the equation to the form $y - k = a(x - h)^2$.
b. Find the vertex.
c. Find the x and y-intercepts.
d. Sketch the graph.
e. By using what you have learned about graphing linear inequalities, shade the region on your graph paper corresponding to the solution set of the *quadratic* inequality

$$y \geq -2x^2 - 12x - 10.$$

9. S. Bones is the doctor in Deathly, Ill., a suburb of Chicago. One day, John Garfinkle comes in with a high fever. Dr. Bones takes a blood sample and finds that it contains 1300 flu viruses per cubic millimeter and is *increasing.* John immediately gets a shot of penicillin. The virus count should continue to increase for awhile, then (hopefully!) level off and go back down. After 5 minutes the virus count is up to 1875, and after 5 more minutes, it is 2400. Assume that the virus count varies quadratically with the number of minutes since the shot.

a. Write an equation expressing the number of viruses per cubic millimeter in terms of the number of minutes since the shot.
b. Dr. Bones realizes that if the virus count ever reaches 4500, John must go to the hospital. Must he go? Explain.
c. According to your model, when will his flu be completely cured?
d. Draw a graph of this quadratic function.
e. Tell the range and domain of this quadratic function.

10. For each of the following equations, calculate the discriminant. Then, without actually solving the equation, tell what *kind* of numbers (rational, irrational, imaginary) will be in the solution set.

a. $3x^2 - 2x - 8 = 0.$
b. $3x^2 - 2x + 8 = 0.$
c. $x^2 - 6x + 2 = 0.$
d. $x^2 - 6x + 9 = 0.$

11. Suppose that a computer runs the following BASIC program. Construct a simulated computer memory and use it to find out what the output of this program will be if the numbers 1, −5, 8 (in that order) are input in Line 10.

```
10   INPUT A,B,C
20   LET Y1 = C
30   LET X = 1
40   LET Y2 = A*X*X + B*X + C
50   IF Y2 >= Y1 THEN 90
60   LET Y1 = Y2
70   LET X = X+1
80   GO TO 40
90   PRINT "VERTEX BETWEEN", X-1, "AND", X
100  END
```

Exponential and Logarithmic Functions

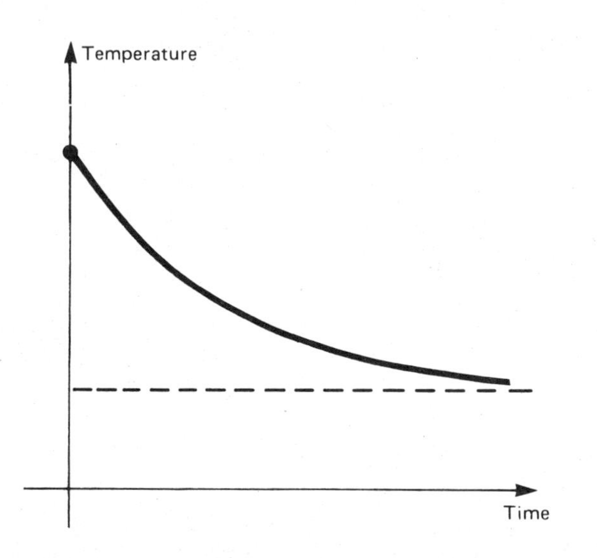

The polynomial functions you have studied so far have had variables raised to constant powers. In this chapter you will encounter exponential functions that have ***constants*** *raised to* ***variable*** *powers. Since variables can be negative numbers or fractions, you must invent meanings for these kinds of exponents. The tool that allows you to operate with such powers is called the "logarithm." The resulting exponential functions are reasonable mathematical models for things from figuring out compound interest to predicting the temperature of a cup of coffee.*

6-1 INTRODUCTION TO EXPONENTIAL FUNCTIONS

A quadratic function might have an equation of the form $y = ax^2$. If the x and the 2 are *reversed*, you get a completely different kind of function.

$$y = a \cdot 2^x$$

To call attention to the fact that the variable, x, is an exponent, this kind of function is called an *exponential* function.

DEFINITION

An ***exponential*** function is a function with an equation of the form

$$\boxed{y = a \cdot b^x,}$$

where a stands for a constant of proportionality, b stands for a positive constant base, and x and y are independent and dependent variables, respectively.

To describe such a function verbally, you may say, "y varies exponentially with x." The effect of the constant a on the graph, and the reason for making b positive will become clear later in this chapter.

Objective:

By plotting points, discover what the graph of an exponential function looks like.

In the exercise that follows, you will accomplish this objective.

Exercise 6-1

1. Suppose that $f = \{(x, y): y = 2^x\}$. Calculate $f(1)$, $f(2)$, $f(3)$, and $f(4)$, and plot the corresponding points on a Cartesian coordinate system.

2. Write the values of $f(4)$, $f(3)$, $f(2)$, and $f(1)$ in a table such as

x	$f(x)$	
4	___	) × ½
3	___	) × ½
2	___	) × ½
1	___	

Observe that $f(3)$ is *half* of $f(4)$, $f(2)$ is *half* of $f(3)$, and so forth. By continuing this pattern, find values of $f(0)$, $f(-1)$, $f(-2)$, and $f(-3)$. Plot these points on your graph.

3. How does the negative x-axis seem to be related to the graph of f?

4. Connect the points on the graph of f with a smooth curve. Since you know no definition for fractional exponents, you must make the curve *dotted*.

5. From your graph, find approximate values (1 decimal place) for $f(2\frac{1}{2})$, $f(1\frac{1}{2})$, and $f(\frac{1}{2})$.

6-2 THE OPERATION OF EXPONENTIATION

If you did Exercise 6-1 correctly you should have gotten a graph such as in Figure 6-2. The points for 0 and negative values of x you got by crossing your fingers and following a pattern. You know no *definitions* for these types of exponents. The points are connected with a dotted line since you know no definition for *fractional* exponents, either. In the following sections you will define these types of exponents so that you can connect the points with a *solid* line.

Before you define negative, zero, and fractional exponents, you should have a good grasp of positive integer exponents. In this section and the next one, you will refresh your memory concerning the definition and properties of the operation *exponentiation*, which means, "raising to powers."

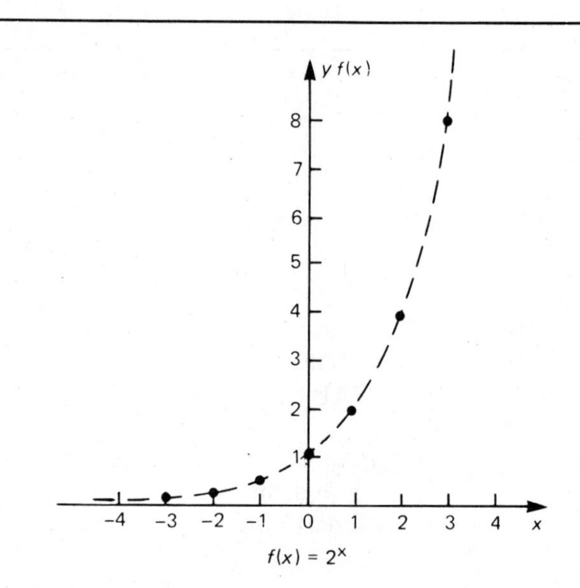

Figure 6-2.

Objective:

Be able to operate with expressions containing positive integer exponents.

From your previous work in mathematics, you should know that

$$x^3 = x \cdot x \cdot x,$$

$$x^7 = x \cdot x \cdot x \cdot x \cdot x \cdot x \cdot x,$$

$$x^1 = x.$$

From these examples, the definition follows.

DEFINITION

Exponentiation for positive integer exponents:

x^n means the *product* of n x's.

The definition of x^n can also be stated, "x used n times as a factor," or, "take n x's and multiply them together." You should *avoid* statements such as, "x^3 means x multiplied by itself 3 times," because as you can see clearly from the example, $x \cdot x \cdot x$ has only *two* "times" signs.

When you combine exponentiation with other operations, it becomes necessary to make some agreements about which one will be done first. In an expression such as $4x^3$, x is the base, not $4x$. If you want $4x$ to be the base, you must use parentheses and write $(4x)^3$. That is,

$$4x^3 = 4(x \cdot x \cdot x),$$

$$(4x)^3 = (4x)(4x)(4x).$$

In an expression such as $-x^4$, x is the base, not $-x$. If you want to raise $-x$ to a power, you must use parentheses and write $(-x)^4$. That is,

$$-x^4 = -(x \cdot x \cdot x \cdot x),$$

$$(-x)^4 = (-x)(-x)(-x)(-x).$$

The above information is summarized in the following agreement.

Agreement: If the base consists of *more than one* symbol, then it must be placed in *parentheses*. For example,

$$(xy)^5,\ (x-y)^5,\ (-x)^5.$$

Otherwise, the exponent applies only to *one* symbol. For example,

$$xy^5 = x(y^5),\ x - y^5 = x - (y^5),\ -x^5 = -(x^5).$$

In the exercise that follows, you will practice using these agreements and the definition of exponentiation.

Exercise 6-2

Evaluate the following expressions subject to the definitions and agreements about exponentiation:

1. a. $(2 \cdot 3)^4$ b. $2 \cdot (3^4)$ c. $2 \cdot 3^4$
2. a. $(3 \div 2)^4$ b. $3 \div (2^4)$ c. $3 \div 2^4$
3. a. $(3 + 2)^4$ b. $3 + (2^4)$ c. $3 + 2^4$
4. a. $(3 - 7)^2$ b. $3 - (7^2)$ c. $3 - 7^2$
5. a. $(4 - 7)^3$ b. $4 - (7^3)$ c. $4 - 7^3$
6. a. $(-6)^2$ b. $-(6^2)$ c. -6^2
7. a. $-2 \cdot 3^4$ b. $(-2 \cdot 3)^4$ c. $-(2 \cdot 3^4)$ d. $-2 \cdot (3^4)$
8. $3 \cdot 4 + 5 \cdot 6^2$
9. $64/2^3 \cdot 4$
10. $4 + 8^2 - 2 \cdot 3^5 + 6(-4)^2$

6-3 PROPERTIES OF EXPONENTIATION

When you started studying algebra, you defined only the operations of addition and multiplication. You assumed *axioms* for these operations such as associativity, closure, identity elements, etc. Then you defined new operations in terms of multiplying and adding. The properties of subtraction and division were proved from the axioms for multiplication and addition. Now you have defined another new operation, exponentiation, which can be thought of as *repeated* multiplication (just like multiplication can be thought of as repeated *addition*).

There are properties of exponentiation that are provable using the axioms and other properties you already know. You should recall these properties from previous mathematics courses. In this section you will be asked to *prove* the properties for specific cases.

Objective:

Be able to *name* and *state* the properties of exponentiation, and supply reasons for the steps in their proofs.

The following exercise lists the properties and shows a proof of an example of each, using specific values for the exponents. By using letters in place of the constant exponents, you can create *general* proofs of the properties. However, the wording of the proofs is a bit untidy. It will be your objective in the exercise to see *why* the property is true, and then *remember* what the property says.

Exercise 6-3

Supply a reason for each step in the following sample proofs. Then state in words what the property says. The first one is done for you to refresh your memory on how to do this.

1. *Product of Two Powers with Equal Bases*— Prove that $x^a \cdot x^b = x^{a+b}$.

Sample Proof:

a.	$x^3 \cdot x^4 = (x \cdot x \cdot x)(x \cdot x \cdot x \cdot x)$	Definition of exponentiation.
b.	$= (x \cdot x \cdot x \cdot x \cdot x \cdot x \cdot x)$	Associativity.
c.	$= x^7$	Definition of exponentiation.

d. $\qquad = x^{3+4}$. $\qquad$ Arithmetic.

e. $\therefore x^3 \cdot x^4 = x^{3+4}$. $\qquad$ Transitivity.

f. Verbal Statement: *When you multiply two powers with equal bases, you add the exponents* (and keep the same base).

2. *Quotient of Two Powers with Equal Bases*—

Prove that $\dfrac{x^a}{x^b} = x^{a-b}$.

Sample Proof:

a. $\dfrac{x^5}{x^3} = \dfrac{x \cdot x \cdot x \cdot x \cdot x}{x \cdot x \cdot x}$

b. $\qquad = \dfrac{x \cdot x \cdot x \cdot x \cdot x}{x \cdot x \cdot x \cdot 1}$

c. $\qquad = \dfrac{(x \cdot x \cdot x)(x \cdot x)}{(x \cdot x \cdot x) \cdot 1}$

d. $\qquad = \dfrac{x \cdot x \cdot x}{x \cdot x \cdot x} \cdot \dfrac{x \cdot x}{1}$

e. $\qquad = 1 \cdot x \cdot x$

f. $\qquad = x^2$.

g. $\therefore \dfrac{x^5}{x^3} = x^2$.

h. Verbal Statement: (Compose your own.)

3. *Power of a Power* – Prove that $(x^a)^b = x^{ab}$.

Sample Proof:

a. $(x^2)^3 = x^2 \cdot x^2 \cdot x^2$

b. $\qquad = x^{2+2+2}$

c. $\qquad = x^6$.

d. $\therefore (x^2)^3 = x^6$.

e. Verbal Statement: (Compose your own.)

4. *Power of a Product* – Prove that $(xy)^a = x^a y^a$.

Sample Proof:

a. $(xy)^3 = xy \cdot xy \cdot xy$

b. $\qquad = (x \cdot x \cdot x)(y \cdot y \cdot y)$ $\qquad$ (Two reasons)

c. $\qquad = x^3 y^3$.

d. $\therefore (xy)^3 = x^3 y^3$.

e. Verbal Statement: (Compose your own.)

5. *Power of a Quotient* – Prove that $\left(\dfrac{x}{y}\right)^a = \dfrac{x^a}{y^a}$.

Sample Proof:

a. $\left(\dfrac{x}{y}\right)^3 = \dfrac{x}{y} \cdot \dfrac{x}{y} \cdot \dfrac{x}{y}$

b. $\qquad = \dfrac{x \cdot x \cdot x}{y \cdot y \cdot y}$

c. $\qquad = \dfrac{x^3}{y^3}$.

d. $\therefore \left(\dfrac{x}{y}\right)^3 = \dfrac{x^3}{y^3}$.

e. Verbal Statement: (Compose your own.)

6. Prove by counter-example that exponentiation does *not* distribute over addition.

7. Without referring to the previous problems, write the equation for each of the following properties:

a. Exponentiation distributes over division.
b. Exponentiation distributes over multiplication.
c. Power of a power.
d. Power of a quotient.
e. Power of a product.
f. Product of two powers with equal bases.
g. Quotient of two powers with equal bases.

8. If you have studied mathematical induction (Appendix B), prove that

a. $x^a \cdot x^b = x^{a+b}$ for all integer values of $b \geq 1$,

b. $(x^a)^b = x^{ab}$ for all integer values of $b \geq 1$,

c. $(xy)^a = x^a y^a$ for all integer values of $a \geq 1$.

6-4 NEGATIVE AND ZERO EXPONENTS

Once you have mastered the properties of positive integer exponents, you are ready to return to the problem of filling in the gaps in the graph of $f(x) = 2^x$. (See Figure 6-4.) In Exercise 6-1 you found a pattern that could be used to calculate 2^0, 2^{-1}, 2^{-2}, and 2^{-3}.

$$2^0 = \frac{1}{2} \cdot 2^1 = \frac{1}{2} \cdot 2 = \underline{1}$$

$$2^{-1} = \frac{1}{2} \cdot 2^0 = \frac{1}{2} \cdot 1 = \underline{1/2}$$

$$2^{-2} = \frac{1}{2} \cdot 2^{-1} = \frac{1}{2} \cdot \frac{1}{2} = \underline{1/4}$$

$$2^{-3} = \frac{1}{2} \cdot 2^{-2} = \frac{1}{2} \cdot \frac{1}{4} = \underline{1/8}$$

In this section you will *define* 0 and negative exponents in a way that is consistent with this pattern, and is also consistent with the properties of exponentiation. Once you know the definitions, you must practice enough to become comfortable using these kinds of exponents.

Objective:

Define negative and zero exponents in such a way that the properties of exponentiation still work, and be able to operate with expressions containing these kinds of exponents.

The property of the quotient of two powers with equal bases states that

$$\frac{x^a}{x^b} = x^{a-b}.$$

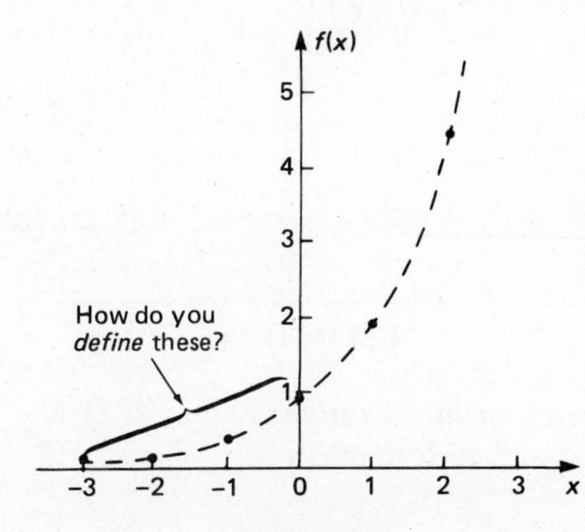

Figure 6-4a.

What would happen if a were *smaller* than b? This property would say, for example, that

$$\frac{x^3}{x^5} = x^{3-5} = x^{-2}.$$

The expression x^{-2} is undefinable using the present definition of exponentiation. (How could you use x as a factor -2 times?) So you are free to *define* x^{-2} in any way you wish! The most reasonable way to define it is in such a way that the properties of exponentiation will still be true for negative exponents. To find out what x^3/x^5 *really* equals, think of reasons for the following steps:

$$\frac{x^3}{x^5} = \frac{x \cdot x \cdot x}{x \cdot x \cdot x \cdot x \cdot x}$$

$$= \frac{x \cdot x \cdot x \cdot 1}{x \cdot x \cdot x \cdot x \cdot x}$$

$$= \frac{(x \cdot x \cdot x) \cdot 1}{(x \cdot x \cdot x)(x \cdot x)}$$

$$= \frac{x \cdot x \cdot x}{x \cdot x \cdot x} \cdot \frac{1}{x \cdot x}$$

$$= 1 \cdot \frac{1}{x^2}$$

$$= \frac{1}{x^2}.$$

Since the *property* gives x^{-2} for an answer, and the *real* answer is $1/x^2$, it seems reasonable to *define*

$$x^{-2} = \frac{1}{x^2}\,.$$

This leads to the following definition in general.

DEFINITION

Exponentiation for negative exponents. The expression x^{-n} is defined to be

$$\boxed{x^{-n} = \frac{1}{x^n}\,.}$$

The desire to make the property of the quotient of two powers with equal bases always work leads to the definition of a zero exponent. For example, the property would say

$$\frac{x^3}{x^3} = x^{3-3} = x^0.$$

The *actual* value of x^3/x^3 is 1, of course. Therefore, the definition $x^0 = 1$ is just the definition which is needed to make the property applicable.

DEFINITION

Exponentiation for 0 exponent:

$$\boxed{x^0 = 1, \text{ provided } x \neq 0.}$$

The expression 0^0 is awkward. If $x = 0$, the property says

$$\frac{x^3}{x^3} = \frac{0^3}{0^3} = 0^{3-3} = 0^0$$

But the following is also true:

$$\frac{x^3}{x^3} = \frac{0^3}{0^3} = \frac{0}{0}\,.$$

So $0^0 = 0/0$. Since division by zero is not defined, the expression 0^0 is not defined either.

Example 1: Simplify $(3x^2y^{-5})^3\,(4x^4y^7)^{-2}$.

Since you have *defined* negative and zero exponents in such a way that the properties of exponentiation still work, you can *use* these properties to simplify expressions involving any integer exponents. Starting with

$$(3x^2y^{-5})^3\,(4x^4y^7)^{-2},$$

you would first *distribute* the exponentiation over the multiplication, getting

$$3^3x^6y^{-15}\cdot 4^{-2}x^{-8}y^{-14},$$

where you raise powers to powers by *multiplying* the exponents. Recalling that 4^{-2} is defined to be $1/4^2$, you get

$$27x^6y^{-15}\cdot\frac{1}{16}x^{-8}y^{-14}.$$

Commuting and associating the constants, x's and y's gives

$$\frac{27}{16}x^{-2}y^{-29}$$

where you multiply powers with equal bases by *adding* exponents.

Example 2: To simplify a fraction such as $\left(\frac{12x^{-3}y^7z^{-4}}{18x^5y^{-2}z^{-6}}\right)^{-3}$, you could first operate inside the parentheses using the multiplication property of fractions to get

$$\left(\frac{12}{18}\cdot\frac{x^{-3}}{x^5}\cdot\frac{y^7}{y^{-2}}\cdot\frac{z^{-4}}{z^{-6}}\right)^{-3}$$

Dividing powers with equal bases by *subtracting* the exponents gives

$$\left(\frac{2}{3}x^{-8}y^9z^2\right)^{-3}$$

Distributing the exponent −3 over the multiplication, and raising powers to powers by multiplying exponents gives

$$\left(\frac{2}{3}\right)^{-3} x^{24}y^{-27}z^{-6}.$$

The fraction $\left(\frac{2}{3}\right)^{-3}$ is equal to $\left(\frac{3}{2}\right)^{3}$, as you will prove in the following exercise. Thus, the answer is

$$\underline{\underline{\frac{27}{8}x^{24}y^{-27}z^{-6}.}}$$

All you need in order to simplify expressions such as these is the properties and definitions of exponentiation. You should have complete faith in these properties, no matter *what* numbers appear as exponents. The following exercise is designed to give you practice operating with various integer exponents, and plotting the graph of $f(x) = \left(\frac{2}{3}\right)^x$.

Exercise 6-4

1. Suppose that $f(x) = \left(\frac{2}{3}\right)^x$.

a. Calculate values of $f(x)$ for each integer value of x from −4 through 4.
b. Plot the points on a Cartesian coordinate system. You must still connect the points with a dotted curve since you know no definition yet for fractional exponents.
c. What relationship does the positive x-axis have to the graph of f?

2. Prove the following properties of exponentiation:

a. $x^{-1} = \frac{1}{x}$. b. $\frac{1}{x^{-n}} = x^n$. c. $\left(\frac{x}{y}\right)^{-n} = \left(\frac{y}{x}\right)^n$.

For Problems 3 through 30, simplify the expression by transforming it to a *product* of powers with positive or negative exponents for the variables, and *no* variables written in denominators.

3. $6x^5 \cdot 3x^{-2}$
4. $7x^{-3} \cdot 8x^9$
5. $5x^{-4} \cdot 2x^{-3}$
6. $9x^{-2} \cdot 6x^{-5}$
7. $11x^7 \cdot 4x^{-12}$
8. $3a^{-11} \cdot 17a^5$
9. $(-2x^2)^3(3x^{-1}y^2)^4$
10. $(4x^3)^2(-2x^{-3}y^{-1})^3$
11. $(3x^2y^{-3})^4(2x^{-4}y^5)^3$
12. $(5a^3b^{-2})^4(5a^{-4}b^{-5})^{-2}$
13. $-3x^4y^5 \cdot (4x^{-2}y)^{-3}$
14. $-4x^2y^7 \cdot (2x^3y^{-5})^{-3}$
15. $(7a^{-5}b^6) \div (21a^4b^{-2})$
16. $(1001x^{-4}y^{-3}) \div (77x^6y^{-7})$
17. $\dfrac{5}{a^{-2}} - \dfrac{3}{a^{-1}}$
18. $\dfrac{7}{x^{-5}} - \dfrac{4}{x^{-1}}$
19. $\dfrac{x^{-1}y^{-4}z}{x^{-2}yz^{-3}}$
20. $\dfrac{r^{-5}st^{-3}}{r^{-1}s^{-5}t}$
21. $\dfrac{8x^{-1}y^5z^{-3}}{4x^2y^{-2}z^5}$
22. $\dfrac{51x^{-4}y^2z^7}{17x^{-4}y^{-2}z^0}$
23. $\dfrac{13x^5y^{-2}z^0}{39xy^{-3}z^2}$
24. $\dfrac{3^4a^{-7}b^3d^{-4}}{43^0a^{-4}b^{-5}c^6}$
25. $\left(\dfrac{18x^4y^{-5}}{27x^3y^6}\right)^2$
26. $\left(\dfrac{75x^6y^{-2}}{25x^{-5}y^{-9}}\right)^3$
27. $\left(\dfrac{81x^{-2}z^3}{162x^3y^5}\right)^{-3}$
28. $\left(\dfrac{43a^{-6}b^5}{129a^4b^{13}}\right)^{-2}$
29. $\left(\dfrac{29x^{11}y^{-15}z^{98}}{37x^{-13}y^{44}z^{-9}}\right)^0$
30. $\left(\dfrac{47x^{51}y^{73}}{37z^5r^{-13}}\right)^{-5} \cdot \left(\dfrac{47x^{51}y^{73}}{37z^5r^{-13}}\right)^5$

6-5 RADICALS AND FRACTIONAL EXPONENTS

By defining exponentiation for zero and negative exponents, you have extended the domain of the exponential function. For example, the graph of $f(x) = 2^x$ looks like that in Figure 6-5. It is now time to extend the domain to include non-integer

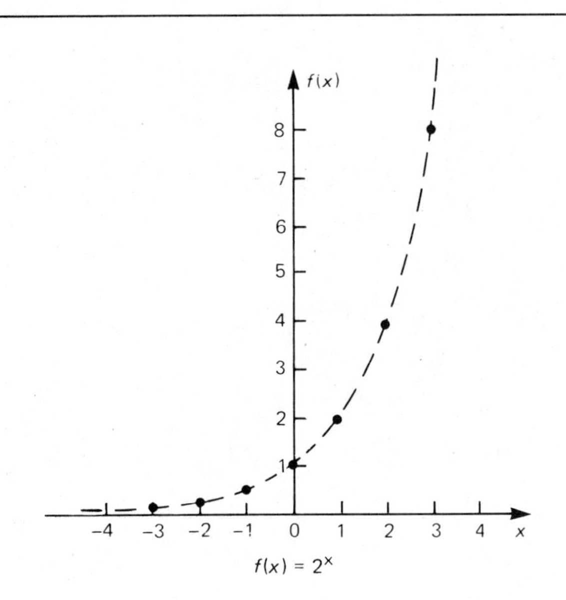

Figure 6-5.

exponents so that you can fill in the "gaps" in the graph.

If you are clever enough, you can define exponentiation for fractional exponents in such a way that the properties you learned for integer exponents will still work. You recall that you defined negative exponents with this consideration in mind, too.

Objective:

Define exponentiation for *fractional* exponents, and be able to operate with expressions containing this type of exponent.

To figure out this clever definition, think of what $a^{1/2}$ could mean. If the property of a power raised to a power still works, then $(a^{1/2})^2$ would be evaluated by *multiplying* the exponents. So

$$(a^{1/2})^2 = a^{1/2 \cdot 2} = a^1 = a.$$

According to the properties of exponentiation, $a^{1/2}$ would have to be a number which when *squared* gives a for an answer. But you recall from your work with quadratics that $\sqrt{a}$ is *also* a number which when squared gives a for an answer. So $a^{1/2}$ should be defined to be $\sqrt{a}$. Similarly, $a^{1/3}$ would be a number which when *cubed* gives a for an answer. Thus, $a^{1/3}$ is a *cube* root of a, written $\sqrt[3]{a}$.

DEFINITION

Definition of nth root: The expression $\sqrt[n]{a}$ means a number which when raised to the nth power gives a for an answer. That is,

$$(\sqrt[n]{a})^n = a.$$

For example, $\sqrt[5]{32} = 2$ because $2^5 = 32$. Also, $\sqrt[4]{81} = 3$ because $3^4 = 81$.

The expression $\sqrt[n]{a}$ is called a *radical*, and is read, "the nth root of a." The word "radical" comes from the same word as "radish," which is, as you well know, a kind of root. The parts of $\sqrt[n]{a}$ also have names:

- n — root index. (It *points out* what root to take.)
- a — radicand.
- $\sqrt{\ }$ — radical sign.
- $\overline{\ \ }$ — a symbol of inclusion (like parentheses) called a "vinculum."

The best way to define $1/n$ as an exponent is now clear.

DEFINITION

Definition of ***reciprocal exponents:***

$$a^{1/n} = \sqrt[n]{a}$$

where n stands for a positive integer.

If the exponent were a fraction such as 3/4, and the properties of exponentiation are still to be true, then

$$a^{3/4} = a^{(1/4)(3)} = (a^{1/4})^3 = (\sqrt[4]{a})^3.$$

This leads to a general definition for fractional exponents.

DEFINITION

Exponentiation for rational exponents:

$$\boxed{a^{m/n} = (\sqrt[n]{a})^m}$$

where m and n stand for integers.

Note that this definition requires there to be an integer for both numerator and denominator of the exponent. Since irrational numbers *cannot* be expressed as a ratio of two integers, you still have no definition for powers with irrational exponents, such as 2^π or $7^{\sqrt{5}}$. Note also that negative values of a sometimes cause trouble (Problem 53, below).

Objective:

Given an expression containing radicals or fractional exponents, be able to *simplify* it.

Example 1: In some cases, $a^{m/n}$ will turn out to be a *rational* number. The properties of exponentiation can be used to simplify the expression. For example,

$8^{2/3} = (2^3)^{2/3}$	Because $8 = 2^3$.
$= 2^{3 \cdot 2/3}$	Property of a power raised to a power.
$= 2^2$	Because $3 \cdot 2/3 = 2$.
$= 4.$	

So $8^{2/3} = 4$. This procedure worked because 8 could be written as a power of 2. The procedure would not have helped for an expression such as $11^{2/3}$ because the base, 11, cannot be expressed as a power. Techniques for finding decimal approximations for such expressions are presented in Section 6-6.

Example 2: If the base is a large number such as 216, it may take some work to find its prime factors. A clever scheme for doing this is illustrated at the right. The procedure might best be described as "upside-down repeated short division." 2 goes into 216 with a quotient of 108. The 108 is written below the 216, as shown in the first step. The process is continued, always dividing by the *smallest prime* which goes into the current quotient. The prime factors of 216 appear listed to the left, outside the division signs. So

First Step:

$$\begin{array}{r} 2\,\underline{)\ 216} \\ 108 \end{array}$$

All Steps:

$$\begin{array}{r} 2\,\underline{)\ 216} \\ 2\,\underline{)\ 108} \\ 2\,\underline{)\ 54} \\ 3\,\underline{)\ 27} \\ 3\,\underline{)\ 9} \\ 3 \end{array}$$

$$216 = 2 \cdot 2 \cdot 2 \cdot 3 \cdot 3 \cdot 3 = 2^3 \cdot 3^3.$$

If you wanted to evaluate $216^{4/3}$, you could write:

$216^{4/3} = (2^3 \cdot 3^3)^{4/3}$	By substitution.
$= (2^3)^{4/3} \cdot (3^3)^{4/3}$	Exponentiation distributes over multiplication.
$= (2^{3 \cdot 4/3}) \cdot (3^{3 \cdot 4/3})$	Power of a power.
$= 2^4 \cdot 3^4$	Because $3 \cdot 4/3 = 4$.
$= 16 \cdot 81$	Arithmetic.
$= 1296$	Arithmetic.
$\therefore 216^{4/3} = 1296.$	Transitivity.

Example 3: Even if an expression turns out to be an irrational number, the techniques illustrated above can sometimes be used to simplify it. For example,

$\sqrt[6]{256} \div \sqrt[4]{64}$	
$= 256^{1/6} \div 64^{1/4}$	Definition of fractional exponents.
$= (2^8)^{1/6} \div (2^6)^{1/4}$	$256 = 2^8$ and $64 = 2^6$.
$= 2^{4/3} \div 2^{3/2}$	Power of a power.
$= 2^{4/3-3/2}$	Quotient of two powers with equal bases.
$= 2^{8/6-9/6}$	Getting common denominator.

$= 2^{-1/6}$ Arithmetic.

$= \dfrac{1}{2^{1/6}}$ Definition of negative exponents.

$= \dfrac{1}{\sqrt[6]{2}}$ Definition of fractional exponents.

The key to success with this simplification technique lies in expressing things as powers with *equal* bases. Then you can use the properties of exponentiation to combine the exponents.

Example 4: Simplify $\left(\dfrac{2a^{2/3}b^{-3/7}c^{-5}}{50a^{-1/3}b^{-5/7}c^{1/4}}\right)^{-1/2}$

$\left(\dfrac{2a^{2/3}b^{-3/7}c^{-5}}{50a^{-1/3}b^{-5/7}c^{1/4}}\right)^{-1/2}$

$= \left(\dfrac{1}{25}ab^{2/7}c^{-21/4}\right)$ Simplifying inside parentheses.

$= \left(\dfrac{1}{25}\right)^{-1/2} a^{-1/2}b^{-1/7}c^{21/8}$ Distributing the $-\frac{1}{2}$.

$= 5a^{-1/2}b^{-1/7}c^{21/8}$. Simplifying $\left(\dfrac{1}{25}\right)^{-\frac{1}{2}}$.

In the exercise that follows you will complete the graph of $f(x) = 2^x$ that you started in Exercise 6-1. Then you will operate with expressions containing fractional and negative exponents so that you can become comfortable using them. In particular, you will be forced to distinguish clearly between fractional and negative exponents.

Exercise 6-5

1. Suppose that $f(x) = 2^x$.

a. Plot values of $f(x)$ for each integer value of x from -3 through 3. The graph should look like Figure 6-5.

b. Using Table I, find decimal approximations for $f(1/3)$ and $f(1/2)$. Plot these points on the graph from part a.

c. Find decimal approximations for $f(2/3)$, $f(4/3)$, $f(5/3)$, $f(3/2)$, and $f(5/2)$. You can use Table I if you realize, for example, that

$$2^{5/3} = (2^5)^{1/3} = 32^{1/3} = \sqrt[3]{32}.$$

Plot the resulting points on the graph from part a.

d. Find decimal approximations for $f(-1/2)$, $f(-1/3)$, $f(-3/2)$, and $f(-5/2)$. You can do this by taking *reciprocals* of values you have already calculated. Plot the resulting points on the graph from part a.

e. Connect the points on the graph with a smooth, solid curve. If any of the points do not seem to lie along this curve, go back and check your calculations.

2. In Problem 1 you connected points on the graph of $f(x) = 2^x$ with a *solid* curve, indicating that 2^x is defined for *all* values of x. Explain why numbers such as $2^{\sqrt{3}}$ and 2^{π} that have *irrational* exponents are *still* undefined using your present definitions of exponentiation.

For Problems 3 through 6 find *exactly* the indicated power of 64.

3. a. $64^{2/3}$ b. $64^{3/2}$ c. $64^{5/6}$ d. $64^{7/6}$

4. a. $64^{-2/3}$ b. $-64^{2/3}$ c. $(-64)^{2/3}$ d. $-64^{-2/3}$

5. a. $64^{-3/2}$ b. $-64^{3/2}$ c. $(-64)^{3/2}$ d. $-64^{-3/2}$

6. a. $64^{-5/6}$ b. $-64^{7/6}$ c. $64^{-7/6}$ d. $-64^{-5/6}$

For Problems 7 through 22, simplify the given power.

7. $32^{6/5}$
8. $32^{-6/5}$
9. $-32^{6/5}$
10. $-32^{-6/5}$
11. $128^{2/7}$
12. $256^{3/4}$
13. $81^{-3/2}$
14. $243^{-4/5}$
15. $343^{2/3}$
16. $36^{3/2}$
17. $-100^{-3/2}$
18. $-1000^{-2/3}$
19. $\left(\frac{9}{49}\right)^{-3/2}$
20. $\left(\frac{256}{625}\right)^{-3/4}$
21. $\left(-\frac{729}{64}\right)^{-2/3}$
22. $\left(-\frac{125}{27}\right)^{-4/3}$

For Problems 23 through 30, find the *exact* value of the radical.

23. $\sqrt[3]{64}$
24. $\sqrt[5]{1024}$
25. $\sqrt[4]{81}$
26. $\sqrt[3]{216}$
27. $\sqrt[3]{1728}$
28. $\sqrt[3]{1331}$
29. $\sqrt[4]{625}$
30. $\sqrt{196}$

For Problems 31 through 44, simplify the radical by first transforming it to exponential form. You may leave the answer in either exponential or radical form, whichever you prefer.

31. $\sqrt{6} \div \sqrt{2}$
32. $\sqrt[3]{88} \div \sqrt[3]{11}$
33. $\sqrt{8} \div \sqrt[4]{32}$
34. $\sqrt[3]{81} \div \sqrt[5]{729}$
35. $\sqrt[3]{128} \cdot \sqrt{32}$
36. $\sqrt[3]{81} \cdot \sqrt[3]{9}$
37. $\sqrt[3]{2^{3.1} \div 2^{0.7}}$
38. $\sqrt[5]{7^{9.8} \div 7^{1.6}}$
39. $\sqrt[5]{10^{4.4} \times 10^{-6.3} \div 10^{-8.1}}$
40. $\sqrt{10^{2.6} \times 10^{0.5} \div 10^{1.5}}$
41. $\sqrt[4]{\sqrt[3]{16}}$
42. $\sqrt{\sqrt[3]{64}}$
43. $\sqrt{\sqrt[3]{36}}$
44. $\sqrt[3]{\sqrt[4]{\sqrt[5]{7}}}$

For Problems 45 through 52, simplify the expression by writing it as a *product* with fractional exponents, and *no* variables written in denominators.

45. $3x^{-1/2} \cdot 4x^{2/3}$
46. $5x^{-2/5} \cdot 6x^{1/8}$
47. $(4x^{-1/2})^3 \div (9x^{1/3})^{-3/2}$
48. $(64x^2)^{-1/6} \div (32x^{5/2})^{-2/5}$
49. $\dfrac{x^{-2/3}y^{5/2}}{x^{-5/3}y^{-1}}$
50. $\dfrac{x^{3/7}y^{-4/5}}{x^{-2/7}y^{3/5}}$
51. $\left(\dfrac{147a^{2/3}b^{-2}}{3a^{3/2}b^{-1/3}}\right)^{-1/2}$
52. $\left(\dfrac{a^{3/2}b^{-4/5}}{32a^0b^{1/5}}\right)^{-2/5}$

53. The following will show you why *negative* bases can be awkward.

a. Evaluate $(-8)^{1/3}$.
b. Evaluate $(-8)^{2/6}$ by squaring first, then taking the sixth root.
c. $1/3 = 2/6$. How do you explain why $(-8)^{1/3} \neq (-8)^{2/6}$?? Can equals *always* be substituted for equals? Later in your mathematical career you will learn De Moivre's Theorem which will help you make some sense out of this awkward situation. For the time being, negative bases are *excluded* when you have fractional exponents, in the same manner that 0 is excluded as a divisor.

6-6 POWERS BY CALCULATOR

You now have a *definition* of exponentiation that works for powers with any rational exponent. You can *evaluate* such powers when the answer "comes out exactly." For example,

$$64^{2/3} = (2^6)^{2/3} = 2^4 = 16.$$

But how do you evaluate powers such as $65^{2/3}$? Since 65 is close to 64, the answer should be close to 16. But *how* close? In the next few sections you will develop a technique with which you can raise *any* positive number to *any* power, and find unknown exponents.

Scientific calculators have been programmed to raise a positive number to any power. In Section 6-11 you will learn how calculators do this. For the time being you will learn how to use a calculator to raise numbers to powers.

Objective:

Given a power with a positive base, such as $65^{2/3}$, use a calculator to find a decimal approximation.

There are three kinds of calculators available which work in basically different ways. The most widely available kind uses "algebraic logic." The order of operations is the same as in algebra. In other words, pressing

3 [+] 4 [X] 5 [=] gives 23.

The multiplication is done first, then the addition. The second kind uses "arithmetic logic." The operations are done in the order they are pressed. On this type of calculator, pressing

3 [+] 4 [X] 5 [=] gives 35.

To tell which type calculator yours is, just try the above example. If the answer is 23, the calculator has algebraic logic. If the answer is 35, it has arithmetic logic.

Changing the order of operations on an algebraic or arithmetic calculator requires either pressing the numbers in a different order or using the parenthesis keys. For instance, to evaluate (3 + 4) X 5 by algebraic calculator requires pressing

[(] 3 [+] 4 [)] [X] 5 [=] which gives 35.

Evaluating 3 + (4 X 5) by arithmetic calculator can be done by pressing

3 [+] [(] 4 [X] 5 [)] [=] which gives 23.

The third type of calculator uses what is called "reverse Polish notation," abbreviated RPN. RPN calculators are most easily distinguished by the fact that there is no [=] key. On this type of calculator the operation key is pressed *after* the two numbers have been entered. To evaluate (3 + 4) X 5 on an RPN calculator you would press

3 [enter] 4 [+] 5 [X] which gives 35.

Pressing 3 [enter] 4 enters the numbers 3 and 4. Pressing [+] adds 3 and 4, giving 7. Pressing 5 [X] enters 5 then multiplies it by the 7 left over from the previous calculation, giving 35. To evaluate 3 + 4 X 5 on an RPN calculator, where the second operation is to be done first, you simply enter all three numbers, then press the operation keys in the reverse order,

3 [enter] 4 [enter] 5 [X] [+]

The X operates on the 4 and 5, which were entered last, giving 20. The + operates on the 3 and 20, giving 23.

Once you know how your calculator operates, you can use it to evaluate powers. Scientific calculators have a key labeled either [y^x] or [x^y]. To evaluate 5^3, you would press

5 [y^x] 3 [=] (for algebraic or arithmetic)

5 [enter] 3 [y^x] (for RPN)

The answer is 125.

Non-integer powers are just as easy to evaluate! To find $47^{1.3}$, you press

47 [y^x] 1.3 [=] (for algebraic or arithmetic)

47 [enter] 1.3 [y^x] (for RPN)

The answer is about 149.1857066.

If the exponent contains an operation, as in $65^{2/3}$, this operation must be done first.

65 [y^x] [(] 2 [÷] 3 [)] [=] (algebraic or arithmetic)

65 [enter] 2 [enter] 3 [÷] [y^x] (RPN)

The answer is about 16.16623563, which is close to 16 as mentioned earlier in this section.

To avoid using parentheses with an algebraic or arithmetic calculator, you can evaluate the exponent first, pressing

2 [÷] 3 [=] [y^x] 65 [$x \leftrightarrow y$] [=]

Pressing the [=] key after $2 \div 3$ carries out this calculation. The $x \leftrightarrow y$ key reverses the two numbers so that the calculator will evaluate $65^{2/3}$ instead of $(2/3)^{65}$.

Example 1: Evaluate $26^{3/2}$. Show that the answer is reasonable.

$$26^{3/2} \approx \underline{132.5745073}$$

Note: Your calculator may round off to fewer places, or give an answer that differs slightly in the last decimal place.

To show that the answer is reasonable, recall that raising to the 3/2 power involves taking a *square* root. The perfect square closest to 26 is 25. Evaluating $25^{3/2}$ as in the previous section gives

$$25^{3/2} = (5^2)^{3/2} = 5^3 = 125.$$

Since 26 is slightly larger than 25, the answer 132.57... is reasonable because it is slightly larger than 125.

Example 2: Evaluate $80^{-3/4}$. Show that the answer is reasonable.

$$80^{-3/4} \approx \underline{0.037383720}$$

Since the exponent is negative you must use the "sign change" key. This key is usually marked [+/−] or [chs]. The sequence of keystrokes is

80 [y^x] [(] 3 [÷] 4 [+/−] [)] [=] (Algebraic)

80 [enter] 3 [enter] 4 [chs] [÷] [y^x] (RPN)

As a check on reasonability, recall that raising to the −3/4 power involves taking a *fourth* root. The fourth power closest to 80 is 81. Evaluating $81^{-3/4}$ gives

$$80^{-3/4} = (3^4)^{-3/4} = 3^{-3} = 1/27 = 0.0370\ldots .$$

Since 0.0373... is close to 0.0370..., the answer is reasonable.

Example 3: Evaluate $72.65 \times 10^{-3.1573}$.

$$72.65 \times 10^{-3.1573} \approx \underline{0.050574968}$$

Most scientific calculators have a 10^x key. Using this key, the sequence of keystrokes is

72.65 [X] 3.1573 [+/−] [10^x] [=] (Algebraic)

72.65 [enter] 3.1573 [chs] [10^x] [X] (RPN)

The exercise that follows is designed to give you practice raising numbers to powers. You will also learn how to find decimal approximations for any radical, and for powers with irrational exponents. The last problem introduces you to the number *e*, a naturally-occurring, transcendental number similar to π.

Exercise 6-6

For Problems 1 through 24, find a decimal approximation for the given expression.

1. 3.57^4
2. 6.43^5
3. 82.4^{-3}
4. 29.7^{-4}
5. 0.521^5
6. 0.368^4
7. $0.00735^{-1/4}$
8. $0.000291^{-1/3}$
9. $437.3^{1.3}$
10. $8657^{0.3}$

11. $0.09528^{0.7}$
12. $0.5743^{5.1}$
13. $2^{1.4}$
14. $5^{1.3}$
15. $3^{-2.6}$
16. $7^{-3.1}$
17. $14.72 \times 10^{1.5639}$
18. $18.24 \times 10^{3.6192}$
19. $384.6 \times 10^{-5.8}$
20. $645.8 \times 10^{-4.3}$
21. $0.00241 \times 10^{4.87}$
22. $0.0918 \times 10^{3.72}$
23. $33^{6/5}$
24. $31^{6/5}$

25. Show that your answer to Problem 23 is *reasonable* by comparing it with the *exact* value of $32^{6/5}$.

26. Show that your answer to Problem 24 is *reasonable* by comparing it with the *exact* value of $32^{6/5}$.

The technique of this section can be used to find *roots* of numbers. For example,

$$\sqrt[3]{0.8326} = 0.8326^{1/3}.$$

After this initial transformation, the problem of finding roots reduces to the now-familiar problem of raising to powers. For Problems 27 through 34, find a decimal approximation for the given radical.

27. $\sqrt[4]{271000}$
28. $\sqrt[3]{67.8}$
29. $\sqrt[7]{92360}$
30. $\sqrt[5]{8432000}$
31. $\sqrt[3]{0.0000519}$
32. $\sqrt[4]{0.00000729}$
33. $\sqrt[5]{1024}$
34. $\sqrt[6]{4096}$

Powers with *irrational* exponents can be found by using a *decimal approximation* for the exponent. For example,

$$5^{\sqrt{7}} = 5^{2.645\ldots} \approx \underline{70.68069398}$$

When you study calculus, you will learn a precise *definition* for powers with irrational exponents.

For Problems 35 through 40, find a decimal approximation for the given power. You can find approximate values of cube roots as above. Also, $\pi =$ 3.14159... which can be found on most scientific calculators.

35. $2^{\sqrt{3}}$
36. $2^{\sqrt{6}}$
37. $23^{\sqrt[3]{2}}$
38. $37^{\sqrt[3]{5}}$
39. 5^{π}
40. π^{π}

41. The expression $\left(1 + \frac{1}{n}\right)^n$ is interesting when n is very large. For instance, if $n = 1000$, the expression is 1.001^{1000}. You would expect it to be *close to 1* because the base is so close to 1. But you would also expect it to be *very large* since the exponent is so large. In this problem you will investigate values of this expression for large values of n.

a. If $n = 10$, the expression equals $\left(1 + \frac{1}{10}\right)^{10}$, or 1.1^{10}. Find a decimal approximation for 1.1^{10}.
b. Using a calculator, find values of the expression for $n = 100$, 1000, 10000, 100000, and 1000000.
c. What number does the expression seem to be approaching as n becomes very large? This number is given the name "e." Like π, it is a naturally-occurring transcendental number. It is used as a base for logarithms in higher mathematics, where it has certain advantages over the base 10.

6-7 SCIENTIFIC NOTATION

Raising numbers to powers can produce numbers that are too large for the calculator to display. If you evaluate

$$365^6$$

most calculators will display two numbers with a space between them, such as

2.3645973 15

This is the way the calculator displays a number in *scientific notation.* It means

$$2.3645973 \times 10^{15},$$

which is a rounded-off version of the exact answer,

2,364,597,285,765,625.

In 2.3645973×10^{15} the decimal point has been moved 15 places to the left so that it is between the first two digits of the number. Multiplying by 10^{15} compensates for this move. The calculator rounds off the decimal number to 2.3645973, or to whatever number is short enough to fit in its display.

The two parts of a number such as 5.38×10^9 or 2.345×10^{-6} in scientific notation are given special names. The factor on the left has the decimal point behind the *first* (non-zero) digit. It is called the *mantissa.* The factor on the right is a power of 10. Its exponent is called the *characteristic.*

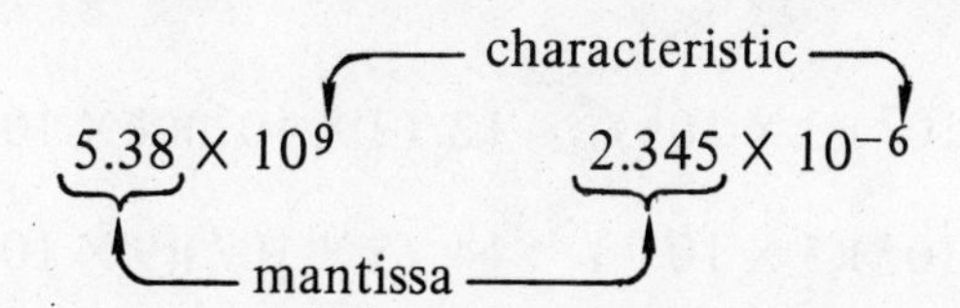

The word *mantissa* comes from a Eutruscan word meaning "makeweight." A makeweight is a small slice of meat the butcher puts with what is already on the scale to *make* the *weight* come out right. From its name, you can readily tell that the mantissa is less significant than the power of 10 which "characterizes" the number in scientific notation.

The number of *significant digits* in a number in scientific notation is the number of digits in the mantissa. For instance,

5.38×10^9 has 3 significant digits,

2.345×10^{-6} has 4 significant digits.

Objectives:

1. Transform numbers to or from scientific notation.
2. Multiply or divide numbers in scientific notation.

Example 1: Write in scientific notation:

a. 3250000
b. 0.007814

a. To get the decimal point to the desired place it must move from the right end of the number (where it is by implication) to the space between the 3 and the 2.

3250000.
6 places

This requires a move of 6 places, as shown above. Therefore,

$3250000 = \underline{3.25 \times 10^6}$

b. To get the decimal point to the desired place it must move 3 places to the right, between the 7 and the 8.

0.007814
3 places

Therefore,

$0.007814 = \underline{7.814 \times 10^{-3}}$

The only tricky part in these transformations is deciding whether the characteristic will be positive or negative. For this purpose it helps to realize that the "=" sign must tell the *truth.* The given number and the answer must really be *equal.* For instance, 3250000 is a *big* number, so its scientific notation form must have a *positive* exponent. 0.007814 is a *small* number (close to 0). So its scientific notation form must have a *negative* exponent, indicating a fraction.

Example 2: Multiply $(3 \times 10^{17})(4 \times 10^{-38})$ mentally. Write the answer in scientific notation.

$(3 \times 10^{17})(4 \times 10^{-38})$

$= 12 \times 10^{-21}$ Multiplying mantissas and adding characteristics.

$= \underline{1.2 \times 10^{-20}}$ $12 = 1.2 \times 10^1$

Example 3: Divide $\dfrac{7.41 \times 10^{13}}{5.941 \times 10^{-8}}$ by calculator.

Round off to the appropriate number of significant digits.

$$\frac{7.41 \times 10^{13}}{5.941 \times 10^{-8}}$$

$= 1.247264... \times 10^{21}$ By calculator (Note 1).

$\approx \underline{1.25 \times 10^{21}}$ Rounding off (Note 2).

Notes:

1. To enter a number in scientific notation you first enter its mantissa. Then you press the "enter exponent" key, usually marked [exp], [eex], or [ee]. The two-digit number you enter will be the exponent. Its sign can be made negative by pressing the sign change key. The sequence of keystrokes is:

   7.41 [exp] 13 [÷] 5.941 [exp] 8 [+/−] [=]
   (algebraic or arithmetic)

   7.41 [eex] 13 [enter] 5.941 [eex] 8 [chs] [÷]
   (RPN)

2. Numbers in scientific notation are usually assumed to be decimal approximations correct to the number of significant digits given. Therefore, the number of significant digits in the answer to a multiplication or division problem should be no more than in the *least* accurate of the numbers. In this case, 7.41×10^{13} has only 3 significant digits. So the answer is rounded off to 1.25×10^{21}.

The following exercise has more problems of this type for you to practice either mentally or by calculator.

Exercise 6-7

For Problems 1 through 10, transform to scientific notation.

1. 372000
2. 57260
3. 0.00261
4. 0.0005423
5. 2001
6. 98.6
7. 0.2024
8. 0.9999
9. 24.7 million
10. 468 million

For Problems 11 through 30, do the operations in your *head,* without using a calculator. Write the answer in scientific notation.

11. $(2 \times 10^7)(3 \times 10^8)$
12. $(2 \times 10^5)(4 \times 10^8)$
13. $(7 \times 10^4)(5 \times 10^{11})$
14. $(5 \times 10^7)(9 \times 10^8)$
15. $(6 \times 10^3)(7 \times 10^{-8})$
16. $(9 \times 10^6)(7 \times 10^{-13})$
17. $(9 \times 10^{-5})(6 \times 10^5)$
18. $(8 \times 10^{-7})(6 \times 10^7)$
19. $(8 \times 10^{-4})(7 \times 10^{-20})$
20. $(4 \times 10^{-5})(7 \times 10^{-10})$
21. $\dfrac{8 \times 10^{15}}{2 \times 10^4}$
22. $\dfrac{8 \times 10^{17}}{4 \times 10^8}$
23. $\dfrac{6 \times 10^4}{3 \times 10^{11}}$
24. $\dfrac{6 \times 10^5}{2 \times 10^{12}}$
25. $\dfrac{7 \times 10^5}{2 \times 10^{-3}}$
26. $\dfrac{9 \times 10^3}{2 \times 10^{-7}}$
27. $\dfrac{4 \times 10^{-3}}{2 \times 10^{-12}}$
28. $\dfrac{8 \times 10^{-9}}{2 \times 10^{-7}}$
29. $\dfrac{6 \times 10^{-8}}{5 \times 10^8}$
30. $\dfrac{8 \times 10^{-4}}{5 \times 10^4}$

For Problems 31 through 46, use a calculator to evaluate the expression. Write the answer in scientific notation, with the appropriate number of significant digits.

31. $(2.05 \times 10^6)(3.12 \times 10^7)$

32. $(3.52 \times 10^4)(2.11 \times 10^7)$

33. $(4.9 \times 10^{17})(1.345 \times 10^{-5})$

34. $(1.3 \times 10^{-5})(4.253 \times 10^{12})$

35. $(8.65 \times 10^3)(2.296 \times 10^{-18})$

36. $(9.06 \times 10^{-9})(5.224 \times 10^3)$

37. $(1.79 \times 10^{-4})(9.7 \times 10^{-9})$

38. $(1.72 \times 10^{-5})(3.6 \times 10^{-11})$

39. $\dfrac{3.75 \times 10^8}{2.93 \times 10^5}$

40. $\dfrac{4.28 \times 10^6}{1.94 \times 10^2}$

41. $\dfrac{4.976 \times 10^{-8}}{8.2 \times 10^5}$

42. $\dfrac{7.732 \times 10^{-7}}{9.63 \times 10^3}$

43. $\dfrac{7.6 \times 10^4}{9.83 \times 10^{-9}}$

44. $\dfrac{1.07 \times 10^4}{8.3 \times 10^{-2}}$

45. $\dfrac{6.802 \times 10^{-7}}{5.1996 \times 10^{-11}}$

46. $\dfrac{7.41 \times 10^{-12}}{6.94 \times 10^{-5}}$

For Problems 47 through 50, find the reciprocal of the given number.

47. 5.87×10^4

48. 2.93×10^5

49. 1.62×10^{-7}

50. 7.8×10^{-9}

For Problems 51 through 56, raise the number to the indicated power.

51. $(2.34 \times 10^5)^2$

52. $(1.29 \times 10^7)^2$

53. $(4.08 \times 10^{-5})^2$

54. $(6.13 \times 10^{-6})^2$

55. $(8.97 \times 10^4)^3$

56. $(7.21 \times 10^5)^3$

57. *Brain Cell Problem.* The average human brain has about 8 billion neurons. There are about 230 million people in the United States. About how many neurons, total, do these people have? Answer in scientific notation.

58. *Heartbeat Problem.* The average life expectancy of women in the United States is about 76 years. The heart beats about once every 0.9 seconds. About how many times does an average woman's heart beat in a lifetime?

59. *Rainfall Problem.* One day it rains one-tenth of an inch. Disappointed over the small amount of rain, Calvin Butterball seeks to find out how much water fell on his 1 square mile of land. He knows that a cubic foot of water weighs 62.4 pounds. How many pounds of water fell on his land?

6-8 INVERSES OF FUNCTIONS

You have learned to evaluate powers by calculator, such as

$$x = 3.7^{2.8}$$

How would you *reverse* the process? Suppose you must find the exponent, x, for which

$$3.7^x = 2.8$$

What keys would you press?

Very often in the field of human endeavor, the way to solve a formerly insoluble problem is to back off and look at it from a different point of view. So you will pause to invent the *inverse* of a function.

DEFINITION

If R is a relation, then the *inverse* of R (abbreviated R^{-1}, and pronounced "R inverse") is the relation obtained by *interchanging* the two coordinates in each ordered pair.

For example, if $R = \{(3, 5), (-2, 1), (4, 5)\}$, then

$$R^{-1} = \{(5, 3), (1, -2), (5, 4)\}.$$

In this case, R is a *function* since each value of the first coordinate has only *one* value of the second coordinate. However, R^{-1} is *not* a function. If the first coordinate is 5, the second coordinate could be either 3 or 4.

Objective:

Given the equation for a function,

a. write the equation for the inverse of the function,
b. transform the inverse equation so that y is in terms of x,
c. plot graphs of the function and its inverse, and
d. tell whether or not the inverse is a function.

Example 1: If $f = \{(x, y): y = 3x + 2\}$, write the equation for the inverse of f.

If the relation to be inverted is specified by an equation, then it is customary to interchange the variables in the *equation* rather than in the sample ordered pair. That is,

$$f^{-1} = \{(x, y): x = 3y + 2\}.$$

In this way, x is still used for the independent variable, and y for the dependent. Since you usually want the dependent variable to be written in terms of the independent one, you can solve the f^{-1} equation for y in terms of x.

$x = 3y + 2$	Equation for f^{-1}.
$x - 2 = 3y$	Subtract 2 from each member.
$\frac{1}{3}x - \frac{2}{3} = y$	Divide each member by 3.
$\therefore y = \frac{1}{3}x - \frac{2}{3}$	Symmetry.

So the inverse of f can be written

$$f^{-1} = \{(x, y): y = \frac{1}{3}x - \frac{2}{3}\}.$$

Since f^{-1} is the name of a function, you could use "$f(x)$" terminology to write

$$f^{-1}(x) = \frac{1}{3}x - \frac{2}{3}.$$

There are two things to observe about what you have done above. First, the symbol $f^{-1}(x)$ is pronounced "f inverse of x" to help distinguish it from the reciprocal of $f(x)$. It should be quite clear from looking at the equations of $f(x)$ and $f^{-1}(x)$ that $f^{-1}(x)$ is *not* equal to the reciprocal of $f(x)$. Second, f^{-1} is a *linear* function since its equation has the form $y = mx + b$. The inverse of any linear function will be another linear function.

Example 2: If $f(x) = |x|$, find $f^{-1}(x)$.

Letting $y = f(x)$, the inverse of the function has the equation

$$x = |y|.$$

Since absolute values are never negative, x must be a non-negative number. But y itself can be positive *or* negative since it is *inside* the absolute value sign. Therefore,

$$y = \pm x \qquad \text{or} \qquad f^{-1}(x) = \pm x.$$

Note that f^{-1} is *not* a function this time because there are *two* different values of y for each positive value of x.

In the exercise which follows, you will find equations for the inverses of some functions. Then you will draw graphs of the functions and their inverses. An interesting graphical property should show up!

Exercise 6-8

1. Suppose that $f = \{(x, y): y = 2x - 6\}$.

a. Write an equation for f^{-1}, and transform it so that y is expressed in terms of x.
b. Find $f(5)$ and $f^{-1}(4)$. Surprising?
c. Show that $f^{-1}(f(x)) = x$, and $f(f^{-1}(x)) = x$.
d. Plot graphs of f and f^{-1} on the *same* Cartesian coordinate system, using the *same* scales for both axes.
e. Is f^{-1} a *function*? Explain.

2. Suppose that $g = \{(x, y): y = x^2\}$.

a. Write an equation for g^{-1} and transform it so that y is expressed in terms of x.
b. Plot graphs of g and g^{-1} on the *same* Cartesian coordinate system, using the *same* scales for both axes.
c. Is g^{-1} a *function*? Explain.

3. Suppose that $h = \{(x, y): y = 2^x\}$.

a. Plot the graph of h using the same scales for both axes.
b. Write the equation for h^{-1}, but do *not* transform it yet.
c. Plot the graph of h^{-1} on the same Cartesian coordinate system. Try picking values of y and calculating x.
d. Is h^{-1} a *function*? Justify your answer.
e. Try to transform the equation for h^{-1} so that y is expressed in terms of x.

4. How are the graph of a function and its inverse related to each other?

6-9 LOGARITHMIC FUNCTIONS

If you invert the function $y = 2^x$, you get the equation

$$x = 2^y.$$

Unfortunately, there is no way to solve this equation for y in terms of x if you use only the algebraic operations, $+, -, \times, \div$, and $\sqrt[n]{\;}$. You may already have concluded this if you worked Problem 3 of Exercise 6-8 above.

When faced with such a situation, mathematicians simply invent a *new* operation to do the job! The word "logarithm" is picked for this new operation. See a dictionary to find out its origin. You write

y is the logarithm of x to the base 2.

This sentence is abbreviated

$$y = \log_2 x.$$

Since y is the exponent in the equation $x = 2^y$, you should realize that a *logarithm* is simply an *exponent*. The graphs of $y = 2^x$ and $y = \log_2 x$ are shown in Figure 6-9 on the next page.

DEFINITION

Definition of ***logarithm***:

$$\boxed{y = \log_b x \text{ means } b^y = x.}$$

The numbers that appear in $y = \log_b x$ have the names:

y is the *logarithm*
b is the *base*
x is the *argument.*

Objective:

Given any *two* of the numbers for logarithm, base, and argument, find the third number.

There is a clever way to remember the definition of logarithm that does not rely on sheer memory. Suppose you are to figure out the meaning of

$$r = \log_s t.$$

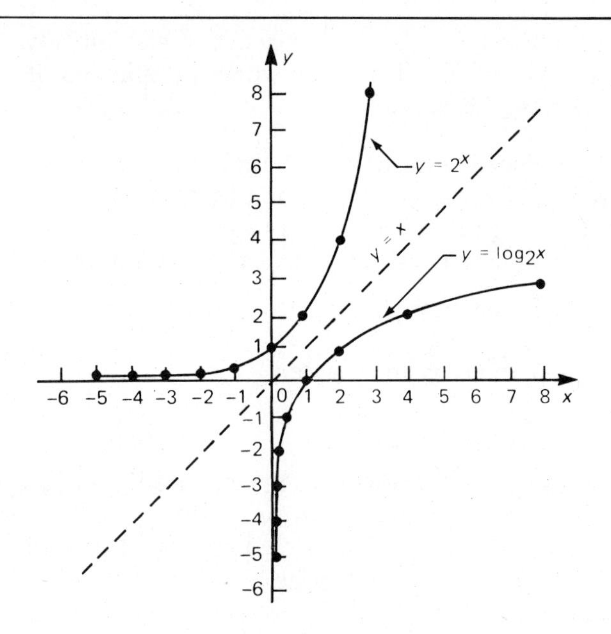

Figure 6-9.

Your thought process would be:

1. The equation is read, "r equals log *to the base* s of t." So s must be the base.

2. The equation is read, "r is a logarithm ...," and a logarithm is an *exponent*. So r is the exponent.

3. The only number left is t. So that must be the answer, and you can write,

$$s^r = t.$$

There are several restrictions on the possible values of the base and the argument of a logarithm. Since fractional powers of negative bases can be imaginary numbers, the base should not be negative. Also, 0 and 1 raised to powers equal 0 and 1, respectively. So the base should not equal these numbers. Once the base is restricted to positive numbers, the power will also equal a positive number, even if the exponent is negative. Therefore, the agrument of a logarithm will always be positive. These restrictions are summarized as follows:

If $y = \log_b x$, then $b > 0$, $b \neq 1$, and $x > 0$.

The following examples show you how to accomplish the objective of this section.

Example 1: Find x if $\log_3 x = -4$.

Since you know no properties of logarithms yet, you simply transform this to a familiar problem using the definition of a logarithm.

$\log_3 x = -4$	Given problem.
$3^{-4} = x$	Definition of a logarithm.
$\frac{1}{3^4} = x$	Definition of negative exponents.
$\underline{\underline{x = \frac{1}{81}}}.$	Symmetry and arithmetic.

Example 2: Find $\log_2 8$.

Let $x = \log_2 8$.	Create an *equation.*
$\therefore 2^x = 8$	Definition of a logarithm.
$2^x = 2^3$	Arithmetic.
$x = 3$	By inspection.
$\therefore \log_2 8 = \underline{\underline{3}}.$	Transitivity.

Example 3: Find x if $\log_x 4 = 2/3$.

$\log_x 4 = 2/3$	Given problem.
$\therefore x^{2/3} = 4$	Definition of a logarithm.
$(x^{2/3})^{3/2} = 4^{3/2}$	Raise each member to the 3/2 power.
$x = (2^2)^{3/2}$	$(x^{2/3})^{3/2} = x$, and $4 = 2^2$.
$x = 2^3$	Power of a power.
$\underline{\underline{x = 8}}.$	Arithmetic.

The exercise which follows is designed to give you enough practice using the definition of a logarithm for you to become thoroughly familiar with the definition.

Exercise 6-9

For Problems 1 through 12, find the argument, x, of the logarithm:

1. $\log_2 x = 3$ 2. $\log_3 x = 2$
3. $\log_3 x = 3$ 4. $\log_5 x = 2$
5. $\log_{1/3} x = 4$ 6. $\log_3 x = -4$
7. $\log_{1/3} x = -4$ 8. $\log_{1/2} x = 7$
9. $\log_{1/2} x = 4$ 10. $\log_4 x = 1/2$
11. $\log_4 x = -1/2$ 12. $\log_{-4} x = 1/2$.

For Problems 13 through 30, find the logarithm, x:

13. $\log_4 16 = x$ 14. $\log_3 81 = x$
15. $\log_3 (1/9) = x$ 16. $\log_5 (1/625) = x$
17. $\log_2 1024 = x$ 18. $\log_{1/2} 8 = x$
19. $\log_{1/3} 243 = x$ 20. $\log_{1/4} (1/16) = x$
21. $\log_{1/5} (1/125) = x$ 22. $\log_5 0 = x$
23. $\log_3 (-9) = x$ 24. $\log_3 3 = x$
25. $\log_3 1 = x$ 26. $\log_1 3 = x$
27. $\log_{10} 1000000000 = x$ 28. $\log_3 3^5 = x$
29. $\log_7 7^9 = x$ 30. $\log_b b^n = x$.

For Problems 31 through 42, find the base, x, of the logarithm:

31. $\log_x 16 = 4$ 32. $\log_x 16 = -4$
33. $\log_x 4 = 1$ 34. $\log_x (1/8) = 3$
35. $\log_x (1/64) = 2$ 36. $\log_x 64 = 3/4$
37. $\log_x 81 = 4/3$ 38. $\log_x (1/25) = 1/2$
39. $\log_x (1/216) = 3/2$ 40. $\log_x (1/64) = -6$
41. $\log_x 2 = 0$ 42. $\log_x 0 = 2$.

6-10 PROPERTIES OF LOGARITHMS

In the previous section you defined a new operation, "taking the logarithm" of a number. Like the operation of "taking the negative" of a number, this is a *unary* operation, meaning it is performed on *one* number. Operations like adding and multiplying are binary operations, performed on two numbers.

Since a logarithm is an exponent, the properties of logarithms should be similar to those of exponents. In this section you will learn and prove three such properties, specifically,

1. Logarithm of a product.
2. Logarithm of a quotient.
3. Logarithm of a power.

Objective:

1. Discover the three properties of logarithms.
2. Prove the three properties.
3. Use the properties to simplify logarithmic expressions.

A calculator can be used to help you discover the properties. Scientific calculators find logs to two different bases. If you enter a number such as 3, then press the [log] key, the calculator displays the base 10, or *common logarithm* of 3.

3 [log] gives $\log_{10} 3$, about 0.477121255

The key labeled [ln] finds the base e, or *natural logarithm* of the argument. The number e is approximately 2.71828, and arises in computations such as Problem 41 of Exercise 6-6.

3 [ln] gives $\log_e 3$, about 1.098612289.

In this text you will use base 10 logs. The symbol $\log x$, written without a base, will be used for base 10 logs. Natural logs are used later, in calculus.

$\log x$ means $\log_{10} x$.
$\ln x$ means $\log_e x$.

Two powers with equal bases may be multiplied by adding their exponents. To see how this property applies to logarithms, you can use a calculator to see how $\log(3 \times 5)$ is related to $\log 3$ and $\log 5$.

$$\log(3 \times 5) = \log 15 \approx 1.176091259$$

$$\log 3 \approx 0.477121255$$
$$\log 5 \approx 0.698970004$$
$$1.176191259 \leftarrow \text{add}$$

Same!

$$\therefore \log(3 \times 5) = \log 3 + \log 5$$

The sequence of keystrokes for the second computation is

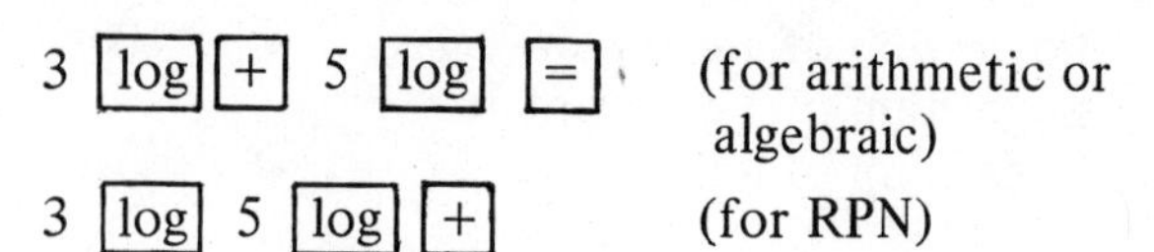

3 [log] [+] 5 [log] [=] (for arithmetic or algebraic)

3 [log] 5 [log] [+] (for RPN)

For the logarithm of a quotient, consider log 12, log 3, and log (12 ÷ 3).

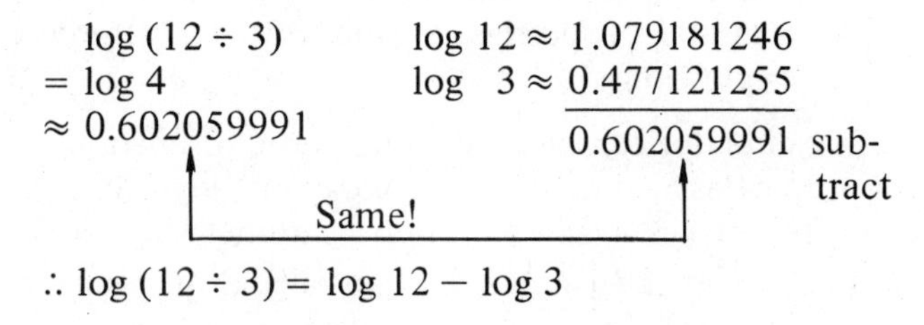

$\log(12 \div 3)$
$= \log 4$
≈ 0.602059991

$\log 12 \approx 1.079181246$
$\log\ \ 3 \approx 0.477121255$
0.602059991 subtract

Same!

$\therefore \log(12 \div 3) = \log 12 - \log 3$

Summarizing the above examples, the log of a product can be found by *adding* the logs of the two factors, and the log of a quotient can be found by *subtracting* the two logs, numerator minus denominator.

For the log of a power, compare log (2^5) with log 2 and 5.

$\log(2^5)$
$= \log 32$
≈ 1.505149978

$\log 2 \approx 0.301029996$
$5(\log 2) \approx 1.505149978$

Same!

$\therefore \log(2^5) = 5(\log 2)$

In other words, the log of a power is the exponent times the log of the base.

The keystrokes for the second calculation are

5 [X] 2 [log] [=] (for arithmetic or algebraic)

5 [enter] 2 [log] [X] (for RPN)

Note: The calculator keeps more significant digits than it displays. This is why 5(log 2) comes out 1.505149978, instead of 1.505149980 which you would get by multiplying 5 by 0.301029996.

The above properties can be proved to work for any valid base and argument. In the following exercise you will supply reasons for these proofs.

Exercise 6-10

For Problems 1 through 6, demonstrate the property by calculator.

1. $\log(3 \times 7) = \log 3 + \log 7$
2. $\log(9 \times 5) = \log 9 + \log 5$
3. $\log(72 \div 8) = \log 72 - \log 8$
4. $\log(51 \div 3) = \log 51 - \log 3$
5. $\log(7^3) = 3(\log 7)$
6. $\log(5^4) = 4(\log 5)$

For Problems 7, 8, and 9, supply reasons for the steps in the proofs. Then state the property in words.

7. *Logarithm of a Product.* Prove that $\log_b(xy) = \log_b x + \log_b y$.

Proof:

a. Let $r = \log_b x$ and $s = \log_b y$.

b. $\therefore b^r = x$ and $b^s = y$.

c. $\therefore b^r b^s = xy$. (Multiplying x by y.)

d. $\therefore b^{(r+s)} = xy$.

e. $\therefore (r+s) = \log_b(xy)$.

f. $\therefore \log_b(xy) = r + s$.

g. $\therefore \log_b(xy) = \log_b x + \log_b y$.

h. Verbal Statement: (Compose your own.)

8. *Logarithm of a Quotient.* Prove that $\log_b \frac{x}{y} = \log_b x - \log_b y$.

Proof:

a. Let $r = \log_b x$ and $s = \log_b y$.

b. $\therefore b^r = x$ and $b^s = y$.

c. $\therefore \frac{b^r}{b^s} = \frac{x}{y}$

d. $\therefore b^{r-s} = \frac{x}{y}$.

e. $\therefore r-s = \log_b \frac{x}{y}$.

f. $\therefore \log_b \frac{x}{y} = r - s$.

g. $\therefore \log_b \frac{x}{y} = \log_b x - \log_b y$.

h. Verbal Statement: (Compose your own.)

9. *Logarithm of a Power.* Prove that $\log_b (x^n) = n(\log_b x)$.

Proof:

a. Let $r = \log_b x$.

b. $\therefore b^r = x$.

c. $\therefore (b^r)^n = x^n$.

d. $b^{rn} = x^n$.

e. $\therefore rn = \log_b (x^n)$.

f. $nr = \log_b (x^n)$.

g. $\therefore \log_b (x^n) = nr$.

h. $\therefore \log_b (x^n) = n(\log_b x)$.

i. Verbal Statement: (Compose your own.)

10. Prove the following:

a. $\log_b 1 = 0$.

b. $\log_b b = 1$.

c. $\log_b 0$ is undefined.

For Problems 11 through 22, use the properties of logarithms to write the expression of a *single* logarithm of a *single* argument.

Example: $\log_7 3 + 2 \log_7 5 = \log_7 3 + \log_7 5^2 = \log_7 75$.

11. $\log_3 5 + \log_3 7$
12. $\log_7 3 + \log_7 8$
13. $\log_2 24 - \log_2 8$
14. $\log_5 12 - \log_5 3$
15. $\log_7 2 + \log_7 5 + \log_7 3$
16. $\log_{11} 6 + \log_{11} 5 + \log_{11} 4$
17. $\log_5 48 - \log_5 12 + \log_5 4$
18. $\log_2 225 - \log_2 5 + \log_2 3$
19. $5 \log_{12} 2$
20. $3 \log_5 4$
21. $3 \log_6 15 - \log_6 25$
22. $4 \log_8 3 - \log_8 6$

For Problems 23 through 30, simplify the expression.

23. $\log_{49} 49^8$
24. $\log_{64} 64^{5/6}$
25. $3 \log_{24} 24^{-2/3}$
26. $5 \log_{37} 37^9$
27. $3^{\log_3 7}$
28. $6^{\log_6 13}$
29. $10^{\log_{10} 9}$
30. $10^{\log_{10} 2001}$

31. With your book closed, *name* and *state* the three properties of logarithms. Then open your book and find out if you are correct. If you missed *any* of them, close your book and write all three of them over again. Keep doing this until you can get all three correct without looking.

6-11 SPECIAL PROPERTIES OF BASE 10 LOGARITHMS

Common logarithms have several important properties which result from the fact that their base is the same as the base of the number system. The first is

$$\boxed{\log 10 = 1}$$

This property follows directly from the definition, and the fact that the base is also 10. Since $10^1 = 10$, you can conclude that $\log 10 = 1$.

The second property concerns the logarithm of a power of 10. For instance,

$\log 10^7$	
$= 7(\log 10)$	Log of a power.
$= 7$	$\log 10 = 1$.

So the log of a power of 10 is simply the *exponent.*

$$\boxed{\log 10^n = n}$$

The third property concerns logarithms of arguments that differ only in the location of the decimal point. For instance,

$$\log 3.456 = 0.53857...$$
$$\log 34.56 = 1.53857...$$
$$\log 345.6 = 2.53857...$$
$$\log 3456. = 3.53857...$$

The logarithms differ only in the *integer* part. The decimal part is the *same.* Scientific notation helps show why this is true.

$\log 345.6$	
$= \log(3.456 \times 10^2)$	Scientific notation.
$= \log 3.456 + \log 10^2$	Log of a product.
$= 0.53857... + 2$	By calculator, and $\log 10^2 = 2$.
$= 2.53857...$	Arithmetic.

Equating the second and the fourth lines, above,

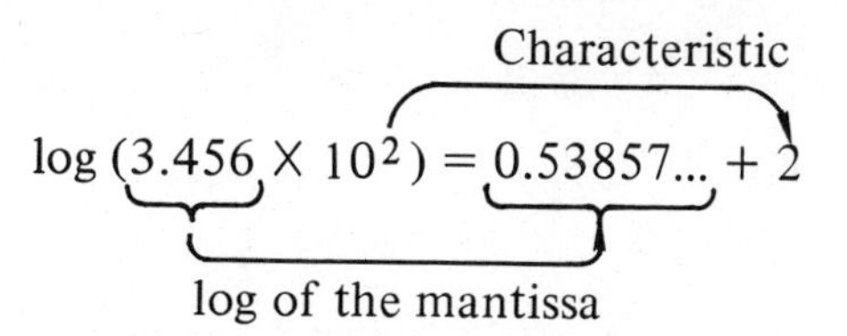

The integer part of the logarithm is the exponent of 10 (the characteristic). The decimal part of the logarithm is the log of the mantissa. These facts lead to the same names being applied to logarithms as to numbers in scientific notation.

$$\log 345.6 = 2.53857...$$

mantissa

characteristic

The mantissa and characteristic properties of common logarithms allowed people to make compact tables of logarithms such as Table II at the end of the book. These tables were useful before there were calculators. A portion of Table II is shown in Figure 6-11a.

Table II. Four-Place Logarithms of Numbers

n	*00*	*10*	*20*	*30*	*40*	*50*	*60*	*70*	*80*	*90*
1.0	0000	0043	0086	0128	0170	0212	0253	0294	0334	0374
1.1	0414	0453	0492	0531	0569	0607	0645	0682	0719	0755
1.2	0792	0828	0864	0899	0934	0969	1004	1038	1072	1106
1.3	1139	1173	1206	1239	1271	1303	1335	[illegible]	[illegible]	4298
[illegible]	4314	4330	4346	[illegible]	[illegible]	[illegible]	[illegible]	4424	4440	4456
2.8	4472	4487	4502	4518	4533	4548	4564	4579	4594	4609
2.9	4624	4639	4654	4669	4683	4698	4713	4728	4742	4757
3.0	4771	4786	4800	4814	4829	4843	4857	4871	4886	4900
3.1	4914	4928	4942	4955	4969	4983	4997	5011	5024	5038

Figure 6-11a.

To find log 31700 using the table, you would reason

$\log 31700$	
$= \log(3.17 \times 10^4)$	In scientific notation.
$= \log 3.17 + 4$	Log of a product, and $\log 10^4 = 4$.

The table contains only mantissas, with the decimal points left off for clarity. The first two digits of the argument, 3.1, are in the column on the left. The next two digits, 70, can be found across the top. The table entry, 5011, means that the logarithm has a mantissa of 0.5011. Finishing the computation,

$$\log 31700 \approx 0.5011 + 4$$
$$= \underline{4.5011}$$

As you can see, finding logarithms was more time-consuming before calculators.

Objective:

Use the special properties of base 10 logarithms to perform various computations.

1. *Antilogs*

The process of evaluating $x = \log 2.79$ is called "finding the logarithm of 2.79." If things are reversed, as in

$$2.79 = \log x,$$

the process of finding x is called "finding the *antilog* of 2.79." Recalling that log by itself implies $\log_{10}$, finding the antilog simply involves raising 10 to the 2.79 power. Therefore,

$$x = 10^{2.79}$$
$$x \approx \underline{616.595}$$

On some calculators the computation is done by pressing the $\boxed{10^x}$ key. On others, two keys, $\boxed{\text{inv}}$ followed by $\boxed{\log}$, must be pressed. The abbreviation inv is for *inverse.* As you know, logarithms and exponentials are inverses of each other.

2. *Powers by Logarithms*

Before there were calculators, it was necessary to evaluate powers with non-integer exponents using logarithms. Since this is still the way calculators find powers internally, it is interesting to see how it is done. To evaluate $3^{2.8}$, the computations are as follows.

$x = 3^{2.8}$	Write an equation.
$\log x = \log 3^{2.8}$	Take log of each member.
$\log x = 2.8 \log 3$	Log of a power.
$\log x \approx 2.8\,(0.4771)$	Table II.
$\log x \approx 1.33588$	Arithmetic.
$x \approx 2.17 \times 10^1$	Find argument from Table II. Characteristic is 1.
$x \approx \underline{21.7}$	Arithmetic.

The calculator pauses briefly while it is evaluating a power since it must do all of these computations. The steps above show you why most calculators will give an "error" message if you try to evaluate a power with a negative base. For $x = (-2)^3$, the calculator tries to take $\log(-2)$, which is undefined.

In the following exercise you will practice finding common logarithms and antilogs by calculator. Then you will use the special properties of common logs to answer some other questions.

Exercise 6-11

For Problems 1 through 10, find the log by calculator.

1. $\log 22.7$
2. $\log 39.6$
3. $\log 0.0548$
4. $\log 0.000921$
5. $\log(3.06 \times 10^5)$
6. $\log(5.79 \times 10^8)$
7. $\log(6.02 \times 10^{23})$
8. $\log(3.00 \times 10^{10})$
9. $\log(1.00 \times 10^{-14})$
10. $\log(6.62 \times 10^{-27})$

For Problems 11 through 20, take the antilog of each member to find the argument, x.

11. $\log x = 2.7745$
12. $\log x = 1.8785$
13. $\log x = 7.5658$
14. $\log x = 9.3444$
15. $\log x = -1.8633$
16. $\log x = -2.8182$
17. $\log x = 2.8633$
18. $\log x = 4.1931$
19. $\log x = -2.8633$
20. $\log x = -4.1931$

21. Find $x = 3.29^{2.5}$ without using a calculator, as shown in the text example. Find the logarithm in Table II. Compare the answer you get with that obtained by calculator.

22. Explain why most scientific calculators give an error message if you attempt to evaluate $(-2)^5$. Explain how the calculator could be used to get the correct answer without giving an error message.

23. ***Powers too Large for Calculators Problem.*** Most calculators cannot give an answer that has an exponent more than 99. However, you can evaluate such powers by calculator if you find the *logarithm* of the answer. Use this clue to do the following.

a. Evaluate 1776^{53}.

b. Evaluate 2001^{97}.

c. Evaluate 0.007^{105}.

d. If 854^{231} were evaluated exactly, how many digits would the answer have?

e. If 0.2^{1000} were evaluated exactly, how many zeros would there be between the decimal point and the first non-zero digit?

24. ***Earthquake Problem.*** The Richter energy number of an earthquake is the base 10 log of the amplitude (i.e., the severity) of the quake vibrations. The Seattle quake of April 29, 1965, measured 7 on the Richter scale. The San Francisco quake of 1906 measured about 8.25 on the Richter scale. To those who know little about logarithms, 8.25 does not sound much more severe than 7. Using what you know about logarithms, tell how many times more severe the San Francisco quake was than the Seattle one.

25. ***Decibel Problem.*** The loudness of sound is measured in decibels. The number of decibels is 10 times the base 10 log of the relative acoustical power of the sound waves. A sound just barely loud enough to be heard is given a relative acoustical power of 1. Calculate to the nearest decibel the loudness of the following sounds:

	Sound	*Relative Acoustical Power*
a.	Threshold of audibility	1
b.	Soft recorded music	4000
c.	Loud recorded music	6.8×10^8
d.	Jet aircraft	2.3×10^{12}
e.	Threshold of pain	1×10^{13}

26. ***pH Problem.*** The strength of an acid solution is measured by its pH (for "*p*ower of *H*ydrogen"). The pH is the *negative* of the common logarithm of the hydrogen ion concentration (in moles per liter).

a. Find the pH of the following solutions:

Solution	*Hydrogen Ion Concentration*
Neutral water	1.0×10^{-7}
Human blood	6.3×10^{-8}
Hydrochloric acid	2.5×10^{-2}
Grandma's lye soap suds	9.2×10^{-12}

b. The pH of ordinary vinegar is 2.8. What is the hydrogen ion concentration?

c. The pH of tomatoes is about 4.2. Find the hydrogen ion concentration of tomatoes. Are tomatoes more or less acidic than neutral water? That is, do tomatoes have a higher hydrogen ion concentration than water, or a lower one?

6-12 EXPONENTIAL EQUATIONS, AND LOGARITHMS WITH OTHER BASES

You have learned how to raise any positive number to any power. You are now ready to *reverse* the process, and find values of unknown exponents. This technique is useful when you are dealing with exponential functions with equations of the form

$$y = a \times b^x,$$

and you wish to calculate x for given values of y.

Objective:

Be able to solve an equation in which the variable appears as an *exponent.*

The basic technique is taking the logarithm of each member of the equation. The property of the log of a power can then be used to get the variable down out of the exponent so that you can solve for it. This technique can also be used to evaluate logarithms to bases *other than* 10.

Secondary Objective:

Be able to find a decimal approximation for the logarithm to *any* positive base ($b \neq 1$) of *any* positive number.

Although logs with other bases are not in the mainstream of this course, they are worth studying. By so doing, you will fill in some more details in your "big picture" of mathematical theory. Base 2 logarithms are used internally by computers since computers "think" in *binary* (base 2) numbers. Base *e* logarithms ($e \approx 2.718$) are used later in mathematical analysis because of a remarkable relationship between the *slope* of the graph and the value of the ordinate.

Example 1: Solve $0.274^{3x} = 1.97$.

Using the fact that if two positive numbers are equal then their logarithms are equal, you can take the log of each member of the equation.

$\log 0.274^{3x} = \log 1.97$	Take log of each member.
$(3x)(\log 0.274) = \log 1.97$	Log of a power.
$x = \dfrac{\log 1.97}{3 \log 0.274}$	Dividing by 3 and by log 0.274.
$x = -0.1746$	By calculator.
$\therefore S = \{-0.1746\}$.	By inspection.

Note that it is better to do all the algebra first, getting x explicitly as an expression involving logarithms, before you do any work on the calculator. In evaluating the expression for x, remember that you must divide by 3 and *divide* by log 0.274.

Example 2: Find an approximation for $\log_7 53$.

The first step is to give $\log_7 53$ a name. Let

$$x = \log_7 53.$$

By the definition of logarithm,

$$7^x = 53.$$

You have now transformed the *new* problem of finding $\log_7 53$ into the "old" problem of solving an exponential equation. Taking $\log_{10}$ of each member, you get

$\log_{10} 7^x = \log_{10} 53$	Take log of each member.
$(x)(\log_{10} 7) = \log_{10} 53$	Log of a power.
$x = \dfrac{\log_{10} 53}{\log_{10} 7}$	Dividing by $\log_{10} 7$.
$x = 2.040$	By calculator.
$\therefore \log_7 53 = \underline{\underline{2.040}}$.	Transitivity.

The answer is reasonable, since $\log_7 49 = \log_7 7^2 = 2$, and the answer is slightly more than 2.

The exercise that follows is designed to give you practice solving exponential equations. You will also get practice evaluating logarithms with bases other than 10. Toward the end of the exercise, you will have the opportunity to prove some less-familiar properties of logarithms, and then use these properties to work problems of the type that often appear on mathematics contests.

Exercise 6-12

For Problems 1 through 14, solve the given equation.

1. $2^x = 3$
2. $3^x = 2$
3. $54.3^x = 17.2$
4. $74.1^x = 2.68$
5. $0.27^x = 5.06$
6. $0.065^x = 13.2$
7. $4.13^{2x} = 987$
8. $8.09^{5x} = 40800$
9. $10^x = 15$
10. $10^{-x} = 72$
11. $3 \times 10^x = 7$
12. $2.43 \times 10^x = 1.84$
13. $91.2 \times 10^{0.3x} = 438$
14. $24.2 \times 10^{0.7x} = 6.42$

For Problems 15 through 24, find decimal approximations for the given logarithms.

15. $\log_4 17.36$
16. $\log_5 63.24$
17. $\log_4 4000$
18. $\log_5 5000$
19. $\log_4 400$
20. $\log_5 500$
21. $\log_4 40$
22. $\log_5 50$
23. $\log_4 4$
24. $\log_5 5$

25. Based on your answers to Problems 17 through 24, do logs with bases other than 10 seem to have the same characteristic-mantissa properties as common logs? Justify your answer.

Problems 26 through 32 present some less-familiar properties of logarithms. Prove the properties.

26. *Change-of-Base:*

$$\log_a x = \frac{\log_b x}{\log_b a}.$$

27. *Direct Proportionality:*

$\log_a x = (c)(\log_b x)$, where c is a *constant.*

28. *Reciprocal:*

$$\log_a b = \frac{1}{\log_b a}.$$

29. *Product of Two Logs:*

$$(\log_a b)(\log_b c) = \log_a c.$$

30. *Log with Power for Base:*

$$\log_{(b^n)} x = \frac{1}{n}\log_b x.$$

31. *Base and Argument are Like Powers:*

$$\log_{(b^n)} (x^n) = \log_b x.$$

32. *Base and Argument are Reciprocals:*

$$\log_{\frac{1}{b}} \frac{1}{x} = \log_b x.$$

33. *Logs to the Base 2* – By Problem 27, preceding, $\log_2 x = (c)(\log_{10} x)$.

a. Find a decimal approximation for the proportionality constant, c.
b. Using a calculator with a memory or a "constant" key, evaluate *quickly*:
   - i. $\log_2 13$.
   - ii. $\log_2 0.63$.
   - iii. $\log_2 1776$.
   - iv. $\log_2 4096$.
   - v. $\log_2 1000000$.

34. *Logs to the Base e – Natural* logarithms have the number e ($e \approx 2.7182818182846$) as a base. Base e logs can be calculated from base 10 logs using the technique of Problem 33.

a. Find the constant c such that $\log_e x = (c)(\log_{10} x)$.

b. Using a calculator with a memory or a "constant" key, evaluate quickly:

   i. $\log_e 2001$.
   ii. $\log_e 7$.
   iii. $\log_e 0.0324$.
   iv. $\log_e 1000000$.
   v. $\log_e 20.08553692$.
   vi. $\log_e \pi$.

Problems 35 through 56 are "tricky" problems involving logarithms of the type that often appear on mathematics contests. You should find the properties of Problems 26 through 32 helpful in answering these. Also, you will need to recall the other properties of logarithms, as well as the definition.

The problems are arranged in *no* special order, so you must decide in each case just what procedure to use. Find *exact* answers unless a decimal approximation is called for.

35. Evaluate $\log_{12} (\log_9 (\log_5 (\log_2 32)))$.

36. Evaluate $(\log_7 5)(\log_3 7)(\log_2 27)(\log_5 2)$.

37. Given only that $\log_{10} 2 = 0.301$, $\log_{10} 3 = 0.477$, and $\log_{10} 7 = 0.845$, find $\log_{10} (1 \cdot 2 \cdot 3 \cdot 4 \cdot 5 \cdot 6 \cdot 7 \cdot 8 \cdot 9 \cdot 10)$ *without* using a calculator.

38. Solve for y:
$(\log_5 x)(\log_x 5x)(\log_{5x} y) = \log_x x^4$.

39. Evaluate $\dfrac{-\log_3 2}{\log_2 (-8)}$.

40. Evaluate $(\log_{27} 32 + \log_{27} \frac{1}{4})(\log_2 9)$.

41. Solve for $n > 1$: $\log_{10} 2 + \log_{10} 4 + \log_{10} 8 + \ldots + \log_{10} 2^n = (\log_{1024} 10)^{-1}$.

42. Simplify: $\log_{1001} 7 + \log_{1001} 11 + \log_{1001} 13$.

43. Solve for x: $(\log_3 x)^2 + \log_3 x^2 + 1 = 0$.

44. Evaluate $\log_7 (7^{\log_7 343})$.

45. Evaluate $\log_2 (\log_3 (\log_4 64))$.

46. Evaluate $\log_7 8 \div \log_7 \frac{1}{8}$.

47. Solve the following system for (x,y):

$$\log_9 x + \log_y 8 = 2.$$

$$\log_x 9 + \log_8 y = \frac{8}{3}.$$

48. Solve for x and y:

$$(\log_3 x)(\log_x 2x)(\log_{2x} y) = \log_x y^2.$$

49. Evaluate $\log_2 56 - \log_4 49$.

50. Given $\log_2 3 = 1.58$, find $\log_{16} 81$ *without* using a calculator.

51. Evaluate $10^{\log_{100} 9}$.

52. Solve for x: $\log_{10} (x^{\log_{10} x}) = 4$.

53. Evaluate $\log_4 (2\sqrt[5]{8}) - \log_8 \sqrt[3]{0.25}$.

54. Evaluate $4 \log_3 \frac{1}{3} + 2 \log_{27} 9$.

55. Evaluate $4^{\log_2 5}$.

56. Solve for x:

$$x^2 \log_{10} 8 - x \log_{10} 5 = 2(\log_2 10)^{-1} - x.$$

6-13 EVALUATION OF EXPONENTIAL FUNCTIONS

At the beginning of the chapter, you defined exponential functions to be functions with equations in the form

$$y = a \cdot b^x,$$

where a and b stand for constants. Now that you know how to raise any positive number to any power using a calculator, you can find values of y for any given values of x. Also, since you have learned how to solve exponential equations, you can find values of x for given values of y. So you are just about ready to use exponential functions as mathematical models.

There is a clever transformation of the above equation which will save you time when you start looking up logarithms. If you let

$$b = 10^k,$$

where k stands for a constant, then the equation becomes

$$y = a \cdot (10^k)^x.$$

Using the property of a power of a power, this becomes

$$y = a \cdot 10^{kx}.$$

Since a, k, x, and y might have decimal points in their numerals, it is advisable to use "X" or "*" for the multiplication sign, rather than "·." So the general equation you will use for an exponential function has the form

$$\boxed{y = a \times 10^{kx}.}$$

Functions of this type may be described verbally by saying, "y varies exponentially with x." The constant k is called the *exponential constant*. The constant a is a constant of proportionality.

Objective:

Given a particular equation of the form $y = a \times 10^{kx}$,

a. find y for given values of x,
b. find x for given values of y, and
c. plot the graph.

The first two parts of the objective are simply applications of the techniques you learned in Sections 6-11 and 6-12. The example below should refresh your memory.

Example 1: If $f(x) = 42.3 \times 10^{0.08x}$,

a. find $f(7)$, and
b. find x if $f(x) = 11.9$.

a. $f(7) = 42.3 \times 10^{(0.08)(7)}$ Substitute 7 for x.

$f(7) \approx \underline{\underline{153.6}}$ By calculator.

b. $11.9 = 42.3 \times 10^{0.08x}$ Substitute 11.9 for $f(x)$.

$\dfrac{11.9}{42.3} = 10^{0.08x}$ Divide by 42.3.

$\log \dfrac{11.9}{42.3} = 0.08x$ Take log of each member.

$\dfrac{\log \frac{11.9}{42.3}}{0.08} = x$ Divide by 0.08.

$\underline{\underline{-6.8849}} \approx x$ By calculator.

The third part of the objective, drawing the graph, can be accomplished by pointwise plotting. You have already calculated two points, namely (7, 153.6) and (−6.8849, 11.9). Another point appears in the equation. If $x = 0$, then

$$\begin{aligned} f(0) &= 42.3 \times 10^{(0.08)(0)} \\ &= 42.3 \times 10^0 \\ &= 42.3. \end{aligned}$$

So the constant a in the equation $y = a \times 10^{kx}$ is equal to the *y-intercept*; the value of y when $x = 0$. Figure 6-13a shows the graph of the function based on these three points. The graph can be extended in the negative x-direction since the x-axis is an *asymptote*.

There is a property of exponential functions that lets you calculate plotting data rapidly, in much the same way as the slope of a linear function lets you plot those graphs rapidly. Suppose that

$$f(x) = 7 \times 10^{0.2x}.$$

If you calculate $f(0), f(5), f(10), f(15), \ldots$, an interesting pattern shows up,

$$f(0) = 7 \times 10^0 = 7,$$

$$f(5) = 7 \times 10^1 = 70,$$

$$f(10) = 7 \times 10^2 = 700,$$

$$f(15) = 7 \times 10^3 = 7000.$$

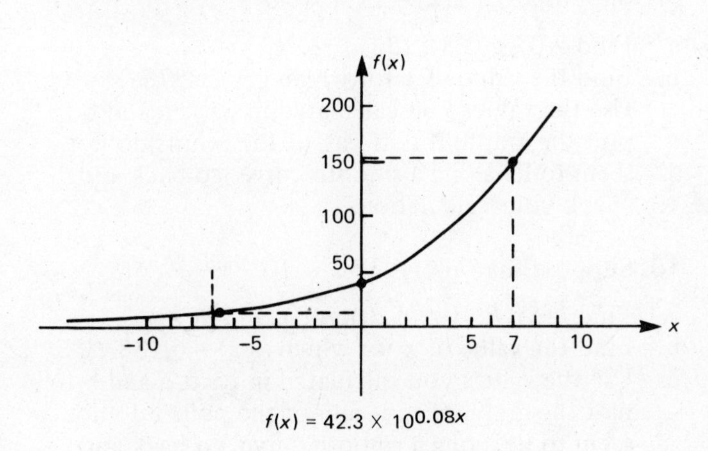

Figure 6-13a.

Every time you *add* 5 to x, you *multiply* $f(x)$ by 10. The reason this happens is not hard to see. Suppose that

$$f(x) = a \times 10^{kx},$$

and you add a constant, c, to x. Then

$f(x + c) = a \times 10^{k(x + c)}$	Substitution.
$f(x + c) = a \times 10^{kx + kc}$	Distributing the k.
$f(x + c) = a \times 10^{kx} \times 10^{kc}$	Product of powers with equal bases.
$f(x + c) = f(x) \times 10^{kc}.$	Substituting $f(x)$ for $a \times 10^{kx}$.

So *adding* the constant c to x causes $f(x)$ to be *multiplied* by the constant 10^{kc}. This property may be stated:

> For exponential functions, *adding* a constant to x causes y to be *multiplied* by a constant.

Note:

> For *linear* functions, *adding* a constant to x *adds* a constant to y.

Example 2: Suppose that $f(x)$ varies exponentially with x, and that $f(3) = 100$ and $f(5) = 80$. Calculate values of $f(x)$ and use them to sketch the graph of f.

By the preceding property you can conclude that *adding* 2 to x causes $f(x)$ to be *multiplied* by 80/100, or 0.8. So you can make a table of values by adding 2 to the previous value of x, and multiplying the previous value of $f(x)$ by 0.8.

x	$f(x)$
3	100
5	80
7	64
9	51.2
11	40.96

You can extend the pattern back the other direction by *subtracting* 2 from x and *dividing* $f(x)$ by 0.8 (or multiplying by 1/0.8, which equals 1.25).

x	$f(x)$
3	100
1	125
−1	156.25
−3	195.3125

The repeated multiplications or divisions are most easily accomplished by a calculator having a memory or a "constant" key.

With these values available, you can plot an accurate graph quite quickly. The result for this example is shown in Figure 6-13b on the next page.

Note that the graph goes *downward* as x increases. This is a feature of exponential functions that have a *negative* value for the exponential constant, k. For such functions, $f(x)$ is said to "decrease exponentially with x." If k is a positive number, $f(x)$ is said to "increase exponentially with x."

The exercise that follows is designed to give you practice finding y for known values of x, finding x for known values of y, and sketching graphs. These

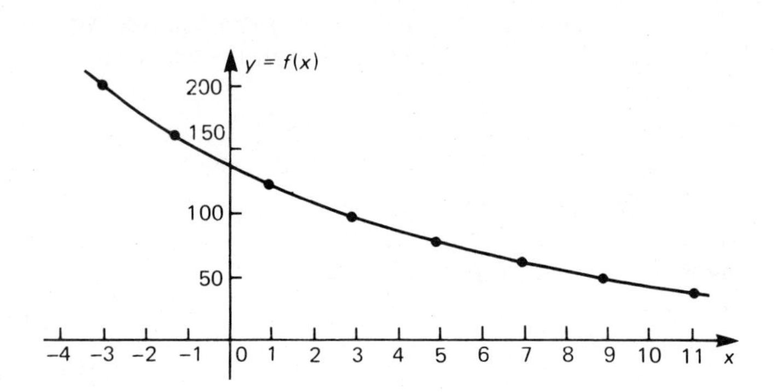

Figure 6-13b.

are the techniques you will need in order to be able to use exponential functions as mathematical models of the real world.

Exercise 6-13

For Problems 1 through 8, assume that f is an *exponential* function with the two given values. Use the property of exponential functions to calculate values of $f(x)$ for two *larger* values of x and for two *smaller* values of x. Then use the calculated and given points to plot the graph.

1. $f(2) = 36$ and $f(5) = 54$.
2. $f(3) = 20$ and $f(7) = 24$.
3. $f(4) = 100$ and $f(6) = 70$.
4. $f(1) = 50$ and $f(3) = 20$.
5. $f(-2) = 1$ and $f(3) = 2$.
6. $f(0) = 30$ and $f(10) = 27$.
7. $f(8) = 25.7$ and $f(11) = 16.8$.
8. $f(-6) = 8.4$ and $f(-1) = 22.7$.
9. Suppose that $g(x) = 21.4 \times 10^{0.6x}$.
   a. Find $g(0)$, $g(2.8)$, and $g(-3.2)$.
   b. Find the value of x for which $g(x) = 578$.
   c. Use the values you calculated in parts a and b to plot the graph of g. If any of the points do not seem to lie along a smooth curve, go back and check your calculations.
10. Suppose that $h(x) = 37.2 \times 10^{-0.05x}$.
    a. Find $h(5)$, $h(10)$, $h(20)$, $h(0)$, and $h(-5)$.
    b. Find the value of x for which $h(x) = 6.138$.
    c. Use the values you calculated in parts a and b to plot the graph of h. If any of the points do not seem to lie along a smooth curve, go back and check your calculations.

6-14 EXPONENTIAL FUNCTIONS AS MATHEMATICAL MODELS

In Section 6-13 you drew graphs of several exponential functions. As shown in Figure 6-14a, the graphs have the x-axis as a horizontal asymptote. Therefore, exponential functions are suitable mathematical models for phenomena in which y increases gradually from small values, or decreases gradually toward zero.

In order to use an exponential function as a model, you must find the particular equation from given points on its graph. The equation can then be used to predict values of y or x.

Objective:

Given two points on the graph of an exponential function,

a. derive the particular equation, and
b. use the function to predict values of y or x.

Example: Suppose that the number of bacteria per square millimeter in a culture in your biology lab is increasing exponentially with time. On Tuesday you find that there are 2000 bacteria per square millimeter. On Thursday, the number has increased to

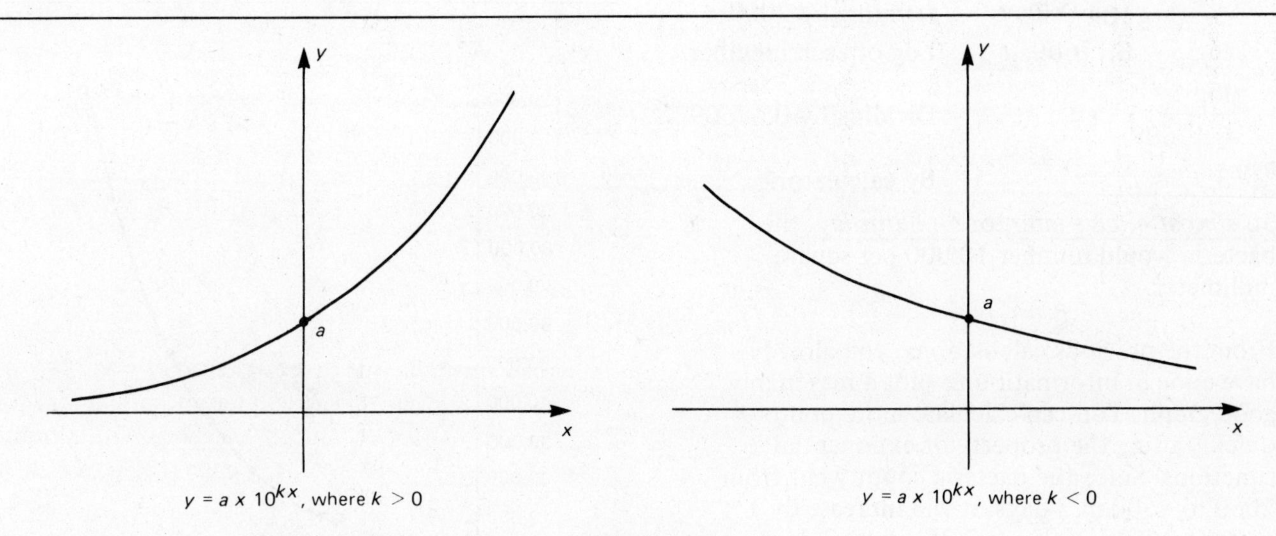

Figure 6-14a.

4500 per square millimeter. Do the following:

a. Derive the particular equation.
b. Predict the number of bacteria per square millimeter on Tuesday of next week.
c. Predict the time when the number of bacteria per square millimeter reaches 10,000.
d. Plot the graph of the function.

a. Let B = number of bacteria per square millimeter.
Let t = number of days since this Tuesday.

The *general* equation is $B = a \times 10^{kt}$.

Substituting (0, 2000) for (t, B) gives

$$2000 = a \times 10^{k(0)}$$

$$2000 = a \times 10^{0}$$

$$2000 = a.$$

On Thursday, $t = 2$. Substituting (2, 4500) for (t, B), and 2000 for a,

$$4500 = 2000 \times 10^{k(2)} \quad \text{Substitution.}$$

$$\frac{4500}{2000} = 10^{2k} \quad \text{Divide by 2000.}$$

$$\log \frac{4500}{2000} = 2k \quad \text{Take log of each member.}$$

$$\frac{\log \frac{4500}{2000}}{2} = k \quad \text{Divide by 2.}$$

$$0.17609... = k$$

Therefore, the particular equation is

$$\underline{\underline{B = 2000 \times 10^{0.17609...t}}}$$

Note: The value of k should be stored, without round-off, in the calculator's memory for use later in the problem.

b. Next Tuesday, $t = 7$. Substituting 7 for t,

$$B = 2000 \times 10^{(0.17609...)(7)} \quad \text{Substitution.}$$

$$\underline{\underline{B \approx 34172 \text{ bacteria/sq. mm.}}} \quad \text{By calculator.}$$

c. If $B = 10{,}000$, then

$$10000 = 2000 \times 10^{0.17609...t} \quad \text{Substitution.}$$

$5 = 10^{0.17609...t}$ Dividing by 2000.

$\log 5 = 0.17609...t$ Log of each member.

$\dfrac{\log 5}{0.17609...} = t$ Dividing by 0.17609...

$3.97 \approx t.$ By calculator.

So almost 4 days later, or on Saturday, the bacteria would number 10,000 per square millimeter.

d. From the previous calculations, you already have enough information to plot a reasonably good graph. You can calculate more points quickly using the property of exponential functions. Since the bacteria count went from 2000 to 4500 in 2 days, it will increase by a *factor* of 4500/2000, or 2.25, every 2 days. Using a calculator, you can rapidly calculate more points.

t	B
0	2000
2	4500
4	10125
6	22781
8	51258
10	115330

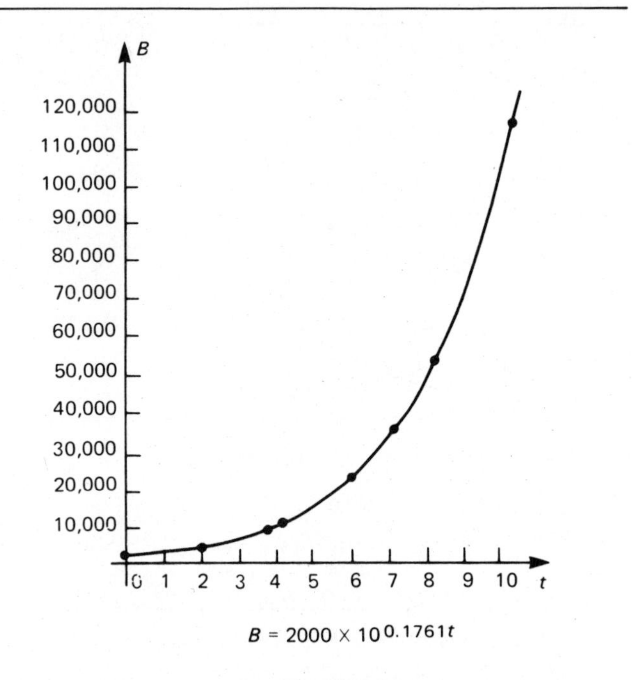

Figure 6-14b.

Figure 6-14b shows a graph of this function. The domain includes only *non-negative* values of t, since it is uncertain whether or not the bacteria were growing *before* this Tuesday.

In the exercise that follows, you will use exponential functions as mathematical models of some real-world phenomena. Toward the end of the exercise, you will use some variations of the general exponential function equation. You will also use *logarithmic* functions as mathematical models.

Exercise 6-14

Problems 1 through 10 are "straightforward" problems similar to the example. Problems 11 through 19 involve variations of exponential functions that will require you to *think* a bit harder.

1. ***Population Problem.*** Assume that the population of the United States is increasing exponentially with time. The 1970 census showed that the population was about 203 million. The 1980 census showed that the population had grown to about 226 million.

   a. By what factor did the population increase from 1970 to 1980? Use the answer and the properties of exponential functions to predict the outcomes of the 1990, 2000, and 2010 censuses.

   b. Plot a graph of population versus time from 1970 through 2010.

   c. Find the particular equation expressing population in terms of the number of years that have elapsed since 1970.

d. Use your equation to predict the population *this* year. Plot the value of the graph of part b. If it does not seem to fit with the other points, go back and check your work.
e. Predict the year in which the population will reach 400 million.
f. According to your mathematical model, what was the population when the Declaration of Independence was signed? Check in an encyclopedia or other source to see how close you came to the *actual* population. Explain any large differences you may observe.

2. ***Rabbit Problem.*** When rabbits were first brought to Australia last century, they had no natural enemies so their numbers increased rapidly. Assume that there were 60,000 rabbits in 1865, and that by 1867 the number had increased to 2,400,000. Assume that the number of rabbits increased exponentially with the number of years that elapsed since 1865.

a. Write the particular equation for this function.
b. How many rabbits would you predict in 1870?
c. According to your model, when was the *first* pair of rabbits introduced into Australia?
d. Based on the properties of exponential functions, why is it appropriate to say that the rabbits "multiplied?"
e. See *The Alien Animals*, by George Laycock (Ballentine Books, 1966) for the eventual outcome of the rabbit problem.

3. ***Car-Stopping Problem.*** Oliver Sudden is driving along a straight, level highway at 64 kilometers per hour (kph) when his car runs out of gas. As he slows down, his speed decreases exponentially with the number of seconds since he ran out of gas, dropping to 48 kph after 10 seconds.

a. Write the particular equation expressing speed in terms of time.
b. Predict Oliver's speed after 25 seconds.

c. At what time will Oliver's speed be 10 kph?
d. Draw the graph of the function for speed in the domain from 0 through the time when Oliver reaches 10 kph.
e. What would the actual speed-time graph look like for *negative* values of time?
f. Explain why this mathematical model would not give reasonable answers at very *large* values of time.

4. ***Milk Spoiling Problem.*** Assume that the number of hours milk stays fresh decreases exponentially with temperature. Suppose that milk in the refrigerator at 0° C will keep for 192 hours, and milk left out in the kitchen at 20° C will keep for only 48 hours.

a. Let h be the number of hours the milk keeps, and T the Celsius temperature. Write the particular equation expressing h in terms of T.
b. Use the equation to show that milk will keep approximately 1/4 as long at 40° C as it will at 20° C.
c. Without any more use of your equation, predict h for temperatures of 60° C and 80° C.
d. Plot the graph of h versus T for values of T from 0 through 80.

5. ***Phoebe's Next Rocket Problem.*** Phoebe Small is out Sunday driving in her rocket ship. She fills up with fuel at the Scorpion Gulch Rocket Fuel Station, and takes off. When she starts the last stage of her rocket, she is going 4230 miles per hour (mph). Ten seconds later she is going 6850 mph. While the last stage is running, you may assume that Phoebe's speed increases exponentially with time.

a. In order to go into orbit, Phoebe must be going 17,500 mph. She took in enough fuel to last for 30 seconds. Will she orbit? Explain.
b. What is the minimum length of time the last stage could run and still get Phoebe into orbit?
c. How long would the last stage have to run to get Phoebe going 25,000 mph so that she could go off to the Moon?

6. ***Compound Interest Problem.*** Banks which compound interest "continuously" use an exponential function to calculate the amount of money you have at any time. Suppose that you put \$1000 into a savings account and find that at the end of one year you have \$1052. Assume that the number of dollars you have in the account increases exponentially with time.

a. Find the particular equation for this exponential function.
b. Predict the amount you will have 10 years after you invested the \$1000.
c. How many years will it take to *double* your investment? That is, when will you have \$2000?

7. ***Car Trade-In Problem.*** A rule-of-thumb used by car dealers is that the trade-in value of a car decreases by 30% each year. That is, the value at the end of any year is 70% of its value at the beginning of that year ("70% of ..." means "0.7 times ...").

a. Suppose that you own a car whose trade-in value is presently \$2350. How much will it be worth 1 year from now? 2 years from now? 3 years from now?
b. Explain how the properties of exponential functions allow you to conclude that the trade-in value varies *exponentially* with time.
c. Write the particular equation expressing the trade-in value of your car as a function of the number of years from the present.
d. In how many years from now should the trade-in value be \$600?
e. If the car is presently 2.7 years old, what was its trade-in value when it was new?
f. The car cost \$7430 when it was new. How do you explain the difference between this number and the answer to part e?

8. ***Air Pressure Problem.*** The pressure of the air in the Earth's atmosphere decreases exponentially with altitude above the surface of the Earth. The pressure at the Earth's surface (sea level) is about 14.7 pounds per square inch (psi) and the pressure at 2000 feet is approximately 13.5 psi.

a. Write the particular equation expressing pressure in terms of altitude.

b. Predict the pressure at

   i. Mexico City (altitude 7500 feet),
   ii. Mount Everest (altitude 29,000 feet),
   iii. where U-2 spy planes fly (80,000 feet),
   iv. the edge of outer space (defined by NASA to be 50 miles up).

c. Human blood at body temperature will boil if the pressure is below 0.9 psi. At what altitude would your blood start to boil if you were in an unpressurized airplane?

d. Recently it has been discovered that the Mediterranean Sea dried up about 6 million years ago, leaving a "valley" 10,000 feet below sea level. What would the air pressure have been at the bottom of this valley?

9. ***Carbon 14 Dating Problem.*** Carbon 14 is an isotope of carbon that is formed when radiation from the Sun strikes ordinary carbon dioxide in the atmosphere. Plants such as trees, which get their carbon dioxide from the atmosphere, therefore contain small amounts of carbon 14. Once a particular part of a plant has been formed, no more new carbon 14 is taken in. The carbon 14 in that part of the plant decays slowly, transmuting into nitrogen 14. Let P be the percent of carbon remaining in a part of a tree that grew t years ago.

a. Write the particular equation expressing P in terms of t. You may assume that the "half-life" of carbon 14 is 5750 years, meaning that of the 100% of carbon 14 present when $t = 0$, only 50% remains when $t = 5750$ years.

b. Christ was crucified about 2000 years ago. If somebody claimed to have a piece of wood from the cross on which he was crucified, what percent of the carbon 14 would you expect to find remaining in this wood?

c. The oldest living trees in the World are the bristlecone pines in the White Mountains of California. 4000 "growth rings" have been counted in the trunk of one of these, meaning that the innermost ring is 4000 years old. What percent of the original carbon 14 would you expect to find in the oldest ring of this tree?

d. A piece of wood believed to have come from Noah's Ark has 48.37% of the carbon 14 remaining. The Great Flood is supposed to have occurred in 4004 B.C. Is this piece of wood *old enough* to have come from Noah's Ark? Justify your answer.

e. Coal is supposed to have been formed from trees which lived 100 million years ago. What percent of the original carbon 14 would you expect to find remaining in coal? Why would carbon 14 dating probably *not* be very good for anything as old as coal?

f. See the article "Carbon 14 and the Prehistory of Europe," by Colin Renfrew in the October, 1971, issue of *Scientific American* for some surprising results of slight inaccuracies in the carbon 14 technique!

10. ***Biological Half-Life Problem.*** You accidentally inhale some mildly poisonous fumes. Twenty hours later you see a doctor. From a blood sample, she measures a poison concentration of 0.00372 milligrams per cubic centimeter (mg/cc), and tells you to come back in 8 hours. On the second visit, she measures a concentration of 0.00219 mg/cc. Let t be the number of hours that have elapsed since your first visit to the doctor, and let C be the concentration of poison in your blood, in mg/cc. Assume that C varies exponentially with t.

a. Write the particular equation for this function.

b. The doctor says you might have had serious body damage if the poison concentration was ever as high as 0.015 mg/cc. Based on your mathematical model, was the concentration ever that high? Justify your answer.

c. You can resume normal activities when the poison concentration has dropped to 0.00010 mg/cc. How long after you inhaled the fumes will you be able to resume normal activities?

d. The "biological half-life" of the poison is the length of time it takes for the concentration to drop to *half* of its present value. Find the biological half-life of this poison.

e. Plot a graph of C versus t from the time you breathed the fumes until the time it is safe to resume normal activities. If you are clever, you can think of a way to get *many* plotting points *quickly*.

Problems 11 through 19 involve exponential functions, but each problem uses them in a slightly different way from that shown in the example for this section. By working these problems, you will demonstrate your ingenuity at applying familiar concepts to unfamiliar problems.

11. ***Coffee Cup Problem.*** After you pour a cup of coffee, it cools off in such a way that the *difference* between the coffee temperature and the room temperature decreases exponentially with time. This model is called "Newton's Law of Cooling." Suppose that you pour a cup of coffee. Three minutes later, you measure its temperature and find that it is 85°C. Five minutes after the first reading, you find that it has cooled to 72°C. The room is at 20°C.

Let c = number of degrees of coffee temperature.
Let D = number of degrees *difference* (coffee minus room).
Let t = number of minutes since first temperature reading.

a. Show that you understand the definitions of D and t by writing the given information as two ordered pairs, (t, D).
b. Write the particular equation expressing D in terms of t.
c. What was t when the coffee was first poured? Substitute this value of t into the equation to find D when the coffee was first poured. Based on your answer, what was the temperature, c, of the coffee when it was first poured?
d. Assume that coffee is "drinkable" when its temperature is at least 55°C. For how long *after it was poured* will the coffee be drinkable?
e. Using the properties of exponential functions, predict the temperature, c, of the coffee for every 5 minutes from $t = 10$ through $t = 60$. You may wish to write a short computer program to do the arithmetic. Don't forget to add room temperature to the values of D you calculate!
f. Plot a graph of the coffee temperature, c, from the time the coffee was poured until $t = 60$. Draw a dotted line at the appropriate place to show the asymptote. If you worked Problem 1.j. in Exercise 2-2, see how well your first thoughts about this real-world situation agree with the exponential model of the present problem.

12. ***Calvin's Mass Problem.*** Calvin Butterball consumes 8000 calories per day, and has a mass of 150 kg. He reads in a health book that he could maintain a more normal mass of 90 kg by reducing his consumption to 2000 calories per day. At time $t = 0$ he starts the diet. When $t = 20$ days, he is down to 141 kg. Assume that the *difference* between his mass and 90 kg decreases exponentially with t.

a. Write the particular equation expressing the difference between Calvin's mass and 90 kg in terms of t.
b. Predict the number of kilograms by which Calvin is above 90 kg for times of 40, 60, 80, and 100 days.
c. Calculate Calvin's mass at each of these times.
d. Plot the graph of Calvin's mass versus time. Show what the graph looks like for values of t less than zero.
e. When should Calvin be 100 kg?
f. According to your model, will Calvin ever actually reach 90 kg? Justify your answer. What action must he take in order to reach 90 kg?
g. If you worked Problem 2.g. in Exercise 2-2, go back and look at your graph to see how well your earlier thoughts agree with this exponential model.

13. ***Car Acceleration Problem.*** Your car is standing still on a straight, level stretch of highway. At time $t = 0$ seconds, you floorboard the gas pedal. After 5 seconds you are going 40 kilometers per hour (kph). Let D be the *difference* between your car's speed and its top speed of 160 kph. Assume that D decreases exponentially with t.

a. Write the particular equation expressing D in terms of t.
b. How fast will you be going 18 seconds after you floorboarded the gas pedal?
c. How long will it take you to reach 150 kph?

d. Plot the graph of *actual* speed (*not D*) versus t. Draw a dotted line at the appropriate place to show the asymptote.
e. If you worked Problem 1.c. of Exercise 2-2, go back and look at your graph to see how well your earlier thoughts agree with this exponential model.
f. Show that the equation for the actual speed, S, has the form

$$S = a(1 - 10^{kt}),$$

where a and k are constants, and S and t are speed and time, respectively.

14. ***Advertising Problem.*** A well-known soft drink company comes out with a new product, Ms. Phizz. Based upon market analysis, they figure that if all potential users of this product were to buy it, they could sell \$300,000 per day worth of the beverage. However, the *actual* sales depend on how much per day they spend on advertising. With *no* advertising, they figure that they will sell *no* Ms. Phizz. With \$40,000 per day spent on advertising, they expect sales of \$100,000 per day. Assume that the *difference* between \$300,000 per day and *actual* sales decreases exponentially with the amount spent per day on advertising.

a. Write the particular equation expressing the *difference* between \$300,000 per day and the actual sales in terms of the amount spent per day on advertizing. It may be more convenient to express both amounts in *thousands* of dollars.
b. Calculate the difference between \$300,000 per day and actual sales if 80, 120, 160, and 200 thousand dollars per day are spent on advertising. Then use the answers to calculate the actual sales for these amounts spent on advertising.
c. Plot the graph of actual sales versus amount spent on advertizing. Draw a dotted line at the appropriate place to indicate the asymptote.
d. The *profit* made on the product is the actual sales *minus* the amount spent on advertising. Use the answers from part b to predict the profit per day if amounts of 0, 40, 80, 120, 160, and 200 thousand dollars per day are spent on advertising.
e. Plot a graph of profit versus amount spent on advertizing. Use the Cartesian coordinate system of part c.
f. What number of dollars per day seems to produce the *maximum* profit? (When you study calculus, you will learn how to *calculate* this maximum value from the equation, *without* having to plot the graph!)
g. What do you suppose is meant by, "The Law of Diminishing Returns?"
h. Show that the equation for actual sales, part b, has the form

$$S = a(1 - 10^{kx}),$$

where a and k are constants, and S and x are the actual sales and the advertizing amounts, respectively.

15. ***Sunlight Below the Water Problem.*** The intensity of sunlight reaching points below the surface of the ocean varies exponentially with the depth of the point below the surface of the water. Suppose that when the intensity at the surface is 1000 units, the intensity at a depth of 2 meters is 60 units.

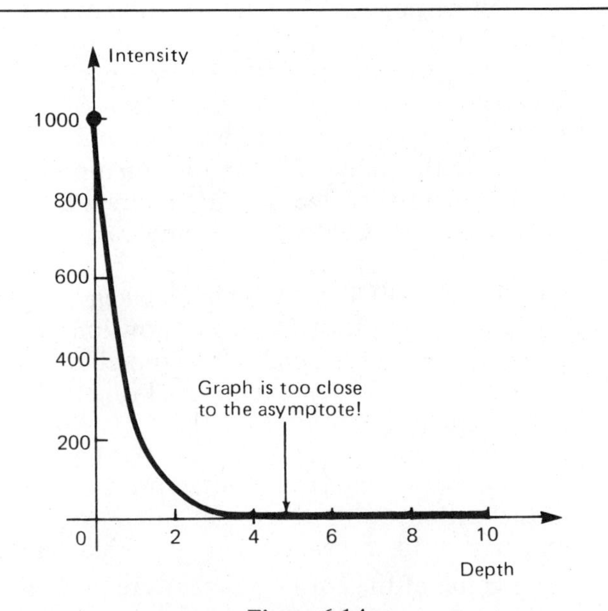

Figure 6-14c.

a. Write the particular equation expressing intensity in terms of depth.
b. Predict the intensity at depths of 4, 6, 8, and 10 meters.
c. Plants cannot grow beneath the surface if the intensity of sunlight is below 0.001 unit. What is the maximum depth at which plants will grow?
d. The intensity of sunlight drops so fast that throughout most of the domain, the graph cannot be distinguished from the asymptote (see Figure 6-14c). Using the given ordered pairs and the results of part b, find the *logarithm* of the intensity for each 2 meters from 0 through 10 meters. Then plot a graph of the *log* of intensity versus depth.
e. What interesting property does the graph from part d seem to have? *Prove* that it has this property.

16. ***Radio Dial Problem.*** You have probably noticed that the distances between markings on most radio dials are *not* uniform. For example, the distance between frequencies of 53 and 60 kilohertz is not much different from the distance between 140 and 160 kilohertz. Assume that the frequency marked on the dial varies exponentially with the distance from the left end of the dial. In this problem you are to figure out how to mark a dial that is to be 12 cm long.

a. The dial is to be 12 cm long, and the lowest and highest frequencies are to be 53 and 160 kilohertz, respectively. Write these pieces of information as ordered pairs, (distance, frequency).
b. Write the particular equation expressing frequency in terms of distance.
c. Transform the equation in part b so that *distance* is expressed in terms of *frequency*. This form of the equation will be more convenient to use when you calculate the distances corresponding to various frequencies.
d. By looking at your equation in part c, think up some appropriate words to describe how *distance* varies with *frequency*.
e. Calculate to the nearest tenth of a millimeter the distances for frequencies of 60, 70, 80, 100, 120, and 140 kilohertz, the frequencies that often appear on radio dials. Since a considerable amount of computation is required, you may want to use a calculator or to write a short computer program. If your computer has only base *e* logarithms, you will need to use the techniques of Section 6-12 (especially Problem 34) to transform to base 10 logs.

f. Use the given points and the results of part e to plot a graph of distance (as the *ordinate*) versus frequency (as the *abscissa*).
g. Use a ruler to make a scale drawing of the radio dial with the frequencies marked in. It should look somewhat like that in the photograph.

17. ***Deep Oil Well Cost Problem.*** Suppose that you are a mathematician for Wells Oil Production, Inc. Your company is planning to drill a well 50,000 feet deep, deeper than anyone has ever drilled before. Your part of the project is to predict the cost of drilling the well.

From previous well records, you ascertain that the price is \$20 per foot for drilling at the surface, and \$30 per foot for drilling at 10,000 feet. Assume that the number of dollars per foot for drilling an oil well increases exponentially with the depth at which the drill is operating.

a. Write the particular equation expressing price per foot in terms of depth. You might find it more convenient to express depth in *thousands* of feet.
b. Predict the price per foot for drilling at depths of 20, 30, 40, and 50 thousand feet.
c. *Carefully* plot the graph of price per foot versus depth. Choose scales that make the graph occupy most of the piece of graph paper.
d. The *area* of the region under the graph (see Figure 6-14d) represents the *total* number of dollars it costs to drill the well. You can see a reason by considering the *units* of this area. The vertical distance is \$/ft, and the horizontal distance is ft., so the area has units

$$\frac{\text{dollars}}{\text{foot}} \times \text{feet},$$

or simply *dollars*. Count the *number of squares* in this region on *your* graph. Estimate fractional squares to the nearest 0.1 unit (see Figure 6-14d). (When you study calculus, you will learn how to *calculate* such areas from the equation, without having to count squares.)
e. Calculate the number of dollars corresponding to each square. For example, if the horizontal spacing is 2000 feet, and the vertical spacing is 5 \$/ft, as in the right-hand sketch of Figure 6-14d, then each square corresponds to

$$(2000)(5) = 10{,}000 \text{ dollars.}$$

f. Calculate the total cost of drilling the well. (Expensive, isn't it!?)

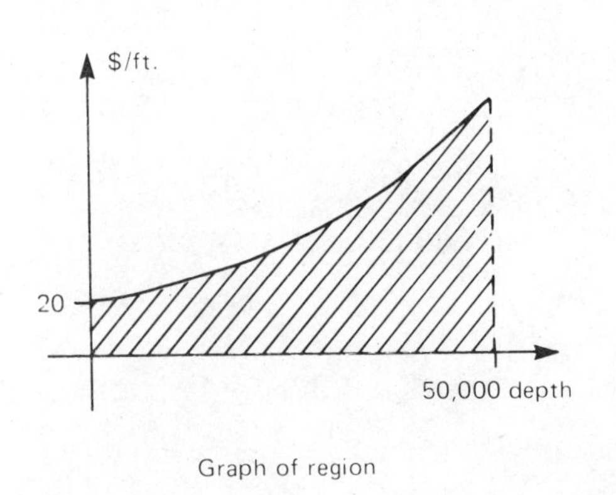

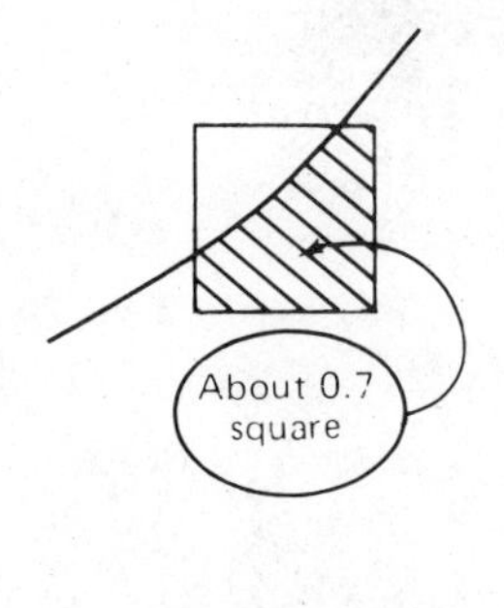

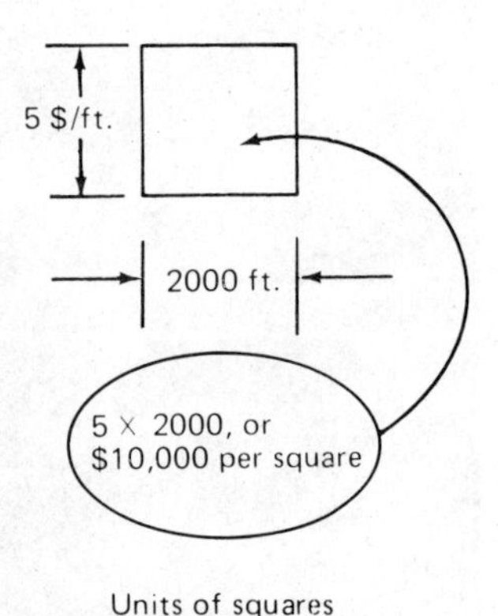

Figure 6-14d.

18. ***Vapor Pressure Problem.*** The vapor pressure of water is the pressure that water vapor (i.e., steam) would exert if it, alone, occupied the space above the water in a closed container (Figure 6-14e). About 200 years ago the French scientists Clausius and Clapeyron found that the vapor pressure varies exponentially with the *reciprocal* of the absolute temperature of the water. That is, if P is the vapor pressure and T is the absolute temperature, then

$$P = a \times 10^{k(1/T)}.$$

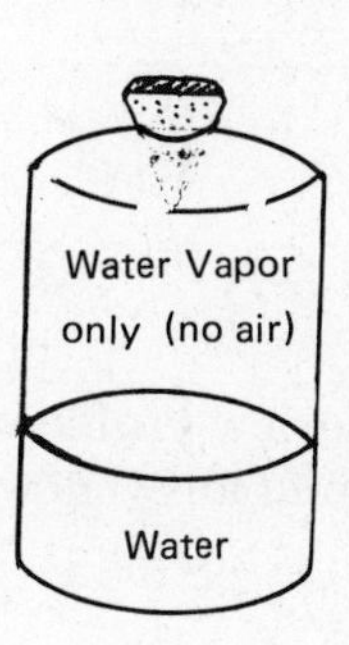

Figure 6-14e.

a. Find the particular equation expressing vapor pressure in terms of absolute temperature if water at 0°C has a vapor pressure of 4.6 millimeters of mercury (mm), and at 100°C, has 760 mm. (The absolute temperature is 273 plus the Celsius temperature.) You must be very clever to figure a way to evaluate the two constants!

b. Use your equation to predict the vapor pressure of water on a hot summer day, 40°C.

c. Water boils when its vapor pressure reaches the pressure of its surrounding atmosphere. At what Celsius temperature would water boil

   i. atop Mt. Everest, where the air is at 220 mm?

   ii. in the reactor of a nuclear power plant, where the pressure is kept at 100,000 mm?

   iii. in the deepest part of the ocean, where the pressure is 800,000 mm?

19. ***Radioactive Brain Tracer Problem.*** Technetium 99m (pronounced "tek-nee´-si-um") is a radioactive isotope used to trace the activity of certain functions in the brain. A small quantity of the isotope is injected. Then the level of radioactivity is measured at various times during the next few hours to see how much technetium remains in the brain. The amount remaining in the brain decreases for two reasons:

> *Physical:* Technetium 99m has a half-life of 6 hours, which means that at the end of any 6-hour period, the amount remaining is only half what it was at the beginning of that period, even if none is eliminated from the brain.
>
> *Biological:* The brain eliminates technetium in such a way that even if it were not radioactive, the amount remaining would decrease exponentially with time.

It is the *biological* half-life that doctors seek to measure. As you work the following parts of the problem you will see how this measurement can be accomplished.

a. Let P be the fraction of technetium that would remain after t hours due to physical (radioactive) decay alone, and let B be the fraction that would be left due to biological activity alone. The actual fraction, F, that remains after t hours is

$$F = PB.$$

Prove that if P and B both vary exponentially with time, then F varies exponentially with time also.

b. A patient is injected with some technetium 99m. Two hours later, only 71.3% ($F = 0.713$) remains. Use this and the fact that $F = 1$ when $t = 0$ to find the particular equation expressing F in terms of t.

c. Use the 6-hour half-life to find the exponential constant in the equation which expresses P in terms of t. Then use the result in an appropriate manner to find the particular equation for B in terms of t.

d. Calculate the *biological* half-life of technetium for this particular patient's brain. You can do this by letting $B = 0.5$ and solving for t.

6-15 CHAPTER REVIEW AND TEST

In this chapter you have learned new concepts such as logarithms, properties of exponentiation, and inverses of functions. These concepts have allowed you to analyze a new type of function, the exponential function. As you study mathematics, it is the *theoretical concepts* that are most important. With a firm grasp of the theory, you can apply your knowledge to *unfamiliar* situations. The review problems that follow are designed to test your knowledge of the new theoretical concepts under controled conditions. The Concepts Test allows you to apply these concepts to the analysis of *another* new type of function. By so doing, you will demonstrate that you have *mastered* the concepts.

The objectives for exponential functions listed throughout the chapter are summarized below:

1. *Be able to operate with powers having negative or non-integer exponents.*
2. *Be able to raise any positive number to any power.*
3. *Given the equation for the function, draw the graph.*
4. *Given points on the graph of the function, find the particular equation.*
5. *Use exponential and related functions as mathematical models.*

Review Problems

The following problems are numbered according to the objectives stated above.

Review Problem 1a. State the following properties, using variables:

i. Power of a power.
ii. Quotient of two powers with the same base.
iii. Exponentiation distributes over multiplication.

b. Evaluate, getting *exact* answers (*no* decimal approximations):

i. -3^2.
ii. 3^{-2}.
iii. 3^0.

c. Simplify:

i. $4x^{-3} \cdot (5x^{-4})^2$.

ii. $\left(\dfrac{12x^4y^{-2}}{4x^5y^2}\right)^{-3}$.

iii. $9x^{-1/2} \cdot 16x^{1/4} \cdot 5x^0$.

Review Problem 2. Evaluate the following, getting *exact* answers where possible:

a. $128^{4/7}$,

b. $\sqrt[3]{512}$,

c. $32^{0.41}$,

d. $0.41^{3.2}$,

e. $\sqrt[5]{1{,}924{,}000{,}000}$.

Review Problem 3. Sketch the graphs of the following exponential functions:

a. $f(x) = 12 \times 2^x$, for $-3 \le x \le 2$.

b. $f(0) = 100$ and $f(5) = 70$, for $0 \le x \le 20$.

Review Problem 4. Find the particular equation for the exponential function described in Problem 3b.

Review Problem 5 – Fog. In a fog, the intensity of light reaching you from the tail lights of the car in front of you varies exponentially with the distance between you and the other car. Suppose that the intensity, I, is given by the equation

$$I = 128 \times 10^{-0.007d},$$

where d is the number of meters between you and the other car.

a. What is the intensity of the tail lights when he is "right at" your car?

b. Predict the intensity when the car is 30 meters in front of you.

c. How far in front of you must the car go to reduce the intensity to $1/2$ its value at $d = 0$? $1/4$ of its value? $1/8$ of its value? $1/16$ of its value?

d. Draw the graph of I versus d.

e. Based on this mathematical model, why do oncoming cars seem to appear out of the fog so *suddenly*?

Concepts Test

For each part of each problem, tell which one(s) of the five objectives of this chapter you used in working that part.

Suppose that f is a function with a general equation of the form

$$f(x) = 21.6x^n,$$

where n is a *constant*, but *not necessarily* an integer. Answer the following questions. For each part of each question, tell by number which one or ones of the five objectives of this chapter you used.

Concepts Test 1. Suppose that $n = 0$. Write an equation for $f(x)$ in its simplest form.

Concepts Test 2. Suppose that $n = 2/3$. Use the definition of fractional exponents to write an equation for $f(x)$ in radical form.

Concepts Test 3. Suppose that $n = -2$.

a. Write the equation for $f(x)$ *without* negative exponents.

b. Find $f(1)$, $f(2)$, $f(3)$, and $f(10)$.

c. What would $f(0)$ equal? Explain.

d. Use the information from parts b and c to draw the graph of f for positive values of x.

e. What relationship does the x-axis have to the graph of f?

f. For *exponential* functions, *adding* 1 to x would cause $f(x)$ to be *multipled* by a *constant* factor. Does this property seem to be true for functions of the form

$$f(x) = 21.6x^n?$$

Justify your answer.

Concepts Test 4. Suppose that $n = 1.4$. Find decimal approximations for

a. $f(56.17)$,

b. $f(0.007)$,

c. x, if $f(x) = 1776$.

Concepts Test 5. Let $f = \{(x, y): y = 21.6x^2\}$.

a. Write the equation for the *inverse* function, f^{-1}. Transform the equation for f^{-1} so that y is expressed in terms of x.

b. Find $f^{-1}(2.4)$.

Concepts Test 6. Suppose that f is a mathematical model of how the mass of a person is related to his or her height. Let x be the number of meters tall, and let $f(x)$ be the number of kg.

a. If a person 1.52 meters tall has a mass of 63 kg, find the particular equation for the function.

b. Predict the mass of a basketball player 2.34 meters tall.

7 Rational Algebraic Functions

In this chapter you will study functions in which there is ***division*** *by the independent variable. The most important mathematical concept is what happens when the variable denominator is close to zero. To investigate this concept, you must refresh your memory about operations with fractions. These operations require you to know various factoring techniques. The mathematical models you will use are called* ***variation*** *functions which relate variables such as the force needed to loosen a bolt and the length of the wrench handle.*

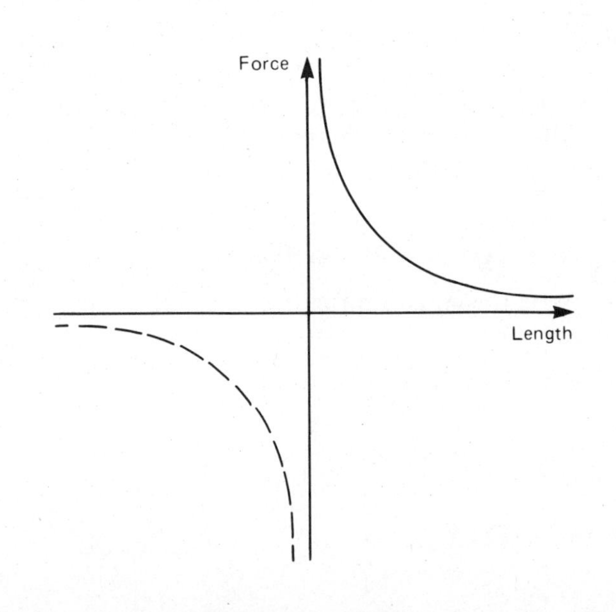

7-1 INTRODUCTION TO RATIONAL ALGEBRAIC FUNCTIONS

You recall that a rational *number* is one that can be written as a ratio of two *integers*. A rational *algebraic expression* is defined by simply changing a few words in this definition.

DEFINITION

A ***rational algebraic expression*** is an expression that can be written as a ratio of two *polynomials.*

Some examples of rational algebraic expressions are

$$\frac{3x+2}{x^2-7x+11},\quad x^2+3x-5$$

(which equals $\frac{x^2+3x-5}{1}$), etc.

The following are *not* called rational algebraic expressions since they contain radicals or non-algebraic operations:

$$\frac{\sqrt{x+3}}{\log x},\quad \frac{2x}{|x-1|},\ \text{etc.}$$

Rational algebraic *functions* are defined by an *equation*, just as other types of functions have been.

DEFINITION

A ***rational algebraic function*** has a general equation of the form

$$f(x)=\frac{P(x)}{Q(x)},$$

where $P(x)$ and $Q(x)$ stand for *polynomials.*

The most interesting thing about rational algebraic functions is what happens to the graphs when you pick a value of x that makes the denominator equal to zero!

Objective:

Discover what the graph of a rational function looks like.

Exercise 7-1

Suppose that

$$f(x)=\frac{x+2}{x^2-x-6}.$$

1. Find $f(0), f(1), f(2), f(4), f(5), f(-1)$, and $f(-3)$.

2. Try to find $f(3)$ and $f(-2)$. What do you notice?

3. What numbers must be *excluded* from the domain of f?

4. Use the points you have already calculated, and any others you think would be helpful, to plot a graph of f. You might try some points close to $x = 3$, such as $f(3.1)$ and $f(2.9)$, or close to $x = -2$, such as $f(-2.1)$ and $f(-1.9)$.

5. Based on your graph, what *two* things might happen to the graph of a rational algebraic function at points excluded from its domain?

7-2 GRAPHS OF RATIONAL FUNCTIONS

In Exercise 7-1 you plotted a graph of the function

$$f(x)=\frac{x+2}{x^2-x-6}.$$

The job of substituting values for x was somewhat tedious, particularly for decimal values of x. If you recall how to factor polynomials and simplify fractions from previous mathematics courses, the job can be made much easier. The denominator will factor, as follows:

$$f(x) = \frac{x + 2}{(x + 2)(x - 3)}.$$

The numerator and denominator now have $(x + 2)$ as a common factor. As long as $x \neq -2$, this factor can be "canceled" as follows:

$$f(x) = \frac{(x + 2) \cdot 1}{(x + 2)(x - 3)}$$ Multiplicative identity.

$$f(x) = \frac{x + 2}{x + 2} \cdot \frac{1}{x - 3}$$ Multiplication property of fractions.

$$f(x) = \frac{1}{x - 3} \text{ and } x \neq -2$$ Multiplicative identity.

Note that substituting values of x into $1/(x - 3)$ is much easier than substituting them into the original expression for $f(x)$. Note also that the statement $x \neq -2$ is necessary in the last step because $1/(x - 3)$ *is* defined when $x = -2$, but $f(x)$ is *not*.

The graph of f is shown in Figure 7-2. It looks exactly like the graph of $y = 1/(x - 3)$ except that the point at $x = -2$ is deleted, leaving a "hole" in the graph. When x is close to 3, the denominator is close to zero. For example,

$$f(3.01) = \frac{1}{3.01 - 3} = 100.$$

The closer x gets to 3, the larger $f(x)$ gets. Since x can never *equal* 3, there is a vertical asymptote at $x = 3$.

Conclusion: At values of x excluded from the domain, there may be either an *asymptote* or a *hole* in the graph.

In order to locate values of x which are excluded from the domain, and to tell whether there is an asymptote or just a hole, you must be able to factor polynomials and simplify rational expressions. The next few sections are designed to refresh your memory on these operational techniques. In Section 7-9 you will pick up the plotting of graphs of rational functions again.

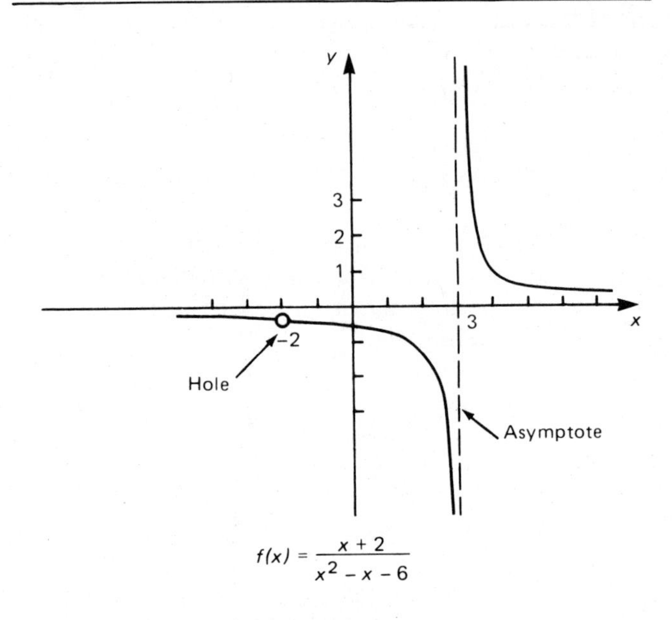

Figure 7-2.

7-3 SPECIAL PRODUCTS

In the previous section, you saw that the job of plotting rational function graphs is much simpler if you can first simplify the rational expression which $f(x)$ equals. To simplify an expression such as

$$\frac{x + 2}{x^2 - x - 6},$$

you must be able to *factor* an expression like $x^2 - x - 6$ into a *product* of two expressions, such as $(x - 3)(x + 2)$. To be able to do this, you should first be good at going in the other direction, multiplying two polynomials together.

Objectives:

1. Given two polynomials, be able to *multiply* them in *one* step, in your head.

2. Given a binomial, be able to *square* it in *one* step, in your head.

Example 1: *Product of Two Binomials* – You will recall from Chapter 1 that two binomials may be multiplied by multiplying each term of one by each term of the other. For example,

$$\begin{aligned}(2x-5)(3x+7) &= (2x)(3x) + (2x)(7) \\ &\quad + (-5)(3x) + (-5)(7) \\ &= 6x^2 + 14x - 15x - 35 \\ &= 6x^2 - x - 35.\end{aligned}$$

Once you understand the procedure, you can shorten the process. The acronym "FOIL" may help you remember to multiply the:

*F*irst terms, *O*utside terms, *I*nside terms, *L*ast terms.

Adding the two x-terms together in your head allows you to write the answer down in *one* step,

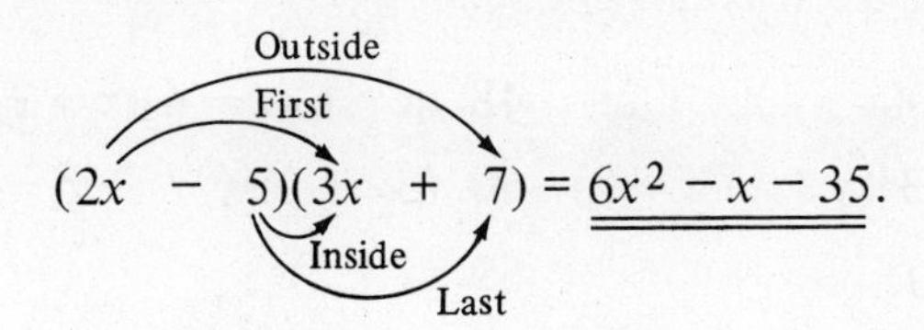

Example 2: *Square of a Binomial* – You may square a binomial by multiplying it by itself, using the FOIL technique. For example,

$$\begin{aligned}(3x-7)^2 &= (3x-7)(3x-7) \\ &= \underline{\underline{9x^2 - 42x + 49}}.\end{aligned}$$

However, there is a quicker way. The middle term, $-42x$, comes from adding $-21x$ and $-21x$, which is equivalent to *doubling* $-21x$. And the $-21x$ is equal to the *product* of the two terms in the binomial, $(3x)(-7)$. So the following *one-step* pattern emerges for squaring a binomial:

a. The first and last terms of the product are the *squares* of the two terms in the binomial.
b. The middle term in the product is *twice* the *product* of the two terms in the binomial.

Note that unwary algebra students sometimes think that squaring distributes over adding and subtracting, and that $(3x-7)^2$ is equal to $9x^2 - 49$. If you are ever in doubt, just write down the polynomial twice, and multiply by the FOIL technique.

Example 3: *Product of Conjugate Binomials* – Binomials of the form

$$a + b \quad \text{and} \quad a - b,$$

which have the *same* terms but an *opposite* sign between the terms, are called *conjugate* binomials. Multiplying two conjugate binomials is very easy because the middle term equals zero. Using the FOIL technique, you find that

$$\begin{aligned}(a+b)(a-b) &= a^2 - ab + ab - b^2 \\ &= a^2 - b^2.\end{aligned}$$

The answer, $a^2 - b^2$, is called a *difference of two squares,* for obvious reasons. If two conjugates are to be multiplied, you can simply *square* the two terms, then *subtract*. For example,

$$(3x - 5y)(3x + 5y) = \underline{\underline{9x^2 - 25y^2}}.$$

Example 4: *Product of Two Polynomials* – An extension of the FOIL technique allows you to multiply in your head polynomials with *several* terms. For example, to multiply

$$(x^3 - 2x^2 + 5x - 7)(x^2 + 4x - 3),$$

you would first multiply x^3 by x^2, getting x^5. Then you would multiply all terms that would give x^4, namely $(x^3)(4x)$ and $(-2x^2)(x^2)$, adding the results in your head to get $2x^4$. The secret is working for a term of *one* given degree at a time. Figure 7-3 shows how you can get the x^2-term by putting your fingers at position ①, multiplying, and getting $6x^2$.

Then you move them to position ②, getting $20x^2$, which you add to the $6x^2$ to get $26x^2$. At position ③ you get $-7x^2$, which, added to the $26x^2$, gives a total of $19x^2$.

$(x^3 - 2x^2 + 5x - 7)(x^2 + 4x - 3) = \ldots + 19x^2 + \ldots$

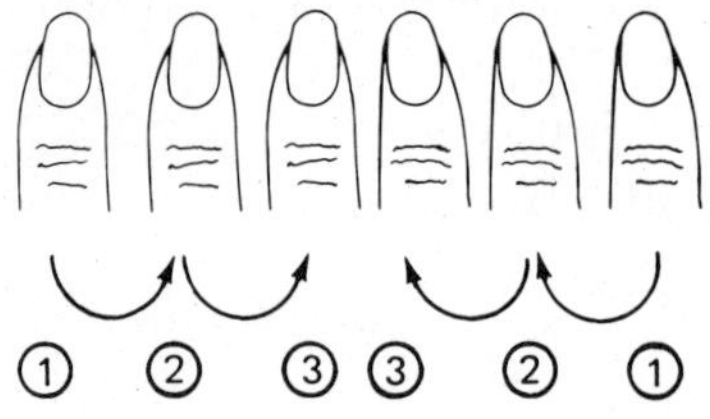

Mental Multiplication of Polynomials

Figure 7-3

Proceeding in the same manner with terms of descending degree gives the complete product,

$$x^5 + 2x^4 - 6x^3 + 19x^2 - 43x + 21.$$

The following exercise is designed to give you practice in multiplying polynomials in your head.

Exercise 7-3

For Problems 1 through 54, multiply the polynomials.

1. $(x + 3)(x - 5)$
2. $(x - 2)(x + 7)$
3. $(x + 6)(x + 1)$
4. $(x + 4)(x + 11)$
5. $(x - 2)(x - 7)$
6. $(x - 9)(x - 1)$
7. $(x - 2)(x + 2)$
8. $(x + 3)(x - 3)$
9. $(2x + 3)(x - 5)$
10. $(3x + 5)(x - 4)$
11. $(x - 4y)(3x - 7y)$
12. $(x - 6y)(2x - 7y)$
13. $(4x + 1)(2x + 5)$
14. $(5x + 1)(3x + 4)$
15. $(7x - 2)(3x + 8)$
16. $(9x - 5)(2x + 7)$
17. $(13x - 2y)(13x + 2y)$
18. $(11x - 3y)(11x + 3y)$
19. $(13x - 2y)(13x - 2y)$
20. $(11x - 3y)(11x - 3y)$
21. $(x + 3)^2$
22. $(x + 4)^2$
23. $(2x - 5)^2$
24. $(3x - 2)^2$
25. $(4x + 7)^2$
26. $(5x - 1)^2$
27. $(11x - 1)^2$
28. $(2x + 7)^2$
29. $(4 - 5x)^2$
30. $(2 - 9x)^2$
31. $(-3x - 5)^2$
32. $(-4x - 3)^2$
33. $(x^2 + 3x - 4)(x + 2)$
34. $(x^2 - 2x + 5)(x - 3)$
35. $(x - 11)(x^2 - 2x + 8)$
36. $(x + 12)(x^2 - 3x - 2)$
37. $(x - y)(x^2 + xy + y^2)$
38. $(x + y)(x^2 - xy + y^2)$
39. $(x^2 - 5x + 7)(2x - 1)$
40. $(x^2 + 2x - 6)(3x - 1)$
41. $(x^2 + 4x - 7)(x^2 - 2x + 5)$
42. $(x^2 - 6x + 8)(x^2 - 7x + 1)$
43. $(x^3 + 5x^2 - 3x - 2)(x^2 - 2x + 1)$
44. $(x^3 - 3x^2 - 4x + 6)(x^2 + 2x + 5)$
45. $(x^2 + 3x - 5)(x^3 + 4x^2 - 3x + 2)$
46. $(x^2 - x + 2)(x^3 + 3x^2 - 5x + 4)$
47. $(x + 5)(x - 1)(x + 2)$
48. $(x - 6)(x - 4)(x + 3)$
49. $(2x - 3)(x + 1)(x - 2)$
50. $(3x + 2)(x - 1)(x - 4)$
51. $(x - 1)^3$
52. $(x + 2)^3$
53. $(x + y)^3$
54. $(x - y)^3$

55. *Mental Multiplication of Two-Digit Numbers* – A two-digit number such as 57 is a compact form of the binomial

$$5(10) + 7.$$

Consequently, you can apply the FOIL technique and multiply two-digit numbers in your *head.* For example, to multiply 57 by 34, your thought process would be

57 × 34 (7 × 4 = 28)	8	7 × 4 = 28. Write down "8" and carry "2."
57 × 34 (5 × 4 = 20, 7 × 3 = 21)	38	5 × 4 = 20. Add the "2" that you carried, getting 22. 7 × 3 = 21, and 21 added to 22 is 43. Write down "3" and carry "4."
57 × 34 (5 × 3 = 15)	= 1938	5 × 3 = 15. Add the "4" that you carried, getting 19.

Multiply the following numbers using the above technique.

a. 23 × 87 b. 37 × 48 c. 56 × 91 d. 89 × 67

56. *Mental Multiplication of Larger Numbers* – The FOIL technique of Problem 55 can be extended to products of numbers with *more* than two digits, just as it was extended to products of polynomials with more than two terms. The thing to remember is that you must work on *one* digit of the answer at a time, starting with the *units* digit. Multiply the following numbers in your head, writing the answer only.

a. 27 × 518 b. 365 × 24
c. 623 × 451 d. 409 × 489

7-4 FACTORING SPECIAL KINDS OF POLYNOMIALS

Factoring a polynomial means writing it as a *product* of two or more other polynomials. There is one simple rule, *termwise multiplication*, by which you can multiply together any two polynomials. Unfortunately, there is no single rule for factoring a polynomial back apart again. In fact, some polynomials are "prime," meaning that they will *not* factor. There are some special cases in which you can recognize a polynomial as the answer to a multiplication problem and thus go backwards to find the factors. In this section your objective is to become proficient at factoring these special kinds of polynomials. In Section 7-6 you will learn a more general technique using the Factor Theorem.

Objective:

Be able to factor special kinds of polynomials that

a. have common factors,
b. have terms that can be grouped,
c. are quadratic trinomials,
d. are differences of two squares.

Example 1: *Common Factors* – The first step in factoring any polynomial is using the distributive property to factor out any factors which are common to all terms of the polynomial. For example, in $4x^5 + 12x^4 - 8x^3$, each term has $4x^3$ as a common factor. Using the distributive property (backwards) you can get

$$4x^5 + 12x^4 - 8x^3 = \underline{\underline{4x^3(x^2 + 3x - 2)}}.$$

It is always possible to factor out *something*, even if it is not a common factor. You can factor various numbers out of $x + y$ and get

$$x + y = x\left(1 + \frac{y}{x}\right)$$

$$x + y = xy\left(\frac{1}{y} + \frac{1}{x}\right)$$

$$x + y = 4\left(\frac{1}{4}x + \frac{1}{4}y\right).$$

This is sometimes called "factoring out a rabbit," because you are pulling out something that was not there to begin with. Factoring out a rabbit is sometimes useful in proving trigonometric identities (Chapter 10). For the time being, however, it is helpful to make an agreement which *avoids* this.

Agreement: A polynomial is *completely factored* when it is written as a *product* of *polynomials* with *integer* coefficients.

Example 2: *Factoring by Grouping* – Sometimes a clever use of the associative property allows you to find common *binomial* factors. For example, the first two and the last two terms in

$$12x^2 + 8ax - 15x - 10a$$

can be associated. Factoring $4x$ from the first two and -5 from the last two gives

$$4x(3x + 2a) - 5(3x + 2a).$$

Now $(3x + 2a)$ is a common factor, and can be factored out, giving

$$\underline{\underline{(3x + 2a)(4x - 5)}}.$$

Example 3: *Quadratic Trinomials* – A quadratic trinomial can either be prime or it can be the product of two linear binomials. For example,

$$x^2 + 4x - 12 = (x - 2)(x + 6),$$

as you can easily see by multiplying $(x - 2)(x + 6)$. However, a small change can produce a polynomial that does not factor. For example, $x^2 + 4x - 11$ has no linear factors with integer coefficients.

To factor $x^2 + 4x - 12$ you must realize that the first term in each factor will be x, so that the product will be x^2. So you write

$$x^2 + 4x - 12 = (x \qquad)(x \qquad),$$

leaving blanks into which to put the second terms. These second terms must be two factors of -12. There are six choices: 1 and -12, -1 and 12, 2 and -6, -2 and 6, 3 and -4, or -3 and 4. In order for the middle term to be $4x$ when you multiply the factors back together, the *sum* of the two factors of -12 must equal $+4$. The only pair which works is -2 and 6. So you fill these two numbers into the blanks you have left and get

$$x^2 + 4x - 12 = \underline{\underline{(x - 2)(x + 6)}}.$$

If the x^2 coefficient is not equal to 1, the factoring process is a bit harder. For example, to factor

$$6x^2 - 7x - 20,$$

the first terms in the factors could be

$$(2x \qquad)(3x \qquad) \text{ or } (x \qquad)(6x \qquad).$$

The second terms must be factors of -20. But these could be ± 1 and ∓ 20, ± 2 and ∓ 10, or ± 4 and ∓ 5. There are 24 possible ways these numbers can be combined! After considerable trial and error, you will find that the combination which gives $-7x$ as the middle term is

$$6x^2 - 7x - 20 = \underline{\underline{(2x - 5)(3x + 4)}}.$$

Students sometimes call this "factoring by groping!"

Example 4: *Difference of Two Squares* – To factor $x^2 - 9$, you seek two factors of -9 that add up to 0. These factors are 3 and -3. So

$$x^2 - 9 = \underline{\underline{(x + 3)(x - 3)}}.$$

Such a quadratic *bi*nomial will factor if both terms are perfect squares, and the sign between them is "–." Such an expression is called a *difference of two squares.* The factors of a difference of two squares will both be the same except for the sign between the terms. Binomials such as $x + 3$ and $x - 3$ are called *conjugate* binomials.

DEFINITION

Two binomials are ***conjugates*** of each other if they are identical except for the signs between the terms.

To factor a difference of two squares quickly, it is necessary only to find the appropriate conjugate binomials. For example,

$$4x^6 - 25 = \underline{\underline{(2x^3 + 5)(2x^3 - 5)}}.$$

Example 5: *Factoring Quadratics by Splitting the Middle Term* – There is a way to organize the trial-and-error search for factors of a quadratic trinomial. The system has the added advantage that it indicates when the trinomial is prime. To see how it works, it is helpful to go backwards, and multiply two linear binomials together.

$$(2x-5)(3x+4) = (2x-5)(3x) + (2x-5)(4) \quad ①$$
$$= 6x^2 - 15x + 8x - 20 \quad ②$$
$$= 6x^2 - 7x - 20 \quad ③$$

The factoring operation consists of *reversing* these steps. You should recognize that from Step ② back, this is a factoring-by-grouping problem. The only difficulty is seeing how to split up the $-7x$ so that you can go from Step ③ back to Step ②. It helps to put a step *between* the first and second lines, in which the multiplication is indicated, but *not* carried out.

$$(2x-5)(3x) + (2x-5)(4)$$
$$= (2\cdot 3)x^2 + (-5\cdot 3)x + (2\cdot 4)x + (-5\cdot 4)$$
$$= \quad 6x^2 \quad - \quad 15x \quad + \quad 8x \quad - \quad 20.$$

The -15 and 8 contain all four coefficients of the binomials, -5, 3, 2, and 4, as factors. But the 6 and -20 *also* contain the -5, 3, 2, and 4 as factors. In other words, $-15 \cdot 8$ will equal $6 \cdot (-20)$, which equals -120. So to split the $-7x$ into two terms, all you have to do is find two factors of $6 \cdot (-20)$ that add up to -7. The search goes as follows:

Factors of −120		*Sum = −7?*	
1,	−120	−119	no
2,	−60	−58	no
3,	−40	−37	no
4,	−30	−26	no
5,	−24	−19	no
6,	−20	−14	no
8,	−15	−7	yes!

So you write $6x^2 - 7x - 20$ as

$$6x^2 + 8x - 15x - 20,$$

which factors by grouping into

$$2x(3x+4) - 5(3x+4)$$
$$= (3x+4)(2x-5).$$

As another example, to factor $8x^2 - 85x + 30$, you seek two factors of $8 \cdot 30$ that add up to -85. Since the sum is negative, both factors must be negative. Since $8 \cdot 30 = 240$, you try:

Factors of 240		*Sum = −85?*		
−1,	−240	−241	no	
−2,	−120	−122	no	Skips over −85!
−3,	−80	−83	no	

Since the sums skip over -85, you know there are *no* factors of 240 that add up to -85. Thus,

$8x^2 - 85x + 30$ is prime.

Example 6: *Discriminant Test for Prime Quadratics* – There is a way to use the discriminant to tell with a *single* calculation whether or not a quadratic trinomial will factor. For example, $6x^2 - 7x - 20$ factors into $(2x-5)(3x+4)$. Writing the quadratic *equation*

$$6x^2 - 7x - 20 = 0,$$

you can solve by factoring to get

$$(2x-5)(3x+4) = 0$$
$$x = \frac{5}{2} \quad \text{or} \quad x = -\frac{4}{3}.$$

Since you have agreed that the factors must contain *integer* coefficients, these solutions will always turn out to be *rational* numbers. Recalling that solutions of quadratic equations are rational if and only if the discriminant, $b^2 - 4ac$, is a perfect square, you can reach the following conclusion:

Conclusion: A quadratic trinomial factors if and only if the discriminant is a perfect square.

For example, for $6x^2 - 7x - 20$, the discriminant is

$$b^2 - 4ac = (-7)^2 - 4(6)(-20) = 49 + 480 = 529 = 23^2.$$

So the equation $6x^2 - 7x - 20 = 0$ has rational solutions, and thus the polynomial $6x^2 - 7x - 20$ *does* factor.

For $x^2 + 4x - 11$, however,

$$b^2 - 4ac = 4^2 - 4(1)(-11) = 16 + 44 = 60,$$

which is not a perfect square. So the equation $x^2 + 4x - 11 = 0$ has irrational solutions, and the polynomial $x^2 + 4x - 11$ does *not* factor.

The following exercise is designed to give you practice in factoring the kinds of polynomials discussed above.

Exercise 7-4

For Problems 1 through 74, factor the polynomial completely.

Common Factors

1. $x^2 + x$
2. $x^3 + x$
3. $2x - 10$
4. $3x - 21$
5. $5x^3 - 50ax$
6. $6x^4 - 8bx^2$
7. $12r^5 + 18r^3x^2$
8. $20s^7t^2 + 15s^3t^4$
9. $8x^9 + 12x^7 - 28x^2$
10. $13m^3n - 26m^2n^2 + 65mn^3$

Factoring by Grouping

11. $2x^3 - 3x^2 - 4x + 6$
12. $10x^3 - 15x^2 + 2x - 3$
13. $12x^3 + 45x^2 + 32x + 120$
14. $2x^3 - 7x^2 - 10x + 35$
15. $18x^3 - 27x^2 + 8x - 12$
16. $10x^3 + 10x^2 + 3x + 3$
17. $24x^3 - 18x^2 + 60x - 45$
18. $64x^3 - 160x^2 + 24x - 60$
19. $8x^2 - 12ax + 6xy - 9ay$
20. $12a^2x - 4a^2 - 9bx + 3b$
21. $7r^2s - 6r^2 - 7s + 6$
22. $a^2c - b^2d + a^2d - b^2c$

Quadratic Trinomials

23. $x^2 + 5x - 14$
24. $x^2 - 9x + 20$
25. $3x^2 + 9x - 30$
26. $2x^2 - 4x - 70$
27. $8x^2 + 6xy + y^2$
28. $10x^2 + 3xy - y^2$
29. $3r^2 + rs - 10s^2$
30. $6p^2 - 11pj - 10j^2$
31. $6x^2 + 11x + 3$
32. $6x^2 + 7x - 20$
33. $6x^2 + 19x - 20$
34. $6x^2 + 37x - 20$
35. $10x^2 - 29x + 21$
36. $4x^2 + 25x - 21$
37. $10x^2 - 29x - 21$
38. $12x^2 - 7x - 12$
39. $5x^2 - 42x + 16$
40. $4x^2 - 5x - 6$
41. $8x^2 + 49x + 6$
42. $21x^2 + 29x - 10$
43. $4x^2 - 12x + 9$
44. $25x^2 + 40x + 16$
45. $x^2 + 16x + 64$
46. $x^2 - 22x + 121$
47. $a(a + 1)x^2 + x - a(a - 1)$
48. $(2a + b)x^2 - (a - b)x - (a + 2b)$

Difference of Two Squares

49. $x^2 - 9$
50. $x^2 - 16$
51. $4x^2 - 25$
52. $25x^2 - 9$
53. $9x^2 - 49$
54. $81x^2 - 4$
55. $x^2 - 1$
56. $36x^2 - 1$
57. $36x^2 - 144$
58. $16x^2 - 64$
59. $5x^2 - 80$
60. $3x^2 - 27$

61. $x^2 - 36y^2$
62. $x^2 - y^2$
63. $x^4 - y^4$
64. $x^8 - y^8$
65. $x^6 - y^4$
66. $p^{10} - q^4$
67. $a^5 - a^3$
68. $s^7 - s^3$
69. $x^4y^2 - x^2y^4$
70. $x^5y^3 - x^3y^5$
71. $(x + 7)^2 - 9$
72. $25 - (x - 3)^2$
73. $36 - (x - 5)^2$
74. $81 - (2x + 7)^2$

For Problems 75 through 84, use the technique of splitting the middle term either to factor the polynomial or to show that it is prime.

75. $12x^2 + 25x + 12$
76. $18x^2 + 27x + 10$
77. $24x^2 - 121x + 5$
78. $12x^2 - 179x - 15$
79. $30x^2 + 41x - 6$
80. $6x^2 - 27x + 20$
81. $35x^2 - 2x - 6$
82. $9x^2 - 56x + 12$
83. $36x^2 - 63x + 20$
84. $12x^2 + 27x + 2$

For Problems 85 through 94, use the discriminant to tell (*without* actually doing the factoring) whether or not the given trinomial will factor. The squares in Table 1, or a square-root calculator will help you determine whether the discriminant is a perfect square.

85. $x^2 - 2x + 5$
86. $x^2 - 2x - 29$
87. $x^2 - 21x + 68$
88. $x^2 - 10x + 74$
89. $18x^2 - 15x + 2$
90. $3x^2 - 15x + 14$
91. $18x^2 - 13x + 3$
92. $3x^2 - 16x + 13$
93. $8x^2 - 79x - 10$
94. $48x^2 - 22x - 15$

Problems 95 through 120 are arranged in *no* particular order. So *you* must decide upon which factoring technique to use.

95. $4x^2 - 16y^2$
96. $a^2b^4 - a^4b^2$
97. $4x^2 - 12xy + 9y^2$
98. $x^2 + 11x + 18$
99. $a^2 + 3a - 88$
100. $7a^2x - 6a^2 - 7x + 6$
101. $30x^2 + 95x + 50$
102. $2x^2 + xy - 3y^2$
103. $x^4 + 5x^3 - 2x^2 - 10x$
104. $7x^4 - 28x^2$
105. $x^4 - x^2 - 12$
106. $15x^2 + 8x - 12$
107. $60x^2 - 68x + 8$
108. $x^4 - 26x^2 + 25$
109. $16x^2 - 35x + 6$
110. $7x^3 - 14x^2 - 3x + 6$
111. $5x^3 + 6x^2 - 45x - 54$
112. $81x^2 + 108x + 36$
113. $x^2 + 3x + 4$
114. $(2x + 3y + a)^2 - (x - y + a)^2$
115. $100 - (x - y)^2$
116. $x^2 - 2x - 7$
117. $35a^2 + 47ab + 6b^2$
118. $343 - 7(x + 3)^2$
119. $(a^2 - b^2 - c^2)^2 - 4b^2c^2$
120. $200x^2 + 510x - 1001$

121. *Factoring Harder Differences of Squares* – The polynomial

$$x^2 + 2x + 1 - y^2$$

can be transformed into a difference of two squares by associating the first *three* terms:

$$\begin{aligned}(x^2 + 2x + 1) - y^2 &= (x + 1)^2 - y^2\\ &= [(x + 1) + y][(x + 1) - y]\\ &= (x + 1 + y)(x + 1 - y).\end{aligned}$$

Using this technique, factor the following:

a. $x^2 + 6x + 9 - y^2$
b. $a^2 - b^2 + 2b - 1$
c. $p^2 - 14p + 49 - 9k^2$
d. $x^2 - 4y^2 + 4y - 1$
e. $x^2 - 10x + 25 - y^2$
f. $r^2 - s^2 - 4s - 4$
g. $x^2 - 4x + 4 - 100a^2$
h. $81 - x^2 + 2xy - y^2$

122. *Factoring by Completing the Square* – The polynomial

$$x^4 + 2x^2 + 9$$

can be transformed into a difference of two squares.

Since x^4 and 9 are both perfect squares, you need only get the correct *middle* term to make a trinomial which is a perfect square. First, you commute the $2x^2$ out of the way, getting

$$x^4 \qquad + 9 + 2x^2.$$

Then you add $6x^2$ to complete the square, and subtract $6x^2$ so that you are actually adding 0:

$$x^4 + 6x^2 + 9 + 2x^2 - 6x^2.$$

Associating the first three and the last two terms, you get

$$(x^2 + 3)^2 - 4x^2 = [(x^2 + 3) + 2x]\,[(x^2 + 3) - 2x]$$
$$= (x^2 + 2x + 3)(x^2 - 2x + 3).$$

Use this technique to factor the following:

a. $x^4 - x^2 + 16$
b. $x^4 + 11x^2 + 36$
c. $x^4 - 19x^2 + 9$
d. $x^4 - 26x^2 + 25$
e. $x^4 + 4$
f. $x^4 - 18x^2 + 1$
g. $x^4 - 19x^2 + 25$
h. $x^4 - 110x^2 + 25$
i. $x^4 + 5x^2 + 9$
j. $x^4 - 13x^2 + 36$

123. How are the binomials $x - y$ and $x + y$ related to each other? How are the binomials $x - y$ and $y - x$ related to each other?

124. It is possible to factor $x - y$ as a difference of two squares:

$$x - y = (\sqrt{x} + \sqrt{y})(\sqrt{x} - \sqrt{y}).$$

The second factor will factor again using fourth roots,

$$x - y = (\sqrt{x} + \sqrt{y})(\sqrt[4]{x} + \sqrt[4]{y})(\sqrt[4]{x} - \sqrt[4]{y}),$$

and so forth, *infinitely*! Since it is desirable in factoring to have as many factors as possible, you are faced with an awkward situation—where do you stop factoring? Explain how your agreement about "completely factored form" gets you out of this awkward situation.

125. You have used the factoring techniques you have learned to factor only expressions containing *variables*. It is important for you to realize that variables represent *numbers*, and that the factoring techniques will work just as well with constants. Find the factors of the following numbers:

a. $2^4 - 1$
b. $3^8 - 1$
c. $2^{16} - 1$
d. $3 \times 2^6 - 2 \times 2^3 - 5$

126. *Factoring Quadratics by Computer* – The following is a flow chart for a computer program to factor quadratic trinomials of the form

$$Ax^2 + Bx + C$$

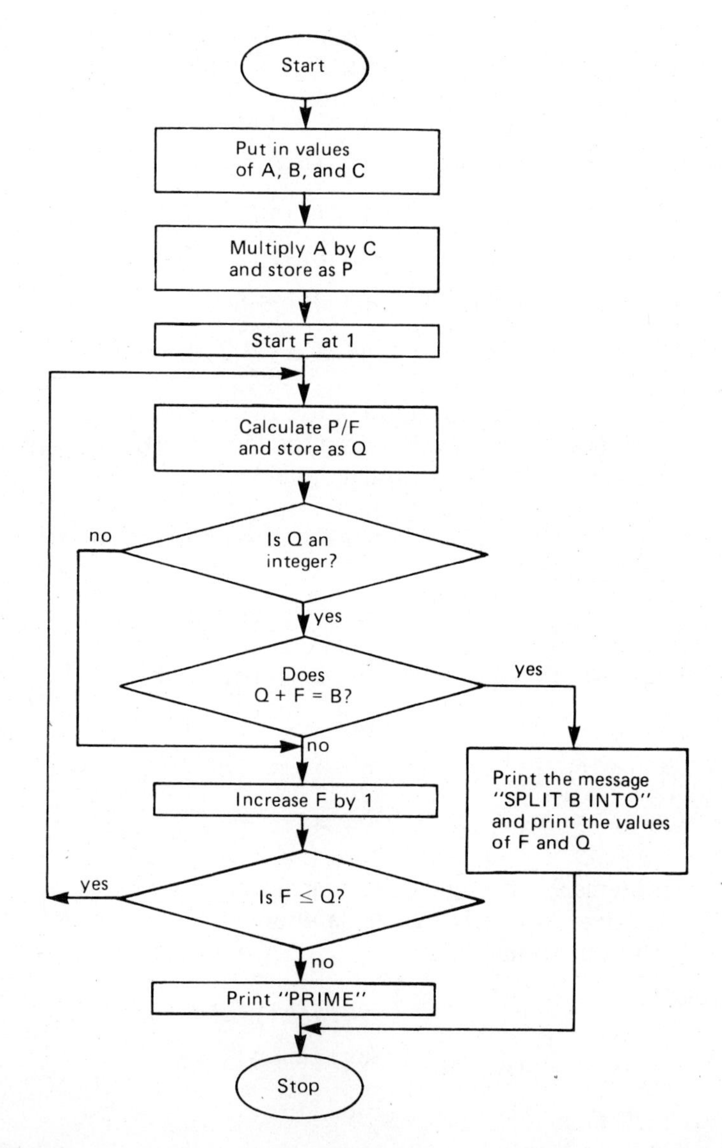

by splitting the middle term. The output of the program is the two numbers to split B into, or the message that the trinomial is prime.

a. Show the simulated computer memory and the output if the following values are put in for A, B, and C:

   i. 5, 14, 8,
   ii. 3, 9, 4.

b. Translate the flow chart into a computer language such as BASIC.
c. Use the program to split the middle term for Problems 75, 77, 79, 81, and 83, above.
d. Use the program to factor the trinomial $2001x^2 - 1336x - 1001$.

127. *LCM, GCF, and Musical Harmony* – The musical notes A and E around middle C have frequencies of 220 and 330 cycles per second (cps), respectively. Whenever you play or sing a note, you also generate "overtones," which are notes with *multiples* of the fundamental frequency. Some overtones of A and E are:

Overtone:	*1 (fund.)*	*2*	*3*	*4*	*5*	*6*	...
Note A:	220	440	660	880	1100	1320	...
Note B:	330	660	990	1320	1650	1980	...

A and E harmonize because many of their overtones (660, 1320, 1980, . . .) have the same frequency. The closer the *least* common multiple (LCM) is to the fundamental frequencies, the better the harmony. You can calculate the LCM by factoring into primes,

$$220 = 2 \cdot 2 \cdot 5 \cdot 11,$$
$$330 = 2 \cdot 3 \cdot 5 \cdot 11,$$

then taking the *union* of the sets of prime factors:

$$\text{LCM} = 2 \cdot 2 \cdot 3 \cdot 5 \cdot 11 = 660.$$

Answer the following questions.

a. A, C, and C sharp have frequencies of 220, 264, and 275 cps, respectively. By factoring these frequencies into primes, find the LCM of A and C, A and C sharp, and C and C sharp.
b. Which pairs of these will harmonize well, and which will not?
c. Is it true that if two notes harmonize well with a third note, then they will harmonize with each other? Explain?
d. Find the greatest common factor (GCF) of 220 and 264, 220 and 275, and 264 and 275. How can the GCF of the frequencies tell you whether or not the notes will harmonize?
e. Show that the product of two numbers equals the product of their LCM and GCF.

(For more information, see "More on the Mathematics of Musical Scales" by Truman Botts in *The Mathematics Teacher*, January, 1974.)

7-5 DIVISION OF POLYNOMIALS

You can determine whether or not one integer is a factor of another by dividing one into the other. Similarly, you can tell whether one polynomial is a factor of another by division. The algorithm (the method) for dividing polynomials is the same as for long division of integers. For example, to divide $3x^3 - 2x^2 - 13x + 10$ by $x - 2$, the procedure is to write

$$x - 2\overline{)3x^3 - 2x^2 - 13x + 10}.$$

Then you divide x into $3x^3$, getting $3x^2$. The $3x^2$ is part of the quotient, and should be placed somewhere convenient, like above the $-2x^2$ term. To find out how much of the original polynomial is accounted for by the $3x^2$, you multiply $3x^2$ by $x - 2$, and subtract the result from the original polynomial. This is complicated to *say*, but easy to *do*, as shown below:

$$\begin{array}{r} 3x^2 \\ x - 2\overline{)3x^3 - 2x^2 - 13x + 10} \\ \underline{3x^3 - 6x^2 } \\ 4x^2 - 13x + 10 \end{array}$$

Subtract.

$-2x^2 - (-6x^2) = 4x^2$.

The $4x^2 - 13x + 10$ is the remainder of the original polynomial which is left to be accounted for by other terms in the quotient. The remaining steps are simply repetitions of the first set of steps, operating each time on the remainder. The whole computation looks like this:

$$
\begin{array}{r}
3x^2 + 4x - 5 \\
x - 2\overline{)3x^3 - 2x^2 - 13x + 10} \\
\underline{3x^3 - 6x^2 \qquad\qquad} \\
4x^2 - 13x + 10 \\
\underline{4x^2 - 8x \qquad} \\
-5x + 10 \\
\underline{-5x + 10} \\
0
\end{array}
$$

The final remainder of 0 shows that $x - 2$ *does* divide evenly into $3x^3 - 2x^2 - 13x + 10$. That is, $x - 2$ *is* a factor of this polynomial. If the remainder were *not* zero, then the one polynomial would *not* divide evenly into the other.

If the polynomial to be divided has some terms "missing" (i.e., with zero coefficients), then spaces should be left for these terms. For example, in $x^3 + 8$, the x^2 and x terms have zero coefficients. To divide $x^3 + 8$ by $x + 2$, you would write

$$x + 2\overline{)x^3 + 0 + 0 + 8},$$

and divide as before. See if you can get the correct quotient, $x^2 - 2x + 4$.

There is a shorter way to divide by a *linear* polynomial, called "synthetic division." If you like, you may look in Section 10-5 to see how it works.

Exercise 7-5

Divide:

1. $x^3 - 3x^2 + 4x + 28$ by $x + 2$
2. $x^3 - 9x^2 + 13x + 15$ by $x - 3$
3. $2x^3 - 3x^2 + 7x - 3$ by $2x - 1$
4. $6x^3 + 2x^2 + 11x - 10$ by $3x - 2$
5. $24x^3 - 35x^2 - 36x + 5$ by $8x - 1$
6. $28x^3 + 30x^2 + 17x - 15$ by $7x - 3$
7. $x^3 + 3x^2 + 3x + 1$ by $x^2 + 2x + 1$
8. $x^3 - 3x^2 + 3x - 1$ by $x^2 - 2x + 1$
9. $x^3 - 6x^2 + 12x - 8$ by $x^2 - 4x + 4$
10. $8x^3 + 12x^2 + 6x + 1$ by $4x^2 + 4x + 1$
11. $x^3 - y^3$ by $x - y$
12. $x^3 + y^3$ by $x + y$
13. $x^4 - 1$ by $x - 1$
14. $x^4 - 1$ by $x + 1$
15. $x^4 + x^2 + 1$ by $x^2 + x + 1$
16. $x^4 + x^2 + 1$ by $x^2 - x + 1$
17. $6x^5 - x^4 + 10x^3 - 14x^2 - 25$ by $3x^2 + 4x + 5$
18. $x^6 - 3x^5 + x^3 + 358x - 357$ by $x^2 + 2x - 3$

7-6 FACTORING HIGHER DEGREE POLYNOMIALS

In Section 7-4 you learned how to factor certain special polynomials. For example, the cubic polynomial

$$x^3 - 2x^2 - 3x + 6,$$

can be factored by grouping. You get

$$x^2(x - 2) - 3(x - 2)$$
$$= (x - 2)(x^2 - 3).$$

The technique worked because the $(x - 2)$ was a *common* factor, and could be factored out.

Unfortunately, this technique will *not* work with a polynomial such as

$$x^3 - 2x^2 - 5x + 6$$

because there is *no* way to group and find common binomial factors.

Objective:

Be able to find *linear* factors for cubic and higher degree polynomials.

One procedure for factoring higher degree polynomials is simply to *guess* a factor, then find out whether you are right by some process such as long division. Obviously, this procedure is extremely tedious. So you seek some systematic procedure for narrowing down the number of factors from which to guess.

Suppose that $P(x) = x^3 - 2x^2 - 5x + 6$, and you wish to find a linear factor. You might guess some of the following:

$$2x - 1, \ x + 5, \ x + 1, \ x - 3.$$

Before you start doing the long division, you can rule out the first two guesses. If $2x - 1$ were a factor, then

$$P(x) = (2x - 1)(\text{some other factor}).$$

The other factor would have to have $\frac{1}{2}x^2$ as one of its terms so that $(2x)(\frac{1}{2}x^2)$ would give the x^3 term in $P(x)$. But you have agreed to look only for factors with *integer* coefficients. Similarly, $x + 5$ is impossible. If

$$P(x) = (x + 5)(\text{some other factor}),$$

then the other factor would have to have $6/5$ as its constant term. This is the only way to get the 6 as the constant term of $P(x)$ when you multiply the two factors back together.

The $x + 1$ and $x - 3$ are possible factors because the x-coefficient equals 1, and the 1 and -3 are each *factors* of the *constant* term, 6. Upon doing the long division, you find that $x + 1$ does *not* divide evenly. But $x - 3$ *does*. So

$$P(x) = (x - 3)(x^2 + x - 2).$$

You should try doing the long division on a piece of scratch paper right now so that you can see why this is right.

There is an easy way to test a linear binomial to see if it is a factor *without* doing the long division. If $x = 3$, then $(x - 3) = 0$. So $P(3)$ is a product with a factor equal to 0. By the Multiplication Property of Zero, $P(3)$ must equal zero. But a product equals zero *only* if one of its factors equals 0. So if by some means or other you find that $P(3) = 0$, you can conclude that $P(x)$ has a factor that is zero when $x = 3$. This factor would be $(x - 3)$ because

$$x - 3 = 0 \text{ when } x \text{ is } 3.$$

The Multiplication Property of Zero and its Converse thus form the basis for the following major theorem:

Factor Theorem: $(x - b)$ is a factor of $P(x)$ if and only if $P(b) = 0$.

The problem of searching for linear factors thus reduces to finding all factors of the *constant* term (positive and negative), and substituting them one at a time for x. Whenever you get 0 for an answer, x *minus* whatever you substituted will be a factor.

Example 1: Find the factors of $P(x) = x^3 + x^2 - 8x - 12$.

The factors of -12 are $\pm 1, \pm 2, \pm 3, \pm 4, \pm 6, \pm 12$.

$P(1) = 1 + 1 - 8 - 12 \neq 0$, so $(x - 1)$ is not a factor.

$P(-1) = -1 + 1 + 8 - 12 \neq 0$, so $(x - (-1)) = (x + 1)$ is not a factor.

$P(2) = 8 + 4 - 16 - 12 \neq 0$, so $(x - 2)$ is not a factor.

$P(-2) = -8 + 4 + 16 - 12 = 0$, so $(x + 2)$ *is* a factor.

The other factor may be found by long division.

$$\begin{array}{r} x^2 - x - 6 \\ x + 2\overline{)x^3 + x^2 - 8x - 12} \\ \underline{x^3 + 2x^2 } \\ -x^2 - 8x - 12 \\ \underline{-x^2 - 2x } \\ -6x - 12 \\ \underline{-6x - 12} \\ 0 \end{array}$$

$\therefore P(x) = (x + 2)(x^2 - x - 6)$.

The second factor may or may not factor again. In this case, it does, giving as the final answer

$$P(x) = (x + 2)(x + 2)(x - 3).$$

Example 2: Factor $P(x) = 2x^3 - x^2 - x - 3$.

If you want to factor a polynomial like $P(x) = 2x^3 - x^2 - x - 3$, where the lead coefficient of $P(x)$ is *not* equal to 1, then a linear factor could be $(2x + 3)$, $(2x - 1)$, etc., where the x-coefficient in the linear factor must be a factor of the lead coefficient (2, in this case). In general, the factor would have the form $(ax - b)$, where a must be a factor of the lead coefficient of $P(x)$, and b must be a factor of the constant term of $P(x)$.

If $(ax - b)$ is a factor of $P(x)$, then $P(x)$ will equal zero whenever $(ax - b)$ equals zero. You find out what x must equal by solving the equation

$$ax - b = 0$$
$$\therefore ax = b$$
$$\therefore x = \frac{b}{a}.$$

So it turns out that $(ax - b)$ is a factor of $P(x)$ if and only if $P\left(\frac{b}{a}\right) = 0$.

Possible values of $\frac{b}{a}$ are $\pm\frac{1}{1}$, $\pm\frac{3}{1}$, $\pm\frac{1}{2}$, and $\pm\frac{3}{2}$.

$P(1) = 2 - 1 - 1 - 3 \neq 0$, so $(x - 1)$ is not a factor.

$P(-1) = -2 - 1 + 1 - 3 \neq 0$, so $(x + 1)$ is not a factor.

$P(3) = 54 - 9 - 3 - 3 \neq 0$, so $(x - 3)$ is not a factor.

$P(-3) = -54 - 9 + 3 - 3 \neq 0$, so $(x + 3)$ is not a factor.

$P(\frac{1}{2}) = \frac{1}{4} - \frac{1}{4} - \frac{1}{2} - 3 \neq 0$, so $(2x - 1)$ is not a factor.

$P(-\frac{1}{2}) = -\frac{1}{4} - \frac{1}{4} - \frac{1}{2} - 3 \neq 0$, so $(2x + 1)$ is not a factor.

$P(\frac{3}{2}) = \frac{27}{4} - \frac{9}{4} - \frac{3}{2} - 3 = 0$, so $(2x - 3)$ *is* a factor!!

By division,

$$\begin{array}{r} x^2 + x + 1 \\ 2x - 3\overline{)\,2x^3 - x^2 - x - 3} \\ \underline{2x^3 - 3x^2} \\ 2x^2 - x - 3 \\ \underline{2x^2 - 3x} \\ 2x - 3 \\ \underline{2x - 3} \\ 0 \end{array}$$

$$\therefore P(x) = (2x - 3)(x^2 + x + 1).$$

In this case, the second factor does *not* factor again.

This is an example of the Rational Root Theorem.

Rational Root Theorem: $(ax - b)$ is a factor of $P(x)$ if and only if $P(b/a) = 0$.

The theorem gets its name from the fact that b/a is a solution, or "root," of the equation $P(x) = 0$.

Example 3: Factor $P(x) = x^3 - 4x^2 + 3x - 2$.

Since the lead coefficient is 1, you need only look for factors of the form $(x - b)$. Possible values of b are ± 1 and ± 2.

$P(1) = 1 - 4 + 3 - 2 \neq 0$, so $(x - 1)$ is not a factor.

$P(-1) = -1 - 4 - 3 - 2 \neq 0$, so $(x + 1)$ is not a factor.

$P(2) = 8 - 8 + 6 - 2 \neq 0$, so $(x - 2)$ is not a factor.

$P(-2) = -8 - 8 - 6 - 2 \neq 0$, so $(x + 2)$ is not a factor.

Since *none* of the factors of the constant term, -2, makes $P(x)$ equal zero, you can conclude that

$P(x)$ has *no* linear factors with integer coefficients.

Sum or Difference of Like, Odd Powers – The Factor Theorem can be used to learn how to factor a sum or difference of two *cubes*, two *fifth* powers, etc. For example, if

$$P(x) = x^5 + c^5,$$

then

$P(-c) = (-c)^5 + c^5$	Definition of $P(-c)$.
$= -c^5 + c^5$	*Negative* number to an *odd* power.
$= 0$	Additive inverses.
$\therefore (x + c)$ is a factor.	Factor Theorem.

By long division, you find that

$$P(x) = (x + c)(x^4 - x^3c + x^2c^2 - xc^3 + c^4).$$

The pattern is easy enough that you can remember it. The first factor, $(x + c)$, is obvious. In the second factor, the powers of x *decrease*, the powers of c *increase*, and the signs *alternate*. The pattern holds for a sum of any two like, *odd* powers. (In Problem 47, you will prove that it does *not* work for a sum of two *even* powers.) For a *difference* of like, odd powers, you must first transform it into a *sum*.

Example 4: Factor $P(x) = x^7 - 128$.

$x^7 - 128 = x^7 + (-128)$	Definition of subtraction.
$= x^7 + (-2)^7$	$-128 = (-2)^7$

$$= (x + (-2))(x^6 - x^5(-2) + x^4(-2)^2 - x^3(-2)^3 + x^2(-2)^4 - x(-2)^5 + (-2)^6)$$

$$= \underline{\underline{(x - 2)(x^6 + 2x^5 + 4x^4 + 8x^3 + 16x^2 + 32x + 64)}}.$$

Note that for $x^n - c^n$, the first factor is $(x - c)$, and all the signs in the second factor turn out to be *positive*.

Appendix B-2 shows ways to shorten the trial-and-error search for factors using Descartes' Rule of Signs and the Upper Bound Theorem.

Exercise 7-6

In Problems 1 through 22, use the Factor Theorem either to factor the polynomial completely or to prove that it has *no* linear factors with integer coefficients.

1. $x^3 + 3x^2 - 18x - 40$
2. $x^3 + 9x^2 + 24x + 20$
3. $x^3 - 10x^2 - 17x + 66$
4. $x^3 + 3x^2 - 6x - 8$
5. $x^3 - x^2 - 5x + 2$
6. $x^3 - 2x^2 - 14x + 3$
7. $x^3 + 3x^2 - 7x + 2$
8. $x^3 + 2x^2 - 9x + 3$
9. $x^4 + 2x^3 - 13x^2 - 14x + 24$
10. $x^4 + 4x^3 - 7x^2 - 34x - 24$
11. $x^4 - 2x^3 - 3x^2 + 4x + 4$
12. $x^4 - 7x^3 + 18x^2 - 20x + 8$
13. $x^5 + 5x^4 + 10x^3 + 10x^2 + 5x + 1$
14. $x^6 - 6x^5 + 15x^4 - 20x^3 + 15x^2 - 6x + 1$
15. $2x^3 + 5x^2 + x - 2$
16. $3x^3 - 2x^2 - 7x - 2$
17. $2x^3 + 3x^2 - x - 1$
18. $3x^3 + 5x^2 - 5x + 1$
19. $12x^3 + 4x^2 - 3x - 1$
20. $18x^3 + 9x^2 - 2x - 1$
21. $12x^3 - 20x^2 - 37x + 30$
22. $18x^3 - 9x^2 - 38x + 24$

For Problems 23 through 46, use the pattern for factoring the sum or difference of like, odd powers, and any other techniques you may need, to factor the polynomial completely.

23. $x^7 + y^7$
24. $x^7 - y^7$
25. $x^{11} - 1$
26. $x^{11} + 1$
27. $x^5 + 32$
28. $x^5 - 32$
29. $a^{13} - b^{13}$
30. $a^{13} + b^{13}$
31. $a^5 + 32b^{10}$
32. $32x^5 - b^{10}$
33. $x^3 - y^3$
34. $x^3 + y^3$
35. $x^3 + 8$
36. $x^3 - 27$
37. $r^3 - s^6$
38. $r^9 + s^6$
39. $x^6 - y^6$
40. $x^{12} - y^{12}$
41. $8a^3 - 64b^3$
42. $27r^3 + 125s^3$

43. $x^4y + xy^4$

44. $a^2b^5 - a^5b^2$

45. $(a - b)^3 - (a + b)^3$

46. $(x + y)^3 + (x - y)^3$

47. *Sum of Two Squares* – Use the Factor Theorem to prove that a *sum* of two squares, $P(x) = x^2 + c^2$, has *no* linear factors.

48. *Factors of* $x^{15} + y^{15}$

a. Factor $x^{15} + y^{15}$ by first considering it to be a sum of two *cubes.* You can get 3 factors.
b. Factor $x^{15} + y^{15}$ by first considering it to be a sum of two *fifth* powers. Again, you can get 3 factors.
c. Since the sets of factors you get in parts *a* and *b* are *not* identical, one or more of the factors must not be prime. By a clever application of techniques you know, factor $x^{15} + y^{15}$ into *four* polynomials with integer coefficients.

49. *Factor Theorem by Computer* – The following is a flow chart for a computer program to search for linear factors of the form $(x - n)$ for cubic polynomials of the form

$$P(x) = Ax^3 + Bx^2 + Cx + D$$

using the Factor Theorem.

a. Show the simulated computer memory and the output if the input values for *A, B, C,* and *D* are:
   i. 1, −1, −5, −3,
   ii. 1, 1, 1, −2.
b. What special name is given to the variable *K*?
c. Translate the flow chart into a computer language such as BASIC.
d. Test your program by finding factors for Problems 1, 3, 5, and 7, above.
e. When your program is working, use it to find factors of $x^3 - 9x^2 - 649x + 2001$.

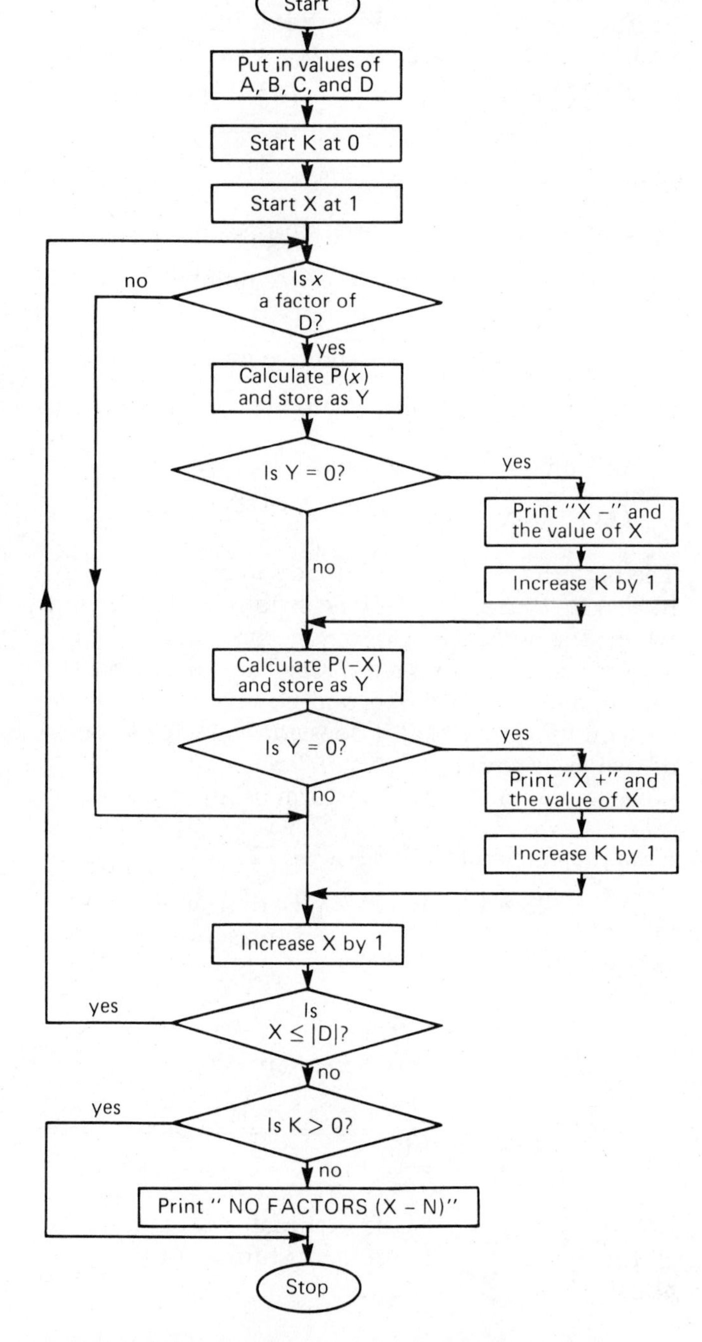

50. Modify the program of Problem 49 so that it searches for linear factors of the form $(ax - b)$, where *a* can be an integer greater than 1. Test your modified program by factoring the polynomials you got in Problems 15, 17, 19, and 21, above.

7-7 PRODUCTS AND QUOTIENTS OF RATIONAL EXPRESSIONS

Now that you have refreshed your memory about factoring polynomials, you are ready to tackle operations with rational expressions. In this section you will multiply and divide them and in the next, you will add and subtract them. When you are comfortable with these operations, you will be ready to return to graphing of rational functions.

Objective:

Be able to multiply or divide several rational expressions and simplify the result.

Multiplication: In Chapter 1 you saw how the Multiplication Property of Fractions can be proved from the Field Axioms. You should recall that this property states:

$$\frac{xy}{ab} = \frac{x}{a} \cdot \frac{y}{b}.$$

Using symmetry to read the property backwards, you get the following technique for multiplying fractions:

Technique: To multiply two fractions, multiply their numerators together and multiply their denominators together.

That is,

$$\boxed{\frac{x}{a} \cdot \frac{y}{b} = \frac{xy}{ab}}.$$

Multiplication of two rational expressions is accomplished by the same technique. It is usually better *not* to carry out the multiplication, but rather to go in the opposite direction by factoring apart the numerators and denominators. For example, to multiply

$$\frac{x^2 + 5x + 6}{x^2 - x - 20} \cdot \frac{x^2 + 3x - 4}{x^2 + x - 2},$$

you could first use the multiplication property of fractions to get

$$\frac{(x^2 + 5x + 6)(x^2 + 3x - 4)}{(x^2 - x - 20)(x^2 + x - 2)}.$$

Each trinomial will factor, giving

$$\frac{(x + 2)(x + 3)(x - 1)(x + 4)}{(x + 4)(x - 5)(x - 1)(x + 2)}.$$

The numerator and denominator have several *common* factors. Commuting the *greatest* common factor to the right leads to a considerable simplification:

$$\frac{(x + 3)(x - 1)(x + 2)(x + 4)}{(x - 5)(x - 1)(x + 2)(x + 4)} \qquad \text{Commutativity.}$$

$$= \frac{x + 3}{x - 5} \cdot \frac{(x - 1)(x + 2)(x + 4)}{(x - 1)(x + 2)(x + 4)} \qquad \text{Multiplication property of fractions.}$$

$$= \underline{\underline{\frac{x + 3}{x - 5}}}. \qquad \text{Multiplicative identity.}$$

It is clear from the answer that x cannot equal 5. It must be understood that the domain of x also excludes 1, −2, and −4, each of which would make the denominator of the original fraction equal to zero.

Division: Division of fractions is carried out by first applying the definition of division,

$$\boxed{\frac{a}{b} \div \frac{c}{d} = \frac{a}{b} \cdot \frac{d}{c}},$$

then multiplying the resulting expressions as above.

Technique: To divide a fraction by a fraction, use the definition of division, then multiply the resulting fractions.

Simplification: The process of eliminating common factors from the numerator and denominator of a rational expression is called "simplification."

DEFINITION

A rational expression is ***simplified*** when its numerator and denominator have no common factors other than 1.

The process of simplification can be done by "canceling."

DEFINITION

Canceling in a fraction means ***dividing*** the numerator and denominator of the fraction by the same common factor.

Canceling may be done rapidly by simply "crossing out" common factors. For example,

$$\frac{\cancel{(x-5)}(x+3)}{(x+2)\cancel{(x-5)}} = \frac{x+3}{x+2}.$$

Note: It is tempting to "cross out the x's," above and get 3/2. Upper classmen sometimes call this error "freshman canceling." You can easily see that this is wrong by substituting a value for x. For example, if $x = 8$, then

$$\frac{x+3}{x+2} = \frac{8+3}{8+2} = \frac{11}{10},$$

which is *not* equal to 3/2. The way to avoid such errors is to remember that canceling is a *division* process. "Canceling x" would mean *dividing* the numerator and denominator by x. Since division *distributes* over addition, you would get

$$\frac{x+3}{x+2} = \frac{1+\frac{3}{x}}{1+\frac{2}{x}},$$

which is more complicated that the original expression!

Special Cases: Specially related binomials such as $x + 3$, $x - 3$, $3 + x$, and $3 - x$ may occur in rational expressions. One case is

$$\frac{x+3}{3+x} = \frac{x+3}{x+3} \qquad \text{Since addition } \textit{commutes}.$$

$$= 1. \qquad \text{Since } n/n = 1.$$

A second case is

$$\frac{x-3}{3-x} = \frac{x-3}{-1(x-3)} \qquad \text{Factoring out } -1.$$

$$= \frac{1}{-1} \qquad \text{Canceling.}$$

$$= \underline{\underline{-1.}} \qquad \text{Positive divided by negative is negative.}$$

An easy way to remember this second case is to realize that $x - 3$ and $3 - x$ are *opposites* of each other and any number divided by its opposite equals -1. For example,

$$\frac{-13}{13} = -1.$$

A third case is

$$\frac{x-3}{x+3}.$$

The two binomials are *conjugates* of each other. But *nothing* can be done to simplify the quotient. For example, if x were equal to 8, the expression would be

$$\frac{x-3}{x+3} = \frac{8-3}{8+3}$$

$$= \frac{5}{11}.$$

Clearly, 5/11 cannot be reduced.

The exercise that follows is designed to allow you to become comfortable with multiplying, dividing, and simplifying rational algebraïc expressions. You should work enough problems so that you can do them quickly and correctly.

Exercise 7-7

1. *Property of the Reciprocal of a Product* – Using the Field Axioms, prove that the reciprocal of a product of two numbers is equal to the product of their reciprocals. That is, prove that

$$\frac{1}{a \cdot b} = \frac{1}{a} \cdot \frac{1}{b}.$$

What restrictions are there on the values of a and b?

2. *Multiplication Property of Fractions* – Use the Property of the Reciprocal of a Product (Problem 1, above) as a lemma to prove the Multiplication Property of Fractions. That is, prove that

$$\frac{x}{a} \cdot \frac{y}{b} = \frac{xy}{ab}.$$

3. Simplify the following as much as possible.

a. $\dfrac{x-5}{5-x}$ b. $\dfrac{x+5}{5+x}$ c. $\dfrac{x-5}{x+5}$ d. $\dfrac{x-5}{x-5}$

4. Simplify the following as much as possible.

a. $\dfrac{r-s}{r+s}$ b. $\dfrac{r-s}{r-s}$ c. $\dfrac{r-s}{s-r}$ d. $\dfrac{r+s}{s+r}$

For Problems 5 through 46, carry out the indicated multiplication and division, and simplify.

5. $\dfrac{2}{x-2} \cdot \dfrac{x^2-4}{4}$

6. $\dfrac{x^2+7x+12}{12} \cdot \dfrac{4}{x+4}$

7. $\dfrac{x^2+4x+3}{5x} \div \dfrac{x+1}{x+5}$

8. $\dfrac{x^2-64}{x^2-16} \div \dfrac{x+8}{x+4}$

9. $\dfrac{x^2+6x}{6} \cdot \dfrac{x^3+6}{x^3+6x^2}$

10. $\dfrac{x^2-4}{2x-4} \cdot \dfrac{2}{x+2}$

11. $\dfrac{x+y}{x-y} \div \dfrac{y+x}{y-x}$

12. $\dfrac{x^2-9}{x^2+x} \div \dfrac{3-x}{x^2-1}$

13. $\dfrac{x^2-49}{49} \cdot \dfrac{7}{7-x}$

14. $\dfrac{x+5}{5-x} \cdot \dfrac{x^2-4x-5}{x^2}$

15. $\dfrac{x^2+2xy+y^2}{x^2-y^2} \div \dfrac{y+x}{y-x}$

16. $\dfrac{(x-y)^2}{y^2+xy} \div \dfrac{y^2-x^2}{x^2+xy}$

17. $\dfrac{x^2-3x-4}{3+x} \cdot \dfrac{x+3}{16-x^2}$

18. $\dfrac{x^2-x}{1-x} \cdot \dfrac{x+2}{x^2-2x}$

19. $\dfrac{x^2-3xy-10y^2}{xy} \div \dfrac{x^2+7xy+10y^2}{x-5y}$

20. $\dfrac{xy}{(x+2y)(x+3y)} \div \dfrac{x+2y}{(x+3y)(xy)}$

21. $\dfrac{x^2+x-2}{x^2-4x-12} \cdot \dfrac{x^2-5x-6}{x^2-2x+1}$

22. $\dfrac{x^2+3x-10}{x^2-7x+6} \cdot \dfrac{x^2+2x-3}{x^2+x-6}$

23. $\dfrac{x^2-7x+12}{x^2-x-6} \div \dfrac{x^2-16}{x^2+x-2}$

24. $\dfrac{x^2-6x+8}{x^2-5x+6} \div \dfrac{x^2-7x+12}{x^2-4x+4}$

25. $(x^2-5x-14) \cdot \dfrac{x+3}{x^2-4x-21}$

26. $\dfrac{x^2-81}{x^2-18x+81} \cdot (x^2-9x)$

27. $\dfrac{(x+5)(x+8)}{5-x} \div (x+8)$

28. $(x^2-x-72) \div \dfrac{x-9}{x+8}$

29. $\dfrac{x^3-1}{x+1} \cdot \dfrac{x^2+2x+1}{x^2+x+1}$

30. $\dfrac{x^3+8}{x-2} \cdot \dfrac{x^2-4x+4}{x^2-2x+4}$

31. $\dfrac{x^4-27x}{x^2-9} \div \dfrac{x^2+3x+9}{x+3}$

32. $\dfrac{x^3+y^3}{(x+y)^3} \div \dfrac{x^2-xy+y^2}{x^2+2xy+y^2}$

33. $\dfrac{3x^2-2xy+6x-4y}{3x^2+xy-2y^2} \cdot \dfrac{x-2}{x^2-4}$

34. $\dfrac{x^2+3ax-4x-12a}{x^2+3x} \cdot \dfrac{x+3}{x^2-ax-12a^2}$

Problems 35 through 38 involve higher degree polynomials, so you must use the Factor Theorem.

35. $\dfrac{x^3 - 4x^2 - 3x + 18}{x^2 + x - 6} \cdot \dfrac{x^2 + 7x + 12}{x^2 + x - 12}$

36. $\dfrac{x^3 - 7x + 6}{x^2 + 4x - 5} \cdot \dfrac{x^2 + 9x + 20}{x^3 - 4x^2 + 5x - 2}$

37. $\dfrac{x^3 + x^2 - 4x - 4}{x^2 - 2x - 15} \div \dfrac{x^2 - 4}{x^2 + 8x + 15}$

38. $\dfrac{x^4 + 8x^3 + 24x^2 + 32x + 16}{x^2 - 4} \div \dfrac{x^2 + 4x + 4}{x^2 - x - 2}$

Problems 39 through 46 involve *more* than two fractions to be multiplied or divided. Recall that by the agreed-upon sequence of operations, $\dfrac{a}{b} \div \dfrac{c}{d} \cdot \dfrac{e}{f}$ is equivalent to $\dfrac{a}{b} \cdot \dfrac{d}{c} \cdot \dfrac{e}{f}$. Carry out the indicated operations and simplify:

39. $\dfrac{x + 3}{x - 7} \div \dfrac{x - 7}{x + 5} \cdot \dfrac{x + 5}{x + 3}$

40. $\dfrac{a + 9}{a - 10} \cdot \dfrac{a + 8}{a + 9} \div \dfrac{a - 10}{a + 8}$

41. $\dfrac{25x^2 - 1}{9x^2 - 4y^2} \div \dfrac{5x - 1}{3x - 2y} \div \dfrac{5x + 1}{3x + 2y}$

42. $\dfrac{x + 5}{x - 7} \div \dfrac{x + 3}{7 - x} \div \dfrac{x + 5}{x + 3}$

43. $\dfrac{x + 2}{x - 4} \div \dfrac{x - 4}{x + 3} \div \dfrac{x - 4}{x + 2}$

44. $\dfrac{xy - y^2}{x^2 - x} \cdot \dfrac{x^2 + xy}{xy - y^2} \div \dfrac{xy + y^2}{x^2 - xy}$

45. $\dfrac{x^4 - y^4}{x^3 - y^3} \cdot \dfrac{x^2 - y^2}{x^3 + y^3} \cdot \dfrac{x^6 - y^6}{x^2 + y^2}$

46. $\dfrac{(r^2 + rx)^2}{(r^2 - rx)^2} \div \dfrac{r + x}{r - x} \div \dfrac{r^3 + x^3}{r^3 - x^3}$

For Problems 47 through 62 you are asked to simplify fractions that have other fractions in their numerators or denominators. This can be accomplished by multiplying the fraction by a clever form of 1. For example, to simplify

$$\frac{3 - \dfrac{6}{x + 5}}{1 + \dfrac{7}{x - 4}},$$

you would multiply by 1 in the form

$$\frac{3 - \dfrac{6}{x + 5}}{1 + \dfrac{7}{x - 4}} \cdot \frac{(x + 5)(x - 4)}{(x + 5)(x - 4)}.$$

Multiplying, and associating in the numerator and denominator,

$$\frac{\left[\left(3 - \dfrac{6}{x + 5}\right)(x + 5)\right](x - 4)}{\left[\left(1 + \dfrac{7}{x - 4}\right)(x - 4)\right](x + 5)}.$$

Distributing and simplifying gives

$$\frac{(3x + 15 - 6)(x - 4)}{(x - 4 + 7)(x + 5)} = \frac{3(x - 3)(x - 4)}{(x - 3)(x + 5)}$$
$$= \frac{3x - 12}{x + 5}.$$

Simplify the following complex fractions:

47. $\dfrac{1 - \dfrac{1}{x^2}}{1 + \dfrac{1}{x}}$

48. $\dfrac{12 + \dfrac{6}{x}}{12 - \dfrac{3}{x^2}}$

49. $\dfrac{x + 3 + \dfrac{2}{x}}{1 - \dfrac{4}{x^2}}$

50. $\dfrac{x - 5 + \dfrac{6}{x}}{1 - \dfrac{9}{x^2}}$

51. $\dfrac{\dfrac{1}{x} - \dfrac{2}{x^2} - \dfrac{3}{x^3}}{\dfrac{9}{x} - x}$

52. $\dfrac{1 - \dfrac{y^2}{x^2}}{1 + \dfrac{y}{x}}$

53. $\dfrac{\dfrac{1}{1 + x}}{1 - \dfrac{1}{1 + x}}$

54. $\dfrac{2x + \dfrac{x}{x - 2}}{2x - \dfrac{x}{x - 2}}$

55. $\dfrac{x-3+\dfrac{12}{x+5}}{x-8+\dfrac{42}{x+5}}$

56. $\dfrac{x+3+\dfrac{5}{x-3}}{x+2+\dfrac{4}{x-3}}$

57. $\dfrac{x^{-2}-y^{-2}}{x^{-1}y^{-1}}$

58. $\dfrac{x^{-4}-y^{-4}}{x^{-2}+y^{-2}}$

59. $\dfrac{1-\dfrac{4}{x+1}}{1-\dfrac{2}{x-1}}$

60. $\dfrac{4-\dfrac{12}{x+2}}{2+\dfrac{4}{x-3}}$

61. $\dfrac{x-2-\dfrac{x^2-5x}{x-3}}{x+\dfrac{3x}{x-3}}$

62. $\dfrac{x+y-\dfrac{x^2+y^2}{x+y}}{x+y-\dfrac{2xy}{x+y}}$

7-8 SUMS AND DIFFERENCES OF RATIONAL EXPRESSIONS

In the previous section you learned how to multiply and divide rational algebraic expressions; in this section you will see how to add and subtract them.

Objective:

Be able to add or subtract several rational expressions and simplify the result.

You recall from Chapter 1 that division distributes over addition. That is,

$$\frac{a+b}{c}=\frac{a}{c}+\frac{b}{c}\,.$$

Using symmetry to read the equation the other way, you can see that two fractions with the same denominator may be added by *adding* their numerators and using their *common* denominator.

Subtraction of fractions with the same denominator is done in the same way, because division distributes over subtraction, too.

$$\frac{a}{c}-\frac{b}{c}=\frac{a-b}{c}\,.$$

Addition or subtraction of fractions with *unlike* denominators may be accomplished by first transforming the fractions so that they have the same denominator. For example, to subtract

$$\frac{7x-1}{x^2-2x-3}-\frac{6x}{x^2-x-2}\,,$$

you would first factor the two denominators, getting

$$\frac{7x-1}{(x-3)(x+1)}-\frac{6x}{(x-2)(x+1)}\,.$$

The LCM of the two denominators is $(x-3)(x-2)(x+1)$. To make the first fraction have this as its denominator, you could multiply it by 1 in the form of $(x-2)/(x-2)$. The second fraction could be multiplied by 1 in the form of $(x-3)/(x-3)$, giving

$$\frac{7x-1}{(x-3)(x+1)}\cdot\frac{x-2}{x-2}-\frac{6x}{(x-2)(x+1)}\cdot\frac{x-3}{x-3}$$

$$=\frac{7x^2-15x+2}{(x-3)(x+1)(x-2)}-\frac{6x^2-18x}{(x-2)(x+1)(x-3)}$$

$$=\frac{7x^2-15x+2-(6x^2-18x)}{(x-3)(x+1)(x-2)}\,.$$

The fractions have now been subtracted, and all that remains is to tidy up the answer. Associating like terms in the numerator, factoring, and canceling gives

$$\frac{x^2+3x+2}{(x-3)(x+1)(x-2)}$$

$$=\frac{(x+1)(x+2)}{(x-3)(x+1)(x-2)}$$

$$=\frac{x+2}{(x-3)(x-2)}\,.$$

The answer may be left in this form or the factors in the denominator may be multiplied together giving

$$\frac{x+2}{x^2-5x+6}.$$

There are several ideas to keep in mind when you add or subtract rational expressions:

1. If any of the expressions can be simplified, do the simplifying *before* adding or subtracting. For example,

$$\frac{x+5}{x^2+2x-15}+\frac{2}{x-3}=\frac{x+5}{(x+5)(x-3)}+\frac{2}{x-3}$$

$$=\frac{1}{x-3}+\frac{2}{x-3}$$

$$=\frac{3}{x-3}.$$

2. Make sure that the common denominator is the *least* common multiple of the denominators or you will carry along a lot of excess numbers that would cancel at the end, anyway.

For example, to add

$$\frac{x-7}{(x+2)(x-1)}+\frac{3}{(x+2)(x+4)},$$

you would use $(x+2)(x-1)(x+4)$ as the common denominator rather than $(x+2)(x-1)(x+2)(x+4)$. The factor $(x+2)$ need appear only *once.*

3. If one denominator has a factor of the form $(a-b)$ and another has a factor of the form $(b-a)$, only *one* of these factors need appear in the common denominator since $a-b$ and $b-a$ are additive inverses of each other. For example, for

$$\frac{3}{x-2}+\frac{4}{2-x}+\frac{1}{x}$$

The LCM would be $(x-2)(x)$ rather than $(x-2)(2-x)(x)$. The middle fraction would be multiplied by $(-x)/(-x)$, giving

$$\frac{4}{2-x}=\frac{4}{2-x}\cdot\frac{-x}{-x}$$ Multiplication property of 1.

$$=\frac{-4x}{(2-x)(-x)}$$ Multiplication property of fractions.

$$=\frac{-4x}{(x-2)(x)}$$ $(-a)(-b)=ab.$

Thus, the middle fraction has $(x-2)(x)$ as its denominator.

4. Keep the denominator in factored form until the very end so that you will have the best chance of canceling.

The exercise which follows is designed to give you practice in adding and subtracting rational expressions, and combining these operations with multiplication and division.

Exercise 7-8

1. *Distributivity of Division over Addition* – Use the Field Axioms, Equality Axioms, and the Definition of Division to prove that division distributes over addition. That is, prove that

$$\frac{a+b}{c}=\frac{a}{c}+\frac{b}{c}.$$

2. The expression $ac+bc$ can be transformed to $(a+b)(c)$ by factoring out a common factor. Explain why addition of the two fractions

$$\frac{a}{c}+\frac{b}{c}$$

can be thought of as "factoring out a common *denominator.*"

For Problems 3 through 42, carry out the indicated addition and subtraction, and simplify.

3. $\frac{x-3}{3} - \frac{x-4}{4}$

4. $\frac{2x-1}{3} - \frac{4x-8}{6}$

5. $\frac{1}{x+1} + \frac{1}{x-1}$

6. $\frac{3}{x-1} + \frac{1}{1-x}$

7. $\frac{6}{2x-3y} - \frac{3}{3y-2x}$

8. $\frac{x}{x+y} + \frac{y}{x-y}$

9. $\frac{2x-1}{x+1} - \frac{2x-1}{x-1}$

10. $\frac{x+3}{x-3} - \frac{x-3}{x+3}$

11. $\frac{a-b}{c-d} - \frac{b-a}{d-c}$

12. $\frac{5}{3(a-b)} + \frac{3}{2(b-a)}$

13. $\frac{1}{x-y} + \frac{2x-y}{x^2-y^2}$

14. $\frac{x}{x^2-y^2} + \frac{y}{y^2-x^2}$

15. $\frac{1}{y-x} + \frac{x}{(x-y)^2}$

16. $\frac{4x}{(x+y)^2} - \frac{4}{x+y}$

17. $\frac{1}{1-2x} - \frac{2}{1-4x^2}$

18. $\frac{3a}{9a^2-4b^2} - \frac{1}{3a+2b}$

19. $\frac{3}{1-x} + \frac{4}{(1-x)^2}$

20. $\frac{2y}{(x-2y)^2} + \frac{1}{x-2y}$

21. $\frac{2}{x-4} - \frac{x+12}{x^2-16}$

22. $\frac{3}{x+5} - \frac{2x-20}{x^2-25}$

23. $\frac{x-y}{x^2-y^2} + \frac{1}{2x+3y}$

24. $\frac{x^2+5x+4}{x+4} - \frac{x^2-5x+6}{x-2}$

25. $\frac{1}{x-2y} - \frac{x^2+4y^2}{x^3-8y^3}$

26. $\frac{x+2}{x^3-1} + \frac{x+1}{x^2+x+1}$

27. $x+2-\frac{x^2+x-6}{x-3}$

28. $2x+5-\frac{x^2+2x-15}{x-3}$

29. $\frac{x+a}{x-a} - \frac{x^2+a^2}{ax-a^2}$

30. $\left(\frac{x+y}{x-y}\right)^2 - 1$

31. $\frac{1}{x+1} - \left(\frac{1}{x-1} - \frac{1}{x^2-1}\right)$

32. $\frac{16x-x^2}{x^2-4} + \frac{2x+3}{2-x} + \frac{3x-2}{x+2}$

33. $\frac{1}{x-1} - \frac{2}{x-2} + \frac{1}{x-3}$

34. $\frac{a}{a^2-b^2} - \frac{1}{3(a-b)} - \frac{1}{3(a+b)}$

35. $\frac{1}{x+y} - \frac{1}{x-y} + \frac{2x}{x^2-y^2}$

36. $\frac{3}{x+6} - \frac{4x}{x^2-36} - \frac{2}{6-x}$

37. $\frac{3}{x^2+x-2} - \frac{5}{x^2-x-6}$

38. $\frac{5}{x^2-3x-4} - \frac{3}{x^2-x-2}$

39. $\frac{3x+13}{x^2-3x-10} - \frac{16}{x^2-6x+5}$

40. $\frac{6}{x^2-7x+12} + \frac{5x+9}{x^2-2x-3}$

41. $\frac{x-2}{x^2-x-2} + \frac{x-4}{x^2-5x+4}$

42. $\frac{x+4}{x^2-3x-28} - \frac{x-5}{x^2+2x-35}$

For Problems 43 through 48 you are asked to *reverse* the process of adding fractions, and split a complicated rational expression apart into a sum of two simpler fractions. For example,

$$\frac{4x+41}{(x-2)(x+5)} = \frac{A}{x-2} + \frac{B}{x+5},$$

where A and B stand for constants. Multiplying both members of this equation by $(x-2)$ gives

$$\frac{4x+41}{x+5} = A + \left(\frac{B}{x+5}\right)(x-2).$$

Substituting 2 for x gives

$$\frac{4\cdot 2+41}{2+5} = A + \left(\frac{B}{x+5}\right)\cdot 0$$

$$7 = A.$$

Similarly, multiplying both members of the original equation by $(x + 5)$, then substituting -5 for x, gives $B = -3$. So

$$\frac{4x+41}{(x-2)(x+5)} = \frac{7}{x-2} - \frac{3}{x+5}.$$

This process is called, "resolving into *partial fractions.*" Resolve the following into partial fractions:

43. $\dfrac{5x-1}{x^2-x-2}$

44. $\dfrac{5x-10}{x^2-x-6}$

45. $\dfrac{3x+18}{x^2+5x+4}$

46. $\dfrac{x-8}{x^2-5x+6}$

47. $\dfrac{4x^2+15x-1}{x^3+2x^2-5x-6}$

48. $\dfrac{-3x^2+22x-31}{x^3-8x^2+19x-12}$

7-9 GRAPHS OF RATIONAL ALGEBRAIC FUNCTIONS, AGAIN

By now you should be comfortable with simplifying rational expressions. If so, you are ready to return to the problem of plotting rational function graphs.

Objective:

Be able to determine what values of x are excluded from the domain of a given rational function, figure out what happens to the graph at these points, and draw the graph.

In Sections 7-1 and 7-2 you plotted the graph of

$$f(x) = \frac{x+2}{x^2-x-6}.$$

This graph is shown in Figure 7-9a. By factoring the denominator you get

$$f(x) = \frac{x+2}{(x+2)(x-3)}.$$

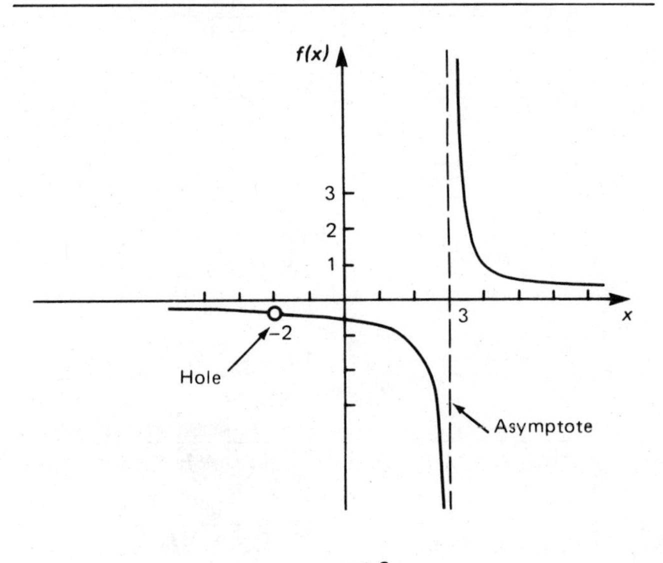

Figure 7-9a.

If $x = -2$ or $x = 3$, $f(x)$ will be *undefined*, since these values of x make the denominator equal *zero.* At $x = 3$, you found a vertical asymptote. At $x = -2$, you found one point deleted from an otherwise "continuous" graph. In order to be able to draw such graphs, you need a technique for locating such "holes" and asymptotes quickly.

If you simply substitute 3 for x in the equation for $f(x)$, you get

$$\frac{5}{0}.$$

Expressions such as $5/0$ are said to be *infinitely large.* You can see why by choosing values of the denominator that are close to zero.

$$\frac{5}{0.01} = 500, \frac{5}{-0.0001} = -50{,}000,$$

$$\frac{5}{0.000001} = 5{,}000{,}000$$

The closer the denominator gets to zero, the bigger the fraction becomes (in absolute value).

If you substitute -2 for x in the equation for $f(x)$, you get the form

$$\frac{0}{0}.$$

The form 0/0 is said to be *indeterminate*. For example,

$$\frac{3x}{x} = 3, \quad \text{if } x \neq 0.$$

When x *does* equal 0, the expression has the form 0/0. So you could conclude that 0/0 is equal to 3! But the same line of reasoning leads to other conclusions, for example,

$$\frac{57x}{x} = 57, \ \frac{x}{\pi x} = \frac{1}{\pi}, \text{ and } \frac{-13x}{x} = -13.$$

Substituting 0 for x in these expressions would lead to the conclusion that 0/0 is also equal to 57, $1/\pi$, and -13! The word "indeterminate" is used since you cannot determine what 0/0 is close to without knowing where the two 0's came from.

These facts allow you to tell how the graph of a rational function behaves when the denominator is 0.

Conclusion: Suppose that substituting c for x makes the denominator of $f(x)$ equal 0.

1. If $f(c)$ has the form $\frac{\text{non-zero}}{\text{zero}}$, then $f(c)$ is *infinitely large,* and there is a vertical asymptote in the graph at $x = c$.

2. If $f(c)$ has the form $\frac{0}{0}$, then $f(c)$ is *indeterminate.*

You must examine the *simplified* fraction to see whether there is a vertical asymptote at $x = c$, or just a deleted point (a "hole") at $x = c$.

Once you have found out what happens to the graph when the denominator equals zero, you need only calculate other points, plot them, and draw the graph.

Example: Draw the graph of

$$f(x) = \frac{x^2 - 1}{x^2 + 2x - 3}.$$

Factoring the numerator and denominator gives

$$f(x) = \frac{(x + 1)(x - 1)}{(x + 3)(x - 1)}.$$

$f(x)$ is *undefined* when $x = -3$ or $x = 1$. If $x = -3$, then $f(x)$ has the form

$$\frac{-4}{0},$$

which is *infinitely large*. So there is a *vertical asymptote* at $x = -3$.

If $x = 1$, $f(x)$ has the form

$$\frac{0}{0},$$

which is *indeterminate.* If the $(x - 1)$ factors are canceled,

$$f(x) = \frac{x + 1}{x + 3}.$$

Substituting $x = 1$ into the simplified fraction gives

$$\frac{1 + 1}{1 + 3} = \frac{2}{4} = \frac{1}{2}.$$

So there is a *deleted point* (a "hole") in the graph at $(1, \frac{1}{2})$.

Values of $f(x)$ for other values of x can be calculated by constructing a table. Choose values of x from below the smallest value of x that makes the denominator zero, to above the largest such value.

x	$x + 1$	$x + 3$	$f(x) = \dfrac{x + 1}{x + 3}$
−6	−5	−3	$\frac{-5}{-3} = 1\frac{2}{3}$
−5	−4	−2	$\frac{-4}{-2} = 2$
−4	−3	−1	$\frac{-3}{-1} = 3$
−3	−2	0	$\frac{-2}{0}$ (infinite)
−2	−1	1	$\frac{-1}{1} = -1$
−1	0	2	$\frac{0}{2} = 0$
0	1	3	$\frac{1}{3}$
1	2	4	$\frac{2}{4}$ (hole at ½)
2	3	5	$\frac{3}{5}$
3	4	6	$\frac{4}{6} = 2/3$
4	5	7	$\frac{5}{7}$

The table is most easily constructed by going *down* the columns. For example, you would find *all* values of $x + 1$ before starting on the values of $x + 3$. The values of $f(x)$ are calculated by simply dividing the numbers in the $x + 1$ column by the numbers in the $x + 3$ column. The graph is shown in Figure 7-9b. Note that as x gets very large, the fraction gets closer and closer to 1. For example,

$$f(1000) = \frac{1001}{1003} \approx 0.998 \approx 1.$$

The following exercise is designed to give you practice in locating critical features of the graphs of rational functions (asymptotes and holes), and sketching the graphs.

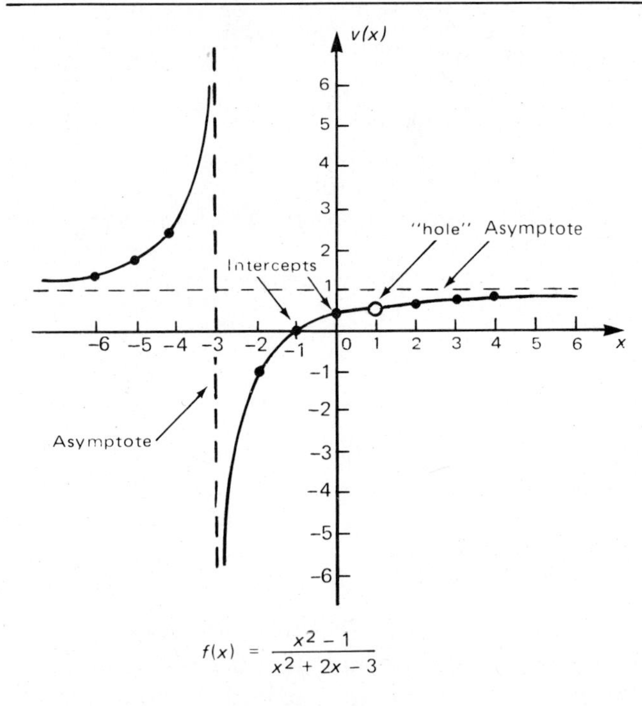

Figure 7-9b.

Exercise 7-9

1. Consider the "expressions" $\frac{0}{7}$, $\frac{7}{0}$, and $\frac{0}{0}$.

   a. Two of these expressions are *undefined. Which* two?
   b. What special words are used to distinguish between the two that are undefined?
   c. The other expression *is* defined. *Which* one? What does it equal?

2. Consider the function $f(x) = \dfrac{5x + 7}{2x + 1}$.

   a. Find $f(100)$, $f(10{,}000)$, and $f(1{,}000{,}000)$. Use a calculator to find decimal approximations for these numbers.
   b. What number does $f(x)$ seem to be approaching as x gets very large?
   c. What does the graph of f do as x becomes very large?

For Problems 3 through 30,

a. find the values of x which must be excluded from the domain,
b. tell which feature, asymptote or hole, there will be at each of the excluded values,
c. calculate some points and draw the graph.

3. $f(x) = \dfrac{4}{x-5}$

4. $f(x) = \dfrac{3}{x+1}$

5. $f(x) = \dfrac{x^2 - 4x + 3}{x^2 - x - 6}$

6. $f(x) = \dfrac{x^2 - x - 6}{x^2 + 3x + 2}$

7. $f(x) = \dfrac{2}{x^2 + 3x - 10}$

8. $f(x) = \dfrac{5}{x^2 - 2x - 8}$

9. $f(x) = \dfrac{x-1}{x^2 + 3x - 4}$

10. $f(x) = \dfrac{x-3}{x^2 - 2x - 3}$

11. $f(x) = \dfrac{x^2 + 2x - 15}{x-3}$

12. $f(x) = \dfrac{x^2 + 6x - 7}{x-1}$

For Problems 13 through 16, it will help if you recall how to plot *quadratic* function graphs.

13. $f(x) = \dfrac{x^3 - 8}{x-2}$

14. $f(x) = \dfrac{x^3 + 1}{x+1}$

15. $f(x) = \dfrac{x^3 - 2x^2 - 5x + 6}{x+2}$

16. $f(x) = \dfrac{x^3 + 3x^2 - 18x - 40}{x+5}$

17. $f(x) = \dfrac{5}{x^2 + 1}$

18. $f(x) = \dfrac{3x + 3}{x^3 + 1}$

19. $f(x) = \dfrac{x+5}{x^3 + 9x^2 + 24x + 20}$

20. $f(x) = \dfrac{x+1}{x^3 + 3x^2 - 6x - 8}$

21. $f(x) = \dfrac{(x-1)(x-4)}{(x-4)(x-3)(x+2)}$

22. $f(x) = \dfrac{(x+2)(x+3)}{(x+1)(x-2)(x+3)}$

23. $f(x) = \dfrac{(x+2)(x-1)}{(x-2)^2(x-1)}$

24. $f(x) = \dfrac{(x-3)(x+2)}{(x+3)^2(x+2)}$

25. $f(x) = \dfrac{(x-2)(x-1)}{(x-2)^2(x-1)}$

26. $f(x) = \dfrac{(x+3)(x+2)}{(x+3)^2(x+2)}$

27. $f(x) = \dfrac{8x}{x^2 - 6x + 5} - \dfrac{x-7}{x^2 + x - 2}$

28. $f(x) = \dfrac{7x+5}{x^2 + 2x - 3} + \dfrac{3x}{x^2 - 3x + 2}$

29. $f(x) = \dfrac{\dfrac{1}{1+x}}{1 - \dfrac{1}{1+x}}$

30. $f(x) = \dfrac{x - 2 - \dfrac{x^2 - 5x}{x-3}}{x + \dfrac{3x}{x-3}}$

31. Plot graphs of the following functions on the *same* set of axes:

a. $f(x) = \dfrac{10}{x^2}$

b. $g(x) = \dfrac{10}{x^2 + 1}$

c. $h(x) = \dfrac{10}{x^2 + 2}$

32. *Computer Program for Evaluating Rational Functions* – The flow chart on the next page is for calculating values of $f(x)$ for each value of x between two given values, L and H, increasing x each time by an amount I.

a. Show the simulated computer memory and the output if the function is

$$f(x) = \frac{x-7}{x^2 - 3x - 10}$$

and the values put in for L, H, and I are -4, 4, and 1, respectively.

b. Write a computer program for this flow chart, and use it to evaluate the function in part a for the given values of L, H, and I.

c. The function in part a reaches a *minimum* value between $x = -2$ and $x = 4$. Between what two integer values of x does this minimum occur?

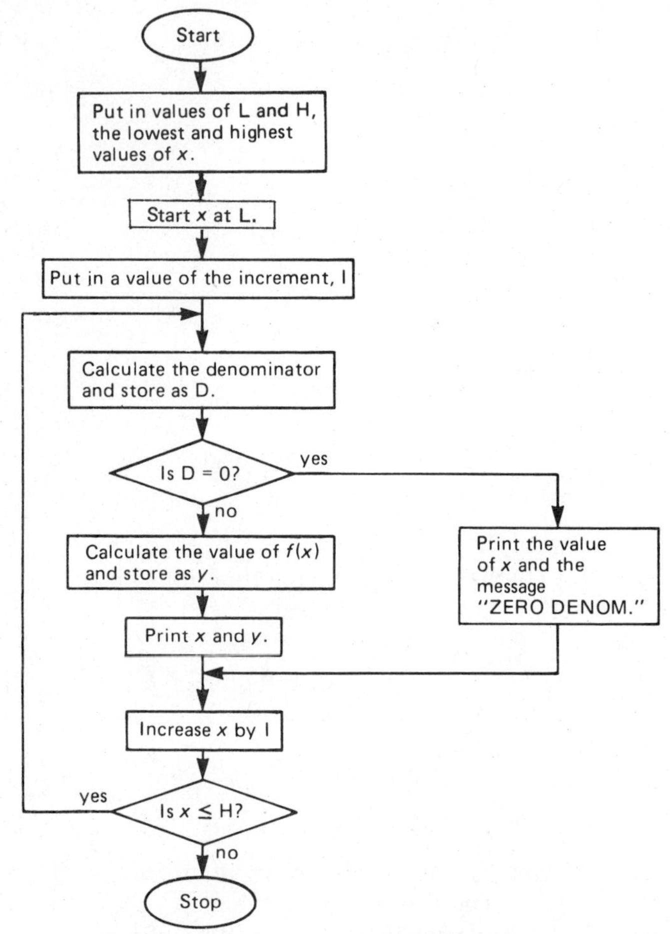

d. Let L and H be the two integers you found in part c. Let $I = 0.1$. Find the two values of x between which the minimum occurs. Repeat the procedure using $I = 0.01$, $I = 0.001$, and $I = 0.0001$ to find a four decimal place approximation for the value of x at which the minimum occurs.

33. ***Man Overboard Problem.*** Suppose that you are traveling with a convoy of ships going at a rate of 8 knots (nautical miles per hour) when a man falls overboard from one of the ships. Six minutes later (0.1 hour), your ship turns around and proceeds at a faster rate, s, to pick him up. Meanwhile, the convoy continues at 8 knots. The number of hours, t, from the time the man falls overboard until your ship returns to the convoy depends on your rate, s.

a. Recalling that distance = rate × time, how far back must you travel to pick up the man?
b. In terms of s, how long will it take you to reach the man from the time you turn around?
c. The rescue takes 12 minutes (0.2 hour). In terms of s, how far away is the convoy when you start the return trip?
d. On the return trip, at what rate are you *catching up* with the convoy?
e. How long will it take you to reach the convoy from the time the rescue is completed?
f. Let t be the total number of hours from the time the man fell overboard until your ship returns to the convoy. Write an equation expressing t in terms of s. Simplify as much as possible.
g. What will t be if s is 20 knots? 16 knots? 10 knots? 7 knots?
h. Graph the function.
i. The maximum speed your ship can go is 22 knots. What is the domain of this function?
j. At what speed must you go to be back in 1 hour?

7-10 FRACTIONAL EQUATIONS

A fractional equation is an equation that has a *variable* in a *denominator*. For example,

$$\frac{x}{x-2} + \frac{2}{x+3} = \frac{10}{x^2 + x - 6}$$

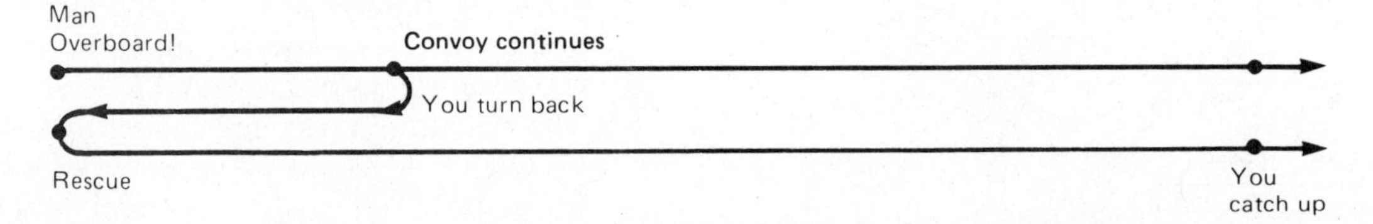

is a fractional equation. An equation such as $\frac{1}{2}x + 7 = 3$ is *not* called a fractional equation, even though it has a fraction in it, because there are no variables in denominators.

Such equations arise, for example, in finding the value of x for a given value of y in a rational algebraic function. Since fractional equations occur in many other places, it is worth spending some time learning how to solve them.

Objective:

Given a fractional equation, be able to solve it.

The first thing you must recognize is that certain values of x might make a denominator equal *zero*. In the above example, denominators are zero when $x = 2$ and when $x = -3$. These values of x must be excluded from the domain. So you write

$$x \neq 2, \ x \neq -3.$$

Next, you must transform the equation to a familiar form. Multiplying both members by an appropriate expression will eliminate the fractions. To see what this expression would be, it helps to factor any denominators. So you write.

$$\frac{x}{x-2} + \frac{2}{x+3} = \frac{10}{(x-2)(x+3)}.$$

The *least common multiple* (LCM) of the denominators is $(x - 2)(x + 3)$. Multiplying both members of the equation by this expression gives

$$(x-2)(x+3)\left(\frac{x}{x-2} + \frac{2}{x+3}\right) = \left(\frac{10}{(x-2)(x+3)}\right)(x-2)(x+3).$$

The denominator on the right cancels immediately. On the left, you must remember to *distribute* the $(x - 2)(x + 3)$ to both terms.

$$(x-2)(x+3)\left(\frac{x}{x-2}\right) + (x-2)(x+3)\left(\frac{2}{x+3}\right) = 10.$$

Canceling and simplifying reduces the equation to

$$(x+3)(x) + (x-2)(2) = 10$$

$$x^2 + 5x - 14 = 0.$$

You could use the Quadratic Formula to solve this quadratic equation, but there is an easier way that sometimes works. Factoring the left member gives

$$(x-2)(x+7) = 0.$$

By the Multiplication Property of Zero and its Converse, you know that the only way this product can be zero is for one of the factors to equal zero. So this equation is equivalent to

$$x - 2 = 0 \quad \text{or} \quad x + 7 = 0,$$

$$x = 2 \quad \text{or} \quad x = -7.$$

Before you write the solution set, you must look back at the excluded values. Since $x \neq 2$, the 2 must be an *extraneous* solution. It satisfies the transformed equation, but not the original one. Such an extraneous solution can occur when you multiply both members of an equation by an expression that can equal zero, as you will demonstrate in Problem 1 of Exercise 7-10. So you write

extraneous

$$\cancel{x = 2} \quad \text{or} \quad x = -7$$

$$\therefore S = \{-7\}.$$

Summarizing the steps involved in solving fractional equations, you

1. write the domain,

2. multiply both members of the equation by the smallest expression needed to eliminate all of the fractions,

3. solve the resulting polynomial equation,

4. discard any extraneous solutions, and

5. write the solution set.

The following exercise is designed to give you practice in solving fractional equations, after exposing you to solving polynomial equations by factoring.

Exercise 7-10

1. Starting with the equation $x = 3$, do the following:

a. Multiply both members by $(x - 4)$.
b. Tell why the transformed equation *is* true when $x = 4$, but the original equation is *not*.
c. What name is given to the solution 4?
d. Multiplying both members of an equation by an expression which can equal zero, such as $(x - 4)$, is called an *irreversible step*. Why do you suppose that this name is used?

2. Starting with the equation $x^2 = 3x$, do the following:

a. Show that the solution set is $S = \{0, 3\}$.
b. Divide both members of the equation by x. What is the solution set of the transformed equation?
c. Why is it more dangerous to *divide* both members of an equation by a variable than it is to *multiply*?

For Problems 3 through 20, solve the equation by factoring.

3. $x^2 - 2x - 3 = 0$
4. $x^2 - x - 20 = 0$
5. $x^2 - 10x - 24 = 0$
6. $x^2 + 5x - 6 = 0$
7. $2x^2 + 7x + 5 = 0$
8. $3x^2 - 10x + 8 = 0$
9. $3x^2 - 5x + 2 = 0$
10. $3x^2 - 7x + 2 = 0$
11. $x^2 + 3x = 0$
12. $x^2 - 5x = 0$
13. $x^2 = 2x$
14. $x^2 = -7x$
15. $(x - 2)(x - 3) = 20$
16. $(x + 3)(x - 1) = 5$
17. $x^3 + 2x^2 - 5x - 6 = 0$
18. $x^3 - 5x^2 - 2x + 24 = 0$
19. $x^3 + x^2 - 6x + 4 = 0$
20. $x^3 + 2x^2 - 6x - 4 = 0$

For Problems 21 through 42, state the domain, then solve the equation. Show the step where you discard any extraneous solutions.

21. $x + \dfrac{x}{x-2} = \dfrac{2}{x-2}$
22. $x + \dfrac{2x}{x-1} = \dfrac{3-x}{x-1}$
23. $\dfrac{x}{x-3} - \dfrac{7}{x+5} = \dfrac{24}{x^2 + 2x - 15}$
24. $\dfrac{x}{x+2} + \dfrac{7}{x-5} = \dfrac{14}{x^2 - 3x - 10}$
25. $\dfrac{3x}{x+4} + \dfrac{4x}{x-3} = \dfrac{84}{x^2 + x - 12}$
26. $\dfrac{4x}{x-1} - \dfrac{5x}{x-2} = \dfrac{2}{x^2 - 3x + 2}$
27. $\dfrac{3}{x-3} + \dfrac{4}{x-4} = \dfrac{25}{x^2 - 7x + 12}$
28. $\dfrac{11x}{x+20} + \dfrac{24}{x} = 11 + \dfrac{88}{x(x+20)}$
29. $\dfrac{x+2}{x-3} + \dfrac{x-2}{x-6} = 2$
30. $\dfrac{3x+2}{x-1} + \dfrac{2x-4}{x+2} = 5$
31. $\dfrac{2}{x+2} - \dfrac{x}{2-x} = \dfrac{x^2+4}{x^2-4}$
32. $\dfrac{x}{x+4} - \dfrac{4}{x-4} = \dfrac{x^2+16}{x^2-16}$
33. $\dfrac{1}{1-x} = 1 - \dfrac{x}{x-1}$
34. $\dfrac{x}{x-1} - \dfrac{2}{1-x^2} = \dfrac{8}{x+1}$
35. $\dfrac{x+3}{2x-3} = \dfrac{18x}{4x^2-9}$
36. $3 - \dfrac{22}{x+5} = \dfrac{6x-1}{2x+7}$
37. $\dfrac{4x}{x^2-9} - \dfrac{x-1}{x^2-6x+9} = \dfrac{2}{x+3}$

38. $\dfrac{x}{x^2 - 2x + 1} = \dfrac{2}{x + 1} + \dfrac{4}{x^2 - 1}$

39. $\dfrac{3x}{x - 2} + \dfrac{2x}{x + 3} = \dfrac{30}{x^2 + x - 6}$

40. $\dfrac{5}{x - 6} - \dfrac{4}{x + 3} = \dfrac{x + 39}{x^2 - 3x - 18}$

41. $\dfrac{x^3 + 3x - 9}{x(x - 3)} - \dfrac{x + 6}{x - 3} = \dfrac{3}{x}$

42. $\dfrac{5x}{x - 5} + \dfrac{4}{x + 6} = \dfrac{54x + 5}{x^2 + x - 30}$

43. *Continued Fractions* – The following is called a *continued fraction* (for obvious reasons!). It equals an *irrational* number. Find that number.

$$2 + \cfrac{1}{2 + \cfrac{1}{2 + \cfrac{1}{2 + \ldots}}}$$

Clue: Let x equal the fraction. Since the pattern continues *forever*, you can write

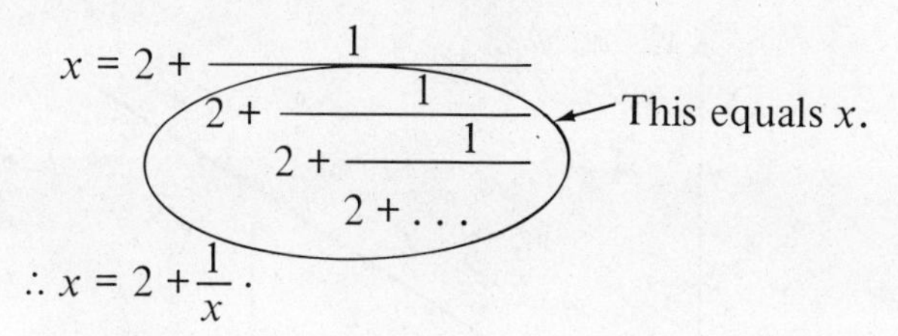

To find the value of this fraction, solve the equation for x. (See Charles G. Moore; *An Introduction to Continued Fractions;* Washington, D.C.; National Council of Teachers of Mathematics; 1964.)

44. Evaluate:

$$3 + \cfrac{1}{2 + \cfrac{1}{3 + \cfrac{1}{2 + \cfrac{1}{3 + \ldots}}}}$$

7-11 VARIATION FUNCTIONS

A relatively simple type of function that is very useful as a mathematical model has an equation in which y is equal to a constant multiplied or divided by a power of x. Functions with equations such as

$$y = 4x^2, \quad y = \frac{13}{x}, \quad y = \frac{0.732}{x^2}, \quad y = 1.9x,$$

are called *variation* functions. If the constant is *multiplied* by the variable, then y varies *directly* with the power of x. If the constant is *divided* by the variable, then y varies *inversely* with the power of x. The constant, such as 1.9 in $y = 1.9x$, is called the *proportionality constant.* The letter k is often used for this constant, after the German word *Konstante,* meaning "constant."

DEFINITION

If k and n are constants, then "y varies directly with the nth power of x" means

$$\boxed{y = kx^n},$$

and "y varies inversely with the nth power of x" means

$$\boxed{y = \frac{k}{x^n}}.$$

Notes:

1. If n is a positive integer, then direct variation functions are special cases of *polynomial* functions (linear, quadratic, etc.), and inverse variation functions are special cases of *rational algebraic* functions (such as those in Section 7-9). Variation functions in which n is *not* an integer will be explored in Chapter 8.
2. The equation $y = kx^n$ can include both direct and inverse variation functions if n is allowed to be *negative.*
3. The words, ". . . varies directly with . . . ," or ". . . varies inversely with . . . ," can be replaced by the words, ". . . is directly porportional to . . . ," or ". . . is inversely proportional to . . . ," respectively.

Some examples of variation function equations and the words that go with them are listed below.

General Equation	*Words*
$y = kx$	y varies directly with x.
$y = kx^2$	y varies directly with the square of x.
$y = kx^3$	y varies directly with the cube of x.
$y = \frac{k}{x}$	y varies inversely with x.
$y = \frac{k}{x^2}$	y varies inversely with the square of x.

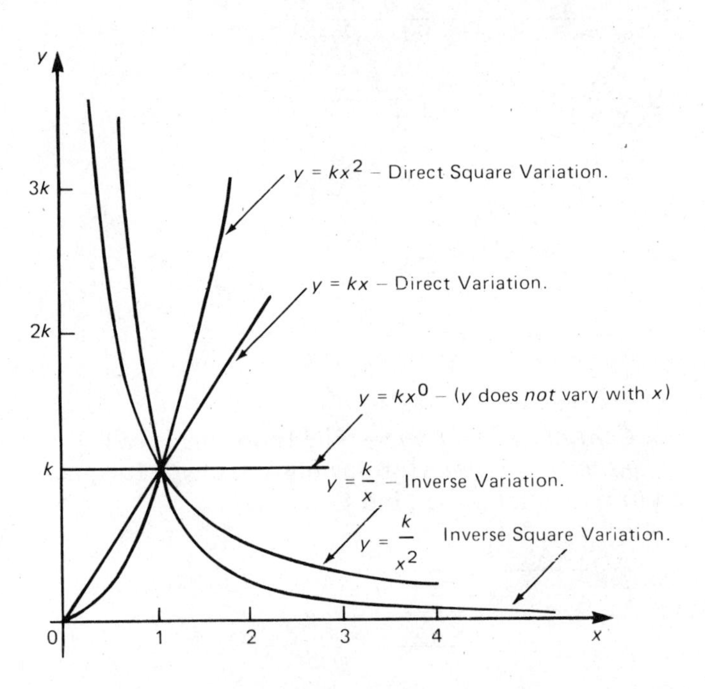

Figure 7-11a.

Objective:

Given a real-world situation,

a. determine which kind of variation function is a reasonable mathematical model,
b. find the particular equation for the function, and
c. predict values of y or x.

One way to tell whether a certain kind of function is a reasonable mathematical model is to compare its graph with the real-world graph. Figure 7-11a shows several kinds of variation function graphs. For *direct* variation functions, y gets *bigger* as x increases. For *inverse* variation functions, y gets *smaller* as x increases. Inverse variation functions have the x- and y-axes as *asymptotes.*

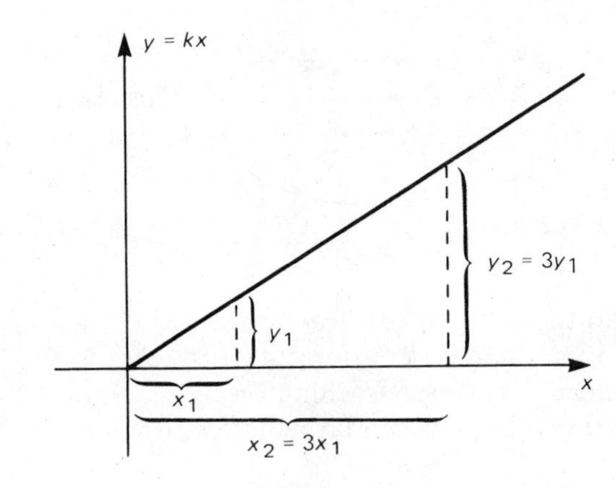

Figure 7-11b.

There is a property of variation functions that lets you tell *which power* of x to use. For $y = kx$, the graph is a straight line through the origin (Figure 7-11b). By the properties of similar triangles, if x_2 is *three* times x_1, then y_2 will also be *three* times y_1. In general, *multiplying* the value of x by some constant causes the value of y to be *multiplied* by the *same* constant.

A similar property holds for higher degree and for inverse variation functions. Suppose that

$$f(x) = 5x^2 \quad \text{and} \quad g(x) = \frac{162}{x^2}.$$

By picking certain values of x and examining the corresponding values of y, an interesting pattern shows up!

	x	$y = f(x)$	
	1	$5 \cdot 1^2 = 5$	
$\times 3$	3	$5 \cdot 3^2 = 45$	$\times 9$
$\times 3$	9	$5 \cdot 9^2 = 405$	$\times 9$
$\times 3$	27	$5 \cdot 27^2 = 3645$	$\times 9$

	x	$y = g(x)$	
	1	$\frac{162}{1^2} = 162$	
$\times 3$	3	$\frac{162}{3^2} = 18$	$\div 9$
$\times 3$	9	$\frac{162}{9^2} = 2$	$\div 9$

For function f, every time x is *multiplied* by 3, y is *multiplied* by 3^2. For function g, every time x is *multiplied* by 3, y is *divided* by 3^2. It is not hard to see why this pattern is true. Suppose that

$$y = f(x) = kx^2.$$

Substituting a constant, c for x gives

$$f(c) = kc^2.$$

Substituting $3c$ for x gives

$f(3c) = k(3c)^2$	Substitution.
$= k(9c^2)$	$(3c)(3c) = 9c^2$.
$= 9(kc^2)$	Commutativity and associativity.
$= 9 \cdot f(c)$.	Because $kc^2 = f(c)$.

This pattern is summarized in the following property:

Property of Variation Functions: If $y = kx^n$, then *multiplying* the value of x by the constant c *multiplies* the value of y by the constant c^n.

If $y = \dfrac{k}{x^n}$, then *multiplying* the value of x by the constant c *divides* the value of y by the constant c^n.

Note: This property is similar to the properties of linear and exponential functions.

> For *linear* functions, *adding* a constant to x *adds* a constant to y.
>
> For *exponential* functions, *adding* a constant to x *multiplies* y by a constant.

> For *variation* functions, *multiplying* x by a constant *multiplies* y by a constant. (Dividing by c^n can be thought of as multiplying by $1/c^n$.)

Once you understand the above properties, you are ready to select the kind of variation function that is appropriate in a given situation.

Example 1: The pressure required to force water through a garden hose depends on the number of gallons per minute (gpm) you want to flow. By experiment, you find that for a flow of 3 gpm, a pressure of 10 pounds per square inch (psi) is required. For 6 gpm, a pressure of 40 psi is required. Predict

a. the pressure required for a flow rate of 12 gpm,
b. the pressure required for a flow rate of 4.2 gpm,
c. the number of gpm you would get with a pressure of 5 psi.

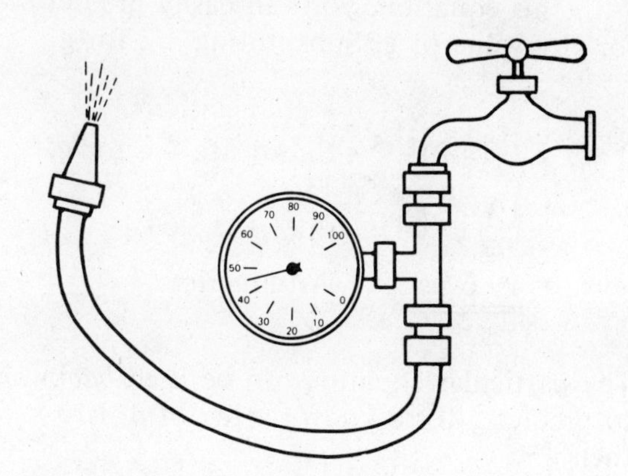

a. Let g = number of gpm.
Let p = number of psi.
You first observe that *doubling* g (from 3 to 6) makes p get *four* times as big (from 10 to 40). Assuming that this pattern continues, you would expect that doubling g again (from 6 to 12) would make p four times bigger again.
So

$$p = 4(40)$$
$$= \underline{\underline{160 \text{ psi}}}.$$

b. Since 4.2 is not a simple multiple of 3, it is helpful to derive a particular equation expressing p in terms of g. Since *doubling* g makes p *four* times as big, and since 4 is equal to 2^2, you can conclude that p varies directly with the *square* of g. The general equation is thus

$$p = kg^2.$$

Substituting one of the given ordered pairs gives

$10 = k(3^2)$	Substitute (3, 10) for (g, p).
$\frac{10}{9} = k.$	Divide by 9.

Thus, the particular equation is $p = \frac{10}{9} g^2$.

With this equation, you can easily predict p for *any* value of g. Substituting 4.2 for g gives

$p = \frac{10}{9}(4.2)^2$	Substitute 4.2 for g.
$= \frac{10}{9}(17.64)$	$4.2^2 = 17.64$
$= 19.6$ psi	Arithmetic.

c. The particular equation can be used *backwards* to predict g for a known value of p. If $p = 5$, then

$5 = \frac{10}{9} g^2$	Substitute 5 for p.
$4.5 = g^2$	Multiply by 9/10.
$2.12 \approx g$	By calculator.

So you would predict about 2.12 gpm.

It is interesting to note from the last part of the example that although 5 psi is only *half* of 10 psi, the flow rate is still over 2/3 of what it was for 10 psi.

Example 2: The intensity of light reaching you from a light bulb depends on how far from the bulb you are standing. Suppose that at 3 meters the intensity is 120 units, and at 6 meters it is 30 units.

a. How does intensity vary with distance? Write the general equation.
b. Find the particular equation.
c. Predict the intensity at 10 meters.
d. Predict the intensity at 12 meters.
e. Predict the distance at which the intensity will be 4 units.
f. Draw the graph.
g. Tell the real-world significance of the asymptotes.

a. Let I = number of units of intensity.
Let d = number of meters distance.

The two given ordered pairs are $(d, I) = (3, 120)$ and $(6, 30)$. From these, you can see that multiplying d by 2 (from 3 to 6) *divides I by 4* (from 120 to 30). Since $4 = 2^2$, I appears to vary inversely with the *square* of d.

Thus, the general equation is $I = \frac{k}{d^2}$.

b. To evaluate the proportionality constant, k, you must substitute one of the given ordered pairs. It is usually better to solve the equation for k *before* you substitute.

$I = \frac{k}{d^2}$	General equation.
$Id^2 = k$	Multiplying both members by d^2.
$(120)(3^2) = k$	Substituting (3, 120) for (d, I).
$1080 = k$	Arithmetic.

$\therefore$ the particular equation is $I = \frac{1080}{d^2}$.

c. The equation is now ready to use for making predictions.

If $d = 10$, then $I = \dfrac{1080}{10^2} = \underline{\underline{10.8 \text{ units}}}$.

d. If $d = 12$, you can calculate I *without* using the equation. The thought process is as follows:

Since 12 is *two* times 6, and the intensity at $d = 6$ is 30 units, the intensity twice as far away will be

$$30 \div (2^2) = 30 \div 4 = \underline{\underline{7.5 \text{ units}}}.$$

This thought process is easy enough to do in your head. It will work whenever the desired value of x is a simple multiple of a *known* value of x.

e. If $I = 4$, then $4 = \dfrac{1080}{d^2}$.

$\therefore d^2 = \dfrac{1080}{4}$	Multiplying by d^2 and dividing by 4.
$d^2 = 270$	Arithmetic.
$d = \sqrt{270}$	Square root of both members.
$d \approx \underline{\underline{16.4 \text{ meters}}}$	By calculator.

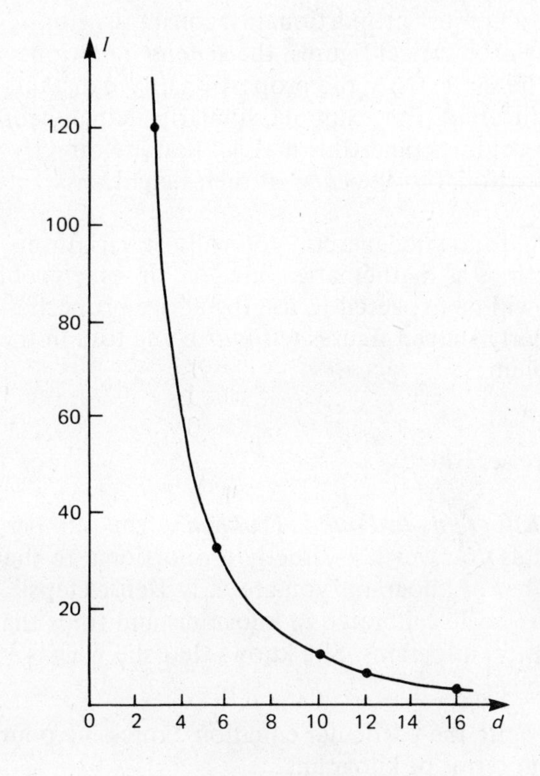

Figure 7-11c.

f. The graph of this function is shown in Figure 7-11c. You can use the given points, the calculated points, and any other points you feel it is advisable to calculate.

g. From the graph it is clear that the illumination drops off very rapidly as you move away from the bulb. The asymptote at the d-axis can be interpreted as telling you that no matter how far you go from the light, there is still *some* illumination. At great distances, however, the illumination would be so small that it would be zero "for all practical purposes."

The asymptote at the I-axis indicates that the intensity at the center of the bulb is *infinite.* This could be true only if all the light came from a single point. Since light comes from the entire bulb, the mathematical model works only in the domain *outside* the bulb.

There is a geometrical property of similarly-shaped objects that helps you to decide upon the proper general equation to use. If two objects are of *similar proportions,* then their linear dimensions, area, and volume vary directly with the first, second, and third power, respectively, of their length. For example, the circumference (a linear dimension), the area, and the volume of a sphere of radius r are given by

$$\text{Circumference} = 2\pi r,$$

$$\text{Area} = 4\pi r^2,$$

$$\text{Volume} = \frac{4}{3}\pi r^3.$$

In these equations, the numbers 2π, 4π, and $(4/3)\pi$ are just proportionality constants. For other geometrical figures, the *general* equations are the *same*. Only the proportionality constants are different. For example, similarly shaped *people* have volumes (and thus masses) that are directly proportional to the *cube* of their height.

In the following exercise you will use variation functions as mathematical models. In some problems you will be expected to use the above properties of similarly shaped figures *without* being told in the problem.

Exercise 7-11

1. ***Kilograms-to-Pounds Problem.*** The number of pounds you weigh is directly proportional to the number of kilograms you are. Kay Dense steps onto a scale calibrated in kilograms and finds that she is 50 kilograms. She knows that she weighs 110 pounds.

a. Write the particular equation expressing pounds in terms of kilograms.
b. How many pounds would a person weigh if the scale read

   i. 100 kilograms?
   ii. 25 kilograms?
   iii. 150 kilograms?

c. How many kilograms would Stan Dupp be if he weighed 165 pounds?
d. How many kilograms are *you*?
e. Plot the graph of this function.
f. What quantity does the proportionality constant represent in the real world?

2. ***Water Pressure Problem.*** When you swim underwater, the pressure in your ears varies directly with the depth at which you swim. At 10 feet, the pressure is about 4.3 pounds per square inch, (psi).

a. Write the particular equation expressing pressure in terms of depth.
b. Predict the pressure at 50 feet.
c. It is unsafe for amateur divers to swim where the pressure is more than 65 psi. How deep can an amateur diver safely swim?
d. Plot the graph of pressure versus depth.

3. ***Wrench Problem.*** The amount of force you must exert on a wrench handle to loosen a rusty bolt depends on how long the wrench handle is. Suppose that for a particular bolt, a wrench 7 inches long would require a force of 270 pounds, and a wrench 21 inches long would require 90 pounds.

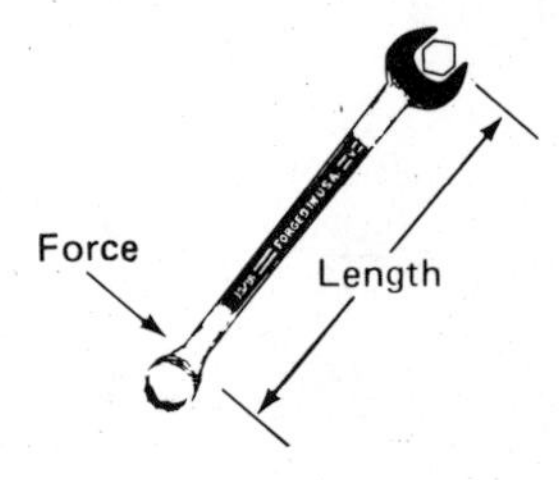

a. How does force vary with length? Write the general equation.
b. Write the particular equation.
c. What are the units of the proportionality constant? (This quantity is called "torque," and is a measure of the amount of "twisting" needed to loosen the bolt.)
d. Find the force needed for wrenches with handles 3, 10, 15, 30, and 60 inches long.
e. How long a wrench would be needed by Sarah Sota, who can exert a force of 300 pounds?
f. How long a wrench would be needed by Sarah's little sister Minnie, who can exert a force of 50 pounds?
g. Plot a graph of force versus handle length in the domain from 3 through 60 inches. Use the answers to parts d, e, and f, above, the given points, and any others you feel are necessary.

4. ***Multi-Story Building Problem.*** The amount of floor space in a building equals the area of each floor multiplied by the number of stories. The area of each floor is equal to the area of land covered by the building. If a building is to contain a *fixed* total floor space, the area of land covered by the

building will depend on how many stories high it is.

a. Show that the number of square meters of land covered by the building *varies inversely* with the number of stories.
b. Show that the proportionality constant equals the total amount of floor space in the building.
c. A building is to contain 12,000 square meters, total, of floor space. Write the particular equation expressing number of square meters of land covered in terms of number of stories.
d. How much land would be covered if the building were
   i. 3 stories high?
   ii. 5 stories high?
   iii. 10 stories high?
e. If the building is to cover no more than 1900 square meters, what is the minimum number of stories it can have?
f. Draw a graph of this function in a suitable domain, if city zoning laws prevent buildings from being over 10 stories high.

5. ***Water Main Problem.*** The number of houses that can be served by a water main is directly proportional to the square of the diameter of the main. This is because the cross-sectional area of the pipe, and thus the number of liters per minute that can flow, varies directly with the square of the diameter. Suppose that the City Waterworks has a 10 cm diameter water main that can supply 50 houses.

a. Write the particular equation for this function.
b. How many houses could be served by a water main of diameter 30 cm? 43 cm? 1 meter?
c. A new subdivision of 1500 houses is being planned. What size water main would be needed?

6. ***Boat's Wake Problem.*** When a boat is going at high speed, most of the power generated by its engine goes into forming the wake. The amount of power used to generate the wake is directly proportional to the seventh power of the speed of the boat.

a. Suppose that a boat going 10 knots (nautical miles per hour) uses 0.1 horsepower for wake generation. Write the particular equation for this function.
b. How much power goes into wake generation when the boat goes 20 knots? 30 knots?
c. Why is it so hard for boats to go very fast?
d. When a boat skims across the water, the proportionality constant in the wake generation function gets much smaller. Why do you suppose that such "hydroplaning" is necessary for a boat that goes 100 knots?

7. ***Gas Law Problem.*** In chemistry you learn Boyle's Law, which states that the volume of a fixed amount of gas (at constant temperature) is inversely proportional to the pressure of the gas.

a. Write the particular equation expressing volume in terms of pressure if a pressure of 46 pounds per square inch (psi) compresses the gas to a volume of 360 cubic feet.
b. Plot a graph of volume versus pressure in the domain from 10 to 100 psi.
c. What pressure would be necessary to compress the gas to a volume of 276 cubic feet?
d. According to your model, could you compress the gas to *zero* volume? How do you tell?
e. Prove that the product of the pressure and the volume is constant. (This is the way Boyle's Law is stated in some chemistry books.)
f. Prove that if (P_1, V_1) and (P_2, V_2) are ordered pairs of pressure and volume, then

$$P_1 V_1 = P_2 V_2 .$$

(This is the way Boyle's Law is stated in other chemistry books.)

8. ***Parkinson's Law Problem.*** In the book *Parkinson's Law* the author, Cyril M. Parkinson, claims (as a joke?) that the amount of time an agency spends discussing an item in its budget is *inversely* proportional to the amount of money involved. Suppose that a university's Board of Regents spent 15 minutes discussing a $1,000,000 item in next year's budget for starting a computer service. According to Parkinson's Law, how much time would they spend discussing

a. faculty salaries, $30,000,000?
b. a new bicycle rack, $1,000?

9. ***Radio Transmitter Problem.*** The strength of a radio signal received from the transmitter varies inversely with the square of your distance from the transmitter.

a. Derive the particular equation expressing strength in terms of distance if the strength is 1000 units at a distance of 2 kilometers.
b. Predict the strength 10 kilometers from the transmitter.
c. Predict the strength 100 meters from the transmitter.
d. From your answer to part c, tell why when you drive past a transmitting station, the signal from that station can sometimes be heard on your car's radio even though the radio is tuned to another station.

10. ***Recycled Can Problem.*** The amount of refund you get for returning recyclable aluminum cans varies directly with the number of cans you collect. In 1983, Pearl Brewery paid 22 cents for each pound of cans (23 cans).

a. Write the particular equation expressing number of cents of refund in terms of number of cans you return.
b. How much refund would you get for 100 cans?
c. If you wanted to earn $100, how many cans would you have to return? Does collecting cans seem to be an easy way to earn a living?

11. ***Lightning Problem.*** In a lightning storm, the time interval between the flash and the bang is directly proportional to the distance between you and the lightning. Answer the following questions:

a. Define the variables you would like to use for the distance and time intervals. Tell which should be dependent.
b. Write the particular equation for this direct variation function, if the thunder clap from

lightning 5 kilometers away takes 15 seconds to reach you.

c. Figure out the units of the constant of proportionality, and from these units figure out what real-world quantity the constant represents.
d. Calculate the times for the thunder sound to reach you from lightning bolts which are 1, 2.5, and 10 kilometers away.
e. Plot a graph of time versus distance.
f. Suppose that you measured a time interval of 29 seconds. Figure out how far away the lightning is.
g. What would be the situation in the real world if you heard the bang at the same instant you see the flash?

12. ***Balloon Temperature Problem.*** The volume of a fixed amount of gas such as air varies directly with its Kelvin temperature (273° + its Celsius temperature). Suppose that a balloon contains 7.5 liters of air at 300° K (about room temperature).

a. Write the particular equation expressing volume in terms of temperature.
b. Use your mathematical model to predict the volume for temperatures of 400, 600, 900, and 1200 degrees Kelvin.
c. Plot the graph of this function.
d. What things in the real world might set upper and lower bounds on the domain of this direct variation function?

13. ***Friction Problem.*** The force needed to overcome friction and drag a person across the floor depends on the person's weight. Manuel Dexterity can drag Gil O'Teen across the floor by pulling with a force of 51 pounds. Gil's brother, Nick, who weighs twice as much, can be dragged across the floor by pulling with a force of 102 pounds.

a. How does force vary with the person's weight? Define variables and write the general equation.
b. Gil weighs 85 pounds. Write the particular equation. (The proportionality constant is called the "coefficient of friction.")
c. Bob Tail weighs 130 pounds. How hard would Manuel have to pull to drag him across the floor?
d. Manuel can drag Kara Vann across the floor by pulling with a force of 65 pounds. How heavy is Kara?
e. Plot the graph of this function.

14. ***Radiotherapy Problem.*** Tumors are sometimes treated by irradiating them with gamma rays from a radioactive source such as cobalt-60. The intensity of radiation you receive depends on how far you are from the source. Suppose that for a particular source, the intensity is 80 mr/hr at 2 meters and 5 mr/hr at 8 meters. ("Mr/hr" stands for "milliroentgens per hour." The word "roentgen" is pronounced "rent′-ken.")

a. How does the intensity vary with distance? Write the general equation.
b. Write the particular equation.
c. What would the intensity be if your distance were
   i. 16 meters?
   ii. 12 meters?
   iii. 10 meters?
   iv. 10 centimeters?
d. At what distance would the intensity be 0.5 mr/hr?
e. Plot the graph of this function.

15. ***Radiant Heat Problem.*** The rate at which a hot object radiates heat varies directly with some power of its Kelvin temperature. (The Kelvin temperature equals the Celsius temperature plus 273°.) By experiment, you find that an electric heater radiates 5 calories per minute when it is at 300° K, and 80 calories per minute when it is at 600° K.

a. At what rate would it radiate heat if the heater were at 1200° K?
b. With what power of the Kelvin temperature does the heat radiation rate vary?

c. Write the particular equation expressing radiation rate for this heater in terms of temperature.
d. Use the equation to predict the radiation rate if the heater is at

   i. 800°K,
   ii. 1000°K.

e. Plot a graph of radiation rate as a function of temperature.

16. ***Diamond Problem.*** A discount diamond house ran an ad in a local newspaper which showed a 3/4 carat diamond for $360 and a 1.5 carat diamond for $1440.

a. Based on these figures, how would you expect the price of a diamond to vary with its weight (number of carats)?

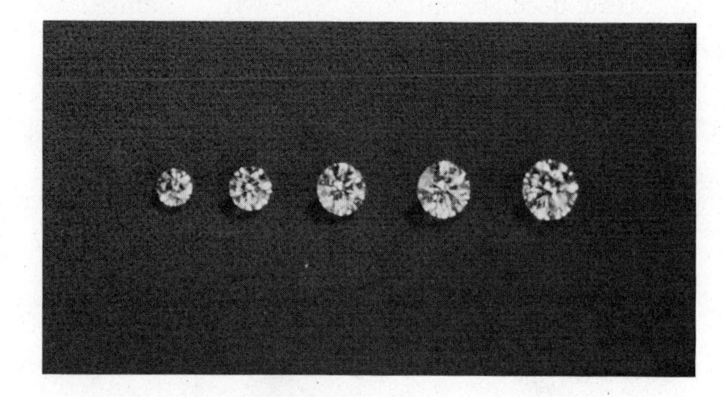

b. Predict the price of

   i. a 3 carat diamond,
   ii. a 1/4 carat diamond.

c. Write the particular equation expressing the number of dollars in terms of the number of carats.
d. The Hope Diamond weighs 44.5 carats. Based on your model, how much should it be worth?
e. Approximately how large a diamond could you get for $2000?
f. The weights of similarly shaped diamonds are directly proportional to the *cubes* of their diameters. Write a general equation expressing this fact.
g. Substitute the expression for weight from part f into the equation for price in part c, and simplify as much as possible. Then tell how the *price* of a diamond varies with its *diameter*.
h. If a particular diamond cost $500, how much would you expect to pay for one with *twice* the diameter?

17. ***Ruby Problem.*** Based on a comparison of prices, you find that the price of a ruby varies directly with the fourth power of its weight. Suppose that a particular ruby is worth $200.

a. How much would you expect to pay for a ruby weighing 10 times as much as the $200 one? What does this tell you about the availability of very large rubies?
b. How much would you expect to pay for a ruby weighing 1/10 as much as the $200 one? What does this tell you about the demand for very small rubies?

18. ***Relative Time Problem.*** Time seems to pass faster when you are old than it did when you were very young. Assume that the length that a particular period of time *seems* to be is inversely proportional to your age. At your present age, a day seems like a day, a week seems like a week, and so forth.

a. How long will a week seem to be when you are *twice* as old as you are now?
b. How long did a week seem to be when you were a *tenth* as old as you are now?
c. A mother 36 years old says to her 3-year-old child, "Don't get those toys out. You have only 5 minutes to play." Based on your model, and in terms of the mother's time scale, how long does that 5 minutes seem to be to the child? Does this result suggest a reason for some types of parent-child conflicts?

19. ***Stopping Distance Problem.*** Your car's stopping distance is the *sum* of the braking distance (the distance you go after you get your foot on the brake pedal) and the reaction distance (the distance you go between the time you realize you need to stop and the time you put your foot on the brake pedal). The braking distance varies directly with the *square* of the car's speed, and the reaction distance varies directly (with the *first* power) with the car's speed.

a. At 30 kph, your reaction distance is 7 meters and your braking distance is 6 meters. Calculate the reaction and braking distances for speeds of 60, 90, 120, 150, and 180 kph.
b. By adding the reaction and braking distances, calculate the total stopping distances for the speeds in part a.
c. Plot the graph of total stopping distance versus speed for speeds in the domain from 0 through 180 kph.
d. Use the given information in part a to write particular equations expressing reaction distance and braking distance in terms of speed.
e. Use the two equations from part d to predict the total stopping distance if you drive 100 kph. Show that this point lies on the graph of part c.
f. If you drive 200 kph, how many football field lengths will it take you to stop your car after you realize that you need to stop?
g. Consult a driver's manual or another source to find out just how accurate your mathematical model is.

20. ***Weight Above and Below the Earth.*** Sometimes a function is defined by *different* equations in different parts of its domain. For example, the weight of an object varies directly with its distance from the center of the Earth when the object is below the Earth's surface, but is inversely proportional to the square of its distance from the center when it is above the surface. (Note that an object *in orbit* is "weightless" only because its weight is exactly counterbalanced by the upward force due to its motion.)

a. Phoebe Small weighs 81 pounds at the Earth's surface. Write particular equations expressing her weight as a function of distance from the center (*two* equations) assuming that the radius of the Earth is 4000 miles.
b. Predict Phoebe's weight 3000, 8000, and 12,000 miles from the center.
c. Draw a graph of this function in the domain from 0 through 12,000 miles.

21. ***Egg Problem.*** According to an ad for South African Airways in a July, 1974, issue of *Time Magazine*, ostrich eggs are served at bush country "braaivleis" (barbecues). Each ostrich egg is claimed to be equivalent to two dozen chicken eggs!

a. If an ostrich egg really is equivalent to two dozen chicken eggs, and a chicken egg is about 6 centimeters long, about how long is an ostrich egg (to the nearest centimeter)? Justify your answer.
b. If a robin's egg is 1 cm long, how many of them would you have to scramble to get the equivalent of one chicken egg?

22. ***Grapefruit Problem.*** Suppose that you are shopping at a supermarket and find that the Texas grapefruits have *twice* the diameter of the Florida ones, but cost *seven* times as much. Recalling that the volumes of similarly shaped solids vary directly with the *cube* of a linear dimension such as a diameter, you instantly determine which kind of grapefruit gives you more for your money. Which one does, and why?

23. ***Large and Small People Problem.*** Two people of different size have the same "build" if they have the same proportions.

a. Write general equations expressing the following quantities in terms of height for people of the same build:
   i. mass,
   ii. skin area,
   iii. belt length.

b. In *Gulliver's Travels*, Gulliver traveled to Brobdingnag, where people are 10 times as tall as normal people, and to Lilliput, where people are 1/10 as tall as normal people. Les Moore, a normal person of average build, is 70 kg, has 25,000 square centimeters of skin, and wears a belt 90 cm long. Estimate the mass, skin area, and belt size of

   i. a Brobdingnagian,
   ii. a Lilliputian.

c. How does the mass of a Lilliputian compare with that of a MacDonald's hamburger patty?

d. The mass a person's legs will support varies directly with the *square* of his or her height because the strength of the legs depends on their cross-sectional *area*. Les Moore's legs can support a total of 200 kg (including his own 70 kg). Based on this information, would a Brobdingnagian's legs be able to support his own mass? Justify your answer.

e. See the article, "On Being the Right Size," by J. B. S. Haldane in *The World of Mathematics,* Volume II, Page 952, for more reasons why people cannot be excessively large or excessively small!

24. ***Airplane Wing Problem.*** John Garfinkle is Chief Mathematician for the Fly-By-Night Aircraft Corporation. His Company is designing the Sopwith Hippopotamus, a larger version of the famous Sopwith Camel and John is called upon to answer the following questions:

a. The weight of an airplane varies directly with the *cube* of the plane's length. Write an equation expressing the weight of an airplane in terms of its length. Evaluate the proportionality constant if a Sopwith Camel is 40 feet long and weighs 3000 pounds.
b. The number of pounds a plane's wings can lift varies directly with the *square* of the plane's length. Write another equation expressing the number of pounds of lift in terms of the plane's length. Evaluate the proportionality constant if a Sopwith Camel's wings can lift 5000 pounds.
c. The Sopwith Hippopotamus is to be 60 feet long. Based on your model, will a Hippopotamus be able to fly? Explain.
d. What is the longest such airplane that would be able to fly?

25. ***Shark Problem.*** In July, 1975, fishermen caught a great white shark off Catalina Island. The shark was 15 feet long, and weighed 2000 pounds.

a. Assuming that all great white sharks have similar proportions, how should the weight of a great white shark vary with its length?
b. Write the particular equation expressing weight in terms of length.
c. Predict the weight of
   i. a baby shark, 2 feet long,
   ii. the shark in the novel *Jaws*, 25 feet long.
d. If you caught a shark weighing 250 pounds, how long would you expect it to be?
e. Fossilized teeth have been found in Florida that resemble those of a great white shark. If this creature had the same porportions as present-day sharks, it would have been 100 feet long. How much would such a shark have weighed?

26. ***Height-Mass Problem.*** Professor Snarff, who considers himself to be of average build, is 91 kg and 194 cm tall.

a. Write the particular equation expressing mass in terms of height for people of Professor Snarff's proportions.
b. Use the equation of part a to calculate *your* mass. What can you conclude if your actual mass is significantly different from what you predicted?
c. Show that a baby is significantly *fatter* in proportion to its height than an adult. You can do this by using the mathematical model to show that the predicted mass of a 50 cm baby is much smaller than its actual 3 to 4 kg.
d. According to *Guinness Book of World Records,* the tallest man who ever lived was Robert P. Wadlow. When he died in 1940 at the age of 22, he was 272 cm tall and had a mass of 196 kg. Was Wadlow fatter or thinner in proportion to his height than Professor Snarff? Justify your answer. (Look at Wadlow's picture in *Guinness* to see if your conclusion is reasonable!)

27. ***Medication Problem.*** The safe dosage of a new medicine is determined by testing it on animals. To predict the safe dosage for humans from the results of animal experiments, scientists assume that the safe dosage is directly proportional to the patient's skin area. This area is, of course, directly proportional to the *square* of the person's height (assuming that patients have the same proportions).

a. Write general equations for safe dosage in terms of skin area, and for skin area in terms of height. Use *different* letters for the two proportionality constants.
b. Combine the two equations in part a to get a general equation expressing safe dosage in terms of height. You can simplify the equation by realizing that the product of two constants is just another constant.
c. Tell in words how safe dosage depends on height.
d. Suppose that a new cold remedy is tested on monkeys 30 cm tall, and that the safe dosage for the monkeys is found to be 1.2 milligrams (mg). Write the particular equation expressing safe dosage in terms of height.
e. Assuming that humans have roughly the same proportions as monkeys, predict the safe dosage of the cold remedy for

    i. a child, 90 cm tall
    ii. an average adult, 170 cm tall
    iii. a very tall adult, 2 meters tall.

28. ***Pizza Problem.*** The price you pay for a pizza may be assumed to be composed of two different terms. One term varies directly with the square of the diameter. It represents the amount you pay for the ingredients in the pizza, and is reasonable because the area (and thus the amount of ingredients) varies directly with the square of the diameter. The other term is a *constant.* It represents the fixed costs such as cooking and serving the pizza, washing the dishes, and making the mortgage payments on the building. The total price you pay is the *sum* of these two terms. Therefore, the general equation would be

$$p = ad^2 + c,$$

where p is the number of cents the pizza costs, d is the number of inches diameter, and a and c are constants. Answer the following questions:

a. A 1983 Pizza Hut menu lists the following prices for Supreme pizzas:

Kind	*Diam. (in.)*	*Price ($)*
Mini	7	3.60
Small	9	5.70
Medium	13	9.30
Large	15	11.25

Use the prices and diameters for Small and Large pizzas to determine the values of the constants a and c. Then write the particular equation expressing price in terms of diameter.

b. Do the prices of the Mini and Medium pizzas seem to agree with this mathematical model? If not, do they seem to be overpriced or underpriced? Justify your answer.

c. Obtain a menu from a pizzeria in your area. For each type of pizza listed, find the particular equation as in part a, using the largest and smallest sizes to determine the constants. Then predict the prices for other sizes of pizza of the same kind. The pizzeria manager might be interested in learning about any pizzas that seem to be particularly overpriced or underpriced!

7-12 CHAPTER REVIEW AND TEST

In this chapter you have studied functions whose equations involve four of the five operations of algebra, namely +, −, ×, and ÷. The domains of these functions must exclude any values of the variable that make a denominator equal zero. You found that when the denominator is close to zero, the graph could

a. approach a vertical asymptote, or
b. approach a deleted point, or "hole."

The analysis of such functions required you to refresh and extend your abilities with elementary algebraic techniques such as factoring and operating with fractions. You used some functions with variables in denominators as mathematical models, and found physical meanings for asymptotes. Finally, you were exposed to some functions that have more than two variables.

The objectives of the chapter may be summarized as follows:

1. *Factor polynomials.*
2. *Add, subtract, multiply, or divide rational expressions.*
3. *Solve fractional equations.*
4. *Graph rational functions.*
5. *Use variation functions as mathematical models.*

The Review Problems below are designed to measure your ability to accomplish these objectives one at a time. The Concepts Test gives you an opportunity to put together several objectives at a time, and to apply your knowledge to new situations.

Review Problems

The problems below are numbered according to the five objectives of this chapter

Review Problem 1. Factor each polynomial completely:

a. $6ax + 15bx$
b. $36r^2 - 81s^2$
c. $2x^2 - 21xy - 36y^2$
d. $2x^3 - 3ax^2 - 18x - 27a$
e. $2x^3 - 13x^2 - 48x + 27$
f. $32x^5 + y^5$
g. $12x^2 + 37x + 8$

Review Problem 2. Carry out the indicated operations and simplify:

a. $\dfrac{2(2x+1)}{x^2+x-6} - \dfrac{2}{x+3} - \dfrac{x}{2-x}$

b. $\dfrac{x^2+4x+4}{x^2-8x+15} \div \dfrac{3+x}{3-x} \cdot \dfrac{x^3-x^2-17x-15}{x^2-4}$

Review Problem 3. The four fractional equations below all look somewhat alike. Yet each illustrates a different example of what might happen when you solve a fractional equation. By solving the equations, show what happens.

a. $\dfrac{x}{x-2} - \dfrac{2}{x+4} = \dfrac{12}{x^2+2x-8}$

b. $\dfrac{x}{x-2} - \dfrac{2}{x+4} = \dfrac{19}{x^2+2x-8}$

c. $\dfrac{x}{x-2} + \dfrac{1}{x+4} = \dfrac{12}{x^2+2x-8}$

d. $\dfrac{1}{x-2} + \dfrac{2}{x+4} = \dfrac{3x}{x^2+2x-8}$

Review Problem 4. Plot the graph of

$$f(x) = \frac{x^2 - 1}{(x-3)^2(x+1)}.$$

Review Problem 5. Suppose that y varies inversely with the cube of x.

a. Write the general equation for this function.
b. Write the particular equation if the function contains the ordered pair (4, 448).
c. Calculate y when x is 0.7.
d. *Without* using your equation, find y when x is 8.

Concepts Test

The following problems require you to combine several concepts of this chapter or to use the concepts to do things you have never done before. Work each problem and tell by number(s), 1 through 5, which one or ones of the objectives of this chapter were used in working each part of each problem.

Concepts Test 1 – One-Question Test on Rational Functions. Suppose that

$$f(x) = \frac{2x^2 - 2x - 4}{x^3 - 4x^2 + x + 6} - \frac{\frac{x}{9} + \frac{1}{3} + \frac{1}{x}}{\frac{x^2}{9} - \frac{3}{x}} + \frac{2-x}{x^2 - 4}.$$

a. Simplify each of the three fractions as much as possible.
b. Carry out the indicated addition and subtraction. Simplify the result as much as possible.
c. Plot the graph of f.
d. Find the value(s) of x for which $f(x) = -5/4$.
e. Find the value(s) of x for which $f(x) = 2$.

Concepts Test 2 – System of Rational Equations. You have learned how to solve systems of equations with two variables when both equations were *linear.* The following system has one equation that is *rational.*

$$y = \frac{x^3 - 4x^2 + x + 6}{x - 2},$$

$$x - y = 5.$$

a. Combine the two equations so that y is eliminated. (There are at least two ways you can do this.)
b. Solve the resulting fractional equation to find the value(s) of x.
c. Find the value(s) of y corresponding to each value of x, and write the solution set.

Concepts Test 3 – Intensity Problem. The *intensity* of light at a certain point is the amount of light that passes through a unit area in a given time. Suppose that Q units of light comes from a light bulb in a given time, and spreads out uniformly in all directions. If you imagine a sphere drawn around the bulb (Figure 7-12a), all the light must pass through this sphere. Thus, the intensity of light on the surface of the sphere is *Q divided by* the area of the sphere.

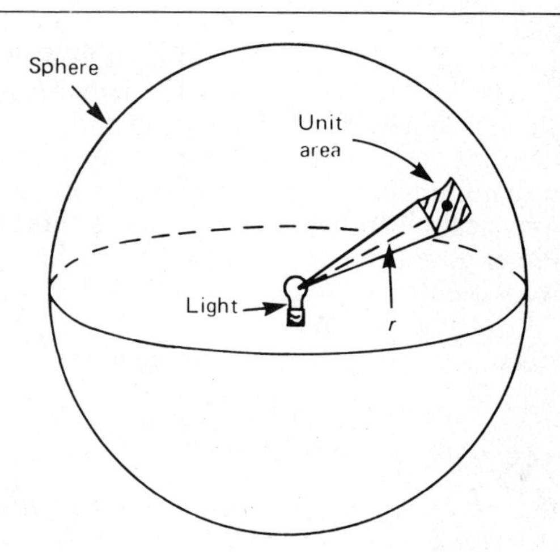

Figure 7-12a.

a. Show that the intensity of light at a distance r from the bulb varies *inversely* with the *square* of r. (Recall that the area of a sphere of radius r is $4\pi r^2$.)

b. What is the proportionality constant for the intensity function?
c. What intensity of light does your hand receive if you hold it 10 centimeters away from a light bulb for which $Q = 200$ watts?
d. If an intensity of 0.005 watts per square centimeter or more will cause damage to your eyes, how far must you stay from a 600 watt bulb to avoid such damage?

Concepts Test 4 – Slope of a Tangent Line. In the mathematical models problems of Section 7-11 you found meanings for asymptotes to graphs. In this problem you will find a meaning for a "hole" in a graph.

a. A *secant* line is a line that cuts a graph in *two* places (Figure 7-12b). Suppose that various secant lines are drawn through the *fixed* point $(2, f(2))$ and the *variable* point $(x, f(x))$. Since the graph of f is *curved*, the slope of the secant line will depend on the value of x. Recalling that slope equals rise/run, write an equation expressing the slope, $m(x)$, in terms of x, $f(x)$, 2, and $f(2)$.
b. Suppose that $f(x) = x^3 - 4x^2 + 5x + 3$. Find $f(2)$. Then use this information to write the particular equation for $m(x)$ in terms of x.
c. If you have been successful so far, the numerator for $m(x)$ should have $(x - 2)$ as a factor. Simplify the fraction.

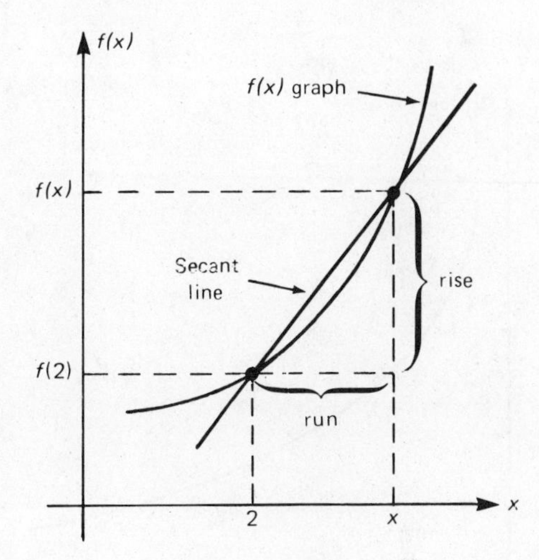

Figure 7-12b.

d. Plot the graph of $m(x)$.
e. $m(x)$ is *undefined* when $x = 2$. What value does $m(x)$ seem to be close to when x is close to 2? What feature does the graph of m have at $x = 2$?
f. When $x = 2$, the secant line becomes a *tangent* line. That is, it touches the $f(x)$ graph at just *one* point. What is the slope of this tangent line? Justify your answer.

Irrational Algebraic Functions

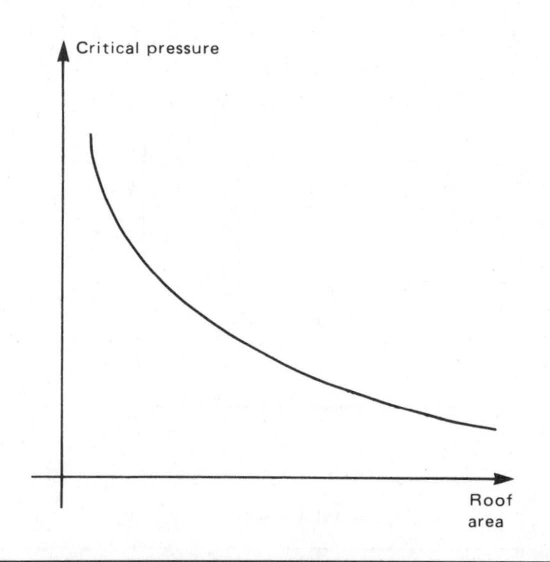

In this chapter you will study the fifth operation of algebra, ***taking nth roots.*** *Your most significant advance in mathematical theory will be in proving that there really are real numbers that cannot be expressed as ratios of two integers. The operations with radical expressions should be easy for you since you already know how to deal with fractional exponents. Applications of radical functions will lead you to some startling conclusions, such as why mammals are the way they are, and where is the safest place to hide during a tornado!*

8-1 EXISTENCE OF IRRATIONAL NUMBERS

You will recall from Chapter 1 that a rational number is a real number that can be expressed as a ratio of two integers. Irrational numbers are defined by negating just one word in this definition.

DEFINITION

An ***irrational number*** is a real number that *cannot* be expressed as a ratio of two integers.

Objective:

Be able to demonstrate that certain radicals represent *irrational* numbers.

It is not a trivial problem to show that there actually *are* real numbers which cannot be expressed as ratios of integers. Pythagoras is reputed to have believed that there was no such thing as an irrational number, and to have spent years looking for two integers whose ratio was exactly $\sqrt{2}$. With what you now know about the properties of exponentiation, you can convince yourself that numbers such as $\sqrt{2}$ really *are* irrational.

Table I lists $\sqrt{2}$ as 1.414. It is easy to prove that this is not exactly correct by squaring 1.414 and getting 1.999396. But there is a more instructive way of showing that 1.414 is not exactly equal to $\sqrt{2}$.

$1.414 = \dfrac{1414}{1000}$ Showing that 1.414 is a rational number.

$1.414 = \dfrac{707}{500}$ Canceling all common factors.

$1.414 = \dfrac{7 \times 101}{2^2 \times 5^3}$ Expressing numerator and denominator as prime factors. (No common factors)

$1.414^2 = \left(\dfrac{7 \times 101}{2^2 \times 5^3}\right)^2$ If $a = b$, then $a^2 = b^2$.

$1.414^2 = \dfrac{7^2 \times 101^2}{2^4 \times 5^6}$ Distributivity of exponentiation over multiplication and division. (*Still* no common factors)

There are no common prime factors in the numerator and denominator, so you cannot do any canceling. Clearly, then, the fraction on the right is *not* an integer. Since the fraction is equal to 1.414^2, this number cannot be an integer, either. So 1.414^2 could not possibly equal 2, meaning that $1.414 \neq \sqrt{2}$. You could repeat the above reasoning for *any* rational number which you think might be $\sqrt{2}$, and you will always come out with that number *not* equal to $\sqrt{2}$.

Two important theorems follow from the above reasoning.

Theorem: If a rational number is not an integer, then its nth power is not an integer.

The proof relies on the fact that squaring, cubing, etc., *distribute* over $\times$ and $\div$. Thus, if the numerator and denominator had no common factors *before* exponentiation, then they will still have none *after* exponentiation.

Suppose you want to tell whether or not $\sqrt[4]{64}$ is a *rational* number. Since $2^4 = 16$ and $3^4 = 81$, it follows that $\sqrt[4]{64}$ is *between* 2 and 3. If $\sqrt[4]{64}$ were a rational number, it would be a *fraction* between 2 and 3. By the above theorem, you also know that no such fraction raised to the fourth power could possibly *equal* the integer 64. So you are forced to conclude that $\sqrt[4]{64}$ is *not* a rational number. That is, $\sqrt[4]{64}$ is *irrational.* This is an example of the following general property:

Theorem: *Irrational radicals* – If $\sqrt[n]{\text{positive integer}}$ is *between* two integers, then it is an *irrational* number.

The following exercise gives you practice using these two theorems to demonstrate that certain radicals represent irrational numbers.

Exercise 8-1

In each one of Problems 1 through 6 there is a decimal approximation for a radical. For each problem,

a. write the decimal approximation as a ratio of two integers,
b. factor the numerator and denominator into primes and do any possible canceling,
c. square or cube the fraction, leaving numerator and denominator factored,
d. tell why you can conclude without further arithmetic that the square or cube of the fraction could not possibly equal the integer under the radical sign (the "radicand"),
e. actually square or cube the decimal approximation to show that the answer is *close* to the radicand.

1. $\sqrt{5} \approx 2.236$
2. $\sqrt{10} \approx 3.162$
3. $\sqrt{22} \approx 4.690$
4. $\sqrt{40} \approx 6.325$
5. $\sqrt[3]{84} \approx 4.380$
6. $\sqrt[3]{4} \approx 1.587$

For Problems 7 through 16, either find the integer value of the given n^{th} root, or prove that it is irrational by finding two consecutive integers whose n^{th} powers are below and above the radicand.

7. $\sqrt[3]{100}$
8. $\sqrt[4]{27}$
9. $\sqrt[5]{53}$
10. $\sqrt[3]{1001}$
11. $\sqrt{3}$
12. $\sqrt{2}$
13. $\sqrt[4]{625}$
14. $\sqrt[7]{128}$
15. $\sqrt[3]{1332}$
16. $\sqrt[5]{1024}$

8-2 SIMPLE RADICAL FORM

In Section 6-5 you learned the relationship between radicals and powers with fractional exponents. For example,

$$\sqrt[3]{128} = 128^{1/3}.$$

Since the operation of taking roots is really just a form of exponentiation, the properties of exponentiation you learned before automatically apply to radicals. Two of these properties are restated here to refresh your memory.

	Radical Form	*Exponential Form*
Root of a product	$\sqrt[n]{ab} = \sqrt[n]{a}\ \sqrt[n]{b}$	$(ab)^{\frac{1}{n}} = a^{\frac{1}{n}}\ b^{\frac{1}{n}}$
Root of a quotient	$\sqrt[n]{\dfrac{a}{b}} = \dfrac{\sqrt[n]{a}}{\sqrt[n]{b}}$	$\left(\dfrac{a}{b}\right)^{\frac{1}{n}} = \dfrac{a^{\frac{1}{n}}}{b^{\frac{1}{n}}}$

These properties can be used to *simplify* radicals. Note that these properties apply only to *non-negative* values of a and b, and that $b \neq 0$ in the second property.

Objective:

Give an expression containing radicals, be able to write it in *simple radical form.*

The first three examples show you exactly what is meant by "simple radical form."

Example 1: Simplify $\sqrt{288}$.

288 is not a perfect square, but it has factors that *are* perfect squares. To find these factors you can use the "upside down short division" process you have learned before.

Divide by perfect squares:

$$\begin{array}{r|l} 4 & 288 \\ 4 & 72 \\ 9 & 18 \\ & 2 \end{array}$$

Stop when the quotient has no more square factors.

So you can write

$$\begin{aligned} \sqrt{288} &= \sqrt{4 \cdot 4 \cdot 9 \cdot 2} & \\ &= \sqrt{4 \cdot 4 \cdot 9}\sqrt{2} & \text{Root of a product.} \\ &= 2 \cdot 2 \cdot 3 \cdot \sqrt{2} & \text{Root of a product.} \\ &= \underline{\underline{12\sqrt{2}}} & \text{Arithmetic.} \end{aligned}$$

The expression $12\sqrt{2}$ is considered to be simpler than $\sqrt{288}$ because the *radicand* is *smaller.*

Example 2: Simplify $\dfrac{1}{\sqrt[3]{2}}$.

At first glance this expression seems to be as simple as possible. However, there is a radical in the *denominator.* Radical denominators sometimes cause difficulty in operations such as adding fractions. By multiplying this expression by a clever form of 1 you can *rationalize* the denominator.

$$\frac{1}{\sqrt[3]{2}} = \frac{1}{\sqrt[3]{2}} \cdot \frac{\sqrt[3]{4}}{\sqrt[3]{4}}$$ Multiply by something that makes the denominator radicand a perfect cube.

$$= \frac{\sqrt[3]{4}}{\sqrt[3]{8}}$$ Multiplication property of fractions, and $\sqrt[n]{a}\ \sqrt[n]{b} = \sqrt[n]{ab}$.

$$= \frac{\sqrt[3]{4}}{2}$$ $\sqrt[3]{8} = 2$.

Note that although the radicand is now larger (4 instead of 2), the expression is considered to be "simpler" because *there are no radicals in the denominator.*

Example 3: Simplify $\sqrt[8]{64}$.

This radical seems to be as simple as possible. 64 is equal to 2^6, and thus has no factors that are perfect 8th powers. However, if you write the radical in *exponential* form, a simplification shows up.

$\sqrt[8]{64} = 64^{1/8}$ Definition of fractional exponents.

$= (2^6)^{1/8}$ $64 = 2^6$.

$= 2^{6/8}$ Power of a power.

$= 2^{3/4}$ Simplifying the fraction.

$= \sqrt[4]{2^3}$ Definition of fractional exponents.

$= \sqrt[4]{8}$ $2^3 = 8$.

This answer is simpler than $\sqrt[8]{64}$ for two reasons. First, the radicand is smaller and second, *the root index is as small as possible.*

From these three examples, the definition of "simple radical form" can be extracted.

DEFINITION

An expression is in ***simple radical form*** if

1. the radicand of an nth root contains no n^{th} powers as factors,
2. the root index is as low as possible, and
3. there are no radicals in the denominator.

Note that the third requirement, "rationalizing the denominator", also makes finding decimal approximations easier if you have no calculator, as you will see if you work Problem 55 in the following exercise. Historically, this is the most important reason for rationalizing denominators. Sometimes a problem is simpler if you rationalize the *numerator,* as you will see if you work Problem 56, below.

The third requirement, rationalizing the denominator, is tricky if the denominator contains two or more terms. For example, if the denominator were $5 + \sqrt{3}$, and you multiplied it by another $5 + \sqrt{3}$, you would get $25 + 2\sqrt{3} + 3$. (The denominator would still contain $\sqrt{3}$). A clever way around this difficulty is shown in Example 4.

Example 4: Simplify $\dfrac{2}{5+\sqrt{3}}$.

$$\frac{2}{5+\sqrt{3}} = \frac{2}{5+\sqrt{3}} \cdot \frac{5-\sqrt{3}}{5-\sqrt{3}}$$ Multiplicative identity. $5 - \sqrt{3}$ is the *conjugate* of $5 + \sqrt{3}$.

$$= \frac{2(5-\sqrt{3})}{5^2 - (\sqrt{3})^2}$$ $(a + b)(a - b) = a^2 - b^2$.

$$= \frac{2(5-\sqrt{3})}{25-3}$$ Arithmetic.

$= \dfrac{2(5-\sqrt{3})}{22}$ More arithmetic.

$= \dfrac{5-\sqrt{3}}{11}$ Canceling.

The procedure is to multiply by the *conjugate* of the denominator.

The exercise which follows is designed to give you practice in transforming irrational algebraic expressions to simple radical form. Some challenging problems come at the end.

Exercise 8-2

For Problems 1 through 38, transform the given expression to simple radical form. You can check your answer by evaluating it and the original expression by calculator.

1. $\sqrt{12} + 2\sqrt{48} + 5\sqrt{147} - 4\sqrt{3}$
2. $3\sqrt{125} - 2\sqrt{80} + \sqrt{405}$
3. $\sqrt[3]{2187} - 2\sqrt[3]{24}$
4. $\sqrt[3]{108} + 10\sqrt[3]{32} + \sqrt[3]{500}$
5. $\dfrac{3}{2\sqrt{2}} + \dfrac{5\sqrt{2}}{4}$
6. $\dfrac{7}{5\sqrt{3}} - \dfrac{8\sqrt{3}}{15}$
7. $\dfrac{12}{\sqrt{6}} + \sqrt{6}$
8. $4\sqrt{5} - \dfrac{15}{\sqrt{5}}$
9. $7\sqrt{3} - \dfrac{12}{\sqrt{3}} + \sqrt{75}$
10. $4\sqrt{5} + \dfrac{35}{\sqrt{5}} - \sqrt{125}$
11. $(\sqrt{7} + \sqrt{2})^2$
12. $(\sqrt{5} - \sqrt{3})^2$
13. $(2\sqrt{5} - 3)^2$
14. $(4 + 3\sqrt{6})^2$
15. $(\sqrt{5}-\sqrt{3})(\sqrt{5}+\sqrt{3})$
16. $(\sqrt{7} - 2)(\sqrt{7} + 2)$
17. $(2\sqrt{3} - \sqrt{2})(2\sqrt{3} + \sqrt{2})$
18. $(3\sqrt{5} + \sqrt{7})(3\sqrt{5} - \sqrt{7})$
19. $(6\sqrt{7} + \sqrt{15})(\sqrt{7} - \sqrt{3})$
20. $(\sqrt{12} - \sqrt{6})(\sqrt{3} + \sqrt{27})$
21. $\dfrac{2}{\sqrt[3]{9}}$
22. $\dfrac{4}{\sqrt[3]{25}}$
23. $\dfrac{4}{\sqrt[4]{32}}$
24. $\dfrac{7}{\sqrt[5]{128}}$
25. $\dfrac{5}{\sqrt[6]{1024}}$
26. $\dfrac{9}{\sqrt[4]{25}}$
27. $\dfrac{7}{\sqrt[4]{49}}$
28. $\dfrac{12}{\sqrt[6]{512}}$
29. $\dfrac{1}{\sqrt{5}-1}$
30. $\dfrac{4}{3-2\sqrt{2}}$
31. $\dfrac{4}{\sqrt{7}+\sqrt{3}}$
32. $\dfrac{57}{5\sqrt{3}-3\sqrt{2}}$
33. $\dfrac{\sqrt{3}-1}{\sqrt{2}-1}$
34. $\dfrac{3\sqrt{3}-1}{3\sqrt{2}-1}$
35. $\dfrac{7\sqrt{2}+3}{7\sqrt{2}-3}$
36. $\dfrac{4\sqrt{7}+3\sqrt{2}}{5\sqrt{2}+2\sqrt{7}}$
37. $\dfrac{1+\dfrac{1}{\sqrt{3}}}{1-\dfrac{1}{\sqrt{3}}}$
38. $\dfrac{1-\dfrac{1}{\sqrt{5}}}{1+\dfrac{1}{\sqrt{5}}}$

For Problems 39 through 42, the denominator is a *tri*nomial. Associating *two* of the three terms gives a *bi*nomial, which can be simplified by multiplying by the conjugate. For example,

$$\frac{1}{2+\sqrt{5}-\sqrt{3}} = \frac{1}{(2+\sqrt{5})-\sqrt{3}} \cdot \frac{(2+\sqrt{5})+\sqrt{3}}{(2+\sqrt{5})+\sqrt{3}}$$ Associating, and multiplying by 1.

$$= \frac{2+\sqrt{5}+\sqrt{3}}{(2+\sqrt{5})^2-3}$$ Carrying out the multiplication.

$$= \frac{2+\sqrt{5}+\sqrt{3}}{6+4\sqrt{5}}$$ Simplify the denominator.

This expression can be simplified as before. Simplify the following:

39. $\dfrac{12}{2+\sqrt{3}-\sqrt{7}}$

40. $\dfrac{\sqrt{3}+\sqrt{2}}{\sqrt{3}+\sqrt{2}-1}$

41. $\dfrac{1}{\sqrt{3}+\sqrt{2}-\sqrt{5}}$

42. $\dfrac{1}{\sqrt{5}+\sqrt{3}+2\sqrt{2}}$

For Problems 43 through 46, the denominator looks like one factor of a sum or difference of two cubes. You recall, for example, that

$$a^3 + b^3 = (a+b)(a^2 - ab + b^2).$$

Using fractional exponents, you can write $a + b$ as

$$a + b = (a^{1/3} + b^{1/3})(a^{2/3} - a^{1/3}b^{1/3} + b^{2/3}).$$

Use this fact to find a clever form of 1 by which to multiply each of the following expressions, and thus rationalize the denominator.

43. $\dfrac{1}{\sqrt[3]{2}-\sqrt[3]{3}}$

44. $\dfrac{1}{\sqrt[3]{4}+\sqrt[3]{5}}$

45. $\dfrac{1}{\sqrt[3]{4}+\sqrt[3]{10}+\sqrt[3]{25}}$

46. $\dfrac{1}{\sqrt[3]{16}-\sqrt[3]{4}+1}$

For Problems 47 through 54, the expression has the form $\sqrt{a \pm \sqrt{b}}$. Expressions like this can be transformed to $\sqrt{x} \pm \sqrt{y}$. For example, letting

$$\sqrt{10+2\sqrt{21}} = \sqrt{x} + \sqrt{y},$$

and squaring each member gives

$$10 + 2\sqrt{21} = x + 2\sqrt{xy} + y.$$

Setting $x + y = 10$ and $2\sqrt{xy} = 2\sqrt{21}$ (from which $xy = 21$), you get a *system* of two equations with two variables. From the first equation, $y = 10 - x$. Substituting this into the second gives

$$x(10 - x) = 21.$$

Distributing the x and getting 0 on the right side gives

$$x^2 - 10x + 21 = 0.$$

Factoring gives $(x-3)(x-7) = 0$, from which $x = 3$ or 7. Substituting these gives $y = 7$ or 3, so that

$$\sqrt{10+2\sqrt{21}} = \underline{\underline{\sqrt{3}+\sqrt{7}.}}$$

Use this technique to express the following as $\sqrt{x} \pm \sqrt{y}$:

47. $\sqrt{4+2\sqrt{3}}$

48. $\sqrt{7+2\sqrt{6}}$

49. $\sqrt{12-6\sqrt{3}}$

50. $\sqrt{17-12\sqrt{2}}$

51. $\sqrt{11+6\sqrt{2}}$

52. $\sqrt{12+2\sqrt{35}}$

53. $\sqrt{32-8\sqrt{15}}$

54. $\sqrt{101-28\sqrt{13}}$

55. a. Find a decimal approximation for $3/\sqrt{2}$ by using $\sqrt{2} \approx 1.414$ and carrying out the long division.
    b. Rationalize the denominator of $3/\sqrt{2}$. Then use $\sqrt{2} \approx 1.414$ to get a decimal approximation.
    c. Why do you suppose that rationalizing the denominator was so important in the days before calculators?

56. a. Solve $x^2 - 1{,}000{,}000x + 1 = 0$ using the quadratic formula.
    b. Show that $\sqrt{b^2 - 4ac}$ is so close to b that the numerator of one solution rounds off to 0.
    c. Rationalize the *numerator* of the solution in part b, and thus get a non-zero decimal approximation of the solution.

8-3 RADICAL EQUATIONS

As the name suggests, a radical equation is an equation with a radical in it. For example,

$$3 = x + \sqrt{x-1}$$

is a radical equation. However, there are equations such as $x + \sqrt{2} = 13$ that have radicals but are *not* radical equations. Radical equations have a *variable* under a radical sign.

DEFINITION

A ***radical equation*** is an equation in which a variable appears under a radical sign.

Objective:

Given a radical equation, be able to find its solution set, discarding any extraneous solutions.

Radical equations can arise in the study of functions when you must find x for a given value of y. For example, if $y = x + \sqrt{x-1}$, then solutions of the radical equation on the preceding page would be values of x for which $y = 3$.

However, radical equations arise in other places too, as you will see in Chapter 9. So it is worth spending time studying them as a background for future work as well as a technique for finding x when you know y.

Example 1: Solve $3 = x + \sqrt{x-1}$.

As with any new problem, the technique is to transform it into an old problem which you already know how to work. In this case, the technique is to transform the equation into a polynomial equation by squaring each member. A preliminary step is necessary in order to *isolate* the radical on one side of the equation.

$3 - x = \sqrt{x-1}$	Subtracting x from each member.
$9 - 6x + x^2 = x - 1$	Squaring each member.
$x^2 - 7x + 10 = 0$	Addition property of equality.
$(x-2)(x-5) = 0$	Factoring.
$x = 2$ or $x = 5$	Setting each factor equal to 0.

Substituting these values for x in the original equation gives you a surprise!

$x = 2$:	$x = 5$:
$3 = 2 + \sqrt{2-1}$	$3 = 5 + \sqrt{5-1}$
$3 = 2 + 1$	$3 = 5 + 2$
$3 = 3$	$3 \neq 7$

So 2 is a solution, but 5 is not! The reason is because squaring each member is an *irreversible* step which leads to an extraneous solution sometimes. The extraneous solution occurred this time because there are two distinct equations,

$$3 - x = \sqrt{x-1} \quad \text{and} \quad 3 - x = -\sqrt{x-1},$$

for which squaring produces the same polynomial equation,

$$9 - 6x + x^2 = x - 1.$$

The 2 is the solution of the equation on the left and the 5 is the solution of the equation on the right. So whenever you square each member of an equation, you must check the solutions and discard any which are extraneous. In this case you would mark 5 as extraneous, and write

$$x = 2 \text{ or } \cancel{x = 5} \quad \leftarrow \text{extraneous}$$

$$S = \{2\}.$$

Example 2: Solve $\sqrt{x^2 + 5x - 6} + \sqrt{x^2 + 3x - 3} = 1$.

In this case the equation contains *two* radicals. So you isolate *one* of them, getting

$$\sqrt{x^2 + 5x - 6} = 1 - \sqrt{x^2 + 3x - 3}.$$

Squaring each member produces

$$x^2 + 5x - 6 = 1 - 2\sqrt{x^2 + 3x - 3} + x^2 + 3x - 3$$

which can be transformed to

$$2x - 4 = -2\sqrt{x^2 + 3x - 3}$$

$$2 - x = \sqrt{x^2 + 3x - 3},$$

with the radical isolated on the right side. Squaring again gives

$$4 - 4x + x^2 = x^2 + 3x - 3$$
$$7 = 7x$$
$$1 = x.$$

Checking by substitution into the original equation shows that 1 is a valid solution.

$x = 1$:

$$\sqrt{1^2 + 5 - 6} + \sqrt{1^2 + 3 - 3} = 1$$
$$\sqrt{0} + \sqrt{1} = 1$$
$$0 + 1 = 1$$
$$1 = 1.$$

Therefore,

$S = \{1\}$.

The steps in transforming and solving radical equations are summarized in the following flow chart.

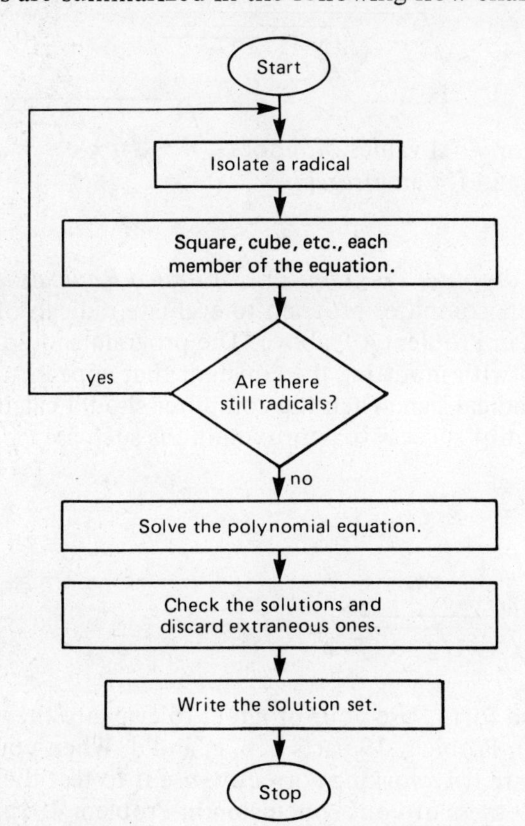

The exercise which follows is designed to give you practice solving radical equations.

Exercise 8-3

Find the solution set of each of the following equations:

1. $\sqrt{2x + 3} = 5$
2. $\sqrt{3x - 5} = 4$
3. $\sqrt[3]{4x - 1} = 3$
4. $\sqrt[3]{x - 2} = 2$
5. $5\sqrt{x - 1} = \sqrt{x + 1}$
6. $\sqrt{x + 14} - \sqrt{3x - 10} = 0$
7. $\sqrt{x - 7} = \sqrt{x} - 7$
8. $\sqrt{x + 5} - 1 = \sqrt{x}$
9. $\sqrt{3x^2 - 4x + 9} = 3$
10. $\sqrt{x^2 - 9} = 4$
11. $x + 5 = \sqrt{x + 5} + 6$
12. $x - 2 = \sqrt{x - 2} + 12$
13. $\sqrt{x - 1} + 3 = x$
14. $\sqrt{-3x - 14} - x = 4$
15. $\sqrt{x - 1} + x = 3$
16. $\sqrt{x + 6} - x = 4$
17. $\sqrt{7 - 3x} + 3 = x$
18. $x - \sqrt{6 - x} = 4$
19. $\sqrt{3x - 11} + 3 = x$
20. $\sqrt{3x + 10} - 4 = x$
21. $\sqrt{22 - 12x} + 5 = 2x$
22. $\sqrt{7 - 6x} + 3x = 2$
23. $\sqrt{x} - \sqrt{7} = \sqrt{x + 7}$
24. $\sqrt{x} + \sqrt{7} = \sqrt{x + 7}$
25. $\sqrt{x + 3} + \sqrt{x - 3} = 3$
26. $\sqrt{x + 4} + \sqrt{x - 4} = 4$
27. $\sqrt{2x + 5} + 2\sqrt{x + 6} = 5$
28. $\sqrt{2x + 25} - 2\sqrt{x + 4} = 1$
29. $x^2 = 21 - \sqrt{x^2 - 9}$
30. $x^2 = 3 - \sqrt{2x^2 - 3}$
31. $\sqrt{x^2 + 3x + 6} - \sqrt{x^2 + 3x - 1} = 1$
32. $\sqrt{x^2 - 4x - 12} - \sqrt{x^2 - 4x - 5} = 1$
33. $\sqrt{x} + \sqrt{x - 7} = \dfrac{21}{\sqrt{x - 7}}$
34. $2\sqrt{x} - \sqrt{4x - 3} = \dfrac{1}{\sqrt{4x - 3}}$
35. $\dfrac{1}{1 - x} + \dfrac{1}{1 + \sqrt{x}} = \dfrac{1}{1 - \sqrt{x}}$

36. $\dfrac{2}{\sqrt{x}+2} - \dfrac{\sqrt{x}}{2-\sqrt{x}} = \dfrac{x+4}{x-4}$

37. $\dfrac{\sqrt{x}}{\sqrt{x}-1} + 3 = \dfrac{1}{\sqrt{x}-1} - 1$

38. $\dfrac{1}{1-\sqrt{x}} = 1 - \dfrac{\sqrt{x}}{\sqrt{x}-1}$

39. $\sqrt[3]{2x^2 - 11x + 14} = 2 - x$
(Remember the Factor Theorem!)

40. $1 - x = \sqrt[3]{(x-1)(x-13)}$
(Remember the Factor Theorem!)

41. $x^2 + \sqrt{x^2 - 5x + 1} = 5x + 1$

42. $x^2 + \sqrt{x^2 + 3x + 5} = 7 - 3x$

For Problems 43 through 48 the transformed equation is a quadratic that does *not* factor. Use the Quadratic Formula to solve the transformed equation. You can check for extraneous solutions by finding decimal approximations for the radicals using a calculator. You can also check the answers *exactly* by simplifying radicals such as $\sqrt{2+\sqrt{3}}$ using the technique of Problems 47 through 54 in Exercise 8-2.

43. $x = 3 + \sqrt{20 - 4x}$

44. $x - \sqrt{15 - 4x} = 4$

45. $2\sqrt{2x} + 1 = \sqrt{4x + 9}$

46. $\sqrt{6x} - 1 = \sqrt{4x + 5}$

47. $\sqrt{2x} - \sqrt{x-3} = \dfrac{2}{\sqrt{x-3}}$

48. $\sqrt{-2x} - \sqrt{5-x} = \dfrac{-3}{\sqrt{5-x}}$

49. *Continued Radicals* – Under certain conditions, radicals such as

$$\sqrt{2+\sqrt{2+\sqrt{2+\sqrt{2+\ldots}}}}$$

that continue forever, can represent *rational* numbers. To find out what number such a radical represents, you can let it equal x and solve the resulting equation. The key to the solution is realizing that x also appears *under the radical sign.*

$$x = \sqrt{2+\underbrace{\sqrt{2+\sqrt{2+\sqrt{2+\ldots}}}}_{\text{This also equals } x.}}$$

So the equation is equivalent to $x = \sqrt{2+x}$. Evaluate the following radicals by solving the appropriate radical equation.

a. $\sqrt{2+\sqrt{2+\sqrt{2+\sqrt{2+\ldots}}}}$

b. $\sqrt{6+\sqrt{6+\sqrt{6+\sqrt{6+\ldots}}}}$

c. $\sqrt{20-\sqrt{20-\sqrt{20-\sqrt{20-\ldots}}}}$

d. $\sqrt{42-\sqrt{42-\sqrt{42-\sqrt{42-\ldots}}}}$

e. For what values of n does $\sqrt{n+\sqrt{n+\sqrt{n+\sqrt{n+\ldots}}}}$ stand for an *integer*?

50. *Computer Program for Continued Radicals* – Write a computer program to evaluate radicals of the form in Problem 49, above. The program should begin with inputting the constant that appears under the radical sign. Then the computer should calculate and print successive approximations such as

$\sqrt{2}$

$\sqrt{2+\sqrt{2}}$

$\sqrt{2+\sqrt{2+\sqrt{2}}}$,

and so forth. Use your program to evaluate the radicals in Problem 49 parts a, b, c, and d. When you are sure it is working correctly, use it to test the correctness of your conclusion in Problem 49, part e.

8-4 IRRATIONAL VARIATION FUNCTIONS

In Chapter 7 you studied problems in which one variable was proportional to an integer power of another variable. In this section you will study variation problems in which the exponent is a *constant,* but *not* necessarily an integer.

Objective:

Given a real-world situation in which one variable is proportional to a non-integer power of another variable, determine the proportionality constant, and use the resulting variation function as a mathematical model.

As an example, the area of an egg's shell is directly proportional to the 2/3 power of the mass of the egg. Suppose that a normal 60 gram chicken egg has a shell area of 28 square centimeters.

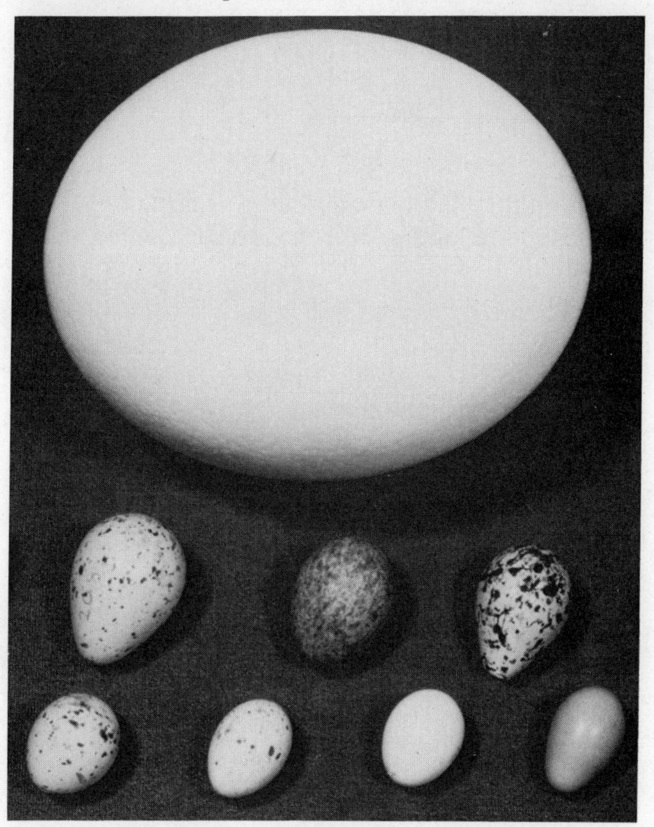

Let S = number of square centimeters.
Let M = number of grams.

So the general equation is

$$S = kM^{2/3}.$$

To find the particular equation, you must calculate k by substituting the known ordered pair, $(M, S) = (60, 28)$.

$$28 = k(60^{2/3})$$

Dividing each member by $60^{2/3}$ gives

$$\frac{28}{60^{2/3}} = k.$$

$$1.8269... = k$$

So the particular equation is

$$\underline{S = 1.8269...M^{2/3}}$$

The value of k should be stored in the calculator's memory for use in the rest of the problem.

The model is now ready to use. For example, to predict the surface area of a 1600 gram ostrich egg, you simply substitute 1600 for M and do the indicated operations.

$$S = 1.827(1600^{2/3})$$

$$S = 249.92...$$

So the ostrich egg has about $\underline{\underline{\text{250 square centimeters}}}$ of shell.

The model can also be used "backwards" to predict the mass of an egg when you know the surface area. Suppose that a lizard's egg has a surface area of 0.6 square centimeters. To find the mass, you would substitute 0.6 for S and solve for M.

$0.6 = 1.827M^{2/3}$ Substitution.

$\frac{0.6}{1.827} = M^{2/3}$ Divide by 1.827.

$\left(\frac{0.6}{1.827}\right)^{3/2} = M.$ Raise each member to the 3/2 power.

$0.188... = M$ By calculator.

So the lizard's egg is about 0.19 grams.

Does doubling the mass cause the surface area to double? Substituting $M = 1$ and $M = 2$ gives:

$M = 1$:

$S = 1.8269...(1^{2/3})$

$S \approx 1.83$

$M = 2$:

$S = 1.8269...(2^{2/3})$

$S \approx 2.90$

Since twice 1.83 is 3.66 and S is only 2.90, doubling the mass does *not* double the area. This fact can be anticipated because M is raised to a power *less* than 1. If S equaled kM^1, then doubling M would double S.

In the exercise which follows, you are to use variation functions with appropriate integer or non-integer exponents as mathematical models for the given situations.

Exercise 8-4

1. ***Ship Power Problem.*** When a ship is travelling at high speed through the water, most of the power generated by the engines goes into formation of the wake (the waves that trail out behind the ship). At these speeds the speed of the ship is proportional to the seventh root of the power being generated by the engines. Suppose that you are on a ship going 30 knots (30 nautical miles per hour) and the engines are generating 45,000 horsepower.

   a. Write the particular equation expressing speed in terms of power.
   b. The engines are capable of producing 90,000 horsepower. How fast would you expect the ship to go if the Captain gives the order, "Full speed ahead!"?
   c. Does doubling the power cause the speed to double?
   d. Why do you suppose that ships do not go much faster than 30 knots?

2. ***Toothpaste Factory Problem.*** A "rule of thumb" used by chemical engineers to estimate the cost of a chemical factory is that the cost is directly proportional to the 0.6 power of the amount of chemical the factory produces per unit time. Suppose that a toothpaste factory which turns out 15 tons per day of toothpaste costs 43 million dollars to build.

   a. Write the particular equation expressing the cost of a toothpaste factory in terms of the number of tons per day it produces.
   b. Predict the cost of building a factory to produce
      i. 500 tons per day,
      ii. 0.07 tons per day.
   c. If you double the manufacturing capacity of a toothpaste factory will the cost be double, more than double, or less than double? Justify your answer.

3. ***River Basin Problem.*** From measurements on many rivers, geographers find that the length of a river that drains a particular "basin" of land is approximately proportional to the 0.6 power of the area of the basin. The Rio Grande is 3034 kilometers long, and drains a basin of about 500,000 square kilometers (see sketch).

   a. Write the particular equation expressing river length in terms of basin area.
   b. The Suwannee River (made famous by Stephen Foster) flows from the Okefenokee Swamp (made famous by Pogo) to the Gulf of Mexico. It drains an area of about 15,000 square kilometers. According to your model, how long is the Suwannee

River? (Check an atlas or encyclopedia to find out how close your prediction is!)

c. The longest river in the world is the 6700 kilometer Nile. Approximately what area of land does the Nile drain?

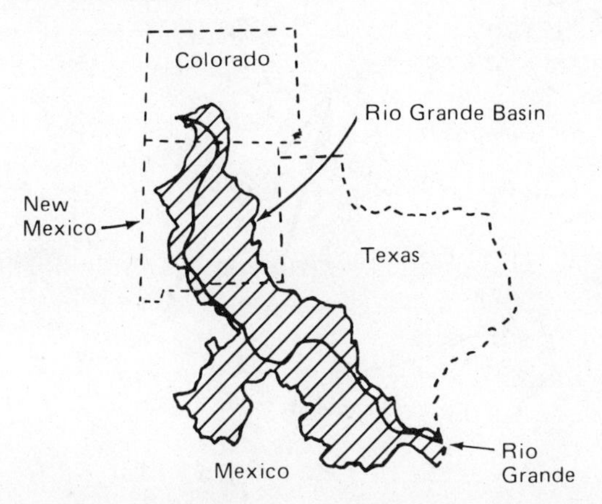

4. ***Home Range Problem.*** According to information on Page 200 of J. M. Emlen's *Ecology: An Evolutionary Approach,* the number of acres to which an animal confines its movements is directly proportional to the 1.41 power of its body mass.

a. Find the particular equation for this function if an 80 kilogram deer confines itself to a range of 2000 acres.

b. Predict the home range of

   i. a 6000 kilogram African elephant,
   ii. a 0.002 kilogram shrew.

c. Suppose that you find evidence of a Tasmanian devil spread over a range of 300 acres. What would you predict the mass of a Tasmanian devil to be?

5. ***Tree Trunk Problem.*** Thomas A. McMahon reports in the July, 1975, issue of *Scientific American* that the base diameter of a tree's trunk varies directly with the 3/2 power of its height.

a. Suppose that you find a young sequoia tree 5 meters tall that has a base diameter of 14.5 centimeters. Write the particular equation expressing the base diameter of a sequoia in terms of its height.

b. What would you expect this tree's diameter to be when it has grown to a height of 50 meters? Does making the height 10 times as big make the diameter *more* than 10 times as big or *less* than 10 times as big?

c. It is difficult to measure the height of a tall tree, especially when it is in a dense forest. But it is relatively easy to measure its base diameter. The largest known sequoia, the General Sherman in California, has a base diameter of 985 centimeters (about the size of a small house). Approximately how tall is the General Sherman?

6. ***Tire Pump Problem.*** If you compress a gas quickly, such as by pushing down the handle of the tire pump on the sketch, the heat generated by compression does not have time to escape, and the gas warms up. In this case, the pressure varies inversely with the 1.4 power of the volume. (In problem 7 of Section 7-11, Boyle's Law said that the pressure varied inversely with the *first* power of the volume, but that was because the temperature remained *constant.*)

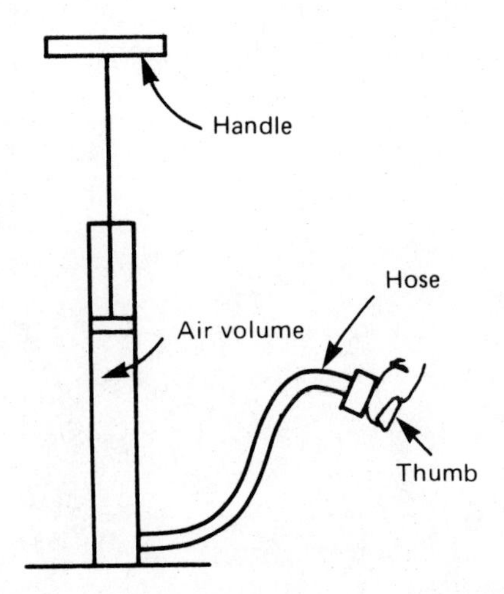

a. Suppose that when the pump handle is fully extended, the volume is 40 cubic inches, and the pressure is 15 pounds per square inch (psi), the normal atmospheric pressure. Write the particular equation expressing pressure in terms of volume.
b. You hold your thumb over the end of the hose, then push down on the handle, reducing the volume to 20 cubic inches. Predict the pressure.
c. Suppose that you connect the hose to a tire containing air at 50 psi. When you push down the handle, the pressure in the pump will increase until it just equals the pressure in the tire. What will the volume of air in the pump be when its pressure just reaches 50 psi?
d. Draw the graph of pressure versus volume in the domain $0 < \text{volume} \leq 40$ cubic inches with the pump connected to the tire. Take into account the fact that when the pressure reaches 50 psi, the air flows into the tire rather than being compressed to a higher pressure.

7. ***Pendulum Problem.*** The period of a pendulum (the length of time it takes to make one complete swing) varies directly with the square root of the length of the pendulum. By experiment, you find that a pendulum 0.3 meters long swings with a period of 0.55 seconds.

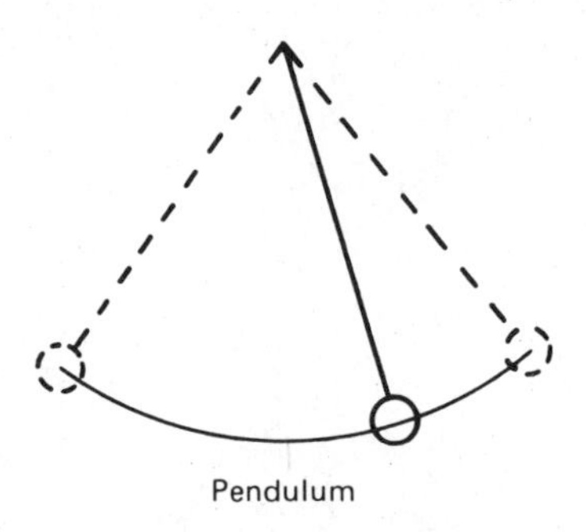

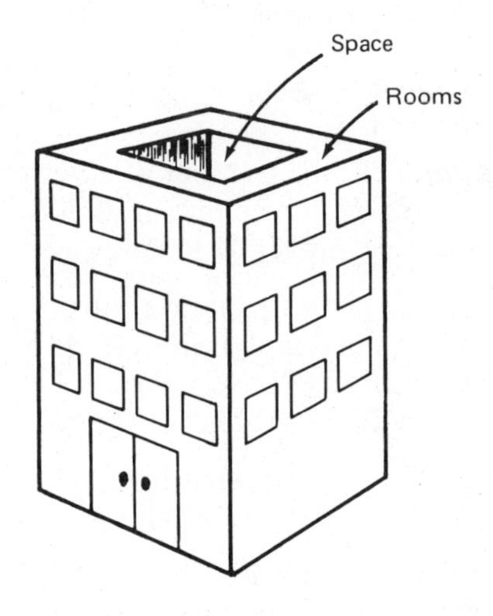

a. Write the particular equation expressing period in terms of length.
b. Grandfather clocks have pendulums that swing with a period of 1 second. How long is the pendulum?
c. Some hotels are constructed with rooms around the outside and empty space above the lobby in the middle. Suppose that a chandelier is suspended by a chain from the top story and hangs all the way down to the lobby. You observe the chandelier swinging with a period of 10 seconds. How tall is the hotel?

8. ***Planetary Period Problem.*** According to Keppler's Third Law, the period of a planet (the length of time it takes to make one revolution about the Sun) varies directly with the 3/2 power of the planet's average distance from the Sun. The Earth is 150 million kilometers from the Sun, and has a period of 1 year.

a. Write the particular equation expressing period in terms of distance.
b. Predict the periods of the following planets:

Planet	*Million km*
Mercury	58
Mars	227.7
Pluto	5913

(Check an encyclopedia or other source to see if you are correct.)

c. The "Asteroid Belt," located about 400 million kilometers from the Sun, is thought to be the path of a planet that disintegrated many centuries ago. What would the period of that planet have been?
d. Suppose that you wanted to build a space station orbiting the Sun every 1/8 of a year. How close to the Sun would it have to be?

9. ***Microwave Oven Problem.*** The number of minutes it takes to cook bacon in a microwave oven depends on how many slices you put in at once. A popular brand of oven specifies 1.75 minutes for 2 slices, and 2.5 minutes for 4 slices.

a. Explain why the number of minutes does *not* vary *directly* with the number of slices.
b. Assume that the number of minutes varies directly with some *power* of the number of slices (not necessarily an integer power). Use the two given ordered pairs to derive the particular equation expressing the number of minutes in terms of the number of slices. (The system of equations you get after substituting the ordered pairs can be solved for the two unknown constants. You must be clever enough to figure out a way!)
c. Where would you set the timer to cook 8 slices? 6 slices? 1 slice?
d. The timer on this oven can be set for as much as 30 minutes. What is the maximum number of slices that could be cooked at once?
e. What are the domain and range of this function?
f. What things in the real world might make the domain even smaller than the domain you wrote in part e?

10. ***Height-Mass Problem, Second Model.*** The masses of similarly shaped people vary directly with the cube of their height (see Problem 26 of Exercise 7-11). However, tall people tend to be thinner in proportion to their height than short people. Assume that the masses of people of average build vary with some power of their height, close to 3 but not quite equal to 3. The following average heights and masses are published for boys and girls:

Height	*Boys*	*Girls*
150 cm	45 kg	41 kg
180 cm	74 kg	67 kg

a. Write the particular equation for boys expressing mass in terms of height. Since there are *two* unknown constants (the exponent and the proportionality constant), you must substitute *both* ordered pairs into the equation and solve the resulting system for the constants. You can make the equations *linear* by taking the logs of both members first.
b. Write another equation for girls, as in part a, above.
c. Predict Sally Forth's mass. She is 173 cm tall. Show that your answer is reasonable by showing that it is *between* 41 and 67 kilograms.
d. Use the equation for boys to predict Professor Snarff's mass. He is 193 cm tall.
e. Professor Snarff is actually 91 kilograms. A person's mass is considered to be "normal" if it is within 5% of that predicted by the model. Is the Professor normal, overweight, or underweight? Justify your answer.
f. Use the appropriate equation to predict *your* mass. Are you within 5% of the predicted mass???

g. What does your model predict for the mass of a newborn baby girl 50 centimeters long? Is this reasonable?

8-5 FUNCTIONS OF MORE THAN ONE INDEPENDENT VARIABLE

In most real-world situations there are many variable quantities. So far you have considered only two at a time, one independent and one dependent variable, with the tacit assumption that the values of the others remain *constant.* If you consider more than two variables, there are several ways they can be interrelated. For example, if there are three variables, then

1. y and z might *both* depend on x,
2. z might depend on *both* x and y,
3. z might depend on y, and y might depend on x.

In this section you will study the second and third kinds, because functions of the first kind are really just *two* functions of the kind you have already studied.

Objective:

Given a real-world situation involving more than two variables, find general and particular equations relating the variables, and use these as mathematical models.

Functions with Several Independent Variables – If you studied linear programming in Section 4-10, you encountered linear functions that had two independent variables. For example, the profit made by an automobile manufacturer depends on both the number of large cars and on the number of small cars that are sold.

Another type of function with more than one independent variable is called a *combined variation function.* For example, the statement, "w varies directly with the square of x, and inversely with the cube of y," means that w is related to x and y by the general equation

$$w = \frac{kx^2}{y^3},$$

where k is, as usual, the constant of proportionality.

The reason for these words can be seen if you hold one independent variable constant. If x is constant, then $w = (\text{constant})/y^3$. Thus, w varies inversely with the cube of y. If y is held constant, then $w = kx^2/(\text{constant})$, which can be written $w = (\text{another constant})(x^2)$. Thus, w varies directly with the square of x.

To find the particular equation for a given situation, you must know one set of corresponding values of the variables. With these values, you can find the proportionality constant. For example, suppose that $w = 12$ when $x = 2$ and $y = 5$. Solving first for k, then substituting,

$$\frac{wy^3}{x^2} = k \quad \text{Multiplying by } y^3 \text{ and dividing by } x^2.$$

$$\frac{(12)(5^3)}{2^2} = k \quad \text{Substituting the given values.}$$

$$375 = k \quad \text{Arithmetic.}$$

$\therefore$ the particular equation is $w = \dfrac{375x^2}{y^3}$.

The equation can then be used for predicting and interpreting, as you have done in past mathematical models problems.

Composite Functions – Functions in which *one* variable depends on a *second* variable, and the second variable depends on a *third* variable are called "composite functions." For example, the statement, "w varies directly with the square of x, and x varies inversely with the cube of y," can be translated into *two* equations

$$w = k_1 x^2 \quad \text{and} \quad x = \frac{k_2}{y^3},$$

where k_1 and k_2 are proportionality constants, *not* necessarily equal to each other. Often it is desirable to express the first variable, w, in terms of the third variable, y. This may be accomplished simply by substituting k_2/y^3 for x in the first equation.

$w = k_1 x^2$ First equation.

$w = k_1\left(\frac{k_2}{y^3}\right)^2$ Substitution.

$w = \frac{k_1 k_2^2}{y^6}$ Properties of exponentiation and fractions.

Since k_1 and k_2 are constants, the quantity $k_1 k_2{}^2$ is also a constant. Calling this third constant k_3, you get

$$w = \frac{k_3}{y^6}.$$

In words, "w varies inversely with the sixth power of y."

Notes:

1. $f(x)$ terminology is sometimes convenient for writing about composite functions. If $w = f(x)$ and $x = g(y)$, then $w = f(g(y))$. You may recall having seen some examples of this terminology in Exercise 4-4, Problem 4.
2. A statement such as, "w depends on x *and* y *and* z," means that there is *one* equation which expresses w in terms of x, y, and z. A statement such as, "w depends on x, x depends on y, and y depends on z," means that there are *several* equations relating *pairs* of variables. You must be able to distinguish between these *combined* variation functions and *composite* functions to interpret the following problems.

Exercise 8-5

1. ***Beam Strength Problem.*** The safe load for a horizontal beam, such as those that hold up the floor in a house, varies directly as the breadth, directly as the square of the depth, and inversely as the length between supports (see Figure 8-5a).

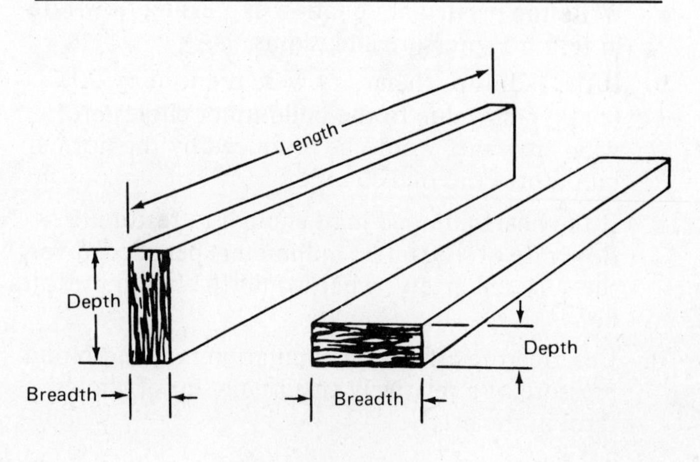

Figure 8-5a.

a. Write an equation expressing safe load in terms of these independent variables if a 2″ × 8″ beam 16 feet long, turned "on edge" so that its breadth is 2″ and its depth is 8″, can support a load of 1000 pounds.
b. If the beam in part a is laid "flat" so that the breadth is 8″ and the depth is 2″, how many pounds can it support?
c. Why are houses built with the floor beams as in part a rather than as in part b?
d. What is the safe load for a beam 8 inches broad, 12 inches deep, and 12 feet long?
e. By what factor is the safe load changed if

   i. the breadth is doubled?
   ii. the depth is doubled?
   iii. the length is doubled?

2. ***Blood Pressure Problem.*** In 1844 the French scientist J. L. Poiseuille found that the rate at which a fluid such as blood flows through small tubes such as arteries and veins varies directly with the product of the pressure acting on the fluid and the fourth power of the radius of the tube. Assume that you have an artery 0.2 centimeters in radius through which blood flows at 400 cubic millimeters per second when acted upon by the normal pressure of 100 units.

a. Write the particular equation expressing flow rate in terms of pressure and radius.

b. If the radius of the artery were reduced by 20% (to 0.16 cm) due to the build-up of cholesterol, what flow rate would be produced by the normal blood pressure of 100 units?

c. If the heart pumped hard enough to restore the flow rate of 400 cubic millimeters per second for the 0.16 cm artery, what would the blood pressure be?

d. Describe the effects on circulation rate and blood pressure of a relatively small build-up of cholesterol in the arteries.

3. ***Spike Heel Problem.*** The pressure exerted on the floor by a person's shoe heel is directly proportional to his or her weight and inversely proportional to the square of the width of his or her heel. Professor Snarff weighs 200 pounds, wears a shoe with a 3 inch wide heel, and exerts a pressure of 24 pounds per square inch (psi) on the floor.

a. Write the particular equation expressing pressure in terms of weight and heel width.

b. Phoebe Small weighs 100 pounds. Plot a graph of the pressure she exerts versus her heel width for heels one inch to 4 inches wide.

c. On the same set of axes, plot two other graphs, one for John Garfinkle who weighs 150 pounds, and one for Professor Snarff who weighs 200 pounds. (The resulting three graphs form a *family* of curves, and the variable which characterizes each curve (weight, in this case) is often called a *parameter*.)

d. In the 1960's, young ladies wore "spike" heels which were very narrow. How much pressure would Phoebe Small's heel exert if it were ¼ inch wide? Surprising??

e. Why do you suppose that commercial airlines did not allow their stewardesses to wear spike heels?

4. ***10-Speed Bike Problem.*** A 10-speed bike changes gears by moving the chain to different size sprockets on the front and back (see Figure 8-5b). The speed at which the bike goes varies directly with the number of revolutions per minute (rpm) you turn the pedals, directly with the number of teeth on the front sprocket, and inversely with the number of teeth on the back sprocket. A standard bike with 70 cm. diameter wheels will go about 7.4 kilometers per hour (kph) if you pedal 40 rpm in the lowest gear (39-tooth front sprocket, 28-tooth back one.)

a. Write an equation expressing kph in terms of the three independent variables.

b. The fastest you can turn the pedals is about 180 rpm. Calculate your top speed in each of the 10 gears.

c. Rank the gears, from "highest" to "lowest."

d. Suppose that you are most comfortable when you pedal 60 rpm. If you want to go 20 kph, which gear should you use to get closest to your most comfortable speed? Justify your answer.

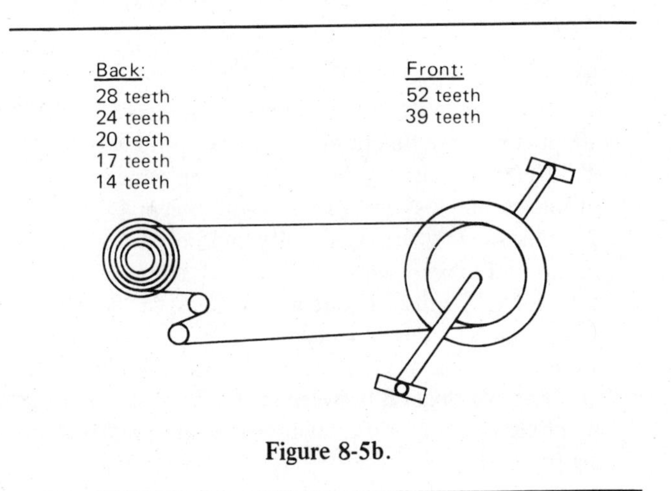

Figure 8-5b.

5. ***Bridge Column Problem.*** A bridge is held up by columns 20 inches in diameter and of varying lengths (see Figure 8-5c). From a Strength of Materials text you find that as you put more and more weight on a column, it can collapse either by *buckling* or by *crushing* (see sketch). The load which will buckle a

column is directly proportional to the fourth power of its diameter, and inversely proportional to the square of its length. The load which will crush a column varies directly with the square of its diameter, and is *independent* of its length (i.e., length does *not* appear in the equation for the crushing load).

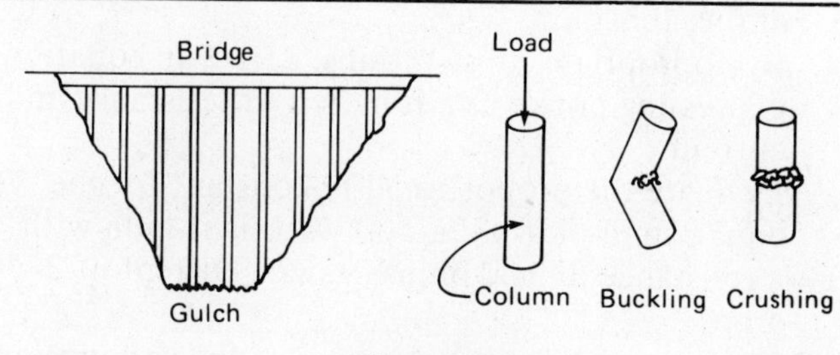

Figure 8-5c.

a. Write an equation expressing the number of tons which will *buckle* a column in terms of the length and diameter of the column. Evaluate the proportionality constant, using your laboratory findings that a 2″ diameter column 3 feet long will buckle under a load of 4 tons.
b. Write another equation expressing the number of tons that will *crush* a column in terms of the diameter of the column, using the fact that the 2″ diameter laboratory column in part a is crushed by a load of 5 tons.
c. Calculate the number of tons load needed to *buckle* 20″ diameter bridge columns which are 20 feet, 30 feet, and 50 feet long.
d. Calculate the number of tons that will *crush* a 20″ diameter column from the bridge.
e. Draw a graph of the number of tons a bridge column will support versus length of the column in the domain from 0 through 50 feet. Remember that the number of tons is either the *buckling* or *crushing* load, whichever is *less.*
f. What conclusions can you make about the way in which *long* columns collapse and the way in which *short* ones collapse?
g. At approximately what length will a bridge column collapse by both buckling *and* crushing? Explain.

6. ***Hospital Room Problem.*** Suppose that a hospital charges you \$110 per day if you stay in a private room, and \$83 per day if you stay in a room for 4 people. Assume that the amount you pay is the sum of a *constant* amount (for the services you receive) and an amount which *varies inversely* with the number of people the room will hold (to pay for your share of the room).

a. Calculate the constant amount you pay for services, and the proportionality constant in the variation function.
b. What does the hospital charge for a room, exclusive of the services to the patient?
c. How much would you expect to pay per day if you stayed in
   i. a semi-private room holding 2 people?
   ii. a ward holding 20 people?

7. ***Reaction to Shock Problem.*** Whether or not you will feel an electric shock depends on how high the current is and on how long the current is applied. According to Weiss' Law, the shock can be felt if the current is greater than or equal to a constant *plus* an amount that varies inversely with the length of time the current is applied. Suppose that you can just feel a current of 30 milliamperes (ma) applied for 2 milliseconds (ms). A current of 20 ma can just be felt when applied for 4 ms.

a. Write the particular *inequality* expressing current that can be felt in terms of time.
b. How many milliamperes can be felt if the current is applied for 10 ms?
c. What is the minimum length of time a current of 100 ma could be applied and still be felt?
d. How low a current could be felt if it were applied for a long period of time?
e. Plot the graph of the inequality in part a.

8. ***Airplane Flight Speed Problem.*** The weights of airplanes of similar shape are directly proportional to the cube of their lengths. In order for the airplane to fly, this weight must be directly proportional to the product of the wing area and the flight speed. The wing area varies directly as the square of the length of the airplane.

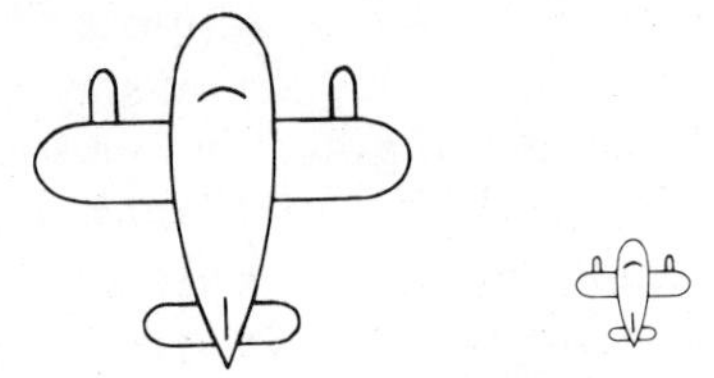

a. Write the general equation for each of the three variation functions above, using different symbols for each proportionality constant.
b. From these equations, derive another equation expressing the speed of an airplane in terms of its length.
c. How does flight speed vary with plane length?
d. How much faster would the plane have to fly if you doubled its length?

9. ***Gas Consumption Problem.*** Assume that the wind resistance acting against a car traveling at high speed is directly proportional to the square of the speed. Assume also that the number of kilometers per liter of gasoline (kpl) is inversely proportional to the wind resistance.

a. Write general equations expressing each of the above relationships. Then use the two equations to get a third equation expressing kpl in terms of speed.
b. How does the number of kpl vary with speed?
c. If you get 8 kpl when you drive 100 kilometers per hour (kph), find the proportionality constant and write the particular equation expressing kpl in terms of kph.
d. Predict the number of kpl for speeds of 120, 140, 160, 180, and 200 kph. Use the results to plot the graph of kpl versus kph.
e. Can you think of a reason other than safety for not driving 200 kph?
f. Do you suppose that your mathematical model would give reasonable answers for kpl at very low speeds? Justify your answer.

10. ***Electric Light Problem.*** There are four variables concerned with the operation of an electric light:

power (watts),	current (amps), and
voltage (volts),	resistance (ohms).

The power is proportional to the product of the voltage and the current. The resistance varies directly with the voltage and inversely with the current.

a. Write general equations for each of the two variations above.
b. By appropriate substitutions, derive an equation expressing power as a function of resistance and current.
c. Evaluate the proportionality constant for the equation in part b if a 400 watt light bulb with a resistance of 100 ohms draws a current of 2 amps.
d. Why do you suppose the sizes of the various units of electrical measurement were chosen the way they were? (Clue: See your answer to part c.)
e. By what factor would the power change if the current were tripled?

11. ***Why Mammals Are the Way They Are.*** Some physical and behavioral characteristics of mammals can be explained by comparing the way their surface area and mass are related. Assume that,

i. the mass of an animal is directly proportional to the *cube* of its length,
ii. the area of its skin is directly proportional to the *square* of its length,
iii. the rate at which it loses heat to its surroundings is directly proportional to its skin area, and
iv. the amount of food it must eat per day is directly proportional to the rate of heat loss.

a. Write general equations for each of the four functions above.
b. By appropriate substitutions, derive a general equation expressing amount of food per day in terms of the animal's mass. Tell how the food consumption varies with mass.
c. Divide both members of the equation in part b by the animal's mass to get an equation for the *fraction* of its mass an animal consumes per day. Tell how this fraction varies with mass.

d. The smallest mammal, the shrew, eats 3 times its mass each day. A shrew has a mass of about 2 grams (about that of a penny!). Find the proportionality constant for the equation in part c.
e. What fraction of its mass would a 6000 *kilogram* elephant eat each day?
f. As the size of the animal decreases, what seems to happen to the food-to-mass ratio? Based upon this result, why do you suppose that there are no mammals smaller than a shrew? What makes a shrew so *mean*? Why do small mammals like shrews have fur, but large ones like elephants do *not*? Apart from the fact that whales have no legs, why is the sea a favorable environment for such large mammals?
g. See Haldane's article, "On Being the Right Size," in James R. Newman's *World of Mathematics*, page 952.

12. ***Tornado Problem.*** When a tornado moves over a house, the sudden decrease in air pressure creates a force that can lift off the roof.

a. The force exerted on the roof equals the pressure times the area of the roof. Write an equation expressing this fact.
b. If the joint between the roof and the walls is strong enough, the roof will not be lifted off. The force a roof will withstand varies directly with the length of the roof-to-walls joint (that is, the perimeter of the room). Write an equation expressing this fact.
c. Write an equation stating that the area of the roof varies directly with the square of the perimeter of the room.
d. Use the equation in parts b and c to show that the force a roof will withstand is directly proportional to the 1/2 power of the roof's area.
e. The "critical pressure" for a room is the pressure at which the force exerted by the tornado just equals the force the roof will withstand. Combine the equations in parts a and d to show that the critical pressure varies *inversely* with the *square root* of the roof's area.
f. If one roof has an area 9 times that of another roof, how does the critical pressure for the first roof compare with the critical pressure of the second?
g. Based on your answer to part f, where is the safest place to hide if a tornado threatens your house?

8-6 CHAPTER REVIEW AND TEST

In this chapter you have studied the last kind of algebraic function, the *irrational* algebraic function. You found that these functions could be analyzed using techniques you had learned before. The properties of exponents are useful for simplifying radicals. Solving radical equations is quite similar to solving fractional equations – both kinds can have *extraneous* solutions. And variation functions with non-integer exponents behave just like other direct and inverse variation functions.

The objectives for this chapter may be summarized as follows:

1. *Show that certain radicals represent* ***irrational*** *numbers.*
2. *Transform expressions to simple radical form.*
3. *Solve radical equations.*
4. *Use as mathematical models variation functions that have*
   a. *non-integer exponents*
   b. *more than one independent variable.*

The Review Problems below have numbers corresponding to these objectives. The Concepts Test has two problems that *combine* these objectives, and require you to recall some techniques you have learned previously.

Review Problems

The following problems are numbered according to the four objectives listed above.

Review Problem 1a. Table I lists $\sqrt{7} \approx 2.646$. Write this approximation as the ratio of two integers, factor the numerator and denominator into primes, and do all possible canceling. Then show that the fraction could not possibly equal $\sqrt{7}$ because when you *square* the factored fraction you do not get an *integer*.

b. Show that $\sqrt[3]{100}$ is an *irrational* number by showing that it is between two consecutive integers.

Review Problem 2. Transform the following to simple radical form:

a. $\sqrt{75} - \sqrt{27}$, b. $2\sqrt{7} + \dfrac{35}{\sqrt{7}}$,

c. $\dfrac{\sqrt{75}}{\sqrt{3} + \sqrt{8}}$, d. $\dfrac{3}{\sqrt[5]{36}}$.

Review Problem 3. Solve the following radical equations:

a. $\sqrt{x+6} + 6 = x$,

b. $\sqrt{3x+1} - \sqrt{2x-10} = 4$,

c. $\sqrt{3x} + \sqrt{2x-1} = \dfrac{5}{\sqrt{2x-1}}$.

Review Problem 4a – Defended Region Problem. An animal will defend the region around its home by attacking intruders that come into the region. According to statistical studies reported in J. M. Emlen's *Ecology: An Evolutionary Approach,* the defended region's area varies directly with the 1.31 power of the animal's body mass.

i. Suppose that a normal 20 kilogram beaver will defend a region of area 300 square meters. Write the particular equation for this function.
ii. Skeletons show that thousands of years ago North American beavers were up to 3.5 meters long (including the tail) and had a mass of 200 kilograms (about the size of an Alaskan brown bear!). How many square meters would such a beaver have defended!
iii. What mass beaver would defend a region of area 50 square meters?

b. Suppose that y varies inversely with the cube of x, and that x varies directly with the square of z and inversely with w.

i. Write general equations expressing y and x in terms of their independent variables.
ii. Write a general equation expressing y in terms of z and w.
iii. Tell in words how y varies with z and w.
iv. Write the particular equation expressing y in terms of z and w if $y = 40$ when $z = 3$ and $w = 6$.

Concepts Test

Work each part of the following problems. Then tell, by number, which one or ones of the objectives of this chapter, listed above, you used for that part.

Concepts Test 1. In this problem you are to plot the graph of the irrational algebraic function

$$f(x) = \frac{1}{3 - \sqrt{x+5}} - 2.$$

In doing so, you will have to use most of the techniques of this chapter, plus some that you have learned before.

a. Prove that the radical $\sqrt{x+5}$ represents a *rational* number when $x = -1$, but an *irrational* number when $x = 1$.

b. What kind of number does $\sqrt{x+5}$ represent when $x < -5$?

c. For what value(s) of x is the denominator of the fraction equal to zero? What do you suppose happens to the graph of f at these values?

d. Write the domain of f.

e. Rationalize the denominator of the fraction in $f(x)$.

f. Write $f(x)$ as a *single* fraction in simple radical form.

g. Find out whether or not there are any x-intercepts by setting the numerator of the fraction in part f equal to zero and solving the resulting radical equation. Remember that to be an x-intercept, the value of x must be in the domain of f.

h. Find decimal approximations for $f(3)$, $f(5)$, and $f(11)$.

i. Find a decimal approximation for the $f(x)$-intercept.

j. Plot the graph of f. Use the information you already know, and calculate any additional points you feel you need. Put dots at any places where the graph ends, and arrows to show places where the graph continues beyond your graph paper.

Concepts Test 2. Suppose that $g(x)$ varies inversely with the cube root of x^2.

a. Write the general equation for $g(x)$.

b. Write the particular equation if $g(8) = 25$.

c. Write the value of $g(5)$ in simple radical form.

d. Find, approximately, the value of x for which $g(x) = 3.47$.

8-7 CUMULATIVE REVIEW, CHAPTERS 6, 7, AND 8

The following exercise may be considered to be a "final exam" covering your work on exponential, rational, and irrational functions. If you are thoroughly familiar with the materials in Chapters 6, 7, and 8, you should be able to work all of these problems in 2 to 3 hours.

Exercise 8-7

1. For each of the following graphs, tell what kind of function it could be, and write the general equation for that kind of function.

a.

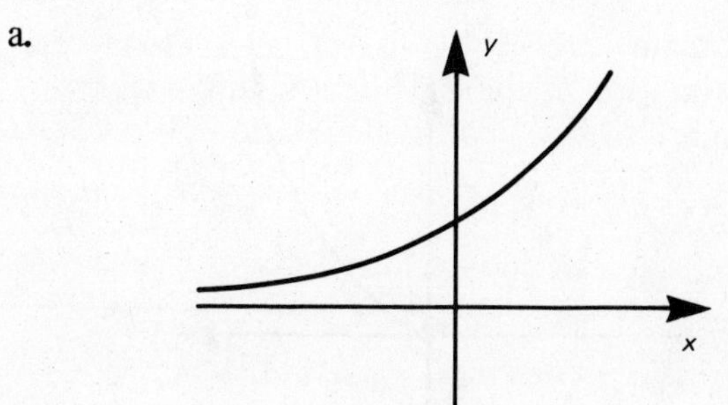

b.

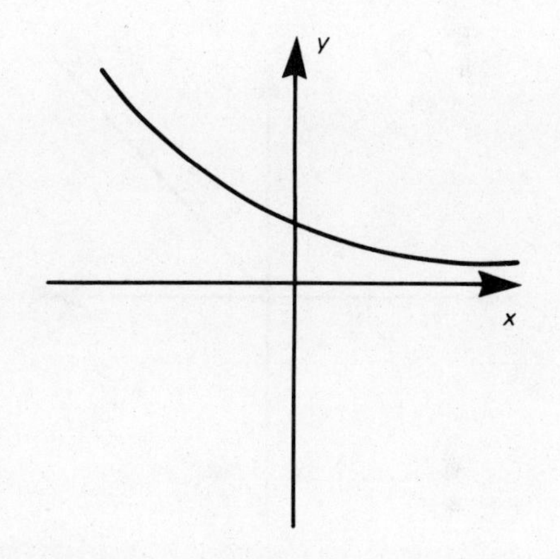

c.

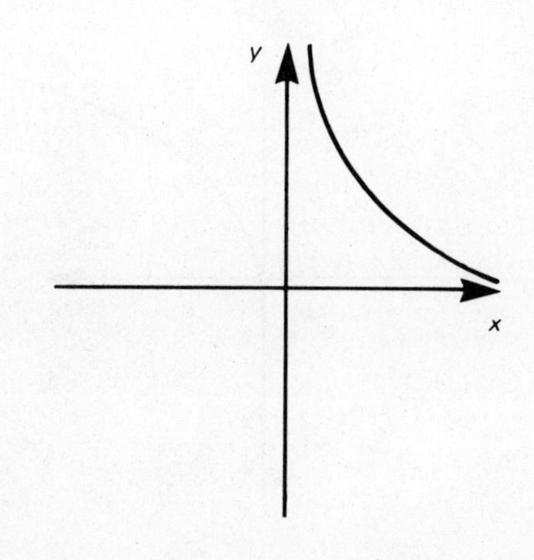

d.

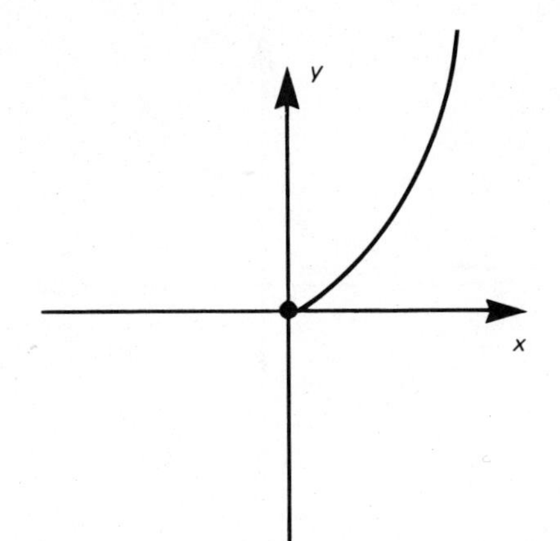

e.

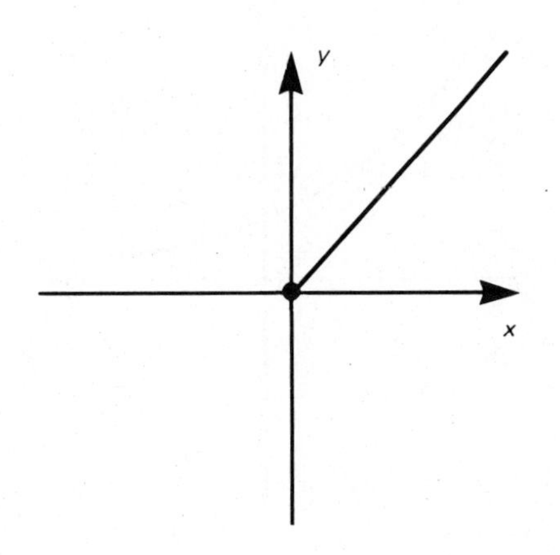

f.

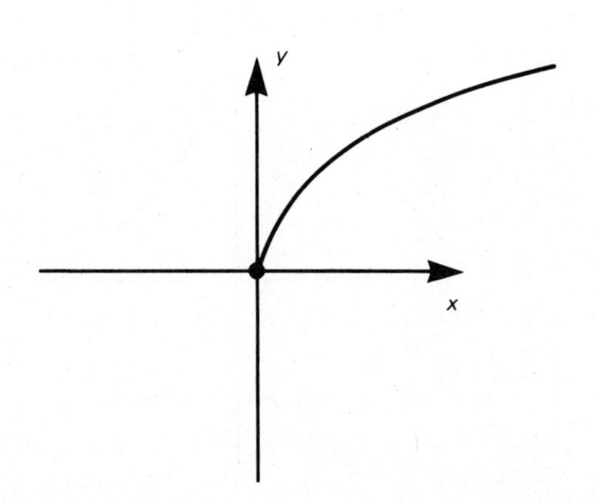

g.

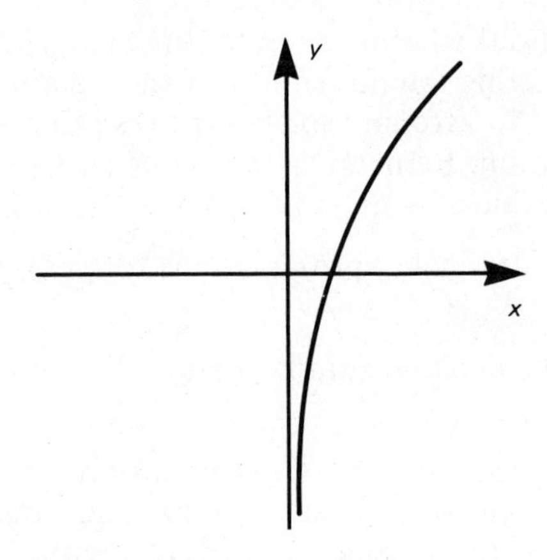

h.

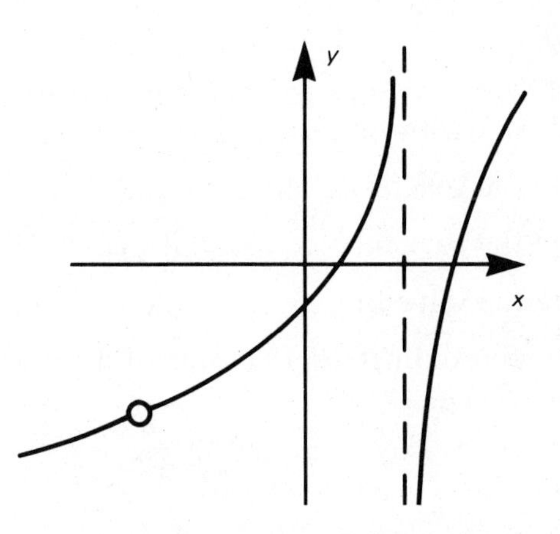

i.

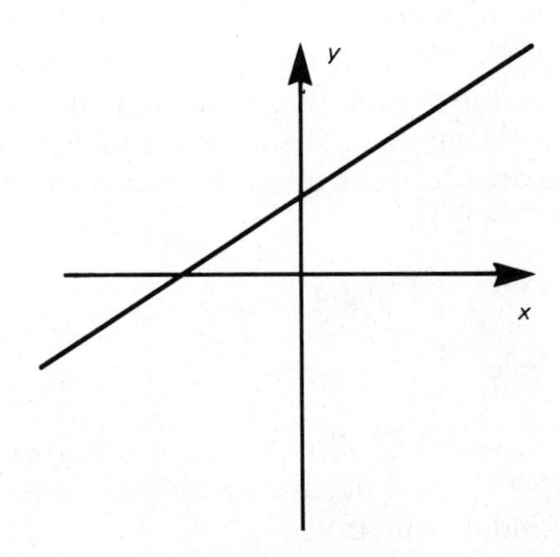

2. Write the general equation for a function that has the indicated property.

a. Doubling x doubles y.
b. Doubling x makes y half as big.
c. Doubling x makes y $1/8$ as big.
d. Doubling x makes y 16 times as big.
e. Adding 7 to x makes y three times as big.

3. Multiply the following and simplify:

a. $(\sqrt{15}+\sqrt{3})(\sqrt{12}-\sqrt{5})$
b. $(\sqrt{2}-\sqrt{7})^2$
c. $(\sqrt{2}-\sqrt{7})(\sqrt{2}+\sqrt{7})$
d. $(\sqrt{x}+5)(\sqrt{x}-7)$
e. $(x+5)(x-7)$
f. $(x^2-5x+4)(x^2+3x-9)$

4. Factor the following polynomials:

a. r^7+s^7 b. $6x^2-71x-120$

c. $x^3+x^2-10x+8$

5. Carry out the indicated operations and simplify:

a. $\dfrac{8}{x^2-16}-\dfrac{5}{x^2+3x-4}$

b. $\dfrac{x^2-4}{x^2+2x-3} \div \dfrac{x^2-7x+10}{(1-x)(x-3)}$

6. Simplify the following expressions (*no* decimal approximations!).

a. $3x^{-1/2} \times 4x^{2/3}$ b. $\sqrt{27} \div \sqrt[3]{81}$

c. $\dfrac{6}{\sqrt{19}-4}$ d. $\dfrac{12}{\sqrt[5]{27}}$

e. $\sqrt[4]{25}$ f. $3\sqrt{28}+\dfrac{21}{\sqrt{7}}$

g. $\log_{27} 81$

h. $\log_3 15+\log_3 6-\log_3 10$

i. $\dfrac{\dfrac{2}{x-1}+3}{\dfrac{13}{x+4}-3}$

j. $\left((16a^{-4}) \div (9a^{-2})\right)^{-1/2}$

7. Solve the following equations:

a. $1-\sqrt{7-x}=x$ b. $x+\dfrac{2}{x-1}=\dfrac{x-3}{1-x}$

8. Consider the expressions $\dfrac{5}{0}$, $\dfrac{0}{0}$, and $\dfrac{0}{5}$.

a. Two of these are *undefined.* Which two? What special words are used to distinguish between these two?
b. The other expression *is* defined. Which one? What does it equal?

9. For the function $f(x)=\dfrac{x^3+x^2-10x+8}{x^2-1}$:

a. Evaluate $f(-10)$, $f(-4)$, $f(-1)$, $f(0)$, $f(1)$, $f(2)$, and $f(10)$.
b. Plot a graph of f showing any holes or asymptotes.

10. Prove that the graph of $f(x)=\dfrac{x-2}{x^2-x+2}$ has *no* "holes" and *no* vertical asymptotes.

11. ***Iodine Problem.*** A radioactive atom "decays" by shooting out something from its nucleus. Once the "thing" is shot out, the atom is no longer radioactive. Radioactive iodine, used to examine people's thyroid glands, decays with a "half-life" of 8.1 days. This means that when 8.1 days is *added* to the time, the amount of iodine remaining is *multiplied* by 1/2.

a. What kind of function would be an appropriate mathematical model relating the amount of iodine remaining to the time that has passed?
b. A hospital receives a shipment of 23.7 millicuries of radioactive iodine. Write the particular equation expressing the amount of iodine remaining as a function of the number of days since they received the iodine.

c. Assume that the hospital uses none of the iodine on patients. Use your equation to predict

  i. the amount of iodine remaining after 1 week,
  ii. the number of days until the amount of iodine has dropped to 0.8 millicuries.

d. Show that your answers to part c are *reasonable* by using the fact that the *half*-life is 8.1 days.
e. The iodine was produced in a nuclear reactor 3.7 days before it was received at the hospital. How many millicuries were there when it was produced?

12. ***Heat Radiation Problem.*** The rate at which you receive heat radiated from a hot sphere such as the Sun varies directly with the fourth power of its absolute temperature, directly with its surface area, and inversely with the square of your distance from the sphere.

a. Write the general equation for this function.
b. The surface area of a sphere varies directly with the square of its radius, and its volume varies directly with the cube of its radius. Write general equations for these two functions.
c. Combine the equations in parts a and b to get an equation expressing the rate at which you receive heat in terms of the sphere's temperature and volume, and the distance between you and the sphere.

13. ***Diesel Engine Problem.*** You recall that when you compress air in a tire pump, the air warms up. A diesel engine works on the same principle. Air in the cylinders is compressed until it is hot enough to ignite the diesel fuel. As the air is compressed, its absolute temperature varies inversely with 0.4 power of its volume.

a. As the air starts being compressed, it is at "room temperature" of 300° K, and occupies 60 cu. cm. volume. Write an equation expressing absolute temperature in terms of volume.
b. What is the temperature when the air is compressed to 4 cu. cm.?
c. The minimum temperature needed to make diesel fuel ignite is 750° K. To what volume must the air be compressed to reach this temperature?

14. The following is a flow chart for calculating values of $f(x)$ for the rational function

$$f(x) = \frac{x^2 - 7x + 10}{x^2 - 8x + 15}\ .$$

Show the simulated computer memory, and find the output.

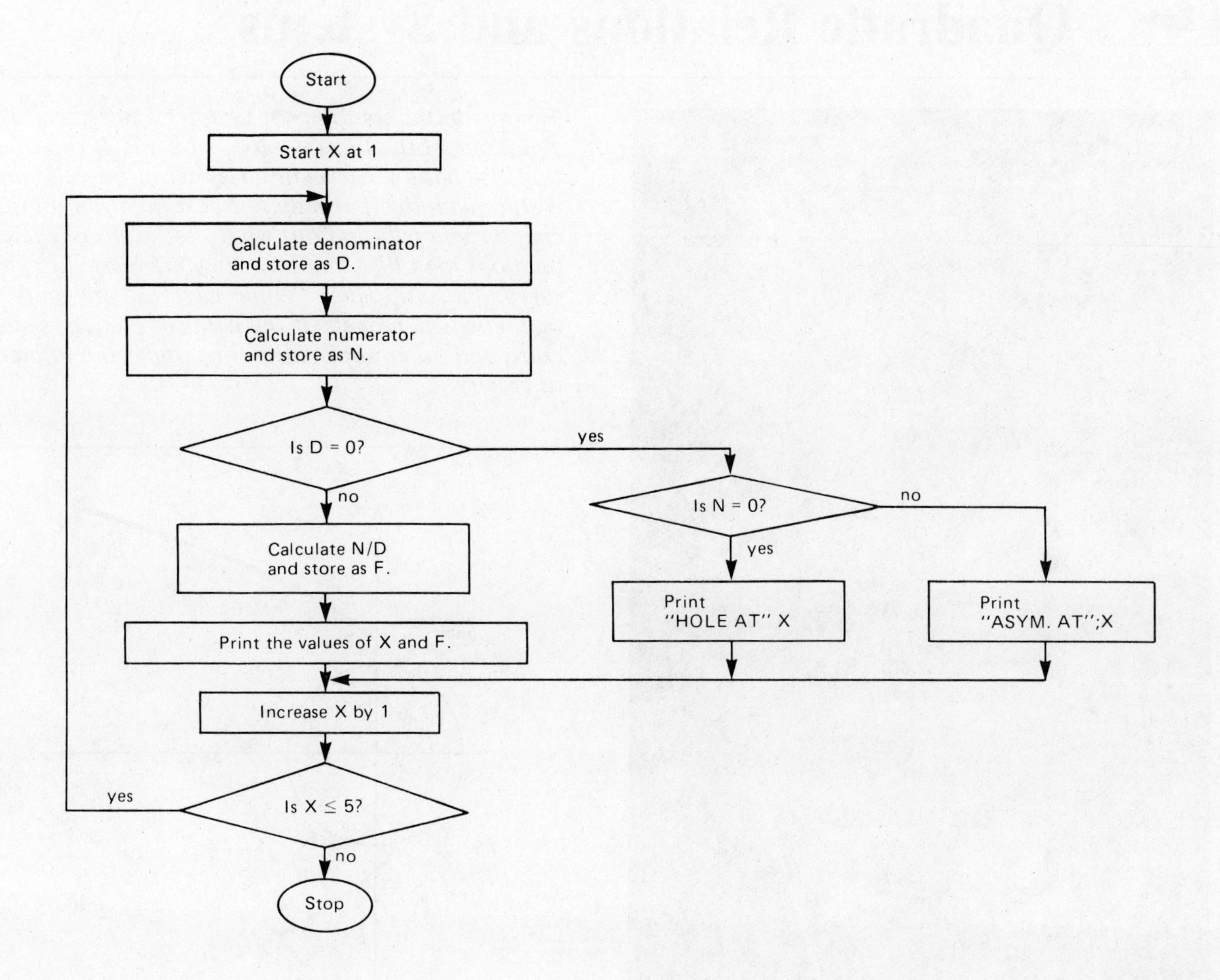
Start
Start X at 1
Calculate denominator and store as D.
Calculate numerator and store as N.
Is D = 0?
yes
no
Is N = 0?
no
yes
Calculate N/D and store as F.
Print "HOLE AT" X
Print "ASYM. AT";X
Print the values of X and F.
Increase X by 1
yes
Is X ≤ 5?
no
Stop

Quadratic Relations and Systems

The quadratic functions of Chapter 5 have general equations of the form $y = ax^2 + bx + c$. *If the equation also has a* y^2*-term or an* xy*-term, the relation is still quadratic, but may not be a function. In this chapter you will find out what the graphs of these relations look like, and how they are related to slices of a cone. These "conic sections" are good mathematical models of the paths of planets, spacecraft, and other objects traveling under the influence of gravity.*

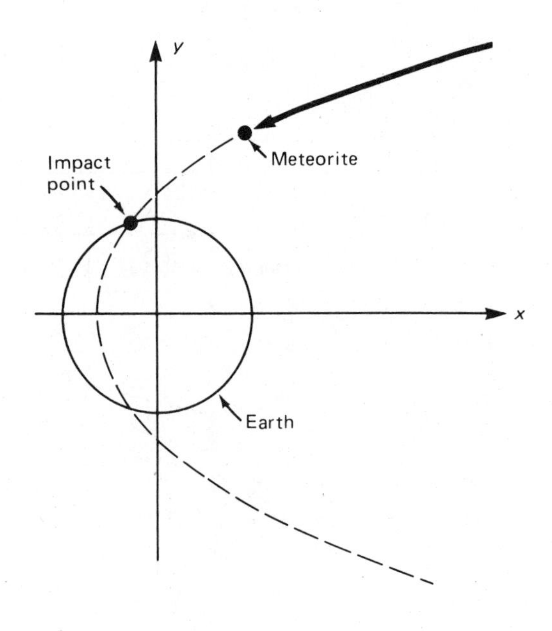

9-1 INTRODUCTION TO QUADRATIC RELATIONS

DEFINITION

A ***quadratic relation*** is a relation specified by an equation or inequality of the form

$$Ax^2 + Bxy + Cy^2 + Dx + Ey + F = 0,$$

where $A, B, C, D, E,$ and F stand for constants, and where the "=" sign may be replaced by an inequality sign.

It will be your objective in the next few sections to determine what the graph of such a relation might look like and how the six constants affect the graph.

To begin this objective, you will plot the graphs of several such relations pointwise, and attempt to discover what geometrical figure each one is.

Exercise 9-1

Plot a graph of each relation. Select values of x and calculate the corresponding values of y until you have enough points to draw a smooth curve. Use a calculator or square root tables to approximate any radicals. When you have finished, see if you can come to any conclusions about the *shape, size,* and *location* of each graph.

1. $\{(x, y): x^2 + y^2 = 25\}$.
2. $\{(x, y): x^2 + y^2 + 6x = 16\}$.
3. $\{(x, y): x^2 + y^2 - 4y = 21\}$.
4. $\{(x, y): x^2 + y^2 + 6x - 4y = 12\}$.

9-2 CIRCLES

In Exercise 9-1 you discovered that the graphs of some quadratic relations look like circles. In this section you will learn *why* this is true, and use the results to sketch the graphs *rapidly*.

Objective:

Given the equation or inequality of a circle or circular region, be able to draw the graph *quickly*.

To accomplish this objective, you will start with the geometric definition of a circle, and use the definition to find out what the equation of a circle looks like.

DEFINITION

A ***circle*** is a set of points in a plane, each of which is equidistant from a *fixed* point called the *center*.

Suppose that a circle of radius r units has its center at the fixed point (h, k) in a Cartesian coordinate system, as shown in Figure 9-2a. If (x, y) is a point on the circle, then the distance between (x, y) and (h, k) is r units, the radius of the circle. This distance can be expressed in terms of the coordinates of the two points using the Distance Formula.

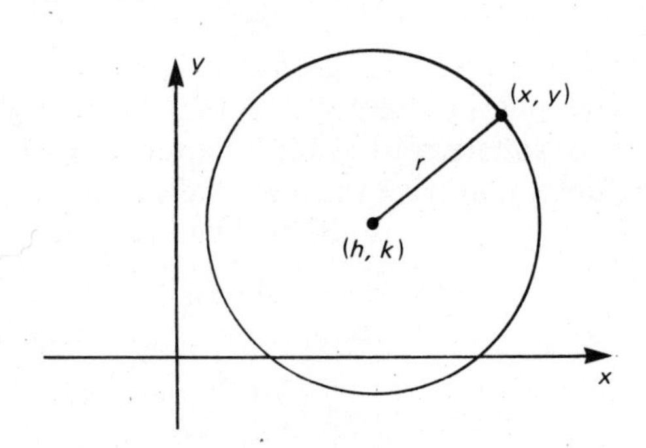

Figure 9-2a.

Background: *The Distance Formula*

Objective:

Express the distance between two points in a Cartesian coordinate system in terms of the coordinates of the two points.

You recall from your work with slopes of linear functions that the rise and run, Δy and Δx, for the line between the points (x_1, y_1) and (x_2, y_2) are

$$\Delta y = y_2 - y_1$$
$$\Delta x = x_2 - x_1 .$$

Figure 9-2b shows that Δy, Δx, and the distance d between the two points are the measures of the sides of a right triangle. By Pythagoras, then,

$$d^2 = (\Delta x)^2 + (\Delta y)^2 \quad \text{——} \quad ① .$$

By substitution,

$$d^2 = (x_2 - x_1)^2 + (y_2 - y_1)^2 \quad \text{——} \quad ② .$$

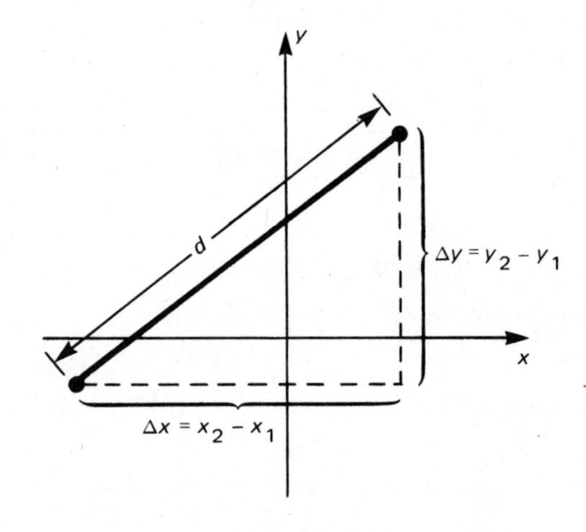

Figure 9-2b.

Taking the positive square root of both members gives

$$d = \sqrt{(x_2 - x_1)^2 + (y_2 - y_1)^2} \quad \text{———} \quad ③ .$$

Equations ①,②, and ③ are different forms of the *distance formula.* Each form has certain advantages for different applications. The formula is, of course, just a fancy form of the Pythagorean Theorem.

Returning to the problem of finding the equation of a circle, the distance r between the points (x, y) and (h, k) in Figure 9-2a can be expressed using form ② of the distance formula as:

$$\boxed{(x - h)^2 + (y - k)^2 = r^2},$$

where the center is at (h, k), and the radius is r. This is called the *general equation* of a circle.

If the center is at the origin, then $(h, k) = (0, 0)$. In this case the equation reduces to

$$\boxed{x^2 + y^2 = r^2.}$$

To show that an equation such as in Problem 4 of Exercise 9-1 is really that of a circle, it is sufficient to transform the equation into the general form, above. This transformation may be accomplished by completing the square.

Example: Graph the relation

$$x^2 + y^2 + 6x - 4y - 12 = 0.$$

Commuting and associating the terms containing x with each other, and the terms containing y with each other, then adding the constant 12 to both members gives

$$(x^2 + 6x \qquad) + (y^2 - 4y \qquad) = 12.$$

The spaces are left inside the parentheses for completing the square, as you did for equations of parabolas in Section 5-3. Adding 9 and 4 on the left to complete the squares, and on the right to obey the Addition Property of Equality gives

$$(x^2 + 6x + 9) + (y^2 - 4y + 4) = 12 + 9 + 4,$$

from which

$$(x + 3)^2 + (y - 2)^2 = 25.$$

Therefore, the graph will be a *circle* of radius 5 units, centered at $(h, k) = (-3, 2)$, as shown in Figure 9-2c. The graph you drew for Problem 4 of Exercise 9-1 should have looked something like this figure. The prior knowledge that the graph is a circle of known center and radius, however, will allow you to plot the graph much more quickly, in accordance with your objective for this section.

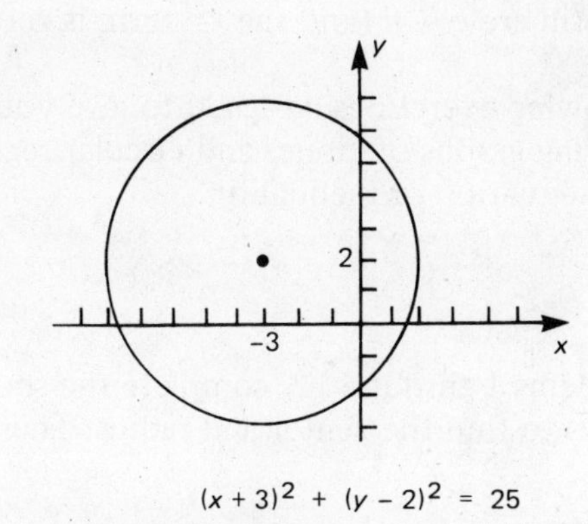

Figure 9-2c.

If the open sentence is an *inequality,* then the graph will be the region *inside* the circle if the symbol is "<," as indicated in Figure 9-2d, or *outside* the circle if the symbol is ">."

An objective stated in Section 9-1 is to determine the effects of the constants A, B, C, D, E, and F on the graph of

$$Ax^2 + Bxy + Cy^2 + Dx + Ey + F = 0.$$

The above leads to a conclusion about the relative sizes of A and C.

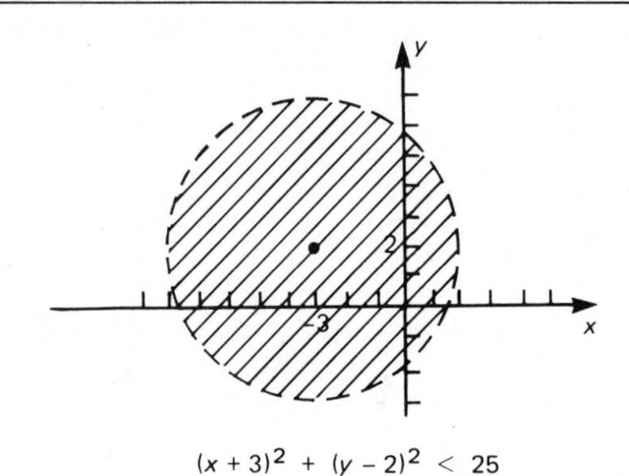

Figure 9-2d.

Conclusion: The graph of a quadratic relation will be a *circle* if the coefficients of the x^2 term and the y^2 term are *equal* (and the xy term is zero).

The following exercise is designed to give you practice drawing graphs of circles and circular regions given the equation or inequality.

Exercise 9-2

For Problems 1 through 12, complete the square (if necessary) to find the center and radius. Then draw the graph.

1. $x^2 + y^2 - 10x + 8y + 5 = 0$
2. $x^2 + y^2 + 12x - 2y + 21 = 0$
3. $x^2 + y^2 - 6x - 4y - 12 > 0$
4. $x^2 + y^2 + 16x + 10y - 11 < 0$
5. $x^2 + y^2 - 8x + 6y - 56 \leq 0$
6. $x^2 + y^2 + 4x - 18y + 69 \geq 0$
7. $x^2 + y^2 + 4x - 5 = 0$
8. $x^2 + y^2 - 14y + 48 = 0$
9. $x^2 + y^2 = 49$
10. $x^2 + y^2 = 4$
11. $x^2 + y^2 = 0$ (Watch for a surprise!)
12. $x^2 + y^2 + 16 = 0$ (Watch for a different surprise!!)

For Problems 13 through 18, write an equation of the circle described.

13. Center at (7, 5), containing (3, −2).
14. Center at (−4, 6), containing (−2, −3).
15. Center at (−9, −2), containing the origin.
16. Center at (5, −4), containing (0, 3).
17. Center at the origin, containing (−6, −8).
18. Center at the origin, containing (−5, 1).
19. Write a general equation for a circle
   a. of radius r, centered at a point on the x-axis,
   b. of radius r, centered at a point on the y-axis,
   c. of radius r, centered at the origin.

20. What do you suppose is meant by a "point circle?" How can you tell from the equation that a circle will be a point circle?

21. Sometimes the graph of a quadratic relation such as you have been plotting turns out to have no points at all! How can you tell from the equation whether or not this will happen?

22. *Introduction to Ellipses* – In this section you have found that the graph of a quadratic relation is a circle if x^2 and y^2 have *equal* coefficients. In this problem you will find out what the graph looks like if the coefficients are *not* equal, but have the same sign. Do the following things for the relation whose equation is

$$9x^2 + 25y^2 = 225.$$

   a. Solve the equation for y in terms of x.
   b. Explain why there are *two* values of y for each value of x between −5 and 5.

c. Explain why there are *no* real values of y for $x > 5$ and for $x < -5$.
d. Calculate values of y for each integer value of x from -5 to 5, inclusive. Use a calculator or square root tables to approximate any radicals which do not come out integers. Round off to one decimal place.
e. Plot the graph. If you have been successful, you should have a *closed* figure called an *ellipse*.

9-3 ELLIPSES

In Section 9-2 you learned that the graph of a quadratic relation is a circle if x^2 and y^2 have the *same* coefficient. If x^2 and y^2 have *different* coefficients, but the *same sign,* then the graph is called an "ellipse." If you worked Problem 22 of Exercise 9-2 you found by pointwise plotting that the graph of

$$9x^2 + 25y^2 = 225$$

is as shown in Figure 9-3a. In this section you will learn properties of ellipses that will allow you to sketch their graphs *quickly*.

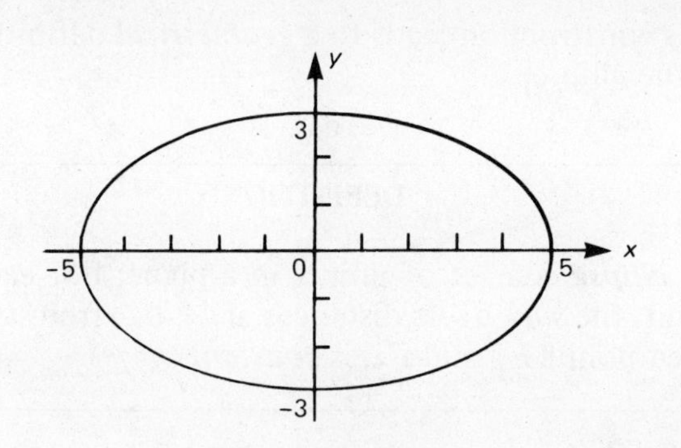

Figure 9-3a.

Various parts of the ellipse are given special names, as shown in Figure 9-3b. All ellipses look rather like flattened circles. Thus, if you know where the

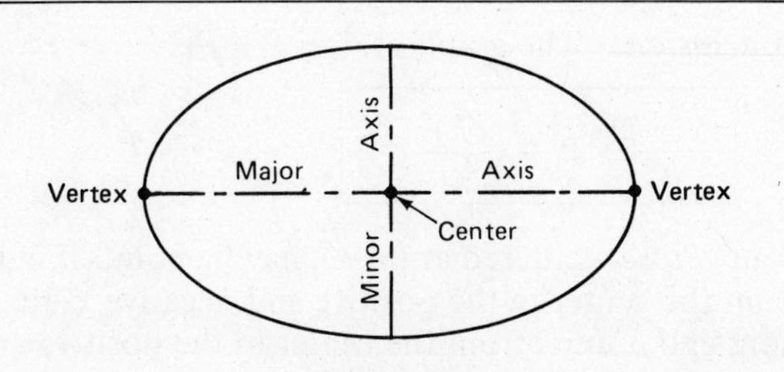

Figure 9-3b.

center is, and how long the major and minor axes are, you can sketch the graph *quickly*. The trick is to transform the equation so that these lengths show up in it. Starting with

$$9x^2 + 25y^2 = 225,$$

you divide by 225 to make the right member equal 1.

$$\frac{9x^2}{225} + \frac{25y^2}{225} = 1 \qquad \text{Divide by 225.}$$

$$\frac{x^2}{25} + \frac{y^2}{9} = 1 \qquad \text{Cancel.}$$

$$\frac{x^2}{5^2} + \frac{y^2}{3^2} = 1 \qquad 25 = 5^2 \text{ and } 9 = 3^2.$$

The "5" under the x^2 is the distance from the center to the graph in the x-direction. The "3" under the y^2 is the distance from the center to the graph in the y-direction. (In Section 9-6 you will *prove* that this is true.) Thus, if

$$\frac{x^2}{a^2} + \frac{y^2}{b^2} = 1,$$

then a and b will be *half* the lengths of the horizontal and vertical axes, respectively. The larger number will be half the *major* axis and the smaller will be half the *minor* axis.

If the center is at (h, k) instead of at the origin, $(0, 0)$, then x and y in the above equation will be replaced by $(x - h)$ and $(y - k)$, as they were for the circle.

Conclusion: The graph of

$$\boxed{\frac{(x-h)^2}{a^2} + \frac{(y-k)^2}{b^2} = 1}$$

is an *ellipse,* centered at (h, k), having points a units from the center in the positive and negative x-direction, and b units from the center in the positive and negative y-direction.

Example 1: Sketch the graph of

$$25x^2 + 9y^2 - 200x + 18y + 184 = 0.$$

To sketch the graph quickly you must transform it to the form above by completing the square. First you subtract 184 from both members and associate what remains on the left, getting

$$(25x^2 - 200x) + (9y^2 + 18y) = -184.$$

Factoring out the coefficients of x^2 and y^2 gives

$$25(x^2 - 8x \qquad) + 9(y^2 + 2y \qquad) = -184.$$

Spaces are left in the parentheses for completing the square. Adding 16 and 1 completes the squares on the left, and requires that $25 \cdot 16$ and $9 \cdot 1$ be added on the right, giving

$$25(x^2 - 8x + 16) + 9(y^2 + 2y + 1)$$
$$= -184 + 25 \cdot 16 + 9 \cdot 1.$$

Writing the left member in terms of perfect squares, and doing the arithmetic on the right gives

$$25(x - 4)^2 + 9(y + 1)^2 = 225.$$

Dividing both members by 225 produces the desired form of the equation,

$$\frac{(x-4)^2}{9} + \frac{(y+1)^2}{25} = 1$$

or

$$\frac{(x-4)^2}{3^2} + \frac{(y+1)^2}{5^2} = 1.$$

So the graph will be an *ellipse;* with the major axis in the y-direction because $b > a$, centered at $(4, -1)$, with critical points ± 3 units from the center in the x-direction and ± 5 units from the center in the y-direction.

The graph may now be drawn quickly by plotting the four critical points, then sketching the ellipse, as shown in Figure 9-3c.

If the "=" sign is replaced by one of the inequalities "<" or ">," then the graph will be the region inside the ellipse or outside the ellipse, respectively.

There is a pair of points associated with an ellipse that has an important geometrical property. Suppose that you tie two pins to a piece of string in such a way that there are 10 cm of string between them. Then you stick the pins at the points $F_1 = (4, 0)$ and $F_2 = (-4, 0)$ in a Cartesian coordinate system that has 1-cm squares (see Figure 9-3d). Placing a pencil as shown in the figure, and keeping the string tight, you can draw a curve. The curve turns out to be an *ellipse,* the same one as in Figure 9-3a!

This construction leads to a *geometrical* definition of the ellipse.

DEFINITION

An ***ellipse*** is a set of points in a plane. For each point, the *sum* of its distances, $d_1 + d_2$, from two fixed points F_1 and F_2, is *constant.*

Each point, F_1 and F_2, is called a *focus.* The name is chosen because sound or light emitted from one focus will be "focused" toward the other one by bouncing off the ellipse. (The plural of focus is "foci," the "c" being pronounced like an "s.")

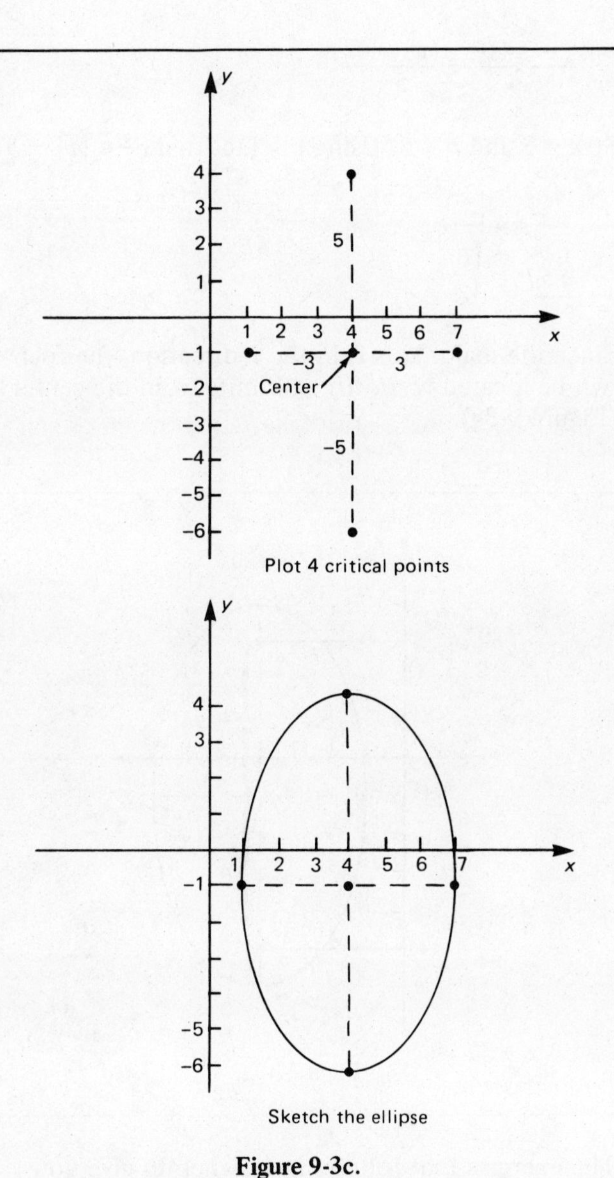

Figure 9-3c.

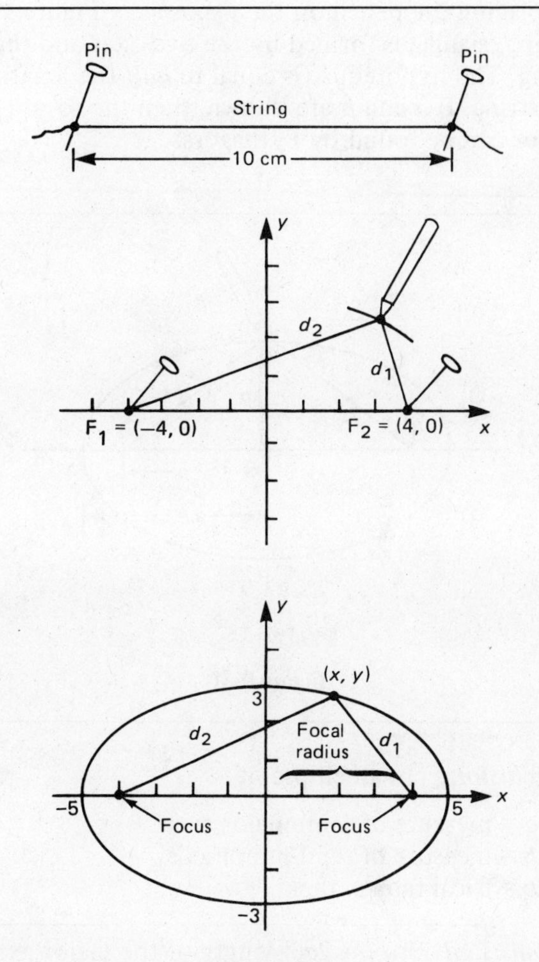

Figure 9-3d.

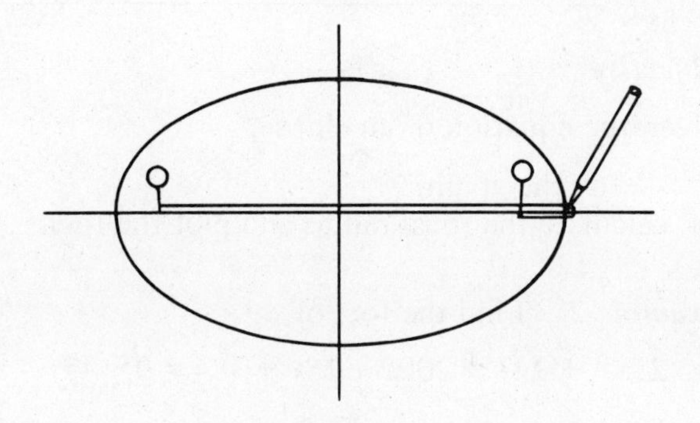

Figure 9-3e.

The distance from the center to a focus is called the *focal radius.* By placing the pencil at two strategic positions, you can find out how to *calculate* the focal radius. Placing it on the x-axis as in Figure 9-3e, you can see that the length of the major axis is equal to the length of the string. That is, the constant sum of the distances, $d_1 + d_2$, equals the length of the major axis.

By placing the pencil on the y-axis as in Figure 9-3f, a right triangle is formed by the two axes and the string. The hypotenuse is equal to *half* the length of the string. If a and b are known, then the focal radius can be found by Pythagoras.

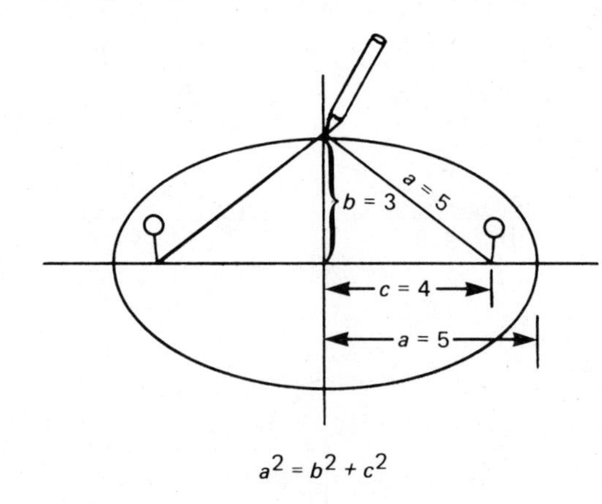

Figure 9-3f.

Conclusion: In an ellipse, if

a = measure of semi-major axis,
b = measure of semi-minor axis,
c = focal radius, then:

$d_1 + d_2 = 2a$ = length of the major axis, and $a^2 = b^2 + c^2$, from which $c^2 = \lvert a^2 - b^2 \rvert$.

Objective:

Given the equation of an ellipse,

a. sketch the graph,
b. calculate the focal radius and plot the foci.

Example 2: Find the foci of

$$25x^2 + 9y^2 - 200x + 18y + 184 = 0.$$

This is the ellipse in Example 1. Completing the square gives

$$\frac{(x-4)^2}{3^2} + \frac{(y+1)^2}{5^2} = 1.$$

So $a = 3$ and $b = 5$. Using the fact that $c^2 = \lvert a^2 - b^2 \rvert$,

$$\begin{aligned} c^2 &= \lvert 3^2 - 5^2 \rvert \\ &= \lvert -16 \rvert \\ &= 16 \\ \therefore c &= 4. \end{aligned}$$

Since the major axis is in the y-direction, the foci will be located *vertically*, ±4 units from the center (Figure 9-3g).

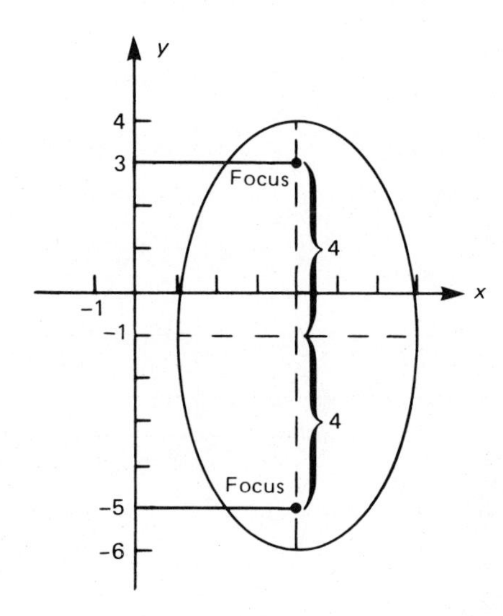

Figure 9-3g.

The exercise that follows is designed to give you practice in sketching graphs of ellipses and finding their foci.

Exercise 9-3

For Problems 1 through 16,

a. Sketch the graph of the relation,
b. Calculate the focal radius and plot the two foci.

1. $4x^2 + 9y^2 - 16x + 90y + 205 = 0$
2. $4x^2 + 36y^2 + 40x - 288y + 532 = 0$
3. $49x^2 + 16y^2 + 98x - 64y - 671 = 0$
4. $25x^2 + 4y^2 - 150x + 32y + 189 = 0$
5. $x^2 + 4y^2 + 10x + 24y + 45 = 0$
6. $16x^2 + y^2 - 128x - 20y + 292 = 0$
7. $25x^2 + 9y^2 + 50x - 36y - 164 < 0$
8. $4x^2 + 36y^2 + 48x + 216y + 324 \geq 0$
9. $16x^2 + 25y^2 - 300y + 500 = 0$
10. $36x^2 + 9y^2 - 216x = 0$
11. $100x^2 + 36y^2 > 3600$
12. $25x^2 + 49y^2 \leq 1225$
13. $12x^2 + y^2 = 48$
14. $x^2 + 6y^2 = 25$
15. $5x^2 + 8y^2 = 77$
16. $11x^2 + 5y^2 = 224$

17. ***Stadium Problem.*** Suppose that you are Chief Mathematician for Ornery & Sly Construction Company. Your company has a contract to build a football stadium in the form of two concentric ellipses, with the field inside the inner ellipse, and the seats between the two ellipses. The seats are in the intersection of the graphs of

$$x^2 + 4y^2 \geq 100 \quad \text{and} \quad 25x^2 + 36y^2 \leq 3600,$$

where each unit on the graph represents 10 meters.

a. Draw a graph of the seating area.
b. From a handbook, you find that the area of an elliptical region is πab, where a and b are the semi-axes, and $\pi \approx 3.14159$. The Engineering Department estimates that each seat occupies 0.8 square meter. What is the seating capacity of the stadium?

18. Show that the equation for an ellipse,

$$\frac{(x-h)^2}{a^2} + \frac{(y-k)^2}{b^2} = 1,$$

reduces to the equation of a *circle* if $a = b$.

19. *Introduction to Hyperbolas.* The ellipses you have plotted in this section have equations in which the x^2 and y^2 terms have the same sign. In this problem you will find out what the graph looks like if they have *opposite* signs. Do the following things for the relation whose equation is

$$9x^2 - 16y^2 = 144.$$

a. Solve the equation for y in terms of x.
b. Explain why there will be *no* real values of y for $-4 < x < 4$.
c. Explain why there will be *two* values of y for each value of x when $x > 4$ or $x < -4$.
d. Make a table of values of y for each integer value of x from 4 to 8, inclusive, finding decimal approximations from square root tables or by calculator, where necessary.
e. Explain why you can use the *same* table of values of y for values of x between -4 and -8.
f. Plot a graph of the relation from $x = -4$ to $x = -8$, and from $x = 4$ to $x = 8$.
g. On the same Cartesian coordinate system, plot graphs of the lines

$$y = \frac{3}{4}x \quad \text{and} \quad y = -\frac{3}{4}x.$$

If your graph in part f is correct, then it should have these two lines as diagonal asymptotes. The graph is called a *hyperbola*, and is a non-closed curve with *two* "branches."

9-4 HYPERBOLAS

In the last section you learned that an ellipse has an equation in which the x^2 and y^2 coefficients are unequal, but have the same sign. If x^2 and y^2 have *opposite* signs, such as

$$9x^2 - 16y^2 = 144,$$

then the graph looks like that in Figure 9-4a, and is called a *hyperbola.* You may have discovered this fact already if you worked Problem 19 in Exercise 9-3.

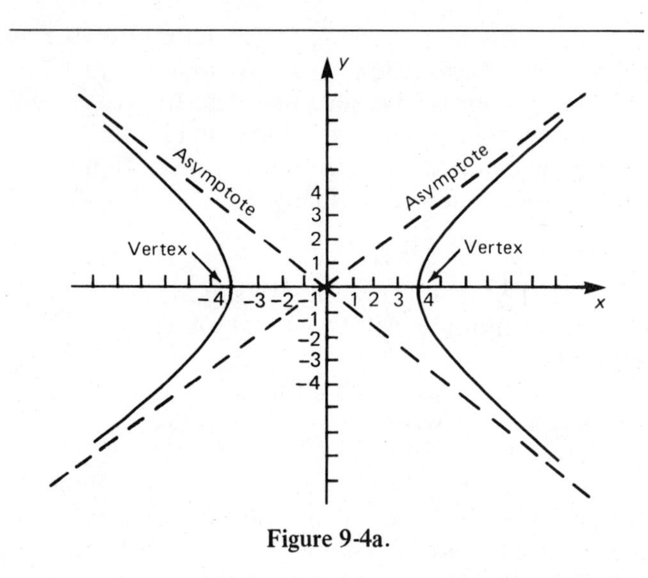

Figure 9-4a.

Objective:

Given the equation of a hyperbola, be able to sketch its graph *quickly*.

Hyperbolas have two disconnected branches. Each branch approaches diagonal asymptotes. You can see *why* by transforming the equation so that y is by itself on the left side.

$-16y^2 = -9x^2 + 144.$	Subtract $9x^2$.
$16y^2 = 9(x^2 - 16)$	Multiply by -1 and factor out 9.
$y = \pm \frac{3}{4}\sqrt{x^2 - 16}$	Divide by 16, then take the square root.

Two observations about the graph may be made from this equation:

1. There are *no* real values of y when x is between -4 and 4 because the radicand, $x^2 - 16$, would be *negative*..This fact explains why the graph is split into two branches.

2. As x becomes larger, the difference between $\sqrt{x^2 - 16}$ and $\sqrt{x^2}$ gets close to zero. For example, if $x = 100$, then

$$\begin{aligned}\sqrt{x^2 - 16} &= \sqrt{10000 - 16}\\ &= \sqrt{9984}\\ &\approx 99.92,\end{aligned}$$

while $\sqrt{x^2} = \sqrt{10000} = 100$. Consequently, the larger x gets, the closer y gets to $\pm \frac{3}{4}\sqrt{x^2}$, or $y \approx \pm \frac{3}{4}x$. The lines $y = \pm \frac{3}{4}x$ are thus *asymptotes* of the graph.

The 4 and 3 in the slopes of the asymptotes can be made to show up in the equation by making the right member equal 1.

$9x^2 - 16y^2 = 144$	Given equation.
$\frac{x^2}{16} - \frac{y^2}{9} = 1$	Divide by 144.
$\frac{x^2}{4^2} - \frac{y^2}{3^2} = 1$	$16 = 4^2$ and $9 = 3^2$.

The vertices are ± 4 units from the center in the x-direction, as shown in the left side of Figure 9-4b. If the signs in the equation are reversed, the hyperbola opens in the y-direction, as shown in the right side of Figure 9-4b. The asymptotes are the same, but the vertices are ± 3 units from the center in the *y-direction.* These two hyperbolas are said to be *conjugates* of each other.

As you recall, ellipses have foci. For any given point on the ellipse, the sum of its distances from the two foci is constant. Hyperbolas also have foci. In this case, the *difference* between the two distances is constant.

DEFINITION

A ***hyperbola*** is a set of points in a plane. For each point (x, y), the *difference* between its distances from two fixed foci is constant.

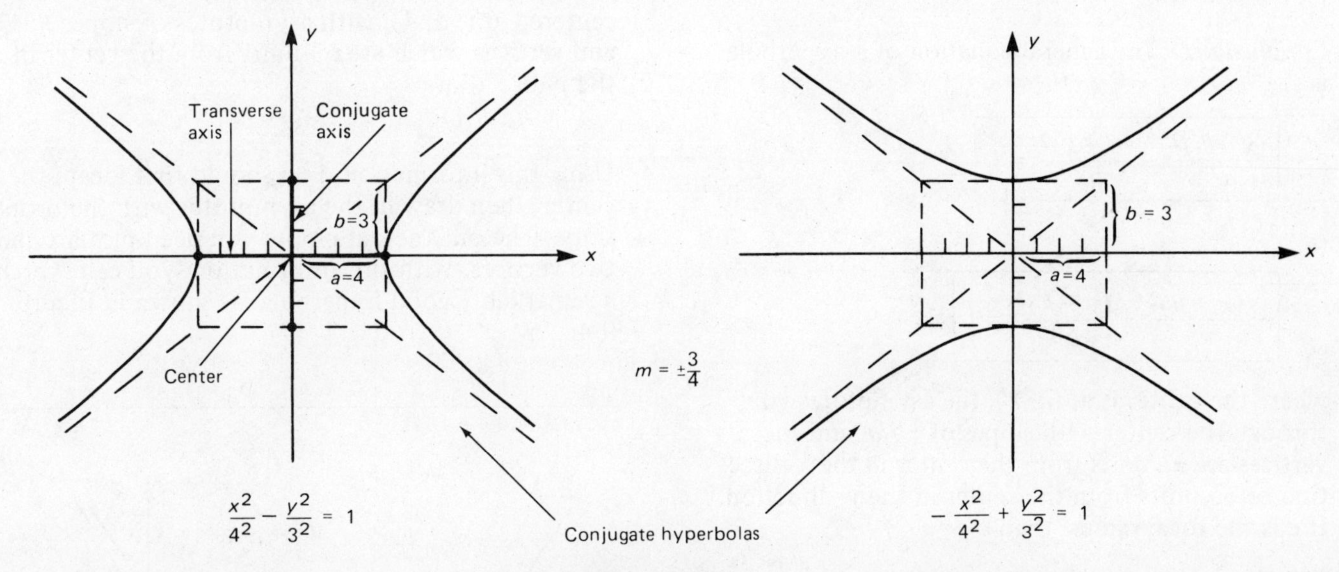

Figure 9-4b.

Figure 9-4c illustrates this definition. For hyperbolas, the *hypotenuse* of the right triangle shown in Figure 9-4c is equal to the focal radius. Letting c stand for focal radius,

$$c^2 = 4^2 + 3^2 = 25.$$
$$\therefore c = 5.$$

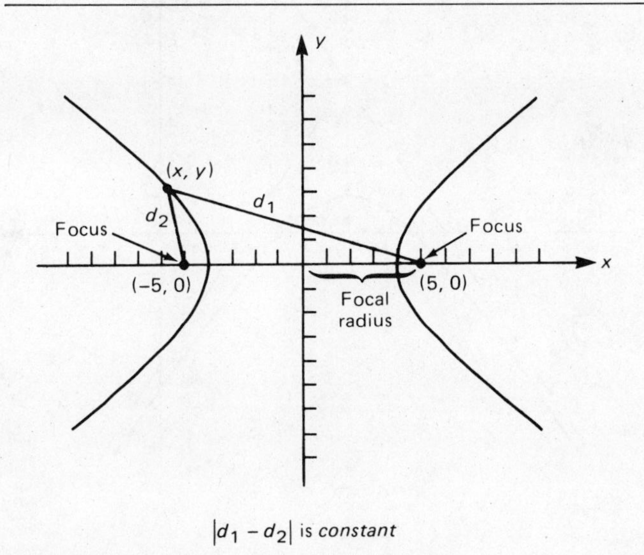

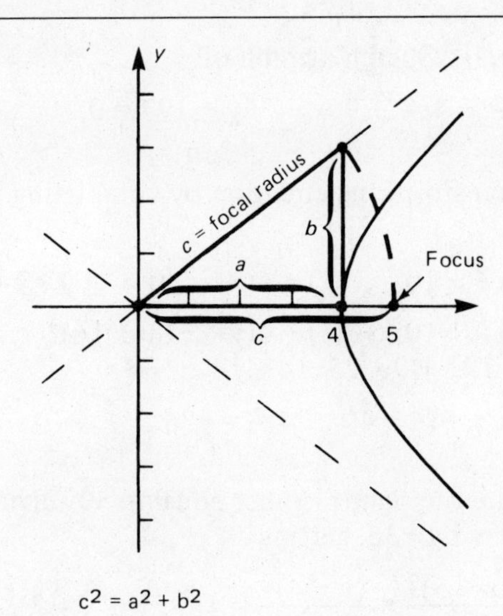

Figure 9-4c.

These observations can be summarized as follows:

Conclusion: The general equation of a hyperbola is

$$\frac{(x-h)^2}{a^2} - \frac{(y-k)^2}{b^2} = 1$$

or

$$-\frac{(x-h)^2}{a^2} + \frac{(y-k)^2}{b^2} = 1,$$

where the center is at (h, k), the asymptotes go through the center with slopes of $\pm b/a$, and the vertices are $\pm a$ units from the center in the x-direction or $\pm b$ units from the center in the y-direction. If c is the focal radius, then $c^2 = a^2 + b^2$.

If the center is at the origin, then $(h, k) = (0, 0)$, and these equations reduce to

$$\frac{x^2}{a^2} - \frac{y^2}{b^2} = 1 \quad \text{or} \quad -\frac{x^2}{a^2} + \frac{y^2}{b^2} = 1.$$

Example: Sketch a graph of

$$9x^2 - 4y^2 + 90x + 32y + 197 = 0.$$

First transform the equation by completing the square.

$$9(x^2 + 10x \quad\quad) - 4(y^2 - 8y \quad\quad) = -197$$

$$9(x^2 + 10x + 25) - 4(y^2 - 8y + 16)$$
$$= -197 + 9 \cdot 25 - 4 \cdot 16$$

$$9(x+5)^2 - 4(y-4)^2 = -36$$

To make the right member equal to +1, divide both members by −36, getting

$$-\frac{(x+5)^2}{2^2} + \frac{(y-4)^2}{3^2} = 1$$

after simplification.

The graph is a *hyperbola,* opening *vertically,* centered at (−5, 4), with asymptotes of slopes ±3/2, and vertices which are ±3 units from the center in the y-direction.

Using this information, you should first locate the center, then draw in the asymptotes with the proper slope. The only actual points you need plot are the two vertices. With this information, you can sketch a remarkably good hyperbola, as shown in Figure 9-4d.

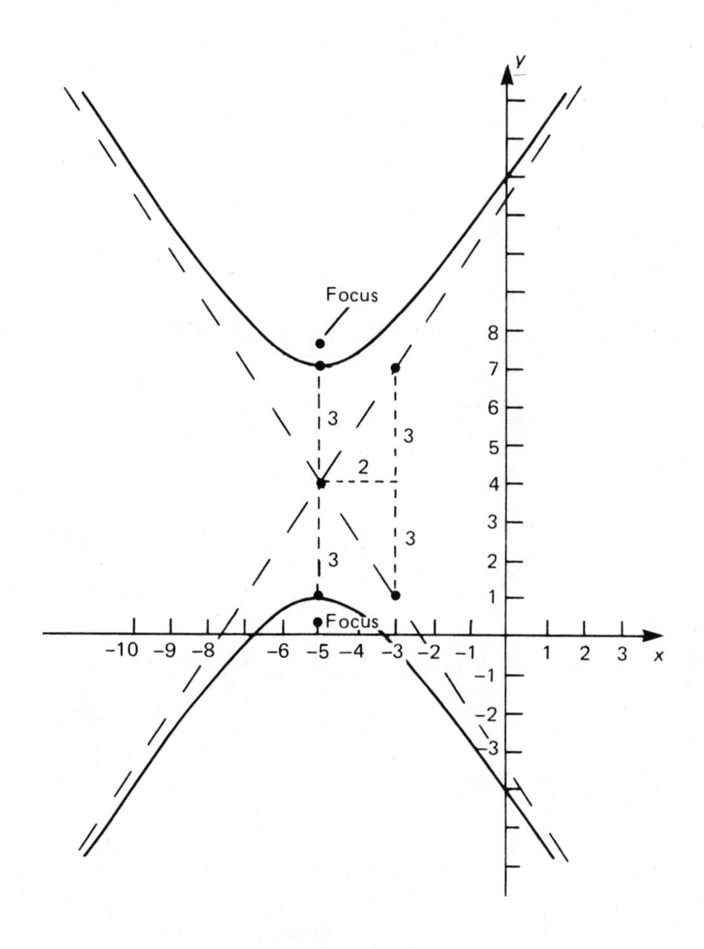

Figure 9-4d.

Since the focal radius, c, is given by $c^2 = a^2 + b^2$, you can locate the foci.

$$c^2 = 2^2 + 3^2$$
$$= 13.$$
$$\therefore c = \sqrt{13} \approx 3.61.$$

The following exercise is designed to give you practice in sketching hyperbolas rapidly.

Exercise 9-4

For each relation:

a. Complete the square (if necessary), find the center, draw the asymptotes, plot the vertices, then sketch the graph.
b. Calculate the focal radius, and plot the foci.

1. $25x^2 - 16y^2 - 100x - 96y - 444 = 0$
2. $4x^2 - 9y^2 + 16x + 108y - 344 = 0$
3. $25x^2 - 9y^2 + 300x - 126y + 684 = 0$
4. $4x^2 - 36y^2 - 40x + 216y - 80 = 0$
5. $x^2 - y^2 + 4x + 16y - 69 = 0$
6. $x^2 - y^2 - 14x - 8y + 37 = 0$
7. $9x^2 - 4y^2 - 54x - 16y - 79 = 0$
8. $x^2 - 4y^2 + 4x + 32y - 96 = 0$
9. $9x^2 - y^2 - 90x + 4y + 302 = 0$
10. $25x^2 - 4y^2 + 200x - 8y + 796 = 0$
11. $16x^2 - 9y^2 + 144 = 0$
12. $25x^2 - 144y^2 - 3600 = 0$
13. $4x^2 - 5y^2 = 16$
14. $27x^2 - 4y^2 = -36$
15. $16x^2 - 3y^2 = -11$
16. $5x^2 - 7y^2 = 17$
17. $9x^2 - y^2 = 0$
18. $4x^2 - y^2 = 0$

9-5 PARABOLAS

You will recall from Chapter 5 that a quadratic function

$$y = ax^2 + bx + c,$$

has a graph called a parabola as shown in Figure 9-5a. This equation can be thought of as that of a quadratic relation with only *one* squared term.

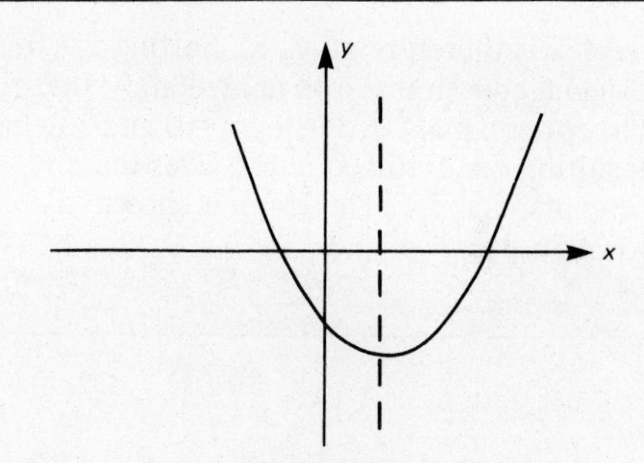

Figure 9-5a.

If y^2 appears instead of x^2, then the equation can be written in the form

$$x = ay^2 + by + c.$$

Since the variables are *interchanged,* the parabola will open in the x-direction instead of in the y-direction. The vertex can be found by transforming the equation to

$$x - h = a(y - k)^2$$

by completing the square, as you have done previously. Finding a few other points will allow you to sketch the graph quickly.

Objective:

Given the equation of a parabola, sketch its graph *quickly*.

Example: Sketch the graph of $x = -2y^2 + 12y - 10$.

First you find the vertex by completing the square.

$x + 10 = -2(y^2 - 6y \quad)$	Add 10, then factor out -2.
$x + 10 - 18 = -2(y^2 - 6y + 9)$	Complete the square on the right, add $(-2)(9)$ on the left.
$x - 8 = -2(y - 3)^2$	Write as a perfect square.

The vertex is therefore at (8, 3). Setting $y = 0$ in the original equation shows immediately that the x-intercept equals -10. Setting $x = 0$ and solving the resulting quadratic equation gives the y-intercepts, 5 and 1. The graph is shown in Figure 9-5b.

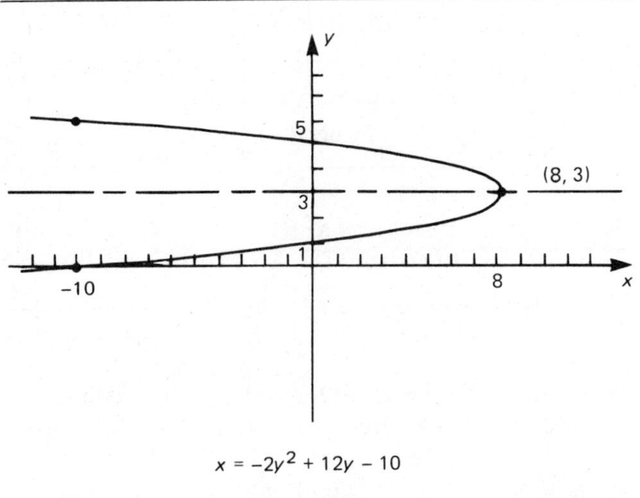

Figure 9-5b.

Like ellipses and hyperbolas, parabolas also have a *geometric* definition. In the next section you will learn how the focus of a parabola is related to the graph.

In the following exercise you will sketch the graphs of parabolas that open in the x-direction as well as in the y-direction.

Exercise 9-5

Find the vertex and intercepts of the following parabolas, and sketch the graph. Find decimal approximations for any radicals you encounter.

1. $x = y^2 - 4y + 3$
2. $x = y^2 + 2y - 3$
3. $x = -3y^2 - 12y - 5$
4. $x = -5y^2 + 30y + 11$
5. $x = \frac{1}{2}y^2 + 3y + 4$
6. $x = \frac{1}{3}y^2 - 2y - 9$
7. $y = -4x^2 + 20x - 16$
8. $y = \frac{1}{5}x^2 + 2x - \frac{11}{5}$
9. $x = \frac{1}{4}y^2$
10. $x = -\frac{1}{10}y^2$

9-6 EQUATIONS FROM GEOMETRICAL DEFINITIONS

You have learned geometrical definitions for circles, ellipses, and hyperbolas in Sections 9-2, 9-3, and 9-4. In this section you will learn a geometrical definition for a parabola. You will also learn that there are *other* geometrical definitions for all four kinds of graphs. Using these definitions, you will derive equations for the graphs.

Objective:

Given the geometrical definition of a set of points, derive an equation relating x and y. If the equation is *quadratic*, tell whether the graph is a circle, ellipse, hyperbola, or parabola.

Circles, ellipses, hyperbolas, and parabolas have a *family* name that comes from their relationship to slices ("sections") of a cone. As shown in Figure 9-6a, a complete cone has two halves, called "nappes." If a plane cuts the cone *parallel* to one of its elements, the section is a *para*bola. If the plane exceeds the parallel position, it will cut *both* nappes, forming a *hyper*bola (from the Greek work "hyperbole," meaning "to exceed"). If the plane does not tilt far enough, one of the nappes will be left out. and the graph will be an *ellipse* (from the Greek

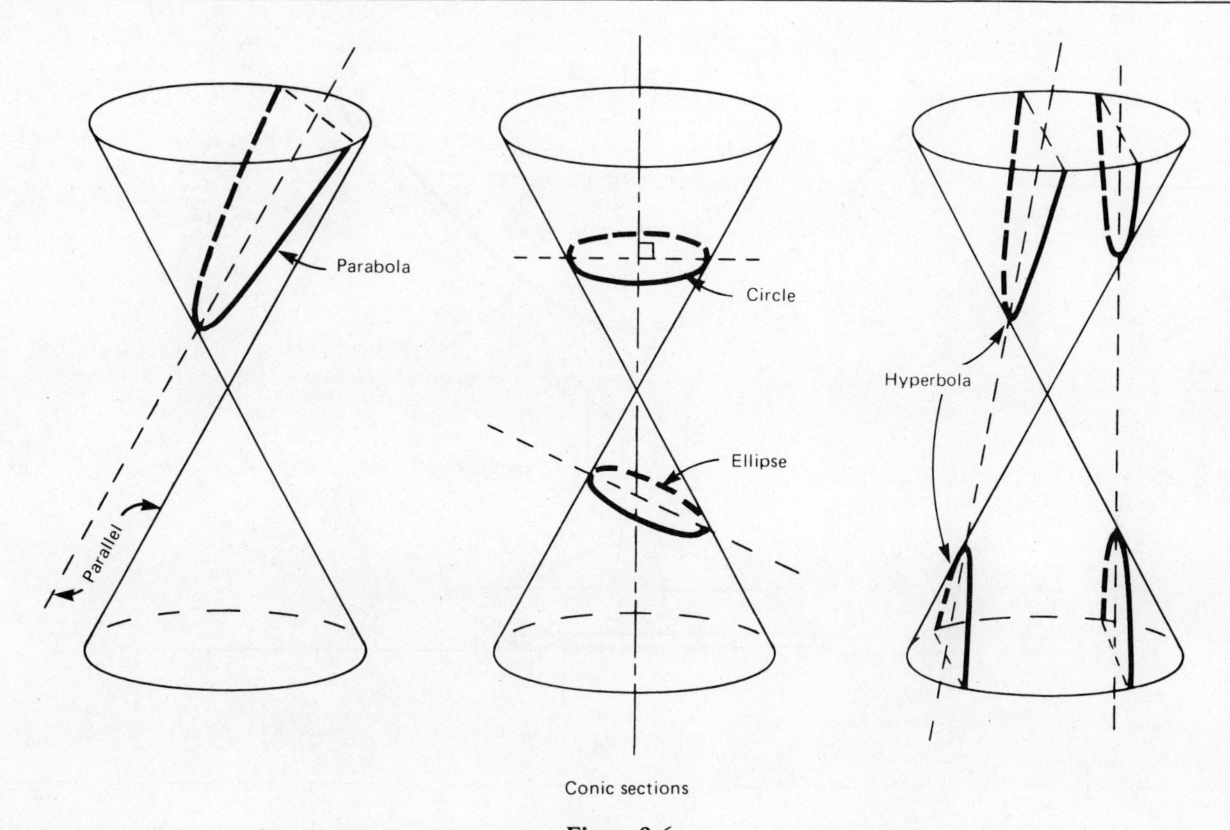

Figure 9-6a.

word "ellipsis," meaning "to leave out"). A circle is formed by a cutting plane perpendicular to the axis of the cone.

Because of the properties stated above, circles, ellipses, hyperbolas, and parabolas are given the name *conic sections.* Perhaps you have seen various conic sections when the cone of light from a lamp is cut by the plane of a wall or ceiling (Figure 9-6b)!

The second part of the objective may now be stated, "If the equation is quadratic, tell which *conic section* the graph is." In order to do this, you should pull together what you have learned about the coefficients of the *squared* terms in the equation of a quadratic relation. The following table should help refresh your memory:

Recognition of Conic Sections from Their Equations

Circle:	x^2 and y^2 terms have *equal* coefficients.
Ellipse:	x^2 and y^2 terms have *unequal* coefficients, but the *same* sign.
Hyperbola:	x^2 and y^2 terms have *opposite* signs.
Parabola:	Equation has only *one* squared term. If the x^2 term is missing, the parabola opens in the x-direction. If the y^2 term is missing, the parabola opens in the y-direction (a quadratic *function*).

Note: All of the above conclusions assume that there is *no* xy-term. If there *is* an xy-term, you must use the discriminant to identify the conic section, as explained in Exercise 9-7, Problem 4.

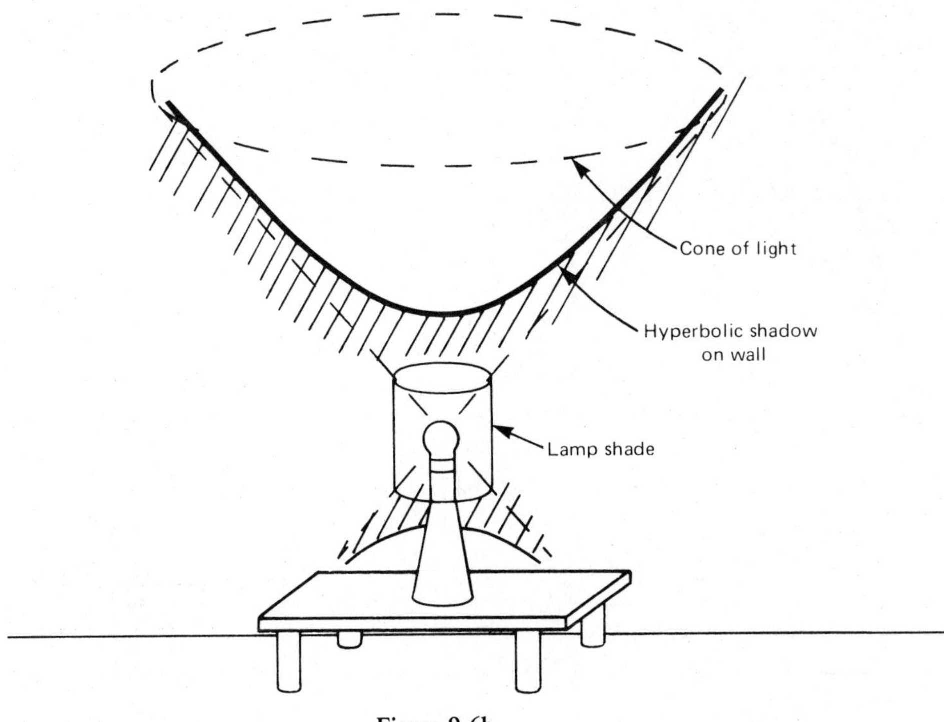

Figure 9-6b.

Example: A *parabola* is defined to be a set of coplanar points, each of which is the *same* distance from a fixed focus as it is from a fixed straight line (called the "directrix"). Find the equation of the parabola whose focus is (2, 3) and whose directrix is the line $y = -7$. Show that the equation has the form $y = ax^2 + bx + c$.

The first step is to draw a picture showing the given focus and directrix (Figure 9-6c). Then pick a point (*any* point!) that could be on the graph, and label it (x, y). By the definition of parabola, above,

$$d_1 = d_2.$$

The point where the segment labeled d_1 meets the line $y = -7$ has coordinates $(x, -7)$. So d_1 and d_2 can be found using the Distance Formula. Substituting the appropriate coordinates into the equation $d_1 = d_2$ gives

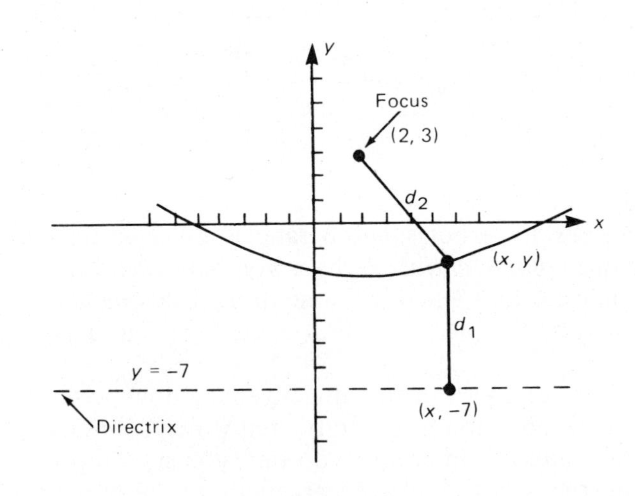

Figure 9-6c.

$$\sqrt{(x-x)^2+(y+7)^2}=\sqrt{(x-2)^2+(y-3)^2}\,,$$

which reduces to

$$|y+7|=\sqrt{(x-2)^2+(y-3)^2}\,.$$

Squaring both members transforms this to a *polynomial* equation,

$$y^2+14y+49=(x-2)^2+(y-3)^2\,.$$

Expanding the binomial squares on the right gives

$$y^2+14y+49=x^2-4x+4+y^2-6y+9.$$

Note that the y^2 terms will cancel, leaving only *one* squared term. The equation can be transformed to

$$20y=x^2-4x-36,$$

or

$$y=\frac{1}{20}x^2-\frac{1}{5}x-\frac{9}{5}\,.$$

Since this last equation has the form $y = ax^2 + bx + c$, you can see that this set of points described geometrically fits your old definition of a parabola as the graph of a quadratic function.

The exercise which follows is designed to give you practice finding equations and identifying graphs from given geometrical definitions.

Exercise 9-6

For Problems 1 through 12, find a *polynomial* equation with *integer* coefficients for the set of coplanar points described. Tell whether or not the graph is a conic section and, if it is, tell *which* conic section.

1. For each point, its distance from the fixed point $(-3, 0)$ is *twice* its distance from the fixed point $(3, 0)$.

2. For each point, its distance from the fixed point $(0, -5)$ is $1/3$ of its distance from the fixed point $(0, 5)$.

3. For each point, its distance from the fixed point $(4, 3)$ is 3 times its distance from the fixed point $(-1, 2)$.

4. For each point, its distance from the fixed point $(-1, -3)$ is $1/2$ of its distance from the fixed point $(5, -8)$.

5. Each point is *twice* as far from the line $x = 3$ as it is from the line $y = -2$.

6. Each point is *three times* as far from the line $y = 5$ as it is from the line $x = 1$.

7. Each point is *equidistant* from the point $(3, -4)$ and the line $y = 2$.

8. Each point is *equidistant* from the point $(2, 5)$ and the line $x = -3$.

9. For each point, its distance from the point $(-3, 1)$ is *half* its distance from the line $y = 4$.

10. For each point, its distance from the point $(-4, -1)$ is $2/3$ times its distance from the line $x = -5$.

11. For each point, its distance from the point $(0, 3)$ is $3/2$ times its distance from the line $y = -3$.

12. For each point, its distance from the point $(4, 0)$ is *twice* its distance from the line $x = -4$.

13. Use the geometric definition of an ellipse (Section 9-3) to show that the ellipse with foci $(4, 0)$ and $(-4, 0)$ and major axis 10 units long is

$$9x^2+25y^2=225.$$

14. Use the geometric definition of hyperbola (Section 9-4) to show that the hyperbola with foci $(5, 0)$

and (−5, 0) and transverse axis (i.e., the distance between vertices) 8 units long is

$$9x^2 - 16y^2 = 144.$$

15. Show that another definition of a circle is a set of points for each of which its distance from the fixed point $(f, 0)$ is c times its distance from the fixed point $(-f, 0)$.

16. Find a polynomial equation of a set of points for each of which the distance from the fixed point $(f, 0)$ is e times its distance from the vertical line $x = -f$, where e stands for a constant. Show that the graph will be an *ellipse* if $0 < e < 1$; a *parabola* if $e = 1$; and a *hyperbola* if $e > 1$.

17. The geometrical definitions in Problem 16 are called the "focus-directrix" definitions of the conic sections. The straight line is called the "directrix," the fixed point is called the "focus," and the constant e is called the "eccentricity." What would be the eccentricity for a *circle*? Where would the directrix be for a circle?

18. A parabola may be thought of as having a *second* focus, just like ellipses and hyperbolas do. Where do you suppose this second focus would be? (Remember the cones and the lamp shade of Figures 9-6a and 9-6b.)

9-7 QUADRATIC RELATIONS – *xy*-TERM

In your investigation of quadratic relations, you have not yet encountered one for which the equation had a non-zero xy-term. In this section you will find out that the graphs are still conic sections, but the xy-term *rotates* the graph and affects its *shape*.

Objective:

Determine by actual plotting what effects the xy-term in the equation of a quadratic relation has on the graph.

The following exercise is designed to accomplish this objective.

Exercise 9-7

1. Plot a graph of $x^2 + y^2 = 9$.

2. The following equations differ from the one in Problem 1 only by the addition of increasingly larger xy-terms:

    i. $x^2 + xy + y^2 = 9$.
    ii. $x^2 + 4xy + y^2 = 9$.

    a. For each relation, select values of x, calculate the corresponding values of y, and plot the graph. Use a calculator or square root tables if necessary.
    b. For each graph, tell which of the conic sections is plotted.
    c. Does the xy-term affect the x and y-intercepts? Explain.

3. The difficulty with the plotting in Problem 2 is the tedious computations. If you have a computer available, these and more complicated quadratic relations can be evaluated and plotted more easily.

    a. For the general quadratic equation

    $$Ax^2 + Bxy + Cy^2 + Dx + Ey + F = 0,$$

    solve for y in terms of x. This can be done by treating x as a "constant" and using the quadratic formula.
    b. Write a computer program into which you can input the six coefficients A, B, C, D, E, and F. The program should calculate values of y for each integer value of x between an input lower value, L, and an input upper value, U. The computer should print out the value of x and the two values of y, or the message NO REAL SOLUTIONS, whichever is appropriate.

4. Use the computer program from Problem 3 to calculate plotting data for each of the following quadratic relations. Calculate values of y for each integer value of x between −10 and 10, inclusive. Then plot the graph of each. Finally,

by looking at the graphs tell which conic section is plotted in each one.

a. $x^2 + y^2 - 10x - 8y + 16 = 0$
b. $x^2 + xy + y^2 - 10x - 8y + 16 = 0$
c. $x^2 + 2xy + y^2 - 10x - 8y + 16 = 0$
d. $x^2 + 4xy + y^2 - 10x - 8y + 16 = 0$

5. Use your graphs from Problems 1, 2, and 4 to answer the following questions:

a. When there is an xy-term in the equation, can you still tell by looking at the coefficients of x^2 and y^2 just which of the conic sections the graph will be?
b. The equation $Ax^2 + Bxy + Cy^2 + Dx + Ey + F = 0$ has a *discriminant*, which equals

$$B^2 - 4AC.$$

For each of the equations in Problems 1 and 4, calculate the discriminant.
c. Show that the *sign* of the discriminant tells the following about the kind of conic section:

Circle or Ellipse–discriminant < 0.
Parabola–discriminant $= 0$.
Hyperbola–discriminant > 0.

6. Prove that the graph of an inverse variation function is a hyperbola.

9-8 SYSTEMS OF QUADRATICS

In Chapter 4 you learned how to solve systems of two linear equations with two variables. The answer tells the point where the two straight-line graphs cross. There are situations in the real world where it is important to know where the graphs of *quadratic* relations cross. For example, the paths of spaceships, planets, comets, and so forth, which travel under the action of gravity, turn out to be circles, ellipses, hyperbolas, or parabolas. Finding the point where a spaceship crosses the path of the planet it is approaching is of fundamental importance! Ships at sea use the LORAN system (for *LO*ng *RA*nge *N*avigation) to find their location by receiving radio signals which locate them on several hyperbolas. The intersection point of the hyperbolas tells the ship's position (See Figure 9-8b).

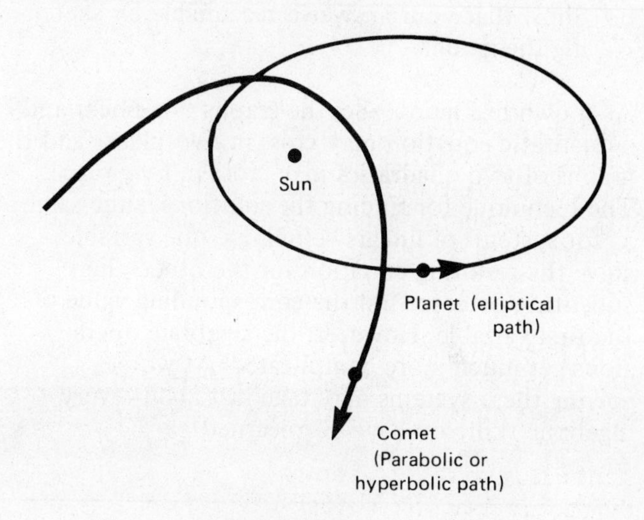

Figure 9-8a.

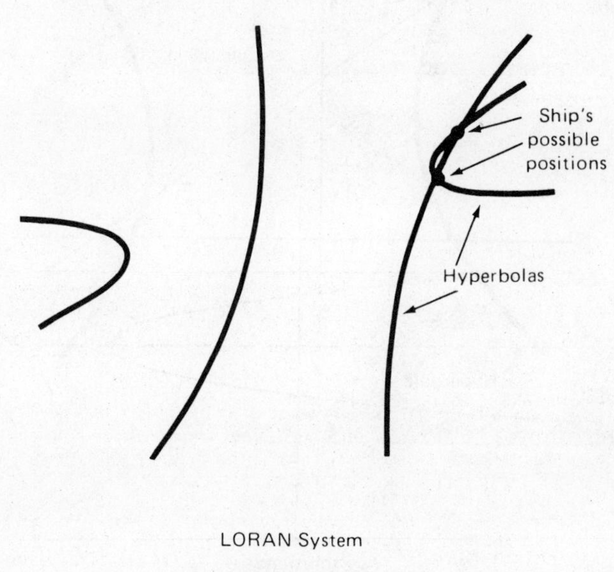

Figure 9-8b.

Objective:

Given a system of two equations in two variables, where at least one equation is quadratic (and none are higher degree), be able to:

a. Calculate the solution set of the system.
b. Show that your answer is reasonable by sketching the graphs.

As shown in Figure 9-8c, the graphs of a linear and a quadratic equation may cross in *two* places and the graphs of two quadratics may cross in *four* places. The technique for finding the solutions is the same as for systems of linears—eliminate one variable, solve the resulting equation for the other, then substitute back to find the corresponding value of the first variable. However, the algebraic operations get much more complicated. At worst, solving these systems may take just about every algebraic skill you have ever learned!

Example 1: Solve the system

$$9x^2 + 32y^2 = 324 \quad \text{——} \quad ① ,$$
$$3x^2 - y^2 = 3 \quad \text{——} \quad ② .$$

Since both equations have *only squared* terms, the problem is almost as easy as solving a system of two linears. You would multiply both members of Equation ② by −3 and add the resulting equation to Equation ①:

$$\begin{aligned} -9x^2 + 3y^2 &= -9 \\ 9x^2 + 32y^2 &= 324 \\ \hline 35y^2 &= 315 \\ y^2 &= 9 \\ y &= \pm 3. \end{aligned}$$

These values of y may be substituted into either of the equations one at a time. Substituting into Equation ② gives

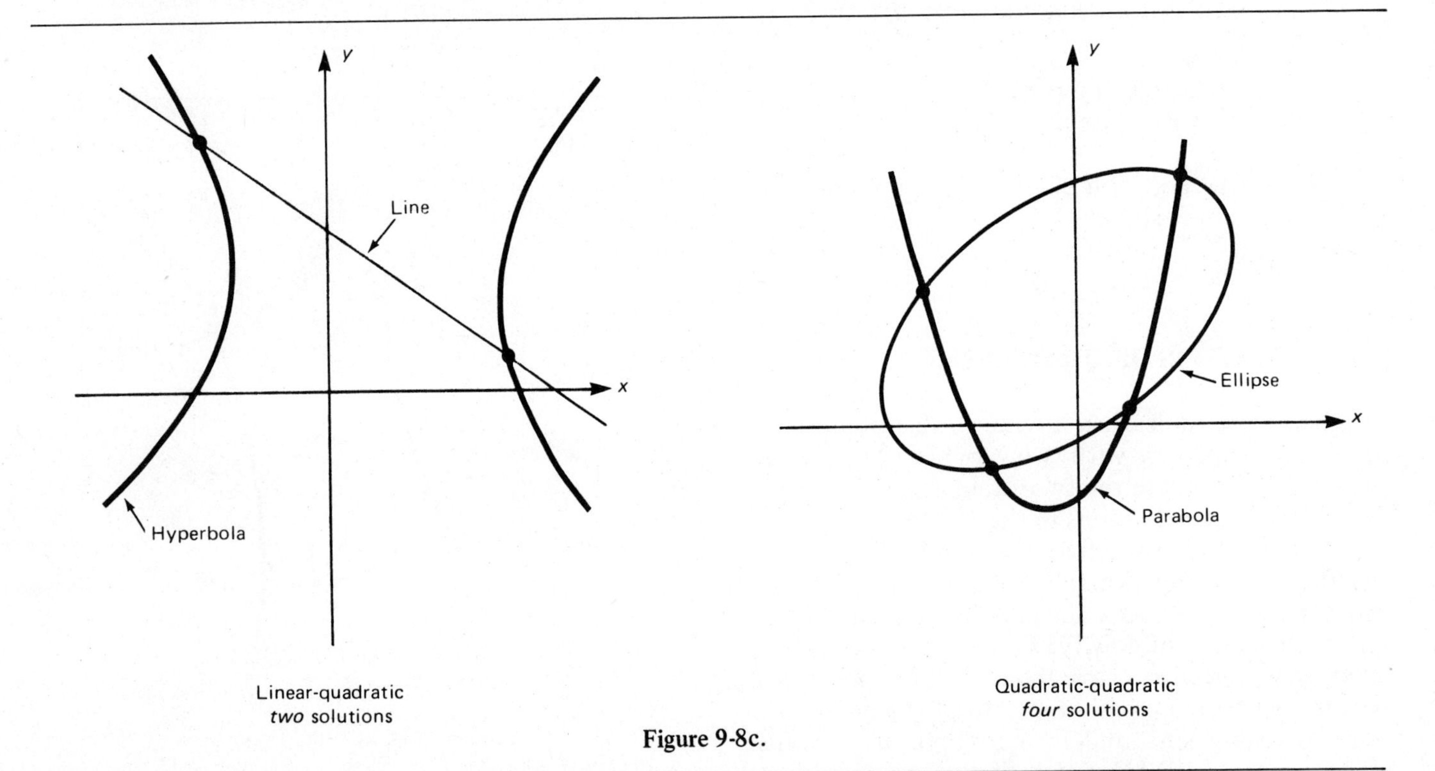

Figure 9-8c.

If $y = 3$,	If $y = -3$,
$3x^2 - 9 = 3$	$3x^2 - 9 = 3$
$3x^2 = 12$	$3x^2 = 12$
$x^2 = 4$	$x^2 = 4$
$x = \pm 2$	$x = \pm 2.$

$\therefore S = \{(2, 3), (2, -3), (-2, 3), (-2, -3)\}.$

The solution set can be checked graphically. Using the plotting techniques of this Chapter, you find that Equation ① gives an ellipse, centered at the origin, with major axis horizontal, semi-major axis $= \sqrt{324/9} = 6$, and semi-minor axis $= \sqrt{324/32} \approx \sqrt{10} \approx 3.2$. Equation ② gives a hyperbola, centered at the origin, opening in the x-direction, with asymptotes of slope $\pm\sqrt{3} \approx \pm 1.7$, and vertices 1 unit from the center. The graphs are shown in Figure 9-8d. You can see that the four points you have calculated are actually the points of intersection.

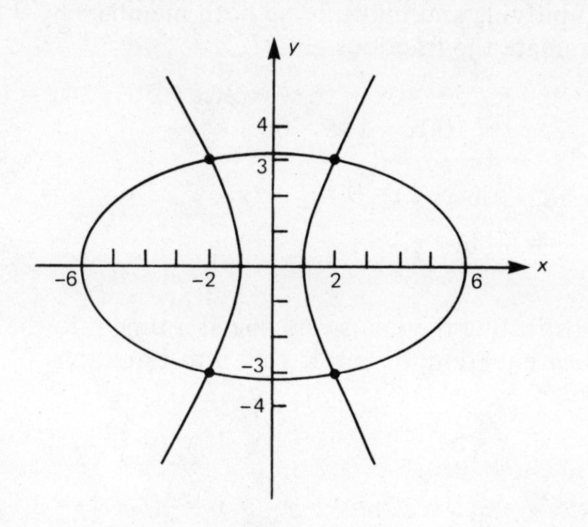

Figure 9-8d.

Example 2: Solve the system

$$7x^2 - 5y^2 + 20y = 3$$
$$21x^2 + 5y^2 = 209$$

using the "substitution" technique.

The linear combination technique used in Example 1 works only in special cases. A more general technique of eliminating a variable, called "substitution," works for any system of quadratics no matter how complicated. Starting with

$$7x^2 - 5y^2 + 20y = 3 \quad \text{——} \ ①,$$
$$21x^2 + 5y^2 = 209 \quad \text{——} \ ②,$$

you could solve Equation ② for x in terms of y, getting

$$21x^2 = 209 - 5y^2$$ Subtracting $5y^2$ from both members.

$$x^2 = \frac{1}{21}(209 - 5y^2) \quad \text{——} \ ③$$ Dividing by 21.

Substituting $\frac{1}{21}(209 - 5y^2)$ for x^2 in Equation ① gives

$$7 \cdot \frac{1}{21}(209 - 5y^2) - 5y^2 + 20y = 3,$$

which reduces to

$$y^2 - 3y - 10 = 0.$$

Factoring and solving for y gives

$$(y - 5)(y + 2) = 0,$$
$$y = 5 \text{ or } y = -2.$$

These values of y may now be substituted, one at a time, into Equation ③ to find x.

If $y = 5$,	If $y = -2$,
$x^2 = \frac{1}{21}(209 - 125)$	$x^2 = \frac{1}{21}(209 - 20)$
$x^2 = 4$	$x^2 = 9$
$x = \pm 2$	$x = \pm 3$

$\therefore S = \{(2, 5), (-2, 5), (3, -2), (-3, -2)\}.$

Example 3: Check the answer to Example 2 by graphing.

By completing the square, Equation ① can be transformed into

$$-\frac{x^2}{\left(\frac{17}{7}\right)} + \frac{(y-2)^2}{\left(\frac{17}{5}\right)} = 1.$$

This is a hyperbola, centered at (0, 2), opening vertically, with asymptotes of slope $\pm\sqrt{17/5}\,/\,\sqrt{17/7} = \pm\sqrt{7/5} \approx \pm 1.2$.

Equation ② can be transformed to

$$\frac{x^2}{\left(\frac{209}{21}\right)} + \frac{y^2}{\left(\frac{209}{5}\right)} = 1\,.$$

This graph is an ellipse with its major axis vertical, and semi-axes of $\sqrt{209/5} \approx \sqrt{42} \approx 6.5$, and $\sqrt{209/21} \approx \sqrt{10} \approx 3.2$.

The graphs are shown in Figure 9-8e. As you can see, the intersection points are (2, 5), (−2, 5), (3, −2), and (−3, −2).

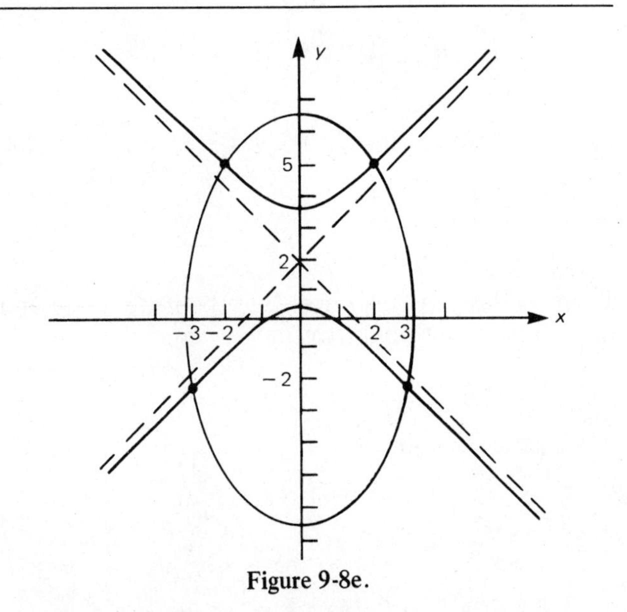

Figure 9-8e.

Example 4: Solve the system

$$x^2 + y^2 - 4x + 2y = 20,$$
$$4x + 3y = 5.$$

If one of the two equations is linear, then the linear combination technique seldom works and you must use substitution.

$$x^2 + y^2 - 4x + 2y = 20 \quad \text{———} \quad ①$$
$$4x + 3y = 5 \quad \text{———} \quad ②$$

Solving ② for y in terms of x gives

$$y = \tfrac{1}{3}(5 - 4x) \quad \text{———} \quad ③$$

Substituting $\frac{1}{3}(5 - 4x)$ for y in Equation ① gives

$$x^2 + [\tfrac{1}{3}(5 - 4x)]^2 - 4x + 2[\tfrac{1}{3}(5 - 4x)] = 20.$$

Simplifying and multiplying both members by 9 to eliminate the fractions gives

$$9x^2 + 25 - 40x + 16x^2 - 36x + 30 - 24x = 180$$
$$25x^2 - 100x - 125 = 0$$
$$x^2 - 4x - 5 = 0$$
$$(x - 5)(x + 1) = 0$$
$$x = 5 \quad \text{or} \quad x = -1.$$

Substituting these answers one at a time into the *linear* equation, or better still, into Equation ③, gives

If $x = 5$, $\qquad y = \frac{1}{3}(5 - 20) = -5$

If $x = -1$, $\qquad y = \frac{1}{3}(5 + 4) = 3$

$\therefore S = \{(5, -5), (-1, 3)\}.$

By completing the square, you can find that the graph of Equation ① is a circle of radius 5, centered at (2, −1). The second graph is a straight line of slope −4/3 and y-intercept 5/3. The graphs are shown in Figure 9-8f.

If both equations have squared and also linear terms, or xy-terms, the job of solving the system becomes much more complicated. Solving one equation for y in terms of x involves a clever use of the quadratic formula. The resulting equation will be *fourth* degree in x, and will have four solutions. Problem 29 in the following exercise leads you stepwise through the solution of one such system.

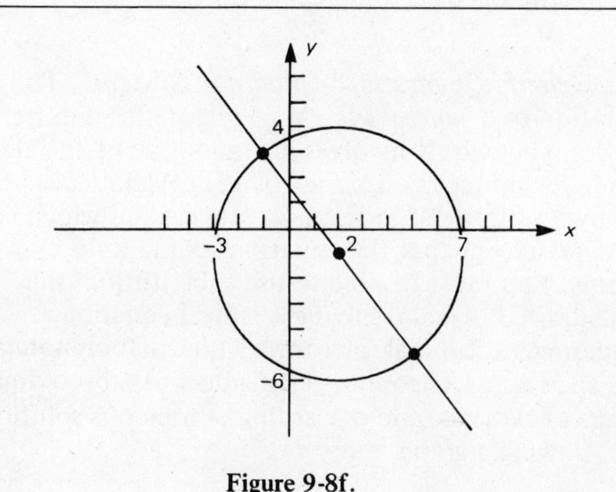

Figure 9-8f.

Exercise 9-8

For Problems 1 through 26,

a. calculate the solution set, and
b. demonstrate that your solutions are right by drawing the graphs.

1. $2x^2 + 5y^2 = 98$
   $2x^2 - y^2 = 2$
2. $x^2 - y^2 = -16$
   $8x^2 - 3y^2 = -3$
3. $x^2 + y^2 = 25$
   $y - x^2 = -5$
4. $4x^2 + y^2 = 100$
   $4x - y^2 = -20$
5. $3x^2 + 7y^2 = 187$
   $3x^2 - 7y = 47$
6. $9x^2 + y^2 = 85$
   $2x^2 - 3y^2 = 6$
7. $x^2 + 2y^2 = 33$
   $x^2 + y^2 + 2x = 19$
8. $5x^2 - 3y^2 = -22$
   $5x^2 - 6y^2 + 12y = -85$
9. $3x^2 - 5y^2 = 22$
   $3x^2 - y^2 - 6x = 8$
10. $x^2 + y^2 + 6x = 16$
    $2x^2 - 3y^2 = 24$
11. $x^2 + y^2 = 64$
    $x^2 + 10y = 100$
12. $x^2 + y^2 = 100$
    $8x^2 + 13y^2 = 1405$
13. $x^2 + y^2 + 8x = -15$
    $9x^2 + 25y^2 = 225$
14. $x^2 - 6y = 34$
    $x^2 + y^2 = 25$
15. $5x^2 + 9y^2 = 161$
    $x^2 - 4y = 4$
16. $20x^2 - 3y^2 + 12y = -16$
    $20x^2 + 3y^2 = 128$
17. $x^2 - 5y^2 = -44$
    $xy = -24$
18. $x^2 + 4y^2 = 68$
    $xy = 8$
19. $16x^2 - 3y^2 = -11$
    $8x - y = -11$
20. $7x^2 + y^2 = 64$
    $x + y = 4$
21. $x^2 + 2y^2 = 33$
    $3x + 2y = -11$
22. $2x^2 - y^2 = 7$
    $2x - 3y = -7$
23. $y = x^2 + 8x + 9$
    $x - y = -3$
24. $y = -2x^2 + 8x + 27$
    $2x + y = 15$
25. $x^2 + 9y^2 - 10x + 36y = 20$
    $x - 3y = 2$
26. $3x^2 - y^2 + 30x + 6y = -63$
    $x - y = -7$
27. Given the equations

    $x^2 - y^2 - 2y = 17$ —— ①
    $7x^2 + 3y^2 - 24y = 139$ —— ②
    $x - y = 9$ —— ③ ,

    a. solve the system of ① and ② ,
    b. solve the system of ① and ③ ,
    c. solve the system of ② and ③ ,
    d. graph all three equations on the same Cartesian coordinate system, showing that your answers for a, b, and c are reasonable.
28. Repeat Problem 27 for the equations

    $8x^2 - 3y^2 + 30y = 80$ —— ①
    $32x^2 + 3y^2 = 140$ —— ②
    $8x - 3y = 10$ —— ③ .

29. ***Meteorite Tracking Problem.*** Suppose that you have been hired by the Palomar Observatory near San Diego. Your assignment is to track incoming meteorites to find out whether or not they will strike the Earth. Since the Earth has a circular cross-section, you decide to set up a Cartesian coordinate system with its origin at the center of the Earth. The equation of the Earth's surface is

$$x^2 + y^2 = 40,$$

where x and y are distances in *thousands* of kilometers.

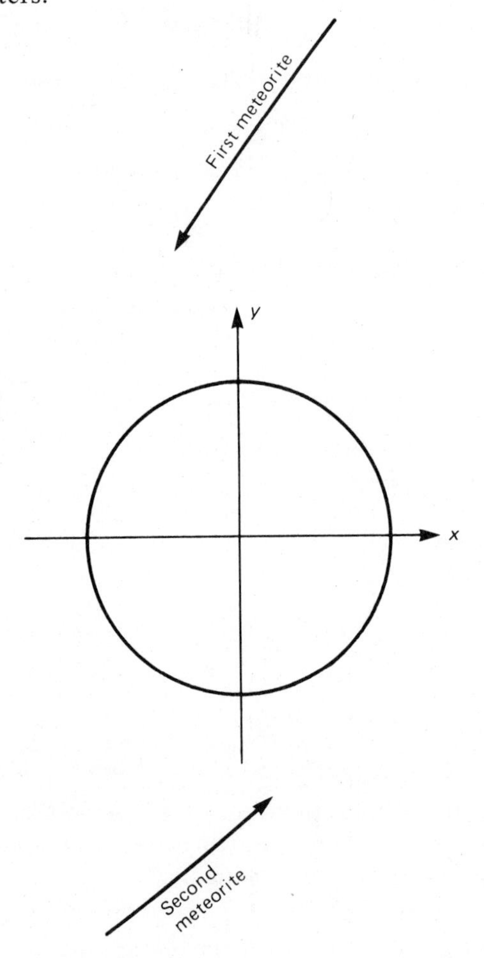

a. The first meteorite you observe is moving along the parabola whose equation is

$$18x - y^2 = -144.$$

Will this meteorite strike the Earth's surface? If so, *where?* If not, how do you tell?

b. The second meteorite is coming in from the lower left along one branch of the hyperbola

$$4x^2 - y^2 - 80x = -340,$$

(see sketch). Will it strike the Earth's surface? If so, *where?* If not, how do you tell?

c. To the nearest 100 kilometers, what is the radius of the Earth?

30. ***General Quadratic – Quadratic System.*** The solution of a *general* system of quadratics is extremely complicated. It involves just about all of the algebraic techniques you know! This problem leads you stepwise through the solution of a system which is general except that the equations contain no xy-terms. You must be able to use substitution, the Quadratic Formula, solving a radical equation, squaring a trinomial, factoring a quartic polynomial by the Factor Theorem, getting decimal approximations of radicals, and discarding extraneous solutions by drawing a graph.

Given the system

$$5x^2 + y^2 + 30x - 6y = -9 \quad \text{——} \quad ①$$
$$x^2 + 3y^2 + 4x - 6y = 21 \quad \text{——} \quad ②\ ,$$

do the following:

a. By considering x to be a "constant," write Equation ① as a quadratic equation in y. Then use the Quadratic Formula to show that

$$y = 3 \pm \sqrt{-5x^2 - 30x}.$$

b. Substitute this value of y into Equation ② and carry out the indicated operations. (Be careful squaring the value of y!) Then collect like terms and simplify as much as possible. You should find that 2 is a common factor of each term.

c. Isolate the remaining radical on one side of the equation, then square both members; all terms should now be polynomial. (Don't worry if some of the coefficients are in the thousands!)

d. Transform the equation so that all terms are on the left and 0 is on the right. You should have a quartic (fourth degree) polynomial on the left, beginning with $49x^4$ and ending with 36.

e. Use the Factor Theorem to show that $(x + 1)$ and $(x + 6)$ are factors of this polynomial. Then find the other polynomial by division.

f. Show that the remaining quadratic factor is *prime.*

g. Solve the equation in part d. Use the Quadratic Formula, where needed, and get decimal approximations for the radicals.

h. Find the values of y corresponding to each value of x from part g (4 values!).

i. Sketch the graphs of the two equations. Then discard the y-values from part h which do not correspond to crossing points.

j. Write the solution set of the system.

9-9 CHAPTER REVIEW AND TEST

In this chapter you have extended your knowledge of quadratics to relations that have a y^2 term in the equation. You have found that the graph could be a circle, ellipse, or hyperbola, as well as the familiar parabola. For each of these conic sections you learned properties that allowed you to sketch the graph quickly. By analysis of geometric properties ("analytic geometry") you were able to derive equations of relations from geometric definitions. Finally, you extended the system-solving of Chapter 4 to systems of quadratic equations.

The specific objectives for this chapter are summarized below:

1. *From the equation of a quadratic relation,*
   a. *tell which conic section the graph will be, and*
   b. *sketch the graph quickly.*

2. *Derive an equation of a set of points from a geometrical definition.*

3. *Solve systems containing quadratic equations with two variables.*

The Review Problems, below, give you a chance to try accomplishing these objectives one at a time. The Concepts Test is designed to see if you are familiar enough with the concepts of the chapter to be able to use them in problems you have never seen before!

Review Problems

The following problems are numbered according to the three objectives, above.

Review Problem 1a. Tell which conic section the graph or the boundary of the region will be.

i. $x^2 + y^2 = 36$ ii. $x^2 + y^2 < 36$
iii. $x^2 + y^2 \geq 36$ iv. $x^2 - y^2 = 36$
v. $x^2 - y^2 = -36$ vi. $x^2 + y^2 = -36$
vii. $x^2 + 4y^2 = 36$ viii. $4x^2 + y^2 = 36$
ix. $4x^2 + y = 36$ x. $4x + y^2 = 36$
xi. $x^2 - 9y^2 + 10x + 54y - 47 = 0$
xii. $x^2 + y^2 + 16x - 22y + 85 = 0$
xiii. $9x^2 + 4y^2 - 54x + 16y - 479 = 0$
xiv. $x - y^2 + 6y - 3 = 0$

b. Transform each equation in part a to a standard form (if necessary), and sketch the graph.

c. Find the focal radius for Equations xi and xiii.

Review Problem 2a. Write the geometric definition of

i. a circle,
ii. an ellipse,
iii. a hyperbola,
iv. a parabola.

b. From the geometric definition, derive a polynomial equation of the ellipse with foci (3, 0) and (−3, 0), and major axis 10 units long.

Review Problem 3a. Solve the following systems:

i. $9x^2 + y^2 - 2y = 80$
$x^2 + y^2 - 10y = 0$

ii. $9x^2 - 16y^2 + 36x - 32y = -195$
$3x + 4y = 15$

b. Show by graphing that your solutions for part a are correct.

Concepts Test

Work the following problems. For each part of each problem, tell by number and letter which one(s) of the three objectives of this chapter you used. Write also "new concept" if you are applying your knowledge to a new situation.

Concepts Test 1. Sketch graphs that illustrate how a system of two quadratic equations with two variables can have

a. exactly four solutions,
b. exactly three solutions,
c. exactly two solutions,
d. exactly one solution,
e. no solutions.

Concepts Test 2. You have learned how to find graphs of quadratic relations from their equations. Using the same properties, *reverse* the process and write equations or inequalities for the following graphs:

a.

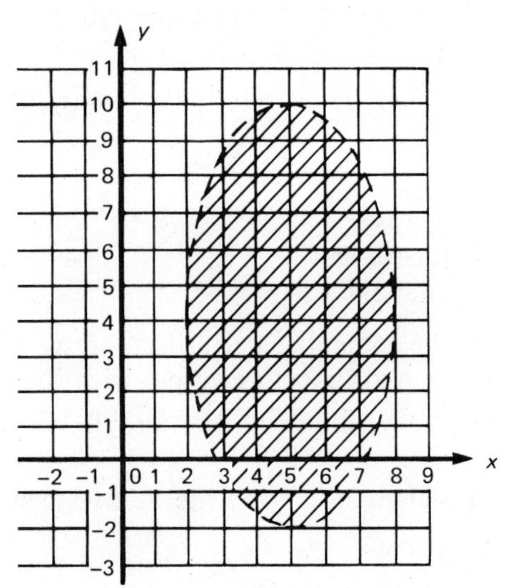

b.

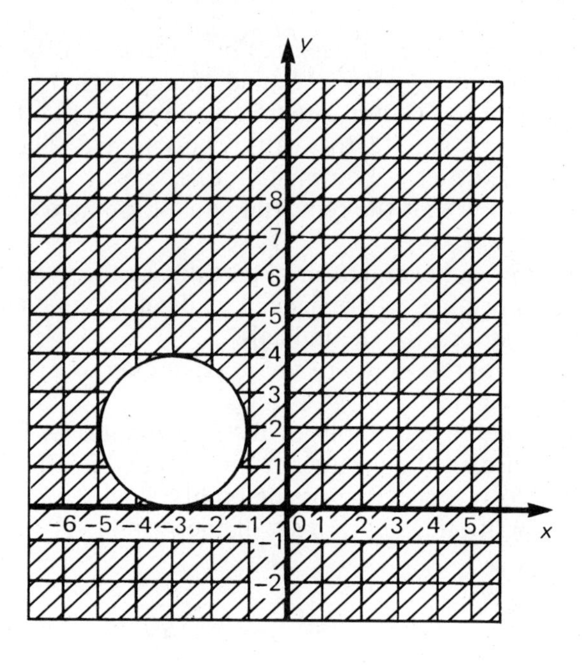

c.

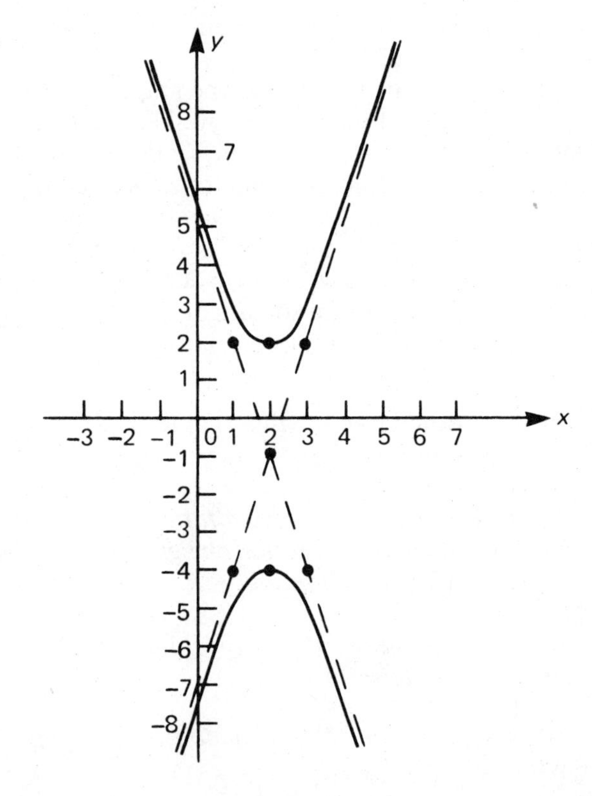

d.

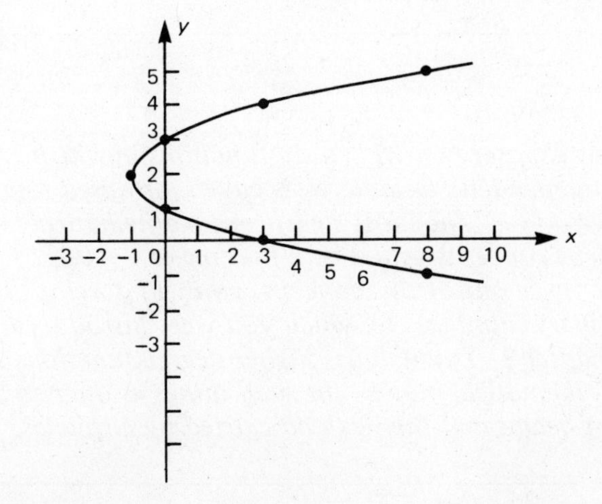

Concepts Test 3. It is possible to write the equation of a parabola in terms of the distance, p, between the focus and the directrix (see sketch).

a. Draw x- and y-axes so that the vertex is at the origin. Then use the definition of a parabola to show that the equation is

$$y = \frac{1}{2p}x^2.$$

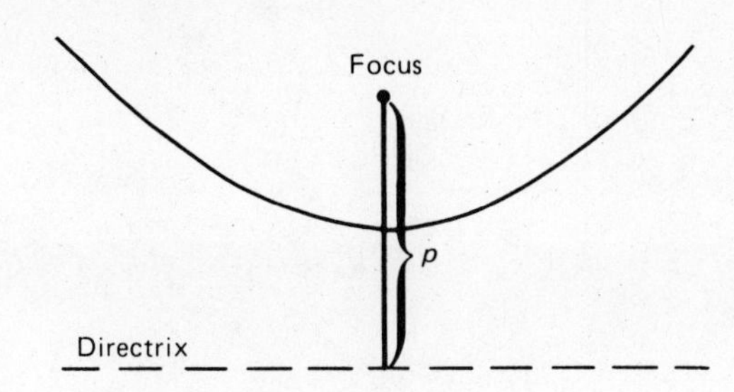

b. For the parabola $y = 3x^2 + 5x - 7$, how far is the focus from the directrix? Justify your answer.

Concepts Test 4. You have graphed inequalities with circles or ellipses as boundaries, but never with hyperbolas. In this problem you will discover what such a graph looks like.

a. The boundary for the graph of $4x^2 - y^2 \leq -16$ is a hyperbola. Draw this hyperbola.

b. To find which side of the boundary line the region lies on, substitute 0 for x, and solve the resulting inequality for y. Remember the properties of order, and the fact that

$$\sqrt{(\text{number})^2} = |\text{number}|.$$

c. Shade the correct region on your graph.

Concepts Test 5. Recalling the definitions of x-intercept and y-intercept, find the intercepts of

$$x^2 + y^2 + 16x - 22y + 85 = 0.$$

Concepts Test 6. Suppose that you are aboard a spaceship. The Earth is at the origin of a Cartesian coordinate system, and the path of your spaceship is the graph of

$$9x^2 + 25y^2 - 72x = 81.$$

a. Which conic section is this path, and how do you tell?

b. Sketch the graph of this path.

c. Show that the Earth is at the focus of this conic section.

d. To determine your position at a particular time, you tune in to the LORAN station and find that you are also on the graph of

$$9x^2 - 15y^2 = 9.$$

Which conic section is this graph, and how do you tell?

e. By solving the system formed by the equations of the two conic sections above, find the *three* possible points at which your spaceship could be located.

f. Draw the graph of the conic section in part d on the same Cartesian coordinate system as in part b, and thus show that your answer to part e is correct.

g. To ascertain at *which* one of the three points you are located, you find that you are also on the graph of

$$3x - 5y = 27.$$

At which of the points are you located? (There is a *clever* way to do this, as well as the longer way.)

Higher Degree Functions and Complex Numbers

*In this chapter you will study functions in which the independent variable, **x**, is **cubed**, or raised to a higher power. The most significant mathematical concept concerns values of **x** that make **y** = 0. To make sense out of this concept, you will develop **imaginary** numbers, to which you were introduced in Chapter 1. You will use higher degree functions as mathematical models for such things as the bending of beams and the payload carried by airplanes.*

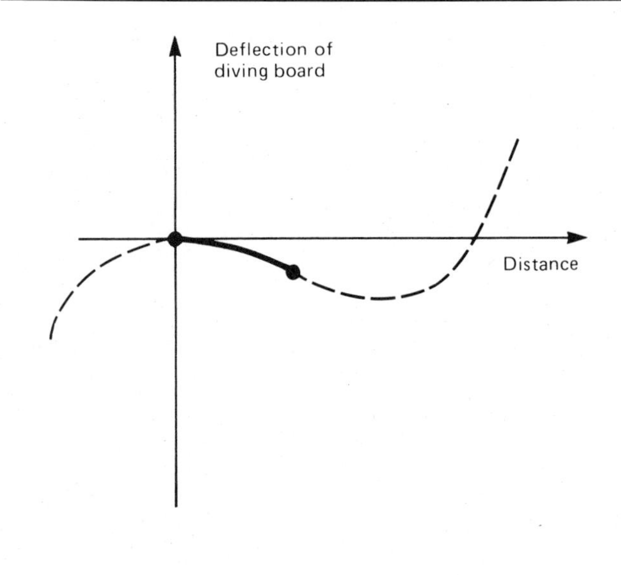

10-1 INTRODUCTION TO HIGHER DEGREE FUNCTIONS

You will recall that a polynomial function is a function with a general equation of the form

y = a polynomial in x.

You studied linear (first degree) and quadratic (second degree) functions in Chapters 3 and 5. In this section you will explore the graph of a cubic function.

Objective:

Discover by pointwise plotting what the graph of a cubic function looks like.

The exercise which follows is designed to allow you to accomplish this objective.

Exercise 10-1

1. Plot a graph of the cubic function

$f(x) = x^3 - 4x^2 - 5x + 14$

in the domain $-3 \leq x \leq 6$. Do this by calculating the value of $f(x)$ for each integer value of x in the domain. For example, $f(6) = 56$. Then plot the resulting points, and connect them with a smooth curve. You may wish to use different scales for the two axes so that the graph fits conveniently onto the graph paper. Be careful with the arithmetic, since even *one* point out of place can make the graph look completely different!

2. From your graph, you should be able to answer the following questions:

a. How many x-intercepts does a cubic function seem to have?
b. How many vertices does a cubic function graph seem to have?
c. Comparing these answers with what you recall about graphs of quadratic functions, how many x-intercepts and vertices would you expect in the graph of a *quartic* (*fourth* degree) function?

3. Plot graphs of

$g(x) = x^3 - 4x^2 - 5x + 24$, and
$h(x) = x^3 - 4x^2 - 5x + 48$.

This is easy if you observe that $g(x) = f(x) + 10$, and $h(x) = f(x) + 34$, where $f(x)$ is defined in Problem 1, above.

a. How do the graphs of g and h compare with the graph of f?
b. How many x-intercepts do the graphs of g and h have?
c. What conclusion can you reach about the number of x-intercepts a cubic function can have?

10-2 IMAGINARY NUMBERS

In Exercise 10-1 you plotted the graphs of three cubic functions. As shown in Figure 10-2, these graphs have the same shape, but are in different vertical positions.

As the constant term in the equation gets larger and larger, the graph rises. The two x-intercepts on the right get closer to each other, merge into one intercept, and then disappear. In this chapter you will find values for the "missing" x-intercepts. You will find that they involve square roots of *negative* numbers. Since the square of any *real* number is *non*-negative, it is necessary for you to invent a new kind of number.

The first step is to define a number whose square is -1.

DEFINITION

i is a number whose square is -1. That is,

$$\boxed{i^2 = -1}\ .$$

Notes:

1. Since $i^2 = -1$, it is customary to write

$$\boxed{i = \sqrt{-1}}\ .$$

2. Square roots of negative numbers are called "imaginary" numbers since when they were proposed several hundred years ago, people could not "imagine" such a number! However, they are no less real than "real" numbers, since both kinds of numbers are inventions of mankind.

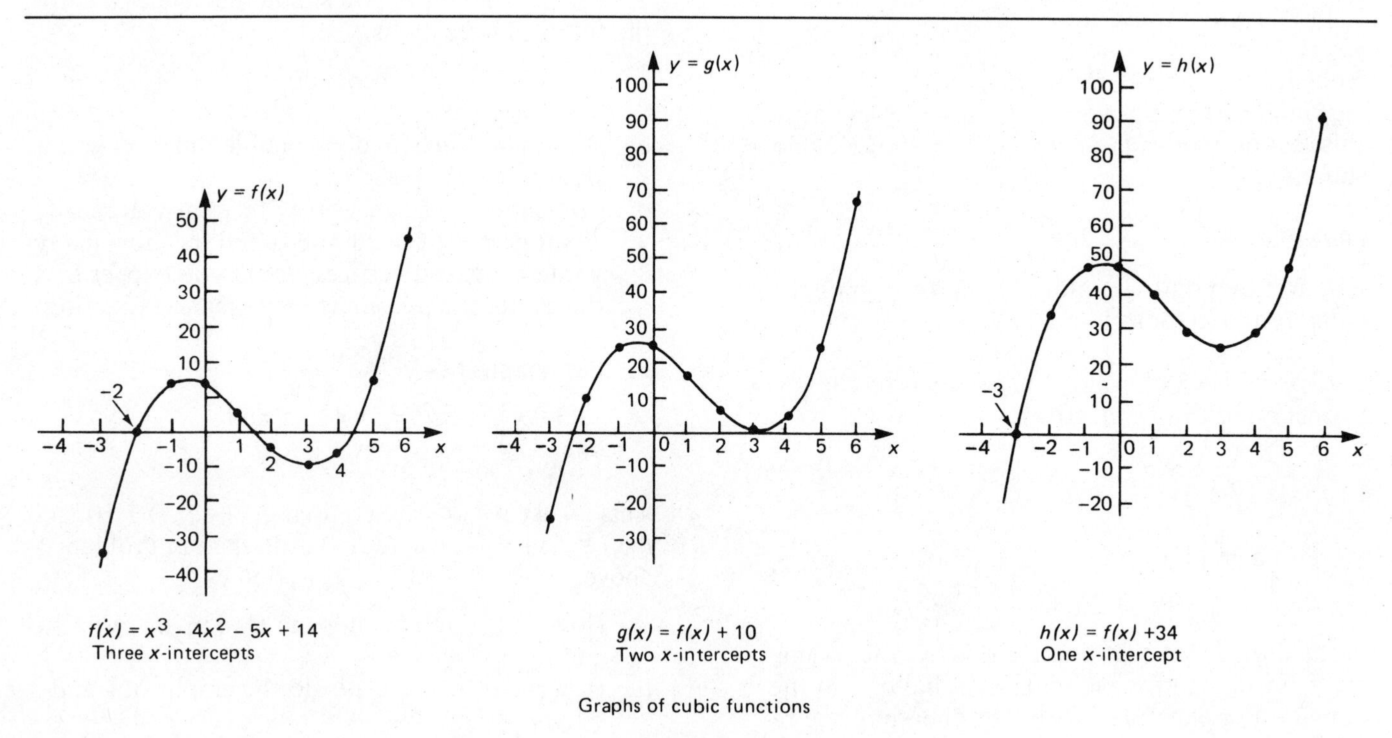

Graphs of cubic functions

Figure 10-2.

3. The number i is called the *unit* imaginary number, just as 1 (or $\sqrt{+1}$) is called the unit *real* number.

The next step is to define square roots of other negative numbers in terms of i. Since a number such as $\sqrt{-25}$ could be written $\sqrt{(-1)(25)}$, or $\sqrt{-1}\,\sqrt{25}$, you simply *define* $\sqrt{-25}$ to equal $i\sqrt{25}$.

DEFINITION

If x is a non-negative real number, then

$$\boxed{\sqrt{-x} = i\sqrt{x}}\,.$$

So any imaginary number is the product of a *real* number and the unit imaginary number i.

Having invented a new kind of number, you must now learn how to compute with it.

Objective:

Be able to perform operations with expressions containing i or square roots of negative numbers.

Powers of i: By definition, $i^2 = -1$. To evaluate other integer powers of i, it is necessary only to use the definition of exponentiation with integer exponents;

$i^3 = i^2 \cdot i = -1 \cdot i = -i$	$\therefore i^3 = -i.$
$i^4 = i^2 \cdot i^2 = (-1)(-1) = 1$	$\therefore i^4 = 1.$
$i^5 = i \cdot i^4 = i \cdot 1 = i$	$\therefore i^5 = i.$
$i^6 = i^2 \cdot i^4 = -1$	$\therefore i^6 = -1.$
$i^7 = i^3 \cdot i^4 = -i$	$\therefore i^7 = -i.$
$i^8 = i^4 \cdot i^4 = 1$	$\therefore i^8 = 1.$
⋮	⋮

Note that any time the exponent is a multiple of 4, the power equals 1. Thus

$$i^{76} = 1,$$

because 76 is a multiple of 4.

Suppose that you wish to evaluate i^{39}. Since $39 = 36 + 3$, and 36 is a multiple of 4,

$$i^{39} = i^{36+3} = i^{36} \cdot i^3 = 1 \cdot i^3 = i^3 = -i.$$

A faster technique is to divide the exponent by 4, getting 9 with a remainder of 3. Throw away the quotient, 9, and keep the remainder, 3. So $i^{39} = i^3$, which equals $-i$.

Products of Imaginary Numbers: To multiply $\sqrt{-12}$ by $\sqrt{-6}$, you would start with the definition of imaginary numbers.

$\sqrt{-12} \cdot \sqrt{-6}$	
$= i\sqrt{12} \cdot i\sqrt{6}$	Definition of $\sqrt{-x}$.
$= i^2 \cdot \sqrt{12} \cdot \sqrt{6}$	Commutativity and associativity.
$= -1 \cdot \sqrt{72}$	$i^2 = -1$ and $\sqrt{a}\sqrt{b} = \sqrt{ab}$.
$= -6\sqrt{2}$	Simple radical form.

Notes:

1. If both negative factors are under the *same* radical sign, you would get

$$\sqrt{(-12)(-6)} = \sqrt{+72} = 6\sqrt{2}.$$

Therefore, the property $\sqrt{ab} = \sqrt{a}\sqrt{b}$ does *not* work if both a and b are *negative* numbers.

2. Observe that the product of two imaginary numbers is a *real* number. So {imaginary numbers} is *not* closed under multiplication.

3. This calculation requires you to assume that multiplication is commutative and associative for imaginary numbers.

Product of a Real and an Imaginary Number: Suppose you are asked to multiply $\sqrt{-12}$ by 5. Assuming commutativity and associativity of multiplication as above, leads to

$5 \cdot \sqrt{-12} = 5 \cdot i\sqrt{12}$	Definition of $\sqrt{-x}$.
$= 5 \cdot i \cdot 2\sqrt{3}$	Simple radical form.
$= 10\,i\sqrt{3}$	Commutativity and associativity.

So the product of a real number and an imaginary number is an *imaginary* number. This fact leads to an interesting conclusion. Assuming that the multiplication property of zero holds for imaginary numbers,

$$0 \cdot i = 0.$$

But $0 \cdot i$ is an *imaginary* number, as shown above. So you are obliged to conclude that zero is also an imaginary number!

Conclusion: Zero is both a real number and an imaginary number.

Division by an Imaginary Number: Suppose you are asked to divide 12 by $\sqrt{-18}$. Since the denominator contains a radical, you can start by rationalizing the denominator.

$\frac{12}{\sqrt{-18}} = \frac{12}{i\sqrt{18}}$	Definition of imaginary numbers.
$= \frac{12}{i\sqrt{18}} \cdot \frac{i\sqrt{2}}{i\sqrt{2}}$	Multiplication by a clever form of 1 to rationalize the denominator.
$= \frac{12i\sqrt{2}}{-6}$	Multiplication property of fractions, and $i^2 = -1$.
$= -2i\sqrt{2}$	Canceling.

Sums of Imaginary Numbers: Suppose you are asked to add $\sqrt{-12}$ and $\sqrt{-75}$. Using the definition of imaginary numbers,

$\sqrt{-12} + \sqrt{-75}$ $= i\sqrt{12} + i\sqrt{75}$	Definition of imaginary numbers.
$= i \cdot 2\sqrt{3} + i \cdot 5\sqrt{3}$	Simplifying the radicals.
$= i\sqrt{3}(2 + 5)$	$i\sqrt{3}$ is a common factor.
$= 7i\sqrt{3}$	Arithmetic and commutativity.

Note: This calculation requires you to assume that multiplication distributes over addition for imaginary numbers. Observe that the answer is an imaginary number. So the set of imaginary numbers is closed under addition.

The exercise which follows is designed to give you practice operating with imaginary numbers. You will also compare some of the properties of imaginary numbers with those of real numbers.

Exercise 10-2

For Problems 1 through 46, carry out the indicated operations and simplify. Leave the answers in *radical* form (i.e., *no* decimal approximations), if necessary. Express all imaginary numbers as power of i.

1. i^5
2. i^7
3. i^{55}
4. i^{25}
5. i^{62}
6. i^{74}
7. i^{300}
8. i^{180}
9. i^0
10. i^{-2}
11. i^{-7}
12. i^{-25}
13. i^{-38}
14. i^{-54}
15. $\sqrt{-16}$
16. $\sqrt{-25}$
17. $\sqrt{-18}$
18. $\sqrt{-48}$
19. $\sqrt{-7}$
20. $\sqrt{-3}$
21. $\sqrt{-54} \times \sqrt{-98}$
22. $\sqrt{-24} \times \sqrt{-72}$
23. $\sqrt{-13} \times \sqrt{26}$
24. $\sqrt{51} \times \sqrt{-17}$
25. $\sqrt{-5} \times \sqrt{-15} \times \sqrt{-12}$
26. $\sqrt{-7} \times \sqrt{-35} \times \sqrt{-20}$
27. $\sqrt{(-3)(-12)}$
28. $\sqrt{(-6)(-18)}$
29. $\frac{4}{\sqrt{-28}}$
30. $\frac{5}{\sqrt{-120}}$
31. $\frac{\sqrt{-3}}{\sqrt{-15}}$
32. $\frac{\sqrt{-22}}{\sqrt{-11}}$
33. $\frac{\sqrt{45}}{\sqrt{-5}}$
34. $\frac{\sqrt{35}}{\sqrt{-7}}$
35. $\frac{\sqrt{-1024}}{\sqrt{96}}$
36. $\frac{\sqrt{-243}}{\sqrt{27}}$
37. $\sqrt{-18} + \sqrt{-32} - \sqrt{-242}$
38. $\sqrt{-12} + \sqrt{-147} - \sqrt{-72}$
39. $5\sqrt{-28} - 4\sqrt{-63}$
40. $4\sqrt{-45} - 7\sqrt{-20}$
41. $2\sqrt{-3} + 5\sqrt{-27} - 7\sqrt{-48}$
42. $11\sqrt{-7} - 5\sqrt{-28} + 3\sqrt{-63}$
43. $\sqrt{-4} + \sqrt{-9} - \sqrt{-16}$

44. $\sqrt{-25} + \sqrt{-144} - \sqrt{-169}$

45. $i + i^2 + i^3 + i^4$

46. $\frac{1}{i} + \frac{1}{i^2} + \frac{1}{i^3} + \frac{1}{i^4}$

47. The set of imaginary numbers is *not* a field with respect to the operations of addition and multiplication. Tell which of the eleven field axioms fail to apply to the set of imaginary numbers.

48. Is $4 + 3i$ a *real* number or an *imaginary* number? Explain.

10-3 COMPLEX NUMBERS

In the previous section you learned how to perform various operations with imaginary numbers. Conspicuously absent from this list of operations was the sum of a real number and an imaginary number.

Addition of numbers like 4 and $\sqrt{-9}$ presents a problem. The 4 can be plotted on a real number line,

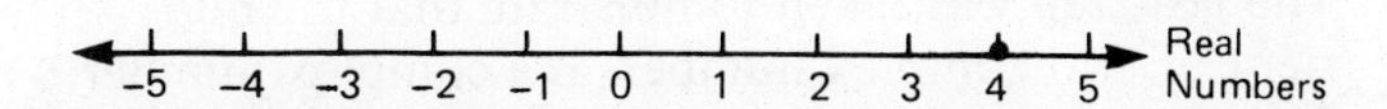

and the $\sqrt{-9}$, or $3i$, can be plotted on an imaginary number line.

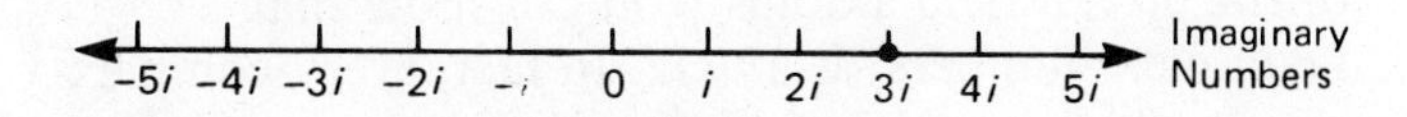

But there is no point on either line corresponding to $4 + 3i$. Since zero is both a real number and an imaginary number, it makes sense to *cross* these two number lines at their origins. The point in the plane with coordinates (4, 3) could then represent the number $4 + 3i$. (See Figure 10-3.) Since this is neither a real number nor an imaginary number, a new name is needed.

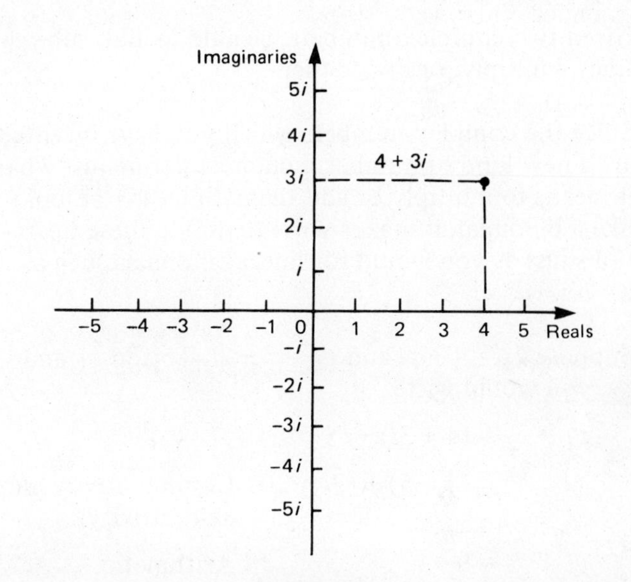

Figure 10-3.

DEFINITION

A ***complex*** number is a number of the form $a + bi$, where a and b stand for *real* numbers, and $i = \sqrt{-1}$.

If $z = a + bi$, then the real number a is called the *real part* of z, and the real number b is called the *imaginary part* of z. The plane determined by the two number lines is called the *complex number plane.* Since the real number line and the imaginary number line are both in this plane, it follows that real and imaginary numbers are also complex numbers. This fact can be deduced from the definition. For example,

$$7 = 7 + 0i$$
$$5i = 0 + 5i$$

Conclusion: {real numbers} and {imaginary} numbers are both *subsets* of {complex numbers}.

Objective:

Given two complex numbers, be able to add, subtract, multiply, or divide them.

Since the complex numbers which you have invented are a new kind of number, you must define just what it means to multiply or add them. Since $a + bi$ looks like a binomial, it makes sense to define these operations just as you would for linear binomials such as $a + bx$.

Suppose $z_1 = 4 + 3i$ and $z_2 = 5 - 2i$. To add z_1 and z_2, you would write

$z_1 + z_2 = (4 + 3i) + (5 - 2i)$	
$= (4 + 5) + (3i - 2i)$	Commutativity and associativity.
$= 9 + i$	Arithmetic.

To subtract z_2 from z_1, you would write

$z_2 - z_1 = (4 + 3i) - (5 - 2i)$	
$= 4 + 3i - 5 + 2i$	Negative of a sum.
$= (4 - 5) + (3i + 2i)$	Commutativity and associativity.
$= -1 + 5i$	Arithmetic.

To multiply z_1 by z_2, you would write

$z_1 z_2 = (4 + 3i)(5 - 2i)$	
$= 20 + 7i - 6i^2$	Product of binomials.
$= 20 + 7i + 6$	$i^2 = -1$.
$= 26 + 7i$	Commutativity and associativity.

Dividing z_1 by z_2 requires a bit of preliminary work. The quotient

$$\frac{4 + 3i}{5 - 2i}$$

has a *radical,* namely i, in the denominator. The technique is to *rationalize* the denominator as you did in Section 8-2. Since $5 - 2i$ is a binomial, the denominator may be rationalized by multiplying by its complex *conjugate,* $5 + 2i$.

DEFINITION

If $\boldsymbol{a + bi}$ is a complex number, then its *complex conjugate* is $\boldsymbol{a - bi}$, and vice versa.

So to divide z_1 by z_2, you would write

$\frac{z_1}{z_2} = \frac{4 + 3i}{5 - 2i} \cdot \frac{5 + 2i}{5 + 2i}$	Multiplication property of 1.
$= \frac{20 + 23i + 6i^2}{25 - 4i^2}$	Multiplication property of fractions.
$= \frac{20 + 23i - 6}{25 + 4}$	$i^2 = -1$.
$= \frac{14 + 23i}{29}$	Arithmetic.
$= \frac{14}{29} + \frac{23}{29} i$	Division distributes over addition.

The last step was taken to illustrate that the quotient of two complex numbers is a complex number. That is, except for division by zero, the set of complex numbers is *closed* under division. In fact, {complex numbers} is also closed under addition, subtraction, and multiplication, as you can readily check by the preceding examples. By checking each of the eleven field axioms, you can show that {complex numbers} is a *field*. In fact, every complex number has an nth root in the set of complex numbers, a property not possessed by the set of real numbers.

The exercise which follows is designed to give you practice operating with complex numbers, and plotting these numbers on a complex number plane. You will also have an opportunity to discover some properties of complex numbers.

Exercise 10-3

For Problems 1 through 8, plot graphs of the given numbers on a complex number plane.

1. $5 + 7i$ 2. $6 + 3i$ 3. $3 - 2i$ 4. $7 - 4i$

5. $-4 + 6i$ 6. $-2 + i$ 7. $-1 - 8i$ 8. $-5 - 2i$

For Problems 9 through 12, find (a) $z_1 + z_2$, (b) $z_1 - z_2$, (c) $z_1 z_2$, and (d) z_1/z_2.

9. $z_1 = 2 + 3i$, $z_2 = 4 - 5i$ 10. $z_1 = 6 + 2i$, $z_2 = 5 - 7i$

11. $z_1 = -1 + 2i$, $z_2 = 6 + 7i$ 12. $z_1 = -3 - i$, $z_2 = 4 + 8i$

13. Find the product of the complex number $a + bi$ and its complex conjugate, and explain why the answer is a *real* number.

14. Use the results of Problem 13 to factor the *sum* of two squares, $x^2 + y^2$.

For Problems 15 through 18, suppose that the *absolute value* of z, $|z|$, is defined to be the distance from the origin to the graph of z on the complex number plane (just the way the absolute value of a real number is its distance from the origin on a real number line). For each problem, plot z on a complex number plane, connect the graph of z to the origin, then find $|z|$ by appropriate use of the Pythagorean Theorem.

15. $z = 4 + 3i$ 16. $z = 2 - 3i$

17. $z = -5 + 12i$ 18. $z = -8 - 15i$

19. Suppose that $z_1 = 3 + 5i$ and $z_2 = 7 + 2i$.

a. Plot the points z_1, z_2, and $z_1 + z_2$ on a complex number plane. Then connect each point to the origin. What do the lengths of these lines represent?
b. Connect z_1 to $z_1 + z_2$. What does the length of this line equal?
c. Explain why the inequality $|z_1 + z_2| \leq |z_1| + |z_2|$ is true for any two complex numbers z_1 and z_2. (Clue: This is called the "Triangle Inequality.")

20. Suppose that $z_1 = 1 + i$ and $z_2 = 2 + 5i$.

a. Find the product $z_1 z_2$.
b. Find $|z_1|$, $|z_2|$, and $|z_1 z_2|$, then try to arrive at some conclusion about how these three absolute values are related to each other.
c. Plot the points z_1, z_2, and $z_1 z_2$ on a complex number plane, then connect each point to the origin. How do the *angles* between the positive real axis and these three lines seem to be related to one another?

21. Draw graphs of i, i^2, i^3, and i^4 on the same complex number plane. What seems to be happening to the graph each time the power of i is increased by one?

22. Suppose that $z = 3 + 4i$.

a. What does iz equal?
b. Plot the points z and iz on a complex number plane, then connect each point to the origin.
c. Show that multiplying z by i has *not* changed its absolute value.
d. Show that multiplying z by i has *rotated* its graph through an angle of 90°.

23. Show that

$$z = \frac{\sqrt{2}}{2}(1 + i)$$

is a *square root* of i by squaring $\frac{\sqrt{2}}{2}(1 + i)$ and showing that you get i for an answer. Also, show that $|z| = 1$.

24. Suppose that $z = -1 + i\sqrt{3}$.

a. Show that $|z| = 2$.
b. Show that z is a *cube root* of 8 by cubing $-1 + i\sqrt{3}$ and showing that you get 8 for an answer.

10-4 COMPLEX SOLUTIONS OF QUADRATIC EQUATIONS

In Section 5-5 you learned how to solve quadratic equations of the form

$$ax^2 + bx + c = 0$$

using the Quadratic Formula. Sometimes a negative number appeared under the radical sign, and you concluded that there were no real solutions. Now you know that such solutions are simply *imaginary* numbers or complex *numbers*. In this section you will practice solving quadratic equations that have complex solutions.

Objectives:

1. Given a quadratic equation with one variable, solve it over the set of complex numbers (i.e., the domain of the variable is the set of complex numbers).

2. Given two numbers, write a quadratic equation having these numbers as its solutions.

3. Given a quadratic polynomial, factor it into two linear factors.

Solution of Equations – To accomplish the first objective, you simply use the Quadratic Formula.

Example 1: Solve $x^2 - 2x + 13 = 0$.

Using the Quadratic Formula, you get

$$x = \frac{2 \pm \sqrt{4 - 4(1)(13)}}{2(1)} = \frac{2 \pm \sqrt{-48}}{2} = \frac{2 \pm 4i\sqrt{3}}{2}$$

$$= 1 \pm 2i\sqrt{3}$$

$$\therefore S = \{1 + 2i\sqrt{3},\ \ 1 - 2i\sqrt{3}\}.$$

Observe that the solutions are *complex conjugates* of each other. This will always happen if the coefficients in the equation are *real* numbers. The imaginary parts of the solutions come from

$$\sqrt{b^2 - 4ac},$$

which is *added* in one of the solutions and *subtracted* in the other.

Conclusion: If a quadratic equation with *real* coefficients has a negative discriminant, then the two solutions will be *complex conjugates* of each other.

This conclusion can be remembered by telling yourself that, "Complex solutions come in conjugate pairs." The reason for the restriction to *real* coefficients will become clear when you work Problem 64 in the following exercise.

Equations from Solutions – You have learned how to solve a quadratic equation by factoring. For example, to solve $x^2 - 5x + 6 = 0$, you would write

$x^2 - 5x + 6 = 0$	Given equation.
$(x - 3)(x - 2) = 0$	Factoring.
$x - 3 = 0$ or $x - 5 = 0$	A product is zero if and only if a factor is zero.
$x = 3$ or $x = 2$	Adding 3 or 2.

To find the equation from its solutions, you simply *reverse* the process. For example, an equation whose solutions are 5 and 7 is

$$(x - 5)(x - 7) = 0.$$

Carrying out the multiplication gives

$$x^2 - 12x + 35 = 0.$$

In general, a quadratic equation whose solutions are s_1 and s_2 is

$$\boxed{(x - s_1)(x - s_2) = 0.}$$

To see how this equation relates to the general quadratic equation, $ax^2 + bx + c = 0$, you can carry out the multiplication.

$$x^2 - s_1x - s_2x + s_1s_2 = 0$$

$$x^2 - (s_1 + s_2)x + s_1s_2 = 0 \qquad \text{——} \qquad ①$$

Dividing both members of $ax^2 + bx + c = 0$ by a gives

$$x^2 + \frac{b}{a}x + \frac{c}{a} = 0 \quad \text{———} \quad ②$$

By comparing Equations ① and ②, you should be able to see that

$$\boxed{s_1 s_2 = \frac{c}{a}} \quad \text{and} \quad \boxed{-(s_1 + s_2) = \frac{b}{a}}.$$

Example 2: Find a quadratic equation with $\frac{5}{3}$ and -2 as its solutions .

$$\frac{b}{a} = -(s_1 + s_2) = -(\tfrac{5}{3} - 2) = -(-\tfrac{1}{3}) = \frac{1}{3}$$

$$\frac{c}{a} = s_1 s_2 = (\tfrac{5}{3})(-2) = -\frac{10}{3}$$

$$\therefore \text{ the equation is } x^2 + \frac{1}{3}x - \frac{10}{3} = 0.$$

If you want the equation to have integer coefficients, you can multiply both members by 3, getting

$$\underline{\underline{3x^2 + x - 10 = 0}}.$$

Example 3: Find a quadratic equation with real-number coefficients if one solution is $2 + 3i$.

Since complex solutions come in *conjugate pairs* when the coefficients are real numbers, the other solution will be $2 - 3i$.

$$\frac{b}{a} = -(s_1 + s_2) = -(2 + 3i + 2 - 3i) = -4$$

$$\frac{c}{a} = s_1 s_2 = (2 + 3i)(2 - 3i) = 4 - 9i^2 = 13.$$

$$\therefore \text{ the equation is } \underline{\underline{x^2 - 4x + 13 = 0}}.$$

Factoring Quadratics – Since factoring can be used to solve quadratic equations, solving quadratic equations can be used to help you factor. You use the fact that if s_1 and s_2 are solutions of the equation, then the equation is

$$(x - s_1)(x - s_2) = 0.$$

Example 4: Factor $x^2 - 2x + 13$.

First you set the polynomial equal to zero, and solve the resulting equation.

$$x^2 - 2x + 13 = 0$$

By the Quadratic Formula, the two solutions of this equation are

$$s_1 = 1 + 2i\sqrt{3} \quad \text{and} \quad s_2 = 1 - 2i\sqrt{3}.$$

Thus, the equation is equivalent to

$$[x - (1 + 2i\sqrt{3})]\,[x - (1 - 2i\sqrt{3})] = 0.$$

Simplifying each factor gives

$$(x - 1 - 2i\sqrt{3})(x - 1 + 2i\sqrt{3}) = 0.$$

By the Transitive Property,

$$x^2 - 2x + 13 = \underline{\underline{(x - 1 - 2i\sqrt{3})(x - 1 + 2i\sqrt{3})}}.$$

Example 5: Factor $5x^2 + 3x - 7$.

In this case, the leading coefficient is not equal to 1. Therefore, you must first factor it out, getting

$$5(x^2 + \frac{3}{5}x - \frac{7}{5}).$$

The polynomial inside the parentheses will factor into $(x - s_1)(x - s_2)$. Solving the equation $5x^2 + 3x - 7 = 0$, you find

$$x = \frac{-3 \pm \sqrt{9 - 4(5)(-7)}}{2(5)}$$

$$= \frac{-3 \pm \sqrt{149}}{10}$$

$$\therefore 5x^2 + 3x - 7 = \underline{\underline{5\left(x - \frac{-3 + \sqrt{149}}{10}\right)\left(x - \frac{-3 - \sqrt{149}}{10}\right)}}.$$

If desired, you can also find decimal approximations for these solutions:

$$\frac{-3 \pm \sqrt{149}}{10} \approx 0.92 \text{ or } -1.52.$$

$$\therefore\ 5x^2 + 3x - 7 \approx \underline{\underline{5(x - 0.92)(x + 1.52)}}.$$

Example 6: Factor $4x^2 + 9$.

This is a *sum* of two squares, previously not factorable. You can factor it by the technique used in Example 5. An easier way is to turn it into an *old* problem by writing it as a *difference* of two squares.

$$4x^2 + 9 = 4x^2 - 9i^2 \qquad \text{Because } i^2 = -1.$$
$$= \underline{\underline{(2 + 3i)(2 - 3i)}}.$$

When you first studied factoring, you ran across some polynomials that were "prime." From Examples 4, 5, and 6, you can see that *any* quadratic polynomial can be factored into linear factors if you are willing to pay the price of using *irrational* or *complex* numbers in the factors.

The exercise that follows is designed to give you practice in finding solutions for given equations, finding equations for given solutions, and factoring quadratics into linear factors.

Exercise 10-4

For Problems 1 through 24, find the solution set of the given equation over the set of complex numbers (i.e., assume that the domain of x is the set of complex numbers).

1. $x^2 - 2x + 2 = 0$
2. $x^2 + 2x + 2 = 0$
3. $x^2 - 4x + 5 = 0$
4. $x^2 - 2x + 5 = 0$
5. $x^2 + 2x + 10 = 0$
6. $x^2 - 6x + 10 = 0$
7. $x^2 - 4x + 29 = 0$
8. $x^2 + 10x + 29 = 0$
9. $x^2 + 4x + 7 = 0$
10. $x^2 - 6x + 11 = 0$
11. $x^2 - 10x + 27 = 0$
12. $x^2 + 4x + 9 = 0$
13. $3x^2 - 4x + 10 = 0$
14. $5x^2 + 2x + 7 = 0$
15. $25x^2 + 10x + 101 = 0$
16. $9x^2 - 12x + 229 = 0$
17. $x^2 + 8x + 7 = 0$
18. $x^2 + 8x + 15 = 0$
19. $x^2 - 6x + 4 = 0$
20. $x^2 - 10x + 22 = 0$
21. $x^2 + 9 = 0$
22. $x^2 + 16 = 0$
23. $4x^2 + 49 = 0$
24. $9x^2 + 25 = 0$

For Problems 25 through 38, find a quadratic equation with the two given numbers as solutions.

25. 2 and -5
26. 4 and -3
27. -3 and -6
28. -5 and -2
29. $2 + i$ and $2 - i$
30. $1 + 2i$ and $1 - 2i$
31. $-3 + 4i$ and $-3 - 4i$
32. $-5 + i$ and $-5 - i$
33. $1 + i\sqrt{5}$ and $1 - i\sqrt{5}$
34. $3 + i\sqrt{2}$ and $3 - i\sqrt{2}$
35. $-5 + 2i\sqrt{3}$ and $-5 - 2i\sqrt{3}$
36. $-6 + 3i\sqrt{5}$ and $-6 - 3i\sqrt{5}$
37. $4 + \sqrt{7}$ and $4 - \sqrt{7}$
38. $5 + \sqrt{3}$ and $5 - \sqrt{3}$

For Problems 39 through 52, factor the given polynomial into *linear* factors.

39. $x^2 - 2x + 5$
40. $x^2 - 4x + 5$
41. $x^2 + 10x + 29$
42. $x^2 + 2x + 10$
43. $x^2 - 6x + 11$
44. $x^2 + 4x + 7$
45. $x^2 - 10x + 22$
46. $x^2 - 6x + 4$
47. $x^2 + 16$
48. $4x^2 + 25$
49. $9x^2 + 121$
50. $36x^2 + 49$
51. $49x^2 + 1$
52. $x^2 + 1$
53. $3x^2 - 4x + 10$
54. $5x^2 + 2x + 7$
55. $25x^2 + 10x + 101$
56. $9x^2 - 12x + 229$
57. Substitute each solution for the equation in Problem 1 into the equation, and thus show that these numbers actually *do* satisfy the equation.

58. Repeat Problem 57 for the equation in Problem 3.

59. Solve the equation you got as the answer to Problem 29, and thus show that the solutions really *are* $2 + i$ and $2 - i$.

60. Repeat Problem 59 for the equation from Problem 31.

61. Multiply together the factors for $x^2 - 2x + 5$ (Problem 39), and thus show that the product actually *is* $x^2 - 2x + 5$.

62. Repeat Problem 57 for $x^2 + 10x + 29$ (Problem 41).

63. *Sum and Product of the Solutions.* Prove directly from the Quadratic Formula that if s_1 and s_2 are the solutions of the equation $ax^2 + bx + c = 0$, then

$$\boxed{s_1 + s_2 = -\frac{b}{a}} \quad \text{and} \quad \boxed{s_1 s_2 = \frac{c}{a}}.$$

64. *Equations with Imaginary-Number Coefficients.*

a. Use the quadratic formula to solve

  i. $x^2 + 3ix - 2 = 0$,

  ii. $ix^2 - 2x - 3 = 0$.

b. Are complex solutions of quadratic equations *always* conjugates of each other? Explain why your answer does *not* contradict the conclusion of this section concerning complex solutions of quadratic equations.

10-5 GRAPHS OF HIGHER DEGREE FUNCTIONS – SYNTHETIC SUBSTITUTION

You are now ready to return to the problem of graphing higher degree functions. The first thing you need is a rapid way to evaluate polynomials so that you can get plotting data quickly. A clever factoring scheme shows how this can be done. For example,

$$P(x) = 3x^4 - 19x^3 - 21x^2 + 51x - 14 \qquad \text{Given polynomial.}$$

$$= (3x - 19)x^3 - 21x^2 + 51x - 14 \qquad \text{Factoring out } x^3.$$

$$= ((3x - 19)x - 21)x^2 + 51x - 14 \qquad \text{Factoring out } x^2.$$

$$= (((3x - 19)x - 21)x + 51)x - 14 \qquad \text{Factoring out } x.$$

The polynomial is said to be in "nested form" since sets of parentheses are nested inside other parentheses.

Suppose, now, that you want to find $P(2)$. Starting with the innermost parentheses, you can use the following sequence of steps:

Start with 3	Start with the highest-degree coefficient.
$3 \times 2 = 6$	Multiply by x.
$6 - 19 = -13$	Add the next coefficient, -19.
$-13 \times 2 = -26$	Multiply the answer by x.
$-26 - 21 = -47$	Add the next coefficient, -21.
$-47 \times 2 = -94$	Multiply the answer by x.
$-94 + 51 = -43$	Add the next coefficient, 51.
$-43 \times 2 = -86$	Multiply the answer by x.
$-86 - 14 = -100$	Add the next coefficient, -14.

The answer is $P(2) = -100$. The virtue of this method is that the same pair of steps, "Multiply by x, then add the next coefficient," is done repeatedly. Such repetitive steps are easy to do on a calculator or to program into a computer, as you will see in Problems 44 and 45 of the following exercise.

There is a convenient way to arrange the steps for pencil-and-paper calculation, too. First, you write the coefficients of $P(x)$, and the value of x to be substituted.

```
 ↙ Value substituted for x
2| 3  -19  -21  51  -14  ←— coefficients of P(x)
 ______________________  ←— space for computation
```

You bring down the first coefficient, 3, multiply it by x, 2, write the answer, 6, under the -19 and *add*. The process is hard to *say* but easy *to do*!

```
2| 3  -19  -21  51  -14
   ↓    6   Add
 ______________________
   3  -13
 Multiply
```

The other steps are done in the same way. The completed computation looks like this:

```
2| 3  -19  -21   51  -14
         6  -26  -94  -86 ←— x times previous answer
 ________________________
   3  -13  -47  -43  -100
                      └─ P(2)
```

This process is called *synthetic substitution.*

The synthetic substitution process is essentially the same as the long division process. Dividing $P(x)$ by $(x - 2)$ gives:

Quotient: $3x^3 - 13x^2 - 47x - 43$,
Remainder: -100.

The remainder and the coefficients of the quotient show up in the synthetic substitution process.

```
2| 3  -19  -21   51   -14
         6  -26  -94   -86
 _________________________
   3. -13  -47  -43  -100
   \_________________/ \___/
   Coefficients of    Remainder
      Quotient
```

For this reason, synthetic substitution is sometimes called "synthetic division." The relationship between the two processes leads to the following major theorem:

Remainder Theorem: If $P(x)$ is a polynomial, then $P(b)$ is equal to the *remainder* when $P(x)$ is divided by $x - b$.

To see *why* this is true, suppose that $P(x)$ is divided by $(x - b)$, giving a quotient $Q(x)$ and a remainder R. Then

$$P(x) = (x - b) \cdot Q(x) + R.$$

Substituting b for x gives

$$P(b) = (b - b) \cdot Q(b) + R$$

$$P(b) = 0 \cdot Q(b) + R$$

$$P(b) = R.$$

The Remainder Theorem can be used as a lemma to prove the Factor Theorem of Section 7-6, as you will see in Problem 43 in the following exercise.

Armed with the synthetic substitution technique, you are ready to tackle the job of graphing higher degree functions.

Objectives:

1. Plot the graph of a given higher degree function by using synthetic substitution to calculate plotting data.

2. Find all values of x that make a polynomial $P(x)$ equal zero.

Example 1: Plot the graph of $P(x) = x^3 - 4x^2 - 5x + 14$, and find all values of x that make $P(x) = 0$.

To draw the graph, you must calculate many points. The table below shows a compact way to do many substitutions into the same polynomial. See if you can figure out how it works! The graph is shown in Figure 10-5a.

x	1	−4	−5	P(x) 14
−3	1	−7	16	−34
−2	1	−6	7	0
−1	1	−5	0	14
0	1	−4	−5	14
1	1	−3	−8	6
2	1	−2	−9	−4
3	1	−1	−8	−10
4	1	0	−5	−6
5	1	1	0	14
6	1	2	7	56

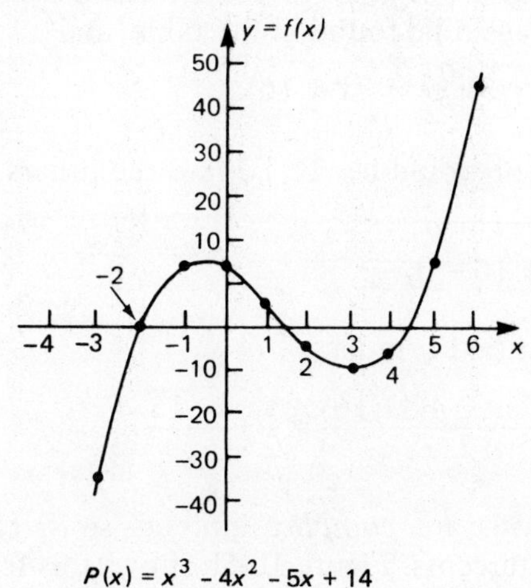

Figure 10-5a.

To find the values of x that make $P(x) = 0$ (that is, the x-intercepts), you have already found that $P(-2) = 0$. By the Factor Theorem, you know that $(x + 2)$ is a factor of $P(x)$. The coefficients of the quotient appear in the synthetic substitution process. Therefore,

$$P(x) = (x + 2)(x^2 - 6x + 7).$$

Setting $P(x) = 0$ gives

$(x + 2)(x^2 - 6x + 7) = 0$

$x + 2 = 0$ or $x^2 - 6x + 7 = 0$ Multiplication Property of 0.

$x = -2$ or

$x = \dfrac{6 \pm \sqrt{36 - 4(1)(7)}}{2}$ Solving for x.

$x = -2$ or $x = 3 \pm \sqrt{2}$ Simplifying.

$x = \underline{\underline{-2}}$ or $x \approx \underline{\underline{4.414, 1.586}}$ $\sqrt{2} \approx 1.414$.

You can see from Figure 10-5a that the graph actually does cross the x-axis at these three points.

Example 2: Plot the graph of $P(x) = x^3 - 4x^2 - 5x + 48$, and find all values of x that make $P(x) = 0$.

Repeating the procedure of Example 1, you will get a table of values which can be plotted as shown in Figure 10-5b. It turns out the $P(-3) = 0$, so that

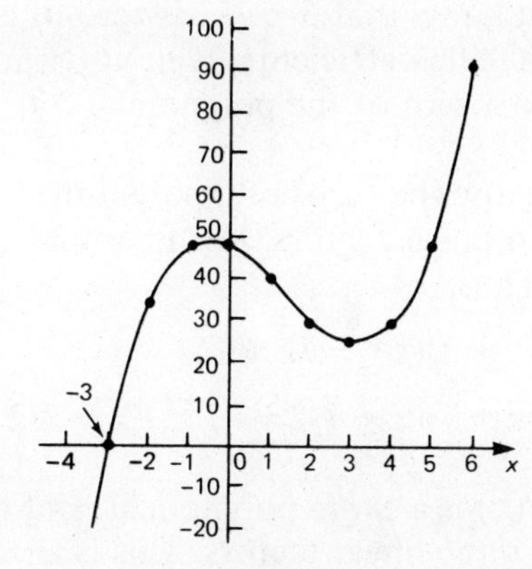

Figure 10-5b.

$(x + 3)$ is a factor of $P(x)$. The coefficients of the quotient can again be found in the table, and

$$P(x) = (x + 3)(x^2 - 7x + 16).$$

One x-intercept would be -3. To find the others you let

$$x^2 - 7x + 16 = 0.$$

By the Quadratic Formula,

$$x = \frac{7 \pm \sqrt{49 - 4(1)(16)}}{2} = \frac{7}{2} \pm \frac{i\sqrt{15}}{2}$$

These values of x are *complex* numbers, so there are *no* other x-intercepts. Figure 10-5b shows this feature. However, these numbers *do* make $P(x)$ equal zero, and this fact leads to a more general name.

DEFINITION

A ***zero*** of a polynomial $P(x)$ is a value of x, real or complex, which makes $P(x) = 0$.

The process of finding zeros is illustrated above. As with quadratics, complex zeros of a polynomial with real coefficients always come in conjugate pairs. Thus, if you know that $5 + 7i$ is a zero of a polynomial with real coefficients, then you know that $5 - 7i$ is also a zero of the polynomial.

Once you know the zeros of a polynomial, you can factor it into linear factors. For Example 2, $P(-3) = 0$. Thus,

$$P(x) = (x + 3)(x^2 - 7x + 16)$$
$$= (x + 3)(x - 7/2 - i\sqrt{15})(x - 7/2 + i\sqrt{15}).$$

Note that $P(x)$ is a *cubic* polynomial, and that it has exactly *three* linear factors. This is an example of an important algebraic theorem.

Theorem: If $P(x)$ is an nth degree polynomial, then $P(x)$ has exactly n linear factors.

This means that $P(x)$ has exactly n zeros, although polynomials such as

$$P(x) = (x - 7)^3 (x - 4)^2$$

would have the zeros 7 and 4 counted 3 and 2 times, respectively. This theorem is a corollary of the Fundamental Theorem of Algebra.

Fundamental Theorem of Algebra: A polynomial $P(x)$ has at least *one* zero, if you allow zeros to be complex numbers.

The exercise which follows is designed to give you practice in analyzing higher degree polynomial functions by finding zeros and intercepts and plotting graphs.

Exercise 10-5

For Problems 1 through 18,

a. Plot the graph of the function by calculating points for the given x-values. You may write a computer program to find points, as in Problem 44.

b. Find *all* the zeros, real and complex. Those not discovered in part a may be found by factoring $P(x)$. (Real zeros may be found by the computer program in Problem 46.)

1. $P(x) = x^3 - 4x^2 + x + 6$, $\quad -3 \le x \le 5$
2. $P(x) = x^3 - x^2 - 4x + 4$, $\quad -4 \le x \le 4$
3. $P(x) = x^3 - 7x^2 + 11x + 3$, $\quad -2 \le x \le 6$
4. $P(x) = x^3 - x^2 - 10x + 10$, $\quad -4 \le x \le 5$
5. $P(x) = x^3 + x^2 - 4x + 6$, $\quad -4 \le x \le 3$
6. $P(x) = x^3 + x^2 - 7x - 15$, $\quad -4 \le x \le 4$
7. $P(x) = x^3 + 2x^2 - 4x - 8$, $\quad -4 \le x \le 3$
8. $P(x) = x^3 - 9x^2 + 24x - 16$, $\quad -1 \le x \le 7$
9. $P(x) = x^3 - 6x^2 + 12x + 19$, $\quad -2 \le x \le 5$
10. $P(x) = x^3 - 6x^2 + 12x - 7$, $\quad -2 \le x \le 5$
11. $P(x) = -2x^3 - 3x^2 + 8x + 12$, $\quad -4 \le x \le 3$

12. $P(x) = 3x^3 - x^2 - 40x + 48, \quad -5 \le x \le 4$
13. $P(x) = 2x^3 + 4x^2 - 17x - 39, \quad -4 \le x \le 4$
14. $P(x) = -2x^3 + 2x^2 + 7x - 10, \quad -3 \le x \le 4$
15. $P(x) = x^4 - x^3 - 11x^2 + 9x + 18, \quad -4 \le x \le 4$
16. $P(x) = x^4 + x^3 - 5x^2 + x - 6, \quad -4 \le x \le 3$
17. $P(x) = x^4 - 5x^3 + 2x^2 + 22x - 20, \quad -3 \le x \le 5$
18. $P(x) = -x^4 - 4x^3 + 12x^2 + 44x - 51, \quad -5 \le x \le 4$

For Problems 19 through 30, find all zeros of the given polynomial.

19. $x^3 + x^2 - x + 15$
20. $x^3 - 6x^2 + 13x - 10$
21. $x^3 + 3x^2 - 6x - 8$
22. $x^3 + 4x^2 + x - 6$
23. $x^3 - 4x^2 + 2x + 4$
24. $x^3 - 5x^2 + 5x + 3$
25. $x^4 + x^3 - 6x^2 - 14x - 12$
26. $x^4 - 8x^2 - 8x + 15$
27. $x^4 + x^3 - 7x^2 - x + 6$
28. $x^4 + 4x^3 - 3x^2 - 14x - 8$
29. $x^4 - 8x^3 + 27x^2 - 38x + 26$

Clues: One of the zeros is $3 + 2i$.
Complex zeros come in conjugate pairs.
It is possible to divide a *quadratic* into a quartic.

30. $x^4 - 12x^3 + 56x^2 - 120x + 100$

Clue: $3 + i$ is one zero.

For Problems 31 through 36, use the Remainder Theorem to find the remainder *quickly* when the polynomial on the left is divided by the linear binomial on the right. You may use synthetic substitution or *direct* substitution, whichever seems more efficient.

31. $2x^3 - 5x^2 + 11x + 6$ by $x - 4$
32. $3x^3 + 7x^2 - 12x + 1$ by $x + 3$
33. $x^5 - 3x^2 + 14$ by $x + 2$
34. $x^4 - 10x^2 + 9$ by $x - 3$
35. $x^{51} + 51$ by $x + 1$
36. $x^{2000} + 2000$ by $x - 1$.

For Problems 37 through 42, use what you have observed about the graphs of higher degree functions and what you know about the Fundamental Theorem of Algebra and its corollary, to sketch graphs of the functions described.

37. Quintic function with exactly 3 real zeros.

38. Sixth degree function with exactly 4 real zeros.

39. Cubic function with exactly two distinct real zeros.

40. Quartic function with no real zeros.

41. Cubic function with no real zeros.

42. Quartic function with exactly five real zeros.

43. *Proof of the Factor Theorem.* Use the Remainder Theorem to prove the Factor Theorem. That is, prove that a polynomial $P(x)$ has a linear factor of the form $(x - b)$ if, and only if, $P(b) = 0$. Remember that there are *two* parts to the proof, an "if" part and an "only if" part!

44. *Synthetic Substitution by Computer.* Write a computer program to carry out repeatedly the steps

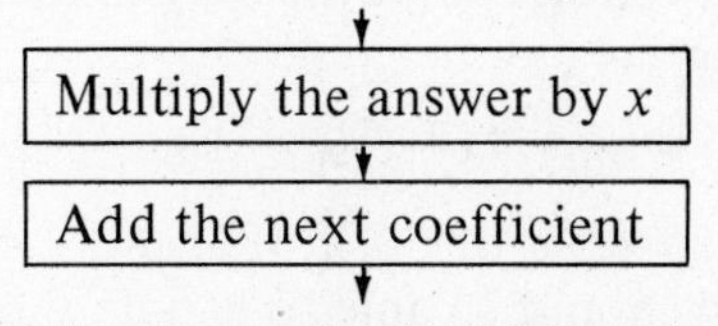

which appear in the synthetic substitution process. The input should be the value of x to be substituted, and the output should be the value of $P(x)$. The coefficients may be stored in a DATA statement and read one at a time as they are needed, using a READ statement. You should also read the *degree* of the polynomial, and use a counter to stop the process when the last coefficient has been read.

45. *Synthetic Division by Computer.* Modify the program in Problem 43 so that the computer prints out the coefficients of the *quotient* polynomial as well as the values of x and $P(x)$.

46. *Zeros by Computer.* If none of the zeros of a higher degree function is rational, it is quite complicated to find the exact values of the zeros. There is a "cubic formula" and a "quartic formula" which you can look up in handbooks such as *CRC Tables.* But they are too cumbersome to be practical. For fifth or higher degree functions, it can be shown that there is *no* general formula for finding zeros, so you seek a technique for finding *decimal approximations* of irrational zeros.

a. Write a computer program which searches for zeros of a function. The program should begin with inputting the coefficients of $P(x)$. Then x should be started at some low value, L, and increased by 1 until a high value, H, is reached. At each step, the computer should see if there is a zero between x and $x + 1$ by checking to see if $P(x)$ and $P(x+1)$ have *opposite* signs. You can do this by checking to see whether $P(x)*P(x+1)$ is *negative.* When a zero has been located, the computer should explore the interval between x and $x+1$ by increments of 0.1, looking for a sign change. Then it should proceed by increments of 0.01, and so forth, until the zero has been determined with the desired amount of accuracy.

b. Test your program on the polynomial of Example 1,

$$P(x) = x^3 - 4x^2 - 5x + 14,$$

which has zeros of -2, 1.586, and 4.414. Your program should be able to handle the *integer* zeros as well as the irrational ones!

c. For the following functions, find all real zeros that lie between -10 and 10.

i. $P(x) = x^3 - 4x^2 + x + 5$

ii. $P(x) = x^3 - 7x^2 + 11x + 3$

iii. $P(x) = x^3 - 9x^2 + 24x - 29$

iv. $P(x) = x^4 - x^3 - 11x^2 + 9x + 10$

v. $P(x) = x^4 + x^3 - 5x^2 + x - 10$

vi. $P(x) = x^4 + 17x^2 + 2x + 30$

d. Explain why your program might *miss* some zeros if they are too close together.

47. *Was Figure 10-2 Really Accurate?* The graph of

$$g(x) = x^3 - 4x^2 - 5x + 24$$

sketched in Figure 10-2 seems to have a vertex on the x-axis. Based on what you now know, was that figure accurate? Explain.

10-6 HIGHER DEGREE FUNCTIONS AS MATHEMATICAL MODELS

Higher degree functions are used as models in two basically different ways. A function may turn out to be polynomial based on theoretical considerations. For example, the shape into which a loaded beam bends is the graph of a polynomial function. The particular function is determined by the way the weight is distributed along the beam and the manner in which the beam is supported.

The second way polynomial functions are applied is to *start out* by assuming that a polynomial function is a reasonable model, and then fit the polynomial graph to measured experimental data points. A model created in this way is called an "empirical" model. For example, if you assume a *cubic* function, then you have assumed an equation of the form

$$y = ax^3 + bx^2 + cx + d,$$

where a, b, c, and d stand for constants. Fitting the model to the data requires finding values of these constants by substituting the values of (x, y) and solving the resulting system of linear equations for a, b, c, and d. This is the same procedure you used for quadratic functions in Section 5-7 and 5-8.

Objective:

Given a real-world situation in which one variable depends on another by a cubic or higher degree function, find the particular equation and use it as a mathematical model.

Since the technique of finding the particular equation is a familiar one, no specific examples are presented here. If you need a refresher on how to do it, see Sections 5-7 and 5-8.

The exercise which follows contains problems of each of the above types. There are also problems in which polynomial functions are used as mathematical models of the *mathematical* world.

Exercise 10-6

1. ***Beam Deflection Problem.*** A horizontal beam 10 meters long has its left end built into a wall, and its right end resting on a support, as shown in Figure 10-6a. The beam is loaded with weight uniformly distributed along its length. As a result, the beam sags downward according to the equation

$$y = -x^4 + 25x^3 - 150x^2$$

where x is the number of meters from the wall to a point on the beam, and y is the number of hundredths of a millimeter from the x-axis to the beam.

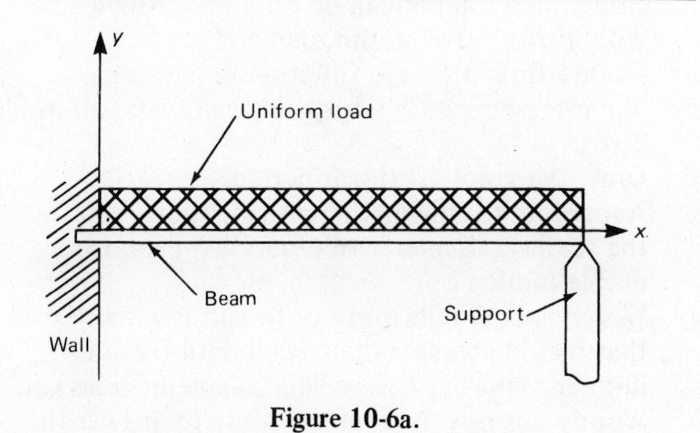

Figure 10-6a.

a. What is an appropriate domain for x?
b. Find the zeros of this function and tell what they represent in the real world.
c. Using all integer values of x in the domain, plot a graph of this function.

2. ***Diving Board Problem.*** When you stand on a diving board (see Figure 10-6b), the amount the board bends, y, below its rest position is a cubic function of x, the distance from the built-in end to the point on the board. Suppose that you measure the following deflections:

x (ft.)	y (thousandth of an inch)
0	0
1	116
2	448
3	972

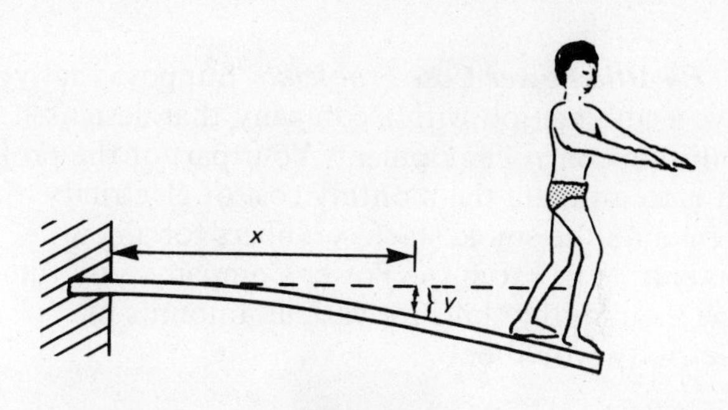

Figure 10-6b.

a. Derive the particular equation expressing y in terms of x.
b. The board is 10 feet long. How far does its tip sag below the horizontal?

3. ***Oil Viscosity Problem.*** The viscosity, or "stickiness" of normal motor oil you use in your car decreases as its temperature increases. The "all-weather" motor oils, however, retain a relatively constant viscosity throughout their range of operating temperatures. Suppose that a chemical lab has found the following viscosities for an all-weather motor oil:

Temperature, °F	Viscosity
100	54
200	50
300	52
400	54

Let V be the number of units of viscosity, and T be the number of *hundreds* of degrees (i.e., $T = 1$ means temperature is 100°). Assume that a *cubic* function is a reasonable model of how V varies with T.

a. Find the particular equation expressing V in terms of T.
b. Predict the viscosity at 0°, 500°, and 600°.
c. Plot a graph of V versus T in the domain $0 \leq T \leq 6$.
d. Does $V = 0$ for any value of T in this domain? Justify your answer.

4. ***Electric Power Cost Problem.*** Suppose that you have a summer job with a company that designs pollution control equipment. Your part of the project is to estimate the monthly cost of electricity to operate the smokestack scrubbers for a new cement plant. From the Power Company, you find that the monthly bills for various amounts of electricity would be

Kilowatt hours (kwh)	Dollars
1,000,000	$20,000
2,000,000	29,000
3,000,000	34,000
4,000,000	41,000

Since you need the cost of *any* amount of power, from 1,000,000 through 4,000,000 kwh, you need an equation expressing cost in terms of kwh. You decide a *cubic* function is reasonable.

a. Let D = the number of thousands of dollars per month, and let k = the number of millions of kwh per month. Write the particular equation expressing D in terms of k.
b. Predict the cost of 1.5 million kwh.
c. According to your model, how much would you pay if you used *no* electricity in a given month? Is this reasonable? Explain.
d. Your Boss wants to know how much electricity could be used without exceeding $35,000 per month. Draw a graph of D versus k, and use the graph to estimate this number.
e. Use the computer program of Problem 43, Exercise 10-5, to get a better answer to part d, above.

5. ***Lumber Problem.*** Woody Forester has the job of figuring out how much lumber can be obtained from various sizes of monkeypuzzle trees. From sawmill records he finds the following numbers of board-feet of lumber can be cut from trees of the given diameters:

Diameter (feet)	Lumber (board-feet)
1	10
2	99
3	324
4	745

He figures that since board-feet is a cubic measure, a *cubic* function would be a reasonable mathematical model.

a. Find the particular equation expressing board-feet in terms of diameter.
b. How much lumber can be obtained from a tree with a trunk five feet in diameter?
c. Woody finds that the function in part a has *one* integer zero. What is that zero? Find all other zeros.
d. Draw the graph of this function.
e. According to this mathematical model, what is the smallest diameter tree that will produce usable lumber?
f. Woody's boss tells him not to cut down any tree that would give less than 200 board-feet of lumber. Approximately what diameter trees can Woody cut down? There is a way to answer this question with almost *no* more work!

6. ***Payload Problem.*** The number of kilograms of "payload" an airplane can carry equals the number of kilograms the wings can lift *minus* the mass of the airplane itself and the mass of the flight crew and their equipment. The lift is directly proportional to the *square* of the plane's length since lift depends on wing *area*. The plane's mass varies directly with the *cube* of its length, since mass depends on *volume*. The mass of the crew and their equipment is *constant*, and does *not* depend on the plane's length.

a. Let L = number of meters long the plane is. Let $P(L)$ = number of kilograms of payload the plane can carry. Write the general equation expressing $P(L)$ in terms of L.
b. Find the particular equation if a plane 20 meters long can lift 2000 kilograms and has a mass of 800 kilograms. The mass of the flight crew and their equipment is 400 kilograms.
c. Calculate $P(L)$ for each 5 meters from $L = 10$ through $L = 50$.
d. Since P is a *cubic* function, it has *three* zeros. Find the zeros, then tell what each one represents in the real world.
e. Use the results of parts c and d to plot the graph of P in a suitable domain.

7. ***Sum of the Squares Problem.*** Let $S(n)$ be the *sum* of the *squares* of the integers from 0 through n. That is,

$$S(n) = 0^2 + 1^2 + 2^2 + 3^2 + \ldots + n^2.$$

a. Find $S(0)$, $S(1)$, $S(2)$, and $S(3)$.
b. $S(n)$ is a *cubic* function of n. Find the particular equation expressing $S(n)$ in terms of n using the ordered pairs from part a.
c. The coefficients in the particular equation are fractions. Factor out the appropriate fraction leaving a polynomial with *integer* coefficients inside the parentheses, then factor this polynomial. (This is the formula for the sum of the squares that you will find in handbooks such as *CRC Tables.*)
d. Use the equation of part d to find $S(4)$ and $S(5)$. Then show that your answers are correct by actually adding the squares of the integers.
e. Find $S(1000)$.
f. If you have studied mathematical induction (Appendix B), prove that your formula gives the right answer for $S(n)$ for *all* integers $n \geq 0$.

8. ***Sum of the Cubes Problem.*** Let $S(n)$ be the *sum* of the *cubes* of the integers from 0 through n. That is,

$$S(n) = 0^3 + 1^3 + 2^3 + 3^3 + \ldots + n^3.$$

a. Find $S(0)$, $S(1)$, $S(2)$, $S(3)$, and $S(4)$. You should find that all of these numbers are perfect squares!
b. $S(n)$ is a *quartic* function of n. Find the particular equation expressing $S(n)$ in terms of n, using the ordered pairs of part a.
c. The coefficients in the particular equation are fractions. Factor out the appropriate fraction, thus leaving a polynomial with *integer* coefficients inside the parentheses.
d. Factor the polynomial inside the parentheses from part c. From your answer, how can you conclude that $S(n)$ is *always* a perfect square, no matter what positive integer n is?
e. Find $S(1000)$. Which is quicker for this calculation—using the formula you have derived, or simply adding up all the cubes of the integers?
f. If you have studied mathematical induction (Appendix B), prove that your formula works for *all* integers $n \geq 0$.

10-7 CHAPTER REVIEW AND TEST

In this chapter you have investigated higher degree polynomial functions. Synthetic substitution allowed you to calculate plotting data quickly. The Fundamental Theorem of Algebra let you conclude that an nth degree function always has exactly n zeros, if you are willing to use complex numbers. You also learned that the familiar Factor Theorem is really just a corollary of the more general Remainder Theorem. Finally, you saw several examples in which higher degree functions are reasonable mathematical models.

The objectives for the chapter may be summarized as follows:

1. *Operate with complex numbers.*
2. *For a given polynomial function, $P(x)$,*
   a. *calculate plotting data and draw the graph,*
   b. *find the zeros by solving the equation $P(x) = 0$,*
   c. *factor $P(x)$ into linear factors.*
3. *Use higher degree functions as mathematical models.*

The Review Problems test your ability to do these things one at a time. The Concepts Test is intended to let you find out how well you can apply one or more of these concepts to new situations.

Review Problems

The following problems are numbered according to the three objectives stated above.

Review Problem 1. Perform the indicated operations and simplify:

a. i^{59} b. i^{-15}

c. i^{500} d. $\sqrt{-18} \times \sqrt{-24}$

e. $\sqrt{(-5)(-35)}$ f. $\dfrac{\sqrt{98}}{\sqrt{-2}}$

g. $\sqrt{-12} + \dfrac{12}{\sqrt{-3}}$ h. $(8 - 2i) - (3 - 11i)$

i. $(12 + i)(12 - i)$ j. $(12 + i)^2$

k. $\dfrac{12 + 9i}{3 - 4i}$

Review Problem 2a. If $P(x) = -2x^3 - x^2 + 6x + 8$, plot the graph of P by finding $P(x)$ for each integer value of x from -3 through 3.

b. i. Solve the equation $2x^2 + 5x + 4 = 0$.
ii. Find a quadratic equation with real coefficients if one of its solutions is $3 + 4i$.
iii. You have learned that if s_1 and s_2 are solutions of a quadratic equation $ax^2 + bx + c = 0$, then $s_1 + s_2 = -\dfrac{b}{a}$. If s_1, s_2, and s_3 are solutions of a *cubic* equation $ax^3 + bx^2 + cx + d$, prove that

$$s_1 + s_2 + s_3 = -\frac{b}{a}.$$

Then find the sum of the solutions of the equation

$$5x^3 + 11x^2 - 13x + 47 = 0.$$

c. Factor the following over the set of complex numbers:
i. $x^2 - 4x + 5$
ii. $25x^2 + 1$

Review Problem 3. A cubic function $P(x)$ contains the ordered pairs $(x, y) = (-1, -10)$, $(2, -7)$, $(3, 2)$, and $(4, 45)$.

a. Find the particular equation.
b. Find $P(-2)$.
c. $P(x)$ has an integer zero. Find that zero.

Concepts Test

Work the following problems. For each part of each problem, tell by number and letter which one(s) of the three objectives of this chapter you used. If you are applying your knowledge to a new concept, write the words, "new concept."

Concepts Test 1. Sketch the graph of

a. a seventh degree function,
b. a quartic function with two distinct real zeros and two complex zeros,
c. a polynomial function with two real zeros at the same point.

Concepts Test 2. Prove that

$$\frac{a - bi}{b + ai} = -i.$$

Concepts Test 3. If $x^7 - 31$ were divided by $x - 2$, what would the remainder be? What theorem allows you to do this *quickly?*

Concepts Test 4. ***Catastrophe Theory Problem*** – According to "catastrophe theory" (see for example *Scientific American,* April, 1976), when a person is under certain kinds of stress, the amount of food eaten is *not* a simple function of how hungry he or she is (see Figure 10-7). As hunger increases, food consumption increases slightly, but not enough to satisfy the hunger. Then at a certain point a "catastrophe" happens, and the person starts "gorging" himself or herself. The hunger decreases, but remains high until a "reverse catastrophe" happens, and the person starts "fasting" again.

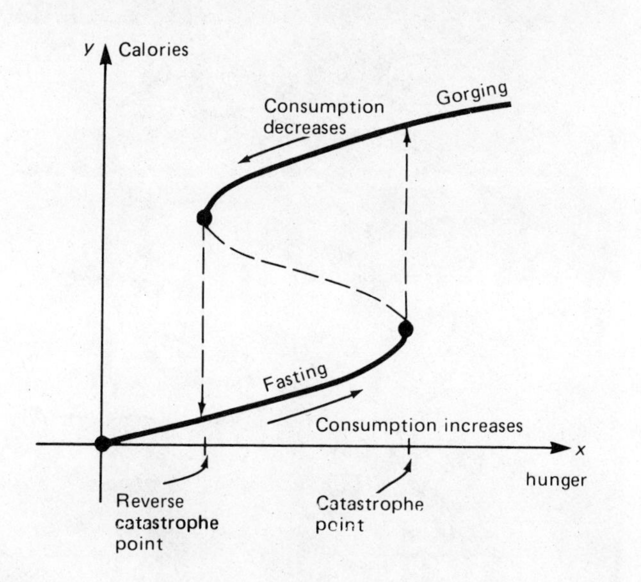

Figure 10-7.

Suppose that Juanita Lott is under this kind of stress. When her hunger is 16 units, she eats only 1000 Calories per day. At the catastrophe point, 20 hunger units, she jumps from 2000 to 5000 Calories per day. The point (0, 0) is also on the graph.

Let x be the number of units of hunger.
Let y be the number of *thousands* of Calories per day.

a. Write the given information as *four* ordered pairs, (x, y).

b. The graph looks like the *inverse* of a cubic function. That is, the general equation would be

$$x = ay^3 + by^2 + cy + d.$$

Based on this assumption, find the particular equation.

c. Explain why this relation is *not* a function.

d. Plot an accurate graph of this function by picking values of y and calculating the corresponding values of x. Use values of y from 0 through 6. You can check the accuracy of your equation by making sure that the given ordered pairs satisfy it.

e. Based on your graph, where does the reverse catastrophe seem to happen? Demonstrate that you are right by showing that there are *exactly two* distinct values of y for this value of x.

f. What is the *range* of this relation?

g. Find the two values of y in the range for which $x = 18$.

h. Show that there is only *one* real value of y when $x = 0$.

i. Juanita's doctor prescribes therapy that relieves the stress. Thereafter, her hunger and food consumption are related by the equation.

$$x = y^3 - 6y^2 + 20y - 15.$$

Plot the graph of this relation. Use values of y from 0 through 4. Does there seem to be a catastrophe under these conditions?

11 Sequences and Series

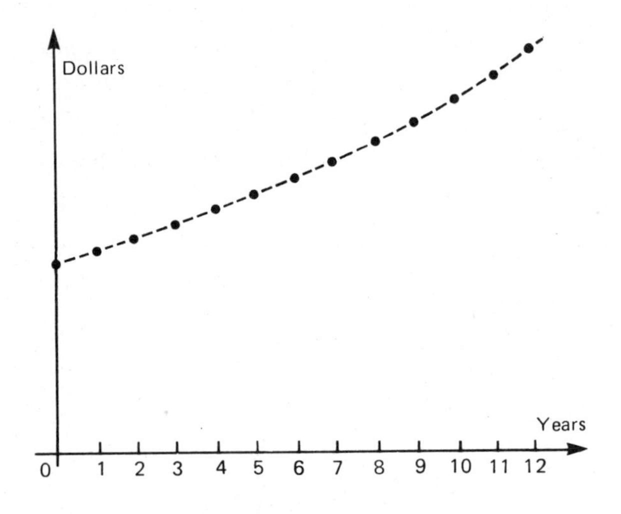

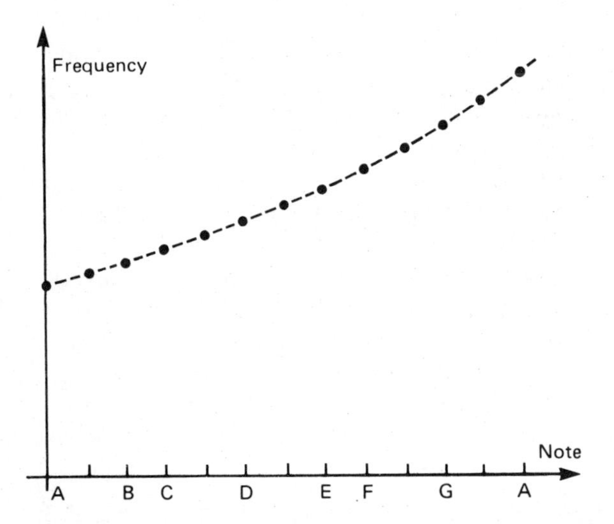

The functions you have studied so far are "continuous" functions. That is, the points can be connected with straight lines or smooth curves, except, perhaps, for a few holes or asymptotes. In this chapter you will study **sequences**, *in which the independent variable jumps from integer to integer, with no values in between. You will find that the same kind of sequence can be used to model things ranging from the amount of money in a savings account to the frequency of notes on a piano.*

11-1 INTRODUCTION TO SEQUENCES

The numbers 3, 5, 7, ..., seem to follow a pattern. Such a set of numbers is commonly called a "sequence," and each of the numbers in the set is called a "term" of the sequence. If you know what the pattern is, you can replace the ellipsis (the "..." punctuation mark) with other terms of the sequence. For example, you might write

3, 5, 7, 9, 11, 13, 15, 17, 19, 21, ...,

assuming that this is the sequence of odd integers beginning with 3. However, it could also be

3, 5, 7, 11, 13, 17, 19, 23, ...,

which is the sequence of odd primes. So you will be asked to discover "a" pattern rather than "the" pattern.

Sequences can be fit into the framework of what you have learned so far by considering them to be *functions*. Each term in a sequence has a *position* or *term number* (1st, 5th, 976th, etc.), and a *value* (3, 11, 1953, etc.). The sequence of odd integers above can be written

term *value*:	3	5	7	9	11	13	15	...,
term *number*:	1	2	3	4	5	6	7	....

For each term *number* there is a unique term *value*. Thus, a sequence can be thought of as a *function*.

DEFINITION

A sequence is a function whose domain is the set of natural numbers (the term numbers), and whose range is the set of term values.

Notes:

1. This definition implies that a sequence has an *infinite* number of terms.

2. The letter n will usually be used for the term number, and the letter t for the function, so that the term value will be $t(n)$. The symbols t_n or a_n are sometimes used for the term values.

Objective:

Given the first few terms of a sequence,

a. discover a pattern,
b. write a few more terms of the sequence,
c. get a formula for $t(n)$,
d. use the formula to calculate other term values, and
e. draw a graph of the sequence.

Example: Do the above five things for the sequence

3, 5, 7, 9, 11, 13, 15, ...

a. An obvious pattern is that the terms increase by 2 each time.

b. Using this pattern, the next few terms are

17, 19, 21, 23, ...

c. Discovering a formula is tricky and will test your ingenuity. What you must do is find some link between the term *number*, n, and the term *value*, $t(n)$. It helps to write the two sets of numbers close to each other.

3,	5,	7,	9,	11,	13,	15,	...	Term *values*.
①	②	③	④	⑤	⑥	⑦	...	Term *numbers*.

In this case the term values seem to be increasing *twice* as fast as the term numbers. If you write the values of $2n$ by the values of $t(n)$, a pattern shows up.

3,	5,	7,	9,	11,	13,	15,	...	← $t(n)$
1	2	3	4	5	6	7		← n
2	4	6	8	10	12	14		← $2n$

The value of $t(n)$ is always *one more than* the value of $2n$. So a formula would be

$$t(n) = 2n + 1.$$

d. If you must calculate $t(n)$ for a *large* value of n, you would use the formula rather than continuing the pattern. For example, if n were 256, then

$$t(256) = 2 \cdot 256 + 1$$
$$= 513.$$

e. The graph may be plotted pointwise, as shown in Figure 11-1a. The dots should *not* be connected with a solid line since the domain contains only *integers*.

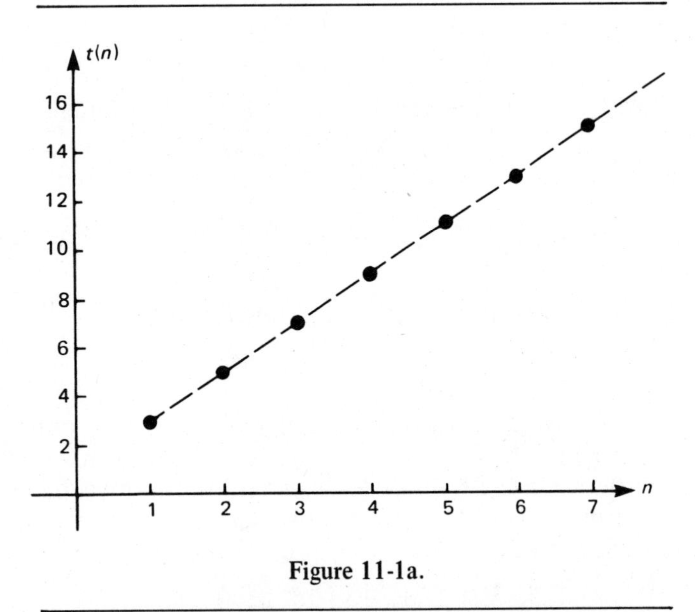

Figure 11-1a.

Not all sequence graphs are linear. For example, the graph of 24, 12, 6, 3, $1\frac{1}{2}$, ..., is shown in Figure 11-1b.

The exercise which follows is designed to give you practice in accomplishing the objectives of this section.

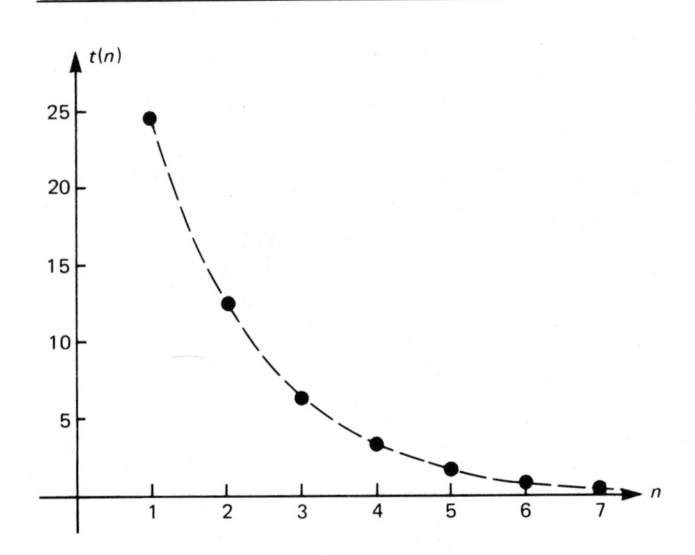

Figure 11-1b.

Exercise 11-1

In Problems 1 through 12, $t(1)$ through $t(6)$ of a sequence are listed. For each sequence, do the following:

a. Draw the graph.
b. Find $t(7)$ and $t(8)$.
c. Figure out a formula for $t(n)$.
d. Calculate $t(100)$.

1. 1, 1/2, 1/3, 1/4, 1/5, 1/6, ...
2. 1, 4, 9, 16, 25, 36, ...
3. 1/3, 1/5, 1/7, 1/9, 1/11, 1/13, ...
4. 1/2, 3/4, 7/8, 15/16, 31/32, 63/64, ...
5. 3, 4, 5, 6, 7, 8, ...
6. 3, 6, 9, 12, 15, 18, ...
7. 3, 6, 12, 24, 48, 96, ...
8. 2, 6, 18, 54, 162, 486, ...

9. 32, −16, 8, −4, 2, −1, ...

10. 1, −1, 1, −1, 1, −1, ...

11. 1, 3/2, 5/4, 7/8, 9/16, 11/32, ...

12. 0, 13/8, 26/27, 39/64, 52/125, 65/216, ...

For Problems 13 through 20, write the next two terms of the sequence, and tell what pattern you used.

13. 1, 2, 4, 7, 11, 16, ...

14. 1, 11, 20, 28, 35, 41, ...

15. 1, 1, 2, 3, 5, 8, 13, ... (A *Fibonacci* sequence.)

16. 1, 2, 6, 24, 120, 720, ... (The sequence of *factorials*.)

17. 1, 3, 6, 10, 15, ... (The sequence of *triangular* numbers.)

18. 1/12, 1/19, 1/26, 1/33, ... (A *harmonic* sequence.)

19. 1, 1, 2, 2, 3, 4, 4, 8, 5, 16, 6, 32, ...

20. 100, 100, 75, 50, 50, 25, 25, ...

21. The sequence of letters *o, t, t, f, f, s, s,* ..., has a mathematical pattern. Demonstrate that you have figured out this pattern by writing the next four letters in the sequence.

11-2 ARITHMETIC AND GEOMETRIC SEQUENCES

Two types of sequences are of special interest mathematically because the formulas are easily derived, and of special interest practically because they fit the real world as mathematical models in many situations. Examples are

3, 10, 17, 24, 31, 38, ... , and
3, 6, 12, 24, 48, 96, ...

In the first sequence the next term is formed by *adding* a constant, 7, to the previous term. In the other, the next term is formed by *multiplying* the previous term by a constant, 2.

DEFINITION

An ***arithmetic*** sequence is a sequence in which one term equals a constant *added* to the preceding term.

The constant for an arithmetic sequence is called the *common difference, d,* because the difference between any two adjacent terms equals this constant. In the sequence above,

$$10 - 3 = 7, \quad 17 - 10 = 7, \quad 24 - 17 = 7,$$

and so forth. The word "arithmetic" is used as an adjective here rather than as a noun, and is pronounced with the stress on the "-met-" syllable instead of the "ith-" syllable.

DEFINITION

A ***geometric*** sequence is a sequence in which each term equals a constant *multiplied* by the preceding term.

The constant for a geometric sequence is called the *common ratio*, *r*, because the ratio of one term to the preceding term is equal to this constant. In the sequence above,

$$6/3 = 2, \quad 12/6 = 2, \quad 24/12 = 2,$$

and so forth.

Arithmetic and geometric sequences are sometimes called "progressions" because the terms "progress" from one to the next in a regular manner. The word "progression" is often used when there is a *finite* number of terms. Sequences have infinite numbers of terms.

Objectives:

1. Given the first few terms of a sequence, tell whether it is arithmetic, geometric, or neither.

2. Given a value of n for a specified arithmetic or geometric sequence, find the value of $t(n)$.

3. Given a value of $t(n)$ for a specified arithmetic or geometric sequence, find the value of n.

Example 1: Is the sequence 4, 7, 10, ... arithmetic, geometric, or neither?

The first objective may be accomplished by seeing if adjacent terms have either a common difference or a common ratio. For the sequence

$$4, 7, 10, \ldots,$$

the differences between adjacent terms are

$$t(2) - t(1) = 7 - 4 = 3, \text{ and}$$
$$t(3) - t(2) = 10 - 7 = 3.$$

Since these terms have a *common* difference, the sequence is *arithmetic*. Each term is formed by *adding* 3 to the preceding term. The terms do *not* have a common ratio, since $10/7 \neq 7/4$, so the sequence is *not* geometric.

Example 2: Is the sequence 3, 6, 12, ... arithmetic, geometric, or neither?

For this sequence the differences are

$$t(2) - t(1) = 6 - 3 = 3, \text{ and}$$
$$t(3) - t(2) = 12 - 6 = 6,$$

so the sequence is *not* arithmetic. But the ratios are

$$t(2)/t(1) = 6/3 = 2, \text{ and}$$
$$t(3)/t(2) = 12/6 = 2.$$

Thus, the sequence is *geometric* because adjacent terms have a common *ratio*. Each term is formed by *multiplying* the preceding term by 2.

Example 3: Is the sequence 2, 6, 24, ... arithmetic, geometric, or neither?

For this sequence the differences are

$$t(2) - t(1) = 6 - 2 = 4, \text{ and}$$
$$t(3) - t(2) = 24 - 6 = 18,$$

and the ratios are

$$t(2)/t(1) = 6/2 = 3 \text{ and}$$
$$t(3)/t(2) = 24/6 = 4,$$

so the sequence is neither arithmetic nor geometric. (This turns out to be a "factorial" sequence where the next term is formed by multiplying the preceding term by a larger number each time.)

Formulas for calculating $t(n)$ for arithmetic and geometric sequences can be found by linking the term number to the term value, as you did in the previous section. The arithmetic sequence

$$3, 10, 17, 24, 31, \ldots,$$

has as a first term $t(1) = 3$, and common difference $d = 7$. The first few terms can be constructed by adding 7 to the preceding term.

$$t(1) = 3$$
$$t(2) = 3 + 7$$
$$t(3) = 3 + 7 + 7 = 3 + (2)(7)$$
$$t(4) = 3 + 7 + 7 + 7 = 3 + (3)(7)$$
$$t(5) = 3 + 7 + 7 + 7 + 7 = 3 + (4)(7).$$

This is always one *less* than the term number.

So the pattern consists of adding $(n - 1)$ common differences to the first term, $t(1)$. Thus, the formula is

$$\boxed{t(n) = t(1) + (n - 1)d} \leftarrow \text{For } \textit{arithmetic} \text{ sequences.}$$

where n is the term number and d is the common difference.

Examination of this formula should reveal to you that an arithmetic sequence is nothing more than a cleverly-disguised *linear* function, because the independent variable n appears to the *first* power. The slope of the function is the common difference d, and the y-intercept would be $t(0)$, which equals $t(1) - d$, if zero were in the domain of the function.

The same procedure gives a formula for $t(n)$ for a geometric sequence. The sequence

$$3, 6, 12, 24, 48, \ldots,$$

has $t(1) = 3$ and common ratio $r = 2$. The first few terms can be constructed by multiplying the preceding term by 2.

$$t(1) = 3$$
$$t(2) = 3 \cdot 2$$
$$t(3) = 3 \cdot 2 \cdot 2 = 3 \cdot 2^2$$
$$t(4) = 3 \cdot 2 \cdot 2 \cdot 2 = 3 \cdot 2^3$$
$$t(5) = 3 \cdot 2 \cdot 2 \cdot 2 \cdot 2 = 3 \cdot 2^4$$

Exponent is always one *less* than the term number.

So the pattern consists of multiplying the first term, $t(1)$, by the common ratio, r, $(n - 1)$ times. Thus, the formula is

$$\boxed{t(n) = t(1) \cdot r^{n-1}}$$ ← For *geometric* sequences.

where n is the term number and r is the common ratio.

Notes:

1. Since the independent variable n appears as an exponent, a geometric sequence is actually just an example of an *exponential* function, which you studied in Chapter 6. The only difference is that the domain of a geometric sequence is positive integers rather than all real numbers.
2. The formulas for arithmetic and geometric sequences are exactly alike. The only difference is the *operation* that is performed. For arithmetic sequences, $(n - 1)$ common differences are *added* to $t(1)$. For geometric sequences, $(n - 1)$ common ratios are *multiplied* by $t(1)$.

With the aid of these two formulas, the second and third objectives of this section can be accomplished.

Example 4: Calculate $t(100)$ for the arithmetic sequence

$$17, 22, 27, 32, \ldots.$$

By subtracting adjacent terms, you find that the common difference, d, equals 5. Substituting in the formula gives

$$\begin{aligned} t(100) &= 17 + (100 - 1)(5) \\ &= 17 + 495 \\ &= \underline{\underline{512}}. \end{aligned}$$

Example 5: Calculate $t(100)$ for the geometric sequence with first term $t(1) = 35$ and common ratio $r = 1.05$.

Since you know $t(1)$, r, and n, you simply substitute these into the formula and get

$$\begin{aligned} t(100) &= 35 \times 1.05^{100-1} \\ &= 35 \times 1.05^{99} \\ &\approx \underline{\underline{4383.375262}}. \end{aligned}$$

Example 6: The number 68 is a term in the arithmetic sequence with $t(1) = 5$ and $d = 3$. *Which* term is it?

In this case you know that $t(n) = 68$, and you must find the term number, n. Using the formula,

$t(n) = t(1) + (n - 1)d$

$68 = 5 + (n - 1)(3)$ Substituting into the formula.

$63 = (n - 1)(3)$ Subtracting 5.

$21 = n - 1$ Dividing by 3.

$\therefore n = \underline{\underline{22}}$ Adding 1 and using symmetry.

Example 7: A geometric sequence has $t(1) = 17$ and $r = 2$. If $t(n) = 34816$, find n.

Substituting in the formula gives

$t(n) = t(1) \cdot r^{n-1}$

$34816 = 17 \times 2^{n-1}$ Substitution.

$2048 = 2^{n-1}$ Dividing by 17.

Since n is an exponent, you could find it by using logarithms. However, the exponent is an *integer*, since it is a term number of a sequence. Therefore, the number 2048 must be a power of 2. By factoring 2048 using "upside-down-repeated-short division" (see Section 6-5) you find that it equals 2^{11}. So

$$2^{11} = 2^{n-1}.$$

The only way this equation can be true is for the *exponents* to be equal. So

$11 = n - 1$ — Equating the exponents.

$\underline{\underline{12}} = n$ — Adding 1.

The exercise which follows is designed to give you practice identifying arithmetic and geometric sequences and finding term values or term numbers for such sequences.

Exercise 11-2

In Problems 1 through 16 tell whether the sequence can be arithmetic or geometric, or if it is neither. If it is arithmetic, find the common difference; if it is geometric, find the common ratio.

1. 7, 12, 17, ...
2. 3, 6, 12, ...
3. 5, 10, 12, ...
4. −1, 0, 1, ...
5. −1, 1, −1, ...
6. −1, 2, −3, ...
7. 25, 50, 100, ...
8. 25, 50, 75, ...
9. 25, 75, 100, ...
10. 2, −4, 8, ...
11. 2, −4, 6, ...
12. 1/2, 1/3, 1/4, ...
13. 1/2, 1/4, 1/8, ...
14. 1/2, 1/4, 0, ...
15. $\sqrt{5}, \sqrt[3]{5}, \sqrt[6]{5}, \ldots$
16. $\sqrt{5}, \sqrt[3]{5}, \sqrt[4]{5}, \ldots$

For Problems 17 through 24, find the specified terms of the indicated arithmetic sequence.

17. 45th term of 2, 5, 8, ...
18. 29th term of 7, 11, 15, ...
19. 51st term of 18, 14, 10, ...
20. 68th term of 95, 92, 89, ...
21. Thirtieth term of 1/3, 1, $1\frac{2}{3}$, ...
22. Seventeenth term of $3\sqrt{2}, 7\sqrt{2}, 11\sqrt{2}, \ldots$
23. $t(64)$, $t(65)$, and $t(66)$ for 8, 11, 14, ...
24. $t(95)$, $t(96)$, and $t(97)$ for 136, 131, 126, ...

For Problems 25 through 36, find the specified terms of the indicated geometric sequence. Use logarithms if you do not have access to a calculator.

25. Seventh term of 2, 6, 18, ...
26. Ninth term of 1, 2, 4, ...
27. Tenth term of 12, 6, 3, ...
28. Eighth term of 54, 18, 6, ...
29. Tenth term of 1, −2, 4, ...
30. Sixth term of 1, −3/2, 9/4, ...
31. 51st term of the sequence for which $t(1) = 7$ and $r = 1.02$
32. 43rd term of the sequence for which $t(1) = 100$ and $r = 1.04$
33. 37th term of the sequence for which $t(1) = 29$ and $r = 0.95$
34. 31st term of the sequence for which $t(1) = 1000$ and $r = 0.92$
35. 28th term of the sequence for which $t(1) = 0.01$ and $r = -3$
36. 64th term of the sequence for which $t(1) = 1$ and $r = -2$

For Problems 37 through 46, find out which term of the given number is in the indicated sequence.

37. 101 in the arithmetic sequence with $t(1) = 5$ and $d = 3$
38. 111 in the arithmetic sequence with $t(1) = 7$ and $d = 4$
39. 13 in the arithmetic sequence with $t(1) = 88$ and $d = -5$

40. 0 in the arithmetic sequence with $t(1) = 57$ and $d = -3$

41. 1536 in the geometric sequence with $t(1) = 3$ and $r = 2$

42. 4374 in the geometric sequence with $t(1) = 2$ and $r = 3$

43. 1 in the geometric sequence with $t(1) = 729$ and $r = 1/3$

44. 27 in the geometric sequence with $t(1) = 1728$ and $r = 1/2$

45. -1215 in the geometric sequence with $t(1) = 5$ and $r = -3$

46. $-170\frac{2}{3}$ in the geometric sequence with $t(1) = 1/3$ and $r = -2$

47. The following is a flow chart for a computer program to calculate and print terms of an arithmetic sequence.

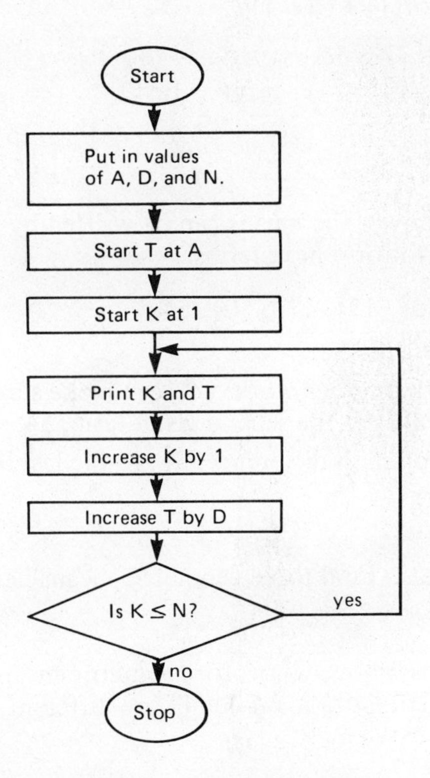

a. Show the simulated computer memory and the output if 47, 6, and 5 are put in for A, D, and N, respectively.
b. Write a computer program for this flow chart.
c. Use the program to print the first 20 terms of the arithmetic sequence whose first term is $7\frac{2}{3}$ and whose common difference is 1/3. (This is the sequence of men's shoe sizes, where n is the size and $t(n)$ is the man's foot length in inches.)

48. Modify the computer program from Problem 47 so that it calculates and prints terms of a *geometric* sequence. Use the modified program to print the first 20 terms of the geometric sequence with $t(1) = 1050$ and $r = 1.05$. (Terms of this sequence represent the number of dollars you would have in a savings account after n years if you invested \$1000 at 5% interest, compounded once a year.)

49. ***Sequences by Computer Graphics.*** Write a computer program to plot the graph of a geometric sequence on the computer's screen. The computer should let you input the first term and the common ratio. Then it should calculate and plot successive terms. You will have to read your computer's instruction manual to find out how to make it do graphics. The resulting program should produce a graph similar to that below for $t(1) = 9$ and $r = -0.95$.

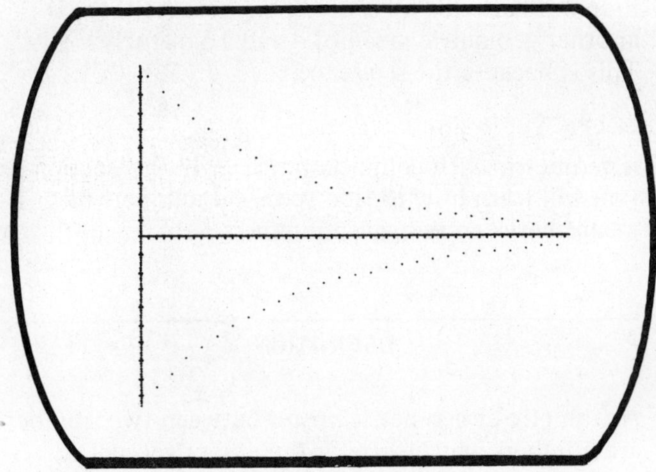

50. What is the difference between a geometric *sequence* and a geometric *progression*?

11-3 ARITHMETIC AND GEOMETRIC MEANS

Suppose that you are asked to find the *average* of two numbers, say 4 and 16. Adding the numbers and dividing by 2 gives 10. The numbers

$4, 10, 16, \ldots$

form an arithmetic sequence, since each pair of adjacent terms has a common difference of 6. The number 10 is called the *arithmetic mean* of 4 and 16, the word "mean" being just another word for "average" or "in between."

There are other ways of inserting numbers between 4 and 16 to form sequences which are either arithmetic or geometric. For example, putting 8 and 12 in between forms

$4, 8, 12, 16, \ldots,$

which is an arithmetic sequence with a common difference of 4. The numbers 8 and 12 are called the *two* arithmetic means between 4 and 16. Inserting just the number 8 between 4 and 16 forms

$4, 8, 16, \ldots,$

which is a *geometric* sequence with a common ratio of 2. So 8 is called a *geometric* mean of 4 and 16. The indefinite article "a" is used here because there is another geometric mean of 4 and 16, namely −8. This is because the sequence

$4, -8, 16, \ldots,$

is geometric with common ratio −2. In this section you will learn how to find specified numbers of means between two given numbers.

DEFINITION

Arithmetic or geometric ***means*** between two numbers are numbers which form arithmetic or geometric sequences with the two given numbers.

Objective:

Given two numbers, be able to find a specified number of arithmetic or geometric means between them.

The key to accomplishing this objective is finding the common difference or common ratio. With this number known, you can use the *definition* of an arithmetic or geometric sequence to write the desired means.

Example 1: Find four arithmetic means between 37 and 54.

The safest way to find these means is to write 37 and 54 with four spaces between them, into which you can write the means.

$37, \_\_\_, \_\_\_, \_\_\_, \_\_\_, 54.$

These will be the first *six* terms of the sequence. So $n = 6$, $t(6) = 54$, and $t(1) = 37$. You can find d by substituting the three known values into the formula.

$$t(6) = t(1) + (6 - 1)d$$
$$54 = 37 + 5d$$
$$17 = 5d$$
$$3\tfrac{2}{5} = d$$

With d known, the means can be written by successively adding d to form next terms.

$37, \underline{40\tfrac{2}{5}}, \underline{43\tfrac{4}{5}}, \underline{47\tfrac{1}{5}}, \underline{50\tfrac{3}{5}}, 54.$

As a check on your work, you can make sure that you actually *get* the second given term, 54, when you add the common difference, $3\tfrac{2}{5}$, to the last mean, $50\tfrac{3}{5}$.

Example 2: Find three geometric means between 3 and 48.

The process above works for geometric means, too. Only the computational details are different. By writing

3, ___, ___, ___, 48,

you realize that there are *five* terms. So $t(5) = 48$, $t(1) = 3$, and $n = 5$. Substituting these into the formula gives

$$t(5) = t(1) \cdot r^{5-1}$$
$$48 = 3 \cdot r^4$$
$$16 = r^4.$$

By taking the *square* root of both members, you get

$$r^2 = 4 \quad \text{or} \quad r^2 = -4.$$

The first clause of the equation has solutions $r = 2$ or $r = -2$. Therefore, two possible sets of means are

3, $\underline{6}$, $\underline{12}$, $\underline{24}$, 48 and 3, $\underline{-6}$, $\underline{12}$, $\underline{-24}$, 48.

The second clause of the equation for r has no real solutions. However, it does have solutions in the set of *imaginary* numbers,

$$r = 2i \quad \text{or} \quad r = -2i,$$

where $i = \sqrt{-1}$. So two other sets of means are

3, $\underline{6i}$, $\underline{-12}$, $\underline{-24i}$, 48 and 3, $\underline{-6i}$, $\underline{-12}$, $\underline{24i}$, 48.

The following exercise is designed to give you practice in finding arithmetic and geometric means. You will also discover some properties of arithmetic and geometric means.

Exercise 11-3

For Problems 1 through 12, find the specified number of *arithmetic* means between the given numbers.

1. Three, between 42 and 70
2. Four, between 55 and 85
3. Six, between −107 and −86
4. Five, between −91 and −67
5. Five, between 23 and −31
6. Six, between 17 and −60
7. Two, between −143 and −215
8. Three, between −257 and −397
9. Three, between 53 and 75
10. Two, between 67 and 90
11. Four, between 123 and 55
12. Five, between 47 and −11

For Problems 13 through 24, find the specified number of *real* geometric means between the given numbers. (There may be *more* than one set of real means!)

13. Two, between 5 and 135
14. Three, between 7 and 112
15. Three, between 81 and 16
16. Two, between 128 and 54
17. Four, between 1/32 and 32
18. Three, between 1/525 and 25/21
19. Two, between 13 and −4459
20. Four, between −7 and 1701
21. Five, between x^5 and x^{17}
22. Two, between x^5 and x^{17}
23. Two, between 13 and 26
24. Three, between 5 and 45

For Problems 25 through 28, find the specified number of geometric means, allowing the common ratio to be a *complex* number.

25. Three, between 2 and 162
26. Three, between 5 and 80
27. Three, between 1 and 16
28. Three, between 2 and 32

29. You know that the formula for the *arithmetic* mean (i.e., the average) of two numbers a and b is $\frac{1}{2}(a + b)$. Prove that the *geometric* mean of a and b is $\sqrt{ab}$.

30. Prove that the geometric mean, m, of two numbers a and b is the *mean proportional* of a and b. That is, show that $a/m = m/b$.

For Problems 31 through 34, find the arithmetic mean and the geometric mean of the given numbers.

31. 2 and 18

32. 3 and 108

33. 2/3 and 24

34. 3/5 and 15

35. From the answers to Problems 31 through 34, you can observe that the arithmetic mean is *larger* than the geometric mean. Prove that this is *always* true for the arithmetic and geometric means of two distinct, positive numbers. The formulas given in Problem 29 should help.

36. Prove that if $a, b, c, d, \ldots$, is a geometric sequence, then $a^2, b^2, c^2, d^2, \ldots$ is also a geometric sequence.

11-4 INTRODUCTION TO SERIES

If you *add* the terms of a sequence, the result is called a *series*. For example, the series which comes from 3, 5, 7, 9, 11, ... , is

$$3 + 5 + 7 + 9 + 11 + \ldots .$$

DEFINITION

A ***series*** is the indicated sum of the terms of a sequence.

Since a sequence has an infinite number of terms, a series can be thought of as a *sum* of an *infinite* number of terms. The sum of *all* of the terms of a series will, therefore, usually be infinite. For this reason it is convenient to study sums of only a *finite* number of terms of a series. For example, the sum of the first four terms of the above series is

$$3 + 5 + 7 + 9,$$

which equals 24. The sum of part of a series is called a *partial sum*. 3 + 5 + 7 + 9 is called the "fourth partial sum" of the above series because it is the sum of the first four terms.

DEFINITION

The ***n^{th} partial sum*** of a series is the sum of the first n terms of that series.

The symbol $S(n)$ will be used to stand for the nth partial sum of a series because the value of the partial sum is clearly the dependent variable in a function whose independent variable is n. For the above series,

$$S(1) = 3$$

$$S(2) = 3 + 5 = 8$$

$$S(3) = 3 + 5 + 7 = 15$$

$$S(4) = 3 + 5 + 7 + 9 = 24,$$

and so fourth. Note that the partial sums themselves form a sequence,

$$3, 8, 15, 24, \ldots .$$

So series can be dealt with by considering sequences of partial sums.

Unfortunately, there are not many series for which there is a convenient formula for calculating $S(n)$. So mathematicians simply invented a symbol telling how to construct the partial sum. The symbol uses the Greek letter Σ ("sigma") for "sum," and is written

$$S(n) = \sum_{k=1}^{n} t(k).$$

The symbol is read, "The sum from $k = 1$ to $k = n$ of $t(k)$," and means $t(1) + t(2) + t(3) + \ldots + t(n)$. The variable k is called the *index*. It does the same thing as the "counter" in a loop of a computer program. For example, if $t(k) = 2 \cdot 3^k$, then

$$S(4) = \sum_{k=1}^{4} 2 \cdot 3^k = 2 \cdot 3^1 + 2 \cdot 3^2 + 2 \cdot 3^3 + 2 \cdot 3^4$$
$$= 6 + 18 + 54 + 162$$
$$= 240.$$

Objectives:

1. Given a partial sum in Σ notation, *evaluate* it by writing all the terms, then adding them.

2. Given the first few terms of a series, write $S(n)$ using Σ notation.

The following exercise is designed to give you practice accomplishing these objectives.

Exercise 11-4

For Problems 1 through 20, evaluate the expression by writing the terms and adding them up.

1. $\sum_{k=1}^{5} 2k + 7$
2. $\sum_{k=1}^{6} 3k - 4$
3. $\sum_{k=1}^{10} k$
4. $\sum_{k=1}^{12} k$
5. $\sum_{k=1}^{4} 1/k$
6. $\sum_{k=1}^{4} 2/k$
7. $\sum_{k=1}^{6} k^2 + 1$
8. $\sum_{k=1}^{5} k^3 - 4$
9. $\sum_{k=1}^{5} (-1)^k (2k + 3)$
10. $\sum_{k=1}^{6} (-1)^k (3k - 2)$
11. $\sum_{k=1}^{5} 2^k$
12. $\sum_{k=1}^{4} 3^k$
13. $\sum_{k=1}^{3} 1.02^k$
14. $\sum_{k=1}^{3} 1.04^k$
15. $\sum_{k=1}^{5} (-1)^{k-1} (2) (3^{k-1})$
16. $\sum_{k=1}^{5} (-1)^{k-1} (3) (2^{k-1})$
17. $\sum_{k=1}^{4} 5 \cdot 3^k$
18. $\sum_{k=1}^{4} 3 \cdot 5^k$
19. $5 \cdot \sum_{k=1}^{4} 3^k$
20. $3 \cdot \sum_{k=1}^{4} 5^k$
21. The answers to Problems 17 and 19 are equal and so are the answers to Problems 18 and 20. What field axiom explains *why* these answers are equal?
22. The Σ notation can be used with numbers other than 1 for the lower value of the index k, and with other variables besides k in the argument. Write down what the following would equal:

    a. $\sum_{k=3}^{5} 4k - 7$ b. $\sum_{k=0}^{10} (-1)^k x^k$

For Problems 23 through 34, write $S(n)$ using sigma notation.

23. $S(10)$ for $1 + 4 + 9 + 16 + 25 + \ldots$
24. $S(30)$ for $1 + 8 + 27 + 64 + 125 + \ldots$
25. $S(100)$ for $1/3 + 1/4 + 1/5 + 1/6 + 1/7 + \ldots$
26. $S(100)$ for $1/3 + 1/5 + 1/7 + 1/9 + 1/11 + \ldots$
27. $S(40)$ for $2 + 6 + 18 + 54 + 162 + \ldots$

28. $S(50)$ for $3 + 6 + 12 + 24 + 48 + \ldots$

29. $S(90)$ for $2 + 6 + 10 + 14 + 18 + \ldots$

30. $S(80)$ for $5 + 8 + 11 + 14 + 17 + \ldots$

31. $S(20)$ for $0 - 10 + 20 - 30 + 40 - \ldots$

32. $S(60)$ for $1 - 1/2 + 1/4 - 1/8 + 1/16 - \ldots$

33. $S(55)$ for $1 \cdot 2 + 2 \cdot 3 + 3 \cdot 4 + 4 \cdot 5 + \ldots$

34. $S(47)$ for $1 \cdot \frac{2}{3} + 2 \cdot \frac{3}{4} + 3 \cdot \frac{4}{5} + 4 \cdot \frac{5}{6} + \ldots$

35. Suppose that $S(n) = \sum_{k=1}^{n} (2k - 1)$. Find $S(n)$ for each value of $n \in \{1, 2, 3, 4, 5, 6, 7\}$. What would you guess might be a *formula* for $S(n)$?

36. Suppose that $S(n) = \sum_{k=1}^{n} \frac{k}{2}(k + 1)$.

a. Write the first five terms of the series.
b. These terms are called *triangular* numbers. Show that each value of $t(k)$ is the number of balls that can be arranged in a triangle with k balls on a side.
c. Find the first five partial sums of the series.
d. These partial sums are called *pyramidal* numbers. Show that each value of $S(n)$ is the number of balls that can be arranged in a triangular pyramid with n balls on each side of the base.

37. Suppose that $S(n) = \sum_{k=1}^{n} (\frac{1}{2})^k$. Find $S(n)$ for each value of $n \in \{1, 2, 3, 4, 5\}$. From what you observe about the sequence of partial sums, what do you suppose happens to $S(n)$ as n becomes very large?

38. The following is a flow chart for calculating partial sums of a series.

a. Show the simulated computer memory and the output for calculating $S(5)$ if the formula for calculating T is $T = K^3$.
b. Write a computer program for this flow chart if the first value of T is 37.2, and each subsequent value of T is 1.01 times the previous value of T.
c. Use the program to find $S(1000)$ for the series in part b.

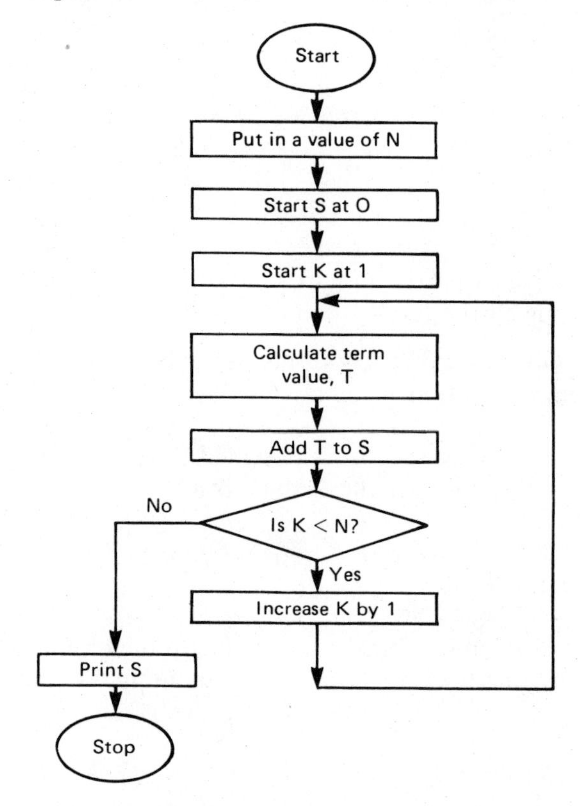

11-5 ARITHMETIC AND GEOMETRIC SERIES

You have learned that a series is what results from adding the terms of a sequence. In this section you will study series which come from adding terms of arithmetic or geometric sequences.

DEFINITION

An ***arithmetic*** or ***geometric*** series is a series which results from adding the terms of an arithmetic or geometric *sequence*, respectively.

In Exercise 11-4 you calculated partial sums of series by writing down each term then adding them up. To find partial sums *quickly*, it is desirable to have a *formula* for $S(n)$.

Objective:

Given an arithmetic or geometric series, be able to calculate $S(n)$, the nth partial sum, *quickly*.

Suppose that you must find the sum of the first 100 terms of the arithmetic series

$$7 + 13 + 19 + 25 + 31 + \ldots .$$

The 100th *term* of the series can be calculated by the formula you already know.

$$t(100) = t(1) + (100 - 1)d.$$

Since $d = 6$ for this series, the formula gives

$$t(100) = 7 + (99)(6)$$
$$= 601.$$

So the *last* few terms of the partial sum are 601, 595, 589, 583, and so forth. The partial sum is

$$S(100) = 7 + 13 + 19 + 25 + 31 + \ldots + 583 + 589 + 595 + 601.$$

An interesting pattern shows up if you add the *first* and *last* terms, the *second* and *next-to-last* terms, and so on.

$$S(100) = 7 + 13 + 19 + 25 + \ldots + 583 + 589 + 595 + 601$$

608
608
608
608

Each pair of terms adds up to 608. And there will be 100/2, or 50 such pairs. Thus, the partial sum is

$$S(100) = 608 + 608 + 608 + 608 + \ldots \text{ (50 terms)}$$
$$= 50(608)$$
$$= 30400.$$

From this work you can find a *formula* for $S(n)$. The 50 is equal to ½ the number of terms, or $n/2$ and the 608 is equal to $t(1) + t(n)$. So the formula is

$$\boxed{S(n) = \frac{n}{2}\Big(t(1) + t(n)\Big)}$$ ← For *arithmetic* series.

The formula was derived with an *even* value of n but it works for odd values of n, too. In this case there would be $(n-1)/2$ pairs, each equal to $t(1) + t(n)$, plus the *middle* term. But the middle term is just the *average* of $t(1)$ and $t(n)$, which equals $\frac{1}{2}(t(1) + t(n))$. Adding this middle term to the sum of the $(n - 1)/2$ pairs gives the same formula.

Carl Friederich Gauss, who lived from 1777 to 1855, is reputed to have discovered this formula at the age of 10. For an interesting account of how he amazed his school master, see James R. Newman's *The World of Mathematics*, Volume I, page 295.

An alternate form of this formula can be found by substituting the quantity $t(1) + (n - 1)d$ for $t(n)$.

$$S(n) = \frac{n}{2}\Big(t(1) + t(n)\Big)$$ Previous formula.

$$S(n) = \frac{n}{2}\Big(t(1) + t(1) + (n - 1)d\Big)$$ Substitution.

$$\boxed{S(n) = \frac{n}{2}\Big(2t(1) + (n - 1)d\Big)}$$ ← For *arithmetic* series.

This formula is more convenient if you know d, but *not* $t(n)$, or if you want to *calculate* n when you know $S(n)$, $t(1)$ and d.

Another clever algebraic trick gives an $S(n)$ formula for *geometric* series. Suppose that you must find

$S(100)$ for the geometric series

$$7 + 21 + 63 + \ldots .$$

In this case, $t(1) = 7$ and the common ratio is $r = 3$. It helps to write the terms in *factored* form.

$$S(100) = 7 + 7 \cdot 3 + 7 \cdot 3^2 + 7 \cdot 3^3 + \ldots + 7 \cdot 3^{98} + 7 \cdot 3^{99}.$$

If you multiply both members of this equation by -3, the *opposite* of the common ratio, then *add* the two equations (as you did with linear systems), you get

$$\begin{array}{rl} S(100) = & 7 + 7 \cdot 3 + 7 \cdot 3^2 + 7 \cdot 3^3 + \ldots + 7 \cdot 3^{99} \\ -3 \cdot S(100) = & \quad - 7 \cdot 3 - 7 \cdot 3^2 - 7 \cdot 3^3 - \ldots - 7 \cdot 3^{99} - 7 \cdot 3^{100} \\ \hline S(100) - 3 \cdot S(100) = & 7 + \quad 0 + \quad 0 + \quad 0 + \ldots + \quad 0 \quad - 7 \cdot 3^{100} \end{array}$$

All but one term in the top equation has its *opposite* in the bottom equation. So all of the terms "drop out" except for the 7 and the $-7 \cdot 3^{100}$. The result is

$$S(100) - 3 \cdot S(100) = 7 - 7 \cdot 3^{100}.$$

Factoring out 7 on the right and $S(100)$ on the left gives

$$(1 - 3) \cdot S(100) = 7(1 - 3^{100}).$$

Dividing by $(1 - 3)$ gives

$$S(100) = 7 \cdot \frac{1 - 3^{100}}{1 - 3}.$$

From this result you can extract a *formula* for $S(n)$. The 7 is equal to $t(1)$. The 3 is the common ratio, r, and the 100 is the number of terms, n. So the formula is

$$\boxed{S(n) = t(1) \cdot \frac{1 - r^n}{1 - r}} \longleftarrow \text{For } geometric \text{ series.}$$

The objective, rapid calculation of $S(n)$, can be accomplished using these formulas.

Example 1: Find the 127th partial sum of the arithmetic series with $t(1) = 17$ and $d = 4$.

From the second form of the arithmetic series formula, you get

$$\begin{aligned} S(127) &= \frac{127}{2}\Big(2 \cdot 17 + (127 - 1)(4)\Big) \\ &= \frac{127}{2}\Big(34 + 504\Big) \\ &= \frac{127}{2}\Big(538\Big) \\ &= \underline{\underline{34163}}. \end{aligned}$$

This calculation is certainly quicker than calculating all 127 terms, then adding them up!

Example 2: Find $S(34)$ for the geometric series with $t(1) = 7$ and $r = 1.03$.

From the formula for geometric series you get

$$S(34) = 7 \cdot \frac{1 - 1.03^{34}}{1 - 1.03}.$$

By calculator, $1.03^{34} \approx 2.731905296$. Thus,

$$\begin{aligned} S(34) &\approx 7 \cdot \frac{1 - 2.731905296}{1 - 1.03} \\ &= 7 \cdot \frac{-1.731905296}{-0.03} \\ &\approx \underline{\underline{404.1112356}}. \end{aligned}$$

The exercise which follows is designed to give you practice finding partial sums of arithmetic and geometric series.

Exercise 11-5

For Problems 1 through 10, find $S(n)$ for the indicated *arithmetic* series by either calculating the terms and adding them up, or by using the formula, whichever you think is quicker.

1. $S(10)$ for $4 + 7 + 10 + \ldots$
2. $S(13)$ for $9 + 20 + 31 + \ldots$
3. $S(20)$ for the series with $t(1) = 15$ and $d = 10$
4. $S(30)$ for the series with $t(1) = 17$ and $d = 8$
5. $S(15)$ for the series with $t(1) = 14$ and $d = -2$
6. $S(18)$ for the series with $t(1) = 29$ and $d = -3$
7. $S(40)$ for the series with $t(1) = 8$ and $t(6) = 38$
8. $S(50)$ for the series with $t(1) = 7$ and $t(9) = 47$
9. $\sum_{k=1}^{60} 3 + 2(k - 1)$
10. $\sum_{k=1}^{70} 4 + 3(k - 1)$

For Problems 11 through 24, find $S(n)$ for the indicated *geometric* series by either calculating the terms and adding them up, or by using the formula, whichever you think is quicker.

11. $S(5)$ for $1 + 2 + 4 + \ldots$
12. $S(6)$ for $2 + 6 + 18 + \ldots$
13. $S(6)$ for $1 - 3 + 9 - \ldots$
14. $S(5)$ for $3 - 6 + 12 - \ldots$
15. $S(10)$ for the series with $t(1) = 5$ and $r = 3$
16. $S(10)$ for the series with $t(1) = 7$ and $r = 2$
17. $S(9)$ for the series with $t(1) = 6$ and $r = -2$
18. $S(9)$ for the series with $t(1) = 5$ and $r = -3$
19. $S(20)$ for the series with $t(1) = 11$ and $r = 1.3$
20. $S(30)$ for the series with $t(1) = 13$ and $r = 1.1$
21. $S(15)$ for the series with $t(1) = 10$ and $t(2) = 9$
22. $S(18)$ for the series with $t(1) = 20$ and $t(2) = 19$
23. $\sum_{k=1}^{6} 2 \cdot 3^{k-1}$
24. $\sum_{k=1}^{7} 3 \cdot 2^{k-1}$
25. ***Partial Sums by Computer.*** Write a program to calculate and print partial sums of a series. The computer should let you input the number of terms. Then it should enter a loop in which the term value is calculated and added to the sum of the previous terms. At each pass through the loop the computer should print the term number, the term value, and the partial sum.
26. For the geometric series $3 + 3/2 + 3/4 + 3/8 + \ldots$, find $S(5)$, $S(10)$, and $S(20)$. What do you notice is happening to $S(n)$ as n becomes very large?

11-6 CONVERGENT GEOMETRIC SERIES

The series $5 + 8 + 11 + 14 + \ldots$ has partial sums that keep getting bigger and bigger as n increases. Consider what happens to the geometric series

$$5 + \frac{5}{2} + \frac{5}{4} + \frac{5}{8} + \frac{5}{16} + \frac{5}{32} + \ldots.$$

The first few partial sums are

$$S(1) = 5$$

$$S(2) = 5 + \frac{5}{2} = 7\frac{1}{2}$$

$$S(3) = 5 + \frac{5}{2} + \frac{5}{4} = 8\frac{3}{4}$$

$$S(4) = 5 + \frac{5}{2} + \frac{5}{4} + \frac{5}{8} = 9\frac{3}{8}$$

$$S(5) = 5 + \frac{5}{2} + \frac{5}{4} + \frac{5}{8} + \frac{5}{16} = 9\frac{11}{16}.$$

The partial sums are getting larger, but seem to be less than 10. To see what happens when n is large, you can use the formula. For example, the 20th partial sum is

$$S(20) = 5 \cdot \frac{1 - (\frac{1}{2})^{20}}{1 - \frac{1}{2}} .$$

By calculator, you can find that $(\frac{1}{2})^{20} \approx 0.0000009537$. Also, $1 - \frac{1}{2} = \frac{1}{2}$. So

$$S(20) \approx 5 \cdot \frac{1 - 0.0000009537}{\frac{1}{2}} .$$

Dividing 5 by ½ (getting 10) and doing the subtraction,

$$\begin{aligned} S(20) &\approx 10(0.9999990463) \\ &= 9.999990463. \end{aligned}$$

This number is *very* close to 10. A graph of the partial sums is shown in Figure 11-6a.

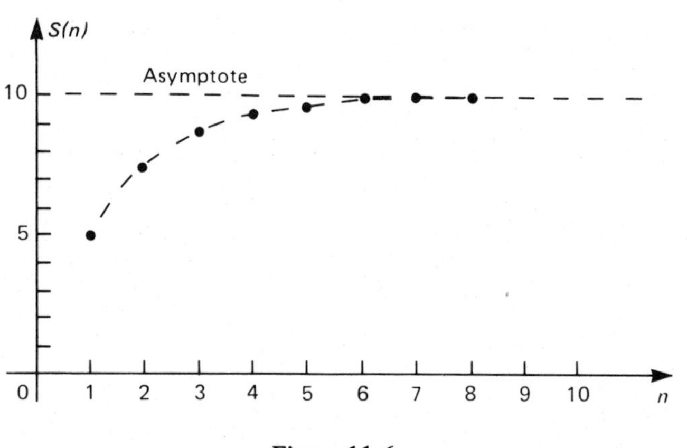

Figure 11-6a.

The horizontal line at 10 is an *asymptote* to the graph. You can see that $S(n)$ will never be larger than 10, because it will always equal 10 multiplied by a number slightly *smaller* than 1. Since the points get closer and closer to 10, the series is said to "converge to 10."

DEFINITION

A series ***converges*** to a number S if the partial sums, $S(n)$, stay arbitrarily close to S as n gets very large.

Two questions now arise:

1. Under what conditions will a geometric series converge?

2. How do you find the number to which it converges?

The above series converges because the term $(\frac{1}{2})^n$ in the formula for $S(n)$ is very close to *zero* when n is large. This will happen as long as the common ratio is a *proper fraction.* That is, r must be between -1 and 1. So the series will converge if $|r| < 1$.

To find the *number* to which the series converges, you simply replace the r^n term with 0 in the formula for $S(n)$, getting

$$t(1) \cdot \frac{1 - r^n}{1 - r} \to t(1) \cdot \frac{1 - 0}{1 - r} = \frac{t(1)}{1 - r} .$$

This number is called the *limit* of the series as n approaches infinity.

Conclusion: A geometric series *converges* if $|r| < 1$. The limit, S, to which it converges is given by

$$\boxed{S = \frac{t(1)}{1 - r}} .$$

Note: Although there is an *infinite* number of terms in the series, you now have a reasonable definition for the sum of *all* the terms. In the above example, if you add any *finite* number of terms, the answer is less than 10. But you could define the sum of *all* the terms to be *equal* to 10. This definition allows you to find *exact* values of repeating decimals, as you will see below.

Objectives:

1. Given a geometric series, tell whether or not it converges. If it does converge, find the limit to which it converges.

2. Given a repeating decimal, write it as a convergent geometric series, and find a *rational* number equal to the decimal.

Example 1: Does the geometric series $15 - 4.5 + 1.35 - \ldots$ converge? If so, to what number does it converge?

The common ratio is

$$r = \frac{-4.5}{15} = -0.3.$$

Since $|r|$ is *less* than 1, the series *does* converge. By the formula,

$$S = \frac{15}{1-(-0.3)}$$

$$= \frac{15}{1.3}$$

$$= 11\tfrac{7}{13}.$$

Example 2: Does the geometric series $2 - 3 + 4.5 - \ldots$ converge? If so, to what number does it converge?

The common ratio is

$$r = \frac{-3}{2} = -1.5.$$

Since $|r|$ is *not* less than 1, the series *does not* converge. Figure 11-6b shows what happens to the partial sums of this series. If a series does not converge, you can say that it "diverges."

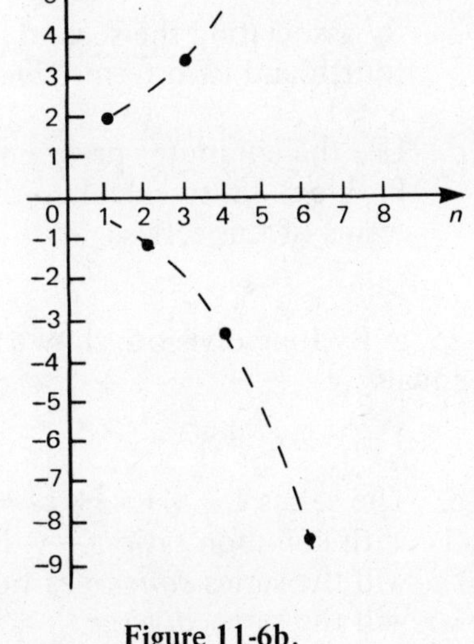

Figure 11-6b.

Example 3: Express 0.324324324... as the ratio of two integers, and simplify.

Let $x = 0.324324324\ldots$.

Since the decimal repeats every *three* digits, you can write

$$x = 0.324 + 0.000324 + 0.000000324 + \ldots .$$

This is a *geometric series*, with $t(1) = 0.324$ and common ratio $r = 0.001$. So x is the sum of *all* the terms of the series. It is reasonable to define this sum to be the number to which the series *converges.* So you can write

$$x = \frac{0.324}{1-0.001}$$

$$= \frac{0.324}{0.999}$$

$$= \frac{324}{999}$$

$$x = \frac{12}{37}.$$

Example 4: Express 0.26717171... as the ratio of two relatively prime integers.

In this case the decimal does not start repeating until after the first few places. So you break off the non-repeating part and express the rest as a convergent geometric series. Letting $x = 0.26717171\ldots$, you can write

$$x = 0.26 + 0.0071 + 0.000071 + 0.00000071 + \ldots$$

$$= 0.26 + \frac{0.0071}{1-0.01}$$

$$= 0.26 + \frac{0.0071}{0.99}$$

$$= \frac{26}{100} + \frac{71}{9900}$$

$$= \frac{2645}{9900}$$

$$= \frac{529}{1980}.$$

The calculations above should reveal to you that a repeating decimal can always be written as a ratio of two integers. This observation leads to the following theorem.

Theorem: If x is a repeating decimal, then x is a rational number.

The *converse* of this theorem is also true. Its proof uses the long division algorithm rather than geometric series, and so will not be presented here.

The exercise which follows is designed to give you practice in accomplishing the two objectives of this section.

Exercise 11-6

For Problems 1 through 10, determine whether or not the indicated geometric series converges. If so, find the value to which it converges.

1. $t(1) = 3$ and $r = 1/5$
2. $t(1) = 5$ and $r = 1/3$
3. $t(1) = 42$ and $r = -3/4$
4. $t(1) = 42$ and $r = -4/3$
5. $t(1) = 18$ and $r = -7/5$
6. $t(1) = 18$ and $r = -5/7$
7. $t(1) = 10$ and $r = 0.1$
8. $t(1) = 100$ and $r = 0.001$
9. $t(1) = 100$ and $t(3) = 1$ (*Two* answers!)
10. $t(1) = 81$ and $t(5) = 1$ (*Two* answers!)

For Problems 11 through 22, write the repeating decimal as a ratio of two integers, and simplify.

11. 0.636363...
12. 0.848484...
13. 0.567567567...
14. 0.990990990...
15. 1.060606...
16. 2.363636...
17. 1.4272727...
18. 1.6454545...
19. 0.4999999...
20. 0.9999999... (Surprising?)
21. 0.012345679012345679012345679 ...
(Clue: Simplifying the fraction is difficult, but obvious.)
22. 0.987654320987654320987654320 ...
(Clue: Use the answers to Problems 20 and 21.)

23. The sequence 1/2, 1/3, 1/4, 1/5, ... is called the *harmonic* sequence. Clearly, $t(n)$ gets closer and closer to zero as n becomes very large. Yet the harmonic *series* 1/2 + 1/3 + 1/4 + 1/5 + ... does *not* converge! It just keeps getting larger and larger as n increases. By associating terms, show that

$$1/2 + 1/3 + 1/4 + 1/5 + 1/6 + 1/7 + 1/8 + \ldots$$
$$> 1/2 + 1/2 + 1/2 + \ldots,$$

and use the result to show that the series "diverges."

24. The *alternating* harmonic series,

$$S = 1/2 - 1/3 + 1/4 - 1/5 + \ldots,$$

does converge.

a. By associating the first and second terms, the third and fourth terms, and so on, show that $S > 0$.
b. By associating the second and third terms, the fourth and fifth terms, and so on, show that $S < 1/2$.
c. Use the computer program of Exercise 11-4, Problem 38, to calculate the sum of the first 1000 terms of this series.

25. a. By long division, show that the fraction $\frac{1}{1-x}$ equals

$$1 + x + x^2 + x^3 + x^4 + \ldots.$$

b. The series $1 + x + x^2 + x^3 + x^4 + \ldots$ is geometric with common ratio $r = x$. For what values of x will the series *converge*? For what values of x will the series *diverge*?
c. Using what you know about convergent geometric series, show that the partial sums of the series in part b really *do* approach $1/(1-x)$ when the series is convergent.

26. Calvin Butterball is taking a test on geometric series. He realizes that the series $1 - 1 + 1 - 1 + 1 - 1 + \ldots$ is geometric, with $t(1) = 1$ and common ratio $r = -1$. So he finds the number the sequence converges to, using the formula

$$S = \frac{t(1)}{1-r} = \frac{1}{1-(-1)} = \frac{1}{2}.$$

a. Write the first few partial sums of this series.
b. Draw a graph of the sequence of partial sums, and thus show Calvin that his answer is not reasonable.
c. Explain to Calvin what error he made.

27. Using the techniques of convergent geometric series, it is possible to show that the series

$$1 + 2(1/3) + 3(1/3)^2 + 4(1/3)^3 + 5(1/3)^4 + \ldots$$

converges to a finite number. Find out what this number equals.

11-7 SEQUENCES AND SERIES AS MATHEMATICAL MODELS

Since a sequence is a function whose domain is a set of integers, the graph of a sequence changes by "jumps" rather than in a smooth, continuous curve. Therefore, sequences are appropriate as mathematical models for real-world phenomena in which the dependent variable changes stepwise rather than continuously. For instance, when you hammer a nail into a board, the distance the nail has gone into the board is a stepwise function of the number of times you hit it with the hammer.

Objective:

Given a situation from the real world in which the dependent variable changes stepwise, use an arithmetic or geometric sequence or series as a mathematical model.

Example 1: Suppose you start pounding a nail into a board. With the first impact the nail moves 20 millimeters (mm); with the second impact it moves 18 mm more. Predict the distance it moves and the total distance it has gone into the board assuming that the distances it moves form

a. an arithmetic sequence,
b. a geometric sequence.

Let n = number of impacts.
Let $t(n)$ = number of mm the nail moves on the nth impact. (See Figure 11-7a.)
Let $S(n)$ = total number of mm the nail has gone after n impacts.

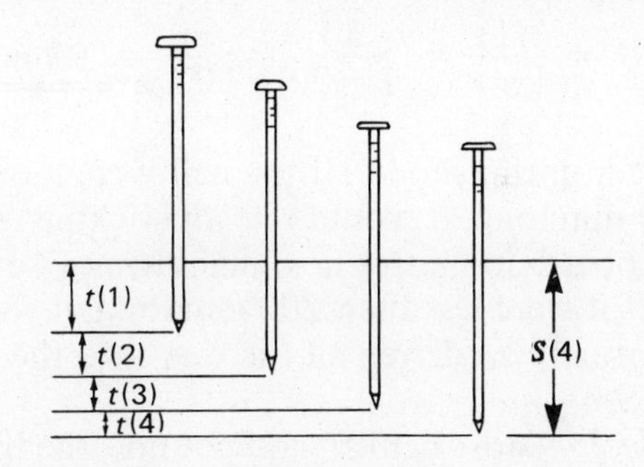

Figure 11-7a.

a. Assuming an *arithmetic* sequence, $d = 18 - 20 = -2$. Therefore,

$$t(n) = 20 + (n - 1)(-2).$$

On the fourth impact, the nail moves

$$t(4) = 20 + (4 - 1)(-2) = 20 - 6 = \underline{\underline{14 \text{ mm}}}.$$

The total distance the nail has gone after four impacts will be the *fourth partial sum* of the corresponding arithmetic series. Since $t(4) = 14$,

$$S(4) = \frac{4}{2}(20 + 14) = \underline{\underline{76 \text{ mm}}}.$$

b. If you assume a *geometric* sequence rather than an arithmetic one, the common ratio would be $r = 18/20 = 0.9$. On the fourth impact, the nail would go

$$t(4) = 20 \times 0.9^{4-1} = \underline{\underline{14.58 \text{ mm.}}}$$

After four impacts, the nail would have gone a a total of

$$S(4) = \frac{20(1 - 0.9^4)}{1 - 0.9} = \underline{\underline{67.78 \text{ mm.}}}$$

Since the common ratio has an absolute value less than 1, the partial sums *converge* to some number as n gets very large. After many impacts, the total distance the nail has gone into the board will approach

$$S = \frac{t(1)}{1-r} = \frac{20}{1-0.9} = \frac{20}{0.1} = \underline{\underline{200 \text{ mm.}}}$$

Based on this model, if the nail were more than 200 mm long, it would remain sticking out of the board no matter how many times you hit it. If it were less than 200 mm long, it would eventually be driven all the way into the board.

Example 2: Most banks "compound" the interest they pay you on savings accounts. This means that at regular intervals (daily, quarterly, yearly, etc.) they add the interest you have earned for that interval to the "principal" which you already had invested. During the next interval, you earn interest on the interest as well as on the original principal. Suppose that a bank pays 5% per year interest, compounded annually (once a year). If you invest a principal of \$100, then at the end of one year the bank pays you 5% of 100, or

$$0.05^* \times 100 = 5$$

dollars. So the new principal for the next compounding period is 100 + 5, or \$105. At the end of the second year, the bank pays you 5% of \$105, or

$$0.05 \times 105 = 5.25$$

dollars, making the new principal equal 105 + 5.25 = 110.25, and so forth. The amount of money you have in the bank at any time is a term in the sequence

100, 105, 110.25, ...

The sequence is geometric, with a common ratio of 1.05, as you could discover by dividing adjacent terms. Some clever factoring shows the same thing more instructively.

Let n = number of years the money has been in the bank.

Let $t(n)$ = number of dollars you have in the bank.

When you make the first deposit, time equals zero. Therefore, it is convenient to start n at 0 rather than 1 as you have been doing up to now. Therefore,

$$t(0) = 100.$$

At the end of one year, when $n = 1$, the interest is added to get the next term. Therefore,

$$\begin{aligned} t(1) &= 100 + 0.05 \times 100 && \text{Adding 5\% interest.} \\ &= 100(1 + 0.05) && \text{Factoring out 100.} \\ &= 100(1.05) && \text{Adding 1 to 0.05.} \end{aligned}$$

To see the pattern more clearly, you should *not* carry out the multiplication yet.

To find $t(2)$, you add the interest for the second year to $t(1)$ (the principal for the first year) getting

$$\begin{aligned} t(2) &= 100(1.05) + 0.05 \times 100(1.05) \\ &\qquad \text{Adding 5\% interest.} \\ &= 100(1.05)(1 + 0.05) \\ &\qquad \text{Factoring out 100(1.05).} \\ &= 100(1.05)(1.05) \\ &\qquad \text{Adding 1 and 0.05.} \\ &= 100(1.05)^2 \\ &\qquad \text{Definition of exponentiation.} \end{aligned}$$

Similarly, $t(3) = 100(1.05)^3$, $t(4) = 100(1.05)^4$, and so on. Since the exponent is always equal to the term number,

$$t(n) = 100(1.05)^n.$$

*"5%" is read, "5 percent." Since "per" means "divided by" and "cent" means "100," 5% means 5 ÷ 100, or 0.05.

This is the formula for the nth term of a geometric sequence with $r = 1.05$ and first term = 100. The exponent is n rather than $n - 1$ because the sequence started with term 0 instead of term 1.

To find the amount of money you would have after 7 years, you would substitute 7 for n and do the calculations.

$$t(7) = 100(1.05)^7$$
$$\approx \underline{\underline{140.71}}.$$

Finding the time at which the amount is $400 involves finding the exponent, n.

$400 = 100\,(1.05)^n$	Substituting 400 for $t(n)$.
$4 = 1.05^n$	Dividing by 100.
$\log 4 = \log 1.05^n$	Log of each member.
$\log 4 = n \log 1.05$	Log of a power.
$\dfrac{\log 4}{\log 1.05} = n$	Dividing by log 1.05.
$28.413... = n$	By calculator.
After 29 years.	

The answer must be rounded *upward* since the amount is still below $400 at the end of 28 years.

Example 3: Suppose you invest $100 in a savings account that pays 5% per year interest, as in Example 2, but the interest is compounded *quarterly* (every 3 months). How much would you have at the end of 7 years?

Let n = number of *quarters* the money has been in the bank.

Let $t(n)$ = number of dollars you have in the bank.

The bank will pay only ¼ of the 5% interest after each compounding period, or 1.25% (0.0125). Using the same reasoning as in Example 1,

$$t(0) = 100$$
$$t(1) = 100(1.0125)$$
$$t(2) = 100(1.0125)^2$$
$$t(3) = 100(1.0125)^3.$$

Again, the term number equals the exponent, so

$$t(n) = 100(1.0125)^n.$$

The amount of money you would have after 7 years would be $t(28)$ because there are 28 quarters in 7 years. Therefore,

$$t(28) = 100(1.0125)^{28}$$
$$\approx 141.60.$$

Comparing this answer with Example 2, you make only 89 cents more in 7 years if the interest is compounded quarterly instead of yearly. If you work Problem 24 in the following exercise, you will discover the *real* advantage of compounding interest more frequently.

The exercise which follows is designed to give you practice using sequences and series as mathematical models. Since the calculations, particularly for geometric sequences, are pretty difficult, it is advisable to use a calculator. This will allow you to spend more time concentrating on setting up the problem, and on interpreting the answers.

Exercise 11-7

1. ***Muscle-Building Problem.*** Suppose that you start an exercise program to build up your biceps. On the first day, your biceps increase by 3 millimeters. The amount of increase on each following day is 0.95 times the amount of increase the day before.

a. Your biceps increase by how much on the tenth day?
b. By how many millimeters, total, have your biceps increased after 10 days?

c. If you continue the exercise program for many days, what number does the total increase in bicep measurement approach?

2. ***Chewing Gum Problem.*** Anne X. Kewse gets caught chewing gum in algebra class. Quickly, she explains to her instructor that she is conducting a practical experiment in geometric sequences and series, since with each chew she gets 0.9 times as much flavor as she did with the preceding chew.

a. If she gets 40 squirts of flavor with the first chew, how many squirts will she get with the tenth chew?
b. What total amount of flavor will she get in the first 10 chews?
c. Her punishment is to chew the gum *forever.* What total amount of flavor will she receive?

3. ***Blue Jeans Problem.*** Assume that whenever you wash a pair of blue jeans, they lose 4% of the color they had just before they were washed.

a. Explain why the percent of the original color *left* in the blue jeans varies geometrically with the number of washings. What is the common ratio?
b. How much of the original color would be left after 10 washings?
c. Suppose that you buy a new pair of blue jeans, and decide to wash them enough times so that only 25% of the original color remains. How many times must you wash them?
d. Explain why an arithmetic sequence would *not* be a reasonable mathematical model for the amount of color left after many washings.

4. ***Bouncing Ball Problem.*** Suppose that you drop a superball from a window 20 meters above the ground. The ball bounces to 90% of its previous altitude with each bounce.

a. How far does the ball travel, up and down, between the first and the second bounce?
b. Show that the numbers of meters the ball travels up and down between bounces are terms of a geometric sequence.
c. Find the number of meters the ball travels up and down between the sixth and seventh bounce.
d. Find the total number of meters the ball travels between the first bounce and the seventh bounce.
e. If the ball continues to bounce in this manner until it comes to rest, how far will it have traveled, up and down, from the time it was dropped from the window?

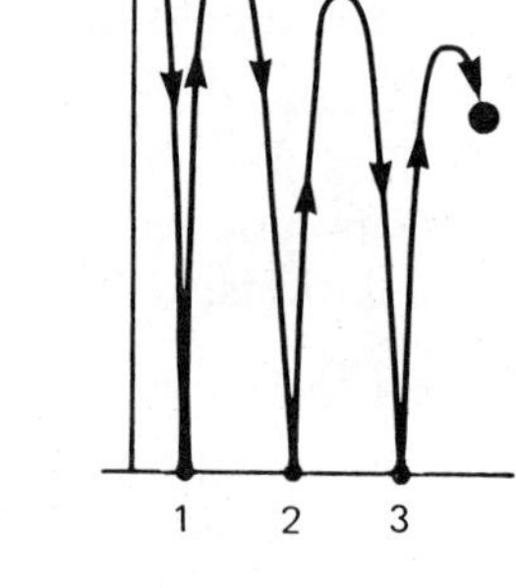

5. ***George Washington Problem.*** Recently, Anna Ward found she is a distant relative of George Washington. When he died in 1799, he left \$1000 in his will, which now belongs to her. The money has been in a savings account at the Old Dominion Bank, where it has been earning interest at 5% per year, compounded annually (once a year). How much does Anna have *this* year? Why do you suppose that there are laws limiting a bank's liability for paying interest on dormant savings accounts?

6. ***Little Brother Problem.*** Your little brother is hard up for cash, and asks you to lend him 10 cents. You get him to agree to pay 5% per day interest, compounded daily. You lend him the money, but you both forget the deal until exactly one year later. Being an expert at geometric sequences, you perform some calculations which show he is deeply in debt.

a. How much does he owe you for a 365-day year?
b. How much *more* would he owe you for a 366-day leap year?

7. ***Manhattan Problem.*** In 1626, Peter Minuit, Governor of the Dutch West India Company, purchased Manhattan Island from the Indians for \$24

worth of beads, cloth, and trinkets. Suppose that the Indians had sold the stuff and invested the $24 in a savings account paying 6% per year interest, compounded annually. How much would they have on deposit *this* year? Surprising??

8. ***Louisiana Purchase Problem.*** In 1803, when Thomas Jefferson was President, the United States purchased about 500,000,000 acres of land from the French for about $15,000,000.

a. If Napoleon had invested this money in a savings account paying 7% per year interest, compounded annually, how much money would he have *now*? How does this amount compare with the current U. S. National Debt?
b. How much per acre did the French get for the land? How much per acre could they afford to pay to buy it back if they had actually kept the money in such a savings account?

9. ***Straight-Line Depreciation Problem.*** The Internal Revenue Service (IRS) assumes that the value of an item which can wear out decreases by a *constant* number of dollars each year. For example, a house "depreciates" by 1/40 of its original value each year.

a. If your house were worth $24,000 originally, by how many dollars does it depreciate each year?
b. What is your house worth after 0, 1, 2, and 3 years?
c. Do these values form an arithmetic or a geometric sequence? What is the common difference or common ratio?
d. Calculate the value of your house at the end of 27 years.
e. According to this model, is your house ever worth *nothing*? Explain.
f. Plot a graph of this sequence using the results of parts b, d, and e. Why do you suppose the IRS calls this model "straight-line" depreciation?

10. ***Piggy Bank Problem.*** Suppose that you put $.50 into an empty piggy bank. One week later you put in $.75. At the end of the next week, you put in $1.00, and so on.

a. What kind of sequence do the individual deposits form?
b. At the end of 76 weeks, how many deposits have you made?
c. What amount do you deposit at the end of the 76th week?
d. What total amount do you have in the piggy bank at the end of 76 weeks?

11. ***Pile Driver Problem.*** A pile driver (see Figure 11-7b) starts driving a piling into the ground. On the first impact, the piling goes 100 cm into the ground. On the second impact, it moves 96 cm more.

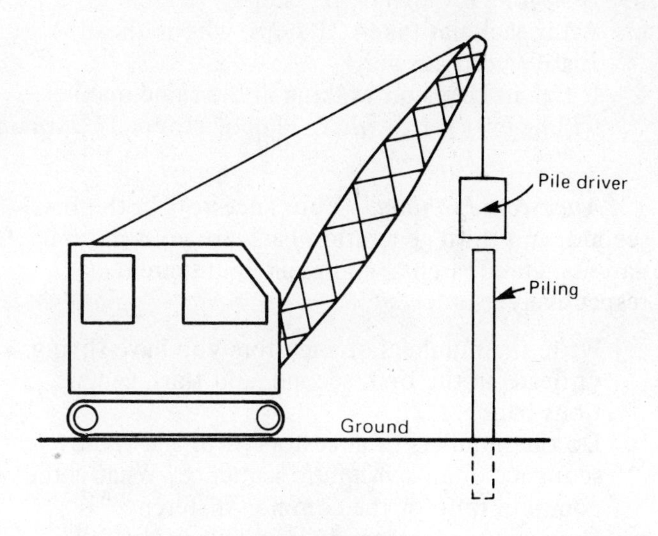

Figure 11-7b.

a. Assuming the distances it moves form an *arithmetic* sequence,

   i. how far will it move on the tenth impact?
   ii. how far into the ground will it be after ten impacts?

b. If you assume that the distances the piling moves with each impact form a *geometric* sequence,

   i. how far will it move on the tenth impact?
   ii. how far into the ground will it be after 10 impacts?
   iii. what is the farthest it can be driven into the ground?

12. ***Calvin and Phoebe Problem.*** Calvin Butterball and Phoebe Small start walking at the same time from the same point and in the same direction. Phoebe starts with a 12 inch step, and increases her stride by ½ inch each step. She goes 21 steps, then stops. Calvin starts with a 36 inch step, and each subsequent step is 90% as long as the preceding one.

a. What kind of sequence do Phoebe's steps follow?
b. What kind of sequence do Calvin's steps follow?
c. How long is Phoebe's last step?
d. How long is Calvin's 21st step?
e. After each has taken 21 steps, who is ahead? Justify your answer.
f. If Calvin keeps on walking in the same manner, will he ever get to where Phoebe stopped? Explain.

13. ***Ancestors Problem.*** Your ancestors in the first, second, and third generation back are your natural parents, grandparents, and great grandparents, respectively.

a. Write the numbers of ancestors you have (living or dead) in the first, second, and third generations back.
b. Do the numbers of ancestors form a *geometric* sequence or an *arithmetic* sequence? What is the common ratio or the common difference?
c. How many ancestors do you have in the 20th generation back?
d. What *total* number of ancestors do you have in the first 20 generations back?

14. ***Month's Pay Problem.*** Your Parents want you to do some work around the house on a regular basis. You get them to agree to pay you \$.01 the first day, \$.02 the second, \$.04 the third, \$.08 the fourth, and so on. At the end of 30 days, how much will they have paid you, total? Surprising?!

15. ***Tree Branch Problem.*** Pine tree branches grow in layers (see sketch). As sunlight hits the top layer, a certain fraction of the light is absorbed by those branches, and the rest goes through to the next layer. 100% of the incoming sunlight reaches the top layer. Suppose you find that 45% of the incoming sunlight reaches the ground after passing through all 5 layers of branches on a particular tree.

a. What fraction of the light reaching one layer makes it through to the next layer?
b. Find the percent of the *original* light that reaches the second, third, fourth, and fifth layers.
c. If a layer of branches gets less than 20% of the original sunlight, the branches in that layer will die and drop off. What is the maximum number of layers of branches you would expect to find on such a tree?

16. ***Musical Scale Problem.*** The musical scale on a piano has 12 notes (counting black keys) from *A* through *G*, then starts over again with *A* (see sketch). The frequence at which the piano strings vibrate is 220 cycles per second (cps) for the *A* below middle *C*, and 440 cps for the *A* above. Each note has a frequency which is one of the 11 positive geometric means between 220 and 440.

a. Find the common ratio.
b. Calculate to one decimal place the frequency of each note between the two A's (220 cps and 440 cps).
c. What is the frequency of the highest note on the piano, the C which is 51 notes above the 220 cps A?
d. What is the frequency of the lowest note on the piano, the A which is 36 notes below the 220 cps A?

17. ***Pineapple Plantation Problem.*** Suppose that you obtain a single pineapple plant of a new and unusual variety. You decide to go into the pineapple-growing business, propagating new plants from this one. You find the following:

i. New pineapple plants are started from cuttings of an old one as follows:

   1 from the crown (see sketch),
   4 from "slips" at the base of the fruit, 3 from "suckers" that grow from the roots, and 3 from sections of the stump.
ii. It takes 2 years for a pineapple to mature, which means that each generation takes 2 years.
iii. You can plant 4000 plants per acre.

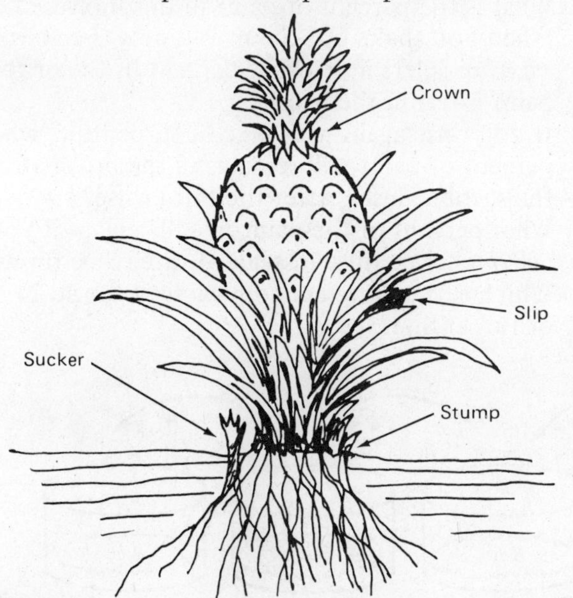

Answer the following questions.

a. Write the number of plants you would have after 0, 1, 2, and 3 generations, assuming that all possible plants live.
b. You have 1000 acres available for planting. How long would it take before all 1000 acres would be covered with pineapple plants?

18. ***Lion Hunting in Africa.*** When asked how to go about catching a lion in Africa, a mathematician replied, "That's easy. You just build a fence across Africa, dividing it in half. Then you find which side of the fence the lion is on, and build another fence, dividing that half in half. You continue the process until the lion is in a corral." Africa has an area of 30,300,000 square kilometers. How many fences would you have to build to trap the lion in a corral less than 10 square meters in area?

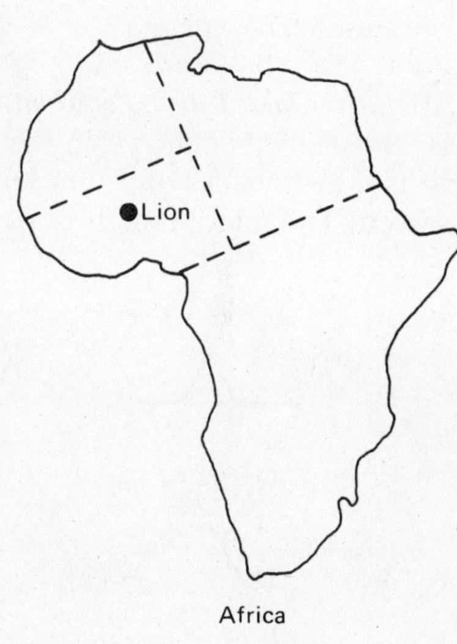

Africa

19. ***Nested Squares Problem.*** A set of nested squares is drawn inside a square of edge 1 unit (see sketch). The corners of the next square are at the midpoints of the sides of the preceding square.

a. Show that the lengths of the sides of the squares form a geometric sequence.
b. Show that the perimeters of the squares form a geometric sequence.
c. Show that the areas of the squares form a geometric sequence.

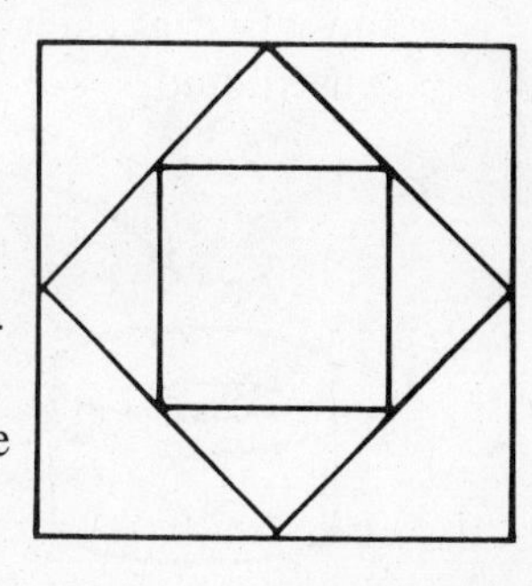

d. Find the area of the tenth square.
e. Find the perimeter of the tenth square.
f. Find the *sum* of the areas of the first 10 squares.
g. Find the sum of the perimeters of the first 10 squares.
h. Show that the sum of the areas of the squares approaches a finite number, and tell what that number is.
i. Does the sum of the perimeters approach a finite number? Explain.

20. ***Snowflake Curve Problem.*** A "snowflake curve" is constructed as shown in the sketch. An equilateral triangle with sides of 1 unit length has each side trisected. The middle sections of each side serve as bases

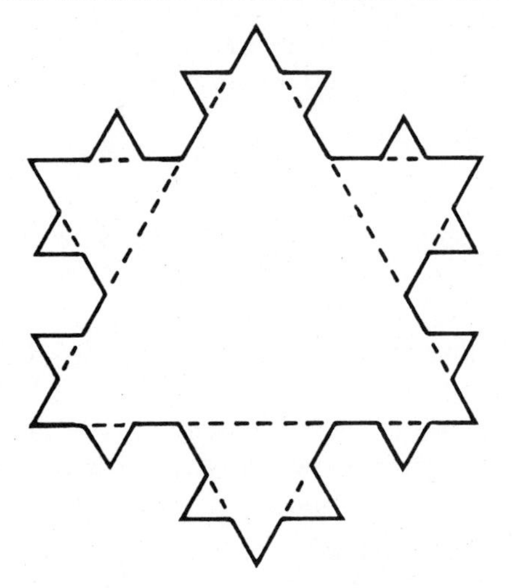

for smaller equilateral triangles. These triangles in turn have their sides trisected, and still smaller equilateral triangles are constructed. The process is carried on "infinitely."

a. Show that as each set of smaller triangles is constructed,
   i. the total area enclosed by the curve is a partial sum of a geometric series, and
   ii. the perimeter of the figure (i.e., the length of the curve) is a term in a geometric sequence.
b. Show that the area enclosed by the snowflake curve approaches a *finite* number, and tell what that number equals.
c. Show that the perimeter of the snowflake curve becomes *infinite* as the number of sides increases. Surprising?
d. See Martin Gardner's article in the December, 1976, issue of *Scientific American* for further information about snowflakes, "flowsnakes," and other "pathological monster" curves.

21. ***Paint Brush Problem.*** When you clean a paint brush, the amount of paint remaining in the brush depends on the number of times you rinse it with turpentine, and on the volume of turpentine you use for each rinse. Assume that a brush retains 2 cubic centimeters (cc) of fluid after it has been shaken out.

a. Before the first rinse, the 2 cc retained in the brush is pure paint. If 8 cc of turpentine is used for the rinse, the total volume of paint and turpentine you are mixing around is $2 + 8 = 10$ cc. What is the percent of paint in this mixture?
b. When you shake out the brush, only 2 cc of the mixture is left in it. What percent of the original paint is left in the brush?
c. If you rinse again with 8 cc of turpentine, what percent of paint still remains in the brush after the second rinse? After the third rinse?
d. What percent of the paint would remain if you use *one* 24 cc rinse instead of three 8 cc rinses?
e. Which is the more economical way to use 24 cc of turpentine?

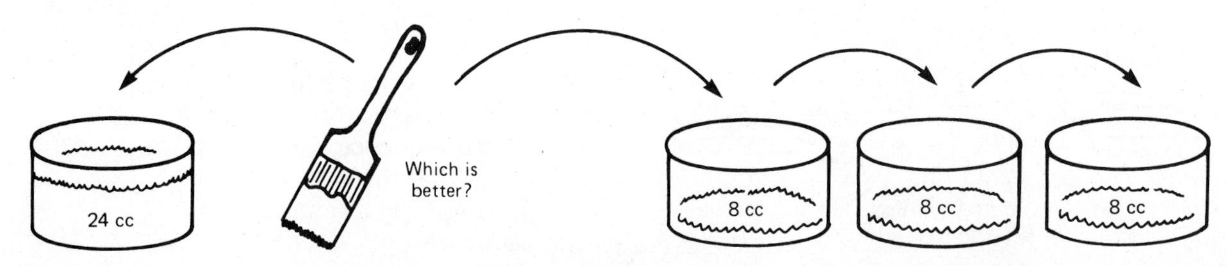

22. ***Geometry Problem.*** Many years ago, the people in Euclid, Ohio, planted a geoma tree. The first year it grew a trunk 2 meters long. The next year it grew two branches at right angles to each other, each 1 meter long. The third year it grew four ½-meter branches, two from each tip of last year's branches, at right angles to each other. Unlike non-Euclidean geoma trees, all the branches and the trunk grew in the same plane (this is a plane geoma tree).

a. Show that the lengths of the branches form a geometric sequence. What is the common ratio?
b. How high will the tree be after 2 years? 3 years? 4 years?
c. Show that the height of the tree is the sum of *two* geometric series. What are the common ratios?
d. If the tree keeps on growing this way forever,
   i. what height will it approach?
   ii. what width will it approach?
   iii. what will the length of a new branch approach?
   iv. what total length of branches will grow each year?
   v. what will the total length of all branches approach?
   vi. how close to the ground will the lowest branches come?

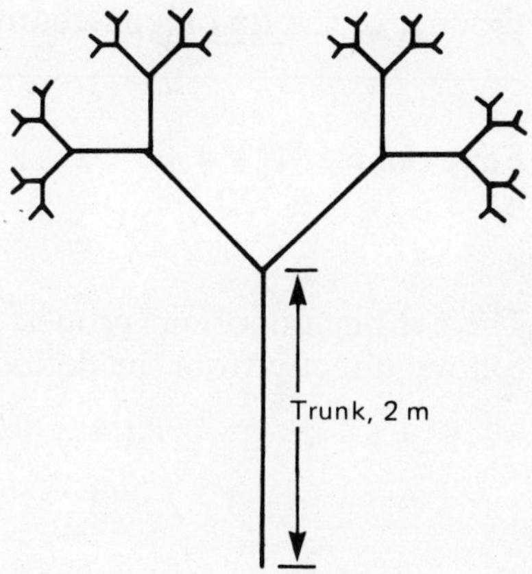

Trunk, 2 m

A Fine Geoma Tree

23. ***Middleman Problem.*** When you purchase food, clothing, etc., in a retail store, the item has passed through many hands before reaching you. For example, the farmer might sell to a trucker, who sells to a wholesaler, who sells to a packager, who sells to a distributor, who sells to a retailer, who sells to you. Each person between you and the farmer is called a "middleman." Suppose that a farmer spends \$.50 raising one pound of beef.

a. If the farmer and the 5 middlemen each make a profit of 30% on what they spent for the pound of beef, how much will you, the consumer, have to pay for the beef?
b. How many cents profit does the farmer make? How many cents profit do the five middlemen make, combined? Surprising?!
c. The farmer and each middleman insist that their profit should be increased to 40%, arguing that the extra 10% increase in price will not hurt the consumer that much. How much would you *actually* pay for the pound of beef under this condition? Does the price really increase by only 10%?

24. ***Interest Compounding Problem.*** Suppose that one bank pays 6¼% per year interest, compounded *annually*, and another bank pays 6% per year interest, compounded *monthly*. You have \$1000 to invest, and wish to know which is the better deal.

a. How much could you withdraw from each account at the end of 10 full years? On this basis, which seems better, a higher interest rate or more frequent compounding?
b. How much could you withdraw from each account at the end of 9 years, 11 months? Remember, if you withdraw before the end of a compounding period, you get *no* interest for that period. Which type of account would be better if you are likely to have to withdraw your money on short notice?
c. What annual interest rate, compounded annually, would you need in order to make the same amount of money at the end of 10 years as you would with 6% compounded monthly?

25. ***Rich Uncle Problem.*** Claire Voyance has a rich uncle who wishes to give Claire \$1000 on her 21st birthday. But there is a string attached. Claire must calculate how much her uncle must invest *now*

in a savings account in order to have \$1000 on the 21st birthday. Claire is 16 years 7 months old now. The savings account pays 6.2% per year interest, compounded quarterly. Since Claire has studied geometric sequences she realizes that the value of the \$1000 if left in the savings account *after* her 21st birthday would be a term in a geometric sequence, with independent variable p being the number of quarters the money has been left. The value of the \$1000 *before* her 21st birthday, therefore, will be given by the *same* sequence, but with the appropriate *negative* value substituted for p. How much should Claire tell her uncle to invest? (Businessmen call this amount the "present value" of the \$1000.)

26. ***Sweepstakes Problem.*** Suppose that you have just won the Readers Digest Sweepstakes. You have a choice of accepting a lump sum of \$20,000 now, or taking \$100 per month for the rest of your life. In either case, you will put the money into a savings account paying 4% per year interest, compounded monthly, and let the interest accumulate. Assume that the Internal Revenue Service charges you no income tax (unlikely, but assume it anyway!).

a. If you accept the \$20,000, the amount you will have after n months is the nth term of a geometric sequence. How much will you have after 10 years? After 50 years?
b. If you accept the \$100 per month, the amount you will have after n months is the nth partial sum of a geometric series. How much will you have after 10 years? After 50 years?
c. How long would it be before the amount you would have from the \$100 per month plan would *exceed* the amount you would have from the \$20,000 lump sum plan?
d. Show that if you can get an annual interest rate of 7%, compounded monthly, the \$100 per month plan will *never* give you as much money as the \$20,000 lump sum plan.
e. Ask your Economics instructor how income tax considerations might make the \$20,000 lump sum plan *less* desirable than your calculations above indicate.

11-8 FACTORIALS

In the next section and in Chapter 12 you will study series and other expressions which contain products of many consecutive integers. In this section you will study a convenient way of writing these products.

DEFINITION

The expression $\boldsymbol{n}!$ (read "n factorial") means the product of the first n consecutive positive integers.

For example, $4! = 4 \cdot 3 \cdot 2 \cdot 1 = 24$, and $6! = 6 \cdot 5 \cdot 4 \cdot 3 \cdot 2 \cdot 1 = 720$.

There is one important property of factorials that follows directly from the definition. For example,

$$\begin{aligned} 5! &= 5 \cdot 4 \cdot 3 \cdot 2 \cdot 1 \\ &= 5 \cdot (4 \cdot 3 \cdot 2 \cdot 1) \\ &= 5 \cdot 4! \end{aligned}$$

So in general,

$$\boxed{n! = n \cdot (n-1)!}\,.$$

In plain English, this property says that you can get the next factorial by multiplying the previous factorial by n. Reversing this pattern, you can get the previous factorial by *dividing*.

$$\begin{aligned} 4! &= 24 && \div 4 \\ 3! &= 6 && \div 3 \\ 2! &= 2 && \div 2 \\ 1! &= 1 && \div 1 \\ 0! &= ? \end{aligned}$$

If the pattern continues backwards, 0! must equal $1 \div 1$, which equals 1. This fact leads to a reasonable *definition* of 0!.

DEFINITION

$0! = 1.$

Objective:

Be able to use the definition of factorials to simplify expressions containing factorials, or to express in factorial form expressions containing products of consecutive integers.

Fractions which have factorials in the numerator and denominator or offer great possibilities for canceling. For example,

$\frac{10!}{7!} = \frac{10\cdot9\cdot8\cdot7\cdot6\cdot5\cdot4\cdot3\cdot2\cdot1}{7\cdot6\cdot5\cdot4\cdot3\cdot2\cdot1}$	Definition of factorials
$= 10\cdot9\cdot8$	Canceling.
$= 720.$	Arithmetic.

Conversely, a product of consecutive integers can be expressed in compact form using factorial notation. For example,

$9\cdot8\cdot7\cdot6$	
$= 9\cdot8\cdot7\cdot6\cdot\frac{5\cdot4\cdot3\cdot2\cdot1}{5\cdot4\cdot3\cdot2\cdot1}$	Multiplication property of 1.
$= \frac{9\cdot8\cdot7\cdot6\cdot5\cdot4\cdot3\cdot2\cdot1}{5\cdot4\cdot3\cdot2\cdot1}$	Multiplication property of fractions.
$= \frac{9!}{5!}$	Definition of factorials.

Once you understand this process, it can be done in *one* step in your head.

The exercise which follows is designed to give you practice working with factorials so that you will be familiar with them when you encounter them in the next section and in the next chapter.

Exercise 11-8

For Problems 1 through 16, simplify the given expression:

1. $3!\,4!$ (i.e., $3!\cdot4!$)
2. $3!\,5!$ (i.e., $3!\cdot5!$)
3. $\frac{8!}{4!}$
4. $\frac{8!}{5!}$
5. $\frac{7!}{0!}$
6. $\frac{8!\,3!}{6!}$
7. $\frac{10!}{5!\,3!}$
8. $\frac{6!}{2!\,4!}$
9. $\frac{5!}{2!\,3!}$
10. $\frac{10!}{7!\,3!}$
11. $(3!)!$
12. $(2!)!$
13. $\frac{(n-1)!}{n!}$
14. $\frac{(n+1)!}{n!}$
15. $\frac{(n+1)!}{(n-1)!}$
16. $\frac{(n+2)!}{n!}$

For problems 17 through 24, write the given expression as a ratio of factorials.

17. $7\cdot6\cdot5\cdot4$
18. $9\cdot8\cdot7\cdot6$
19. $20\cdot19\cdot18\cdot17\cdot16\cdot15$
20. $35\cdot34\cdot33\cdot32$
21. $20\cdot19\cdot18\cdot12\cdot11\cdot10$
22. $100\cdot99\cdot98\cdot50\cdot49\cdot48$
23. $\frac{30\cdot29\cdot28\cdot27\cdot26\cdot25}{6!}$
24. $\frac{43\cdot42\cdot41\cdot40}{4!}$
25. For what values of n will $n!$ be evenly divisible by 9?
26. For what values of n will $n!$ end in zero, when multiplied out?

27. 100! ends with a string of zeros. How many zeros are there at the end of 100!?

28. a. Write a computer program for calculating values of $n!$ for given values of n. The program should begin with inputting a value of n. You can start a variable P (for "product") at 1, then set up a loop in which P is multiplied by successive integers 2, 3, 4, ... , up through n. The output should be the value of n and the value of $n!$ at each pass through the loop.

b. Modify the program so that it *adds* the *logs* of the integers, then takes the antilog, thus allowing you to calculate numbers like 100!.

For Problems 29 through 32, evaluate the given sum.

29. $\sum_{k=0}^{4} k!$

30. $\sum_{k=0}^{4} \frac{(k+1)!}{k!}$

31. $\sum_{k=1}^{4} \frac{(k-1)!}{k!}$

32. $\sum_{k=1}^{5} \frac{(k+1)!}{(k-1)!}$

33. You have studied quadratic functions, exponential functions, and now factorials, each of which gets much bigger as the independent variable increases. Make a table of n^2, 2^n, and $n!$ for the integers $n = 0$ through 10. Tell which one of these functions grows the *fastest.*

34. In the sequence for which $t(n) = (n-1)! + 1$, certain ones of the terms are divisible by n and others are not. For example, $t(5) = 25$, and 25 is divisible by 5. But $t(6) = 121$, and 121 is *not* divisible by 6. Write the values of $t(1)$ through $t(11)$. Then try to determine what kind of number n is when $t(n)$ is divisible by n. (This property is known as Wilson's Theorem.)

35. The series specified by $S(n) = \sum_{k=0}^{n} \frac{1}{k!}$ has partial sums which get closer and closer to an interesting number as n gets very large.

a. Show that the ninth term of the series (i.e., $k = 8$) is so small that it contributes nothing to the first four decimal places.
b. Find a four decimal place approximation for $S(7)$. (This number is called "e," and is used as the base for "natural" logarithms.)

36. The function specified by the series

$$f(x) = 1 - \frac{x^2}{2!} + \frac{x^4}{4!} - \frac{x^6}{6!} + \dots,$$

converges to a *rational* number when x is certain multiples of the irrational number π.

a. Using a calculator or computer, find values of the first, second, third, ... , tenth partial sum of $f(\pi)$. Use the decimal approximation 3.141592654 for π. To what number does the series seem to be converging?
b. Find an approximation for $f(\pi/2)$. Use enough terms to find the rational number to which the series converges.
c. To what rational number does $f(\pi/3)$ converge?

11-9 INTRODUCTION TO BINOMIAL SERIES

You recall how to square a binomial. For example,

$$(a+b)^2 = a^2 + 2ab + b^2.$$

You can *cube* a binomial by using the above results. You get

$$\begin{aligned}(a+b)^3 &= (a+b)^2 (a+b)\\ &= (a^2 + 2ab + b^2)(a+b)\\ &= a^3 + 2a^2b + ab^2 + a^2b + 2ab^2 + b^3\\ &= a^3 + 3a^2b + 3ab^2 + b^3.\end{aligned}$$

The answer is a *series* of terms. It is called a *binomial* series because it comes from expanding a binomial raised to a power. There are several patterns which show up in this series. If you know the patterns, you will eventually be able to raise a binomial to a power *mentally*, in one step. In this section you will try to discover some of these patterns.

Objective:

Discover patterns followed by the signs, exponents, and coefficients in a binomial series.

The following exercise is designed to lead you to some of these discoveries.

Exercise 11-9

1. Expand the binomial power $(a + b)^4$ into a binomial series. This is most easily done by observing that $(a + b)^4 = (a + b)^3 \, (a + b)$, and using the series for $(a + b)^3$ from above.

2. Expand $(a + b)^5$ as a binomial series using the results of Problem 1, above.

3. If you expand $(a + b)^6$, you will get

$$(a + b)^6 = a^6 + 6a^5b + 15a^4b^3 + 20a^3b^3 + 15a^2b^4 + 6ab^5 + b^6.$$

By observing the patterns in this binomial series and in your answers to Problems 1 and 2, tell

a. the pattern followed by the powers of a,
b. the pattern followed by the powers of b,
c. the degree of each term,
d. the number of terms that will be in the series, and
e. any pattern you see in the coefficients as you look at the series from both ends.

4. The pattern followed by the powers of a and b is easy to see. The pattern of the coefficients is more difficult. By arranging the coefficients for the various powers of $(a + b)$ in a *triangle*, one pattern shows up.

$(a+b)^0$. 1
$(a+b)^1$. 1 1
$(a+b)^2$. 1 2 1
$(a+b)^3$. 1 3 3 1
$(a+b)^4$. 1 4 6 4 1
$(a+b)^5$. 1 5 10 10 5 1
$(a+b)^6$. 1 6 15 20 15 6 1

This triangular array of numbers is called Pascal's Triangle after the French mathematician Blaise Pascal, who lived from 1623 to 1662. Tell what pattern relates the numbers in one row of Pascal's Triangle to the numbers in the *preceding* row. Then demonstrate that you *understand* the pattern by finding the coefficients which will be in the row for $(a + b)^7$.

5. Suppose someone tells you that the coefficients of $(a + b)^9$ are 1 9 36 84 126 126 84 36 9 1. Use what you have discovered to expand $(a + b)^{10}$ as a binomial series in *one* step.

6. Suppose you wish to expand a binomial with a "–" sign between the terms, such as $(x - y)^6$. Observe that $(x - y)^6$ equals $[x + (-y)]^6$. Then use the pattern for the binomial series for $(a + b)^6$ from Problem 3, above, to expand $(x - y)^6$ as a binomial series. What is the *only* difference in the series when the sign of the binomial is "–" instead of "+"?

7. You have defined a "series" to be a sum of an *infinite* number of terms. Yet the binomial series of this section appears to have only a finite number of terms. What do you suppose all the rest of the terms of a binomial series must equal?

11-10 THE BINOMIAL FORMULA

In Exercise 11-9 you raised binomials to powers. For example,

$$(a + b)^6 = a^6 + 6a^5b + 15a^4b^2 + 20a^3b^3 + 15a^2b^4 + 6ab^5 + b^6.$$

By examining such "binomial series," you can find several patterns. For the series from expanding $(a + b)^n$:

1. There are $n + 1$ terms.

2. Each term is n^{th} degree.

3. The powers of a start at a^n, and the exponent *decreases* by 1 each term; the powers of b start at b^0 and *increase* by 1 each term.

4. The coefficients are *symmetrical* with respect to the ends.

5. The coefficients can be calculated from the *previous row* of Pascal's Triangle by *adding* pairs of adjacent coefficients in that previous row (see Exercise 11-9, Problem 4).

In this section you will learn how to calculate *any* coefficient in *any* binomial series *without* having to know the previous row in Pascal's Triangle.

Objectives:

1. Given a binomial power, expand it as a binomial series in *one* step.

2. Given a binomial power of the form $(a + b)^n$, find term number k, or find the term which contains b^r, where k and r are integers from 0 through n.

There is a pattern to the coefficients which is difficult to discover, but is easy to remember and use once you have learned it. Consider the binomial series

$$(a+b)^8 =$$

$$\underset{①}{a^8} + \underset{②}{8a^7b} + \underset{③}{28a^6b^2} + \underset{④}{56a^5b^3} + \underset{⑤}{70a^4b^4}$$

$$+ \underset{⑥}{56a^3b^5} + \underset{⑦}{28a^2b^6} + \underset{⑧}{8ab^7} + \underset{⑨}{b^8}.$$

The numbers in circles are the term numbers. If you multiply the coefficient of a term by the exponent of a over the term number, you get the coefficient of the *next* term! For example,

$$8\cdot\frac{7}{2} = 28,$$

$$28\cdot\frac{6}{3} = 56,$$

$$\underset{\text{Coefficient}}{56}\cdot\frac{\overset{\text{Exponent of } a}{5}}{\underset{\text{Term number}}{4}} = 70, \xleftarrow{\text{Next coefficient}}$$

In general,

$$\frac{(\text{coefficient})\ (a - \text{exponent})}{\text{term number}} = \text{next coefficient.}$$

A still more interesting pattern shows up if you do *not* simplify the coefficients as you calculate them. For example, the first term of $(a + b)^8$ is

① a^8.

Using the pattern, the second term is

② $\frac{8}{1}a^7b$.

Multiplying the coefficient, 8/1, by the exponent of a over the term number, 7/2, allows you to write term 3.

③ $\frac{8}{1}\cdot\frac{7}{2}a^6b^2$.

The fourth and fifth terms may be calculated in the same way.

④ $\frac{8}{1}\cdot\frac{7}{2}\cdot\frac{6}{3}a^5b^3$ ⑤ $\frac{8}{1}\cdot\frac{7}{2}\cdot\frac{6}{3}\cdot\frac{5}{4}a^4b^4$.

By now you should be able to see the pattern. The coefficients can be written as *factorials*. For example, the coefficient of the fourth term is

$\frac{8}{1}\cdot\frac{7}{2}\cdot\frac{6}{3} = \frac{8\cdot7\cdot6}{1\cdot2\cdot3}$	Multiplication property of fractions.
$= \frac{8\cdot7\cdot6}{1\cdot2\cdot3}\cdot\frac{5}{5}$	Multiplicative identity.
$= \frac{8!}{3!\,5!}$	Multiplication property of fractions, and definition of factorials.

The three numbers 8, 3, and 5 show up as exponents at various places, as shown below.

$$(a+b)^8 = \ldots + \frac{8!}{3!\,5!}a^5b^3 + \ldots .$$

The 8 in the numerator is the exponent n, to which $(a + b)$ is raised. The 3 and 5 in the denominator are the exponents of a and b in the term you are looking for.

With this knowledge, you can now accomplish the objective of this section.

Example 1: Find the term in $(a - b)^{17}$ which contains b^{11}.

Your reasoning should be as follows:

1. The power of a must be a^6, since the exponents of a and b must add up to 17.

2. Therefore, the term is $\frac{17!}{11!\,6!} a^6 (-b)^{11}$, which equals

$-\frac{17!}{11!\,6!} a^6 b^{11}$ since $-b$ is raised to an *odd* power.

3. If asked to, you can do the arithmetic, getting

$$\frac{17!}{11!\,6!} = \frac{17 \cdot 16 \cdot 15 \cdot 14 \cdot 13 \cdot 12 \cdot 11!}{11! \cdot 6 \cdot 5 \cdot 4 \cdot 3 \cdot 2}$$

$= 12376$, so the term is $-12376a^6 b^{11}$.

Example 2: Find the 8th term of $(a + b)^{12}$.

Your reasoning should be as follows:

1. The 8th term is the one that has b^7, since the 1st term has b^0.

2. Therefore, the 8th term has $a^5\, b^7$, since the exponents must add up to 12.

3. The term is thus $\frac{12!}{7!\,5!} a^5 b^7$, which equals $792\, a^5 b^7$.

Example 3: Find the 5th term of $(r^3 - 2s)^{10}$.

Your reasoning should be as follows:

1. Instead of a and b, the terms of the binomial are r^3 and $-2s$.

2. The 5th term will have $(-2s)^4$.

3. The power of r^3 will be $(r^3)^6$, since the exponents of r^3 and $-2s$ must add up to 10.

4. The term is therefore $\frac{10!}{4!\,6!} (r^3)^6 (-2s)^4$.

5. This can be simplified to $210r^{18} (16s^4)$ which equals $3360r^{18}s^4$.

Note that the exponents no longer add up to 10 after the simplification is done.

Example 4: Find the term containing b^r in $(a + b)^n$.

Using the same reasoning you used in the previous examples, you can immediately write the answer:

$$\text{term} = \frac{n!}{r!\,(n-r)!} a^{n-r} b^r.$$

This formula is called the *Binomial Formula* since it is used to calculate terms in a binomial series.

Example 5: Expand $(r^2 - 3s)^7$ as a binomial series.

This is most easily accomplished by using the "pattern" to find the coefficients. The first term is

$\underset{①}{(r^2)^7}$.

Multiplying the coefficient, 1, by the r^2 exponent, 7, and dividing by the term number, 1, gives 7. Thus the series begins

$\underset{①}{(r^2)^7} + \underset{②}{7(r^2)^6 (-3s)} + \ldots$.

Working with the second term, you multiply the coefficient 7 by 6/2 (the r^2 exponent over the term number), getting 21. So the series continues

$$(r^2)^7 + 7(r^2)^6(-3s) + 21(r^2)^5(-3s)^2 + \dots .$$
① ② ③

Using the same reasoning for following terms gives the complete series,

$$(r^2 - 3s)^7 = (r^2)^7 + 7(r^2)^6(-3s) + 21(r^2)^5(-3s)^2 + 35(r^2)^4(-3s)^3 + 35(r^2)^3(-3s)^4 + 21(r^2)^2(-3s)^5 + 7(r^2)(-3s)^6 + (-3s)^7.$$

The objective, writing the binomial series in one step, has thus been accomplished. If you must do the arithmetic and simplify, you get

$$(r^2 - 3s)^7 = \underline{\underline{r^{14} - 21r^{12}s + 189r^{10}s^2 - 945r^8s^3 + 2835r^6s^4 - 5103r^4s^5 + 5103r^2s^6 - 2187s^7}}.$$

Note that the symmetrical pattern of the coefficients and the constant degree of the terms may disappear after you have done the simplifying.

Exercise 11-10

For Problems 1 through 20:

a. Expand the given binomial power as a binomial series.
b. Simplify the resulting series as much as possible.

1. $(x + y)^5$
2. $(a + b)^8$
3. $(p - m)^7$
4. $(z - w)^9$
5. $(a + 2)^4$
6. $(x - 2)^6$
7. $(2x - 3)^5$
8. $(3a + 2)^4$
9. $(x^2 + y^3)^6$
10. $(a^3 - b^2)^5$
11. $(2a - b^4)^3$
12. $(4r + s^2)^3$
13. $(x + x^{-1})^8$
14. $(x^2 - x^{-2})^{10}$
15. $(x^{½} - x^{-½})^4$
16. $(x^{¼} + x^{-¼})^8$
17. $(1 + i)^6$
18. $(1 - i)^4$
19. $(1 - i)^5$
20. $(1 + i)^7$

Note: $i = \sqrt{-1}$.

For Problems 21 through 40, find the term with the specified power in the expansion of the given binomial power. You may leave the coefficient in factorial form unless instructed otherwise.

21. $(x + y)^8$, y^5
22. $(p + j)^{11}$, j^4
23. $(r + s)^{13}$, s^7
24. $(m + q)^{10}$, q^5
25. $(p - j)^{15}$, j^{11}
26. $(c - d)^{19}$, d^{15}
27. $(a - b)^{21}$, b^{16}
28. $(x - y)^{25}$, y^{20}
29. $(e - f)^{58}$, f^{23}
30. $(g - h)^{64}$, h^{39}
31. $(x + y)^{73}$, x^{48}
32. $(a + b)^{81}$, a^{19}
33. $(x^3 + y^2)^{15}$, y^{12}
34. $(x^3 + y^2)^{29}$, y^{10}
35. $(x^3 - y^2)^{13}$, x^{18}
36. $(x^3 - y^2)^{24}$, x^{30}
37. $(x^3 + y^2)^{42}$, y^{15}
38. $(x^3 + y^2)^{107}$, y^{77}
39. $(3x + 2y)^8$, y^5
40. $(3x + 2y)^7$, y^4

For Problems 41 through 50, find the specified term in the expansion of the given binomial power. You may leave the coefficient in factorial form unless instructed otherwise.

41. $(j + k)^{34}$, 17th term
42. $(p + k)^{51}$, 19th term
43. $(r - q)^{15}$, 12th term
44. $(a - b)^{17}$, 8th term
45. $(a - b)^{12}$, 13th term
46. $(x - y)^{19}$, 20th term
47. $(3x^2 - 2y^3)^7$, 4th term
48. $(3x^2 - 2y^3)^8$, 6th term
49. $(x^3 - 5)^6$, 3rd term (Simplify the answer.)
50. $(r^4 - 7)^5$, 3rd term (Simplify the answer.)

51. Suppose that $(r + s)$ is raised to some positive integer power, and one term in the binomial series is $27132\, r^{13} s^6$.

a. *Which* term is it?
b. To what *power* was $(r + s)$ raised?
c. What is the *next* term? (Simplify the coefficient.)
d. How *many* terms are there in this series?

52. Suppose you are told that $3274\, r^{17} s^{66}$ is a term in a binomial series, and you are asked to find the *next* term. Can you do it? If so, *how*? If not, *why* not?

53. Find a decimal approximation for 1.02^{10} by writing it as $(1 + 0.02)^{10}$ and calculating the first five terms of the resulting binomial series.

54. The pattern of multiplying the coefficient by the exponent and then dividing by the term number works for *non-integer* exponents, too. Use this fact to expand

$$(a + b)^{3/2}$$

as a binomial series. Tell how many terms there will be in this binomial series. Then tell why a binomial series has a *finite* number of terms when the exponent is an integer.

55. Expand the binomial $(1 + 1)^5$ as a binomial series, simplify each term, then add up the terms. Next, show how you could have gotten the answer in a much easier way. Demonstrate that you understand the significance of what you have done by finding the *sum* of the *coefficients* of the binomial series for $(a + b)^{57}$ in *one* step.

11-11 CHAPTER REVIEW AND TEST

In this chapter you have studied functions whose domain is a set of integers rather than the set of all real numbers. Sometimes you can figure out a formula for such sequences of numbers just by observing a pattern. In particular, you were able to derive general formulas for arithmetic and geometric sequences, and for partial sums of their corresponding series. You used sequences and series as mathematical models for things in the real world that change by "steps" rather than changing "continuously." Finally, you learned about factorials and binomial series, mainly as background for work with probability in the next chapter.

The objectives for the chapter may be summarized as follows:

1. *Given the first few terms of a sequence,*
   a. *discover a pattern,*
   b. *write the next few terms,*
   c. *draw the graph,*
   d. *derive a formula,*
   e. *calculate* ***any*** *specified term.*
2. *Calculate any specified term in a given arithmetic or geometric sequence, or in a binomial series.*
3. *Find partial sums of arithmetic or geometric series.*
4. *Find arithmetic and geometric means.*
5. *Use sequences or series as mathematical models.*

The Review Problems below are meant to be a relatively straight-forward test of your ability to accomplish these objectives one at a time. The Concepts Test requires you to use several techniques on the same problem, and to apply your knowledge to new situations.

Review Problems

The following problems are numbered according to the five objectives listed above.

Review Problem 1. For the sequence, 2, 6, 12, 20, 30, 42, 56, ...:

a. Write the next two terms, telling what pattern you used.
b. Is the sequence arithmetic, geometric, or neither? Justify your answer.

c. Draw the graph of the first six terms of the sequence.
d. Write an equation expressing $t(n)$ in terms of n.
e. Use the equation to calculate $t(700)$.

Review Problem 2a. i. Find $t(56)$ for the arithmetic sequence with $t(1) = 237$ and common difference of -7.

ii. An arithmetic sequence has $t(1) = 164$ and common difference of -9. If $t(n) = -484$, find n.

b. i. Find $t(31)$ for the geometric sequence with $t(1) = 17$ and common ratio of 1.2.

ii. A geometric sequence has $t(1) = 11$ and common ratio of 3. If $t(n) = 24057$, find n.

c. For the binomial series from expanding $(h - p)^{10}$,

i. find the term that has p^7,
ii. find the seventh term.

Review Problem 3a. Find $S(200)$ for the arithmetic series $6 + 13 + 20 + \ldots$.

b. i. Find $S(20)$ for the geometric series $100 + 96 + 92.16 + \ldots$.

ii. Find the number to which the above geometric series converges as n becomes very large.

c. Evaluate $\sum_{k=1}^{5} k!$.

Review Problem 4a. Insert five arithmetic means between 17 and 63.

b. Find the geometric mean of 4 and 100 (*two* possible answers).

Review Problem 5 – Swinging Problem. When a person is swinging, the "amplitude" of the swing increases with each pump (see sketch). According to *Scientific American* magazine (April, 1977, page 60), the amplitude increases by *adding* a constant when the person pumps sitting down, and by *multiplying* by a constant when the person pumps standing up. Suppose that when you get on a swing, it has an amplitude of 3 centimeters (i.e., $t(0) = 3$). Assume that the amplitude increases by adding 2 centimeters each pump when you are sitting, and by multiplying by 1.2 each pump when you are standing.

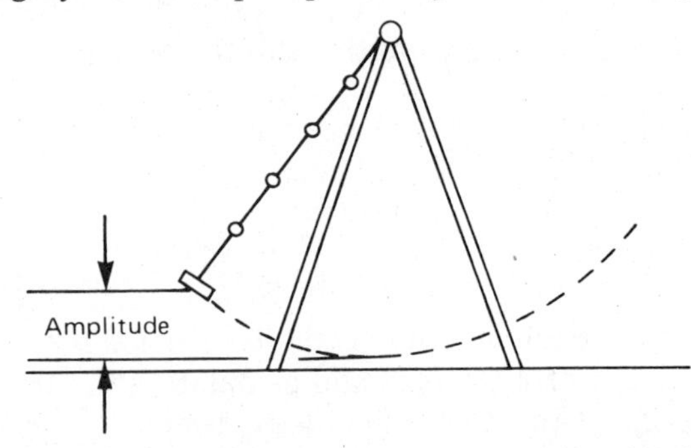

a. If you are sitting, what is your amplitude after 10 pumps? After 20 pumps?
b. If you are standing, what is your amplitude after 10 pumps? After 20 pumps?
c. What is the minimum number of pumps it takes to make the amplitude equal 25 centimeters when you are sitting? Standing?
d. What is the minimum number of pumps it takes to make the amplitude equal 75 centimeters when you are sitting? Standing?
e. On the same Cartesian coordinate system, draw graphs of amplitude versus number of pumps for sitting and for standing. You should have enough plotting data from the above information and calculations.
f. According to your mathematical model, what would happen if you started with an amplitude of 0 and tried to swing standing up?
g. Which way of pumping increases your amplitude faster, sitting or standing? Justify your answer.

Concepts Test

Work each of the following problems. Then tell by number and letter which one or ones of the five objectives for this chapter you used. If you are applying your knowledge to a new concept, write, "new concept."

Concepts Test 1. The following is a geometric sequence that extends infinitely far in *both* directions.

$$\ldots, \frac{128}{243}, \frac{64}{81}, \frac{32}{27}, \frac{16}{9}, \frac{8}{3}, 4, \ldots \longleftarrow t(n)$$

$$(-2) \quad (-1) \quad (0) \quad (1) \quad (2) \quad (3) \longleftarrow n$$

a. Find the common ratio, r.
b. Show that $t(n) = t(0) \cdot r^n$.
c. Find decimal approximations for $t(20)$ and $t(-10)$.
d. Find the *sum* of the terms from $t(1)$ through $t(20)$.
e. Show that the sum $t(0) + t(1) + t(2) + \ldots$ *diverges,* but that the sum $t(0) + t(-1) + t(-2) + \ldots$ *converges.* Find the number to which this second sum converges.
f. Plot the graph of the above terms, $t(-2)$ through $t(3)$.
g. If n could be *any* real number, rather than just an integer, what kind of function would $t(n)$ be called?

Concepts Test 2 – Bode's Law Problem. In 1776, Johann Titus discovered that the distances of the planets from the Sun are proportional to the terms of a rather simple sequence:

4	7	10	16	28	52	100
Mercury	Venus	Earth	Mars	Ceres (asteroid)	Jupiter	Saturn
(1)	(2)	(3)	(4)	(5)	(6)	(7)

a. Describe the pattern followed by the terms of this sequence. (Planet 1, Mercury, does *not* fit this pattern.) Then use this pattern to find the term for Uranus, planet number 8.
b. Planets 9 and 10, Neptune and Pluto, have distances corresponding to the numbers 305 and 388. Are these numbers terms of the sequence? Justify your answer.
c. Find a formula for $t(n)$, the term value, in terms of n, the planet number. For what values of n is the formula valid?
d. See *Scientific American* magazine, July, 1977, page 128, to see why this is called "Bode's" law.

Concepts Test 3 – Office Building Problem. Suppose that you are responsible for predicting the cost of constructing a new multi-story office building. The cost per square meter of floor space for constructing the higher stories increases because it is more difficult to build them.

a. You find that the first story costs \$400 per square meter, and the fifth story costs \$500 per square meter. Assuming that the costs per square meter form an arithmetic sequence, find the common difference.
b. What is the cost per square meter for the second, third, and fourth stories?
c. The building is to be 48 stories high. How much per square meter will the top story cost?
d. What is the total cost per square meter for building all 48 stories?
e. If each story is to have 1000 square meters of floor space, what is the total cost of the building?

Concepts Test 4. Let ${}_nC_r$ be the *coefficient* of the b^r term in the binomial series from expanding $(a + b)^n$. For example, in $(a + b)^5$, the term with b^2 is $10a^3b^2$. So ${}_5C_2 = 10$.

a. Write ${}_{17}C_9$ in terms of factorials.
b. Find ${}_3C_0$, ${}_3C_1$, ${}_3C_2$, and ${}_3C_3$, then *add* these four numbers.
c. Find $\sum_{r=0}^{4} C_r$.
d. Based on your answers to parts b and c, what is a formula for the *sum* of the coefficients in the binomial series that comes from expanding $(a + b)^n$? Show that your formula gives the correct answers for $(a + b)^2$, $(a + b)^1$, and $(a + b)^0$.

Concepts Test 5. Prove that the partial sum, $S(n)$, of a geometric series is equal to a *constant* plus a term in a *different* geometric sequence.

Probability, and Functions of a Random Variable

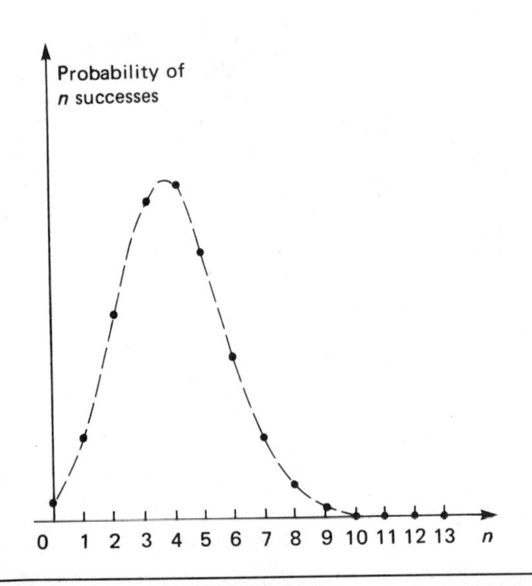

An occurrence is said to happen "at random" if there is **no** *way of telling for sure what the outcome will be on any particular occasion. In this chapter you will make a precise definition of the* ***probability*** *that a random occurrence comes out a certain way. The mathematical concepts were originally invented for analyzing gambling problems. You will use the same concepts for problems ranging from genetics and insurance to the reliability of complex systems such as power plants and rocket engines. Such problems will tie together much of the mathematics you have learned so far!*

12-1 INTRODUCTION TO PROBABILITY

You have used statements such as, "I shall probably be home by 10:00 o'clock." Mathematicians attempt to make the word "probably" have a more precise meaning by attaching *numbers* to it. For example, the probability of rain may be 30%. The idea for doing this arose from the analysis of gambling games several hundred years ago.

Suppose that two dice are rolled, a black one and a white one. The possible outcomes are shown in Figure 12-1.

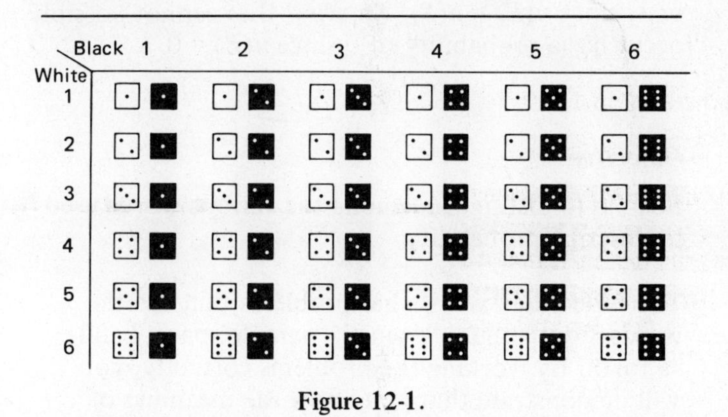

Figure 12-1.

There are *five* ways the total on the dice could equal 6.

Since each outcome is *equally likely*, you would expect that in many rolls of the dice, the total would be 6 roughly 5/36 of the time. This number, 5/36, is called the *probability* of rolling a total of 6.

In the following exercise, you will find the probabilities of other such events.

Exercise 12-1

A pair of dice is rolled, one black and one white. Find the probability of each of the following events:

1. The total is 10.
2. The total is at least 10.
3. The total is less than 10.
4. The total is at most 10.
5. The total is 7.
6. The total is 2.
7. The total is between 3 and 7, inclusive.
8. The total is between 3 and 7.
9. The total is between 2 and 12, inclusive.
10. The total is 13.
11. The numbers are 2 and 5.
12. The black die has 2 and the white die has 5.
13. The black die has 2 or the white die has 5.

12-2 WORDS ASSOCIATED WITH PROBABILITY

So that people may talk more efficiently about probability, various things are assigned *names*, with precise meanings. For the dice-rolling example in Section 12-1, the names would be as follows:

The act of rolling the dice is called a *random experiment.* The word "random" lets you know that there is *no* way of telling beforehand just how any one roll is going to come out.

Each way in which the dice can come up, such as

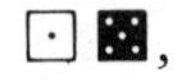

is called an *outcome*. Outcomes are *equally likely* results of a random experiment.

An *event* is a *set* of outcomes. For example, the event, "The total on the two dice is 6," is the five-element set

{⚀ ⚄, ⚁ ⚃, ⚂ ⚂, ⚃ ⚁, ⚄ ⚀}.

Two events are usually *not* equally likely. The only condition under which they would be equally likely is if they contained the same number of outcomes.

The set of *all* outcomes is called the *sample space.* The sample space for the dice roll is the set of all 36 outcomes in the table of Section 12-1.

The *probability* of an event may now be defined more precisely.

DEFINITION

The ***probability*** of an event is

$$\text{Probability} = \frac{\text{Number of outcomes in the event}}{\text{Number of outcomes in the sample space}}.$$

Since the probability of an event depends on the number of outcomes in that event, a form of "$f(x)$" terminology can be used. Letting P stand for "probability," you can write

$$P(\text{The total is 6}) = \frac{5}{36}.$$

Note that the argument inside the parentheses may be a *set* rather than simply a number. The "$f(x)$" terminology can be used for the numbers of elements in the event and in the sample space. Letting n stand for "number," the numbers of elements in Event E and in the sample space, S, would be $n(E)$ and $n(S)$, respectively.

In general, therefore, the probability of Event E is

$$\boxed{P(E) = \frac{n(E)}{n(S)}}.$$

Note: All probabilities are numbers between 0 and 1. An event that is *certain* to occur has a probability of 1, since $n(E) = n(S)$. An event that cannot possibly occur has a probability of 0 since $n(E) = 0$.

Objective:

Be able to distinguish among the various words used to describe probability.

The following exercise has problems that use the words, "outcome," "event," "sample space," and so forth. By working the problems correctly, you will demonstrate that you know the meanings of these words.

Exercise 12-2

1. A card is drawn at random from a normal 52-card deck.

   a. What name is given to the act of drawing the card?

b. How many outcomes are there in the sample space?
c. How many outcomes are there in the event, "The card is a face card?"
d. Calculate *P*(The card is a face card).
e. Calculate *P*(The card is black).
f. Calculate *P*(The card is an Ace).
g. Calculate *P*(The card is between 3 and 7, inclusive).
h. Calculate *P*(The card is the 8 of clubs).
i. Calculate *P*(The card belongs to the deck).
j. Calculate *P*(The card is a Joker).

2. A penny, a nickel, and a dime are flipped at the same time. Each coin can come out either heads (*H*) or tails (*T*).

a. What name is given to the act of flipping the coins?
b. There are eight elements in the sample space (for example, *HHH, THT,* and so forth). List all eight outcomes.
c. How many outcomes are there in the event, "Exactly two of the coins are heads?"
d. Calculate *P*(*HHT*).
e. Calculate *P*(Exactly two heads).
f. Calculate *P*(At least two heads).
g. Calculate *P*(Penny and nickel are tails).
h. Calculate *P*(Penny or nickel are tails).
i. Calculate *P*(None is tails).
j. Calculate *P*(Zero, one, two or three heads).
k. Calculate *P*(Four heads).

12-3 TWO COUNTING PRINCIPLES

The difficulty with calculating probabilities is counting up the number of elements in an event or in the sample space. For example, if ten people line up at random, and you want to find the probability that you will be next to your best friend, the sample space contains over three *million* outcomes! Obviously, it is impractical to list all of the outcomes and count them as you did in Exercise 12-1 and 12-2.

Objective:

Be able to determine the number of outcomes in an event or sample space *without* listing and counting them.

There are two principles of counting which form the basis for accomplishing this objective. One applies when Event *A* and Event *B* are *both* performed. The other applies when *either* Event *A* or Event *B* is performed, but not both.

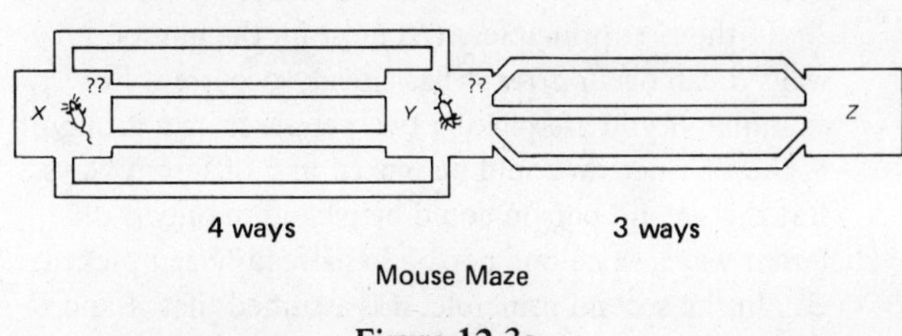

Mouse Maze

Figure 12-3a.

Suppose that a mouse is placed in Room *X* of a maze (See Figure 12-3a). There are four ways it can get to Room *Y*, and three ways it can get from *Y* to *Z*. For each *one* way of getting to *Y*, there are *three* ways of getting to *Z*. So there is a total of

$$4 \times 3 \text{ ways,}$$

or 12 ways it could go to *Y and then* to *Z*.

If the mouse were placed in Room *Y*, it could go *either* to *X or* to *Z*. Since there are 4 ways of getting to *x* and 3 ways of getting to *Z*, there is a total of

$$4 + 3 \text{ ways,}$$

or seven ways it could go to *X or* to *Z*.

These examples illustrate the two counting principles.

Principle 1: If $n(A)$ and $n(B)$ are the numbers of ways that Events *A* and *B* can occur, respectively, then

$$\boxed{n(A \text{ and then } B) = n(A) \times n(B)}\ .$$

Principle 2: If $n(A)$ and $n(B)$ are the numbers of ways that Events A and B can occur, respectively, then

$$\boxed{n(A \text{ or } B) = n(A) + n(B)}\ .$$

Notes:

1. An easy way to remember these two principles is that you *multiply* the numbers of ways when you do one *and then* the other. You *add* the numbers of ways when you do one *or* the other. If you ever forget, you can think of the mouse in the maze!
2. In the first principle, $n(B)$ must be the number of ways B can occur *after* A has already occurred. For example, if you are picking two people from a group of 5, the first person could be picked in 5 different ways. But the second person could be picked in only 4 different ways, since one person has already been picked.
3. In the second principle, it is assumed that A and B *cannot both* occur. For instance, the mouse cannot go to both rooms X and Z at the same time if it starts at Room Y. Such events are said to be "mutually exclusive." In Problem 17 of the following exercise, you will see what has to be done if Events A and B *could* both occur.

The following exercise is designed to let you see if you can distinguish between these two counting principles.

Exercise 12-3

1. Ward Robe has 15 pairs of slacks and 23 shirts. In how many different ways could he select a slacks-and-shirt combination?

2. Natalie Attired has 20 dresses and 17 pants outfits. In how many different ways could she select a dress or a pants outfit to wear?

3. A salesman has 7 customers in Denver and 13 customers in Reno. In how many different ways could he telephone

   a. a customer in Denver and then a customer in Reno?
   b. a customer in Denver or a customer in Reno?

4. Sally visits the pet store. There are 37 dogs and 15 cats. In how many ways could she select

   a. a dog or a cat?
   b. a dog and a cat?

5. A pizza establishment offers 12 kinds of meat topping (pepperoni, sausage, etc.) and 5 kinds of vegetable topping (onions, green peppers, etc.). In how many different ways could you select

   a. a meat topping or a vegetable topping?
   b. a meat topping and a vegetable topping?

6. A first grade class has 13 girls and 11 boys. In how many different ways could the teacher select

   a. a boy and a girl to go to the office?
   b. a boy or a girl to go to the office?

7. A reading list contains 11 novels and 5 mysteries. In how many different ways could a student select

   a. a novel or a mystery?
   b. a novel and then a mystery?
   c. a mystery and then another mystery?

8. A submarine practices attacking a convoy that has 20 cargo ships and 5 escort vessels. In how many different ways could it attack

   a. a cargo ship and then an escort vessel?
   b. a cargo ship or an escort vessel?
   c. a cargo ship and then another cargo ship?

9. The menu at Valerio's lists seven kinds of salad, eleven entrees, and nine kinds of dessert. How many different salad-entree-dessert meals could you select? (Meals are considered to be "*different*" if any *one* thing is different.)

10. Admiral Motors manufactures cars with five different body styles, uses eleven different colors of paint, and has six different interior colors. Suppose that the Admiral Motors dealer for whom you work wants to order one of each possible variety of car to display in the showroom. Show your boss that the plan would be impractical because so many cars would have to be ordered.

11. Using only the letters in LOGARITHM:

a. In how many ways could you pick a vowel or a consonant?
b. In how many ways could you pick a vowel and a consonant?
c. How many different three-letter "words" (such as "ORL," "HLG," "AOI," etc.) could you make, using each letter no more than once in any given word? (There are *three* events; A, "Select the first letter;" B, "Select the second letter;" and C, "Select the third letter." Find $n(A)$, $n(B)$, and $n(C)$, then figure out what to do with these three numbers.)

12. Using only the letters in SEQUOIA:

a. In how many ways could you select a vowel and a consonant?
b. In how many ways could you select a vowel or a consonant?
c. How many different four-letter "words" could you form using no letter more than once in any given word? (See Problem 11, above, for a hint!)

13. There are 10 students in a class, and 10 chairs numbered 1 through 10.

a. In how many different ways could a student be selected to occupy Chair Number 1?
b. After somebody has been seated in Chair Number 1, how many different ways are there of seating someone in Chair Number 2?
c. In how many different ways could Chairs 1 and 2 be filled?
d. If Chairs 1 and 2 are already occupied with two of the students, in how many ways could Chair 3 be filled?
e. In how many different ways could Chairs 1, 2, and 3 be filled?
f. In how many different ways could all ten chairs be filled?

14. Nine people on a baseball team are trying to decide who will play which position.

a. In how many different ways could they select a person to be pitcher?
b. After someone has already been selected as pitcher, how many different ways could they select someone else to be catcher?
c. In how many different ways could they select a pitcher and a catcher?
d. After the pitcher and catcher have been selected, in how many ways could they select a first baseman?
e. In how many different ways could they select a pitcher, a catcher, and a first baseman?
f. In how many different ways could all nine positions be filled? Surprising?!

15. Many states use car license plates that have six characters. Some of these states use 2 letters and a number from 1 through 9999. Others use 3 letters and a number from 1 through 999.

a. Which of these two plans allows there to be *more* possible license plates? How *many* more?
b. How many license plates could there be using *either* 2 letters and 4 digits *or* 3 letters and 3 digits?
c. There are about 100,000,000 motor vehicles in the United States. Would it be possible to have a *national* license plate program using the scheme of part b? Explain.

16. Telephone numbers in the United States and Canada have three groups of digits which must meet certain requirements:

i. Area Code – 3 digits, the first of which is *not* 0 or 1, and the second of which *must* be 0 or 1.
ii. Exchange – 3 digits, first and second cannot be 0 or 1.
iii. Line Number – 4 digits, not all zeros.

a. How many possible area codes are there?
b. How many possible exchanges are there?
c. How many possible line numbers are there?
d. How many valid 10-digit phone numbers can there be?
e. What is the probability that a 10-digit number dialed *at random* is a valid phone number?

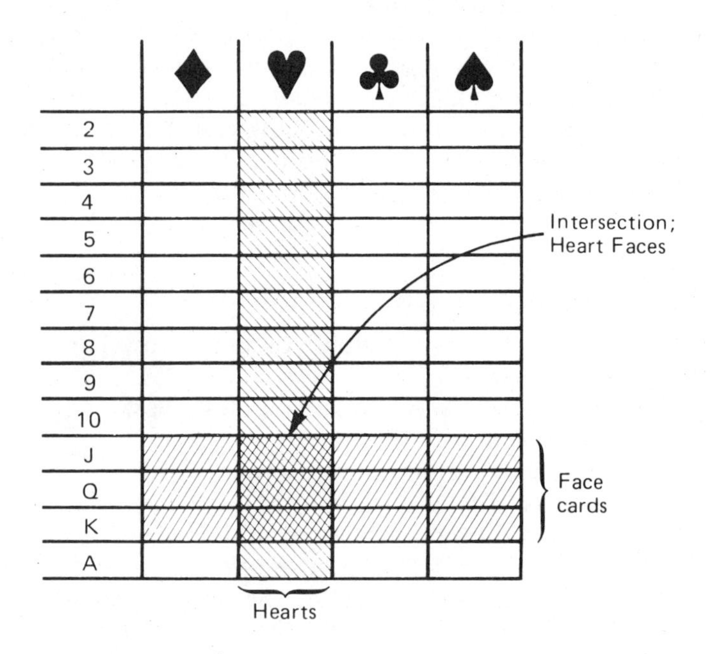

Figure 12-3b.

17. *Overlapping Events* – Suppose you draw one card from a normal 52-card deck, and ask, "In how many ways could it be a Heart *or* a Face Card?" As shown in Figure 12-3b, n(Heart) = 13 and n(Face) = 12. But simply adding 13 and 12 gives the *wrong* answer. The cards in the *intersection* (those that are Hearts *and* Faces) have been counted *twice*. An easy way to get the right answer is to *subtract* the number that are Hearts and Faces from the 13 + 12. That is,

$$n(\text{Heart } or \text{ Face}) = n(\text{Heart}) + n(\text{Face}) - n(\text{Heart } and \text{ Face}).$$

In general, the number of ways Events A *or* B could occur when they are *not* mutually exclusive is

$$\boxed{n(A \text{ or } B) = n(A) + n(B) - n(A \cap B)}\ .$$

Show that you understand this principle by working the following problems:

a. 20 girls are on the basketball team. 17 are over 16 years old, 12 are over 170 centimeters tall, and 9 are both over 16 years old and over 170 centimeters tall. How many of the girls are over 16 years old or over 170 centimeters tall?
b. Researchers discover 37 techniques used in working mathematics problems, and 29 techniques used in working physics problems. 21 of the techniques are used in both fields. If Wanda Learnit studies both physics and mathematics, how many different techniques must Wanda learn?
c. The library has 463 books dealing with science and 592 fiction books. Of these, 37 are science fiction books. How many books do they have that are science or fiction?
d. A jewelry store has 544 necklaces and 215 pieces containing puka shells. 129 of the pieces are puka shell necklaces. How many of the pieces are necklaces *or* contain puka shells?
e. The Senior Class has 367 girls, and 425 students with brown hair. 296 of the girls have brown hair. In how many different ways could you select a girl or a brownhaired student from the Senior Class?

12-4 PROBABILITIES OF VARIOUS PERMUTATIONS

One of the more tedious counting problems is counting the number of different ways of *arranging* things. For example, the three letters "ABC" can be arranged in six different ways:

ABC ACB BAC BCA CAB CBA.

But the ten letters "ABCDEFGHIJ" can be arranged in more than three *million* different ways!

An arrangement of letters, objects, or any elements from a given set is called a *permutation*. In this section you will learn how to calculate the probability that certain kinds of permutations will occur if you select elements at random from a set.

DEFINITION

A ***permutation*** is an ***arrangement*** of some or all of the elements from a given set in a definite order.

Objective:

Given a description of a desired permutation, find the probability of getting that permutation if an arrangement is selected at random.

The first step toward accomplishing this objective is developing an efficient method of counting numbers of permutations.

Example 1: In how many different ways could you arrange 3 books on a shelf if you have 7 books to choose from?

The process of selecting an arrangement of three books can be divided into three acts.

A – Select a book to go in the first position (Figure 12-4a).

B – Then select another book to go in the second position.

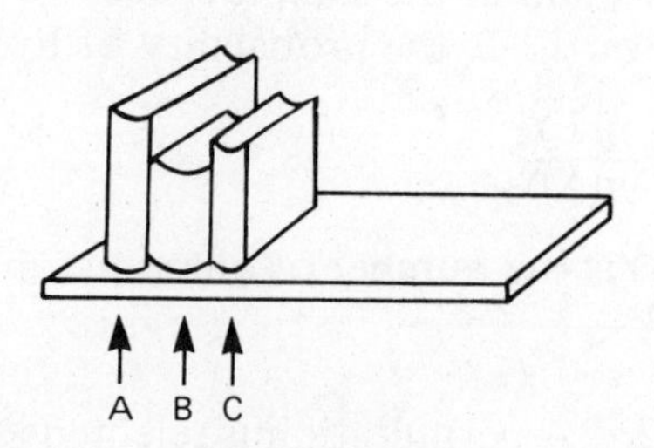

Figure 12-4a.

C – Finally, select another book to go in the third position.

The numbers of ways of doing each act are

$n(A) = 7$, since any of the 7 books can occupy the first position.

$n(B) = 6$, since only 6 books remain to choose from for the second position.

$n(C) = 5$, since only 5 books remain to choose from for the third position.

A convenient way to organize the calculations is to write three spaces representing the positions of the three books. You can pick a variable such as n to represent the number of ways of selecting all three books. You would write

$$n = \_\ \_\ \_\ .$$

Then you would fill in the number of ways of selecting each book.

$$n = \underline{7}\ \underline{6}\ \underline{5}\ .$$

Finally, you would *multiply* the numbers of ways since all three acts are performed *in succession.*

$$n = \underline{7} \cdot \underline{6} \cdot \underline{5}$$
$$= \underline{\underline{210}}\ .$$

Once you understand the technique for calculating numbers of permutations, you can use it to find numbers of permutations in a given event of a random experiment, and in the sample space. As you recall from Section 12-2, the probability of Event E is

$$P(E) = \frac{n(E)}{n(S)}$$

where $n(S)$ is the number of outcomes in the sample space.

Example 2: A permutation is selected at random from the letters SEQUOIA. What is the probability that it has Q in the fourth position, and ends with a vowel?

To answer this question, you need to know the number of outcomes in the sample space, and the number of outcomes in the event, "The fourth letter is Q and the last letter is a vowel." Since the event is hard to analyze all at once, you break it down into three smaller events.

Event A – Select Q for the fourth position.
Event B – Select a vowel for the last position.
Event C – Select the five other letters.

Let $n(E)$ stand for the number of outcomes in the event, "The fourth letter is Q and the last letter is a vowel." You would write spaces for each letter, as you did in Example 1, above.

$n(E) = \underline{\ }\ \underline{\ }\ \underline{\ }\ \underline{1}\ \underline{\ }\ \underline{\ }\ \underline{\ }$	Write 1 in the fourth position, since only Q can go there.
$n(E) = \underline{\ }\ \underline{\ }\ \underline{\ }\ \underline{1}\ \underline{\ }\ \underline{\ }\ \underline{5}$	Write 5 in the last position since any of the 5 vowels can go there.
$n(E) = \underline{5}\ \underline{4}\ \underline{3}\ \underline{1}\ \underline{2}\ \underline{1}\ \underline{5}$	Fill in the remaining 5 letters in the remaining 5 spaces.
$n(E) = \underline{5}\cdot\underline{4}\cdot\underline{3}\cdot\underline{1}\cdot\underline{2}\cdot\underline{1}\cdot\underline{5}$	Multiply, since the events are done *in sequence*.
$= 600$	Do the arithmetic.

Therefore, there are 600 permutations that have Q in the fourth position and end in a vowel.

The fourth position is said to be a *fixed* position, since there is only *one* letter that can go there. The last position is said to be a *restricted* position because more than one letter can go there, but not all of the letters.

To calculate $n(S)$, the number of permutations in the sample space, you again write seven spaces. Then you write the number of ways to fill each space, as you did above.

$$n(S) = \underline{7}\cdot\underline{6}\cdot\underline{5}\cdot\underline{4}\cdot\underline{3}\cdot\underline{2}\cdot\underline{1}$$
$$= 5040.$$

So there are 5040 possible permutations if *none* of the spaces is fixed or restricted.

To calculate the *probability* that the fourth letter is Q and the last is a vowel, you simply recall the definition of probability.

$P(E) = \frac{n(E)}{n(S)}$	Definition of probability.
$= \frac{600}{5040}$	Substitution.
$= \underline{\underline{\frac{5}{42}}}$	Canceling.

If desired, this fraction can be expressed as a *percent*. You do the division, and multiply by 100.

$$P(E) \approx \underline{\underline{12\%}}\ .$$

The following problems are designed to give you practice in finding numbers of permutations, and in calculating probabilities of various permutations.

Exercise 12-4

Problems 1 through 10 involve calculating numbers of permutations.

1. In how many ways could you arrange the following numbers of books on a shelf?

a. 4 books from a set of 9 books?

b. 3 books from a set of 12 books?
c. 5 books from a set of 8 books?
d. All 7 books from a set of 7 books?

2. Fran Tick takes a 10-problem algebra test. The problems may be worked in any order.

a. In how many orders could she work all 10 problems?
b. In how many orders could she work any 7 of the 10 problems?

3. The Hawaiian alphabet has twelve letters. How many permutations could be made using

a. 2 different letters?
b. 4 different letters?
c. all 12 letters, once each?

4. How many permutations of the 26 letters in the English alphabet could be made using

a. 2 different letters?
b. 3 different letters?
c. 4 different letters?

5. Suppose that prestige license plates are made using exactly 4 of the 26 letters in the alphabet. How many different prestige plates could be made if all 4 letters are different?

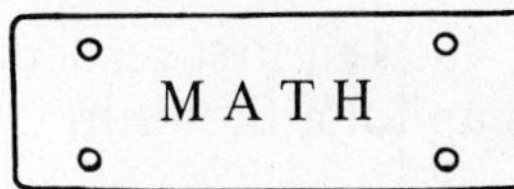

6. Triangles are usually named by placing a different letter at each vertex. In how many different ways could a given triangle be named?

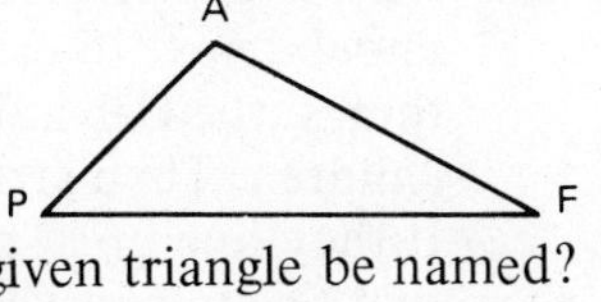

7. Fifteen people try out for a baseball team. In how many different ways could they select

a. the pitcher and the catcher?
b. the three outfielders, *after* the pitcher and catcher have been selected?
c. the First, Second, and Third Basemen and the Shortstop, *after* the pitcher, catcher, and three outfielders have been selected?

8. Tom, Dick, and Harry each draw two cards from a normal deck of 52 cards, and do not replace them. Tom goes first.

a. In how many orders could Tom draw his two cards?
b. In how many orders could Dick draw his two cards, *after* Tom has drawn his?
c. In how many orders could Harry draw his two cards, *after* Tom and Dick have drawn theirs?

9. The parking lot at Gypsum Bank has three spaces labeled "Vice-President." However, there are seven vice-presidents. In how many ways could these spaces be occupied by the vice-presidents' cars?

10. Professor Snarff says, "You may work these six problems in any order you choose." There are 100 students in the class. Is it possible for all 100 students to work the problems in a different order? Explain.

Problems 11 through 20 involve finding the probability of getting a certain kind of permutation if a permutation is selected at random.

11. A 6-letter permutation is selected at random from the letters NIMBLE.

a. How many permutations are possible?
b. How many of these permutations begin with M?
c. What is the probability that the permutation begins with *M*?
d. What is the percent probability in part c?
e. What is the probability that the word is "NIMBLE?"

12. A 5-letter permutation is selected at random from the letters GRATE.

a. How many permutations are possible?
b. How many of these permutations begin with *G*?
c. What is the probability that the permutation begins with *G*?
d. What is the percent probability in part c?
e. What is the probability that the permutation is GREAT?

13. A 6-letter permutation is selected at random from the letters NIMBLE, as in Problem 11. What is the probability that

a. the third letter is *I* and the last letter is *B*?
b. the second letter is a vowel and the third is a consonant?
c. the second and third letters are both vowels?
d. the second letter is a consonant and the last letter is *E*?
e. the second letter is a consonant and the last letter is *L*?

14. A 5-letter permutation is selected at random from the letters GRATE, as in Problem 12. What is the probability that

a. the second letter is *T* and the last letter is *G*?
b. the second letter is a vowel and the third is a consonant?
c. the second and third letters are both consonants?
d. the second letter is a consonant and the last letter is *E*?
e. the second letter is a consonant and the last letter is *R*?

15. Nine people try out for the nine positions on a baseball team.

a. In how many ways could the positions be filled if there are no restrictions on who plays which position?
b. In how many ways could the positions be filled if Fred must be the pitcher, but the other eight can take any positions?
c. If the positions are selected at random, what is the probability that Fred will be the pitcher?
d. What is the percent probability in part c?

16. Eleven girls try out for the eleven positions on the varsity soccer team.

a. In how many ways could the eleven positions be filled if there are no restrictions on who plays which position?
b. In how many ways could the positions be filled if Mabel must be the goal keeper?
c. If the positions are selected at random, what is the probability that Mabel will be goal keeper?
d. What is the percent probability in part c?

17. Nine people try out for the nine positions on the baseball team, as in Problem 15. If the players are selected at random for the positions, what is the probability that

a. Fred, Mike, or Joe is pitcher?
b. Fred, Mike, or Joe is pitcher, and Sam or Paul is first baseman?
c. Fred, Mike, or Joe is pitcher, Sam or Paul plays first base, and Bob is catcher?

18. Eleven girls try out for the eleven positions on the varsity soccer team, as in Problem 16. If the players are selected at random for the positions, what is the probability that

a. Mabel, Sue, or Diedra is goal keeper?
b. Mabel, Sue, or Diedra is goal keeper, and Alice or Phyllis is center forward?
c. Mabel, Sue, or Diedra is goal keeper, Alice or Phyllis is center forward, and Bea is left fullback?

19. Ten first-graders line up for a fire drill.

a. How many possible arrangements are there?
b. How many of these arrangements have Calvin and Phoebe next to each other? (Clue: Arrange *nine* things, the Calvin-Phoebe pair and the eight other children. Then arrange Calvin and Phoebe.)
c. If they line up at random, what is the probability that Calvin and Phoebe will be next to each other?

20. The ten digits, 0, 1, 2, 3, ..., 9, are arranged at random with no repeats. What is the probability that the numeral thus formed represents

a. a number greater than 5 billion?
b. an *even* number greater than 5 billion? (Clue: There are *two* cases to consider, "first digit odd" or "first digit even.")

21. *Permutations with Repeated Elements* – The word "TRUSSES" has 7 letters. But there are *less* than 7! permutations since rearranging the 3 *S*'s does *not* produce a different permutation. Since there are 3! ways of arranging the 3 *S*'s, only 1/3! or 1/6, of the 7! permutations are actually different. So the number of permutations is

$$\frac{7!}{3!} = 840.$$

If several letters are repeated, such as in MISSISSIPPI, then the number of permutations would be

$$\frac{11!}{4!\ 4!\ 2!} = 34650,$$

because there are 4 identical *I*'s, 4 identical *S*'s, and 2 identical *P*'s. Answer the following questions:

a. Find the number of different permutations of the letters in:

   i. FREELY.
   ii. BUBBLES.
   iii. LILLY.
   iv. MISSISSAUGA.
   v. HONOLULU.
   vi. HAWAIIAN.

b. In how many different ways could seven first-graders line up if April, Mae, June, and Julie are quadruplets, and are considered to be identical?

c. Nine pennies are lying on a table. Five are "heads" and four are "tails." In how many ways such as "HHTHTTHHT" could the coins be lined up, considering only "heads" and "tails?"

22. *Circular Permutations* – In Figure 12-4b, the letters ABCD are arranged in a circle. Though these

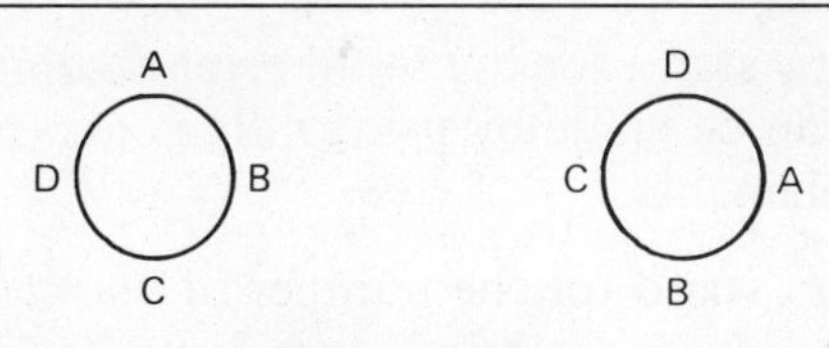

Figure 12-4b.

may appear to be different permutations, they are considered to be the *same* since each of the four letters has the same position *with respect to the others.* An easy way to calculate the number of different "circular permutations" of n elements is to *fix* the position of one of them, and then arrange the other $(n - 1)$ elements with respect to it (Figure 12-4c). Thus, for the letters ABCD, the number of circular permutations would be

$$1 \cdot 3 \cdot 2 \cdot 1 = 6.$$

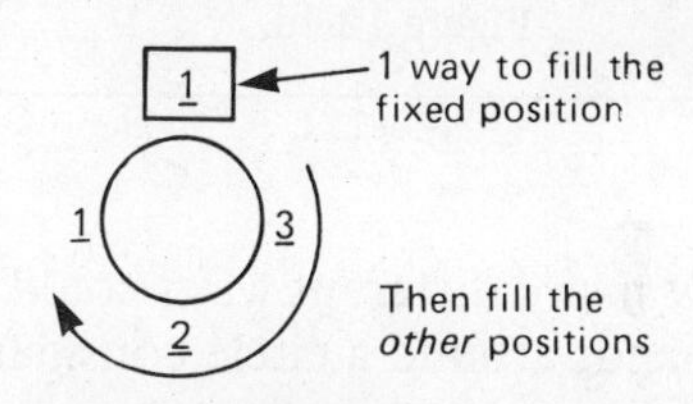

Figure 12-4c.

a. How many different circular permutations could be made from

   i. ABCDE,
   ii. QLMTXN,
   iii. LOGARITHM.

b. In how many different ways could King Arthur's twelve knights be seated around the Round Table?

c. Four boys and four girls sit around the merry-go-round.

   i. In how many different ways could they be arranged if boys and girls come alternately?
   ii. If they seat themselves at random, what is the probability that boys and girls alternate?

d. If you are concerned only with which elements come *between* other elements, then two different circular permutations would be "the same," a clockwise one and a counterclockwise one

(Figure 12-4d). So there would be only *half* as many permutations as calculated above.

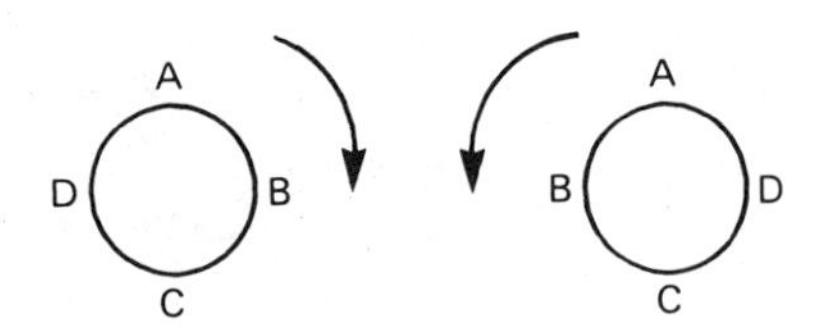

The same "betweenness" property

Figure 12-4d.

i. In how many different ways could 7 people hold hands around a circle considering only who is *between* whom?
ii. In how many different ways could 5 keys be arranged on a key ring considering only which key is *between* which?

12-5 PROBABILITIES OF VARIOUS COMBINATIONS

There are 24 different 3-letter "words" that can be made from the 4 letters ABCD. These are listed below.

ABC	ACB	BAC	BCA	CAB	CBA
ABD	ADB	BAD	BDA	DAB	DBA
ACD	ADC	CAD	CDA	DAC	DCA
BCD	BDC	CBD	CDB	DBC	DCB

Only *four* different combinations.

Since these "words" are *arrangements* of letters in a definite order, each one is a *permutation* of four elements taken three at a time.

Suppose that you are concerned only with *which* letters appear in the "word," not with the order in which they appear. For instance, ADC and DAC would be considered to be the same, since they have the same three letters. Each different group is called a three-letter *combination* of the letters ABCD.

DEFINITION

A ***combination*** of elements of a set is a ***subset*** of those elements, without regard to how the elements are arranged.

Objective:

Be able to calculate the number of combinations containing r elements that can be made from a set that has n elements.

From the table above, you can see that for every *one* combination, there are *six* possible permutations. So the total number of combinations equals the total number of permutations *divided by 6*. That is,

$$\text{number of combinations} = \frac{24}{6} = 4.$$

The "6" in the denominator is the number of permutations that can be made of each 3-letter combination.

Conclusion:

Number of combinations

$$= \frac{\text{total number of permutations}}{\text{number of permutations of each } one \text{ combination}}.$$

The conclusion stated above can be used to accomplish the objective of this section. It helps to define symbols for numbers of combinations and permutations, and derive some formulas.

Let ${}_nC_r$ stand for the number of *combinations* that can be made by using r elements from a set of n elements.

Let ${}_nP_r$ stand for the number of *permutations* that can be made by using r elements from a set of n elements.

Notes:

1. $_nC_r$ and $_nP_r$ are pronounced, "n, C, r" and "n, P, r," respectively. Words such as "number of combinations of n elements taken r at a time" are often used.
2. In the example, above, $_4C_3 = 4$ and $_4P_3 = 24$.
3. The symbols C_r^n, $C(n, r)$, and $\binom{n}{r}$ are sometimes used for $_nC_r$.
4. The symbols P_r^n and $P(n, r)$ are sometimes used for $_nP_r$.

The conclusion above can now be written in symbols.

$$_nC_r = \frac{_nP_r}{_rP_r}.$$

For example,

$$_4C_3 = \frac{_4P_3}{_3P_3} = \frac{24}{6} = 4.$$

It is possible to write $_nP_r$ in terms of *factorials*. For example, by the technique of Section 12-4, you can write

$$_9P_4 = \underline{9}\cdot\underline{8}\cdot\underline{7}\cdot\underline{6}\,.$$

Multiplying by a "clever" form of 1 gives

$$_9P_4 = 9\cdot8\cdot7\cdot6\cdot\frac{5\cdot4\cdot3\cdot2\cdot1}{5\cdot4\cdot3\cdot2\cdot1}$$
$$= \frac{9!}{5!}.$$

The "9" in the numerator is the total number of elements in the set, n. The "5" in the denominator is the number of elements *not* used in the permutation, that is, $(n-r)$. In general,

$$\boxed{_nP_r = \frac{n!}{(n-r)!}}\,. \quad \text{Example:} \quad \boxed{_9P_4 = \frac{9!}{5!}}\,.$$

As a special case, $_nP_n$ is given by

$$_nP_n = \frac{n!}{(n-n)!} = \frac{n!}{0!} = \frac{n!}{1} = n!.$$

$$\boxed{_nP_n = n!}\,. \quad \text{Example:} \quad \boxed{_3P_3 = 3!}\,.$$

Using this information, $_nC_r$ can also be written in terms of factorials.

$$_nC_r = \frac{_nP_r}{_rP_r}$$

$$_nC_r = \frac{\frac{n}{(n-r)!}}{r!}$$

$$\boxed{_nC_r = \frac{n!}{r!\,(n-r)!}}\,. \quad \text{Example:} \quad \boxed{_7C_3 = \frac{7!}{3!\,4!}}\,.$$

The formula is easier to remember by the example than by using n and r, as shown below.

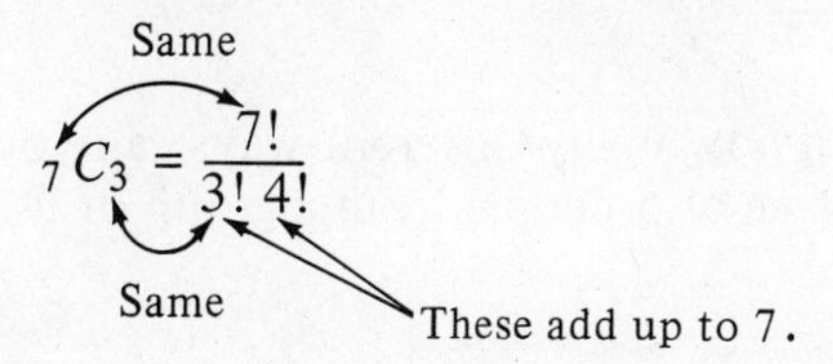

Note: You should recognize the expression for $_7C_3$ as being equal to a coefficient in a *binomial series* (see Section 11-10). If $(a + b)$ is raised to the 7th power, then $_7C_3$ is the coefficient of the term that has b^3. That is,

$$(a + b)^7 = \ldots + {_7C_3}\, a^4\, b^3 + \ldots$$

With this background you are now equipped to calculate numbers of combinations.

Example 1: Evaluate $_9C_6$.

By the combinations formula,

$$_9C_6 = \frac{9!}{6!\,3!}.$$

If you have a calculator with a factorial key, the computation is straightforward, and the answer is 84. If not, the computations are still not too difficult.

$$_9C_6 = \frac{9\cdot8\cdot7\cdot6!}{6!\cdot3\cdot2\cdot1}$$ Properties of factorials.

$$= \frac{9\cdot8\cdot7}{3\cdot2\cdot1}$$ Canceling the 6!'s.

$$= 3\cdot4\cdot7$$ Further canceling.

$$= \underline{\underline{84}}$$ Arithmetic.

Example 2: Evaluate $_9P_6$.

By the permutations formula,

$$_9P_6 = \frac{9!}{3!}$$ The "3" is $9 - 6$.

$$= \frac{9\cdot8\cdot7\cdot6\cdot5\cdot4\cdot3!}{3!}$$ Properties of factorials.

$$= \underline{\underline{60480}}$$ Arithmetic.

Example 3: In how many different ways can you form a committee of 5 people from a group of 9 people?

Committees are "different" only if different people are on the committee. It does not matter how the people are arranged. So the answer is a number of *combinations* rather than a number of permutations. Letting n(5 people) be the number of committees.

$$n(\text{5 people}) = {}_9C_5$$

$$= \frac{9!}{5!\,4!}$$

$$= \underline{\underline{126.}}$$

Example 4: If a committee of five is selected at random from a group of 9 people (6 women and 3 men), what is the probability that it will have:

a. Eileen and Ben (two of the 9 people)?
b. Exactly 3 women and 2 men?
c. At least 3 women?

By the definition of probability,

$$P(\text{Event}) = \frac{n(\text{Event})}{n(\text{Sample Space})}.$$

For this problem, the sample space is the set of all possible 5-member committees. In Example 3 you found that the number of possible 5-member committees is $_9C_5$, which equals 126.

$$\therefore n(\text{Sample Space}) = 126.$$

To answer the three questions, you need to know the numbers of outcomes in each of the 3 events described.

a. To find $n(E \text{ and } B)$, you can divide the act of selecting a committee into two simpler acts:

   1. Select Eileen and Ben. . . . . 1 possible way.
   2. Select the other 3 people from the remaining 7 people. . . . . $_7C_3$ possible ways.

   Since these two acts are performed *in succession,* the total number of ways is found by *multiplying.*

   $$n(E \text{ and } B) = 1 \cdot {}_7C_3$$

   $$= 1 \cdot \frac{7!}{3!\,4!}$$

   $$= 35.$$

   $$\therefore P(E \text{ and } B) = \underline{\underline{\frac{35}{126}}}, \quad \text{or about } 28\%$$

b. To find the number of 3-woman, 2-man committees, you must realize that people are being selected from two different groups. So you should divide the act of selecting a committee into two simpler acts.

   1. Select the 3 women . . . . $_6C_3$ possible ways.
   2. Select the 2 men. . . . . . . $_3C_2$ possible ways.

   Since the acts must *both* be performed, *in succession,* the total number of ways, $n(3W, 2M)$, is found by *multiplying.*

   $$n(3W, 2M) = {}_6C_3 \cdot {}_3C_2$$

   $$= \frac{6!}{3!\,3!} \cdot \frac{3!}{2!\,1!}$$

$= 20 \cdot 3$

$= 60$

$\therefore P(3W, 2M) = \frac{60}{126}$, or about 48%.

c. If the committee has *at least* 3 women, then it could have 3 women *or* 4 women *or* 5 women. In each case, the remainder of the committee consists of *men*. The act of selecting a committee can be divided into 3 simpler acts:

1. Select a 3-woman, 2-man committee . . . . $n(3W, 2M)$ ways.
2. Select a 4-woman, 1-man committee . . . . $n(4W, 1M)$ ways.
3. Select a 5-woman, 0-man committee . . . . $n(5W, 0M)$ ways.

Since these are "either-or" acts, the total number of ways is found by *adding*.

$$n(\text{at least } 3W) = n(3W, 2M) + n(4W, 1M) + n(5W, 0M)$$

Each of the numbers on the right may be calculated as in part b, above.

$$n(\text{at least } 3W) = {}_6C_3 \cdot {}_3C_2 + {}_6C_4 \cdot {}_3C_1 + {}_6C_5 \cdot {}_3C_0$$

$$= 20 \cdot 3 + 15 \cdot 3 + 6 \cdot 1$$

$$= 111.$$

$\therefore P(\text{at least } 3W) = \frac{111}{126}$, or about 88%.

The following exercise is designed to give you practice finding numbers of combinations and permutations by using the formulas, and calculating the probability that a certain event will occur if a combination or permutation is selected at random.

Exercise 12-5

For Problems 1 through 12, calculate the indicated number of combinations.

1. ${}_5C_3$ 2. ${}_6C_4$ 3. ${}_8C_5$ 4. ${}_9C_3$
5. ${}_6C_5$ 6. ${}_{10}C_4$ 7. ${}_{11}C_4$ 8. ${}_{11}C_5$
9. ${}_{10}C_{10}$ 10. ${}_{100}C_{100}$ 11. ${}_{10}C_0$ 12. ${}_{100}C_0$

For Problems 13 through 16, calculate the indicated number of permutations.

13. ${}_6P_4$ 14. ${}_9P_3$ 15. ${}_{11}P_5$ 16. ${}_{15}P_8$

Problems 17 through 28 involve finding various numbers of combinations.

17. A committee of five is to be selected from the 30 students in a Freshman English class. In how many ways could the committee be composed?

18. Twelve people apply to go on a biology field trip but there is room in the car for only five of them. In how many different ways could the group making the trip be composed?

19. Seven people come to an evening bridge party. Since only four people can play bridge at any one time, they decide to play as many games as it takes to use every possible foursome *once.* How many games would have to be played? Could all of these games be played in *one* evening?

20. A well-known donut dealer has 34 varieties of donuts. Suppose that they decide to make sample boxes containing six different donuts each. How many different sample boxes could they make? Would it be practical to stock one of each kind?

21. Ann Jellik goes to the toy store, where her mother will let her buy any three different toys. There are 1000 toys to choose from. How many different selections could Ann make?

22. The Supreme Court justices have a hand-shaking ritual that they conduct just before each session. Each of the justices shakes hands with every other justice. How many hand-shakes will there be in this ritual?

23. Horace Holmsley has a breakfast of scrambled eggs, bacon, sausage, grits, hash browns, and toast. How many different combinations of these foods could he put on his fork if he uses

a. three ingredients?
b. four ingredients?
c. three ingredients or four ingredients?
d. all six ingredients?

24. A well-known chain of pizza parlors has eleven different kinds of topping they can put on their pizzas. How many different kinds of pizza could they make using

a. three of the toppings?
b. five of the toppings?
c. three of the toppings or five of the toppings?
d. all eleven of the toppings?

25. A normal deck of cards has 52 cards.

a. How many different 5-card poker hands could be formed?
b. How many different 13-card bridge hands could be formed?

26. The diagonals of a convex polygon are made by combining the vertices two at a time. However, some of the combinations are *sides* rather than diagonals. How many diagonals are there in a convex

a. pentagon (5 sides)?
b. decagon (10 sides)?
c. n-gon (n sides)? Simplify your answer as much as possible.

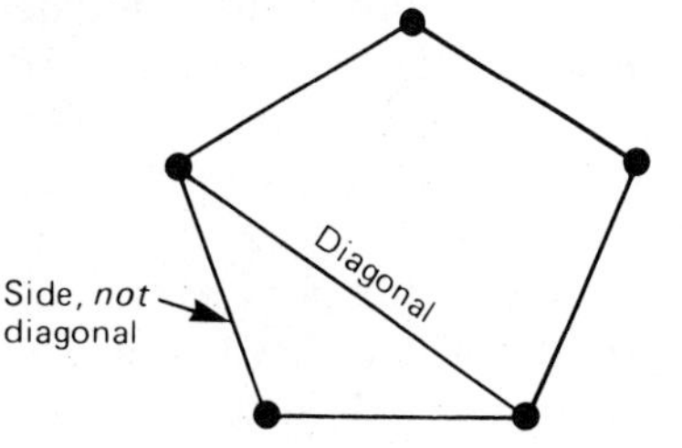

27. A set has 10 elements. How many subsets are there that contain

a. 2 of the elements?
b. 5 of the elements?
c. 8 of the elements? Explain the relationship between this answer and the answer in part a.

28. A set has five elements.

a. How many subsets of this set are there that contain
   i. 1 element?
   ii. 2 elements?
   iii. 3 elements?
   iv. 4 elements?
   v. 5 elements?
   vi. no elements?

How many subsets?

b. How many subsets are there altogether? What relationship does this number have to the number 5?

Problems 29 through 36 involve finding the probability that a certain event will occur if a combination is selected at random.

29. Ten first-graders, 6 boys and 4 girls, are playing on the playground. Miss Twiddle selects a group of 5 at random. What is the probability that the group has

a. 3 boys and 2 girls?
b. 2 boys and 3 girls?
c. 3 boys and 2 girls, or 2 boys and 3 girls?
d. Ella Quence, one of the girls?

30. Charlie Brown has 13 socks in his drawer, 7 blue and 6 green. He selects 5 socks at random. What is the probability that he gets

a. 2 blue and 3 green?
b. 3 blue and 2 green?
c. 2 blue and 3 green, or 3 blue and 2 green?
d. the one sock that has a hole in it?

31. In a group of 15 people, 6 are left-handed and the rest are right-handed. If 7 people are selected at random from this group, what is the probability that

a. 3 are left-handed and 4 are right-handed?
b. all are right-handed?
c. all are left-handed?
d. Harry and Peg Legg, two of the left-handers, are selected?

32. The varsity croquet team, 4 boys and 8 girls, travels to an out-of-town game. Their coach, Miss Teak, will take 7 of them in her station wagon. If they get into cars at random, what is the probability that Miss Teak's car has

a. 2 boys and 5 girls?
b. all girls?
c. all boys?
d. Peter Doubt and Manuel Dexterity, two of the boys?

33. Three cards are selected at random from a group of 7. Two of the cards have been marked with winning numbers.

a. What is the probability that exactly 1 of the 3 cards has a winning number?
b. What is the probability that at least 1 of the 3 cards has a winning number?
c. What is the probability that *none* of the 3 cards has a winning number?
d. What relationship is there between the answers to parts b and c?

34. Kay Paso, who is 3 years old, tears the labels off all 10 of the soup cans on her mother's shelf. Her mother knows that there were 2 cans of tomato soup and 8 cans of vegetable. She selects 4 cans at random.

Tomato?
Vegetable?

a. What is the probability that exactly 1 of the 4 cans is tomato?
b. What is the probability that at least 1 of the 4 cans is tomato?
c. What is the probability that *none* of the 4 cans is tomato?
d. What relationship exists between the answers to parts b and c?

35. Sata Light Company tests a sample of 5 of every 100 light bulbs produced to make sure they work.

a. How many different ways could a 5-bulb sample be taken from the 100 bulbs?
b. To check on the Quality Control Department, the Chief Engineer puts 2 defective bulbs in with 98 good ones. How many ways could a sample of 5 of these 100 bulbs have *at least* one defective bulb?
c. What is the probability that Quality Control will discover at least 1 of the defective bulbs?
d. Do you think Sata Light Company has an adequate sampling program? Explain.

36. An ordinary deck of playing cards has 4 suits, with 13 cards of each suit. In many games, each of 4 players is dealt 13 cards at random.

a. What is the probability that such a hand has
   i. exactly 5 spades?
   ii. exactly 3 clubs?
   iii. exactly 5 spades and 3 clubs?
   iv. exactly 5 spades, 3 clubs, and 2 diamonds?
b. Which is more probable, getting all 4 aces or getting all 13 cards of the same suit? Justify your answer.

12-6 PROPERTIES OF PROBABILITY

In the preceding sections you have learned how to use the definition of probability,

$$P(\text{Event}) = \frac{n(\text{Event})}{n(\text{Sample Space})},$$

to calculate probabilities. You simply counted the numbers of elements in the event and in the sample space, then divided. In this section you will learn some properties that will allow you to calculate probabilities *without* having to go all the way back to the definition.

Objective:

Given $P(A)$ and $P(B)$, the probabilities of Events A and B, be able to calculate

a. $P(A \text{ and then } B)$,
b. $P(A \text{ or } B)$,
c. $P(\text{not } A)$ and $P(\text{not } B)$.

Suppose that you draw two cards in succession from a normal 52-card deck, without replacing the first card before you draw the second. What is the probability that both cards are black?

There are 52 ways to choose the first card, and 51 ways to choose the second card *after* the first has been chosen. So the sample space contains

$$n(S) = 52 \cdot 51 = 2652$$

outcomes. There are 26 ways the first card could be black. After the first black has been drawn, there are only 25 ways that the second could be black. So

$$n(\text{both black}) = 26 \cdot 25 = 650.$$

Therefore,

$$P(\text{both black}) = \frac{650}{2652} = \frac{25}{102}.$$

It is more instructive, however, *not* to simplify. That is,

$$P(\text{both black}) = \frac{26 \cdot 25}{52 \cdot 51}$$

$$= \frac{26}{52} \cdot \frac{25}{51} \quad \text{Multiplication property of fractions.}$$

The 26/52 is the probability that the first card is black, and the 25/51 is the probability that the second card is black, *after* you have already drawn one black card. From this example, you can extract the following property of probability:

$$\boxed{P(A \text{ and then } B) = P(A) \cdot P(B)}.$$

Notes:

1. $P(B)$ must be the probability that B occurs *after* A has already occurred. If $P(B)$ does *not* depend on whether or not A has occurred, then A and B are said to be *independent* events.
2. This property corresponds exactly to the first counting principle, $n(A \text{ and then } B) = n(A) \cdot n(B)$.

The second part of the objective can be illustrated by the act of drawing a single marble at random from a bag containing 13 red ones, 17 black ones, and 11 green ones. What is the probability that the marble is *either* red *or* black?

The sample space contains 41 outcomes since there are 41 marbles in the bag. The event "red or black" contains

$$n(\text{red or black}) = n(\text{red}) + n(\text{black})$$
$$= 13 + 17$$
$$= 30$$

outcomes. The numbers of ways are *added* since these are "either-or" events. Therefore,

$$P(\text{red or black}) = \frac{30}{41}.$$

Again, it is more instructive *not* to simplify.

$$P(\text{red or black}) = \frac{13 + 17}{41} \quad \text{Definition of probability.}$$

$$= \frac{13}{41} + \frac{17}{41} \quad \text{Division distributes over addition.}$$

$$= P(\text{red}) + P(\text{black}).$$

In general,

$$\boxed{P(A \text{ or } B) = P(A) + P(B)}.$$

Notes:

1. Events A and B must be "mutually exclusive" for this property to be true. That is, it must be impossible for A and B *both* to occur. If they *can* both occur, then $P(A \text{ or } B) = P(A) + P(B) - P(A \cap B)$, where $P(A \cap B)$ is the probability that A and B *both* occur.
2. Again, this property of probability corresponds exactly to the second counting principle, $n(A \text{ or } B) = n(A) + n(B)$.

The third part of the objective can be accomplished using the "either-or" property above.

Let $P(A)$ = the probability that Event A occurs.
Let $P(\text{not } A)$ = the probability that Event A does *not* occur.

Then

$$P(A \text{ or not } A) = P(A) + P(\text{not } A).$$

But $P(A \text{ or not } A) = 1$, because one of the two events, A or not A, is *certain* to happen. And the probability of an event that is certain to happen equals 1. Therefore,

$$P(A) + P(\text{not } A) = 1,$$

from which

$$\boxed{P(\text{not } A) = 1 - P(A)}\ .$$

These three properties can be used to calculate probabilities *without* having to go all the way back to the definition.

Example: Calvin and Phoebe visit the Children's Ward at the hospital. The probability that Calvin will catch mumps as a result of the visit is $P(C) = 0.13$, and the probability that Phoebe will catch mumps is $P(Ph) = 0.07$. Find the probability that

a. both catch mumps,
b. Calvin does not catch mumps,
c. Phoebe does not catch mumps,
d. Calvin and Phoebe both do not catch mumps,
e. at least one of them catches mumps.

a. $P(C \text{ and } Ph) = P(C) \cdot P(Ph)$ — Property of probability.
$= 0.13 \times 0.07$ — Substitution.
$= \underline{\underline{0.0091}}$ — Arithmetic.

b. $P(\text{not } C) = 1 - P(C)$ — Property of probability.
$= 1 - 0.13$ — Substitution.
$= \underline{\underline{0.87}}$ — Arithmetic.

c. $P(\text{not } Ph) = 1 - P(Ph)$ — Property of probability.
$= 1 - 0.07$ — Substitution.
$= \underline{\underline{0.93}}$ — Arithmetic.

d. $P(\text{not } C \text{ and not } Ph)$
$= P(\text{not } C) \cdot P(\text{not } Ph)$ — Property of probability.
$= 0.87 \times 0.93$ — From parts b and c.
$= \underline{\underline{0.8091}}$ — Arithmetic.

e. If Event A is "neither one catches mumps," then "not A" is the event, "at least one *does* catch mumps." You have already calculated P(neither one) in part d. Therefore,

$P(\text{at least } 1) = 1 - P(\text{not } C \text{ and not } Ph)$ — Property of probability.
$= 1 - 0.8091$ — From part d.
$= \underline{\underline{0.1909}}$ — Arithmetic.

The following exercise is designed to give you practice using these three properties of probability. In the next section you will use the properties to come to some rather startling conclusions about how probability relates to binomial series.

Exercise 12-6

1. ***Calculator Components Problem*** . The "heart" of a pocket calculator is one or more "chips," each of which contains several thousand components. These chips are mass produced, and have a fairly

high probability of being defective. Suppose that a particular kind of calculator uses two chips. Chip A has a probability of 70% (0.7) of being defective, and Chip B has a probability of 80% of being defective. If one chip of each kind is selected at random, what is the probability that

a. both are defective?
b. A is not defective?
c. B is not defective?
d. neither chip is defective?
e. at least one chip is defective?

2. ***Grade Problem.*** Kara Vann is a good student. She figures that her probability of making A is 0.92 for algebra and 0.88 for history. What is her probability of making

a. A in algebra and history?
b. no A in algebra?
c. no A in history?
d. A in neither algebra nor historv?
e. at least one A?

3. ***Car Breakdown Problem.*** You drive on a long vacation trip. The probability you will have a flat tire is 0.1, and the probability of engine trouble is 0.05. What is the probability you will have

a. no flat tire?
b. no engine trouble?
c. no flat tire and no engine trouble?
d. both a flat tire and engine trouble?
e. at least one, either a flat tire or engine trouble?

4. ***Menu Problem.*** Doc Worker is a regular customer at the Waterfront Coffee Shop. The manager has figured that Doc's probability of ordering ham is 0.8; and eggs, 0.65. What is the probability that

a. he does not order ham?
b. he does not order eggs?
c. he orders neither ham nor eggs?
d. he orders ham and eggs?
e. he orders at least one, either ham or eggs?

Eggs? Ham?

5. ***Traffic Light Problem.*** Two traffic lights on Broadway operate independently. Your probability of being stopped at the first one is 0.4 and your probability of being stopped at the second one is 0.7. What is your probability of being stopped at

a. both lights?
b. neither light?
c. the first but not the second?
d. the second but not the first?
e. exactly one of the lights?

6. ***Visiting Problem.*** The Dover children, Eileen and Ben, are away at college. They visit home on random weekends, Eileen with a probability of 0.2 and Ben with a probability of 0.25. On any given weekend, what is the probability that

a. both will visit?
b. neither will visit?
c. Eileen will visit but Ben will not?
d. Ben will visit but Eileen will not?
e. exactly one will visit?

7. ***Back-Up System Problem.*** Vital systems such as electric power distribution systems have "back-up" components in case one component fails. Suppose that two generators each have a probability of 98% (0.98) of working. The system will continue to work as long as *at least one* of the generators works. What is the probability that the system will continue to operate?

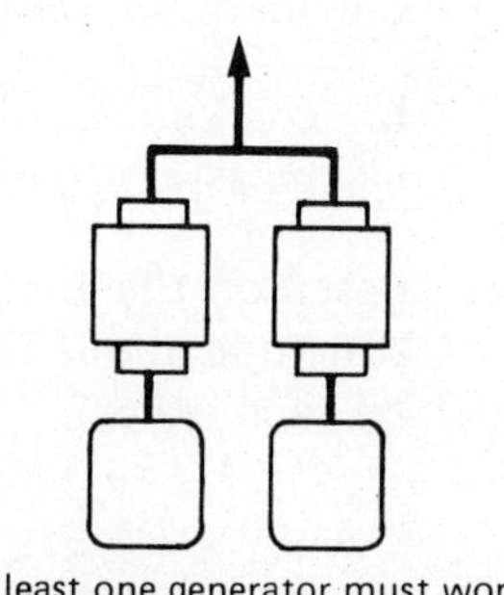
At least one generator must work

8. ***Hide-and-Seek Problem.*** The Katz brothers, Bob and Tom, are hiding in the cellar. If either one sneezes, they will reveal their hiding place and be found. Bob's probability of sneezing is 0.6 and Tom's probability is 0.7. What is the probability that at least one sneezes?

9. ***Basketball Problem.*** The three basketball teams from Lowe High each play on Friday night. The probabilities that they will win are: varsity, 0.7; junior varsity, 0.6; and freshman, 0.8. What is the probability that

a. all three win?
b. all three lose?
c. at least one team wins?
d. the varsity wins and the other two lose?

10. ***Another Grade Problem.*** Terry Torrey has the following probabilities of passing various courses: Humanities, 90%; Speech, 80%; and Latin, 95%. What is his probability of

a. passing all 3?
b. failing all 3?
c. passing at least 1?
d. passing exactly 1?

11. ***Spaceship Problem.*** Fly-By-Night Spaceship Company produces a booster rocket that has 1000 vital parts. If any *one* of these parts fails, the booster will crash, so they design each part with a reliability of 99.9%, meaning that the probability of an individual part working is 0.999.

a. What is the probability that all 1000 parts work, and the booster does not crash? Surprising?
b. What is the minimum reliability needed for each part to insure that there is a 90% probability of all 1000 parts working?

12. ***Silversword Problem.*** The silversword is a rare plant that grows only atop the 10,000 foot high Haleakela volcano in Maui, Hawaii. The seeds have only a small probability of germinating, but if enough are planted, there is a fairly good chance of getting a new plant. Suppose that the probability of any one seed germinating is 0.004.

a. What is the probability that a seed will *not* germinate?

b. If 100 seeds are planted, what is the probability that

   i. none will germinate?
   ii. at least one will germinate?

c. If 1000 seeds are planted, what is the probability that at least one will germinate?

13. ***Football Plays Problem.*** Backbay Polytechnic Institute's quarterback selects pass plays and run plays randomly. His probability of selecting a pass on first down is 0.4. If he selects a pass on first down, his probability of selecting another pass on second down is 0.3. Otherwise, his probability of passing on second down is 0.8. What is his probability of passing on

a. first down and second down?
b. first down but not second down?
c. second down but not first down?
d. neither first down nor second down?

14. ***Operation Problem.*** Ann Teak must undergo two operations. The first has a 70% probability of success. If it succeeds, the second operation has a 90% probability of success; if not, the second operation has only a 40% probability of success. What is the probability that

a. both succeed?
b. both fail?
c. the first succeeds and the second fails?
d. the first fails and the second succeeds?

15. ***Measles and Chicken Pox Problem.*** Suppose that a child has a probability of 0.12 of catching measles and a probability of 0.2 of catching chicken pox in any one given year.

a. If these events are *independent* of each other, what is the probability that he or she will get *both* diseases in a given year?
b. Suppose that statistics show the following probabilities for getting both diseases in the same year:

   P(measles, then chicken pox) = 0.006.
   P(chicken pox, then measles) = 0.18.

   Calculate the probability of getting

   i. chicken pox *after* measles,
   ii. measles *after* chicken pox.

c. Based on the answers to part b, what could you conclude about the effects of the two diseases on each other?

16. ***Airplane Engine Problem.*** Wing and Prayer Aircraft Corporation manufactures a twin-engine plane. Laboratory tests indicate that the probability of any one engine failing during a particular flight is 0.03.

a. If the engines operate independently, what is the probability that *both* fail?
b. Flight records reveal that the probability of both engines failing during a particular flight is actually 0.006. What is the probability that the second engine will fail *after* the first has already failed?
c. Based on your answer to part b, do the engines actually seem to operate independently? Explain.

17. ***Combinations and Binomial Series Problem.*** A group of five students compete for National Merit Scholarships.

a. Calculate the numbers of ways that 0, 1, 2, 3, 4, or 5 of them could win a scholarship.
b. Add the answers to part a, and thus show that there are 32 ways, total, for 0 through 5 of them to win scholarships.
c. Expand $(1 + 1)^5$ as a binomial series. Simplify each term, but do not carry out the addition. How does this series relate to parts a and b, above?
d. Show a very easy way to answer part b directly, without using the results of part a.
e. Demonstrate that you understand the significance of pard d by calculating *quickly* the number of ways 0 through 10 students could win a scholarship.

18. ***Combinations and Powers of 2 Problem.*** In Problem 17, above, you found that

$${}_5C_0 + {}_5C_1 + {}_5C_2 + {}_5C_3 + {}_5C_4 + {}_5C_5 = 2^5.$$

In general,

$$\boxed{\sum_{r=0}^{n} {}_nC_r = 2^n}.$$

a. Show that this property works for $n = 3$ and $n = 4$.
b. Use this property to find *quickly*
   i. the number of combinations of 0 or 1 or 2 or ... or 10 people that could be made from a set of 10 people,
   ii. the number of subsets of a 12-element set,
   iii. the number of combinations of 10 people taken at least one at a time; at least two at a time.

12-7 FUNCTIONS OF A RANDOM VARIABLE

Suppose that you conduct the random experiment of flipping a coin. The coin is bent, so that the probability of "heads" on any one flip is only 0.4. If you flip the coin 5 times, what is $P(3T, 2H)$, the probability that exactly two of the outcomes are "heads" and the other three are "tails?"

To answer this question, it helps to look at the simpler event $P(TTTHH)$, the probability of 3 tails and 2 heads *in that order*.

$$P(TTTHH) = P(T) \cdot P(T) \cdot P(T) \cdot P(H) \cdot P(H).$$

You are told that $P(H) = 0.4$. Since T is the only other possible outcome on any one flip,

$$\begin{aligned} P(T) &= 1 - 0.4 \\ &= 0.6. \end{aligned}$$

Therefore,

$$P(TTTHH) = 0.6^3 \times 0.4^2.$$

There are 10 possible outcomes that have exactly 2 heads. They are

HHTTT, HTHTT, HTTHT, HTTTH, THHTT, THTHT, THTTH, TTHHT, TTHTH, TTTHH.

The 10 equals the number of ways of selecting a *group* of 2 of the 5 flips to be "heads." But this is just the number of *combinations* of 5 elements taken 2 at a time, ${}_5C_2$. So

$$\begin{aligned} P(3T, 2H) &= {}_5C_2 \times 0.6^3 \times 0.4^2 \\ &= \frac{5!}{2!\,3!} \times 0.6^3 \times 0.4^2. \end{aligned}$$

You should recognize this as a term in the *binomial series* that comes from expanding

$$(0.6 + 0.4)^5.$$

In general,

> If b is the probability that Event B happens, and a is the probability that Event B does *not* happen, then
>
> $$\begin{aligned} P(B \text{ happens } x \text{ times out of } n) &= {}_nC_x \cdot a^{n-x}\, b^x \\ &= \frac{n!}{(n-x)!\,x!}\, a^{n-x}\, b^x \\ &= \text{term with } b^x \text{ in the binomial series } (a+b)^n. \end{aligned}$$

Once you see the pattern, you can quickly write the probabilities of other events.

Let x = number of times the coin is "heads" in 5 flips.

Let $P(x)$ = probability that it is "heads" x times.

Therefore,

$$\begin{aligned} P(0) &= {}_5C_0 \times 0.6^5 \times 0.4^0 = 1 \times 0.6^5 \times 0.4^0 \\ &= 0.07776 \end{aligned}$$

$$\begin{aligned} P(1) &= {}_5C_1 \times 0.6^4 \times 0.4^1 = 5 \times 0.6^4 \times 0.4^1 \\ &= 0.2592 \end{aligned}$$

$$P(2) = {}_5C_2 \times 0.6^3 \times 0.4^2 = 10 \times 0.6^3 \times 0.4^2$$
$$= 0.3456$$

$$P(3) = {}_5C_3 \times 0.6^2 \times 0.4^3 = 10 \times 0.6^2 \times 0.4^3$$
$$= 0.2304$$

$$P(4) = {}_5C_4 \times 0.6^1 \times 0.4^4 = 5 \times 0.6^1 \times 0.4^4$$
$$= 0.0768$$

$$P(5) = {}_5C_5 \times 0.6^0 \times 0.4^5 = 1 \times 0.6^0 \times 0.4^5$$
$$= 0.01024$$

A calculator is helpful for the last step.

As a check on the answers, you should realize that x is *certain* to take on one of the values 0 through 5. So $P(0 \text{ or } 1 \text{ or } 2 \text{ or } 3 \text{ or } 4 \text{ or } 5)$ must equal 1, or 100%. Adding the probabilities,

$$P(0) + P(1) + P(2) + P(3) + P(4) + P(5)$$
$$= 0.07776 + 0.2592 + 0.3456 + 0.2304 + 0.0768 + 0.01204$$
$$= 1.00000,$$

which shows that the answers are reasonable.

The independent variable x is called a *random* variable since you cannot be sure what value x will have on any one run of the random experiment. The dependent variable $P(x)$ is the *probability* that the value is x. So P is a *function* of a random variable. The graph of this function is shown in Figure 12-7.

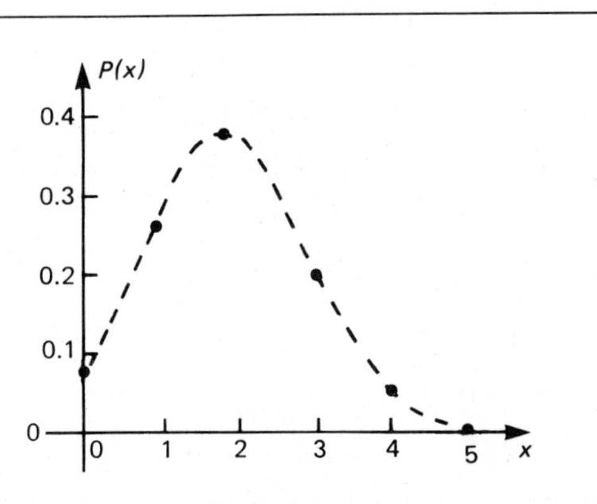

Figure 12-7.

The function P shows how the total probability, 1.00000, is "distributed" among the possible values of x. This function of a random variable is often called a *probability distribution.* Since this particular distribution has probabilities that are terms of a binomial series, it is called a *binomial* distribution.

Binomial distributions occur when you perform a random experiment repeatedly, and each time there are only *two* possible outcomes (e.g., heads or tails, boy or girl, win or lose, yes or no).

Once you have found the probability distribution, you can use the properties of probability to calculate the probabilities of related events. For example, if the bent coin is flipped 5 times, as above, then the probability of getting *at least two* heads is

$$P(x \geq 2) = P(2) + P(3) + P(4) + P(5)$$
$$= 0.3456 + 0.2304 + 0.0768 + 0.01024$$
$$= \underline{\underline{0.66504}}\ ,$$

which is about 2/3, or 67%.

The following exercise starts with some binomial distribution problems. The later problems let you analyze some probability distributions that are *not* binomial.

Exercise 12-7

Note: You may want to write the computer program of Problem 19 *before* you work Problems 1 through 11.

1. ***Heredity Problem.*** If a dark-haired mother and father have a particular type of genes, they have a 1/4 probability of having a light-haired baby.
   a. What is their probability of having a dark-haired baby?
   b. If they have 3 babies, calculate $P(0)$, $P(1)$, $P(2)$, and $P(3)$, the probabilities of having exactly 0, 1, 2, and 3 *dark*-haired babies, respectively.
   c. Show that your answers to part b are reasonable by finding their sum.
   d. Plot the graph of the probability distribution, P.

2. ***Multiple Choice Test Problem.*** A short multiple choice test has 4 questions. Each question has 5 choices, exactly *one* of which is right. Willie Makitt has not studied for the test, so he guesses at random.

a. What is his probability of guessing any one answer right? Wrong?
b. Calculate his probabilities of guessing 0, 1, 2, 3, and 4 answers right.
c. Perform a calculation that shows your answer to part b is reasonable.
d. Plot the graph of the probability distribution in part b.
e. Willie passes the test if he gets at least 3 answers right. What is his probability of passing?

3. ***Thumbtack Problem.*** If you flip a thumbtack, it can come out either "up" or "down" (see sketch). Suppose that the probability of "up" on any one flip is 0.7.

a. If the tack is flipped 4 times, find the probabilities that it is "up" exactly 0, 1, 2, 3, and 4 times.
b. Plot the graph of this probability distribution.
c. Which is more probable, more than 2 "ups" or at most 2 "ups?" Justify your answer.

4. ***Traffic Light Problem.*** Three widely-separated traffic lights on U.S. 1 operate independently of each other. The probability that you will be stopped at any one of them is 40%.

a. Calculate the probability that you will make all 3 lights "green."
b. Calculate the probabilities that you will be stopped at exactly *one*, exactly *two,* and *all three* lights.
c. Plot the graph of this probability distribution.
d. Which is more probable, being stopped at more than one light or at one or less lights? Justify your answer.

5. ***Bull's-Eye Problem.*** Mark Wright can hit the bull's-eye with his 22 rifle 30% of the time. He fires 5 shots.

a. Calculate his probabilities of making 0, 1, 2, 3, 4, and 5 bull's-eyes.
b. Plot the graph of this probability distribution.
c. Calculate the probability that he will make at least 2 bull's-eyes.

6. ***Dice Problem.*** You roll a die 6 times. Let x be the number of times the die comes up 1 or 2, and let $P(x)$ be the probability it comes out that way x times.

a. On any one roll, what is the probability that the die *will* come up 1 or 2? Will *not* come up 1 or 2?
b. Calculate $P(x)$ for each value of x in the domain.
c. Plot the graph of P.
d. You win the game if the die comes up 1 or 2 at least twice. What is your probability of winning?
e. Since the probability of getting a 1 or 2 is 1/3, you might assume that the probability of getting a 1 or 2 on *two* of the six rolls is also 1/3. Is this assumption true or false? Justify your answer.

7. ***Color-Blindness Problem.*** Statistics show that about 5% of all males are color-blind. Suppose that 20 males are selected at random. Let x be the number who are color-blind, and let $P(x)$ be the probability that x of them are color-blind.

a. Calculate $P(0)$, $P(1)$, $P(2)$, and $P(3)$.
b. Plot a graph of this probability distribution.
c. What is the probability that at least 4 of the 20 people are color-blind?

8. ***Operation Problem.*** Suppose that for a certain kind of surgery, the probability that an individual will survive is 98%. If 100 people have the operation:

a. Calculate the probabilities of 0, 1, 2, 3, and 4 people dying.
b. Plot the graph of this probability distribution. Show how the graph would look for larger values of the random variable.

c. What is the probability that at least one of the 100 people dies?

9. ***Eighteen-Wheeler Problem.*** Large tractor-trailer trucks usually have 18 tires. Suppose that the probability of any one tire blowing out on a cross-country trip is 0.03.

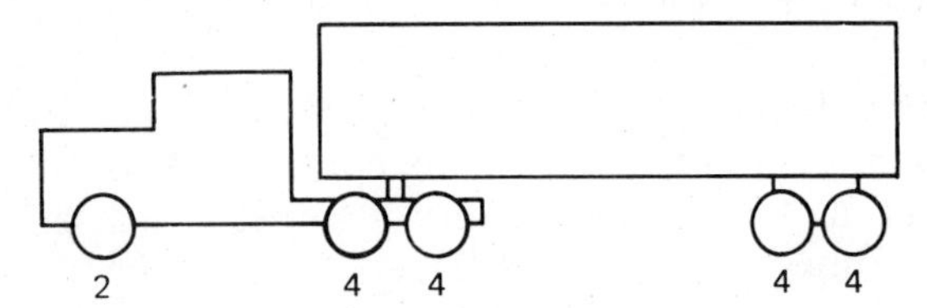

a. What is the probability that any one tire does *not* blow out?
b. What is the probability that

   i. none of the 18 tires blows out?
   ii. exactly one tire blow‘ out?
   iii. exactly two tires blow out?
   iv. more than two tires blow out?

c. If the trucker wants to have a 95% probability of making the trip without a blowout, what must the reliability of each tire be? That is, what is the probability that any one tire will blow out?

10. ***Perfect Solo Problem.*** Clara Nett plays a musical solo. She is quite good, and figures that her probability of playing any one note right is 99%. The solo has 60 notes.

a. What is her probability of

   i. getting every note right?
   ii. making exactly one mistake?
   iii. making exactly two mistakes?
   iv. making at least two mistakes?
   v. making more than two mistakes?

b. What must be Clara's probability of getting any one note right if she wants to have a 95% probability of getting all 60 notes right?

11. ***Airplane Engine Problem.*** A Boeing 707 has 4 jet engines. Assume that the probability of any one engine failing during flight is 0.1. (It is not nearly that high, but assume it anyway!)

Which is safer?

a. What is the probability that a given engine does *not* fail?
b. Calculate the probabilities that 0, 1, 2, 3, and 4 of the engines fail during a given flight.
c. Show that the probabilities in part b add up to 1.
d. If the plane will keep flying as long as *no more than one* engine fails, what is the probability that the plane keeps flying?
e. A Boeing 727 has only 3 jet engines. If the plane will keep flying with no more than 1 engine out, and the probability of any one engine failing is the same as in part a, above, what is the probability that a 727 keeps flying?
f. Based on these calculations, which is safer, a 3-engine plane or a 4-engine plane?

12. ***World Series Problem.*** Suppose that the Dodgers and Yankees are in the World Series of baseball. From their season's records, you predict that the Dodgers have a probability of 0.6 of beating the Yankees in any particular game. In order to win the *Series,* a team must win *four* games.

a. What is the probability that the Yankees beat the Dodgers in any particular game?
b. What is the probability that

   i. the Dodgers win all the first four games?
   ii. the Yankees win all the first four games?

c. For the Dodgers to win the Series in exactly 5 games, they must win exactly three of the first four games, then win the fifth game. What is the probability that the Series goes exactly 5 games, and

   i. the Dodgers win?
   ii. the Yankees win?

d. What is the probability that the Series lasts
   i. exactly 4 games?
   ii. exactly 5 games?
e. Recalling, as in part c, that the winner of the Series must win the *last* game, calculate the probability that
   i. the Dodgers win in 6 games.
   ii. the Dodgers win in 7 games.
   iii. the Yankees win in 6 games.
   iv. the Yankees win in 7 games.
f. What is the most probable length of the Series, 4, 5, 6, or 7 games?

Problems 13 through 18 involve probability distributions *other* than binomial distributions.

13. ***Another Dice Problem.*** Suppose that a random experiment consists of rolling two dice, a black one and a white one, as in the example of Section 12-1.

a. Plot graphs of the probability distributions for the following random variables. You may count outcomes from Figure 12-1.
   i. x is the *sum* of the numbers on the two dice.
   ii. x is the *difference*, black die minus white die.
   iii. x is the *absolute value* of the difference between the numbers on the two dice.
b. For each of the probability distributions in part a, find the most probable value of x.

14. ***Proper Divisors Problem.*** An integer from 1 through 10 is selected at random. Let x be the number of proper divisors the integer has. (A "proper divisor" of a number is an integer *less* than the number that divides *exactly* into the number. For example, 12 has five proper divisors, 1, 2, 3, 4, and 6.)

a. List the proper divisors and the *number* of proper divisors for each integer from 1 through 10.
b. For each possible value of x, tell how many of the integers from 1 through 10 *have* that number of proper divisors.
c. Let $P(x)$ be the probability that the integer selected at random has x proper divisors. Calculate $P(x)$ for each value of x in the domain.
d. Plot the graph of the probability distribution P. You may wish to leave the points *un*connected.

15. ***First Girl Problem.*** Eva and Paul Lution decide to keep having babies until they have a girl. They know that the probability of having a girl on any single birth is 0.5.

a. Let x be the number of babies they have, and let $P(x)$ be the probability that the xth baby is the *first* girl. $P(1)$ will be 0.5. $P(2)$ is the probability that the first is *not* a girl, and the second *is* a girl. Calculate $P(2)$, $P(3)$, and $P(4)$.
b. Plot the graph of P. Show what happens as x becomes very large.
c. Besides being called a probability distribution, what other special kind of function is P?
d. Show that the *sum* of the values of $P(x)$ approaches 1 as x becomes very large.

16. ***Target Practice Problem.*** A single bullet is loaded into one of the six chambers in a revolver, and the cylinder is spun around so that the bullet is in a random position. The revolver is aimed at a target and the trigger is pulled. Let x be a random variable equal to the number of times the trigger is pulled before the revolver goes off. Let $P(x)$ be the probability that it goes off on the xth pull.

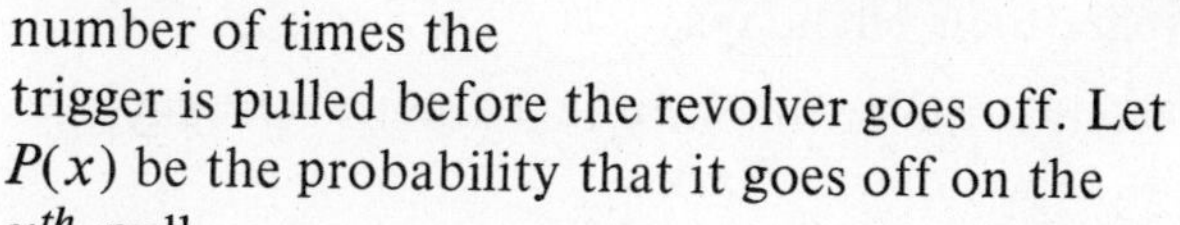

a. Find $P(1)$.
b. $P(2)$ is the probability the gun does *not* go off on the first pull, then *does* go off on the second pull. Find $P(2)$.
c. Find $P(3)$, $P(4)$, $P(5)$, and $P(6)$.
d. What is the significance of the fact that the sum of $P(1)$ through $P(6)$ equals 1?

e. Find $P(1)$ through $P(6)$ if the cylinder is spun between each pull, so that the bullet is *always* in a random position.
f. Plot a graph of $P(x)$ versus x using the values from part e.
g. Show that the sum of the probabilities calculated as in part e approaches 1 as x becomes very large.

17. ***Lucky Card Problem.*** An index card is marked with a lucky number, then shuffled with 9 other blank index cards. Three people play a game in which each, in turn, draws a card at random. If it is not the marked card, it is replaced, the cards are shuffled again, and the next player draws at random. The winner is the first player to draw the marked card. Let $P(x, y)$ be the probability that the xth player (1, 2 or 3) wins on his y^{th} turn.

a. What is $P(1, 1)$?
b. $P(2, 1)$ is the probability that Player 1 does *not* win on his first turn, and Player 2 *does* win on his first turn. Find $P(2, 1)$ and $P(3, 1)$.
c. Show that $P(1, 1), P(1, 2), P(1, 3), \ldots$ is a geometric sequence.
d. The probability that Player 1 wins is the sum of *all* the terms of the corresponding geometric series. Find his probability of winning.
e. Find the second and third players' probabilities of winning.
f. Show that your answers to parts d and e are reasonable based on their relative sizes and their sum.

18. ***Same Birthday Problem.*** A group of students compare their birthdays.

a. What is the probability that John's birthday is *not* the same as Mark's?
b. If John and Mark have different birthdays, what is the probability that Fred's birthday is not the same as John's and not the same as Mark's?
c. What is the probability that John and Mark have different birthdays, *and* that Fred has one different from both?
d. Using the pattern you observe in part c, predict the probability that a group of 10 students will all have different birthdays. If you have access to a calculator or computer, get a decimal approximation for this answer.
e. What is the probability that in a group of 10 students, at least two have the same birthday (i.e., *not* all of them have different birthdays)?
f. Write a computer program to print a table showing the probability that at least two people in a group have the same birthday as a function of the number of people in the group. Run the program for groups from 2 through 60 people.
g. Plot the graph of the probability of at least two people with the same birthday versus the number of people in the group. You may use the output of the computer program in part f for plotting data.
h. From your graph or computer output, find how many people must be in the group to make the probability of at least two birthdays the same equal

   i. 50%, and
   ii. 99%.

   Surprising?

19. ***Computer Program for Binomial Distributions.*** Write a computer program to calculate $P(x)$ for each value of x in a binomial distribution. The input should be n, the number of times the experiment is run, and b, the probability of success on any one run. The appropriate binomial is $(a + b)^n$, where $a = 1 - b$. Note that x will equal the exponent of b in the series. The output should be a table of values such as

x	$P(x)$
0	0.216
1	0.432
2	0.288
3	0.064

You should find that the coefficients are most easily calculated using the pattern

$$\frac{\text{(coefficient)(exponent of } a)}{\text{term number}} = \text{next coefficient,}$$

which you recall from Section 11-10. Also, since the powers of a *drop* one each time, and the powers of b *increase* one, you can simply divide the previous term by a and multiply it by b. Debug your program by using $n = 3$ and $b = 0.4$, and showing that the output is the above table of values.

12-8 MATHEMATICAL EXPECTATION

Suppose that you play a dice game for money. You pay 50 cents, then roll a die. The payoffs are

Roll a 6: Win $1.00 (and get the 50 cents back).
Roll a 2 or 4: Win 10 cents (and get the 50 cents back).
Roll an odd number: Win nothing (and lose the 50 cents).

Since each outcome, 1, 2, 3, 4, 5, and 6, is equally likely, you would "expect" to get each number *once* in 6 rolls. (You probably *won't*, but that is what you would "expect" on the average, if you rolled *many* times.) If this did happen, your winnings would be

Number	*Cents Won*
1	−50
2	10
3	−50
4	10
5	−50
6	100
Total:	−30

Since there were 6 rolls, and you *lost* a total of 30 cents, you would expect your average winnings per roll to be

$$\frac{-30}{6} = -5 \text{ cents per roll.}$$

The number "−5 cents per roll" is called your *mathematical expectation,* or the *expected value* of your winnings. If you play the game many times (thousands!), you would expect to lose about 5 cents per roll, on the average.

Objective:

Be able to calculate the mathematical expectation of a given random experiment.

An important pattern shows up if you do *not* carry out the addition of the payoffs in the above experiment. Letting E stand for mathematical expectation,

$$E = \frac{-50 + 10 - 50 + 10 - 50 + 100}{6}$$ Adding the 6 payoffs and dividing by 6.

$$= \frac{3(-50) + 2(10) + 1(100)}{6}$$ Commuting, associating, and factoring.

$$= \frac{3(-50)}{6} + \frac{2(10)}{6} + \frac{1(100)}{6}$$ Division distributes over addition.

$$= \frac{3}{6}(-50) + \frac{2}{6}(10) + \frac{1}{6}(100)$$ Properties of fractions.

The 3/6 is the probability of getting an odd number, and the −50 is the payoff for getting an odd number. The 2/6 and 1/6 are the probabilities of getting a 2 or 4, and of getting a 6, respectively, and the 10 and 100 are the respective payoffs. So the mathematical expectation can be calculated by multiplying *probability* × *payoff* for each event, then *adding* the results.

DEFINITION

The ***mathematical expectation*** of a random experiment is

$$E = (\text{probability})(\text{payoff}) + (\text{probability})(\text{payoff}) + \ldots + (\text{probability})(\text{payoff}),$$

for all possible mutually exclusive events in the experiment.

Note: This definition can be stated compactly using Σ terminology (see Section 11-4):

$$E = \sum_{k=1}^{n} \text{Probability}(k) \times \text{Payoff}(k)$$

for the n mutually exclusive events in the experiment.

Example: The basketball toss at Playland costs you 50 cents to play. You shoot three balls. If you make no baskets, you win nothing (and lose your 50 cents, of course!). If you make just one basket, you win a paper hat worth 5 cents. For two baskets you win a stuffed animal worth 60 cents. Making all three baskets wins you a doll worth 2.50. The basket hoop is small, so your probability of making any one shot is only 0.3. What is your mathematical expectation for this game?

Let $P(x)$ be your probability of making x baskets. Your probability of *missing* on any one shot is $1 - 0.3 = 0.7$. Therefore,

$$P(0) = {}_3C_0 \times (0.7)^3(0.3)^0 = 1 \times 0.343 = 0.343$$
$$P(1) = {}_3C_1 \times (0.7)^2(0.3)^1 = 3 \times 0.147 = 0.441$$
$$P(2) = {}_3C_2 \times (0.7)^1(0.3)^2 = 3 \times 0.063 = 0.189$$
$$P(3) = {}_3C_3 \times (0.7)^0(0.3)^3 = 1 \times 0.027 = 0.027$$

Total probability 1.000

The payoffs for each event are found by subtracting the 50 cent "admission fee" from the amount you win.

x	*Payoff*
0	$0 - 50 = -50$
1	$5 - 50 = -45$
2	$60 - 50 = 10$
3	$250 - 50 = 200$

By the definition of mathematical expectation,

$$E = (0.343)(-50) + (0.441)(-45) + (0.189)(10) + (0.027)(200)$$
$$= -17.15 - 19.845 + 1.89 + 5.4$$
$$= -29.705.$$

So on the average, you would expect to *lose* about 30 cents per game. (This is the way amusement parks make money!)

Once you understand the computations, you can arrange the work compactly in table form. The table for the above example would look like this:

x	$P(x)$	Payoff	$P(x) \times$ Payoff
0	0.343	−50	−17.15
1	0.441	−45	−19.845
2	0.189	10	1.89
3	0.027	200	5.4
Totals:	1.000		−29.705

$\therefore E = -29.705.$

Adding up the values of $P(x)$ and seeing that you get exactly 1 gives you a check against possible errors.

The following problems are designed to give you practice finding the mathematical expectation of various random experiments. The ability to estimate expected payoffs is the major reason people are interested in calculating probabilities.

Exercise 12-8

1. ***Card Draw Problem.*** You pay 26 cents, and draw a card at random from a normal 52-card deck. The payoffs are as follows:

   Ace: Get $1.56.
   Face card: Get 65 cents.
   Any other card: Get nothing.

   a. What is your mathematical expectation for this game?
   b. In the long run, would you expect to gain money or lose money playing this game? How much money?

2. ***Uranium Fission Problem.*** When a uranium atom splits ("fissions"), it releases 0, 1, 2, 3, or 4 neutrons. Assume that the probabilities of these various numbers are:

Number	*P(Number)*
0	0.05
1	0.2
2	0.25
3	0.4
4	0.1

a. What is the mathematically expected number of neutrons released per fission?
b. The number of neutrons released in a fission must be an *integer*. How do you explain the fact that the mathematically expected number is *not* an integer?

3. ***Archery Problem.*** An expert marksman at archery has the following probabilities of hitting various rings on the target (see sketch):

Color	*Probability*	*Points*
Gold	0.20	9
Red	0.36	7
Blue	0.23	5
Black	0.14	3
White	0.07	1

9 7 5 3 1

a. What is her mathematical expectation on any one shot?
b. In a National Round, she shoots 48 arrows. What would you expect her score to be?

4. ***Calvin's Grade Problem.*** Calvin Butterball's Father offers to pay him \$90 if he makes all A's, or to pay him \$10 for each A he makes. However, he must decide in advance which offer to accept. Calvin is a good student, and estimates his probabilities of making A's to be

Algebra – 0.9
English – 0.7
Chemistry – 0.8
Spanish – 0.6

a. Calculate his mathematical expectation if he chooses \$10 per A.
b. Calculate his probability of making all A's.
c. Calculate his mathematical expectation if he chooses \$90 for making all A's.
d. Which offer should Calvin choose?

5. ***Seed Germination Problem.*** A package of seeds for an exotic tropical plant states that the probability of any one seed germinating is 80%. You plant 4 of the seeds.

a. Find the probabilities that exactly 0, 1, 2, 3, and 4 of the seeds germinate.
b. Find the mathematically expected number of seeds that will germinate.

6. ***Batting Average Problem.*** Milt Famey has a baseball batting average of 300, which means that his probability of getting a hit at any one time at bat is 0.3. Suppose that Milt comes to bat 5 times during a game.

a. Calculate the probabilities that he gets exactly 0, 1, 2, 3, 4, and 5 hits.
b. What is Milt's mathematically expected number of hits in these 5 at-bats?

7. ***Expectation of a Binomial Experiment.*** Suppose that you conduct a random experiment that has a binomial probability distribution. Suppose that the probability of success on any one trial is 0.4. Let $P(x)$ be the probability of x successes in 5 trials.

a. Calculate $P(x)$ for all values of x in the domain.
b. Find the mathematically expected value of x.
c. Show that the mathematical expectation of x is equal to 0.4 (the probability of success on any *one* trial) times 5 (the total number of trials).
d. If the probability of success on one trial is b, and the probability of failure on one trial is a, prove that in 5 trials, the expected value of x is $5b$.
e. From what you have observed above, what do you suppose the mathematically expected value of x would be in n trials, if the probability of success on any one trial is b?

f. If you plant 100 seeds, each of which has a probability of 0.71 of germinating, how many seeds would you expect to germinate?

8. ***Dollar Bill Problem.*** You play a game in which a dollar bill is selected at random. You win the bill if all 8 digits in its serial number are different. What is your mathematical expectation, if you pay 5 cents each time to play the game?

9. ***Dice Game Problem.*** You pay a dollar and roll a die 3 times. If the outcome is "Ace" (that is, 1), at least two of the three times, you get back $10.00. Otherwise you get back nothing. What is your mathematical expectation for this game?

10. ***Multiple Choice Test Problem.*** Suppose that you are taking your College Board tests. You answer all the questions you can, and have some time left over, so you decide to guess at the answers to the rest of the questions.

a. Each question is multiple choice with 5 choices. If you guess at random, what is the probability of getting an answer right?
b. What is the probability of getting an answer wrong?
c. When Educational Testing Service grades your paper, they give you 1 point if the answer is right, and subtract 1/4 point if the answer is wrong. What is your mathematically expected score on any question for which you guess at random?
d. Suppose you can eliminate two of the choices you know are wrong, and you randomly guess among the other three. What is your mathematically expected score on the question?
e. If you can eliminate three choices, and randomly guess between the other two, what is your mathematically expected score?
f. Based on your answers, is it really worthwhile guessing answers on a multiple choice test?

11. ***Bad Egg Problem.*** An egg salesman has 5 dozen eggs that he will sell for $1.00 per dozen. Before they can be sold, they must pass an inspection. Three eggs are selected at random. If all three are good, the inspection is passed. If exactly one is bad, the 5 dozen are rejected, and the salesman loses his cost of $.60 per dozen. If two or more are bad, he loses his $.60 per dozen, and must pay a fine of $100 for trying to sell such inferior products. Suppose that exactly two eggs *are* bad. What is the salesman's mathematically expected profit?

12. ***Accident/Illness Insurance Problem.*** Ripov's Insurance Company has an accident/illness policy that pays $500 if you get ill during the year, $1000 if you have an accident, and $6000 if you both get ill and have an accident. For this policy, you pay $100 per year premium. One of your friends who has studied acturial science tells you that your probability of becoming ill in any one year is 0.05, and your probability of having an accident is 0.03.

a. What is your probability of
   i. becoming ill and having an accident?
   ii. becoming ill and not having an accident?
   iii. not becoming ill, but having an accident?
   iv. not becoming ill and not having an accident?
b. What is your mathematical expectation for this policy?

13. ***Nuclear Reactor Problem.*** When a uranium atom inside the reactor of a nuclear power plant is hit by a neutron, it splits, or "fissions," releasing energy. It also releases 1, 2, 3, or 4 new neutrons. The mathematically expected number of new neutrons is about 2.3 per fission.

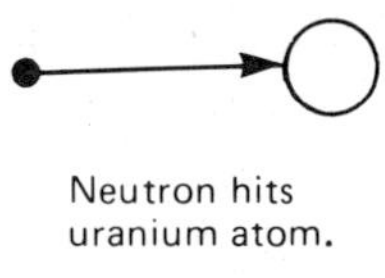
Neutron hits uranium atom.

a. Suppose that there are 100 neutrons in the reactor initially. If all of these neutrons cause fissions, how many neutrons would you expect there to be after this first "generation" of fissions?

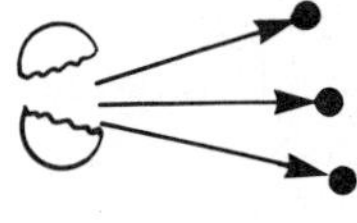
Atom splits. New neutrons are released.

b. If all of the new neutrons from the first "generation" of fissions strike other uranium atoms and cause them to fission, what is the mathematically expected number of neutrons after 2 generations? 3 generations? 4 generations?
c. If each generation takes 1/1000 of a second, how many neutrons would you expect there to be after 1 second? Surprising?! This is what makes atomic bombs *explode*!
d. Not all of the neutrons from one generation actually do cause fissions in the next generation. Some leak out of the reactor, some are captured by atoms other than uranium, and some that are captured by uranium do not cause fissions. Assume that

$P(\text{leaking}) = 0.36$
$P(\text{other atom capture}) = 0.2$
$P(\text{non-fission capture}) = 0.15.$

Calculate the probability that *none* of these things happens, and thus the neutron *does* cause a fission in the next generation.
e. Use the probability of part c and the payoff of 2.3 new neutrons per fission to calculate the expected number of neutrons in the second generation produced by 1 neutron in the first generation.
f. If you start with one neutron in the reactor as in parts d and e, and each generation takes 1/1000 second, as in part c, how many neutrons would you expect to have at the end of one second? Would this reactor explode like a bomb?
g. Why can you say that the number of neutrons in the reactor is "increasing exponentially" with time?

14. ***Life Insurance Problem.*** Functions of random variables are useful as mathematical models in the insurance business. Some of the highest-paid mathematicians are the actuaries, who figure out what rates you should pay for various types of insurance policies. The following is a portion of the *Commissioners 1958 Standard Ordinary* mortality table. The table shows the probability, $P(x)$, that a person alive on his x^{th} birthday will die before he reaches age $x + 1$.

Age, x	$P(x)$
15	0.00146
16	0.00154
17	0.00162
18	0.00169
19	0.00174
20	0.00179

A group of 10,000 fifteen-year-olds gets together to form their own life insurance company. For a premium (i.e., a payment) of $2.00 per year, they agree to pay $1000 to the family of anyone in the group who dies while he is 15 through 20 years old.

a. Calculate $D(15)$, the number expected to die while they are 15. This should be rounded off to an integer, for obvious reasons!
b. Calculate $A(16)$, the number expected to be alive on their sixteenth birthday. (Subtract the number who die from 10,000).
c. Calculate $D(16)$. Use the answer to part b, and round off to an integer.
d. Calculate $A(x)$ and $D(x)$ for ages 17 through 20. Present the results in table form by adding more columns to the above mortality table.
e. For each x, calculate $I(x)$ and $O(x)$, the income from the $2.00 premiums and the outgo from the $1000 death payments, respectively. You realize of course, that if a person dies, he no longer pays the $2.00 per year premium! Present the results as additional columns in the table.
f. Calculate $NI(x)$, the net income per year, for each year by subtracting the total paid out from the total premiums taken in. Why does $NI(x)$ *decrease* each year? (2 reasons!)
g. On the average, how much does the company expect to earn per year? Would this be enough to pay a person full time to operate the company?

15. ***Another Life Insurance Problem.*** A group of 10,000 people, each now 55 years old, is to be insured as in Problem 14, above. Upon the death of the insured person, his survivors receive $1000, at

any age 55 through 59. You are to calculate the annual premium which should be charged for each policy. Another protion of the mortality table, as in Problem 14, is:

Age, x	*P(x)*
55	0.01300
56	0.01421
57	0.01554
58	0.01700
59	0.01859

a. Calculate $D(x)$ and $A(x)$ for $x = 55$ through 59, as in Problem 14.
b. Calculate $O(x)$ for each year, the amounts to be paid out for $1000 death benefits.
c. A part-time administrator is to be paid $5000 per year to operate the insurance program. Calculate the *total* expenses (death benefits plus administration costs) for the five-year period.
d. Calculate the *number* of premiums which will be paid during the five years. This will be 10,000 the first year, and will decrease each subsequent year.
e. Divide the total number of premiums into the total expenses to find the annual premium which should be charged.
f. Why is the premium for this insurance program so much higher than the $2.00 per year for the program in Problem 14?

12-9 CHAPTER REVIEW AND TEST

In this chapter you have analyzed functions in which the independent variable takes on *random* values. The dependent variable is the *probability* that a certain value occurs. For this purpose you made a precise definition of "probability" as the ratio of favorable outcomes to total outcomes. At first you calculated probabilities from this definition by calculating numbers of combinations or permutations and dividing. Then you found properties that would let you calculate probabilities using other known probabilities. You found that some probability distributions are closely related to binomial series. You concluded by using probability distributions to calculate mathematically expected "payoffs" of random experiments.

The objectives of this chapter may be summarized as follows:

1. *Calculate the number of outcomes in an event or sample space.*
2. *Calculate the probability of a given event.*
3. *Calculate the mathematical expectation of a random experiment.*

The Review Problems, which follow, are numbered according to these three objectives. The Concepts Test requires you to apply all of these concepts to an analysis of the game of BINGO.

Review Problems

The problems below are numbered according to the three objectives of this chapter.

Review Problem 1a. Baskin Robbins has 20 flavors of ice cream and 11 flavors of sherbet. In how many different ways could you select

i. a scoop of ice cream and then a scoop of sherbet?
ii. a scoop of ice cream or a scoop of sherbet?

b. The senior class has 100 students. In how many different ways could they select

i. a group of 4 students to be officers?
ii. a president, vice-president, secretary, and treasurer?

c. Calculate:

i. ${}_8P_3$.
ii. ${}_8C_3$.

Review Problem 2a. Find the probability that a random arrangement of the letters LUCIFER will begin with R and end with a consonant.

b. Five marbles are selected from a bag, without replacement. The bag originally contained 7 red marbles and 9 blue ones. What is the probability that

i. exactly 3 are red?
ii. at least 3 are red?

c. Mr. Rhee's car has a probability of 70% of starting, and Ms. Rhee's car has an 80% probability of starting. What is the probability that

i. neither car will start?
ii. both cars will start?
iii. either both cars or neither car will start?
iv. exactly one of the cars will start?

d. Flicka Bick's cigarette lighter has a 60% probability of lighting on any one flick. She flicks it 4 times. Let $P(x)$ be the probability that it lights exactly x of those times.

i. Find $P(x)$ for each value of x in the domain.
ii. Plot the graph of P.
iii. Find the probability that it lights at least half of the times.

Review Problem 3. Professor Snarff determines his students' averages by "weighting" each test a different amount. Test 1 counts 10% of the grade. Tests 2, 3, and 4 count 20% each. The Final Exam counts 30%.

a. Suppose that Nita B. Topaz makes grades of 72, 86, 93, 77, and 98 on the five tests, in that order. What is her average?
b. The average in part a is called a "weighted" average. Explain why the process of finding a weighted average is exactly the same as finding the mathematical expectation of a probability distribution.

Concepts Test

The following questions concern the game of BINGO. Each player receives a card as shown in the sketch. The B-column contains exactly 5 of the integers 1 – 15, arranged in any order. The I, N, G, and 0 columns contain the integers from 16 – 30, 31 – 45, 46 – 60, and 61 – 75, respectively, the N-column having only 4 integers. Answer the following questions. For each part of each question, tell by number(s) which one(s) of the three objectives of this chapter you used on that part.

B	I	N	G	O
3	19	45	52	67
11	22	37	53	75
7	16	X	49	66
2	30	38	60	68
13	18	41	47	72

Concepts Test 1. These questions concern the BINGO card itself.

a. How many different groups of 4 integers could there be for the N-column?
b. How many different groups of 5 integers could there be for the B-column?
c. How many different arrangements of 5 of the 15 integers could there be for the B-column?
d. After 5 integers have been selected for the B-column, in how many different ways could these 5 be arranged?
e. If integers are selected at random for the B-column, what is the probability that all are *even*? All are *odd*?
f. To avoid the possibility of two different people winning a game with the same 5 numbers, no two cards have the same group of numbers in the same column. What is the maximum possible number of cards that satisfy this requirement? (The answer is obvious, but requires careful thought.)

Concepts Test 2. The game is played by selecting integers at random, without replacement, and seeing if they are on your card. What is the probability that

a. the first 4 numbers selected for the N-column are all on your card?
b. the first 5 numbers selected for the B-column are all on your card?
c. the first 10 numbers selected for the B-column include the 5 on your card?

Concepts Test 3. Suppose that you are one of 5 people playing BINGO.

a. Since each person is equally likely to win any one given game, what is the probability that you win? That you lose?
b. If the 5 people play 4 games, find the probabilities that you win 0, 1, 2, 3, and all 4 of these games.
c. Plot the graph of the probability distribution in part b.
d. Which is more likely, that you win no games or that you win at least 2 games? Justify your answer.
e. Use the definition of mathematical expectation and the probability distribution of part b to find the expected number of games you will win. Show that this number equals the number of games played, multiplied by your probability of winning any one game.
f. The five of you decide on the following payoffs:

Win 0 games, pay \$2.00.
Win 1 game, break even. (Pay nothing, win nothing.)
Win 2 games, win \$3.00.
Win 3 games, win \$5.00.
Win all 4 games, win \$100.00.

What is *your* mathematically expected number of dollars for playing the 4 games?

12-10 CUMULATIVE REVIEW, CHAPTERS 9 THROUGH 12

The following exercise may be considered to be a "final exam" covering quadratic relations, higher degree functions, complex numbers, sequences and series, and probability. If you are thoroughly familiar with these concepts, as presented in Chapters 9 through 12, you should be able to work all of the problems in about 2 hours.

Exercise 12-10

1. Write an equation in the form $Ax^2 + Bxy + Cy^2 + Dx + Ey + F = 0$ for each of the quadratic relations whose graph is sketched in Figure 12-10.

a.

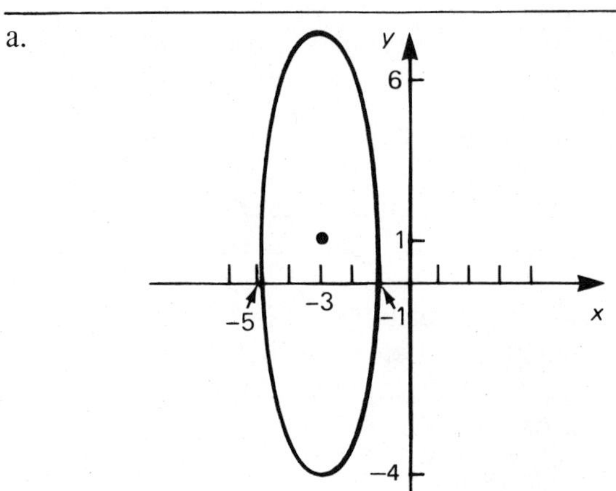

b.

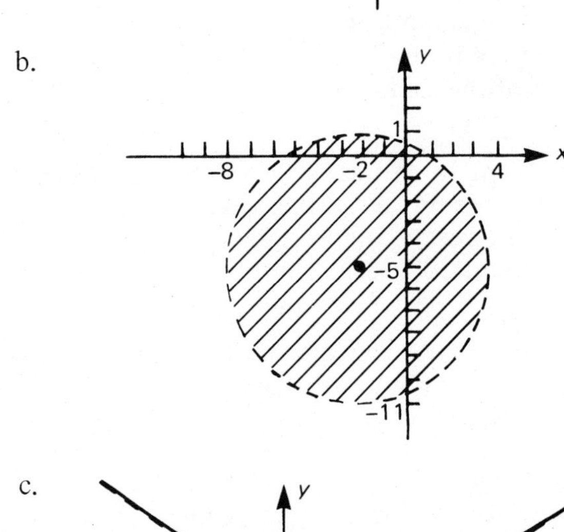

c.

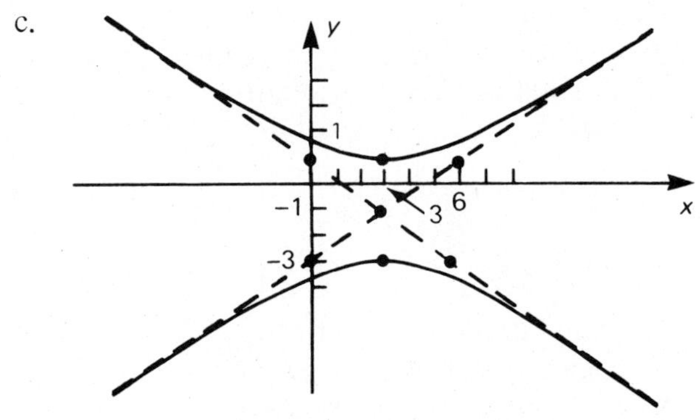

d.

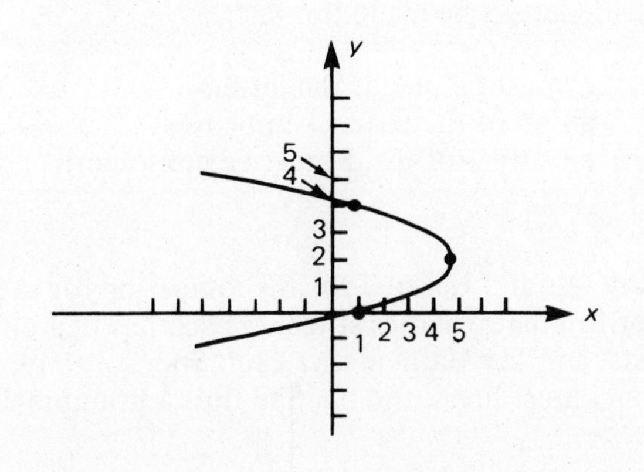

Figure 12-10.

2. Solve the system:

$$x^2 + 2y^2 = 33$$

$$x^2 + y^2 + 2x = 19$$

3. For the cubic function $f(x) = x^3 - 7x^2 + 11x + 3$:

a. Show that 3 is a real zero of $f(x)$.
b. Find all the other zeros of $f(x)$, real and complex.
c. Plot a graph of f in the domain $-2 \leq x \leq 6$.

4. The "payload" an airplane will carry is defined to be the difference between the weight the plane's wings will lift and the actual weight of the plane. A Sopwith Camel is 40 feet long, can lift 5000 pounds, and weighs 3000 pounds.

a. The lift varies directly with the square of the length, and the weight varies directly with the cube of the length. Write two equations, expressing lift and weight, respectively, in terms of length.
b. Combine the two equations from part a to get an equation expressing payload in terms of length.
c. Calculate the payload for airplanes similar in shape to a Sopwith Camel 10, 20, 30, 50, and 60 feet long.
d. Find the zeros of the function in part b, and tell what each of them represents in the real world.
e. Plot the graph of payload versus length, using a suitable domain.
f. Approximately what length plane shaped like a Sopwith Camel can carry the *maximum* payload?

5. Simplify the following expressions:

a. i^{55}.

b. $\dfrac{6}{\sqrt{-24}}$.

c. $\dfrac{10}{1 - 2i}$.

6. Find a quadratic equation with one variable and real-number coefficients if one of the solutions is $3 + 4i$.

7. An arithmetic sequence has term $t(1) = 3$, and $t(9) = 7$.

a. Find the common difference.
b. Write the seven arithmetic means between 3 and 7.
c. Find $t(101)$, and the 101st term.
d. Find $S(101)$, the 101st partial sum.

8. A geometric series has first term $t(1) = 8$ and common ratio $r = -1/2$.

a. Find $s(4)$, the fourth partial sum.
b. Find the sum of *all* the terms. That is, find the number which $S(n)$ approaches as n becomes very large.

9. Find the term that contains b^3 in the binomial series that comes from expanding $(a - b)^7$.

10. Annie Moore invests $500 in a savings account that pays 8% per year interest, compounded quarterly (4 times a year). She knows that the amount

of money she has after n quarters is a term in a geometric sequence. How much will she have after 5 years?

11. Using the 9 letters ABCDEFGHI:

a. In how many different ways can an arrangement be made using

   i. any 5 of these letters?
   ii. all 9 of these letters?
   iii. all 9 of these letters, with a consonant first and G last?

b. In how many different ways can a group of 5 of these letters be made if

   i. F must be one of the letters?
   ii. any 5 of the letters can be used?
   iii. exactly 3 of the 5 must be consonants?

12. Kay Oss and Hezzy Tate are competing for the title of top mathematics student. Each has tied on all tests, and the Mathematics Club Sponsor, Miss Fortune, must break the tie. She flips a thumbtack

five times. The winner is the one who correctly guesses the number of times the tack will land point-up. Kay selects 3 and Hezzy selects 4. If the probability of "point-up" on any one flip is 0.6, who is more likely to win, Kay or Hezzy? Justify your answer.

13. Plutonium is used as a fuel in some nuclear power plants. When a plutonium atom fissions ("splits"), it emits 0, 1, 2, 3, or 4 new neutrons. Assume that the probabilities are

Number	*P(Number)*
1	0.1
2	0.15
3	0.5
4	0.25

a. Calculate $P(0)$.
b. What is the expected number of neutrons per fission?
c. Neutrons come only in integer quantities. How do you explain the fact that the expected number is *not* an integer?

Trigonometric and Circular Functions

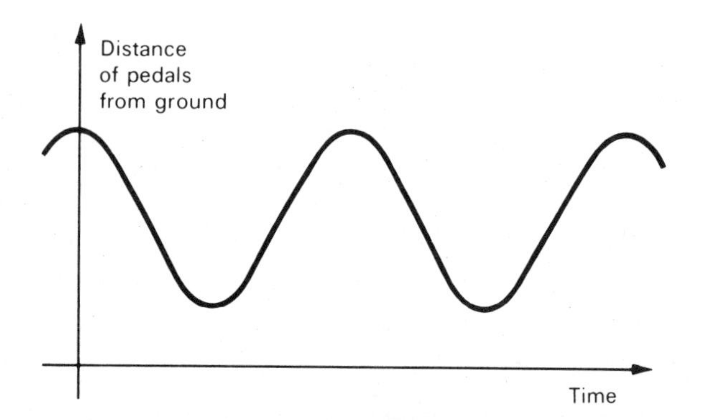

The position of the pedal on a bicycle is a ***periodic*** *function of how far the cyclist has traveled. This means that the dependent variable takes on the same set of values over and over again as the independent variable increases. In this chapter you will invent "trigonometric" and "circular" functions whose graphs are periodic. Your ultimate objective will be to write the particular equation of such a function from information about its graph. This way you can use the function as a mathematical model of periodic phenomena.*

13-1 INTRODUCTION TO PERIODIC FUNCTIONS

A function is said to be "periodic" if the dependent variable takes on the same set of values over and over again as the independent variable changes. In this section you will sketch reasonable graphs of real-world situations, as you did in Section 2-2. Some will be periodic and others will not.

Objective:

Given a situation from the real world with two related variables,

a. sketch a reasonable graph showing how those two variables are related,
b. tell whether or not the function is periodic.

The following exercise is designed to see if you can accomplish this objective. If you have difficulty, reread Section 2-2.

Exercise 13-1

For each of the following,

a. sketch a reasonable graph,
b. tell whether or not the function graphed is periodic.

1. The depth of the water at the beach depends on the time of day due to the motion of the tides.

2. The distance required to stop your car depends on how fast you were going when you applied the brakes.

3. The temperature of a cup of coffee depends on how long it has been since the coffee was poured.

4. As you breathe, the volume of air in your lungs depends on time.

5. A gymnast is jumping up and down on a trampoline. Her distance from the floor depends on time.

6. The distance you go depends on how long you have been going (at a constant speed).

7. As you ride the Ferris wheel at the amusement park, your distance from the ground depends on how long you have been riding.

8. The average temperature for any particular day (averaged over many years) depends on the day of the year.

9. A pendulum swings back and forth in a grandfather clock. The distance from the end of the pendulum to the left side of the clock depends on time.

10. A straight line starts along the positive x-axis and rotates counterclockwise around and around the origin of a Cartesian coordinate system. The *slope* of the line depends on the number of degrees through which the line has been rotated.

13-2 MEASUREMENT OF ARCS AND ROTATION

In the last section you sketched graphs of "periodic" functions which repeat themselves at regular intervals. One of the simplest examples of a periodic phenomenon is rotation. In this section you will invent ways to specify how far an object has rotated and in which direction.

Suppose that an object is going around a circular path. For simplicity, suppose that the circle has a radius of 1 unit (a "unit circle"), as in Figure 13-2a. If the object starts at the point $(u, v) = (1, 0)$, then its position may be specified in one of two different ways:

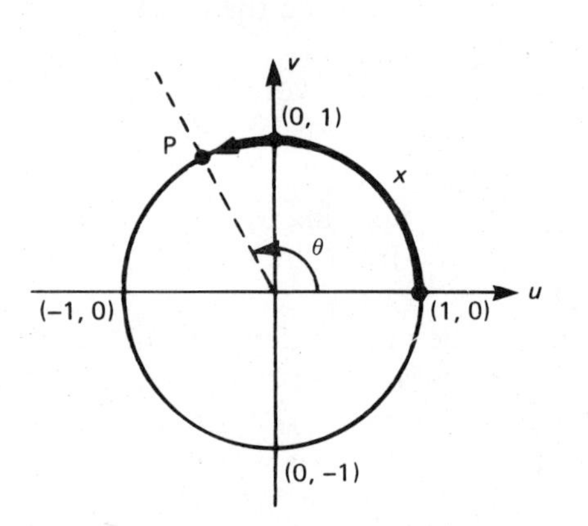

Figure 13-2a.

1. The distance, x, the point has traveled around the curved path.

2. The number of degrees, θ, through which the point has rotated.

Since the point could travel many times around the circle, the angle θ could be more than 360°.

Objective:

Given the angle measure, θ, or the arc length, x, sketch the position of the point P on the unit circle.

Notes:

1. The letters u and v are used for the coordinate system since x and y will be used later for independent and dependent variables.
2. The symbol θ is the Greek letter "theta." Greek letters are often used for angles. Some more frequently used ones are:

α alpha	θ theta
β beta	ϕ phi
γ gamma	ω omega

3. The letter θ may be used for the angle *itself* rather than the measure of the angle. In this case the measure of θ in degrees would be written

$$m^\circ(\theta).$$

For the sake of brevity, the symbol θ is often used interchangeably for the angle or for its measure, provided no confusion will result.

A directed angle or arc beginning at the positive u-axis and measured *counterclockwise* is said to be in *standard position.* Thus, if θ or x is *negative,* the angle or arc would be measured in the *clockwise* direction. Several arcs and angles in standard position are shown in Figure 13-2b.

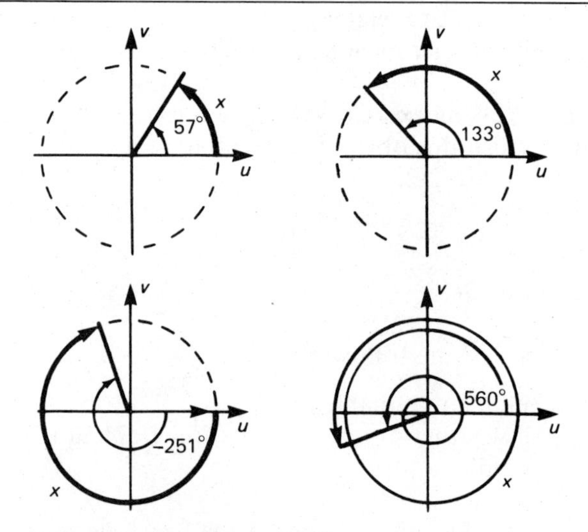

Figure 13-2b.

Two angles or arcs in standard position are said to be *coterminal* if they terminate ("end") at the same place. Coterminal arcs or angles differ by an integral number of revolutions. One complete revolution corresponds to 360° or to 2π units of arc length. Therefore, angles θ and ϕ, or arcs x and w, are coterminal if and only if

$$\boxed{\phi = \theta + 360n^\circ \quad \text{and} \quad w = x + 2\pi n} \quad ,$$

where n stands for an integer. Figure 13-2c shows three coterminal angles.

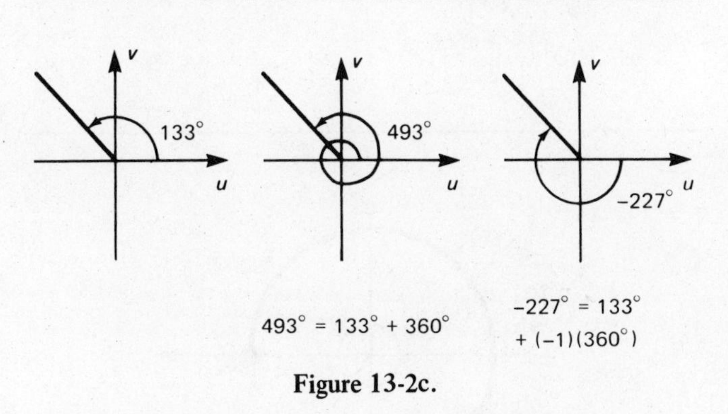

Figure 13-2c.

In order to draw an angle or arc in standard position, it helps to find the measure of the acute angle between the terminal side and the u-axis. This angle is called the *reference angle.* Figure 13-2d shows reference angles for various values of θ.

If θ terminates in Quadrant I, $\theta_{ref} = \theta$.
If θ terminates in Quadrant II, $\theta_{ref} = 180° - \theta$.
If θ terminates in Quadrant III, $\theta_{ref} = \theta - 180°$.
If θ terminates in Quadrant IV, $\theta_{ref} = 360° - \theta$.

Note: It is usually easier to draw a picture and *figure out* a formula than it is to memorize the formula.

With the help of coterminal and reference angles, you are now ready to accomplish the objective of this section.

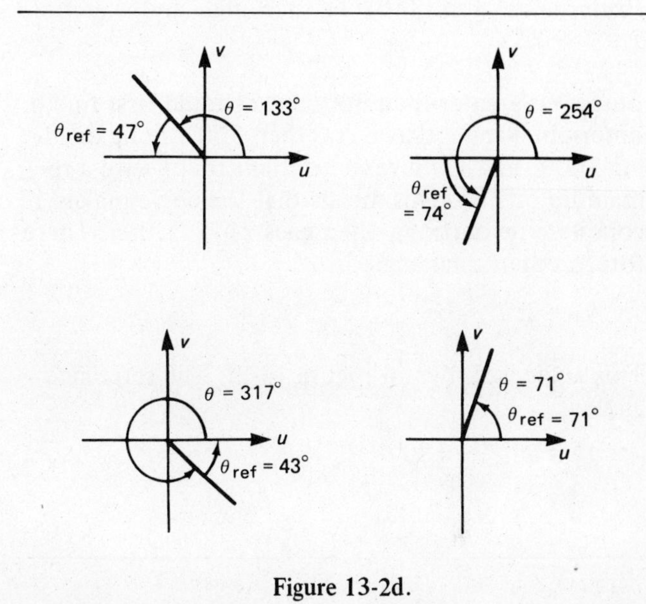

Figure 13-2d.

Example 1: Find the reference angle and sketch $\theta = 156°$.

Since 156° is between 90° and 180°, the angle terminates in Quadrant II. Therefore,

$$\theta_{ref} = 180° - 156° = 24°.$$

To draw the sketch, you go *back* 24° from the negative u-axis, as shown in Figure 13-2e. A rough estimate of 24° is enough for this sketch.

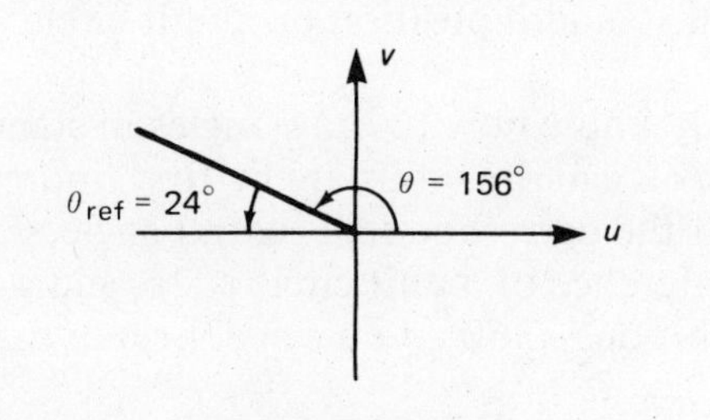

Figure 13-2e.

Example 2: Find the reference angle and sketch $\theta = 4897°$.

Since θ is greater than 360°, you should first find a coterminal angle that is *less* than 360°. Long dividing 360 into 4897 gives a quotient of 13 with a remainder of 217. This means that the angle makes 13 complete revolutions, then goes 217° farther. Therefore, a coterminal angle is

$$\theta_c = 217°.$$

This angle is shown in Figure 13-2f. The reference angle is

$$\theta_{ref} = 217° - 180° = 37°.$$

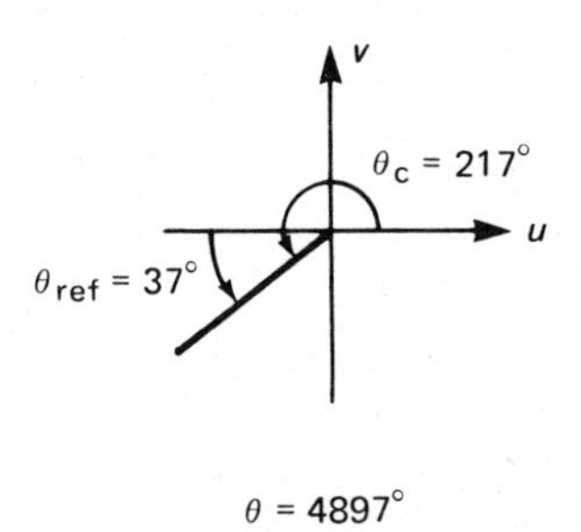

Figure 13-2f.

So θ terminates in the third quadrant, 37° beyond the negative u-axis.

Example 3: Sketch an arc of a unit circle measuring $x = 5\pi/3$ in standard position on a unit circle.

You already know how to draw angles in standard position. You can draw this arc by first finding the measure of the corresponding central angle, θ. Since the circumference of a unit circle is 2π, and a complete revolution is 360°, you can conclude that

$$\theta = \frac{360}{2\pi}x,$$

or, after canceling,

$$\boxed{\theta = \frac{180}{\pi}x}.$$

Similarly,

$$\boxed{x = \frac{\pi}{180}\theta}.$$

Therefore, for this example,

$$\theta = \frac{180}{\pi} \cdot \frac{5\pi}{3} = 300°.$$

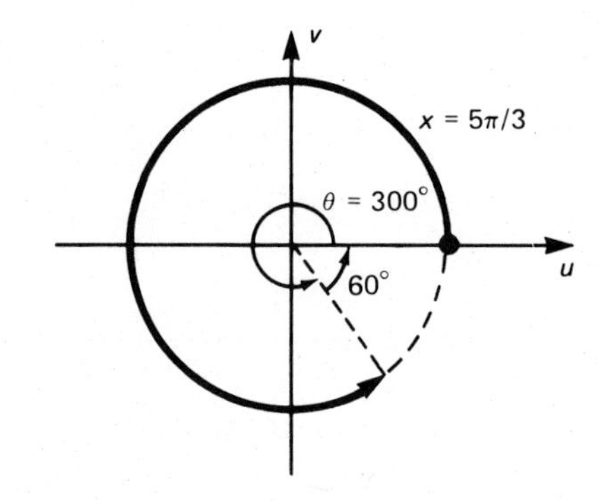

Figure 13-2g.

Since θ terminates in Quadrant IV, the reference angle is

$$\theta_{ref} = 360° - 300° = 60°.$$

After you have drawn the corresponding angle, you simply sketch an arc as shown in Figure 13-2g.

The following exercise is designed to give you practice with sketching arcs and angles in standard position.

Exercise 13-2

For Problems 1 through 12, find the measure of the reference angle, then draw the angle in standard position. Mark the reference angle on your sketch.

1. 137°	2. 259°	3. 342°
4. 54°	5. 412°	6. 591°
7. −186°	8. −303°	9. 5481°
10. 7321°	11. −2746°	12. −3814°

For Problems 13 through 24, sketch the indicated arc of a unit circle in standard position.

13. $\pi/3$	14. $2\pi/5$	15. π
16. $3\pi/2$	17. $7\pi/6$	18. 0.8π
19. $-\pi/4$	20. $-2\pi/3$	21. $7\pi/2$
22. 4.7π	23. -9.8π	24. $-11\pi/3$

13-3 DEFINITIONS OF TRIGONOMETRIC AND CIRCULAR FUNCTIONS

In the previous section you drew arcs and angles in standard position around a unit circle. The relationships among the angle measure, θ, the arc length, x, and the coordinates u and v are shown in Figure 13-3a.

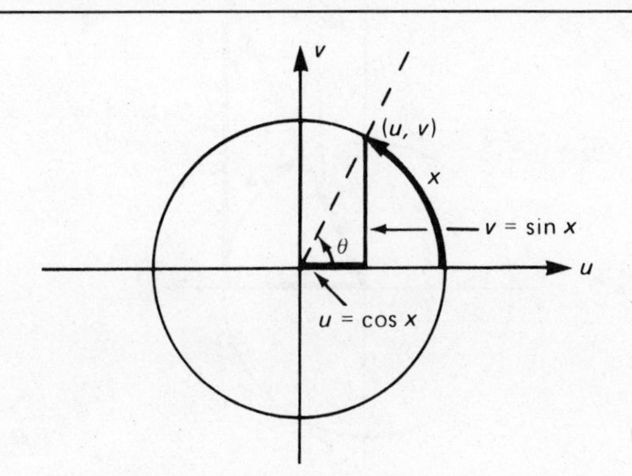

Figure 13-3a.

Since there is one and only one ordered pair (u, v) for each angle or arc, u and v are *functions* of θ or x. These functions are given special names.

DEFINITION

Sine Function: $v = \sin\theta = \sin x.$

Cosine Function: $u = \cos\theta = \cos x.$

Notes:

1. The symbols $\cos x$ and $\sin x$ are forms of "$f(x)$" terminology. The letters "cos" and "sin" are the *names* of the functions, and the independent variable θ or x is called the *argument.* As was true with $\log x$, the parentheses around the argument are usually omitted unless they are needed for clarity.
2. These functions are called *circular* functions if the argument is an arc length, and *trigonometric* functions if the argument is an angle measure. The latter name comes from "trigon" meaning triangle, and "metry" meaning measurement.
3. The name "sine" comes from the Latin word "sinus," which is a mistranslation of an Arabic word meaning "bowstring." The v-distance to the terminal point resembles half of a bowstring! The prefix "co-" in cosine comes from the word "complement," as you will see later.

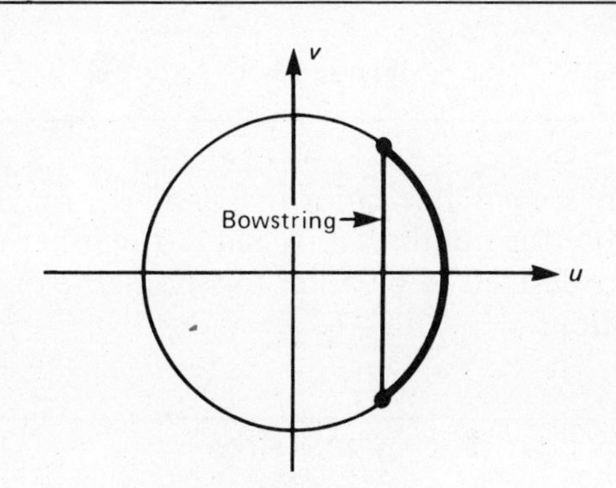

Figure 13-3b.

The sine and cosine are also equal to the *ratios* of sides of a right triangle. The two triangles in Figure 13-3c are similar, so their corresponding sides are proportional. Therefore,

$$\frac{v_1}{r} = \frac{v}{1} = v = \sin\theta, \text{ and}$$

$$\frac{u_1}{r} = \frac{u}{1} = u = \cos\theta.$$

In general, if $P(u, v)$ is *any* point on the terminal side of θ, and r is the distance from the origin to P, then

$$\boxed{\sin\theta = \frac{v}{r} \text{ and } \cos\theta = \frac{u}{r}}\ .$$

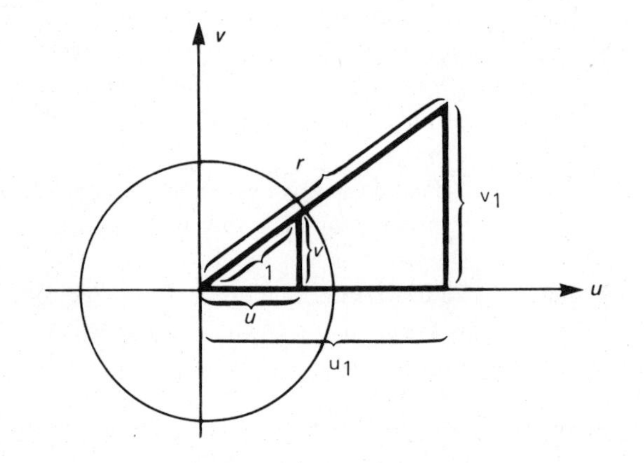

Figure 13-3c.

There are six possible ratios that can be formed using any two of the numbers u, v, and r. The other four ratios are defined to be other trigonometric or circular functions.

DEFINITION

tangent: $\tan\theta = \tan x = \dfrac{v}{u}$.

cotangent: $\cot\theta = \cot x = \dfrac{u}{v}$.

secant: $\sec\theta = \sec x = \dfrac{r}{u}$.

cosecant: $\csc\theta = \csc x = \dfrac{r}{v}$.

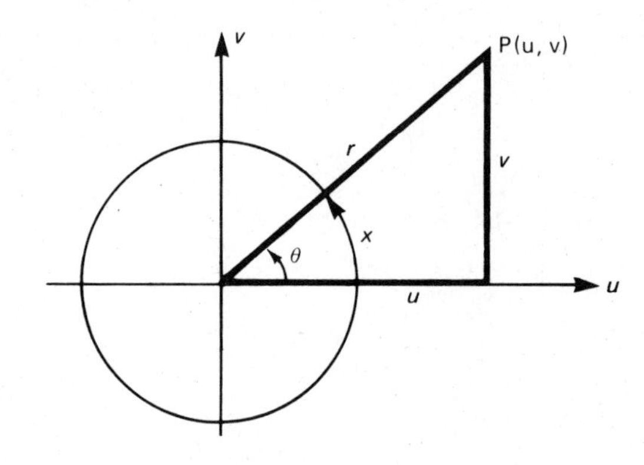

Notes:

1. Again, these are called *trigonometric* functions if the argument is an angle measure, and *circular* functions if the argument is an arc length.
2. The names "tangent" and "secant" come from the facts that tan x and sec x equal the lengths of the tangent and secant lines, respectively, as shown in Figure 13-3d.

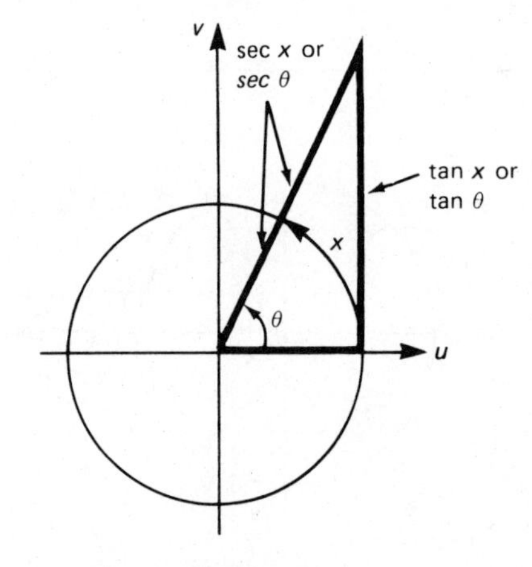

Figure 13-3d.

These trigonometric ratios provide an easy way to figure out the values of the six functions for certain special angles or arcs. If $\theta = 60°$, as in the left hand sketch of Figure 13-3e, then u, v, and r form sides of a $30° - 60°$ right triangle. From geometry, you will recall that the hypotenuse of such a triangle is *twice* the shorter leg. Therefore, if $u = 1$, then $r = 2$. By the Pythagorean Theorem, $v = \sqrt{3}$. In the middle sketch of Figure 13-3e, $\theta = 45°$. If $u = 1$, then $v = 1$ also, because the triangle is isosceles. By Phythagoras, $r = \sqrt{2}$. In the right sketch of Figure 13-3e, $\theta = 30°$. Reversing the triangle in the left sketch, $u = \sqrt{3}$, $v = 1$, and $r = 2$. From these three sketches, you can use the ratios to write the function values.

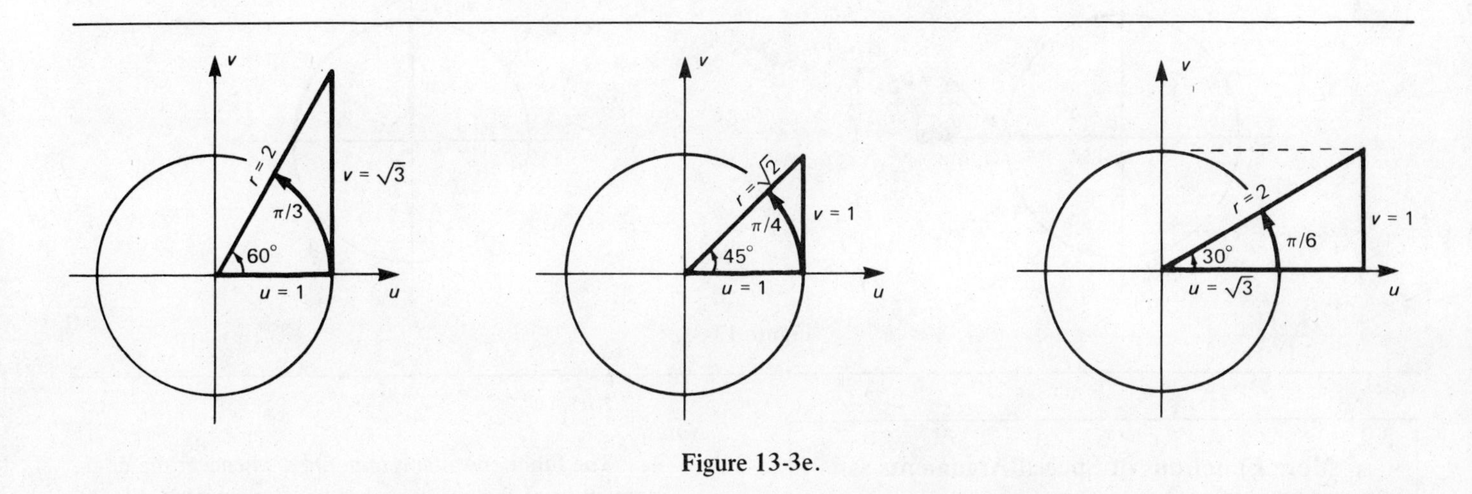

Figure 13-3e.

Functions of Special Arguments

Function	Ratio	$\theta = 30°$ $x = \pi/6$	$\theta = 45°$ $x = \pi/4$	$\theta = 60°$ $x = \pi/3$
$\sin\theta$ $\sin x$	$\frac{v}{r}$	$\frac{1}{2}$	$\frac{1}{\sqrt{2}} = \frac{\sqrt{2}}{2}$	$\frac{\sqrt{3}}{2}$
$\cos\theta$ $\cos x$	$\frac{u}{r}$	$\frac{\sqrt{3}}{2}$	$\frac{1}{\sqrt{2}} = \frac{\sqrt{2}}{2}$	$\frac{1}{2}$
$\tan\theta$ $\tan x$	$\frac{v}{u}$	$\frac{1}{\sqrt{3}} = \frac{\sqrt{3}}{3}$	1	$\sqrt{3}$
$\cot\theta$ $\cot x$	$\frac{u}{v}$	$\sqrt{3}$	1	$\frac{1}{\sqrt{3}} = \frac{\sqrt{3}}{3}$
$\sec\theta$ $\sec x$	$\frac{r}{u}$	$\frac{2}{\sqrt{3}} = \frac{2\sqrt{3}}{3}$	$\sqrt{2}$	2
$\csc\theta$ $\csc x$	$\frac{r}{v}$	2	$\sqrt{2}$	$\frac{2}{\sqrt{3}} = \frac{2\sqrt{3}}{3}$

If the angle or arc terminates at a quadrant boundary, the function values are even easier to find. For example, if $\theta = 90°$, then the point to pick is $(u, v) = (0, 1)$. If $\theta = 0°$, then the point to pick is $(u, v) = (1, 0)$. In both cases, $r = 1$. The arcs and angles are shown in Figure 13-3f, and the function values are given in the following table.

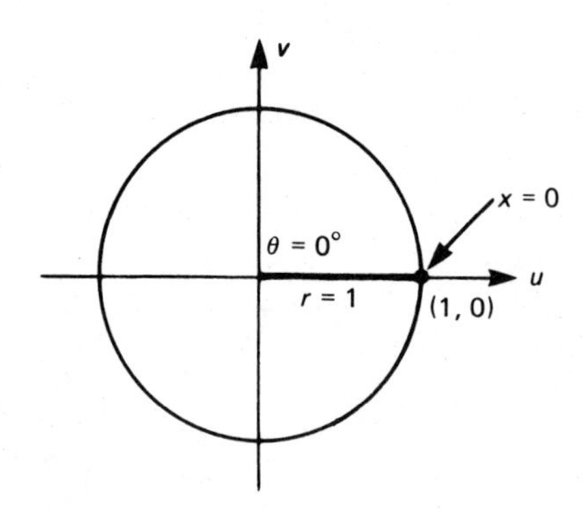

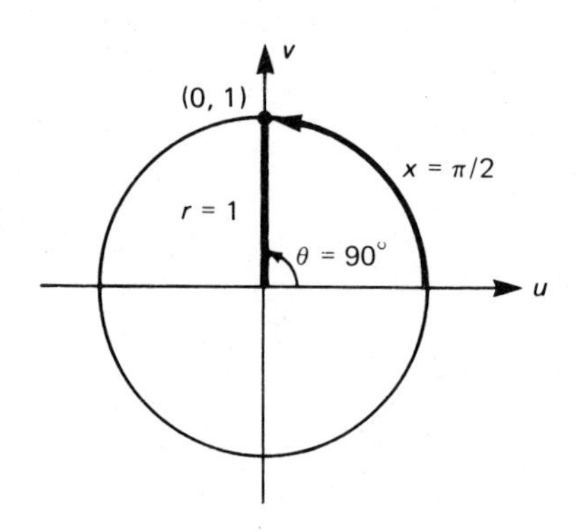

Figure 13-3f.

More Functions of Special Arguments

Function	Ratio	$\theta = 0°$ $x = 0°$	$\theta = 90°$ $x = \pi/2$
$\sin \theta$ $\sin x$	$\frac{v}{r}$	0	1
$\cos \theta$ $\cos x$	$\frac{u}{r}$	1	0
$\tan \theta$ $\tan x$	$\frac{v}{u}$	0	Undefined
$\cot \theta$ $\cot x$	$\frac{u}{v}$	Undefined	0
$\sec \theta$ $\sec x$	$\frac{r}{u}$	1	Undefined
$\csc \theta$ $\csc x$	$\frac{r}{v}$	Undefined	1

Notes:

1. The function value is *undefined* whenever the denominator of the ratio equals zero. For example, tan 90° would be 1/0, and division by zero is undefined.
2. There is a clever way to remember the function values by writing them in *radical* form.

θ	0°	30°	45°	60°	90°	
$\sin \theta$	$\frac{\sqrt{0}}{2}$	$\frac{\sqrt{1}}{2}$	$\frac{\sqrt{2}}{2}$	$\frac{\sqrt{3}}{2}$	$\frac{\sqrt{4}}{2}$	
$\cos \theta$	$\frac{\sqrt{4}}{2}$	$\frac{\sqrt{3}}{2}$	$\frac{\sqrt{2}}{2}$	$\frac{\sqrt{1}}{2}$	$\frac{\sqrt{0}}{2}$	
$\tan \theta$	$\sqrt{\frac{0}{4}}$	$\sqrt{\frac{1}{3}}$	$\sqrt{\frac{2}{2}}$	$\sqrt{\frac{3}{1}}$	$\sqrt{\frac{4}{0}}$	(undefined)

With the help of reference angles, you can find the functions of *any* size angle or arc, provided the angle is a multiple of 30° or 45°.

Objective:

Find *exact* values of the six trigonometric or circular functions if the angle measure is a multiple of 30° or 45°.

Example 1: Find the six trigonometric functions of 315°.

Since 315° terminates in Quadrant IV, its reference angle is

$$\theta_{ref} = 360° - 315°$$
$$= 45°.$$

Therefore, an isosceles right triangle can be drawn as shown in Figure 13-3g for which $u = 1$, $v = -1$, and $r = \sqrt{2}$. The value of v is *negative* in Quadrant IV. Using the ratios,

$$\sin 315° = \frac{v}{r} = \frac{-1}{\sqrt{2}} = -\frac{\sqrt{2}}{2} \qquad \cot 315° = \frac{u}{v} = \frac{1}{-1} = -1$$

$$\cos 315° = \frac{u}{r} = \frac{1}{\sqrt{2}} = \frac{\sqrt{2}}{2} \qquad \sec 315° = \frac{r}{u} = \frac{\sqrt{2}}{1} = \sqrt{2}$$

$$\tan 315° = \frac{v}{u} = \frac{-1}{1} = -1 \qquad \csc 315° = \frac{r}{v} = \frac{\sqrt{2}}{-1} = -\sqrt{2}$$

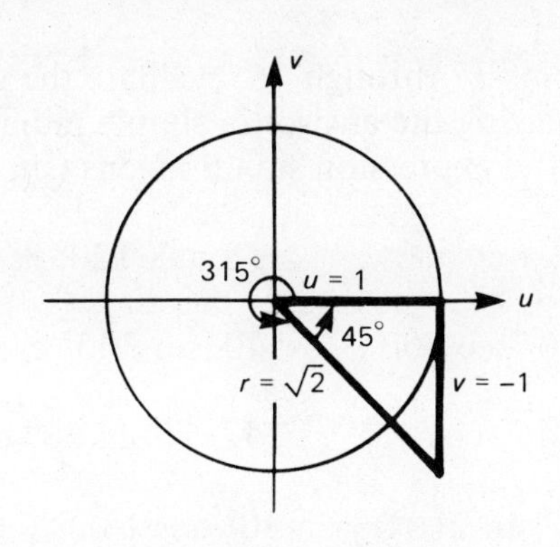

Figure 13-3g.

Example 2: Find the six circular functions of $4\pi/3$.

If $x = 4\pi/3$, then

$$\theta = \frac{180}{\pi} \cdot \frac{4\pi}{3}$$
$$= 240°.$$

Therefore,

$$\theta_{ref} = 240° - 180°$$
$$= 60°.$$

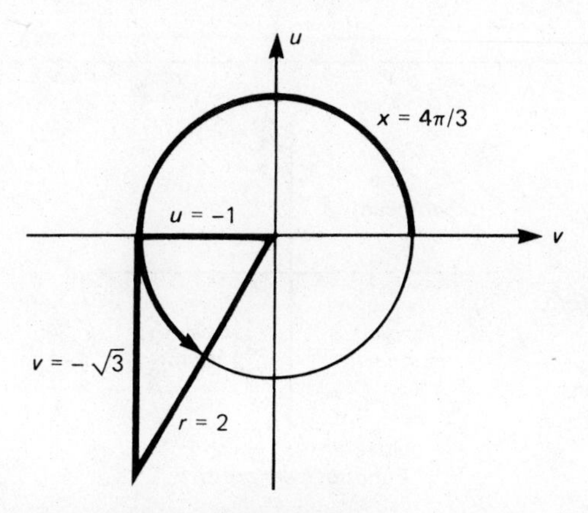

Figure 13-3h.

So you can draw a right triangle for which $u = -1$, $v = -\sqrt{3}$, and $r = 2$, as shown in Figure 13-3h. Using the ratios,

$$\sin\frac{4\pi}{3} = \frac{v}{r} = -\frac{\sqrt{3}}{2} \qquad \cot\frac{4\pi}{3} = \frac{u}{v} = \frac{-1}{-\sqrt{3}} = \frac{\sqrt{3}}{3}$$

$$\cos\frac{4\pi}{3} = \frac{u}{r} = -\frac{1}{2} \qquad \sec\frac{4\pi}{3} = \frac{r}{u} = -2$$

$$\tan\frac{4\pi}{3} = \frac{v}{u} = \frac{-\sqrt{3}}{-1} = \sqrt{3} \qquad \csc\frac{4\pi}{3} = \frac{r}{v} = \frac{2}{-\sqrt{3}} = -\frac{2\sqrt{3}}{3}$$

From the above examples you should be able to see that the *absolute value* of a function is equal to the same function of its *reference angle.* For example,

$$|\sin\theta| = \sin\theta_{ref},$$

or

$$\sin\theta = \pm\sin\theta_{ref}.$$

The proper sign to pick depends on the signs of u and v in the quadrant where θ terminates. As shown in

Figure 13-3i, *all* the functions are positive in the *first* quadrant. The *sine* and its reciprocal are positive in the *second* quadrant. The *tangent* and its reciprocal are positive in the *third* quadrant. The cosine and its reciprocal are positive in the *fourth quadrant.*

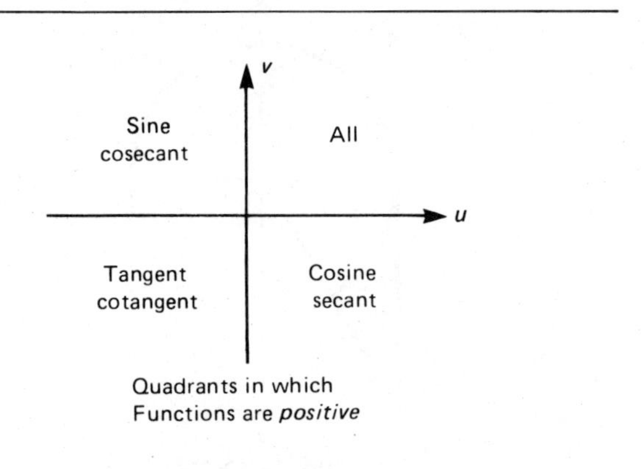

Figure 13-3i.

The following exercise has two purposes: First, you are to demonstrate your knowledge of the *definitions* of the trigonometric and circular functions and second, you are to get practice in finding *special* values of these functions. These special values will come up frequently throughout your future mathematical career, and you should be comfortable using them.

Exercise 13-3

For Problems 1 through 6, find exact values of the six trigonometric functions of the given angle.

1. $60°$
2. $135°$
3. $-315°$
4. $330°$
5. $180°$
6. $-270°$

For Problems 7 through 12, find exact values of the six circular functions of the given argument.

7. $5\pi/4$
8. $2\pi/3$
9. $5\pi/6$
10. $-7\pi/4$
11. $-5\pi/2$
12. $-\pi$

For Problems 13 through 36, find the exact value of the given trigonometric or circular function. You should try doing this *quickly,* either from memory or by visualizing the diagram in your head.

13. $\sin 180°$
14. $\sin 225°$
15. $\cos 240°$
16. $\cos 120°$
17. $\tan 315°$
18. $\tan 270°$
19. $\cot 0°$
20. $\cot 300°$
21. $\sec 150°$
22. $\sec 0°$
23. $\csc 45°$
24. $\csc 330°$
25. $\sin \frac{\pi}{3}$
26. $\sin \frac{11\pi}{6}$
27. $\cos \frac{3\pi}{4}$
28. $\cos \frac{\pi}{4}$
29. $\tan \pi$
30. $\tan \frac{3\pi}{4}$
31. $\cot \frac{7\pi}{6}$
32. $\cot \pi$
33. $\sec 2\pi$
34. $\sec \frac{\pi}{2}$
35. $\csc \frac{4\pi}{3}$
36. $\csc \frac{2\pi}{3}$

For Problems 37 through 76, evaluate the given expression, leaving the answer in simple radical form. Note that the expression $\sin^2 \theta$ means $(\sin \theta)^2$.

37. $\sin 30° + \cos 60°$
38. $\tan 120° + \cot (-30°)$
39. $\tan 300° \sec 300°$
40. $\sin 300° \csc 300°$
41. $12 \sin 45° \cos 45°$
42. $20 \sin 60° \cos 240°$
43. $\cos 45° \sin 210° - \sin 30° \cos 135°$
44. $\cos 180° \cos 45° - \sin 180° \sin 45°$
45. $\tan 30° \cot 30° + \tan 60° \cot 60°$
46. $\sec 60° \tan 135° - \cot 60° \sin 60°$
47. $\cos^2 60° + \sin^2 60°$
48. $\cos^2 150° + \sin^2 150°$
49. $\cot^2 330° - \csc^2 330°$

50. $\tan^2 240° - \sec^2 240°$

51. $\cos^2 45° - \sin^2 135°$ 52. $\sin^2 150° + \cos^2 30°$

53. $\dfrac{\sec 30°}{\cos 30°}$ 54. $\dfrac{\sin 120°}{\cos 120°}$

55. $\sin^2 30° + \cos^2 30° + \tan^2 30° - \sec^2 30°$

56. $\sin^2 30° + \cos^2 150° + \tan^2 60°$

57. $\sin \frac{\pi}{2} + 6 \cos \frac{\pi}{3}$ 58. $\sin \frac{\pi}{3} + 6 \cos \frac{\pi}{4}$

59. $\csc \frac{\pi}{2} \sin \frac{\pi}{2}$ 60. $4 \sin \frac{4\pi}{3} \cos \frac{4\pi}{3}$

61. $4 \sin \frac{\pi}{3} \cos \frac{\pi}{3}$ 62. $\sin \frac{\pi}{6} \csc \frac{\pi}{6}$

63. $\sin \frac{2\pi}{3} \cos \frac{5\pi}{6} - \cos \frac{2\pi}{3} \sin \frac{5\pi}{6}$

64. $\sin \frac{2\pi}{3} \cos \frac{\pi}{6} + \cos \frac{2\pi}{3} \sin \frac{\pi}{6}$

65. $\sec \frac{\pi}{4} \sin \frac{\pi}{4} - \tan \frac{3\pi}{4} \csc \frac{\pi}{3}$

66. $\sec \frac{\pi}{3} \cos \frac{\pi}{3} + \tan \frac{\pi}{3} \cot \frac{\pi}{3}$

67. $\cos^2 \pi + \sin^2 \pi$ 68. $\cos^2 \frac{2\pi}{3} + \sin^2 \frac{2\pi}{3}$

69. $\tan^2 \frac{\pi}{6} - \csc^2 \frac{\pi}{6}$ 70. $\csc^2 \pi - \tan^2 \pi$

71. $\cos^2 \frac{3\pi}{4} - \sin^2 \frac{\pi}{3}$ 72. $\sin^2 \frac{7\pi}{6} + \cos^2 \frac{\pi}{4}$

73. $\dfrac{\cos \frac{5\pi}{3}}{\sin \frac{5\pi}{3}}$ 74. $\dfrac{\cos \frac{\pi}{4}}{\sec \frac{\pi}{4}}$

75. $\tan \frac{\pi}{6} \cot \frac{\pi}{3} + \tan \frac{\pi}{4}$ 76. $\tan^2 \frac{2\pi}{3} \left(1 - \tan^2 \frac{7\pi}{6}\right)$

77. Find all values of θ from 0° through 360° for which

a. $\sin \theta = 0$ b. $\cos \theta = 0$ c. $\tan \theta = 0$
d. $\cot \theta = 0$ e. $\sec \theta = 0$ f. $\csc \theta = 0$

78. Find all values of θ from 0° through 360° for which

a. $\sin \theta = 1$ b. $\cos \theta = 1$ c. $\tan \theta = 1$
d. $\cot \theta = 1$ e. $\sec \theta = 1$ f. $\csc \theta = 1$

79. Find all values of x from 0 through 2π for which

a. $\sin x = 1$ b. $\cos x = 1$ c. $\tan x = 1$
d. $\cot x = 1$ e. $\sec x = 1$ f. $\csc x = 1$

80. Find all values of x from 0 through 2π for which

a. $\sin x = 0$ b. $\cos x = 0$ c. $\tan x = 0$
d. $\cot x = 0$ e. $\sec x = 0$ f. $\csc x = 0$

81. *Radians* – Dividing a complete revolution into 360° was done because there are approximately 360 days in a year. A more natural unit of angular measure is called the "radian." The radian measure of an angle is defined to be equal to the length of the corresponding arc of a unit circle. Using $m°(\theta)$ and $m^R(\theta)$ for the degree and radian measures of angle θ, respectively, if $m°(\theta) = 120$, then

$$x = \frac{\pi}{180} \cdot 120$$

$$= 2\pi/3$$

$$\therefore \quad m^R(\theta) = 2\pi/3.$$

As you can see from Figure 13-3j, the *radian* measure of an angle is equal to the *arc length, x,* of the unit circle.

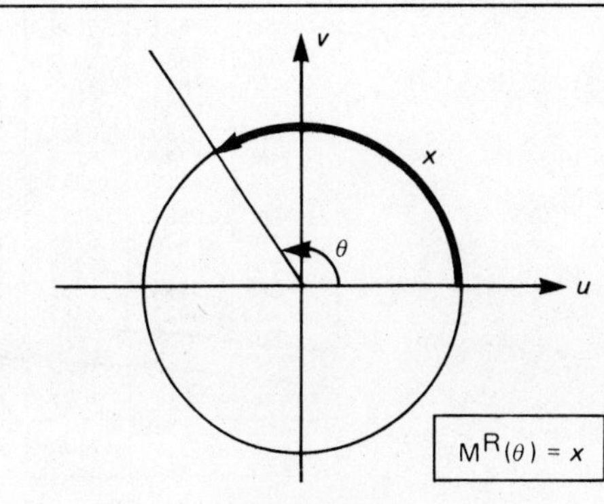

Figure 13-3j.

Find the radian measure of each of the following angles.

a. 30° b. 45° c. 60°
d. 90° e. 150° f. 180°
g. 300° h. −270° i. 3000°
j. −1080°

13-4 APPROXIMATE VALUES OF TRIGONOMETRIC AND CIRCULAR FUNCTIONS

Exact values of the trigonometric and circular functions can be found for certain "special" arguments such as 30° and 45°. Unfortunately, most values of the functions cannot be expressed exactly, even using radicals. For these "transcendental" numbers, the best you can hope for are decimal approximations. These may be found by calculator (which uses convergent series like those in Section 11-6 internally), or by tables at the back of this book.

Objective:

Given a value of the argument, find a decimal approximation for the trigonometric or circular function by calculator or from Tables III or IV.

Table III lists trigonometric functions in degrees and *minutes.* (There are 60 minutes in a degree.)

Example 1: Find tan 37° 40′.

Figure 13-4a shows a portion of Table III. You simply look in the first column, labeled "$m(\theta)$," find 37° 40′, and read across to the column headed "tan θ." The answer is tan 37° 40′ = 0.7720.

To find tan 37° 40′ by calculator, you must first convert the minutes to decimal degrees.

$$\tan 37° 40' = \tan 37.666667°$$
$$= 0.7719589447.$$

Table III. Trigonometric Functions and Degrees-to-Radians

	$m(\theta)$ Degrees	$m(\theta)$ Radians	sin θ	csc θ	tan θ	cot θ	sec θ	cos θ			
	36° 00′	.6283	.5878	1.701	.7265	1.376	1.236	.8090	.9425	54° 00′	
	10′	.6312	.5901	1.695	.7310	1.368	1.239	.8073	.9396	50′	← *Example 3:*
	20′	.6341	.5925	1.688	.7355	1.360	1.241	.8056	.9367	40′	
	30′	.6370	.5948	1.681	.7400	1.351	1.244	.8039	.9338	30′	
	40′	.6400	.5972	1.675	.7445	1.343	1.247	.8021	.9308	20′	
	50′	.6429	.5995	1.668	.7490	1.335	1.249	.8004	.9279	10′	
	37° 00′	.6458	.6018	1.662	.7536	1.327	1.252	.7986	.9250	53° 00′	
	10′	.6487	.6041	1.655	.7581	1.319	1.255	.7969	.9221	50′	← *Example 2:*
	20′	.6516	.6065	1.649	.7627	1.311	1.258	.7951	.9192	40′	← *Example 2:*
	30′	.6545	.6088	1.643	.7673	1.303	1.260	.7934	.9163	30′	
Example 1: →	40′	.6574	.6111	1.636	.7720	1.295	1.263	.7916	.9134	20′	
	50′	.6603	.6134	1.630	.7766	1.288	1.266	.7898	.9105	10′	
	38° 00′	.6632	.6157	1.624	.7813	1.280	1.269	.7880	.9076	52° 00′	
	10′	.6661	.6180	1.618	.7860	1.272	1.272	.7862	.9047	50′	
	20′	.6690	.6202	1.612	.7907	1.265	1.275	.7844	.9018	40′	
	30′	.6720	.6225	1.606	.7954	1.257	1.278	.7826	.8988	30′	
	40′	.6749	.6248	1.601	.8002	1.250	1.281	.7808	.8959	20′	
	50′	.6778	.6271	1.595	.8050	1.242	1.284	.7790	.8930	10′	
	39° 00′	.6807	.6293	1.589	.8098	1.235	1.287	.7771	.8901	51° 00′	
	10′	.6836	.6316	1.583	.8146	1.228	1.290	.7753	.8872	50′	
	20′	.6865	.6338	1.578	.8195	1.220	1.293	.7735	.8843		
	30′	.6894	.6361	1.572	.8243	1.213	1.296	.7716			
	40′	.6923	.6383	1.567	.8292	1.206	1.299		.7941	30′	
	50′							.7112	.7912	20′	
				1.418			1.410	.7092	.7883	10′	
	45° 00′	.7854	.7071	1.414	1.000	1.000	1.414	.7071	.7854	45° 00′	
			cos θ	sec θ	cot θ	tan θ	csc θ	sin θ	$m(\theta)$ Radians	$m(\theta)$ Degrees	

Figure 13-4a.

Example 2: Find cos 52° 43′.

This problem is tricky for at least two reasons. First, angles over 45° appear on the *right* side of Table III, going *up* (Figure 13-4a). Therefore, you must use the column headings at the *bottom* of the page rather than at the top. Second, 52° 43′ is *between* two table entries, so you must *interpolate,* as you did with logarithms in Section 6-10. Referring to Figure 13-4a, your thought process would be:

1. The two entries in the table are .6065 and .6041.

2. The difference (ignoring the decimal point) is −24.

3. 43′ is .3 of the way from 40′ to 50′, so you assume that the cosine is .3 of the way from .6065 to .6041.

4. .3 *of* −24 means .3 × (−24), which equals −7.2, or about −7.

5. The −7 is added to the function of the smaller angle, .6065, to give the answer cos 52° 43′ = .6058.

Great care is required when interpolating in the trigonometric tables for two reasons:

1. The cofunctions (cos, cot, csc) have values that *decrease* as the angle increases from 0° to 90°. Therefore the difference between two entries in the table is *negative.*

2. Angles between 45° and 90° are read *up* the table, so the function of the smaller angle appears *below* the function of the larger angle.

By calculator,

$$\cos 52° \, 43' = \cos 52.716667°$$
$$= 0.60575698.$$

Example 3: Find sec 126° 10′.

For angles outside Quadrant I, you simply find the same function of the *reference* angle. If θ = 126° 10′, then

$$\theta_{ref} = 180° - 126° \, 10'$$
$$= 179° \, 60' - 126° \, 10'$$
$$= 53° \, 50'.$$

Since 126° 10′ terminates in Quadrant II, its secant is *negative.* Therefore,

$$\sec 126° \, 10' = -\sec 53° \, 50'$$
$$= -1.695.$$

To find sec 126° 10′ by calculator, you must use the fact that sec θ = 1/cos θ.

$$\sec 126° \, 10' = \sec 126.16667°$$
$$= 1/\cos 126.16667°$$
$$= 1/(-0.59013610)$$
$$= -1.6945244.$$

Example 4: Find cot 1.073.

Figure 13-4b shows a portion of Table IV.

Table IV (Concluded)

Real Number x or $m^R(\theta)$	$m^o(\theta)$	sin x or sin θ	csc x or csc θ	tan x or tan θ	cot x or cot θ	sec x or sec θ	cos x or cos θ
1.05	60°10′	0.8674	1.153	1.743	0.5736	2.010	0.4976
1.06	60°44′	.8724	1.146	1.784	.5604	2.046	.4889
Example 4: → 1.07	61°18′	.8772	1.140	1.827	.5473	2.083	.4801
Example 4: → 1.08	61°53′	.8820	1.134	1.871	.5344	2.122	.4713
1.09	62°27′	.8866	1.128	1.917	.5216	2.162	[illegible]
1.10	63°02′	0.8912	1.122	1.965	0.5090		
1.11	[illegible]	[illegible]	1.116	2.014			

Figure 13-4b.

Your reasoning should be as follows:

1. The two entries in Table IV are 0.5473 and 0.5344.

2. The difference (ignoring the decimal point) is −129.

3. 1.073 is .3 of the way from 1.07 to 1.08, so the cotangent is about .3 of the way from 0.5473 to 0.5344.

4. .3 *of* −129 means (.3) (−129), which equals −38.7, or about −39.

5. The −39 is *added* to the function of the *smaller* argument, 0.5473, to give the answer, $\underline{\underline{\text{cot } 1.073 = 0.5434}}$.

By putting your calculator in the "radians" mode (Problem 81 of Exercise 13-3), you find that

$$\begin{aligned}\cot 1.073 &= 1/\tan 1.073\\ &= 1/1.8401141\\ &= \underline{\underline{0.54344457}}.\end{aligned}$$

Problems 1 through 8, below, are designed to lead you through progressively more involved uses of Table III. The remaining problems give you practice finding function values using Table III, Table IV, or a calculator.

Exercise 13-4

For Problems 1 through 8, cover up the answer at the end of the problem, then look up the function using Table III. If you do not get the answer listed, then read the hints and try looking in the table again.

1. sin 19° 10′

Hints: Find 19° 10′ in the left column, then find the column labeled "sin θ" at the *top* of the table.

Answer: .3283

2. cos 26° 40′

Hints: Be sure to use the column headings at the *top* of the table. If you got .4488 for an answer, then you used the column headings at the bottom.

Answer: .8936

3. tan 48° 50′

Hints: Use the angles in the *right* column and the headings at the *bottom* of the page. If you got .8744 or .9057, then you are using the column headings at the top instead of the bottom. If you got 1.104 then you are reading *down* the right-hand column instead of up, and found 47° 50′ instead of 48° 50′.

Answer: 1.144

4. sin 34° 23′

Hints: The entries in the table are .5640 and .5664. The difference is 24; .3 × 24 = 7.2, which rounds off to 7. The answer is found by adding 7 to 5640, the function of the *smaller* angle.

Answer: .5647

5. cos 25° 47′

Hints: The entries in the table are .9013 and .9001. Ignoring the decimal point, the difference is −12; .7 × (−12) = −8.4, which rounds off to −8. Add −8 to the function of the *smaller* angle.

Answer: .9005

6. sin 68° 59′

Hints: Remember that you are reading *up* the *right* column to find the angle, and that the function names are at the *bottom* of the page. The table entries of .9325 and .9336 have a difference of 11; .9 × 11 = 9.9, which rounds off to 10. Add 10 to the function of the *smaller* angle.

Answer: .9335

7. cos 79° 53′

Hints: Remember that you are reading angles *up* the *right* column, that you need the table headings at the *bottom* of the page, and that the cosine *decreases* as the angle gets larger. The difference between the table entries of .1765 and .1736 is −29; .3 × (−29) = −8.7, which rounds off to −9. Add −9 to the function of the smaller angle.

Answer: .1756

8. cos 156° 20′

Hints: The reference angle is 180° −156° 20′ = 23° 40′. cos 23° 40′ is found in the tables to be .9159. Since cos θ is defined to be u/r, and u is *negative* in Quadrant II, cos 156° 20′ will be negative.

Answer: −.9159

For Problems 9 through 26, find the trigonometric function value

a. using Table III,
b. using a calculator.

9. sin 34° 20′ 10. cos 29° 10′

11. sec 54° 50′ 12. csc 47° 40′

13. tan 15° 30′ 14. cot 81° 00′

15. sin 75° 23′ 16. cos 21° 46′

17. sec 52° 37′ 18. csc 2° 52′

19. tan 85° 18′ 20. cot 17° 4′

21. sin 241° 29′ 22. cos 116° 43′

23. sec (−81° 12′) 24. csc (−253° 51′)

25. tan 352° 6′ 26. cot 196° 34′

For Problems 27 through 38, find the circular function value

a. using Table IV,
b. using a calculator.

27. sin 0.354 28. cos 0.216

29. tan 0.428 30. cot 0.245

31. sec 0.693 32. csc 0.934

33. cos 1.254 34. sin 1.031

35. cot 1.432 36. tan 1.249

37. csc 1.111 38. sec 1.047

For Problems 39 through 50, the argument is outside the range $0 \leq x \leq \pi/2$. Find the indicated function value either by calculator or by first finding the reference *arc* and using Table IV. For example, if $x = 2$, then $x_{\text{ref}} = \pi - 2 \approx 3.142 - 2 = 1.142$.

39. sin 2 40. cos 3

41. tan 3.5 42. cot 4

43. sec 5 44. csc 6

45. cos 6 46. sin 5

47. cot 4.5 48. tan 3

49. csc 4 50. sec 2

For Problems 51 through 56, use a calculator to demonstrate that the equation is a true statement. Note that $\cos^2 x$ means $(\cos x)^2$.

51. $\cos^2 0.34 + \sin^2 0.34 = 1$

52. $\cos^2 1.96 + \sin^2 1.96 = 1$

53. $\cos^2 3 - \sin^2 3 = \cos 6$

54. $\cos^2 5 - \sin^2 5 = \cos 10$

55. $\cos 5 \sin 5 = \frac{1}{2} \sin 10$

56. $\cos 3 \sin 3 = \frac{1}{2} \sin 6$

13-5 GRAPHS OF TRIGONOMETRIC AND CIRCULAR FUNCTIONS

You have defined six new functions and know how to find their values. The next step in the study of a new kind of function is to see what the graphs look like.

As you have done with other functions, you will first draw the graphs accurately by pointwise plotting. Then you will identify features that will allow you to *sketch* the graphs *quickly*.

Objective:

Be able to draw graphs of the six trigonometric or circular functions

a. *accurately,* by pointwise plotting, and
b. *quickly,* by finding certain "critical points."

Start by drawing the graphs of $y = \cos x$ and $y = \sin x$. Since the values of $\cos x$ repeat themselves each 2π units of x, it is convenient to mark off the x-axis in multiples of π. For both graphs, y will range from -1 to 1. The easiest values to plot are those for x equals multiples of $\pi/2$. These "critical" values will be either 1, 0, or -1. You can recall them quickly by thinking of the unit circle. The points are plotted in Figure 13-5a.

To see what the graph looks like *between* these points, recall the *special* values

$$\cos\frac{\pi}{6} = \frac{\sqrt{3}}{2} \approx 0.87 \qquad \sin\frac{\pi}{6} = \frac{1}{2} = 0.5$$

$$\cos\frac{\pi}{4} = \frac{\sqrt{2}}{2} \approx 0.71 \qquad \sin\frac{\pi}{4} = \frac{\sqrt{2}}{2} \approx 0.71$$

$$\cos\frac{\pi}{3} = \frac{1}{2} = 0.5 \qquad \sin\frac{\pi}{3} = \frac{\sqrt{3}}{2} \approx 0.87$$

Figure 13-5b shows these values plotted on an expanded portion of Figure 13-5a.

The remainder of each graph consists of repetitions of these curves, as shown in Figure 13-5c.

The graphs are called *sinusoids.* The prefix "sinus-" is pronounced like what gets stopped up when you have a cold. The suffix "-oid" means "like."

Certain features of these graphs are given special names, as shown in Figure 13-5d.

DEFINITIONS

A ***cycle*** of a periodic function is a portion of the graph from one point to the point at which the graph starts repeating itself.

The ***period****, p,* of a periodic function is the change in x corresponding to one cycle. That is, $f(x + p) = f(x)$ for all values of x.

The ***amplitude*** of a sinusoid is the distance from its axis to a high point or a low point.

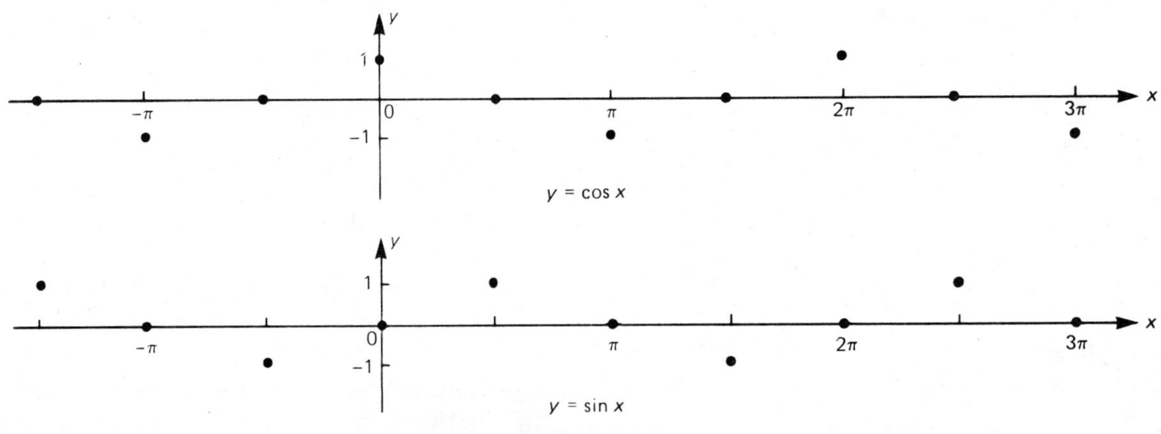

Figure 13-5a.

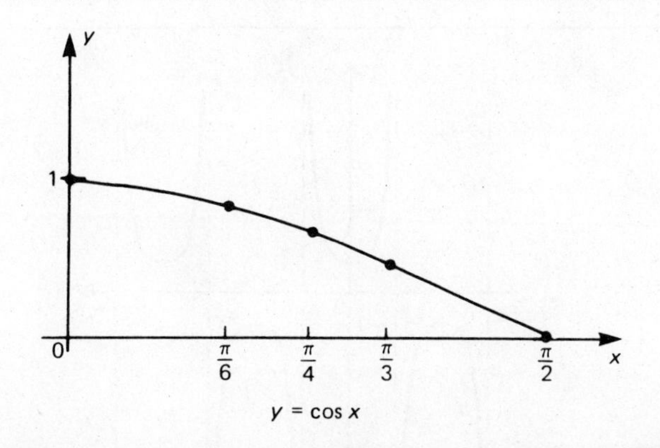

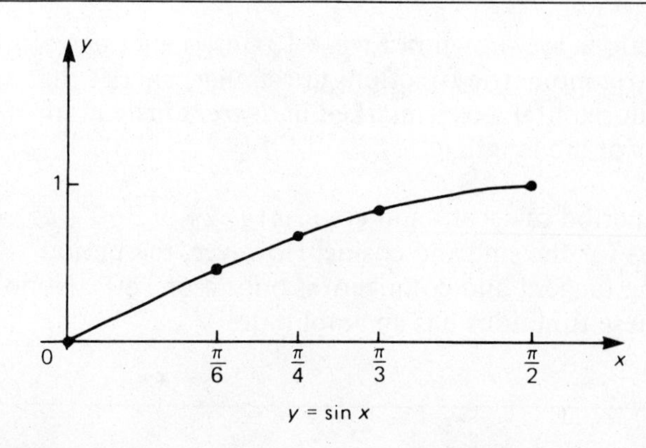

Figure 13-5b.

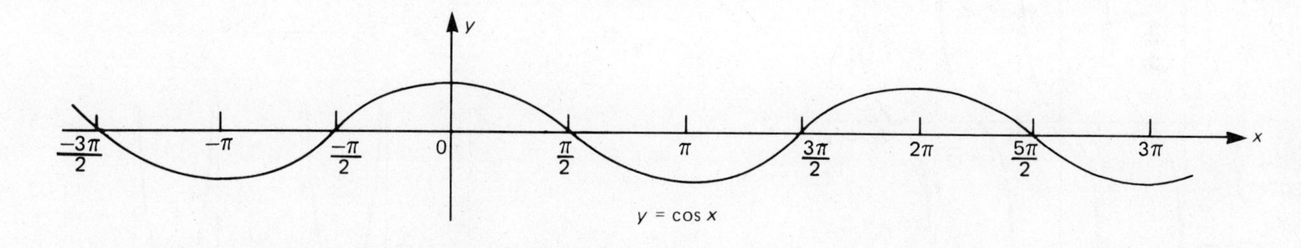

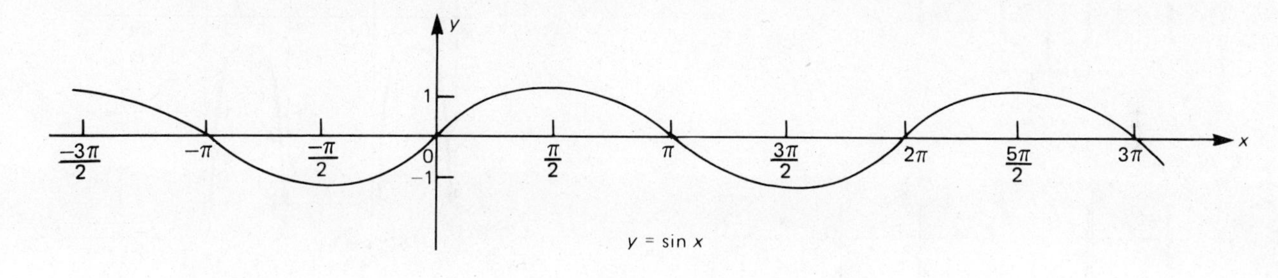

Figure 13-5c.

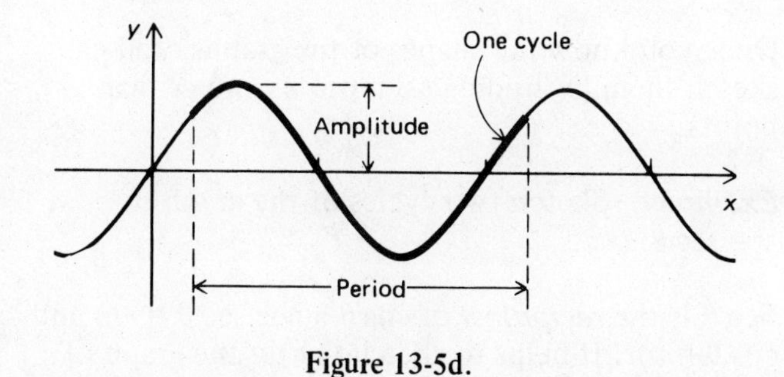

Figure 13-5d.

For both sine and cosine, the amplitude is 1 and the period is 2π (for circular functions) or 360° (for trigonometric functions).

The graphs of the other four functions can also be drawn by pointwise plotting. Since the function is *undefined* when the denominator of the ratio equals zero, there will be a *vertical asymptote* at each such value of x or θ. The graphs of the four circular functions are shown in Figure 13-5e. The graphs of the trigonometric functions are similar, except that the horizontal axis is marked in *degrees* instead of units of arc length.

The period of secant and cosecant is 2π or $360°$, as it was for the sine and cosine. However, the period of the tangent and cotangent is only π or $180°$. None of these functions has an amplitude.

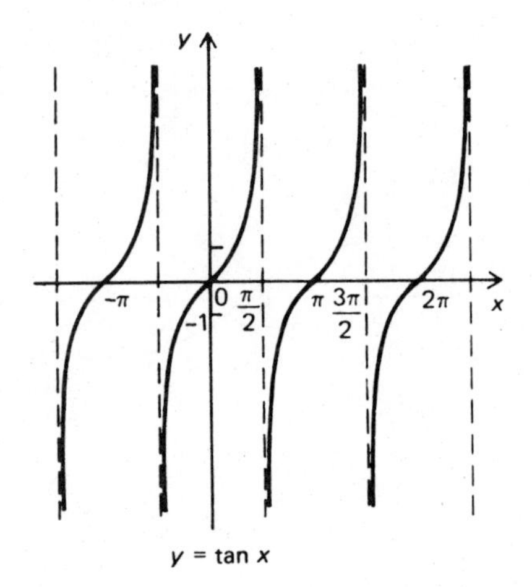

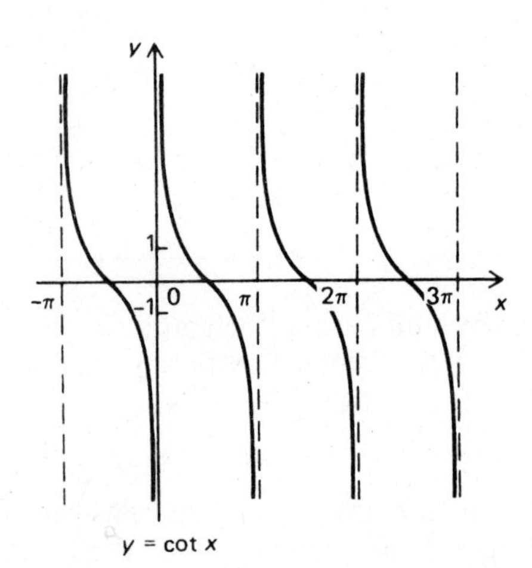

Figure 13-5e.

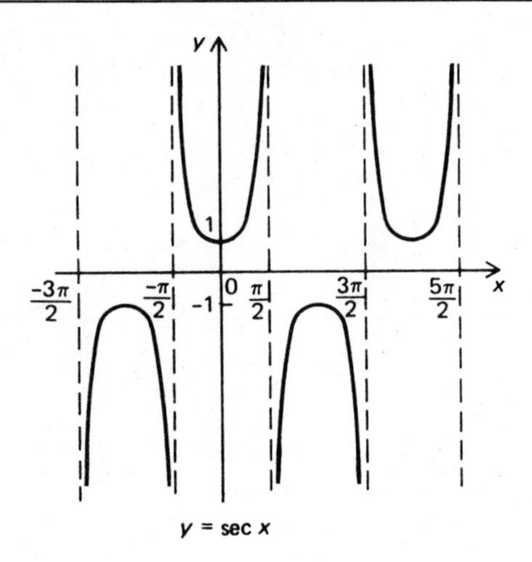

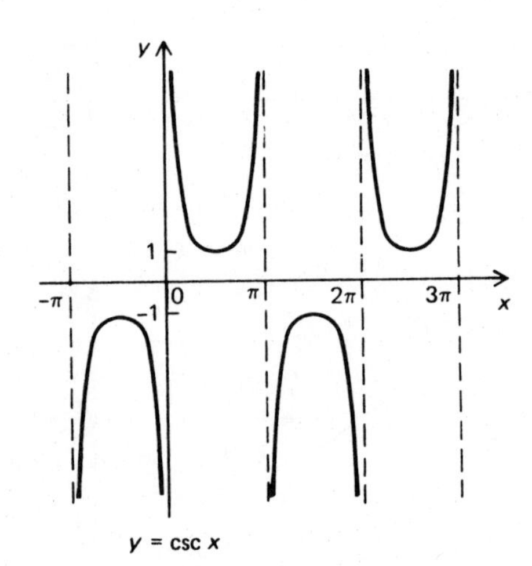

Figure 13-5f.

Once you know the shapes of the graphs, you can sketch them by finding asymptotes and critical points.

Example: Sketch two cycles of the graph of $y = \sec \theta$.

Sec θ is the *reciprocal* of $\cos \theta$ since $\sec \theta = r/u$ and $\cos \theta = u/r$. It helps to sketch lightly the graph of $y = \cos \theta$ first, as in the left side of Figure 13-5g.

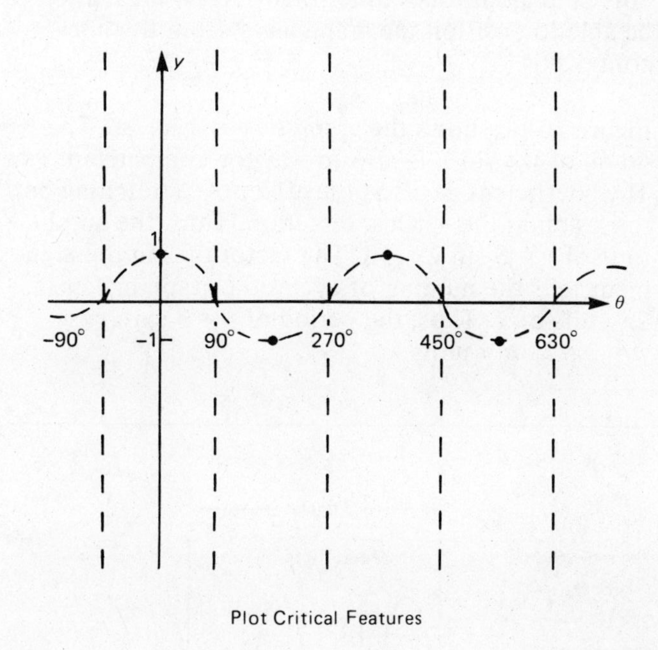

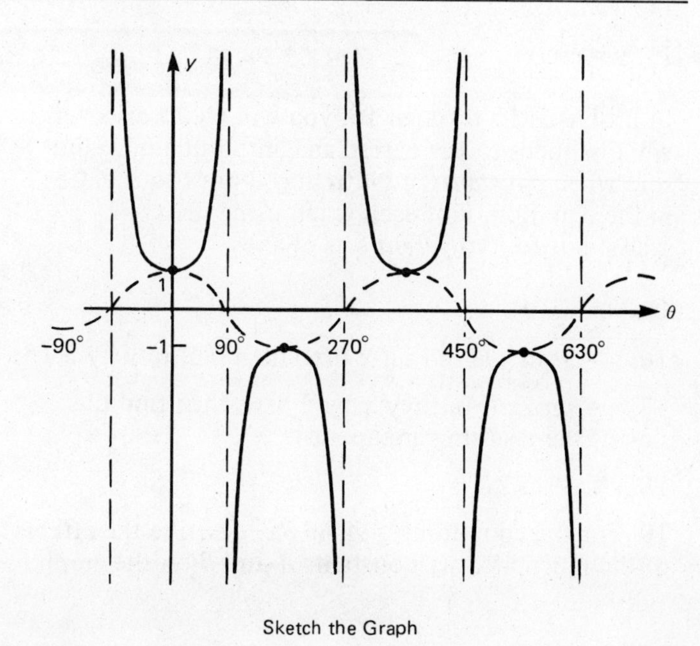

Figure 13-5g.

Each place $\cos\theta = 0$, the secant graph will have a vertical asymptote, so you sketch these. Critical points occur halfway between each asymptote, where $\sec\theta = \pm 1$. With the knowledge of what the graph looks like, you can sketch it as shown in the right side of Figure 13-5g.

The following exercise is designed to let you find out the exact shapes of the graphs of the six functions by pointwise plotting. Then you will use critical features to sketch these graphs quickly.

Exercise 13-5

1. Carefully plot the graphs of the six trigonometric functions for each 10° from $\theta = 0°$ to $\theta = 90°$. You should choose scales so that the graphs are fairly *large*. It is *not* necessary to use the same scale for both the y-axis and the θ-axis. Connect the points with smooth curves.

2. Carefully plot the graphs of the six circular functions for each 0.2 units from $x = 0$ to $x = \pi/2$. Choose as large a scale as possible, and use the *same* scale for both axes. Connect the points with smooth curves.

For Problems 3 through 14, sketch *quickly* two complete cycles of the trigonometric or circular function graph. Do this by locating critical features such as high or low points and asymptotes. After you have sketched the graph, look back at this section to make sure you are correct. If not, then redraw your graphs.

3. $y = \sin\theta$
4. $y = \cos\theta$
5. $y = \tan\theta$
6. $y = \cot\theta$
7. $y = \sec\theta$
8. $y = \csc\theta$
9. $y = \cos x$
10. $y = \sin x$

11. $y = \cot x$ 12. $y = \tan x$

13. $y = \csc x$ 14. $y = \sec x$

In Problems 15 through 19, you will try to discover what happens to the period and amplitude of a sinusoid when *constants* are introduced at various places in the equation. Plot each graph using the *same* scales, but four *different* sets of axes.

15. $y = \sin x$

16. $y = 3 \sin x$ (Find $\sin x$ *first,* then multiply by 3.)

17. $y = \sin 2x$ (Multiply x by 2 *first,* then find the sine of the resulting argument.)

18. $y = 3 \sin 2x$

19. For the equation $y = A \sin Bx$, describe the effects of the multiplicative constants A and B on the graph.

13-6 GENERAL SINUSOIDAL GRAPHS

Many periodic phenomena have graphs that look like sinusoids. For example, the time of sunrise as a function of the day of the year has a graph that looks like Figure 13-6a.

In order to use sinusoidal functions as mathematical models for these phenomena, you must be able to graph sinusoids that have periods other than 2π or 360° and amplitudes other than 1. You must also be able to position the graph away from the horizontal axis.

Figure 13-6b shows the graph of $y = 3 \sin 2x$. The graph of $y = \sin x$ is also drawn for comparison. The coefficient of 3 has the effect of "stretching out" the graph in the vertical direction. Thus, the amplitude of $y = 3 \sin 2x$ is 3. The factor of 2 in the argument tells the number of cycles the graph makes in 2π units of x. Thus, the period of $y = 3 \sin 2x$ is $p = 2\pi/2$, or π units.

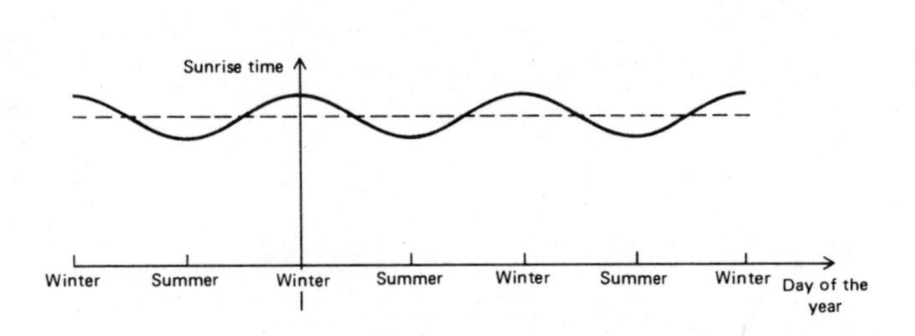

Figure 13-6a.

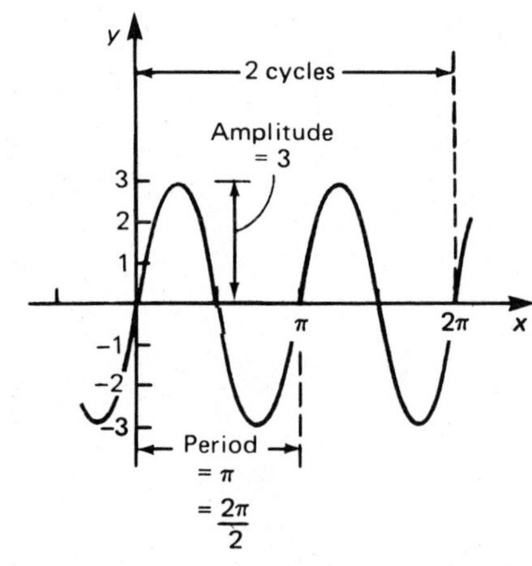

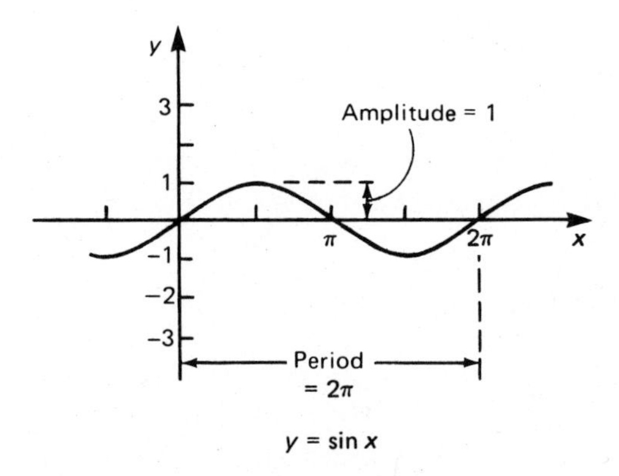

Figure 13-6b.

Suppose that you are to graph

$$y = \cos\left(x - \frac{\pi}{3}\right).$$

To find critical points on the graph, you need to find values of x that make the argument $(x - \pi/3)$ equal to 0, $\pi/2$, π, $3\pi/2$, 2π, and so forth. Setting the argument equal to 0,

$$x - \frac{\pi}{3} = 0$$

$$x = \frac{\pi}{3}.$$

Since $\cos 0 = 1$, the graph is a *high* point when $x = \pi/3$. Similarly, setting the argument equal to $\pi/2$ gives

$$x - \frac{\pi}{3} = \frac{\pi}{2}$$

$$x = \frac{5\pi}{6}.$$

Since $\cos(\pi/2) = 0$, the graph crosses the axis at $x = 5\pi/6$. By similar reasoning, you can get other critical points and draw the graph as shown in Figure 13-6c. The graph is *congruent* to that of $y = \cos x$, but is *displaced* $\pi/3$ units to the right.

Conclusion: If $y = \cos(x - D)$, then the graph is congruent to that of $y = \cos x$, but is *displaced D* units in the x-direction.

The constant D is called the *phase displacement.*

Now, suppose that you must graph

$$y = 3 + \cos x.$$

Each point will be 3 units *higher* than the corresponding point on the graph of $y = \cos x$, as shown in Figure 13-6d.

Conclusion: If $y = C + \cos x$, then the *sinusoidal axis* (see Figure 13-6d) is C units above the x-axis.

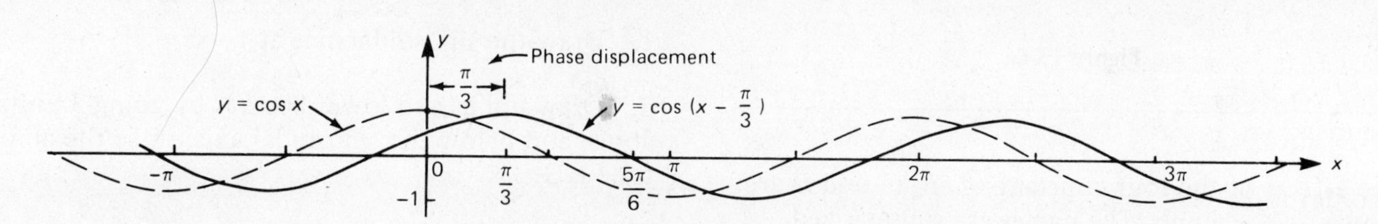

Figure 13-6c.

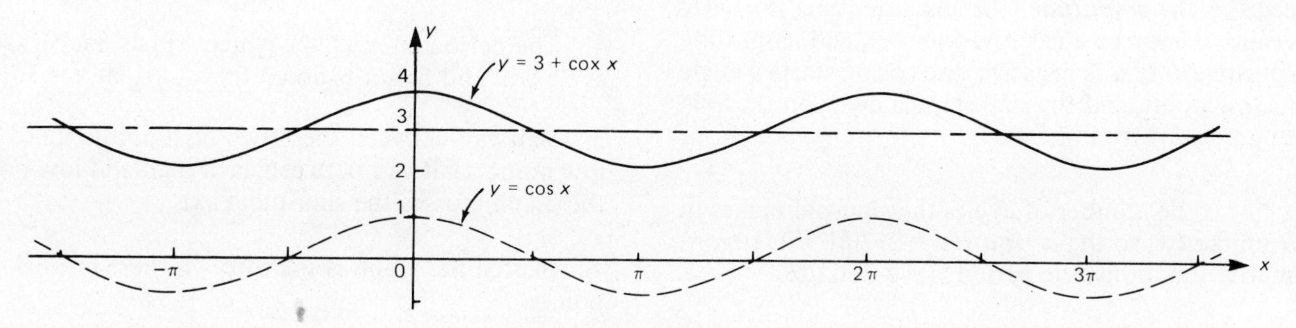

Figure 13-6d.

Drawing together the above conclusions, the *general* sinusoidal equation is

$$y = C + A \cos B(x - D).$$

A graph of this equation is shown in Figure 13-6e.

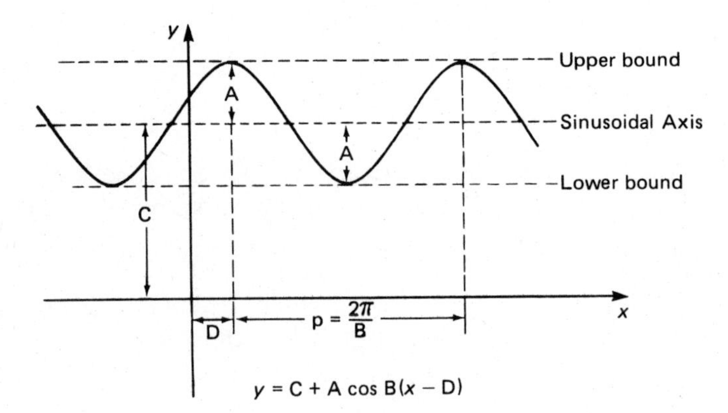

Figure 13-6e.

The effects of the four constants A, B, C, and D are shown on the graph. The names are summarized as follows.

1. $|A|$ is the ***amplitude*** (the absolute value is needed because A may be a negative number, and amplitude is positive). If A is negative, the cosine starts a cycle at a *low* point, and the sine starts a cycle on the axis, but going *downward.*

2. $|B|$ is the number of cycles the sinusoid makes in 2π units of x, so the ***period*** is $p = 2\pi/|B|$. For trigonometric functions, the period is $p = 360°/|B|$.

3. C is the ***vertical shift***. It may be positive or negative.

4. D is the ***phase displacement***. It, too, may be positive or negative.

Note: A cycle "starts" when the argument is 0, or a multiple of 2π. Since $\cos 0 = 1$ and $\sin 0 = 0$, the cosine graph starts a cycle at a *high* point and the sine function starts a cycle at a *midpoint*, going *up*.

These conclusions allow you to accomplish the following objective.

Objective:

Given the equation of a particular sinusoid, sketch the graph *quickly*.

Example 1: Sketch the graph of

$$y = 5 + 3 \cos \frac{1}{4}(x + \pi).$$

An efficient stepwise procedure for drawing the graph is:

1. Draw the sinusoidal axis at $y = 5$.

2. Draw upper and lower bounds by going 3 units above and below the sinusoidal axis, since the amplitude is 3.

3. Find the starting point of a cycle at $x = -\pi$, the phase displacement. Cosine starts a cycle at a *high* point.

4. The period is $2\pi/(1/4)$, which equals 8π. So the cycle will end 8π units down the x-axis, at $x = 7\pi$.

5. Halfway between these two high points there is a low point. Halfway between each high and low point the graph crosses the sinusoidal axis.

6. Sketch the graph going through these critical points.

The graph is shown in Figure 13-6f.

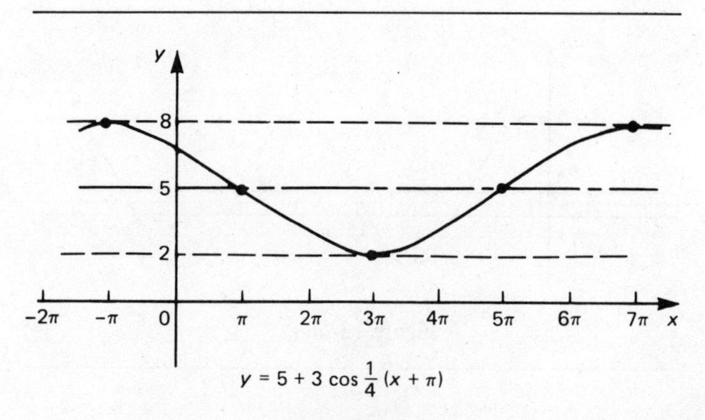

Figure 13-6f.

Example 2: Sketch the graph of
$y = -2 + 4 \sin \frac{\pi}{5}(x - 3)$.

The period is

$$p = \frac{2\pi}{\frac{\pi}{5}} = 10.$$

Notice that when B is a multiple of π, the π's will cancel, and the period will *not* be a multiple of π. In this case, the x-axis can be marked in more familiar units.

The sinusoidal axis is at $y = -2$. The upper and lower bounds are 4 units above and below this sinusoidal axis. A cycle starts at $x = 3$, the phase displacement. Since this is the *sine* function, the graph starts at a *midpoint*, going *up*. The end of this cycle is at $x = (3 + 10) = 13$. Again, there are critical points each 1/4 cycle. The graph is shown in Figure 13-6g.

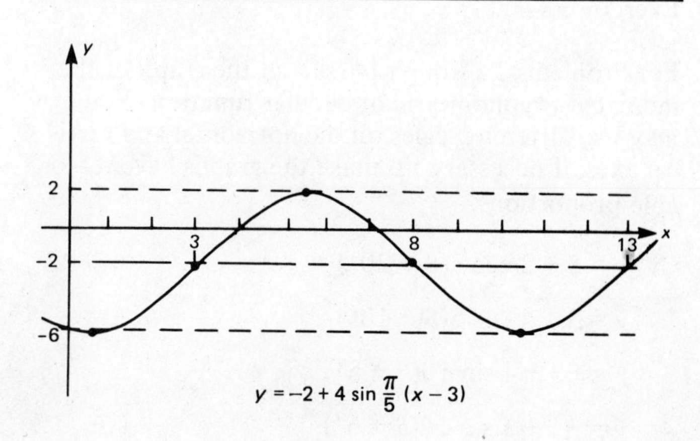

Figure 13-6g.

Example 3: Sketch the graph of the trigonometric function $y = -5 + 7 \cos 30(\theta + 4°)$.

The only way the trigonometric function graphs differ from the circular functions is that the horizontal axis is labeled in degrees. The period is

$$p = \frac{360°}{30} = 12°.$$

Since the phase displacement is $-4°$, the graph starts with a high point at $\theta = -4°$, and has its next high point at $\theta = (-4 + 12)° = 8°$. The sinusoidal axis is at $y = -5$, and the amplitude is 7. The graph is shown in Figure 13-6h.

The exercise that follows is designed to give you practice in sketching the graphs of sinusoids.

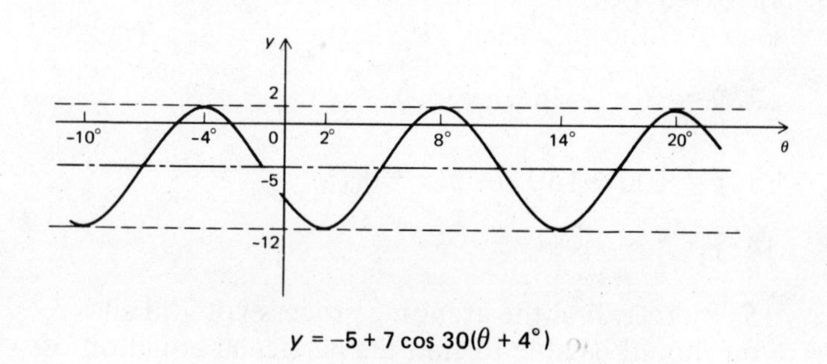

Figure 13-6h.

Exercise 13-6

For Problems 1 through 14, sketch the graph of the indicated trigonometric or circular function. You may use different scales on the horizontal and vertical axes, if necessary, to make the graphs have reasonable proportions.

1. $y = 5 + 2 \cos 3(\theta - 20°)$
2. $y = 3 + 4 \cos 5(\theta - 10°)$
3. $y = -3 + 4 \sin 10(\theta + 5°)$
4. $y = -1 + 3 \sin 12(\theta + 6°)$
5. $y = 3 + 2 \cos \frac{1}{5}(x - \pi)$
6. $y = 7 + 3 \cos \frac{1}{4}(x - 3\pi)$
7. $y = -4 + 5 \sin \frac{2}{3}(x + \frac{\pi}{2})$
8. $y = -5 + 4 \sin \frac{1}{3}(x + \frac{\pi}{2})$
9. $y = -10 + 20 \cos \frac{\pi}{3}(x - 1)$
10. $y = 2 + 6 \cos \frac{\pi}{4}(x - 3)$
11. $y = 3 + 5 \sin \frac{\pi}{4}(x - 3)$
12. $y = -6 + 7 \sin \frac{\pi}{4}(x - 2)$
13. $y = 100 + 150 \cos \pi(x + 0.7)$
14. $y = 1 + \cos \pi(x + 0.3)$
15. By *reversing* the graphing process you did above, you should be able to find the particular equation of a sinusoid from a given graph. Demonstrate that you can do this by writing the particular equation of the following sinusoid:

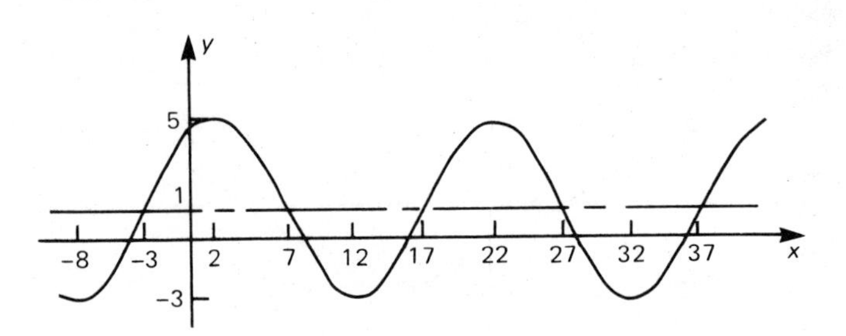

Figure 13-6i.

13-7 EQUATIONS OF SINUSOIDS FROM THEIR GRAPHS

Now that you know the effects of the four constants A, B, C, and D on the graph of the general equation $y = C + A \cos B(x - D)$, you are ready to use such sinusoidal functions as mathematical models. In order to do this, you must be able to write the *particular* equation that has the right period, amplitude, phase, and vertical placement.

Objective:

Given the graph of a sinusoidal function or information about the graph, write the particular equation.

Example: Write the particular equation of the sinusoid in Figure 13-7a.

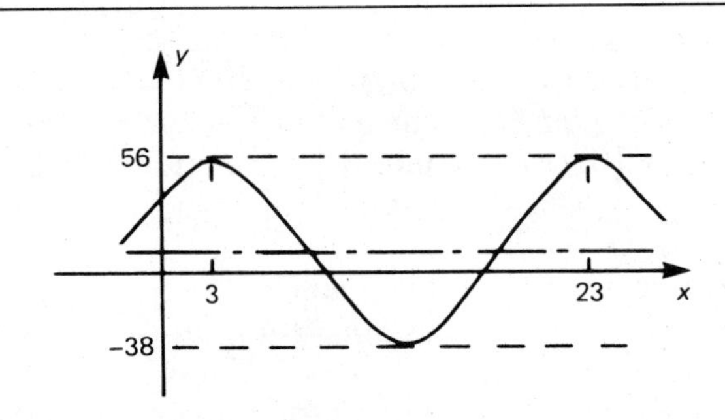

Figure 13-7a.

Your reasoning should be as follows:

1. It is easier to use the *cosine* function since a cycle starts at a *high* point. So the general equation is

$$y = C + A \cos B(x - D).$$

2. One cycle starts at $x = 3$ and ends at $x = 23$. So the period is $p = 23 - 3 = 20$. Therefore,

$$B = \frac{2\pi}{p} = \frac{2\pi}{20} = \frac{\pi}{10}.$$

3. The sinusoidal axis is halfway between the upper bound, 56, and the lower bound, −38. So the vertical shift is the *average* of 56 and −38. Therefore,

$$C = \frac{1}{2}(56 + (-38)) = \frac{1}{2}(18) = 9.$$

4. The amplitude is the distance between the sinusoidal axis and the upper bound. Therefore,

$$A = 56 - 9 = 47.$$

5. Using the cosine function, the phase displacement is 3. Therefore, $D = 3$.

6. Having found the four constants, you can write the equation:

$$\underline{\underline{y = 9 + 47 \cos \frac{\pi}{10}(x - 3).}}$$

If the horizontal axis had been labeled in *degrees*, the function would be *trigonometric*, and the variable would (probably) have been θ. In this case, the period would be 20°, and the value of B would be

$$B = \frac{360°}{20°} = 18.$$

The equation would be $\underline{\underline{y = 9 + 47 \cos 18(\theta - 3°)}}$.

The exercise that follows is designed to give you practice finding particular equations from the graph or from information about the graph.

Exercise 13-7

For Problems 1 through 14, write the particular equation of the sinusoid sketched.

1.

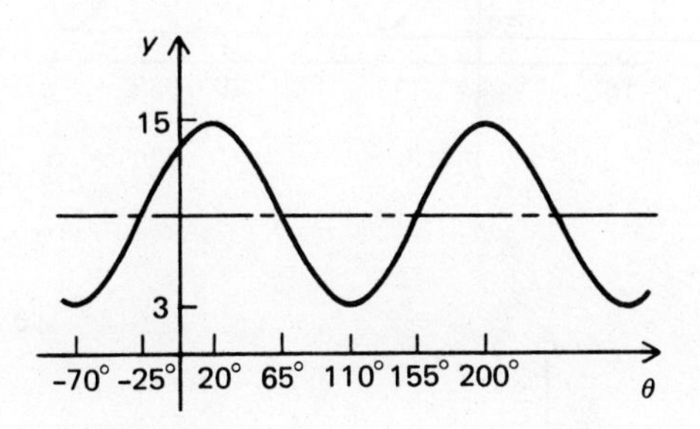

2.

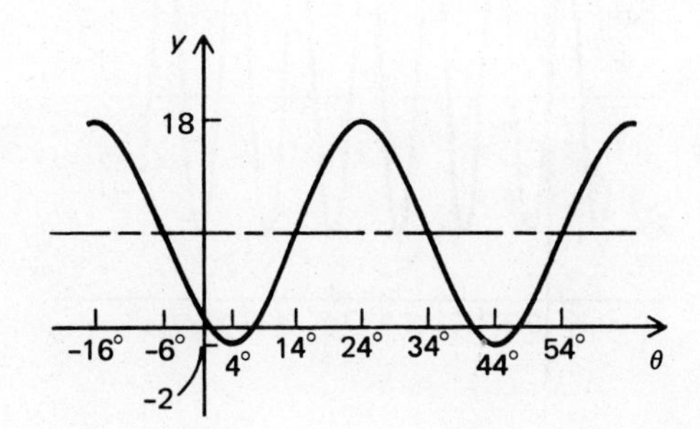

3.

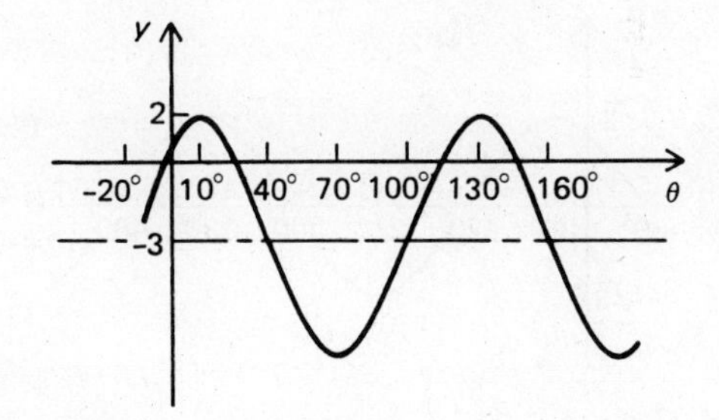

4.

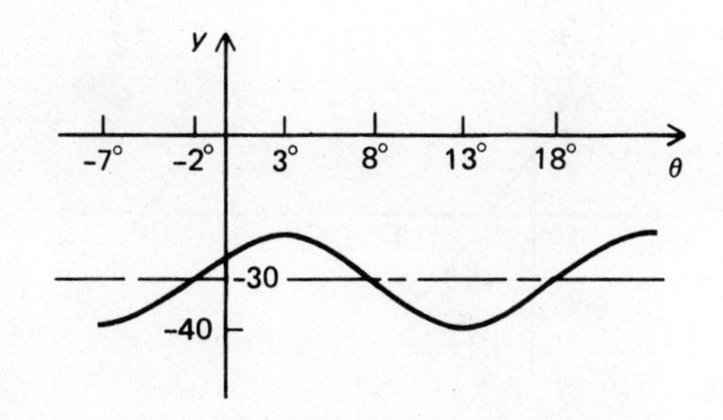

5.

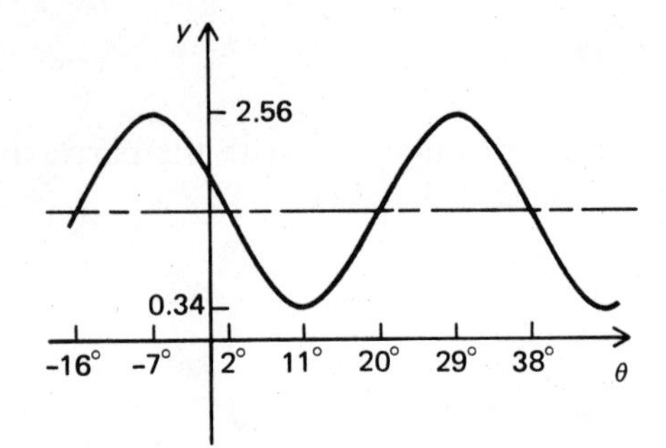

6.

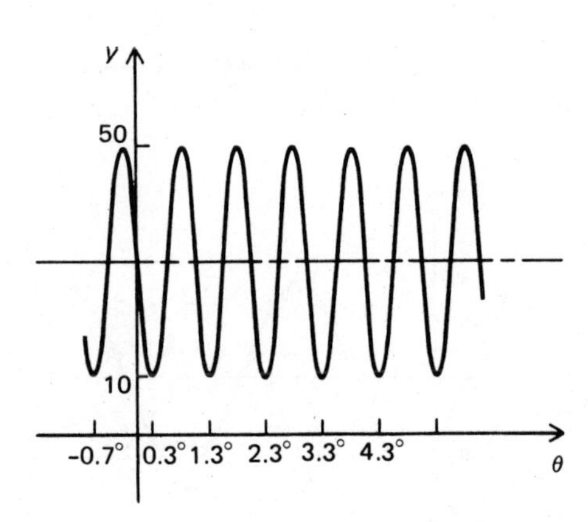

7.

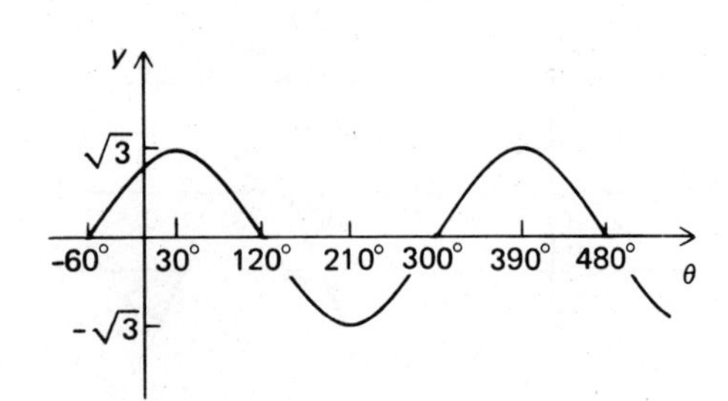

8.

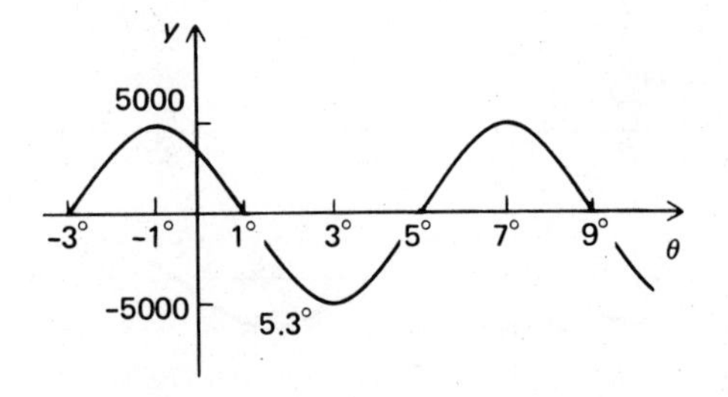

9.

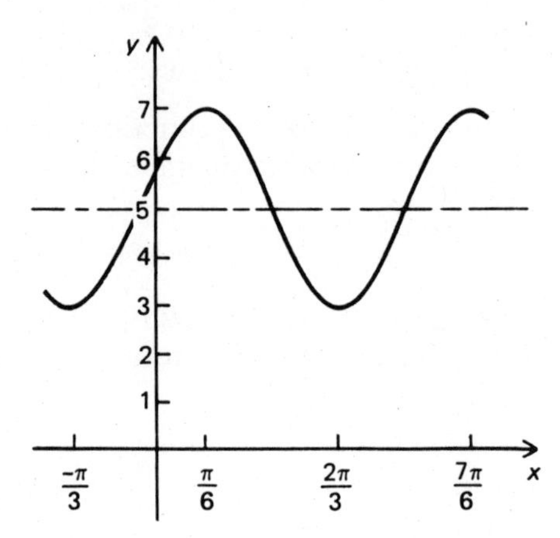

10.

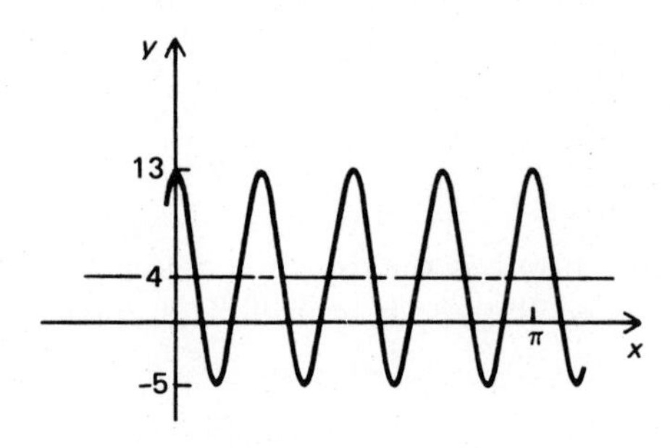

11.

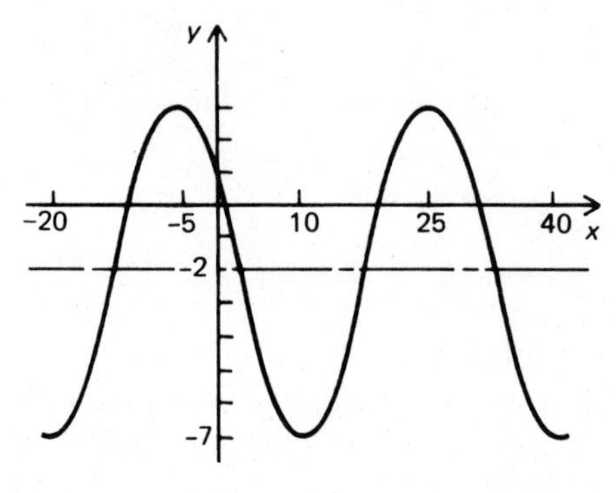

12.

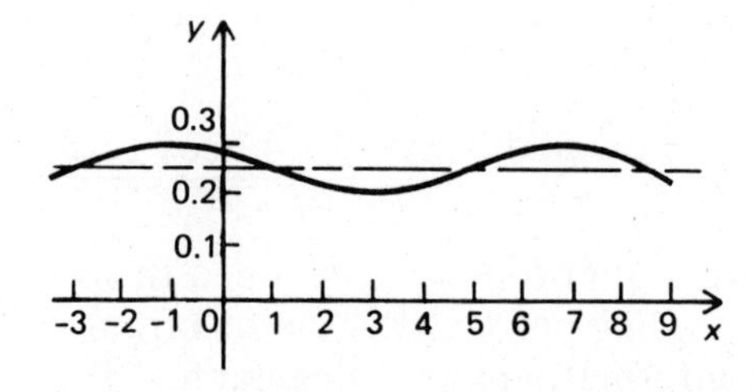

13.

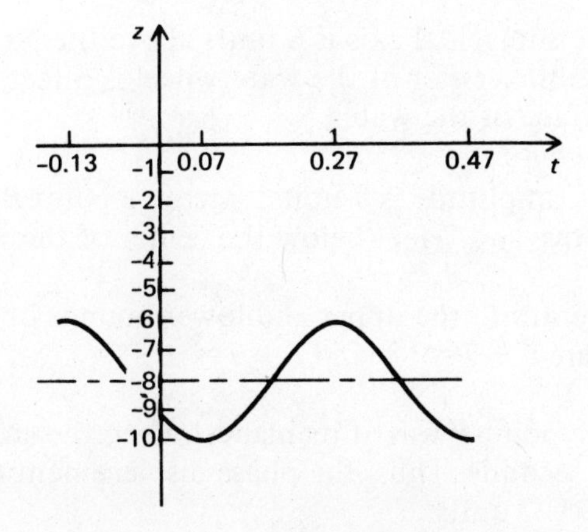

14.

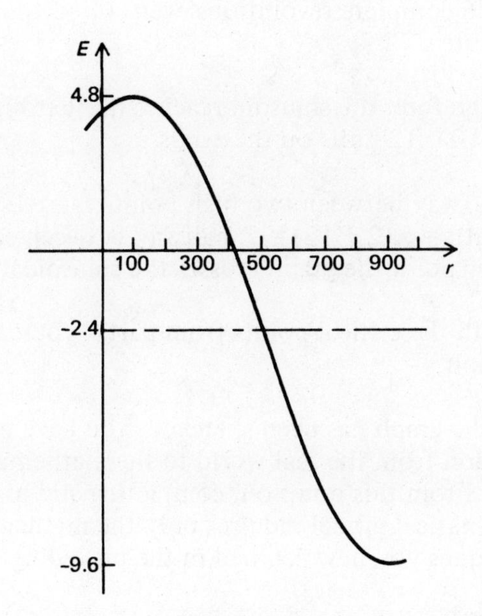

For Problems 15 through 18, sketch the graph of the sinusoid described. Then write the particular equation.

15. Period = 12, amplitude = 7, phase displacement (for the cosine) = 5, vertical displacement = 3.

16. Period = 0.1, amplitude = 8, phase displacement (for the cosine) = 0.02, vertical displacement = −5.

17. Low point at $(x, y) = (2, -1)$, next high point at $(5, 4)$.

18. High point at $(r, s) = (-5, -3)$, next low point at $(-1, -13)$.

13-8 SINUSOIDAL FUNCTIONS AS MATHEMATICAL MODELS

In Section 13-1 you found several real-world situations in which a dependent variable repeated its values at regular intervals as the independent variable changed. For example, the volume of air in your lungs varies periodically with time as you breathe. A reasonable sketch of the graph of this function is shown in Figure 13-8a.

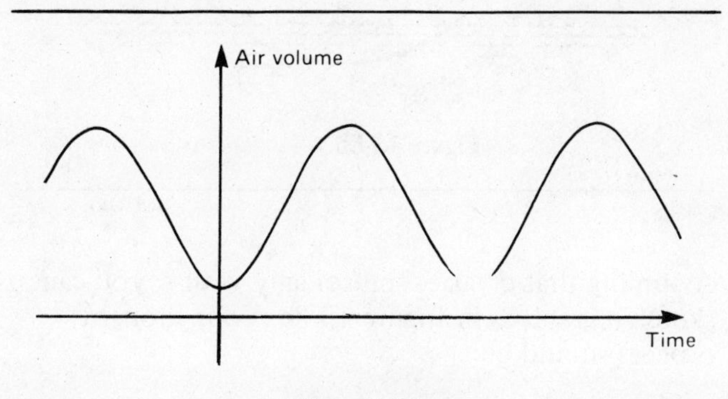

Figure 13-8a.

Since the graph looks like a sinusoid, a sine or cosine function would be a reasonable mathematical model. You now know how to write an equation for a sinusoidal function with any given period, amplitude, phase, and axis location. This is the technique you will use for mathematical modeling.

Objective:

Given a situation from the real world in which something varies sinusoidally, derive an equation and use it as a mathematical model to make predictions and reach conclusions about the real world.

Example: Suppose that the waterwheel in Figure 13-8b rotates at 6 revolutions per minute (rpm). You start your stopwatch. Two seconds later, point P on the rim of the wheel is at its greatest height. You are to model the distance d of point P from the surface of the water in terms of the number of seconds t the stopwatch reads.

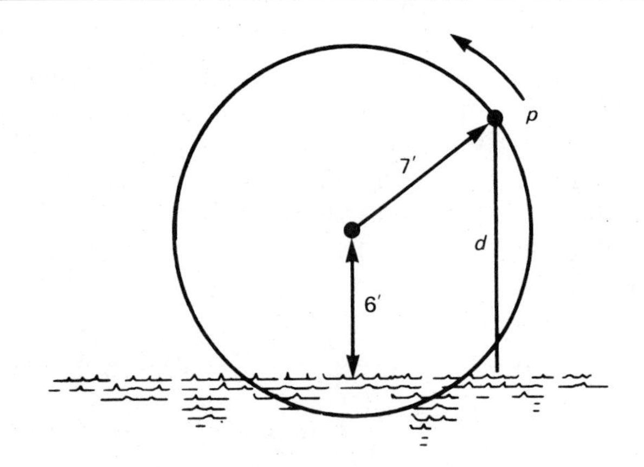

Figure 13-8b.

Assuming that d varies sinusoidally with t, you can sketch a graph as in Figure 13-8c. Your thought process should be:

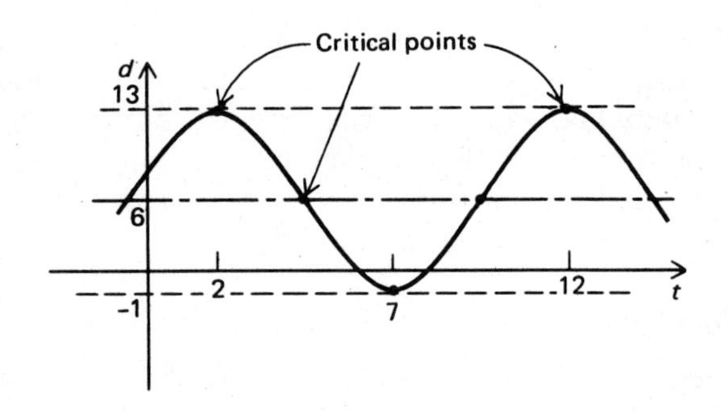

Figure 13-8c.

1. The sinusoidal axis is 6 units above the t-axis, because the center of the waterwheel is 6 feet above the surface of the water.

2. The amplitude is 7 units, since the point P goes 7 feet above and 7 feet below the center of the wheel.

3. Therefore, the upper and lower bounds of the graph are $6 + 7 = 13$, and $6 - 7 = -1$.

4. The point P was at its highest when the stopwatch read 2 seconds. Thus, the phase displacement (for the cosine) is 2 units.

5. The period is 10 seconds, since the waterwheel makes 6 complete revolutions every 60 seconds (1 minute).

6. Therefore, the sinusoid reaches its next high point at $2 + 10 = 12$ units on the t-axis.

7. Halfway between two high points there is a low point at $t = \frac{1}{2}(2 + 12) = 7$; halfway between each high and low point the graph crosses the sinusoidal axis.

8. With the critical points from part 7 you can sketch the graph.

Once the graph has been sketched, you have made the transition from the real world to the mathematical world. From this point on, completing and using the mathematical model requires only the mathematical techniques you have learned in the preceding sections.

From the graph, the four constants in the sinusoidal equation are

$A = 7,$
$B = 2\pi/\text{period} = 2\pi/10 = \pi/5,$
$C = 6,$
$D = 2.$

The equation is therefore

$$d = 6 + 7 \cos \frac{\pi}{5}(t - 2).$$

The equation can be used to make predictions of d for given values of t. For example, to find out how far P is from the water when $t = 5.5$, you would simply substitute 5.5 for t in the equation and carry out the indicated operations.

$d = 6 + 7 \cos \frac{\pi}{5}(5.5 - 2)$	Substitution.
$= 6 + 7 \cos \frac{3 \cdot 5\pi}{5}$	Arithmetic.
$= 6 + 7 \cos 0.7\pi$	More arithmetic.
$\approx 6 + 7(-0.5878)$	By calculator or by the methods of Section 13-4.
$= 1.8854$	

So the point is about 1.9 feet above the water when $t = 5.5$.

The exercise that follows is designed to give you practice using sinusoidal functions as mathematical models. In doing so, you will use most of the trigonometric techniques you have learned so far, as well as a lot of algebraic techniques you have learned in the past.

Exercise 13-8

1. ***Ferris Wheel Problem.*** As you ride the Ferris wheel, your distance from the ground varies sinusoidally with time. When the last seat is filled and thé Ferris wheel starts, your seat is at the position shown in Figure 13-8d. Let t be the number of seconds that have elapsed since the Ferris wheel started. You find that it takes you 3 seconds to reach the top, 43 feet above the ground, and that the wheel makes a revolution once every 8 seconds. The diameter of the wheel is 40 feet.

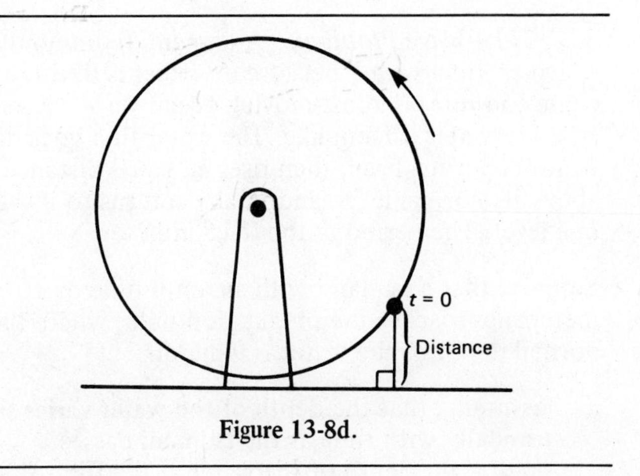

Figure 13-8d.

a. Sketch a graph of this sinusoid.
b. What is the lowest you go as the Ferris wheel turns, and why is this number greater than zero?
c. Write the particular equation of this sinusoid.
d. Predict your height above the ground when
   i. $t = 6$.
   ii. $t = 4\ 1/3$,
   iii. $t = 9$,
   iv. $t = 0$.

2. ***Bouncing Spring Problem.*** A weight attached to the end of a long spring is bouncing up and down (Figure 13-8e). As it bounces, its distance from the floor varies sinusoidally with time. You start a stopwatch. When the stopwatch reads 0.3 second, the weight first reaches a high point 60 centimeters above the floor. The next low point, 40 centimeters above the floor, occurs at 1.8 seconds.

Figure 13-8e.

a. Sketch a graph of this sinusoidal function.
b. Write the particular equation expressing distance from the floor in terms of the number of seconds the stopwatch reads.
c. Predict the distance from the floor when the stopwatch reads 17.2 seconds.
d. What was the distance from the floor when you started the stopwatch?

3. ***Tidal Wave Problem.*** A tsunami (commonly called a "tidal wave" because its effect is like a rapid change in tide) is a fast-moving ocean wave caused by an underwater earthquake. The water first goes down from its normal level, then rises an equal distance above its normal level, and finally returns to its normal level. The period is about 15 minutes.

Suppose that a tsunami with an amplitude of 10 meters approaches the pier at Honolulu, where the normal depth of the water is 9 meters.

a. Assuming that the depth of the water varies sinusoidally with time as the tsunami passes, predict the depth of the water at the following times after the tsunami first reaches the pier:

   i. 2 minutes.
   ii. 4 minutes.
   iii 12 minutes.

b. According to your model, what will the *minimum* depth of the water be? How do you interpret this answer in terms of what will happen in the real world?
c. The "wavelength" of a wave is the distance a crest of the wave travels in one period. It is also equal to the distance between two adjacent crests. If a tsunami travels at 1200 kilometers per hour, what is its wavelength?
d. If you were far from land on a ship at sea and a tsunami was approaching your ship, what would you *see*? Explain.

4. ***Tarzan Problem.*** Tarzan is swinging back and forth on his grapevine. As he swings, he goes back and forth across the river bank, going alternately over land and water (Figure 13-8f.). Jane decides to mathematically model his motion and starts her stopwatch. Let t be the number of seconds the stopwatch reads and let y be the number of meters Tarzan is from the river bank. Assume that y varies sinusoidally with t, and that y is positive when Tarzan is over water and negative when he is over land.

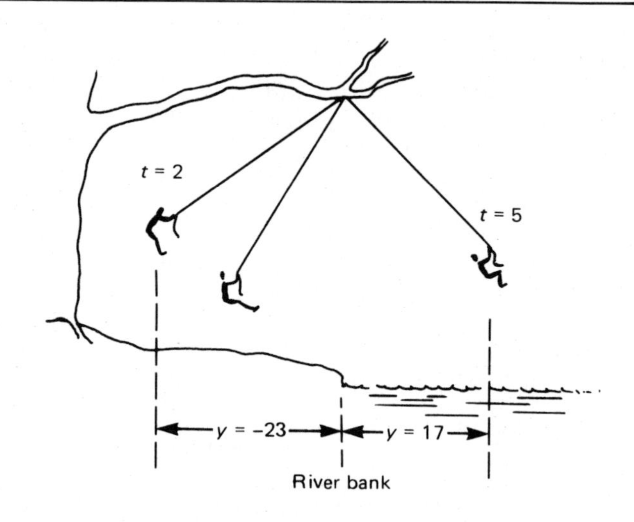

Figure 13-8f.

Jane finds that when $t = 2$, Tarzan is at one end of his swing, where $y = -23$. She finds that when $t = 5$ he reaches the other end of his swing and $y = 17$.

a. Sketch a graph of this sinusoidal function.
b. Write the particular equation expressing Tarzan's distance from the river bank in terms of t.
c. Predict y when

   i. $t = 2.8$,
   ii. $t = 6.3$,
   iii. $t = 15$.

d. Where was Tarzan when Jane started the stopwatch?

5. ***Pebble-in-the-Tire Problem.*** As you stop your car at a traffic light, a pebble becomes wedged between the tire treads. When you start off, the distance of the pebble from the pavement varies sinusoidally with the distance you have traveled. The period is, of course, the circumference of the wheel. Assume that the diameter of the wheel is 24 inches.

a. Sketch a graph of this function.
b. Write the particular equation of this function. Be sure to use a *circular* function. It is possible to get

a form of the equation that has *zero* phase displacement.

c. Predict the distance from the pavement when you have gone 15 inches.

6. ***Spaceship Problem.*** When a spaceship is fired into orbit from a site such as Cape Canaveral, which is not on the equator, it goes into an orbit that takes it alternately north and south of the equator. Its distance from the equator is approximately a sinusoidal function of time.

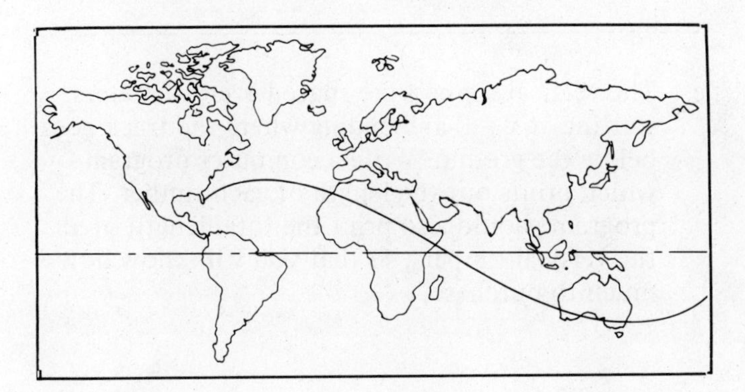

Figure 13-8g.

Suppose that a spaceship is fired into orbit from Cape Canaveral. Ten minutes after it leaves the Cape, it reaches its farthest distance *north* of the equator, 4000 kilometers. Half a cycle later it reaches its farthest distance *south* of the equator (on the other side of the Earth, of course!), also 4000 kilometers. The spaceship completes an orbit once every 90 minutes.

Let y be the number of kilometers the spaceship is *north* of the equator (you may consider distances south of the equator to be negative). Let t be the number of minutes that have elapsed since liftoff.

a. Sketch a complete cycle of the graph of y versus t.
b. Write the particular equation expressing y in terms of t.
c. Use your equation to predict the distance of the spaceship from the equator when
   i. $t = 25$,
   ii. $t = 41$,
   iii. $t = 163$.
d. Calculate the distance of Cape Canaveral from the equator by calculating y when $t = 0$.
e. See if you can find how far Cape Canaveral *really* is from the equator to see if the model gives reasonably accurate answers.

7. ***Electric Current Problem.*** The electricity supplied to your house is called "alternating current" because the current varies sinusoidally with time (Figure 13-8h.). The frequency of the sinusoid is 60 cycles per second. Suppose that at time $t = 0$ seconds the current is at its maximum, $i = 5$ amperes.

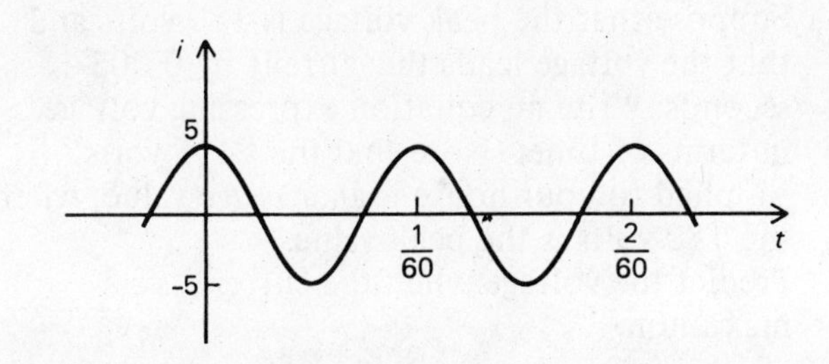

Figure 13-8h.

a. Write an equation expressing current in terms of time.
b. What is the current when $t = 0.01$?

8. ***Electrical Voltage Problem.*** The alternating electrical current supplied to your house (see Problem 7) is created by an alternating electrical potential, or "voltage," which also has a frequency of 60 cycles per second. For reasons you will learn when you study electricity, the voltage usually reaches a peak

slightly before the current does (Figure 13-8i). Since the voltage peak occurs *before* the current peak, the voltage is said to "lead" the current, or the current "lags" the voltage. Leading corresponds to a *negative* phase displacement, and lagging corresponds to a *positive* phase displacement.

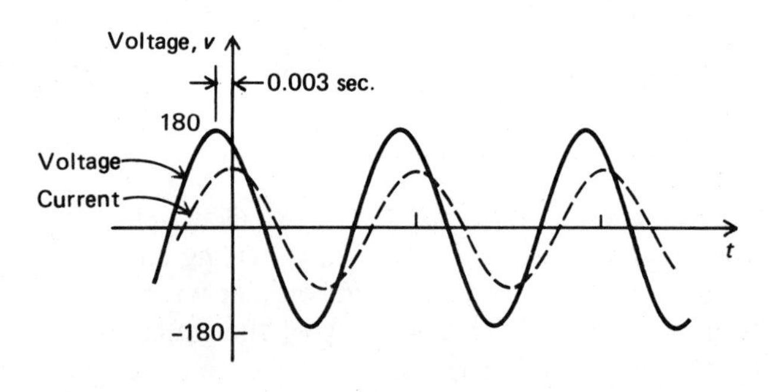

Figure 13-8i.

a. Suppose that the peak voltage is 180 volts and that the voltage leads the current by 0.003 seconds. Write an equation expressing voltage in terms of time. (Note that the "115 volts" supplied to your house is an *average* value, whereas the 180 volts is the peak value.)
b. Predict the voltage when the current is a maximum.

9. ***Roller Coaster Problem.*** A portion of a roller coaster track is to be built in the shape of a sinusoid (Figure 13-8j). You have been hired to calculate the lengths of the vertical timber supports to be used.

a. The high and low points on the track are separated by 50 meters horizontally and by 30 meters vertically. The low point is 3 meters below the ground. Letting y be the number of meters the track is above the ground and x be the number of meters horizontally from the high point, write the particular equation expressing y in terms of x.
b. How long is the vertical timber at the high point? At $x = 4$ meters? At $x = 32$ meters?

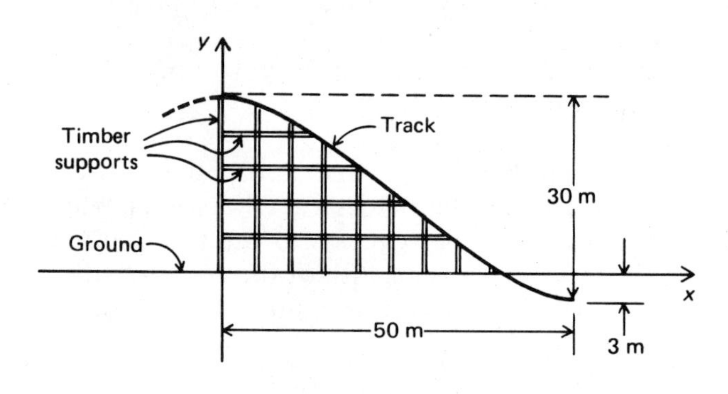

Figure 13-8j.

c. The vertical timbers are spaced every 2 meters, starting at $x = 0$ and ending where the track goes below the ground. Write a computer program which prints out the length of each timber. The program should also print the total length of all the vertical timbers, so that you will know how much to purchase.

10. ***Rock Formation Problem.*** An old rock formation is warped into the shape of a sinusoid. Over the centuries, the top has eroded away, leaving the ground with a flat surface from which various layers of rock are cropping out (Figure 13-8k).

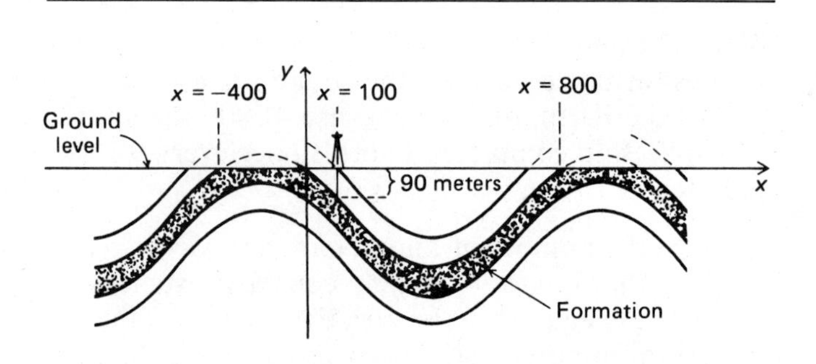

Figure 13-8k.

Since you have studied sinusoids, the geologists call upon you to predict the depth of a particular formation at various points. You construct an x-axis along the ground and a y-axis at the edge of an outcropping, as shown. A hole drilled at $x = 100$ meters shows that the top of the formation is 90 meters deep at that point.

a. Write the particular equation expressing the y-coordinate of the formation in terms of x.
b. If a hole were drilled to the top of the formation at $x = 510$, how deep would it be?
c. What is the maximum depth of the top of the formation, and what is the value of x where it reaches this depth?
d. How high above the present ground level did the formation go before it eroded away?
e. The geologists decide to drill holes to the top of the formation every 50 meters from $x = 50$ through $x = 750$, searching for valuable minerals. Write a computer program to print the depth and the cost of each hole if drilling costs $75 per meter of depth. The program should also print the total cost of drilling the holes.

11. ***Sunrise Problem.*** Assume that the time of sunrise varies sinusoidally with the day of the year. Let t be the time of day that the Sun rises, and let d be the number of the day of the year, starting with $d = 1$ on January 1. To calculate the constants in the equation, recall that the period is 365 days. The amplitude and axis location can be calculated from the times of sunrise on the longest and shortest days of the year (i.e., June 21 and December 21). You may find these times for various cities in an almanac (they are 5:34 a.m. and 7:24 a.m. CST for San Antonio). The phase displacement will, of course, be related to the day number on which the sinusoid reaches its maximum, if you use cosine.

a. Sketch a graph of this sinusoid. You may neglect daylight-saving time.
b. Write an equation for this function.
c. Calculate the time of sunrise for your city *today*. Check your answer with today's newspaper to see how close your model is to the actual sunrise time.

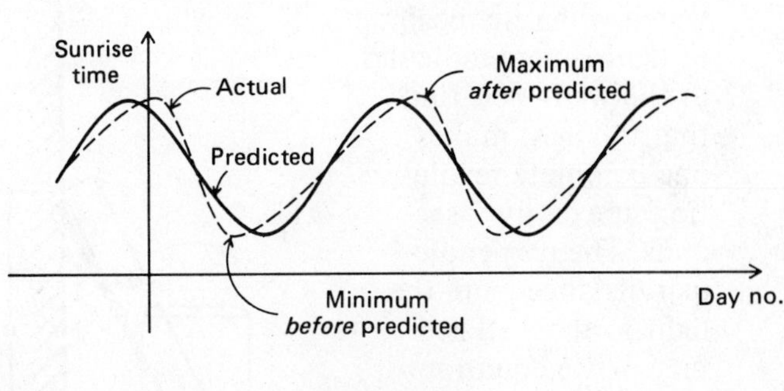

Figure 13-8l.

d. Predict the time of sunrise on your birthday, taking daylight-saving time into account if necessary.
e. The difficulty with parts c and d is the tedious computations involved. To ease the computations, write a computer program that will predict the sunrise time for all days of the year which you input at the beginning of the program.
f. Modify the program from part e so that it calculates the sunrise time for each day, starting with $d = 1$, and searches for the first day on which the sun rises at 6:07 a.m. (or earlier).
g. If you plot a graph of predicted and actual sunrise times versus d, you will find something like Figure 13-8l, where the maximum occurs *after* the predicted maximum, but the minimum occurs *before* the predicted minimum. From what you have learned about how the Earth orbits the Sun, think of a reason why the actual sunrise times differ from the predicted ones in this manner.

12. ***Rotating Beacon Problem.*** A police car pulls up alongside a long brick wall. While the police are away from the car, the red beacon light continues to rotate, shining a spot of red light that moves along the wall (Figure 13-8m). As you watch, you decide to make a mathematical model of the position of the light spot as a function of time.

a. You start your stopwatch when the beam of light is perpendicular to the wall. You find that the light makes one complete revolution in exactly 2 seconds. The perpendicular distance from the light to the wall is 6 feet. Write equations expressing the distances d and L in terms of the number of seconds t, that the stopwatch reads.

b. Calculate d and L when t equals

   i. 0.1,
   ii. 0.3,
   iii. 0.5,
   iv. 0.8.

c. Sketch graphs of d and L versus time t.

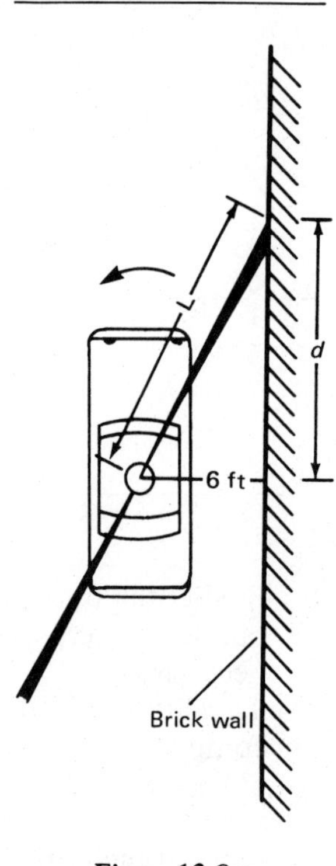

Figure 13-8m.

13-9 INVERSE CIRCULAR FUNCTIONS

In the problems of the last section you used functions with equations of the form

$$y = C + A \cos B(x - D).$$

After getting the particular equation, you used it to find values of y for given values of x. But how would you find x for a given value of y? In this section and the next, you will develop the mathematical techniques to do this job. In Section 13-11 you will apply the techniques to real-world problems.

From Section 6-8 you should recall that the *inverse* of a function is the relation you get by interchanging the independent and dependent variables. For example, if

$$f = \{(x, y) : y = \cos x\},$$

then the *inverse* of f (abbreviated f^{-1} and pronounced, "f inverse") is

$$f^{-1} = \{(x, y) : x = \cos y\}.$$

There is no way to transform the equation $x = \cos y$ so that y is expressed in terms of x. So you simply say that

$$y = \cos^{-1} x.$$

The symbol $\cos^{-1} x$ is pronounced "cosine inverse of x." It means, "The number whose cosine is x." Note that $\cos^{-1} x$ does *not* mean $1/\cos x$. The negative exponent is used for the *function* inverse of $\cos x$, not for the multiplicative inverse.

The geometrical meaning of $\cos^{-1} x$ can be seen by drawing a unit circle, as in Figure 13-9a. If $x = \cos y$, then y is the *arc* whose *cosine* is x. Because of this fact, $\cos^{-1} x$ is often written

$$\cos^{-1} x = \arccos x.$$

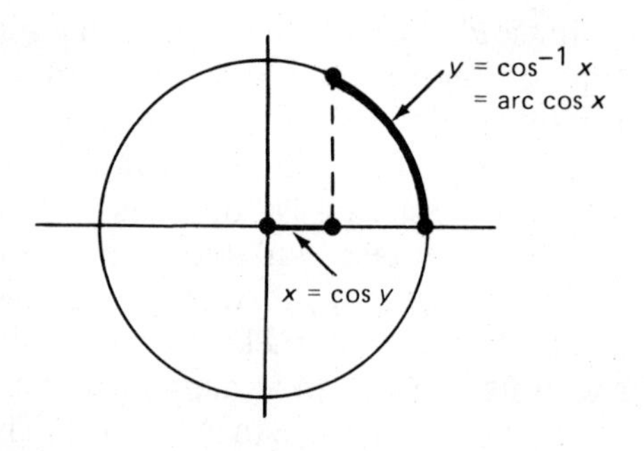

Figure 13-9a.

Since *any* arc coterminal with y has x as its cosine, there are *many* values of $\arccos x$. Thus, the inverse cosine is *not* a function. So arccos is called an inverse circular *relation.* This and other inverse circular relations are defined as follows.

DEFINITIONS

$y = \arcsin x = \sin^{-1} x$ means $x = \sin y$.
$y = \arccos x = \cos^{-1} x$ means $x = \cos y$.
$y = \arctan x = \tan^{-1} x$ means $x = \tan y$.
$y = \text{arccot}\, x = \cot^{-1} x$ means $x = \cot y$.
$y = \text{arcsec}\, x = \sec^{-1} x$ means $x = \sec y$.
$y = \text{arccsc}\, x = \csc^{-1} x$ means $x = \csc y$.

The next thing you do after inventing a new kind of relation is draw the graphs. To draw

$y = \arcsin x,$

for example, you would simply plot

$x = \sin y.$

This is most easily done by picking values of y and finding x. The graph is shown in Figure 13-9b. Since x and y are *reversed* for the inverse relation, everything that happened along the x-axis for $y = \sin x$ will happen along the y-axis for $y = \arcsin x$.

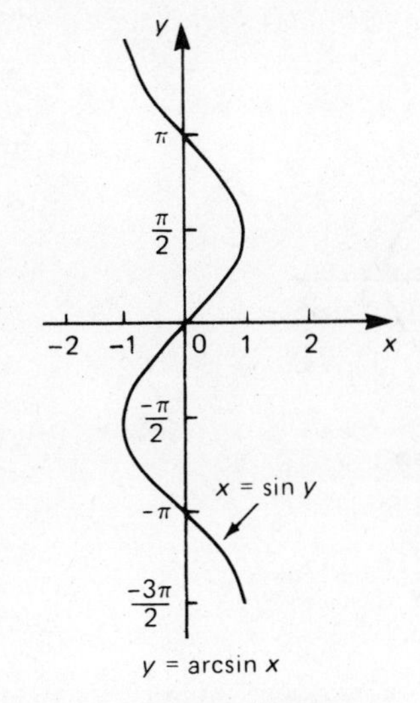

Figure 13-9b.

Figure 13-9b makes it clear that $y = \arcsin x$ is *not* a function. There are *many* values of y for each value of x. By picking only *part* of the graph in Figure 13-9b, it is possible to get a function graph. Several attempts are shown in Figure 13-9c with comments below. The first one is still not a function. The second one is a function but does not use the entire domain $-1 \le x \le 1$. The third sketch uses the entire domain but is not in one continuous piece. The fourth one is continuous and uses the entire domain, but is located away from the origin. The last sketch is the most desirable.

Picking part of the graph is accomplished by specifying what you want the *range* of the function to be. The last graph in Figure 13-9c is specified by

$y = \arcsin x$ *and* $-\pi/2 \le y \le \pi/2$.

This function is called an *inverse circular function.* If the range were $-90° \le y \le 90°$, it would be an *inverse trigonometric function.* To distinguish between the relation and the function, a *capital* letter is usually used for the function.

$y = \text{Arcsin}\, x = \text{Sin}^{-1} x$ means $x = \sin y$ *and* $-\pi/2 \le y \le \pi/2$.

The ranges of the other five inverse circular functions are found by the same reasoning process. The graph must meet the following requirements:

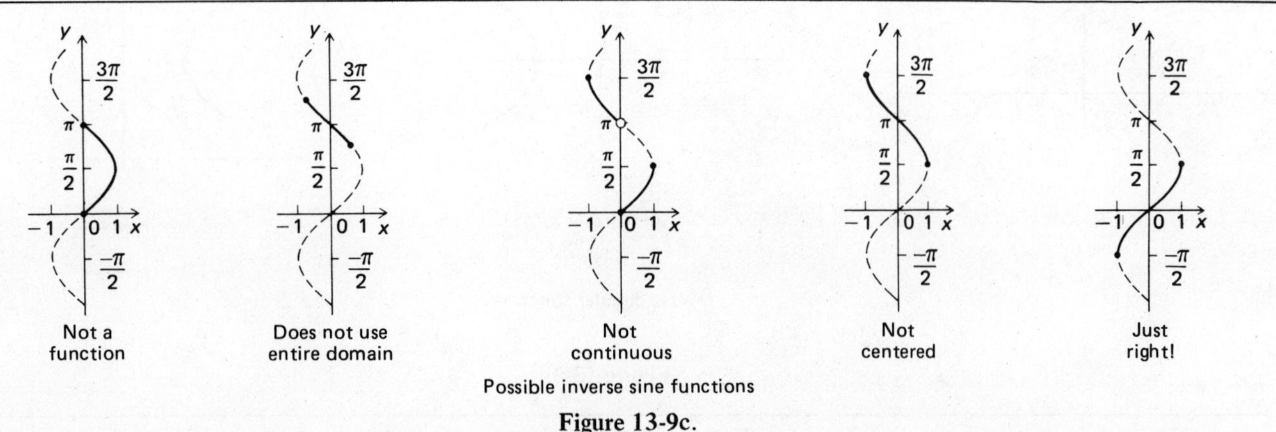

Possible inverse sine functions

Figure 13-9c.

1. It must be a *function* graph.
2. It must use the *entire domain* of the corresponding inverse circular relation.
3. It must be *continuous* if possible.
4. It must be as close as possible to the x-axis.
5. If there is a choice of two branches meeting the above requirements, it must be the *positive* branch (i.e., $y \geq 0$).

The graphs of the inverse circular functions are shown in Figure 13-9d.

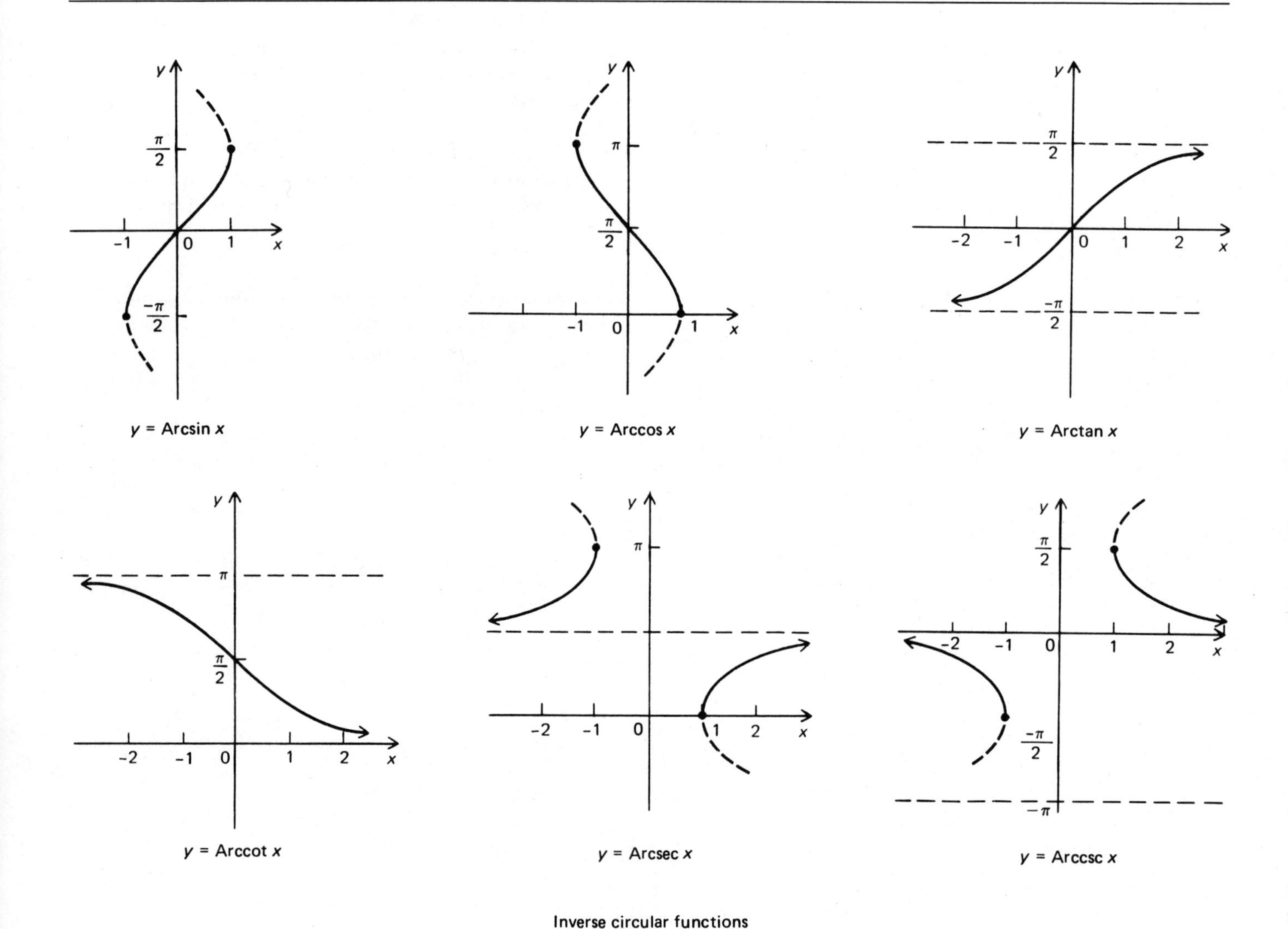

Figure 13-9d.

From these graphs you should be able to read off the ranges of the inverse circular functions. These ranges are tabulated below.

Ranges of Inverse Circular Functions

$y = \text{Arcsin } x = \text{Sin}^{-1} x,$	$-\pi/2 \le y \le \pi/2$
$y = \text{Arccos } x = \text{Cos}^{-1} x,$	$0 \le y \le \pi$
$y = \text{Arctan } x = \text{Tan}^{-1} x,$	$-\pi/2 < y < \pi/2$
$y = \text{Arccot } x = \text{Cot}^{-1} x,$	$0 < y < \pi$
$y = \text{Arcsec } x = \text{Sec}^{-1} x,$	$0 \le y \le \pi$ *and* $y \ne \pi/2$
$y = \text{Arccsc } x = \text{Csc}^{-1} x,$	$-\pi/2 \le y \le \pi/2$ *and* $y \ne 0$

Note: The inverse *trigonometric* functions are similarly defined. The range of y is in degrees rather than in units of arc length. For example, if $y = \text{Arccos } x$ is an inverse trigonometric function, then y is in the range $0° \le y \le 180°$.

The most important thing for you to observe about these ranges appears if you draw the arc (or angle) on a uv-graph, as in Figure 13-9e. The arc or angle is always in either Quadrants I and II or in Quadrants I and IV, depending on the function. The only differences are whether the endpoints are included and where there is a "hole" in the middle.

The next thing to do after inventing and graphing new functions is to learn their properties and become familiar with them.

Objectives:

1. Be able to draw the graphs of the inverse circular or trigonometric functions or relations.

2. Be able to simplify expressions containing inverse circular or trigonometric functions or relations.

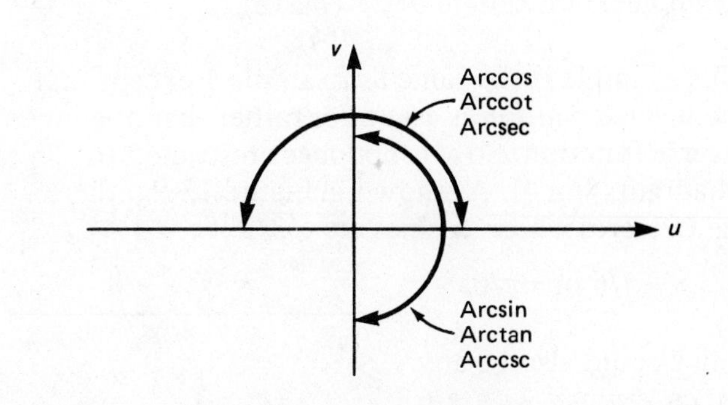

Figure 13-9e.

The first objective is accomplished as described above. The following examples illustrate the second objective.

Example 1: Evaluate $\text{Cos}^{-1}(\sqrt{3}/2)$.

Let $x = \text{Cos}^{-1}(\sqrt{3}/2)$.

By the definition of Cos^{-1},

$$\cos x = \sqrt{3}/2, \text{ and } 0 \le x \le \pi.$$

From the functions of special arguments which you learned in Section 13-3, you should recognize that $\cos \pi/6 = \sqrt{3}/2$. Therefore,

$$x = \underline{\underline{\pi/6}}.$$

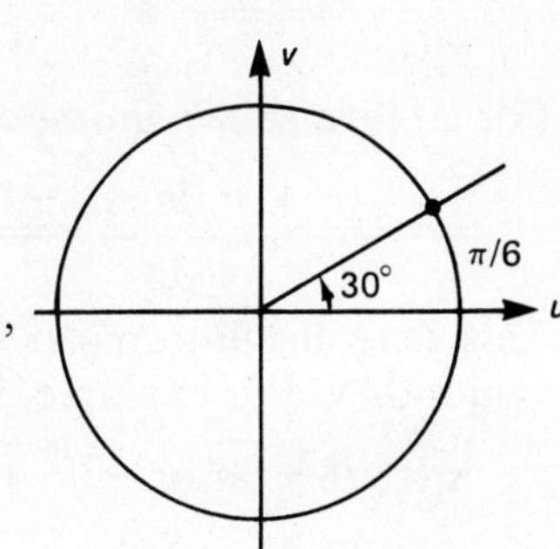

Figure 13-9f.

For an inverse trigonometric function,

$$x = \underline{\underline{30°}}.$$

These answers are illustrated in Figure 13-9f.

Example 2: Evaluate $\cos^{-1}(\sqrt{3}/2)$.

This example is the same as Example 1 except that the inverse *relation* is asked for rather than the inverse function. So x is no longer restricted to Quadrants I or II. As shown in Figure 13-9g, there are two arcs whose cosines are $\sqrt{3}/2$. So

$$x = \pi/6 \text{ or } -\pi/6.$$

But x could also be any arc *coterminal* with $\pi/6$ or $-\pi/6$. Coterminal arcs differ by multiples of 2π. Letting n stand for an integer, any multiple of 2π has the form $2\pi n$. So in general,

$$x = \pi/6 + 2\pi n$$

$$\text{or } -\pi/6 + 2\pi n.$$

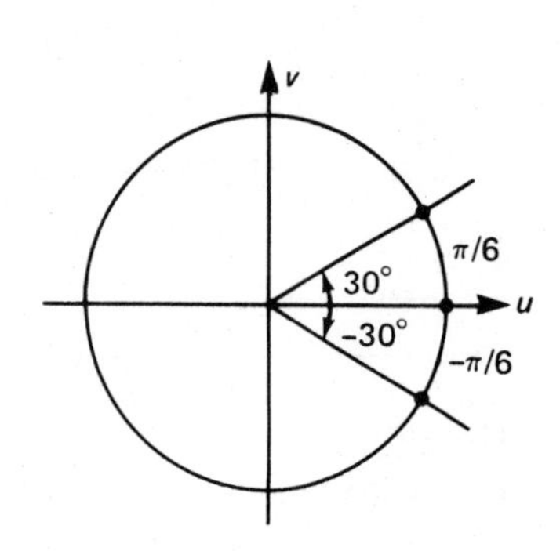

Figure 13-9g.

For an inverse *trigonometric* relation, x would be

$$x = 30° + 360n° \text{ or } -30° + 360n°.$$

This general solution can be used to find particular values of x. For example, if $n = 3$, then

$$x = \pi/6 + 6\pi \text{ or } -\pi/6 + 6\pi$$
$$= 37\pi/6 \text{ or } 35\pi/6.$$

If $n = -1$, then

$$x = \pi/6 - 2\pi \text{ or } -\pi/6 - 2\pi$$
$$= -11\pi/6 \text{ or } -13\pi/6.$$

Example 3: Evaluate $\tan(\text{Sin}^{-1} 3/5)$.

A picture helps. Since $\text{Sin}^{-1} 3/5$ means "the angle whose sine is 3/5," you can draw that angle in standard position. Figure 13-9h shows the angle with $v = 3$ and $r = 5$. By Pythagoras, $u = 4$. Therefore,

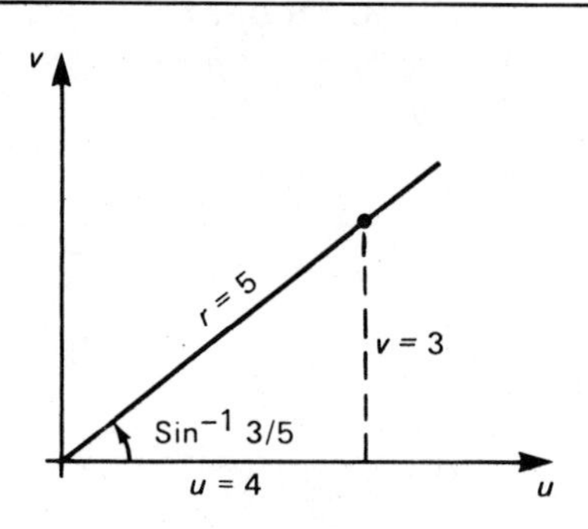

Figure 13-9h.

$$\tan(\text{Sin}^{-1} 3/5) = 3/4,$$

because $\tan x = v/u$.

Example 4: Evaluate $\sin(\text{Sin}^{-1} 5/7)$.

The uv-graph for $\text{Sin}^{-1} 5/7$ is shown in Figure 13-9i. Here, $v = 5$ and $r = 7$. Since you are looking for the *sine* of this angle,

$$\sin(\text{Sin}^{-1} 5/7) = 5/7$$

because $\sin x = v/r$.

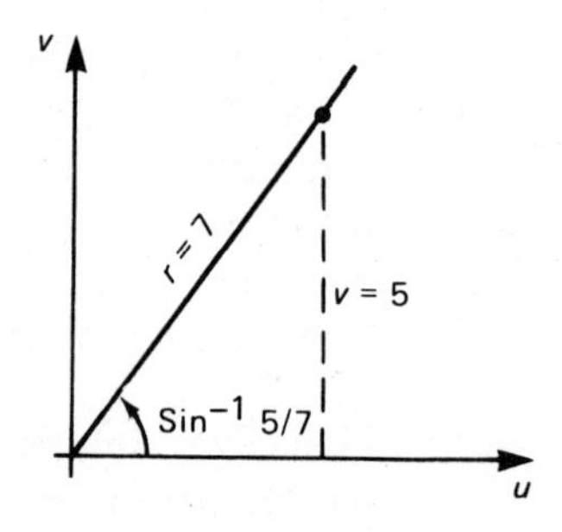

Figure 13-9i.

The result of Example 4 is really not surprising. Translated into words, the problem says, "Find the sine of the angle whose sine is 5/7." It is important because it illustrates a property that applies to inverses of *any* kind of function.

Property: $f(f^{-1}(x)) = x$ for all x in the appropriate domain.

In fact, this property is sometimes used as a *definition* of inverse functions. Beware that you do not use the property for x in an inappropriate domain. For example,

$$\sin(\text{Sin}^{-1} 15) \neq 15,$$

because there is no arc or angle whose sine is 15.

Example 5: Evaluate $\text{Cos}^{-1}(\cos 60°)$.

$\text{Cos}^{-1}(\cos 60°) = \text{Cos}^{-1}(1/2)$ because $\cos 60° = 1/2$.

$= 60°$ because $\cos 60° = 1/2$.

Here again the answer is the number you started with. This will work for all appropriate values. But beware of such things as $\text{Cos}^{-1}(\cos 315°)$, which equals $\text{Cos}^{-1}(\sqrt{2}/2)$, which equals 45° and *not* 315°.

Property: $f^{-1}(f(x)) = x$ for all x in the appropriate domain.

You will prove these two properties in Problems 97 and 98 of the following exercise. You will also get practice working with inverse trigonometric and circular functions and relations.

Exercise 13-9

For Problems 1 through 6, sketch the graph of the inverse circular relation *without* looking at Figure 13-9d. After you have drawn the graph, check the figure to make sure you are correct.

1. $y = \text{arcsin } x$
2. $y = \text{arccos } x$
3. $y = \text{arctan } x$
4. $y = \text{arccot } x$
5. $y = \text{arcsec } x$
6. $y = \text{arccsc } x$

For Problems 7 through 12, sketch the graph of the inverse circular function *without* looking at Figure 13-9d. After you have drawn the graph, check the figure to make sure you are correct.

7. $y = \text{Arccos } x$
8. $y = \text{Arcsin } x$
9. $y = \text{Arccot } x$
10. $y = \text{Arctan } x$
11. $y = \text{Arccsc } x$
12. $y = \text{Arcsec } x$

For Problems 13 through 36, find *exact* values of the inverse *circular* functions or relations.

13. $\text{arctan } 1$
14. $\tan^{-1} 1$
15. $\text{Tan}^{-1} 1$
16. $\text{Arctan } 1$
17. $\text{arcsin}(-1/2)$
18. $\text{Arccos}(-1)$
19. $\text{Sec}^{-1} 2$
20. $\cot^{-1}\sqrt{3}$
21. $\text{arccsc } 2$
22. $\cos^{-1} 0$
23. $\text{Arccos } 2$
24. $\text{Arcsec } 0$
25. $\text{arccot}\sqrt{3}$
26. $\text{Arcsin}(-1)$
27. $\text{Sin}^{-1}\sqrt{2}/2$
28. $\sin^{-1}(-\sqrt{2}/2)$
29. $\cos^{-1}(-1)$
30. $\text{Csc}^{-1} 1$
31. $\text{Arccot } 0$
32. $\text{Sec}^{-1} 1$
33. $\text{arcsec}(-2/\sqrt{3})$
34. $\text{arccot}(-\sqrt{3}/3)$
35. $\text{Csc}^{-1}(-\sqrt{2})$
36. $\text{arccsc } 2/\sqrt{3}$

For Problems 37 through 60, find *exact* values of the inverse *trigonometric* functions or relations.

37. $\cos^{-1}\sqrt{2}/2$
38. $\text{Cot}^{-1} 1$
39. $\text{Arcsin}(-1)$
40. $\text{Cos}^{-1} 1$
41. $\text{arccot}(-1)$
42. $\text{Sin}^{-1} 1$
43. $\text{Csc}^{-1}\sqrt{2}$
44. $\text{Arcsec } 2$
45. $\text{arcsec } 1/2$
46. $\text{Arcsin } 2$
47. $\text{arcsin } 1/2$
48. $\tan^{-1}\sqrt{3}$
49. $\cos^{-1} 1$
50. $\text{arccsc}(-1)$
51. $\text{Arctan}(-\sqrt{3})$
52. $\text{Csc}^{-1}(-2/\sqrt{3})$
53. $\csc^{-1}(-2/\sqrt{3})$
54. $\text{Arcsec}(-1)$
55. $\text{Arcsec}(-2)$
56. $\cot^{-1}(-\sqrt{3})$

57. Arccos $(-1/2)$
58. Arctan (-1)
59. $\tan^{-1} \sqrt{3}/3$
60. arccos $(-\sqrt{2}/2)$

For Problems 61 through 72, use tables or a calculator to find the following inverse circular functions correct to 3 decimal places.

61. Arccos 0.4531
62. Sin^{-1} 0.7753
63. Sec^{-1} 1.233
64. Arccsc 2.647
65. Arccot 2.331
66. Arctan 1.872
67. Tan^{-1} (-0.4375)
68. Cot^{-1} (-4.011)
69. Arcsin (-0.9692)
70. Cos^{-1} (-0.7573)
71. $\text{Csc}^{-1}(-1.873)$
72. Arcsec (-4.502)

For Problems 73 through 96, find the *exact* value of the expression.

73. tan $(\text{Cos}^{-1}\ 4/5)$
74. cos (Arctan 4/3)
75. sin $(\text{Tan}^{-1}\ 5/12)$
76. sec (Arcsin 15/17)
77. cos (Arcsin $(-8/17)$)
78. cot $(\text{Csc}^{-1}\ (-13/12))$
79. sec (Arccos 2/3)
80. sin $(\text{Cot}^{-1}\ 4)$
81. cot $(\text{Sin}^{-1}\ (-\sqrt{2}/2))$
82. tan (Arcsec $(-\sqrt{2})$)
83. csc (Arccot 3)
84. csc $(\text{Tan}^{-1}\ 1/2)$
85. csc (Arccsc 5)
86. sin $(\text{Sin}^{-1}\ 2/3)$
87. cos $(\text{Sin}^{-1}\ 2)$
88. tan (Arcsec 0)
89. Arctan $(\tan \pi/6)$
90. Cos^{-1} $(\cos 30°)$
91. Sin^{-1} $(\sin(-17°))$
92. Arcsec $(\sec(-1.5))$
93. Sec^{-1} $(\sec 210°)$
94. Arccsc $(\csc 2\pi/3)$
95. Arccot $(\tan 79°)$
96. Sin^{-1} $(\cos 17°)$

97. Prove that $f^{-1}(f(x)) = x$ by letting $y = f(x)$, using the definition of f^{-1} and using a clever substitution.

98. Prove that $f(f^{-1}(x)) = x$ by letting $y = f^{-1}(x)$, using the definition of f^{-1} and using a clever substitution.

99. Suppose that you have been hired by I.M.F. Computer Corporation to do some programming using inverse circular functions. Since many computer languages such as BASIC and FORTRAN have only the Arctan function available, I.M.F. calls upon you to find other inverse circular functions in terms of Arctangents.

a. Let $y = \text{Arcsin}\ x$. Write an equation expressing x in terms of y.
b. Figure 13-9j is a uv-graph showing an arc of length y in standard position on a unit circle. The coordinate v equals the x of part a, above. What does u equal in terms of x?

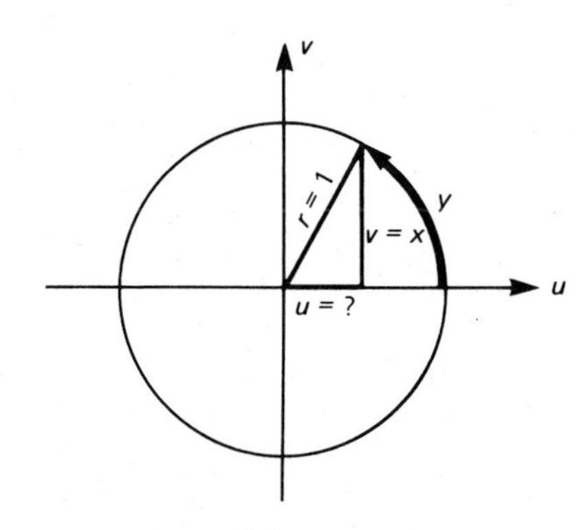

Figure 13-9j.

c. With the help of the sketch, write an equation expressing $\tan y$ in terms of x.
d. Transform the equation from part c so that y is expressed as an Arctangent. Explain why you can use the Arctangent *function* here instead of the arctangent relation.
e. Demonstrate that your equation in part d gives correct answers by using it to calculate Arcsin 0.5 and by comparing the answer with the actual value of Arcsin 0.5.

100. Repeat Problem 99 to get an equation for Arccos x as an Arctangent. Explain what you must do to get the right value of Arccos x if $x < 0$.

101. Show that there is a very easy way to find Arccsc x in terms of an Arcsine. Then use the results of Problem 99 to write an equation expressing Arcsec x as an Arctangent function.

102. Use the techniques of Problems 100 and 101 to derive equations expressing Arccot x and Arcsec x in terms of Arctangents.

103. Write a computer program and use it to print a short table of values of Arcsin x, Arccos x, Arctan x, and Arccot x for each 0.1 unit from $x = -1$ through $x = 1$. The formulas of Problems 99 through 102 should help if your computer has only the Arctangent function available.

104. a. Draw graphs of $y = \tan x$, for $-\pi/2 < x < \pi/2$, $y = \text{Tan}^{-1} x$ on the same set of axes, using the same scale for both axes.

b. Draw the line $y = x$ on this set of axes and tell what the relationship is among this line and the graphs of part a.

c. Explain why the relationship in part b would *not* be true for the *trigonometric* tangent function $y = \tan \theta$, for $-90° < \theta < 90°$.

105. a. Carefully plot the portions of $y = \sin x$ and $y = \text{Sin}^{-1} x$ for each 0.1 unit from $x = 0$ through $x = 0.8$. You may find values by calculator or from Table IV. Use the same scales on both axes, and make the scale large enough to show the separation between the two graphs.

b. Draw the line $y = x$ on the same set of axes. Does this line intersect either of the graphs at any points besides the origin?

c. What is the slope of the line *tangent* to each graph at $x = 0$?

d. Plot another graph of the *trigonometric* sine function $y = \sin x$ for values of x from 0° through 0.8°, using the *same* scale for both axes. Does the line $y = x$ seem to have the same relationship to the trigonometric sine function as it does to the circular sine function? Explain.

13-10 EVALUATION OF INVERSE RELATIONS

In Section 13-8 you used sinusoidal functions with the general equation

$$y = C + A \cos B(x - D)$$

as mathematical models. You derived the particular equation from information about the real world, then used the equation to find values of y for given values of x.

It is just as important to be able to find x when you know y. The difficulty is that there are *many* values of x for the *same* value of y. In this section you will learn how inverse circular or trigonometric relations can be used to help out.

Objective: Given a circular function with equation

$$y = C + A \cos B(x - D) \text{ or } y = C + A \sin B(x - D),$$

or a trigonometric function with equation

$$y = C + A \cos B(\theta - D) \text{ or } y = C + A \sin B(\theta - D),$$

find the values of x or θ for a given value of y.

Example 1: For the trigonometric function

$$y = 2 + 3 \cos 4(\theta + 31°),$$

find the first three positive values of θ for which $y = 1.5$.

Substituting 1.5 for y gives

$1.5 = 2 + 3 \cos 4(\theta + 31°)$	Substitution.
$-0.5 = 3 \cos 4(\theta + 31°)$	Subtracting 2.
$-0.1667 = \cos 4(\theta + 31°)$	Dividing by 3.
$\arccos(-0.1667) = 4(\theta + 31°)$	Taking arccos of both members.

By calculator, the value of Arccos (−0.1667) is about 99° 36′. The same thing can be found by using Table III to get the reference angle, 80° 24′, then subtracting this number from 180°. So the values of the arccosine relation are

$\pm 99° \; 36' + 360n°$,

as indicated in Figure 13-10a.

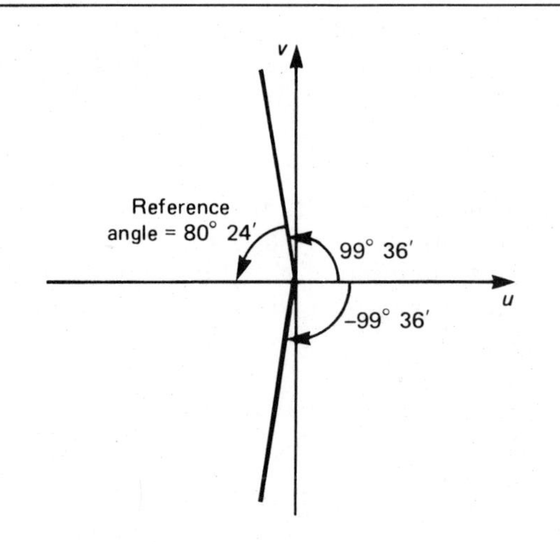

Figure 13-10a.

Substituting these values for the left member of the equation gives

$\pm 99° \; 36' + 360n° = 4(\theta + 31°)$	
$\pm 24° \; 54' + 90n° = \theta + 31°$	Dividing by 4.
$-7° \; 54' + 90n°$ or $-55° \; 54' + 90n° = \theta$	Subtracting 31°.
$\theta = -7°54', 82°6', 172°6', \ldots$ or $-55°54', 34°6', 124°6', \ldots$	Substituting $n = 0, 1, 2, \ldots$.

So the first three positive values of θ are

$\underline{\underline{\theta = 34° \; 6', 82° \; 6', \text{ and } 124° \; 6'.}}$

Example 2: For the circular function

$y = 2 + 3 \sin 4(x + 0.5)$,

find the least positive value of x for which $y = 4.1$.

Substituting 4.1 for y and using symmetry,

$2 + 3 \sin 4(x + 0.5) = 4.1$	Substitution.
$3 \sin 4(x + 0.5) = 2.1$	Substracting 2.
$\sin 4(x + 0.5) = 0.7$	Dividing by 3.
$4(x + 0.5) = \arcsin 0.7$	arcsin of both members.

From Table IV or by calculator, Arcsin 0.7 is approximately 0.7754. But arcsin 0.7 could terminate in Quadrant I or Quadrant II since sine is positive in these quadrants. As shown in Figure 13-10b, another value of arcsin 0.7 is

$$\begin{aligned} \arcsin 0.7 &\approx \pi - 0.7754 \\ &\approx 3.1416 - 0.7754 \\ &= 2.3662. \end{aligned}$$

So in general,

$$\arcsin 0.7 \approx 0.7754 + 2\pi n \text{ or } 2.3662 + 2\pi n.$$

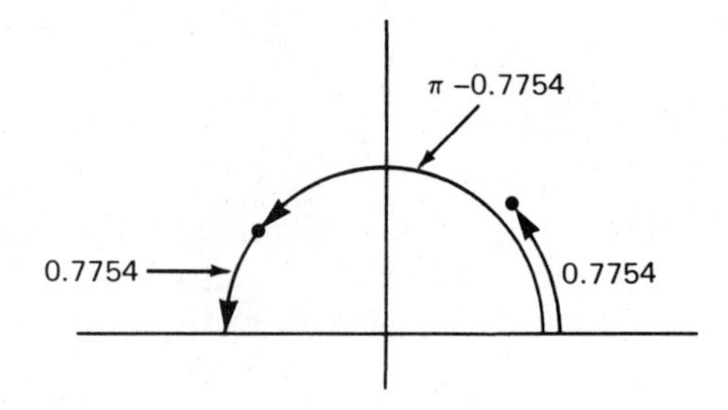

Figure 13-10b.

Returning to the original problem,

$4(x + 0.5) \approx 0.7754 + 2\pi n$ or $2.3662 + 2\pi n$	Substitution.
$x + 0.5 \approx 0.1938 + \frac{\pi}{2} n$ or $0.5915 + \frac{\pi}{2} n$	Dividing by 4.

$x \approx -0.3062 + \frac{\pi}{2}n$ or $0.0915 + \frac{\pi}{2}n$ — Subtracting 0.5.

Substituting 0 for n gives

$x \approx -0.3062$ or 0.0915.

So the least positive value of x is about $\underline{\underline{0.0915}}$.

Example 3: For the circular function

$$y = 2 + 3 \cos \frac{\pi}{4}(x - 0.6),$$

a. Transform the equation so that x is expressed in terms of y.
b. Find the third positive value of x for which $y = -0.7$.

a. $2 + 3 \cos \frac{\pi}{4}(x - 0.6) = y$ — Symmetry.

$\cos \frac{\pi}{4}(x - 0.6) = \frac{1}{3}(y - 2)$ — Subtracting 2 and dividing by 3.

$\frac{\pi}{4}(x - 0.6) = \arccos \frac{1}{3}(y - 2)$ — arccos of both members.

$x - 0.6 = \frac{4}{\pi} \arccos \frac{1}{3}(y - 2)$ — Multiplying by $4/\pi$.

$\underline{\underline{x = 0.6 + \frac{4}{\pi} \arccos \frac{1}{3}(y - 2)}}$ — Adding 0.6.

b. To find x when $y = -0.7$, you simply substitute -0.7 for y and carry out the indicated operations. All of the algebra has already been done.

$x = 0.6 + \frac{4}{\pi} \arccos \frac{1}{3}(-0.7 - 2)$ — Substitution.

$= 0.6 + \frac{4}{\pi} \arccos(-0.9)$ — Arithmetic.

$\approx 0.6 + \frac{4}{\pi}(\pm 2.6906 + 2\pi n)$ — Arccos 0.9 ≈ 0.4510.

$\approx 0.6 \pm 3.4257 + 8n$ — Arithmetic, and $\pi \approx 3.1416$.

$= 4.0257 + 8n$ or $-2.8257 + 8n$ — Arithmetic.

Letting $n = 0$ and then $n = 1$ gives

$x \approx 4.0257, -2.8257, 12.0257,$ or 5.1743

So the third positive value of x is about $\underline{\underline{5.1743}}$.

In the following exercise you will get practice finding x or θ for given values of y. This technique will be used for real-world problems in the following section.

Exercise 13-10

For Problems 1 through 10

a. Transform the equation so that θ or x is in terms of y.
b. Find the first three positive values of θ or x for which $y = 5$.
c. By sketching the graph of the *given* function, show that your answers are reasonable.

1. $y = 3 + 4 \cos 2(\theta + 10°)$
2. $y = -1 + 12 \cos 3(\theta - 10°)$
3. $y = 6 + 2 \cos \frac{1}{4}(x - 3\pi)$
4. $y = 7 + 4 \cos \frac{1}{3}(x - 5\pi)$
5. $y = -3 + 10 \cos 4(\theta - 40°)$
6. $y = 4 + 3 \cos 2(\theta + 20°)$
7. $y = 1 + 5 \sin \frac{\pi}{2}(x + 0.7)$
8. $y = -2 + 8 \sin \frac{2\pi}{3}(x - 1)$
9. $y = 1 + 3 \cos \pi(x + 0.2)$
10. $y = 2 + 2 \cos 2\pi(x - 0.1)$

For Problems 11 and 12, you must apply the techniques of this section to the inverse tangent and secant functions.

11. Suppose that $y = 2 + \frac{1}{2} \tan \frac{\pi}{4}(x - 0.6)$.

a. Transform the equation so that x is expressed in terms of y.
b. Find the smallest positive value of x for which

   i. $y = 2.5$,
   ii. $y = 0$.

c. Show that your answers are right by sketching the graph.

12. Suppose that $y = 4 + 2 \sec 3(\theta + 20°)$.

a. Transform the equation so that θ is expressed in terms of y.
b. Find the least positive value of θ for which

   i. $y = 6$,
   ii. $y = 3$,
   iii. $y = 0$.

c. Show that your answers are right by sketching the graph.

13-11 INVERSE CIRCULAR RELATIONS AS MATHEMATICAL MODELS

In the previous section you learned how to find values of x for known values of y in the equation

$$y = C + A \cos B(x - D).$$

In this section you will apply the techniques you learned to problems from the real world.

Objective:

Given a situation from the real world in which y varies sinusoidally with x, derive the particular equation and use it to find x for a given value of y.

Example: A water wheel 14 feet in diameter is rotating as shown in Figure 13-11a. You start a stopwatch and observe the motion of point P on the rim of the wheel. When the stopwatch reads 2 seconds, P is at its maximum distance from the surface of the water. When the stopwatch reads 7 seconds, P is at its maximum depth below the water.

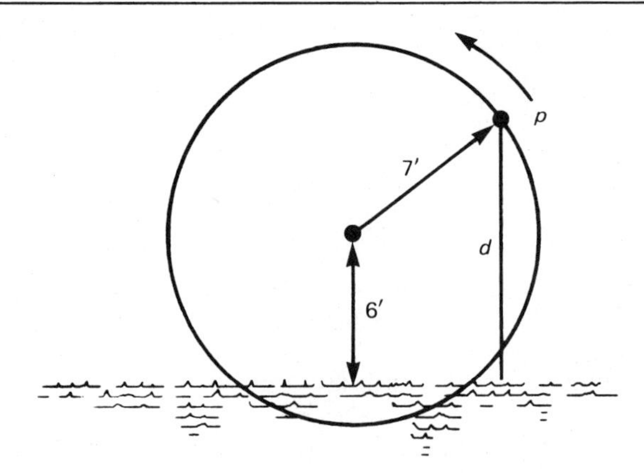

Figure 13-11a.

a. Find the particular equation expressing the distance of P from the surface as a function of stopwatch reading.
b. Find the time at which P first *emerges* from the water.

a. The first part of this problem is similar to the example in Section 13-8. By sketching the graph (Figure 13-11b) you can derive the following equation:

$$\underline{\underline{d = 6 + 7 \cos \frac{\pi}{5}(t - 2),}}$$

where d is feet above the surface of the water, and t is the number of seconds the stopwatch reads.

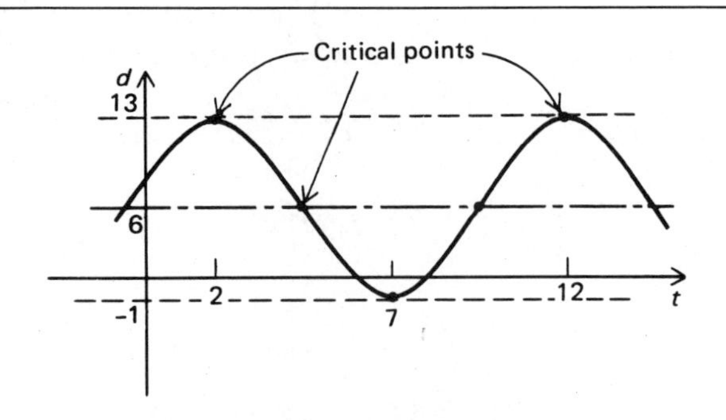

Figure 13-11b.

b. Point P emerges from the water when $d = 0$. Setting $d = 0$ gives

$6 + 7\cos\frac{\pi}{5}(t-2) = 0$	Substitution and symmetry.
$\cos\frac{\pi}{5}(t-2) = -\frac{6}{7}$	Subtracting 6 and dividing by 7.
$\frac{\pi}{5}(t-2) = \arccos\left(-\frac{6}{7}\right)$	arccos of both members.

Since $6/7 \approx 0.8571$, the reference arc (Table IV or calculator) is 0.541. Since cosine is *negative,* the arc must terminate in Quadrant II or III. So $\text{Arccos}\,(-6/7) \approx \pi - 0.541 \approx 2.600$. Consequently, $\arccos\,(-6/7) \approx \pm\, 2.600 + 2\pi n$. Substituting these values for arccos $(-6/7)$ gives

$\frac{\pi}{5}(t-2) \approx \pm 2.600 + 2\pi n$	Substituting for arccos $(-6/7)$.
$t - 2 \approx \pm 4.14 + 10n$	Multiplying by $\frac{5}{\pi}$.
$t \approx 6.14 + 10n$ or $-2.14 + 10n$	Adding 2.
$t \approx 6.14, 16.14, \ldots,$ or $-2.14, 7.86, \ldots$	Substituting integers for n.

To find out which of these is the particular value of t desired, you must return to the real world. From Figure 13-11b, you can see that P *emerges* from the water the *second* time $d = 0$. So the particular value of t required is $\underline{\underline{t = 7.86}}$.

In the following exercise you will find values of the independent variable for given values of the dependent one. Some of the problems are continuations of those in Exercise 13-8.

Exercise 13-11

1. ***Ferris Wheel Problem.*** For Problem 1 in Exercise 13-8, find the second time your seat in the Ferris wheel is 18 feet above the ground.

2. ***Bouncing Spring Problem.*** For Problem 2 in Exercise 13-8, predict the first positive value of time at which the weight is 59 centimeters above the floor.

3. ***Tidal Wave Problem.*** For Problem 3 in Exercise 13-8, between what two times is there no water at the pier?

4. ***Tarzan Problem.*** For Problem 4 in Exercise 13-8, find the least positive value of time t for which Tarzan is directly over the river bank (i.e., $y = 0$).

5. ***Pebble-in-the-Tire Problem.*** For Problem 5 in Exercise 13-8, find the first two distances you have traveled when the pebble in the tire tread is 11 inches above the pavement.

6. ***Spaceship Problem.*** For Problem 6 in Exercise 13-8, find the first time at which the spaceship is 1600 kilometers *south* of the equator.

7. ***Another Spaceship Problem.*** A spacecraft is in an elliptical orbit around the Earth (Figure 13-11c). At time $t = 0$ hours, it is at its apogee (highest point) $d = 1000$ kilometers above the Earth's surface. Fifty minutes later, it is at its perigee $d = 100$ kilometers above the surface.

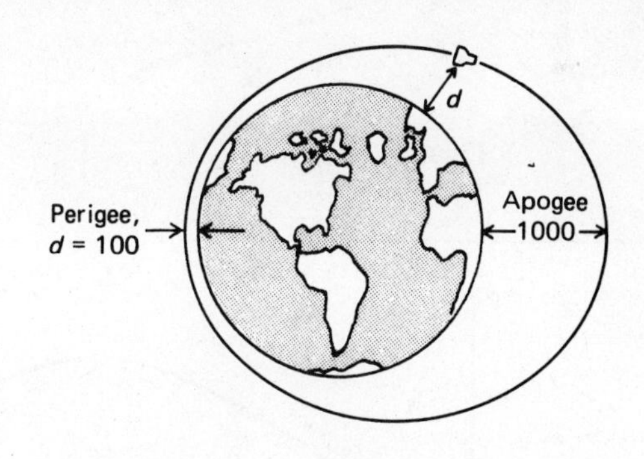

Figure 13-11c.

a. Assuming that d varies sinusoidally with time, write the particular equation expressing d in terms of t.
b. Solve the equation for t in terms of d.
c. Predict the first three positive values of t for which the spacecraft is 200 kilometers from the surface.

d. In order to transmit information back to Earth, the spacecraft must be within 700 kilometers of the surface. For how many consecutive minutes will the spacecraft be able to transmit?

8. ***Tide Problem.*** At a certain point on the beach, a post sticks out of the sand, its top being 76 centimeters above the beach (Figure 13-11d). The depth of the water at the post varies sinusoidally with time due to the motion of the tides. Using the techniques of Section 13-8, you find that the depth d in centimeters is

$$d = 40 + 60 \cos \frac{\pi}{6} (t - 2),$$

where t is the time in hours since midnight.

a. Sketch a graph of the sinusoid.
b. Solve the equation for t in terms of d.

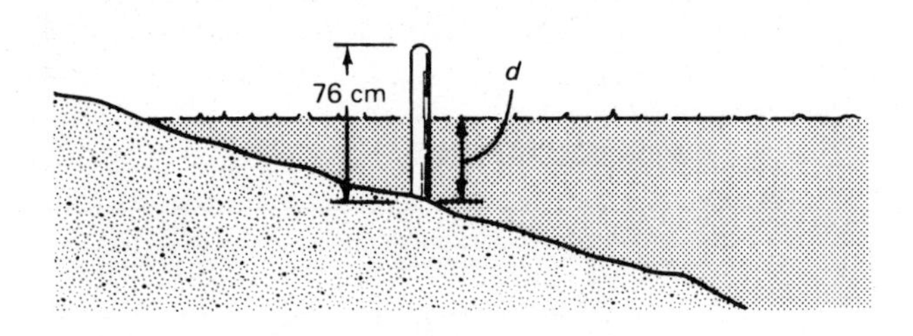

Figure 13-11d.

c. What is the earliest time of day at which the water level is just at the top of the post?
d. At the time you calculated in part c, is the post just going under water or just emerging from the water? Explain.
e. When d is negative, the tide is completely out and there is no water at the post. Between what times will the entire post be out of the water?

9. ***Tunnel Problem:*** Scorpion Gulch & Western Railway is preparing to build a new line through Rolling Mountains. They have hired you to do some calculations for tunnels and bridges needed on the line (Figure 13-11e).

You set up a Cartesian coordinate system with its origin at the entrance to the tunnel through Bald Mountain. Your surveying crew finds that the mountain rises 250 meters above the level of the track and that the next valley goes down 50 meters below the level of the track. The cross section of the mountain and valley is roughly sinusoidal with a horizontal distance of 700 meters from the top of the mountain to the bottom of the valley.

a. Write the particular equation expressing the vertical distance y from the track to the surface of the mountain or valley in terms of the horizontal distance x from the tunnel entrance. This can be done by finding the constants A, B, and C from the distances given. Finding D requires that you

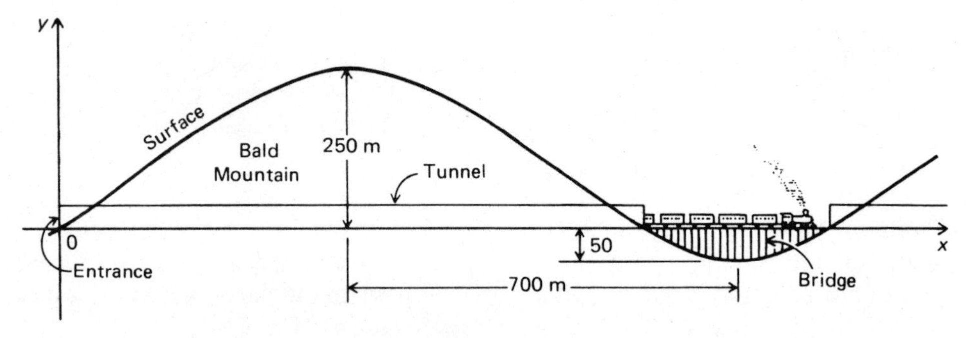

Figure 13-11e.

substitute the other constants and the ordered pair (0, 0) into the equation and solve for D.

b. Solve the equation from part a for x in terms of y.
c. How long will the tunnel be?
d. How long will the bridge be?
e. The company thinks it might be cheaper to build the line if the entire project is raised to $y = 20$, thus making the tunnel shorter and the bridge longer. Find the new values of x at the ends of the tunnel and bridge. Then find the new lengths of each.

10. ***Roller Coaster Problem.*** A sinusoidal roller coaster track is to be built with a high point $h = 27$ meters at a horizontal distance $d = 0$ meters. It has a low point $h = -3$ meters at $d = 50$ meters (Figure 13-11f).

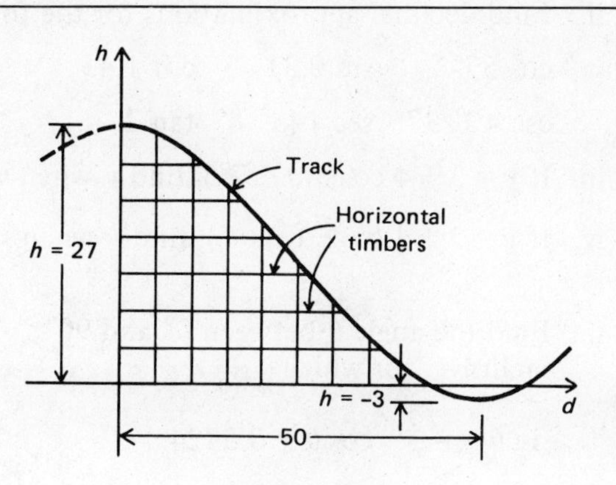

Figure 13-11f.

a. Write the particular equation expressing h in terms of d.
b. Solve the equation of part a for d in terms of h.
c. The lengths of the horizontal timbers used to build the roller-coaster supporting structure are values of d. Calculate the length of the horizontal timber that is 4 meters above the ground.
d. Write a computer program to print the lengths of each of the horizontal timbers. The first one is 2 meters above the ground, and they are spaced 2 meters apart. The program should also find the *sum* of the lengths of the timbers, so that the builders will know how much material to order. If your computer has only the Arctangent function available, you can use the results of Problem 100 in Section 13-9 to express Arccosine as an Arctangent.

11. ***Sun Elevation Problem.*** The "angle of elevation" of an object above you is the angle between a horizontal line and the line of sight between you and the object, as shown in Figure 13-11g. After the Sun rises, its angle of elevation increases rapidly at first, then more slowly, reaching a maximum near noontime. Then the angle decreases until sunset. The next day the phenomenon repeats itself.

Assume that when the Sun is up, its angle of elevation E varies sinusoidally with the time of day. Let t be the number of hours that has elapsed since midnight last night. Assume that the amplitude of this sinusoid is 60°, and the maximum angle of elevation occurs at 12:45 p.m. Assume that at this time of year the sinusoidal axis is at $E = -5°$. The period is, of course, 24 hours.

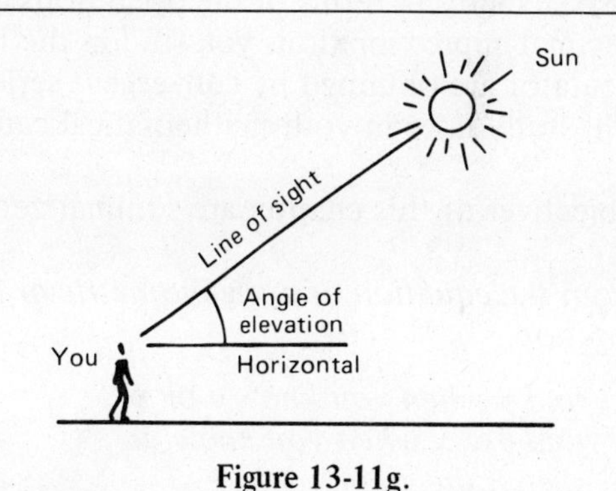

Figure 13-11g.

a. Sketch a graph of this function.
b. What is the real-world significance of the t-intercepts?
c. What is the real-world significance of the portion of the sinusoid which is *below* the t-axis?
d. Predict the angle of elevation at 9:27 a.m., at 2:30 p.m.

e. Predict the time of sunrise.
f. As you know, the maximum angle of elevation increases and decreases with the changes of season. Also, the times of sunrise and sunset change with the seasons. What *one* change could you make in your mathematical model that would allow you to use it for predicting the angle of elevation of the Sun at *any* time on *any* day of the year?

13-12 CHAPTER REVIEW AND TEST

In this chapter you have been introduced to a new kind of function, the trigonometric or circular function. The most significant thing about these functions is that they are *periodic*, repeating themselves regularly as the independent variable changes. As was the case with exponential and logarithmic functions, the trigonometric and circular functions are *transcendental*. This means that except for certain special values such as $\sin \pi/3$, their values cannot be expressed exactly in terms of the operations of algebra. The decimal approximations you find in the tables or by calculator are obtained by convergent series which you will study later in your mathematical career.

The objectives of this chapter are summarized below:

1. *From the equation of a trigonometric or circular function,*
   a. *find* **y** *when you know* θ *or* **x**.
   b. *find* θ *or* **x** *when you know* **y**.
   c. *sketch the graph.*
2. *From the graph of a trigonometric or circular function, find the particular equation.*
3. *Be able to evaluate and graph inverse trigonometric and circular functions and relations.*
4. *Use circular or inverse circular functions or relations as mathematical models of periodic phenomena in the real world.*

The four problems in the chapter review, below, are numbered according to these objectives. The Concepts Test requires you to decide which objective is being tested in each part of each problem.

Review Problems

The following problems are numbered according to the four objectives of this chapter.

Review Problem 1a. i. Find *exact* values of the following:

$\sin 60°$ $\quad\cos \frac{\pi}{4}$ $\quad\tan \frac{\pi}{6}$

$\sec 180°$ $\quad\csc 225°$ $\quad\cot 0$

ii. Find decimal approximations for the following:

$\cos 57°$ $\quad\sin 0.31$ $\quad\cot 1.43$

$\csc 42°\,37'$ $\quad\sec 143°\,8'$ $\quad\tan 2$

iii. If $y = 3 + 4\cos 5(\theta - 7°)$, find y when $\theta = 2°$.

iv. If $y = 3 + 4\cos \frac{\pi}{5}(x - 7)$, find y when $x = 2$.

b. i. Find the angle θ between 0° and 90°, inclusive, for which

$\sin \theta = \frac{\sqrt{2}}{2}$ $\quad\cos \theta = 0.5724$

$\sec \theta = 2$ $\quad\csc \theta = 1/2$

$\tan \theta$ is undefined

$\cot \theta = 0.2056$

ii. Find the real number x between 0 and $\pi/2$, inclusive, for which

$\sin x = \frac{\sqrt{3}}{2}$ $\quad\cos x = 2$

$\sec x = 1$ $\quad\csc x = 1.34$

$\tan x = 1$

$\cot x$ is undefined

iii. If $y = 3 + 4 \cos 5(\theta - 7°)$, find the first positive value of θ for which $y = 1$.

iv. If $y = 3 + 4 \cos \frac{\pi}{5}(x - 7)$, find the first positive value of x at which $y = 1$.

c. Sketch two complete cycles of the graph of:

i. $y = 3 + 4 \cos 5(\theta - 7°)$

ii. $y = 3 + 4 \cos \frac{\pi}{5}(x - 7)$

iii. $y = \tan \theta$

iv. $y = \sec x$

Review Problem 2. Write the particular equation of each sinusoid sketched in Figure 13-12a.

a.

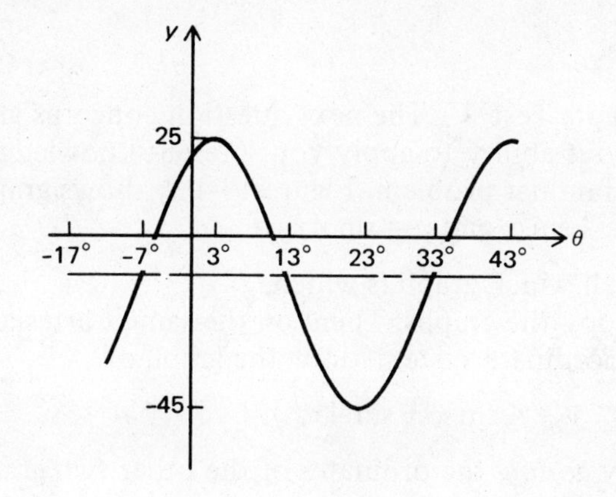

b.

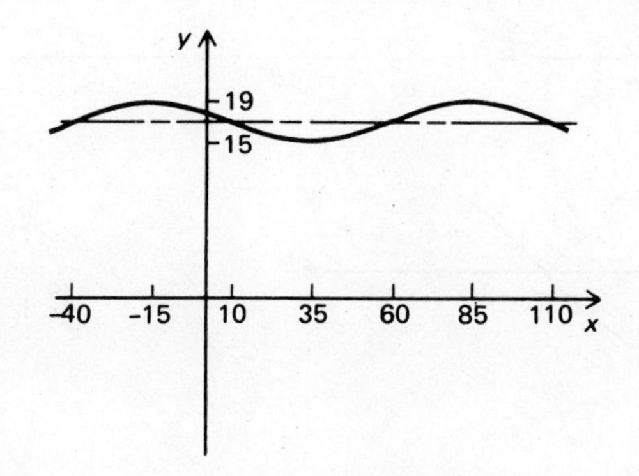

Figure 13-12a.

Review Problem 3a. Find exact values of the following, if possible. Otherwise, find decimal approximations.

Arcsin 0.5	arccos (−1)	$\text{Tan}^{-1} 0.3241$
$\sec^{-1} \sqrt{2}$	Csc^{-1} 0.5	arccot (−1)

b. Transform the following equations so that θ or x is expressed in terms of y. Then find the first three positive values of θ or x for which $y = 1$.

i. $y = 5 + 4 \cos 3(\theta - 77°)$,

ii. $y = 6 + 2 \tan \frac{\pi}{5}(x - 3)$.

c. Sketch the graph of

i. $y = \text{Arccot}\, x$,

ii. $y = \sin^{-1} x$.

Review Problem 4 – Porpoising Problem. Assume that you are aboard a submarine, submerged in the Pacific Ocean. At time $t = 0$, you make contact with an enemy destroyer. Immediately, you start "porpoising" (going deeper and shallower). At time $t = 4$ minutes, you are at your deepest, $y = -1000$ meters. At time $t = 9$ minutes, you next reach your shallowest, $y = -200$ meters. Assume that y varies sinusoidally with t for $t \geq 0$.

a. Sketch the graph of y versus t.
b. Write the particular equation expressing y in terms of t.
c. Your submarine is "safe" when it is below $y = -300$ meters. At time $t = 0$, was your submarine safe? Justify your answer.
d. Between what two non-negative times is your submarine first safe?

Concepts Test

Work each of the following problems. Then tell which of the preceding objectives you used in working that problem.

You apply for a job with Y. O. Ming Mining Company. Since your work will involve using circular functions,

Mr. Ming poses some problems for you to solve during your interview to see how useful you would be to his company.

Concepts Test 1. The first question concerns exact values. Find

a. $\sin \frac{5\pi}{3}$

b. $\cos 180°$

c. $\tan \frac{3\pi}{4}$

d. $\sec 120°$

e. $\csc 0$

f. $\text{Cos}^{-1}(-\sqrt{2}/2)$

g. $\text{Arccot}\,(-\sqrt{3})$

h. arcsec 2

i. arcsin 2 (Your résumé is *shredded* if you miss *this* one!)

j. $\tan\,(\text{Arccot}\,\frac{2}{5})$

$m(\theta)$	covers θ	
23′ 00′	0.6093	67° 00′
10′	0.6066	50′
20′	0.6039	40′
30′	0.6013	30′
40′	0.5986	20′
50′	0.5959	10′
24° 00′	0.5933	66° 00′
10′	0.5906	50′
20′	0.5880	40′
30′	0.5853	30′
40′	0.5827	20′
50′	0.5800	10′
25° 00′	0.5774	65° 00′
	vers θ	$m(\theta)$

Concepts Test 2. Another question concerns the use of tables. There are other functions of angles called "versine" and "coversine." The abbreviations are "vers θ" and "covers θ," respectively. The tables are constructed just like Table III at the back of this book. Use the adjacent portion of the table to find

a. vers 66° 52′

b. θ, if covers $\theta = 0.6045$.

c. By looking at Table III, see if you can figure out what vers θ and covers θ *really* are!

Concepts Test 3. The next question concerns graphs, and your ability to apply your present knowledge to an unfamiliar problem. Figure 13-12b shows graphs of $y = 4 \sin x$ and $y = \sin 4x$.

a. Tell which graph is which.

b. Copy the graphs. Then, on the same Cartesian coordinate system, draw the graph of

$$y = 4 \sin x + \sin 4x$$

by *adding* the ordinates of the other two graphs.

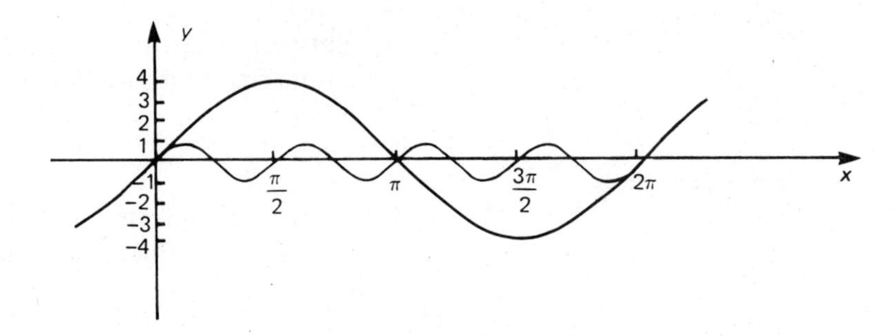

Figure 13-12b.

For example, at a value of x for which the "small" graph has $y = 1$, the desired graph is 1 unit *above* the "big" graph.

Concepts Test 4. The last question concerns your knowledge of graphs. Sketch:

a. $y = \csc \theta$,
b. $y = \text{Cos}^{-1} x$.

Concepts Test 5. Mr. Ming is satisfied with your performance and assigns you to the Uranium Mining Project (UMP). A layer of ore beneath the ground has surfaces that are sinusoidal in cross section, as shown Figure 13-12c. UMP plans to drill a vertical mine shaft through the ore layer and then dig a horizontal tunnel, again going through the ore layer. Your job is to find out how far this horizontal tunnel goes through the ore.

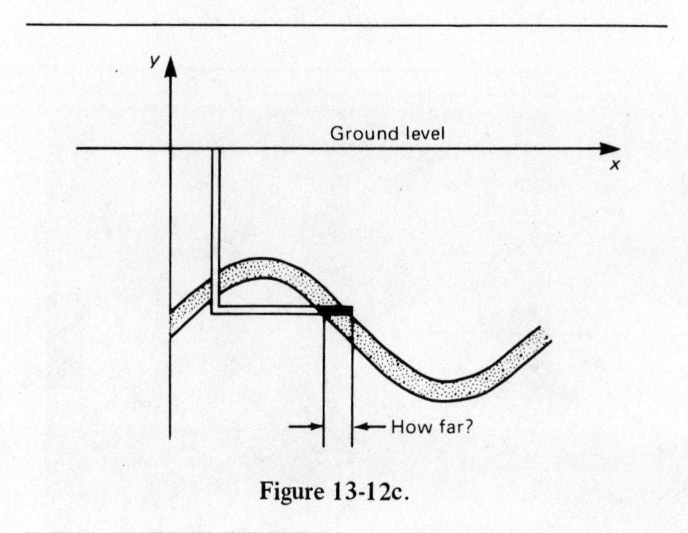

Figure 13-12c.

You set up a Cartesian coordinate system with the x-axis at gound level and x and y in meters. You find that the top surface of the ore layer has a high point at $(x, y) = (80, -40)$, and that the graph next crosses the sinusoidal axis at $(x, y) = (205, -100)$. The bottom surface of the ore layer is 20 meters below the top surface for every value of x.

a. Write the particular equations of the top and bottom surfaces.

b. The vertical shaft is dug at $x = 50$. At what depth will it first reach the ore layer?

c. Transform the equations of the top and bottom surfaces so that x is expressed in terms of y.

d. How far will the horizontal shaft go through the formation if it is dug at a depth of

i. $y = -90$?
ii. $y = -170$?

Properties of Trigonometric and Circular Functions

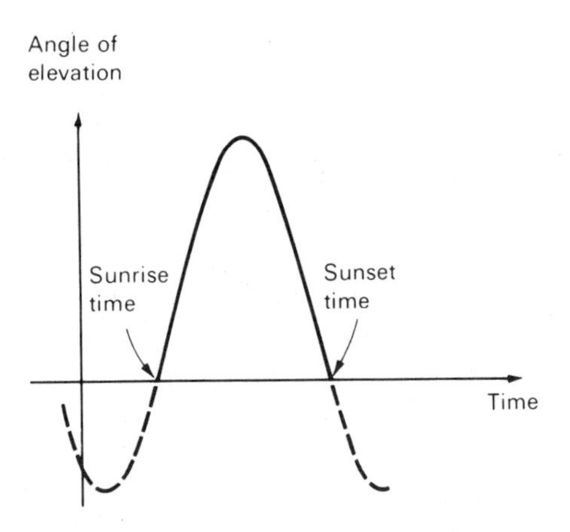

The angle of elevation of the Sun varies sinusoidally with the time of day. Predicting the time of sunrise involves finding $\mathbf{x}$ *in an equation such as* $4 + 3\cos \pi(\mathbf{x} - 2) = 0$. *In order to solve more complicated trigonometric equations such as*

$$4 \sin (\mathbf{x} + 75^\circ) \cos (\mathbf{x} - 75^\circ) = 1,$$

you must learn some ***properties*** *that will allow you to* ***transform*** *the left member to a simpler expression. Since the properties of trigonometric and circular functions are virtually the same, the word "trigonometric" function will be used for both kinds unless it is necessary to distinguish between them.*

14-1 THREE PROPERTIES OF TRIGONOMETRIC FUNCTIONS

When you defined the six trigonometric functions, you probably observed that certain ones were *reciprocals* of others. For example, $\sec x = 1/\cos x$. In this section you will become familiar with this property and two other kinds that also come directly from the definitions. The variable x will be used for the argument whether the function is trigonometric or circular.

Objective:

Be able to use the three types of properties below to transform a given expression to a specified, equivalent form, possibly simplifying the expression.

1. *Reciprocal Properties:* By the definitions of the trigonometric functions,

$$\tan x = \frac{v}{u} \quad \text{and} \quad \cot x = \frac{u}{v},$$

where u and v are the abscissa and ordinate, respectively, of a point on the terminal side of the angle or arc x (see Figure 14-1). Therefore, $\tan x$ and $\cot x$ are *reciprocals* of each other. Similarly, $\sin x$ and $\csc x$ are reciprocals of each other, and $\cos x$ and $\sec x$ are reciprocals of each other. In summary:

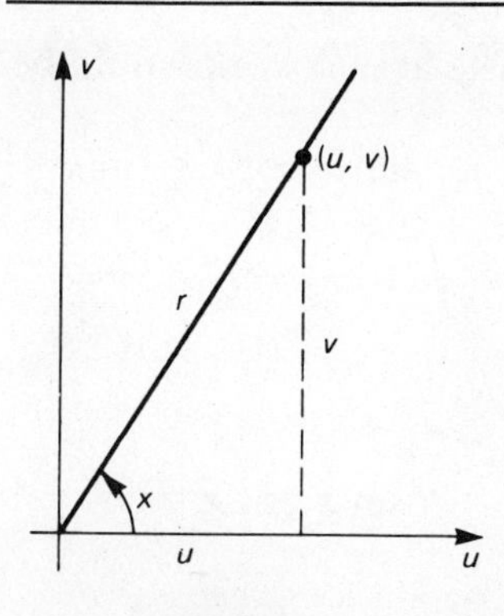

Figure 14-1.

$$\boxed{\cot x = \frac{1}{\tan x} \quad \csc x = \frac{1}{\sin x} \quad \sec x = \frac{1}{\cos x}}.$$

These are known as the *reciprocal* properties. Since the product of a number and its reciprocal equals 1, these properties may also be written:

$$\boxed{\tan x \cot x = 1 \quad \sin x \csc x = 1 \quad \cos x \sec x = 1}.$$

2. *Quotient Properties:* The expression $\sin x/\cos x$ can be written using the definitions of the trigonometric functions as:

$$\frac{\sin x}{\cos x} = \frac{\frac{v}{r}}{\frac{u}{r}} \qquad \text{Definition of } \sin x \text{ and } \cos x.$$

$$= \frac{v}{r} \cdot \frac{r}{u} \qquad \text{Definition of division.}$$

$$= \frac{v}{u} \qquad \text{Multiplication property of fractions and canceling.}$$

But v/u is defined to be $\tan x$ (Figure 14-1). So by transitivity:

$$\boxed{\tan x = \frac{\sin x}{\cos x}}.$$

This is called a *quotient* property because $\tan x$ is expressed as a quotient. Since $\cot x$ is the reciprocal of $\tan x$, the quotient can be inverted to give a second quotient property:

$$\boxed{\cot x = \frac{\cos x}{\sin x}}.$$

There is another form of these two quotient properties that is sometimes useful. The $\sin x$ and $\cos x$ can be replaced by using the reciprocal properties to give:

$$\tan x = \frac{\sin x}{\cos x} \qquad \text{Quotient property.}$$

$$= \frac{\frac{1}{\csc x}}{\frac{1}{\sec x}}$$ Reciprocal properties.

$$= \frac{1}{\csc x} \cdot \frac{\sec x}{1}$$ Definition of division.

$$= \frac{\sec x}{\csc x}$$ Multiplication property of fractions.

$$\therefore \boxed{\tan x = \frac{\sec x}{\csc x}}$$ Transitivity.

Again using the reciprocal property, $\cot x = 1/\tan x$:

$$\boxed{\cot x = \frac{\csc x}{\sec x}}\ .$$

These two are usually not considered to be new properties, but rather to be alternate forms of the two quotient properties.

3. *Pythagorean Properties:* The numbers u, v, and r in Figure 14-1 are the legs and hypotenuse of a right triangle. By the Pythagorean Theorem,

$$u^2 + v^2 = r^2.$$

Dividing both members by r^2 gives

$$\frac{u^2}{r^2} + \frac{v^2}{r^2} = 1.$$

Since $u/r = \cos x$ and $v/r = \sin x$, it follows that

$$(\cos x)^2 + (\sin x)^2 = 1.$$

It is customary to drop the parentheses and to write "$(\cos x)^2$" as "$\cos^2 x$." The exponent is placed in a position where it cannot possibly be mistaken for x^2. So this property is usually written:

$$\boxed{\cos^2 x + \sin^2 x = 1}\ .$$

By dividing both members of $u^2 + v^2 = r^2$ by u^2, you get

$$\boxed{1 + \tan^2 x = \sec^2 x}\ ,$$

and by dividing both members by v^2, you get

$$\boxed{\cot^2 x + 1 = \csc^2 x}\ .$$

These three properties are called the *Pythagorean* properties, since they come from the Pythagorean Theorem.

You are now prepared to accomplish the objective of transforming given expressions to equivalent, simpler forms.

Example 1: Suppose you are asked to transform $\sin x \cot x$ to $\cos x$. The thought process you would go through is as follows:

1. Neither factor in $\sin x \cot x$ has $\cos x$ in it.

2. Therefore, either $\sin x$ or $\cot x$ should be *replaced* by something that *does* have $\cos x$ in it.

3. Sin x has no convenient forms that have $\cos x$ in them.

4. But by the quotient properties, $\cot x = \cos x/\sin x$.

The actual work would be done as follows:

$$\sin x \cot x = \sin x \cdot \frac{\cos x}{\sin x}$$ Quotient properties.

$$= \cos x$$ Multiplication property of fractions and canceling $\sin x$.

$$\therefore \sin x \cot x = \cos x$$ Transitivity.

Example 2: Suppose that you are asked to transform the expression $\sin x \sec x \cot x$ into 1. The thought process is:

1. Since the answer is 1, there must be some canceling that can be done.

2. Canceling requires *fractions,* which can be obtained either from *reciprocal* properties or from *quotient* properties.

3. By the reciprocal properties, $\sec x = 1/\cos x$. By the quotient properties, $\cot x = \cos x/\sin x$.

The actual work you would write down is as follows:

$\sin x \sec x \cot x$

$= \sin x \cdot \dfrac{1}{\cos x} \cdot \dfrac{\cos x}{\sin x}$ Reciprocal and quotient properties.

$= \dfrac{\sin x \cos x}{\cos x \sin x}$ Multiplication property of fractions.

$= 1$ Canceling, or $n/n = 1$.

$\therefore \sin x \sec x \cot x = 1$ Transitivity.

There are often several ways the transformation can be done. In the preceding example, you could have written

$\sin x \sec x \cot x$

$= \sin x \cdot \dfrac{1}{\cos x} \cdot \cot x$ Reciprocal properties.

$= \dfrac{\sin x}{\cos x} \cdot \cot x$ Multiplication property of fractions.

$= \tan x \cot x$ Quotient properties.

$= 1$ Reciprocal properties.

$\therefore \sin x \sec x \cot x = 1$ Transitivity.

Example 3: Suppose you are asked to transform the expression $\cos^2 x - \sin^2 x$ into the equivalent expression $1 - 2\sin^2 x$. Your thought process would be:

1. The answer has no cosines in it. So you must get *rid* of the $\cos^2 x$.

2. Since the expression involves *squares* of functions, the Pythagorean properties should be helpful.

3. The Pythagorean property that has cosines in it is $\cos^2 x + \sin^2 x = 1$, from which $\cos^2 x = 1 - \sin^2 x$.

The actual steps in the transformation would be

$\cos^2 x - \sin^2 x$

$= (1 - \sin^2 x) - \sin^2 x$ Pythagorean properties.

$= 1 - \sin^2 x - \sin^2 x$ Associativity.

$= 1 - 2\sin^2 x$ Adding like terms.

$\therefore \cos^2 x - \sin^2 x = 1 - 2\sin^2 x$ Transitivity.

In the exercise that follows, you will get practice using the reciprocal, quotient, and Pythagorean properties to transform expressions.

Exercise 14-1

For Problems 1 through 26, transform the expression on the left to the one on the right.

1. $\cos x \tan x$ to $\sin x$
2. $\csc x \tan x$ to $\sec x$
3. $\sec x \cot x \sin x$ to 1
4. $\csc x \tan x \cos x$ to 1
5. $\sin^2 \theta \sec \theta \csc \theta$ to $\tan \theta$
6. $\cos^2 A \csc A \sec A$ to $\cot A$
7. $\tan A + \cot A$ to $\csc A \sec A$
8. $\sin \theta + \cot \theta \cos \theta$ to $\csc \theta$
9. $\csc x - \sin x$ to $\cot x \cos x$
10. $\sec \theta - \cos \theta$ to $\sin \theta \tan \theta$
11. $\tan x (\sin x + \cot x \cos x)$ to $\sec x$
12. $\cos x (\sec x + \cos x \csc^2 x)$ to $\csc^2 x$
13. $(1 + \sin B)(1 - \sin B)$ to $\cos^2 B$
14. $(\sec x - 1)(\sec x + 1)$ to $\tan^2 x$
15. $(\cos \phi - \sin \phi)^2$ to $1 - 2\cos \phi \sin \phi$
16. $(1 - \tan \phi)^2$ to $\sec^2 \phi - 2\tan \phi$

17. $(\tan n + \cot n)^2$ to $\sec^2 n + \csc^2 n$

18. $(\cos k - \sec k)^2$ to $\tan^2 k - \sin^2 k$

19. $\dfrac{\csc^2 x - 1}{\cos x}$ to $\cot x \csc x$

20. $\dfrac{1 - \cos^2 x}{\tan x}$ to $\sin x \cos x$

21. $\dfrac{\sec^2 \theta - 1}{\sin \theta}$ to $\tan \theta \sec \theta$

22. $\dfrac{1 + \cot^2 \theta}{\sec^2 \theta}$ to $\cot^2 \theta$

23. $\dfrac{\sec A}{\sin A} - \dfrac{\sin A}{\cos A}$ to $\cot A$

24. $\dfrac{\csc B}{\cos B} - \dfrac{\cos B}{\sin B}$ to $\tan B$

25. $\dfrac{1}{1 - \cos C} + \dfrac{1}{1 + \cos C}$ to $2 \csc^2 C$

26. $\dfrac{1}{\sec D - \tan D} + \dfrac{1}{\sec D + \tan D}$ to $2 \sec D$

27. There are quite a few properties in which the number 1 occurs. By appropriate algebra, if necessary, write *six* trigonometric expressions, each of which equals 1.

28. Use the Pythagorean properties to write expressions equivalent to:

a. $\sin^2 x$ b. $\cos^2 x$ c. $\tan^2 x$
d. $\cot^2 x$ e. $\sec^2 x$ f. $\csc^2 x$

29. Write equations expressing each of the six trigonometric functions in terms of $\sin x$.

30. Write equations expressing each of the six trigonometric functions in terms of $\cos x$.

14-2 TRIGONOMETRIC IDENTITIES

A *trigonometric* open sentence is (obviously!) an open sentence that contains trigonometric functions. For example,

$$\sin x = 1/2$$

is a trigonometric equation. The solution set of such an open sentence is the set of all values of the argument x that makes the sentence true. If x is the degree measure of an angle, then the only solutions of the above open sentence are 30°, 150°, or any angle coterminal with these. Such an equation is called a *conditional* equation, because it is true only under certain conditions.

Some open sentences are true under *all* conditions. For example,

$$\cos^2 x = 1 - \sin^2 x$$

is true for *every* value of x. Such an equation is called an *identity*, because the two members are "identical" to each other. (Actually, the two members are *equivalent* expressions.)

Objective:

Given a trigonometric equation, prove that it is an *identity*.

There are two purposes for learning how to prove *identities*.

1. To learn the relationships among the functions.
2. To learn to transform one trigonometric expression to another equivalent form, usually simplifying it.

To accomplish the objective without defeating these purposes, the following agreement will be made:

Agreement: To prove that an equation is an identity, start with one member and transform it into the other.

Note that this is exactly what you were doing in the previous section! The only thing that is new is that you are free to pick *either* member to start with.

Example 1: Prove that $(1 + \cos x)(1 - \cos x) = \sin^2 x$.

Proof:

$(1 + \cos x)(1 - \cos x)$	Start with the more complicated member.
$= 1 - \cos^2 x$	Do the obvious algebra.
$= \sin^2 x$	Look for familiar expressions.
$\therefore (1 + \cos x)(1 - \cos x) = \sin^2 x$, Q.E.D.	Transititivy.

Notes:

1. It is tempting to *start* with the given equation and then work on *both* members until you have reduced the equation to an obviously true statement, such as "$\cos x = \cos x$." What this actually does is prove the *converse* of what you were asked to prove. That is, "*If* the identity is true, *then* the reflexive property is true." This is circular reasoning. It is dangerous because you might actually "prove" a *false* identity by taking an irreversible step, such as squaring both members.
2. The letters "Q.E.D." at the end stand for the Latin words *quod erat demonstrandum,* which means, "which was to be demonstrated."

Example 2: Prove that $\cot x + \tan x = \csc x \sec x$.

Proof:

$\cot x + \tan x$	Pick a member to work on.
$= \dfrac{\cos x}{\sin x} + \dfrac{\sin x}{\cos x}$	Answer has only *one* term, so try adding fractions. The fractions must be *created* first.
$= \dfrac{\cos^2 x + \sin^2 x}{\sin x \cos x}$	Find common denominator and add the fractions.
$= \dfrac{1}{\sin x \cos x}$	Familiar Pythagorean property.
$= \dfrac{1}{\sin x} \cdot \dfrac{1}{\cos x}$	Answer has *two* factors, so *make* two factors.
$= \csc x \sec x$	Familiar reciprocal properties.
$\therefore \cot x + \tan x = \csc x \sec x$, Q.E.D.	Transititivy.

Example 3: Prove that $\dfrac{\sin x}{1 + \cos x} = \dfrac{1 - \cos x}{\sin x}$.

Proof:

$\dfrac{\sin x}{1 + \cos x}$	
$= \dfrac{\sin x}{1 + \cos x} \cdot \dfrac{1 - \cos x}{1 - \cos x}$	Pick a member to work on and multiply it by a clever form of 1.
$= \dfrac{\sin x(1 - \cos x)}{1 - \cos^2 x}$	Do the obvious albegra, but *don't* destroy the "$1 - \cos x$," because you want it in the answer.
$= \dfrac{\sin x(1 - \cos x)}{\sin^2 x}$	Familiar Pythagorean property.
$= \dfrac{1 - \cos x}{\sin x}$	Do the obvious canceling.
$\therefore \dfrac{\sin x}{1 + \cos x} = \dfrac{1 - \cos x}{\sin x}$, Q.E.D.	Transitivity.

Note that there could be *two* reasons for picking the form of "1" used in the first step. It has the *conjugate* of $1 + \cos x$ in its denominator. Or it has $1 - \cos x$ in its numerator, an expression you *want* in the answer.

Example 4: Prove that $\csc \theta \cos^2 \theta + \sin \theta = \csc \theta$.

Proof:

$\csc \theta \cos^2 \theta + \sin \theta$	Pick the more complicated member.
$= \csc \theta \left(\cos^2 \theta + \dfrac{\sin \theta}{\csc \theta}\right)$	If you want $\csc \theta$ as a factor of the answer, then *factor it out!*

$= \csc\theta\,(\cos^2\theta + \sin\theta \cdot \sin\theta)$	Familiar reciprocal property.
$= \csc\theta\,(\cos^2\theta + \sin^2\theta)$	Obvious algebra.
$= \csc\theta$	Familiar Pythagorean property.
$\therefore \csc\theta\cos^2\theta + \sin\theta = \csc\theta$, Q.E.D.	Transitivity.

Factoring out the $\csc\theta$ in the second line of the proof is sometimes called "factoring out a rabbit," because you are reaching in and pulling out a common factor that wasn't there!

Note that the reasons written for the steps in the above examples are reasons you *chose to do* the particular step, rather than mathematical reasons why the steps are true. From these examples, certain useful techniques emerge that help guide your throught process as you attempt to prove identities. These steps are summarized below for your convenience.

Steps in Proving Identities

1. Pick the member you wish to work with and write it down. Usually it is easier to start with the more complicated member.
2. Look for *algebraic* things to do.
   a. If there are two terms and you want only one,
      i. add fractions,
      ii. factor something out.
   b. Multiply by a clever form of 1
      i. to multiply a numerator or denominator by its conjugate,
      ii. to get a desired expression in numerator or denominator.
   c. Do any obvious algebra such as distributing, squaring, or multiplying polynomials.
3. Look for *trigonometric* things to do.
   a. Look for familiar trigonometric expressions like $1 - \cos^2 x$, $\cos x \sec x$, or $\sin x/\cos x$.
   b. If there are *squares* of functions, think of Pythagorean properties.
   c. Reduce the number of different functions, transforming them to the ones you want in the answer.
4. Keep looking at the answer to make sure you are headed in the right direction.

The exercise that follows is designed to give you practice proving identities, so that you may become more familiar with the properties of the trigonometric functions and may gain practice transforming one expression into another simpler form.

Exercise 14-2

In Problems 1 through 34, prove that each equation is an identity.

1. $\sec x(\sec x - \cos x) = \tan^2 x.$
2. $\tan x(\cot x + \tan x) = \sec^2 x.$
3. $\sin x(\csc x - \sin x) = \cos^2 x.$
4. $\cos x(\sec x - \cos x) = \sin^2 x$
5. $\csc^2\theta - \cos^2\theta\csc^2\theta = 1$
6. $\cos^2\theta + \tan^2\theta\cos^2\theta = 1$
7. $(\sec\theta + 1)(\sec\theta - 1) = \tan^2\theta$
8. $(1 + \sin\theta)(1 - \sin\theta) = \cos^2\theta$
9. $\sec^2 A + \tan^2 A\sec^2 A = \sec^4 A$
10. $\cot^2 A\csc^2 A - \cot^2 A = \cot^4 A$
11. $\cos^4 t - \sin^4 t = 1 - 2\sin^2 t$
12. $\sec^4 t - \tan^4 t = 1 + 2\tan^2 t$
13. $\dfrac{1}{\sin x\cos x} - \dfrac{\cos x}{\sin x} = \tan x$
14. $\dfrac{\sec x}{\sin x} - \dfrac{\sin x}{\cos x} = \cot x$
15. $\dfrac{\sin x}{\csc x} + \dfrac{\cos x}{\sec x} = 1$

16. $\dfrac{1}{\sec^2 x} + \dfrac{1}{\csc^2 x} = 1$

17. $\dfrac{1}{1 + \cos s} = \csc^2 s - \csc s \cot s$

18. $\dfrac{1}{1 - \sin r} = \sec^2 r + \sec r \tan r$

19. $\dfrac{\cos x}{\sec x - 1} - \dfrac{\cos x}{\tan^2 x} = \cot^2 x$

20. $\dfrac{\sin x}{1 - \cos x} + \dfrac{1 - \cos x}{\sin x} = 2 \csc x$

21. $\dfrac{\sec x}{\sec x - \tan x} = \sec^2 x + \sec x \tan x$

22. $\dfrac{1 + \sin x}{1 - \sin x} = 2 \sec^2 x + 2 \sec x \tan x - 1$

23. $\sin^3 z \cos^2 z = \sin^3 z - \sin^5 z$

24. $\sin^3 z \cos^2 z = \cos^2 z \sin z - \cos^4 z \sin z$

25. $\sec^2 \theta + \csc^2 \theta = \sec^2 \theta \csc^2 \theta$

26. $\sec \theta + \tan \theta = \dfrac{1}{\sec \theta - \tan \theta}$

27. $\dfrac{1 - 3 \cos x - 4 \cos^2 x}{\sin^2 x} = \dfrac{1 - 4 \cos x}{1 - \cos x}$

28. $\dfrac{\sec^2 x - 6 \tan x + 7}{\sec^2 x - 5} = \dfrac{\tan x - 4}{\tan x + 2}$

29. $\dfrac{\sin^3 A + \cos^3 A}{\sin A + \cos A} = 1 - \sin A \cos A$

30. $\dfrac{\sec^3 B - \cos^3 B}{\sec B - \cos B} = \sec^2 B + 1 + \cos^2 B$

31. $\csc^6 x - \cot^6 x = 1 + 3 \csc^2 x \cot^2 x$

32. $(2 \sin x + 3 \cos x)^2 + (3 \sin x - 2 \cos x)^2 = 13$

33. $\dfrac{1 + \sin x + \cos x}{1 + \sin x - \cos x} = \dfrac{1 + \cos x}{\sin x}$

34. $\dfrac{1 + \sin x + \cos x}{1 - \sin x + \cos x} = \dfrac{1 + \sin x}{\cos x}$

35. *Graphs of $1 + \tan^2 x$ and $\sec^2 x$:* In this problem you will demonstrate by graphing that the identity

$$1 + \tan^2 x = \sec^2 x$$

is reasonable.

a. Draw an auxiliary graph of $y = \tan x$.
b. On the same Cartesian coordinate system, draw a graph of $y = \tan^2 x$ by *squaring* the ordinates of the graph from part a.
c. Draw a graph of $y = \sec^2 x$ the same way.
d. By comparing the two graphs, show that the graph of $y = \sec^2 x$ is always 1 unit above the graph of $y = \tan^2 x$.

36. *Introduction to Odd and Even Functions:* In this problem you will learn a property possessed by some algebraic functions and all six trigonometric functions, namely, "oddness" and "evenness." Suppose that functions f_1, f_2, f_3, and f_4 are defined as follows:

$$f_1(x) = x, \quad f_2(x) = x^2, \quad f_3(x) = x^3, \quad f_4(x) = x^4.$$

a. For each function, find $f(-3)$, $f(-2)$, $f(2)$, and $f(3)$.
b. A function is called an *even* function if $f(-x) = f(x)$ for all values of x. From your answers in part a, tell which of the above functions are even functions.
c. A function is called an *odd* function if $f(-x) = -f(x)$ for all values of x. From your answers to part a, tell which of the above functions are odd functions.
d. Why do you suppose the names "odd" and "even" were picked to describe the properties in parts b and c?
e. Write down the values of the six trigonometric functions of $-30°$. Then decide which of the trigonometric functions satisfy the requirements of an *odd* function and which ones satisfy the requirements of an *even* function. Write your answers in a form such as

$$\sin(-x) = \sin x \quad \text{or} \quad \sin(-x) = -\sin x,$$

whichever is correct.

14-3 PROPERTIES INVOLVING FUNCTIONS OF MORE THAN ONE ARGUMENT

The Pythagorean, quotient, and reciprocal properties of Section 14-1 involve *one* argument only. In this section you will learn properties in which more than one argument appears. For example, the argument may be composed of a *sum* of two numbers, such as $\cos(x - D)$, which you encountered when you worked with sinusoids.

The operation cos does *not* distribute over addition or subtraction. That is, $\cos(x - D)$ does *not* equal $\cos x - \cos D$. You can prove this by substituting angles such as $x = 60°$ and $D = 90°$ and showing that the two expressions are not equal. In this section you will learn how to express functions of sums or differences of two angles in terms of functions of the angles themselves.

Objectives:

1. Be able to express functions of $-x$ in terms of functions of x.
2. Be able to express $\cos(A - B)$, $\cos(A + B)$, $\sin(A - B)$, and $\sin(A + B)$ in terms of $\sin A$, $\cos A$, $\sin B$, and $\cos B$.
3. Be able to express $\tan(A - B)$ and $\tan(A + B)$ in terms of $\tan A$ and $\tan B$.

1. *Functions of* $-x$: Suppose that $f(x) = x^5$. Then:

$f(-x) = (-x)^5$	Definition of f.
$f(-x) = -x^5$	Negative number raised to odd power.
$f(-x) = -f(x)$	Substitution.

But if $f(x) = x^4$, then

$f(-x) = (-x)^4$	Definition of f.
$f(-x) = x^4$	Negative number raised to even power.
$f(-x) = f(x)$	Definition of f.

So when the exponent is odd, $f(-x) = -f(x)$; when the exponent is even, $f(-x) = f(x)$. This property of odd and even exponents leads to a general definition of odd and even functions.

DEFINITION

If $f(-x) = f(x)$, then f is called an ***even function.***
If $f(-x) = -f(x)$, then f is called an ***odd function.***

The names "odd" and "even" carry over to functions that do not involve exponents. If you worked Problem 36 in the previous section, you discovered that the trigonometric functions possess these properties also.

Figure 14-3a shows two arcs on a unit circle, one with measure x, the other with measure $-x$. From the picture, you can see that if (u, v) is the endpoint of the arc x, then $(u, -v)$ is the endpoint of the arc $-x$. By the definition of circular functions, $\cos x = u$, and $\cos(-x) = u$ also. Therefore,

$$\cos(-x) = \cos x,$$

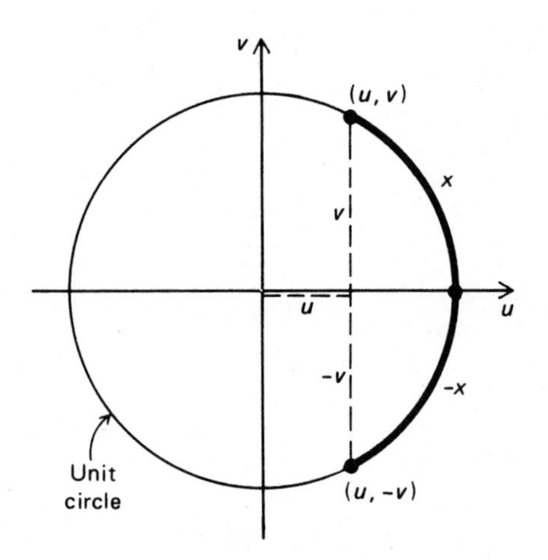

Figure 14-3a.

which means that cosine is an *even* function. Similarly, $\sin x = v$ and $\sin(-x) = -v$, which implies that

$$\sin(-x) = -\sin x,$$

and sine is an *odd* function. Each function whose definition involves v will be an odd function, since the ordinate of $-x$ is $-v$. For example,

$$\tan(-x) = \frac{-v}{u} = -\frac{v}{u} = -\tan x.$$

The cosine and its reciprocal, the secant, which involve only u, are the only even functions. The properties are summarized below.

Odd and Even Function Properties

$\cos(-x)$	$= \cos x$	even function
$\sin(-x)$	$= -\sin x$	odd function
$\tan(-x)$	$= -\tan x$	odd function
$\cot(-x)$	$= -\cot x$	odd function
$\sec(-x)$	$= \sec x$	even function
$\csc(-x)$	$= -\csc x$	odd function

Note that these properties also hold for the trigonometric functions. For example, $\sin(-\theta) = -\sin\theta$ and $\cos(-\theta) = \cos\theta$.

2. *Functions of Complementary Arcs:* You recall from Section 13-4 that the *co*sine of an angle equals the *sine* of its *complement.* For example, the complement of 76° is 90° − 76°, or 14°. So cos 76° = sin 14°.

Using the definitions of the circular functions, it is easy to see why this property is true. Figure 14-3b shows an arc x with endpoint (u, v). The *complement* of x is $\pi/2 - x$. If this arc is moved into standard position, as in the right-hand sketch, its endpoint will be (v, u).

Consequently,

$$\cos\left(\frac{\pi}{2} - x\right) = v,$$

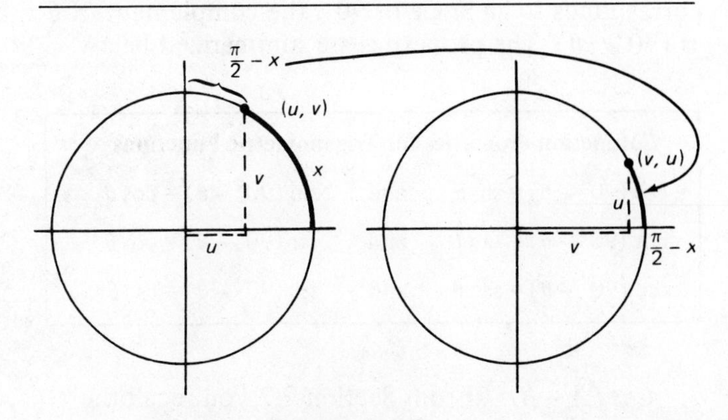

Figure 14-3b.

the abscissa of the endpoint of $(\pi/2 - x)$. But $v = \sin x$, as shown in the left-hand sketch of Figure 14-3b. Therefore,

$$\cos\left(\frac{\pi}{2} - x\right) = \sin x.$$

By similar reasoning,

$$\sin\left(\frac{\pi}{2} - x\right) = \cos x.$$

The cofunction properties for tangent and secant are similarly derived and are summarized below.

Cofunction Properties for Circular Functions

$$\cos\left(\frac{\pi}{2} - x\right) = \sin x \quad \text{and} \quad \sin\left(\frac{\pi}{2} - x\right) = \cos x$$

$$\cot\left(\frac{\pi}{2} - x\right) = \tan x \quad \text{and} \quad \tan\left(\frac{\pi}{2} - x\right) = \cot x$$

$$\csc\left(\frac{\pi}{2} - x\right) = \sec x \quad \text{and} \quad \sec\left(\frac{\pi}{2} - x\right) = \csc x$$

The cofunction properties for the trigonometric functions are virtually the same. Since an arc of $\frac{\pi}{2}$

corresponds to an angle of 90°, the complement of θ is $(90° - \theta)$. The properties are summarized below.

Cofunction Properties for Trigonometric Functions

$\cos(90° - \theta) = \sin\theta$	and	$\sin(90° - \theta) = \cos\theta$
$\cot(90° - \theta) = \tan\theta$	and	$\tan(90° - \theta) = \cot\theta$
$\csc(90° - \theta) = \sec\theta$	and	$\sec(90° - \theta) = \csc\theta$

3. *Cos (A – B):* From Section 9-2 you recall the Distance Formula for finding the distance between two points in a Cartesian coordinate system. It is really just a special case of the Pythagorean Theorem. The distance between points (x_1,y_1) and (x_2,y_2) in Figure 14-3c is given by

$$d^2 = (\Delta x)^2 + (\Delta y)^2,$$

from which

$$d^2 = (x_2 - x_1)^2 + (y_2 - y_1)^2.$$

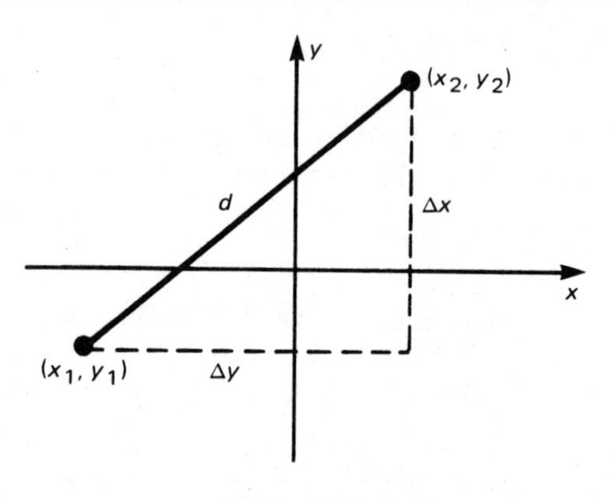

Figure 14-3c.

With this piece of background information, you are now ready to derive a formula for $\cos(A - B)$ in terms of sines and cosines of A and B. Figure 14-3d shows two arcs of measures A and B in standard position on a unit circle. Their terminal points have coordinates $(\cos A, \sin A)$ and $(\cos B, \sin B)$, respectively. The arc between these two terminal points has measure $(A - B)$. Figure 14-3e shows the arc of measure $(A - B)$ moved around the circle into standard position. In this position, the coordinates of its terminal point are $(\cos(A-B), \sin(A-B))$.

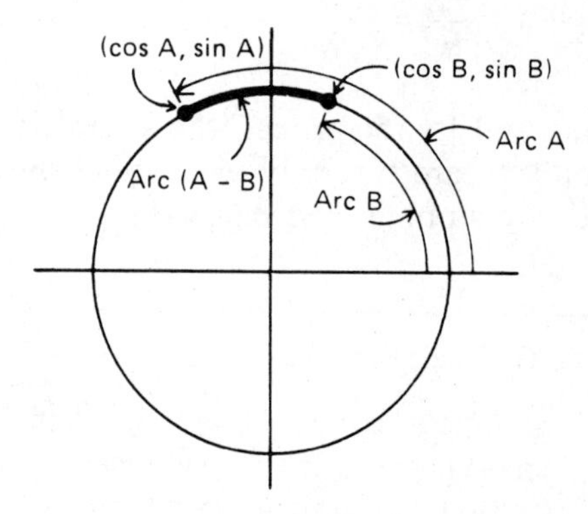

Figure 14-3d.

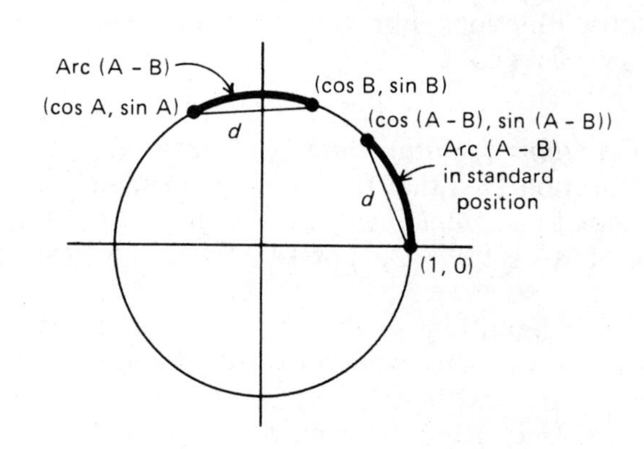

Figure 14-3e.

The length d of the chord for this arc is simply the distance between two points in a Cartesian coordinate system. The Distance Formula can be used to calculate d in two different ways. When the arc $(A - B)$ is in its original position,

$$d^2 = (\cos A - \cos B)^2 + (\sin A - \sin B)^2.$$

Doing the indicated squaring gives

$$d^2 = \cos^2 A - 2 \cos A \cos B + \cos^2 B + \sin^2 A - 2 \sin A \sin B + \sin^2 B.$$

You have learned to think of Pythagorean properties whenever you see squares of functions. Commuting and associating $\cos^2 A + \sin^2 A$ and $\cos^2 B + \sin^2 B$ allows you to use the Pythagorean properties to replace each expression with the number 1. Therefore,

$$d^2 = 2 - 2 \cos A \cos B - 2 \sin A \sin B.$$

When the arc is rotated into standard position, the distance formula can be applied again, giving

$$d^2 = (\cos (A - B) - 1)^2 + (\sin (A - B) - 0)^2.$$

Upon carrying out the indicated squaring, this becomes

$$d^2 = \cos^2 (A - B) - 2 \cos (A - B) + 1 + \sin^2 (A - B).$$

Associating the $\cos^2 (A - B) + \sin^2 (A - B)$ and using the Pythagorean properties gives

$$d^2 = 2 - 2 \cos (A - B).$$

Note that this form of d^2 contains the desired $\cos (A - B)$. The other form of d^2 above contains sines and cosines of A and B. Using the transitive property to equate these expressions gives

$$2 - 2 \cos (A - B) = 2 - 2 \cos A \cos B - 2 \sin A \sin B.$$

Subtracting 2 and then dividing by -2 gives

$$\boxed{\cos (A - B) = \cos A \cos B + \sin A \sin B}\,.$$

The property is called a "composite argument" property, because the argument is composed of measures of two arcs or angles.

4. *Composite Argument Properties for cos (A+B), sin (A – B), sin (A + B):* The way to solve a new problem is to turn it into an old problem. To derive a formula for $\cos (A + B)$, you can first transform the argument into a *difference*, getting

$$\cos (A + B) = \cos (A - (-B)).$$

Using the composite argument property you have just derived, this becomes

$$\cos A \cos (-B) + \sin A \sin (-B).$$

Using the odd-even properties on this, you get

$$\cos A \cos B - \sin A \sin B.$$

$$\therefore \quad \boxed{\cos (A + B) = \cos A \cos B - \sin A \sin B}\,.$$

The expression $\sin (A - B)$ can be transformed to a cosine using the cofunction properties. Considering A and B to be angles,

$$\sin (A - B) = \cos (90° - (A - B)) \quad \text{Cofunction property.}$$

$$= \cos ((90° - A) + B) \quad \text{Associativity.}$$

The last expression is now a *cosine* of a *sum*. Using the above composite argument property gives

$$\sin (A - B) = \cos (90° - A) \cos B - \sin (90° - A) \sin B.$$

Applying the cofunction properties again gives:

$$\boxed{\sin (A - B) = \sin A \cos B - \cos A \sin B}\,.$$

Similar reasoning gives:

$$\boxed{\sin (A + B) = \sin A \cos B + \cos A \sin B}\,.$$

5. *Composite Argument Properties for tan $(A+B)$ and tan $(A-B)$:* The expression tan $(A+B)$ can be written in terms of functions of A and B with the aid of the quotient properties:

$$\tan(A+B)$$

$$= \frac{\sin(A+B)}{\cos(A+B)} \quad \text{Quotient property.}$$

$$= \frac{\sin A \cos B + \cos A \sin B}{\cos A \cos B - \sin A \sin B} \quad \text{Composite argument properties.}$$

It is possible to transform this into terms of tan A and tan B *alone.* Since tangents have cosines for denominators, you seek a way to get cos A and cos B as denominators. A clever way to do this is simply to *factor out* cos A cos B (i.e., "factor out a rabbit!").

$$\tan(A+B)$$

$$= \frac{\cos A \cos B \left(\frac{\sin A \cos B}{\cos A \cos B} + \frac{\cos A \sin B}{\cos A \cos B}\right)}{\cos A \cos B \left(\frac{\cos A \cos B}{\cos A \cos B} + \frac{\sin A \sin B}{\cos A \cos B}\right)} \quad \text{Factoring.}$$

$$= \frac{\frac{\sin A}{\cos A} + \frac{\sin B}{\cos B}}{1 - \frac{\sin A \sin B}{\cos A \cos B}} \quad \text{Canceling.}$$

$$= \frac{\tan A + \tan B}{1 - \tan A \tan B} \quad \text{Quotient properties.}$$

$$\therefore \boxed{\tan(A+B) = \frac{\tan A + \tan B}{1 - \tan A \tan B}} \quad \text{Transitivity.}$$

Writing tan $(A-B)$ as tan $(A+(-B))$ and using the odd-even properties gives:

$$\boxed{\tan(A-B) = \frac{\tan A - \tan B}{1 + \tan A \tan B}}.$$

Note: The composite argument properties are also called the "Addition Formulas."

The following exercise is designed to help you learn the odd-even, cofunction, and composite argument properties by using them.

Exercise 14-3

For Problems 1 through 6, show by substituting 60° for A and 90° for B that:

1. $\cos(A+B) \neq \cos A + \cos B$
2. $\sin(A+B) \neq \sin A + \sin B$
3. $\tan(A-B) \neq \tan A - \tan B$
4. $\cot(A-B) \neq \cot A - \cot B$
5. $\sec(A+B) \neq \sec A + \sec B$
6. $\csc(A-B) \neq \csc A - \csc B$

For Problems 7 through 12, demonstrate that the given property really works by substituting:

a. $A = 60°, \quad B = 30°$
b. $A = 2\pi/3, \quad B = \pi/6$

7. $\cos(A-B) = \cos A \cos B + \sin A \sin B$
8. $\cos(A+B) = \cos A \cos B - \sin A \sin B$
9. $\sin(A-B) = \sin A \cos B - \cos A \sin B$
10. $\sin(A+B) = \sin A \cos B + \cos A \sin B$
11. $\tan(A-B) = \dfrac{\tan A - \tan B}{1 + \tan A \tan B}$
12. $\tan(A+B) = \dfrac{\tan A + \tan B}{1 - \tan A \tan B}$

For Problems 13 through 16, prove that the given equation is an identity.

13. $\cos(x - 90°) = \sin x$
14. $\sin(x - \pi/2) = -\cos x$
15. $\tan(x - \pi/2) = -\cot x$
16. $\sec(x - 90°) = \csc x$

17. Use what you recall about sinusoids and phase displacements from Section 13-6 to do the following:

a. Draw a graph of $y = \cos(x - 90°)$.
b. Draw a graph of $y = \sin x$.

c. Explain from your graphs why the equation in Problem 13 is *reasonable.*

18. Use what you recall about sinusoids and phase displacements from Section 13-6 to do the following:

a. Draw a graph of $y = \sin(x - \pi/2)$.
b. Draw a graph of $y = \cos x$.
c. Explain how your graphs show that the equation in Problem 14 is *reasonable.*

19. Show that giving a sinusoid a phase displacement of 180° has the same effect as turning it over. That is, show that $\cos(x - 180°) = -\cos x$.

20. Show that giving a sinusoid a phase displacement of −180° has the same effect as giving it a phase displacement of +180°. That is, show that $\cos(x + 180°) = \cos(x - 180°)$.

For Problems 21 through 26, assume that A and B are in standard position and that $\sin A = 1/2$, $\cos A > 0$, $\tan B = 3/4$, and $\sin B < 0$. Draw A and B in standard position, then find the following:

21. $\cos(A - B)$
22. $\sin(A - B)$
23. $\sin(A + B)$
24. $\cos(A + B)$
25. $\tan(A - B)$
26. $\tan(A + B)$

The composite argument properties may be used to find *exact* values of functions of 15°. Use a clever choice of A and B, (such as 30°, 45°, 60°, 90°, etc.) and any of the properties you need to find exact values of the following. Express the answers in simple radical form.

27. cos 15°
28. sin 15°
29. tan 15°
30. cot 15°
31. sec 15°
32. csc 15°

Use the cofunction properties and the answers to Problems 27 through 32 to find *exact* values of the following, leaving the answers in simple radical form.

33. sin 75°
34. cos 75°
35. cot 75°
36. tan 75°
37. csc 75°
38. sec 75°

Use a calculator or square-root table and the answers to Problems 27 through 38 to find decimal approximations for the following. Then look in Table III or use a calculator to make sure you are right.

39. cos 15°
40. sin 15°
41. tan 15°
42. tan 75°
43. csc 75°
44. sec 75°

For Problems 45 through 50, prove that the given equation is an identity.

45. $\sin(x + 60°) - \cos(x + 30°) = \sin x$

46. $\sin(x + 30°) + \cos(x + 60°) = \cos x$

47. $\tan(x + \pi/4) + 1 = \sqrt{2}\cos x \sec(x + \pi/4)$

48. $\sqrt{2}\cos(x - \pi/4) = \cos x + \sin x$

49. $(\cos A \cos B - \sin A \sin B)^2 + (\sin A \cos B + \cos A \sin B)^2 = 1$

50. $\sin(3x/7)\cos(4x/7) + \cos(3x/7)\sin(4x/7) = \sin x$

The composite argument properties have sums of *two* angles or arcs. Similar properties for sums of *three* angles can be derived by first associating two of the angles. For Problems 51 and 52, write the given expression in terms of $\sin A$, $\sin B$, $\sin C$, $\cos A$, $\cos B$, and $\cos C$.

51. $\cos(A + B + C)$
52. $\sin(A + B + C)$

Problems 53 through 56 let you discover for yourself some of the properties you will learn in the next section.

The composite argument properties can be used to express functions of *twice* an angle or arc in terms of functions of that angle or arc. For example, $\sin 2A = \sin (A + A)$, which is a function of a composite argument. Use this fact to derive "double argument" properties, expressing Problems 53 through 56 in terms of functions of A.

53. $\sin 2A$
54. $\cos 2A$
55. $\tan 2A$
56. $\cot 2A$

14-4 MULTIPLE ARGUMENT PROPERTIES

The composite argument properties can be used to derive properties expressing functions of *twice* an angle or arc in terms of functions of that angle or arc.

Objective:

Be able to express $\sin 2A$, $\cos 2A$, and $\tan 2A$ in terms of $\sin A$, $\cos A$, and $\tan A$.

1. *Double Argument Property for sin 2A:* Recognizing that $\sin 2A = \sin (A + A)$, you can write:

$\sin 2A = \sin (A + A)$

$= \sin A \cos A + \cos A \sin A$ — Composite argument properties.

$= 2 \sin A \cos A$ — Adding "like terms."

$\therefore \boxed{\sin 2A = 2 \sin A \cos A}$ — Transitivity.

2. *Double Argument Property for cos 2A:* Using the above reasoning:

$\cos 2A = \cos (A + A)$

$= \cos A \cos A - \sin A \sin A$ — Composite argument properties.

$= \cos^2 A - \sin^2 A$ — Definition of exponentiation.

$\therefore \boxed{\cos 2A = \cos^2 A - \sin^2 A}$ — Transitivity.

Note that this property looks a lot like the Pythagorean property $\cos^2 A + \sin^2 A = 1$. In fact, the Pythagorean property can be used to transform the double argument property to two other forms.

$\cos 2A = \cos^2 A - \sin^2 A$ — Double argument property.

$= (1 - \sin^2 A) - \sin^2 A$ — Pythagorean property.

$= 1 - 2 \sin^2 A$ — Associativity.

$\therefore \boxed{\cos 2A = 1 - 2 \sin^2 A}$ — Transitivity.

In this form, $\cos 2A$ is expressed in terms of $\sin A$ *alone*. Cos $2A$ can also be expressed in terms of $\cos A$ alone.

$\cos 2A = \cos^2 A - \sin^2 A$ — Double argument property.

$= \cos^2 A - (1 - \cos^2 A)$ — Pythagorean property.

$= 2 \cos^2 A - 1$ — Commutativity and associativity.

$\therefore \boxed{\cos 2A = 2 \cos^2 A - 1}$ — Transitivity.

3. *Double Argument Property for tan 2A:* A double argument property for $\tan 2A$ may be derived the same way as for $\sin 2A$ and $\cos 2A$.

$\tan 2A = \tan (A + A)$

$= \dfrac{\tan A + \tan A}{1 - \tan A \tan A}$ — Composite argument properties.

$= \dfrac{2 \tan A}{1 - \tan^2 A}$ — Associativity and definition of exponentiation.

$\therefore \boxed{\tan 2A = \dfrac{2 \tan A}{1 - \tan^2 A}}$ — Transitivity.

The exercise that follows is intended to give you enough practice using these properties so that you will *learn* them. You will also use them to derive properties for *higher* multiples of angles. In the last problems you will derive properties of *inverse* circular functions.

Exercise 14-4

1a. Recalling what you have learned about graphing sinusoids in Section 13-6, draw a graph of $y = \cos 2x$.

b. Draw another graph of $y = 2 \cos x$ using the *same* scales as in part a.

c. Based on your graphs, explain why $\cos 2x$ is *not* equal to $2 \cos x$.

2. Repeat Problem 1 using $y = \sin 2x$ and $y = 2 \sin x$.

For Problems 3 through 8, show that the double argument property really works by substituting the given measures of angles or arcs into the formula and showing that you get the right answer.

3. $\sin 2A,\ A = 30°$
4. $\cos 2A,\ A = 30°$
5. $\cos 2A,\ A = \pi/4$
6. $\sin 2A,\ A = \pi/4$
7. $\tan 2A,\ A = 60°$
8. $\tan 2A,\ A = \pi/4$

For Problems 9 through 12, calculate $\sin 2A$, $\cos 2A$, and $\tan 2A$ for the angle or arc described.

9. $\sin A = 3/5$, A terminates in Quadrant I
10. $\cos A = -3/5$, A terminates in Quadrant II
11. $\tan A = -3/4$, A terminates in Quadrant IV
12. $\tan A = 4/3$, A terminates in Quadrant III

For Problems 13 through 22, prove that the given equation is an identity.

13. $\sin 2x = \dfrac{2 \tan x}{1 + \tan^2 x}$
14. $\sec 2x = \dfrac{\sec^2 x}{2 - \sec^2 x}$
15. $\cos 2\phi = \dfrac{1 - \tan^2 \phi}{1 + \tan^2 \phi}$
16. $\sin 2\phi = 2 \cot \phi \sin^2 \phi$
17. $\dfrac{\cos 2D}{\cos D - \sin D} = \cos D + \sin D$
18. $(1 + \tan x) \tan 2x = \dfrac{2 \tan x}{1 - \tan x}$
19. $\tan r = \dfrac{1 - \cos 2r}{\sin 2r}$
20. $\tan y = \dfrac{\sin 2y}{1 + \cos 2y}$
21. $\sin^2 \theta = \frac{1}{2}(1 - \cos 2\theta)$
22. $\cos^2 \theta = \frac{1}{2}(1 + \cos 2\theta)$

The composite argument properties can be combined with the double argument properties to derive triple, quadruple, etc., argument properties. For example, $\sin 3x = \sin (2x + x)$, which is a sine of a composite argument. For Problems 23 through 26, derive multiple argument properties for the given functions.

23. $\sin 3x$ in terms of $\sin x$ alone.
24. $\cos 3x$ in terms of $\cos x$ alone.
25. $\cos 4x$ in terms of $\cos x$ alone.
26. $\sin 4x$ in terms of $\sin x$ and $\cos x$.

The double argument properties work whenever one of the arguments is twice as large as the other. For Problems 27 through 34, write equations expressing:

27. $\tan 14x$ in terms of $\tan 7x$.
28. $\cot 14x$ in terms of $\cot 7x$.
29. $\cos 6x$ in terms of $\sin 3x$ and $\cos 3x$.
30. $\sin 6x$ in terms of $\sin 3x$ and $\cos 3x$.
31. $\cos 10x$ in terms of $\cos 5x$ *alone.*
32. $\cos 10x$ in terms of $\sin 5x$ *alone.*
33. $\cos x$ in terms of $\cos \frac{1}{2}x$ *alone.*
34. $\cos x$ in terms of $\sin \frac{1}{2}x$ *alone.*

Problems 35 and 36 allow you to discover something about the next section.

35. Use the answer to Problem 33 to write an equation expressing cos ½x in terms of cos x.

36. Use the answer to Problem 34 to write an equation expressing sin ½x in terms of cos x.

In Problems 37 through 40 you will learn some properties of *inverse* circular functions. For example, there is a *cofunction* property,

$$\text{Arccos } x = \pi/2 - \text{Arcsin } x.$$

This equation states, "The arc whose cosine is x is the *complement* of the arc whose sine is x." The geometrical meaning can be seen in Figure 14-4. A proof is given below.

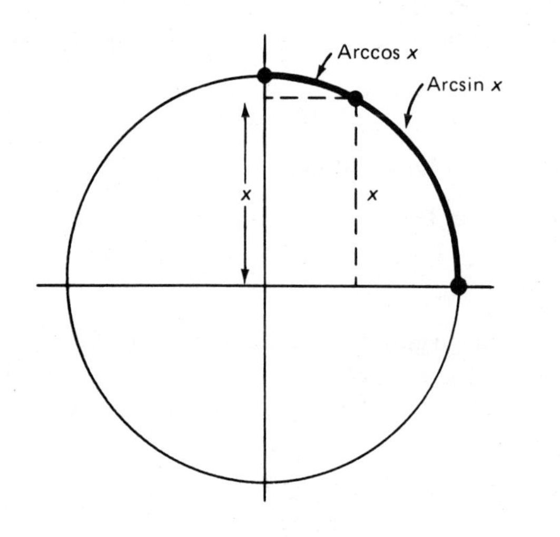

Figure 14-4.

Let y = Arcsin x

$\therefore x = \sin y$, and
$-\pi/2 \leq y \leq \pi/2$ — Definition of Arcsin.

$\therefore x = \cos(\pi/2 - y)$ — Cofunction property.

$\therefore \text{arccos } x = \pi/2 - y$ — arccos of both members.

The arccos *relation* can be replaced with the Arccos *function* if $(\pi/2 - y)$ is always in the range of Arccos. Operating with the inequality $-\pi/2 \leq y \leq \pi/2$, you get

$\pi/2 \geq -y \geq -\pi/2$ — Multiplying by -1.

$\pi \geq (\pi/2 - y) \geq 0$ — Adding $\pi/2$.

Since the interval from 0 to π *is* the range of Arccos,

$\text{Arccos } x = \pi/2 - y$ — Definition of Arccos.

$\therefore \text{Arccos } x = \pi/2 - \text{Arcsin } y$ — Substitution.

For Problems 37 through 40, use similar reasoning to derive the indicated property.

37. Prove the *cofunction* property $\text{Csc}^{-1}\ x = \pi/2 - \text{Sec}^{-1}\ x$.

38. Prove the *cofunction* property $\text{Cot}^{-1}\ x = \pi/2 - \text{Tan}^{-1}\ x$.

39. Prove the *reciprocal* property Arctan x = Arccot $1/x$, and tell for which values of x the property is true.

40. Prove the *reciprocal* property Arcsec x = Arccos $1/x$, and tell for which values of x the property is true.

For Problems 41 through 54, use the appropriate properties to find the *exact* value of the expression.

41. cos (2 Arctan 4/3)
42. cos (2 Arcsin 1/3)
43. sin (2 Arcsec 3)
44. tan (2 Arctan 1/2)
45. tan (2 Arctan 5)
46. sin (2 Arccot 2/3)
47. $\sin(\text{Tan}^{-1}\ 1/2 + \text{Tan}^{-1}\ 1/3)$
48. $\cos(\text{Sec}^{-1}\ 3/2 - \text{Cos}^{-1}\ 1/5)$
49. sin ($\pi/2$ − Arccos 11/13)
50. csc (90° − Arcsec 19)
51. cos (Arcsec 1000)
52. $\csc(\pi/2 - \text{Tan}^{-1}\ (-3))$
53. $\tan(\text{Cot}^{-1}\ 4)$
54. $\sin(\text{Csc}^{-1}\ (-5))$

14-5 HALF-ARGUMENT PROPERTIES

Objective:

Be able to express $\cos \frac{1}{2}x$, $\sin \frac{1}{2}x$, and $\tan \frac{1}{2}x$ in terms of functions of x.

The cosine double argument property in the forms

$$\cos 2A = 2\cos^2 A - 1$$

and

$$\cos 2A = 1 - 2\sin^2 A$$

is useful for deriving other properties. For example, starting with the first form, you can isolate $\cos^2 A$:

$\cos 2A = 2\cos^2 A - 1$	Double argument property.
$1 + \cos 2A = 2\cos^2 A$	Adding 1 to both members.
$\frac{1}{2}(1 + \cos 2A) = \cos^2 A$	Dividing by 2.
$\therefore \boxed{\cos^2 A = \frac{1}{2}(1 + \cos 2A)}$	Symmetry.

This equation expresses a *power* of a function in terms of a function of a *multiple* angle or arc. Solving $\cos 2A = 1 - 2\sin^2 A$ for $\sin^2 A$ gives:

$$\boxed{\sin^2 A = \frac{1}{2}(1 - \cos 2A)}\ .$$

These two properties are also interesting because the argument A on the left is *half* the argument $2A$ on the right. Letting $A = \frac{1}{2}x$ and substituting gives

$$\cos^2 \tfrac{1}{2}x = \tfrac{1}{2}(1 + \cos x)$$

and

$$\sin^2 \tfrac{1}{2}x = \tfrac{1}{2}(1 - \cos x).$$

Taking the square root of both members of each equation gives:

$$\boxed{\begin{aligned} \cos \tfrac{1}{2}x &= \pm\sqrt{\tfrac{1}{2}(1 + \cos x)} \\ \sin \tfrac{1}{2}x &= \pm\sqrt{\tfrac{1}{2}(1 - \cos x)} \end{aligned}}\ .$$

These are called *half-argument* properties for sine and cosine, because they express $\cos \frac{1}{2}x$ and $\sin \frac{1}{2}x$ in terms of $\cos x$. The ambiguous sign $\pm$ is determined by the quadrant in which $\frac{1}{2}x$ terminates (*not* by where x terminates!). For example, if $x = 120°$ then $\frac{1}{2}x = 60°$, which terminates in Quadrant I. So,

$$\sin \tfrac{1}{2}x = +\sqrt{\tfrac{1}{2}(1 - \cos x)}.$$

But if $x = 480°$, which is coterminal with $120°$ as shown in Figure 14-5, then $\frac{1}{2}x = 240°$. Since $240°$ terminates in Quadrant III,

$$\sin \tfrac{1}{2}x = -\sqrt{\tfrac{1}{2}(1 - \cos x)}.$$

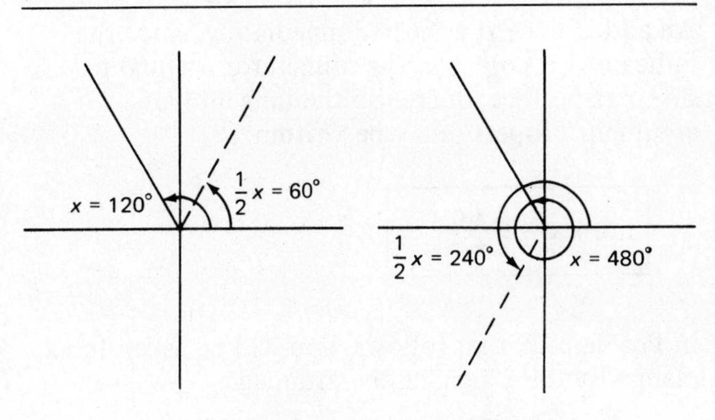

Figure 14-5.

The half-argument property for the tangent is obtained from the sine and cosine half-argument properties by using the quotient property.

$\tan \frac{1}{2}x = \dfrac{\sin \frac{1}{2}x}{\cos \frac{1}{2}x}$	Quotient property.
$= \dfrac{\pm\sqrt{\frac{1}{2}(1 - \cos x)}}{\pm\sqrt{\frac{1}{2}(1 + \cos x)}}$	Half-argument properties.
$= \pm\sqrt{\dfrac{1 - \cos x}{1 + \cos x}}$	Properties of radicals and fractions.
$\therefore \boxed{\tan \frac{1}{2}x = \pm\sqrt{\dfrac{1 - \cos x}{1 + \cos x}}}$	Transitivity.

This tangent half-argument property can be simplified by rationalizing the denominator.

$$\tan \tfrac{1}{2}x = \pm\sqrt{\frac{1-\cos x}{1+\cos x}\cdot\frac{1+\cos x}{1+\cos x}}$$ Multiplication property of 1.

$$= \pm\sqrt{\frac{1-\cos^2 x}{(1+\cos x)^2}}$$ Multiplication property of fractions.

$$= \pm\frac{\sqrt{1-\cos^2 x}}{|1+\cos x|}$$ $\sqrt{n^2} = |n|$.

$$= \pm\frac{\sqrt{1-\cos^2 x}}{1+\cos x}$$ $1+\cos x \geq 0$ for all x.

An added benefit appears immediately, since the radicand $1-\cos^2 x$ in the numerator is equal to $\sin^2 x$, a perfect square. So the tangent half-argument property may be written:

$$\boxed{\tan \tfrac{1}{2}x = \frac{\sin x}{1+\cos x}}.$$

In Problem 29 that follows, you will be asked to explain why the ± sign can be dropped.

A still more interesting form of the property can be obtained by rationalizing the *numerator* instead of the denominator.

$$\boxed{\tan \tfrac{1}{2}x = \frac{1-\cos x}{\sin x}}.$$

Since the denominator in this last form has only *one* term, it is useful for proving identities that involve fractions. In Problem 30 of the following exercise, you will be asked to prove this form of the tangent half-argument property.

The exercise that follows is designed to give you enough practice using the half-argument properties so that you will *learn* them.

Exercise 14-5

1. Use what you recall about graphing sinusoids from Section 13-6 to:

a. Draw a graph of $y = \cos \frac{1}{2}x$.
b. Draw a graph of $y = \frac{1}{2}\cos x$.
c. Explain from your graphs why $\cos \frac{1}{2}x \neq \frac{1}{2}\cos x$.

2. Repeat Problem 1 using $y = \sin \frac{1}{2}x$ and $y = \frac{1}{2}\sin x$.

3a. Draw a graph of $y = \cos x$.

b. Using this as an auxiliary graph, *square* ordinates to get a graph of $y = \cos^2 x$.
c. The graph of $y = \cos^2 x$ is a *sinusoid*. From your sketch in part b, figure out its period, amplitude, and axis location. Then write an equation of the sinusoid in the form $y = C + A\cos Bx$.
d. Use the properties of this section to demonstrate that your equation is correct.

4. Repeat Problem 3 using $y = \sin^2 x$. You must be clever enough to get an equation where the phase displacement is 0.

For Problems 5 through 14, verify that the half-argument properties actually work by substituting the given measure of the angle or arc into the formula and showing that you get the right answer.

5. $\cos \frac{1}{2}x$, $x = 60°$
6. $\sin \frac{1}{2}x$, $x = 60°$
7. $\sin \frac{1}{2}x$, $x = \pi/2$
8. $\cos \frac{1}{2}x$, $x = \pi/2$
9. $\tan \frac{1}{2}x$, $x = 2\pi/3$
10. $\tan \frac{1}{2}x$, $x = \pi$
11. $\cos \frac{1}{2}x$, $x = 420°$
12. $\sin \frac{1}{2}x$, $x = 420°$
13. $\sin \frac{1}{2}x$, $x = -60°$
14. $\cos \frac{1}{2}x$, $x = -60°$

For Problems 15 through 20, calculate $\sin \frac{1}{2}x$, $\cos \frac{1}{2}x$, and $\tan \frac{1}{2}x$ for the angle described.

15. $\cos x = 3/5$, $0° < x < 90°$
16. $\cos x = 3/5$, $270° < x < 360°$
17. $\cos x = -3/5$, $180° < x < 270°$

18. $\cos x = -3/5, \quad 90° < x < 180°$

19. $\cos x = 3/5, \quad 630° < x < 720°$

20. $\cos x = -3/5, \quad 450° < x < 540°$

For Problems 21 through 28, prove that the given equation is an identity.

21. $\tan \frac{1}{2}x + \cot \frac{1}{2}x = 2 \csc x$

22. $\tan x \tan \frac{1}{2}x = \sec x - 1$

23. $\dfrac{2 \tan \frac{1}{2}x}{1 + \tan^2 \frac{1}{2}x} = \sin x$

24. $\tan \frac{1}{2}x\,(2 \cot x + \tan \frac{1}{2}x) = 1$

25. $\dfrac{\cos \frac{1}{2}\theta - \sin \frac{1}{2}\theta}{\cos \frac{1}{2}\theta + \sin \frac{1}{2}\theta} = \dfrac{\cos \theta}{1 + \sin \theta}$

26. $\dfrac{\cos \frac{1}{2}\phi + \sin \frac{1}{2}\phi}{\cos \frac{1}{2}\phi - \sin \frac{1}{2}\phi} = \sec \phi + \tan \phi$

27. $\tan \frac{1}{2}A = \csc A - \cot A$

28. $\tan \left(\dfrac{\pi}{4} + \dfrac{x}{2}\right) = \sec x + \tan x$

29. The property

$$\tan \tfrac{1}{2}x = \frac{\sin x}{1 + \cos x} \quad \text{comes from} \quad \frac{\pm\sqrt{1 - \cos^2 x}}{1 + \cos x}.$$

By considering the quadrants in which x and $\frac{1}{2}x$ may terminate, explain why the ambiguous sign $\pm$ disappears in this property.

30. Prove that

$$\tan \tfrac{1}{2}x = \pm\sqrt{\frac{1 - \cos x}{1 + \cos x}}$$

can be transformed to

$$\tan \tfrac{1}{2}x = \frac{1 - \cos x}{\sin x},$$

and explain what happens to the $\pm$ sign.

Problems 31 and 32 allow you to discover something about the next section.

31. The right-hand members of the composite argument properties for $\sin(A + B)$ and $\sin(A - B)$ are *conjugates* of each other. When you add or subtract conjugates, one of the two terms drops out, leaving twice the other term.

a. Write an equation expressing $\sin(A + B) + \sin(A - B)$ in terms of functions of A and B.
b. Write an equation expressing $\sin(A + B) - \sin(A - B)$ in terms of functions of A and B.

32. Repeat Problem 31 for $\cos(A + B)$ and $\cos(A - B)$.

Problem 33 is meant to challenge the most diligent students!

33. *Sin 18°, etc.:* You have learned how to find exact values of functions of multiples of 15°. It is possible to find exact values for functions of certain other angles, too. In this problem, you will combine trigonometric properties with algebraic techniques and a little ingenuity to find the exact value of sin 18°.

a. Use the double argument property for sine to write an equation expressing sin 72° in terms of sin 36° and cos 36°.
b. Transform the equation in part a so that sin 72° is expressed in terms of sin 18° and cos 18°. You should find that the *sine* form of the double argument property for cos 36° is best.
c. You recall by the cofunction property that sin 72° = cos 18°. Replace sin 72° in your equation from part b with cos 18°. Then simplify the resulting equation. If you have done everything correctly, the cos 18° should disappear from the equation, leaving a *cubic* (third degree) equation in sin 18°.
d. Solve the equation in part c for sin 18°. It may help to let $x = \sin 18°$ and solve for x. If you transform the equation so that the right member is 0, you should find that $(2x - 1)$ is a factor of the left member. The other factor may be found by long division. To solve the equation, recall the multiplication property of zero and the Quadratic Formula.

e. You should get *three* solutions for the equation in part d. Only *one* of these could possibly be sin 18°. *Which* one?

f. A pattern shows up for some exact values of $\sin\theta$:

$$\sin 15° = \frac{\sqrt{6}-\sqrt{2}}{4}$$

$$\sin 18° = \frac{\sqrt{5}-1}{4} = \frac{\sqrt{5}-\sqrt{1}}{4}$$

$$\sin 30° = \frac{1}{2} = \frac{2}{4} = \frac{\sqrt{4}-\sqrt{0}}{4}$$

See if you can figure out what the pattern is!!

14-6 SUM AND PRODUCT PROPERTIES

The composite argument properties for sine are

$$\sin(A+B) = \sin A \cos B + \cos A \sin B,$$

$$\sin(A-B) = \sin A \cos B - \cos A \sin B.$$

By *adding* respective members of the two equations, you get:

$$\boxed{\sin(A+B) + \sin(A-B) = 2\sin A \cos B}\ .$$

This property is of interest because the left member is a *sum* of two sines, and the right member is a *product* of a sine and a cosine.

Objectives:

1. Be able to transform a *sum* (or difference) of two sines or two cosines into a *product* of sines and cosines.

2. Be able to transform a *product* of two sines, two cosines, or a sine and cosine into a *sum* (or difference) of two sines or cosines.

By subtracting respective members of the composite argument properties for sine, above, you get:

$$\boxed{\sin(A+B) - \sin(A-B) = 2\cos A \sin B}\ .$$

The other two "sum and product" properties needed to accomplish the objectives come from adding and subtracting the composite argument properties for the cosine:

$$\cos(A+B) = \cos A \cos B - \sin A \sin B$$

$$\cos(A-B) = \cos A \cos B + \sin A \sin B$$

$$\boxed{\cos(A+B) + \cos(A-B) = 2\cos A \cos B}\ .$$

By subtracting instead of adding:

$$\boxed{\cos(A+B) - \cos(A-B) = -2\sin A \sin B}\ .$$

Example 1: Express 2 sin 13° cos 48° as a sum.

Using the first form of the sum and product properties,

$2\sin 13° \cos 48°$

$= \sin(13° + 48°) + \sin(13° - 48°)$	Sum and product properties.
$= \sin 61° + \sin(-35°)$	Arithmetic.
$= \underline{\underline{\sin 61° - \sin 35°}}$	Sine is an *odd* function.

If you want sin 13° cos 48° instead of 2 sin 13° cos 48°, you can simply divide both members by 2, getting

$$\underline{\underline{\sin 13° \cos 48° = \tfrac{1}{2}\sin 61° - \tfrac{1}{2}\sin 35°}}\ .$$

Example 2: Express cos 47° + cos 59° as a *product*.

The sum of two cosines property is

$$\cos(A+B) + \cos(A-B) = 2\cos A \cos B.$$

So you let the arguments $A + B = 47°$ and $A - B = 59°$, and solve the resulting *system* of equations for A and B.

$A + B = 47°,$

$A - B = 59°.$

Adding the respective members of these equations gives

$2A = 47° + 59° = 106°$

$A = \frac{1}{2}(106°) = 53°,$

and subtracting the respective members gives

$2B = 47° - 59° = -12°$

$B = \frac{1}{2}(-12°) = -6°.$

$\therefore \cos 47° + \cos 59°$

$= 2 \cos 53° \cos(-6°)$ Substitution.

$= \underline{\underline{2 \cos 53° \cos 6°}}$ Cosine is an *even* function.

The procedure of Example 2 can be done in *general*, letting

$A + B = x,$

$A - B = y.$

The solutions are $A = \frac{1}{2}(x + y)$ and $B = \frac{1}{2}(x - y)$. Substituting these values into the sum and product properties gives alternate forms in which a *sum* is expressed as a *product*.

$\sin x + \sin y = 2 \sin \frac{1}{2}(x + y) \cos \frac{1}{2}(x - y)$
$\sin x - \sin y = 2 \cos \frac{1}{2}(x + y) \sin \frac{1}{2}(x - y)$
$\cos x + \cos y = 2 \cos \frac{1}{2}(x + y) \cos \frac{1}{2}(x - y)$
$\cos x - \cos y = -2 \sin \frac{1}{2}(x + y) \sin \frac{1}{2}(x - y)$

These properties are difficult to remember correctly, since they are all so similar to one another. Therefore, it is usually more reliable for you to remember how you derived them from the composite argument properties (which you *should* remember by now!). Then you can *derive* the sum and product properties whenever you need them.

In the following exercise you will use these properties to transform sums to products, and vice versa.

Exercise 14-6

For Problems 1 through 12, transform the indicated product to a *sum* (or difference) of sines or cosines of *positive* arguments.

1. $2 \sin 41° \cos 24°$
2. $2 \cos 73° \sin 62°$
3. $2 \cos 53° \cos 49°$
4. $2 \sin 29° \sin 16°$
5. $2 \cos 3.8 \sin 4.1$
6. $2 \cos 2 \cos 3$
7. $2 \sin 4.6 \sin 7.2$
8. $2 \sin 1.8 \cos 6.2$
9. $2 \sin 3x \cos 5x$
10. $2 \sin 8x \sin 2x$
11. $2 \cos 4x \cos 7x$
12. $2 \cos 11x \sin 9x$

For Problems 13 through 24, transform the indicated sum (or difference) into a *product* of sines and cosines of *positive* arguments.

13. $\cos 46° + \cos 12°$
14. $\cos 56° - \cos 24°$
15. $\sin 54° + \sin 22°$
16. $\sin 29° - \sin 15°$
17. $\cos 2.4 - \cos 4.4$
18. $\sin 1.8 + \sin 6.4$
19. $\sin 2 - \sin 6$
20. $\cos 3.2 + \cos 4.8$
21. $\sin 3x + \sin 9x$
22. $\sin 9x - \sin 11x$
23. $\cos 8x - \cos 10x$
24. $\cos 5x + \cos 13x$

For Problems 25 through 36, prove that the given equation is an identity.

25. $\cos x - \cos 5x = 4 \sin 3x \sin x \cos x$
26. $\sin 5x + \sin 3x = 4 \sin 2x \cos 2x \cos x$
27. $\dfrac{\sin 3x + \sin x}{\sin 3x - \sin x} = \dfrac{2 \cos^2 x}{\cos 2x}$
28. $\dfrac{\sin 5x + \sin 7x}{\cos 5x + \cos 7x} = \tan 6x$
29. $\sin x + \sin 2x + \sin 3x = \sin 2x(1 + 2 \cos x)$
30. $\cos x + \cos 2x + \cos 3x = \cos 2x(1 + 2 \cos x)$

31. $1 + \cos x = \frac{1}{2} + \frac{\sin \frac{3}{2}x}{2 \sin \frac{1}{2}x}$

32. $1 + \cos x + \cos 2x = \frac{1}{2} + \frac{\sin \frac{5}{2}x}{2 \sin \frac{1}{2}x}$

33. $\sin (x + y) \sin (x - y) = \sin^2 x - \sin^2 y$

34. $\sin (x + y) \sin (x - y) = \cos^2 y - \cos^2 x$

35. $\cos (x + y) \cos (x - y) = \cos^2 x - \sin^2 y$

36. $\sin (x + y) \cos (x - y) = ½ \sin 2x + ½ \sin 2y$

14-7 LINEAR COMBINATION OF COSINE AND SINE WITH EQUAL ARGUMENTS

If r and s are variables, then an expression such as $3r - 5s$ is called a *linear combination* of r and s. In this section you will study expressions such as $7 \cos x + 2 \sin x$, which are linear combinations of a cosine and a sine with equal arguments. These expressions differ from those of the previous section because the sum involves both a sine and a cosine. The sum and product properties have sums of two cosines or two sines.

Objective:

Be able to express $A \cos x + B \sin x$ in the form $C \cos (x - D)$, where A, B, C, and D stand for constants.

Starting with the expression $C \cos (x - D)$ and using the composite argument properties, you can write

$$C \cos (x - D) = C(\cos x \cos D + \sin x \sin D).$$

Upon distributing, commuting, and associating, this equation becomes

$$C \cos (x - D) = (C \cos D) \cos x + (C \sin D) \sin x.$$

Since D is a constant, $\cos D$ and $\sin D$ are also constants. The objective may be accomplished by making

$$A = C \cos D,$$
$$B = C \sin D.$$

By a clever combination of algebra and trigonometry, C and D can be expressed in terms of A and B. Squaring and then adding gives

$$A^2 = C^2 \cos^2 D$$
$$B^2 = C^2 \sin^2 D$$

$$\begin{aligned} A^2 + B^2 &= C^2 \cos^2 D + C^2 \sin^2 D \\ &= C^2 (\cos^2 D + \sin^2 D) \\ &= C^2 \end{aligned}$$

$$\therefore C = \boxed{\sqrt{A^2 + B^2}}.$$

The positive square root is used simply as a matter of convenience. Once C is known, D is the argument that satisfies:

$$\boxed{\cos D = \frac{A}{C} \quad \text{and} \quad \sin D = \frac{B}{C}}.$$

Example: Transform $3 \cos 2x - 4 \sin 2x$ to the form $C \cos (2x - D)$.

First, observe that the sine and cosine have the *same* argument, $2x$. Also, $A = 3$ and $B = -4$. Therefore,

$$C = \sqrt{3^2 + (-4)^2} = 5.$$

Consequently,

$$\cos D = \frac{3}{5} \quad \text{and} \quad \sin D = -\frac{4}{5}$$

From Table III, the reference angle is approximately 53°. Since $\cos D$ is positive and $\sin D$ is negative, D must terminate in Quadrant IV. Therefore

$D \approx 360° - 53° = 307°$. In summary,

$$3 \cos 2x - 4 \sin 2x = 5 \cos (2x - 307°).$$

Note that the two terms on the left have graphs that are sinusoids with the same period (180°) but different amplitudes (3 and 4). The expression on the right is another sinusoid with the *same* period (180°), but a *different* amplitude (5) and phase angle (307°). It is this property that explains why two sound waves of the same pitch will add together to form another sound wave of the same pitch. Only the amplitude and phase are changed. Note also that the amplitudes do not simply add together. This explains why a choir of 100 members all singing the same note is not 100 times as loud as only one member singing the note.

In the following exercise you will get practice expressing linear combinations of sine and cosine in terms of a single cosine.

Exercise 14-7

For Problems 1 through 10, transform the given expression to the form $C \cos (x - D)$, assuming that the functions are:

a. trigonometric,
b. circular.

1. $\cos x + \sqrt{3} \sin x$
2. $\sqrt{3} \cos x + \sin x$
3. $5 \cos x - 5 \sin x$
4. $\sqrt{2} \cos x - \sqrt{2} \sin x$
5. $5 \sin x - 12 \cos x$
6. $4 \sin x - 3 \cos x$
7. $-15 \cos 3x - 8 \sin 3x$
8. $-12 \cos 7x - 5 \sin 7x$
9. $(\sqrt{6} + \sqrt{2}) \cos x + (\sqrt{6} - \sqrt{2}) \sin x$
10. $0.6561 \cos x + 0.7547 \sin x$

For Problems 11 through 14, sketch at least one cycle of the graph of the given equation.

11. $y = 5\sqrt{3} \cos 2x - 5 \sin 2x$
12. $y = 6 \cos 3x + 6 \sin 3x$
13. $y = 4 \cos \pi x + 4 \sin \pi x$
14. $y = -\cos \frac{\pi}{6} x + \sqrt{3} \sin \frac{\pi}{6} x$

For Problems 15 through 24, transform the expression to the form $C \cos (x - D)$, where C and D are constants. Express D exactly as an *Arctangent* function. If D terminates in Quadrant II or III, you should use the *opposite* of the appropriate Arctangent.

15. $3 \cos x + 4 \sin x$
16. $5 \cos x + 12 \sin x$
17. $5 \cos x + 6 \sin x$
18. $3 \cos x + 8 \sin x$
19. $0.6 \cos x - 0.8 \sin x$
20. $0.8 \cos x - 0.6 \sin x$
21. $-4 \cos x - 2 \sin x$
22. $-8 \cos x - 15 \sin x$
23. $-2 \cos x + 7 \sin x$
24. $-\cos x + 4 \sin x$

14-8 SIMPLIFICATION OF TRIGONOMETRIC EXPRESSIONS

The ultimate objective of this chapter is for you to be able to solve trigonometric equations. Doing this requires that you be able to transform a trigonometric expression to an equivalent, specified form, possibly simplifying it. Unfortunately, merely stating "simplify" a given trigonometric expression is not enough. As with any mathematical expression, "simple" means "simpler to use in subsequent work." For example, if $\sin 3x + \sin 5x$ appears in an equation, the equation may be easier to solve if the expression is transformed to a product. But in more advanced mathematics such as calculus, it may be simpler to use if left as a sum. It is for this reason that the desired form of the answer will always be specified.

In this section you will be given the desired algebraic form (sum, product, etc.) or trigonometric form (sines, double arguments, etc.), but not the answer itself. The transformations will give you skills you must have for solving trigonometric equations in the next section. There, *you* must decide which form is most desirable.

Objective:

Given a trigonometric expression, transform it to a specified algebraic or trigonometric form.

For your convenience, the properties you have learned so far are summarized below. You should, of course, try to work the problems *without* reference to the table.

Summary of Properties of Trigonometric Functions

1. *Reciprocal*

$\cot x = 1/\tan x$ or $\tan x \cot x = 1$

$\sec x = 1/\cos x$ or $\cos x \sec x = 1$

$\csc x = 1/\sin x$ or $\sin x \csc x = 1$

2. *Quotient*

$\tan x = \sin x/\cos x = \sec x/\csc x$

$\cot x = \cos x/\sin x = \csc x/\sec x$

3. *Pythagorean*

$\cos^2 x + \sin^2 x = 1$

$1 + \tan^2 x = \sec^2 x$

$\cot^2 x + 1 = \csc^2 x$

4. *Odd-Even*

$\sin(-x) = -\sin x$ (odd) $\cot(-x) = -\cot x$ (odd)

$\cos(-x) = \cos x$ (even) $\sec(-x) = \sec x$ (even)

$\tan(-x) = -\tan x$ (odd) $\csc(-x) = -\csc x$ (odd)

5. *Cofunction*

$\cos(90° - x) = \sin x$ or $\cos(\pi/2 - x) = \sin x$

$\cot(90° - x) = \tan x$ or $\cot(\pi/2 - x) = \tan x$

$\csc(90° - x) = \sec x$ or $\csc(\pi/2 - x) = \sec x$

6. *Composite Argument*

$\cos(A - B) = \cos A \cos B + \sin A \sin B$

$\cos(A + B) = \cos A \cos B - \sin A \sin B$

$\sin(A - B) = \sin A \cos B - \cos A \sin B$

$\sin(A + B) = \sin A \cos B + \cos A \sin B$

$\tan(A - B) = \dfrac{\tan A - \tan B}{1 + \tan A \tan B}$

$\tan(A + B) = \dfrac{\tan A + \tan B}{1 - \tan A \tan B}$

7. *Double Argument*

$\sin 2x = 2 \sin x \cos x$

$\cos 2x = \cos^2 x - \sin^2 x = 1 - 2\sin^2 x = 2\cos^2 x - 1$

$\tan 2x = \dfrac{2 \tan x}{1 - \tan^2 x}$

8. *Half Argument*

$\sin \tfrac{1}{2}x = \pm\sqrt{\tfrac{1}{2}(1 - \cos x)}$

$\cos \tfrac{1}{2}x = \pm\sqrt{\tfrac{1}{2}(1 + \cos x)}$

$\tan \tfrac{1}{2}x = \pm\sqrt{\dfrac{1 - \cos x}{1 + \cos x}} = \dfrac{\sin x}{1 + \cos x} = \dfrac{1 - \cos x}{\sin x}$

9. *Sum and Product*

$2 \cos A \cos B = \cos(A + B) + \cos(A - B)$

$2 \sin A \sin B = -\cos(A + B) + \cos(A - B)$

$2 \sin A \cos B = \sin(A + B) + \sin(A - B)$

$2 \cos A \sin B = \sin(A + B) - \sin(A - B)$

$\cos x + \cos y = 2 \cos \tfrac{1}{2}(x + y) \cos \tfrac{1}{2}(x - y)$

$\cos x - \cos y = -2 \sin \tfrac{1}{2}(x + y) \sin \tfrac{1}{2}(x - y)$

$\sin x + \sin y = 2 \sin \tfrac{1}{2}(x + y) \cos \tfrac{1}{2}(x - y)$

$\sin x - \sin y = 2 \cos \tfrac{1}{2}(x + y) \sin \tfrac{1}{2}(x - y)$

10. *Linear Combination of Sine and Cosine*

$A \cos x + B \sin x = C \cos (x - D)$,

where

$C = \sqrt{A^2 + B^2}$, and $\cos D = \dfrac{A}{C}$ and $\sin D = \dfrac{B}{C}$.

Exercise 14-8

For Problems 1 through 26, transform the expression on the left of the comma so that it involves *only* the trigonometric functions or algebraic form on the right. Note that you should work *all* these problems, rather than just the odd- or even-numbered ones.

1. $\sin 2x$, $\sin x$ *and* $\cos x$
2. $\sin^2 x$, $\cos 2x$
3. $\sin^2 x$, $\cos x$
4. $\cos^2 x$, $\sin x$
5. $\cos^2 x$, $\cos 2x$
6. $\cos 2x$, $\cos x$
7. $\cos 2x$, $\sin x$
8. $\cos 2x$, $\cos x$ *and* $\sin x$
9. $\tan 2x$, $\tan x$
10. $\sin \frac{1}{2}x$, $\cos x$
11. $\cos \frac{1}{2}x$, $\cos x$
12. $\tan \frac{1}{2}x$, $\cos x$
13. $\tan \frac{1}{2}x$, $\cos x$ *and* $\sin x$
14. $\sin 3x$, $\sin x$
15. $\cos 4x$, $\cos 2x$
16. $\cos 6x$, $\cos 3x$
17. $\sin x \cos x$, $\sin 2x$
18. $\sin x \cos y$, sum of sines or cosines
19. $\cos x \cos y$, sum of sines or cosines
20. $\cos x + \cos y$, product of sines and/or cosines
21. $\sin x + \sin y$, product of sines and/or cosines
22. $\sin x + \cos x$, *single* cosine
23. $\sin 3x \sin 7x$, sum of sines or cosines of positive multiples of x
24. $\sin 3x + \sin 7x$, product of sines and/or cosines of positive multiples of x
25. $\sqrt{3} \cos x - \sin x$, cosine with phase displacement
26. $-4 \cos x - 4 \sin x$, cosine with phase displacement

For Problems 27 through 30, a definition of "simple form" is given. Use the definition to "simplify" the expressions.

27. Simple form involves no multiple or composite arguments. Simplify:

a. $\cos (x - 37°)$
b. $\cos (x + y + z)$
c. $\cos 3x$

28. Simple form consists of *one* term, which may be composed of several factors. Simplify:

a. $\cos 37° \cos x + \sin 37° \sin x$
b. $\cos 37° + \cos x$
c. $\cos x + \cos 2x + \cos 3x$

29. Simple form involves *no* products or powers of trigonometric functions, but may involve functions of multiple arguments. That is, the expression should be *linear* (first degree) in functions of multiple arguments. Simplify:

a. $\sin x \cos x$ b. $\cos^2 x$
c. $\cos^2 x \sin x$ d. $\sin^4 x$

30. Simple form involves a *single* cosine or sine term (no products, powers or sums of functions!), but may involve multiple or composite arguments. Simplify:

a. $\cos x + \sin x$ b. $\cos x \sin x$
c. $\cos x - \sin x$ d. $\sin x - \cos x$

Problem 31 is designed to let you discover how to solve a trigonometric equation.

31. Given the equation $\sin 3x - \sin x = 0$:

a. Use the sum and product properties to transform the left member to a product of sines and/or cosines.
b. Use the multiplication property of zero to set each factor on the left in part a, equal to zero.
c. Solve the equations in part b for all values of the arguments that make the equations true. Remember that if θ is such a value, then all angles coterminal with θ will also make the equation true (i.e., $\theta + 360n°$).
d. If the argument in part c is a multiple of x, solve for x.
e. Find all values of x in the domain $0° \leq x < 360°$ that make the equation true. There should be *six*!

14-9 TRIGONOMETRIC EQUATIONS

When you were working with sinusoids in Section 13-6, you had equations of the form

$$y = C + A \cos B(x - D),$$

where, A, B, C, and D stand for constants. In Section 13-10, when you substituted a value of y and calculated the corresponding values of x, you were actually solving a trigonometric equation. In this section you will solve other types of trigonometric equations.

DEFINITION

A ***solution*** of a trigonometric equation is a value of the variable in the argument that makes the equation true.

That is, if an expression such as $\cos 3(x - 5)$ appears in the equation, a solution is a value of x and not a value of the entire argument $3(x - 5)$.

Objective:

Given a trigonometric equation and a domain of the variable, find all solutions of the equation in the domain.

Example 1: Solve $2 \sin x - 1 = 0$ for $x \in$ {real numbers of degrees}.

$2 \sin x - 1 = 0$	Given equation.
$2 \sin x = 1$	Adding 1 to both members.
$\sin x = ½$	Dividing by 2.

Recalling functions of special angles, you know that $\sin 30°$ equals ½. So does $\sin 150°$, since 150° has a reference angle of 30° and terminates in Quadrant II, where the sine is positive. Any angles coterminal with 30° and 150° are also solutions. Thus

$$x = 30° + 360n° \quad \text{or} \quad x = 150° + 360n°,$$

so that

$$S = \{30° + 360n°, 150° + 360n°\}.$$

This is known as the *general* solution of the equation. Particular solutions can be obtained by substituting *integers* for n. For instance, if the domain were $0° \leq x < 360°$, then

$$S = \{30°, 150°\},$$

which you get by letting $n = 0$. If the function had been *circular* instead of trigonometric, the general solution would have been

$$S = \{\pi/6 + 2\pi n, 5\pi/6 + 2\pi n\}.$$

If the domain had been $0 \leq x < 2\pi$, the particular solution would have been

$$S = \{\pi/6, 5\pi/6\}.$$

Finally, if the domain had been $180° \leq x \leq 360°$, then the particular solution would have been.

$$S = \phi,$$

since there are *no* solutions in this interval.

From this example, you can see that the solution set depends on the domain. A compact way of writing a set of numbers such as $180° \leq x \leq 360°$ is $[180°, 360°]$. This is read, "the closed interval between 180° and 360°." Other interval notations are as follows:

Interval Notation

Written	Meaning	Name
$x \in [180°, 360°]$	$180° \leq x \leq 360°$	Closed interval
$x \in (180°, 360°)$	$180° < x < 360°$	Open interval
$x \in [180°, 360°)$	$180° \leq x < 360°$	Half-open interval
$x \in (180°, 360°]$	$180° < x \leq 360°$	Half-open interval

Example 2: Solve $\sin 2x \cos x + \cos 2x \sin x = 1$ for $x \in [0, 2\pi)$.

$\sin 2x \cos x + \cos 2x \sin x = 1$	Given equation.
$\sin (2x + x) = 1$	Composite argument properties.
$\sin 3x = 1$	Adding "like" terms."
$3x = \pi/2 + 2\pi n$	$\sin \pi/2 = 1$
$x = \pi/6 + (2\pi/3)n$	Dividing by 3.
$x = \pi/6, 5\pi/6, 3\pi/2$	Picking those integer values of n for which $x \in [0, 2\pi)$.
$\therefore S = \{\pi/6, 5\pi/6, 3\pi/2\}$	

The most difficult part of solving an equation is getting started. The thing that gets you started this time is recognizing a familiar property, the composite argument property, that can be used on the left-hand member.

Example 3: Solve $\cos 2\theta - \cos 4\theta = \sqrt{3} \sin 3\theta$ for $\theta \in [0°, 360°)$.

This problem is complicated because there are three different arguments, 2θ, 4θ, and 3θ. Luckily, the number of arguments can be reduced by applying the sum and product properties to the left member, getting

$$-2 \sin 3\theta \sin (-\theta) = \sqrt{3} \sin 3\theta.$$

This has the advantage of reducing the number of *functions,* too. Now everything is in terms of sines. Since sin is an *odd* function,

$$2 \sin 3\theta \sin \theta = \sqrt{3} \sin 3\theta.$$

At this point it is tempting to divide both members by $\sin 3\theta$. But you recall that dividing by a variable that can equal zero might *lose* solutions. So you resist the temptation and try something else.

$2 \sin 3\theta \sin \theta - \sqrt{3} \sin 3\theta = 0$	Subtracting $\sqrt{3} \sin 3\theta$.
$\sin 3\theta (2 \sin \theta - \sqrt{3}) = 0$	Factoring.
$\sin 3\theta = 0$ or $2 \sin \theta - \sqrt{3} = 0$	Multiplication property of 0.
$\sin 3\theta = 0$ or $\sin \theta = \sqrt{3}/2$	Addition and multiplication properties of equality.

$$\therefore 3\theta = 0° + 360n°, 180° + 360n°;$$

$$\text{or } \theta = 60° + 360n°, 120° + 360n°$$

$$\therefore \theta = 0° + 120n°, 60° + 120n°, 60° + 360n°,$$

$$\text{or } 120° + 360n°$$

$$\therefore S = \{0°, 60°, 120°, 180°, 240°, 300°\}$$

Again, the solution set is determined by picking all values of n that give solutions in the domain.

Example 4: Solve $\cos (x - 57°) = -1$ for $x \in (-180°, 180°)$.

The composite argument $(x - 57°)$ suggests use of the composite argument properties. However, the equation is already in the form of *one* function of *one* argument equal to a *constant*. Thus, the composite argument properties would take you in the wrong direction! So you simply write

$x - 57° = 180° + 360n°$	$\cos 180° = -1$.
$x = 237° + 360n$	Adding 57° to both members.
$x = -123°$	Letting $n = -1$.

This is the only solution in the domain.

$\therefore S = \{-123°\}$

Example 5: Solve $\cos^2 x + \sin x + 1 = 0$ for $x \in [-90°, 270°)$.

There are two different functions, sine and cosine. $\text{Cos}^2\, x$ is easily transformed to sines using the Pythagorean properties. This transformation will *reduce* the number of functions and thus simplify the equation.

$1 - \sin^2 x + \sin x + 1 = 0$	Pythagorean properties.
$\sin^2 x - \sin x - 2 = 0$	Commutativity, associativity and multiplication by -1.
$(\sin x - 2)(\sin x + 1) = 0$	Factoring.
$\therefore \sin x = 2$ or $\sin x = -1$	Multiplication property of zero.

The equation $\sin x = 2$ has no real solutions.

$\therefore \sin x = -1$	Only other choice.
$x = 270° + 360n°$	$\sin 270° = -1$.

The only solution in the domain occurs when $n = -1$, for which $x = -90°$. The 270° you get when $n = 0$ is out of the domain, since the interval is *open* at the upper end.

$\therefore S = \{-90°\}$.

Example 6: Solve $\dfrac{\sin x}{1 + \cos x} = 1$ for $x \in (0°, 360°)$.

There are two ways of approaching this problem. The first is to eliminate the fraction by multiplying both members by $1 + \cos x$.

$\sin x = 1 + \cos x$.

This equation contains two different functions. Since it is easier to transform *squares* of sines into cosines, you can square both members.

$\sin^2 x = 1 + 2\cos x + \cos^2 x$	Squaring both members.
$1 - \cos^2 x = 1 + 2\cos x + \cos^2 x$	Pythagorean properties.
$0 = 2\cos x + 2\cos^2 x$	Addition property of equality.
$0 = \cos x(1 + \cos x)$	Division by 2, then factoring.
$\cos x = 0$ or $1 + \cos x = 0$	Multiplication property of zero.
$\cos x = 0$ or $\cos x = -1$	Subtracting 1.
$x = 90° + 360n°, 270° + 360°$ or $180° + 360n°$	

The solutions in the domain are 90°, 270°, and 180°. However, substituting these into the original equation reveals that only 90° works. The 180° is an *extraneous* solution introduced by multiplying both members by $1 + \cos x$, and the 270° is another extraneous solution introduced by squaring both members. Your next step should look something like this:

extraneous

$x = 90°, \cancel{270°}, \cancel{180°}$

$\therefore S = \{90°\}$.

You should remember that whenever you square both members or multiply by a variable that can equal zero, you must check *all* the solutions. Any which do not satisfy the original equation should be marked "extraneous" and should not be put in the solution set.

A clever application of trigonometric properties can sometimes avoid steps that give extraneous solutions. In this example, $\sin x/(1 + \cos x)$ is recognizable as the half-argument property for tangent. Therefore, the original equation becomes

$\tan \frac{1}{2}x = 1$	Half-argument properties.
$\frac{1}{2}x = 45° + 180n°$	$\tan 45° = 1$, and period of tangent is 180°.
$x = 90° + 360n°$	Multiplying by 2.

$\therefore S = \{90°\}$

The preceding examples were selected to show you many of the useful techniques that exist for solving trigonometric equations and some of the pitfalls you might stumble into. The following table summarizes these techniques.

Techniques for Solving Trigonometric Equations

1. Get an equation (or equations) in which *one* function of *one* argument equals a *constant.* Some ways are:
   a. Reduce the number of different arguments (Examples 2 and 3).
   b. Reduce the number of different functions (Examples 5 and 6).
   c. Do any obvious algebra (Example 1).
   d. *Do not divide* both members by a variable (Example 3).
   e. Use any obvious trigonometric properties (Examples 2, 6).
   f. Get a product equal to zero (Examples 3, 5, 6).

2. Get the *general* solution by finding the argument.
   a. If it is a *special* angle, write the *exact* value.
   b. If it is *not* a special angle, use the tables.

3. Do whatever algebra you need to find the *variable* in the argument (Examples 2, 3, 4, and 6).

4. Write the solution set.
   a. Find all solutions in the domain by picking integer values of n in the general solution.
   b. Check for extraneous solutions if you have multiplied by a variable.

The following exercise will give you practice solving trigonometric equations.

Exercise 14-9

For Problems 1 through 40, solve the equation in the indicated domain.

1. $\tan x + \sqrt{3} = 0$,	$x \in [0°, 360°)$
2. $2 \cos x + \sqrt{3} = 0$,	$x \in [0°, 360°)$
3. $2 \sin (x + 47°) = 1$,	$x \in [0°, 360°)$
4. $\sec (x + 81°) = 2$,	$x \in [0°, 360°)$
5. $4 \cos^2 x = 1$,	$x \in [-180°, 180°]$
6. $4 \sin^2 x = 3$,	$x \in [-180°, 180°]$
7. $2 \sin x \cos x = \sqrt{2} \cos x$,	$x \in$ {real numbers of degrees}
8. $\tan x \sec x = \tan x$,	$x \in$ {real numbers of degrees}
9. $\tan x - \sqrt{3} = 2 \tan x$,	$x \in$ {real numbers}
10. $\cos x + 2 = 3 \cos x$,	$x \in$ {real numbers}
11. $2 \sin^2 x + \sin x = 0$,	$x \in (-180°, 180°)$
12. $\tan^2 x + \tan x = 0$,	$x \in [-90°, 90°)$
13. $2 \cos^2 x - 5 \cos x + 2 = 0$,	$x \in [0, 2\pi)$
14. $2 \sec^2 x - 3 \sec x - 2 = 0$,	$x \in [0, 2\pi)$
15. $\sin^2 x + 5 \sin x + 6 = 0$,	$x \in [0°, 360°)$
16. $4 \csc^2 x + 4 \csc x + 1 = 0$,	$x \in [0, 360°)$

17. $\tan^2 x - \sec x - 1 = 0$, $\quad x \in [-\pi, \pi)$

18. $3 - 3 \sin x - 2 \cos^2 x = 0$, $\quad x \in [-\pi, \pi]$

19. $1 - \cos x = -\sin x$, $\quad x \in [-180°, 180°)$

20. $\dfrac{1 + \cos x}{\sin x} = -1$, $\quad x \in [-180°, 180°)$

21. $4 \sin x \cos x = \sqrt{3}$, $\quad x \in [0, 2\pi)$

22. $\sin x = \sin 2x$, $\quad x \in [0, 2\pi)$

23. $\dfrac{\sin (90° - x)}{\sin x} = -\sqrt{3}$, $\quad x \in (-270°, 270°)$

24. $\tan (90° - x) = -1$, $\quad x \in (-180°, 180°)$

25. $\sin 2x \cos 64° + \cos 2x \sin 64° = \sqrt{3}/2$, $\quad x \in [0°, 360°)$

26. $\cos 3x \cos 12° - \sin 3x \sin 12° = ½$, $\quad x \in [-120°, 120°)$

27. $\cos 4x - \sin 2x = 0$, $\quad x \in (-90°, 90°)$

28. $\cos 4x - \sin 2x = 1$, $\quad x \in [-90°, 90°)$

29. $\cos 3x + \cos 5x = 0$, $\quad x \in (-90°, 90°)$

30. $\sin 5x + \sin 7x = 0$, $\quad x \in [-90°, 90°)$

31. $\cos x - \sqrt{3} \sin x = 1$, $\quad x \in (0, 2\pi]$

32. $\sin x - \sqrt{3} \cos x = 1$, $\quad x \in [-\pi, \pi]$

33. $\dfrac{\tan 10x + \tan 50°}{1 - \tan 10x \tan 50°} = \dfrac{\sqrt{3}}{3}$, $\quad x \in (0°, 90°)$

34. $\tan x - \tan 10° = 1 + \tan x \tan 10°$, $\quad x \in [-180°, 180°]$

35. $\tan ½x + 1 = \cos x$, $\quad x \in [0, 4\pi]$

36. $2 \cos^2 ½x - 2 = 2 \cos x$, $\quad x \in [-\pi, \pi)$

37. $2 \cos (x + 30°) \cos (x - 30°) = 1$, $\quad x \in [-180°, 180°]$

38. $4 \sin (x + 75°) \cos (x - 75°) = 1$, $\quad x \in [-180°, 180°)$

39. $\cos^2 ½x - ½ \cos x = ½$; $\quad x \in$ {real numbers}

40. $\sin x \tan ½x = 1 - \cos x$, $\quad x \in$ {real numbers}

For Problems 41 through 50, solve the equation in the domain $\theta \in [0, 360°)$ or $x \in [0, 2\pi)$, obtaining approximate values by calculator or from Tables III or IV.

41. $5 \sec^2 \theta + 2 \tan \theta - 8 = 0$

42. $3 \tan^2 \theta - 5 \sec \theta - 9 = 0$

43. $3 \cos \theta - 4 \sin \theta = 1$

44. $5 \cos \theta + 12 \sin \theta = 13$

45. $3 \sin^2 x - \sin x = 0$

46. $4 \cos^2 x + \cos x = 0$

47. $\sin^2 x + \sin x - 1 = 0$

48. $\cos^2 x - 2 \cos x - 2 = 0$

49. $4 \sin (-x) = 3$

50. $5 \cos (-x) = 2$

For Problems 51 through 53, solve the equation for the indicated values of x. Note that this is what you did with sinusoids in Section 13-10.

51. $3 + 4 \cos 2(x - 10°) = 5$, $\quad x \in [0°, 360°)$.

52. $4 + 3 \cos 2(x - 10°) = 5$, $\quad x \in [0°, 360°)$.

53. $4 + 3 \cos 2(x - 1) = 2$, $\quad x$ is the smallest positive real number satisfying the equation.

14-10 CHAPTER REVIEW AND TEST

In this chapter you have learned properties with which you can *transform* trigonometric expressions to forms that look simpler or are simpler to use in a given problem. The properties are particularly useful for solving triognometric equations.

The objectives of this chapter may be summarized as follows:

1. *Use the quotient, reciprocal, and Pythagorean properties to transform expressions containing functions with the* ***same*** *argument.*

2. *Use the odd-even properties to transform functions with **negative** arguments.*
3. *Use the composite argument properties and co-function properties to transform functions whose arguments contain a **sum**.*
4. *Use the double argument and half-argument properties to transform to functions of **half** or **twice** the argument.*
5. *Use the sum and product properties to express sums of sines or cosines as products of sines and cosines, and vice versa.*
6. *Use the linear combination of sinusoid properties to transform to a cosine with a phase displacement.*
7. *Transform a trigonometric expression using any combination of the above properties.*
8. *Solve trigonometric equations.*

This section contains Review Problems and a Concepts Test. The six review problems are numbered according to the first six of the objectives above. By working these problems, you can be sure you know what to do when you are told the objective being tested. The Concepts Test consists of expressions to transform and equations to solve. These problems will measure your ability to draw together the necessary techniques when you are *not* told which objective is being tested.

Review Problems

The following problems are numbered according to the objective being tested.

Review Problem 1a. Express $\tan x$ and $\cot x$ in terms of $\sin x$ and $\cos x$.

b. Express $\tan x$ and $\cot x$ in terms of $\sec x$ and $\csc x$.
c. Write three different equations in which a product of two trigonometric functions equals 1.
d. Write equations expressing $\sin^2 x$ in terms of $\cos x$, $\tan^2 x$ in terms of $\sec x$, and $\csc^2 x$ in terms of $\cot x$.

Review Problem 2. Express $\sin(-x)$, $\cos(-x)$, $\tan(-x)$, $\cot(-x)$, $\sec(-x)$, and $\csc(-x)$ as the *same* function of (positive) x.

Review Problem 3. Write equations expressing:

a. $\sin(x+y)$ in terms of sines and cosines of x and y
b. $\cos(x+y)$ in terms of sines and cosines of x and y
c. $\sin(x-y)$ in terms of sines and cosines of x and y
d. $\cos(x-y)$ in terms of sines and cosines of x and y
e. $\tan(x-y)$ in terms of $\tan x$ and $\tan y$
f. $\tan(x-y)$ in terms of $\tan x$ and $\tan y$.

Review Problem 4. Write equations expressing:

a. $\sin 2x$ in terms of sines and cosines of x
b. $\cos 2x$ in terms of

i. sines and cosines of x
ii. $\sin x$
iii. $\cos x$

c. $\tan 2x$ in terms of $\tan x$
d. $\sin \frac{1}{2}x$ in terms of $\cos x$
e. $\cos \frac{1}{2}x$ in terms of $\cos x$
f. $\tan \frac{1}{2}x$ in terms of

i. $\sin x$ and $\cos x$, with a single-term denominator
ii. $\sin x$ and $\cos x$, with a single-term numerator
iii. $\cos x$ alone

Review Problem 5a. Write the following as *sums* of functions:

i. $\sin x \sin y$
ii. $\cos x \cos y$
iii. $\sin x \cos y$
iv. $\cos x \sin y$

b. Write the following as *products* of functions:

i. $\sin x + \sin y$
ii. $\cos x + \cos y$
iii. $\sin x - \sin y$
iv $\cos x - \cos y$

Review Problem 6. Express $5 \cos x - 7 \sin x$ as a cosine with a phase displacement.

Concepts Test

The following problems require you to draw upon your knowledge of the trigonometric function properties in Objectives 1 through 6. The problems are either transforming expressions or solving equations, as in Objectives 7 and 8. Work each problem. Then tell by number which of Objectives 1 through 8 you used in that problem. Since there may be several ways of working the problem, your list of objectives may differ from that in the answers at the back of the book.

Concepts Test 1. Solve the following equations in the indicated domain.

a. $\sin 2x = \cos x$, $x \in [\pi/4, 3\pi/4]$

b. $\tan 2(x + 41°) = 1$, $x \in$ {real numbers of degrees}

c. $\sqrt{3} \sin x - \cos x = 1$, $x \in$ {real numbers}

d. $\csc x \tan x \cos x = 1$, $x \in [-\pi, \pi]$

e. $\sin x + \sin 3x + \sin 5x = 0$, $x \in [0°, 360°]$

f. $\sin x \cos 37° = \cos x \sin 37°$, $x \in (-180°, 180°)$

Concepts Test 2. Prove that the following are identities.

a. $\sin(x + y)\sin(x - y) = \sin^2 x - \sin^2 y$

b. $\cot^2 x - \cos^2 x = \cot^2 x \cos^2 x$

c. $\dfrac{\sin x + \sin y}{\cos x + \cos y} = \tan\dfrac{1}{2}(x + y)$

Concepts Test 3. Find *exact* values of:

a. $\cos 15°$

b. $\sin\frac{\pi}{8}$

Concepts Test 4a. Transform $\cos 2(x + y)$ to functions of x and y.

b. Transform $\sin 58° + \sin 33°$ to a *single* term.

c. Transform $\sin^2 x \cos x$ to a *sum* of functions of *multiples* of x, which involves *no* products of functions.

d. Transform $\cos 38x$ to

i. functions of $19x$,
ii. functions of $76x$.

e. Simplify $\sin 37° \cos 53° + \cos 37° \sin 53°$.

f. Transform $\cos 23° + \sin 23°$ to a *multiple* of a *cosine* of a *positive* angle.

Triangle Problems

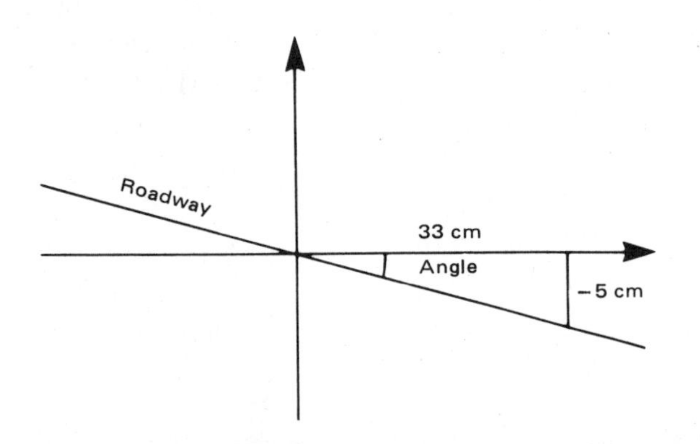

The word "trigonometry" means "triangle measuring." In this chapter you will use the trigonometric functions for their original purpose. In the first section you will solve right triangle problems using just the ***definitions*** *of the six functions. Then you will learn properties that will allow you to find unknown sides, angles, and the area of* ***any*** *triangle. The techniques you will develop are used, for example, by surveyors to measure irregularly-shaped tracts of land, or for finding the slope of a hill.*

15-1 RIGHT TRIANGLE PROBLEMS

You will recall from your previous studies that the six trigonometric functions are defined in terms of the coordinates (u, v) of a point on the terminal side of an angle in standard position (Figure 15-1a). Together with the distance r from the origin to (u, v), the numbers u and v can be thought of as lengths of the sides of a right triangle. If the angle measure and one side are known, the other two sides can be found. Similarly, the lengths of any two sides can be used to find the measures of the acute angles.

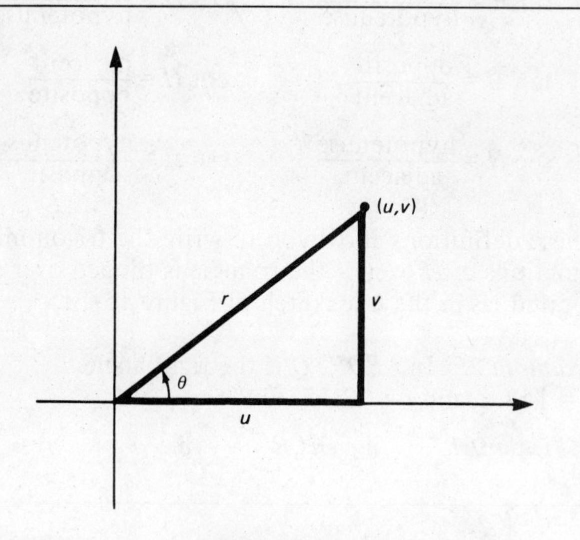

Figure 15-1a.

Objective:

Given two sides or a side and an angle of a right triangle, find measures of the other sides and angles.

Example 1: Suppose you have been assigned the job of measuring the height of the local water tower. Climbing makes you dizzy, so you decide to do the whole job at ground level. From a point 47.3 meters from the base of the tower, you find that you must look up at an angle of 53° to see the top of the tower (Figure 15-1b.). How high is the tower?

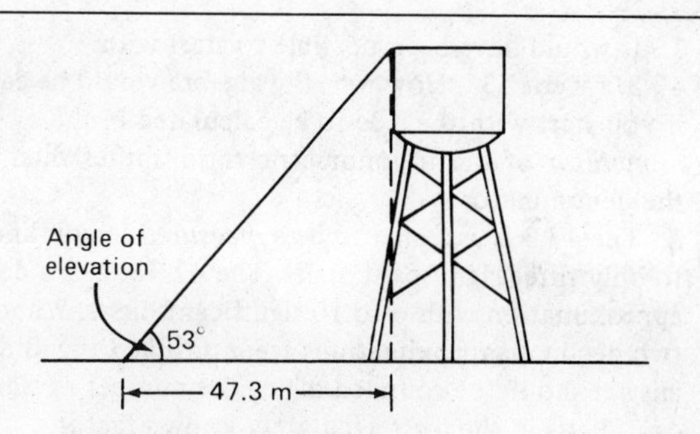

Figure 15-1b.

From the picture, you can find the right triangle sketched in Figure 15-1c. Placing the 53° angle in standard position, you can see that $u = 47.3$, and you must find v. By the definition of tangent,

$$\frac{v}{47.3} = \tan 53°.$$

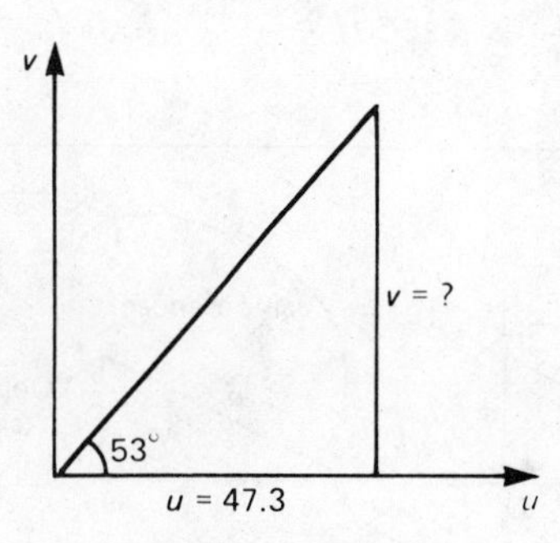

Figure 15-1c.

Multiplying by 47.3 then doing the indicated operations, you get

$$v = 47.3 \tan 53^\circ$$
$$= 62.769...$$

So the tower is approximately 62.8 meters high.

Notes:

1. It would have been possible to start with $47.3/v = \cot 53^\circ$. However, the algebra would be easier if you start with the side to be calculated in the *numerator* of the trigonometric ratio, rather than in the denominator.
2. The 47.3 is assumed to be a *measured* length known to only three significant digits. The 62.769... is a decimal approximation with 6 to 10 significant digits. Whenever two decimal approximations are *multiplied* together, the answer should be rounded off to the number of significant digits in the *least* accurately known factor.

It is possible to use the definitions of the trigonometric functions *without* placing the angle in standard position. For this purpose, it helps to observe where the sides u, v, and r are with respect to the angle.

Figure 15-1d shows ΔHDJ (triangle HDJ) with acute angle H placed in standard position and right angle D on the u-axis. By the definitions of the six trigonometric functions:

$$\sin H = \frac{v}{r} \qquad \sec H = \frac{r}{u} \qquad \cot H = \frac{u}{v}$$

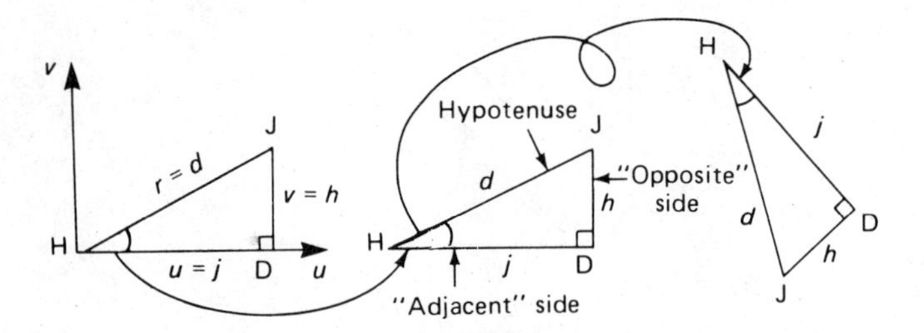

Figure 15-1d.

$$\tan H = \frac{v}{u} \qquad \cos H = \frac{u}{r} \qquad \csc H = \frac{r}{v}\,.$$

It is customary to name the length of the side *opposite* an angle with the same *lower* case letter as the name of the angle. So the side opposite Angle H has measure h, and so forth. Using this lettering scheme, $\sin H = h/d$, since $h = v$ and $d = r$. But d and h are properties of the triangle itself, not the coordinate system. So it is possible to write the trigonometric functions of the angles of a right triangle *without* reference to the coordinates u, v, and r. The *hypotenuse* is one of the sides that includes Angle H. The other side is called the *adjacent* side, the word "adjacent" meaning "next to." Using these words, the definitions of the six trigonometric functions for Angle H of a right triangle become:

$$\sin H = \frac{\text{opposite}}{\text{hypotenuse}} \qquad \cos H = \frac{\text{adjacent}}{\text{hypotenuse}}$$

$$\tan H = \frac{\text{opposite}}{\text{adjacent}} \qquad \cot H = \frac{\text{adjacent}}{\text{opposite}}$$

$$\sec H = \frac{\text{hypotenuse}}{\text{adjacent}} \qquad \csc H = \frac{\text{hypotenuse}}{\text{opposite}}$$

These definitions allow you to write the trigonometric functions of H even if the triangle is flipped over and rotated, as in the last sketch of Figure 15-1d.

Example 2: In ΔPQR, Q is the right angle, $q = 147.6$, and $r = 72.15$. Find:

i. $m\angle P$ ii. $m\angle R$ iii. p

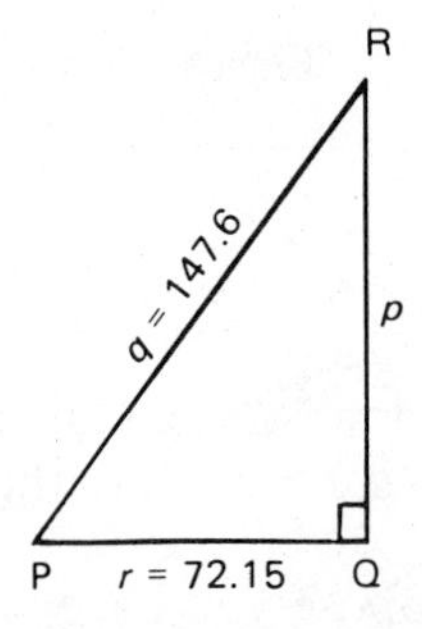

Figure 15-1e.

i. $\cos P = \frac{r}{q}$ — $\cos P = \frac{\text{adjacent}}{\text{hypotenuse}}$.

$= \frac{72.15}{147.6}$ — Substitution.

$= 0.4888...$ — By calculator.

$\therefore m\angle P = 60.7368...^\circ$ — By calculator.

$\approx 60^\circ\ 44'$ — Converting to minutes.

Note: Save 60.7368... in memory, without round-off, for use in the next part of the problem.

ii. $m\angle R = 180^\circ - 90^\circ - 60^\circ 44' = \underline{29^\circ 16'}$

iii. $\frac{p}{r} = \tan P$ — $\tan P = \frac{\text{opposite}}{\text{adjacent}}$.

$p = r \tan P$ — Multiplication by r.

$= 72.15 \tan P$ — Substitution.

$= 128.763...$ — Recalling P from memory.

$= \underline{128.8}$ — Rounding off to four significant digits.

The exercise that follows is designed to give you practice finding unknown side and angle measures of right triangles.

Exercise 15-1

For Problems 1 through 8, find the other side and angle measures for the given right triangle:

	Name	*Right angle*	*Given data*
1.	ABC	A	$m\angle B = 34^\circ$, $c = 14.7$
2.	ABC	B	$m\angle C = 71^\circ$, $b = 36.8$
3.	LMN	M	$m\angle N = 47^\circ 32'$, $m = 3.465$
4.	HPJ	H	$m\angle J = 29^\circ 51'$, $j = 4651$
5.	XYZ	X	$x = 35$, $y = 27$
6.	KLM	L	$k = 5.2$, $m = 3.3$
7.	RST	T	$s = 9.85$, $t = 47.3$
8.	UVW	W	$u = 439.8$, $v = 641.2$

9. ***Ladder Problem.*** You lean a ladder 6.7 meters long against the wall. It makes an angle of 63° with the level ground. How high up is the top of the ladder?

10. ***Flagpole Problem.*** You must order a new rope for the flagpole. To find out what length of rope is needed, you observe that the pole casts a shadow 11.6 meters long on the ground. The angle of elevation of the sun is 36° 50′ (Figure 15-1f). How tall is the pole?

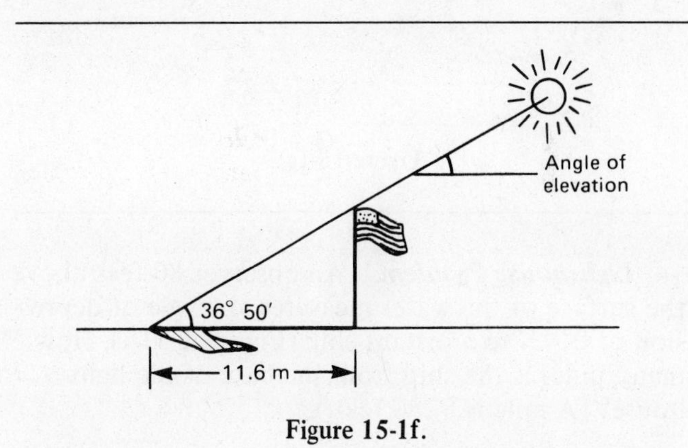

Figure 15-1f.

11. ***Cat Problem.*** Your cat is trapped on a tree branch 6.5 meters above the ground. Your ladder is only 6.7 meters long. If you place the ladder's tip on the branch, what angle will the ladder make with the ground?

12. ***Observation Tower Problem.*** The tallest free-standing structure in the world is the 553-meter tall CN Tower in Toronto, Ontario. Suppose that at a certain time of day it casts a shadow 1100 meters long on the ground. What is the angle of elevation of the sun at that time of day?

13. ***Moon Crater Problem.*** Scientists estimate the heights of features on the moon by measuring the lengths of the shadows they cast on the moon's surface. From a photograph, you find that the shadow cast on the inside of a crater by its rim is 325 meters long (Figure 15-1g). At the time the

photograph was taken, the sun's angle of elevation from this place on the moon's surface was 23°37′. How high does the rim rise above the inside of the crater?

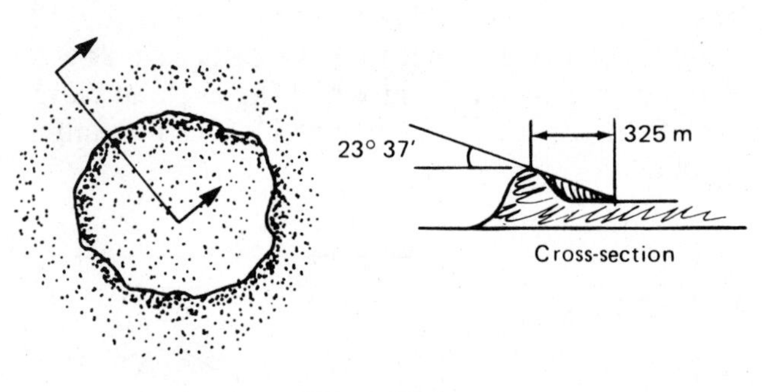

Figure 15-1g.

14 ***Lighthouse Problem.*** An observer 80 feet above the surface of the water measures an angle of depression of 0°42′ to a distant ship (Figure 15-1h). How many miles is the ship from the base of the lighthouse? (A mile is 5280 feet.)

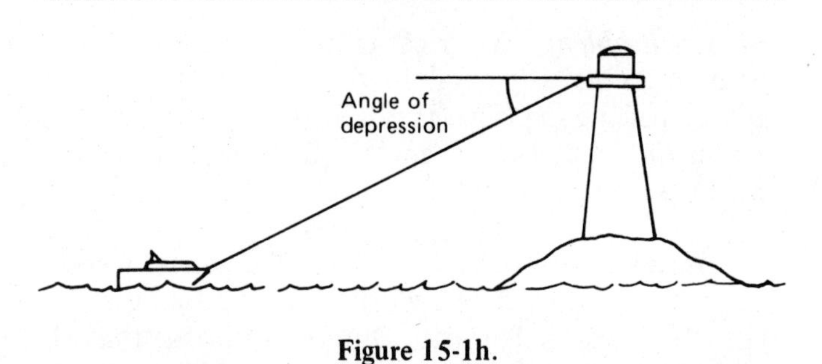

Figure 15-1h.

15. ***Airplane Landing Problem.*** Commercial airliners fly at an altitude of about 10 kilometers. They start descending toward the airport when they are still far away, so that they will not have to dive at a steep angle.

a. If the pilot wants the plane's path to make an angle of 3° with the ground, how far from the airport must he start descending?
b. If he starts descending 300 kilometers from the airport, what angle will the plane's path make with the horizontal?

16. ***Radiotherapy Problem.*** A beam of gamma rays is to be used to treat a tumor known to be 5.7 centimeters beneath the patient's skin. To avoid damaging a vital organ, the radiologist moves the source over 8.3 centimeters (Figure 15-1i).

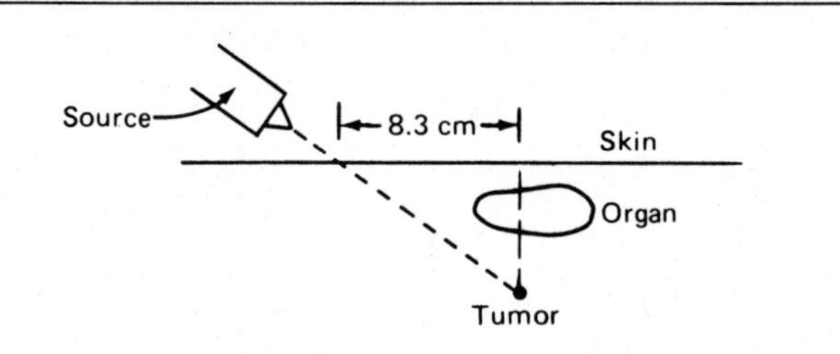

Figure 15-1i.

a. At what angle to the patient's skin must the radiologist aim the gamma ray source to hit the tumor?
b. How far will the beam have to travel through the patient's body before reaching the tumor?

17. ***Triangular Block Problem.*** A block bordering Market Street is a right triangle (Figure 15-1j). You start walking around the block, taking 125 paces on Market Street and 102 paces on Pine Street.

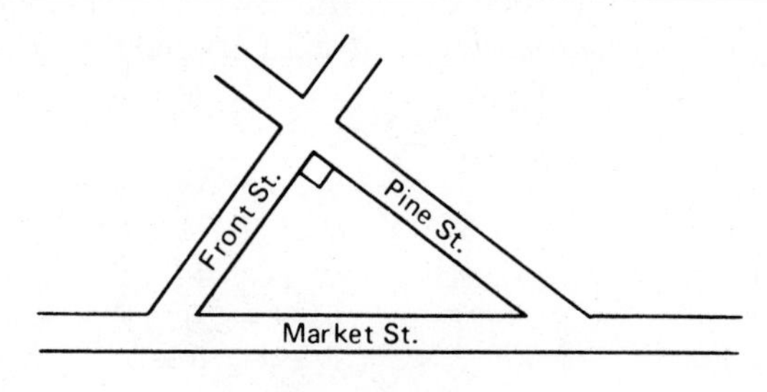

Figure 15-1j.

a. At what angle do Pine and Market Streets intersect?
b. How many paces must you take on Front Street to complete the trip?

18. ***Cable Car Problem.*** Wendy Uptmore is waiting for the cable car in the 600 block of Powell Street in San Francisco. Since the street seems to be so steep, she decides to find out what angle it makes with the horizontal. On the wall of a house, she measures horizontal and vertical distances of 33 centimeters and 5 centimeters, respectively (Figure 15-1k).

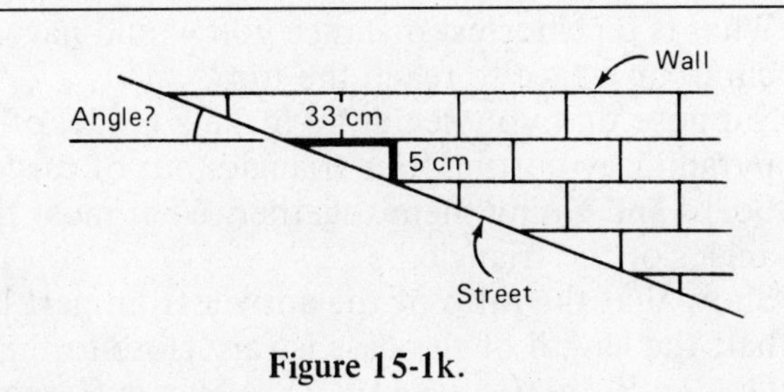

Figure 15-1k.

a. What angle does Powell Street make with the horizontal?
b. While she waited, Wendy went up to the top of the block, counting 101 paces. She is tall and figures each pace is 1 meter long. How many meters did she go *vertically*?
c. If Powell Street had been level instead of slanted, how many paces would Wendy have to go to walk the 600 block of Powell Street? Surprising?!

19. ***Surveying Problem.*** When surveyors measure land that slopes significantly, the distance which is measured will be *longer* than the *horizontal* distance which must be drawn on the map. Suppose that the distance from the top edge of the Cibolo Creek bed to the edge of the water is 37.8 meters (Figure 15-1l). The land slopes downward at 27°36′ to the horizontal.

a. What is the horizontal distance from the top of the bank to the edge of the creek?
b. How far is the surface of the creek below the level of the surrounding land?

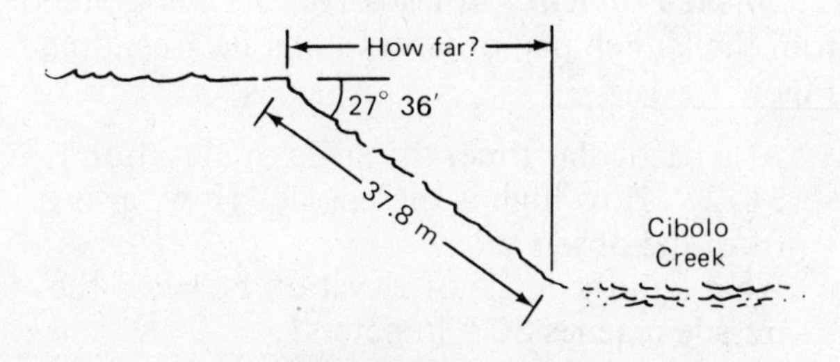

Figure 15-1l.

20. ***Grand Canyon Problem.*** From a point on the North Rim of the Grand Canyon, a surveyor measures an angle of depression of 1°18′ to a point on the South Rim (Figure 15-1m). From an aerial photograph, he determines that the horizontal distance between the two points is 10 miles. How many feet is the South Rim below the North Rim? (A mile is 5280 feet.)

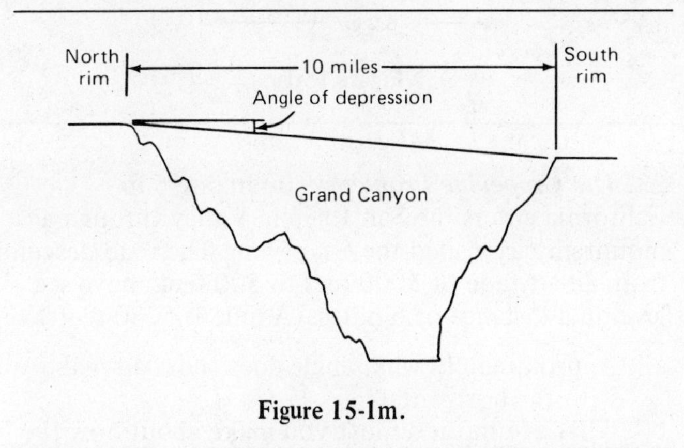

Figure 15-1m.

21. ***Submarine Problem.*** A submarine at the surface of the ocean makes an emergency dive, its path making an angle of 21° with the surface.

a. If it goes for 300 meters along its downward path, how deep will it be? What horizontal distance is it from its starting point?

b. How many meters must it go along its downward path to reach a depth of 1000 meters?

22. ***Missile Problem.*** An observer 5.2 kilometers from the launch pad observes a missile ascending (Figure 15-1n).

a. At a particular time, the angle of elevation is 31°27′. How high is the missile? How far is it from the observer?
b. What will the angle of elevation be when the missile reaches 30 kilometers?

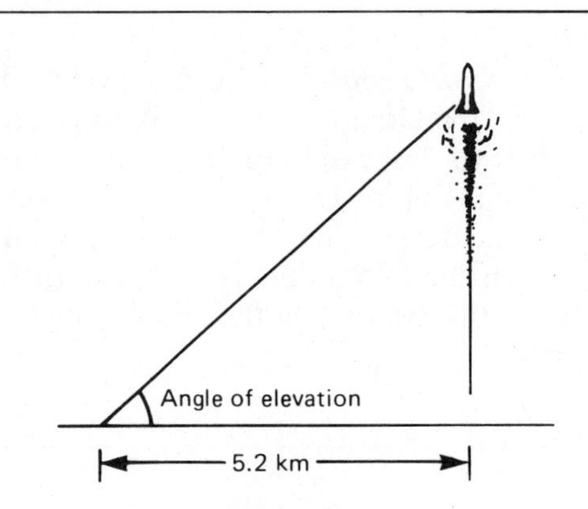

Figure 15-1n.

23. ***The Grapevine Problem.*** Interstate 5 in California enters the San Joaquin Valley through a mountain pass called the Grapevine. The road descends from an altitude of 3000 feet to 500 feet above sea level in a distance of 6 miles. (A mile is 5280 feet.)

a. Approximately what angle does the road make with the horizontal?
b. What assumption must you make about how the road slopes?

24. ***Pyramid Problem.*** The Great Pyramid of Cheops in Egypt has a square base 230 meters on each side. The faces of the pyramid make an angle of 51°50′ with the horizontal (Figure 15-1o).

a. How tall is the pyramid?

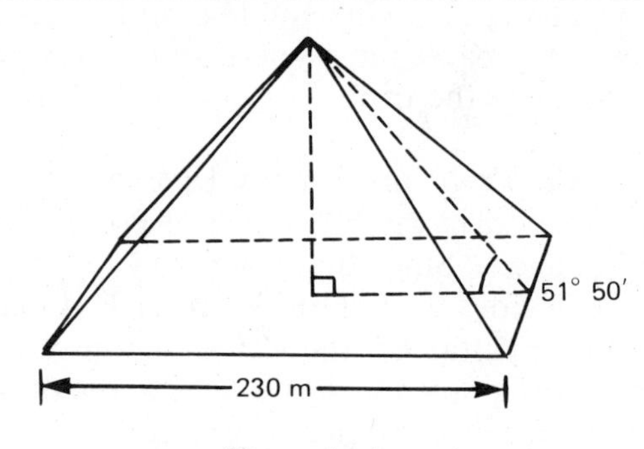

Figure 15-1o.

b. What is the shortest distance you would have to climb up a face to reach the top?
c. Suppose that you decide to make a model of the pyramid by cutting four triangles out of cardboard and gluing them together. What must the angles of the triangle be?
d. Show that the ratio of the answer from part b to half the length of the base is very close to the Golden Ratio, $(\sqrt{5} + 1)/2$. (See Martin Garner's article in the June 1974 issue of *Scientific American* for other startling relationships among the dimensions of the pyramid.)

15-2 OBLIQUE TRIANGLES – LAW OF COSINES

You have learned to find unknown measures in right triangles. Now you must learn how to do the same things for *oblique* triangles, which do *not* have a right angle.

Objectives:

1. Given two sides and the included angle, find the length of the third side of the triangle.

2. Given three sides of a triangle, find the measure of a specified angle.

Suppose that the lengths of two sides, b and c, of ΔABC are known, and also the measure of the included angle A (Figure 15-2a). The length of the third side, a, is to be found.

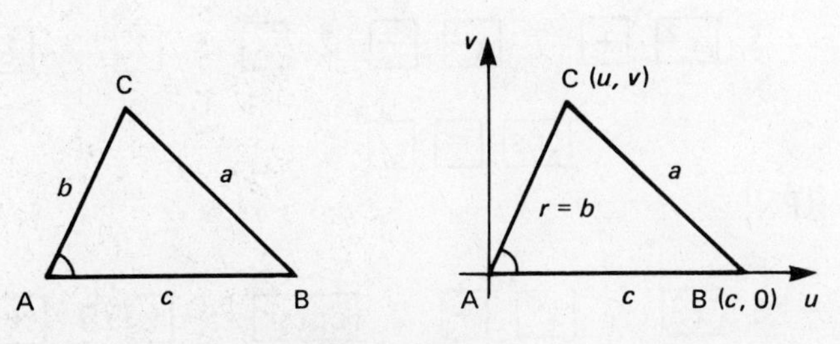

Figure 15-2a.

If you construct a uv-coordinate system with Angle A in standard position, as in the second sketch of Figure 15-2a, then a becomes a distance between two points in a Cartesian coordinate system. These are the points $B(c, 0)$ and $C(u, v)$. By the Distance Formula,

$$a^2 = (u - c)^2 + (v - 0)^2.$$

In order to get a^2 in terms of b, c, and $m\angle A$, all you need to do is observe that A is the *angle* and b is the *radius* to point $C(u, v)$. By the definitions of cosine and sine,

$$\frac{u}{b} = \cos A \quad \text{and} \quad \frac{v}{b} = \sin A.$$

Multiplying both members of each equation by b gives

$$u = b \cos A \quad \text{and} \quad v = b \sin A.$$

Substituting these values for u and v into the distance formula above gives

$$a^2 = (b \cos A - c)^2 + (b \sin A - 0)^2.$$

Upon expanding the squares on the right, you get

$$a^2 = b^2 \cos^2 A - 2bc \cos A + c^2 + b^2 \sin^2 A.$$

Associating the $\sin^2 A$ and $\cos^2 A$ terms and then factoring out b^2 gives

$$a^2 = b^2 (\cos^2 A + \sin^2 A) - 2bc \cos A + c^2.$$

The Pythagorean properties may now be applied to give

$$a^2 = b^2 - 2bc \cos A + c^2,$$

which is usually written:

$$\boxed{a^2 = b^2 + c^2 - 2bc \cos A} \longleftarrow \text{Law of cosines.}$$

This equation is called the Law of Cosines because the *cosine* of an angle appears in it. Note that if $m\angle A = 90°$, then $\cos A = 0$. The Law of Cosines thus reduces to

$$a^2 = b^2 + c^2,$$

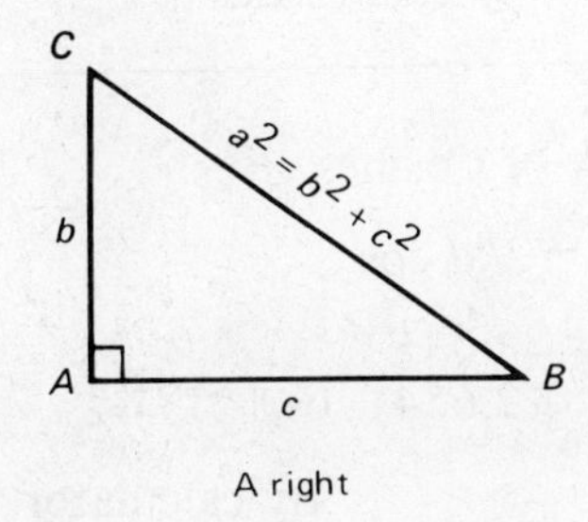

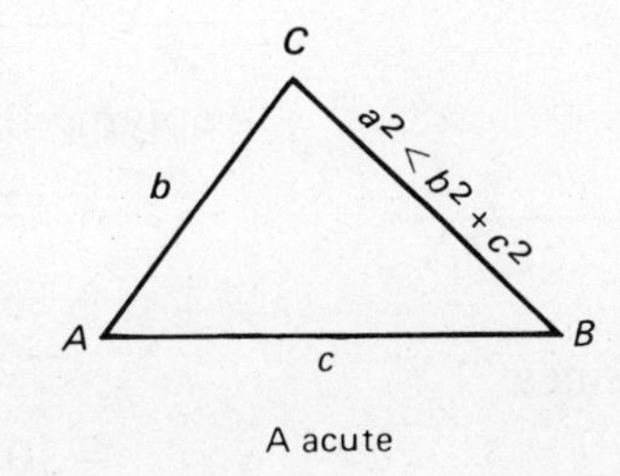

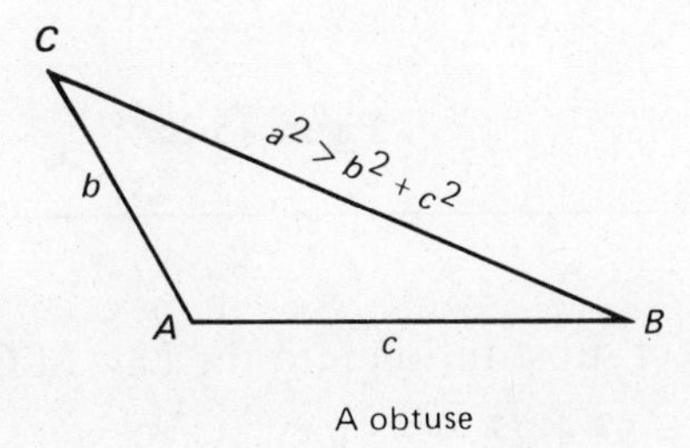

Law of Cosines and the Pythagorean Theorem

Figure 15-2b.

which is, of course, the Pythagorean Theorem! In fact, the Law of Cosines simply says that $2bc \cos A$ is what you must *subtract* from the Pythagorean $a^2 = b^2 + c^2$ in order to get the proper value for a^2 when A is not a right angle. If A is obtuse then $\cos A$ is negative. So subtracting $2bc \cos A$ actually *adds* a number to $b^2 + c^2$. The situation is illustrated in Figure 15-2b.

You should not jump to the conclusion that the Law of Cosines gives an easy way to *prove* the Pythagorean Theorem. Doing so would involve circular reasoning, because Pythagoras was used to *derive* the Law of Cosines!

Accomplishing the first objective, finding the third side from two sides and the included angle, is illustrated in Examples 1 and 2. Accomplishing the second objective, finding an angle from three given sides, is illustrated in Examples 3 and 4.

Example 1: In $\triangle ABC$, if $b = 5$, $c = 7$, and $m\angle A = 39°$, find a (Figure 15-2c).

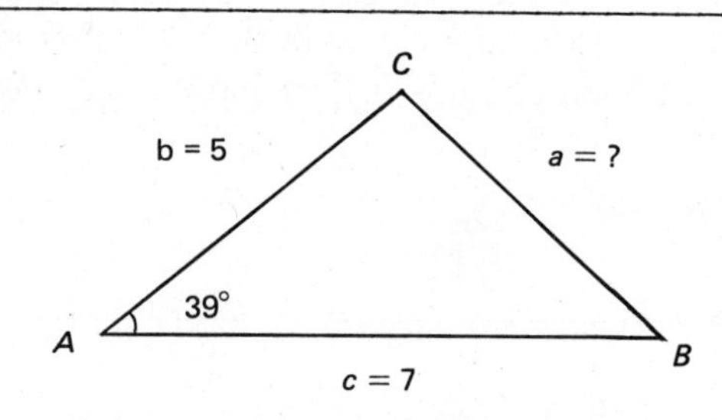

Figure 15-2c.

By direct substitution into the Law of Cosines:

$a^2 = 5^2 + 7^2 - 2(5)(7)\cos 39°$	
$= 19.5997...$	By calculator.
$\therefore a = 4.42716...$	Taking the square root.
$\approx \underline{\underline{4.427}}$	Rounding off.

The sequence of keystrokes is long, but straightforward.

Arithmetic logic:

5 [x^2] [+] 7 [x^2] [−] 2 [×] 5 [×] 7 [×]

39 [cos] [=] [√]

RPN:

5 [x^2] 7 [x^2] [+] 2 [enter] 5 [×] 7 [×]

39 [cos] [×] [−] [√]

Example 2: In $\triangle KSD$, $m\angle S = 127°42'$, $k = 15.78$, and $d = 2.654$. Find s.

You should recognize that the Law of Cosines is *independent* of the letters you use to express it. All that matters is that you know two sides and the *included* angle. A picture such as Figure 15-2d will help.

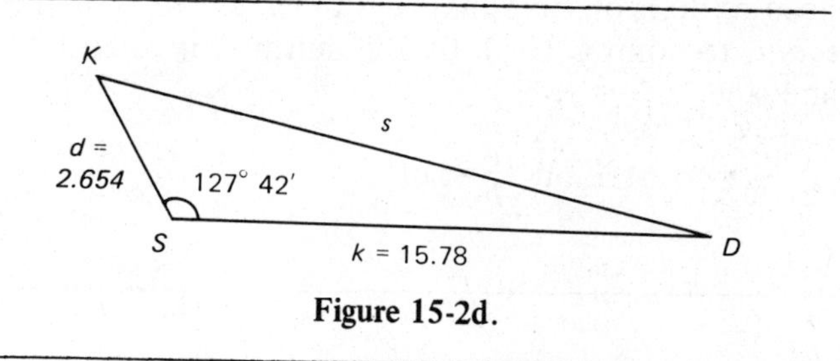

Figure 15-2d.

Applying the Law of Cosines gives:

$s^2 = k^2 + d^2 - 2kd \cos S$	
$= 15.78^2 + 2.654^2 - 2(15.78)(2.654)\cos 127° 42'$	Substitution.
$= 307.27...$	By calculator.
$\therefore s = 17.5292...$	Taking the square root.
$\approx \underline{17.53}$	Rounding off to 4 significant digits.

Note: A *product* of two decimal approximations is rounded off to the least number of *significant digits* in any factor. A *sum* of two decimal approximations is rounded off to the least number of *decimal places* in any term. You can see why if you will write each factor or term with an "x" for each digit beyond the last one you know. Wherever you add or multiply by an unknown digit, place another x. Where an x first appears in the answer is the place to which you must round off.

Example 3: In $\triangle XYZ$, $x = 3$, $y = 7$, and $z = 9$. Find $m\angle Z$.

Again, drawing a picture will help you figure out how to set up the Law of Cosines (Figure 15-2e). Since Angle Z is to be measured, and Sides x and y *include* Angle Z, you use the Law of Cosines in the form

$$z^2 = x^2 + y^2 - 2xy \cos Z.$$

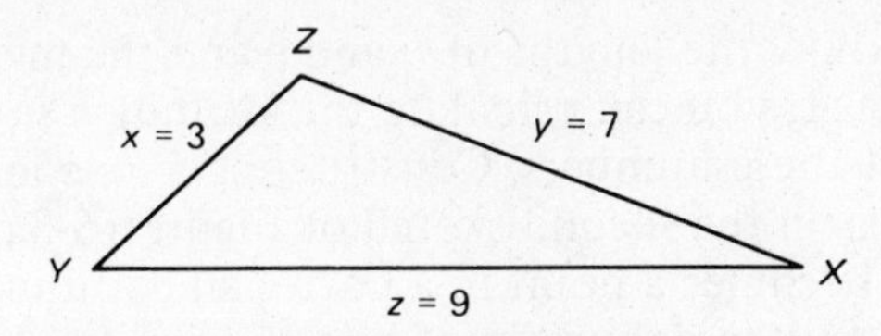

Figure 15-2e.

Since x, y, and z are known and $m\angle Z$ is to be found, the equation can be solved for cos Z, giving

$$\cos Z = \frac{x^2 + y^2 - z^2}{2xy}.$$

Substituting the given information yields

$$\cos Z = \frac{3^2 + 7^2 - 9^2}{2(3)(7)}$$

$$= \frac{-23}{42} \qquad \text{Arithmetic.}$$

$$= -0.54761... \qquad \text{By calculator.}$$

$$\therefore m\angle Z = 123.203...^\circ \qquad \text{By calculator.}$$

$$\approx 123^\circ\ 12' \qquad \text{Converting to minutes.}$$

Example 4: In $\triangle XYZ$, $x = 3$, $y = 7$, and $z = 11$. Find $m\angle Z$.

At first glance, this problem seems equivalent to Example 3. Using the Law of Cosines as before gives:

$$\cos Z = \frac{3^2 + 7^2 - 11^2}{2(3)(7)}$$

$$= \frac{-63}{42} \qquad \text{Arithmetic.}$$

$$= -1.5 \qquad \text{Arithmetic.}$$

But there can be *no* such angle Z, because cosines must be between -1 and 1, inclusive. The reason is clear when you compare the lengths of the three sides (Figure 15-2f). Sides x and y add up to 10, which is *less* than the length 11 of the third side z, so there can be no such triangle! The Law of Cosines automatically detects this situation, if you have overlooked it at first, by giving a cosine *outside* the range $-1 \le \cos\theta \le 1$.

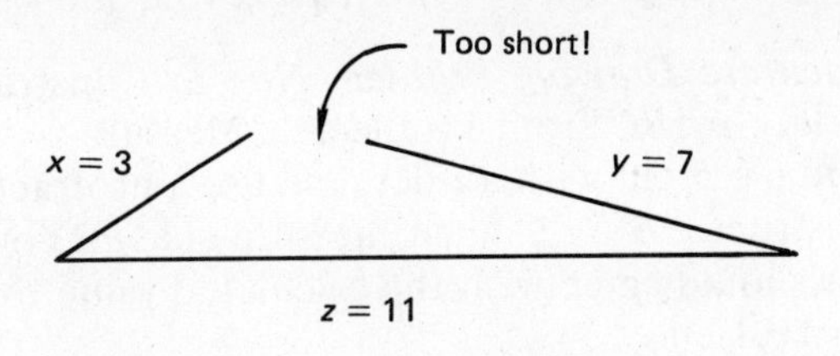

Figure 15-2f.

The exercise that follows is designed to give you practice using the Law of Cosines to find the third side given two sides and the included angle, or to find an angle given three sides.

Exercise 15-2

For Problems 1 through 6, find the length of the side *opposite* the given angle.

1. In $\triangle ABC$, $b = 4$, $c = 5$, and $m\angle A = 51°$.
2. In $\triangle ABC$, $a = 7$, $c = 9$, and $m\angle B = 34°$.
3. In $\triangle PQR$, $p = 3$, $q = 2$, and $m\angle R = 138°$.
4. In $\triangle HJK$, $h = 8$, $j = 6$, and $m\angle K = 172°$.
5. In $\triangle DEF$, $d = 36.2$, $f = 49.8$, and $m\angle E = 67°40'$.
6. In $\triangle BAD$, $a = 2.897$, $d = 5.921$, and $m\angle B = 119°\ 23'$.

For Problems 7 through 14, find the measure of the specified angle.

7. $m\angle A$ in $\triangle ABC$, if $a = 2$, $b = 3$, and $c = 4$.
8. $m\angle C$ in $\triangle ABC$, if $a = 5$, $b = 6$, and $c = 8$.
9. $m\angle T$ in $\triangle BAT$, if $b = 6$, $a = 7$, and $t = 12$.
10. $m\angle E$ in $\triangle PEG$, if $p = 12$, $e = 20$, and $g = 16$.
11. $m\angle Y$ in $\triangle GYP$, if $g = 7$, $y = 5$, and $p = 13$.
12. $m\angle N$ in $\triangle GON$, if $g = 8$, $o = 3$, and $n = 12$.
13. $m\angle O$ in $\triangle NOD$, if $n = 1475$, $o = 2053$, and $d = 1428$.
14. $m\angle Q$ in $\triangle SQR$, if $s = 1504$, $q = 2465$, and $r = 1953$.
15. ***Accurate Drawing Problem No. 1.*** Construct accurately $\triangle ABC$ from Problem 1. Measure $b = 4$ cm, and $c = 5$ cm with a ruler, and use a protractor to construct $\angle B$ 51°. Then measure side a. Your answer should agree with the calculated value to within ±0.1 cm.
16. ***Accurate Drawing Problem No. 2.*** Construct accurately $\triangle ABC$ from Problem 8. Draw side c, 8 cm, as the base. Use a compass to mark off arcs of 5 cm and 6 cm from the two ends of side c. Where the arcs intersect will be point C. Then measure angle C with a protractor. The measured value should agree with the calculated value to within ±1°.

15-3 AREA OF A TRIANGLE

Objective:

Given the measures of two sides and the included angle, find the area of the triangle.

From geometry you recall that the area of a triangle is half the product of the base and the altitude. For $\triangle ABC$ in Figure 15-3a,

$$\text{Area} = \tfrac{1}{2}bh.$$

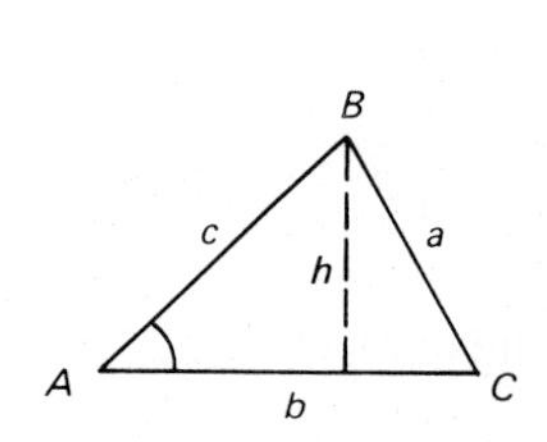

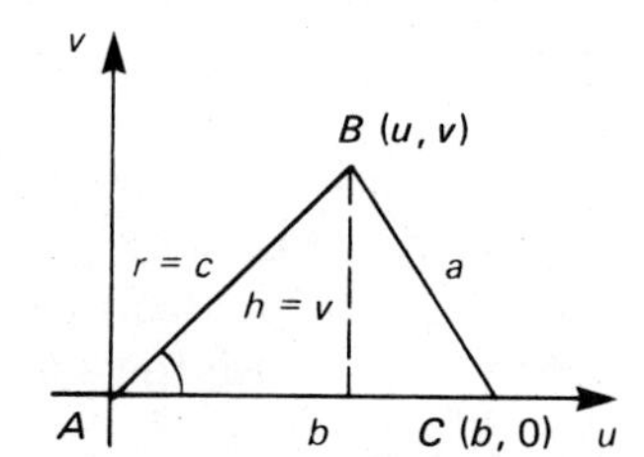

Figure 15-3a.

If you know the lengths of b and c and the measure of Angle A, you can calculate the altitude h in terms of these numbers. Constructing a uv-coordinate system as in the second sketch of Figure 15-3a, $B(u, v)$ becomes a point in a Cartesian coordinate system. By the definition of sine,

$$\frac{v}{r} = \sin A.$$

Multiplying both members of this equation by r gives

$$v = r \sin A.$$

Since $h = v$ and $c = r$, you can substitute these and get

$$h = c \sin A.$$

Substituting this value of h into the area equation gives:

$$\boxed{\text{Area} = \tfrac{1}{2}bc \sin A}$$

Example 1: Find the area of $\triangle ABC$ if $b = 13$, $c = 15$, and $m\angle A = 71°$.

Using the above equation,

$$\text{Area} = \tfrac{1}{2}(13)(15)\sin 71°$$

$= 92.1880...$ By calculator.

$\approx \underline{\underline{92.19}}$ Assuming 13 and 15 are *exact* and rounding off to four significant digits.

Example 2: Find the area of $\triangle HPJ$ if $h = 5$, $p = 7$, and $j = 11$.

In order to use the area equation, you must first know one of the angles. Suppose you decide to calculate $m\angle J$. Drawing a picture as in Figure 15-3b and applying the Law of Cosines as you did in Section 15-2,

$$j^2 = h^2 + p^2 - 2hp\cos J.$$

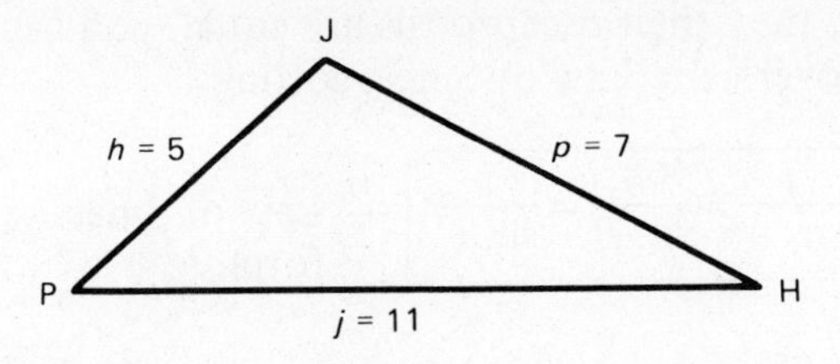

Figure 15-3b.

Solving for cos J gives:

$$\cos J = \frac{h^2 + p^2 - j^2}{2hp}$$

$= \dfrac{25 + 49 - 121}{2(5)(7)}$ Substitution.

$= -0.671428...$ By calculator.

$\therefore m\angle J = 132.177...°$ By calculator.

This value should be saved in memory, without round-off, for use next. Applying the area equation gives:

$\text{Area} = \tfrac{1}{2}hp\sin J$

$= \tfrac{1}{2}(5)(7)\sin J$ Substitution.

$= 12.9687...$ Recalling 132.177...° from memory.

$\approx \underline{\underline{12.97}}$ Rounding off to four significant digits.

The following exercise is designed to give you practice finding areas of specified triangles.

Exercise 15-3

Find the area of each triangle.

1. $\triangle ABC$, if $a = 5$, $b = 9$, and $m\angle C = 14°$.
2. $\triangle ABC$, if $b = 8$, $c = 4$, and $m\angle A = 67°$.
3. $\triangle RST$, if $r = 4.8$, $t = 3.7$, and $m\angle S = 43°\,10'$.
4. $\triangle XYZ$, if $y = 34.19$, $z = 28.65$, and $m\angle X = 138°\,27'$.
5. $\triangle MAP$, if $m = 6$, $a = 9$, and $p = 13$.
6. $\triangle ABX$, if $a = 5$, $b = 12$, and $x = 13$.

7. You may recall from geometry that Heron's Formula can be used to find the area in *one* computation. This formula is

$$\text{Area} = \sqrt{s(s-a)(s-b)(s-c)}\,,$$

where s (for "semiperimeter") equals half the perimeter of the triangle. Use Heron's Formula to find the area of ΔMAP in Problem 5.

8. Use Heron's Formula to find the area of ΔABX in Problem 6.

15-4 OBLIQUE TRIANGLES – LAW OF SINES

The Law of Cosines may be used directly when you know two sides and the included angle. If you know only *one* side length, the Law of Cosines cannot be used. In this section you will find out what can be done in this case to calculate other side measures.

Objective:

Given the measure of an angle, its opposite side, and one other angle measure, calculate the length of another side.

In the previous section you learned that the area of a triangle such as ΔABC in Figure 15-4a is

$$\text{Area} = \tfrac{1}{2}bc \sin A.$$

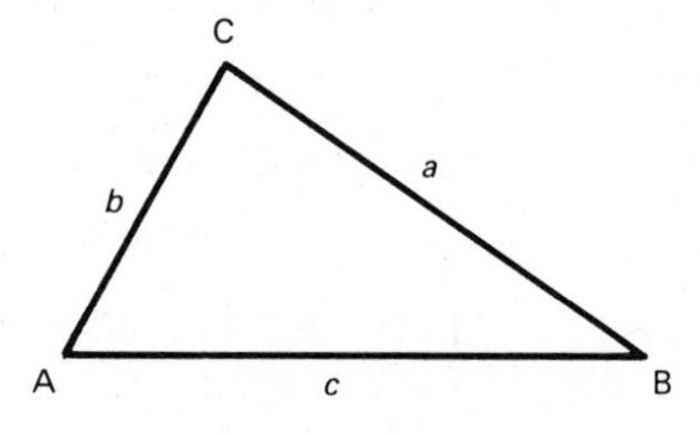

Figure 15-4a.

The area is also equal to $\frac{1}{2}ac \sin B$ and to $\frac{1}{2}ab \sin C$, since the area is *constant* no matter which sides and angle you use to measure it. Setting these expressions equal to each other gives

$$\tfrac{1}{2}bc \sin A = \tfrac{1}{2}ac \sin B = \tfrac{1}{2}ab \sin C.$$

Multiplying all three members by 2 eliminates the fraction $\frac{1}{2}$ each time it occurs. Dividing all three members by abc produces a rather startling simplification.

$$\frac{bc \sin A}{abc} = \frac{ac \sin B}{abc} = \frac{ab \sin C}{abc}\,.$$

In each member, the coefficients of the sine cancel, giving:

$$\boxed{\frac{\sin A}{a} = \frac{\sin B}{b} = \frac{\sin C}{c}} \quad \leftarrow \text{Law of Sines, first form.}$$

This relationship is called the Law of Sines for reasons that should be obvious! Since a is opposite Angle A, and so forth, the law actually says,

> "Within any given triangle, the ratio of the sine of an angle to the length of its opposite side is *constant.*"

Since you know that if two non-zero numbers are equal, then their reciprocals are equal, you can "flip over" the Law of Sines, getting:

$$\boxed{\frac{a}{\sin A} = \frac{b}{\sin B} = \frac{c}{\sin C}} \quad \leftarrow \text{Law of Sines, second form.}$$

The following examples show you how the Law of Sines may be used to accomplish the objective of this section. Watch out for surprises!

Because of the different combinations of sides and angles that might be given in a triangle, it is convenient to revive some terminology that you may recall from geometry. The abbreviation "SAS" stands for "side, angle, side." This means that as you go around the perimeter of the triangle, you are given the length of a side, the measure of the next angle, and the length of the next side. Thus, "SAS" is equivalent to knowing two sides and the included angle. Similar meanings are attached to ASA, AAS, SSA, and SSS.

Example 1: Given AAS, find the other sides.

In $\triangle ABC$, $m\angle B = 64°$, $m\angle C = 38°$ and $b = 9$. Find c and a.

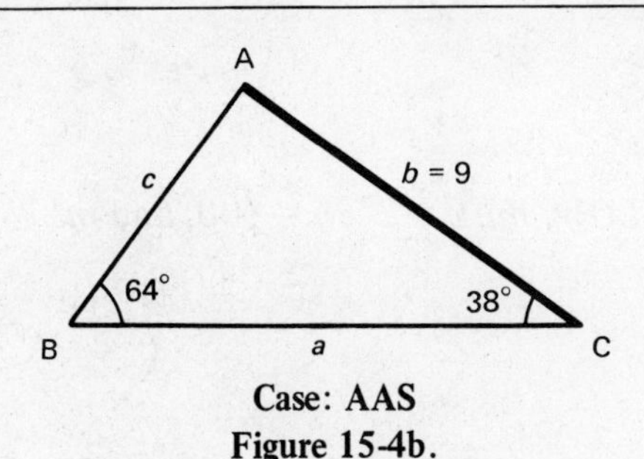

Case: AAS
Figure 15-4b.

The first thing to do is draw a picture, as in Figure Figure 15-4b. Using part of the Law of Sines,

$$\frac{c}{\sin C} = \frac{b}{\sin B}.$$

You pick two of the three members of the Law of Sines equation in such a way that the quantity you are looking for is in the *numerator* of the left member and the *known* side and opposite angle are in the *right* member. Multiplying each member by sin C isolates the quantity you seek on the left side, giving:

$$c = \frac{b \sin C}{\sin B}$$

$$= \frac{9 \sin 38°}{\sin 64°} \quad \text{Substitution.}$$

$$= 6.16487... \quad \text{By calculator.}$$

$$\approx \underline{6.165} \quad \text{Rounding off to four significant digits.}$$

To find a, you must first find the measure of its opposite angle A. Since the sum of the three angle measures is 180°,

$$m\angle A = 180° - 38° - 64° = 78°.$$

Using the appropriate form of the Law of Sines:

$$\frac{a}{\sin A} = \frac{b}{\sin B}$$

$$\therefore a = \frac{b \sin A}{\sin B} \quad \text{Multiplication by sin } A.$$

$$= \frac{9 \sin 78°}{\sin 64°} \quad \text{Substitution.}$$

$$= 9.79460... \quad \text{By calculator.}$$

$$\approx \underline{9.795} \quad \text{Correct to four significant digits.}$$

Example 2: Given ASA, find another side.

In $\triangle ABC$, $a = 8$, $m\angle B = 64°$, and $m\angle C = 38°$. Find c.

Drawing a picture as in Figure 15-4c, you immediately notice that you do *not* know the measure of Angle A, opposite the given side. To use the Law of Sines, you must know a side and the opposite angle. Again using the fact that the sum of the angle measures is 180°,

$$m\angle A = 180° - 64° - 38° = 78°.$$

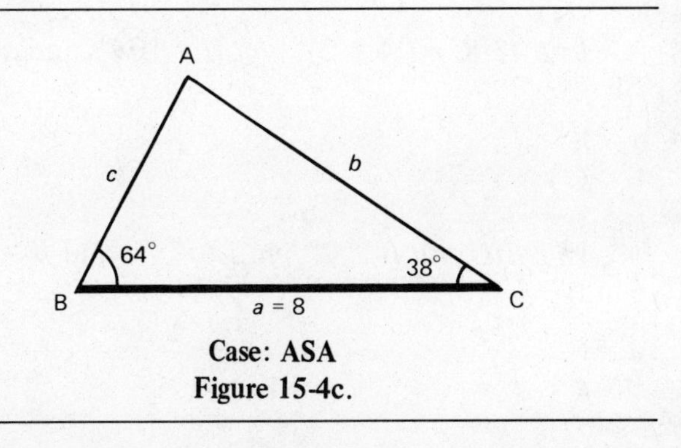

Case: ASA
Figure 15-4c.

Using the appropriate part of the Law of Sines:

$$\frac{c}{\sin C} = \frac{a}{\sin A}$$

$\therefore c = \dfrac{a \sin C}{\sin A}$ Multiplication by sin C.

$= \dfrac{8 \sin 38^\circ}{\sin 78^\circ}$ Substitution of known information.

$= 5.03532...$ By calculator.

$\approx \underline{5.035}$ Correct to four significant digits.

The exercise that follows is designed to give you practice using the Law of Sines to find unknown side lengths when you know one side and its opposite angle.

Exercise 15-4

1. In ΔABC, $m\angle A = 52^\circ$, $m\angle B = 31^\circ$, and $a = 8$. Find:

a. b
b. c

2. In ΔPQR, $m\angle P = 13^\circ$, $m\angle Q = 133^\circ$, and $q = 9$. Find:

a. p
b. r

3. In ΔAHS, $m\angle A = 27^\circ$, $m\angle H = 109^\circ$, and $a = 120$. Find:

a. h
b. s

4. In ΔBIG, $m\angle B = 2^\circ$, $m\angle I = 79^\circ$, and $b = 20$. Find:

a. i
b. g

5. In ΔPAF, $m\angle P = 28^\circ$, $f = 6$, and $m\angle A = 117^\circ$. Find:

a. a
b. p

6. In ΔJAW, $m\angle J = 48^\circ$, $a = 5$, and $m\angle W = 73^\circ$. Find:

a. j
b. w

7. In ΔALP, $m\angle A = 85^\circ$, $p = 30$, and $m\angle L = 87^\circ$. Find:

a. a
b. l

8. In ΔLOW, $m\angle L = 2^\circ$, $o = 500$, and $m\angle W = 3^\circ$. Find:

a. l
b. w

9. ***Law of Sines for Angles Problem.*** The Law of Sines can be used to find an unknown *angle* measure. However, the technique is risky! In this problem you will find out *why*.

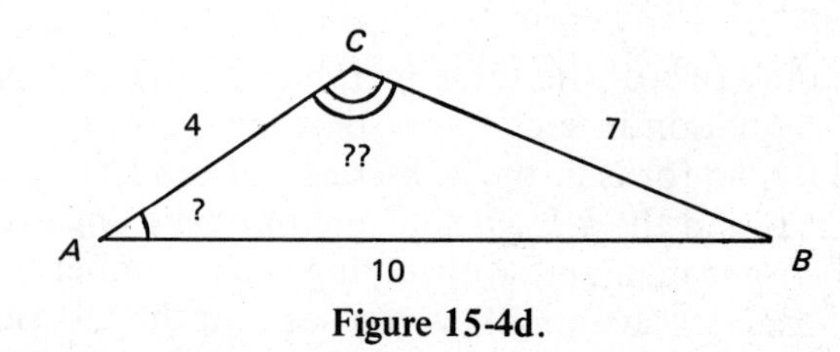

Figure 15-4d.

Triangle ABC has sides of 4, 7, and 10 units, as shown in Figure 15-4d.

a. Use the Law of Cosines to find $m\angle A$, as you did in Section 15-2.
b. Use the answer to part (a) in the Law of *Sines* to find $m\angle C$.
c. Find $m\angle C$ again, directly from the side lengths, using the Law of Cosines.

d. Your answers to parts (b) and (c) probably do not agree! If not, and if you have made no computation errors, your error is in interpreting the results of the Law of Sines in part (b). Use the fact that there is also an *obtuse* angle whose sine is the same as in part (b) to correct your answer to part (b).
e. Explain why it is dangerous to use the Law of Sines to find an angle measure, but is *not* dangerous to use the Law of Cosines.

10. ***Accurate Drawing Problem.*** Using ruler, protractor, and a sharp pencil, draw a triangle with base 10.0 centimeters, and angles of 40° and 30° at the ends of the base. Measure the side opposite the 30° angle. Then calculate its length by the Law of Sines. Your measured value should be within ±0.1 centimeter of the calculated value.

15-5 THE AMBIGUOUS CASE

You may recall from geometry that a triangle is not necessarily determined uniquely by the measures of two sides and a non-included angle. This case is called SSA (side, side, angle), and is illustrated in Figure 15-5a.

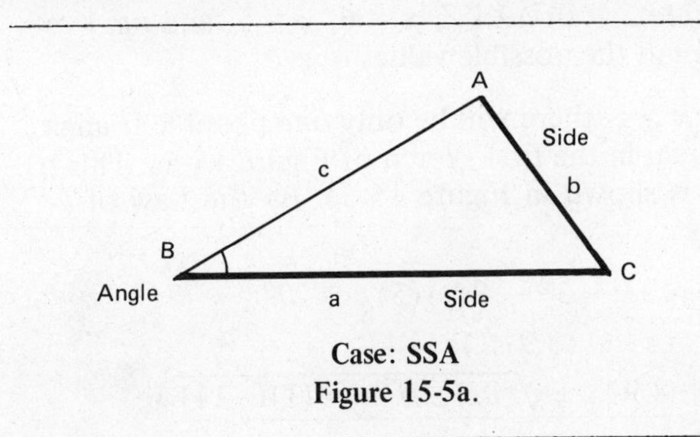

Case: SSA
Figure 15-5a.

There are *four* ways a triangle ABC would come out if you knew the lengths of a and b and the measure of Angle B. To see why, it is helpful to start *constructing* the triangle. Figure 15-5b shows a and Angle B constructed.

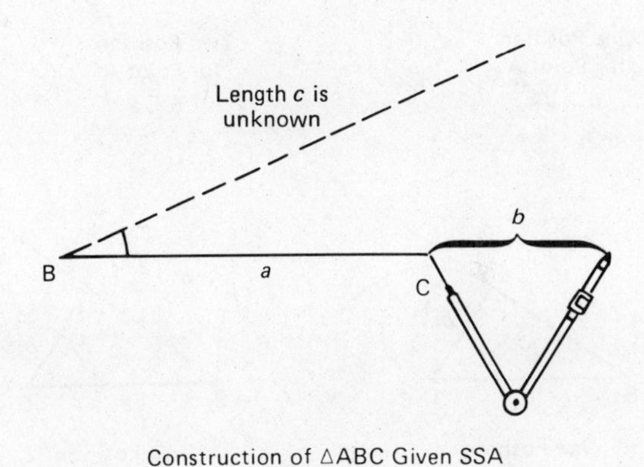

Construction of △ABC Given SSA
Figure 15-5b.

Since c is not given, you simply draw a long line in the correct direction, making an angle of $m\angle B$ with a. To complete the triangle, you place the point of a compass at Point C, open it to length b, which *is* given, and draw an arc. Wherever the compass arc cuts the dotted line in Figure 15-5b is the correct position for Point A. Figure 15-5c on the next page shows the four possible ways Point A might come out. There are either *two, one,* or *no* possible triangles when you are given SSA. For this reason, SSA is sometimes called the "ambiguous case." *Ambiguous* means "two or more possible meanings."

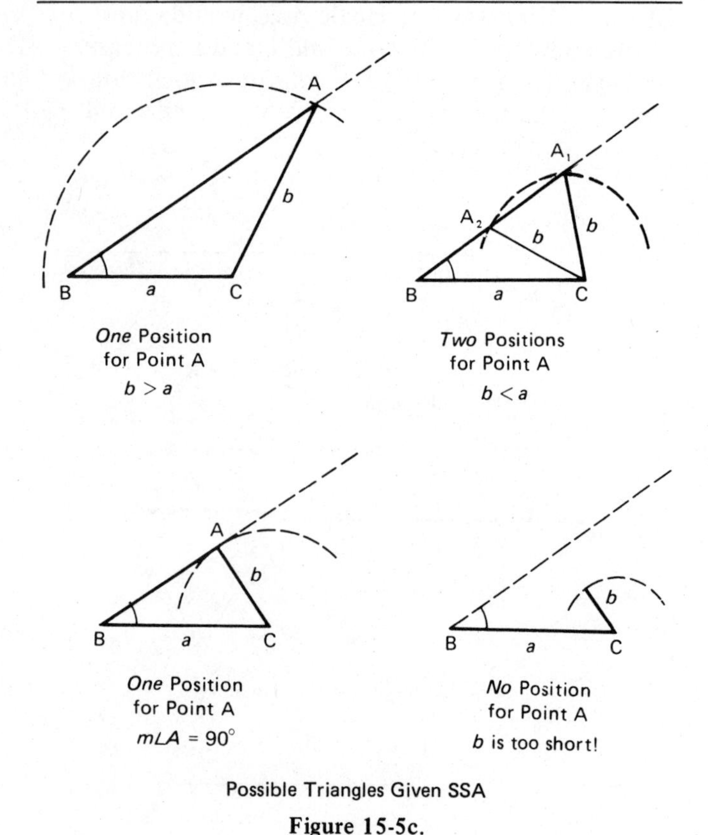

Possible Triangles Given SSA

Figure 15-5c.

Objective:

Given SSA, determine whether or not there are possible triangles, and if so, find the other side length and angle measures.

Example 1: In ΔXYZ, $x = 4$, $y = 5$, and $m\angle Y = 27°$. Find the possible values of z.

Since $x < y$, there can be two possible triangles, as shown in Figure 15-5d. The Law of Sines cannot be used directly to find z, since $m\angle Z$ is unknown. The Law of Cosines can be used, however, because there are two side lengths known. The process is complicated by the fact that z is not opposite the known angle. You write:

$$4^2 = z^2 + 5^2 - 2(z)(5)\cos 27°$$

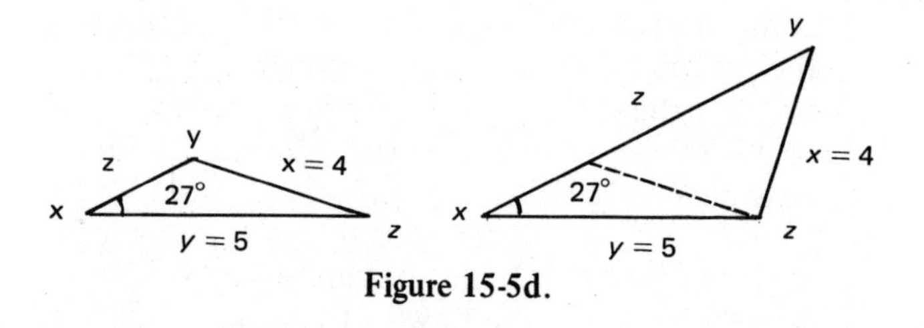

Figure 15-5d.

This is a quadratic equation in the variable z. Making the left member equal 0, and using the Quadratic Formula gives

$$0 = z^2 - (10\cos 27°)z + 25 - 16$$

$$0 = z^2 - (8.91...)z + 9$$

$$z = \frac{8.91... \pm \sqrt{(8.91...)^2 - 4(5)(9)}}{2(1)}$$

$$z = \frac{8.91... \pm 6.58...}{2}$$

$$z \approx \underline{7.75 \text{ or } 1.16}$$

When you evaluate the Quadratic Formula, it helps to evaluate the radical first. Then save its value in memory for use when you calculate the second value of z.

Example 2: In ΔXYZ, $x = 6$, $y = 5$, and $m\angle Y = 27°$. Find the possible values of z.

Since $x > z$, there will be only *one* possible triangle, as shown in the first sketch of Figure 15-5c. This triangle is shown in Figure 15-5e. By the Law of Cosines,

$$6^2 = z^2 + 5^2 - 2(z)(5)\cos 27°$$

$$0 = z^2 - (8.91...)z - 11$$

$$z = \frac{8.91... \pm \sqrt{(8.91...)^2 - 4(1)(-11)}}{2}$$

$$z = 10.009... \text{ or } -1.099...$$

$$z \approx \underline{10.01}$$

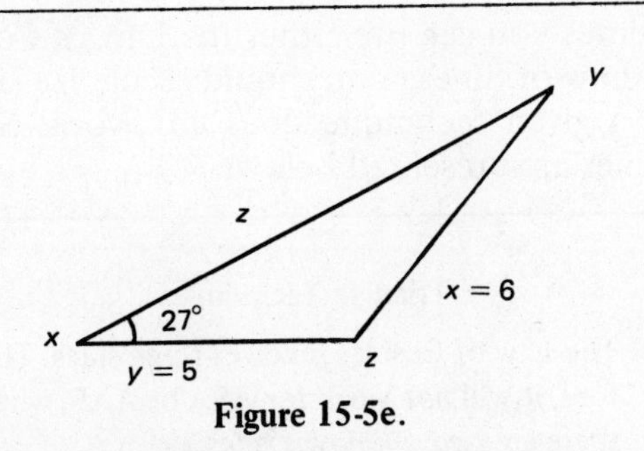

Figure 15-5e.

The negative value of z confirms that there is only *one* possible triangle.

Example 3: In ΔXYZ, $x = 2$, $y = 5$, and $m\angle X = 27°$. Find the possible values of z.

This example is the same as Examples 1 and 2 except for the value of x. By the Law of Cosines,

$$2^2 = z^2 + 5^2 - 2(z)(5)\cos 27°$$

$$0 = z^2 - (8.91...)z + 21$$

$$z = \frac{8.91... \pm\sqrt{(8.91...)^2 - 4(1)(21)}}{2(1)}$$

$$z = \frac{8.91... \pm \sqrt{-4.61...}}{2}$$

No solutions because of the negative radicand.

∴ There is *no triangle.*

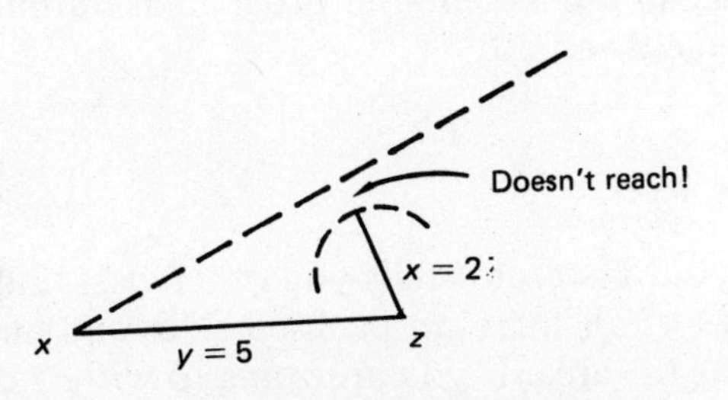

Figure 15-5f.

The radicand will be negative whenever the side opposite the given angle is too short to reach the other side, as shown in Figure 15-5f.

In the following exercise you will analyze more triangles of the case SSA.

Exercise 15-5

For Problems 1 through 8, find the possible lengths of the indicated side.

1. In ΔABC, $m\angle B = 34°$, $a = 4$, and $b = 3$. Find c.
2. In ΔXYZ, $m\angle X = 13°$, $x = 12$, and $y = 5$. Find z.
3. In ΔABC, $m\angle B = 34°$, $a = 4$, and $b = 5$. Find c.
4. In ΔXYZ, $m\angle X = 13°$, $x = 12$, and $y = 15$. Find z.
5. In ΔABC, $m\angle B = 34°$, $a = 4$, and $b = 2$. Find c.
6. In ΔXYZ, $m\angle X = 13°$, $x = 12$, and $y = 60$. Find z.
7. In ΔRST, $m\angle R = 130°$, $r = 20$, and $t = 16$. Find s.
8. In ΔOBT, $m\angle O = 170°$, $o = 19$, and $t = 11$. Find b.

The Law of Sines can be used to find an angle in the SSA case. But you must be careful because there can be two different angles, one acute and the other obtuse, which have the same sine. For Problems 9 through 12, determine beforehand whether there can be two triangles or just one. Then find the possible values of the indicated angle measure.

9. In ΔABC, $m\angle A = 19°$, $a = 25$, and $c = 30$. Find $m\angle C$.
10. In ΔHDJ, $m\angle H = 28°$, $h = 50$, and $d = 20$. Find $m\angle D$.
11. In ΔXYZ, $m\angle X = 58°$, $x = 9.3$, and $z = 7.5$. Find $m\angle Z$.
12. In ΔBIG, $m\angle B = 110°$, $b = 1000$, and $g = 900$. Find $m\angle G$.
13. ***Accurate Drawing Problem No. 1.*** Triangles ABC in Problems 1, 3, and 5 differ only in the length of side b. Draw side a 4 centimeters long as the base. Then construct Angle B of measure 34° at one end of the base.

a. Use a compass to mark off the two possible triangles if $b = 3$ centimeters, as in Problem 1. Measure the two possible values of c. Your answers should be within ± 0.1 centimeter of the calculated values.
b. Use a compass to mark off $b = 5$ centimeters, and in Problem 3. Measure the value of c, and confirm that it agrees with the calculated value. Then extend segment $\overline{AB}$ beyond angle B. Find the point on this segment where the 5 centimeter arc cuts it. Show that the distance between this point and B equals the *negative* value of c that is discarded in working Problem 3.
c. Use a compass to draw an arc of radius $b = 2$ centimeters, as in Problem 5. Show that this arc *misses* the other side of Angle B, and thus that there is *no* possible triangle.

14. ***Accurate Drawing Problem No. 2.*** Repeat Problem 13 for Triangles XYZ in Problems 2, 4, and 6.

15-6 GENERAL SOLUTION OF TRIANGLES

You have learned the techniques necessary for analyzing oblique triangles when you are told which to use, the Law of Sines or the Law of Cosines. Before you tackle the real-world problems in Section 15-7, you must be sure that you can select the appropriate technique when you are *not* told which one to use.

Objective:

Given SSS, SAS, ASA, AAS, or SSA, be able to select the appropriate technique and to calculate the other side and angle measures and the area of the triangle.

Sometimes you can use either the Law of Cosines or the Law of Sines. You should recognize situations where a given technique does *not* work. Some guidelines are presented below.

Triangle Techniques

1. The Law of Cosines involves *three* sides. Therefore, it will *not* work for ASA or AAS, where there are *two unknown sides.*
2. The law of Sines involves the ratio of the sine of an angle to the length of its *opposite* side. Therefore, it will *not* work where *no* angle is known (SSS) or where only *one* angle is known, but *not* its opposite side (SAS).
3. The Law of Sines should *not* be used to find *angle* measures unless you know in advance whether the angle is obtuse or acute.
4. The area formula requires you to know SAS. If you do not know two sides and the included angle, you must first find them. Heron's Formula (Exercise 15-3, Problem 7) will work if you know SSS.

The following exercise requires you to select the appropriate technique and then work the problem. Your instructor may assign you only certain parts of each problem. Since working all parts of the problems requires much computation, you may wish to write and use a computer program as outlined in Problems 29 and 30.

Exercise 15-6

Unless your instructor tells you otherwise, find the measures of all three unspecified sides and angles as well as the area. If you are going to write a computer program to do this, go immediately to Problem 29 and then come back to Problems 1 through 28.

Case	*a*	*b*	*c*	*A*	*B*	*C*
1. SAS	3	4	———	———	———	71°40′
2. SAS	8	5	———	———	———	32°10′
3. SAS	30	60	———	———	———	23°50′
4. SAS	18	40	———	———	———	82°30′
5. SAS	100	210	———	———	———	113°20′
6. SAS	2000	1700	———	———	———	142°00′
7. SSS	8	9	7	———	———	———
8. SSS	4	3	2	———	———	———
9. SSS	3	6	4	———	———	———
10. SSS	18	10	9	———	———	———
11. SSS	3	9	4	———	———	———
12. SSS	18	8	9	———	———	———
13. ASA	———	———	400	143°10′	8°20′	———
14. ASA	———	———	30	122°50′	15°00′	———
15. ASA	———	———	50	11°30′	27°40′	———
16. ASA	———	———	17	84°20′	87°30′	———
17. AAS	6	———	———	56°20′	64°30′	———
18. AAS	10	———	———	139°10′	38°40′	———
19. SSA	7	5	———	25°50′	———	———
20. SSA	10	6	———	31°10′	———	———
21. SSA	5	7	———	25°50′	———	———
22. SSA	6	10	———	31°10′	———	———
23. SSA	5	7	———	126°40′	———	———
24. SSA	10	6	———	144°50′	———	———
25. SSA	7	5	———	126°40′	———	———
26. SSA	3	10	———	31°10′	———	———
27. SSA	3	5	———	36°52.19386′	———	———
28. SSA	5	13	———	22°37.19189′	———	———

29. *Computer Solution of Triangles:* In this problem you will write a computer program to find unspecified sides, angles, and area of a given triangle. Since there are five different ways the data might be given and a different sequence of computations for each case, the program is divided into blocks. Your class may wish to divide into seven groups, each group responsible for one of the tasks outlined below. By writing the program in this manner, you will get experience in the way really big programs are written, such as those used to get spacecraft to the Moon. The length of such programs makes them impossible for one person to write in a reasonable length of time.

Specific tasks are given below. (If you are using a language other than BASIC, you may need to modify some of the instructions.)

Task 1: Write a "menu." This part of the program directs the activities of all other parts. It should first print on the screen a way for you to select the case, such as

```
ENTER CASE
1 FOR SSS
2 FOR SAS
3 FOR SSA
4 FOR AAS
5 FOR ASA
WHICH?
```

Depending on what the person using the program enters, the menu should direct the computer to the proper subroutine described below. Upon returning from the subroutine, this portion of the program should direct the computer to find the area, convert angles in radians (as the computer calculates them) to angles in degrees and minutes, and print the results in a form such as

```
CASE: SAS
S1 = 3
S2 = 4
S3 = 4.177
A1 = 42D, 59M
A2 = 65D, 21M
A3 = 71D, 40M
AREA = 5.695
```

Task 2: Write a subroutine for the case SSS. It should begin with a statement letting the user input the three lengths. The screen should display a message such as

```
TYPE S1, S2, S3
```

The computer should then calculate the three angles, and return to the menu. The subroutine should also be able to detect when the three given sides are impossible for a triangle.

Task 3: Write a subroutine for the case SAS. It should first print a message such as

```
TYPE S1, D3, M3 S2
```

where S1 and S2 are two side lengths, and D3 and M3 are the degrees and minutes in Angle 3, included between sides 1 and 2. The subroutine should calculate the third side length and the measures of the other two angles, then return to the menu.

Task 4: Write a subroutine for the case SSA. Since this is the ambiguous case, the subroutine should allow the computer to detect when there are two possible triangles, or no triangle, and return an appropriate message to the menu so that it will know what to do.

Task 5: Write a subroutine for the case AAS, as above.

Task 6: Write a subroutine for the case ASA, as above.

Task 7: Write any special functions needed. If you use ordinary BASIC, you will have COS(X), SIN(X), and ATN(X) available. These are the *circular* cosine, sine, and Arctangent, respectively, where the argument or Arctangent is in *radians.* So you will need to define

a. FNR(X) for converting X degrees to radians.
b. FND(X) and FNM(X) for converting X radians to degrees and minutes, respectively.
c. FNC(X) and FNS(X) for finding arccos X

and arcsin X, respectively, from ATN(X). The properties of Exercise 13-9, Problem 99 and 100, should be helpful.

30. Debug the program from Problem 29 by using it to work the *odd*-numbered problems above. If your program can solve each triangle correctly, you may consider it to be completely debugged.

15-7 VECTORS

A *vector quantity* is something that has *direction* as well as magnitude (size). Velocity is an example. When you are traveling, it is important to know in what direction as well as how fast! A quantity that has *no* direction, such as volume, is called a *scalar* quantity. The *scalar* quantity "speed" and a direction combine to form the *vector* quantity "velocity."

In this section you will use directed line segments to represent vector quantities. The *length* of the line segment represents the *magnitude* (size) of the vector quantity, and the direction of the line segment represents the direction of the vector quantity. An arrowhead is used to distinguish the head of a vector from its tail (Figure 15-7a).

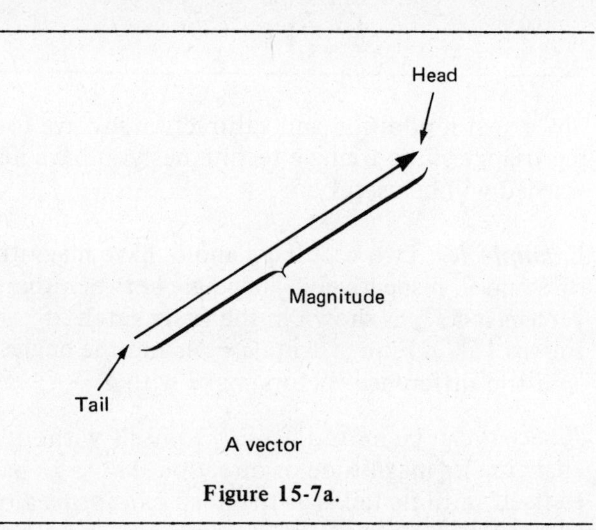

Figure 15-7a.

If a variable is used to represent a vector, you put a small arrow over the top of it, such as $\vec{x}$, to distinguish it from a scalar. The three vectors in Figure 15-7b are considered to be *equal* since they have the same length and the same direction.

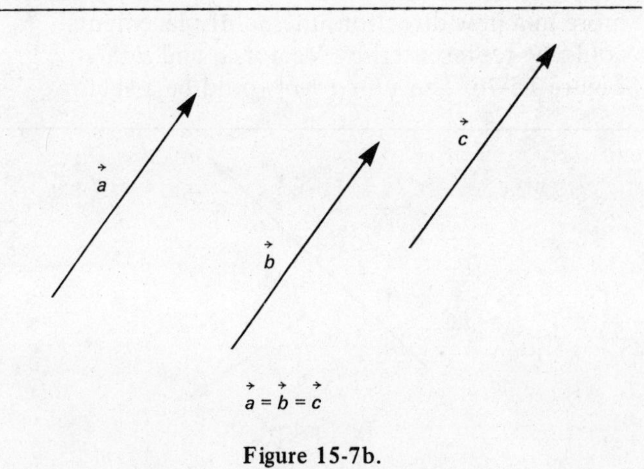

Figure 15-7b.

The above discussion leads to the following definitions.

DEFINITIONS

1. A ***vector*** is a directed line segment.

2. Two vectors are ***equal*** if and only if they have the same magnitude and the same direction.

3. The ***absolute value*** of a vector is its length or magnitude.

Having invented a new kind of mathematical quantity, you must now find out how to operate with it.

Objective:

Given two vectors, be able to *add* them or *subtract* them.

Vector Addition: Vectors are added by placing the tail of one vector at the head of the other, as shown in Figure 15-7c. The *sum* of the vectors is defined to be the vector that goes from the tail of the first vector to the head of the last one. This definition arises from vectors in the real world. If you walk 20 meters in a certain direction, then turn and walk 13 meters more in a new direction, these "displacements" could be represented by Vectors $\vec{a}$ and $\vec{b}$ in Figure 15-7c. The sum $\vec{a} + \vec{b}$ would be a vector representing your net displacement from the starting point. A displacement of $\vec{a} + \vec{b}$ would produce the same result as a displacement of $\vec{a}$ followed by another displacement of $\vec{b}$. For this reason, the sum of two vectors is often called the *resultant* vector.

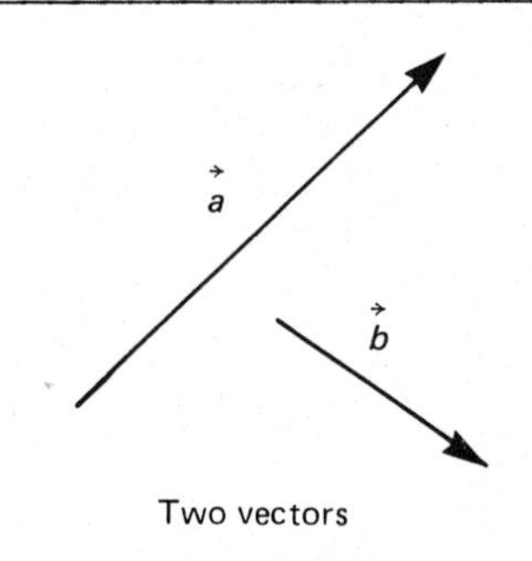

Two vectors

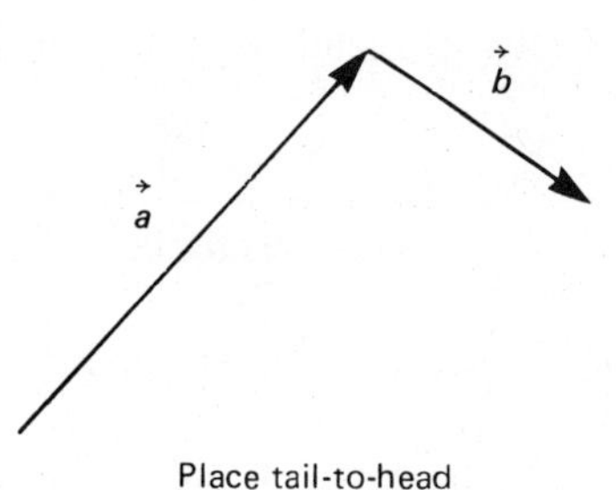

Place tail-to-head

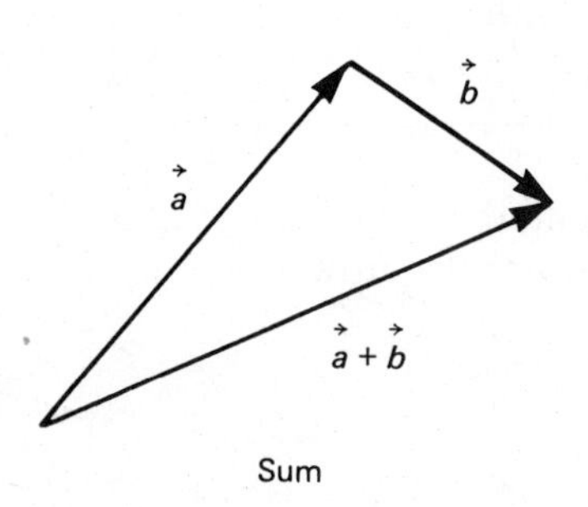

Sum

Figure 15-7c.

Vector Subtraction: The opposite of a number x, $-x$, is the same distance from the origin as x, but in the opposite direction. Similarly, the opposite of vector $\vec{b}, -\vec{b}$, is defined to be a vector having the *same magnitude* as $\vec{b}$ but pointing in the *opposite direction* (Figure 15-7d).

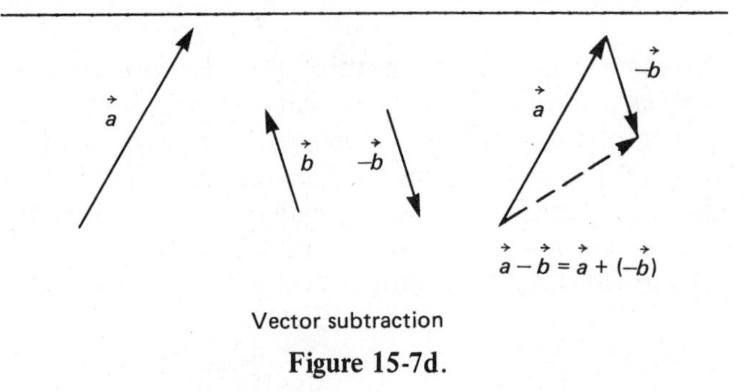

Vector subtraction

Figure 15-7d.

The definition of vector subtraction follows directly from the definition of subtraction for real numbers.

DEFINITION

$$\vec{a} - \vec{b} = \vec{a} + (-\vec{b})$$

Since vector addition and subtraction involve forming triangles, the triangle techniques you have just learned will be useful.

Example 1: Two vectors, $\vec{a}$ and $\vec{b}$, have magnitudes of 5 and 9, respectively. The angle between the vectors is 53°, as shown in the first sketch of Figure 15-7e. Find $|\vec{a} + \vec{b}|$, $|\vec{a} - \vec{b}|$, and the angles these sum and difference vectors make with $\vec{a}$.

A vector can be moved *parallel* to itself without changing its magnitude or direction. Moving $\vec{b}$ parallel to itself until its tail is at the head of $\vec{a}$ forms a triangle

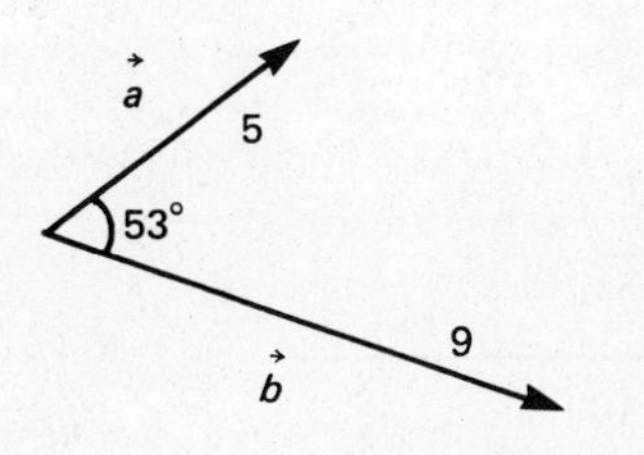

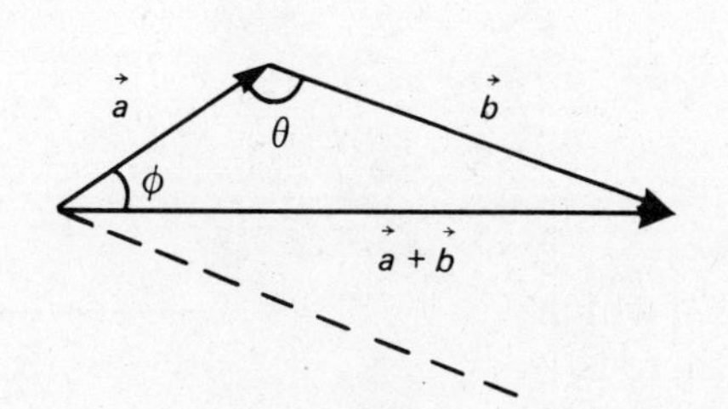

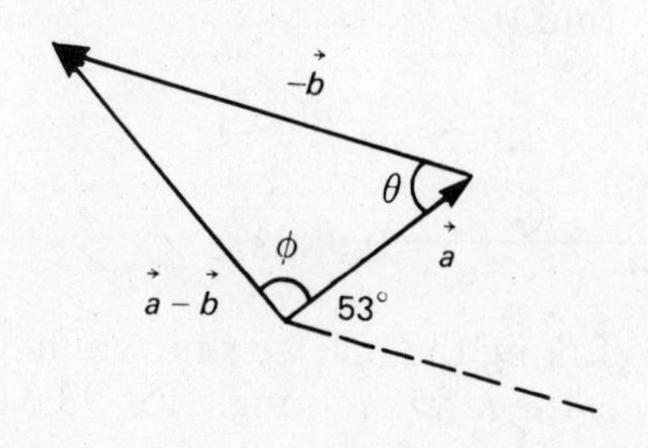

Figure 15-7e.

(Figure 15-7e). Since θ and the 53° angle are supplementary,

$$\theta = 180° - 53° = 127°.$$

Vectors $\vec{a}$ and $\vec{b}$ are two sides of a triangle with included angle θ and third side $\vec{a} + \vec{b}$. By the Law of Cosines,

$|\vec{a} + \vec{b}|^2 = 5^2 + 9^2 - 2(5)(9)\cos 127°$

$= 160.163...$ — By calculator.

$\therefore |\vec{a} + \vec{b}| = 12.655...$ — Taking the square root.

$\approx \underline{\underline{12.66}}$ — Rounding off.

The 12.655... should be stored, without round-off, in the calculator's memory for use in the next part of the problem.

To find the angle ϕ that $\vec{a} + \vec{b}$ makes with $\vec{a}$, you may use the Law of Cosines as in Section 15-2.

$$\cos\phi = \frac{5^2 + (12.655...)^2 - 9^2}{2(5)(12.655...)}$$

$= 0.8230...$ — By calculator.

$\therefore \phi = 34.607...°$ — Taking inverse cosine.

$\approx \underline{\underline{34°36'}}$ — Transforming to degrees and minutes.

To find $\vec{a} - \vec{b}$, you simply turn $\vec{b}$ around in the opposite direction and slide its tail to the head of $\vec{a}$, as shown in the third sketch of Figure 15-7e. Angle θ between $\vec{a}$ and $-\vec{b}$ is now an alternate interior angle of the 53° angle. So θ is also 53°. Applying the Law of Cosines:

$|\vec{a} - \vec{b}|^2 = 5^2 + 9^2 - 2(5)(9)\cos 53°$

$= 51.836...$ — By calculator.

$\therefore |\vec{a} - \vec{b}| = 7.199...$ — Taking the square root.

$\approx \underline{\underline{7.200}}$ — Rounding off.

Using the Law of Cosines to find ϕ:

$$\cos\phi = \frac{5^2 + (7.199...)^2 - 9^2}{2(5)(7.199...)}$$

$= 0.0578...$ — By calculator.

$\therefore \phi = 93.315...°$ — Taking inverse cosine.

$\approx \underline{\underline{93°19'}}$ — Transforming to degrees and minutes.

Note: The Law of Sines could have been used to find ϕ.

$$\sin \phi = \frac{9 \sin 53^\circ}{7.119...}$$

$$= 0.9983... .$$

Taking the inverse sine would give 86.684...°, which is the *reference* angle for 93.315...°. Since there is no easy way to tell whether ϕ is obtuse or acute, it is preferable to use the Law of Cosines technique. The sign of $\cos \phi$ tells that ϕ is obtuse.

Vectors can be used to find displacements of objects that move on the earth's surface. Navigators commonly measure angles clockwise from north as shown in Figure 15-7f, rather than counterclockwise from the positive x-axis. The direction thus determined is called a *bearing*. The figure shows a vector with a bearing of 250 degrees.

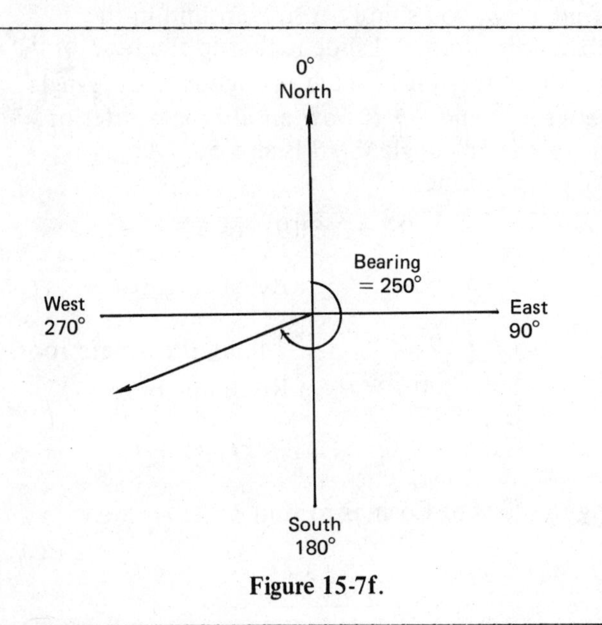

Figure 15-7f.

Example 2: An object moves 90 meters due south (bearing 180 degrees), then turns and moves 40 more meters along a bearing of 250 degrees (Figure 15-7g).

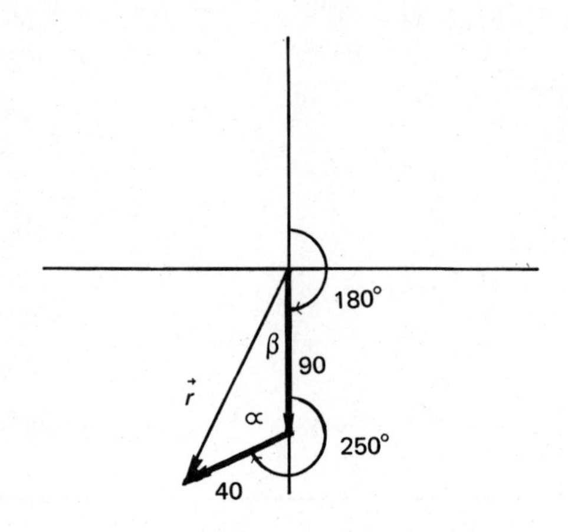

Figure 15-7g.

a. Find the resultant of these two displacement vectors.
b. What is the bearing from the ending point back to the starting point?

a. The resultant, $\vec{r}$, goes from the tail of the first vector to the head of the last.

$$\text{Angle } \alpha = 360^\circ - 250^\circ = 110^\circ.$$

By the Law of Cosines,

$$|\vec{r}|^2 = 90^2 + 40^2 - 2(90)(40) \cos 110^\circ$$

$$= 12162.54...$$

$$\therefore |\vec{r}| = 110.283...$$

By the Law of Cosines,

$$\cos \beta = \frac{90^2 + (110.283...)2 - 40^2}{2(90)(110.283...)}$$

$$= 0.9401...$$

$$\therefore \beta = 19.927...^\circ$$

From Figure 15-7g,

$$\text{Bearing} = 180^\circ + 19.927...^\circ$$

$$= 199.927...^\circ$$

Vector is 110.3 at 199.9°

b. To find the bearing from the end point to the starting point, all you need realize is that it points the opposite direction.

$$\text{Bearing} = 199.9° + 180° = 379.9°.$$

Since this is greater than 360°, you subtract 360° (one full revolution) getting

$$\underline{\underline{\text{Bearing} = 19.9°}}$$

The following exercise gives you practice adding and subtracting vectors.

Exercise 15-7

For Problems 1 through 4, find $|\vec{a} + \vec{b}|$, $|\vec{a} - \vec{b}|$, and the angle that each of these resultant vectors makes with $\vec{a}$, (Figure 15-7h).

1. $|\vec{a}| = 7$, $|\vec{b}| = 11$, $\theta = 73°$
2. $|\vec{a}| = 8$, $|\vec{b}| = 2$, $\theta = 41°$
3. $|\vec{a}| = 9$, $|\vec{b}| = 20$, $\theta = 163°$
4. $|\vec{a}| = 10$, $|\vec{b}| = 30$, $\theta = 122°$

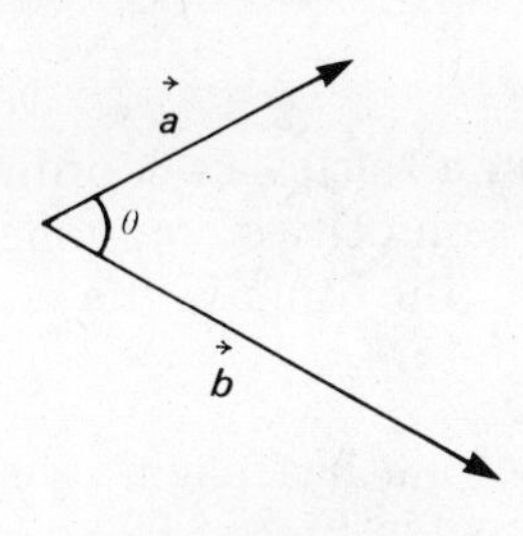

Figure 15-7h.

For Problems 5 through 8, do the following:

a. Find the resultant of the two given displacements. Express the answer as a distance and a bearing (clockwise from north) from the starting point to the ending point.
b. Tell the bearing from the ending point back to the starting point.
c. Draw the vectors on graph paper, using ruler and protractor, and thus show that your answers are correct to 0.1 unit of length and 1 degree of angle.

5. 11 units north (0°) followed by 5 units along a bearing of 70°.
6. 8 units east (90°) followed by 6 units along a bearing of 210°.
7. 6 units west (270°) followed by 14 units along a bearing of 110°.
8. 4 units south (180°) followed by 9 units along a bearing of 320°.

15-8 VECTORS–RESOLUTION INTO COMPONENTS

Sometimes it is important to reverse the addition process and express a single vector as the sum of two other vectors. For example, if a pilot knows the plane's air speed and angle of climb, the rate of climb and the ground velocity can be calculated (Figure 15-8a). For this purpose you must be able to resolve, or break up, a vector into components whose sum is the original vector. Two pieces of background information are helpful.

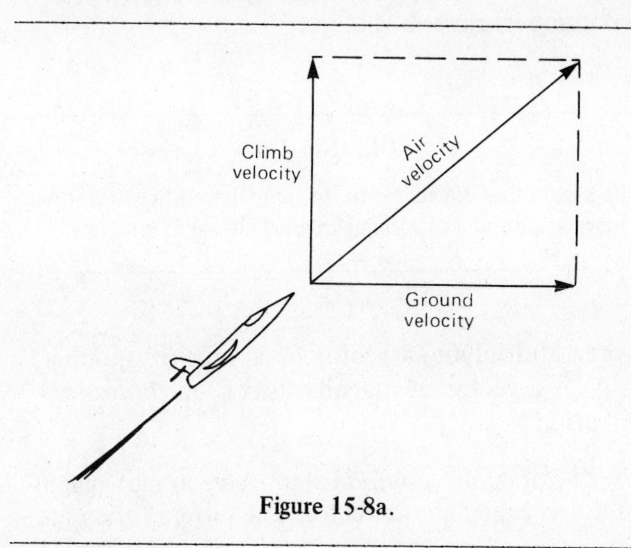

Figure 15-8a.

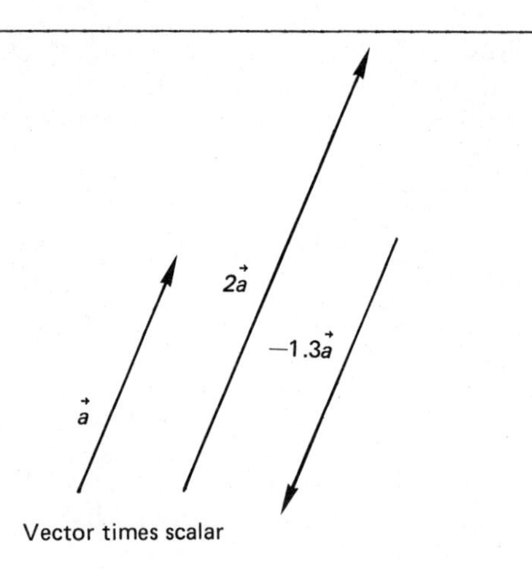

Figure 15-8b.

Vector times scalar: When you add a real number x to itself, you get

$$x + x = 2x.$$

It is reasonable to say that when you add a *vector* to itself, you get twice that vector. That is,

$$\vec{a} + \vec{a} = 2\vec{a}.$$

The answer $2\vec{a}$ is a vector in the *same* direction, but with 2 times the magnitude. The reasoning leads you to define what you mean by a scalar (i.e., a real number) times a vector.

DEFINITION

The *product* $x\vec{a}$ is a vector in the *same* direction as $\vec{a}$ but with a magnitude equal to x times the magnitude of $\vec{a}$.

Note: Multiplying a vector by a *negative* number, x, gives a vector of magnitude $|x|$, but *opposite* direction.

Unit vectors and components: A vector of magnitude 1 is called a *unit* vector. The letters $\vec{i}$ and $\vec{j}$ are used for unit vectors in the x- and y-directions, respectively. Any vector in the x-direction can be written as a scalar multiple of $\vec{i}$ and any vector in the y-direction can be written as a scalar multiple of $\vec{j}$. Figure 15-8c shows vector $\vec{v}$ which is the sum of $4\vec{i}$ and $3\vec{j}$. These two perpendicular vectors are called *components* of $\vec{v}$.

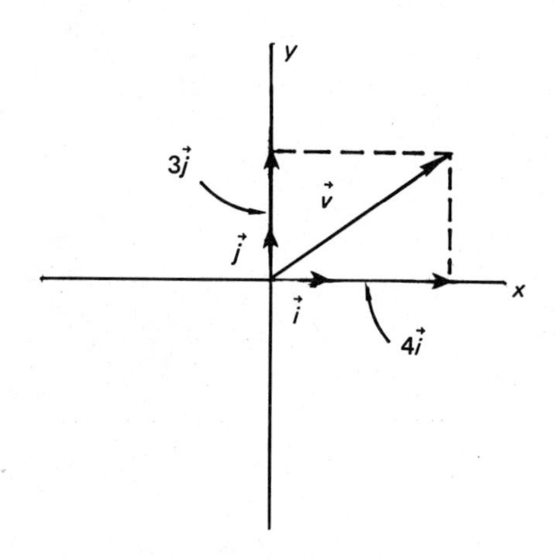

Figure 15-8c.

Objective:

Given a vector in a Cartesian coordinate system, express it as the sum of two component vectors, one in the x-direction, the other in the y-direction.

Example 1: Vector $\vec{a}$ has magnitude 3 and direction 143°, as shown in Figure 15-8d. Resolve $\vec{a}$ into horizontal and vertical components.

By definition of sine and cosine,

$$\frac{x}{3} = \cos 143° \text{ and } \frac{y}{3} = \sin 143°.$$

$$\therefore\ x = 3 \cos 143° = 2.3959...$$

$$y = 3 \sin 143° = 1.8054...$$

$$\therefore\ \vec{a} \approx -\ 2.396\,\vec{i} + 1.8054\,\vec{j}$$

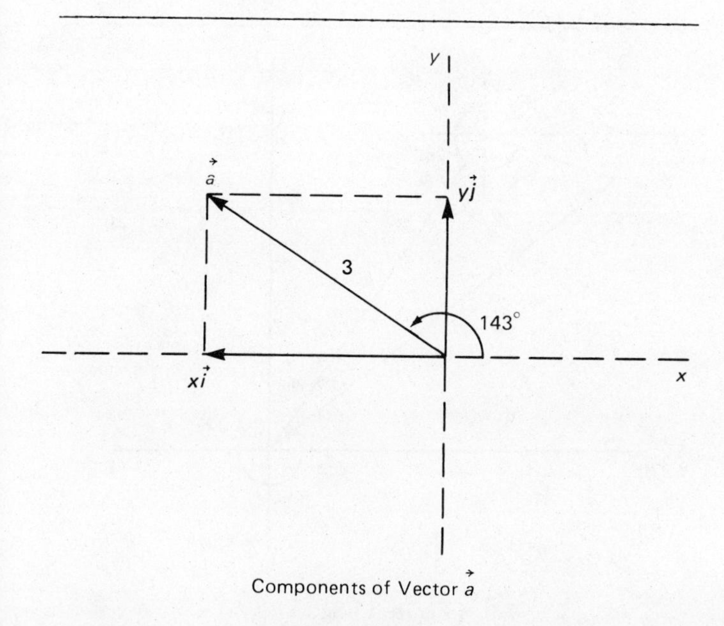

Figure 15-8d.

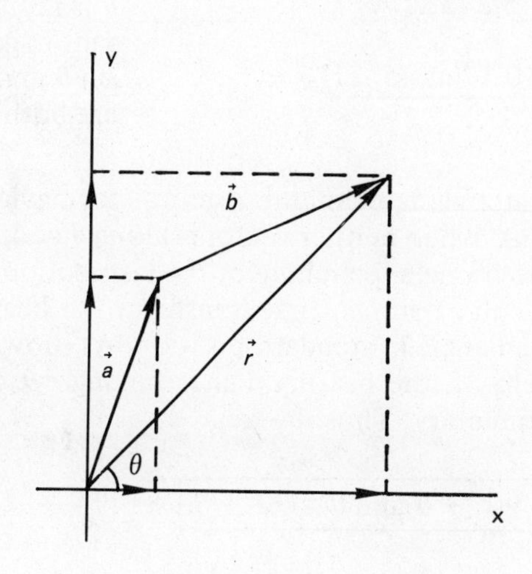

Figure 15-8e.

From Example 1 you can conclude that if θ is the angle in standard position for vector $\vec{v}$, then

$$\boxed{\vec{v} = (|\vec{v}| \cos \theta)\,\vec{i} + (|\vec{v}| \sin \theta)\,\vec{j}}$$

Components give an easy way to add two vectors. As shown in Figure 15-8e, if $\vec{r}$ is the resultant of $\vec{a}$ and $\vec{b}$, then the components of $\vec{r}$ are the sums of the components of $\vec{a}$ and $\vec{b}$. Since the two horizontal components have the same direction, they can be added simply by adding their magnitudes. The same is true for the vertical components.

Example 2: If $\vec{a}$ has magnitude 5 and direction 70°, and $\vec{b}$ has magnitude 6 and direction 25° (Figure15-8e), find the resultant, $\vec{r}$,

a. as the sum of two components,
b. as a magnitude and direction.

a. $\vec{r} = \vec{a} + \vec{b}$

$= (5 \cos 70°)\,\vec{i} + (5 \sin 70°)\,\vec{j}$ $+ (6 \cos 25°)\,\vec{i} + (6 \sin 25°)\,\vec{j}$ — Resolving $\vec{a}$ and $\vec{b}$ into components.

$= (5 \cos 70° + 6 \cos 25°)\,\vec{i}$ $+ (5 \sin 70° + 6 \sin 25°)\,\vec{j}$ — Collecting like terms.

$= 7.1479...\,\vec{i} + 7.2341...\,\vec{j}$ — By calculator.

$\approx \underline{7.15\,\vec{i} + 7.23\,\vec{j}}$ — Rounding off.

b. The more precise values of x and y should be stored in the calculator's memory or written down for use in this part of the problem.

$|\vec{r}| = \sqrt{(7.147...)^2 + (7.234...)^2}$ — By the Pythagorean Theorem.

$= 10.169...$ — By calculator.

$\tan \theta = \dfrac{7.234...}{7.147...} = 1.012...$ — Definition of tangent.

$\therefore \theta = 45.343...°$

$\therefore \vec{r} \approx 10.17$ at $45°21'$

θ is in Quadrant I since $\sin\theta$ and $\cos\theta$ are both positive.

The component technique is useful for navigation problems. When neither vector is along a coordinate axis, the triangle technique of the last section can be difficult. You can first transform the bearing, β, into an angle in standard position. As shown in Figure 15-8f, the bearing β and the angle θ are complementary. Thus,

$$\boxed{\beta = 90° - \theta \text{ and } \theta = 90° - \beta}$$

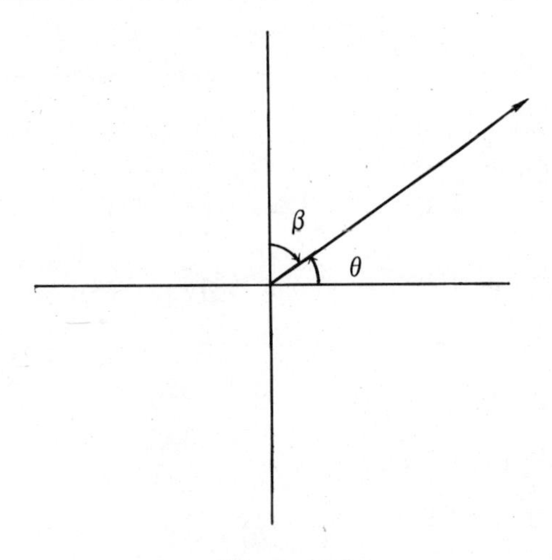

Figure 15-8f.

If this relationship produces a negative value of θ or β, a positive coterminal angle can be found by adding 360°.

Example 3: A ship sails for 20 miles on a bearing of 325°, then turns and sails on a bearing of 250° for 7 more miles. Find its displacement vector, $\vec{d}$, from the starting point (Figure 15-8g).

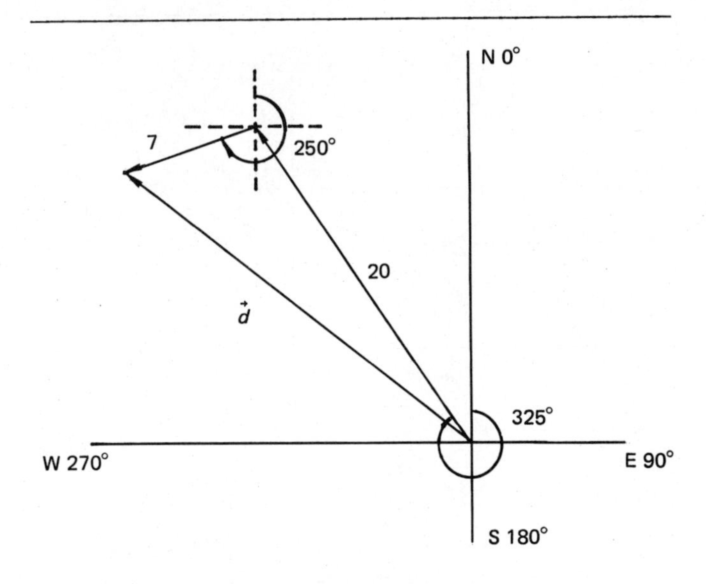

Figure 15-8g.

The two angles are

$$\theta_1 = 90° - 325° = -235°$$
$$\theta_2 = 90° - 250° = -160°$$

Adding 360° to each gives coterminal angles of positive measure.

$$\theta_1 = 125°$$
$$\theta_2 = 200°$$

$$\therefore \vec{d} = (20\cos 125° + 7\cos 200°)\vec{i} + (20\sin 125° + 7\sin 200°)\vec{j}$$

Adding the components.

$$\vec{d} = -18.049...\vec{i} + 13.988...\vec{j}$$

This vector can be transformed to a distance and bearing.

$$|\vec{d}| = \sqrt{(-18.049...)^2 + (13.988...)^2}$$
$$= 22.835...$$

$$\tan\theta = \frac{13.988...}{-18.049...}$$
$$= {}^{-}0.7750...$$

$\therefore \theta = 142.223...°$

$\therefore \beta = 90° - 142.223...°$

$= -52.223...°$

Adding 360° gives

$\beta = 307.776...°$

$\therefore \vec{d} \approx 22.84$ miles at 307° 47′

Note: This problem can also be worked using the bearings themselves, without first finding θ. The main thing to remember is that the quadrants will be numbered *clockwise*, starting from northeast.

Example 4: A ship sails at a speed of 20 knots (nautical miles per hour) on a bearing of 325°. The water has a current of 7 knots along a bearing of 250°. Find the ship's resultant velocity vector, $\vec{v}$.

$\vec{v} \approx 22.84$ knots at 307° 47′

Note: This problem is the same, mathematically, as Example 3. The mathematics is independent of what physical quantity the vectors represent.

The following exercise gives you practice resolving vectors into components, and adding vectors by adding their components.

Exercise 15-8

For problems 1 through 4, resolve the vector into horizontal and vertical components.

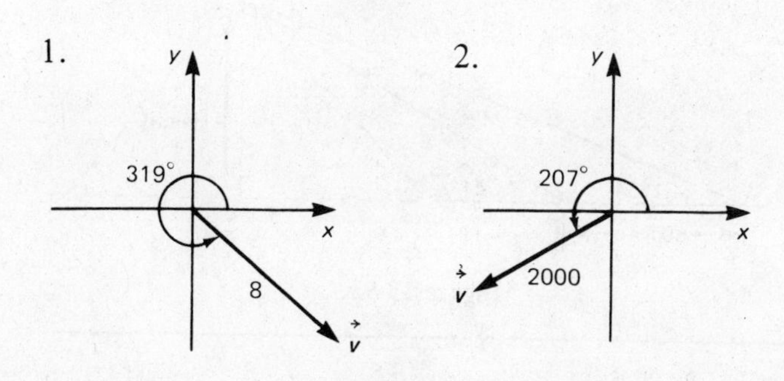

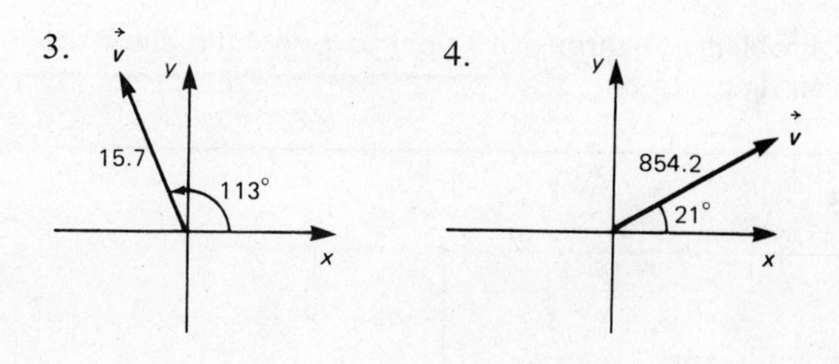

5. If $\vec{r} = 21$ units at $\theta = 70°$ and $\vec{s} = 40$ units at $\theta = 120°$, find $\vec{r} + \vec{s}$

a. as a sum of two components,
b. as a magnitude and a direction.

6. If $\vec{u} = 12$ units at $\theta = 160°$ and $\vec{v} = 8$ units at 310°, find $\vec{u} + \vec{v}$.

a. as a sum of two components,
b. as a magnitude and direction.

7. A ship sails 50 miles on a bearing of $\beta = 20°$, then 30 miles further on a bearing of $\beta = 80°$. Find the resultant displacement vector as a distance and bearing.

8. A plane flies 30 miles on a bearing of $\beta = 200°$, then turns and flies 40 miles on a bearing of $\beta = 10°$. Find the resultant displacement vector as a distance and bearing.

9. A plane flies 200 miles per hour (mph) along a bearing of 320°. The air is moving with a wind speed of 60 mph along a bearing of 190°. Find the plane's resultant velocity (speed and bearing) by adding these two velocity vectors.

10. A scuba diver swims 100 feet per minute along a bearing of 170°. The water is moving with a current of 30 feet per minute along a bearing of 115°. Find the diver's resultant velocity (speed and bearing) by adding these two velocity vectors.

Problems 11 through 13 refer to $\vec{a}$, $\vec{b}$, and $\vec{c}$, shown in figure 15-8h.

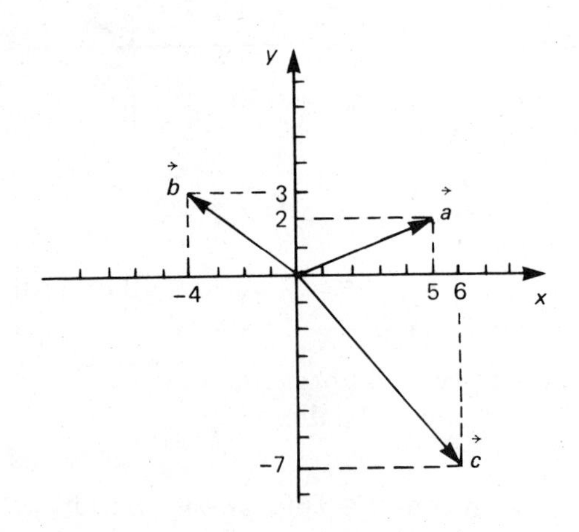

Figure 15-8h.

11a. On a piece of graph paper, draw $\vec{a} + \vec{b}$ by placing the tail of $\vec{b}$ at the head of $\vec{a}$.

b. On the same Cartesian coordinate system, draw $\vec{b} + \vec{a}$ by placing the tail of $\vec{a}$ at the head of $\vec{b}$.

c. How does your picture illustrate the fact that vector addition is *commutative*?

12. Illustrate that vector addition is *associative* by drawing $(\vec{a} + \vec{b}) + \vec{c}$ and $\vec{a} + (\vec{b} + \vec{c})$.

13. Draw $\vec{a} + (-\vec{a})$ on a Cartesian coordinate system. What is the *magnitude* of $\vec{a} + (-\vec{a})$? Does it make sense to assign a *direction* to this vector? Why do you suppose this vector is called the "zero vector"?

14. How can you conclude that {vectors} is *closed* under addition? Why is the zero vector necessary to insure closure?

15. How can you conclude that {vectors} is closed under multiplication by a scalar? Is the zero vector necessary to insure closure in this case? Explain.

15-9 REAL-WORLD TRIANGLE PROBLEMS

Throughout this chapter you have been developing the computational skills you need to work real-world problems involving measurement of triangles. Each problem in the following exercise requires you to identify one or more triangles, right or oblique, and then apply the appropriate technique to find the side, angle, or area you seek. You may use the computer program of Section 15-6.

Exercise 15-9

1. ***Swimming Problem 1.*** You swim at 3 km/h with your body perpendicular to a stream with a current of 5 km/h. Your actual velocity is the vector sum of the stream's velocity and your swimming velocity. Find your actual velocity.

2. ***Swimming Problem 2.*** You swim at 4 km/h with your body perpendicular to a stream. But because the water is moving, your actual velocity vector makes an angle of 34° with the direction you are heading.

a. How fast is the current?
b. What is the magnitude of your *actual* velocity?

3. ***Mountain Height Problem.*** A surveying crew is given the job of measuring the height of a mountain (Figure 15-9a). From a point on level ground, they measure an angle of elevation to the top of 21° 34′.

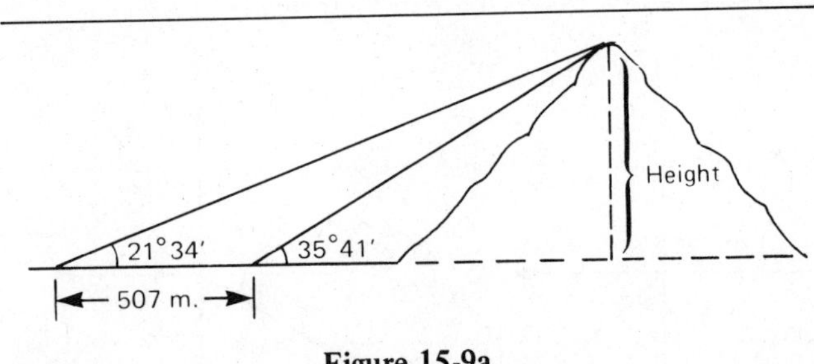

Figure 15-9a.

They move 507 meters closer and find the angle is now 35° 41′. How high is the mountain? (You may need to calculate some other numbers first!)

4. ***Harbor Problem.*** As a ship sails into harbor, the navigator sights a buoy at an angle of 15° to the path of the ship (Figure 15-9b). The ship sails 1300 meters further and finds that the buoy now makes an angle of 29°.

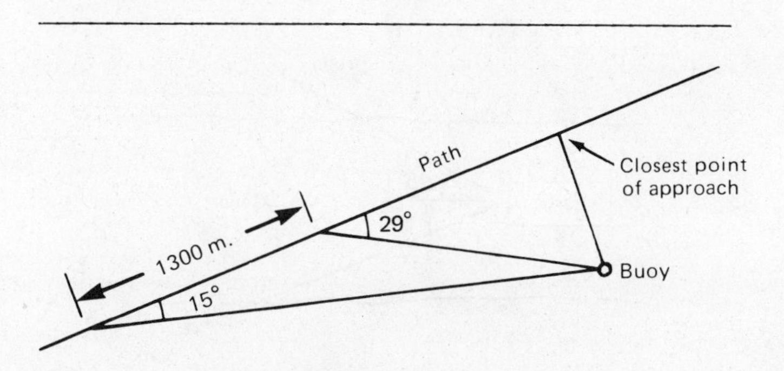

Figure 15-9b.

a. How far is the ship from the buoy at the second sighting?
b. What is the closest the ship will come to the buoy?
c. How far must the ship go from the second sighting point to this closest point of approach?
d. When the ship has gone 7000 meters beyond the second sighting point, what will be the angle from the bow of the ship to the line-of-sight with the buoy?

5. ***Missile Problem.*** An observer 2 kilometers from the launching pad observes a vertically ascending missile at an angle of elevation of 21°. Five seconds later, the angle has increased to 35°.

a. How far did the missile travel during the 5-second interval?
b. What was its average speed during this interval?
c. If it keeps going vertically at the same average speed, what will its angle of elevation be 15 seconds after the *first* sighting?

6. ***Oil Well Problem:*** An oil well is to be located on a hillside that slopes at 10° (Figure 15-9c). The desired rock formation has a dip of 27° to the horizontal in the same direction as the hill slope. The well is located 3200 feet downhill from the nearest edge of the outcropping rock formation. How deep will the driller have to go to reach the top of the formation?

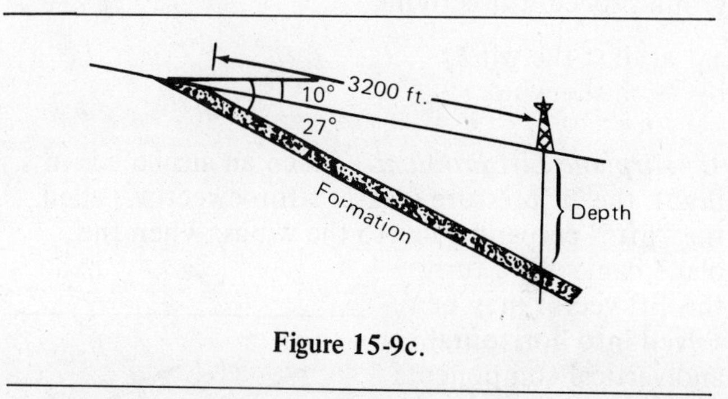

Figure 15-9c.

7. ***Visibility Problem:*** Suppose you are aboard a jet destined for Hawaii. The pilot announces that your altitude is 10 kilometers. Since you have nothing to do but stare at the Pacific Ocean, you decide to calculate how far away the horizon is. You draw a sketch as in Figure 15-9d and realize that you must calculate an *arc length.* You recall that the radius of the Earth is about 6400 kilometers. How far away is the horizon along the Earth's curved surface? Surprising?

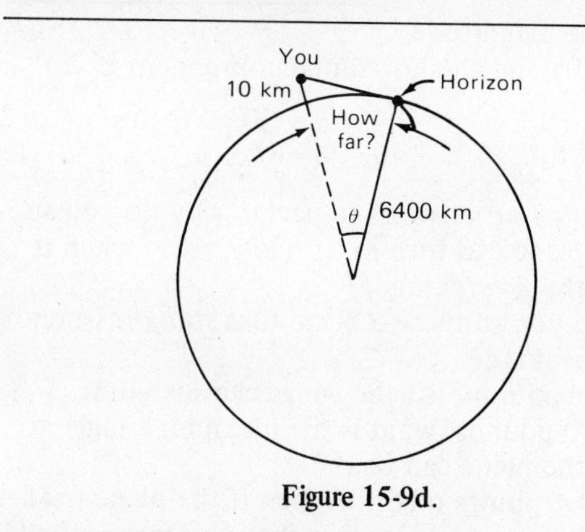

Figure 15-9d.

8. ***Airplane Velocity Problem:*** An airplane is flying through the air at a speed of 500 km/h. At the same time, the air is moving with respect to the ground at an angle of 23° to the plane's path through the air with a speed of 40 km/h (i.e., the wind speed is 40 km/h). The plane's ground speed is the magnitude of the *vector sum* of the plane's air speed and the air's speed with respect to the ground. Find the plane's ground speed if it is flying

a. against the wind,
b. with the wind.

9. ***Airplane Lift Problem:*** When an airplane is in flight, the air pressure creates a force vector, called the "lift," perpendicular to the wings. When the plane banks for a turn, this lift vector may be resolved into horizontal and vertical components. The vertical component has magnitude equal to the plane's weight (this is what holds the plane up), and the horizontal component "pushes" the plane into its curved path. Suppose that a jet plane weighing 500,000 pounds banks at an angle θ (Figure 15-9e).

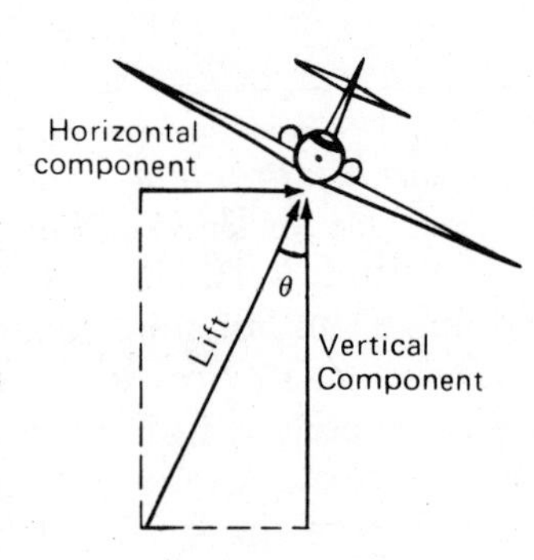

Figure 15-9e.

a. Find the magnitude of the lift and the horizontal component if

   i. $\theta = 10°$, ii. $\theta = 20°$.
   iii. $\theta = 30°$, iv. $\theta = 0°$.

b. Based on your answers to part a, why do you suppose a plane can turn in a *smaller* circle when it banks at a *greater* angle?
c. Why do you suppose a plane flies *straight* when it *is not* banking?
d. If the maximum lift the wings can sustain is 600,000 pounds, what is the maximum angle at which the plane can bank?
e. What *two* things might happen if the plane tried to bank at an angle *steeper* than this maximum?

10. ***Canal Barge Problem.*** Freda Pulliam and Yank Hardy are on opposite sides of a canal, pulling a barge with tow ropes (Figure 15-9f). Freda exerts a force of 50 pounds at 20° to the canal, and Yank pulls at an angle of 15° with just enough force so that the resultant force vector is directly along the canal. Find the number of pounds with which Yank must pull and the magnitude of the resultant vector.

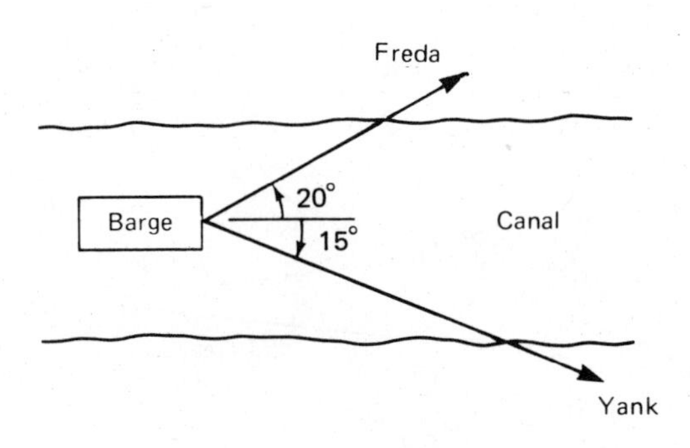

Figure 15-9f.

11. ***Detour Problem.*** Suppose that you are the pilot of a commercial airliner. You find it necessary to detour around a group of thunder-showers (Figure 15-9g). You turn at an angle of 21° to your original path, fly for a while, turn, and intercept your original path at an angle of 35°, 70 kilometers from where you left it.

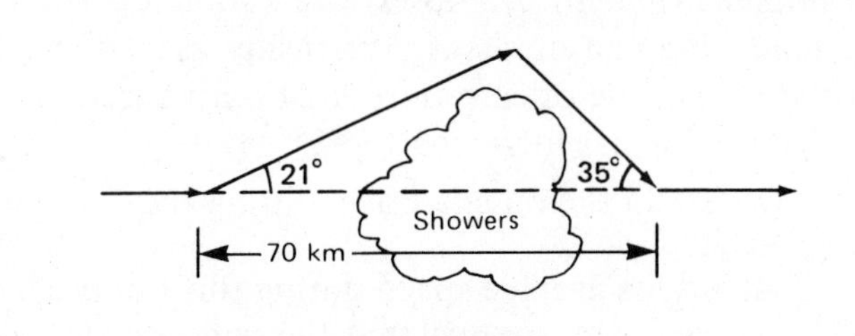

Figure 15-9g.

a. How much further did you have to go because of the detour?
b. What area is enclosed by this triangle?

12. ***Surveying Problem 1.*** A surveyor measures the three sides of a triangular field and gets 114, 165, and 257 meters.

a. What is the measure of the largest angle of the triangle?
b. What is the area of the field?

13. ***Surveying Problem 2.*** A field has the shape of a quadrilateral that is *not* a rectangle. Three sides measure 50, 60, and 70 meters, and two angles measure 127° and 132° (Figure 15-9h).

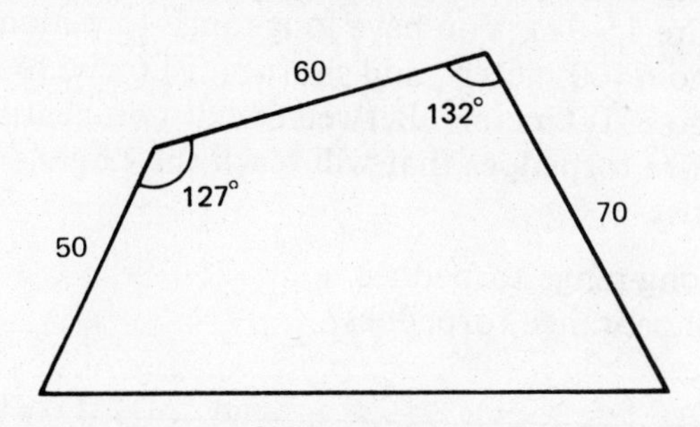

Figure 15-9h.

a. By dividing the quadrilateral into two triangles, find its area. You may have to find some intermediate sides and angles first.
b. Find the length of the fourth side.
c. Find the measures of the other two angles.

14. ***Surveying Problem 3.*** Surveyors can find the area of an irregularly shaped tract of land by taking "field notes." These notes consist of the length of each side and information for finding each angle measure. Then, starting at one vertex, the tract is divided into triangles. For the first triangle, two sides and the included angle are known (Figure 15-9i), so its area can be calculated. To calculate the area of the second triangle, you must recognize that one of its sides is also the *third* side of the *first* triangle, and one of its angles is an angle of the polygon (147° Figure 15-9i) *minus* an angle of the first triangle. By calculating this side and angle and using the next side of the polygon (15 in Figure 15-7i), you can calculate the area of the second triangle. The areas of the remaining triangles are calculated in the same manner. The area of the tract is the *sum* of the areas of the triangles.

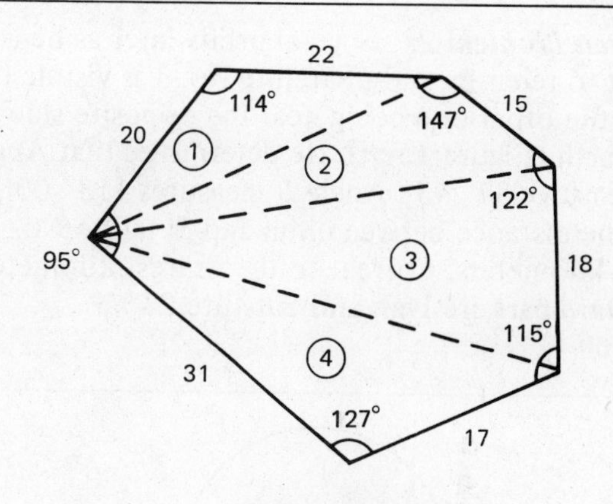

Figure 15-9i.

a. Write a computer program for calculating the area of a tract, using the technique described above. The input should be the sides and angles of the polygon, and the output should be the area of the tract. As a check on your program, you can have it print out the intermediate sides and angles.
b. Use your program to show that the area of the tract in Figure 15-9i is 1029.69 square units.
c. Show that the last side of the polygon is calculated to be 30.6817 units, which is close to the measured value of 31.
d. The polygon in Figure 15-9i is called a *convex polygon*, because none of its angles measure more

than 180°. Explain why your program might give *wrong* answers if the polygon were *not* convex.

For Problems 15 through 18, suppose that the country of Parah has just launched two satellites. The government of Noya sends aloft its most self-reliant astronaut, Ivan Advantage, to observe the satellites.

15. ***Ivan Problem 1.*** As Ivan approaches the two satellites, he finds that one of them is 8 kilometers from him and the other is 11 kilometers, and the angle between the two (with Ivan at the vertex) is 120°. How far apart are the satellites?

16. ***Ivan Problem 2.*** Several orbits later as he is about to re-enter, only Satellite No. 1 is visible to Ivan, the other one being near the opposite side of the Earth (Figure 15-9j). He determines that Angle A measures 37° 43′, Angle B measures 113° 00′, and the distance between him and Satellite No. 1 is 4362 kilometers. Correct to the nearest kilometer, how far apart are Ivan and Satellite 2?

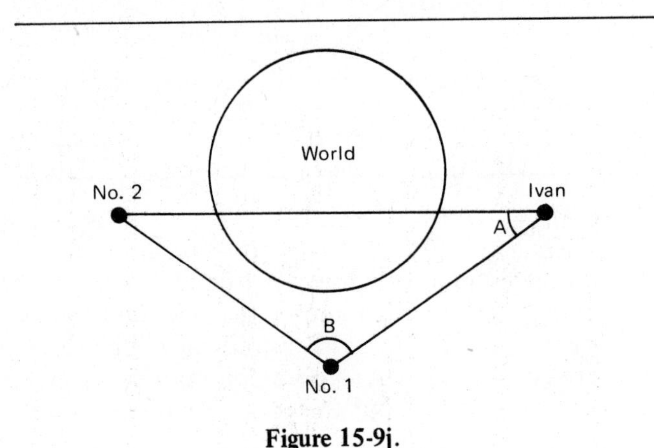

Figure 15-9j.

17. ***Ivan Problem 3.*** Three ships are assigned to rescue Ivan as his spacecraft plunges into the ocean. The ships are at the vertices of a triangle with sides of 5, 7, and 10 kilometers.

a. Find the measure of the largest angle of this triangle.
b. Find the area of ocean in the triangular region bounded by the three ships.

18. ***Ivan Problem 4.*** To welcome their returning hero, the Noyacs give Ivan a parade. The parade goes between the cities of Om, Mann, and Tra. These cities are at the vertices of an equilateral triangle. The roads connecting them are straight, level, and direct, and the parade goes at a constant speed with no stops. From Om to Mann takes 80 minutes, from Mann to Tra takes 80 minutes, but from Tra back to Om takes 1 hour and 20 minutes. How do you explain the discrepancy in times?

19. ***Torpedo Problem.*** Suppose that you are Torpedo Officer aboard the U.S.S. Skipjack. Your submarine is conducting torpedo practice off the Florida coast. The target is 7200 meters from you on a bearing of 276° and is steaming on a course of 68° (Figure 15-9k). You have long-range torpedoes that will go 6400 meters, and short-range torpedoes that will go 3200 meters. Between what two bearings can you fire torpedoes that will reach the target's path if you use

a. long-range torpedoes,
b. short-range torpedoes?

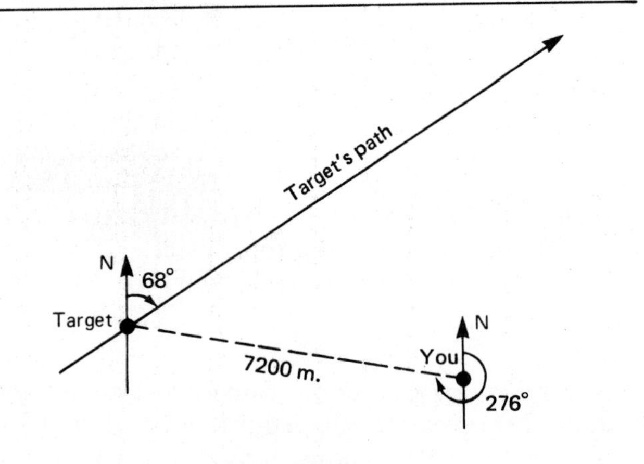

Figure 15-9k.

20. ***Alligator Problem.*** Calvin Butterball is swimming in Lake Rancid when he spots two alligators. He tells you that his distance to Alligator 1 is 30 meters, the distance between the alligators is 20 meters, and the angle at Calvin is 58° (Figure 15-71).

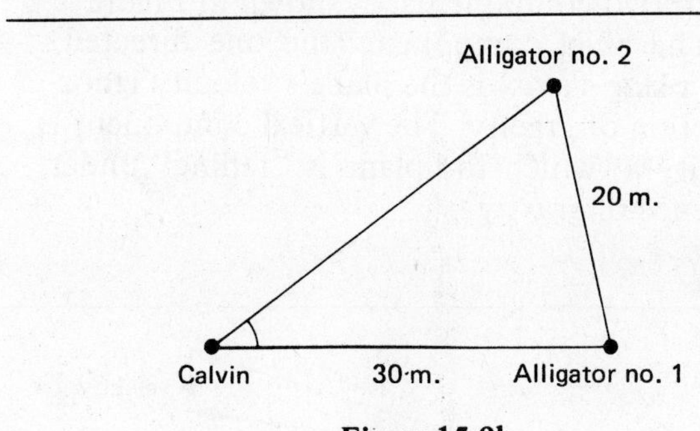

Figure 15-9l.

a. Show Calvin that he must have made a mistake in measurement, since there is no such triangle.
b. Find the *two* possible distances between Calvin and Alligator 2 using the *correct* angle, 28°.

15-10 CHAPTER REVIEW AND TEST

The objectives for this chapter may be summarized as follows:

1. *Solve right triangle problems, given*

   a. *two sides,*
   b. *one side and one acute angle.*

2. *Solve oblique triangle problems, given*

   a. *two sides and the included angle,*
   b. *three sides,*
   c. *one side, the opposite angle, and one other side or angle.*

3. *Find the area of a triangle, given two sides and the included angle measure.*

4. *Add and resolve vectors.*

5. *Solve real-world problems involving triangles.*

The Review Problems below are designed to measure your ability to accomplish these objectives when you are *told* which one is being tested. The Concepts Test is designed to measure how well you can apply all of these techniques to the analysis of a single real-world problem, selecting whichever technique is appropriate for the part you are working on.

Review Problems

The following problems are numbered according to the five objectives listed above.

Review Problem 1a. Right triangle XYZ has hypotenuse of length $y = 14.7$ centimeters and one leg of length $z = 8.3$ centimeters. Find x and the measures of the acute angles.

b. Right triangle BJF has $m\angle B = 39° \ 54'$ and hypotenuse $f = 19$ kilometers. Find b, j, and $m\angle J$.

Review Problem 2a. Triangle FUN has $u = 14$, $n = 13$, and $m\angle F = 145° \ 20'$. Find f, $m\angle U$, and $m\angle N$.

b. Triangle GYM has $g = 7$, $y = 8$, and $m = 13$. Find the measures of the three angles.

c. i. Triangle BAS has $m\angle B = 123°$, $m\angle S = 56°$, and $a = 10$ millimeters. Find the lengths of the other two sides. Surprising?

   ii. Triangle TWO has $o = 20$ meters, $t = 12$ meters, and $m\angle T = 31°$. Find the two possible values of w.

Review Problem 3. Find the area of ΔFUN in Review Problem 2a.

Review Problem 4. Vectors $\vec{a}$ and $\vec{b}$, which have magnitudes 6 and 10, respectively, make an angle of 174° with each other, as shown in Figure 15-10a on the next page. Find the magnitude of $\vec{a} - \vec{b}$ and the angle that this difference vector makes with $\vec{a}$ when placed tail-to-tail.

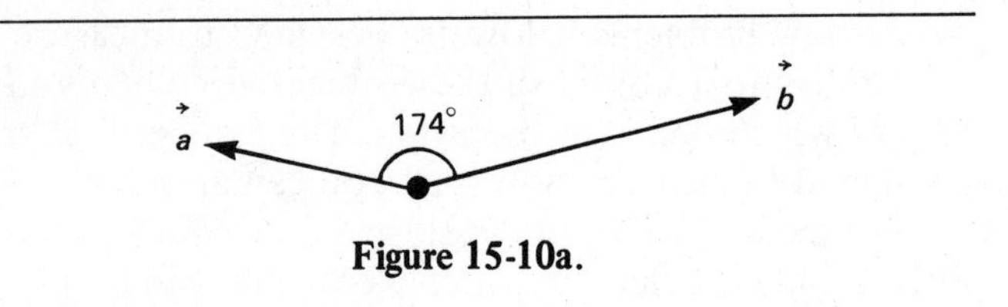

Figure 15-10a.

Review Problem 5. The rotor on a helicopter creates an upward force vector. To move forward, the pilot tilts the helicopter forward. The vertical component of the force vector (the "lift") holds the helicopter up, and the horizontal component (the "thrust") makes it move forward.

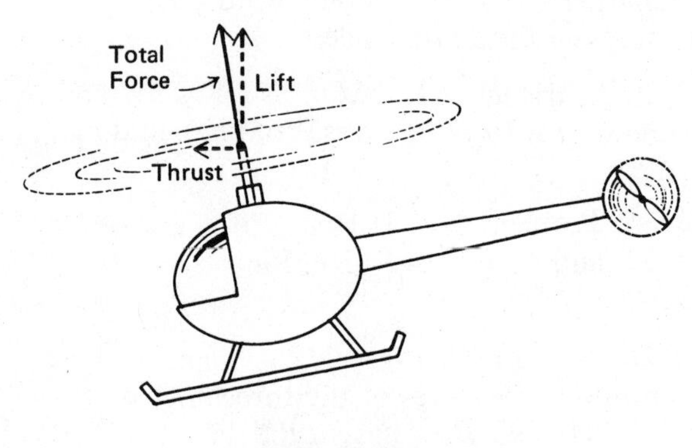

Figure 15-10b.

Suppose that a helicopter tilts forward at a sufficient angle to generate a thrust of 400 pounds. The helicopter weighs 3000 pounds, so the lift has a magnitude of 3000 pounds.

a. What angle does the helicopter make with the ground?
b. What total force must the rotor generate?

Concepts Test

Work each part of the following problem. Then tell which one or ones of the five objectives of the chapter you used in working that part of the problem.

As a jet plane takes off, its path makes a fairly steep angle to the ground. The plane itself makes an even steeper angle. Its velocity vector may be resolved into two components, as shown in Figure 15-10c. The axial component (the one directed along the plane's axis) is the plane's velocity ignoring the action of gravity. The vertical component is the velocity at which the plane is "falling" under the influence of gravity.

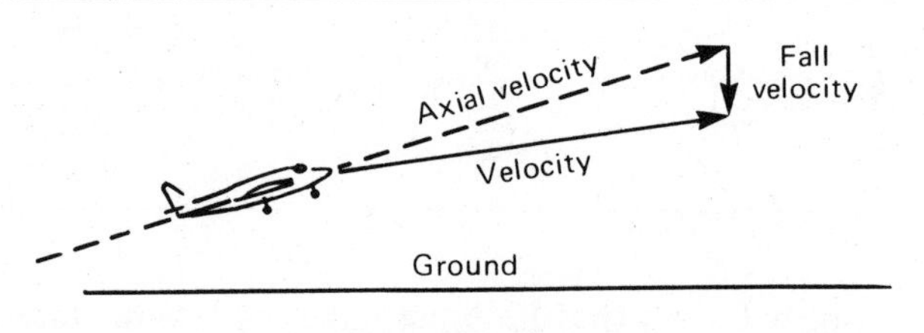

Figure 15-10c.

Concepts Test 1: A plane's velocity vector is 250 km/h at an angle of 10° to the ground. The plane's axis makes an angle of 15° with the ground.

a. Find the speed in the axial direction.
b. Find the speed at which the plane is falling.
c. Find the area of the triangle formed by the three vectors.

Concepts Test 2: If the plane maintains an angle of 15° with the ground, and the fall vector stays the same as in Problem 1:

a. What is the minimum speed the plane can go without going downward? (That is, the velocity vector must be *horizontal.*)
b. At this minimum speed, what will the axial velocity vector equal?

Concepts Test 3: The plane increases its speed to 700 km/h. The navigator determines the axial speed to be 702 km/h and the fall speed to be the same as in

Problem 1. What angle does the plane's path make with the ground?

15-11 CUMULATIVE REVIEW, CHAPTERS 13, 14, AND 15

The following exercise may be considered to be a final examination that tests your ability to use all you have learned about trigonometric and circular functions. If you are thoroughly familiar with the concepts, you should be able to work all of the problems in about two hours.

Suppose you have joined the Navy and are stationed aboard the nuclear submarine Seawolf. Your boat has been assigned to conduct torpedo target practice in the Gulf of Mexico. Answer the following questions.

1. As you approach the practice area, you travel a displacement vector of 6 kilometers along a bearing of 22°. Then you turn and go a displacement vector of 15 kilometers along a bearing of 82°.

a. Add these two vectors to find your resultant displacement.
b. What is the area of ocean enclosed by these three vectors?

2. While you wait for the target ships to arrive, you steam submerged in a circle of diameter 100 kilometers, making a complete revolution every 20 hours. The closest you come to the coastline is 30 kilometers (see Figure 15-11).

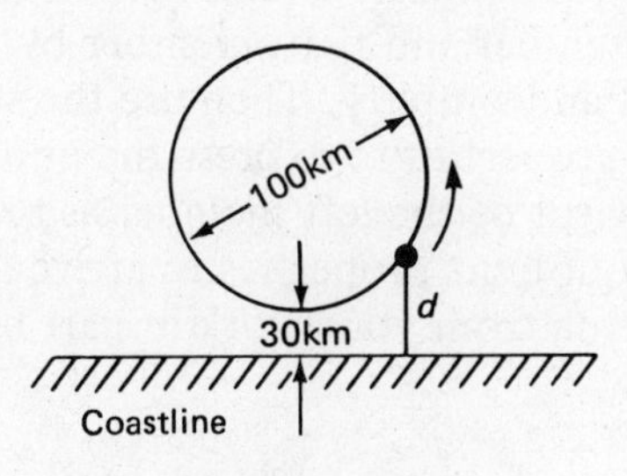

Figure 15-11.

a. Your distance, d, from the coastline varies sinusoidally with time, t. When $t = 2$ hours, you are at your maximum distance from the coastline. Write an equation expressing d in terms of t.
b. Predict your distance from the coastline when $t = 9$.
c. Transform the equation from part c so that t is expressed in terms of d.
d. What are the first three positive values of t for which $d = 123$?

3. On the fourth day of your patrol, your electronic gear picks up four strange signals:

i. $y_1 = \sec^2 x \sin^2 x + \tan^4 x$

ii. $y_2 = \dfrac{\sin^2 x}{\cos^4 x}$

iii. $y_3 = \cos^2 x$

iv. $y_4 = \cos 3x \cos 5x$

a. Prove that the expressions for y_1 and y_2 are identical.
b. Before your computer can analyze signals iii and iv, number iii must be expressed in terms of cos $2x$ and number iv must be expressed as a sum of sines or cosines. Perform these two transformations.

4. The signals are coming from three unmanned target ships, the Hic, the Haec, and the Hoc. Your sonar detects a sound coming from Hoc's engines. The sound is described by the equation

$$y = 5 + 3 \cos \frac{\pi}{4}(x - 1)$$

Sketch one complete cycle of the graph of this function.

5. You can fire torpedoes at any angle θ to the ship's axis in the interval $[-180°, 180°]$, but $\theta \neq 0$. (Firing them straight ahead might give away your position.) However, the best angles are those in the solution set of the trigonometric equation

$$\cos 4\theta - \cos 2\theta = 0.$$

Solve this equation in the given domain.

6. In preparation for firing torpedoes, you must evaluate the following. Leave the answer in *exact* form using π or radicals, if necessary, unless otherwise specified. Assume circular functions unless degrees or θ is specified.

a. $\cos 150°$ b. $\sin \frac{5\pi}{4}$

c. $\tan \theta$, if θ terminates in Quadrant IV, and $\cos \theta = 7/11$

d. csc 3, to 5 decimal places.

e. $\theta = \text{Arcsin}(1/2)$ f. $x = \text{Arccos}(-1)$

g. $x = \text{Tan}^{-1}(-\sqrt{3})$ h. $x = \text{Sec}^{-1}(-2/\sqrt{3})$

i. $\sin(\cos^{-1}(-2/3))$

j. $\cos(\cos(\cos 2))$, to 5 decimal places

7. The last piece of information you need is sin 0.06 correct to *twelve* decimal places. Unfortunately, you have no calculator accurate enough. From a handbook of mathematical tables, you find that $\sin x$ is given by a "Taylor series," as follows:

$$\sin x = x - \frac{x^3}{3!} + \frac{x^5}{5!} - \frac{x^7}{7!} + \frac{x^9}{9!} - \frac{x^{11}}{11!} + \dots .$$

Calculate sin 0.06 correct to 12 decimal places using as many terms as you need of the Taylor series for $\sin x$. Show that your answer, when rounded off, agrees with the value of sin 0.6 on your calculator.

8. Your first torpedo sinks the Hic. The Haec and Hoc leave the practice area along a bearing of $\text{Arctan}(-\sqrt{3})$.

a. Find exactly the inverse *circular* $\text{Arctan}(-\sqrt{3})$.
b. Find exactly the inverse *trigonometric* $\text{Arctan}(-\sqrt{3})$.
c. What are the values of $\arctan(-\sqrt{3})$? (Trig or circular.)
d. What is the value of $\sin(\text{Arctan } 2)$?
e. Sketch a graph of $y = \text{Arctan } x$.

9. On the trip home you dig out an old trigonometry text and see if you can apply your knowledge to a new situation.

You find stated, without proof, the *Law of Tangents*, which says that in any triangle ABC,

$$\frac{\tan \frac{1}{2}(A - B)}{\tan \frac{1}{2}(A + B)} = \frac{a - b}{a + b}$$

You decide to prove the Law of Tangents.

a. Use the Law of Sines and the addition property of equality to show

i. $$\frac{\sin A + \sin B}{\sin B} = \frac{a + b}{b}$$

and

ii. $$\frac{\sin A - \sin B}{\sin B} = \frac{a - b}{b}$$.

b. Divide equation ii by equation i, left member by left member and right member by right member, and simplify. Then use the sum and product properties to express the numerator and denominator of the left member as *products*.
c. Use the quotient properties to arrive at the Law of Tangents from your work in part b.

Final Examination

In this book you have studied various kinds of algebraic and trigonometric functions and relations. For each kind, you have ***defined*** *it,* ***graphed*** *it, found its* ***properties****, and used it as a* ***mathematical model****. The examination below tests your ability to do these four things with the functions you have studied. If you are thoroughly familiar with the concepts, you should be able to work the examination in about three hours.*

I. DEFINITIONS

1. Usually, a function is defined by a *general equation.* Write the general equation for each of the following:

a. A linear function.
b. A direct variation function.
c. y is directly proportional to the 1.6 power of x.
d. y varies directly with x and inversely with the square of z.
e. A quadratic function.
f. A quartic function.
g. An exponential function.
h. A rational algebraic function.
i. A hyperbola opening in the y-direction.
j. The probability of Event E.
k. A sinusoidal function.
l. The inverse secant function.

2. For sequences and series, the definition is usually a *pattern* followed by the terms. You had to *figure out* a general equation from the pattern.

a. Write the definition of a *geometric* sequence.
b. Write the definition of an *arithmetic* sequence.
c. Write the general equation for $t(n)$, the nth term, of a

   i. geometric sequence,
   ii. arithmetic sequence.

d. Figure out a general equation for $t(n)$ in terms of n for the sequence 0, 3, 8, 15, 24, 35, ...

II. GRAPHS

3. For each of the following graphs, tell what kind of function or relation it could be.

a.

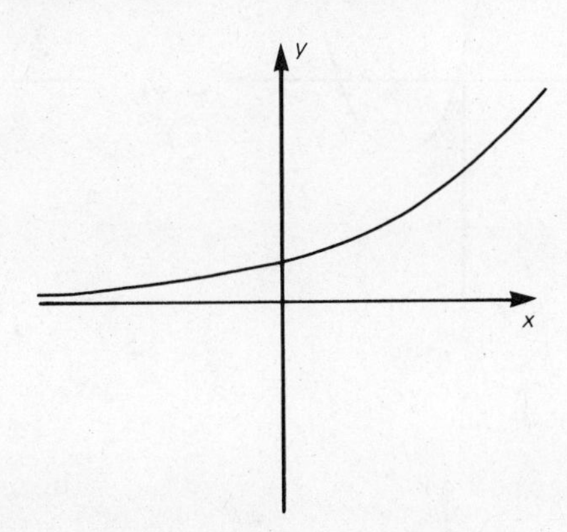

b.

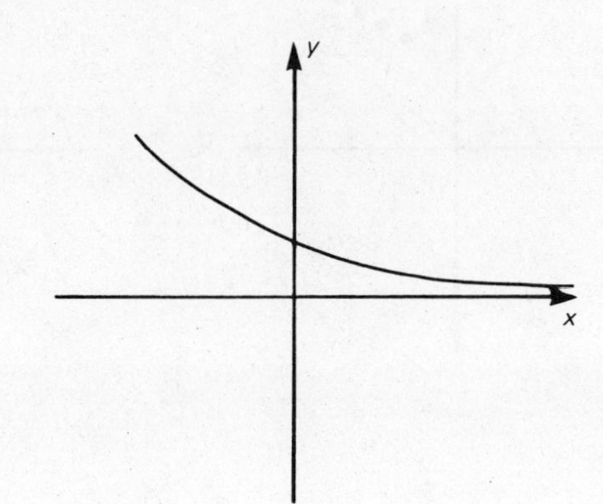

c.

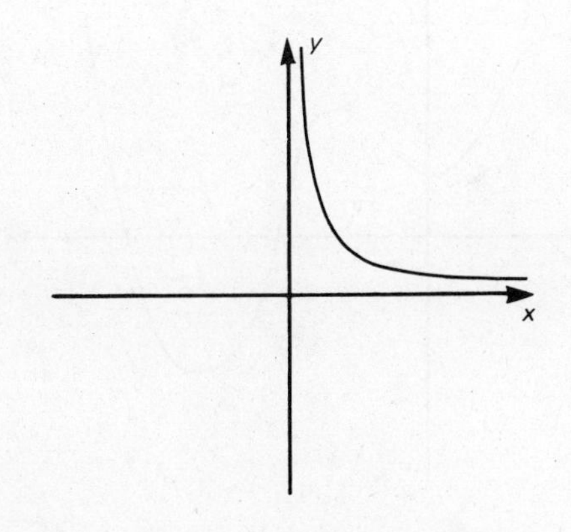

d.

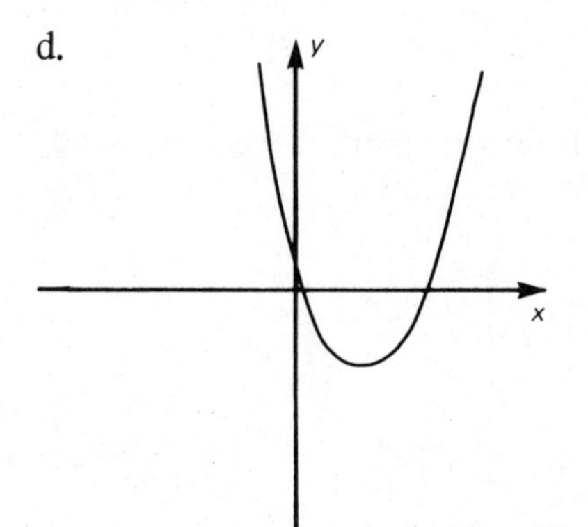

e.

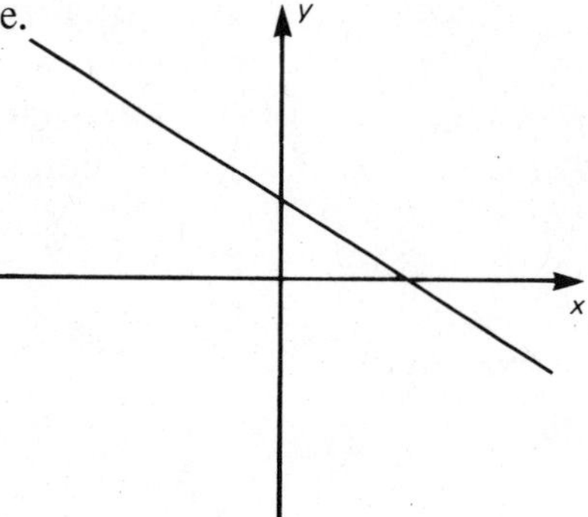

f.

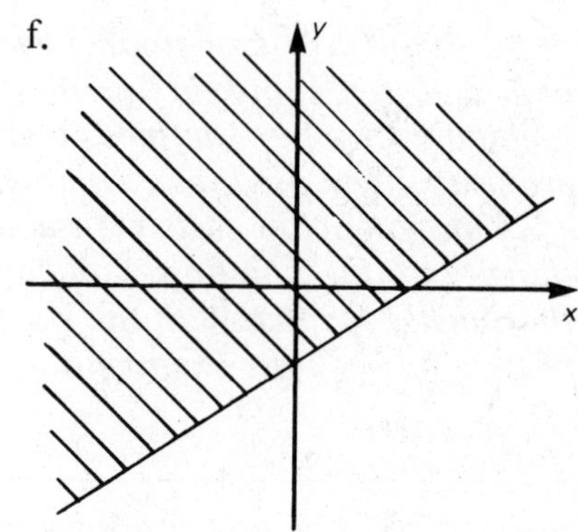

g.

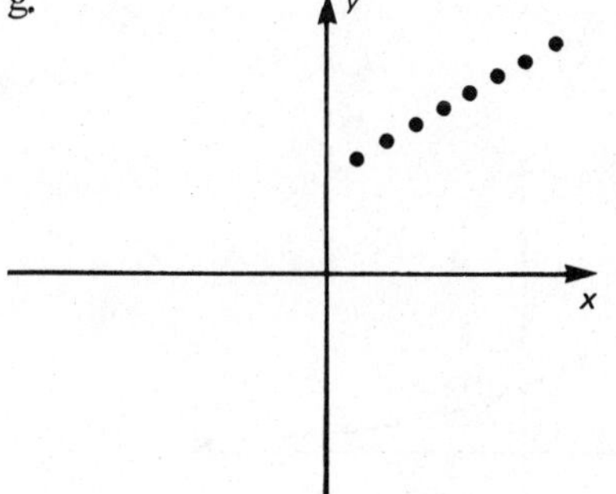

h.

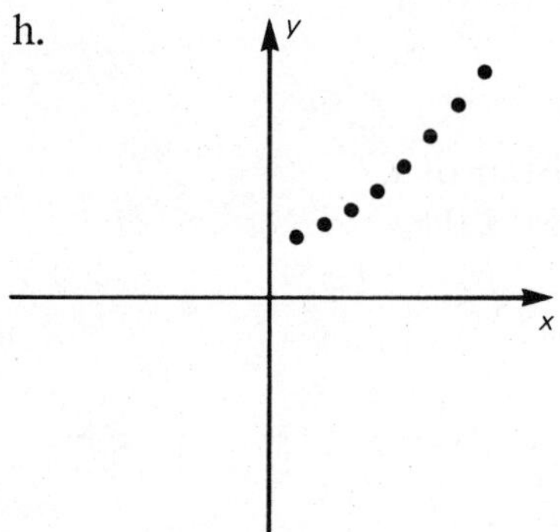

i.

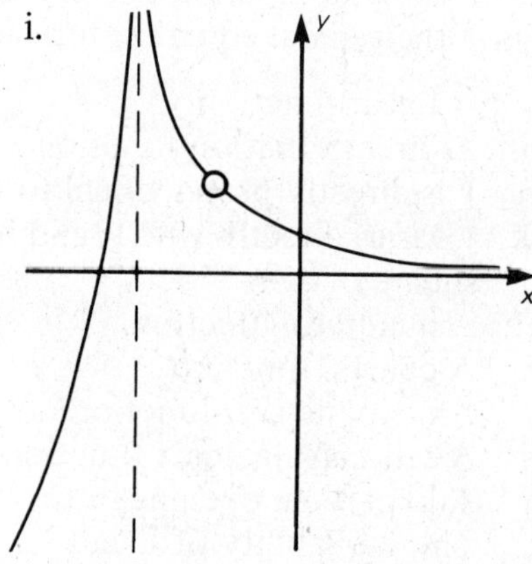

j.

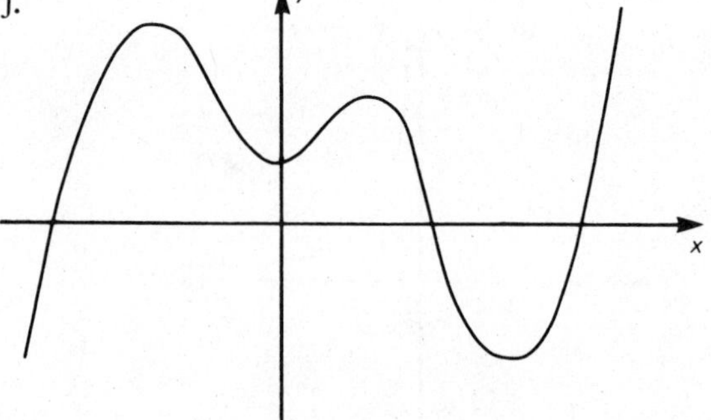

k.

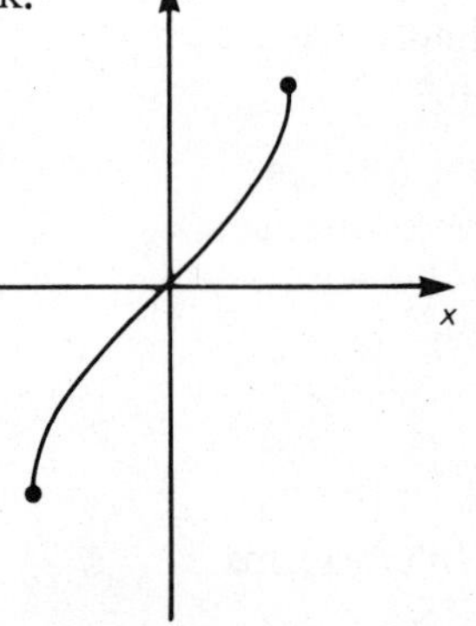

l.

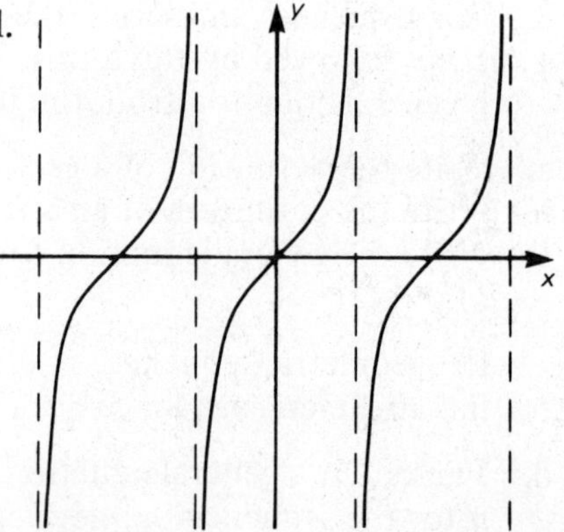

III. PROPERTIES

4. At the beginning of the course you learned fundamental properties, called *axioms,* on which all of algebra and trigonometry is based. Tell which axiom is illustrated by each equation. No abbreviations, please! Write out the *full* name.

a. $x \cdot (y \cdot z) = (x \cdot y) \cdot z$

b. $x \cdot (y \cdot z) = x \cdot (y \cdot z)$

c. $x \cdot (y + z) = xy + xz$

d. If $x < y$ and $y < z$, then $x < z$.

e. $x < y$, $x = y$, or $x > y$.

f. $x \cdot \frac{1}{x} = 1$

g. $x \cdot 1 = x$

5. From the axioms and definitions, you have proved other properties that allow you to write expressions in different forms. Complete each equation below. Do not write in the book unless you own it.

a. $x^{3/4} =$ ____________

b. $(x^3)^4 =$ ____________

c. $(xy)^4 =$ ____________

d. $(x + y)^3 =$ ____________

e. $x^3 + y^3 =$ ____________

f. $\log (xy) =$ ________

g. $\log (x^y) =$ ________

h. If $r = \log_s t$, then $t =$ __________

i. $n(A \text{ or } B) =$ _______

j. $\sqrt{x^2} =$ ____________

k. If $rx^2 + sx + t = 0$, then $x =$ ___________

l. $\cos 2x =$ __________ (in terms of $\sin x$)

m. $\tan (x + y) =$ _______ (in terms of $\tan x$ and $\tan y$)

n. $\sin^2 x =$ ________ (in terms of $\cos x$)

IV. MATHEMATICAL MODELS

A. Selecting a Function

6. Often you can tell what kind of function to use as a mathematical model by looking at the real-world graph. For instance, the distance you are standing from the fireplace and how hot you feel are related.

a. Sketch a reasonable graph showing how these two variables are related.
b. Based on your graph, what kind of function would make a reasonable mathematical model?

7. Sometimes a kind of function can be selected based on known properties of the real-world variables. For example, when a taxi driver first turns on the meter, it reads a certain number of cents. Each time the meter clicks, you owe an additional number of cents. The total you owe depends on the number of clicks.

a. Sketch a reasonable graph of this function.
b. What kind of function would be a reasonable mathematical model? Tell *why* this kind of function would be reasonable.

B. Getting the Particular Equation

8. Once you have selected a kind of function, you must get the particular equation that fits the given ordered pairs. For each of the following, find the particular equation.

a. y varies inversely with the cube of x, and (3, 50) is on the graph.
b. y varies exponentially with x, and (0, 73) and (3, 146) are on the graph.
c. y varies quadratically with x, and (1, 8), (4, 5), and (−3, −16) are on the graph.
d. y varies sinusoidally with x. The ordered pair (−3, 2) is a low point, and (5, 11) is the next high point.

C. Using the Mathematical Model

The following problems test your ability to use various techniques associated with functions and relations.

9. For the system of inequalities

$4x + 9y \geq 13$

$3x - 6y > -20$

a. Use determinants to find the intersection of the boundary lines.
b. Plot the graph of the solution set of the system of inequalities.

10. If $f(x) = \dfrac{x-9}{\sqrt{x}+3}$,

a. Express $f(x)$ in simple radical form.
b. Find $f(45)$, and express the answer in simple radical form.
c. Set $f(x) = -5$, and solve the resulting radical equation. Show that the solution you get is *extraneous.*

11a. Tell which conic section the graph of each of the following will be:

i. $4x^2 + 4y^2 - 40x + 6y = -93$

ii. $4x^2 - 4y^2 - 40x + 6y = -93$

iii. $4x^2 + 4y - 40x + 6y = -93$

iv. $4x^2 + y^2 - 40x + 6y = -93$

b. For the equation in iv, complete the square, find the center, and sketch the graph.

12a. If $P(x)$ is a quintic polynomial with real-number coefficients, and if $3 + 2i$ is a complex zero of $P(x)$, what is another complex zero of $P(x)$?

b. If $f(x) = 3x^4 - 5x^3 + 7x^2 - 4x + 9$, use synthetic substitution to find $f(2)$.
c. If $3x^4 - 5x^3 + 7x^2 - 4x + 9$ is divided by $x - 2$, what is

i. the quotient?
ii. the remainder?

d. Name the theorem that allows you to get the answer to part c. ii. *quickly*.

13. The following is a computer program in BASIC:

```
10 INPUT N, R
20 LET U = N−R+1
30 LET P = 1
40 FOR F = U TO N
50 LET P = P*F
60 NEXT F
70 PRINT N; "P"; R; "="; P
80 END
```

a. Show the simulated computer memory and the output if 10 and 3 are put in for N and R, respectively, in Step 10.
b. What kind of problem would the program be used for?

14a. For the arithmetic series $7 + 10 + 13 + \ldots$, find $t(1000)$ and $S(1000)$.

b. For the geometric series with first term 300 and common ratio 1.03, find $t(100)$ and $S(100)$.

15a. Expand $(x - 2i)^5$ as a binomial series, and simplify. Assume that $i = \sqrt{-1}$.

b. Find the term of the binomial series from $(r - s)^{37}$ that has s^{15}. You may leave the coefficient in factorial form.
c. Find two geometric means between 5 and 50. Get a decimal approximation for the common ratio and for each mean.

16. A triangular tract of land has sides of 10 meters and 15 meters, and an included angle of 120°. Find *exact* values of

a. the length of the third side,
b. the area of the triangle.

D. Putting It All Together

The following problems require you to use many of the things you have learned in the *same* problem.

17. ***Skin Diving Problem.*** When you skin dive, nitrogen from the air you breathe dissolves in your blood. According to Henry's Law, the concentration of nitrogen in your blood varies linearly with the depth at which you are swimming. At 20 feet, the concentration is about 190 milligrams of nitrogen per liter of blood. At 30 feet, the concentration is about 225 milligrams per liter.

a. Write the particular equation expressing nitrogen concentration in terms of depth.
b. Predict the nitrogen concentration at 100 feet.
c. If too much nitrogen dissolves in your blood, you will get "nitrogen narcosis," the rapture of the depths. Assume that nitrogen narcosis starts to set in when the nitrogen concentration reaches 680 milligrams per liter. How deep can you safely dive?
d. According to your mathematical model, how much nitrogen is normally dissolved in your blood when you are *out* of the water? What part of the model tells you this?

18. ***Probability Distribution Problem.*** A group of four people is to be selected at random from a class containing 5 girls and 4 boys. Answer the following questions.

a. In how many different ways could such a group be selected?
b. In how many different ways could the group have

   i. all girls?
   ii. 3 girls and 1 boy?
   iii. 2 girls and 2 boys?
   iv. 1 girl and 3 boys?
   v. all boys?

c. Find the probability for each event in part b.
d. Show that the sum of the probabilities in part c equals 1, and tell the significance of this fact.
e. Let x be the number of girls in the group. Let $P(x)$ be the probability that the group has x girls. Plot the graph of $P(x)$ versus x.
f. In how many different ways could a line be formed having exactly 2 boys and 2 girls?
g. In how many different ways could all 9 people line up if boys and girls must come alternately?

Appendix A
Review of Computer Programming

There are two kinds of computation that require a lot of time. One kind is problems that involve difficult steps, but need to be done only *once*. The second is problems in which the *same* sequence of steps is repeated *many* times. For the one-time computation, a calculator is usually the best tool to use. For the repetitive computation, a computer is usually more efficient.

Problems requiring repetitive computations appear at various places throughout the text. This appendix is designed to provide you with enough background to work such problems. The actual computer language you will use depends, of course, on what type of computing equipment you have available. This appendix concentrates on the process of making a *flow chart*, which can be translated into *any* computer language. The last section shows how this translation can be accomplished with BASIC, an easily-learned computer language adapted to computer time-sharing facilities.

Specific objectives are:

1. Given a mathematical problem, write a flow chart which breaks the problem down into a sequence of steps simple enough to be handled by a computer.

2. Given a flow chart, be able to tell what will be the output if the steps are carried out.

3. Given a computer program in BASIC, be able to tell what the output will be if the steps are carried out.

4. Given a flow chart for a mathematical problem, be able to write a computer program in BASIC.

A-1 READING FLOW CHARTS

A computer program is essentially a set of simple instructions which tell the computer what steps to take, and in what sequence to take them. For your purposes, you may assume that a computer can do only the following things:

1. Take a number which you give it, and store that number at a specified location in its memory.

2. Print the value of a number stored at a specified place in its memory.

3. Perform arithmetic using values of numbers stored in its memory, and store the answer at a specified place in its memory.

4. Compare the values of two numbers stored in its memory, and decide which step in the program to perform next, based upon this comparison.

Suppose that you wish to teach a computer to average three numbers. The first thing you would have to tell the computer is the values of the three numbers you wish to average. Then you would tell it to add the three numbers and divide by 3, and put the answer somewhere in its memory. Finally, you would tell it to print the answer. These steps could be arranged in a sequence of boxes with arrows connecting them to show in what sequence the steps are to be performed. The result is called a "flow chart," and is shown in Figure A-1a.

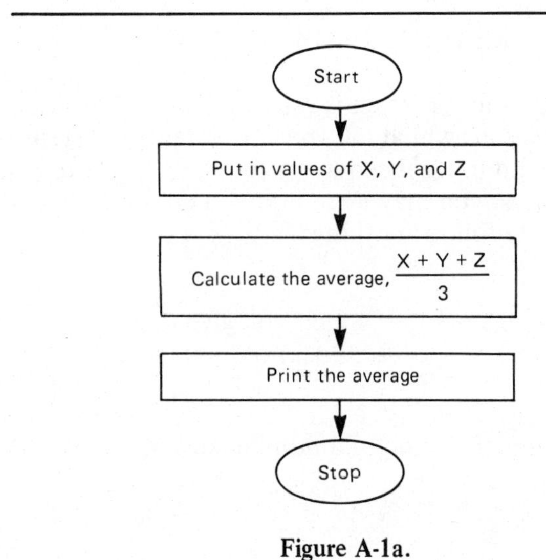

Figure A-1a.

The "Start" and "Stop" boxes are made a different shape since they have only *one* path leading from them, rather than two (one "in" and one "out") as for the rectangular boxes.

Suppose that this program is to be used for averaging grades, and it is desired to print out a message such as, "Your grade is A," when the average is 90 or more. The computer must make a *decision* whether or not to print out the message, depending on how big the average is. For this purpose there must be a *branch* in the flow chart. It is customary to use a different shaped box for a branch instruction since there will be *two* ways out of the box instead of just one. A diamond-shaped box is often used, as shown.

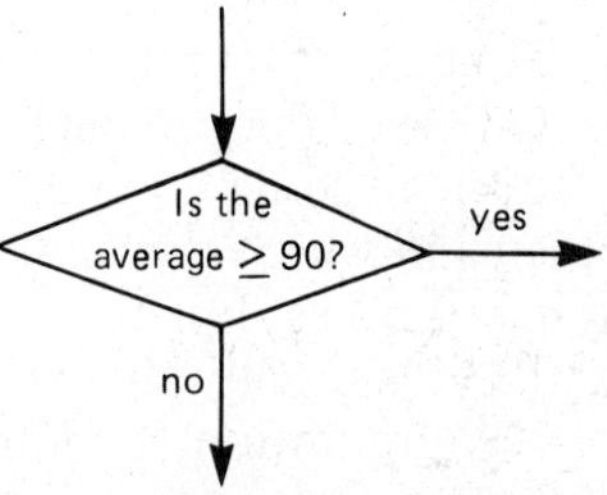

The incoming path comes to the *top* vertex of the diamond. The "yes" and "no" outgoing paths can come from any two of the other three vertices, depending only on what is convenient in making the flow chart tidy. The modified flow chart would be as shown in Figure A-1b.

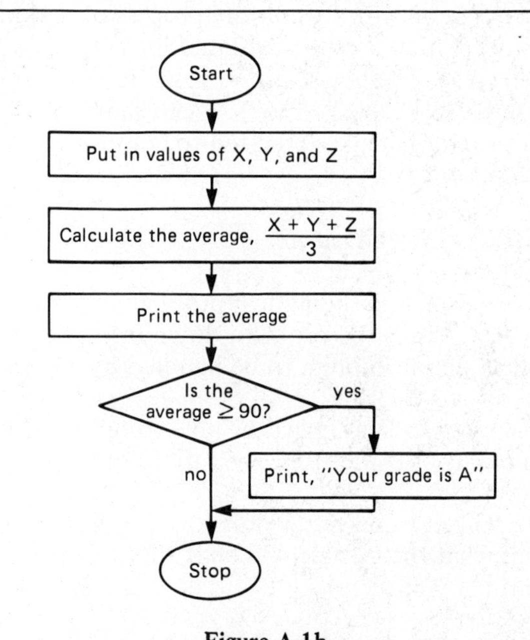

Figure A-1b.

Finally, suppose that there are *many* sets of three numbers to be averaged. Rather than starting and stopping the computer for each set of numbers, it is possible to have the computer *loop* back to the beginning of the program until all sets have been averaged. To accomplish this looping, a branch is put in the flow chart just before the "Stop" instruction telling the computer to return to the top of the flow chart if there are more sets to be averaged.

The flow chart for this modified program is shown in Figure A-1c. You recognize that the heart of the flow chart is unchanged. However, the flow chart is made more complicated by the introduction of two new variables. One of them, N, you put in at the beginning to tell how many sets of numbers to average. The other, C, tells how many times you have looped through the program. Such a variable is called a "counter," for obvious reasons. When $C = N$, you are finished, and can stop.

Before you try writing a flow chart, it helps to practice *reading* some; in this section you will get this practice. In the next section, you will be expected to write flow charts yourself from the statement of a problem.

Objective:

Given a flow chart, go through it step by step as a computer would, and determine what information will be printed by the computer.

In order to read a flow chart such as in Figure A-1c, you should understand how a computer's memory works. You may think of the memory as simply rows and rows of boxes, each of which can hold *one* number. When a computer is going to run a program, it first assigns one box to each variable it finds named in the program. For example, if the program contains the variable X, then the computer assigns

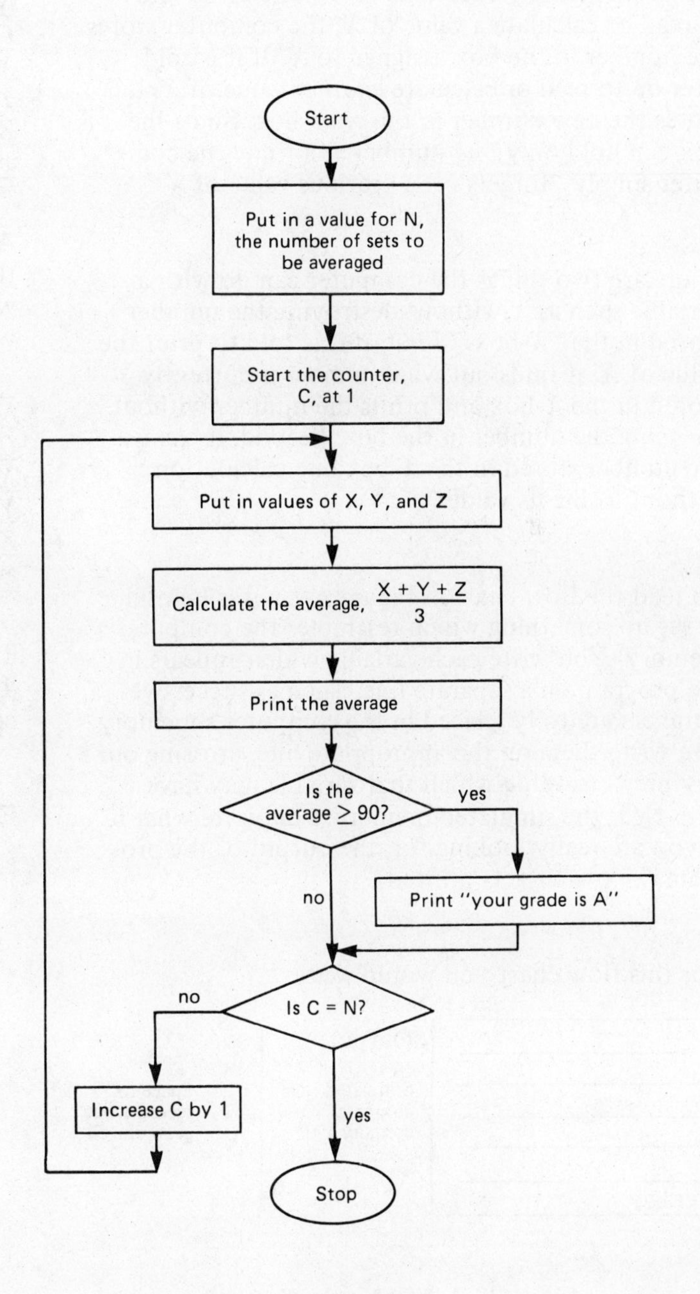

Figure A-1c.

one of its memory locations to X. When it is told to read or calculate a value of X, the computer stores the number in the box assigned to X. If it is told later on to read or calculate *another* value of X, it stores the new number in the *same* box. Since the box can hold only *one* number at a time, the computer simply "forgets" the previous value of X.

There are two things the computer can do with a variable such as X without destroying the number stored in the "X-box." First, if it is told to print the value of X, it finds out what number is currently stored in the X-box and prints the number without changing the number in the box. Second, it can use the number stored in the X-box in a calculation without losing its value.

To read the flow chart in Figure A-1c, it is helpful to set up something which resembles the computer's memory. You write each variable which appears in the program on a separate line. Then as successive numbers are to be placed in the computer's memory, you write them on the appropriate line, crossing out any previous value which that variable may have had. Near the simulated memory, you write what it is you are really looking for, the output of the program (i.e., what gets printed).

For this flow chart you would write:

Simulated computer memory	Space to record what gets printed
N	Output:
C	
X	
Y	
Z	
average	

If you were to put in 4 for N at the first step, and 73, 82, and 80 for X, Y, and Z at the third step, then follow the sequence of steps as they occur, the "memory" would wind up as follows when you first reach the "Is $C = N$" step:

Memory	Output:
N 4	78.33
C 1	
X 73	
Y 82	
Z 80	
average 78.33	

At the "Increase C by 1" step, you would *cross out* the original 1 on the C line and insert 2. If the new values of X, Y, and Z are 95, 86, and 98, the memory would look like:

Memory	Output:
N 4	78.33
C ~~1~~ 2	93.00 YOUR GRADE IS A
X ~~73~~ 95	
Y ~~82~~ 86	
Z ~~80~~ 98	
average ~~78.33~~ 93.00	

The fourth time through, the value of C would be 4, and when the step "Is $C = N$?" is reached, the computer would branch to the "Stop" instruction.

Exercise A-1

1. Find the output:

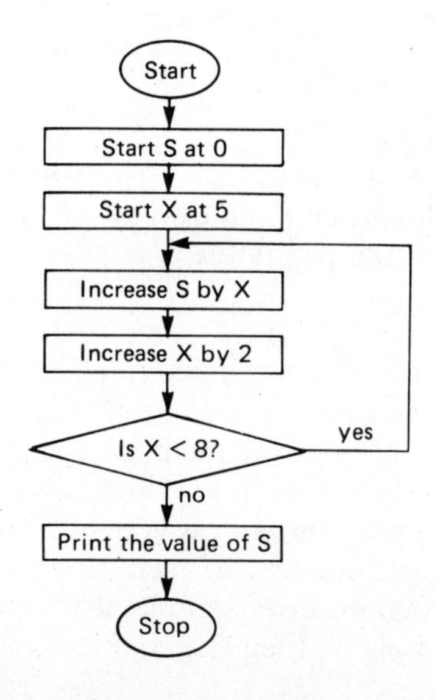

2. Find the output:

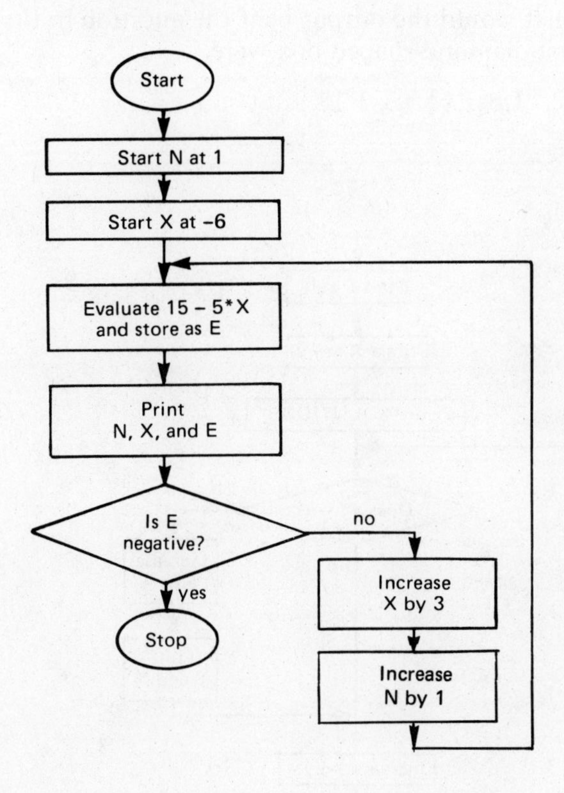

Note: The asterisk in the third rectangular box means "multiply."

3. Find the output if:

a. 4 is put in for N.
b. 7 is put in for N.

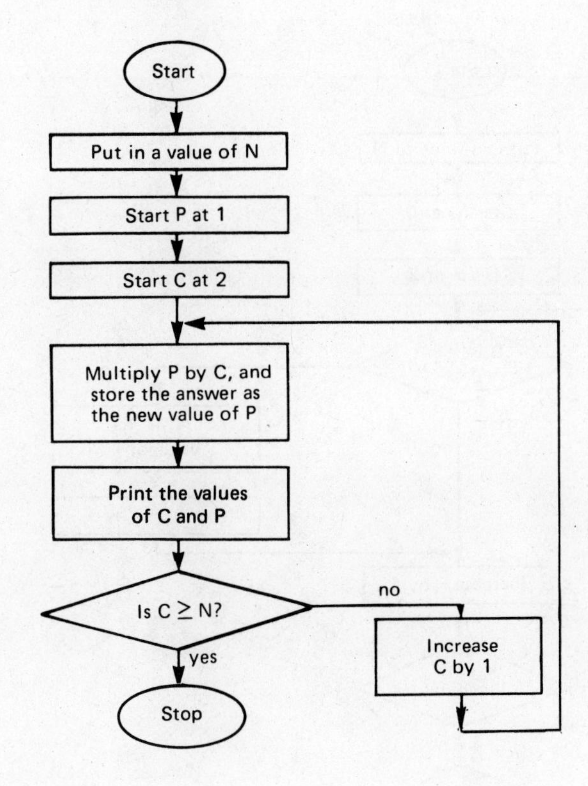

4. Find the output if:

a. 6 is put in for N.
b. 7 is put in for N.

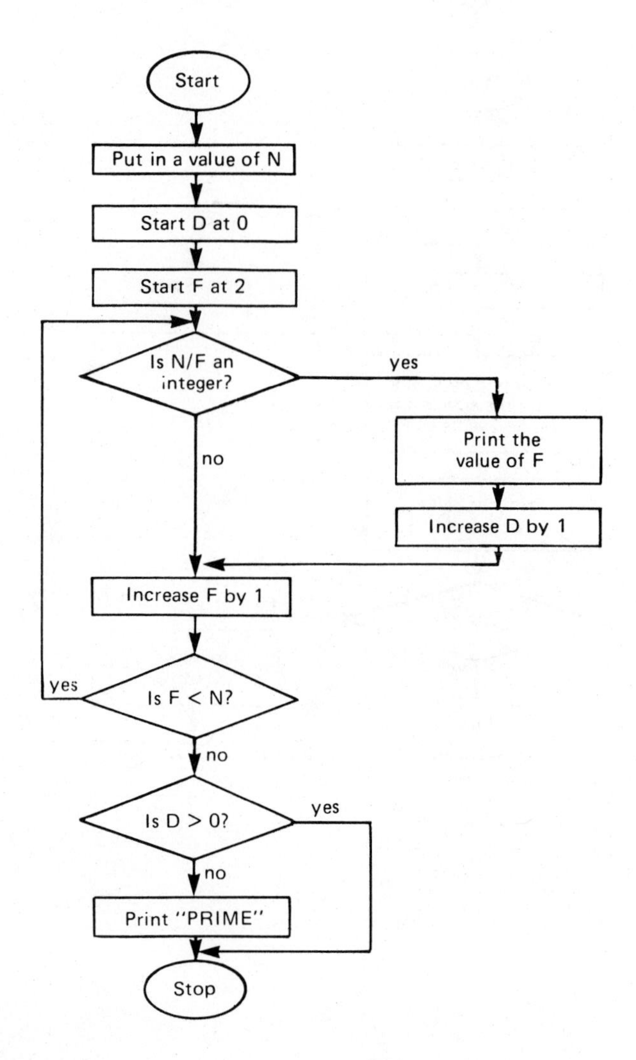

5a. Find the output.

b. What would the output be if the question in the first diamond-shaped box were,

"Does $x^3 = x + 1$?"

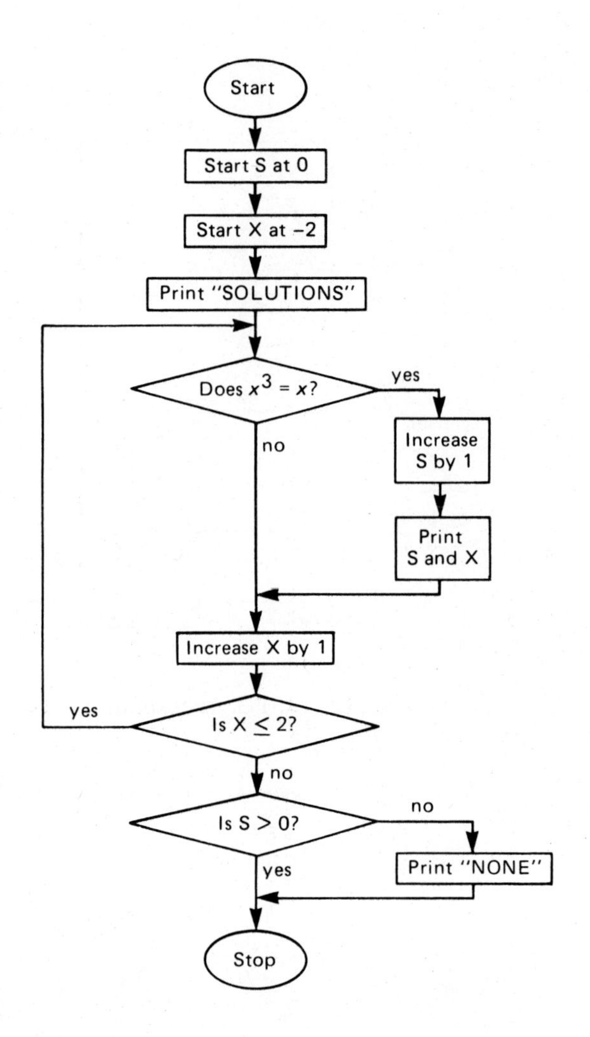

6a. Find the output if 64 is put in for N.

b. Find the output if 34 is put in for N.

c. Tell in your own words what type of problem this flow chart is useful for.

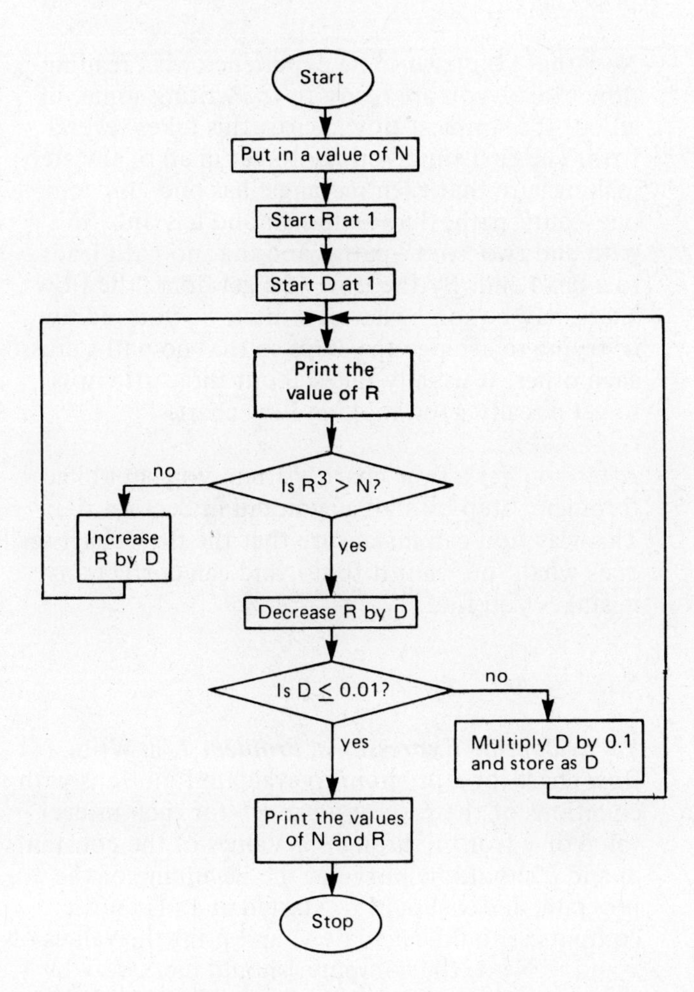

7. Find the output if 4 is put in for N.

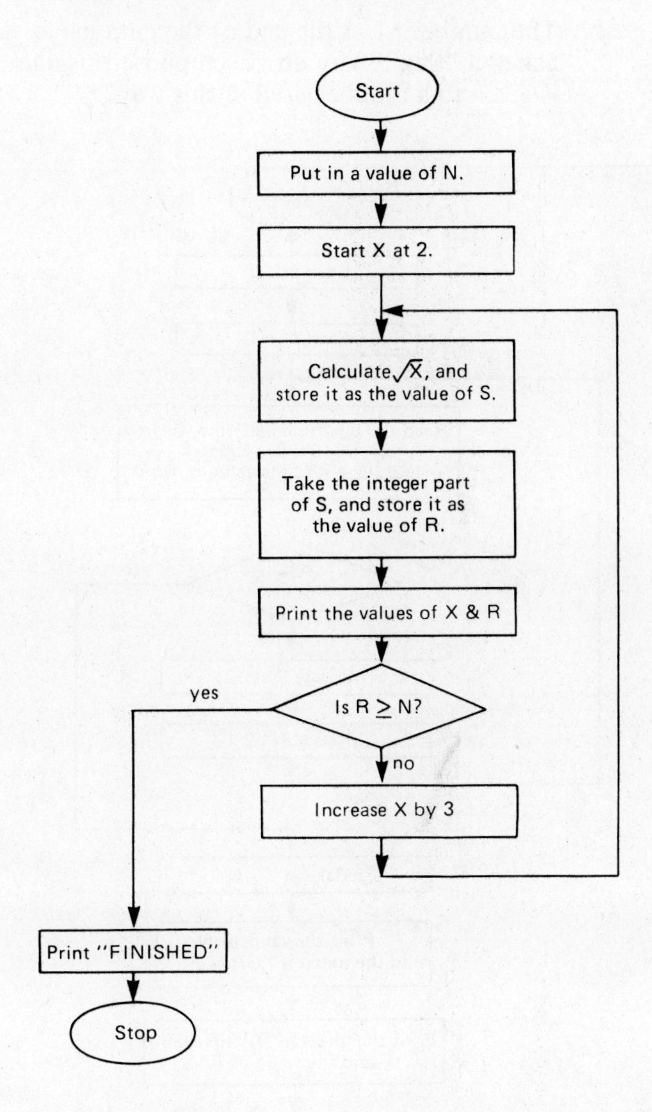

8a. Find the output.

b. The number −1 at the end of the data list is called a "flag." Why do you suppose this name is given to a number used in this way?

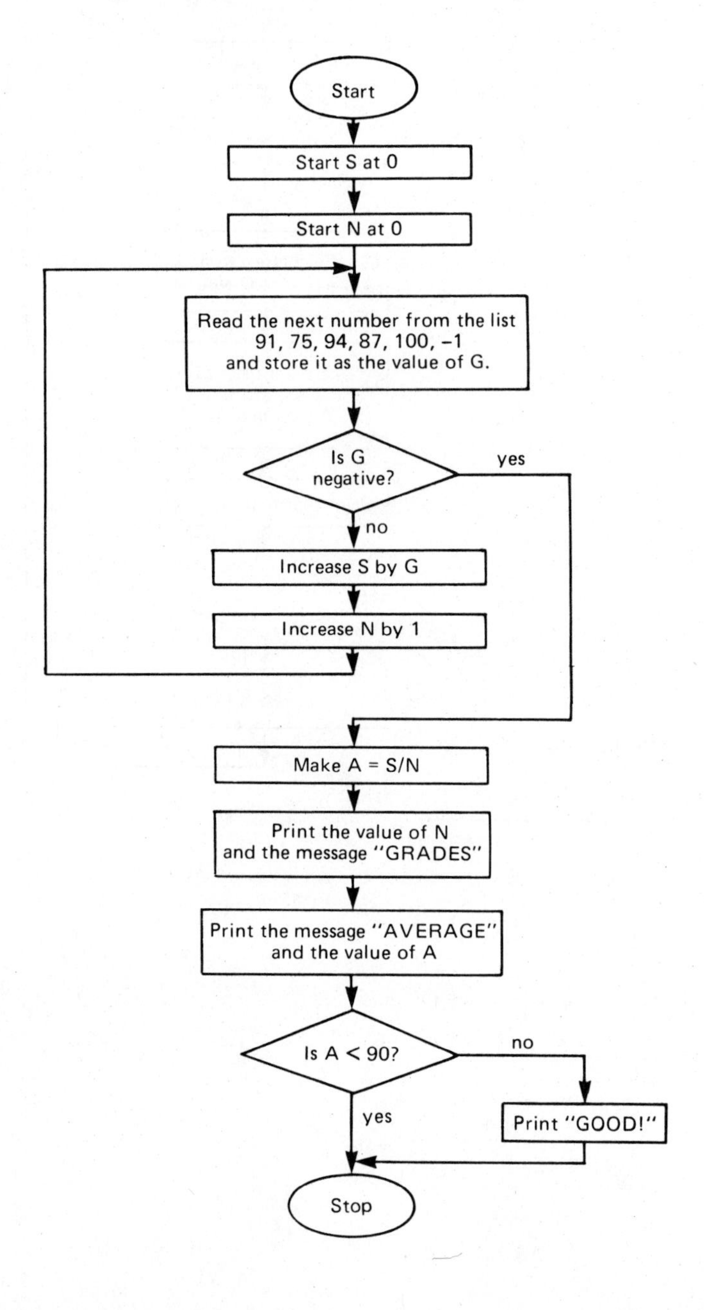

A-2 WRITING FLOW CHARTS

Objective:

Given a problem involving repetitive steps, write a flow chart.

Now that you have some experience with reading flow charts, you are ready to try writing some. In all but the simplest flow charts, this takes several tries. The first time you try to get in all of the steps, making sure that each rectangle has one "in" and one "out" path, that each diamond has one "in" path and two "out" paths, and that no path leads to a dead end. By the time you get done, the flow chart may be somewhat of a mess. So you redraw it, trying to arrange the steps so that no paths cross each other. It usually takes about three attempts to get a really good looking flow chart.

After you get a flow chart written, you should go through it step by step as you did in Section A-1. This way you can make sure that the flow chart really does what you want it to do, and can correct any mistakes you find.

Exercise A-2

1. ***Evaluating Expressions, Problem 1.*** a. Write a flow chart for a program to evaluate functions with equations of the form $y = mx + b$ for each integer value of x from 1 through 5. Values of the constants m and b should be put in at the beginning of the program, and x should be started at 1. Then the computer should calculate y, and print the values of x and y. Next, the computer should increase x by 1, and check to see if x is less than or equal to 5. If so, it should go back and repeat the calculating and printing steps. If not, it should stop.

b. Test your flow chart by putting in 7 for m and 11 for b, then showing the simulated computer memory and the output.

2. ***Evaluating Expressions, Problem 2.*** a. Write a flow chart for a program to evaluate functions with

a general equation of the form $y = ax^2 + bx + c$ for each integer value of x from 1 through 5. The values of the constants a, b, and c should be put in at the beginning of the program, and x should be started at 1. Then the computer should calculate the value of y, print the values of x and y, and increase x by 1. If the new value of x is still less than or equal to 5, the computer should loop back and calculate and print as before. If not, it should stop.

b. Test your flow chart by putting in 2 for a, −11 for b, and 7 for c, then showing the simulated memory and the output.

3. ***Evaluating Expressions, Problem 3.*** a. Write a flow chart for a program to evaluate functions with a general equation of the form $y = a \times b^x$. Values of the constants a and b should be put in at the beginning of the program. Then the computer should evaluate y and print x and y for each integer value of x from 1 through 5.

b. Test your flow chart by putting in 3 for a and 2 for b, then showing the simulated computer memory and the output.

4. ***Evaluating Expressions, Problem 4.*** a. Write a flow chart for a program to evaluate functions with a general equation of the form $y = \sqrt{a - bx}$. The values of a and b should be put in at the beginning of the program. Then y should be evaluated for each integer value of x, starting at $x = 1$. The output should be the values of x and y. Before the computer attempts to take a square root, the value of $a - bx$ should be checked to be sure it is not negative. The computer should stop when $a - bx$ becomes negative for the first time.

b. Test the flow chart by putting in 10 for a and 3 for b, then showing the simulated memory and the output.

5. ***Evaluating Expressions, Problem 5.*** a. Write a flow chart which will find values of the fraction

$$\frac{x-4}{(x-3)(x-5)}$$

for all integer values of x from 1 to 10, inclusive. The program should print the value of x and the value of the fraction for each x. Before doing a calculation with a new value of x, the flow chart should first check to see that the denominator is not equal to zero. If it *is* equal to zero, the flow chart should print the message, "Zero denominator when x =," and then print the value of x.

6. ***Evaluating Expressions, Problem 6.*** a. Write a flow chart for a program to calculate and print N values of a variable Y. The first two values of Y and the value of N should be input at the beginning of the program. Then the computer should divide the *second* value of Y by the *first* value of Y and store the answer as R. The third value of Y is found by multiplying the second value of Y by R. The fourth value of Y is found by multiplying the third value of Y by R, and so forth. The computer should stop when a total of N values of Y have been calculated and printed.

b. Test your flow chart by putting in 5 for N, and 1000 and 800 for the first two values of Y, then showing the simulated memory and the output.

7. ***Sum of the Even Numbers Problem.*** a. Write a flow chart to find the *sum* of the even integers from 14 through N. The value of N should be input at the beginning of the program. Then a variable S (for "sum") should be started at 0 and a variable T (for "term") should be started at 14. A loop should be set up in which T is added to S, then T is increased by 2 (to get the next even integer). The looping should stop when $T = N$, and this value of T has been added to S. At each pass through the loop, the values of T and S should be printed.

b. Test the flow chart by putting in 20 for N, and showing the simulated computer memory and the output.

8. ***Sum of the Odd Numbers Problem.*** a. Write a flow chart for a program to find the sum of N consecutive odd integers, starting at the odd integer X.

Values of N and X should be put in at the beginning of the program, and a variable S (for "sum") should be started at 0. Then a loop should be set up that adds X to S, and then increases X to the next odd integer. The looping should stop when N integers have been added together. At each pass through the loop the values of X and S should be printed.

b. Test your flow chart by putting in 6 for N and 11 for X, then showing the simulated computer memory and the output.

9. ***Sum of the y-Values Problem.*** a. Write a flow chart for a program to calculate N values of y for a function whose general equation is $y = mx + b$, and find the *sum* of these y-values. A value of N should be put in at the beginning of the program along with values of the constants m and b. Then x should be started at 1 and a variable S (for "sum") should be started at 0. A loop should be set up in which the value of y is calculated, the values of x and y are printed, and the value of y is added to the sum of the previous values, S. When N values of y have been calculated, printed, and summed, the looping should stop, and the computer should print the message, "SUM =" and the value of S.

b. Test the flow chart by putting in 5, 7, and 11 for N, m, and b, respectively, and showing the simulated computer memory and the output.

10. ***Sum of the Data Problem.*** a. Write a flow chart for a program to read numbers one at a time from a list such as

11, 37, 22, 64, 51, 73, −1, 29, 44, 36, 16, ... ,

and find their sum. The computer should stop reading numbers when it first encounters a *negative* number. The negative number is just a "flag" that signals the end of the data, and is *not* to be added to the sum of the data. The output should be the *sum* of the numbers read and the *number* of numbers that were read *before* encountering the flag.

b. Test the flow chart by showing the simulated computer memory and the output.

11. ***Data Search Problem.*** a. Suppose that you have a list of numbers such as

37, 41, 92, 76, 81, 42, 25, 21, 52, 71, 94, 105.

Write a flow chart for a program to read numbers one at a time from such a list, and stop at the first one which is evenly divisible by 7. The flow chart should have the computer print out this number (42 for the above example) as well as its position in the list (6th in the above example). The position can be determined by using a variable as a counter, and increasing its value by 1 each time another number is read. (More sophisticated versions of programs like this are commonly used by computers for such things as locating a given name in a list of airplane passengers, or identifying a person by his fingerprints from millions of fingerprints on file.)

b. Write the simulated memory and show the output for this flow chart. You should get the output 6 and 42.

c. Modify the flow chart in part a, above, so that it prints the value and position of *all* numbers in the given list which are divisible by 7. This can be accomplished by asking the question, "Are there any more numbers in the list?" before the "stop" instruction, and looping back to the "read" instruction if the answer is "Yes."

12. ***Data Round-Off Problem.*** a. Write a flow chart for a program to round off data from decimal form to the nearest integer. The computer should read each number from a list of data, round it off, then print it. 1 should be printed by the first number, 2 by the second, and so forth. The program should stop when the first negative number is read from the list.

b. Test your flow chart using the data list 37.8, 15.2, 9.785, 17.33, 0.48, 99.9, −1.

13. ***Product of the Integers Problem.*** a. Write a flow chart for a program to start a variable P at 1, then multiply it by 2, multiply that answer by 3, and so forth. The flow chart should stop the multiplication when P first becomes larger than the value of a

variable L which you put in at the beginning of the flow chart. Then it should print the final value of P as well as the value of the last factor used in calculating P.

b. Test the flow chart by putting in 100 for L and showing that the output is the numbers 120 and 5.

c. Test the flow chart again by putting in 6 for L and showing that the output is 24 and 4.

14. ***Sum of the Cubes Problem.*** a. Write a flow chart for a program to find the *sum, S,* of the *cubes* of the first n counting numbers.

$$S = 1^3 + 2^3 + 3^3 + \ldots + n^3,$$

for each integer value of n from 1 through 6. The output at each step should be the value of n, the value of S, and the value of the *square root* of S.

b. Show the simulated memory and the output of the program. What interesting fact seems to be true about each value of S?

15. ***Bank Balance Problem.*** a. Write a flow chart for a program to determine and print your bank balances. Your beginning balance should be input at the beginning of the program, along with the number of transactions (checks or deposits) to be made. Then the transactions should be read one at a time from a list of data, and *added* to the balance. This means, of course, that the deposits you make will be *positive* numbers and the checks you write will be *negative* numbers. Each time a transaction is read, the computer should print the amount of that transaction, and the balance *after* that transaction. If your balance ever becomes *negative*, the computer should stop reading transactions and print the message "OVERDRAWN BY," and the amount by which your account is overdrawn.

b. Test the flow chart by starting with a balance of \$153, and using the transactions –50, 13, 22, –100, 47, –30, –99, 110, –33, 71. Use 0 to flag the end of the data.

16. ***Telephone Call Problem.*** Write a flow chart for "making a telephone call." It should contain such instructions as "Dial the number," "Listen for a ring," "Did someone answer?", "Should I let it ring another time?", "Complete the conversation," and "Hang up." You should take into account such things as whether or not you know the phone number, whether or not the line is busy, and what happens if nobody is home.

A-3 THE BASIC COMPUTER LANGUAGE

The flow charts in Sections A-1 and A-2 break down problems into a sequence of steps which are simple enough to be handled by a computer. Before a computer can carry out these steps, however, the instructions must be written in a language which it can "understand." One such language is called BASIC. A set of instructions written in a computer language is called a *program*. The act of putting the instructions into the computer is called *programming* the computer. The act of having the computer perform the instructions is called *running* the program, or *executing* the program.

Objectives:

1. Given a flow chart, write the corresponding BASIC program.

2. Given a BASIC program, go through it step by step as the computer would do, and determine the output.

3. Given a BASIC program, feed it to the computer and have the computer execute the program.

The names of most computer languages are written in capital letters since they are acronyms composed of the first letters of other words. For example, BASIC comes from Beginners All-purpose Symbolic Instruction Code. BASIC is popular because it is easily learned by students who are familiar with algebra. The instructions look like algebraic statements, use the same sequence of operations as in algebra (multiplication before addition, etc.), and are easily derived from flow charts such as those in the previous sections. Also, BASIC is available on many computers, particularly with time-sharing systems.

The treatment of BASIC which follows is meant to give you enough information to translate the flow charts in the previous sections. You should consult your library for books on more extended versions of BASIC or for other languages such as FORTRAN, ALGOL, APL, etc.

Putting Information into the Computer

Putting information into the computer is accomplished by an *INPUT* statement or by *READ* and *DATA* statements. Upon encountering the statement.

```
10 INPUT X, Y, Z
```

the computer will print a ? then wait for you to type in three numbers. If you then type

```
73, 82, 80
```

the computer will store 73 as the value of *X*, 82 as the value of *Y*, and 80 as the value of *Z*.

Notes:

a. The number 10 at the left of the statement is called the "line number" or the "statement number." Every statement in BASIC must have a number so that the computer can tell the sequence in which the statements are to be executed. It starts with the statement having the lowest number and proceeds in numerical order unless it is given an instruction to branch to a different statement number.

b. Statements in BASIC are written with capital letters since these are all that are available for most computers.

Putting in numbers can be accomplished more rapidly with the *READ* and *DATA* statements. If the computer encounters the statement

```
50 READ X, Y, Z
```

then it looks for a DATA statement somewhere in the program. If it finds the statement

```
100 DATA 73,82,80,95,86,98,74,53,89,66,100,72
```

then it would store 73 as the value of *X*, 82 as the value of *Y*, and 80 as the value of *Z*. If the computer later *returns* to the same READ statement, it will use the *next* three numbers in the DATA statement, namely 95 for *X*, 86 for *Y*, and 98 for *Z*. The previous values of *X*, *Y*, and *Z* will be "forgotten."

Getting Information out of the Computer

Getting information out of the computer is accomplished by a *PRINT* statement. If the computer encounters the statement

```
70 PRINT X,Y,Z
```

then it prints whatever numbers are *currently* stored in its memory as the values of *X*, *Y*, and *Z*. It prints only the numbers, such as

```
95    86    98
```

without telling that these are values of *X*, *Y*, and *Z*. If you want the computer to print a *message*, you use a PRINT statement with quotation marks around the message, such as

```
80 PRINT "YOUR GRADE IS A."
```

Printing values of variables can be combined with printing messages. For example, if 95 has been stored as the value of *X*, then the statement

```
90 PRINT "X = "; X
```

would cause the computer to print

$X = 95.$

If 7 has been stored as the value of *N*, then the statement

```
120 PRINT "THE NUMBER"; N; "IS PRIME."
```

would cause the computer to print

```
THE NUMBER 7 IS PRIME.
```

The semicolon in the PRINT statement instead of the comma is used to make the computer leave less space between the things being printed.

Calculating Values of Variables

Calculating values of variables is accomplished with the LET statement. For example,

```
20 LET C = 1
```

would cause 1 to be stored as the value of C. The statement

```
30 LET A = (X + Y + Z)/3
```

would tell the computer to do the arithmetic on the right side of the = sign using the numbers currently stored in its memory for X, Y, and Z, and store the answer as the value of the variable A on the left. Notice that the expression on the right must all be written on *one* line. The parentheses around the $X + Y + Z$ tell the computer to do the addition *before* dividing by 3. Like any good algebra student, the computer has been taught that $X + Y + Z/3$ would mean to divide Z by 3 before adding X and Y.

The statement

```
60 LET C = C + 1
```

would instruct the computer to take the present value of C, add 1 to it, and then store the answer as the *new* value of C, "forgetting" the previous value. This is the way you could increase (or decrease) the value of a variable in BASIC.

Note: The = sign in a LET statement in BASIC is *not* the same as the = sign in algebraic equations. Since the variable on the left simply tells the computer where to store the answer, the computer would not recognize a statement such as LET $C + 1 = C$. In other words, the = sign in a LET statement does not possess the symmetric property of the = in an algebraic equation.

Operations in BASIC

Each *operation* in a BASIC expression requires an operation sign. The computer will not recognize expressions such as x^2 or $3x$. The operation signs are as follows:

Sign	Operation	Example	Meaning
+	Addition	X+Y+Z	$x + y + z$
−	Subtraction	X−Y	$x - y$
*	Multiplication	X*Y	xy
/	Division	X/Y	$x \div y$ or $\frac{x}{y}$
↑	Exponentiation	X↑Y	x^y

The sequence of operations is the same as that used in algebra.

a. What is in parentheses is done first.
b. Exponentiation is done next.
c. Multiplication and division are then done in the order in which they occur, from left to right.
d. Addition and subtraction are done last, from left to right in the order in which they occur.

Branching

Branching may be accomplished either by the GO TO statement or by the IF ... THEN statement. Upon encountering the statement

```
200 GO TO 10
```

somewhere in a program, the computer would go to the statement numbered 10 and continue running the program from that point.

If the computer encounters the statement

```
60 IF A >= 90 THEN 85
```

it would first check to see whether the number currently stored as the value of A were greater than or equal to 90. If so, it would go to statement 85 and continue running the program from there. If not, it would continue with the next statement in the program without branching. The following relationships can be used in the IF ... THEN statement:

= equal

< less than

> greater than

<= less than or equal to

>= greater than or equal to

<> not equal to

Looping

Looping repeatedly through certain steps in a program can be accomplished by a combination of IF ... THEN, GO TO, and LET statements. However, an easier way has been provided, the FOR ... TO ... STEP and the NEXT statements. If the computer encounters the set of statements

```
10 FOR X = 4 TO 100 STEP 2
 .   .
 .   .
 .   .
40 NEXT X
 .   .
 .   .
 .   .
```

it would start X at 4, and carry out all the statements up to NEXT X. Then it would increase X by 2 because of the ... STEP 2 in the statement, and *repeat* all the steps between the FOR and the NEXT statements. It would keep looping through these statements with X = 6, 8, 10, 12, ..., until it had been through them with X equal to 100. Then it would leave the loop and continue with the statement right under the NEXT statement. If the STEP portion of the statement is left off, the computer assumes a step of 1.

Functions in BASIC

Functions such as "absolute value" and "square root" are available in BASIC. For example, if the computer encounters the statement

```
60 LET Y = SQR(X)
```

then it takes the square root of the number stored in the X memory location and stores the answer as the new value of Y. Some functions available in BASIC are:

ABS(X)	Absolute value of X
SQR(X)	Square root of X
INT(X)	Greatest integer less than or equal to X. This function has the effect of "chopping off" the decimal part of X, leaving only the integer part. For example, INT(3.9) is equal to 3.
RND(X)	Picks a number at random from the interval between 0.00000 and 1.00000.
SIN(X)	Circular sine of X (i.e., X is in radians).
COS(X)	Circular cosine of X (i.e., X is in radians).
ATN(X)	Circular arctangent of X.
LOG(X)	Natural logarithm of X (i.e., base e log).
EXP(X)	Exponential function, e^x.

Subscripted Variables

Subscripted variables may be used if you want the computer to remember *all* values which a variable has taken on. As you recall, when you assign a new value to a variable, the computer "forgets" the previous value. For a computer to remember previous values, it is necessary to use a different memory location for each value. For example, if the computer encounters the statement

```
10 DIM G(100)
```

at the beginning of a program, then it will save 100 memory spaces for the variable G as shown in the sketch, instead of just one.

G(1)
G(2)
G(3)
.
.
.
G(99)
G(100)

The letters DIM stand for the word "dimension," meaning how long a block of spaces must be saved for the variable. If the computer later on encounters the statements

```
50 FOR K = 1 TO 10 STEP 1
60 READ G(K)
70 DATA 73,82,80,95,86,98,74,53,89,66,100,72
80 NEXT K
```

then it would start K at 1, read the first number, 73, from the DATA list, and store 73 as the value of G(1). Then it would loop back with K equal to 2, read 82 from the DATA list, and store 82 as the value of $G(2)$. The computer would read only the first 10 numbers from the DATA list, ignoring the rest. Variables $G(11)$ through $G(100)$, for which the computer has reserved space, would remain empty.

END Statement

The *last statement* of a BASIC program must be an END instruction. The statement is written

```
200 END.
```

The statement number must be the *largest* of any in the program so that the END instruction will really come last.

The following exercise contains programs written in BASIC for you to read and determine the output. You simply go through the program step by step, recording numbers in a simulated memory as in Section A-1. When you have become familiar with the instructions, you will actually translate some flow charts into BASIC.

Exercise A-3

For Problems 1 through 10, figure out what the computer will print if it runs the BASIC program. Set up a simulated computer memory as for the flow charts in Section A-1.

1.
```
10   LET S=0
20   FOR X=−2 TO 2
30   LET Y=3−5*X
40   LET S=S+Y
50   NEXT X
60   IF Y<=0 THEN 80
70   PRINT "POSITIVE Y"
80   PRINT Y;S
90   END
```

2.
```
10   FOR X=1 TO 10
20   LET Y=X↑2−6*X
30   LET Z=ABS (Y)
40   IF Z > 8 THEN 80
50   NEXT X
60   PRINT "NEVER WAS"
70   GO TO 90
80   PRINT X,Z
90   END
```

3a. Put in 1 for N.

b. Put in 5 for N.

```
10   INPUT N
20   LET S=0
30   FOR X=−1 TO N
40   LET Y=1+3*X
50   LET S=S+Y
60   IF S>25 THEN 100
70   NEXT X
80   PRINT "S NEVER OVER 25"
90   GO TO 110
100  PRINT "S > 25 WHEN X=";X
110  END
```

4. Put in 5 for N in Step 20.

```
10   LET S=0
20   INPUT N
30   FOR K=1 TO N
40   READ X
50   DATA 85,91,53,100,97
60   LET S=S+X
70   NEXT K
80   LET A=S/N
90   PRINT A
100  END
```

5.
```
10   LET C=1
20   READ X,Y
30   DATA 2,5,8,−3,4,9
40   LET E=X+Y↑2
50   LET D=2
60   IF INT(E/D)=E/D THEN 120
70   IF D>=5 THEN 100
80   LET D=D+1
90   GO TO 60
100  PRINT E;"NEVER"
110  GO TO 130
120  PRINT E;"DIVISIBLE BY";D
130  LET C=C+1
140  IF C<3 THEN 20
150  END
```

6a. Put in 12 for N.

b. Put in 11 for N.

```
10   LET D=0
20   INPUT N
30   LET Q=N
40   LET F=2
50   IF Q/F = INT(Q/F) THEN 90
60   LET F=F+1
70   IF F<= SQR(N) THEN 50
80   GO TO 130
90   PRINT F
100  LET Q=Q/F
110  LET D=D+1
120  IF Q>1 THEN 50
130  PRINT N; "HAS"; D; "FACTORS"
140  END
```

7. Put in 3 for M and −11 for B.

```
10   INPUT M, B
20   LET Y1=B
30   FOR X=1 TO 10
40   LET Y2=M*X+B
50   IF Y2*Y1<=0 THEN 100
60   LET Y1=Y2
70   NEXT X
80   PRINT "NO ROOTS"
90   GO TO 110
100  PRINT "ROOTS BETWEEN";X-1;
     "AND";X
110  END
```

8. Put in 4 for N in Step 20.

```
10   DIM G(100)
20   INPUT N
30   LET C=1
40   READ G(C)
50   DATA 73,82,80,95,86,98,74,
51   DATA 53,89,66,100,72
60   LET C=C+1
70   IF C>3*N THEN 90
80   GO TO 40
90   FOR C=1 TO 3*N STEP 3
100  LET A=(G(C)+G(C+1)+G(C+2))/3
110  PRINT (C+2)/3,A
120  IF A<90 THEN 140
130  PRINT "YOUR GRADE IS A."
140  NEXT C
150  END
```

9.
```
10   DIM X(20)
20   DATA 82,97,59,74,91,86,
21   DATA 74,93,100,68,−1
30   LET N=0
40   READ X(N)
50   IF X(N)<0 THEN 80
60   LET N=N+1
70   GO TO 40
80   LET S=0
90   FOR I=1 TO N
100  LET S=S+X(I)
110  NEXT I
120  LET A=S/N
130  PRINT"AVGE OF";N;"NOS.=";A
140  END
```

10.
```
10   DIM X(20)
20   DATA 82,97,59,74,91, 86,−1
30   LET N=1
40   READ X(N)
50   IF X(N)<0 THEN 80
60   LET N=N+1
70   GO TO 40
```

```
80    PRINT"RANK SCORE"
90    FOR I=1 TO N−1
100   FOR K=I+1 TO N
110   IF X(I)>=X(K) THEN 150
120   LET H=X(I)
130   LET X(I)=X(K)
140   LET X(K)=H
150   NEXT K
160   PRINT I;X(I)
170   NEXT I
180   END
```

Problems 11 and 12 contains programming errors. List all the errors you can find.

11.
```
10    LET X = 0
20    LET B = 7
30    LET Y=MX + B
40    LET X+1 = X
50    GO TO 30
60    STOP
```

12.
```
10.   LET S=0
15.   LET C=0
20.   READ X
30.   LET S=S+X
40.   LET C=C+1
50.   IF C≤N THEN GO TO 20
60.   A=S/N
70.   PRINT"AVGE OF";N;"NOS.=";S
```

For Problems 13 through 36, write a BASIC program for the problem indicated.

13. Exercise A-1, Problem 1

14. Exercise A-1, Problem 2

15. Exercise A-1, Problem 3

16. Exercise A-1, Problem 4

17. Exercise A-1, Problem 5

18. Exercise A-1, Problem 6

19. Exercise A-1, Problem 7

20. Exercise A-1, Problem 8

21. Exercise A-2, Problem 1

22. Exercise A-2, Problem 2

23. Exercise A-2, Problem 3

24. Exercise A-2, Problem 4

25. Exercise A-2, Problem 5

26. Exercise A-2, Problem 6

27. Exercise A-2, Problem 7

28. Exercise A-2, Problem 8

29. Exercise A-2, Problem 9

30. Exercise A-2, Problem 10

31. Exercise A-2, Problem 11

32. Exercise A-2, Problem 12

33. Exercise A-2, Problem 13

34. Exercise A-2, Problem 14

35. Exercise A-2, Problem 15

Problems 36 and 37 are intended to allow you to start from scratch and write a BASIC program from a stated problem. You should, of course, write a flow chart first, then translate the flow chart into BASIC. You can either go through the program manually to see if it works, or you can put it on the computer and let the computer "debug" it for you!

36. ***Computer Game Problem.*** Write a program that allows you and the computer to play a game. You first put in a number to be your score. This must be an integer less than 21. The computer starts its score, S, at 0. Then it picks random numbers between 1 and 11 and adds them one at a time to its score. The computer wins if its score becomes greater than or equal to your score, but less than or equal to 21. You win if the computer goes over 21. The function RND(X) can be used to pick a random number between 0 and 1. If you multiply this random number by 11, you get a random number between 0 and 11. Taking the greatest integer of this result gives a random integer from 0 through 10. Adding 1 to this answer produces a random integer from 1 through 11. Test your program by putting 17 in for your score, then using 3, 7, 4, and 11 for the random numbers the computer adds to its score.

37. ***Coin Flipping Problem.*** Write a BASIC program to conduct a coin-flipping game. Variables H and T stand for the numbers of heads and tails, respectively, flipped during the game. The computer will flip a coin at random by using the function RND(X) to select a random number between 0 and 1. If RND(X) < 0.5, assume the flip was tails. If RND(X) ≥ 0.5, assume the flip was heads. The game ends when either H or T is 5 *larger* than the other, or when the coin has been flipped 1000 times, whichever comes first. The output should be *which* won, heads or tails, and the total number of flips used. If 1000 flips is reached without a winner, the game is a draw.

Appendix B Additional Topics

There are several topics which are sometimes made a part of advanced algebra, but which would interrupt the mainstream of the text. These topics, Descartes' Rule of Signs, operations with matrices, and mathematical induction, are presented in this appendix for those instructors who choose to present them.

B-1 OPERATIONS WITH MATRICES

In Sections 4-3, 4-7, and 4-8 you were introduced to determinants and matrices. A matrix is a rectangular array of numbers such as those below.

$$\begin{bmatrix} 2 & 5 & 3 \\ -1 & 4 & -2 \end{bmatrix} \quad \begin{bmatrix} 5 & 1 \\ 7 & 3 \\ 2 & -4 \end{bmatrix} \quad \begin{bmatrix} 9 & 7 & 1 & 3 \end{bmatrix} \quad \begin{bmatrix} 2 \\ -5 \end{bmatrix}$$

2×3 matrix, 3×2 matrix, 1×4 matrix, 2×1 matrix

The first matrix above has an *order* of "2 by 3" since it has 2 *rows* and 3 *columns.* The number of rows is written first and the number of columns second. Each number in a matrix is called an *element.* Two matrices (the plural of matrix) are equal if and only if they are the same order and have their corresponding elements equal.

A determinant is a number produced by doing operations on the elements of a square matrix, as explained in Chapter 4. So a determinant is a number whereas a matrix is an array of numbers.

Addition and Subtraction

Two matrices of the same order can be added or subtracted by adding or subtracting the corresponding elements. For instance,

$$\begin{bmatrix} 5 & 2 & 7 \\ 1 & 3 & 9 \end{bmatrix} + \begin{bmatrix} 4 & 6 & 8 \\ 5 & 1 & 3 \end{bmatrix} = \begin{bmatrix} 9 & 8 & 15 \\ 6 & 4 & 12 \end{bmatrix}$$

$$\begin{bmatrix} 6 & 5 \\ 2 & 3 \end{bmatrix} - \begin{bmatrix} 4 & 7 \\ 5 & 2 \end{bmatrix} = \begin{bmatrix} 1 & -2 \\ -3 & 1 \end{bmatrix}$$

Multiplication by a Scalar

A matrix can be multiplied by a scalar (a number) by multiplying each element of the matrix by that scalar. For instance,

$$5\begin{bmatrix} 2 & 3 & 4 \\ 6 & 1 & 7 \end{bmatrix} = \begin{bmatrix} 10 & 15 & 20 \\ 30 & 5 & 35 \end{bmatrix}$$

Multiplication of Two Matrices

Multiplying a matrix by another matrix is complicated to explain, but easy to do once you learn the pattern. The order in which you write the two factors is important. The number of *columns* in the *left* matrix must equal the number of *rows* in the *right* one, as shown below. You begin by multiplying the elements in the first row of the left matrix by the corresponding elements in the first column of the right matrix.

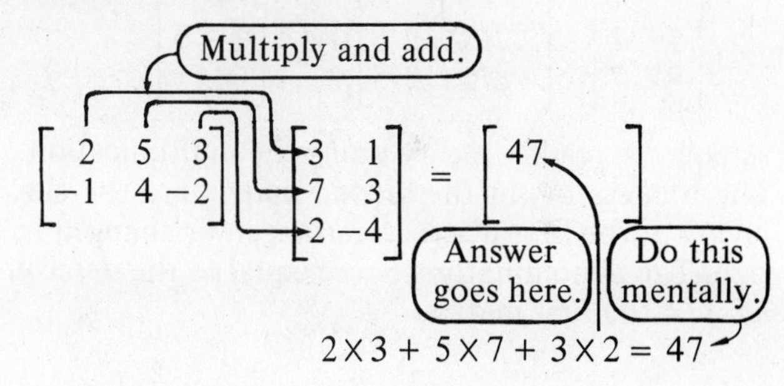

An efficient way to do this is to slide your left index finger across the row and your right index finger down the column, multiplying and adding mentally. The answer goes in the first row, first column of the product matrix.

The procedure is repeated for each possible pairing of rows in the left matrix and columns in the right one. The result goes in the product matrix at the row you used from the left matrix and the column you used from the right matrix. The completed product is shown on the next page.

$$\begin{bmatrix} 2 & 5 & 3 \\ -1 & 4 & 2 \end{bmatrix} \begin{bmatrix} 3 & 1 \\ 7 & 3 \\ 2 & -4 \end{bmatrix} = \begin{bmatrix} 47 & 5 \\ 29 & 3 \end{bmatrix}$$

You should try doing the above multiplication to make sure you understand the process.

Identities and Inverses

The matrix

$$\begin{bmatrix} 1 & 0 & 0 \\ 0 & 1 & 0 \\ 0 & 0 & 1 \end{bmatrix}$$

is the *multiplicative identity* for 3 X 3 matrices. The identity for a square matrix has 1's along the *main diagonal* and 0's everywhere else. Multiplying a square matrix by the identity matrix leaves the matrix unchanged.

If the product of two square matrices is the identity matrix, then those two matrices are *inverses* of each other. For instance,

$$\begin{bmatrix} 3 & 2 \\ 8 & 7 \end{bmatrix} \begin{bmatrix} 7/5 & -2/5 \\ -8/5 & 3/5 \end{bmatrix} = \begin{bmatrix} 1 & 0 \\ 0 & 1 \end{bmatrix}$$

as you can readily see by doing the multiplication. The numerators in the second matrix are the elements of the first matrix, rearranged or changed in sign. The denominators, 5, are equal to the *determinant* of the first matrix.

$$\det \begin{bmatrix} 3 & 2 \\ 8 & 7 \end{bmatrix} = \begin{vmatrix} 3 & 2 \\ 8 & 7 \end{vmatrix} = 3 \cdot 7 - 2 \cdot 8 = 5$$

An easy way to write the inverse of a 2 X 2 matrix is to interchange the top left and bottom right elements, reverse the signs of the other two elements, and multiply the resulting matrix by the reciprocal of the determinant. That is,

$$\text{If } M = \begin{bmatrix} a & b \\ c & d \end{bmatrix}, \text{ then } M^{-1} = \frac{1}{\det M} \begin{bmatrix} d & -b \\ -c & a \end{bmatrix}$$

Adjoint of M, adj M

The last matrix is called the *adjoint* of matrix M, abbreviated adj M. The inverse of M, as indicated above, is abbreviated M^{-1}.

Finding the inverse of a higher-order square matrix is explained at the end of this section.

Matrix Solution of a Linear System

A system of linear equations such as

$$\begin{aligned} ax + by &= c \\ dx + ey &= f \end{aligned}$$

can be written as a product of matrices,

$$\begin{bmatrix} a & b \\ d & e \end{bmatrix} \begin{bmatrix} x \\ y \end{bmatrix} = \begin{bmatrix} c \\ f \end{bmatrix}$$

or more briefly,

$$CV = A,$$

where C is the coefficient matrix, V is the variables matrix, and A is the answers matrix. The variables matrix can be isolated by multiplying each member of the equation by the inverse of C.

$$\begin{aligned} C^{-1}CV &= C^{-1}A \\ V &= C^{-1}A \end{aligned}$$

For example, to solve

$$\begin{aligned} 5x - y &= 7 \\ 2x + 3y &= -1 \end{aligned}$$

you would write

$$\begin{bmatrix} 5 & -1 \\ 2 & 3 \end{bmatrix} \begin{bmatrix} x \\ y \end{bmatrix} = \begin{bmatrix} 7 \\ -1 \end{bmatrix}$$

$$\begin{bmatrix} x \\ y \end{bmatrix} = \frac{1}{17} \begin{bmatrix} 3 & 1 \\ -2 & 5 \end{bmatrix} \begin{bmatrix} 7 \\ -1 \end{bmatrix}$$

$$\begin{bmatrix} x \\ y \end{bmatrix} = \frac{1}{17} \begin{bmatrix} 20 \\ -19 \end{bmatrix}$$

$$\therefore x = \frac{20}{17} \text{ and } y = -\frac{19}{17}$$

$$S = \{(20/17, -19/17)\}$$

Inverse of a Higher Order Matrix

As for 2×2 matrices, the inverse of a higher-order square matrix is given by

$$M^{-1} = \frac{1}{\det M} \cdot \text{adj } M$$

For higher-order matrices the adjoint matrix must be found by a more general procedure. For example, let

$$M = \begin{bmatrix} 2 & 3 & 4 \\ 5 & 1 & 2 \\ 6 & 8 & 7 \end{bmatrix}$$

First, find the *transpose* of M, written M^T, which is the matrix formed by interchanging the rows and the columns of M.

$$M^T = \begin{bmatrix} 2 & 5 & 6 \\ 3 & 1 & 8 \\ 4 & 2 & 7 \end{bmatrix}$$

The elements of the adjoint matrix are found by evaluating the *minor* determinant for each element of M^T. For instance, the minor of the 2 in the upper left corner is the determinant

$$\begin{vmatrix} 1 & 8 \\ 2 & 7 \end{vmatrix} \quad \text{which equals } -9.$$

Each of these minor determinants is given a sign according to the following pattern:

$$\begin{matrix} + & - & + \\ - & + & - \\ + & - & + \end{matrix}$$

Since the sign in the upper left corner is +, the -9 remains -9. Had the sign been $-$, the -9 would have changed to $+9$. The minor determinants along with their signs shown above are called *cofactors* of the elements in the matrix. Evaluating the other eight cofactors produces the answer,

$$\text{adj } M = \begin{bmatrix} -9 & 11 & 2 \\ -23 & -10 & 16 \\ 34 & 2 & -13 \end{bmatrix}$$

You should try evaluating a few of the cofactors above to make sure you know the pattern. In general, the adjoint of a square matrix is the *matrix of cofactors of the transpose* of that matrix. The pattern for a 2×2 matrix is a special case of this more general pattern.

Evaluating the determinant of M as in Section 4-7 gives $\det M = 49$. Therefore, the inverse of M is

$$M^{-1} = \frac{1}{49} \begin{bmatrix} -9 & 11 & 2 \\ -23 & -10 & 16 \\ 34 & 2 & -13 \end{bmatrix}$$

The $1/49$ can be distributed to each element in the matrix or can be left factored out, whichever is more convenient.

The following exercise has problems to make you familiar with matrix operations.

Exercise B-1

For Problems 1 through 16 perform the indicated operations.

1. $\begin{bmatrix} 3 & 5 \\ -2 & 4 \\ 7 & 1 \end{bmatrix} + \begin{bmatrix} -5 & 8 \\ 2 & 6 \\ -7 & 10 \end{bmatrix}$

2. $\begin{bmatrix} -4 & 7 & 11 \\ 13 & 5 & -2 \end{bmatrix} + \begin{bmatrix} -9 & 3 & 7 \\ 1 & 0 & 14 \end{bmatrix}$

3. $\begin{bmatrix} 7 & 9 & 2 & 5 \end{bmatrix} - \begin{bmatrix} 4 & 8 & 2 & 3 \end{bmatrix}$

4. $\begin{bmatrix} 5 & 7 & -4 \\ 10 & 0 & -2 \\ 11 & -3 & 12 \end{bmatrix} - \begin{bmatrix} 4 & 5 & -7 \\ 6 & -5 & -8 \\ 4 & -11 & 3 \end{bmatrix}$

5. $7 \begin{bmatrix} 2 & 8 \\ -4 & 1 \end{bmatrix} + 3 \begin{bmatrix} -5 & 1 \\ 2 & -6 \end{bmatrix}$

6. $4 \begin{bmatrix} -8 & 5 & 3 \end{bmatrix} - 2 \begin{bmatrix} -5 & -1 & 7 \end{bmatrix}$

7. $\begin{bmatrix} 5 & 2 & 1 \\ 4 & -3 & 8 \end{bmatrix} \begin{bmatrix} 2 & -3 \\ 5 & -1 \\ 4 & -1 \end{bmatrix}$

8. $\begin{bmatrix} 5 & -8 \\ 7 & 1 \end{bmatrix} \begin{bmatrix} 4 & 7 \\ -3 & 9 \end{bmatrix}$

9. $\begin{bmatrix} 4 & 7 \\ 5 & 3 \\ 2 & -1 \end{bmatrix} \begin{bmatrix} 6 & 8 \\ 3 & -6 \end{bmatrix}$

10. $\begin{bmatrix} -2 & 3 & 5 \end{bmatrix} \begin{bmatrix} 1 & 4 \\ 7 & -3 \\ -1 & -5 \end{bmatrix}$

11. $\begin{bmatrix} 2 & 4 & -3 \\ 5 & 1 & 2 \\ -1 & 3 & 4 \end{bmatrix} \begin{bmatrix} -1 & 3 & 1 \\ 2 & 4 & 3 \\ 1 & 0 & 2 \end{bmatrix}$

12. $\begin{bmatrix} -1 & 3 & 1 \\ 2 & 4 & 3 \\ 1 & 0 & 2 \end{bmatrix} \begin{bmatrix} 2 & 4 & -3 \\ 5 & 1 & 2 \\ -1 & 3 & 4 \end{bmatrix}$

13. $\begin{bmatrix} 1 & 0 & 0 \\ 0 & 1 & 0 \\ 0 & 0 & 1 \end{bmatrix} \begin{bmatrix} 1 & 4 & 7 \\ 2 & 5 & 8 \\ 3 & 6 & 9 \end{bmatrix}$

14. $\begin{bmatrix} 10 & 20 \\ 30 & 40 \end{bmatrix} \begin{bmatrix} 1 & 0 \\ 0 & 1 \end{bmatrix}$

15. $\begin{bmatrix} 7 & 4 \\ 5 & 3 \end{bmatrix} \begin{bmatrix} 3 & -4 \\ -5 & 7 \end{bmatrix}$

16. $\begin{bmatrix} 2 & 3 & 4 \\ 5 & 1 & 2 \\ 6 & 5 & 7 \end{bmatrix} \begin{bmatrix} -3 & -1 & 2 \\ -23 & -10 & 16 \\ 19 & 8 & -13 \end{bmatrix}$

For Problems 17 through 22 find the inverse of the matrix, or show that it has no inverse because its determinant is zero.

17. $\begin{bmatrix} 5 & 7 \\ 3 & 2 \end{bmatrix}$

18. $\begin{bmatrix} 6 & -1 \\ -10 & 8 \end{bmatrix}$

19. $\begin{bmatrix} -2 & 1 \\ \frac{3}{2} & -\frac{1}{2} \end{bmatrix}$

20. $\begin{bmatrix} 0.2 & -0.3 \\ -0.4 & 1.1 \end{bmatrix}$

21. $\begin{bmatrix} 6 & 3 \\ 8 & 4 \end{bmatrix}$

22. $\begin{bmatrix} -10 & 5 \\ 6 & -3 \end{bmatrix}$

For Problems 23 and 24, find the adjoint matrix.

23. $\begin{bmatrix} 4 & -3 \\ 1 & 5 \end{bmatrix}$

24. $\begin{bmatrix} -2 & 8 \\ 5 & 6 \end{bmatrix}$

For Problems 25 through 30 solve the system by writing it in matrix form, then multiplying each member by the inverse of the coefficient matrix.

25. $5x + 2y = 11$
$x + y = 4$

26. $x - y = -11$
$7x + 4y = -22$

27. $6x - 7y = 47$
$2x + 5y = -21$

28. $8x + 3y = 41$
$6x + 5y = 39$

29. $4x - 3y = 11$
$5x - 6y = 9$

30. $3x + 4y = 18$
$9x + 6y = 17$

For Problems 31 through 34, find the inverse.

31. $\begin{bmatrix} 3 & 7 & -1 \\ 4 & 1 & -5 \\ -2 & 3 & 1 \end{bmatrix}$

32. $\begin{bmatrix} 2 & -6 & -3 \\ 1 & -3 & 5 \\ 0 & 4 & -2 \end{bmatrix}$

33. $\begin{bmatrix} 5 & -1 & 4 \\ 0 & 3 & -2 \\ 1 & 4 & 3 \end{bmatrix}$

34. $\begin{bmatrix} 5 & 7 & 2 \\ 1 & 4 & 3 \\ 2 & 8 & 6 \end{bmatrix}$

35. Show that matrix multiplication is *not* commutative by showing that

$$\begin{bmatrix} 2 & 3 \\ 4 & 5 \end{bmatrix} \begin{bmatrix} 6 & 7 \\ 8 & 9 \end{bmatrix} \neq \begin{bmatrix} 6 & 7 \\ 8 & 9 \end{bmatrix} \begin{bmatrix} 2 & 3 \\ 4 & 5 \end{bmatrix}$$

36. Show that the set of matrices is *not* closed under multiplication by showing two matrices whose product is not a matrix.

37. For real numbers, the converse of the multiplication property of zero states that a product can be zero only if one of the factors is zero. Show that the corresponding property for matrix multiplication is *false* by finding two 2×2 matrices whose product is the *zero matrix* (each element equals 0), but for which no element of either matrix is 0. (Clue: Find a matrix whose determinant is 0, and multiply it by its adjoint matrix.)

38. Prove that if the elements of a 3×3 matrix are consecutive integers, starting at the upper left corner, then the determinant of the matrix is 0.

B-2 DESCARTES' RULE OF SIGNS, AND THE UPPER BOUND THEOREM

In Section 7-6 you learned the Factor Theorem, which states that $(x - c)$ is a factor of a polynomial $P(x)$ if and only if $P(c) = 0$. The number c is called a *zero* of $P(x)$, as you learned in Section 10-5, or a *root* of the equation $P(x) = 0$. In this section you will learn properties that shorten the trial-and-error search for factors or zeros of a polynomial.

Substituting a positive value for x into a polynomial such as

$$P(x) = 3x^4 + 5x^3 + x^2 + 8x + 6$$

which has only positive coefficients always produces a positive answer. So such a polynomial could not possibly have any zeros that are positive. If the polynomial has some positive and some negative coefficients, then there can be positive zeros. Rene Descartes is credited with discovering that it is the number of *reversals* of sign as you go from term to term that sets a limit on the number of positive zeros. For instance,

$$P(x) = x^5 - 7x^4 - 5x^3 + 11x^2 - 2x + 5$$

Sign reversals

has four sign reversals. Descartes' Rule of Signs is as follows:

Descartes' Rule of Signs: The number of positive zeros of a polynomial is less than or equal to the number of sign reversals in $P(x)$. The number of negative zeros is less than or equal to the number of sign reversals in $P(-x)$. In both cases, the number of zeros has the same parity (odd or even) as the number of reversals.

So the polynomial above could have 4, 2, or 0 positive zeros. Since

$$P(-x) = -x^5 - 7x^4 + 5x^3 + 11x^2 + 2x + 5$$

One sign reversal

there is exactly *one* negative zero. By the corollary of the Fundamental Theorem of Algebra in Section 10-5, $P(x)$ has exactly 5 zeros. So there are three possible combinations of zeros for this polynomial.

positive	negative	complex
4	1	0
2	1	2
0	1	4

Recall that complex zeros of polynomials with real coefficients always come in conjugate pairs.

A proof of Descarte's Rule can be found in texts such as L. E. Dickson; *New First Course in the Theory of Equations*; John Wiley & Sons, Inc.; 1939.

A related theorem concerns the maximum and minimum possible values of real zeros. If you divide a polynomial by $(x - c)$, and the signs of the quotient coefficients and the remainder are all the same, then no real zeros are greater than c.

Upper Bound Theorem: For a positive number c, if $P(x)$ is divided by $(x - c)$ and the resulting quotient and remainder have no sign reversals, then $P(x)$ has no real zeros greater than c.

A lower bound can be found by determining an upper bound for the zeros of $P(-x)$, or equivalently, by finding a number c such that the signs of the quotient and remainder *alternate.*

The bounds are most easily found by the synthetic substitution technique of Section 10-5. In Example 1 of that section you plotted $P(x) = x^3 - 4x^2 - 5x + 14$.

From the graph on the following page you can see that there are two positive zeros (x-intercepts), and one negative zero, thus confirming Descartes' Rule of Signs.

$P(x) = x^3 - 4x^2 - 5x + 14$ Two reversals

$P(-x) = -x^3 - 4x^2 + 5x + 14$ One reversal

x	1	-4	-5	$P(x)$ 14	
-3	1	-7	16	-34	
-2	1	-6	7	0	Signs alternate. -3 is a lower bound.
-1	1	-5	0	14	
0	1	-4	-5	14	
1	1	-3	-8	6	
2	1	-2	-9	-4	No sign reversals. 5 is an upper bound.
3	1	-1	-8	-10	
4	1	0	-5	-6	
5	1	1	0	14	
6	1	2	7	56	

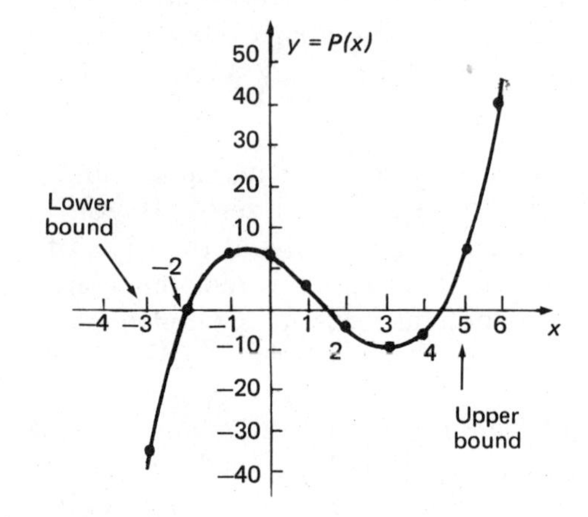

$P(x) = x^3 - 4x^2 - 5x + 14$

Three real zeros

Figure B-2a.

From the table you can see that if x is 5 or more, the signs of the quotient and remainder are all +. So 5 is an upper bound for the zeros. If x is -3 or less, the signs of the quotient and remainder alternate. So -3 is a lower bound for the zeros. These facts are confirmed by the x-intercepts of the graph.

Exercise B-2

For Problems 1 through 10, find the possible numbers of positive, negative, and complex (imaginary) zeros.

1. $P(x) = x^3 - 5x^2 + 3x + 7$
2. $P(x) = x^3 + 4x^2 - 7x + 2$
3. $P(x) = x^3 - 5x^2 + 3x - 1$
4. $P(x) = x^3 - x^2 + 4x - 6$
5. $P(x) = x^4 - 2x^3 - x + 1$
6. $P(x) = x^4 + 5x^2 + 2x - 11$
7. $P(x) = x^4 + x^3 - 5x^2 + x - 6$
8. $P(x) = x^4 - x^3 - 11x^2 + 9x + 8$
9. $P(x) = x^8 + 3x^6 + 4x^2 + 5$
10. $P(x) = x^5 + 4x^3 + 2x$ (Think!)

For Problems 11 through 14, find the least integer that is an upper bound for the zeros and the greatest integer that is a lower bound. Plot the part of the graph for values of x between these two integers, inclusive.

11. $P(x) = x^3 - 2x^2 - 5x + 3$
12. $P(x) = x^3 + 2x^2 + 4x - 5$
13. $P(x) = 3x^3 - 5x^2 + 7x + 4$
14. $P(x) = 2x^3 + x^2 - 8x - 3$
15. The trinomial $x^2 + 2x + 8$ has no positive zeros because there are no sign reversals. If this trinomial is multiplied by $(x - c)$, then the resulting polynomial, $P(x) = (x - c)(x^2 + 2x + 8)$, has exactly one positive zero, namely c (provided c is a positive number).
    a. Find a value of c for which $P(x)$, when multiplied out, has exactly one sign reversal.
    b. Find a value of c for which $P(x)$ has exactly three sign reversals.
    c. Explain why $P(x)$ could not have exactly two sign reversals.
16. The trinomial $x^2 - 3x + 15$ has two sign reversals.
    a. According to Descartes' rule, how many positive zeros could it have?
    b. How many positive zeros does it actually have? Justify your answer.

c. Use Descartes' rule to show that the trinomial has no negative zeros.

d. If c is a positive number, then the polynomial $P(x) = (x + c)(x^2 - 3x + 15)$ has exactly one negative zero, namely $-c$. Find a value of c for which $P(-x)$, when multiplied out, has exactly three sign reversals.

e. Explain why $P(-x)$ in part (d) could not possibly have exactly two sign reversals.

17. According to Descartes' Rule, what can be said about the number of zeros of polynomials that begin and end as follows:

a. $x^{14} \ldots + 8$ b. $x^{15} \ldots + 8$

c. $x^9 \ldots -4$ d. $x^8 \ldots -4$

18. Based on the answers to Problem 17, write a conclusion concerning the parity (odd or even) of the number of sign reversals in a polynomial whose first term is positive.

19. Find the real zeros of the following. Then use Descartes' rule to prove that there are no other real zeros.

a. $P(x) = x^5 - 32$ b. $P(x) = x^5 + 32$

c. $P(x) = x^6 - 64$ d. $P(x) = x^{10} + 1$

20. The polynomial $P(x) = 2x^3 - 8x^2 + 2x + 13$ has two sign reversals, and thus could have 0 or 2 positive zeros.

a. Show that although $P(2)$ and $P(3)$ are positive, there is a zero *between* $x = 2$ and $x = 3$. Draw a graph.

b. Use the Upper Bound Theorem to show that 4 is an upper bound for the zeros of $P(x)$.

c. Explain why 3 is also an upper bound for the zeros, even though the quotient $P(x)/(x-3)$ does have sign reversals. How is this fact consistent with the Upper Bound Theorem?

21. Write a computer program to find an upper bound for the zeros of a polynomial up to 5th degree using the Upper Bound Theorem. The program should allow you to input the six coefficients. Then the computer should divide by $(x-1)$, $(x-2), \ldots,$ until all signs of the quotient and remainder are alike. The output of the program should be the quotient coefficients and the remainder at each pass through the loop. When the upper bound is reached, the computer should print its value along with an appropriate message. Test your program by showing that 9 is the least integer upper bound for the zeros of $P(x) = x^5 - 7x^4 - 9x^3 + x^2 - 99x - 11$.

B-3 MATHEMATICAL INDUCTION

You recall that the distributive axiom states that

$$a(x_1 + x_2) = ax_1 + ax_2.$$

In other words, multiplication distributes over a sum of *two* terms. You then simply assume that multiplication distributes over sums of three terms, four terms, etc. For example,

$$a(x_1 + x_2 + x_3 + x_4 + x_5)$$
$$= ax_1 + ax_2 + ax_3 + ax_4 + ax_5.$$

Although it *seems* reasonable that this property should be true for any number of terms, mathematicians prefer to be able to *prove* properties based upon axioms. The technique used to prove such properties as the extended distributive property is called *mathematical induction.*

Objective:

Given a sequence of statements, $S_1, S_2, S_3, S_4, \ldots, S_n$, prove that each of the statements is true for *any* positive integer n.

In the above example, the statements would be

S_2: $a(x_1 + x_2) = ax_1 + ax_2$

S_3: $a(x_1 + x_2 + x_3) = ax_1 + ax_2 + ax_3$

S_4: $a(x_1 + x_2 + x_3 + x_4) = ax_1 + ax_2 + ax_3 + ax_4$

.
.
.

S_n: $a(x_1 + x_2 + ... + x_n) = ax_1 + ax_2 + ... + ax_n$

(The statement S_1: $a(x_1) = ax_1$ is trivial, and so is omitted.)

Unfortunately, the extended distributive property cannot be proved using the Field Axioms alone. Another axiom must be introduced. Although there are several axioms which will do the job, the most easily understood is the Well-ordering Axiom.

Well-ordering Axiom: Any non-empty set of positive integers contains a *least* element.

The truth of the axiom should be obvious to you. The name comes from the fact that a set is said to be "well-ordered" if it has a least element. The reason for restricting the axiom to non-empty sets is that the empty set has no elements at all. Thus, it could not possibly have a least element.

Proof of the extended distributive property is possible with the aid of this axiom. It is done by assuming that the property is *false*, and showing that this leads you to a *contradiction* (an impossible conclusion).

***Example 1:* Extended Distributive Property of Multiplication over Addition.** Prove that $a(x_1 + x_2 + ... + x_n) = ax_1 + ax_2 + ... + ax_n$ for *any* positive integer n.

Proof:

By the Distributive Axiom, $a(x_1 + x_2) = ax_1 + ax_2$. So Statement S_2 is true.

Assume that there is an integer n for which the property is false. Let F be the set of integers n for which S_n is false. So F is a non-empty set of positive integers. By the Well-ordering Axiom, F has a least element. Let ℓ be the least element of F.

Because S_ℓ is false, you know that

$$a(x_1 + x_2 + ... + x_{\ell-1} + x_\ell) \neq ax_1 + ax_2 + ... + ax_{\ell-1} + ax_\ell.$$

Because ℓ is the *least* element of F, $S_{\ell-1}$ must be *true*. That is,

$$a(x_1 + x_2 + ... + x_{\ell-1}) = ax_1 + ax_2 + ... + ax_{\ell-1}.$$

A contradiction can now be reached by starting with the expression

$$a(x_1 + x_2 + ... + x_{\ell-1} + x_\ell)$$

and *associating* terms to get

$$a[(x_1 + x_2 + ... + x_{\ell-1}) + x_\ell].$$

The expression in brackets now has only *two* terms. The ordinary distributive axiom lets you distribute a to both terms, giving

$$a(x_1 + x_2 + ... + x_{\ell-1}) + ax_\ell.$$

By substituting the above for $a(x_1 + x_2 + ... + x_{\ell-1})$ you get

$$ax_1 + ax_2 + ... + ax_{\ell-1} + ax_\ell.$$

By the transitive property, the first expression equals the last, so

$$a(x_1 + x_2 + ... + x_{\ell-1} + x_\ell) = ax_1 + ax_2 + ... + ax_{\ell-1} + ax_\ell,$$

which directly contradicts the above *in*equality! The only thing that could have gone wrong was assuming at the beginning that the theorem was false. So the assumption was wrong, the theorem is *true*, and

$$a(x_1 + x_2 + ... + x_n) = ax_1 + ax_2 + ... + ax_n$$

for *any* integer value of $n \geq 2$. Q.E.D.

Once you understand the process, the proof may be shortened a great deal. All you need to do is to show that assuming one of the statements is true implies that the next statement is true, and that one of the statements actually is true. These two ideas combined are called the induction principle.

Induction Principle

If you can show that:

1. *Assuming* that one of the statements is true implies that the *next* statement is true, and
2. One of the statements actually *is* true,

then you can conclude that *all* of the statements are true, from the one that actually is true on up.

The proof of the extended distributive property can be condensed as follows. Proofs done in this manner are said to be done by *Mathematical Induction.*

Example 1 (by Mathematical Induction): Prove that $a(x_1 + x_2 + ... + x_n) = ax_1 + ax_2 + ... + ax_n$ for *any* positive integer n.

Proof:

Anchor: $a(x_1 + x_2) = ax_1 + ax_2$ by the Distributive Axiom, so the theorem is actually true when $n = 2$.

Induction Hypothesis: Assume the theorem is true for $n = k$. That is,

$$a(x_1 + x_2 + ... + x_k) = ax_1 + ax_2 + ... + ax_k.$$

Demonstration for $n = k+1$: If $n = k+1$, then

$$a(x_1 + x_2 + ... + x_k + x_{k+1})$$

$$= a[(x_1 + x_2 + ... + x_k) + x_{k+1}]$$
Associativity.

$$= a(x_1 + x_2 + ... + x_k) + ax_{k+1}$$
By the anchor.

$$= ax_1 + ax_2 + ... + ax_k + ax_{k+1}$$
By the Induction Hypothesis.

Conclusion: Since 1., assuming the property is true for $n = k$ implies that it is true for $n = k+1$, and 2., it is actually true for $n = 2$, you can conclude that

$$a(x_1 + x_2 + ... + x_n) = ax_1 + ax_2 + ... + ax_n$$

for *all* integers $n \geq 2$. Q.E.D.

If you have been very alert, you may have detected one weakness in the above proof. You are trying to extend one of the field axioms, distributivity, to sums of more than two terms. But without ever stating it, you have used in the proof an extended *associative* property and an extended *transitive* property!

The proof of the extended associative property is a bit tricky, and is presented below as Example 2. *You* will prove the extended transitive property in Exercise B-3. First, you must define just what it is you mean by a sum of n terms.

DEFINITION

$$x_1 + x_2 + x_3 + x_4 + ... + x_{n-1} + x_n$$
$$= (...(((x_1 + x_2) + x_3) + x_4) + ... + x_{n-1}) + x_n.$$

In plain English, you associate the first two terms and get an answer. Then you add the third term to this answer to get another answer, and so on. The extended associative property which you will prove says that you get the same answer if you remove all of the *inner* parentheses. That is, you can associate the first $n-1$ terms.

Example 2: Extended Associative Property of Addition: Prove that $x_1 + x_2 + x_3 + ... + x_{n-1} + x_n$

$$= (x_1 + x_2 + x_3 + ... + x_{n-1}) + x_n$$

for all integers $n \geq 3$.

Proof:

Anchor: For $n = 3$, $x_1 + x_2 + x_3 = (x_1 + x_2) + x_3$ by definition.

Induction Hypothesis: Assume that for k terms, you can associate $x_1 + x_2 + x_3 + \ldots + x_{k-1} + x_k$

$= (x_1 + x_2 + x_3 + \ldots + x_{k-1}) + x_k$.

Demonstration for $n = k+1$: If there are $k+1$ terms, then

$$x_1 + x_2 + x_3 + \ldots + x_{k-1} + x_k + x_{k+1}$$

$= ((\ldots((x_1 + x_2) + x_3) + \ldots + x_{k-1}) + x_k)$

$+ x_{k+1}$ Definition of a sum of $k+1$ terms.

$= ((x_1 + x_2 + x_3 + \ldots + x_{k-1}) + x_k)$

$+ x_{k+1}$ Definition of a sum of $k-1$ terms.

$= (x_1 + x_2 + x_3 + \ldots + x_{k-1} + x_k) + x_{k+1}$
By Induction Hypothesis.

Conclusion:

$$\therefore x_1 + x_2 + x_3 + \ldots + x_{n-1} + x_n$$

$$= (x_1 + x_2 + x_3 + \ldots + x_{n-1}) + x_n$$

for *all* integers $n \geq 3$. Q.E.D.

Example 3: Induction is useful for proving that formulas for series (see Chapter 11) are true for *any* finite number of terms. For example, the sum of the first n terms in a geometric series

$$a + ar + ar^2 + ar^3 + \ldots + ar^{n-1}$$

is equal to $\dfrac{a(1 - r^n)}{1 - r}$, where $a = t(1)$.

Proof:

Anchor: For $n = 1$, the series consists of just *one* term, a. The formula gives

$$\frac{a(1 - r^1)}{1 - r},$$

which is equal to a. So the formula is correct for $n = 1$.

Induction Hypothesis: Assume that the formula is true for $n = k$. That is,

$$a + ar + ar^2 + \ldots + ar^{k-1} = \frac{a(1 - r^k)}{1 - r}.$$

Demonstration for $n = k+1$: If there are $k+1$ terms, then

$a + ar + ar^2 + \ldots + ar^{k-1} + ar^k$

$= (a + ar + ar^2 + \ldots + ar^{k-1}) + ar^k$ Associativity.

$= \dfrac{a(1 - r^k)}{1 - r} + ar^k$ Induction Hypothesis.

$= \dfrac{a(1 - r^k) + ar^k(1 - r)}{1 - r}$ Adding fractions.

$= \dfrac{a - ar^k + ar^k - ar^{k+1}}{1 - r}$ Distributivity.

$= \dfrac{a - ar^{k+1}}{1 - r}$ Adding like terms.

$= \dfrac{a(1 - r^{k+1})}{1 - r}$ Factoring out a.

$\therefore a + ar + ar^2 + \ldots + ar^k + ar^{k+1}$

$= \dfrac{a(1 - r^{k+1})}{1 - r}$ Transitivity.

Conclusion: Since assuming that the formula is true for $n = k$ implies that it is also true for $n = k+1$, and since the formula actually *is* true for $n = 1$, you can conclude that

$$a + ar + ar^2 + \ldots + ar^{n-1} = \frac{a(1 - r^n)}{1 - r}$$

for *all* integers $n \geq 1$ Q.E.D.

You should be careful not to read too much into the conclusion of an induction proof. The proof is good only for any *finite* value of n. For example, in Problem 4, below, you will prove the extended closure property for addition. This says that the sum of any finite number of terms equals a real number.

If there is an *infinite* number of terms, the sum may or may not equal a real number. For example,

$$1 + 1/2 + 1/4 + 1/8 + ... + 1/2^n + ...$$

is equal to 2, which *is* a real number. But

$$1 + 1/2 + 1/3 + 1/4 + ... + 1/n + ...$$

is *larger* than any real number. When you study analysis or calculus, you will learn about the concepts of limits, which will allow you to prove things about sums of infinite numbers of terms.

In the exercise which follows, you will use mathematical induction to prove other extended field, equality, and order axioms, properties of exponentiation and logarithms, formulas for sequences and series, and some interesting properties of numbers.

Exercise B-3

Prove the following theorems either

a. directly from the Well-Ordering Axiom, or
b. by Mathematical Induction.

You may use the conclusions of the examples or of any *prior* problem in the exercise to help you prove the one you are working on.

1. *Extended Associativity for Multiplication* – The product $a_1a_2a_3...a_{n-1}a_n$ is defined in the same way as the *sum* of n terms (see the above text material). Prove that for $n \geq 3$,

$$a_1a_2a_3...a_{n-1}a_n = (a_1a_2a_3...a_{n-1})a_n.$$

2. *Extended Commutativity for Multiplication* – Prove that you can commute a factor in a product past *any* number of other factors. That is, prove for any $n \geq 2$,

$$a_1a_2a_3...a_{n-1}a_n = a_na_1a_2a_3...a_{n-1}.$$

3. *Extended Closure under Multiplication* – Prove that the product of *any* finite number of real numbers is a real number. That is, if $a_1, a_2, a_3, ..., a_n$ are real numbers, then $a_1a_2a_3...a_n$ is a real number, for *any* integer $n \geq 2$.

4. *Extended Closure under Addition* – Prove that the sum of *any* finite number of real numbers is a real number. That is, if $a_1, a_2, a_3, ..., a_n$ are real numbers, then $a_1 + a_2 + a_3 + ... + a_n$ is a real number, for *any* integer $n \geq 2$.

5. *Extended Transitivity for Equality* – Prove that in a chain of equalities, the first number equals the last number, no matter how many "=" signs come between. That is, if $a_1 = a_2 = a_3 = ... = a_n$, then $a_1 = a_n$, for any integer $n \geq 3$.

6. *Extended Transitivity for Order* – Prove that in a chain of inequalities with "<," the first number is less than the last number, no matter how many "<" signs come between. That is, if $a_1 < a_2 < a_3 < ... < a_n$, then $a_1 < a_n$ for any integer $n \geq 3$.

7. *Exponentiation Distributes over Multiplication* – Exponentiation with positive integer exponents may be defined as follows:

$$x^1 = x$$

$$x^n = x^{n-1} \cdot x, \quad \text{for} \quad n > 1.$$

Use this definition to prove inductively that $(ab)^n = a^n b^n$ for all integers $n \geq 1$.

8. *Exponentiation Distributes over Division* – Prove that

$$\left(\frac{a}{b}\right)^n = \frac{a^n}{b^n} \quad \text{for } any \text{ integer } n \geq 1.$$

9. *Product of Two Powers with the Same Base* – Prove that a product of two powers with the same base may be evaluated by *adding* the exponents. That is, prove that

$$a^r a^n = a^{r+n}$$

for all integers $n \geq 1$.

10. *Power of a Power* – Prove that a power of a power may be evaluated by *multiplying* the exponents. That is, prove that

$$(a^r)^n = a^{rn}$$

for *any* integer $n \geq 1$.

11. *n^{th} power of a Number Larger than 1* – Prove that if $x > 1$, then $x^n > x^{n-1}$ for all integers $n \geq 1$.

12. *n^{th} Power of a Number Between 0 and 1* – Prove that if $0 < x < 1$, then $x^n < x^{n-1}$ for all integers $n \geq 1$.

13. *Logarithm of a Product of n Factors* – Prove that the log of a product is the sum of the logs of the factors, no matter how many factors there are. That is, for any $n \geq 2$, prove that

$$\log(x_1 x_2 x_3 \ldots x_n)$$
$$= \log x_1 + \log x_2 + \log x_3 + \ldots + \log x_n.$$

14. *Logarithm of a Power* – Prove that the log of a number raised to a power is the exponent times the log of the number. That is,

$$\log(x^n) = n \log x$$

for *any* integer $n \geq 1$. The property of the log of a product should be helpful as a lemma.

15. *Absolute Value of a Product* – Prove that the absolute value of a product equals the product of the absolute values of the factors. That is, for all integers $n \geq 2$,

$$|a_1 a_2 a_3 \ldots a_n| = |a_1||a_2||a_3|\ldots|a_n|.$$

The anchor, $|a_1 a_2| = |a_1||a_2|$, can be proved by recalling that $|x| = \sqrt{x^2}$, and using the properties of products of square roots.

16. *Absolute Value of a Sum (Polygonal Inequality)* – Prove that the absolute value of a sum is less than or equal to the sum of the absolute values of the terms. That is, for all integers $n \geq 2$,

$$|a_1 + a_2 + a_3 + \ldots + a_n| \leq |a_1| + |a_2| + |a_3|$$
$$+ \ldots + |a_n|.$$

The anchor, $|a_1 + a_2| \leq |a_1| + |a_2|$, is called the Triangle Inequality. It is quite tricky to prove. You may find that the inequality $-|x| \leq x \leq |x|$, and the fact that if $-r \leq s \leq r$, then $|s| \leq r$, are helpful.

17. *Upper Bound for the nth Prime* – Let p_n be the nth prime number. That is, $p_1 = 2$, $p_2 = 3$, $p_3 = 5$, $p_4 = 7$, $p_5 = 11$, and so forth. Bertrand's Postulate (actually a proven theorem) states that there is at least one prime between p_n and $2p_n$. For example, $p_4 = 7$, so $2p_4 = 14$. The prime 11 (and also the prime 13) is between 7 and 14. Use Bertrand's Postulate to prove that

$$p_n < 2^n$$

for all integers $n \geq 2$.

18. *Formula for Triangular Numbers* – The numbers 1, 3, 6, 10, 15, 21, ..., are called "triangular numbers" because these are the numbers of balls that can be arranged in various sized equilateral triangles (see sketch). As you can see from the sketch, the nth triangular number is formed by adding n to the previous triangular number. Prove by induction that the nth triangular number, T_n, is given by

$$T_n = \frac{n}{2}(n + 1).$$

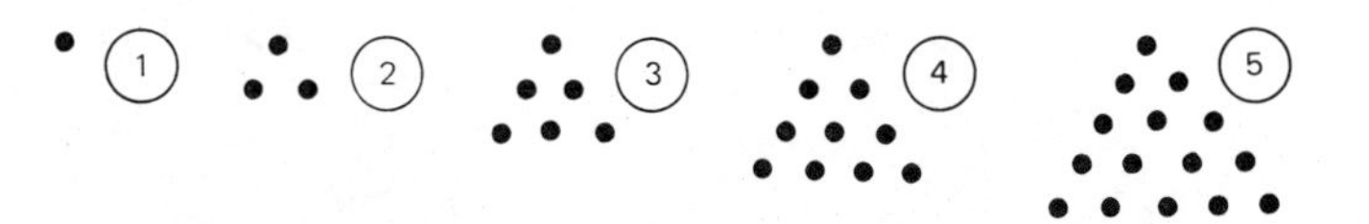

19. *Formula for Pyramidal Numbers* – In the "olden days" when armies and navies used cannon balls, the balls were often stacked in triangular pyramids. Each layer in the stack was an equilateral triangle, with one less cannon ball per side than the layer below. It was important to be able to determine the number of cannon balls in a stack *quickly*.

You realize that this number is just the *sum* of the first n triangular numbers if the stack has n cannon balls in the side of the bottom layer. Prove by induction that the nth "pyramidal number," P_n, is

$$P_n = \frac{n}{6}(n+1)(n+2).$$

20. *Formula for Pentagonal Numbers* – The numbers 1, 5, 12, 22, 35, 51, 70, 92, ..., are called "pentagonal numbers" because these are the numbers of balls that can be arranged in regular pentagons (see sketch).

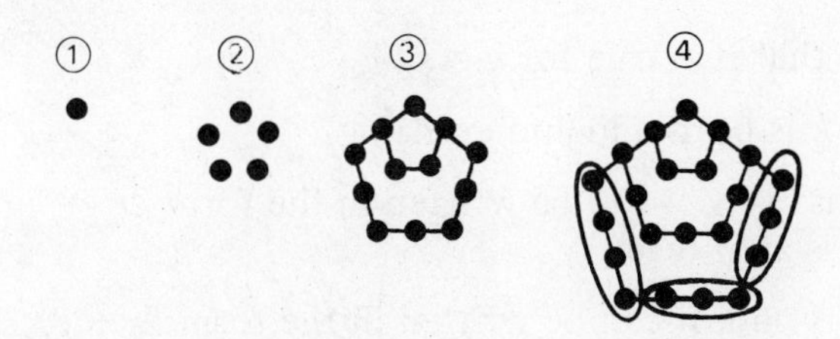

As you can see from the sketch, the fourth pentagonal number is formed from the third one by adding three rows of dots, one containing 4 dots, and the other two containing 3 dots each. In general, if P_n is the nth pentagonal number, then

$$P_{n+1} = P_n + (n+1) + 2n = P_n + (3n + 1).$$

Prove by induction that for all integers $n \geq 1$,

$$P_n = \frac{n}{2}(3n - 1).$$

21. *Sum of the First n Positive Integers* – Prove that the sum of the first n positive integers is

$$1 + 2 + 3 + ... + n = \frac{n}{2}(n + 1).$$

22. *Sum of the Squares of the First n Positive Integers* – Prove that the sum of the squares of the first n positive integers is

$$1^2 + 2^2 + 3^2 + ... + n^2 = \frac{1}{6}(n)(n + 1)(2n + 1).$$

23. *Sum of the Cubes of the First n Positive Integers* – Prove that the *sum* of the *cubes* of the first n positive integers is equal to the *square* of the *sum* of the first n positive integers. That is,

$$1^3 + 2^3 + 3^3 + ... + n^3 = (1 + 2 + 3 + ... + n)^2.$$

The results of Problem 21 should be helpful!

24. *Sum of the First n Odd Positive Integers is a Perfect Square* – Prove that for all $n \geq 1$,

$$1 + 3 + 5 + ... + (2n - 1) = n^2.$$

25. *Sum of Reciprocals of Pairwise Products* – Prove for all $n \geq 1$,

$$\frac{1}{1 \cdot 2} + \frac{1}{2 \cdot 3} + \frac{1}{3 \cdot 4} + ... + \frac{1}{n(n+1)} = \frac{n}{n + 1}.$$

26. *Sum of Reciprocals of Powers of 2* – Prove for all $n \geq 1$ that

$$\frac{1}{2^1} + \frac{1}{2^2} + \frac{1}{2^3} + ... + \frac{1}{2^n} = 1 - \frac{1}{2^n}.$$

27. *Factorials Get Large Faster than Exponentials* – Prove that $n! > 3^n$ whenever n is sufficiently large. Note that to get an anchor, you may have to try several values of n, because the conclusion is *false* when n is close to 1.

28. *Exponentials Get Large Faster than Powers* – Prove that $2^n > n^2$ whenever n is sufficiently large. As in Problem 27, you will have to search for the lowest value of n for which the theorem is true. The algebra in the demonstration for $n = k+1$ is rather difficult. You may need to know that $2^k > 2k + 1$ in order to arrive at the desired conclusion.

29. *n^{th} Term of an Arithmetic Sequence* – You recall that the nth term of an arithmetic sequence is formed by adding a constant, d, to the preceding

term. If $t(n)$ is the nth term of an arithmetic sequence, prove that

$$t(n) = t(1) + (n-1)d$$

for all integer values of $n \geq 1$.

30. *n^{th} Term of a Geometric Sequence* – You recall that the nth term of a geometric sequence is formed by multiplying the preceding term by a constant, r. If t_n is the nth term of a geometric sequence, prove that

$$t(n) = t(1) \cdot r^{n-1}$$

for all integer values of $n \geq 1$.

31. *n^{th} Partial Sum of an Arithmetic Series* – You recall that the nth partial sum of a series is the sum of the first n terms. You also recall that an arithmetic series has as its $n{+}1^{st}$ term

$$t(n+1) = t(n) + d$$

where d is a constant called the common difference. Prove by induction that $S(n)$, the nth partial sum, of an arithmetic series, is equal to

$$S(n) = \frac{n}{2}(2t(1) + (n-1)d)$$

for all integer values of $n \geq 1$.

32. *Term of a Binomial Series* – Prove that the term containing b^r in the expansion of the binomial $(a + b)^n$ is

$$\frac{n!}{r!\,(n-r)!}\, a^{n-r} b^r$$

for all integers $n \geq 2$. Note that after assuming it is true for $n = k$, you will use the fact that $(a + b)^{k+1} = (a + b)^k (a + b)$. The term with b^r in $(a + b)^{k+1}$ comes from only *two* terms in the expansion of $(a + b)^k$; the one with b^r, which gets multiplied by a, and the one with b^{r-1}, which gets multiplied by b. You will need to be very clever in finding the common denominator when you add the two fractions!!

33. *Sum of Cosines of Odd Multiples of x* – Prove that

$$\cos x + \cos 3x + \cos 5x + \ldots + \cos (2n-1)x = \frac{\sin 2nx}{2 \sin x}$$

for all integers $n \geq 1$.

34. *Every Positive Integer is both Odd and Even?* – Calvin Butterball supplies the following "proof" that every positive integer is both odd and even.

Proof:

Assume that it is true for $n = k$.

That is, k is both odd and even.

Since k is even, it can be written in the form $2r$, where r is an integer.

Since k is odd, it can be written in the form $2s + 1$, where s is an integer.

So $k = 2r = 2s + 1$ by transitivity.

If $n = k{+}1$, then $k + 1 = 2r + 1 = 2s + 2$.

Since $k + 1 = 2r + 1$, $k + 1$ is *odd*.

Since $k + 1 = 2s + 2 = 2(s + 1)$, $k + 1$ is *even*.

$\therefore$ $k{+}1$ is both odd and even.

Therefore, *any* positive integer n is both odd and even, Q.E.D.

Where did Calvin go wrong in his proof?

35. *Primes from Polynomials??* – The polynomial $n^2 - n + 41$ is interesting because when you substitute numbers for n, you get *primes.*

n	$n^2 - n + 41$	
1	41	(prime)
2	43	(prime)
3	47	(prime)
4	53	(prime)
5	61	(prime)
6	71	(prime)

Either prove by induction that $n^2 - n + 41$ is prime for *all* integers $n \geq 1$, or find a counter-example in which it is *not* prime.

36. *Every Number Equals Every Other Number???* – Phoebe Small supplies the following "proof" that every number is equal to every other number:

Prove that if $a_1, a_2, a_3, \ldots, a_n$ are real numbers, then

$a_1 = a_2 = a_3 = \ldots = a_n$ for *all* integers $n \geq 1$.

Proof:

Let S be a set containing n numbers.

Anchor: If $n = 1$, then S contains only *one* number, a_1. By the Reflexive Axiom, $a_1 = a_1$, which anchors the induction.

Induction Hypothesis: Suppose that whenever S contains k elements, $a_1 = a_2 = a_3 = \ldots = a_k$.

Demonstration for n = k+1: Suppose that S contains $k+1$ numbers. If a_1 is removed from the set, then S contains only k numbers, $a_2, a_3, a_4, \ldots, a_k$, and a_{k+1}.

By the Induction Hypothesis, $a_2 = a_3 = a_4 = \ldots = a_k = a_{k+1}$. If a_1 is put back, and a_{k+1} is removed from the set, S again contains only k numbers, and thus $a_1 = a_2 = a_3 = \ldots = a_k$.

Therefore, by the Extended Transitive Property (Problem 5),

$a_1 = a_2 = a_3 = \ldots = a_k = a_{k+1}$.

Conclusion: Therefore, $a_1 = a_2 = a_3 = \ldots a_n$ for *all* integers $n \geq 1$, Q.E.D.

What did Phoebe do wrong in her proof? (Note: This "proof" is sometimes used to show that all people in a room have the same sex!)

Tables

Table I. Squares and Square Roots, Cubes and Cube Roots

N	N^2	$\sqrt{N}$	$\sqrt{10N}$	N	N^2	$\sqrt{N}$	$\sqrt{10N}$
1.0	1.00	1.000	3.162	5.5	30.25	2.345	7.416
1.1	1.21	1.049	3.317	5.6	31.36	2.366	7.483
1.2	1.44	1.095	3.464	5.7	32.49	2.387	7.550
1.3	1.69	1.140	3.606	5.8	33.64	2.408	7.616
1.4	1.96	1.183	3.742	5.9	34.81	2.429	7.681
1.5	2.25	1.225	3.873	6.0	36.00	2.449	7.746
1.6	2.56	1.265	4.000	6.1	37.21	2.470	7.810
1.7	2.89	1.304	4.123	6.2	38.44	2.490	7.874
1.8	3.24	1.342	4.243	6.3	39.69	2.510	7.937
1.9	3.61	1.378	4.359	6.4	40.96	2.530	8.000
2.0	4.00	1.414	4.472	6.5	42.25	2.550	8.062
2.1	4.41	1.449	4.583	6.6	43.56	2.569	8.124
2.2	4.84	1.483	4.690	6.7	44.89	2.588	8.185
2.3	5.29	1.517	4.796	6.8	46.24	2.608	8.246
2.4	5.76	1.549	4.899	6.9	47.61	2.627	8.307
2.5	6.25	1.581	5.000	7.0	49.00	2.646	8.367
2.6	6.76	1.612	5.099	7.1	50.41	2.665	8.426
2.7	7.29	1.643	5.196	7.2	51.84	2.683	8.485
2.8	7.84	1.673	5.292	7.3	53.29	2.702	8.544
2.9	8.41	1.703	5.385	7.4	54.76	2.720	8.602
3.0	9.00	1.732	5.477	7.5	56.25	2.739	8.660
3.1	9.61	1.761	5.568	7.6	57.76	2.757	8.718
3.2	10.24	1.789	5.657	7.7	59.29	2.775	8.775
3.3	10.89	1.817	5.745	7.8	60.84	2.793	8.832
3.4	11.56	1.844	5.831	7.9	62.41	2.811	8.888
3.5	12.25	1.871	5.916	8.0	64.00	2.828	8.944
3.6	12.96	1.897	6.000	8.1	65.61	2.846	9.000
3.7	13.69	1.924	6.083	8.2	67.24	2.864	9.055
3.8	14.44	1.949	6.164	8.3	68.89	2.881	9.110
3.9	15.21	1.975	6.245	8.4	70.56	2.898	9.165
4.0	16.00	2.000	6.325	8.5	72.25	2.915	9.220
4.1	16.81	2.025	6.403	8.6	73.96	2.933	9.274
4.2	17.64	2.049	6.481	8.7	75.69	2.950	9.327
4.3	18.49	2.074	6.557	8.8	77.44	2.966	9.381
4.4	19.36	2.098	6.633	8.9	79.21	2.983	9.434
4.5	20.25	2.121	6.708	9.0	81.00	3.000	9.487
4.6	21.16	2.145	6.782	9.1	82.81	3.017	9.539
4.7	22.09	2.168	6.856	9.2	84.64	3.033	9.592
4.8	23.04	2.191	6.928	9.3	86.49	3.050	9.644
4.9	24.01	2.214	7.000	9.4	88.36	3.066	9.695
5.0	25.00	2.236	7.071	9.5	90.25	3.082	9.747
5.1	26.01	2.258	7.141	9.6	92.16	3.098	9.798
5.2	27.04	2.280	7.211	9.7	94.09	3.114	9.849
5.3	28.09	2.302	7.280	9.8	96.04	3.130	9.899
5.4	29.16	2.324	7.348	9.9	98.01	3.146	9.950
5.5	30.25	2.345	7.416	10	100.00	3.162	10.000

Table I (Concluded)

N	N^3	$\sqrt[3]{N}$	$\sqrt[3]{10N}$	$\sqrt[3]{100N}$	N	N^3	$\sqrt[3]{N}$	$\sqrt[3]{10N}$	$\sqrt[3]{100N}$
1.0	1.000	1.000	2.154	4.642	5.5	166.375	1.765	3.803	8.193
1.1	1.331	1.032	2.224	4.791	5.6	175.616	1.776	3.826	8.243
1.2	1.728	1.063	2.289	4.932	5.7	185.193	1.786	3.849	8.291
1.3	2.197	1.091	2.351	5.066	5.8	195.112	1.797	3.871	8.340
1.4	2.744	1.119	2.410	5.192	5.9	205.379	1.807	3.893	8.387
1.5	3.375	1.145	2.466	5.313	6.0	216.000	1.817	3.915	8.434
1.6	4.096	1.170	2.520	5.429	6.1	226.981	1.827	3.936	8.481
1.7	4.913	1.193	2.571	5.540	6.2	238.328	1.837	3.958	8.527
1.8	5.832	1.216	2.621	5.646	6.3	250.047	1.847	3.979	8.573
1.9	6.859	1.239	2.668	5.749	6.4	262.144	1.857	4.000	8.618
2.0	8.000	1.260	2.714	5.848	6.5	274.625	1.866	4.021	8.662
2.1	9.261	1.281	2.759	5.944	6.6	287.496	1.876	4.041	8.707
2.2	10.648	1.301	2.802	6.037	6.7	300.763	1.885	4.062	8.750
2.3	12.167	1.320	2.844	6.127	6.8	314.432	1.895	4.082	8.794
2.4	13.824	1.339	2.884	6.214	6.9	328.509	1.904	4.102	8.837
2.5	15.625	1.357	2.924	6.300	7.0	343.000	1.913	4.121	8.879
2.6	17.576	1.375	2.962	6.383	7.1	357.911	1.922	4.141	8.921
2.7	19.683	1.392	3.000	6.463	7.2	373.248	1.931	4.160	8.963
2.8	21.952	1.409	3.037	6.542	7.3	389.017	1.940	4.179	9.004
2.9	24.389	1.426	3.072	6.619	7.4	405.224	1.949	4.198	9.045
3.0	27.000	1.442	3.107	6.694	7.5	421.875	1.957	4.217	9.086
3.1	29.791	1.458	3.141	6.768	7.6	438.976	1.966	4.236	9.126
3.2	32.768	1.474	3.175	6.840	7.7	456.533	1.975	4.254	9.166
3.3	35.937	1.489	3.208	6.910	7.8	474.552	1.983	4.273	9.205
3.4	39.304	1.504	3.240	6.980	7.9	493.039	1.992	4.291	9.244
3.5	42.875	1.518	3.271	7.047	8.0	512.000	2.000	4.309	9.283
3.6	46.656	1.533	3.302	7.114	8.1	531.441	2.008	4.327	9.322
3.7	50.653	1.547	3.332	7.179	8.2	551.368	2.017	4.344	9.360
3.8	54.872	1.560	3.362	7.243	8.3	571.787	2.025	4.362	9.398
3.9	59.319	1.574	3.391	7.306	8.4	592.704	2.033	4.380	9.435
4.0	64.000	1.587	3.420	7.368	8.5	614.125	2.041	4.397	9.473
4.1	68.921	1.601	3.448	7.429	8.6	636.056	2.049	4.414	9.510
4.2	74.088	1.613	3.476	7.489	8.7	658.503	2.057	4.431	9.546
4.3	79.507	1.626	3.503	7.548	8.8	681.472	2.065	4.448	9.583
4.4	85.184	1.639	3.530	7.606	8.9	704.969	2.072	4.465	9.619
4.5	91.125	1.651	3.557	7.663	9.0	729.000	2.080	4.481	9.655
4.6	97.336	1.663	3.583	7.719	9.1	753.571	2.088	4.498	9.691
4.7	103.823	1.675	3.609	7.775	9.2	778.688	2.095	4.514	9.726
4.8	110.592	1.687	3.634	7.830	9.3	804.357	2.103	4.531	9.761
4.9	117.649	1.698	3.659	7.884	9.4	830.584	2.110	4.547	9.796
5.0	125.000	1.710	3.684	7.937	9.5	857.375	2.118	4.563	9.830
5.1	132.651	1.721	3.708	7.990	9.6	884.736	2.125	4.579	9.865
5.2	140.608	1.732	3.733	8.041	9.7	912.673	2.133	4.595	9.899
5.3	148.877	1.744	3.756	8.093	9.8	941.192	2.140	4.610	9.933
5.4	157.464	1.754	3.780	8.143	9.9	970.299	2.147	4.626	9.967
5.5	166.375	1.765	3.803	8.193	10	1000.000	2.154	4.642	10.000

Table II. Four-Place Logarithms of Numbers

n	*00*	*10*	*20*	*30*	*40*	*50*	*60*	*70*	*80*	*90*
1.0	0000	0043	0086	0128	0170	0212	0253	0294	0334	0374
1.1	0414	0453	0492	0531	0569	0607	0645	0682	0719	0755
1.2	0792	0828	0864	0899	0934	0969	1004	1038	1072	1106
1.3	1139	1173	1206	1239	1271	1303	1335	1367	1399	1430
1.4	1461	1492	1523	1553	1584	1614	1644	1673	1703	1732
1.5	1761	1790	1818	1847	1875	1903	1931	1959	1987	2014
1.6	2041	2068	2095	2122	2148	2175	2201	2227	2253	2279
1.7	2304	2330	2355	2380	2405	2430	2455	2480	2504	2529
1.8	2553	2577	2601	2625	2648	2672	2695	2718	2742	2765
1.9	2788	2810	2833	2856	2878	2900	2923	2945	2967	2989
2.0	3010	3032	3054	3075	3096	3118	3139	3160	3181	3201
2.1	3222	3243	3263	3284	3304	3324	3345	3365	3385	3404
2.2	3424	3444	3464	3483	3502	3522	3541	3560	3579	3598
2.3	3617	3636	3655	3674	3692	3711	3729	3747	3766	3784
2.4	3802	3820	3838	3856	3874	3892	3909	3927	3945	3962
2.5	3979	3997	4014	4031	4048	4065	4082	4099	4116	4133
2.6	4150	4166	4183	4200	4216	4232	4249	4265	4281	4298
2.7	4314	4330	4346	4362	4378	4393	4409	4424	4440	4456
2.8	4472	4487	4502	4518	4533	4548	4564	4579	4594	4609
2.9	4624	4639	4654	4669	4683	4698	4713	4728	4742	4757
3.0	4771	4786	4800	4814	4829	4843	4857	4871	4886	4900
3.1	4914	4928	4942	4955	4969	4983	4997	5011	5024	5038
3.2	5051	5065	5079	5092	5105	5119	5132	5145	5159	5172
3.3	5185	5198	5211	5224	5237	5250	5263	5276	5289	5302
3.4	5315	5328	5340	5353	5366	5378	5391	5403	5416	5428
3.5	5441	5453	5465	5478	5490	5502	5514	5527	5539	5551
3.6	5563	5575	5587	5599	5611	5623	5635	5647	5658	5670
3.7	5682	5694	5705	5717	5729	5740	5752	5763	5775	5786
3.8	5798	5809	5821	5832	5843	5855	5866	5877	5888	5899
3.9	5911	5922	5933	5944	5955	5966	5977	5988	5999	6010
4.0	6021	6031	6042	6053	6064	6075	6085	6096	6107	6117
4.1	6128	6138	6149	6160	6170	6180	6191	6201	6212	6222
4.2	6232	6243	6253	6263	6274	6284	6294	6304	6314	6325
4.3	6335	6345	6355	6365	6375	6385	6395	6405	6415	6425
4.4	6435	6444	6454	6464	6474	6484	6493	6503	6513	6522
4.5	6532	6542	6551	6561	6571	6580	6590	6599	6609	6618
4.6	6628	6637	6646	6656	6665	6675	6684	6693	6702	6712
4.7	6721	6730	6739	6749	6758	6767	6776	6785	6794	6803
4.8	6812	6821	6830	6839	6848	6857	6866	6875	6884	6893
4.9	6902	6911	6920	6928	6937	6946	6955	6964	6972	6981
5.0	6990	6998	7007	7016	7024	7033	7042	7050	7059	7067
5.1	7076	7084	7093	7101	7110	7118	7126	7135	7143	7152
5.2	7160	7168	7177	7185	7193	7202	7210	7218	7226	7235
5.3	7243	7251	7259	7267	7275	7284	7292	7300	7308	7316
5.4	7324	7332	7340	7348	7356	7364	7372	7380	7388	7396

Table II (Concluded)

n	00	10	20	30	40	50	60	70	80	90
5.5	7404	7412	7419	7427	7435	7443	7451	7459	7466	7474
5.6	7482	7490	7497	7505	7513	7520	7528	7536	7543	7551
5.7	7559	7566	7574	7582	7589	7597	7604	7612	7619	7627
5.8	7634	7642	7649	7657	7664	7672	7679	7686	7694	7701
5.9	7709	7716	7723	7731	7738	7745	7752	7760	7767	7774
6.0	7782	7789	7796	7803	7810	7818	7825	7832	7839	7846
6.1	7853	7860	7868	7875	7882	7889	7896	7903	7910	7917
6.2	7924	7931	7938	7945	7952	7959	7966	7973	7980	7987
6.3	7993	8000	8007	8014	8021	8028	8035	8041	8048	8055
6.4	8062	8069	8075	8082	8089	8096	8102	8109	8116	8122
6.5	8129	8136	8142	8149	8156	8162	8169	8176	8182	8189
6.6	8195	8202	8209	8215	8222	8228	8235	8241	8248	8254
6.7	8261	8267	8274	8280	8287	8293	8299	8306	8312	8319
6.8	8325	8331	8338	8344	8351	8357	8363	8370	8376	8382
6.9	8388	8395	8401	8407	8414	8420	8426	8432	8439	8445
7.0	8451	8457	8463	8470	8476	8482	8488	8494	8500	8506
7.1	8513	8519	8525	8531	8537	8543	8549	8555	8561	8567
7.2	8573	8579	8585	8591	8597	8603	8609	8615	8621	8627
7.3	8633	8639	8645	8651	8657	8663	8669	8675	8681	8686
7.4	8692	8698	8704	8710	8716	8722	8727	8733	8739	8745
7.5	8751	8756	8762	8768	8774	8779	8785	8791	8797	8802
7.6	8808	8814	8820	8825	8831	8837	8842	8848	8854	8859
7.7	8865	8871	8876	8882	8887	8893	8899	8904	8910	8915
7.8	8921	8927	8932	8938	8943	8949	8954	8960	8965	8971
7.9	8976	8982	8987	8993	8998	9004	9009	9015	9020	9025
8.0	9031	9036	9042	9047	9053	9058	9063	9069	9074	9079
8.1	9085	9090	9096	9101	9106	9112	9117	9122	9128	9133
8.2	9138	9143	9149	9154	9159	9165	9170	9175	9180	9186
8.3	9191	9196	9201	9206	9212	9217	9222	9227	9232	9238
8.4	9243	9248	9253	9258	9263	9269	9274	9279	9284	9289
8.5	9294	9299	9304	9309	9315	9320	9325	9330	9335	9340
8.6	9345	9350	9355	9360	9365	9370	9375	9380	9385	9390
8.7	9395	9400	9405	9410	9415	9420	9425	9430	9435	9440
8.8	9445	9450	9455	9460	9465	9469	9474	9479	9484	9489
8.9	9494	9499	9504	9509	9513	9518	9523	9528	9533	9538
9.0	9542	9547	9552	9557	9562	9566	9571	9576	9581	9586
9.1	9590	9595	9600	9605	9609	9614	9619	9624	9628	9633
9.2	9638	9643	9647	9652	9657	9661	9666	9671	9675	9680
9.3	9685	9689	9694	9699	9703	9708	9713	9717	9722	9727
9.4	9731	9736	9741	9745	9750	9754	9759	9763	9768	9773
9.5	9777	9782	9786	9791	9795	9800	9805	9809	9814	9818
9.6	9823	9827	9832	9836	9841	9845	9850	9854	9859	9863
9.7	9868	9872	9877	9881	9886	9890	9894	9899	9903	9908
9.8	9912	9917	9921	9926	9930	9934	9939	9943	9948	9952
9.9	9956	9961	9965	9969	9974	9978	9983	9987	9991	9996

Table III. Trigonometric Functions and Degrees-to-Radians

$m(\theta)$									
Degrees	*Radians*	*sin θ*	*csc θ*	*tan θ*	*cot θ*	*sec θ*	*cos θ*		
0°00′	.0000	.0000	Undef.	.0000	Undef.	1.000	1.0000	1.5708	90°00′
10′	.0029	.0029	343.8	.0029	343.8	1.000	1.0000	1.5679	50′
20′	.0058	.0058	171.9	.0058	171.9	1.000	1.0000	1.5650	40′
30′	.0087	.0087	114.6	.0087	114.6	1.000	1.0000	1.5621	30′
40′	.0116	.0116	85.95	.0116	85.94	1.000	.9999	1.5592	20′
50′	.0145	.0145	68.76	.0145	68.75	1.000	.9999	1.5563	10′
1°00′	.0175	.0175	57.30	.0175	57.29	1.000	.9998	1.5533	89°00′
10′	.0204	.0204	49.11	.0204	49.10	1.000	.9998	1.5504	50′
20′	.0233	.0233	42.98	.0233	42.96	1.000	.9997	1.5475	40′
30′	.0262	.0262	38.20	.0262	38.19	1.000	.9997	1.5446	30′
40′	.0291	.0291	34.38	.0291	34.37	1.000	.9996	1.5417	20′
50′	.0320	.0320	31.26	.0320	31.24	1.001	.9995	1.5388	10′
2°00′	.0349	.0349	28.65	.0349	28.64	1.001	.9994	1.5359	88°00′
10′	.0378	.0378	26.45	.0378	26.43	1.001	.9993	1.5330	50′
20′	.0407	.0407	24.56	.0407	24.54	1.001	.9992	1.5301	40′
30′	.0436	.0436	22.93	.0437	22.90	1.001	.9990	1.5272	30′
40′	.0465	.0465	21.49	.0466	21.47	1.001	.9989	1.5243	20′
50′	.0495	.0494	20.23	.0495	20.21	1.001	.9988	1.5213	10′
3°00′	.0524	.0523	19.11	.0524	19.08	1.001	.9986	1.5184	87°00′
10′	.0553	.0552	18.10	.0553	18.07	1.002	.9985	1.5155	50′
20′	.0582	.0581	17.20	.0582	17.17	1.002	.9983	1.5126	40′
30′	.0611	.0610	16.38	.0612	16.35	1.002	.9981	1.5097	30′
40′	.0640	.0640	15.64	.0641	15.60	1.002	.9980	1.5068	20′
50′	.0669	.0669	14.96	.0670	14.92	1.002	.9978	1.5039	10′
4°00′	.0698	.0698	14.34	.0699	14.30	1.002	.9976	1.5010	86°00′
10′	.0727	.0727	13.76	.0729	13.73	1.003	.9974	1.4981	50′
20′	.0756	.0756	13.23	.0758	13.20	1.003	.9971	1.4952	40′
30′	.0785	.0785	12.75	.0787	12.71	1.003	.9969	1.4923	30′
40′	.0814	.0814	12.29	.0816	12.25	1.003	.9967	1.4893	20′
50.	.0844	.0843	11.87	.0846	11.83	1.004	.9964	1.4864	10′
5°00′	.0873	.0872	11.47	.0874	11.43	1.004	.9962	1.4835	85°00′
10′	.0902	.0901	11.10	.0904	11.06	1.004	.9959	1.4806	50′
20′	.0931	.0929	10.76	.0934	10.71	1.004	.9957	1.4777	40′
30′	.0960	.0958	10.43	.0963	10.39	1.005	.9954	1.4748	30′
40′	.0989	.0987	10.13	.0992	10.08	1.005	.9951	1.4719	20′
50′	.1018	.1016	9.839	.1022	9.788	1.005	.9948	1.4690	10′
6°00′	.1047	.1045	9.567	.1051	9.514	1.006	.9945	1.4661	84°00′
10′	.1076	.1074	9.309	.1080	9.255	1.006	.9942	1.4632	50′
20′	.1105	.1103	9.065	.1110	9.010	1.006	.9939	1.4603	40′
30′	.1134	.1132	8.834	.1139	8.777	1.006	.9936	1.4573	30′
40′	.1164	.1161	8.614	.1169	8.556	1.007	.9932	1.4544	20′
50′	.1193	.1190	8.405	.1198	8.345	1.007	.9929	1.4515	10′
7°00′	.1222	.1219	8.206	.1228	8.144	1.008	.9925	1.4486	83°00′
10′	.1251	.1248	8.016	.1257	7.953	1.008	.9922	1.4457	50′
20′	.1280	.1276	7.834	.1287	7.770	1.008	.9918	1.4428	40′
30′	.1309	.1305	7.661	.1317	7.596	1.009	.9914	1.4399	30′
40′	.1338	.1334	7.496	.1346	7.429	1.009	.9911	1.4370	20′
50′	.1367	.1363	7.337	.1376	7.269	1.009	.9907	1.4341	10′
8°00′	.1396	.1392	7.185	.1405	7.115	1.010	.9903	1.4312	82°00′
10′	.1425	.1421	7.040	.1435	6.968	1.010	.9899	1.4283	50′
20′	.1454	.1449	6.900	.1465	6.827	1.011	.9894	1.4254	40′
30′	.1484	.1478	6.765	.1495	6.691	1.011	.9890	1.4224	30′
40′	.1513	.1507	6.636	.1524	6.561	1.012	.9886	1.4195	20′
50′	.1542	.1536	6.512	.1554	6.435	1.012	.9881	1.4166	10
9°00′	.1571	.1564	6.392	.1584	6.314	1.012	.9877	1.4137	81°00′
		cos θ	*sec θ*	*cot θ*	*tan θ*	*csc θ*	*sin θ*	*Radians*	*Degrees*
								$m(\theta)$	

Table III (Continued)

$m(\theta)$		$\sin\theta$	$\csc\theta$	$\tan\theta$	$\cot\theta$	$\sec\theta$	$\cos\theta$		
Degrees	*Radians*								
9°00′	.1571	.1564	6.392	.1584	6.314	1.012	.9877	1.4137	81°00′
10′	.1600	.1593	6.277	.1614	6.197	1.013	.9872	1.4108	50′
20′	.1629	.1622	6.166	.1644	6.084	1.013	.9868	1.4079	40′
30′	.1658	.1650	6.059	.1673	5.976	1.014	.9863	1.4050	30′
40′	.1687	.1679	5.955	.1703	5.871	1.014	.9858	1.4021	20′
50′	.1716	.1708	5.855	.1733	5.769	1.015	.9853	1.3992	10′
10°00′	.1745	.1736	5.759	.1763	5.671	1.015	.9848	1.3963	80°00′
10′	.1774	.1765	5.665	.1793	5.576	1.016	.9843	1.3934	50′
20′	.1804	.1794	5.575	.1823	5.485	1.016	.9838	1.3904	40′
30′	.1833	.1822	5.487	.1853	5.396	1.017	.9833	1.3875	30′
40′	.1862	.1851	5.403	.1883	5.309	1.018	.9827	1.3846	20′
50′	.1891	.1880	5.320	.1914	5.226	1.018	.9822	1.3817	10′
11°00′	.1920	.1908	5.241	.1944	5.145	1.019	.9816	1.3788	79°00′
10′	.1949	.1937	5.164	.1974	5.066	1.019	.9811	1.3759	50′
20′	.1978	.1965	5.089	.2004	4.989	1.020	.9805	1.3730	40′
30′	.2007	.1994	5.016	.2035	4.915	1.020	.9799	1.3701	30′
40′	.2036	.2022	4.945	.2065	4.843	1.021	.9793	1.3672	20′
50′	.2065	.2051	4.876	.2095	4.773	1.022	.9787	1.3643	10′
12°00′	.2094	.2079	4.810	.2126	4.705	1.022	.9781	1.3614	78°00′
10′	.2123	.2108	4.745	.2156	4.638	1.023	.9775	1.3584	50′
20′	.2153	.2136	4.682	.2186	4.574	1.024	.9769	1.3555	40′
30′	.2182	.2164	4.620	.2217	4.511	1.024	.9763	1.3526	30′
40′	.2211	.2193	4.560	.2247	4.449	1.025	.9757	1.3497	20′
50′	.2240	.2221	4.502	.2278	4.390	1.026	.9750	1.3468	10′
13°00′	.2269	.2250	4.445	.2309	4.331	1.026	.9744	1.3439	77°00′
10′	.2298	.2278	4.390	.2339	4.275	1.027	.9737	1.3410	50′
20′	.2327	.2306	4.336	.2370	4.219	1.028	.9730	1.3381	40′
30′	.2356	.2334	4.284	.2401	4.165	1.028	.9724	1.3352	30′
40′	.2385	.2363	4.232	.2432	4.113	1.029	.9717	1.3323	20′
50′	.2414	.2391	4.182	.2462	4.061	1.030	.9710	1.3294	10′
14°00′	.2443	.2419	4.134	.2493	4.011	1.031	.9703	1.3265	76°00′
10′	.2473	.2447	4.086	.2524	3.962	1.031	.9696	1.3235	50′
20′	.2502	.2476	4.039	.2555	3.914	1.032	.9689	1.3206	40′
30′	.2531	.2504	3.994	.2586	3.867	1.033	.9681	1.3177	30′
40′	.2560	.2532	3.950	.2617	3.821	1.034	.9674	1.3148	20′
50′	.2589	.2560	3.906	.2648	3.776	1.034	.9667	1.3119	10′
15°00′	.2618	.2588	3.864	.2679	3.732	1.035	.9659	1.3090	75°00′
10′	.2647	.2616	3.822	.2711	3.689	1.036	.9652	1.3061	50′
20′	.2676	.2644	3.782	.2742	3.647	1.037	.9644	1.3032	40′
30′	.2705	.2672	3.742	.2773	3.606	1.038	.9636	1.3003	30′
40′	.2734	.2700	3.703	.2805	3.566	1.039	.9628	1.2974	20′
50′	.2763	.2728	3.665	.2836	3.526	1.039	.9621	1.2945	10′
16°00′	.2793	.2756	3.628	.2867	3.487	1.040	.9613	1.2915	74°00′
10′	.2822	.2784	3.592	.2899	3.450	1.041	.9605	1.2886	50′
20′	.2851	.2812	3.556	.2931	3.412	1.042	.9596	1.2857	40′
30′	.2880	.2840	3.521	.2962	3.376	1.043	.9588	1.2828	30′
40′	.2909	.2868	3.487	.2994	3.340	1.044	.9580	1.2799	20′
50′	.2938	.2896	3.453	.3026	3.305	1.045	.9572	1.2770	10′
17°00′	.2967	.2924	3.420	.3057	3.271	1.046	.9563	1.2741	73°00′
10′	.2996	.2952	3.388	.3089	3.237	1.047	.9555	1.2712	50′
20′	.3025	.2979	3.356	.3121	3.204	1.048	.9546	1.2683	40′
30′	.3054	.3007	3.326	.3153	3.172	1.049	.9537	1.2654	30′
40′	.3083	.3035	3.295	.3185	3.140	1.049	.9528	1.2625	20′
50′	.3113	.3062	3.265	.3217	3.108	1.050	.9520	1.2595	10′
18°00′	.3142	.3090	3.236	.3249	3.078	1.051	.9511	1.2566	72°00′
		$\cos\theta$	$\sec\theta$	$\cot\theta$	$\tan\theta$	$\csc\theta$	$\sin\theta$	*Radians*	*Degrees*
								$m(\theta)$	

Table III (Continued)

$m(\theta)$ Degrees	$m(\theta)$ Radians	$\sin\theta$	$\csc\theta$	$\tan\theta$	$\cot\theta$	$\sec\theta$	$\cos\theta$		
18°00′	.3142	.3090	3.236	.3249	3.078	1.051	.9511	1.2566	72°00′
10′	.3171	.3118	3.207	.3281	3.047	1.052	.9502	1.2537	50′
20′	.3200	.3145	3.179	.3314	3.018	1.053	.9492	1.2508	40′
30′	.3229	.3173	3.152	.3346	2.989	1.054	.9483	1.2479	30′
40′	.3258	.3201	3.124	.3378	2.960	1.056	.9474	1.2450	20′
50′	.3287	.3228	3.098	.3411	2.932	1.057	.9465	1.2421	10′
19°00′	.3316	.3256	3.072	.3443	2.904	1.058	.9455	1.2392	71°00′
10′	.3345	.3283	3.046	.3476	2.877	1.059	.9446	1.2363	50′
20′	.3374	.3311	3.021	.3508	2.850	1.060	.9436	1.2334	40′
30′	.3403	.3338	2.996	.3541	2.824	1.061	.9426	1.2305	30′
40′	.3432	.3365	2.971	.3574	2.798	1.062	.9417	1.2275	20′
50′	.3462	.3393	2.947	.3607	2.773	1.063	.9407	1.2246	10′
20°00′	.3491	.3420	2.924	.3640	2.747	1.064	.9397	1.2217	70°00′
10′	.3520	.3448	2.901	.3673	2.723	1.065	.9387	1.2188	50′
20′	.3549	.3475	2.878	.3706	2.699	1.066	.9377	1.2159	40′
30′	.3578	.3502	2.855	.3739	2.675	1.068	.9367	1.2130	30′
40′	.3607	.3529	2.833	.3772	2.651	1.069	.9356	1.2101	20′
50′	.3636	.3557	2.812	.3805	2.628	1.070	.9346	1.2072	10′
21°00′	.3665	.3584	2.790	.3839	2.605	1.071	.9336	1.2043	69°00′
10′	.3694	.3611	2.769	.3872	2.583	1.072	.9325	1.2014	50′
20′	.3723	.3638	2.749	.3906	2.560	1.074	.9315	1.1985	40′
30′	.3752	.3665	2.729	.3939	2.539	1.075	.9304	1.1956	30′
40′	.3782	.3692	2.709	.3973	2.517	1.076	.9293	1.1926	20′
50′	.3811	.3719	2.689	.4006	2.496	1.077	.9283	1.1897	10′
22°00′	.3840	.3746	2.669	.4040	2.475	1.079	.9272	1.1868	68°00′
10′	.3869	.3773	2.650	.4074	2.455	1.080	.9261	1.1839	50′
20′	.3898	.3800	2.632	.4108	2.434	1.081	.9250	1.1810	40′
30′	.3927	.3827	2.613	.4142	2.414	1.082	.9239	1.1781	30′
40′	.3956	.3854	2.595	.4176	2.394	1.084	.9228	1.1752	20′
50′	.3985	.3881	2.577	.4210	2.375	1.085	.9216	1.1723	10′
23°00′	.4014	.3907	2.559	.4245	2.356	1.086	.9205	1.1694	67°00′
10′	.4043	.3934	2.542	.4279	2.337	1.088	.9194	1.1665	50′
20′	.4072	.3961	2.525	.4314	2.318	1.089	.9182	1.1636	40′
30′	.4102	.3987	2.508	.4348	2.300	1.090	.9171	1.1606	30′
40′	.4131	.4014	2.491	.4383	2.282	1.092	.9159	1.1577	20′
50′	.4160	.4041	2.475	.4417	2.264	1.093	.9147	1.1548	10′
24°00′	.4189	.4067	2.459	.4452	2.246	1.095	.9135	1.1519	66°00′
10′	.4218	.4094	2.443	.4487	2.229	1.096	.9124	1.1490	50′
20′	.4247	.4120	2.427	.4522	2.211	1.097	.9112	1.1461	40′
30′	.4276	.4147	2.411	.4557	2.194	1.099	.9100	1.1432	30′
40′	.4305	.4173	2.396	.4592	2.177	1.100	.9088	1.1403	20′
50′	.4334	.4200	2.381	.4628	2.161	1.102	.9075	1.1374	10′
25°00′	.4363	.4226	2.366	.4663	2.145	1.103	.9063	1.1345	65°00′
10′	.4392	.4253	2.352	.4699	2.128	1.105	.9051	1.1316	50′
20′	.4422	.4279	2.337	.4734	2.112	1.106	.9038	1.1286	40′
30′	.4451	.4305	2.323	.4770	2.097	1.108	.9026	1.1257	30′
40′	.4480	.4331	2.309	.4806	2.081	1.109	.9013	1.1228	20′
50′	.4509	.4358	2.295	.4841	2.066	1.111	.9001	1.1199	10′
26°00′	.4538	.4384	2.281	.4877	2.050	1.113	.8988	1.1170	64°00′
10′	.4567	.4410	2.268	.4913	2.035	1.114	.8975	1.1141	50′
20′	.4596	.4436	2.254	.4950	2.020	1.116	.8962	1.1112	40′
30′	.4625	.4462	2.241	.4986	2.006	1.117	.8949	1.1083	30′
40′	.4654	.4488	2.228	.5022	1.991	1.119	.8936	1.1054	20′
50′	.4683	.4514	2.215	.5059	1.977	1.121	.8923	1.1025	10′
27°00′	.4712	.4540	2.203	.5095	1.963	1.122	.8910	1.0996	63°00′
		$\cos\theta$	$\sec\theta$	$\cot\theta$	$\tan\theta$	$\csc\theta$	$\sin\theta$	$m(\theta)$ Radians	$m(\theta)$ Degrees

Table III (Continued)

$m(\theta)$ Degrees	$m(\theta)$ Radians	$\sin\theta$	$\csc\theta$	$\tan\theta$	$\cot\theta$	$\sec\theta$	$\cos\theta$		
27° 00′	.4712	.4540	2.203	.5095	1.963	1.122	.8910	1.0996	63° 00′
10′	.4741	.4566	2.190	.5132	1.949	1.124	.8897	1.0966	50′
20′	.4771	.4592	2.178	.5169	1.935	1.126	.8884	1.0937	40′
30′	.4800	.4617	2.166	.5206	1.921	1.127	.8870	1.0908	30′
40′	.4829	.4643	2.154	.5243	1.907	1.129	.8857	1.0879	20′
50′	.4858	.4669	2.142	.5280	1.894	1.131	.8843	1.0850	10′
28° 00′	.4887	.4695	2.130	.5317	1.881	1.133	.8829	1.0821	62° 00′
10′	.4916	.4720	2.118	.5354	1.868	1.134	.8816	1.0792	50′
20′	.4945	.4746	2.107	.5392	1.855	1.136	.8802	1.0763	40′
30′	.4974	.4772	2.096	.5430	1.842	1.138	.8788	1.0734	30′
40′	.5003	.4797	2.085	.5467	1.829	1.140	.8774	1.0705	20′
50′	.5032	.4823	2.074	.5505	1.816	1.142	.8760	1.0676	10′
29° 00′	.5061	.4848	2.063	.5543	1.804	1.143	.8746	1.0647	61° 00′
10′	.5091	.4874	2.052	.5581	1.792	1.145	.8732	1.0617	50′
20′	.5120	.4899	2.041	.5619	1.780	1.147	.8718	1.0588	40′
30′	.5149	.4924	2.031	.5658	1.767	1.149	.8704	1.0559	30′
40′	.5178	.4950	2.020	.5696	1.756	1.151	.8689	1.0530	20′
50′	.5207	.4975	2.010	.5735	1.744	1.153	.8675	1.0501	10′
30° 00′	.5236	.5000	2.000	.5774	1.732	1.155	.8660	1.0472	60° 00′
10′	.5265	.5025	1.990	.5812	1.720	1.157	.8646	1.0443	50′
20′	.5294	.5050	1.980	.5851	1.709	1.159	.8631	1.0414	40′
30′	.5323	.5075	1.970	.5890	1.698	1.161	.8616	1.0385	30′
40′	.5352	.5100	1.961	.5930	1.686	1.163	.8601	1.0356	20′
50′	.5381	.5125	1.951	.5969	1.675	1.165	.8587	1.0327	10′
31° 00′	.5411	.5150	1.942	.6009	1.664	1.167	.8572	1.0297	59° 00′
10′	.5440	.5175	1.932	.6048	1.653	1.169	.8557	1.0268	50′
20′	.5469	.5200	1.923	.6038	1.643	1.171	.8542	1.0239	40′
30′	.5498	.5225	1.914	.6128	1.632	1.173	.8526	1.0210	30′
40′	.5527	.5250	1.905	.6168	1.621	1.175	.8511	1.0181	20′
50′	.5556	.5275	1.896	.6208	1.611	1.177	.8496	1.0152	10′
32° 00′	.5585	.5299	1.887	.6249	1.600	1.179	.8480	1.0123	58° 00′
10′	.5614	.5324	1.878	.6289	1.590	1.181	.8465	1.0094	50′
20′	.5643	.5348	1.870	.6330	1.580	1.184	.8450	1.0065	40′
30′	.5672	.5373	1.861	.6371	1.570	1.186	.8434	1.0036	30′
40′	.5701	.5398	1.853	.6412	1.560	1.188	.8418	1.0007	20′
50′	.5730	.5422	1.844	.6453	1.550	1.190	.8403	.9977	10′
33° 00′	.5760	.5446	1.836	.6494	1.540	1.192	.8387	.9948	57° 00′
10′	.5789	.5471	1.828	.6536	1.530	1.195	.8371	.9919	50′
20′	.5818	.5495	1.820	.6577	1.520	1.197	.8355	.9890	40′
30′	.5847	.5519	1.812	.6619	1.511	1.199	.8339	.9861	30′
40′	.5876	.5544	1.804	.6661	1.501	1.202	.8323	.9832	20′
50′	.5905	.5568	1.796	.6703	1.492	1.204	.8307	.9803	10′
34° 00′	.5934	.5592	1.788	.6745	1.483	1.206	.8290	.9774	56° 00′
10′	.5963	.5616	1.781	.6787	1.473	1.209	.8274	.9745	50′
20′	.5992	.5640	1.773	.6830	1.464	1.211	.8258	.9716	40′
30′	.6021	.5664	1.766	.6873	1.455	1.213	.8241	.9687	30′
40′	.6050	.5688	1.758	.6916	1.446	1.216	.8225	.9657	20′
50′	.6080	.5712	1.751	.6959	1.437	1.218	.8208	.9628	10′
35° 00′	.6109	.5736	1.743	.7002	1.428	1.221	.8192	.9599	55° 00′
10′	.6138	.5760	1.736	.7046	1.419	1.223	.8175	.9570	50′
20′	.6167	.5783	1.729	.7089	1.411	1.226	.8158	.9541	40′
30′	.6196	.5807	1.722	.7133	1.402	1.228	.8141	.9512	30′
40′	.6225	.5831	1.715	.7177	1.393	1.231	.8124	.9483	20′
50′	.6254	.5854	1.708	.7221	1.385	1.233	.8107	.9454	10′
36° 00′	.6283	.5878	1.701	.7265	1.376	1.236	.8090	.9425	54° 00′
		$\cos\theta$	$\sec\theta$	$\cot\theta$	$\tan\theta$	$\csc\theta$	$\sin\theta$	$m(\theta)$ Radians	$m(\theta)$ Degrees

Table III (Concluded)

$m(\theta)$ Degrees	$m(\theta)$ Radians	$\sin\theta$	$\csc\theta$	$\tan\theta$	$\cot\theta$	$\sec\theta$	$\cos\theta$		
36° 00′	.6283	.5878	1.701	.7265	1.376	1.236	.8090	.9425	54° 00′
10′	.6312	.5901	1.695	.7310	1.368	1.239	.8073	.9396	50′
20′	.6341	.5925	1.688	.7355	1.360	1.241	.8056	.9367	40′
30′	.6370	.5948	1.681	.7400	1.351	1.244	.8039	.9338	30′
40′	.6400	.5972	1.675	.7445	1.343	1.247	.8021	.9308	20′
50′	.6429	.5995	1.668	.7490	1.335	1.249	.8004	.9279	10′
37° 00′	.6458	.6018	1.662	.7536	1.327	1.252	.7986	.9250	53° 00′
10′	.6487	.6041	1.655	.7581	1.319	1.255	.7969	.9221	50′
20′	.6516	.6065	1.649	.7627	1.311	1.258	.7951	.9192	40′
30′	.6545	.6088	1.643	.7673	1.303	1.260	.7934	.9163	30′
40′	.6574	.6111	1.636	.7720	1.295	1.263	.7916	.9134	20′
50′	.6603	.6134	1.630	.7766	1.288	1.266	.7898	.9105	10′
38° 00′	.6632	.6157	1.624	.7813	1.280	1.269	.7880	.9076	52° 00′
10′	.6661	.6180	1.618	.7860	1.272	1.272	.7862	.9047	50′
20′	.6690	.6202	1.612	.7907	1.265	1.275	.7844	.9018	40′
30′	.6720	.6225	1.606	.7954	1.257	1.278	.7826	.8988	30′
40′	.6749	.6248	1.601	.8002	1.250	1.281	.7808	.8959	20′
50′	.6778	.6271	1.595	.8050	1.242	1.284	.7790	.8930	10′
39° 00′	.6807	.6293	1.589	.8098	1.235	1.287	.7771	.8901	51° 00′
10′	.6836	.6316	1.583	.8146	1.228	1.290	.7753	.8872	50′
20′	.6865	.6338	1.578	.8195	1.220	1.293	.7735	.8843	40′
30′	.6894	.6361	1.572	.8243	1.213	1.296	.7716	.8814	30′
40′	.6923	.6383	1.567	.8292	1.206	1.299	.7698	.8785	20′
50′	.6952	.6406	1.561	.8342	1.199	1.302	.7679	.8756	10′
40° 00′	.6981	.6428	1.556	.8391	1.192	1.305	.7660	.8727	50° 00′
10′	.7010	.6450	1.550	.8441	1.185	1.309	.7642	.8698	50′
20′	.7039	.6472	1.545	.8491	1.178	1.312	.7623	.8668	40′
30′	.7069	.6494	1.540	.8541	1.171	1.315	.7604	.8639	30′
40′	.7098	.6517	1.535	.8591	1.164	1.318	.7585	.8610	20′
50′	.7127	.6539	1.529	.8642	1.157	1.322	.7566	.8581	10′
41° 00′	.7156	.6561	1.524	.8693	1.150	1.325	.7547	.8552	49° 00′
10′	.7185	.6583	1.519	.8744	1.144	1.328	.7528	.8523	50′
20′	.7214	.6604	1.514	.8796	1.137	1.332	.7509	.8494	40′
30′	.7243	.6626	1.509	.8847	1.130	1.335	.7490	.8465	30′
40′	.7272	.6648	1.504	.8899	1.124	1.339	.7470	.8436	20′
50′	.7301	.6670	1.499	.8952	1.117	1.342	.7451	.8407	10′
42° 00′	.7330	.6691	1.494	.9004	1.111	1.346	.7431	.8378	48° 00′
10′	.7359	.6713	1.490	.9057	1.104	1.349	.7412	.8348	50′
20′	.7389	.6734	1.485	.9110	1.098	1.353	.7392	.8319	40′
30′	.7418	.6756	1.480	.9163	1.091	1.356	.7373	.8290	30′
40′	.7447	.6777	1.476	.9217	1.085	1.360	.7353	.8261	20′
50′	.7476	.6799	1.471	.9271	1.079	1.364	.7333	.8232	10′
43° 00′	.7505	.6820	1.466	.9325	1.072	1.367	.7314	.8203	47° 00′
10′	.7534	.6841	1.462	.9380	1.066	1.371	.7294	.8174	50′
20′	.7563	.6862	1.457	.9435	1.060	1.375	.7274	.8145	40′
30′	.7592	.6884	1.453	.9490	1.054	1.379	.7254	.8116	30′
40′	.7621	.6905	1.448	.9545	1.048	1.382	.7234	.8087	20′
50′	.7650	.6926	1.444	.9601	1.042	1.386	.7214	.8058	10′
44° 00′	.7679	.6947	1.440	.9657	1.036	1.390	.7193	.8029	46° 00′
10′	.7709	.6967	1.435	.9713	1.030	1.394	.7173	.7999	50′
20′	.7738	.6988	1.431	.9770	1.024	1.398	.7153	.7970	40′
30′	.7767	.7009	1.427	.9827	1.018	1.402	.7133	.7941	30′
40′	.7796	.7030	1.423	.9884	1.012	1.406	.7112	.7912	20′
50′	.7825	.7050	1.418	.9942	1.006	1.410	.7092	.7883	10′
45° 00′	.7854	.7071	1.414	1.000	1.000	1.414	.7071	.7854	45° 00′
		$\cos\theta$	$\sec\theta$	$\cot\theta$	$\tan\theta$	$\csc\theta$	$\sin\theta$	$m(\theta)$ Radians	$m(\theta)$ Degrees

Table IV. Circular Functions and Radians-to-Degrees

Real Number x or $m^R(\theta)$	$m^o(\theta)$	sin x or sin θ	csc x or csc θ	tan x or tan θ	cot x or cot θ	sec x or sec θ	cos x or cos θ
0	0°	0	Undef.	0	Undef.	1	1
0.01	0° 34′	0.0100	100.0	0.0100	100.0	1.000	1.000
.02	1° 09′	.0200	50.00	.0200	49.99	1.000	0.9998
.03	1° 43′	.0300	33.34	.0300	33.32	1.000	0.9996
.04	2° 18′	.0400	25.01	.0400	24.99	1.001	0.9992
0.05	2° 52′	0.0500	20.01	0.0500	19.98	1.001	0.9988
.06	3° 26′	.0600	16.68	.0601	16.65	1.002	.9982
.07	4° 01′	.0699	14.30	.0701	14.26	1.002	.9976
.08	4° 35′	.0799	12.51	.0802	12.47	1.003	.9968
.09	5° 09′	.0899	11.13	.0902	11.08	1.004	.9960
0.10	5° 44′	0.0998	10.02	0.1003	9.967	1.005	0.9950
.11	6° 18′	.1098	9.109	.1104	9.054	1.006	.9940
.12	6° 53′	.1197	8.353	.1206	8.293	1.007	.9928
.13	7° 27′	.1296	7.714	.1307	7.649	1.009	.9916
.14	8° 01′	.1395	7.166	.1409	7.096	1.010	.9902
0.15	8° 36′	0.1494	6.692	0.1511	6.617	1.011	0.9888
.16	9° 10′	.1593	6.277	.1614	6.197	1.013	.9872
.17	9° 44′	.1692	5.911	.1717	5.826	1.015	.9856
.18	10° 19′	.1790	5.586	.1820	5.495	1.016	.9838
.19	10° 53′	.1889	5.295	.1923	5.200	1.018	.9820
0.20	11° 28′	0.1987	5.033	0.2027	4.933	1.020	0.9801
.21	12° 02′	.2085	4.797	.2131	4.692	1.022	.9780
.22	12° 36′	.2182	4.582	.2236	4.472	1.025	.9759
.23	13° 11′	.2280	4.386	.2341	4.271	1.027	.9737
.24	13° 45′	.2377	4.207	.2447	4.086	1.030	.9713
0.25	14° 19′	0.2474	4.042	0.2553	3.916	1.032	0.9689
.26	14° 54′	.2571	3.890	.2660	3.759	1.035	.9664
.27	15° 28′	.2667	3.749	.2768	3.613	1.038	.9638
.28	16° 03′	.2764	3.619	.2876	3.478	1.041	.9611
.29	16° 37′	.2860	3.497	.2984	3.351	1.044	.9582
0.30	17° 11′	0.2955	3.384	0.3093	3.233	1.047	0.9553
.31	17° 46′	.3051	3.278	.3203	3.122	1.050	.9523
.32	18° 20′	.3146	3.179	.3314	3.018	1.053	.9492
.33	18° 54′	.3240	3.086	.3425	2.919	1.057	.9460
.34	19° 29′	.3335	2.999	.3537	2.827	1.061	.9428
0.35	20° 03′	0.3429	2.916	0.3650	2.740	1.065	0.9394
.36	20° 38′	.3523	2.839	.3764	2.657	1.068	.9359
.37	21° 12′	.3616	2.765	.3879	2.578	1.073	.9323
.38	21° 46′	.3709	2.696	.3994	2.504	1.077	.9287
.39	22° 21′	.3802	2.630	.4111	2.433	1.081	.9249
0.40	22° 55′	0.3894	2.568	0.4228	2.365	1.086	0.9211
.41	23° 29′	.3986	2.509	.4346	2.301	1.090	.9171
.42	24° 04′	.4078	2.452	.4466	2.239	1.095	.9131
.43	24° 38′	.4169	2.399	.4586	2.180	1.100	.9090
.44	25° 13′	.4259	2.348	.4708	2.124	1.105	.9048
0.45	25° 47′	0.4350	2.299	0.4831	2.070	1.111	0.9004
.46	26° 21′	.4439	2.253	.4954	2.018	1.116	.8961
.47	26° 56′	.4529	2.208	.5080	1.969	1.122	.8916
.48	27° 30′	.4618	2.166	.5206	1.921	1.127	.8870
.49	28° 04′	.4706	2.125	.5334	1.875	1.133	.8823

Table IV (Continued)

Real Number or $m^R(\theta)$	$m^o(\theta)$	sin x or sin θ	csc x or csc θ	tan x or tan θ	cot x or cot θ	sec x or sec θ	cos x or cos θ
0.50	28°39′	0.4794	2.086	0.5463	1.830	1.139	0.8776
.51	29°13′	.4882	2.048	.5594	1.788	1.146	.8727
.52	29°48′	.4969	2.013	.5726	1.747	1.152	.8678
.53	30°22′	.5055	1.978	.5859	1.707	1.159	.8628
.54	30°56′	.5141	1.945	.5994	1.668	1.166	.8577
0.55	31°31′	0.5227	1.913	0.6131	1.631	1.173	0.8525
.56	32°05′	.5312	1.883	.6269	1.595	1.180	.8473
.57	32°40′	.5396	1.853	.6410	1.560	1.188	.8419
.58	33°14′	.5480	1.825	.6552	1.526	1.196	.8365
.59	33°48′	.5564	1.797	.6696	1.494	1.203	.8309
0.60	34°23′	0.5646	1.771	0.6841	1.462	1.212	0.8253
.61	34°57′	.5729	1.746	.6989	1.431	1.220	.8196
.62	35°31′	.5810	1.721	.7139	1.401	1.229	.8139
.63	36°06′	.5891	1.697	.7291	1.372	1.238	.8080
.64	36°40′	.5972	1.674	.7445	1.343	1.247	.8021
0.65	37°15′	0.6052	1.652	0.7602	1.315	1.256	0.7961
.66	37°49′	.6131	1.631	.7761	1.288	1.266	.7900
.67	38°23′	.6210	1.610	.7923	1.262	1.276	.7838
.68	38°58′	.6288	1.590	.8087	1.237	1.286	.7776
.69	39°32′	.6365	1.571	.8253	1.212	1.297	.7712
0.70	40°06′	0.6442	1.552	0.8423	1.187	1.307	0.7648
.71	40°41′	.6518	1.534	.8595	1.163	1.319	.7584
.72	41°15′	.6594	1.517	.8771	1.140	1.330	.7518
.73	41°50′	.6669	1.500	.8949	1.117	1.342	.7452
.74	42°24′	.6743	1.483	.9131	1.095	1.354	.7385
0.75	42°58′	0.6816	1.467	0.9316	1.073	1.367	0.7317
.76	43°33′	.6889	1.452	.9505	1.052	1.380	.7248
.77	44°07′	.6961	1.437	.9697	1.031	1.393	.7179
.78	44°41′	.7033	1.422	.9893	1.011	1.407	.7109
.79	45°16′	.7104	1.408	1.009	.9908	1.421	.7038
0.80	45°50′	0.7174	1.394	1.030	0.9712	1.435	0.6967
.81	46°25′	.7243	1.381	1.050	.9520	1.450	.6895
.82	46°59′	.7311	1.368	1.072	.9331	1.466	.6822
.83	47°33′	.7379	1.355	1.093	.9146	1.482	.6749
.84	48°08′	.7446	1.343	1.116	.8964	1.498	.6675
0.85	48°42′	0.7513	1.331	1.138	0.8785	1.515	0.6600
.86	49°16′	.7578	1.320	1.162	.8609	1.533	.6524
.87	49°51′	.7643	1.308	1.185	.8437	1.551	.6448
.88	50°25′	.7707	1.297	1.210	.8267	1.569	.6372
.89	51°00′	.7771	1.287	1.235	.8100	1.589	.6294
0.90	51°34′	0.7833	1.277	1.260	0.7936	1.609	0.6216
.91	52°08′	.7895	1.267	1.286	.7774	1.629	.6137
.92	52°43′	.7956	1.257	1.313	.7615	1.651	.6058
.93	53°17′	.8016	1.247	1.341	.7458	1.673	.5978
.94	53°51′	.8076	1.238	1.369	.7303	1.696	.5898
0.95	54°26′	0.8134	1.229	1.398	0.7151	1.719	0.5817
.96	55°00′	.8192	1.221	1.428	.7001	1.744	.5735
.97	55°35′	.8249	1.212	1.459	.6853	1.769	.5653
.98	56°09′	.8305	1.204	1.491	.6707	1.795	.5570
.99	56°43′	.8360	1.196	1.524	.6563	1.823	.5487
1.00	57°18′	0.8415	1.188	1.557	0.6421	1.851	0.5403
1.01	57°52′	.8468	1.181	1.592	.6281	1.880	.5319
1.02	58°27′	.8521	1.174	1.628	.6142	1.911	.5234
1.03	59°01′	.8573	1.166	1.665	.6005	1.942	.5148
1.04	59°35′	.8624	1.160	1.704	.5870	1.975	.5062

Table IV (Concluded)

Real Number x or $m^R(\theta)$	$m^o(\theta)$	*sin x or sin θ*	*csc x or csc θ*	*tan x or tan θ*	*cot x or cot θ*	*sec x or sec θ*	*cos x or cos θ*
1.05	60°10′	0.8674	1.153	1.743	0.5736	2.010	0.4976
1.06	60°44′	.8724	1.146	1.784	.5604	2.046	.4889
1.07	61°18′	.8772	1.140	1.827	.5473	2.083	.4801
1.08	61°53′	.8820	1.134	1.871	.5344	2.122	.4713
1.09	62°27′	.8866	1.128	1.917	.5216	2.162	.4625
1.10	63°02′	0.8912	1.122	1.965	0.5090	2.205	0.4536
1.11	63°36′	.8957	1.116	2.014	.4964	2.249	.4447
1.12	64°10′	.9001	1.111	2.066	.4840	2.295	.4357
1.13	64°45′	.9044	1.106	2.120	.4718	2.344	.4267
1.14	65°19′	.9086	1.101	2.176	.4596	2.395	.4176
1.15	65°53′	0.9128	1.096	2.234	0.4475	2.448	0.4085
1.16	66°28′	.9168	1.091	2.296	.4356	2.504	.3993
1.17	67°02′	.9208	1.086	2.360	.4237	2.563	.3902
1.18	67°37′	.9246	1.082	2.427	.4120	2.625	.3809
1.19	68°11′	.9284	1.077	2.498	.4003	2.691	.3717
1.20	68°45′	0.9320	1.073	2.572	0.3888	2.760	0.3624
1.21	69°20′	.9356	1.069	2.650	.3773	2.833	.3530
1.22	69°54′	.9391	1.065	2.733	.3659	2.910	.3436
1.23	70°28′	.9425	1.061	2.820	.3546	2.992	.3342
1.24	71°03′	.9458	1.057	2.912	.3434	3.079	.3248
1.25	71°37′	0.9490	1.054	3.010	0.3323	3.171	0.3153
1.26	72°12′	.9521	1.050	3.113	.3212	3.270	.3058
1.27	72°46′	.9551	1.047	3.224	.3102	3.375	.2963
1.28	73°20′	.9580	1.044	3.341	.2993	3.488	.2867
1.29	73°55′	.9608	1.041	3.467	.2884	3.609	.2771
1.30	74°29′	0.9636	1.038	3.602	0.2776	3.738	0.2675
1.31	75°03′	.9662	1.035	3.747	.2669	3.878	.2579
1.32	75°38′	.9687	1.032	3.903	.2562	4.029	.2482
1.33	76°12′	.9711	1.030	4.072	.2456	4.193	.2385
1.34	76°47′	.9735	1.027	4.256	.2350	4.372	.2288
1.35	77°21′	0.9757	1.025	4.455	0.2245	4.566	0.2190
1.36	77°55′	.9779	1.023	4.673	.2140	4.779	.2092
1.37	78°30′	.9799	1.021	4.913	.2035	5.014	.1994
1.38	79°04′	.9819	1.018	5.177	.1931	5.273	.1896
1.39	79°38′	.9837	1.017	5.471	.1828	5.561	.1798
1.40	80°13′	0.9854	1.015	5.798	0.1725	5.883	0.1700
1.41	80°47′	.9871	1.013	6.165	.1622	6.246	.1601
1.42	81°22′	.9887	1.011	6.581	.1519	6.657	.1502
1.43	81°56′	.9901	1.010	7.055	.1417	7.126	.1403
1.44	82°30′	.9915	1.009	7.602	.1315	7.667	.1304
1.45	83°05′	0.9927	1.007	8.238	0.1214	8.299	0.1205
1.46	83°39′	.9939	1.006	8.989	.1113	9.044	.1106
1.47	84°13′	.9949	1.005	9.887	.1011	9.938	.1006
1.48	84°48′	.9959	1.004	10.98	.0910	11.03	.0907
1.49	85°22′	.9967	1.003	12.35	.0810	12.39	.0807
1.50	85°57′	0.9975	1.003	14.10	0.0709	14.14	0.0707
1.51	86°31′	.9982	1.002	16.43	.0609	16.46	.0608
1.52	87°05′	.9987	1.001	19.67	.0508	19.70	.0508
1.53	87°40′	.9992	1.001	24.50	.0408	24.52	.0408
1.54	88°14′	.9995	1.000	32.46	.0308	32.48	.0308
1.55	88°49′	0.9998	1.000	48.08	0.0208	48.09	0.0208
1.56	89°23′	.9999	1.000	92.62	.0108	92.63	.0108
1.57	89°57′	1.000	1.000	1256	.0008	1256	.0008
$\frac{\pi}{2}$	90°	1	1	Undef.	0	Undef.	0

Glossary

Algebraic expression (p. 9): An expression containing no operations on variables other than +, −, ×, ÷, or $\sqrt[n]{\ }$.

Amplitude (p. 394): The distance from the axis of a periodic function to a high or low point.

Antilogarithm (p. 150): The argument of a logarithm.

Argument (p. 143, 383): The variable or expression on which a function operates. For $\log(3x + 5)$, the expression $(3x + 5)$ is the argument.

Asymptote (p. 31): A line which a graph gets closer and closer to, but never touches, as x or y become very large.

Axiom (p. 3): A property accepted without proof, used as a starting point for a mathematical system.

BASIC (p. 519): Acronym for "Beginners All-purpose Symbolic Instruction Code," an easily-learned computer language.

Cancellation in fractions (p. 194): Dividing the numerator and the denominator by the same common factor.

Characteristic (p. 145, 154): The integer part of a base-10 logarithm, or exponent of 10 for a number in scientific notation.

Circular function (p. 383-388): A function (sin, cos, tan, cot, sec, csc) whose independent variable, x, is a real number representing the length of an arc of a unit circle.

Combination (p. 350): A subset of the elements in a given set, without regard to the order in which these elements are arranged.

Conic sections (p. 265): Circles, ellipses, parabolas, and hyperbolas (and occasionally lines or points) formed by the intersection of a plane and a cone.

Conjugate binomials (p. 179): Binomials of the form $a + b$ and $a - b$, where the only difference is the sign of the second term.

Constant function (p. 43): A function whose general equation is y = a constant.

Converge (p. 316): To get closer and closer to, without necessarily reaching.

Cubic function (p. 279): A function whose general equation is $y = ax^3 + bx^2 + cx + d$, where a, b, c, and d stand for constants, and $a \neq 0$.

Degree of a polynomial (p. 10): The maximum number of variables that are multiplied together in any one term of the polynomial.

Dependent variable (p. 30): The variable appearing *second* in an ordered pair.

Descartes' Rule of Signs (p. 531): The number of positive zeros of a polynomial is less than or equal to the number of sign reversals in $P(x)$. The number of negative zeros is less than or equal to the number of sign reversals in $P(-x)$. In both cases, the number of zeros has the same parity (odd or even) as the number of reversals.

Determinant (p. 69, 81): A square array of numbers evaluated according to specific rules explained in the text.

Discriminant (p. 109): The quantity $b^2 - 4ac$, where a, b, and c are coefficients of the quadratic equation $ax^2 + bx + c = 0$.

Domain of a function (p. 32): The set of permissible values of the independent variable.

Ellipse (p. 256): A set of points in a plane, for each of which the *sum* of its distances from two fixed points (foci) is equal to a constant.

Equivalent equations (p. 12): Equations that have the same solution set.

Equivalent expressions (p. 8): Expressions that stand for equal numbers, no matter what permissible value is substituted for the variable(s).

Evaluating an expression (p. 6): Substituting numbers for the variables, and thus finding the *value* of the expression.

Exponential function (p. 131): A function whose general equation is $y = a \times b^x$ or $y = a \times 10^{kx}$, where a, b, and k stand for constants.

Extraneous solution (p. 13, 205): A number that satisfies an equation which has been transformed, but does not satisfy the original equation.

Function (p. 35): A relation in which each value of the independent variable takes on a *unique* value of the dependent variable.

General equation (p. 49): An equation such as $y = ax^2 + bx + c$, in which letters (a, b, c) are used to stand for the constants.

Greatest common factor, GCF (p. 187): The GCF of integers x and y is the product of the numbers in the *intersection* of the sets of prime factors of x and y.

Higher degree function (p. 279): A function whose general equation is $y = P(x)$, where $P(x)$ is a cubic or higher degree polynomial.

Hyperbola (p. 260): A set of points in a plane, for each of which the *difference* of its distances from two fixed points (foci) is constant.

Independent variable (p. 30): The variable appearing *first* in an ordered pair.

Indeterminate expression (p. 201): An expression such as 0/0 or 0^0, which could take on *any* value, depending on what expression gives the "0."

Infinitely large (p. 201): Larger than any real number.

Intercept (p. 45, 78): A value of a variable when all other variables in the equation equal zero.

Intersection (p. 344): The intersection of sets A and B, written $A \cap B$, is the set of elements that are in both A and B.

Interval (p. 457): A set of all the real numbers between (and sometimes including) two given numbers.

Irrational algebraic function (p. 233): A function whose general equation is $y = f(x)$, where $f(x)$ is an expression containing a root of a variable, and perhaps the operations +, −, ×, and ÷, but no other operations on a variable.

Irreversible step (p. 13): A transformation of an equation, such as multiplying both sides by an expression that can equal 0, which when reversed may not produce the original equation uniquely.

Least common multiple, LCM (p. 187): The LCM of integers x and y is the product of the numbers in the *union* of the sets of prime factors of x and y.

Linear (p. 10): First degree.

Linear combination (p. 66): A linear combination of two functions f and g

is an expression of the form $af(x) + bg(x)$, where a and b stand for constants.

Linear function (p. 43): A function whose general equation is $y = mx + b$, where m and b stand for constants, and $m \neq 0$.

Mantissa (p. 145, 154): The decimal part of a base-10 logarithm, or factor multiplied by the power of 10 for numbers in scientific notation.

Mathematical model (p. 53): A representation in the mathematical world of some phenomenon in the real world. In this text, a mathematical model consists of a *function* or *relation* specifying how two or more variables are related.

Matrix (p. 84): A rectangular array of numbers representing such things as the coefficients in a system of equations.

Open sentence (p. 65): An equation or inequality containing at least one variable.

Parabola (p. 266): A set of points in a plane for each of which its distance from a fixed line (the directrix) is equal to its distance from a fixed point (the focus) not on the line.

Partial sum (p. 310): The sum of the first n terms of a series.

Particular equation (p. 49): An equation such as $y = 3x + 7$, where particular numbers (3 and 7 in this case) appear as the constants.

Permutation (p. 345): An arrangement of some or all of the elements from a given set.

Phase displacement (p. 399): The value of the independent variable of a function when the argument of the function is equal to zero.

Polynomial (p. 9): An expression containing no operations other than +, −, or × performed on the variable(s).

Polynomial function (p. 41): A function whose general equation is $y = P(x)$, where $P(x)$ is a polynomial in x.

Prime number (p. 301): One of the numbers 2, 3, 5, 7, 11, . . . , which have no integer factors besides 1 and itself.

Quadratic equation (p. 105): An equation (with one variable) of the form $ax^2 + bx + c = 0$, where a, b, and c stand for constants, and $a \neq 0$.

Quadratic Formula (p. 108): If $ax^2 + bx + c = 0$, then $x = \dfrac{-b \pm \sqrt{b^2 - 4ac}}{2a}$.

Quadratic function (p. 99): A function whose general equation is $y = ax^2 + bx + c$, where a, b, and c stand for constants, and $a \neq 0$.

Quartic function (p. 279): A function whose general equation is $y = ax^4 + bx^3 + cx^2 + dx + e$, where a, b, c, d, and e stand for constants, and $a \neq 0$.

Quintic function (p. 293): A function whose general equation is $y = ax^5 + bx^4 + cx^3 + dx^2 + ex + f$, where a, b, c, d, e, and f stand for constants, and $a \neq 0$.

Radian (p. 389): A unit of angular measure equal to $1/(2\pi)$ of a complete revolution.

Range of a function (p. 32): The set of numbers that are actually used as values of the dependent variable.

Rational algebraic expression (p. 177): An expression that can be written as the ratio of two polynomials.

Rational Root Theorem (p. 190): $(ax - b)$ is a factor of $P(x)$ if and only if $P(b/a) = 0$.

Reference angle (p. 381): The smallest positive angle between the terminal side of an angle in standard position and the x-axis.

Relation (p. 30): A set of ordered pairs, triples, etc.

Relatively prime (p. 317): Two integers are relatively prime if they have no common prime factors.

Sequence (p. 301): A function whose independent variable is the term number, and whose dependent variable is the term value.

Series (p. 310): The indicated sum of the terms of a sequence.

Sinusoidal function (p. 398): A function whose general equation is $y = C + A \cos B(x - D)$, or $y = C + A \sin B(x - D)$, where A, B, C, and D stand for constants.

Slope (p. 44): The constant m in the linear function equation $y = mx + b$. Also equal to rise/run.

Solution set (p. 11): The set of all values of the variable that satisfy a given equation.

System (p. 65): A set of equations or inequalities containing the same variables.

Terms (p. 10): Parts of an expression separated by "+" or "−" signs.

Transcendental number (p. 1): A number that cannot be expressed exactly by performing a finite number of algebraic operations (+, −, ×, ÷, and $\sqrt[n]{\ }$) on integers.

Trichotomy (p. 18): The Comparison Axiom; For any two real numbers x and y, exactly *one* of the following is true: $x < y$, $x = y$, or $x > y$.

Trigonometric function (p. 383): A function (sin, cos, tan, cot, sec, csc) whose independent variable is an *angle measure*, usually in degrees or radians.

Union: The union of sets A and B, written A∪B, is the set of elements that are in at least one of sets A or B.

Unique (p. 3): "There is a unique number . . . ," means both that there *is* a number and that there is only *one* number of the kind described.

Upper Bound Theorem (p. 531): For a positive number c, if $P(x)$ is divided by $(x - c)$ and the resulting quotient and remainder have no sign reversals, then $P(x)$ has no zeros greater than c.

Variable (p. 6): A letter used to stand for an unspecified number in a given set of numbers (the domain).

Variation function (p. 207): A function whose general equation is $y = kx^n$, where k and n stand for constants.

Vector (quantity), (p. 485): A directed line segment used to represent a quantity that has both magnitude (size) and direction, such as force or velocity.

Zero of a function (p. 292): A value of x (real or complex) that makes $y = 0$.

Index of Problem Titles

Accident/Illness Insurance Problem, 370
Accurate Drawing Problems, 474, 479, 481, 482
Achievement Test Problem, 95
Advertising Problem, 170
Air Pressure Problem, 166
Aircraft Problem, 92
Airplane Engine Problems, 360, 364
Airplane Flight Speed Problem, 241
Airplane Landing Problem, 468
Airplane Lift Problem, 496
Airplane Velocity Problem, 496
Airplane Wing Problem, 219
Alligator Problem, 499
Ancestors Problem, 324
Archery Problem, 369
Artillery Problem, 120

Back-Up System Problem, 359
Bad Egg Problem, 370
Balloon Temperature Problem, 215
Bank Balance Problem, 514
Barley Problem, 123
Basketball Problem, 359
Bathtub Problem, 119
Batting Average Problem, 369
Beam Deflection Problem, 295
Beam Strength Problem, 239
Biological Half-Life Problem, 168
Blood Pressure Problem, 239
Blue Jeans Problem, 322
Boat's Wake Problem, 213
Bode's Law Problem, 337
Bouncing Ball Problem, 322
Bouncing Spring Problems, 407, 423
Brain Cell Problem, 147
Bridge Column Problem, 241
Bull's-Eye Problem, 363

Cable Car Problem, 469
Calculator Components Problem, 357
Calorie Consumption Problem, 57
Calvin and Phoebe Problem, 324
Calvin Butterball's Gasoline Problem, 121
Calvin's Grade Problem, 369
Calvin's Mass Problem, 169
Camping Trip Problem, 96
Canal Barge Problem, 496
Car Acceleration Problem, 169
Car Breakdown Problem, 358
Car Insurance Problem, 120
Car-Stopping Problem, 165
Car Trade-In Problem, 166
Carbon 14 Dating Problem, 167
Card Draw Problem, 368
Cat Problem, 467
Catastrophe Theory Problem, 299
CB Radio Manufacturing Problem, 92
Celsius-to-Fahrenheit Temperature Conversion, 59
Charles's Gas Law, 59
Chewing Gum Problem, 322
Coal Problem, 94
Coffee Cup Problem, 169
Coin Flipping Problem, 521
Color-Blindness Problem, 363
Color TV Problem, 121
Combinations and Binomial Series Problem, 360
Combinations and Powers of 2 Problem, 361
Compound Interest Problem, 166
Computer Game Problem, 521
Computer Program for Binomial Distributions, 366
Computer Solution of Triangles, 484
Computer Time Problem, 60
Cops and Robbers Problem, 75
Cost of Operating a Car Problem, 120
Cost of Owning a Car Problem, 56
Cricket Problem, 56

Data Round–Off Problem, 513
Data Search Problem, 513
Decibel Problem, 156
Deep Oil Well Cost Problem, 172
Defended Region Problem, 244
Detour Problem, 496
Diamond Problem, 216
Dice Game Problem, 370
Dice Problems, 363, 365
Diesel Car vs. Gasoline Car Problem, 75
Diesel Engine Problem, 248
Direct Variation, Pancake Problem, 59
Diving Board Problem, 295
Dollar Bill Problem, 370

Earthquake Problem, 156
Egg Problem, 217
Eighteen-Wheeler Problem, 364
Electric Current Problem, 409
Electric Light Problem, 240
Electric Power Cost Problem, 296
Electrical Voltage Problem, 409
Evaluating Expressions Problems, 511-512
Expectation of a Binomial Experiment, 369

Feedlot Problem, 93
Ferris Wheel Problems, 407, 423
First Girl Problem, 365
Flagpole Problem, 467
Fog Problem, 173
Football Plays Problem, 360
Football Problem, 121
Friction Problem, 215

Gas Consumption Problem, 240
Gas Law Problem, 214
Gas Tank Problem, 58
Gateway Arch Problem, 122
General Quadratic–Quadratic System, 274
Geometry Problem, 327
George Washington Problem, 322
Grade Problems, 358, 359
Grand Canyon Problem, 469
Grapefruit Problem, 217
Grapevine Problem, 470

Hamburger Problem, 76
Harbor Problem, 495
Heartbeat Problem, 147
Heat Radiation Problem, 248
Height-Mass Problems, 220, 237
Heredity Problem, 362
Hide-and-Seek Problem, 359
Hippopotamus Problem, 59
Home Range Problem, 235
Hospital Room Problem, 242

Intensity Problem, 222
Interest Compounding Problem, 327
Iodine Problem, 247
Ivan Problems, 498

Kilograms-to-Pounds Problem, 212

Ladder Problem, 467
Large and Small People Problem, 217
Law of Sines for Angles Problem, 478
Life Insurance Problems, 371
Lighthouse Problem, 468
Lightning Problem, 214
Linear Depreciation Problem, 58
Lion Hunting in Africa, 325
Little Brother Problem, 322
Louisiana Purchase Problem, 323
Lucky Card Problem, 366
Lumber Problem, 296

Man Overboard Problem, 204
Manhattan Problem, 322
Measles and Chicken Pox Problem, 360
Medication Problem, 220
Menu Problem, 358
Meteorite Tracking Problem, 274
Microwave Oven Problem, 237
Middleman Problem, 327
Milk Problem, 56
Milk Spoiling Problem, 166
Missile Problems, 470, 495
Month's Pay Problem, 324
Moon Crater Problem, 467
Mountain Height Problem, 494
Multi-Story Building Problem, 212
Multiple Choice Test Problems, 363, 370
Muscle-Building Problem, 321
Musical Scale Problem, 324

Nested Squares Problem, 325
Nuclear Reactor Problem, 370

Observation Tower Problem, 467
Office Building Problems, 126, 337
Oil Viscosity Problem, 295
Oil Well Problem, 495
Operation Problems, 360, 363
Paint Brush Problem, 326
Park Clean-Up Problem, 91
Parkinson's Law Problem, 214
Partial Sums by Computer, 315
Payload Problem, 297
Pebble-in-the-Tire Problems, 408, 423
Pedalboat Problem, 75
Pendulum Problem, 236
Perfect Solo Problem, 364
pH Problem, 156
Phoebe Small's Rocket Problems, 119, 166
Piggy Bank Problem, 323
Pile Driver Problem, 323
Pineapple Plantation Problem, 325
Pizza Problem, 220
Planetary Period Problem, 237
Population Problem, 164
Porpoising Problem, 427
Powers too Large for Calculators Problem, 156
Probability Distribution Problem, 503
Product of the Integers Problem, 513
Proper Divisors Problem, 365
Pyramid Problem, 470
Rabbit Problem, 165
Radiant Heat Problem, 215
Radio Dial Problem, 171
Radio Transmitter Problem, 214
Radioactive Brain Tracer Problem, 173
Radiotherapy Problems, 215, 468
Rainfall Problem, 147
Reaction Time Problem, 56
Reaction to Shock Problem, 242
Recycled Can Problem, 214
Relative Time Problem, 216
Rich Uncle Problem, 327
River Basin Problem, 234
Rock Formation Problem, 410
Roller Coaster Problems, 410, 425
Rotating Beacon Problem, 411
Ruby Problem, 216
Same Birthday Problem, 366
Seed Germination Problem, 369
Sequences by Computer Graphics, 307
Shark Problem, 218
Ship Power Problem, 234
Shoe Size Problem, 58
Silversword Problem, 359
Skin Diving Problem, 502
Snowflake Curve Problem, 326
Spaceship Problems, 122, 359, 409, 423
Speed on a Hill Problem, 57
Spike Heel Problem, 240
Stadium Problem, 259
Stopping Distance Problem, 217
Straight-Line Depreciation Problem, 323
Submarine Problem, 469
Sum of the Cubes Problems, 297, 514
Sum of the Data Problem, 513
Sum of the Even Numbers Problem, 512
Sum of the Odd Numbers Problem, 512
Sum of the Squares Problem, 297
Sum of the y-values Problem, 513
Sun Elevation Problem, 425
Sunlight Below the Water Problem, 170
Sunlight Problem, 147
Sunrise Problem, 411
Surveying Problems, 469, 497
Suspension Bridge Problem, 124
Sweepstakes Problem, 328
Swimming Problems, 494
Swinging Problem, 336
Target Practice Problem, 365
Tarzan Problems, 408, 423
Telephone Call Problem, 514
10-Speed Bike Problem, 241
Terminal Velocity Problem, 58
Thermal Expansion Problem, 57
Thoroughbreds and Quarter Horses Problem, 93
Thumbtack Problem, 363
Tidal Wave Problems, 408, 423
Tide Problem, 424
Tire Pump Problem, 236
Toothpaste Factory Problem, 234
Tornado Problem, 243
Torpedo Problem, 498
Traffic Light Problems, 358, 363
Tree Branch Problem, 324
Tree Trunk Problem, 235
Triangular Block Problem, 468
Tunnel Problem, 424
Uranium Fission Problem, 369
Vapor Pressure Problem, 173
Visibility Problem, 495
Visiting Problem, 358
Vitamin Problem, 91
Water Main Problem, 213
Water Pressure Problem, 212
Weight Above and Below the Earth, 217
Why Mammals Are the Way They Are, 242
World Series Problem, 364
Wrench Problem, 212

General Index

AAS, 477
Abscissa, 30
Absolute value, 7
in inequalities, 16
of a complex number, 285
of a perfect square, 105, 107
of a product, 526
of a real number, 7
of a sum, 538
of a vector, 485
Addition formulas, 442
Addition of Matrices, 527
Addition Property of Equality, 12, 19
Addition Property of Order, 14
Addition Property of Zero, 4
Additive identity, 3, 4
Additive inverse, 3, 4
Adjacent side, 466
Adjoint matrix, 528, 529
Alegbraic expression, 9
Ambiguous case, 479-481
Amplititude, 394, 400
Angle, coterminal, 381
of depression, 468
directed, 380
of elevation, 425
negative, 380
phase, 453
reference, 381
Antilogarithm, 150
Applications, (see mathematical models)
Arc measurement, 380-382
Area, of an ellipse, 259
of a triangle, 474
under a graph, 172
Argument, of a logarithm, 149
of a trigonometric function, 383
Arithmetic means, 308
Arithmetic sequence, 303-306
formula for, 304
nth term of, 539
Arithmetic series, 312-315
formula for the sum of, 313
partial sums of, 313, 540
Arrangement, 345
ASA, 477
Associativity, 3, 4
Asymptote, 31, 32
Axiom, of equality, 18
field, 3
of order, 18
Axis of symmetry, 100

Base, e as, 158
of logarithms, 149
of powers, 10
BASIC, 519
Binomial(s), 10
coefficients, 332
conjugate, 179
distribution, 361
Formula, 331-334
products, 10-11, 179
series, 330-334, 351, 540
series and combinations, 360
Braces, 8
Brackets, 8
Branch, 510, 521

Cancellation, in fractions, 194
Property of Equality for Addition, 20
Property of Equality for Multiplication, 21
Cartesian coordiante system, 29
Characteristic, 145, 154
Charles, Jacques, 59
Circles, 252-254, 265
Circular function(s), 383-422, 431-455
argument of, 383
graphs of, 393-401
inverses of, 412-417, 421
linear combination of, 452-453, 455
properties of, 454
values of, 385-387, 390-392
Circular permutation, 349
Closed interval, 457
Closure, 3, 4, 537
Coefficient, 10
Cofactor, 529
Cofunction Properties, 439-440
Column of a matrix, 527
Combinations, 350-353
and binomial series, 360
formula for, 351
Common difference, 303
Common factors, 181
Common logarithms, 154
Common ratio, 303
Commutativity, 3, 4
Comparison Axiom, 18
Completing the square, 102-103, 253
factoring by, 185-186
Complex conjugate, 284, 286
Complex number plane, 283
Complex numbers, 1, 283-284
absolute value of, 285
conjugate, 284, 286
operations with, 284
Components of a vector, 489-493
Composite Argument Properties, 441-442
Composite function, 238-239
Compound interest, 320
Computer program(s), 519-523
for arithmetic sequences, 307
for binomial distribution, 366
for continued radicals, 232
for evaluating rational functions, 203-204
for the Factor Theorem, 192
for factorials, 330
for factoring quadratic expressions, 186-187
for linear function equations, 52
for partial sums of series, 312
for solution of triangles, 484-485
for solving quadratic equations, 110
for solving systems of linear equations 71-72
for synthetic division, 294
for synthetic substitution, 293
for transforming matrices, 85-86
for zeros of a function, 294
Conclusion, 19
Conic sections, 264-265
Conjugate binomials, 179
Conjugate of a complex number, 284
Conjugate hyperbolas, 261
Constant(s), 6, 10
Constant of proportionality, 207
Continued fractions, 207
Continued radicals, 232
Converge, 316
Convergent geometric series, 315
Converse of a statement, 19
Coordinates, 29
Corollary, 20
Cosecant function, 384
Cosine function, 383-384
Cosines, Law of, 471
Cotangent function, 384
Coterminal angles, 381
Counter, 510
Counting numbers, 2
Counting principles, 341
Cramer's Rule, 70, 83
Critical points, 422
Cubic polynomial, 10
function, 279
Cumulative Reviews
Chapters 1-5, 126-129
Chapters 6-8, 245-249
Chapters 9-12, 374-377
Chapters 13-15, 501-502
Final Examination, 503-508
Cycle, 394

Degree of a polynomial, 10
Delta, 45
Dependent equations, 67, 71, 79
Dependent variable, 30
Descartes, Rene, 29
Descartes' Rule of Signs, 191, 531
Determinants, minor, 82, 529
second-order, 69-70
third order, 81-83
Difference of like powers, 190-191
Difference of two squares, 179, 182
Digits, Facing 1
Direct variation, 208
Directed angles, 380
coterminal, 381
standard position of, 380
Discriminant, 109
of a conic section, 269
test for prime quadratics, 183
Distance formula, 252-253
Distributivity, 3, 4

Divergent geometric series, 317
Division, 6
synthetic, 294
by zero, 200
Domain, 6
of a function, 32
Double Argument Properties, 444

e, 156
as a base, 158
Eccentricity, 268
Element of a matrix, 527
Ellipses, 254-258, 265
Empty set, 12
Equations, 11-13
dependent, 67, 79
equivalent, 12
exponential, 156-157
extraneous solutions of, 13
fractional, 204-206
general, 49
of horizontal lines, 51
inconsistent, 67, 79
linear, 29, 43, 47-52, 76-78
of linear functions, 43, 47-52
particular, 49
quadratic, 105-109, 114-116
radical, 229-231
solution set of, 11
solving, 11-13
systems of, 65-90, 269-275
trigonometric, 456-459
of vertical lines, 51
Equivalent equations, 12
Equivalent expressions, 8
Equivalent systems, 66
Evaluating expressions, 6
Even function, 438
Even number, 1
Event(s), 340
independent, 356
outcome of, 340
overlapping, 344
probability of, 340
Expanding a binomial power, 330
Explicit, 54
Exponent, 10
fractional, 137
irrational, 156
negative, 135
zero, 135
Exponential constant, 160
Exponential equations, 156-157
Exponential functions, 131-140
base of, 131
evaluation of, 159-162
graphs of, 131, 132, 160-162
as mathematical models, 162-172
Exponentiation, 6
for irrational exponents, 144
for negative and zero exponents, 136-137
for positive integer exponents, 132
properties of, 133-134, 537
for rational exponents, 137-140
Expressions, 5, 6
equivalent, 8
evaluating, 6
factoring, 178
rational algebraic, 177
simplifying, 8, 136-137, 139-140
Extraneous solutions, 13, 205
Extrapolation, 59

Factor Theorem, 189, 293
Factorials, 328-329
sequence of, 303
Factoring, by completing the square, 185-186
by grouping, 182
difference of two squares, 182, 185
higher-degree polynomials, 188-191
polynomials, 181-191
quadratic trinomials, 182, 183, 186, 287
a rabbit, 303, 436
Fibonacci sequence, 303
Field Axioms, 3-5
Flag, 513
Flow charts, 509-516
Fractional equations, 204-206
Fractional exponents, 139
Fractions, partial, 199-200
continued, 207
Multiplication Property of, 23, 195
Focal radius, of an ellipse, 257
of a hyperbola, 261
Focus, of an ellipse, 256
of a parabola, 266
of a hyperbola, 261
FOIL technique, 179
Functions, 34-37
circular, 383-388
combined variation, 238
composite, 238-239
constant, 43
cubic, 279
exponential, 131-140
even, 437, 438
f(x) terminology for, 73
graphs of, 35-37, 44-47
higher degree, 279, 289-295
inverses of, 148-149
irrational algebraic, 233-234
linear, 43-55
logarithmic, 149-159
odd, 437, 438
periodic, 379
polynomial, 41
probability distribution, 362
quadratic, 96, 99-119
of a random variable, 361-362
rational algebraic, 177-178
sequence, 301-307
sinusoidal, 398
with several independent variables, 238
trigonometric, 383-388
variation, 207-212, 233-234, 238-239
zeros of, 292
Fundamental Theorem of Algebra, 292
f(x) terminology, 72

Gauss, Carl Friederich, 313
GCF, 187
General equation, 49
Geometric mean(s), 308
Geometric sequences, 303-306
nth term of, 528
Geometric series, 312-315
convergent, 315-318
formula for sums of, 314
Graphs,
of functions, (see the kind of function.)
number-line, 15-17
reasonable real-world, 30-33, 379
three-variable equations, 77-78
two-variable equations, 29
two-variable inequalities, 86-87
Greatest common factor, 187
Greek letters, 380
Gulliver, 218

Half-argument Properties, 447
Harmonic sequences, 303, 318
Heron's Formula, 476
Hippopotamus, 59, 92, 219
Higher degree function, 279
"Hole" in a graph, 200
Horizontal line, 51
Hyperbolas, 259-263, 265
Hypotenuse, 466
Hypothesis, 19

i, 280
powers of, 281
square root of, 285
Identities, trigonometric, 434-436
Identity element, 3, 4
Identity matrix, 528
If and only if statements, 20
Imaginary numbers, 1, 280-282
Included angle, 472
Inconsistent equations, 67, 71, 79
Independent equations, 67, 71, 79
Independent variable, 30
Indeterminate, 201
Induction, mathematical, 533-537
Inequalities, 14-17
linear, 86-87
quadratic, 128, 254, 256, 277
systems, 86-87
Infinite geometric series, (See convergent geometric series.)
Infinitely large, 201
Input, 520

General Index

Integers, 1
Intercept, 45, 78
  form of an equation, 52
Interpolation, 59, 391
Intersection, of curves, 269
  of lines, 66
  of planes, 79
  of regions, 86
  of sets, 344
Intervals, open and closed, 457
Inverse, 4
  additive, 3, 5
  circular functions, 412
  of a function, 147
  of a matrix, 528
  multiplicative, 4
  trigonometric functions, 415
  variation, 208
Irrational, algebraic functions, 233
  exponents, 144
  numbers, 1, 225
  radicals, 225
Irreversible step, 13

Lagging phase, 410
Law, of Cosines, 471
  of Diminishing Returns, 170
  of Sines, 476
  of Tangents, 502
LCM, 187
Leading phase, 410
Least common multiple, 187
Lemma, 20
Like, odd powers, 191
Limit of a series, 316
Linear combination, 66
  of cosine and sine, 452
  for solving systems, 66, 80, 271
Linear function(s), 43-56
  equations of, 43, 47-52
  intercepts of, 45, 52
  as mathematical models, 53-56
  slope of, 44
Linear inequalities, 86-87
Linear polynomials, 10
Linear programming, 88-90
Linear systems
  2-variable equations, 66, 69
  2-variable inequalities, 86
  3-variable equations, 79, 81
  4-variable equations, 81, 83
  5-variable equations, 86
Logarithms, 149-159
  argument of, 149
  base of, 149
  base of, 149
  base e, 151, 158
  base 10, 154-156
  characteristic of, 154
  mantissa of, 154
  natural, 151, 158-159
  powers by, 155
  properties of, 151-153, 158
Loop, 510, 522

Major axis, 255
Mantissa, 145, 156
Mathematical expectation, 367-368
Mathematical induction, 533-537
Mathematical models, 53
  exponential functions as, 162-164
  higher degree functions as, 294-295
  inverse circular relations as, 422-423
  linear functions as, 53-56
  quadratic functions as, 117-119
  sequences and series as, 319-321
  sinusoidal functions as, 405-407
  systems of linear functions as, 73-74
  variation functions as, 207-212, 233-234, 238-239
Matrix, 84, 527-530
Mean proportional, 310
Means, 308
Memory, computer, 507
Mental multiplication, 180-181
Method of detached coefficients, (See Matrix)
Minor axis, 255
Minor determinant, 82, 529
Monomial, 10
Multiplication of matrices, 527-528
Multiplication Property,
  of Equality, 12, 21
  of Fractions, 23, 195
  of Negative One, 22
  of One, 4
  of Order, 15, 27
  of Zero, 12, 21
Multiplicative identity, 3, 4
Multiplicative inverse, 3, 4
Multiple Argument Properties, 444

Natural numbers, 1
Negative numbers, 1
Negative of a sum, 23
nth partial sum, 310
nth root, 138
Null set, (See Empty set.)
Number, kinds of, Facing 1
Number line, 1

Oblique triangles, 470-474, 476-482
Odd number, 1
Odd function, 438
Open interval, 457
Open sentence, 65
Operations, 3
  sequence of, 7
Opposite of a real number, 4
Opposite side, 466
Optimum point, 90
Order, 15, 18, 26
  of a matrix, 527
Ordered pair, 29
Ordered triple, 76
Ordinate, 29
Origin, 29
Outcome, 340
Output, 507
Overlapping events, 344

Parabolas, 99-102, 263-266, 277
  directrix, 266, 277
  focus, 266, 277
  general equation, 99, 263
  vertex form, 102, 264
Paragraph proof, 19
Parallel lines, 50
Partial fractions, 200
Partial sum, 310
Particular equation, 49
Pascal's Triangle, 331, 332
Pentagonal numbers, 539
Periodic functions, 379
  cycle of, 394
  period of, 394, 400
Permutations, 345-346
  circular, 349
  formula for, 351
  with repeated elements, 349
Perpendicular lines, 126-127
Phase displacement, 399, 400
Phase angle, 453
Point-slope form of an equation, 47-48
Polygonal Inequality, 538
Polynomial(s), 9-11
  common factors, 181
  constant, 10
  cubic, 10
  degree of, 10
  division of, 187-188
  factoring, 181-186, 188-191
  function, 41
  higher degree, 188, 279
  linear, 10
  multiplying, 10-11, 179-180
  numerical coefficient, 10
  prime, 183
  quadratic, 10
  quartic, 10
  quintic, 10
  terms of, 10
  zeros of, 292
Positive numbers, 1
Postulate, (See Axiom)
Powers, 10
  with negative base, 141
Prime, number, 301
  polynomial, 183
Proportionality constant, 207
Probability, 339-372
  counting principles of, 341-342, 344
  definition of, 340

distribution, 362
properties of, 355-357
Progression, 303
Properties,
of equality, 18
of exponentiation, 133
of logarithms, 151, 158
of numbers, 3
of order, 15, 18, 26
provable from axioms, 17
of trigonometric and circular functions, 430
of variation functions, 209
Pyramidal numbers, 312, 527
Pythagorean Properties, 432

Q.E.D., 19, 435
Quadratic equations, 105-109
completing the square, 105-106
discriminant of, 109
one-variable, 105-109
Quadratic Formula, 108-109
sum and product of the solutions, 289
systems of, 269-273
Quadratic Formula, 108-109
Quadratic functions, 96, 99-119
equations of, 114-116
evaluating, 111-113
graphs of, 100, 114-116
as mathematical models, 117-119
vertex form, 102, 110, 264
x-intercepts of, 104-105
Quadratic relations, 251-277
inequalities, 128, 254, 256, 277
systems of, 269-273
xy-term, 268
Quartic, 10
function, 279
Quintic, 10
function, 293
Quotient Properties, 431

Radians, 389
Radical equations, 229-231
Radicals, 1, 107, 137-138, 226-228
continued, 232
irrational, 225
simplifying, 226-228
Radicand, 138
Random experiment, 340
Random variable, 362
Range of a function, 32
Rational algebraic expresessions, 177
products and quotients of, 193-194
simplifying, 193-194
sums and differences of, 197-198
Rational algebraic functions, 177-178, 200-202
Rational exponents, 139
Rational numbers, 1
Rational Root Theorem, 190
Rationalizing, the denominator, 227
the numerator, 229
Real number line, 1
Real numbers, 1
Field Axioms for, 3-5
properties of, 17-23
Real part of a complex number, 283
Reciprocal, 4
exponents, 138
of a product, 21, 195
Properties, of trig functions, 431
Reference angle, 381
Reflexive Property, 18
Relations, 30-33
domain of, 32
graphs of, 30-33
inverses of, 142
Linear, 86-87
quadratic, 251
range of, 32
Relatively prime, 317
Remainder Theorem, 290
Repeating decimals, 317-318
Replacement set, 32
Reversal, of sign, 531
Right triangles, 465-467
Rise, 45
Root of an equation, 190, 531
of a number, 138
index, 138
Rotation, 380-382
Row of a matrix, 527
Run, 45

Sample space, 340
SAS, 477
Scalar, 484
Scientific notation, 144-147
Secant function, 384
Secant line, 223
Semiperimeter, 476
Sequence(s), 301-306
arithmetic, 303-306
of factorials, 303
Fibonacci, 303
geometric, 303-306
harmonic, 303, 318
of operations, 7
of triangular numbers, 303
Series, 310-318
arithmetic, 312-315
binomial, 330-334
convergent, 315
geometric, 312-315
harmonic, 318
partial sums of, 310
Sigma notation, 310
Sign reversal, 531
Significant digits, 145
Simple radical form, 227
Simplified,
rational expressions, 194
trigonometric expressions, 453
Simulated computer memory, 507
Simultaneous equations (see Systems of equations)
Sin 18°, 449
Sine function, 383-384
Sines, Law of, 476
Sinusoids, 394, 398-403
Slope, 44
Formula, 45
–intercept form, 43, 48
of parallel lines, 50
of perpendicular lines, 126-127
of a tangent line, 223
Solving, an equation, 12
a triangle, 482
Solution, of an equation, 11
set, 11
of a system, 65
of a trigonometric equation, 456
Space coordinates, 77
Special arguments, 386
Special products, 178-180
Splitting the middle term, 183
Square of a binomial, 103, 179
Square roots, 107, 281-282
SSA, 477
SSS, 477
Standard position of an arc or angle, 380
Subscripted variable, 517
Substitution into sums and products, 19
Subtraction, 6
of matrices, 527
Sum of like powers, 190-191
Sum and Product Properties, 450
Summation, 310-311
Symmetry, 18
Synthetic division, 294
Synthetic substitution, 289-290, 293
Systems of linear equations, 65-90
dependent, 67-68, 71, 81
equivalent, 65, 84
five-variable, 86
four-variable, 81, 83
inconsistent, 67-68, 71, 81
independent, 67-68
matrix solution of, 83-85
solution set of, 65
solving by linear combination, 66, 80
solving by determinants, 70, 81-83
three-variable, 80
two-variable, 65-74
Systems of linear inequalities, 65, 86-87
Systems of quadratic equations, 269-273

Tables, 542-553
Tangent function, 384
Tangent line, slope of, 223
Tangents, Law of, 497
Taylor series, 497
Term(s), 10
of a binomial series, 333
of a sequence or series, 301-306

Trace, 77-78
Transcendental numbers, 1, 151, 390, 426
Transitivity, 18, 26, 537
Transpose of a matrix, 529
Triangle Inequality, 538
Triangles, 465-494
  area of, 474-475
  general solution of, 482
  oblique, 470-474, 476-481
  right, 465-467
Triangular numbers, 303, 312, 538
Trichotomy, 18
Trigonometric equations, 456-459
Trigonometric functions, 383-422, 431-459
  argument of, 383
  cofunction properties of, 439-440, 454
  composite argument properties of, 440-442, 454
  double argument properties of, 444-445, 454
  graphs of, 393-401
  half-argument properties of, 447-448, 454
  identities, 434-436
  inverse, 413, 415-416, 419-421
  linear combination of, 452-453, 455
  odd and even properties of, 438-439, 454
  Pythagorean properties of, 432, 454
  quotient properties of, 431, 454
  reciprocal properties of, 431, 454
  sum and product properties of, 450-451, 454
  values of, 385-387, 390-392
Trinomials, 10

Unique, 3
Unit circle, 380
Unit imaginary number, 281
Upper Bound Theorem, 191, 531
Upside-down division, 139

Variable, 6
  dependent, 30, 73
  domain of, 6, 12
  independent, 30
  random, 362
  subscripted, 522
Variation functions, 207-212, 238-239
  combined, 238
  direct and inverse, 208
  irrational, 233-234
  property of, 209
Vectors, 484-487
Vertex,
  of an ellipse, 255
  of a hyperbola, 260
  of a parabola, 100, 110
Vertex form, 102, 110, 264
Vertical line, 46, 51
Vertical shift, 400
Vinculum, 8

Wavelength, 408

x-axis, 29, 77
x-intercept, 45, 78, 104
xy-trace, 77-78

y-axis, 29, 77
y-intercept, 45, 78

z-axis, 77
Zero, division by, 200
Zero factorial, 329
Zeros of a function, 292
Zero matrix, 530
Zero of a polynomial, 292, 530
z-intercept, 78

Answers to Selected Problems

CHAPTER 1
PRELIMINARY INFORMATION

Exercise 1-1, page 2; Sets of Numbers

1. See Chart facing page 1. 2. e.g.: a. 71 b. 3 c. –58 d. 17.23 e. –9-2/3 f. 5-3/8 g. $\sqrt{3}$ h. $\sqrt{-9}$ i. 56.732 j. 1027 k. 19 l. π 5. {natural nos.}, since "counting" was probably the first thing people did with nos. 7. Yes. 2.718 = 2718/1000 which *is* a ratio of 2 integers. 9. 0

Exercise 1-2, page 5; The Field Axioms

1. a. Additive identity element is 0. b. Multiplicative identity element is 1. 3. See chart in text. 5. The answer to the *same* addition problem or multiplication problem always comes out the *same*. It isn't one thing today and something else tomorrow. 7. *Calvin* is right. He *associated* $3(x + 4)(x + 7) = [3(x + 40)](x + 7)$, and got $(3x + 12)(x + 7)$. Phoebe wrote $[3(x + 4)] \cdot [3(x + 7)]$, thus "distributing" mult. over mult. 9. a. Associativity for + b. Closure under $\cdot$ c. Commutativity for + d. Commutativity for $\cdot$ e. Distributivity of $\cdot$ over + f. Additive identity g. Additive inverses h. Multiplicative identity i. Multiplicative inverses

Exercise 1-3, page 8; Variables and Expressions

1. 47 3. 10 5. 8 7. 7 9. 16 11. a. 7 b. –13 13. a. 1 b. 14 15. a. –17 b. 18 17. a. 0 b. 21 19. a. 2 b. 27 21. a. –5 b. 40 23. a. 1 b. 11 25. $4 - x$ 27. $21x - 42$ 29. $4x + 17$ 31. 5 33. $32 - 6x$ 35. $3x + 3$ 37. $2y^2 - 2xy$ 39. x 42. Reflexive Axiom.

Exercise 1-4, page 11; Polynomials

1. Yes. Quartic trinomial 3. Yes. Quadratic binomial 5. No. ÷ by variable 7. Yes. Linear binomial 9. No. ÷ by variable 11. Yes. Constant monomial 13. Yes. Four-term cubic 15. Yes. Quintic binomial 17. No. $\sqrt{\ }$ of variable 19. Yes. Linear binomial 21. Yes. 10th degree binomial 23. Yes. Monomial, *no* degree 25. $x^2 + 4x - 21$ 27. $2x^2 + 7x - 4$ 29. $6x^2 - 37x + 56$ 31. $4x^2 - 20x + 25$ 33. $4x^2 - 20x + 25$

Exercise 1-5, page 13; Equations

1. {–5} 3. ϕ (–4 is not positive) 5. {0} 7. a. {–11/3} b. ϕ (–11/3 is not an integer.) 9. a. {–4} (+4 is out of domain.) b. {4, –4} 11. a. {1/3, –1/3} b. ϕ (1/3 and –1/3 are not integers) 13. {–3, 2/3} 15. {–1} (5/2 is not an integer.) 17. {2/3} (–3 is not positive.) 19. {–3/2, 5, –1} 21. {0, 1/2, –4} 23. {7, –7} 25. ϕ 27. {2, –8} 29. {3, –5/2} 31. {2, 16/9} 33. a. S = {–7, 5/3} for *both* equations. b. *Equivalent*. 35. a. $5 + 3 < 7 + 3 \Rightarrow 8 < 10$; true b. $5 + (-8) < 7 + (-8) \Rightarrow -3 < -1$; true c. If $a < b$, then $a + c < b + c$, and if $a > b$, then $a + c > b + c$.

Exercise 1-6, page 17; Inequalities

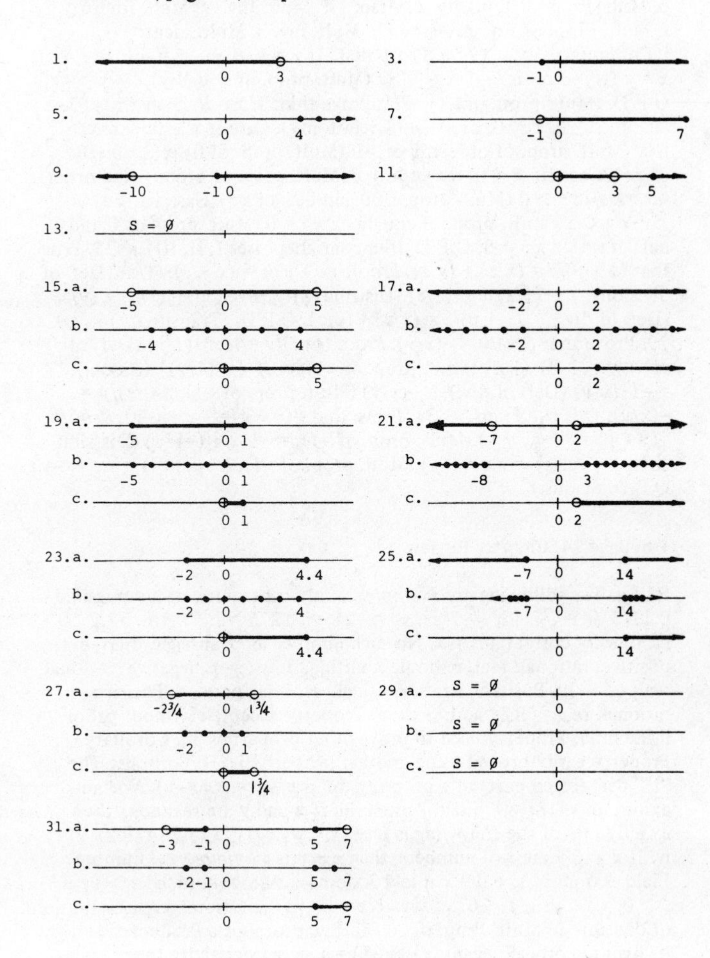

Exercise 1-7, page 20; Properties Provable from Axioms

1-7 #1. Check the index for these properties. 3. An axiom is *assumed* to be true without proof. Other properties are provable from the axioms. 5. Reflexive axiom 9. If $xz = yz$ *and* $z \neq 0$, then $x = y$. 10. The two properties *cannot* be combined using "if and only if" since the hypothesis of the second one is *not* exactly the same as the conclusion of the first one. 11. Prove that $-(-x) = x$. *Proof:* $-(-x) + (-x) = 0$ (Additive inverses), $x + (-x) = 0$ (Additive inverses), $0 = x + (-x)$ (Symmetry), $\therefore -(-x) + (-x) = x + (-x)$ (Transitivity), $\therefore -(-x) = x$, Q.E.D. (Converse of Add. Prop. of Equality) 13. a. Mult. Inverses

b. Mult. Prop. of Equality c. Assoc. d. Mult. Inv. e. Mult. Identity f. Mult. Prop. of Eq. g. Assoc. h. Mult. inv. i. Mult. ident. j. Commutativity 19. a. Prove that: If $x = y$, then $-x = -y$. b. *Proof:* $x = y$ (Hypothesis), $-1 \cdot x = -1 \cdot y$ (Mult. prop. of equality), $\therefore -x = -y$, Q.E.D. (Mult. prop. of -1) 21. Prove that: If $x \in R$, then $x^2 \geq 0$. *Proof:* $x > 0, x < 0$, or $x = 0$ (Trichotomy). Case I: $x > 0$: $\therefore x \cdot x > 0 \cdot x$ (Mult. prop. of order), $x \cdot x > 0$ (Mult. prop. of 0), $x^2 > 0$ (Def. of x^2). Case II: $x < 0$: $\therefore x \cdot x > 0 \cdot x$ (Mult. prop. of order – the order *reverses*), $x^2 > 0$ (Mult. prop. of 0 and def. of x^2). Case III: $x = 0$: $\therefore x \cdot x = 0 \cdot x$ (Mult. prop. of equality), $x^2 = 0$ (Mult. prop. of 0 and def. of x^2), $\therefore x^2 \geq 0$, Q.E.D. (Summarizing cases I, II, III). 23. Prove that $(x + y)/z = (x/z) + (y/z)$. *Proof:* $(x + y)/z = (x + y) \cdot (1/z)$ (Def. of division), $= x \cdot (1/z) + y \cdot (1/z)$ (Distributivity, $\cdot$ over +), $= (x/z) + (y/z)$ (Def. of div.), $\therefore (x + y)/z = (x/z) + (y/z)$, Q.E.D. (Transitivity) 29. Prove that $(-x)/y = -(x/y)$. *Proof:* $(-x)/y = (-x) \cdot (1/y)$ (Def. of div.), $= (-1 \cdot x) \cdot (1/y)$ (Mult. prop. of -1), $= -1 \cdot [x \cdot (1/y)]$ (Assoc.), $= -1 \cdot (x/y)$ (Def. of div.), $= -(x/y)$ (Mult. prop. of -1), $\therefore (-x)/y = -(x/y)$, Q.E.D. (Trans.) 31. Prove that $-(x + y) = -x + (-y)$. *Proof:* $-(x + y) = -1 \cdot (x + y)$ (Mult. prop. of -1), $= -1 \cdot x + (-1 \cdot y)$ (Distributivity, $\cdot$ over +), $= -x + (-y)$ (Mult. prop. of -1), $\therefore -(x + y) = -x + (-y)$, Q.E.D. (Trans.)

1-8, page 24; Chapter Review

R1. a. The following are *examples*. Many other answers are possible. i. 11/3 ii. $-\sqrt{7}$ iii. $\sqrt{-7}$ iv. π v. -24 vi. 1275 vii. 37 viii. 37.2 ix. 0 (*only* correct ans.) x. No such number b. i. Integer, digit, even, positive, rational, real, natural, counting. i. Integer, negative, rational, real, odd. iii. Positive, irrational, real. iv. Imaginary. v. Positive, rational, real. R2. a. i. Axiom: Property accepted without proof ii. Lemma: Property used to prove other properties iii. Corollary: Property easily proved using another property iv. Hypothesis: The "if" part (*given* part) of a property b. i. $x - y = x + (-y)$. *Not* an axiom ii. $x \cdot 1/x = 1$. Field Axiom iii. If x and y are real nos., then exactly *one* of the following is true: $x < y, x = y, x > y$. Axiom iv. If x and y are real numbers, then $x + y$ is a *unique*, *real* number. Field Axiom v. $x + 0 = x$. Field Axiom vi. If $x > y$, then $xz > yz$ if $z > 0, xz < yz$ if $z < 0, xz = yz$ if $z = 0$. *Not* an axiom. c. i. Definition of division ii. Mult. Prop. of -1 iii. Reciprocal of a Product iv. -1 is its own reciprocal because $(-1)(-1) = 1$ v. Associativity for $\cdot$ vi. Commutativity for $\cdot$ vii. Associativity for $\cdot$ viii. Def. of div. ix. Mult. Prop. of -1 x. Transitivity for = d. Prove that $a(b + c + d) = ab + ac + ad$. *Proof:* $a(b + c + d) = a(b + (c + d))$ (Associativity for +), $= ab + a(c + d)$ (Distributive Axiom), $= ab + ac + ad$ (Distributive Axiom), $\therefore a(b + c + d) = ab + ac + ad$ (Transitivity for =) R3. a. i. Cubic binomial ii. Not a polynomial. Division by a variable. iii. Not a polynomial. Square root of a variable. iv. Quartic binomial. v. Quintic monomial vi. Quadratic trinomial b. i. 9 ii. 8 iii. 21 iv. $3x^2 - 17x - 56$ v. $39x - 121$ c. Evaluations for $x = 5$, then $x = -4$: i. 7; -20 ii. 0; 18 iii. 76; 67 R4. a. i. $\{-6\}$ ii. ϕ (11/5 is *not* an integer.) iii. $\{-3, 2/3\}$ iv. $\{3/2\}$ (-3 is *not* positive.) v. $\{9, -9\}$

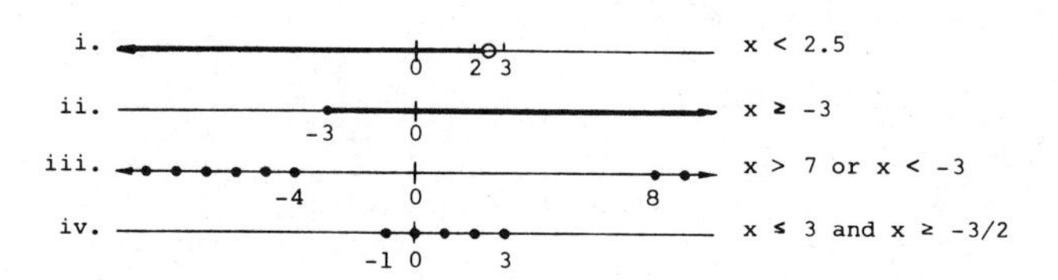

CHAPTER 2
FUNCTIONS AND RELATIONS

Exercise 2-1, page 29; Graphs of Equations with Two Variables

3. $y = -1$
4. $x = 2, y = -3; x = -1, y = -5; x = -4, y = -7$ 5. See graph in Problem 3.

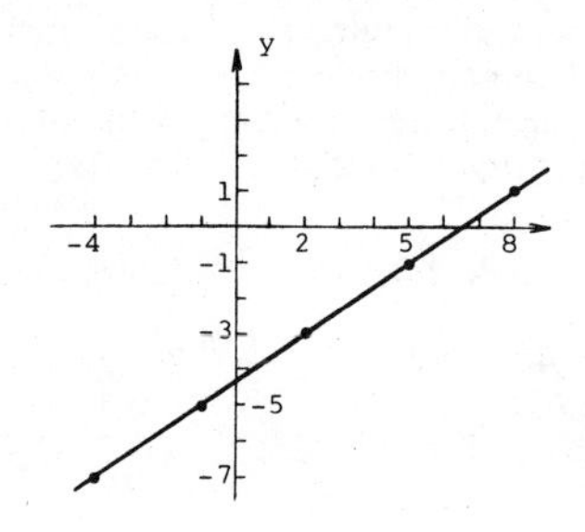

Exercise 2-2, page 33; Mathematical Relations

Note: The following graphs are *reasonable* rather than "right." You may find ways to draw graphs that are even *more* reasonable!

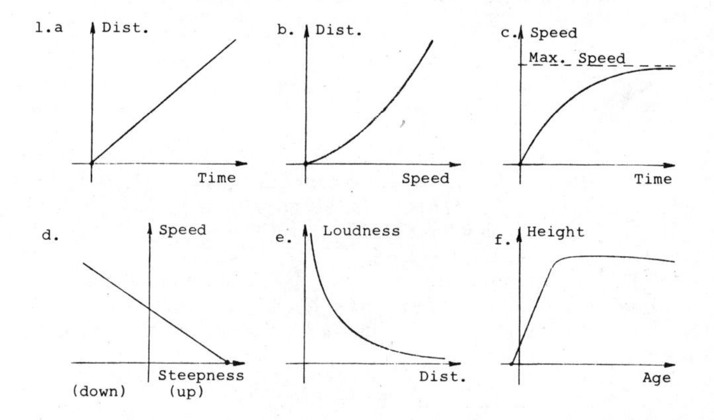

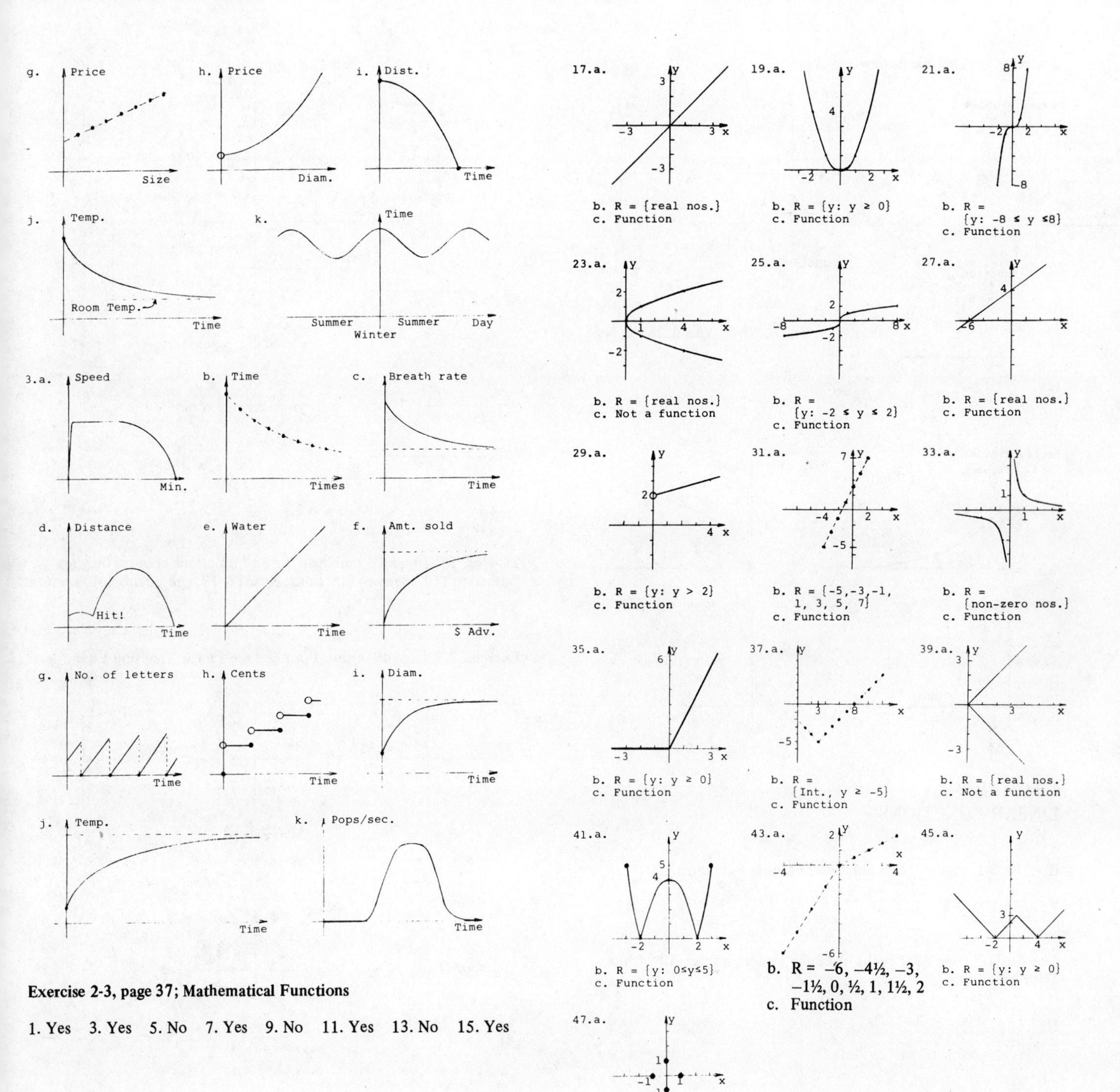

Exercise 2-3, page 37; Mathematical Functions

1. Yes 3. Yes 5. No 7. Yes 9. No 11. Yes 13. No 15. Yes

2-4, page 39; Chapter Review

Review Problem 1

a.

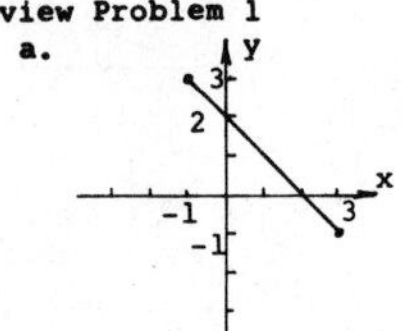

R = {y: $-1 \le y \le 3$}
Function

b.

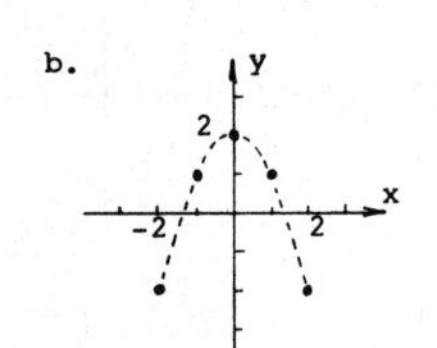

R = {-2, 1, 2}
Function

c.

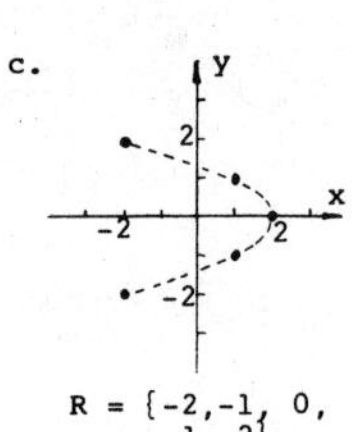

R = {-2, -1, 0, 1, 2}
Not a function

d.

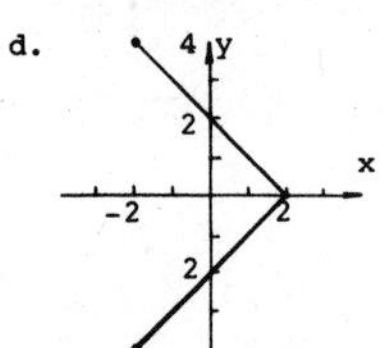

R2. a. Yes b. No c. Yes d. No e. No f. Yes

Review Problem 3

a.

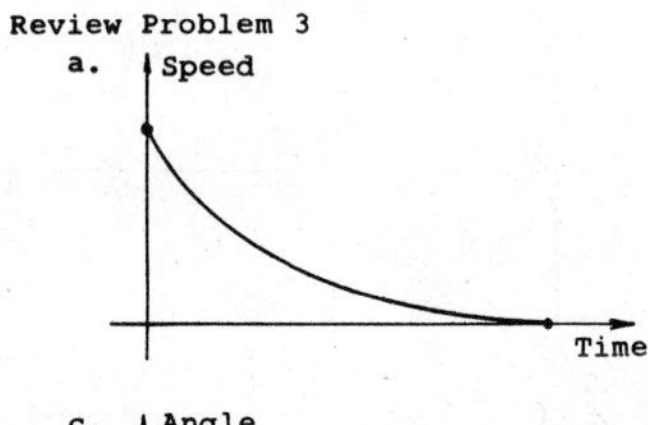

b.

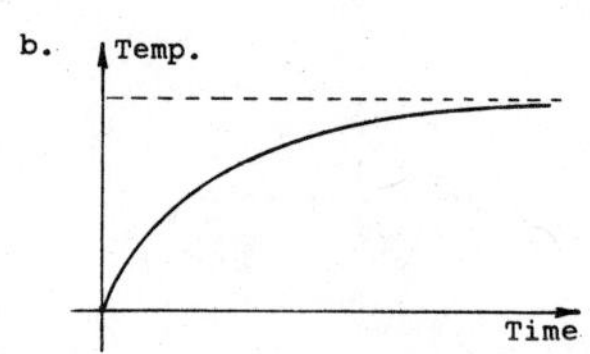

c.

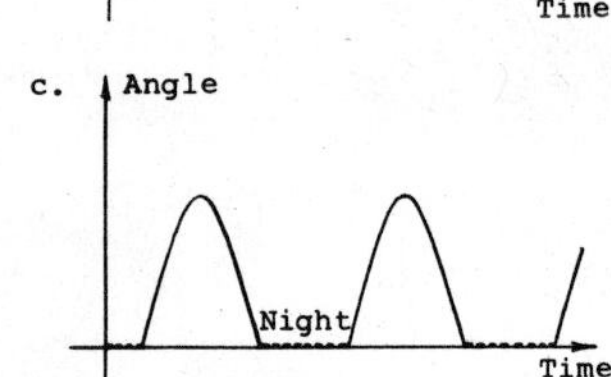

d.

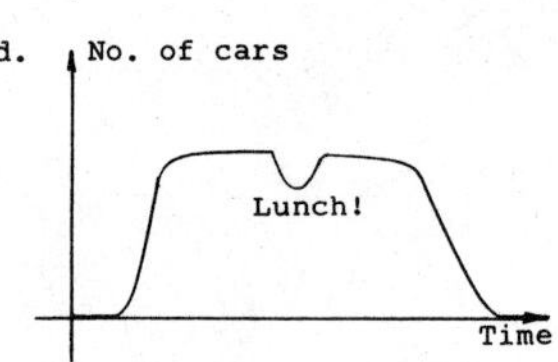

CHAPTER 3
LINEAR FUNCTIONS

Exercise 3-1, page 43; Introduction to Linear Functions

Read Section 3-2.

Exercise 3-2, page 47; Graphs of Linear Function from All within Their Equations

1.

$y = \frac{3}{5}x + 3$

3.

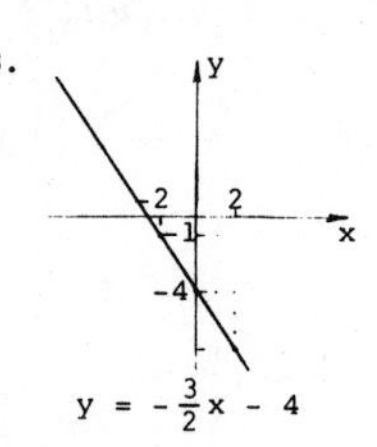

$y = -\frac{3}{2}x - 4$

5.

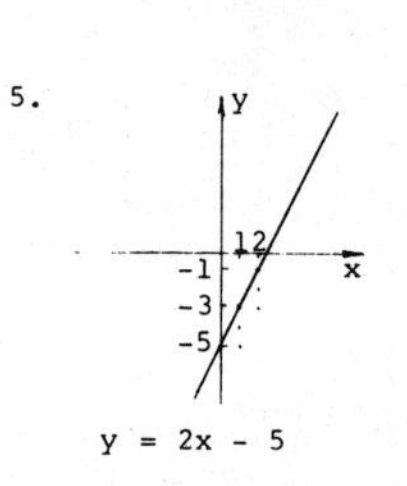

$y = 2x - 5$

7.

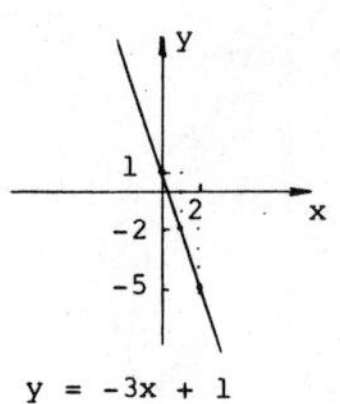

$y = -3x + 1$

9.

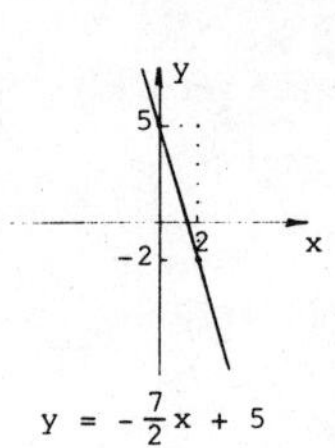

$y = -\frac{7}{2}x + 5$

11.

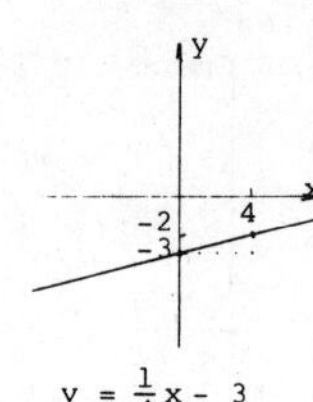

$y = \frac{1}{4}x - 3$

13.

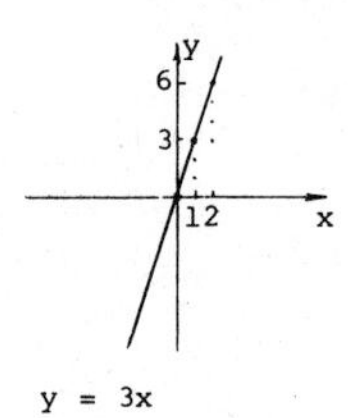

$y = 3x$

15.

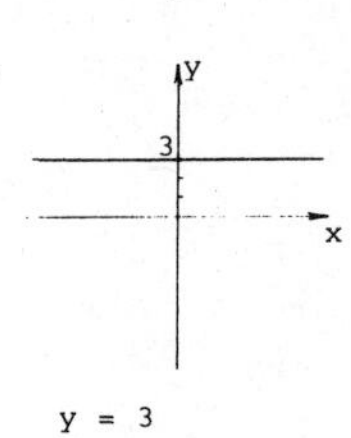

$y = 3$

17.

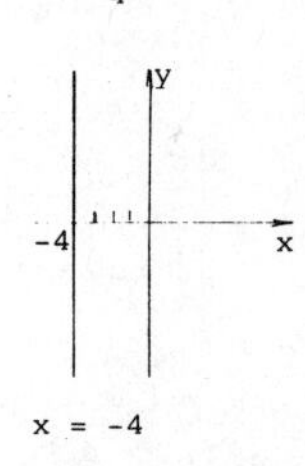

$x = -4$

19.

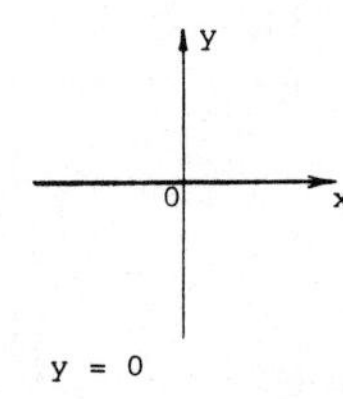

$y = 0$

21. Nos. 15, 16, and 19 are not "linear" functions because they are "constant" (0 · degree) functions. Nos. 17, 18, and 20 are not *functions.*

Exercise 3-3, page 49; Other Forms of the Linear Function Equation

1.a.

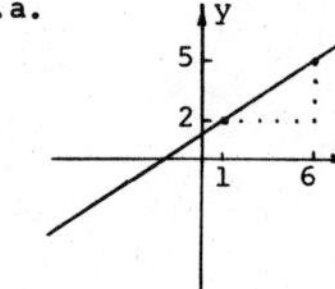

b. $y = \frac{3}{5}x + \frac{7}{5}$

c. $3x - 5y = -7$

3.a.

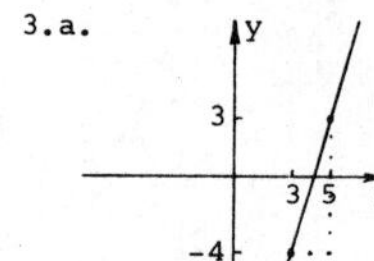

b. $y = \frac{7}{2}x - \frac{29}{2}$

c. $7x - 2y = 29$

5.a.

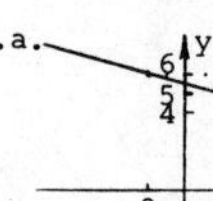

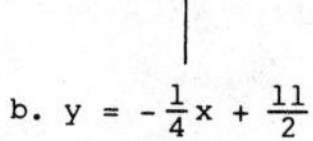

b. $y = -\frac{1}{4}x + \frac{11}{2}$

c. $x + 4y = 22$

7.a.

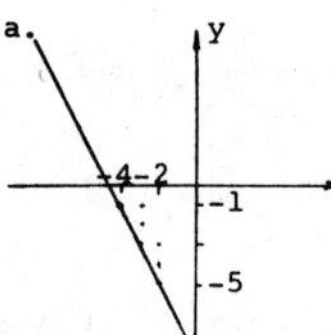

b. $y = -2x - 9$

c. $2x + y = -9$

9.a.

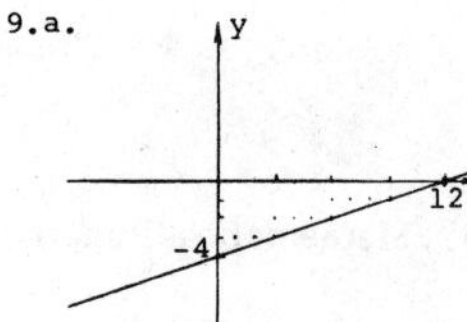

b. $y = \frac{1}{3}x - 4$

c. $x - 3y = 12$

11. $y - 7 = -3(x - 5)$ 13. $y - 5 = (9/13)(x + 2)$

Exercise 3-4, page 51; Equations of Linear Functions from Their Graphs

1. a. $y = -5x + 21$ b. $y = -5x + 21$ c. $5x + y = 21$ 3. a. $y - 7 = 11(x - 3)$ b. $y = 11x - 26$ c. $11x - y = 26$ 5. a. $y + 5 = -6(x - 4)$ b. $y = -6x + 19$ c. $6x + y = 19$ 7. a. $y - 7 = (3/2)(x - 1)$ b. $y = (3/2)x + (11/2)$ c. $3x - 2y = -11$ 9. a. $y + 4 = (6/7)(x - 2)$ b. $y = (6/7)x - (40/7)$ c. $6x - 7y = 40$ 11. a. $y - 8 = 7(x - 5)$ b. $y = 7x - 27$ c. $7x - y = 27$ 13. a. $y - 8 = -(2/3)(x - 5)$ b. $y = -(2/3)x + (34/3)$ c. $2x + 3y = 34$ 15. a. $y - 0 = -(2/3)(x - 5)$ b. $y = -(2/3)x + (10/3)$ c. $2x + 3y = 10$ 17. a. $y = 0.315x$ b. $y = 0.315x$ b. $y = 0.315x + 0$ c. $315x - 1000y = 0$ 19. a. $y = 9$ b. $y = 0x + 9$ c. $0x + y = 9$ 21. a. $x = -8$ b. (can't be done) c. $x + 0y = -8$ 23. All slopes $= -1$, so particular eqn. is $x + y = 8$. 25. All slopes $= 3/2$, so particular eqn. is $3x - 2y = -5$. 27. Intercept Form a. If $x = 0$, then $(0/a) + (y/b) = 1$, so $y = b$. If $y = 0$, then $(x/a) + (0/b) = 1$, so $x = a$. $\therefore$ a and b are x- and y-intercepts, respectively. b. i. $5x + 3y = 15$ ii. $y = -(5/3)x + 5$ c. $(x/3) - (y/12) = 1$. x-int. is 3, y-int. is -12.

Exercise 3-5, page 56; Linear Functions as Mathematical Models

1. Milk Problem a. Let c = no. of cents for carton of milk. Let q = no. of quarts carton holds. $c = 38q + 19$ b. i. $q = 38$ cents ii. $c = \$4.75$ iii. c = about 59 cents c. 4.5 quarts

d.

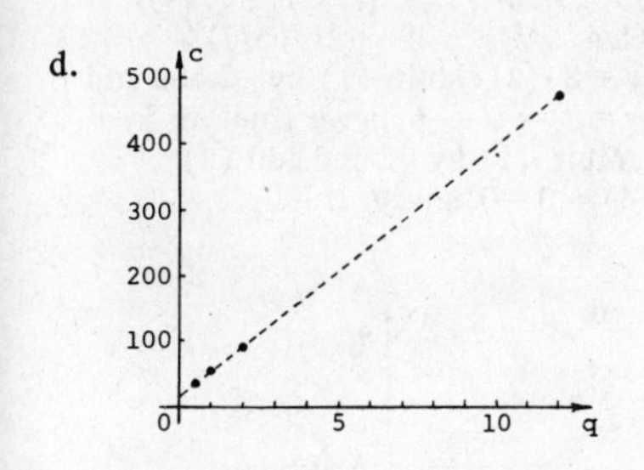

e. 19 cents for carton, handling, etc. f. Slope is *cents per quart*. That's how much the milk itself costs.

3. Cricket Problem a. Let c = no. of chirps/min. Let t = no. of degrees. $c = 4t - 160$ b. $t = 90$, 200 chirps/min. $t = 100$, 240 chirps/min.

c.

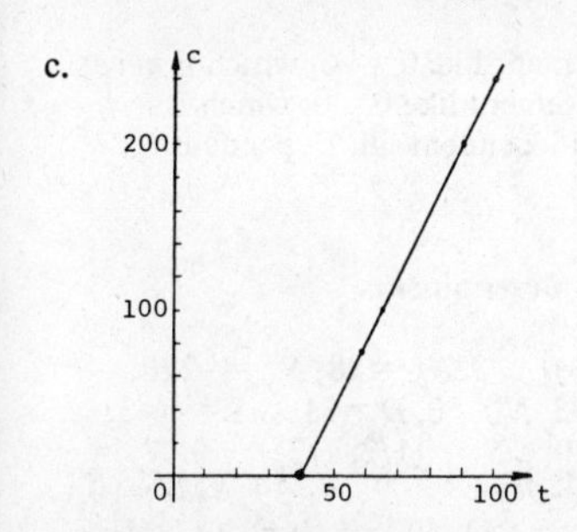

d. $t = 40$. Below 40°F, cricket can't chirp! e. Domain: $t \geq 40$ f. Chirping rate intercept would equal -160. But $t = 0$ is out of the domain. So there is *no* real-world significance. g. $t = (1/4)c + 40$ h. i. 70° ii. 115°

5. Speed on a Hill Problem a. Let a = no. of degrees. Let s = no. of kph. $s = -11a + 132$ b. 209 kph c. 3° (Up, because $a > 0$) d. $s = 132$, meaning your top speed is 132 kph on the *level*. e. $a = 12$. So the steepest hill you can go up is less than 12°.

f.

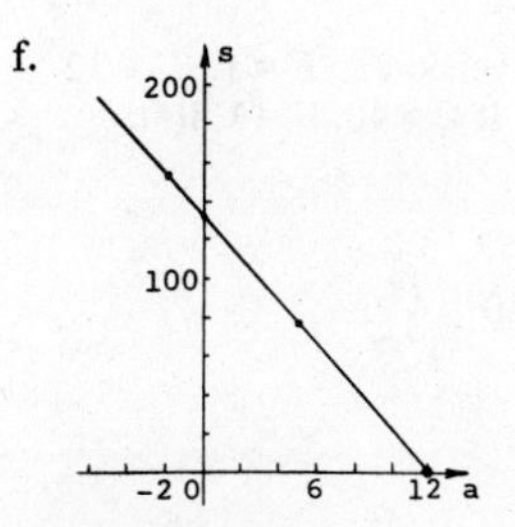

7. Calorie Consumption Problem a. The rate is a *constant* 30 calories per degree. Let c = no. of calories. Let T = no. of °Celsius. $c = -30T + 3630$ b. i. 2130 cal. ii. 5130 cal. c. 121°C (Unlikely to get this hot!)

d.

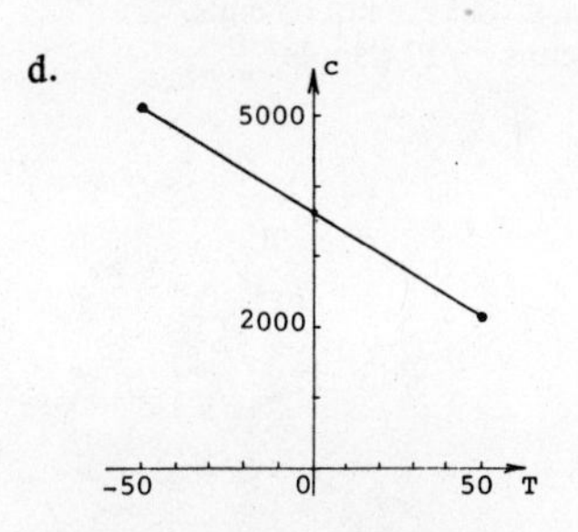

9. Terminal Velocity Problem a. d varies linearly with t because your *velocity* is *constant*. b. $d = -60t + 4500$ c. 75 seconds d. 4500 meters e. You had not yet reached your terminal velocity before $t = 15$.

f.

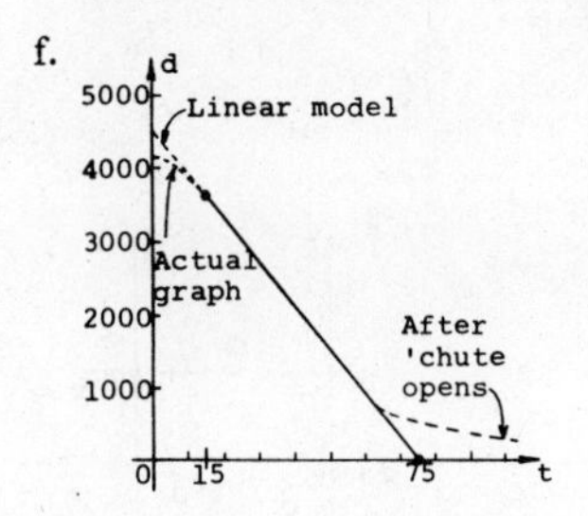

g. 60 meters/sec. 216 km/hr.

11. Shoe Size Problem. a. Let S = shoe size. Let L = no. of inches long. $S = 3L - 22$ b. 14 c. 14-2/3 inches

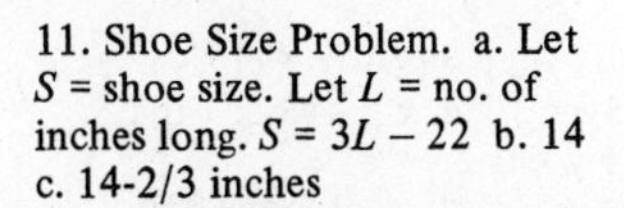

d. Note: Shoes come only in ½ sizes.

13. Celsius-to-Fahrenheit Temperature Conversion a. $F = 1.8C + 32$ b. $C = (5/9)(F - 32)$ c. 2948°F d. 37°C e. If $C = 40$, $F = 1.8(40) + 32 = 104$. So it is *hot*! f. −459.4°F

g.

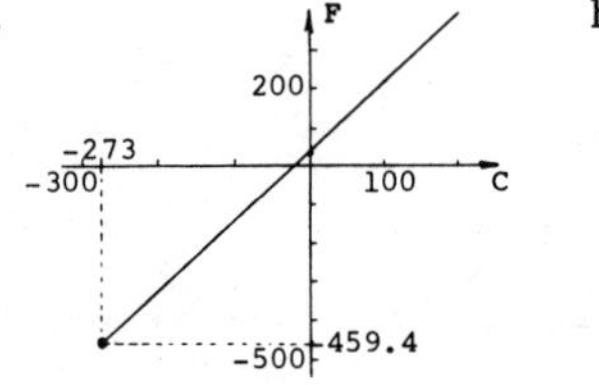

h. $C = -40$

15. Direct Variation, Pancake Problem a. Let c = no. of cups. Let p = no. of people. $c = 0.7p$ b. 35 cups c. 17 people

d.

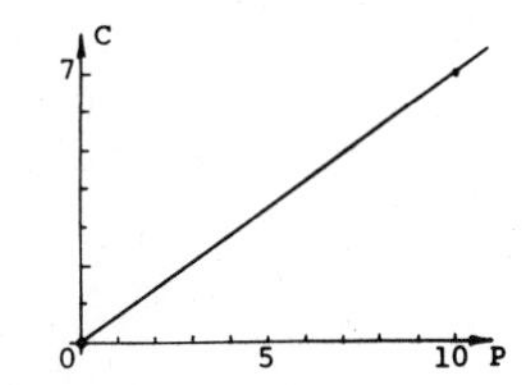

Exercise 3-6, page 60; Chapter Review

R1. a.

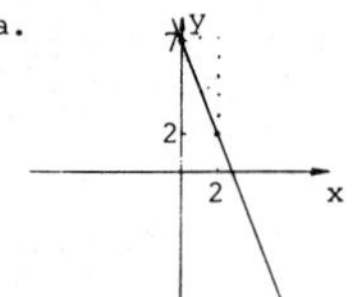

b.

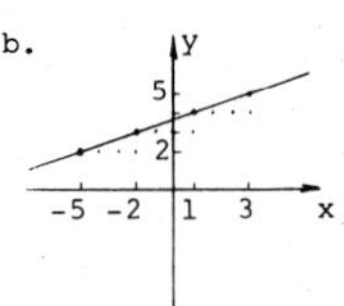

c. d.

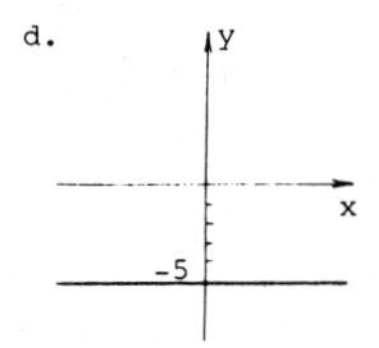

e.

R2. a. $y = (10/3)x - 41/3$, $10x - 3y = 41$ b. $y = (2/9)x + (17/9)$, $2x - 9y = -17$ c. $y = (6/5)x - 6$, $6x - 5y = 30$ d. $x = -13$, $x + 0y = -13$ e. $y = \pi$, $0x + y = \pi$ f. $y = 0$, $0x + y = 0$

R3. Computer Time Problem a. Let n = no. of times it loops. Let t = no. of seconds. $t = 0.06n + 2$ b. $n = 30$, 3.8 sec. $n = 10{,}000$, 602 sec. c. 350 times d. If $n = 0$, $t = 2$, so the t-intercept tells you it takes 2 seconds. e. m is in *seconds per loop,* so it takes 0.06 seconds for one loop.

f.

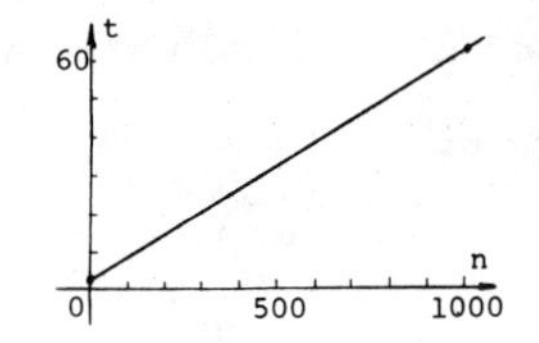

CHAPTER 4
SYSTEMS OF LINEAR EQUATIONS AND INEQUALITIES

Exercise 4-1, page 65; Systems of Two Linear Equations with Two Variables

See Section 4-2.

Exercise 4-2, page 67; Solution of Systems by Linear Combination

1. $\{(1, 3)\}$ 3. $\{(2, -5)\}$ 5. $\{(-3, 0)\}$ 7. $\{(1/2, 1/3)\}$ 9. $\{(-1, -2)\}$ 11. $\{(4, -2)\}$ 13. $\{(1, -2)\}$ 15. $\{(3,2)\}$ 17. $\{(1, -3)\}$ 19. $\{(-2, 0)\}$ 21. $\{(13/3, 19/9)\}$ 23. $\{(22/7, -5/14)\}$ 25. $\{(139/33, 173/33)\}$ 27. $\{(1/4, 5/3)\}$ 29. $\{(2/3, 5/7)\}$

31. a. i. $3x + 2y = 7$ (1), $6x + 4y = 8$ - (2). Mult. (1) by −2 and add (2): $(-6x - 4y = -14) + (6x + 4y = 8) \Rightarrow 0 = -6$, *never* true! ii. $3x + 2y = 7$ - (1), $6x + 4y = 14$ - (2). Mult. (1) by −2 and add (2): $(-6x - 4y = -14) + (6x + 4y = 14) \Rightarrow 0 = 0$, *always* true!

b. i.

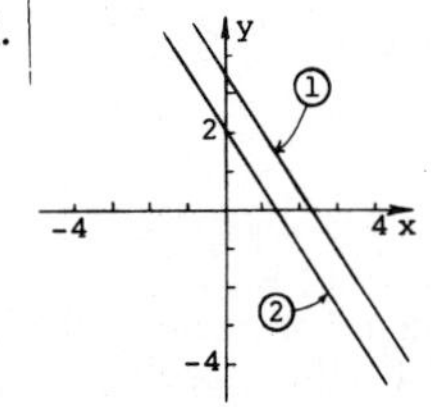

ii.

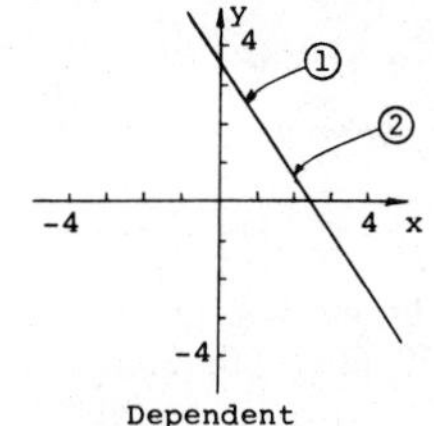

c. Inconsistent equations give a statement like $0 = -6$, which is *never* true. Dependent equations give a statement like $0 = 0$, which is *always* true. d. i. Inconsistent ii. Independent iii. Dependent iv. Inconsistent 33. $S = \{(2, 3)\}$

Exercise 4-3, page 71; Second-Order Determinants

1. $N_x = 3$, $N_y = 9$, $D = 3$, $\therefore S = \{(1, 3)\}$ 3. $N_x = 88$, $N_y = -220$, $D = 44$, $\therefore S = \{(2, -5)\}$ 5. $N_x = -93$, $N_y = 0$, $D = 31$, $\therefore S = \{(-3, 0)\}$ 7. $N_x = -15$, $N_y = -10$, $D = -30$, $\therefore S = \{(1/2, 1/3)\}$ 9. $N_x = -97$, $N_y = -194$, $D = 97$, $\therefore S = \{(-1, -2)\}$ 11. $N_x = -39$, $N_y = -19$, $D = -9$, $\therefore S = \{(13/3, 19/9)\}$ 13. $N_x = -44$, $N_y = 5$, $D = -14$, $\therefore S = \{(22/7, -5/14)\}$ 15. $N_x = -139$, $N_y = -173$, $D = -33$, $\therefore S = \{(139/33, 173/33)\}$

17. a.

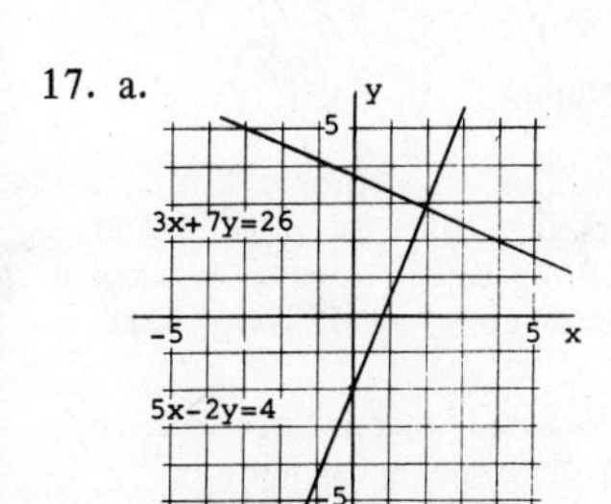

b. $S = \{(1\text{-}39/41, 2\text{-}36/41)\}$ c. Although the solutions *looked* like integers on the graph, they actually were *not* integers!

19. a. (See graphs for Problem 31. b., Exercise 4-2.) b. Equations (i) are *inconsistent*. Equations (ii) are *dependent*. c. $D = 0$ d. i. $N_x = 12$ ii. $N_x = 0$. If x = non-zero/zero, equations are *inconsistent*. If x = zero/zero, equations are *dependent*. e. i. $N_x = -92, N_y = 115, D = 0$; ∴ Eqns. are *inconsistent*. ii. $N_x = 220, N_y = 115, D = 240$; ∴ Eqns. are *independent*. iii. $N_x = 0, N_y = 0, D = 0$; ∴ Eqns. are *dependent*. iv. $N_x = 24, N_y = 30, D = 0$; ∴ Eqns. are *inconsistent*.

Exercise 4-4, page 74; $f(x)$ Terminology and Systems as Models

1. a. $f(7) = 32$ b. $f(-4) = -1$ c. $f(0) = 11$ d. $f(0.5) = 12.5$ e. $g(5) = 31$ f. $g(-3) = 7$ g. $g(0) = 1$ h. $g(1/3) = 1\text{-}4/9$ 3. a. $f(5)/g(5) = 26/31$ b. $g(2)/f(2) = 7/17$ c. $f(6)/g(3) = 29/13$ d. $g(1)/g(0) = 3/1 = 3$
5. Cops and Robbers Problem a. $f(12) = 9; f(4) = 3; f(-8) = -6$ b. $g(7) = 4; g(15) = 20$ c. Willie catches up after 8 minutes, at 6 km from the truck stop. d. 5 minutes after Robin passed

e.

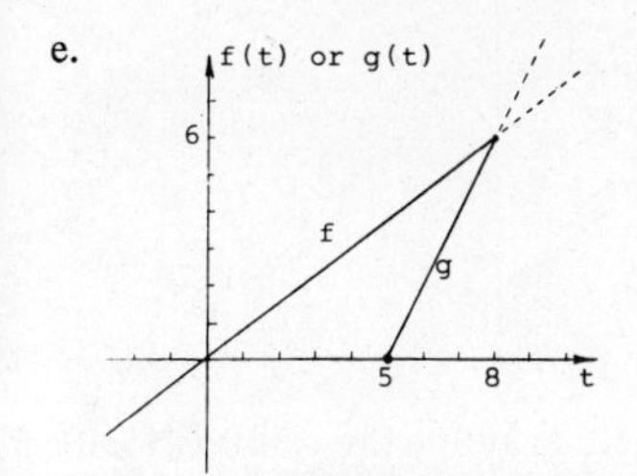

f. km per minute g. Robin: 45 kph; Willie: 120 kph

7. Diesel Car vs. Gasoline Car Problem. a. $f(d) = 7d$ b. $f(1000) = 7000$, $f(10{,}000) = 70{,}000; f(100{,}000) = 700{,}000$ c. \$7000. Wow! d. $g(d) = 2.5d + 250{,}000$ e. $g(1000) = 252{,}500; g(10{,}000) = 275{,}000$; $g(100{,}000) = 500{,}000$

f.

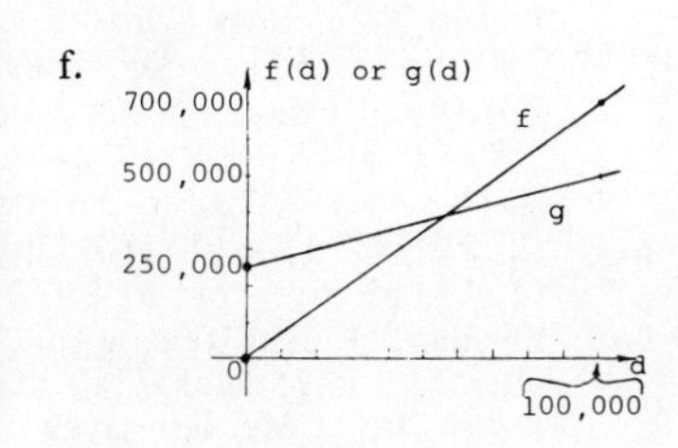

g. (i) Diesel, for many thousands, because g graph is *below* f graph. (ii) Gasoline, for few thousands, because f graph is *below* g graph. h. Break even at 55,555-5/9, or about 56,000 miles

Exercise 4-5, page 78; Linear Equations with Three or More Variables

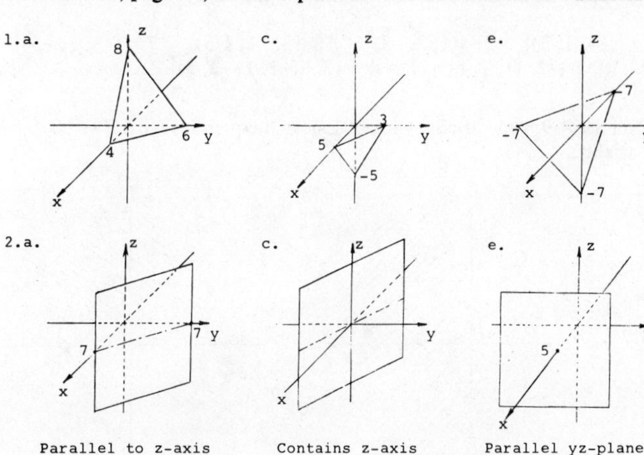

3. a. w-int. = 6; x-int. = 14; y-int. = −10-1/2; z-int. = 7 b. $3x - 4y + 6z = 42$ 5. a. Line. Represents all ordered triples satisfying *both* eqns. *Infinite* no. b. Three planes can meet at a *unique* point.

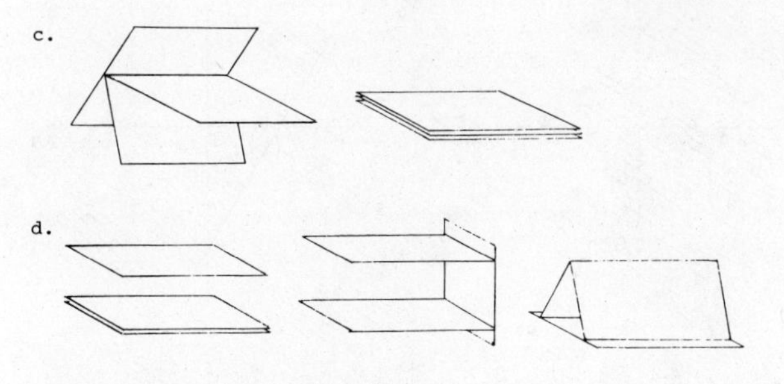

e. No. Sometimes there is an *infinite* number, and sometimes *none.*

Exercise 4-6, page 80; Systems of Linear Equations with Three or More Variables

1. $\{(-4, 1, 3)\}$ 3. $\{(1, 2, -3)\}$ 5. $\{(2, -1, 1/3)\}$ 7. $\{(-1, 1, -5)\}$ 9. $\{(3, 2\text{-}1/2, 1)\}$ 11. $\emptyset$ (inconsistent) 13. $\{(-1, 1, -2, 3)\}$ 15. $\{(12, 24, 36)\}$ 17. Watusi runs 12 mph. Ubangi can't run! ($U = 0$) Pigmy runs 18 mph.

Exercise 4-7, page 83; Higher Order Determinants

1. $D = -55, N_x = 220, N_y = -55, N_z = -165, S = \{(-4, 1, 3)\}$ 3. $D = -55, N_x = -55, N_y = -110, N_z = 165, S = \{(1, 2, -3)\}$ 5. $D = -660$, $N_x = -1320, N_y = 660, N_z = -220, S = \{(2, -1, 1/3)\}$ 7. $D = 27$, $N_x = -27, N_y = 27, N_z = -135, S = \{(-1, 1, -5)\}$ 9. $D = 18, N_x = 54$, $N_y = 45, N_z = 18, S = \{(3, 2\text{-}1/2, 1)\}$ 11. $D = 0, N_x = 329$, $N_y = -282, N_z = -47, S = \emptyset$ 13. $D = 116, N_w = -116, N_x = 116$, $N_y = -232, N_z = 348$, ∴ $S = \{(-1, 1, -2, 3)\}$ 15. $D = -118, N_x = -180$, $N_y = -16, N_z = 134$. ∴ $S = \{(1\text{-}31/59, 8/59, -1\text{-}16/59)\}$

Exercise 4-8, page 85; Matrix Solution of Linear Systems

1. $\{(-4, 1, 3)\}$ 3. $\{(1, 2, -3)\}$ 5. $\{(2, -1, 1/3)\}$ 7. $\{(-1, 1, -5)\}$ 9. $\{(3, 2\text{-}1/2, 1)\}$ 11. $\therefore S = \emptyset$ 13. $\{(-1, 1, -2, 3)\}$

Exercise 4-9, page 88; Systems of Linear Inequalities with Two Variables

1.

3.

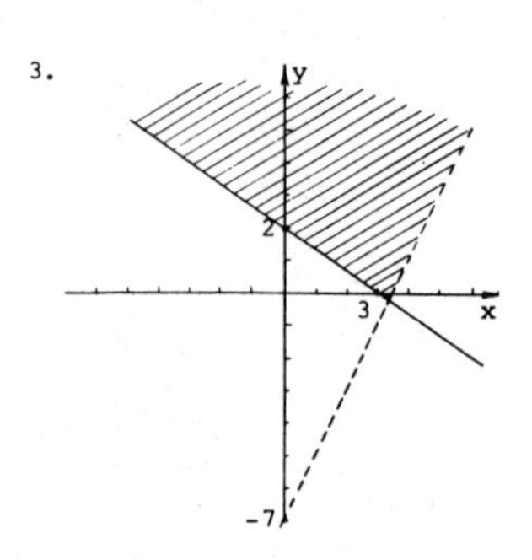

5.

7.

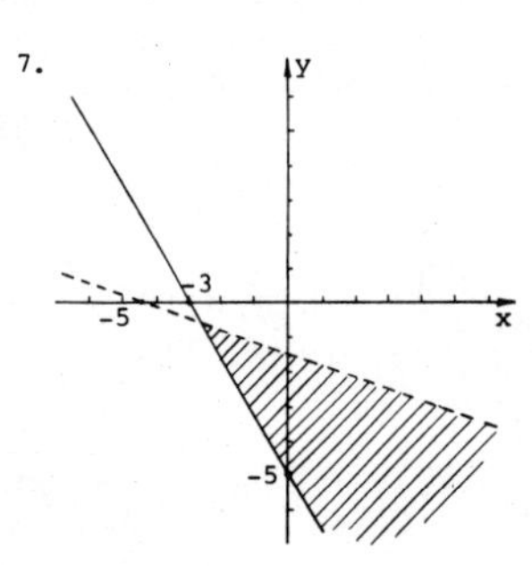

9.

11.

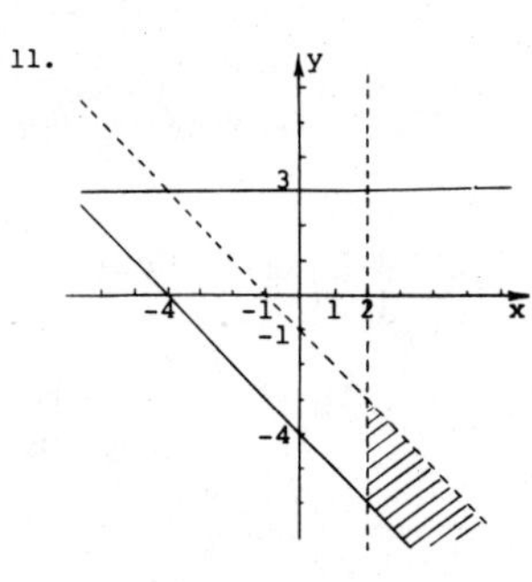

13.

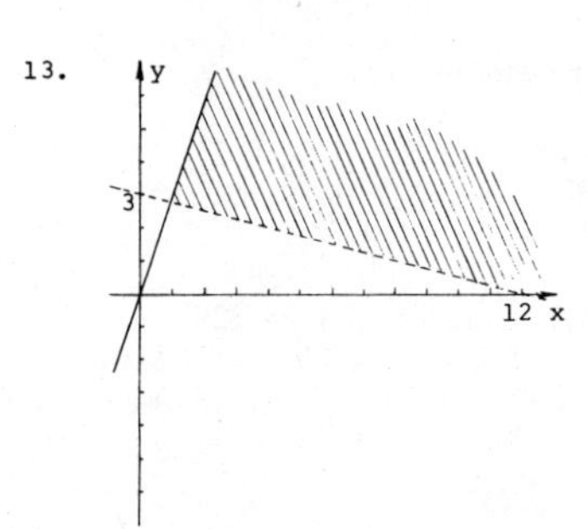

15.

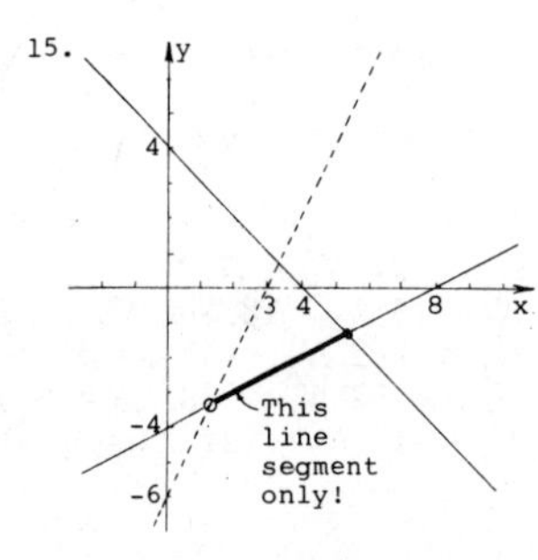

Exercise 4-10, page 91; Linear Programming

Park Clean-Up Problem 1. a. Let x = no. of old members. Let y = no. of new members. Let D = no. of dollars. $D = 10x + 8y$ b. D depends on x and y c. (i) $x \geqslant 0, y \geqslant 0$ (ii) $x \leqslant 9, y \leqslant 8$ (iii) $x + y \geqslant 6, x + y \leqslant 15, \therefore y \geqslant -x + 6, y \geqslant -x + 15$ (iv) $y \geqslant 3$ (v) $y \geqslant (1/2)x, y < 3x$

d.

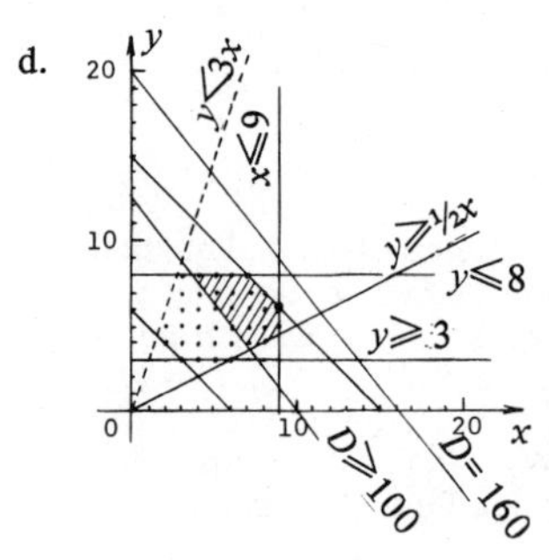

e. "Feasible" means "possible," or "capable of being carried out," f. No. There are no points on y-axis (no old members) in the feasible region g. $10x + 8y \geqslant 100$. h. $10x + 8y = 160$. *Not* feasible to make \$160 since the line *misses* the feasible region. i. 9 old members and 6 new members. D = \$138. j. 2 old members and 4 new members. D = \$52 k. Use 7 old members and 8 new members. D = \$166

3. Aircraft Problem a. Let H = no. of Hippos per day. Let C = no. of Camels per day. i. $H \leqslant 7, C \leqslant 11$ ii. $H + C \leqslant 12$ iii. $H \leqslant 2C$ iv. $100C + 200H > 1000$

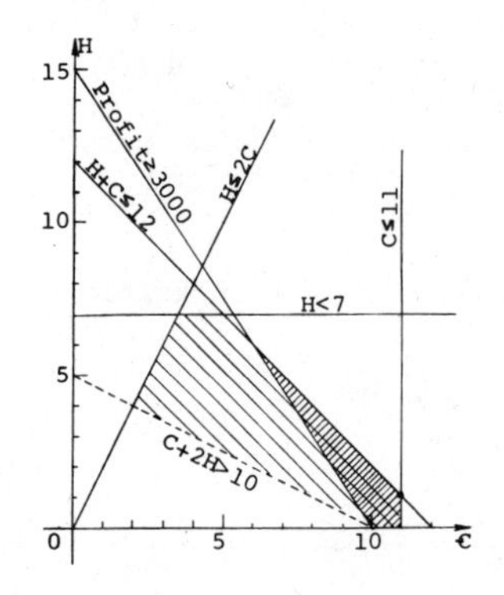

b. Profit = $300\,C + 200\,H$. If Profit $\geqslant 3000$, then $300\,C + 200\,H \geqslant 3000$. See graph. c. 11 Camels and 1 Hippo. Profit = \$3500 5. Feedlot Problem a. i. $25x + 10y \geqslant 250$, protein ii. $12x + 10y \geqslant 200$, minerals iii. $20x + 32y \geqslant 512$, starch iv. $12x + 48y \geqslant 480$, bulk v. $100x + 100y \leqslant 3000$, total

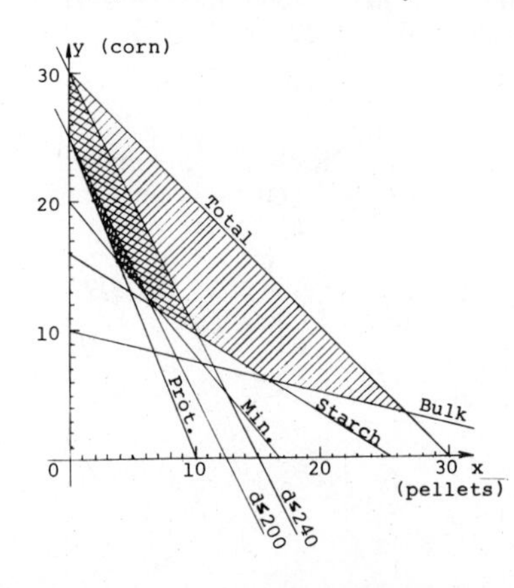

b. All corn *is* feasible since feasible region touches y-axis. All pellets is *not* feasible since feas. reg. does *not* touch x-axis. c. $d = 16x + 8y$ d. i. $16x + 8y \leqslant 240$ ii. $16x + 8y \leqslant 200$ e. Points (2, 20), (3, 18), (4, 16), and (5, 14) all give cost of \$192 per day. f. Min. feas. all-corn diet is 25 sacks/day. $\therefore d = 200$. $\therefore$ Mr. Err saves 200–192, or \$8 per day. g. $x = 3\text{-}11/13$, $y = 15\text{-}5/13$. $\therefore d \approx 184.61$. $\therefore$ Mr. Err saves \$5.39 per day.

7. Coal Problem a. Pikkit: $7\text{-}x$, Weedies: $5\text{-}y$, Treadwell: $4\text{-}z$ b. $z = 10\text{-}x\text{-}y$ c. Treadwell: $x + y - 6$ d. i. $x \geqslant 0$ ii. $y \geqslant 0$ iii. $x \leqslant 7$ iv. $y \leqslant 5$ v. $10\text{-}x\text{-}y \leqslant 4$ vi. $x + y \leqslant 10$

e.

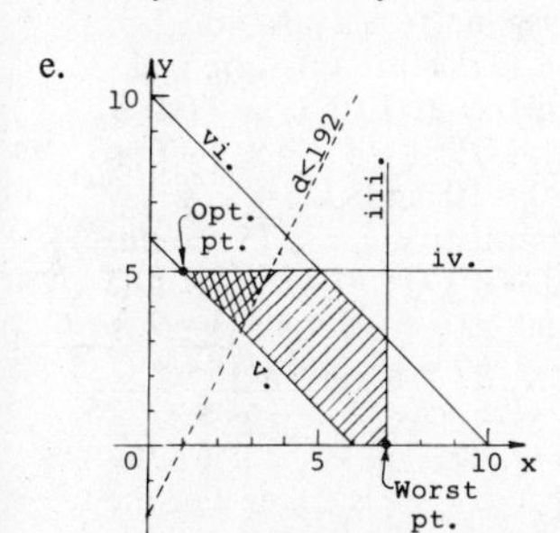

f. $d = 6x - 3y + 186$ g. $y > 2x - 2$
h. Minimum cost at (1, 5); d = \$177
Maximum cost at (7, 0); d = \$228

Exercise 4-11, page 95; Chapter Review

R1. a. $S = \{(3, -4)\}$ b. $D = 29, N_x = 7, N_y = -39, \therefore S = \{(7/29, -39/29)\}$ R2. a. $w(t) = 16 + 0.7t, c(t) = 24 + 0.3t$ b. $w(5) = 19.5$, $c(5) = 25.5$ c. \$20

d.

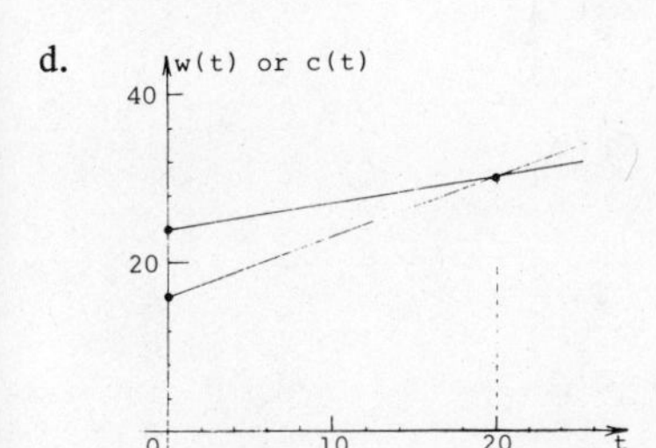

R3. $D = 40, N_x = -120, N_y = 40, N_z = 80, \therefore S = \{(-3, 1, 2)\}$

R4.

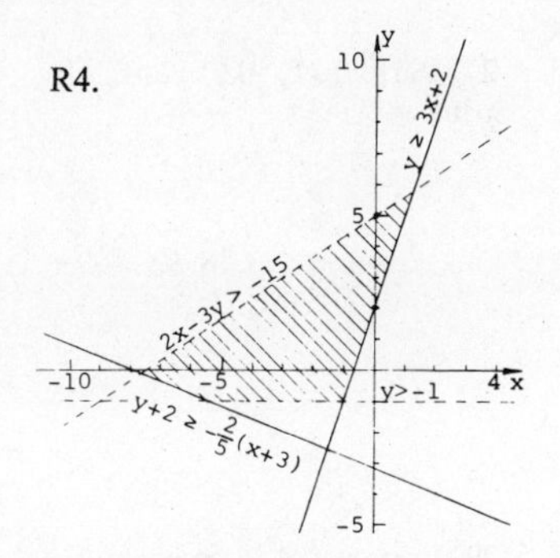

R5. Achievement Test Problem. Let x = no. of 7-point questions answered. Let y = no. of 5-point questions answered. i. $x \leqslant 10, y \leqslant 16$ (also $x \geqslant 0, y \geqslant 0, x, y, \epsilon$ {integers} ii. $x + y \leqslant 20$ iii. $x + y \geqslant 5$ iv. $y \leqslant 2x$ v. $y > (1/2)(x - 5)$ a. Let p = total no. of points. $p = 7x + 5y$. Try $p \geqslant 100$. $7x + 5y \geqslant 100$. From graph, optimum point is (10, 10). b. 120 c. 29

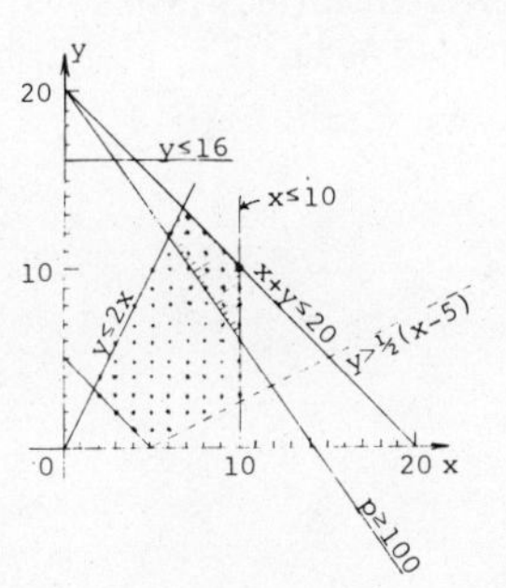

CHAPTER 5
QUADRATIC FUNCTIONS

Exercise 5-1, page 99; Introduction to Quadratic Functions

See Section 5-2.

Exercise 5-2, page 101; Quadratic Function Graphs

1.

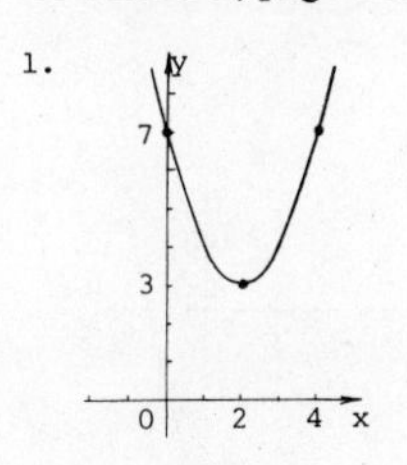

3.

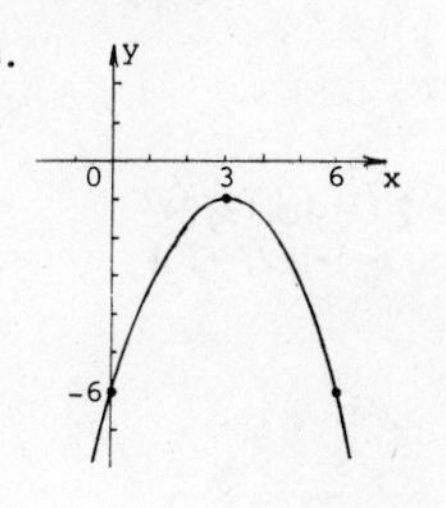

5.

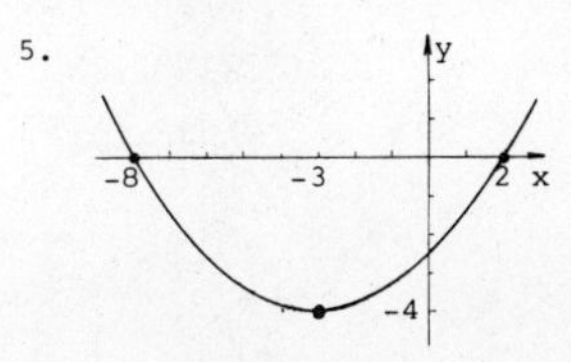

7. a. $y = 3x^2 - 24x + 55$ b. $x = 0, y = 55$; $x = 1, y = 34$; $x = 2, y = 19$; $x = 3, y = 10$; $x = 4, y = 7$; $x = 5, y = 10$; $x = 6, y = 19$ c. Vertex: (4, 7) d. 4 and 7 are as shown: $y - 7 = 3(x-4)^2$ e. By symmetry, $y = 34$ when $x = 7$, and $y = 55$ when $x = 8$.

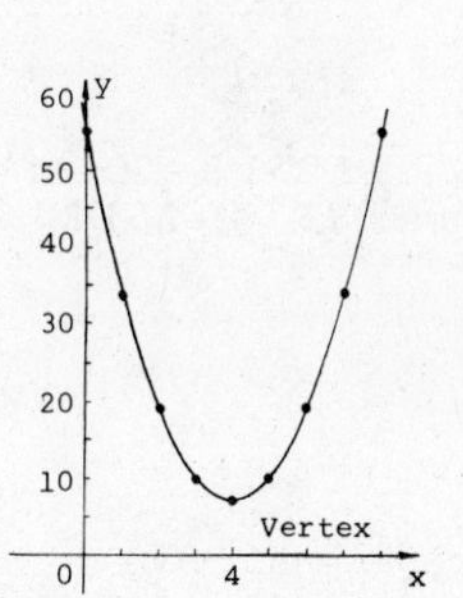

f. $y - 2 = (1/2)(x-6)^2$

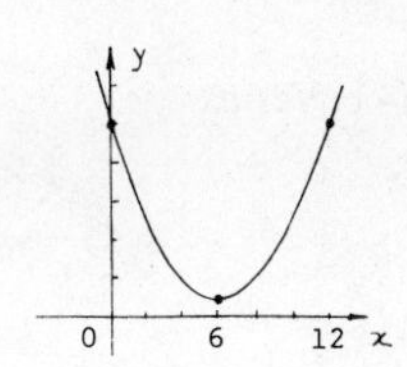

Exercise 5-3, page 104; Completing the Square to Find the Vertex

1. a. $y - 2 = (x+3)^2$ b. Vertex: (−3, 2) c.

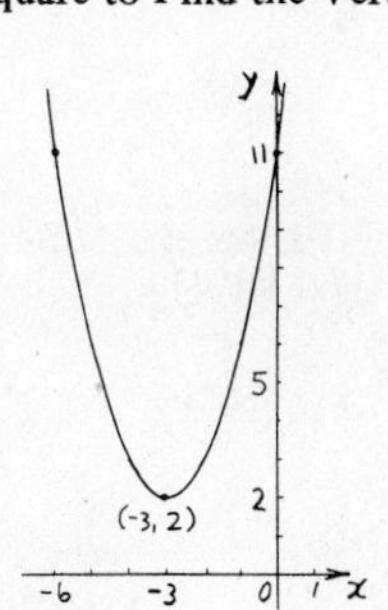

3. a. $y + 31 = 3(x-4)^2$ b. Vertex: (4, –31) c.

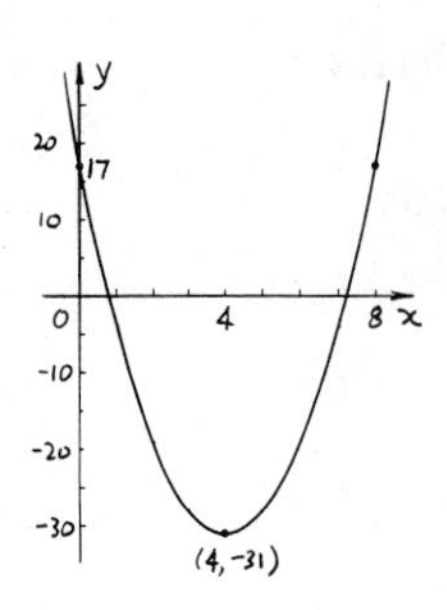

5. a. $y + 20 = 4(x+3/2)^2$
b. Vertex: (–1-1/2, –20) c.

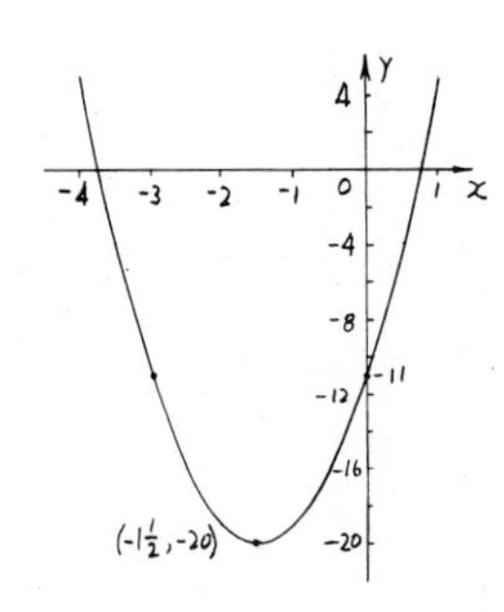

7. a. $y + 27\text{-}1/8 = 2(x-11/4)^2$ b. Vertex: (2-3/4, –27-1/8) c.

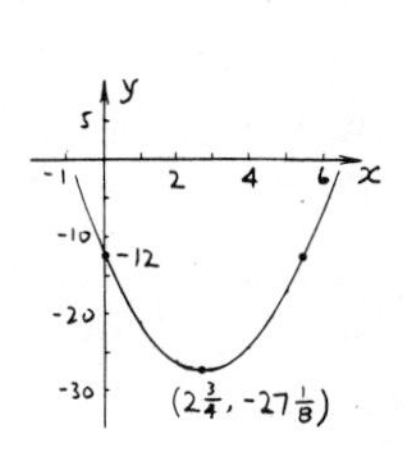

9. a. $y - 96 = -5(x+3)^2$ b. Vertex: (–3, 96) c.

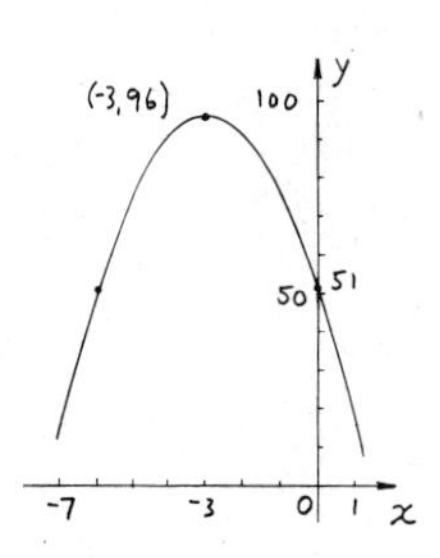

11. a. $y - 1\text{-}1/4 = -(x-1/2)^2$ b. Vertex: (1/2, 1-1/4) c.

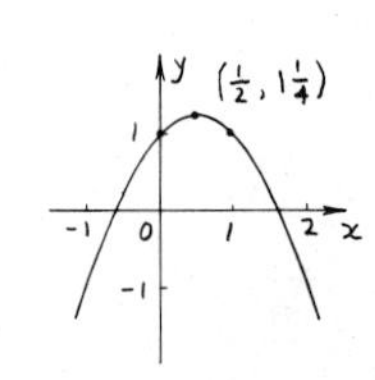

Exercise 5-4, page 107; Quadratic Equations with One Variable

1. a. $3^2 = 9$, ∴ 3 is a square root of 9. b. $(-3)^2 = 9$, ∴–3 is a square root of 9. c. (Pos. no.)^2 is a pos. no. (Neg. no.)^2 is a pos. no. $(0)^2 = 0$. All real nos. are pos., neg., or 0 (Trichotomy). ∴ no real number can be $\sqrt{-9}$. d. (i) 5 (ii) –6 (iii) –9 (iv) 10 (v) –7 (vi) 8 (vii) no real number. 2. a. $\sqrt{36} = 6$, and $\sqrt{9} \times \sqrt{4} = 3 \times 2 = 6$, so $\sqrt{36} = \sqrt{9} \times \sqrt{4}$ (Transitivity). b. $\sqrt{100} = 10$, and $\sqrt{25} \times \sqrt{4} = 5 \times 2 = 10$, so $\sqrt{100} = \sqrt{25} \times \sqrt{4}$ (Transitivity). c. $\sqrt{75}$ should equal $\sqrt{25} \times \sqrt{3}$, which equals $5\sqrt{3}$. d. (i) $2\sqrt{3}$ (ii) $3\sqrt{2}$ (iii) $5\sqrt{2}$ (iv) $3\sqrt{3}$ 3. a. $\sqrt{16+9} = \sqrt{25} = 5$, and $\sqrt{16} + \sqrt{9} = 4 + 3 = 7$, so $\sqrt{16+9} \neq \sqrt{16} + \sqrt{9}$. b. $\sqrt{144+25} = \sqrt{169} = 13$, and $\sqrt{144} + \sqrt{25} = 12 + 5 = 17$, so $\sqrt{144+25} \neq \sqrt{144} + \sqrt{25}$. c. $\sqrt{289-64} = \sqrt{225} = 15$, and $\sqrt{289} - \sqrt{64} = 17 - 8 = 9$, so $\sqrt{289-64} \neq \sqrt{289} - \sqrt{64}$. 4. a. $\sqrt{3^2} = \sqrt{9} = 3$; b. $\sqrt{(-3)^2} = \sqrt{9} = 3$ c. $\sqrt{x^2}$ is either x or the additive inverse of x, whichever is non-negative. This is the definition of absolute value (Sect. 1-3). 5. {2, 1} 7. {–3, –4} 9. {4, –2} 11. $\{-2 + \sqrt{7}, -2 - \sqrt{7}\}$ 13. {1/2, –5} 15. {0, 7/3} 17. {3/2} 19. ∅, for $x \in R$ 21. {2/3, –1} 23. $\{2 + \sqrt{2}, 2 - \sqrt{2}\}$

25. Vertex: (3, –1) y-int. = 8; x-int = 4, 2

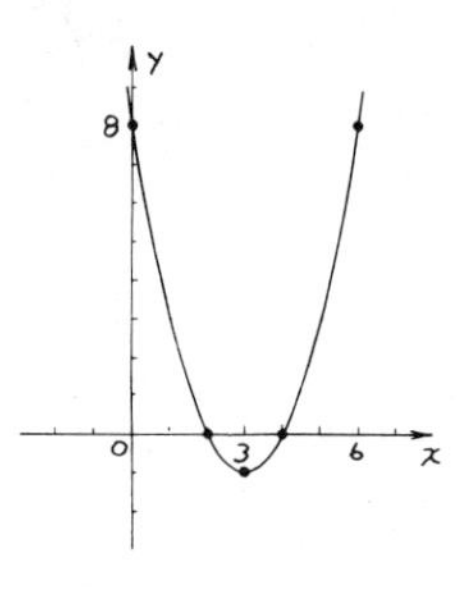

27. Vertex: (1, –16) y-int. –15; x-int. = 5, –3

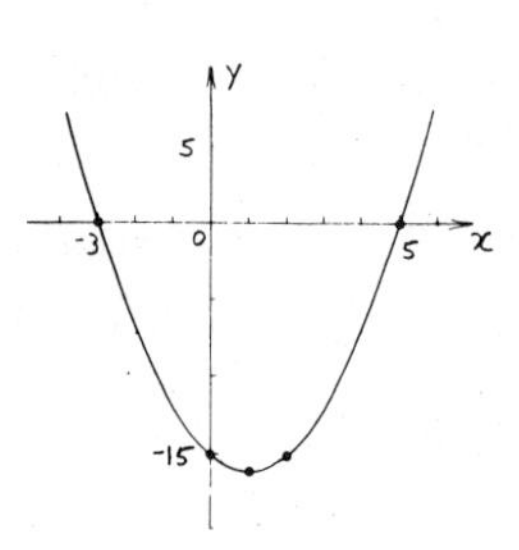

29. Vertex: (–1, 4) y-int. = 3; x-int. = 1, –3

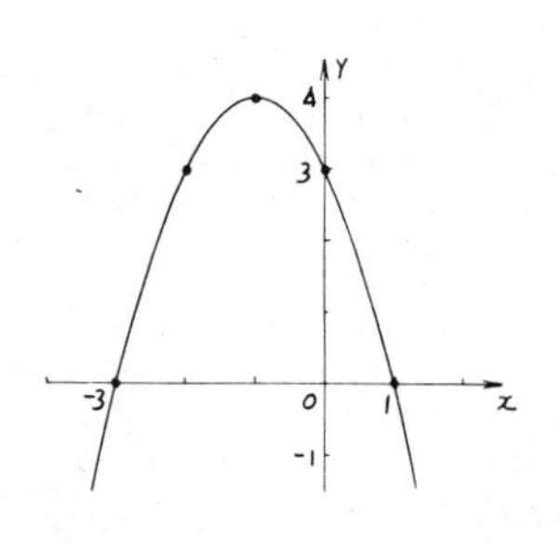

31. Vertex: (–1-3/4, –3-1/8) y-int. = 3; x-int. = –1/2, –3

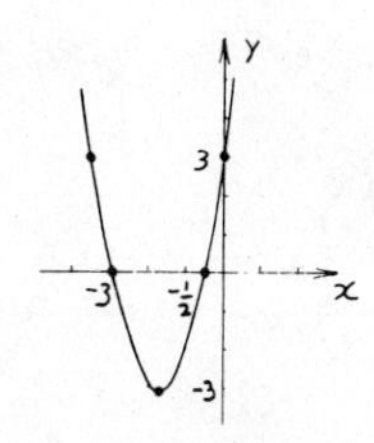

33. Vertex: (1/2, 0) y-int. = –1; x-int. = 1/2

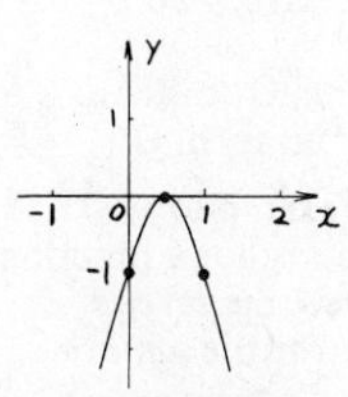

35. Vertex: (–1, 4) y-int. = 5; *no* x-int,

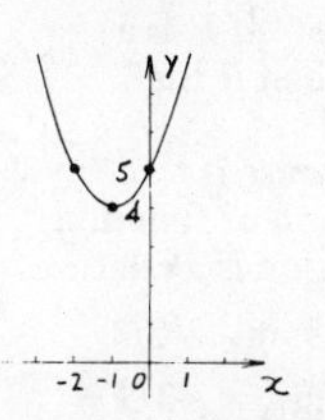

37. Vertex: (–1, –6); y-int. = –5; x-int. = $-1 \pm \sqrt{6} \approx 1.4, -3.4$

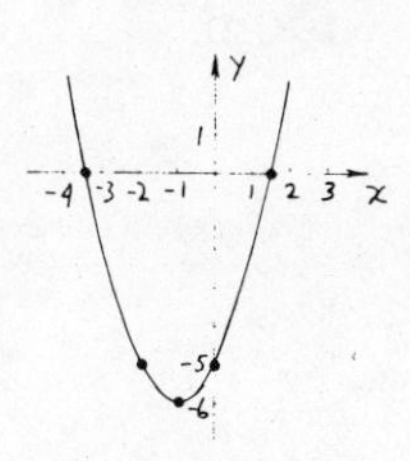

Exercise 5-5, page 109; The Quadratic Formula

Note: Problems 1 through 20 are the same as Problems 5 through 24 in Exercise 5-4. 1. $S = \{2, 1\}$ 3. $S = \{-3, -4\}$ 5. $S = \{4, -2\}$ 7. $S = \{-2 + \sqrt{7}, -2 - \sqrt{7}\}$ 9. $S = \{1/2, -5\}$ 11. $S = \{7/3, 0\}$ 13. $S = \{3/2\}$ 15. $S = \phi$, for $x \in R$ 17. $S = \{2/3, -1\}$ 19. $S = \{2 + \sqrt{2}, 2 - \sqrt{2}\}$ 21. The Discriminant a. i. 49 ii. 0 iii. –48 iv. 28 b. i. Rational ii. Rational iii. Imaginary iv. Irrational c. *Solutions:* Imaginary Nos., *Discriminant: Negative; Solutions:* Real Nos., *Discriminant: Positive or 0; Solutions:* Irrational Nos., *Discriminant: Not a perfect square; Solutions:* Rational Nos., *Discriminant: Perfect Square; Solutions:* Equal Nos., *Discriminant: 0; Solutions:* Unequal Nos., *Discriminant: Positive* 22. Discriminant Practice a. –47; imaginary b. 109; real, irrational c. 49; real, rational, unequal d. 0; real, rational, equal e. 81; real, rational, unequal f. 24; real, irrational g. 1; real, rational, unequal h. –3; imaginary i. 0; real, rational, equal j. –4; imaginary 24. *Quick Vertex Practice* a. (–3, 2) b. (4, –31) c. (–1-1/2, –20) d. (2-3/4, –27-1/8) e. (–3, 96) f. (1/2, 1-1/4) 26. *Untidy Quadratic Equations* a. {.3269, –1.44859} b. {2.49537, .222122} c. {–30.9905, – 60.0733} d. {2.5, –1.4} e. no solution f. {2000.99, 0.0126987}

Exercise 5-6, page 113; Evaluating Quadratic Functions

1. a. $f(-3) = 14$ b. $x = 0.4$ or -2 c. 0.628 or –2.228 3. a. $h(-9) = -199$ b. $x = 1$ or 0.5 c. Disc. = –71. No x-intercepts 5. a. –2 or –6 b. $-4 \pm \sqrt{3}$ c. –3 or –5 d. –4 e. No values of x f. 0 or –8 7. $y = 5$: disc. = 97, yes; $y = -3$: disc. = –31, no 9. $y = 1$: disc. = –31, no $y = -5$: disc. = –79, no 11. $y = 4$: disc. = –11, no; $y = -3$: disc. = 73, yes 13. $y = 7$: disc. = 40, yes; $y = 0$: disc. = 68, yes

Exercise 5-7, page 116; Equations of Quadratic Functions from Their Graphs

1. $y = 3x^2 - 2x + 5$ 3. $y = -4x^2 + 11x - 3$ 5. $y = 0.2x^2 + 0.7x + 1.3$ 7. $y = 0.4x^2$ 9. $y = (-1/2)x^2 + 7x - 1$ 11. $y = 2x^2 - 5x$ 13. $y = 2x^2 + 16x + 35$ 15. Since (5, 2) and (5, –7) are on the same vertical line, the relation is not a *function.*

Exercise 5-8, page 119; Quadratic Functions as Mathematical Models

1. Phoebe Small's Rocket Problem a. $d = 3t^2 - 78t + 500$ b. d-intercept is 500, so Phoebe was 500 km away when she *started* firing her rocket engine. c. If $t = 15$, $d = 5$ km. If $t = 16$, $d = 20$ km. So Phoebe appears to be pulling away when $t = 16$. d. Vertex is at (13, –7), meaning Phoebe crashed into the surface sometime before $t = 13$.

e.

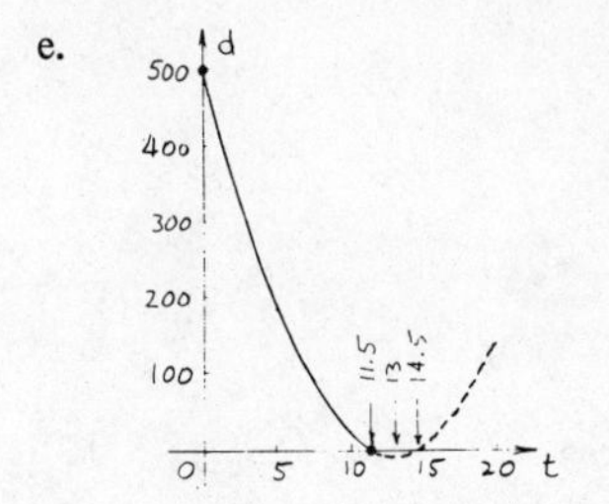

f. Model is reasonable until $d = 0$. If $d = 0$, then $t \approx 14.5$ or 11.5. ∴ Domain = $\{t: 0 \leqslant t \leqslant 11.5\}$

3. Car Insurance Problem. a. Let t = no. of years old. Let A = no. of accidents/100 million km. $A = 0.4t^2 - 36t + 1000$ b. 680 accidents/100 million km c. $t = 16 \Rightarrow A = 526.4$. $t = 70 \Rightarrow A = 440$. ∴ 70-year-olds appear to be safer. d. Vertex is at $t = 45$. Since $a > 0$, parabola opens *up*, and *minimum* is at $t = 45$. So 45-year-olds appear to be safest. e. $A = 830 \Rightarrow t = 85$ or 5. Since insurance applies only to *licensed* drivers, domain is $\{t: 16 \leqslant t \leqslant 85\}$
5. Artillery Problem a. $y = -80x^2 + 120x + 610$ b. i. 530 meters ii. 610 meters c. 3.3 km or –1.8 km d. Vertex is at $y = 655$. ∴ plane is above vertex and there is *no* danger.

7. Color TV Problem

a.

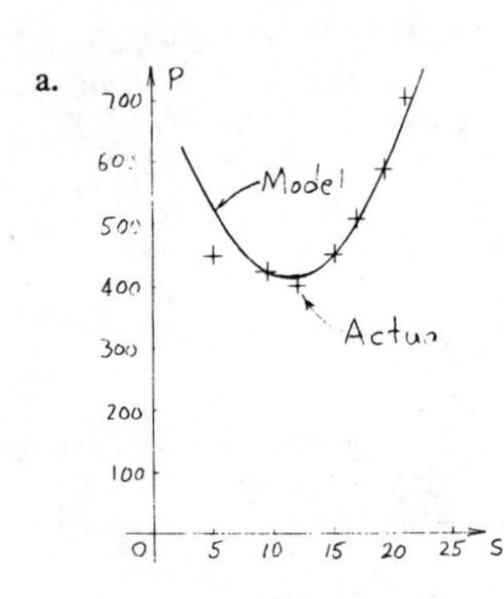

b. Let p = no. of \$. Let s = no. of inches. $p = (8/3)s^2 - (60\text{-}2/3)s + 760$ c. \$840 d. 5″: 523-2/3; 12″: 416; 17″: 499-1/3; 21″: 662 e. (See graph.) f. Model is reasonable, except at $s = 5$. g. It is harder to make very small sets than it is to make medium-size sets. Also, not very many small sets will be sold. For these reasons, price is higher for very small sets.

9. Spaceship Problem a. $y = 0.03x^2 - 7$ b. If $x = 100$, $y = 293$. ∴ graph does *not* contain (100, 295), and a mid-course maneuver will be necessary.

11. Barley Problem a. Let s = no. of hundreds of thousands of seeds per acre. Let B = no. of bushels per acre harvested. $B = -0.5s^2 + 12s$ c. 64 d. Yes, by planting *10 or 14 hundred thousand seeds per acre.* e. Plant *12 hundred thousand seeds/acre.* f. Yes, Planting 24 hundred thousand seeds/acre gives *no* barley.

g.

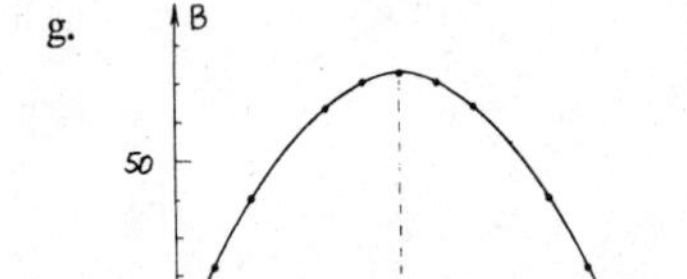

Exercise 5-9, page 125; Chapter Review

R1 a. $x = 0 \Rightarrow y = -5$; $x = 4 \Rightarrow y = -21$; $x = -2 \Rightarrow y = -33$. b. $y = 0 \Rightarrow x = 1$ or $-5/3$; $y = 1 \Rightarrow x$ = no real solution; $y = -2 \Rightarrow x = (-4 \pm \sqrt{7})/(-3)$ c. (1-1/3, 1/3).

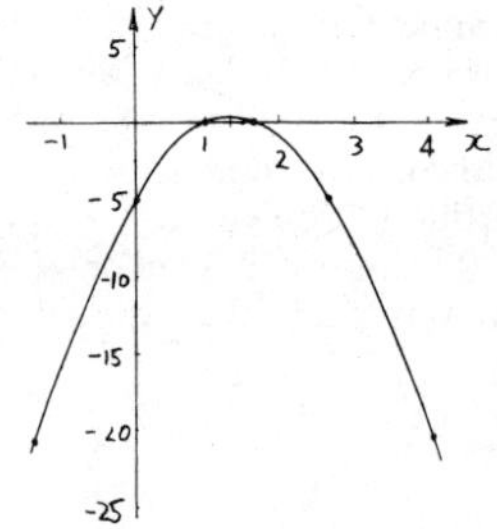

R2. $y = 2x^2 - 4x + 1$ R3. d. Let d = no. of meters above water. Let t = no. of sec. since dive. $d = -5t^2 + 9t + 20$ b. Diving board is 20 m from surface, since this is d when $t = 0$. c. 24.05 m d. about 3.1 sec.

e.

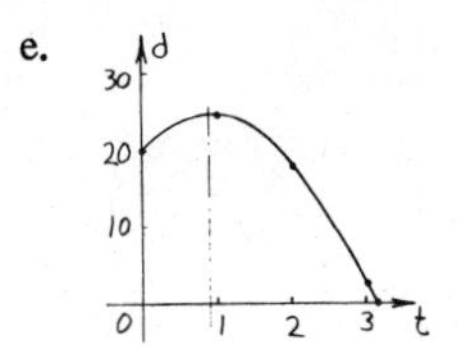

Domain ends when he hits the water, $d = 0$.

Exercise 5-10, page 126

1. a. C b. P c. C d. B e. C or B f. P g. C h. N i. C j. N k. N l. P m. P n. C
2. a. Graph. (example) The relation is not a function because there are *several* values of y for the *same* value of x.

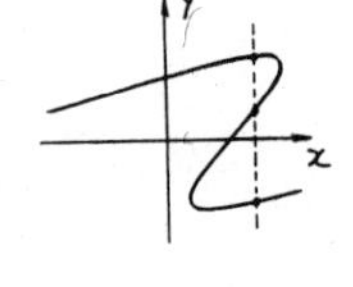

   b. Converse: "If R is a function, then R is a relation."
   c. The converse is *true.* By the definition of function, every function is a relation.
3. a. $m_A = 2/3$, $m_B = 2/3$, $m_C = 2/3$, $m_D = -3/2$
   b. A∥C, A⊥D, C⊥D.
   c. Graph.

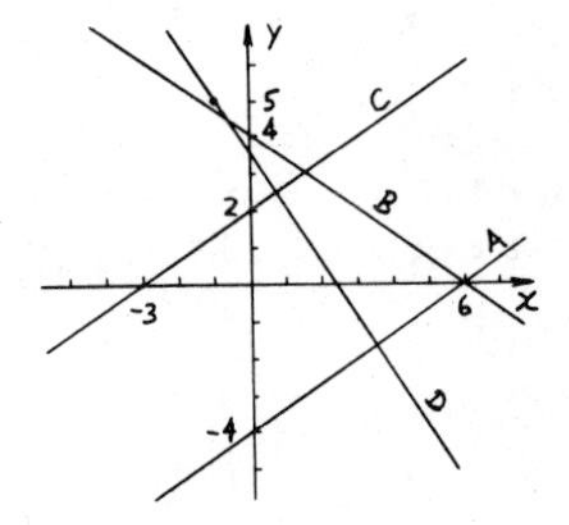

4. a.

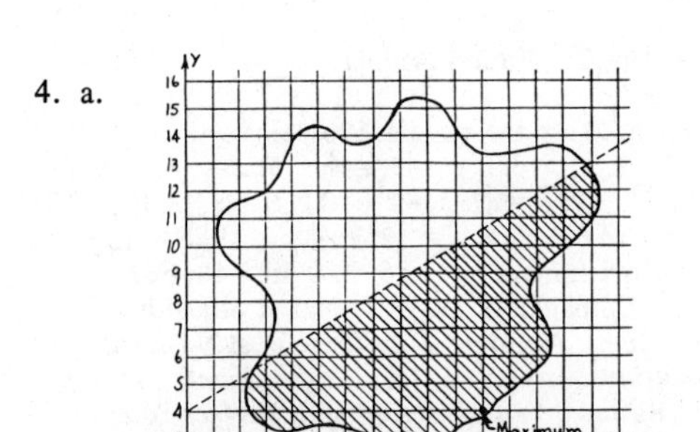

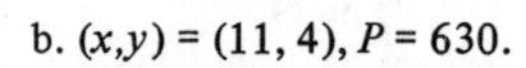
b. $(x,y) = (11, 4)$, $P = 630$.

5. Def. of Division. no. no.
   Multiplicative Inverses. yes. yes.
   Transitivity. yes. no.

6.

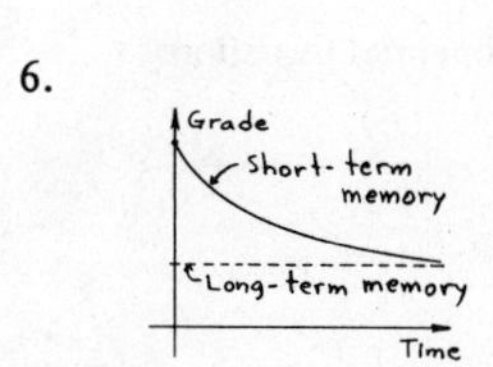

8. d., e.

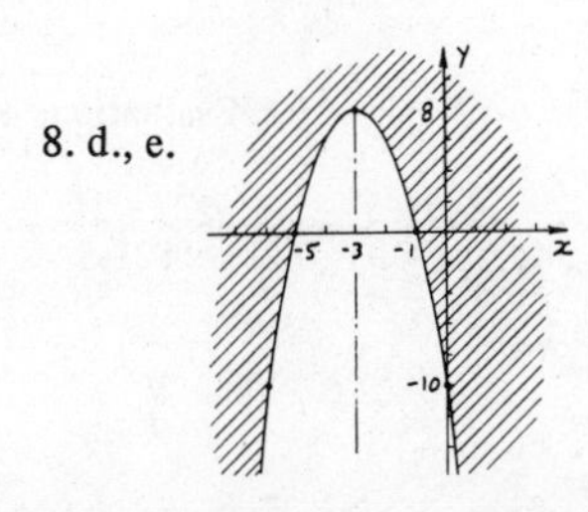

7. a. H = 4A + 19 b. 83 in. c. 19 inches at birth. d. Constant rate of 4 in./year. e. You stop growing. 8. a. $y - 8 = -2(x + 3)^2$ b. $(-3, 8)$ c. $y = -10$; $x = -5$ or -1 d., e. Graph above 9. a. $V = t^2 + 120t + 1300$ b. John must go c. 130 min. d. Graph below e. Domain: t: $0 \leq t \leq 130$ Range: $\{V: 0 \leq V \leq 4900\}$ 10. a. 100 Rational b. -92 Imaginary. c. 28 Irrational d. 0. Rational

9.

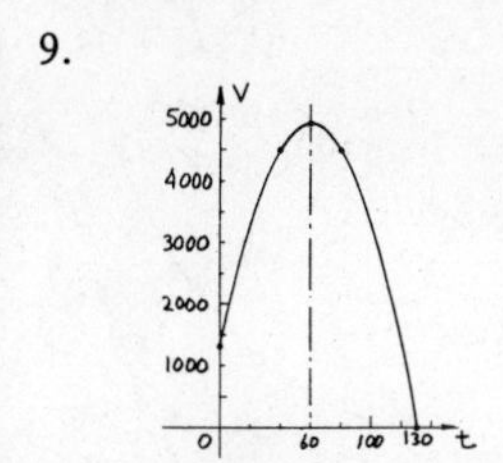

11.

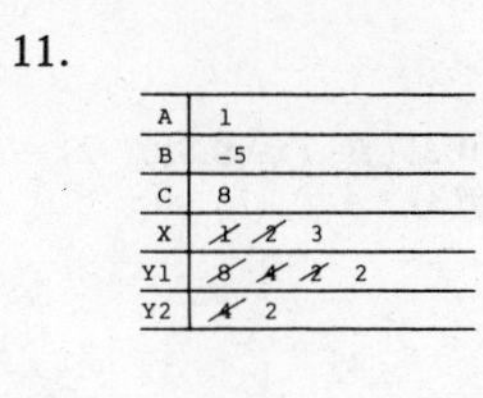

A	1
B	-5
C	8
X	~~1~~ ~~2~~ 3
Y1	~~8~~ ~~4~~ ~~2~~ 2
Y2	~~4~~ 2

CHAPTER 6
EXPONENTIAL AND LOGARITHMIC FUNCTIONS

Exercise 6-1, page 131; Introduction to Exponential Functions

1. $f(1) = 2$; $f(2) = 4$; $f(3) = 8$; $f(4) = 16$ 2. $f(0) = 1$; $f(-1) = 1/2$; $f(-2) = 1/4$; $f(-3) = 1/8$ 3. *asymptote.* 4. See Figure 6-2. 5. $f(2\text{-}1/2) \approx 5.7$; $f(1\text{-}1/2) \approx 2.8$; $f(1/2) \approx 1.4$

Exercise 6-2, page 133; The Operation of Exponentiation

1. a. 1296 b. 162 c. 162 3. a. 625 b. 19 c. 19 5. a. -27 b. -339 c. -339 7. a. -162 b. 1296 c. -162 d. -162 9. 32

Exercise 6-3, page 133; Properties of Exponentiation

1. See text. 3. a. Def. of exponentiation b. Product of powers with same base c. Arithmetic d. Transitivity for = e. When you raise a power to a power, you *multiply* the exponents. 6. Example: $(2 + 3)^2 = 5^2 = 25$, and $2^2 + 3^2 = 4 + 9 = 13$. So $(2 + 3)^2 \neq 2^2 + 3^2$. 7. a. $(x/y)^a = x^a/y^a$ b. $(xy)^a = x^a y^a$ c. $(x^a)^b = x^{ab}$ d. $(x/y)^a = x^a/y^a$ e. $(xy)^a = x^a y^a$ f. $x^a \cdot x^b = x^{a+b}$ g. $x^a/x^b = x^{a-b}$

Exercise 6-4, page 137; Negatives and Zero Exponents

1. a. $f(-4) = 81/16 = 5.06$; $f(-3) = 27/8 = 3.38$; $f(-2) = 9/4 = 2.25$; $f(-1) = 3/2 = 1.5$; $f(0) = 1$; $f(1) = 2/3 = 0.67$; $f(2) = 4/9 = 0.44$; $f(3) = 8/27 = 0.30$; $f(4) = 16/81 = 0.20$

b.

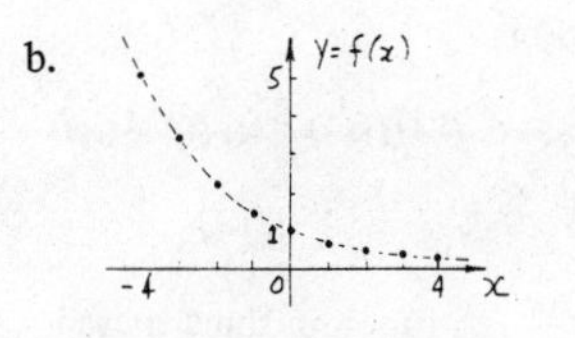

c. Positive x-axis is an *asymptote.*

3. $18x^3$ 5. $10x^{-7}$ 7. $44x^{-5}$ 9. $-648x^2y^8$ 11. $648x^{-4}y^3$ 13. $(-3/64)x^{10}y^2$ 15. $(1/3)a^{-9}b^8$ 17. $5a^2 - 3a$ 19. $xy^{-5}z^4$ 21. $2x^{-3}y^7z^{-8}$ 23. $(1/3)x^4yz^{-2}$ 25. $(4/9)x^2y^{-22}$ 27. $8x^{15}y^{15}z^{-9}$ 29. 1

Exercise 6-5, page 140; Radicals and Fractional Exponents

1. a.

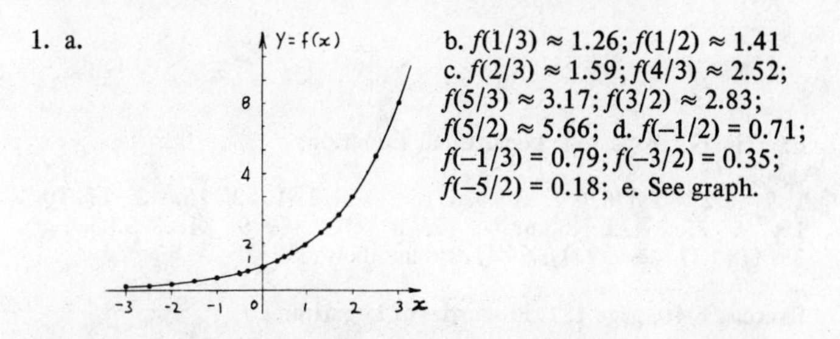

b. $f(1/3) \approx 1.26$; $f(1/2) \approx 1.41$ c. $f(2/3) \approx 1.59$; $f(4/3) \approx 2.52$; $f(5/3) \approx 3.17$; $f(3/2) \approx 2.83$; $f(5/2) \approx 5.66$; d. $f(-1/2) = 0.71$; $f(-1/3) = 0.79$; $f(-3/2) = 0.35$; $f(-5/2) = 0.18$; e. See graph.

3. a. 16 b. 512 c. 32 d. 128 5. a. 1/512 b. -512 c. imag. no. d. $-1/512$ 7. 64 9. -64 11. 4 13. 1/729 15. 49 17. $-1/1000$ 19. 343/17 21. 16/81 23. 4 25. 3 27. 12 29. 5 31. $\sqrt{3}$ 33. $2^{1/4}$ 35. $2^{29/6}$ 37. $2^{0.8}$ 39. $10^{1.24}$ 41. $\sqrt[3]{2}$ or $2^{1/3}$ 43. $\sqrt[3]{6}$ or $6^{1/3}$ 45. $12x^{1/6}$ 47. $1728x^{-1}$ 49. $xy^{7/2}$ 51. $(1/7)a^{5/12}b^{5/6}$ 53. a. $(-8)^{1/3} = -2$ b. $(-8)^{2/6} = ((-8)^2)^{1/6} = 2$ c. Since $-2 \neq 2$, you *cannot* necessarily substitute equals for equals when the base is *negative.*

Exercise 6-6, page 143; Powers by Calculator

1. 162.432 3. 1.78738×10^{-6} 5. 0.0383873 7. 3.41529 9. 2710.27 11. 0.192886 13. 2.63902 15. 0.0574757 17. 539.272 19. 6.09549×10^{-4} 21. 178.656 23. 66.4074 25. $36^{6/5} = 64$; $33^{6/5} \approx 66.4$, which is *slightly larger* than 64. 27. 22.8161 29. 5.12100 31. 0.0373011 33. 4 35. 3.32200 37. 51.9600 39. 156.993 41. a. 2.594 b. $n = 100$: 2.704813815; $n = 1000$: 2.716923842; $n = 10000$: 2.718145918; $n = 100000$: 2.718254646; $n = 1000000$: 2.718281828 c. 2.718281828...

Exercise 6-7, page 146; Scientific Notation

1. 3.72×10^5 3. 2.61×10^{-3} 5. 2.001×10^3 7. 2.024×10^{-1} 9. 2.47×10^7 11. 6×10^{15} 13. 3.5×10^{16} 15. 4.2×10^{-4} 17. 5.4×10^1 19. 5.6×10^{-23} 21. 4×10^{11} 23. 2×10^{-7} 25. 3.5×10^8 27. 2×10^9 29. 1.2×10^{-16} 31. 6.40×10^{13} 33. 6.6×10^{12} 35. 1.99×10^{-14} 37. 1.7×10^{-12} 39. 1.28×10^3 41. 6.1×10^{-14} 43. 7.7×10^{12} 45. 1.308×10^4 47. 1.70×10^{-5} 49. 6.17×10^6 51. 5.48×10^{10} 53. 1.66×10^{-9} 55. 7.22×10^{14} 57. Brain Cell Problem. 2×10^{18}

Exercise 6-8, page 149; Inverses of Functions

1. a. $y = (1/2)x + 3$ b. $f(5) = 4; f^{-1}(4) = 5$ c. $f^{-1}(f(x)) = x; f(f^{-1}(x)) = x$

d.

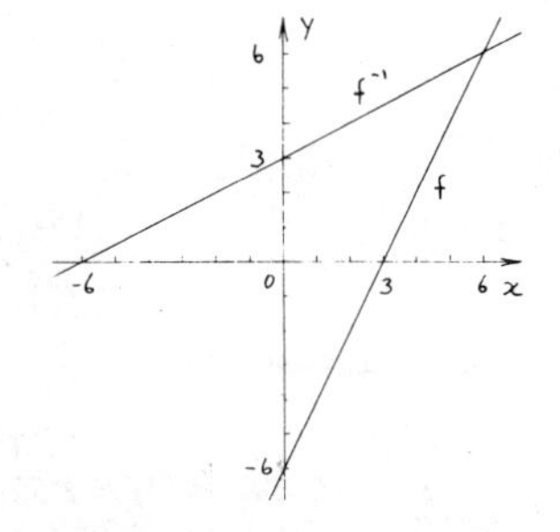

e. f^{-1} *is* a function since there is only *one* y-value for each x-value. It is a *linear* function.

Exercise 6-9, page 151; Logarithmic Functions

1. 8 3. 27 5. 1/81 7. 81 9. 1/16 11. 1/2 13. 2 15. −2 17. 10 19. −5 21. 3 23. No number 25. 0 27. 9 29. 9 31. 2 33. 4 35. 1/8 37. 27 39. 1/36 41. No number

Exercise 6-10, page 151; Properties of Logarithms

1. Both equal 1.3222193 3. Both equal 0.9542425 5. Both equal 2.5352941 7. a. (def. of r & s) b. Def. of logarithm c. multiplying x by y d. Prod. of powers w. same base e. Def. of logarithm f. Symmetry g. Subst. for r & s h. When you *multiply* the arguments, you can *add* the logs. 11. $\log_3 35^3$ 13. $\log_2 3$ 15. $\log_7 30$ 17. $\log_5 16$ 19. $\log_{12} 32$ 21. $\log_6 135$ 23. 8 25. −2 27. 7 29. 9 31. log of a product: $\log_b (xy) = \log_b x + \log_b y$; log of a quotient: $\log_b (x/y) = \log_b x - \log_b y$; log of a power: $\log_b (x^y) = y \cdot \log_b x$

Exercise 6-11, page 155; Special Properties of Base 10 Logarithms

1. 1.3560259 3. −1.2612194 5. 5.4857214 7. 23.779596 9. −14 11. 594.97676 13. 3.67959×10^7 15. 0.0136993 17. 729.96157 19. 1.36993×10^{-3} 21. By tables, $x \approx 19.6$. By calculator, $x \approx 19.633139$ 24. Earthquake Problem for San Francisco, severity $= 1.78 \times 10^8$. For Seattle, severity $= 10^7$. ∴ San Francisco quake was 17.8 *times* as severe as Seattle. 26. pH Problem a. pH = 7 b. pH = 7.2 c. pH = 1.6 d. pH = 11.04

Exercise 6-12, page 158; Exponential Equations, and Logarithm with Other Bases

1. {1.585} 3. {0.7122} 5. {−1.238} 7. {2.431} 9. {1.1761} 11. {0.3680} 13. {2.272} 15. 2.0588 17. 5.9829 19. 4.3219 21. 2.6610 23. 1 25. They do *not* have the same characteristic-mantissa property. For example, $\log 4000 \neq 1 + \log 400$. $\log 5000 \neq 1 + \log 500$. 26. Let $y = \log_a x$. Then $a^y = x$, $\log_b a^y = \log_b x$, $y \cdot \log_b a = \log_b x$, $y = \log_b x / \log_b a$, Q.E.D. 33. a. $c \approx 3.321928096$ b. i. 3.700439722 ii. −0.666576266 iii. 10.79441587 iv. 12 (exactly) v. 19.93156858 35. Undefined 37. 6.559 39. Undefined. 41. 4 43. 1/3 45. 0 47. {(27, 64), (3, 4)} 49. 3 51. 3 53. 46/45 55. 25

Exercise 6-13, page 162; Evaluations of Exponential Functions

1. $f(-4) = 16; f(-1) = 24; f(2) = 36; f(5) = 54; f(8) = 81; f(11) = 121.5$

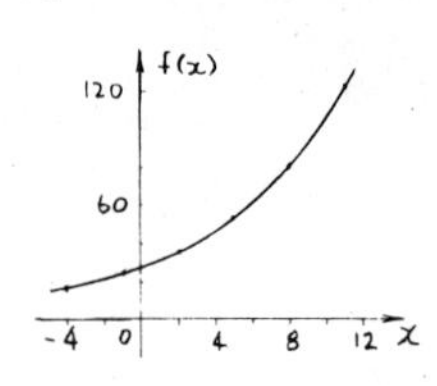

3. $f(0) = 204.1; f(2) = 142.9; f(4) = 100; f(6) = 70; f(8) = 49; f(10) = 34.3$

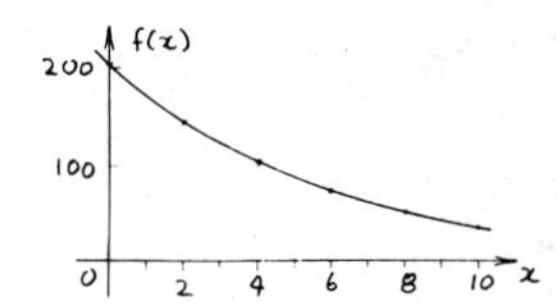

5. $f(-12) = 0.25; f(-7) = 0.5; f(-2) = 1; f(3) = 2; f(8) = 4; f(13) = 8$

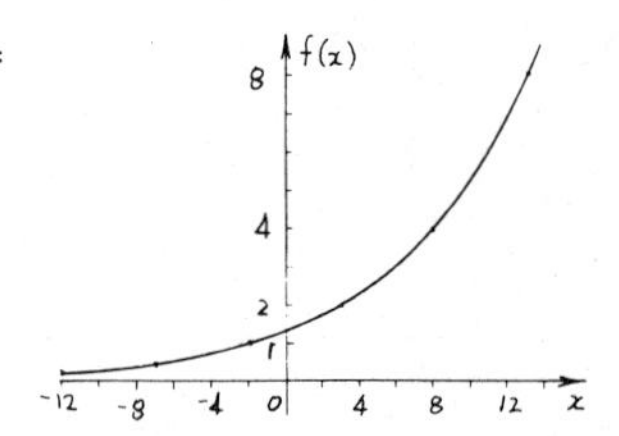

7. $f(2) = 60.1; f(5) = 39.3; f(8) = 25.7; f(11) = 16.8; f(14) = 11.0; f(17) = 7.2$

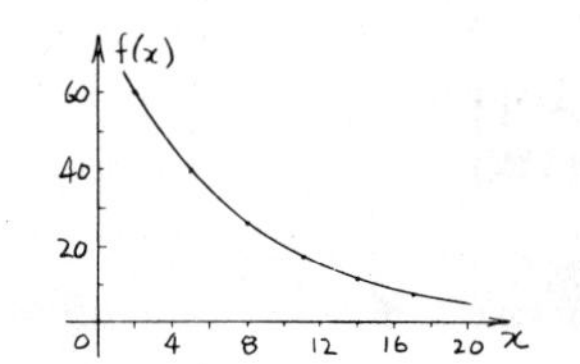

9. a. $g(0) = 21.4; g(2.8) = 1024$; c. $g(-3.2) = 0.26$ b. $x = 2.39$

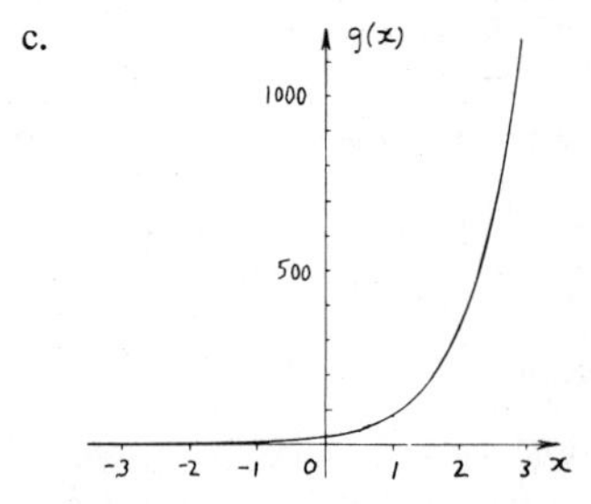

Exercise 6-14, page 164; Exponential Functions as Mathematical Models

Note: Answers are by calculator, *without* rounding off at intermediate steps, and are correct to the number of digits shown. Answers by 4-place logs, or with round-off at intermediate steps may differ slightly from those shown.

1. Population Problem. b.
a. Population increases by a factor of $226/203 \approx$ 1.1133. 1990: 251.6 million; 2000: 280.1 million; 2010: 311.8 million.
c. Let p = no. of millions of people. Let t = no. of years since 1970. $p = 203 \times 10^{0.00466124t}$
d. Answers will vary. For 1984, $p = 235.9$ 3. Early in 2023. f. 25 million. 1790 census showed only 4 million, meaning population grew more rapidly between 1776 and 1970 than predicted.

3. Car-Stopping Problem a. Let s = no. of kph. Let t = no. of seconds since car ran out of gas. $s = 64 \times 10^{-0.0125t}$ b. $s \approx 31$ kph c. About 65 seconds

d. and e.

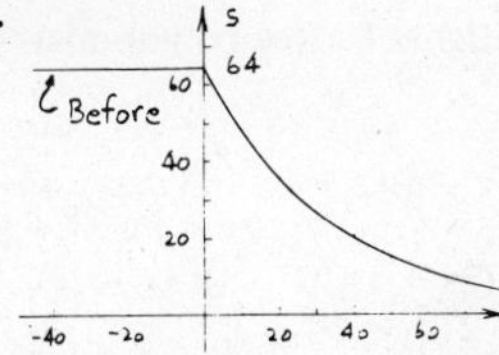

f. For very large values of t, model says he is still moving slightly. But he is actually stopped.

5. Phoebe's Next Rocket Problem. a. Let s = no. of mph. Phoebe is going. Let t = no. of sec. since she fired last stage. $t = 0, s = 4230$; $t = 10, s = 6850$; $t = 20, s = 11093$; $t = 30, s = 17964$. So speed is *more* than 17500 at $t = 30$, and she *will* orbit. b. Equation is $s = 4230 \times 10^{0.02094t}$. Minimum time to get Phoebe into orbit is 29.46 sec. c. 36.86 sec.
7. Car Trade-In Problem. a. 0 years: \$2350; 1 year: \$1645; 2 years: \$1151.50; 3 years: \$806.05 b. Adding 1 to time *multiplies* the value by 0.7. c. Let V = \$ trade-in value. Let t = no. of years from present. $V = 2350 \times 10^{-0.1549t}$ d. About 3.8 years e. \$6156 f. The difference, 7430 –6156 = 1274, is mostly the dealer's profit on sale of the car.
9. Carbon 14 Dating Problem a. Let P = % of carbon 14 remaining. Let t = no. of years since the carbon 14 was absorbed. $P = 100 \times 10^{-0.00005235t}$ b. 78.58% c. 61.75% (61.74 by calc.) d. 6025 years ago. If sample were taken in 1980, year would be 4045 B.C. So wood *is* old enough. e. If t = 100 million = 100,000,000, then $P = 10^{-5233}$ This is *so* small that there would be *no* detectable carbon 14 left after 100 million years.
11. Coffee Cup Problem a. $(t, D) = (0, 65), (5, 52)$ b. $D = 65 \times 10^{-0.01938t}$ c. 94.3°C d. 16.87 min. e. $t = 10, D = 41.6, T = 61.6$; $t = 15, D = 33.3, T = 53.3$; $t = 20, D = 26.6, T = 46.6$; $t = 25, D = 21.3$, $T = 41.3$; $t = 30, D = 17.0, T = 370$; $t = 35, D = 13.6, T = 33.6$; $t = 40, D = 10.9, T = 30.9$; $t = 45, D = 8.7, T = 28.7$; $t = 50, D = 7.0$, $T = 27.0$; $t = 55, D = 5.6, T = 25.6$; $t = 60, D = 4.5$; $T = 24.5$

f.

13. Car Acceleration Problem a. Let S = no. of kph. Let $D = 160 - S$. Let t = no. of sec. $D = 160 \times 10^{-0.02499t}$ b. 103.2 kph c. About 48 sec

d.

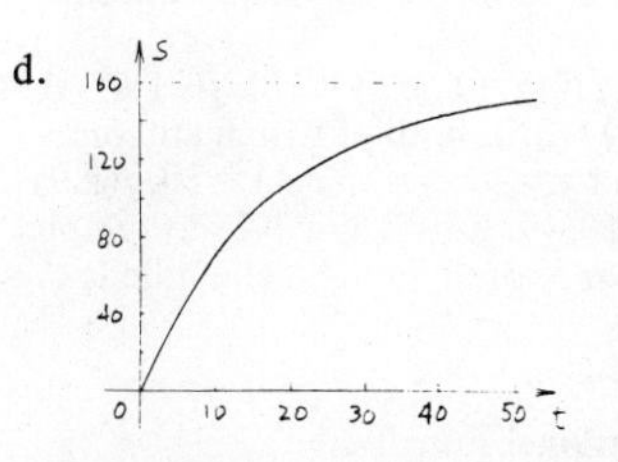

e. See Exercise 2-2, Problem 1c.
f. $S = 160\ (1 - 10^{-0.02499t})$

15. Sunlight Below the Water Problem. a. Let I = no. of units intensity. Let d = no. of meters deep. $I = 1000 \times 10^{-0.6109d}$ b. $d = 0, I = 1000$, $\log I = 3.00$; $d = 2, I = 60, \log I = 1.78$; $d = 4, I = 3.6, \log I = 0.56$; $d = 6$, $I = 0.216, \log I = 0.67$; $d = 8, I = 0.01296, \log I = -1.89$; $d = 10, I = 0.0007776$; $\log I = -3.11$ c. About 9.8 meters

d.

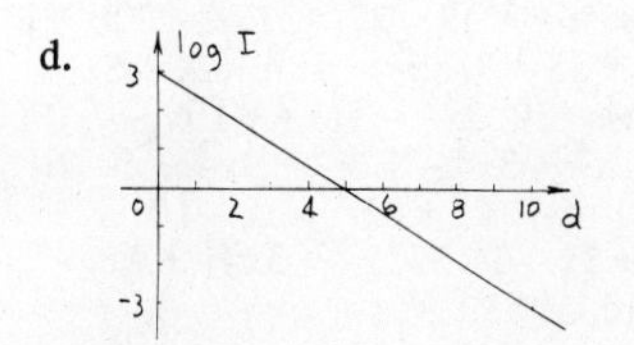

e. Graph is a *straight line*!

Exercise 6-15, page 173; Chapter Review

R. 1. a. i. $(x^a)^b = x^{ab}$ ii. $x^a/x^b = x^{a-b}$ iii. $(xy)^a = x^a y^a$ b. i. –9 i. 1/9 iii. 1 c. i. $100x^{-11}$ i. $(1/27)x^3y^{12}$ iii. $720x^{-1/4}$ R2. a. 16 b. 8 c. 4.141 d. 0.05766 e. 71.92

R3. a.

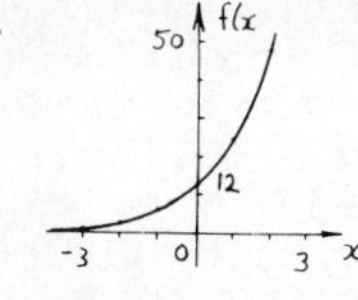

b.

R4. $f(x) = 100 \times 10^{-0.03098x}$ R5. a. 128 b. 78.92 c. 1/2; $d = 43$, $I = 64$; 1/4: $d = 86, I = 32$; 1/8: $d = 129, I = 16$; 1/16: $d = 172, I = 8$

d.

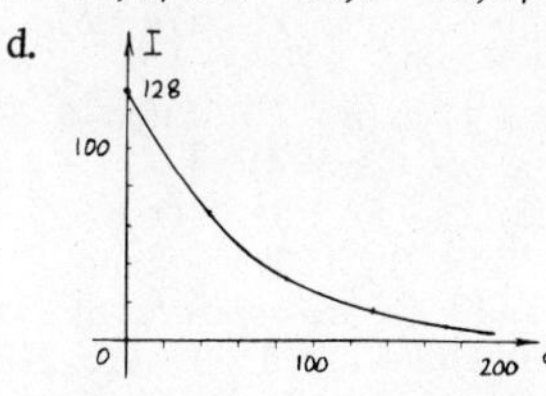

e. Each *equal* change in distance, 43 feet in this problem, makes the lights seem *twice* as bright, no matter how far away the car is.

CHAPTER 7
RATIONAL ALGEBRAIC FUNCTIONS

Exercise 7-1, page 177; Introduction to Rational Algebraic Functions

1. $f(0) = -1/3, f(1) = -1/2, f(2) = -1, f(4) = 1, f(5) = 1/2, f(-1) = -1/4, f(-3) = -1/6$ 2. $f(3) = 5/0$ and $f(-2) = 0/0$, both of which are *undefined.* 3. Domain excludes $x = 3$ and $x = -2$. 4. $f(3.1) = 10, f(2.9) = -10, f(-2.1) = -10/51, f(-1.9) = -10/49$. See Figure 7-2a for graph. 5. Graph might have a *vertical asymptote* or a "hole." (The hole is called a "removable discontinuity.")

Exercise 7-2, page 177; Graphs of Rational Functions

(No exercises.)

Exercise 7-3, page 180; Special Products

1. $x^2 - 2x - 15$ 3. $x^2 + 7x + 6$ 5. $x^2 - 9x + 14$ 7. $x^2 - 4$ 9. $2x^2 - 7x - 15$ 11. $3x^2 - 19xy + 28y^2$ 13. $8x^2 + 22x + 5$ 15. $21x^2 + 50x - 16$ 17. $169x^2 - 4y^2$ 19. $169x^2 - 52xy + 4y^2$ 21. $x^2 + 6x + 9$ 23. $4x^2 - 20x + 25$ 25. $16x^2 + 56x + 49$ 27. $121x^2 - 22x + 1$ 29. $16 - 40x + 25x^2$ 31. $9x^2 + 30x + 25$ 33. $x^3 + 5x^2 + 2x - 8$ 35. $x^3 - 13x^2 + 30x - 88$ 37. $x^3 - y^3$ 39. $2x^3 - 11x^2 + 19x - 7$ 41. $x^4 + 2x^3 - 10x^2 + 34x - 35$ 43. $x^5 + 3x^4 - 12x^3 + 9x^2 + x - 2$ 45. $x^5 + 7x^4 + 4x^3 - 27x^2 + 21x - 10$ 47. $x^3 + 6x^2 + 3x - 10$ 49 $2x^3 - 5x^2 - x + 6$ 51. $x^3 - 3x^2 + 3x - 1$ 53. $x^3 + 3x^2y + 3xy^2 + y^3$ 55. a. 2001 b. 1776 c. 5096 d. 5963

Exercise 7-4, page 184; Factoring Special Kinds of Polynomials

1. $x(x + 1)$ 3. $2(x - 5)$ 5. $5x(x^2 - 10a)$ 7. $6r^3(2r^2 + 3x^2)$ 9. $4x^2(2x^7 + 3x^5 - 7)$ 11. $(2x - 3)(x^2 - 2)$ 13. $(4x + 15)(3x^2 + 8)$ 15. $(2x - 3)(9x^2 + 4)$ 17. $3(4x - 3)(2x^2 + 5)$ 19. $(2x - 3a)(4x + 3y)$ 21. $(7s - 6)(r + 1)(r - 1)$ 23. $(x + 7)(x - 2)$ 25. $3(x + 5)(x - 2)$ 27. $(4x + y)(2x + y)$ 29. $(3r - 5s)(r + 2s)$ 31. $(3x + 1)(2x + 3)$ 33. $(x + 4)(6x - 5)$ 35. $(2x - 3)(5x - 7)$ 37. $(2x - 7)(5x + 3)$ 39. $(x - 8)(5x - 2)$ 41. $(8x + 1)(x + 6)$ 43. $(2x - 3)^2$ 45. $(x + 8)^2$ 47. $(ax - a + 1)(ax + x + a)$ 49. $(x + 3)(x - 3)$ 51. $(2x + 5)(2x - 5)$ 53. $(3x + 7)(3x - 7)$ 55. $(x + 1)(x - 1)$ 57. $36(x + 2)(x - 2)$ 59. $5(x + 4)(x - 4)$ 61. $(x + 6y)(x - 6y)$ 63. $(x^2 + y^2)(x + y)(x - y)$ 65. $(x^3 + y^2)(x^3 - y^2)$ 67. $a^3(a + 1)(a - 1)$ 69. $x^2y^2(x + y)(x - y)$ 71. $(x + 7 + 3)(x + 7 + 3) = (x + 10)(x + 4)$ 73. $(6 + (x - 5))(6 - (x - 5)) = (x + 1)(11 - x)$ 75. $(4x + 3)(3x + 4)$ 77. $(x - 5)(24x - 1)$ 79. $(15x - 2)(2x + 3)$ 81. Prime 83. $(3x - 4)(12x - 5)$ 85. D = –16. Won't factor 87. $D = 169 = 13^2$. Will factor 89. $D = 81 = 9^2$. Will factor 91. D = –47. Won't Factor 93. $D = 6561 = 81^2$. Will factor 95. $4(x + 2y)(x - 2y)$ 97. $(2x - 3y)^2$ 99. $(a + 11)(a - 8)$ 101. $5(3x + 2)(2x + 5)$ 103. $x(x + 5)(x^2 - 2)$ 105. $(x^2 + 3)(x + 2)(x - 2)$ 107. $4(15x - 2)(x - 1)$ 109. $(x - 2)(16x - 3)$ 111. $(5x + 6)(x + 3)(x - 3)$ 113. Prime 115. $(10 + x - y)(10 - x + y)$ 117. $(7a + b)(5a + 6b)$ 119. $(a + b + c)(a - b + c)(a + b - c)(a - b + c)$ 121. a. $(x + 3 + y)(x + 3 - y)$ b. $(a + b - 1)(a - b + 1)$ c. $(p - 7 + 3k)(p - 7 - 3k)$ d. $(x + 2y - 1)(x - 2y + 1)$ 122. a. $(x^2 + 3x + 4)(x^2 - 3x + 4)$ b. $(x^2 + x + 6)(x^2 - x + 6)$ c. $(x^2 + 5x + 3)(x^2 - 5x + 3)$ d. $(x + 5)(x - 1)(x - 5)(x + 1)$ e. $(x^2 + 2x + 2)(x^2 - 2x + 2)$ 123. $x - y$ and $x + y$ are *conjugates* of each other. $x - y$ and $y - x$ are *opposites* of each other. 125. a. $2^4 - 1 = (2^2 + 1)(2 + 1)(2 - 1) = 5 \cdot 3 \cdot 1$ c. $2^{16} - 1 = (2^8 + 1)(2^4 + 1)(2^2 + 1)(2 + 1)(2 - 1) = 257 \cdot 17 \cdot 5 \cdot 3 \cdot 1$

127. LCM, GCF, and Musical Harmony a. *A:* $220 = 2 \cdot 2 \cdot 5 \cdot 11$; *C:* $264 = 2 \cdot 2 \cdot 2 \cdot 3 \cdot 11$; *C#:* $275 = 5 \cdot 5 \cdot 11$; LCM (*A, C*) $= 2 \cdot 2 \cdot 2 \cdot 3 \cdot 5 \cdot 11 = 1320$; LCM (*A, C#*) $= 2 \cdot 2 \cdot 5 \cdot 5 \cdot 11 = 1100$; LCM (*C, C#*) $= 2 \cdot 2 \cdot 2 \cdot 3 \cdot 5 \cdot 5 \cdot 11 = 6600$ b. *A* and *C*, *A* and *C#* harmonize. *C* and *C#* do *not*. c. No. *C* and *C#* both harmonize with *A*, but *not* with each other. d. GCF(220, 264) $= 2 \cdot 2 \cdot 11 = 44$; GCF (220, 275) $= 5 \cdot 11 = 55$; GCF(264, 275) = 11. A *high* GCF indicates *good* harmony. A *low* GCF indicates *poor* harmony. e. (e.g.,) For 220 and 264, LCM × GCF = 1320 × 44 = 58080 220 × 264 = 58080

Exercise 7-5, page 188; Division of Polynomials

1. $x^2 - 5x + 14$ 3. $x^2 - x + 3$ 5. $3x^2 - 4x - 5$ 7. $x + 1$ 9. $x - 2$ 11. $x^2 + xy + y^2$ 13. $x^3 + x^2 + x + 1$ 15. $x^2 - x + 1$ 17. $2x^3 - 3x^2 + 4x - 5$

Exercise 7-6, page 191; Factoring Higher Degree Polynomials

1. $(x + 2)(x - 4)(x + 5)$ 3. $(x - 2)(x + 3)(x - 11)$ 5. $(x + 2)(x^2 - 3x + 1)$ 7. Prime 9. $(x - 1)(x + 2)(x - 3)(x + 4)$ 11. $(x + 1)^2(x - 2)^2$ 13. $(x + 1)^5$ 15. $(2x - 1)(x + 2)(x + 1)$ 17. $(2x + 1)(x^2 + x - 1)$ 19. $(2x + 1)(3x + 1)(2x - 1)$ 21. $(2x + 3)(2x - 5)(3x - 2)$ 23. $(x + y)(x^6 - x^5y + x^4y^2 - x^3y^3 + x^2y^4 - xy^5 + y^6)$ 25. $(x - 1)(x^{10} + x^9 + x^8 + x^7 + x^6 + x^5 + x^4 + x^3 + x^2 + x + 1)$ 27. $(x + 2)(x^4 - 2x^3 + 4x^2 - 8x + 16)$ 29. $(a - b)(a^{12} + a^{11}b + a^{10}b^2 + a^9b^3 + a^8b^4 + a^7b^5 + a^6b^6 + a^5b^7 + a^4b^8 + a^3b^9 + a^2b^{10} + ab^{11} + b^{12})$ 31. $(a + 2b^2)(a^4 - 2a^3b^2 + 4a^2b^4 - 8ab^6 + 16b^8)$ 33. $(x - y)(x^2 + xy + y^2)$ 35. $(x + 2)(x^2 - 2x + 4)$ 37. $(r - s^2)(r^2 + rs^2 + s^4)$ 39. $(x + y)(x^2 - xy + y^2)(x - y)(x^2 + xy + y^2)$ 41. $8(a - 2b)(a^2 + 2ab + 4b^2)$ 43. $xy(x + y)(x^2 - xy + y^2)$ 45. $[(a - b) - (a + b)]\ [(a - b)^2 + (a - b) + (a + b)^2] = -2b(3a^2 + b^2)$

Exercise 7-7, page 195; Products and Quotients of Rational Expressions

1. See Exercise 1-7. Problem 13. 3. a. –1 b. 1 c. $(x - 5)/(x + 5)$ d. 1 5. $(x + 2)/2$ 7. $(x + 3)(x + 5)/5x$ 9. $(x^3 + 6)/6x$ 11. –1 13. $-(x + 7)$ 7 15. –1 17. $-(x + 1)/(4 + x)$ 19. $(x - 5y)^2/(xy(x + 5y))$ 21. $(x +$ $(x - 1)$ 23. $(x - 1)/(x + 4)$ 25. $x + 2$ 27. $(x + 5)/(5 - x)$ 29. $x^2 - 1$ 31. x 33. $1/(x + y)$ 35. $(x + 2)(x - 3)/(x - 2)$ 37. $(x + 1)(x + 5)/(x - 5)$ 39. $(x + 5)^2/(x - 7)^2$ 41. 1 43. $(x + 2)^2(x + 3)/(x - 4)^3$ 45. $(x^2 - y^2)^2$ 47. $(x - 1)/x$ 49. $(x^2 + x)/(x - 2)$ 51. $-(x + 1)/(x^2(3 + x))$ 53. $1/x$ 55. $(x + 3)/(x - 2)$ 57. $(y^2 - x^2)/(xy)$ 59. $(x - 1)/(x + 1)$ 61. $6/x^2$

Exercise 7-8, page 198; Sums and Differences of Rational Expressions

1. See Exercise 1-7, Problem 23 3. $x/12$ 5. $2x/(x^2 - 1)$ 7. $9/(2x - 3y)$ 9. $(2 - 4x)/(x^2 - 1)$ 11. 0 13. $3x/(x^2 - y^2)$ 15. $y/(x - y)^2$ 17. $-1/(1 + 2x)$ 19. $(7 - 3x)/(1 - x)^2$ 21. $1/(x + 4)$ 23. $(3x + 4y)/((x + y)(2x + 3y))$ 25. $2xy/(x^3 - 8y^3)$ 27. $-2x/(x - 3)$ 29. $-x/a$ 31. $-1/(x^2 - 1)$ 33. $2/((x - 1)(x - 2)(x - 3))$ 35. $2/(x + y)$ 37. $-2/((x - 1)(x - 3))$ 39. $(3x + 9)/((x + 2)(x - 1))$ 41. $2x/(x^2 - 1)$ 43. $2/(x + 1) + 3/(x - 2)$ 45. $5/(x + 1) - 2/(x + 4)$ 47. $2/(x + 1) + 3/(x - 2) - 1/(x + 3)$

Exercise 7-9, page 202; Graphs of Rational Algebraic Functions, Again

1. a. 7/0 and 0/0 are undefined. b. 7/0 is *infinite* and 0/0 is *indeterminate.* c. 0/7 = 0.

3. a. $x \neq 5$ b. asymptote c. $f(1) = -1$; $f(3) = -2$; $f(4) = -4$; $f(7) = 2$; $f(9) = 1$

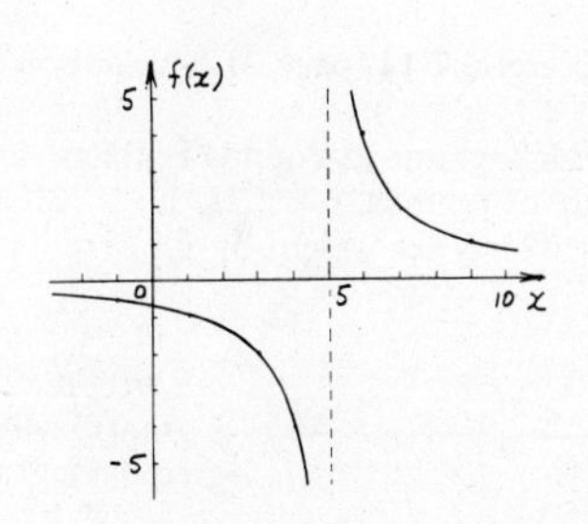

5. a. $f(x) = (x - 1)/(x + 2)$
b. $x \neq -2$ (asymp.), $x \neq 3$ (hole)
c. $f(-5) = 2$; $f(-4) = 2.5$; $f(-3) = 4$; $f(-1) = -2$; $f(0) = -0.5$; $f(1) = 0$; $f(2) = 1/4$, $f(3) = \boxed{2/5}$, $f(4) = 1/2$

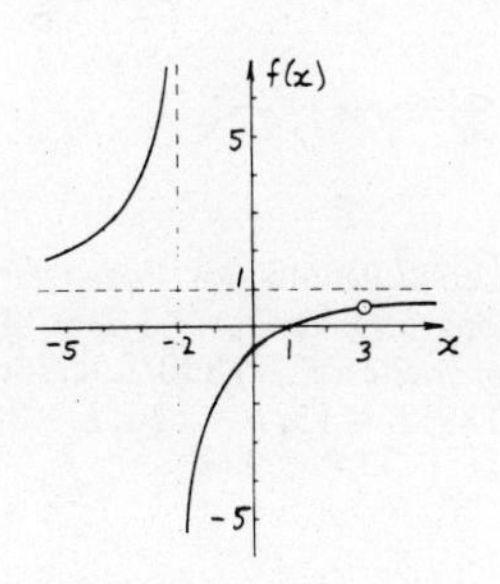

7. $f(x) = 2/((x + 5)(x - 2))$ a, b. $x \neq -5$ (asymp.), $x \neq 2$ (asymp.) c. $f(-6) = 1/4$; $f(-4) = -1/3$; $f(-3) = -1/5$; $f(-2) = -1/6$; $f(-1) = -1/6$; $f(0) = -1/5$; $f(1) = -1/3$; $f(3) = 1/4$

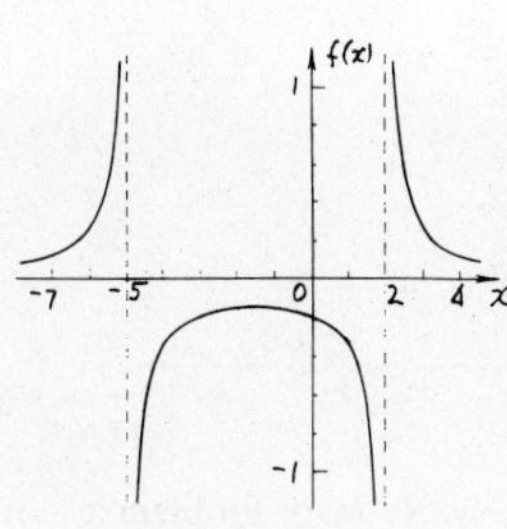

9. $f(x) = 1/(x + 4)$ a, b: $x \neq -4$ (asymp.), $x \neq 1$ (hole) c. $f(-6) = -1/2$; $f(-5) = -1$; $f(-3) = 1$; $f(-2) = 1/2$; $f(-1) = 1/3$; $f(0) = 1/4$; $f(1) = \boxed{1/5}$

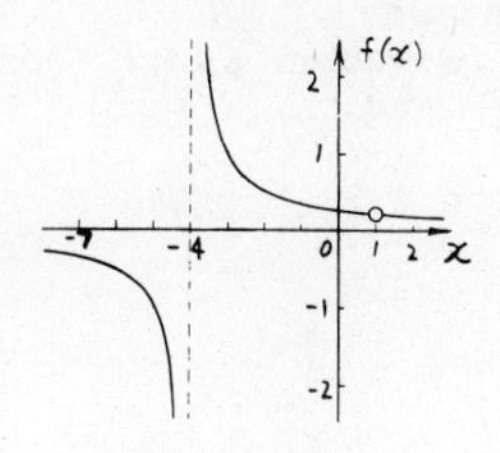

11. $f(x) = x + 5$ a, b: $x \neq 3$ (hole)
c. Linear funct. $m = 1$, $b = 5$

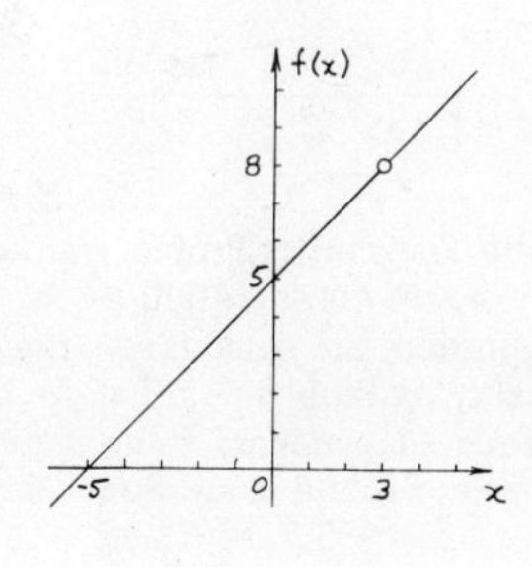

13. $f(x) = x^2 + 2x + 4$ a, b: $x \neq 2$ (hole) c. Quad. Funct., Vertex: $(-1, 3)$, y-int. = 4

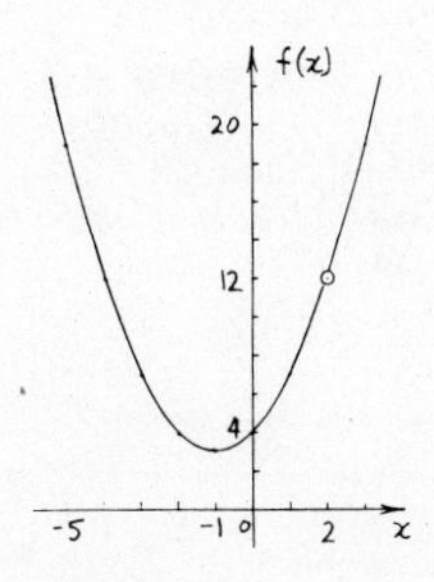

15. $f(x) = x^2 - 4x + 3$ a, b: $x \neq -2$ (hole) c. Quad. Funct., Vert: $(2, -1)$

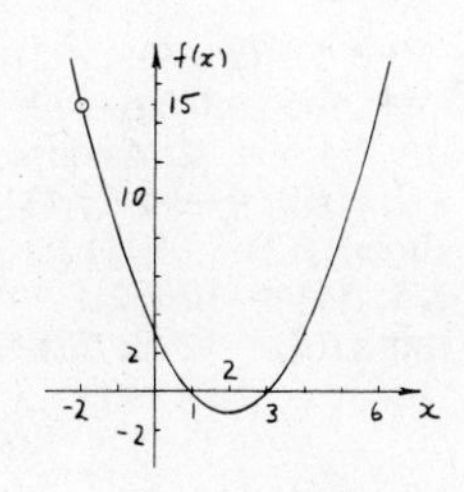

17. $f(x) = 5/(x^2 + 1)$ a, b: no excluded values. c. $f(-3) = 1/2$; $f(-2) = 1$; $f(-1) = 5/2$; $f(0) = 5$; $f(1) = 5/2$; $f(2) = 1$; $f(3) = 1/2$

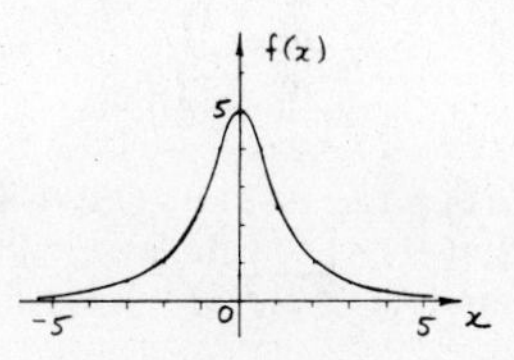

19. $f(x) = 1/(x + 2)^2$ a, b: $x \neq -2$ (asymp.), $x \neq -5$ (hole) c. $f(-5) = \boxed{1/9}$; $f(-4) = 1/4$; $f(-3) = 1$; $f(-1) = 1$; $f(0) = 1/4$; $f(1) = 1/9$

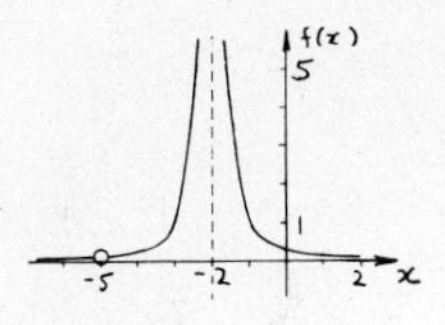

For Problems 21 through 30, parts a and b are included in the list of values.

21. $f(x) = (x - 1)/((x - 3)(x + 2))$; $f(-5) = -1/4$; $f(-4) = -5/14$; $f(-3) = -2/3$; $x = -2$: Asymp; $f(-1) = 1/2$; $f(0) = 1/6$; $f(1) = 0$; $f(2) = -1/4$; $x = 3$: Asymp.; $f(4) = \boxed{1/2}$ (hole); $f(5) = 2/7$

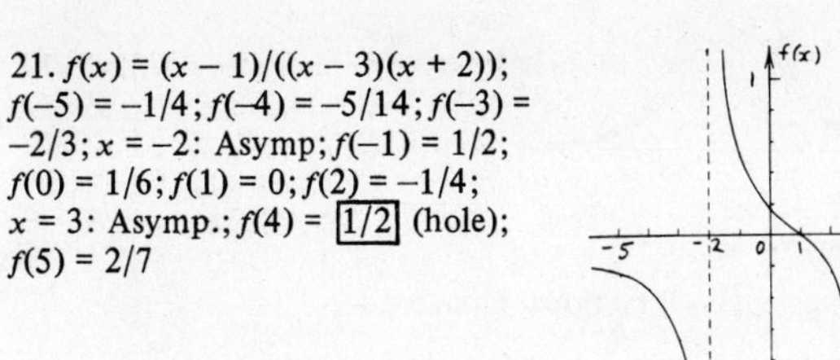

23. $f(x) = (y + 2)/(x - 2)^2$; $f(-5) = 3/49$; $f(-4) = -1/18$; $f(-3) = -1/25$; $f(-2) = 0$; $f(-1) = 1/9$; $f(0) = 1/2$; $f(1) = \boxed{3}$ hole; $x = 2$: Asymp.; $f(3) = 5$; $f(4) = 3/2$; $f(5) = 7/9$

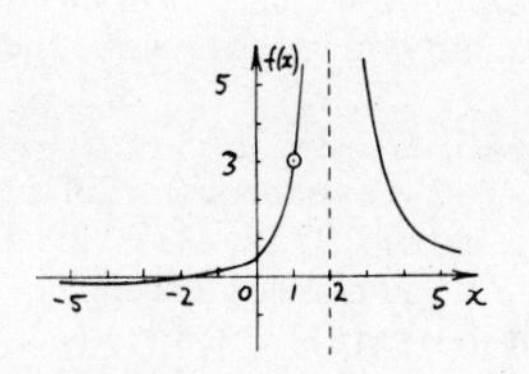

25. $f(x) = 1/(x-2); f(-3) = -1/5;$ $f(-2) = -1/4; f(-1) = -1/3; f(0) =$ $-1/2; f(1) =$ $\boxed{-1}$ (hole); $x = 2$: Asymp.; $f(3) = 1; f(4) = 1/2; f(5) =$ $1/3; f(6) = 1/4; f(7) = 1/5.$

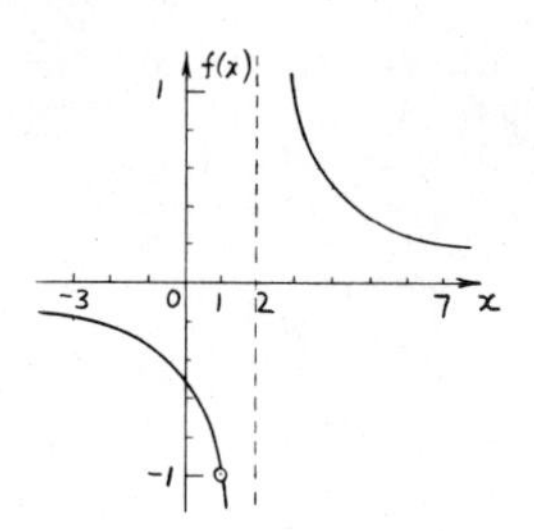

27. $f(x) = 7(x+5)/((x-5)(x+2));$ $f(-6) = -7/44; f(-5) = 0;$ $f(-4) =$ $7/18; f(-3) = 7/4; x = -2$: Asymp.; $f(-1) = -4\text{-}2/3; f(0) = -3\text{-}1/2; f(1) =$ $\boxed{-3\text{-}1/2}$ (hole); $f(2) = -4\text{-}1/12;$ $f(3) = -5\text{-}3/5; f(4) = -10\text{-}1/2;$ $x = 5$: Asymp.; $f(6) = 9\text{-}5/8; f(7) =$ $4\text{-}2/3.$

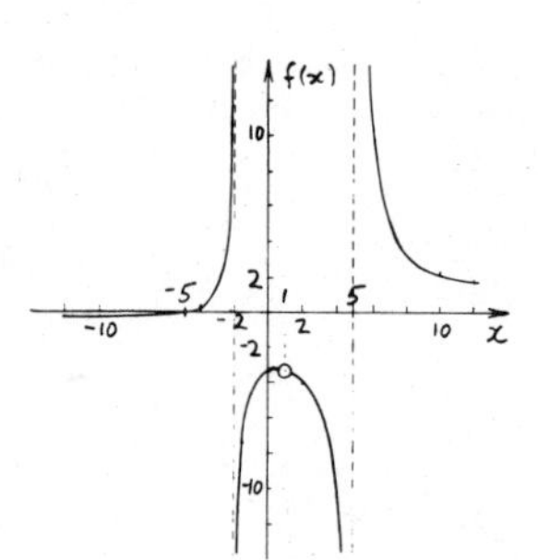

29. $f(x) = 1/x; f(-3) = -1/3; f(-2) =$ $-1/2; f(-1) =$ $\boxed{-1}$ (hole); $x = 0$: Asymp.; $f(1) = 1; f(2) = 1/2; f(3) =$ $1/3.$

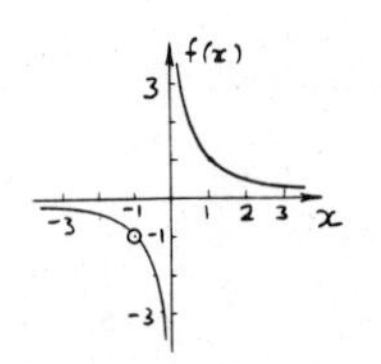

31.

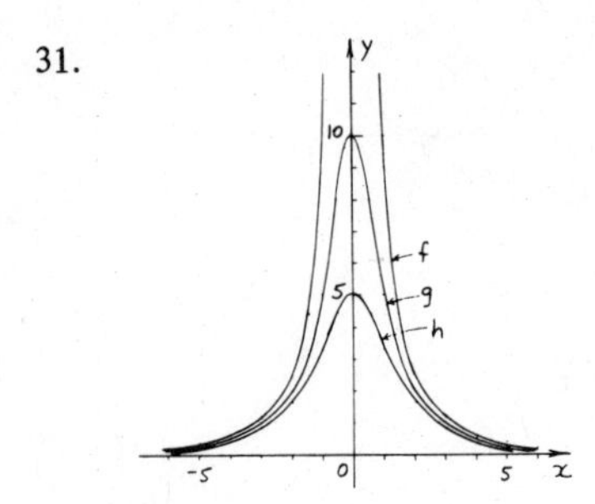

Exercise 7-10, page 206; Fractional Equations

1. a. $x(x-4) = 3(x-4)$ b. Both members of the transformed equation equal *zero* when $x = 4$. c. 4 is an *extraneous* solution. d. You can *multiply* both members of an equation by an expression that can equal zero, but you cannot *reverse* the process and *divide* by zero. 3. $\{3, -1\}$ 5. $\{12, -2\}$ 7. $\{-5/2, -1\}$ 9. $\{2/3, 1\}$ 11. $\{0, -3\}$ 13. $\{0, 2\}$ 15. $\{7, -2\}$ 17. $\{-1, 2, -3\}$ 19. $\{1, -1+\sqrt{5}, -1-\sqrt{5}\}$ 21. $s = \{-1\}$(2 is extraneous) 23. $S = \{-1\}$(3 is extr.) 25. $S = \emptyset$ (−4 and 3 are extr.) 27. $S = \{7\}$ 29. $S = \{14/3\}$ 31. $S = \emptyset$ (2 is extr.) 33. $S = \{$all real nos. except 1$\}$ 35. $S = \{3\}$(3/2 is extr.) 37. $S = \{5\}$(−3 is extr.) 39. $S = \phi$ (−3, 2 are extr.) 41. $S = \{-2\}$(0, 3 are extr.) 43. *Continued Fractions* $1 + \sqrt{2}$

Exercise 7-11, page 212; Variation Functions

1. Kilograms-to-Pounds Problem a. Let k = no. of kilograms. Let p = no. of pounds. $p = 2.2k$ b. i. 220 lb ii. 55 lb iii. 330 lb c. 75 kg d. (Divide your weight by 2.2)

e.

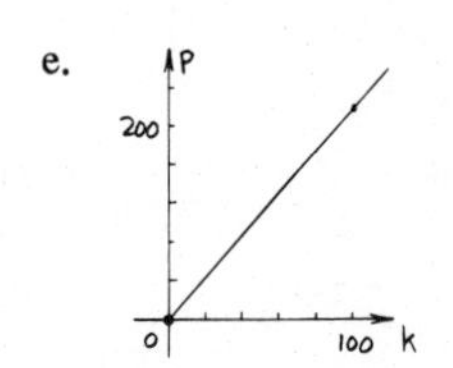

f. $m = p/k$. So m has units "pounds per kilogram," which means that a kilogram is equivalent to 2.2 pounds.

3. Wrench Problem a. Force varies inversely with length, since tripling the length makes force 1/3 as big. b. Let F = no. of pounds. Let L = no. of inches. $F = 1890/L$ c. inch-pounds d. $L = 3, F = 630; L = 10, F = 189; L = 15, F = 126; L = 30, F = 63; L = 60, F = 31.5$ e. 6.3 inches f. 37.8 inches

g.

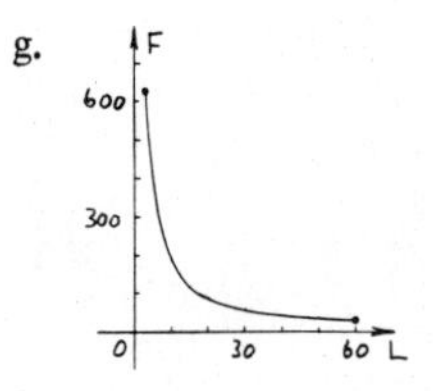

5. Water Main Problem a. Let h = no. of houses. Let d = no. of cm diameter. $h = 0.5\,d^2$ b. $d = 30, h = 450; d = 43, h = 924; d = 100, h = 5000$ c. 55 cm main

7. Gas Law Problem a. Let P = no. of psi. Let V = no. of ft^3. $V = 16560/P$

b.

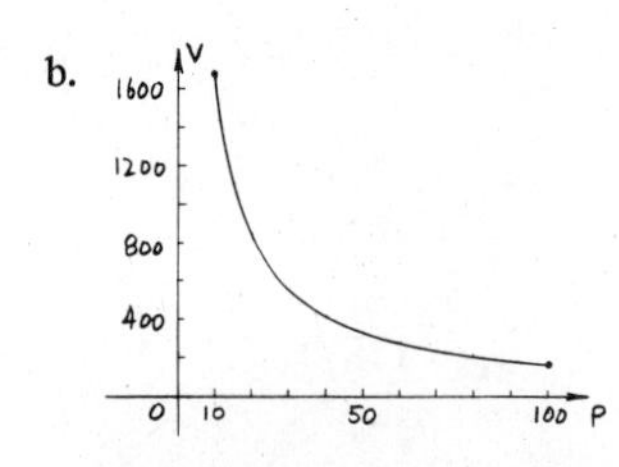

c. 60 psi d. V cannot be zero because K/P is never zero. e. $V = K/P$ (given); $\therefore PV = k$ (mult. by P) $\therefore PV$ is a constant, Q.E.D. f. $P_1V_1 = k$ and $P_2V_2 = k$, by part (e). $\therefore P_1V_1 = P_2V_2$ by transitivity, Q.E.D.

9. Radio Transmitter Problem a. Let S = no. of units signal strength. Let d = no. of km. $s = 4000/d^2$ b. 40 units c. 400,000 units d. Close to the station, the signal is *so* strong it can *drown out* other stations.

11. Lightning Problem. a. Let d = no. of km (independent). Let t = no. of sec. (dependent) b. $t = 3d$ c. k is *sec per km*, the reciprocal of the speed of sound in air. So it takes 3 sec. for sound to travel 1 km. $d = 1, t = 3; d = 2.5, t = 7.5; d = 10, t = 30$

e. 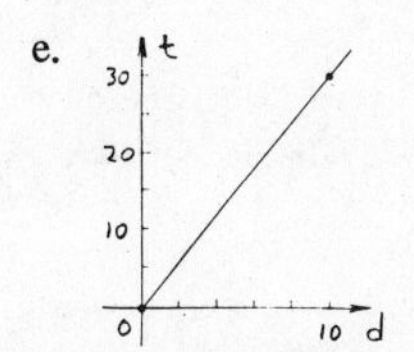

f. 9-2/3 km g. If $t = 0$, then so does d. So the lightning strikes *you*!

13. Friction Problem a. Force varies *directly* with weight, since doubling the weight *doubles* the force. b. Let F = no. of pounds force. Let W = no. of pounds weight $F = 0.6\ W$ c. 78 pounds d. 108 pounds

e. 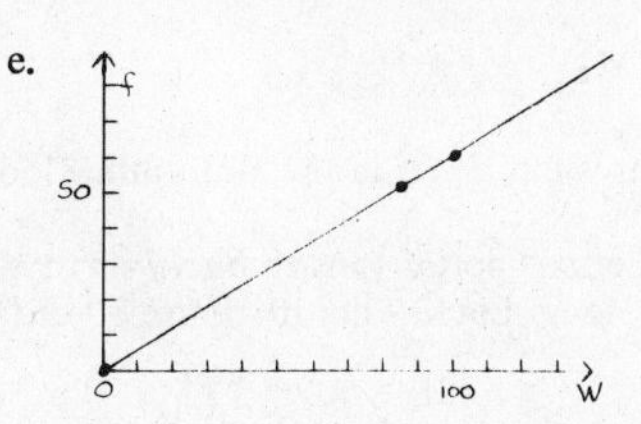

15. Radiant Heat Problem a. Let H = no. of calories per min. Let T = no. of degrees K. 1280 cal. per min. b. H varies directly with the 4th power of T. c. $H = (5/300^4)T^4$ d. i. 252.8 cal/min. ii. 617.3 cal./min.

e. 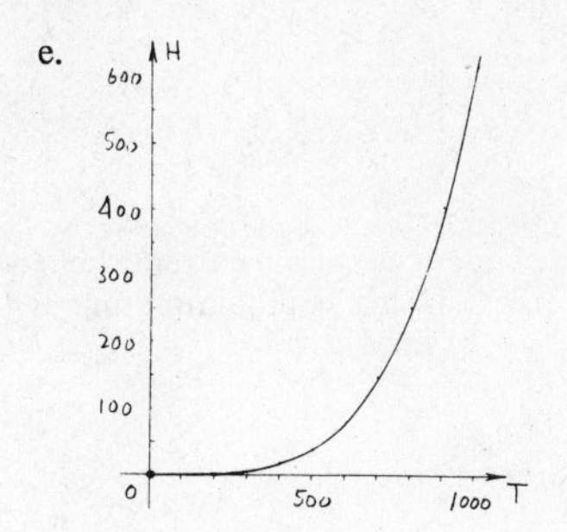

17. Ruby Problem a. $2,000,000. Large rubies are very scarce! b. 20 cents. Small rubies aren't much in demand.
19. Stopping Distance Problem a., b. Let s = no. of kph. Let r = no. of meters reaction distance. Let b = no. of meters braking distance. Let t = total no. of meters stopping distance. $s = 30, r = 7, b = 6, t = 13; s = 60, r = 14, b = 24, t = 38; s = 90, r = 21, b = 54, t = 75; s = 120, r = 28, b = 96, t = 124; s = 150, r = 35, b = 150, t = 185; s = 180, r = 42, b = 216, t = 258;$

c. 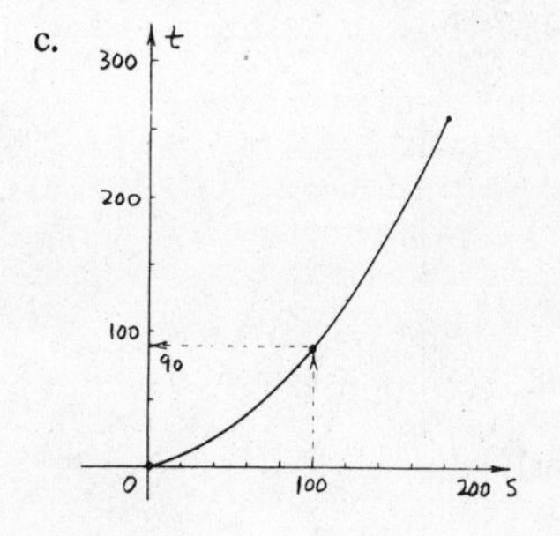

d. $r = (7/30)s$; $b = (1/150)s^2$ e. 90 m f. $b \approx 3$ football fields g. (See driver's manual.)

21. Egg Problem a. Twice the length implies 2^3 or 8 times the volume. Three times the length implies 3^3 or 27 times the volume. 27 ≈ two dozen. ∴ length ≈ 3 times as big, or 3 × 6 = 18 cm b. 216 robin's eggs
23. Large and Small People Problem a. Let h = height. Let m = no. of kg mass. Let s = no. of cm^2 skin area. Let b = no. of cm belt length. i. $m = k_1h^3$ ii. $s = k_2h^2$ iii. $b = k_3h$ b. i. Brobdingnagian: mass = 70,000, skin area = 2,500,000, belt size = 900; ii. Lilliputian: mass = 0.07, skin area = 250, belt size = 9 c. Lilliputian is 70 grams, which is about 0.15 pounds, or slightly less than a MacDonald's hamburger patty. d. Brobdingnagian's legs can support $(200)(10^2) = 20{,}000$ kg. But his mass is 70,000 kg. So he could *not* support his own mass!
25. Shark Problem a. Weight varies directly with *cube* of length. b. Let W = no. of pounds. Let L = no. of feet. $W = 0.5926L^3$ c. i. 5 lb. ii. 9259 lb. d. 7.5 feet e. 592,600 lb. Wow!
27. Medication Problem a. Let d = mg safe dosage, Let A = skin area. Let h = cm height. $d = k_1A$, $A = k_2h^2$ b. $d = k_3h^2$ c. Safe dosage varies directly with the square of the height. d. $d = 0.001333\ h^2$ e. i. $h = 90, d = 10.8$; ii. $h = 170, d = 38.5$; ii. $h = 200, d = 53.3$

Exercise 7-12, page 221; Chapter Review

R1. a. $3x(2a + 5b)$ b. $9(2r + 3s)(2r - 3s)$ c. $(2x + 3y)(x - 12y)$ d. $(2x + 3a)(x + 3)(x - 3)$ e. $(x + 3)(2x - 1)(x - 9)$ f. $(2x + y)(16x^4 - 8x^3y + 4x^2y^2 - 2xy^3 + y^4)$ g. Prime R2. a. $(x + 2)/(x - 2)$ b. $-(x + 2)(x + 1)/(x - 2)$ R3. a. $S = \phi$ (−4 and 2 are both extraneous.) b. $S = \{-5, 3\}$ c. $S = \{-7\}$ (2 is extraneous.) d. S = {all real nos. except 2 and −4}

R4. 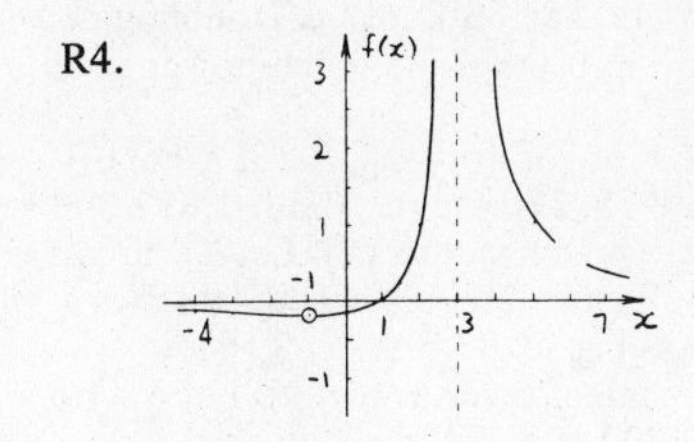

R5. a. $y = k/x^3$ b. $y = 28672/x^3$ c. 83592 d. 56

CHAPTER 8
IRRATIONAL ALGEBRAIC FUNCTIONS

Exercise 8-1, page 226; Existence of Irrational Numbers

1. a. 2236/1000 b. $(13 \cdot 43)/(2 \cdot 5^3)$ c. $(13^2 \cdot 43^2)/(2^2 \cdot 5^6)$ d. No common factors. Not an integer. e. 4.999696 3. a. 469/100 b. $(7 \cdot 67)/(2^2 \cdot 5^2)$ c. $(7^2 \cdot 67^2)/(2^4 \cdot 5^4)$ d. No common factors. Not an integer. e. 21.9961 5. a. 438/100 b. $(3 \cdot 73)/(2 \cdot 5^2)$ c. $(3^3 \cdot 73^3)/(2^3 \cdot 5^6)$ d. No common factors. Not an integer. e. 84.027672 7. $4^3 = 64$, $5^3 = 125$. ∴ $4 < \sqrt[3]{100} < 5$. ∴ $\sqrt[3]{100}$ is irrational. 9. $2^5 = 32$, $3^5 = 243$. ∴ $2 < \sqrt[5]{53} < 3$. ∴ $\sqrt[5]{53}$ is irrational. 11. $1^2 = 1$, $2^2 = 4$, ∴ $1 < \sqrt{3} < 2$. ∴ $\sqrt{3}$ is irrational. 13. $5^4 = 625$, ∴ $\sqrt[4]{625} = 5$ 15. $11^3 = 1331$, $12^3 = 1728$, ∴ $11 < \sqrt[3]{1332} < 12$. ∴ $\sqrt[3]{1332}$ is irrational.

Exercise 8-2, page 228; Simple Radical Form

1. $41\sqrt{3}$ 3. $5\sqrt[3]{3}$ 5. $2\sqrt{2}$ 7. $3\sqrt{6}$ 9. $8\sqrt{3}$ 11. $9 + 2\sqrt{14}$ 13. $29 - 12\sqrt{5}$ 15. 2 17. 10 19. $42 - 6\sqrt{21} + \sqrt{105} - 3\sqrt{5}$ 21. $2\sqrt[3]{3}/3$ 23. $\sqrt[4]{8}$ 25. $5\sqrt[3]{2}/4$ 27. $\sqrt{7}$ 29. $(\sqrt{5} + 1)/4$ 31. $\sqrt{7} - \sqrt{3}$ 33. $\sqrt{6} + \sqrt{3} - \sqrt{2} - 1$ 35. $(107 + 42\sqrt{2})/89$ 37. $2 + \sqrt{3}$ 39. $2\sqrt{3} + 3 + \sqrt{21}$ 41. $(3\sqrt{2} + 2\sqrt{3} + \sqrt{30})/12$ 43. $-(\sqrt[3]{3} + \sqrt[3]{6} + \sqrt[3]{9})$ 45. $(\sqrt[3]{5} - \sqrt[3]{2})/3$ 47. $\sqrt{3} + 1$ 49. $3 - \sqrt{3}$ 51. $3 + \sqrt{2}$ 53. $2\sqrt{5} - 2\sqrt{3}$ 55. a. $3/\sqrt{2} \approx 3/1.414 \approx 2.1216$ b. $3/\sqrt{2} = 3\sqrt{2}/2 \approx (3 \times 1.414)/2 = 4.242/2 = 2.121$ c. With rational denom., you can use *short* division, which is easier than long division.

Exercise 8-3, page 231; Radical Equations

1. {11} 3. {7} 5. {13/12} 7. ϕ (16 is extraneous) 9. {0, 4/3} 11. {4}(−1 is extraneous.) 13. {5} (2 is extr.) 15. {2} (5 is extr.) 17. ϕ (1, 2 are extr.) 19. {4, 5} 21. ϕ (3/2, 1/2 are extr.) 23. ϕ (0 is extr.) 25. {13/4} 27. {−2}(58 is extr.) 29. $\{3\sqrt{2}, -3\sqrt{2}\}$ (5, −5 are extr.) 31. {2, −5} 33. {16} 35. {1/4} 37. ϕ (1 is extr.) 39. {−1, 2, 3} 41. {0, 5} ($(5 \pm \sqrt{37})/2$ are extr.) 43. $\{1 + 2\sqrt{3}\}$ ($1 - 2\sqrt{3}$ is extr.) 45. $\{3 - \sqrt{5}\}$ ($3 + \sqrt{5}$ is extr.) 47. $\{2 + \sqrt{5}\}$ ($2 - \sqrt{5}$ is extr.) 49. Continued Radicals a. 2 c. 4 e. If n is a *product* of *consecutive integers,* then the radical stands for an integer.

Exercise 8-4, page 234; Irrational Variation Functions

1. Ship Power Problem a. Let p = no. of horsepower. Let s = no. of knots. $s = 6.492p^{1/7}$ b. A bit more than 33 knots c. Doubling power *nowhere near* doubles the speed! d. It takes too much power to go faster than 30 knots.
3. River Basin Problem a. Let L = no. of km long. Let A = no. of thousands of km^2. $L = 72.88A^{0.6}$ b. 370 km c. About 1,872,000 km^2
5. Tree Trunk Problem a. Let h = no. of meters tall. Let d = no. of cm diameter. $d = 1.30 \times h^{3/2}$ b. 459 cm. ∴ Making height 10 times as big makes diameter about 32 times as big. c. 83.2 m
7. Pendulum Problem a. Let p = no. of seconds period. Let L = no. of meters long. $p = 1.004\sqrt{L}$ b. 0.992 m c. 99.2 m
9. Microwave Oven Problem a. Time does *not* vary directly with the number of slices because doubling the number of slices (2 to 4) *less* than doubles the number of minutes (1.75 to 2.5). b. Let t = no. of minutes. Let s = no. of slices. $t = 1.225 \times s^{0.5146}$ c. $s = 8, t = 3.57$; $s = 6, t = 3.08$; $s = 1, t = 1.225$ d. 500 slices e. Domain: $0 < s < 500$, Range: $0 \leq t \leq 30$ f. The oven probably would not hold 500 slices of bacon!

Exercise 8-5, page 239; Functions of More Than One Independent Variable

1. Beam Strength Problem a. Let w = no. lb. safe load. Let b = no. of inches broad. Let d = no. of inches deep. Let L = no. of feet long. $w = 125\,bd^2/L$ b. 250 lb c. Beams will support more weight when they are "on edge." d. 12,000 lb e. (i.) twice (ii.) 4 times (iii.) half.
3. Spike Heel Problem a. Let p = no. of psi pressure. Let d = no. of inches wide heel is. Let w = no. of lb. person weighs. $p = 1.08w/d^2$

b., c

	w=100	w=150	w=200
d	p	p	p
1	108	162	216
2	27	40.5	54
3	12	18	24
4	6.75	10.125	13.5

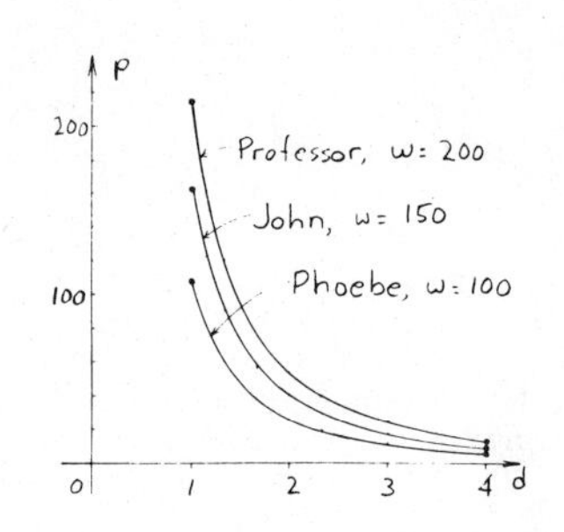

d. 1728 psi e. Such high pressures put *dents* in the aluminum floors of airplanes!
5. Bridge Column Problem a. Let b = no. of tons to *buckle.* Let c = no. of tons to crush. Let L = no. ft. long. Let d = no. in. diam. $b = 9d^4/(4L^2)$ b. $c = (5/4)d^2$

c. $L = 20, b = 900$; $L = 30, b = 400$; $L = 50, b = 144$ d. 500 tons f. Long columns *buckle*; short columns *crush* g. 26.8 ft.

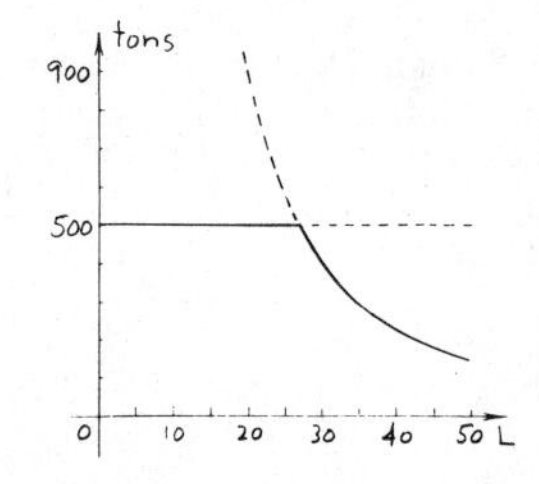

7. Reaction to Shock Problem. a. Let c = no. of ma current. Let t = no. of ms time. $c \geq 10 + (40/t)$ b. $c \geq 14$ ma c. Minimum time is 4/9 ms. d. 10 ma.

e.

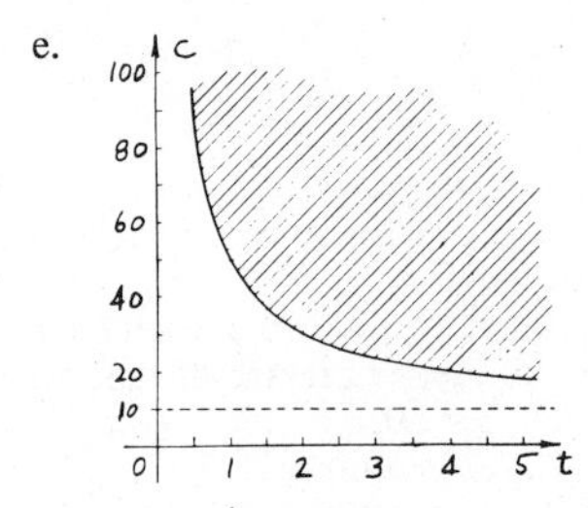

9. Gas Consumption Problem a. Let w = wind resistance. Let S = no. of kph. Let g = no. of km/liter. $g = k/s^2$ b. *kpl* varies inversely with the square of the speed. c. $g = 80000/S^2$

d. $s = 120, g = 5.56$; $s = 140, g = 4.08$; $s = 160, g = 3.125$; $s = 180, g = 2.47$; $s = 200, g = 2.00$ e. Driving 200 kph uses *4 times* as much gas as driving 100 kph! f. At very low speeds, the gas is used primarily to overcome friction and heat losses. So this model does *not* apply at low speeds. The actual graph resembles the one sketched.

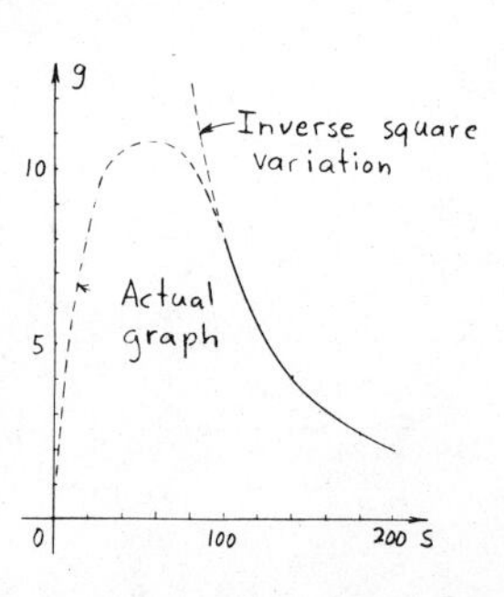

11. Why Mammals Are the Way They Are a. Let m = no. of g mass of animal. Let L = length of animal. Let A = area of skin. Let H = rate of heat loss. Let F = no. of g food eaten per day. i. $m = k_1L^3$ i. $A = k_2L^2$ iii. $H = k_3A$ iv. $F = k_4H$ b. $F = k_5m^{2/3}$. F varies directly with the 2/3 power of m. c. $F/m = k_5m^{-1/3}$; F/m varies inversely with the 1/3 power of m. d. $F/m = 3.78m^{-1/3}$ e. $F/m = 0.0208$. So an elephant eats about 2% of its mass per day f. F/m ratio *increases* as mass decreases. There can be no mammals smaller than a shrew because they would have to eat faster than their digestive system could handle it. Shrews are mean because they must continually seek food. Small animals have fur to reduce heat loss, and thus food consumption. Large animals have *no* fur to help dissipate *excess* heat. Very large animals use sea-water to help dissipate excess heat.

Exercise 8-6, page 243; Chapter Review

R1 a. $2.646 = (3^37^2)/(2^25^3) \therefore 2.646^2 = (3^67^4)/(2^43^6)$, which is *not* an integer since the denominator does not cancel. b. $4^3 = 64$, $5^3 = 125$. $\therefore 4 < \sqrt[3]{100} < 5$. Since $\sqrt[3]{100}$ is not an integer, it must be irrational. R2 a. $2\sqrt{3}$ b. $7\sqrt{7}$ c. $2\sqrt{6} - 3$ d. $\sqrt[5]{216}/2$ R3 a. $\{10\}$(3 is extraneous.) b. $\{5, 133\}$ c. $\{3/2\}$(-12 gives imaginary nos.) R4a. Defended Region i. Let m = no. of kg mass of beaver. Let A = no. of sq. m. defended region. $A = 5.926m^{1.31}$ ii. 6125 sq. m. iii. approximately 5 kg R4b. i. $y = k_1/x^3$ $x = k_2z^2/w$ ii. $y = k_3w^3/z^6$ iii. y varies directly with the cube of w, and inversely with the sixth power of z. iv. $y = 135w^3/z^6$

Exercise 8-7, page 245; Cumulative Review, Chapters 6, 7, and 8

1. a. Exponential, $y = a \times 10^{kx}$ $(k > 0)$ b. Exponential, $y = a \times 10^{kx}$ $(k < 0)$ c. Inverse variation, $y = k/x^n$ d. Direct n^{th} power variation, $y = kx^n$ $(n > 1)$ e. Direct variation, $y = kx$ f. Direct n^{th} power variation, $y = kx^n$ $(0 < n < 1)$ g. Logarithmic, $y = \log_b x$ (Could also be rational algebraic.) h. Rational algebraic, $y = P(x)/Q(x)$, where $P(x)$ and $Q(x)$ are polynomials. i. Linear, $y = mx + b$ 2. a. $y = kx$ b. $y = k/x$ c. $y = k/x^3$ d. $y = kx^4$ e. $y = a \times 10^{kx}$ 3. a. $6\sqrt{5} + 6 - 5\sqrt{3} - \sqrt{15}$ b. $9 - 2\sqrt{14}$ c. -5 d. $x - 2\sqrt{x} - 35$ e. $x^2 - 2x - 35$ f. $x^4 - 2x^3 - 20x^2 + 57x - 36$ 4. a. $(r + s)(r^6 - r^5s + r^4s^2 - r^3s^3 + r^2s^4 - rs^5 + s^6)$ b. $(2x + 3)(3x - 40)$ c. $(x - 1)(x - 2)(x + 4)$ 5. a. $3/((x - 4)(x - 1))$ b. $-(x + 2)(x - 3)/((x + 3)(x - 5))$ 6. a. $12x^{1/6}$ b. $\sqrt[6]{3}$ c. $2\sqrt{19} + 8$ d. $4\sqrt[3]{9}$ e. $\sqrt{5}$ f. $9\sqrt{7}$ g. $4/3$ h. 2 i. $-(x + 4)/(x - 1)$ j. $(3/4)|a|$ 7. a. $S = \{-2\}$(3 is extraneous.) b. $S = \{-1\}$(1 is extraneous.) 8. a. 5/0 is *infinite*. 0/0 is *indeterminate*. b. 0/5 = 0

9. $f(x) = (x - 2)(x + 4)/(x + 1)$. $f(-10) = -8$; $f(-4) = 0$; $x = -1$ Asymp; $f(0) = \boxed{-8}$ (hole); $f(1) = -2.5$; $f(2) = 0$; $f(10) = 10.2$

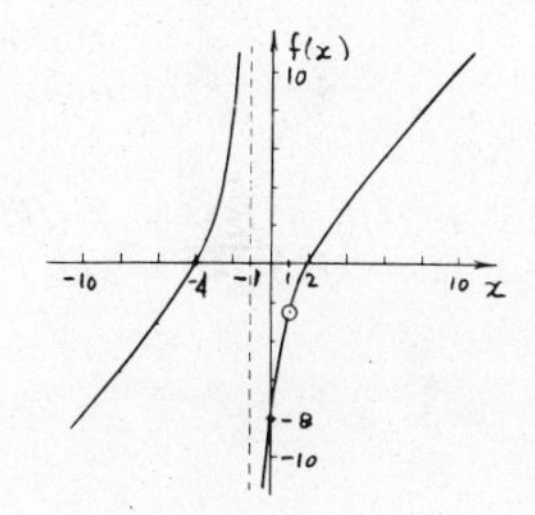

10. $x^2 - x + 2 = 0 \Rightarrow 1^2 - 4(1)(2) = -7 < 0 \Rightarrow$ no solns. $\therefore (x - 2)/(x^2 - x + 2)$ *never* has 0 denom.

11. Iodine Problem a. *Exponential* funct. b. Let m = no. of millicuries. Let d = no. of days since they received it. $m = 23.7 \times 10^{-0.037164d}$ c. i. 13.02 millicuries ii. 39.6 days d. 1 week is just less than 8.1 days, and the amt. remaining is just over (1/2) (23.7) = 11.85. 39.6 days is very near 40.5 days, which is 5 half-lives. $23.7\,(1/2)^5 = 23.7 \div 32 \approx 0.74$, which is slightly less than 0.8. e. 32.53 mc.

12. *Heat Radiation* a. Let h = rate of heat rad. Let T = ° abs temp. Let S = Surface Area. Let d = dist. away. $h = k_1T^4S/d^2$ b. $S = k_2r^2$ and $V = k_3r^3$. $h = k_4T^4V^{2/3}/d^2$

13. *Diesel Engine* a. Let T = no. of °K. Let V = no. of cu. cm. $T = 1543/V^{0.4}$ b. 886.3° K c. 6.07 cu cm

14.

X	1	2	3	4	5	6
D	8	3	0	−1	0	
N	4	0	−2	−2	0	
F	0.5		0			

Output:

1	0.5
2	0
ASYMP. AT 3	
4	2
HOLE AT 5	

(Computer programs will vary depending on language used.)

CHAPTER 9
QUADRATIC RELATIONS AND SYSTEMS

Exercise 9-1, page 251; Introduction to Quadratic Relations

Graphs are all *circles* of radius *5 units*. Adding on x-term moves the graph in the x-direction. Adding a y-term moves the graph in the y-direction. See Section 9-2.

Exercise 9-2, page 254; Circles

1. $(x - 5)^2 + (y + 4)^2 = 36$; center: $(5, -4)$, $r = 6$

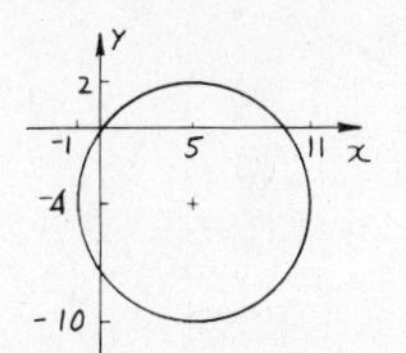

3. $(x - 3)^2 + (y - 2)^2 > 25$; center: $(3, 2)$, $r = 5$

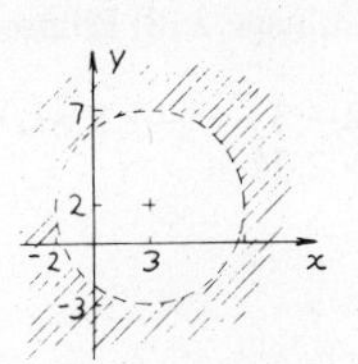

5. $(x - 4)^2 + (y + 3)^2 \leq 81$; center: $(4, -3)$, $r = 9$

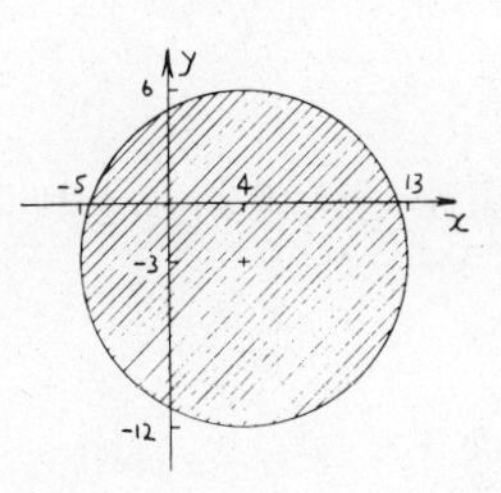

7. $(x + 2)^2 + y^2 = 9$; center: $(-2, 0)$, $r = 3$

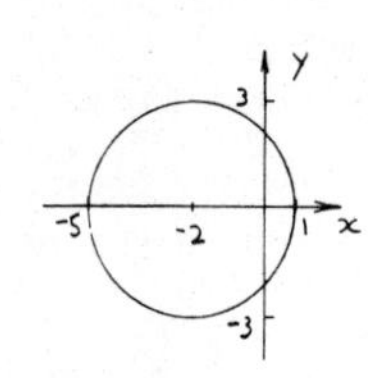

9. $x^2 + y^2 = 49$; center: $(0, 0)$, $r = 7$

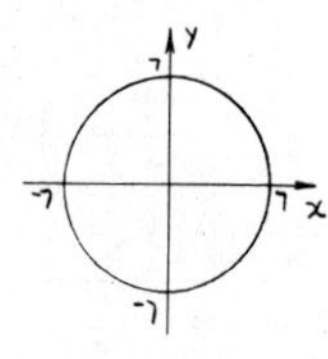

11. $x^2 + y^2 = 0$; center $(0, 0)$; $r = 0$

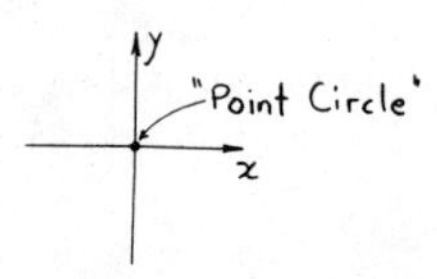

13. $(x - 7)^2 + (y - 5)^2 = 65$ 15. $(x + 9)^2 + (y + 2)^2 = 85$ 17. $x^2 + y^2 = 100$ 19. a. $(x - h)^2 + y^2 = r^2$ b. $x^2 + (y - k)^2 = r^2$ c. $x^2 + y^2 = r^2$ 21. If $(x - h)^2 + (y - k)^2 < 0$, r^2 will be < 0, so r is *imaginary*, and there is *no* circle. 22. *Introduction to Ellipses* a. $9x^2 + 25y^2 = 225 \Rightarrow y = \pm (3/5)\sqrt{25 - x^2}$

e.

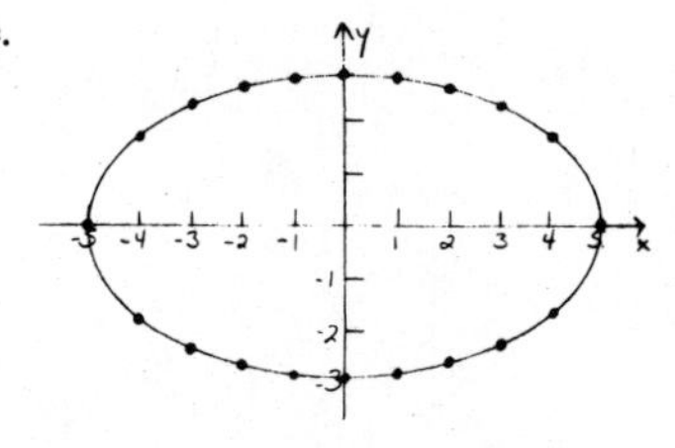

Exercise 9-3, page 258; Ellipses

1. a. $(x - 2)^2/3^2 + (y + 5)^2/2^2 = 1$
b. $c = \sqrt{5} \approx 2.24$

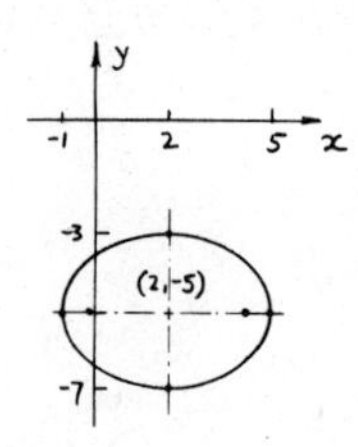

3. a. $(x + 1)^2/4^2 + (y - 2)^2/7^2 = 1$
b. $c = \sqrt{33} \approx 5.7$

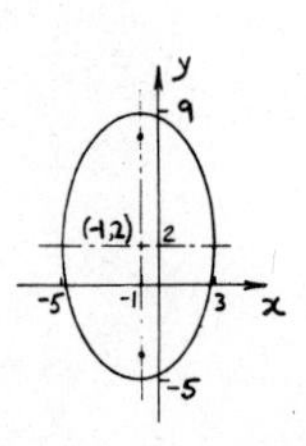

5. a. $(x + 5)^2/4^2 + (y + 3)^2/2^2 = 1$
b. $c = \sqrt{12} \approx 3.5$

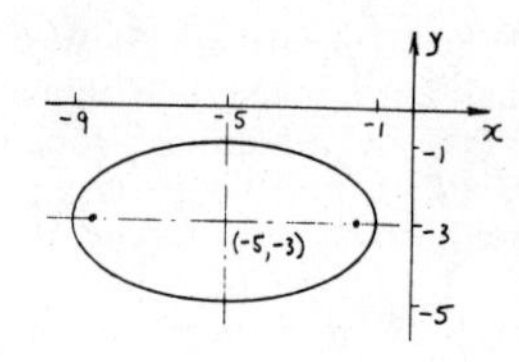

7. a. $(x + 1)^2/3^2 + (y - 2)^2/5^2 < 1$
b. $c = 4$

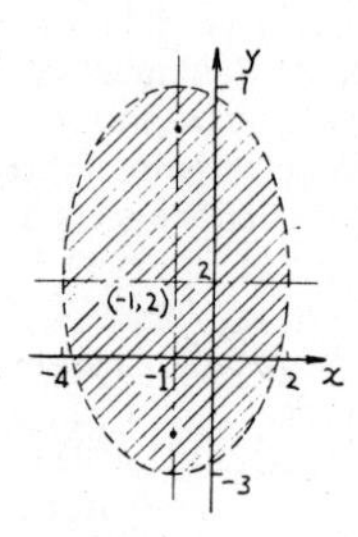

9. a. $x^2/5^2 + (y - 6)^2/4^2 = 1$
b. $c = 3$

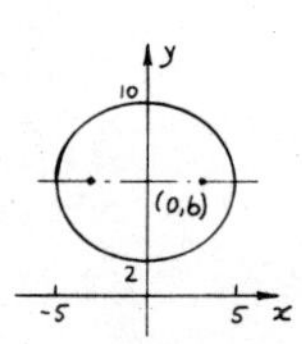

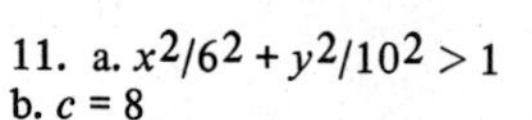

11. a. $x^2/6^2 + y^2/10^2 > 1$
b. $c = 8$

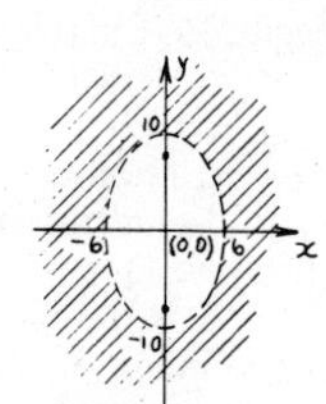

13. a. $x^2/2^2 + y^2/(\sqrt{48})^2 = 1$; $a = 2$, $b \approx 6.9$ b. $c = \sqrt{44} \approx 6.6$

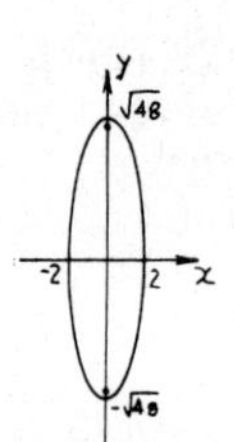

15. a. $x^2/(\sqrt{77/5})^2 + y^2/(\sqrt{77/8})^2 = 1$ b. $c = \sqrt{231/40} \approx 2.40$

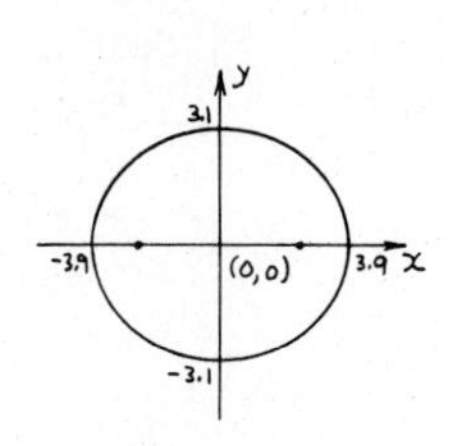

Exercise 9-4, page 263; Hyperbolas

1. a. $(x-2)^2/4^2 - (y+3)^2/5^2 = 1$
b. $c = \sqrt{41} \approx 6.4$

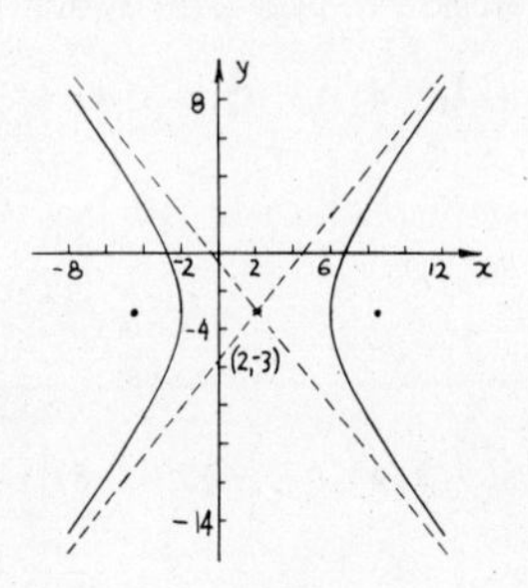

3. a. $-(x+6)^2/3^2 + (y+7)^2/5^2 = 1$
b. $c = \sqrt{34} \approx 5.8$

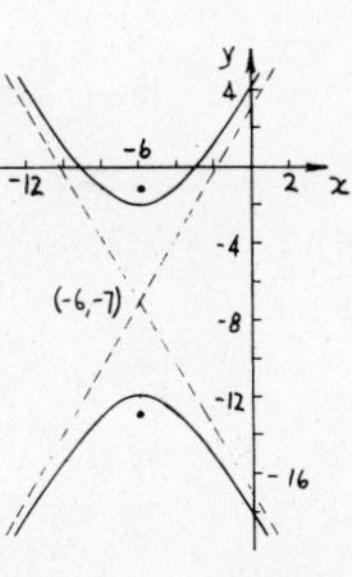

5. a. $(x+2)^2/3^2 - (y-8)^2/3^2 = 1$
b. $c = \sqrt{18} \approx 4.2$

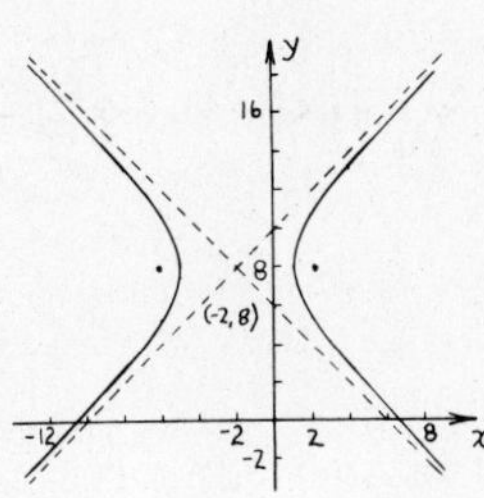

7. a. $(x-3)^2/4^2 - (y+2)^2/6^2 = 1$
b. $c = \sqrt{52} \approx 7.2$

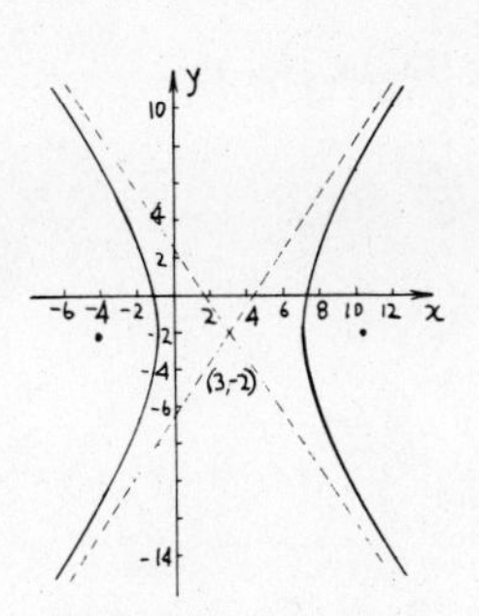

9. a. $-(x-5)^2/3^2 + (y-2)^2/9^2 = 1$
b. $c = \sqrt{90} \approx 9.5$

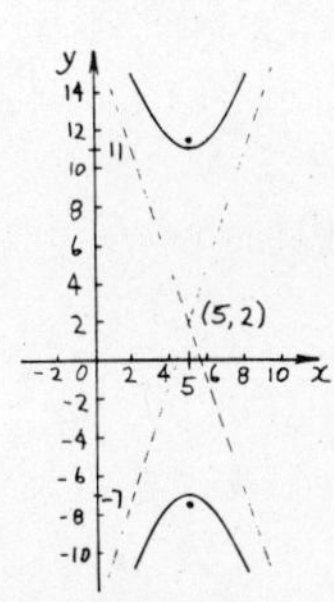

11. a. $-x^2/3^2 + y^2/4^2 = 1$; $m = \pm 4/3$ b. $c = 5$

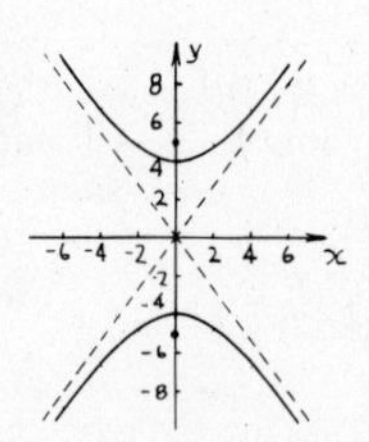

13. a. $x^4/2^2 - y^2/(16/5) = 1$, $m \approx \pm 0.90$ b. $c \approx 2.7$

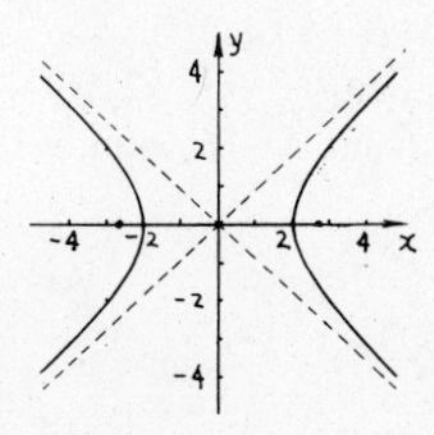

15. a. $-x^2/(11/16) + y^2/(11/3) = 1$; $b \approx 1.91$, $m \approx \pm 2.3$, b. $c \approx 2.1$

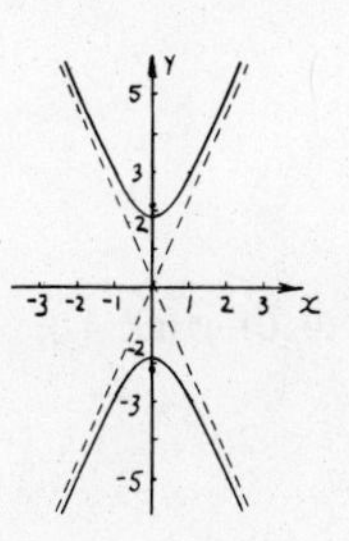

17. $9x^2 - y^2 = 0$; $a = b = c = 0$; $m = \pm 3$

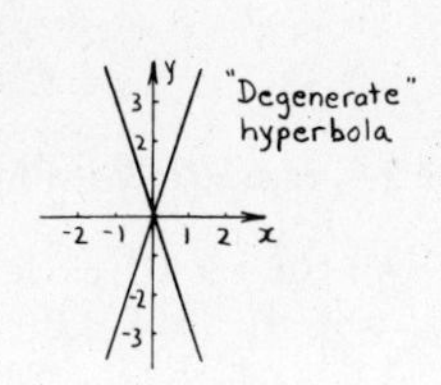

Exercise 9-5, page 264; Parabolas

1. $x + 1 = (y-2)^2$; Vertex: $(-1, 2)$; y-int. = 1, 3; x-int. = 3

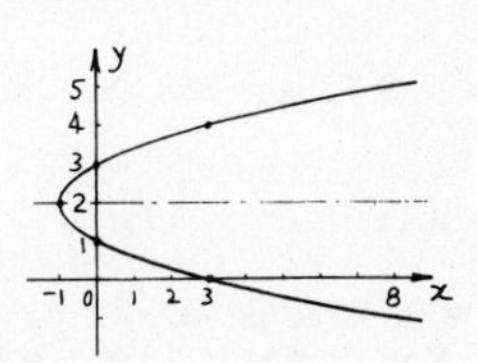

3. $x - 7 = -3(y+2)^2$; Vertex: $(7, -2)$; y-int. $\approx -0.5, -3.5$; x-int. = -5

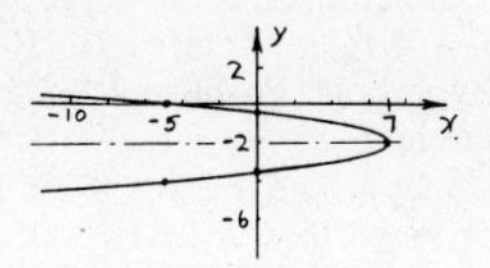

5. $x + (1/2) = (1/2)(y + 3)^2$; Vertex: $(-1/2, -3)$; y-int. $= -2, -4$; x-int. $= 4$

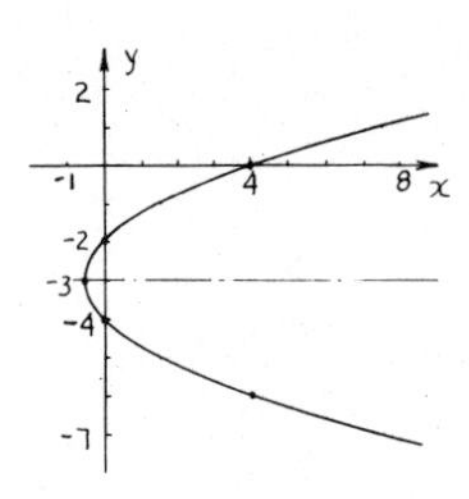

7. $y - 9 = -4(x - (5/2))^2$; Vertex: $(5/2, 9)$; y-int. $= -16$; x-int. $= 1, 4$

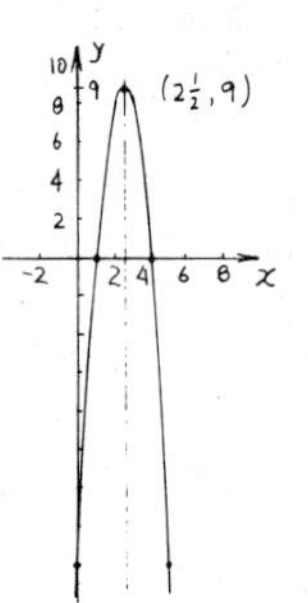

9. Vertex: $(0, 0)$; y-int. $= 0$; x-int. $= 0$

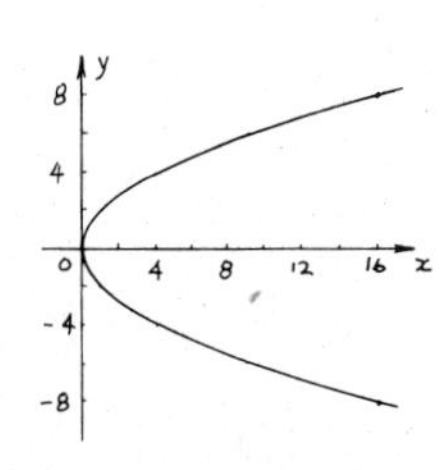

Exercise 9-6, page 267; Equations from Geometric Definitions

1. $x^2 + y^2 - 10x + 9 = 0$; circle 3. $4x^2 + 4y^2 + 13x - 15y + 10 = 0$; circle 5. $x^2 - 4y^2 - 6x - 16y - 7 = 0$; hyperbola (degenerate) 7. $x^2 - 6x + 12y + 21 = 0$; parabola 9. $4x^2 + 3y^2 + 24x + 24 = 0$; ellipse 11. $4x^2 - 5y^2 - 78y - 45 = 0$; hyperbola 15. $(1 - c^2)x^2 + (1 - c^2)y^2 - 2f(1 + c^2)x + (1 - c^2)f^2 = 0$; graph is a *circle* since x^2 and y^2 *both* have $(1 - c^2)$ as a coefficient. 17. For a circle, $e = 0$. But the radius is 0 unless the directrix is infinitely far away.

Exercise 9-7, page 268; Quadratic Relations – xy = Term

3. a. $Ax^2 + Bxy + Cy^2 + Dx + Ey + F = 0$; $\therefore y = [-(Bx + E) \pm \sqrt{(Bx + E)^2 - 4(C)(Ax^2 + Dx + F)}]/(2C)$. 4. a. $x^2 + y^2 - 10x - 8y + 16 = 0$; circle b. $x^2 + xy + y^2 - 10x - 8y + 16 = 0$; ellipse c. $x^2 + 2xy + y^2 - 10x - 8y + 16 = 0$; parabola d. $x^2 + 4xy + y^2 - 10x - 8y + 16 = 0$; hyperbola 5. a. You *cannot* still tell just by the x^2 and y^2 terms. All examples in Probs. 1 and 4 have *equal* x^2 and y^2 coefficients, but some are circles, some are ellipses, some parabolas, and some hyperbolas. b. Problem 1: (i) $B^2 - 4AC = -4$; circle (ii) $B^2 - 4AC = -3$; ellipse (iii) $B - 4AC = 12$; hyperbola. Problem 4: a. $x^2 + y^2 - 10x - 8y + 16 = 0$: $B^2 - 4AC = -4$; circle b. $x^2 + xy + y^2 - 10x - 8y + 16 = 0$; $B^2 - 4AC = -3$; ellipse c. $x^2 + 2xy + y^2 - 10x - 8y + 16 = 0$; $B^2 - 4AC = 0$; parabola d. $x^2 + 4xy + y^2 - 10x - 8y + 16 = 0$; $B^2 - 4AC = 12$; hyperbola c. Circles and ellipses have $B^2 - 4AC < 0$; parabola has $B^2 - 4AC = 0$; hyperbolas have $B^2 - 4AC > 0$.

Exercise 9-8, page 273; Systems of Quadratics

1. a. $\{(3, 4), (3, 4), (-3, 4), (-3, -4)\}$ b.

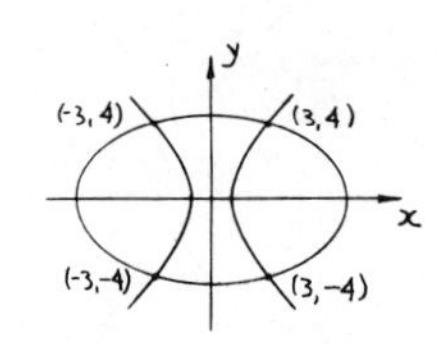

3. a. $\{(3, 4), (-3, 4), (0, -5)\}$ b.

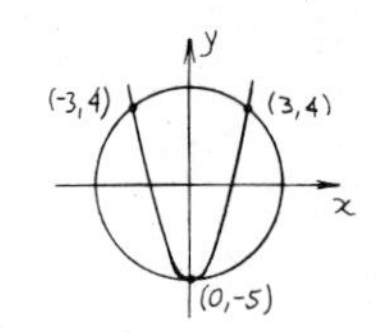

5. a. $\{(5, 4), (-5, 4), (2, -5), (-2, -5)\}$ b.

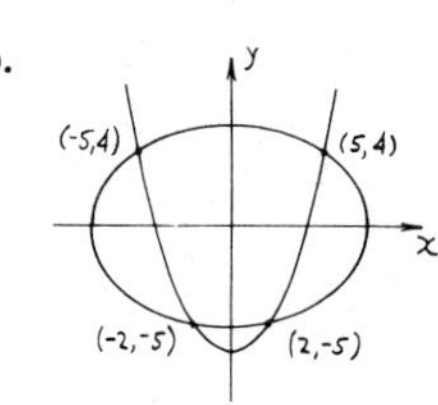

7. a. $\{(1, 4), (1, -4), (-5, 2), (-5, -2)\}$ b.

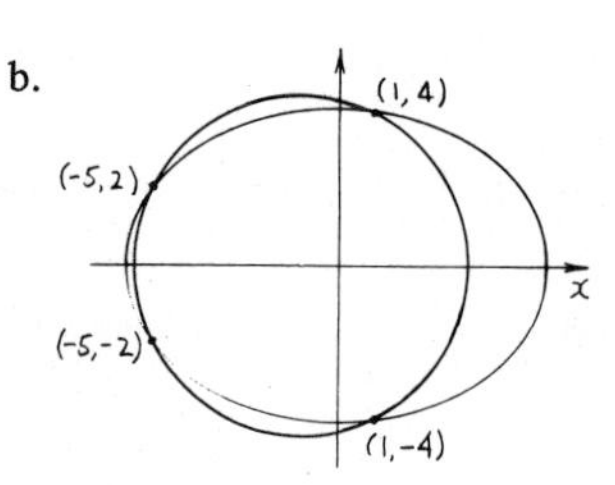

9. a. $\{(3, 1), (3, -1)\}$ b.

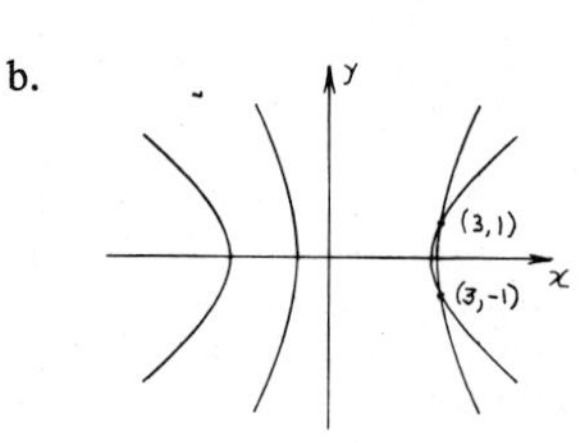

11. a. ϕ b.

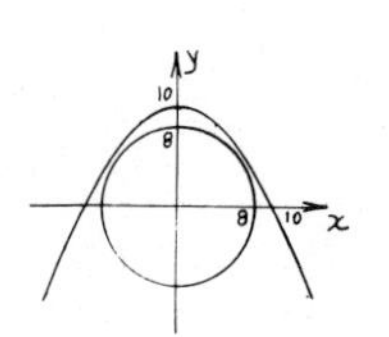

13. a. $\{(-5, 0)\}$ b.

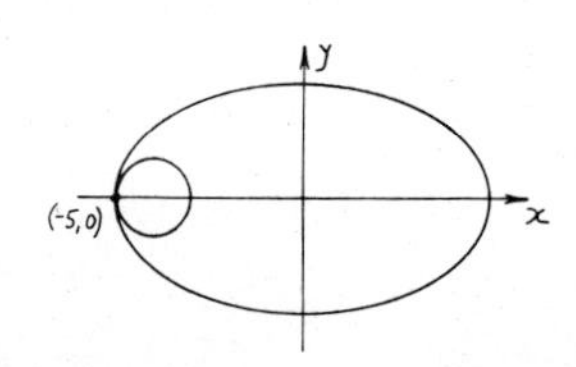

15. a. $\{(4, 3), (-4, 3)\}$ b.

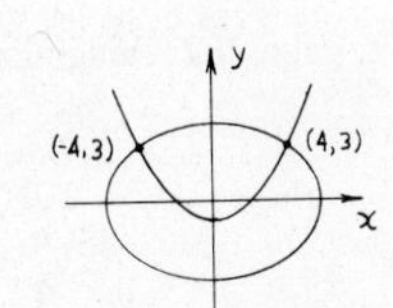

17. a. $\{(-6, 4), (6, -4)\}$ b.

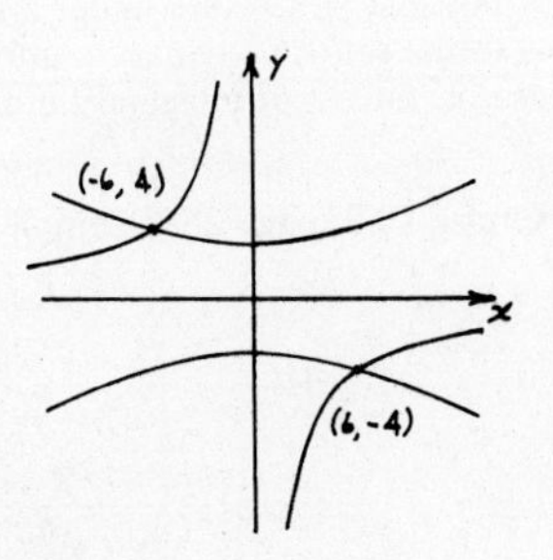

19. a. $\{(-1, 3), (-2, -5)\}$ b.

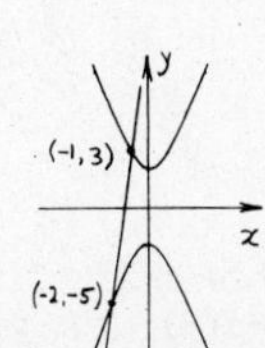

21. a. $\{(-5, 2), (-1, -4)\}$ b.

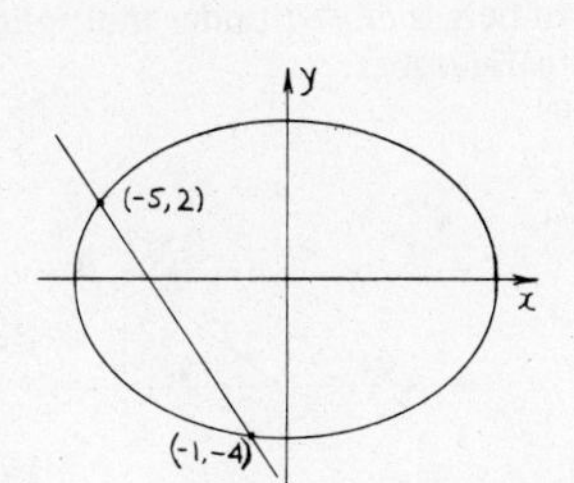

23. a. $\{(-1, 2), (-6, -3)\}$ b.

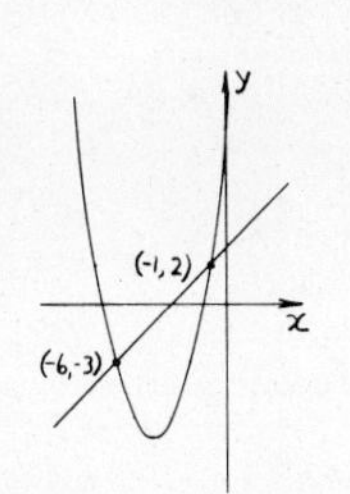

25. a. $\{(5, 1), (-4, -2)\}$ b.

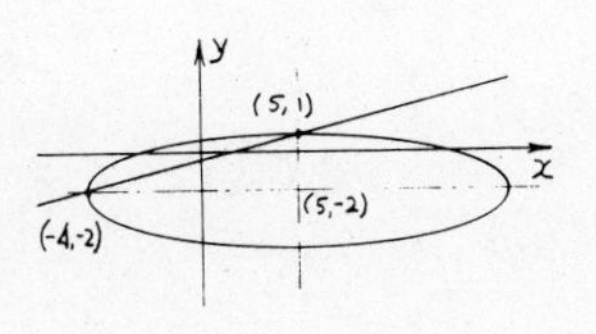

27. a. $\{(5, 2), (-5, 2), (4, -1), (-4, -1)\}$ b. $\{(5, -4)\}$ c. ϕ

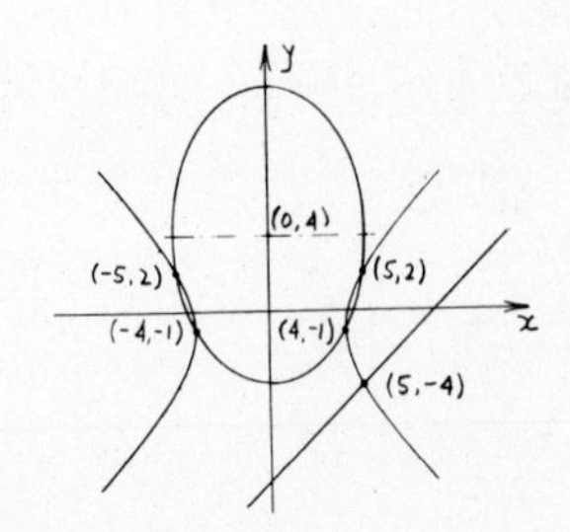

Exercise 9-9, page 275; Chapter Review

R1. a. i. Circle ii. Circle iii. Circle iv. Hyperbola v. Hyperbola vi. (Empty set) vii. Ellipse viii. Ellipse ix. Parabola x. Parabola xi. Hyperbola xii. Circle xiii. Ellipse xiv. Parabola

b. i. $x^2 + y^2 = 6^2$ ii. $x^2 + y^2 < 6^2$ iii. $x^2 + y^2 \geq 6^2$

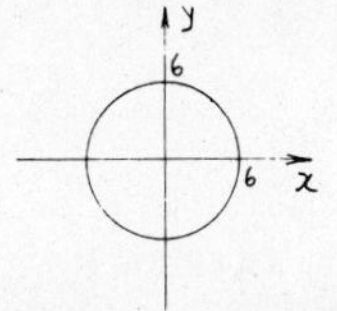

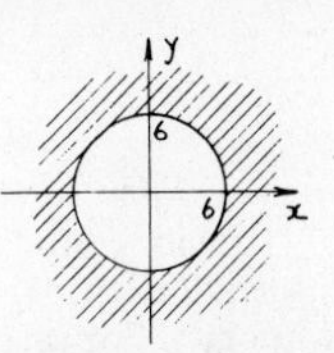

iv. $x^2/6^2 - y^2/6^2 = 1$ v. $-x^2/6^2 + y^2/6^2 = 1$

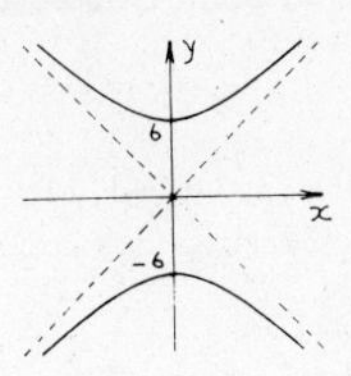

vi. $x^2 + y^2 = -36$ (No graph)
vii. $x^2/6^2 + y^2/3^2 = 1$ viii. $x^2/3^2 + y^2/6^2 = 1$ ix. $y = -4x^2 + 36$

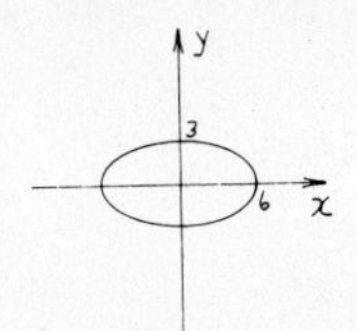

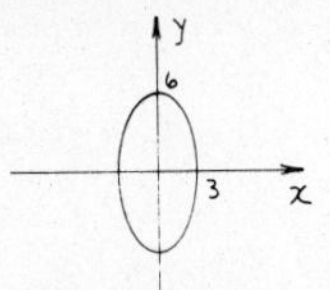

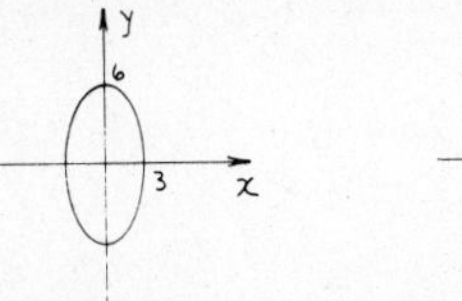

x. $x = (-1/4)y^2 + 9$ xi. $-(x + 5)^2/3^2 + (y - 3)^2/1^2 = 1$ xii. $(x + 8)^2 + (y - 11)^2 = 10^2$

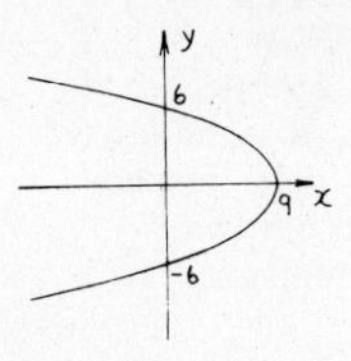

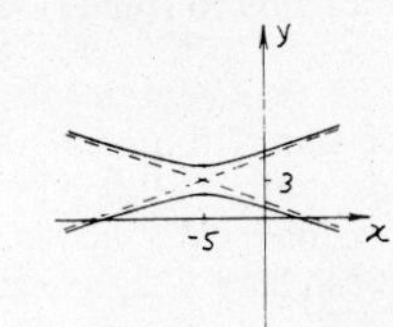

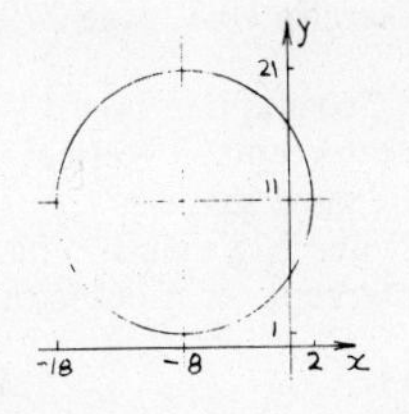

xiii. $(x-3)^2/8^2 + (y+2)^2/12^2 = 1$

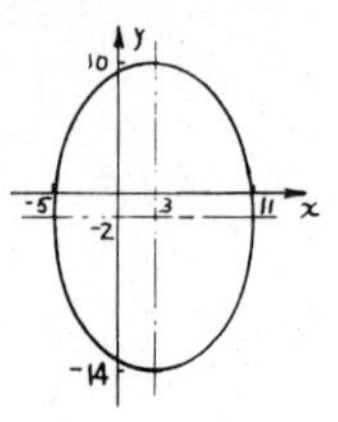

xiv. $x = y^2 - 6y + 3 \Rightarrow x + 6 = (y-3)^2$

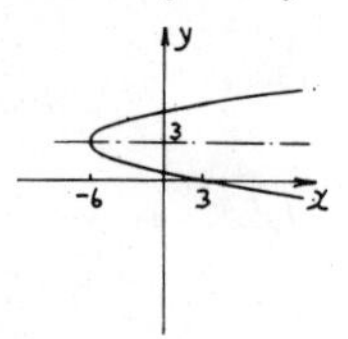

c. xi. $\sqrt{10}$ xiii. $\sqrt{80}$ R2. a. i. A circle is a set of points each of which is equidistant from a fixed point (the center). ii. An ellipse is a set of points. For each point, the sum of its distances from two fixed points (foci) is constant. iii. A hyperbola is a set of points. For each point, the difference between its distances from two fixed points (foci) is constant. iv. A parabola is a set of points, each of which is equidistant from a fixed point (focus) and a fixed straight line (directrix). b. $16x^2 + 25y^2 = 400$

R3. a. i. $\{(0, 10), (3, 1), (-3, 1)\}$ ii. $\{(1, 3)\}$

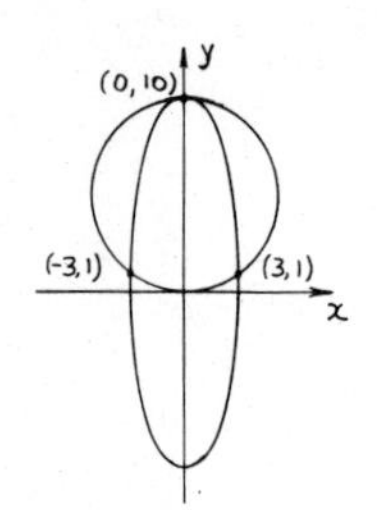

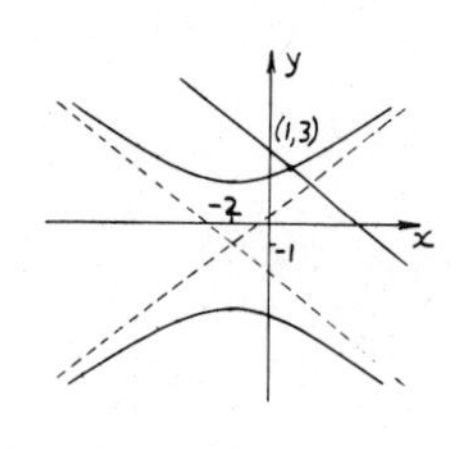

CHAPTER 10
HIGHER DEGREE FUNCTIONS AND COMPLEX NUMBERS

Exercise 10-1, page 279; Introduction to Higher Degree Functions

1. See graph in Figure 10-2a. 2. a. 3 x-intercepts b. 2 vertices c. 4 x-intercepts, 3 vertices 3. See graphs in Figure 10-2a. a. g and h have the same shape as f, but are raised up by 10 and by 34 units, respectively. b. g seems to have *two* x-intercepts. h seems to have *one* x-intercept. c. A cubic function can have 1, 2, or 3 x-intercepts.

Exercise 10-2, page 282; Imaginary Numbers

1. i. 3. $-i$. 5. -1. 7. 1. 9. 1. 11. i. 13. -1. 15. $4i$. 17. $3i\sqrt{2}$. 19. $i\sqrt{7}$. 21. $-42\sqrt{3}$. 23. $13i\sqrt{2}$. 25. $-30i$. 27. 6. 29. $-2i\sqrt{7}/7$. 31. $\sqrt{5}/5$. 33. $-3i$. 35. $4i\sqrt{6}/3$. 37. $-4i\sqrt{2}$. 39. $-2i\sqrt{7}$. 41. $-11i\sqrt{3}$. 43. i. 45. 0. 47. The set of imaginary numbers does *not* possess: a. Closure under Multiplication, since $i \times i = -1$ and -1 is *not* in the set of imaginary numbers. b. Multiplicative Identity, since 1 is *not* in the set of imaginary numbers.

Exercise 10-3, page 285; Complex Numbers

1, 3, 5, 7

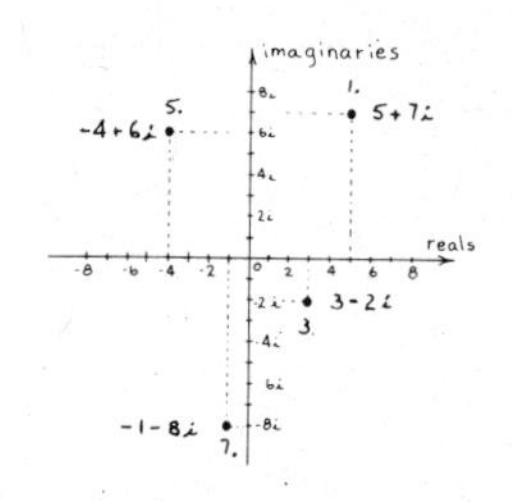

9. a. $6 - 2i$ b. $-2 + 8i$ c. $23 + 2i$ d. $(-7/41) + (22/41)i$ 11. a. $5 + 9i$ b. $-7 - 5i$ c. $-20 + 5i$ d. $(8/85) + (19/85)i$ 13. $(a + bi)(a - bi) = a^2 - b^2i^2 = a^2 + b^2$. Since a and b are real numbers, and the set of real numbers is *closed* under multiplication and addition, $a^2 + b^2$ is also a real number.

15.

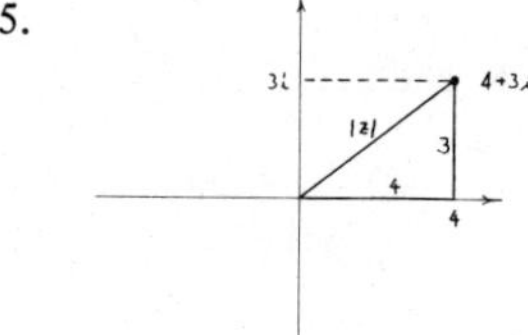

$|z|^2 = 4^2 + 3^2 = 25$.
$\therefore |z| = 5$.

17.

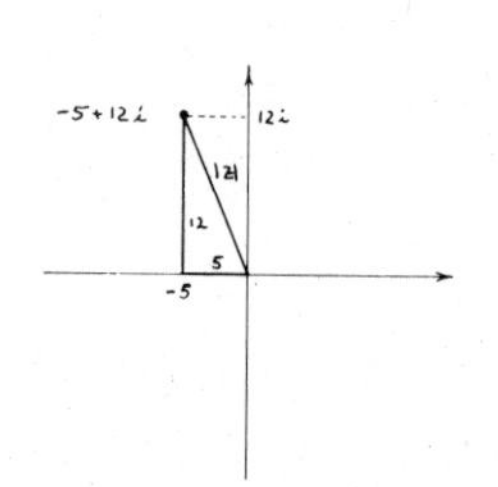

$|z|^2 = 5^2 + 12^2 = 169$. $\therefore |z| = 13$.

19.

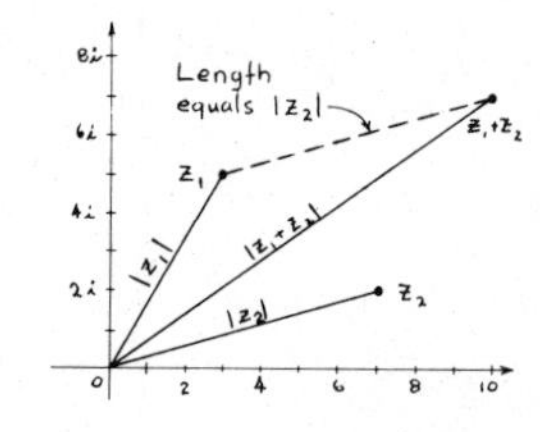

a. Lengths of lines equal absolute values of numbers. b. Length = $|z_2|$. See graph. c. $|z_1 + z_2| \le |z_1| + |z_2|$ because each is the length of the side of a triange, and no side of a triangle can be longer than the sum of the other two lengths.

21. 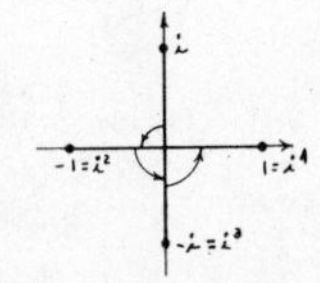

Raising the power of i by 1 *rotates* the line 90°.

23. $z = (\sqrt{2}/2)(1 + i)$, $z^2 = ((\sqrt{2}/2)(1 + i))^2 = (2/4)(1 + 2i + i^2) = (1/2)(1 + 2i - 1) = i$. So $(\sqrt{2}/2)(1 + i) = \sqrt{i}$, Q.E.D. $|z| = \sqrt{(\sqrt{2}/2)^2 + (\sqrt{2}/2)^2} = \sqrt{(1/2) + (1/2)} = 1$, Q.E.D.

Exercise 10-4, page 288; Complex Solutions of Quadratic Equations

1. $\{1 + i, 1 - i\}$ 3. $\{2 + i, 2 - i\}$ 5. $\{-1 + 3i, -1 - 3i\}$ 7. $\{2 + 5i, 2 - 5i\}$ 9. $\{-2 + i\sqrt{3}, -2 - i\sqrt{3}\}$ 11. $\{5 + i\sqrt{2}, 5 - i\sqrt{2}\}$ 13. $\{(2 + i\sqrt{26}/3), (2 - i\sqrt{26}/3)\}$ 15. $\{(-1/5) + 2i, (-1/5) - 2i\}$ 17. $\{-1, -7\}$ 19. $\{3 + \sqrt{5}, 3 - \sqrt{5}\}$ 21. $\{3i, -3i\}$ 23. $\{7i/2, -7i/2\}$ 25. $x^2 + 3x - 10 = 0$ 27. $x^2 + 9x + 18 = 0$ 29. $x^2 - 4x + 5 = 0$ 31. $x^2 + 6x + 25 = 0$ 33. $x^2 - 2x + 6 = 0$ 35. $x^2 + 10x + 37 = 0$ 37. $x^2 - 8x + 9 = 0$ 39. $(x - 1 - 2i)(x - 1 + 2i)$ 41. $(x + 5 - 2i)(x + 5 + 2i)$ 43. $(x - 3 - i\sqrt{2})(x - 3 + i\sqrt{2})$ 45. $(x - 5 - \sqrt{3})(x - 5 + \sqrt{3})$ 47. $(x + 4i)(x - 4i)$ 49. $(3x + 11i)(3x - 11i)$ 51. $(7x + i)(7x - i)$ 53. $3(x - (2 + i\sqrt{26})/3)(x - (2 - i\sqrt{26}/3)$ 55. $(5x + 1 + 10i)(5x + 1 - 10i)$

Exercise 10-5, page 292; Graphs of Higher Degree Functions – Synthetic Substitution

1. a. $x = -3, y = -60; x = -2, y = -20; x = -1, y = 0; x = 0, y = 6; x = 1, y = 4; x = 2, y = 0; x = 3, y = 0; x = 4, y = 10; x = 5, y = 36$ b. Zeros: $-1, 2, 3$

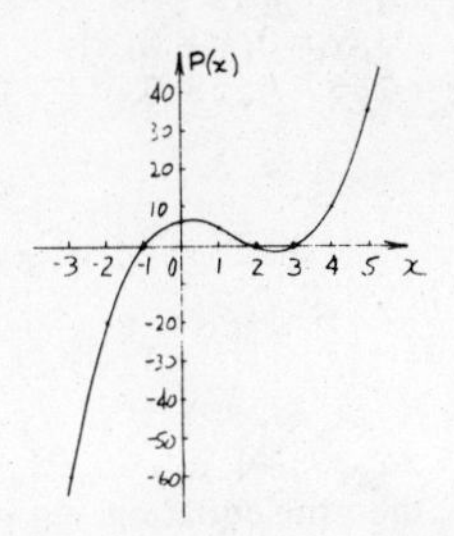

3. a. $x = -2, y = -55; x = -1, y = -16; x = 0, y = 3; x = 1, y = 8; x = 2, y = 5; x = 3, y = 0; x = 4, y = -1; x = 5, y = 8; x = 6, y = 33$ b. Zeros: $3, 2 + \sqrt{5}, 2 - \sqrt{5}$

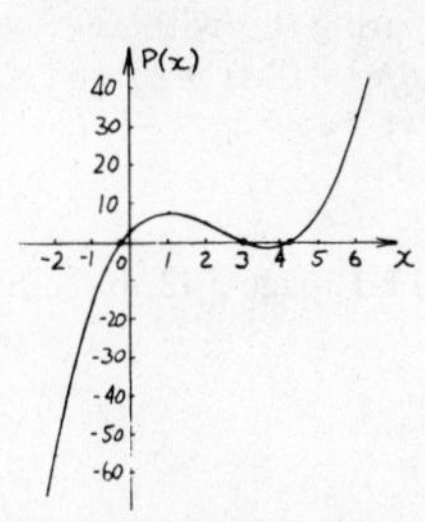

5. a. $x = -4, y = -26; x = -3, y = 0; x = -2, y = 10; x = -1, y = 10; x = 0, y = 6; x = 1, y = 4; x = 2, y = 10; x = 3, y = 30$ b. Zeros: $-3, 1 + i, 1 - i$

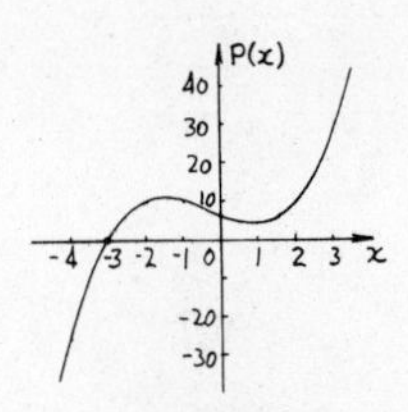

7. a. $x = -4, y = -24; x = -3, y = -5; x = -2, y = 0; x = -1, y = -3; x = 0, y = -8; x = 1, y = -9; x = 2, y = 0; x = 3, y = 25$ b. Zeros: $-2, -2, 2$

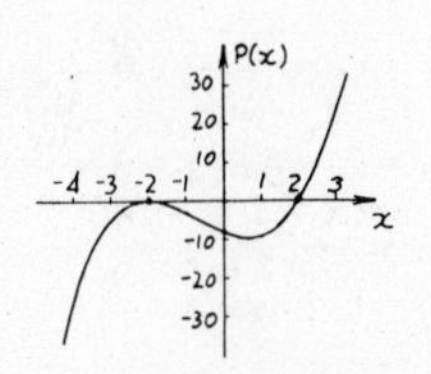

9. a. $x = -2, y = -37; x = -1, y = 0; x = 0, y = 19; x = 1, y = 26; x = 2, y = 27; x = 3, y = 28; x = 4, y = 35; x = 5, y = 54$; b. Zeros: $-1, (7/2) + (3i\sqrt{3}/2), (7/2) - (3i\sqrt{3}/2)$

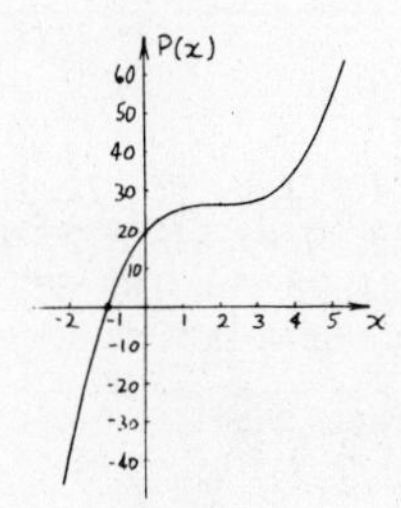

11. a. $x = -4, y = 60; x = -3, y = 15; x = -2, y = 0; x = -1, y = 3; x = 0, y = 12; x = 1, y = 15; x = 2, y = 0; x = 3, y = -45$ b. Zeros: $-2, 2, -3/2$

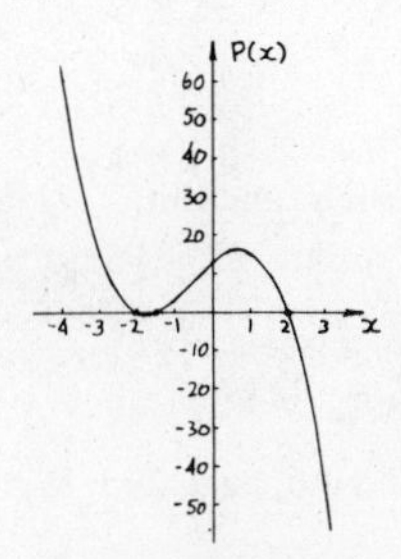

13. a. $x = -4, y = -35; x = -3, y = -6; x = -2, y = -5; x = -1, y = -20; x = 0, y = -39; x = 1, y = -50; x = 2, y = -41; x = 3, y = 0; x = 4, y = 85$; b. Zeros: $3, (-5 \pm i)/2$

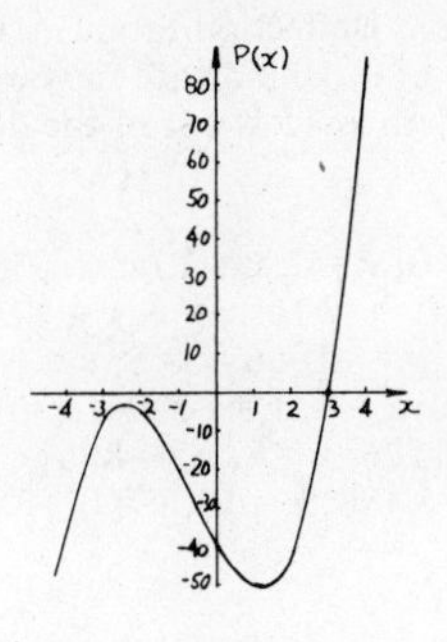

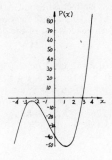

15. $x = -4, y = 126; x = -3, y = 0; x = -2, y = -20; x = -1, y = 0; x = 0, y = 18; x = 1, y = 16; x = 2, y = 0; x = 3, y = 0; x = 4, y = 70$ b. Zeros: $-3, -1, 2, 3$

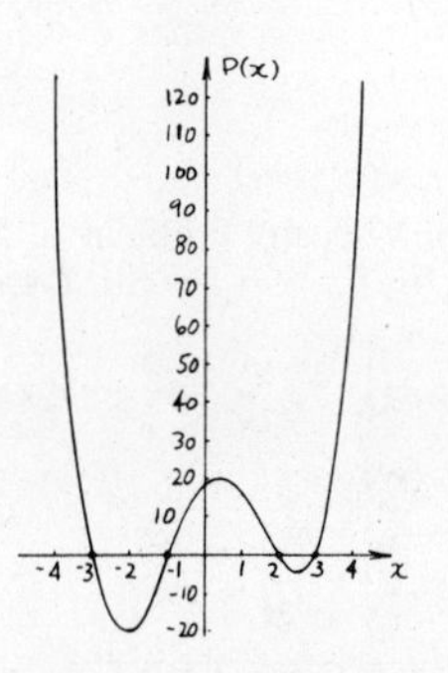

17. a. $x = -3, y = 148; x = -2, y = 0; x = -1, y = -34; x = 0, y = -20; x = 1, y = 0; x = 2, y = 8; x = 3, y = 10; x = 4, y = 36; x = 5, y = 140$
b. Zeros: $-2, 1, 3 + i, 3 - i$

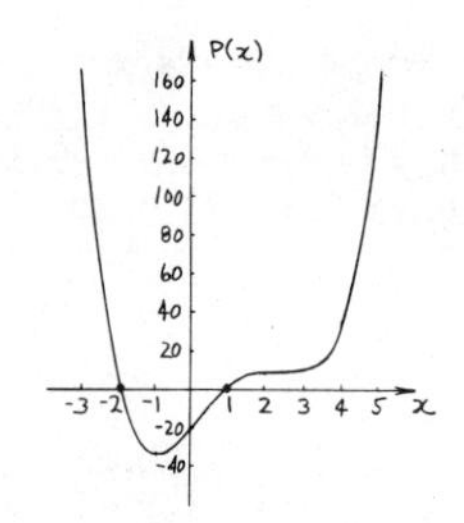

19. $-3, 1 + 2i, 1 - 2i$ 21. $2, -1, -4$ 23. $2, 1 + \sqrt{3}, 1 - \sqrt{3}$ 25. $-2, 3, -1 + i, -1 - i$ 27. $1, -1, 2, -3$ 29. $3 + 2i, 3 - 2i, 1 + i, 1 - i$ 31. 98 33. -30 35. 50 37–42 are *examples* only. There are other satisfactory examples.

37.

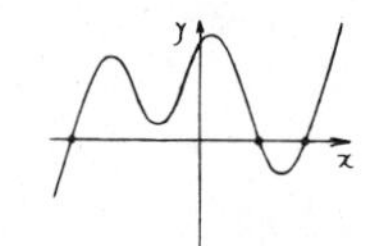

39.

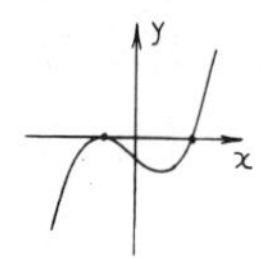

41. No such function.

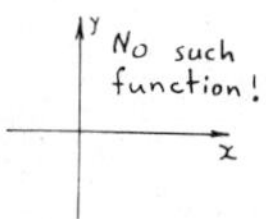

Exercise 10-6, page 295; Higher Degree Functions as Mathematical Models

1. Beam Deflection Problem a. $0 \leq x \leq 10$ b. Zeros: 0, 0, 10, 15. Zeros at 0, 10 indicate the beam is *not* deflected where the *supports* are. Zero at 15 is out of the domain, and is thus *meaningless.*

c. $x = 0, y = 0; x = 1, y = -126; x = 2, y = -416; x = 3, y = -756; x = 4, y = -1056; x = 5, y = -1250; x = 6, y = -1296; x = 7, y = -1176; x = 8, y = -896; x = 9, y = -486; x = 10, y = 0$

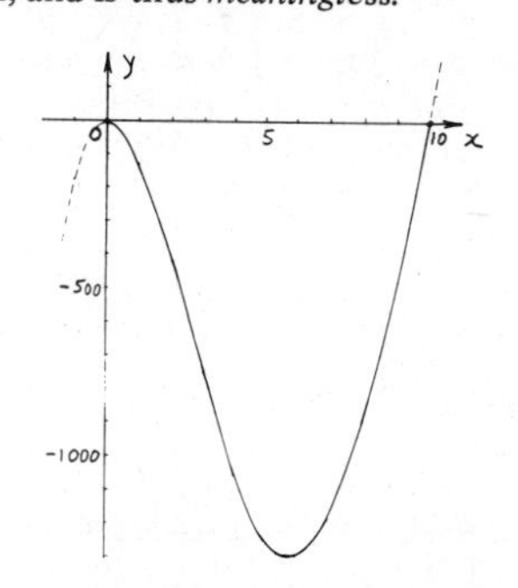

3. Oil Viscosity Problem a. $V = -T^3 + 9T^2 - 24T + 70$ b. $T = 0 \Rightarrow V = 70; T = 5 \Rightarrow V = 50; T = 6 \Rightarrow V = 34$

c.

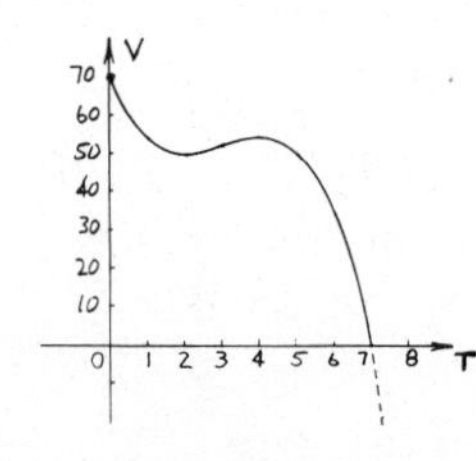

d. $V = -(T - 7)(T^2 - 2T + 10)$. If $V = 0$, then $T = 7$ or $T^2 - 2T + 10 = 0$. For $T^2 - 2T + 10 = 0$, discriminant $= 4 - 40 = -36 < 0$. $\therefore$ There are *no* real zeros of V for $0 \leq T \leq 6$.

5. Lumber Problem a. Let D = no. of feet diameter. Let $B(D)$ = no. of board feet. $B(D) = 10D^3 + 8D^2 - 5D - 3$ b. 1422 board feet c. $-1, 0.1 + \sqrt{0.31}, 0.1 - \sqrt{0.31}$

d.

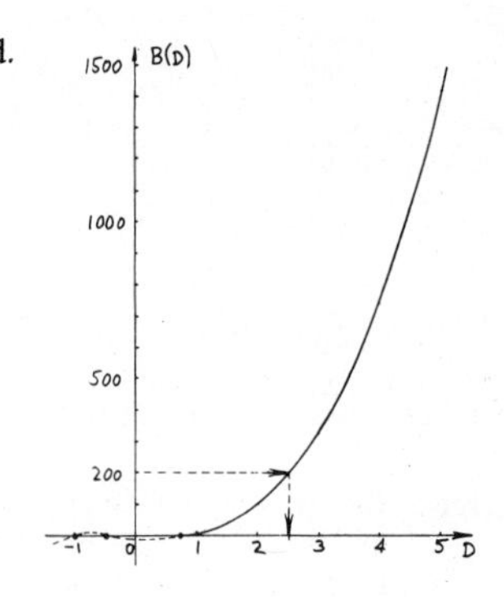

e. $0.1 + \sqrt{0.31} \approx 0.657$ feet, f. $D \approx 2.5$ feet
7. Sum of the Squares Problem a. $S(0) = 0, S(1) = 1, S(2) = 5, S(3) = 14$ b. $S(n) = (1/3)n^3 + (1/2)n^2 + (1/6)n$ c. $S(n) = (n/6)(2n + 1)(n + 1)$ d. $S(4) = 30$ both ways. $S(5) = 55$ both ways. e. 333,833,500

Exercise 10-7, page 297; Chapter Review

R1. a. $-i$ b. i c. 1 d. $-12\sqrt{3}$ e. $5\sqrt{7}$ f. $-7i$ g. $-2i\sqrt{3}$ h. $5 + 9i$ i. 145 j. $143 + 24i$ k. $3i$

2. a. $x = -3, y = 35; x = -2, y = 8; x = -1, y = 3; x = 0, y = 8; x = 1, y = 11; x = 2, y = 0; x = 3, y = -37$

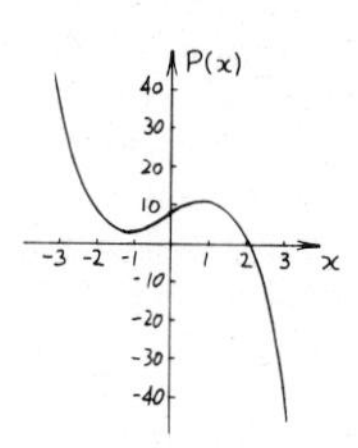

b. i. $\{(-5 \pm i\sqrt{7})/4\}$ ii. $x^2 - 6x + 25 = 0$ iii. If s_1, s_2, and s_3 are solutions, then the equation can be written $(x - s_1)(x - s_2)(x - s_3) = 0$. Expanding gives $x^3 - (s_1 + s_2 + s_3)x^2 + (s_1s_2 + s_1s_3 + s_2s_3)x - s_1s_2s_3 = 0$. Comparing this with $ax^3 + bx^2 + cx + d = 0$, or $x^3 + (b/a)x^2 + (c/a)x + (d/a) = 0$, $s_1 + s_2 + s_3 = -b/a$, Q.E.D. For $5x^3 + 11x^2 - 13x + 47 = 0$, $s_1 + s_2 + s_3 = -11/5$ R3. a. $P(x) = 3x^3 - 10x^2 + 2x + 5$ b. -63 c. 1

Exercise 11-1, page 302; Introduction to Sequences

1. a.

b. $t(7) = 1/7, t(8) = 1/8$ c. $t(n) = 1/n$ d. $t(100) = 1/100$

3. a.

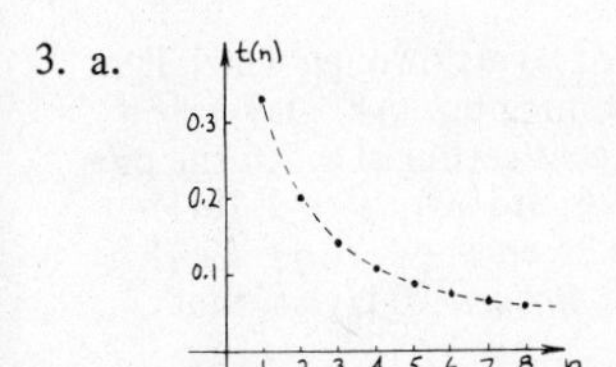

b. $t(7) = 1/15$, $t(8) = 1/17$ c. $t(n) = 1/(2n + 1)$ d. $t(100) = 1/201$

5. a.

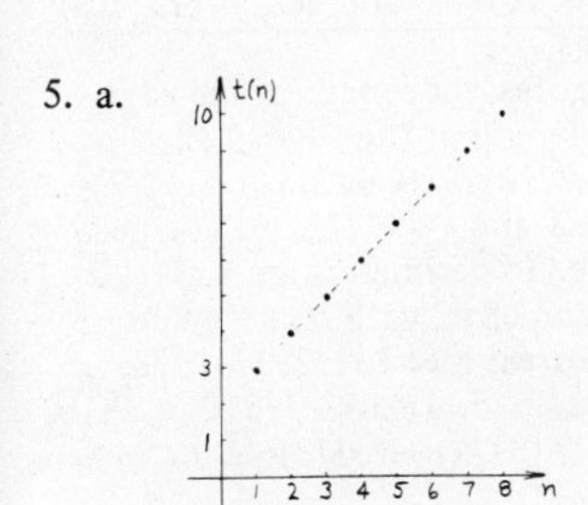

b. $t(7) = 9$, $t(8) = 10$ c. $t(n) = 2 + n$ d. $t(100) = 102$

7. a.

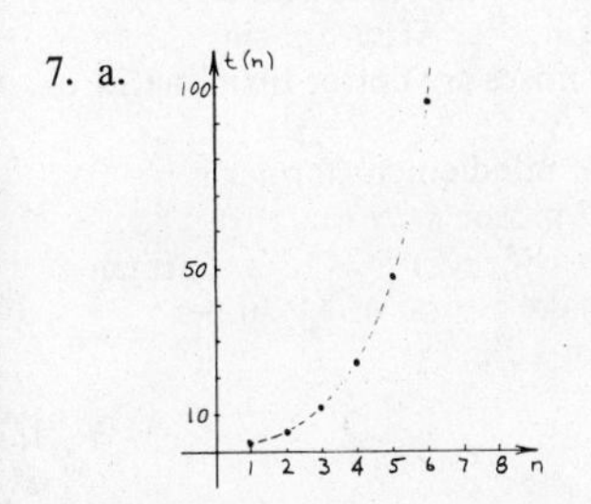

b. $t(7) = 192$, $t(8) = 384$ c. $t(n) = 3 \times 2^{n-1}$ d. $t(100) = 3 \times 2^{99} \approx 1.901459 \times 10^{30}$

9. a

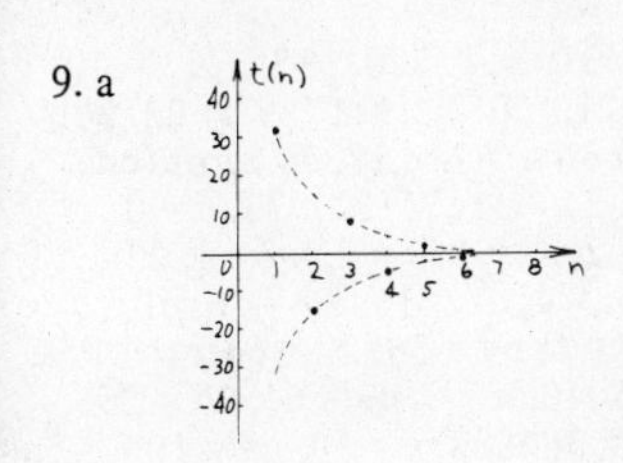

b. $t(7) = 1/2$, $t(8) = -1/4$ c. $t(n) = -64 \times (-2)^{-n}$ *or* $t(n) = -(-2)^{-n+6}$ d. $t(100) = -2^{94} \approx -5.0481098 \times 10^{-29}$

11. a.

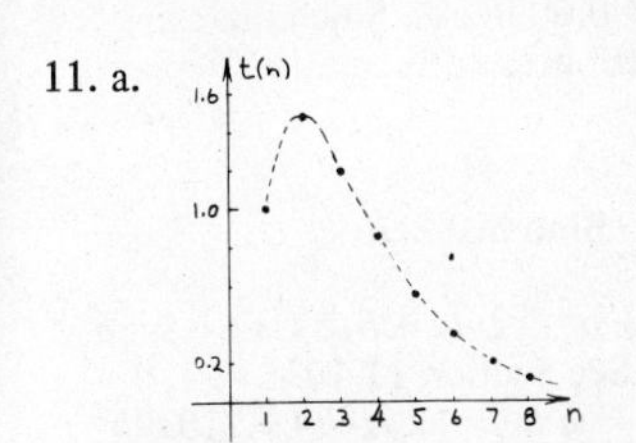

b. $t(7) = 13/64$, $t(12) = 15/128$ c. $t(n) = (2n - 1)/2^{(n-1)}$ d. $t(100) = 199/2^{99} \approx 3.1396664 \times 10^{-28}$

13. 22, 29. Add *1 more* each time to get next term. 15. 21, 34. Add preceding 2 terms to get next term. 17. 21, 28. Starting with 1, add 2, then 3, then 4, and so forth. 19. 7, 64. *Two* sequences put together: 1, 2, 3, 4, 5, 6, 7,... and 1, 2, 4, 8, 16, 32, 64.

Exercise 11-2, page 306; Arithmetic and Geometric Sequences

1. Arithmetic. $d = 5$ 3. Neither 5. Geometric, $r = -1$ 7. Geometric $r = 2$ 9. Neither 11. Neither 13. Geometric. $r = 1/2$ 15. Geometric. $r = 5^{-1/6}$ 17. 134 19. –182 21. 19-2/3 23. 197, 200, 203 [For Problems 25–35, approximate answers by 4-place logs are in parentheses.] 25. 1458 27. 3/128 29. –512 31. 18.84 (18.84) 33. 4.576 (4.576) 35. -7.626×10^{10} (-7.615×10^{10}) 37. 33 39. 16 41. 10 43. 7 45. 6

47. a.

A	47
D	6
N	5
T	~~47~~ ~~53~~ ~~59~~ ~~65~~ ~~71~~ 77
K	~~1~~ ~~2~~ ~~3~~ ~~4~~ ~~5~~ 6

Output:

1	47
2	53
3	59
4	65
5	71

b. and c. (Program and output will vary depending on computer language used.)
50. "Progression" and "sequence" mean the same thing. "Progression" is sometimes used if there is only a *finite* number of terms.

Exercise 11-3, page 309; Arithmetic and Geometric Means

1. 42, 49, 56, 63, 70 3. –107, –104, –101, –98, –95, –92, –89, –86 5. 23, 14, 5, –4, –13, –22, –31 7. –143, –167, –191, –215 9. 53, 58-1/2, 64, 69-1/2, 75 11. 123, 109-2/5, 95-4/5, 82-1/5, 68-3/5, 55 13. 5, 15, 45, 135 15. 81, 54, 36, 24, 16 *or* 81, –54, 36, –24, 16 17. 1/32, 1/8, 1/2, 2, 8, 32 19. 13, –91, 637, –4459 21. $x^5, x^7, x^9, x^{11}, x^{13}, x^{15}, x^{17}$ *or* $x^5, -x^7, x^9, -x^{11}, x^{13}, -x^{15}, x^{17}$; $r = \pm x^2$

23. 13, $13\sqrt[3]{2}$, $13\sqrt[3]{4}$, 26 25. 2, 6, 18, 54, 162; 2, –6, 18, –54, 162; 2, $6i$, –18, $-54i$, 162; 2, $-6i$, –18, $54i$, 162; $r = \pm 3$; $\pm 3i$ 27. 1, 2, 4, 8, 16; 1, –2, 4, –8, 16; 1, $2i$, –4, $-8i$, 16; 1, $-2i$, –4, $8i$, 16; $r = \pm 2$, $\pm 2i$ 29. The sequence is a, m, b. Dividing adjacent terms, $b/m = m/a$, $ab = m^2$, $m = \sqrt{ab}$, Q.E.D. (Also, $m = -\sqrt{ab}$.) 31. $AM = 10$, $GM = \pm 6$ 33. $AM = 12\text{-}1/3$, $GM = \pm 4$

Exercise 11-4, page 311; Introduction to Series

1. 9 + 11 + 13 + 15 + 17 = 65 3. 1 + 2 + 3 + 4 + 5 + 6 + 7 + 8 + 9 + 10 = 55 5. (1/1) + (1/2) + (1/3) + (1/4) = 25/12 7. 2 + 5 + 10 + 17 + 26 + 37 = 97 9. – 5 + 7 – 9 + 11 – 13 = –9 11. 2 + 4 + 8 + 16 + 32 = 62 13. 1.02 + 1.0404 + 1.061208 = 3.121608 15. 2 – 6 + 18 – 54 + 162 = 122 17. 15 + 45 + 135 + 405 = 600 19. 5(3 + 9 + 27 + 81) = 600 21. The 5 and 3 have been *factored out* of each term in Problems 19 and 20. Axiom is *Distributivity*.

23. $\sum_{k=1}^{10} k^2$ 25. $\sum_{k=1}^{100} \frac{1}{(k+2)}$ 27. $\sum_{k=1}^{40} 2 \times 3^{k-1}$ 29. $\sum_{k=1}^{90} 2 + (k-1)(4)$

31. $\sum_{k=1}^{20} (-1)^{k-1}(k-1)(10)$ 33. $\sum_{k=1}^{55} (k)(k+1)$

35. 1, 4, 9, 16, 25, 36, 49; $S(n) = n^2$ 37. As n becomes very large, $S(n)$ gets closer and closer to 1.

Exercise 11-5, page 315; Arithmetic and Geometric Series

1. 175 3. 2200 5. 0 7. 5000 9. 3720 [For Problems 13–24, approximate answers by 4-place logs are in parentheses. 11. 31 13. –182 15. 147620 17. 1026 19. 6931.82 (6915.) 21. 79.4109 (79.44) 23. 728 25. $S(5) = 5.8125$; $S(10) = 5.994140625$; $S(20) \approx 5.999994278$. As n becomes very large, $S(n)$ gets very close to 6.

Exercise 11-6, page 318; Convergent Geometric Series

1. 3-3/4 3. 24 5. Diverges 7. 11-1/9 9. $r = 0.1$: $S = 111\text{-}1/9$; $r = -0.1$: $S = 90\text{-}10/11$ 11. 7/11 13. 21/37 15. 35/33 17. 157/110 19. 1/2 21. Fraction is 12345679/999999999. By dividing numerator into denominator, answer is 1/81. 23. $S = (1/2) + (1/3) + (1/4) + (1/5) + (1/6) + (1/7) + (1/8) + (1/9) + (1/10) + (1/11) + (1/12) + (1/13) + (1/14) + (1/15) + (1/16) + \ldots = (1/2) + [(1/3) + (1/4)] + [(1/5) + (1/6) + (1/7) + (1/8)] + [(1/9) + (1/10) + (1/11) + (1/12) + (1/13) + (1/14) + (1/15) + (1/16)] + \ldots > 1/2 + [(1/4) + (1/4)] + [(1/8) + (1/8) + (1/8) + (1/8)] + [(1/16) + (1/16) + (1/16) + (1/16) + (1/16) + (1/16) + (1/16) + (1/16)] + \ldots = (1/2) + (1/2) + (1/2) + (1/2)$; $+ \ldots \therefore S > (1/2) + (1/2) + (1/2) + \ldots$, which clearly diverges! 26. a. Long division gives $1 + x + x^2 + x^3 + x^4 + \ldots$. b. Converges for $-1 \le x < 1$, diverges for $x \ge 1$ or $x < -1$ c. $S = t(1)/(1 - r) = 1/(1 - x)$, Q.E.D. 27. 9/4

Exercise 11-7, page 321; Sequences and Series as Mathematical Models

1. Muscle-Building Problem a. Let n = no. of days. Let $t(n)$ = no. of mm on n^{th} day. $t(10) = 1.89$mm b. $S(10) = 24.076$mm $\approx$ 24mm c. S = 60mm
3. Blue Jeans Problem a. Losing 4% is equivalent to *keeping* 96%. So color left after each washing is 0.96 *times* color before, which implies *geometric* sequence. b. Let n = no. of washings; $t(n)$ = % of original color left. $\therefore t(10) \approx 66.48\%$ c. 34 washings d. Arithmetic sequence would predict a *negative* amount of color after enough washings.
5. George Washington Problem Let m = no. of \$ after t years. $\therefore m = 1000\,(1.05)^t$. For 1980, $t = 181$, m = 6, 843, 261.43 The amount builds up so rapidly (over \$300,000 from 1979 to 1980), that a bank could go broke if there were no such laws.
7. Manhattan Problem Let t = no. of years since 1626. Let m = no. of dollars. $\therefore m = 24 \times 1.06^t$ For 1980, $t = 354$, m = 21, 801, 558, 740 Tens of *billions* of dollars! *Very* surprising!!

9. Straight-Line Depreciation Problem a. Depreciation = \$600 b. 0 years: \$24,000; 1 year: \$23,400; 2 years: \$22,800; 3 years: \$22,200 c. Arithmetic sequence, $d = -600$ d. \$7,800 e. 40 years

f.

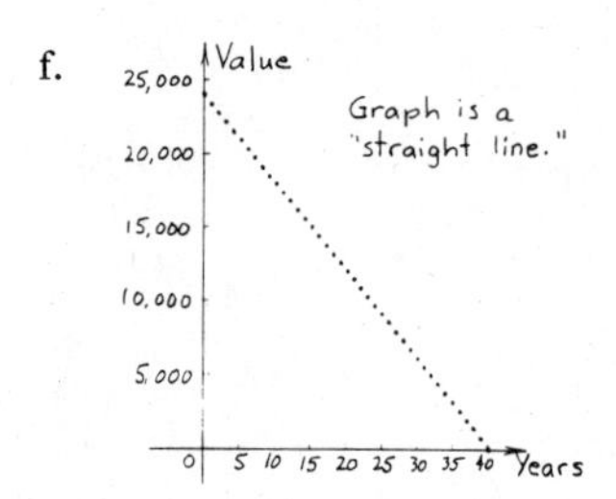

11. Pile Driver Problem a. $d = -4$. i. $t(10) = 64$ cm ii. $S(10) = 820$ cm b. $r = 0.96$ i. $t(10) = 69.25$ cm ii. $S(10) = 837.92$ cm iii. $S = 2500$ cm
13. Ancestors Problem a. Let n = no. of generations back. Let $t(n)$ = no. of ancestors in n^{th} generation back. Let $S(n)$ = total no. of ancestors n generations back. $t(1) = 2$, $t(2) = 4$, $t(3) = 8$. b. Sequence is *geometric*, with $r = 2$. c. $t(20) = 1{,}048{,}576$ d. $S(20) = 2{,}097{,}150$
15. Tree Branch Problem a. Let n = no. of layers through which light has come. Let $t(n)$ = % of original light coming through n layers. r = 0.8524 b. Percent *reaching* a layer is percent getting *through* the preceding layer. 2nd layer: $n = 1$, $t(n) = 85.24$; 3rd layer: $n = 2$, $t(n) = 72.66$; 4th layer: $n = 3$, $t(n) = 61.93$; 5th layer: $n = 4$, $t(n) = 52.79$ c. $t(n) < 20$; $n > 10.077$. Light can come *through* 10 layers, thus *reaching* layer 11.
17. Pineapple Plantation Problem a. $t(0) = 1$; $t(1) = 11$; $t(2) = 121$; $t(3) = 1331$ (Note: *Old* plant *dies*.) b. 14 years
19. Nested Squares Problem a. By Pythagoras, the length of the edge of an inside square is $(1/2)\cdot\sqrt{2}$ times the length of the edge of the square outside. So the lengths are: $1, 1\cdot(\sqrt{2}/2), 1\cdot(\sqrt{2}/2)^2, 1\cdot(\sqrt{2}/2)^3, \ldots$, which is a geometric sequence with $t(1) = 1$ and $r = \sqrt{2}/2$. b. The perimeter is the sum of the edges, or 4 $\times$ edge. So perimeters are $4, 4\cdot(\sqrt{2}/2), 4\cdot(\sqrt{2}/2)^2, 4\cdot(\sqrt{2}/2)^3, \ldots$, which is geometric, with $t(1) = 4$ and $r = \sqrt{2}/2$. c. Area = (length of edge)^2. So areas are: 1, 1/2, 1/4, 1/8, ..., which is geometric, with $t(1) = 1$ and $r = 1/2$. d. $A(10) = 1/512$ e. $P(10) = \sqrt{2}/8$ f. $S_a(10) = 1\text{-}511/512$ g. $S_p(10) = (31/4) + (31/8)\sqrt{2}$ h. $S_a = 2$ i. $S_p = 8 + 4\sqrt{2}$
21. Paint Brush Problem a. Percent paint left = 20% b. Still 20%. c. After 2nd rinse, 20% of 20% remains, i.e., 4%. After 3rd rinse, .2 $\times$ 4% = 0.8% d. 7.69% e. Three 8 oz. rinses are better than one 24 oz. rinse.
23. Middleman Problem a. Let n = no. of middlemen (farmer is $n = 0$). Let $t(n)$ = no. of cents per pound the nth person *sells* for. $t(0) = 65$; $t(1) = 84.5$; $t(2) = 110$; $t(3) = 143$; $t(4) = 186$; $t(5) = 241$. b. Farmer makes 15 cents per pound. Middlemen make a total of \$1.76 per pound! c. \$3.76 per pound. 56%, not just 10%!
25. Rich Uncle Problem \$769.91

Exercise 11-8, page 329; Factorials

1. 144 3. 1680 5. 5040 7. 5040 9. 10 11. 720 13. $1/n$ 15. $(n + 1)(n)$ 17. $7!/3!$ 19. $20!/14!$ 21. $(20!\ 12!)/(17!\ 9!)$ 23. $30!/(6!\ 24!)$ 25. $n!$ is evenly divisible by 9 for all $n \ge 6$. 27. 24 zeros (one for each time 5 appears as a factor). 29. 34 31. 25/12 33. $n = 0$, $n^2 = 0$, $2^n = 1$, $n! = 1$; $n = 1$, $n^2 = 1$, $2^n = 2$, $n! = 1$; $n = 2$, $n^2 = 4$, $2^n = 4$, $n! = 2$; $n = 3$, $n^2 = 9$, $2^n = 8$, $n! = 6$; $n = 4$, $n^2 = 16$, $2^n = 16$, $n! = 24$; $n = 5$, $n^2 = 25$, $2^n = 32$, $n! = 120$; $n = 6$, $n^2 = 36$, $2^n = 64$, $n! = 720$; $n = 7$, $n^2 = 49$, $2^n = 128$, $n! = 5040$; $n = 8$, $n^2 = 64$, $2^n = 256$, $n! = 40320$; $n = 9$, $n^2 = 81$, $2^n = 512$, $n! = 362880$; $n = 10$, $n^2 = 100$, $2^n = 1024$, $n! = 3628800$ 35. a. $t(8) = 0.0000248$. Since there are *zeros* in the first 4 places, this term contributes nothing in these places. b. $S(7) \approx 2.7183$

Exercise 11-9, page 331; Introduction to Binomial Series

1. $(a + b)^4 = a^4 + 4a^3b + 6a^2b^2 + 4ab^3 + b^4$ 2. $(a + b)^5 = a^5 + 5a^4b + 10a^3b^2 + 10a^2b^3 + 5ab^4 + b^5$ 3. a. See Section 11-10. 4. 1, 7, 21, 35, 35, 21, 7, 1. 5. $a^{10} + 10a^9b + 45a^8b^2 + 120a^7b^3 + 210a^6b^4 + 252a^5b^5 + 210a^4b^6 + 120a^3b^7 + 45a^2b^8 + 10ab^9 + b^{10}$ 6. $(x - y)^6 = x^6 - 6x^5y + 15x^4y^2 - 20x^3y^3 + 15x^2y^4 - 6xy^5 + y^6$. The only difference is that the signs *alternate*. 7. The rest of the terms are all *zeros*.

Exercise 11-10, page 334; The Binomial Formula

1. $(x + y)^5 = x^5 + 5x^4y + 10x^3y^2 + 10x^2y^3 + 5xy^4 + y^5$ 3. $(p - m)^7 = p^7 - 7p^6m + 21p^5m^2 - 35p^4m^3 + 35p^3m^4 - 21p^2m^5 + 7pm^6 - m^7$ 5. $(a + 2)^4 = a^4 + 8a^3 + 24a^2 + 32a + 16$ 7. $(2x - 3)^5 = 32x^5 - 240x^4 + 720x^3 - 1080x^2 + 810x - 243$ 9. $(x^2 + y^3)^6 = x^{12} + 6x^{10}y^3 + 15x^8y^6 + 20x^6y^9 + 15x^4y^{12} + 6x^2y^{15} + y^{18}$ 11. $(2a - b^4)^3 = 8a^3 - 12a^2b^4 + 6ab^8 - b^{12}$ 13. $(x + x^{-1})^8 = x^8 + 8x^6 + 28x^4 + 56x^2 + 70 + 56x^{-2} + 28x^{-4} + 8x^{-6} + x^{-8}$ 15. $(x^{1/2} - x^{-1/2})^4 = x^2 - 4x + 6 - 4x^{-1} + x^{-2}$ 17. $(1 + i)^6 = -8i$ 19. $(1 - i)^5 = -4 + 4i$ 21. $(8!/3!\,5!)x^3y^5$ 23. $(13!/6!\,7!)r^6s^7$ 25. $(-15!/4!\,11!)p^4j^{11}$ 27. $(21!/5!\,16!)a^5b^{16}$ 29. $(-58!/35!\,23!)e^{35}f^{23}$ 31. $(73!/48!\,25!)x^{48}y^{25}$ 33. $(15!/9!\,6!)x^{27}y^{12}$ 35. $(-13!/6!\,7!)x^{18}y^{14}$ 37. No such term 39. $(8!/3!\,5!)(3x)^3(2y)^5$ 41. $(34!/18!\,16!)i^{18}k^{16}$ 43. $(-15!/4!\,11!)r^4q^{11}$ 45. b^{12} 47. $(-7!/4!\,3!)(3x^2)^4(2y^3)^3$ 49. $375x^{12}$ 51. a. 7th term b. 19th power c. $50388r^{12}s^7$ d. 20 terms 53. a. $1^{10} + 10 \times 1^9 \times 0.02 + 45 \times 1^8 \times 0.02^2 + 120 \times 1^7 \times 0.02^3 + 210 \times 1^6 \times 0.02^4$ b. $1 + 0.2 + 0.018 + 0.00096 + 0.0000336 = 1.2189936$ c. 5th and following terms have only 0's in the first 4 decimal places. d. Difference is 0.000,000,8 e. Error is 0.0000656%. 55. $(1 + 1)^5 = 1^5 + 5 \times 1^4 \times 1 + 10 \times 1^3 \times 1^2 + 10 \times 1^2 \times 1^3 + 5 \times 1 \times 1^4 + 1^5 = 1 + 5 + 10 + 10 + 5 + 1 = 32$. Easier way: $(1 + 1)^5 = 2^5 = 32$. Sum of coefficients for $(a + b)^{57}$ is 2^{57}

Exercise 11-11, page 335; Chapter Review

R1. a. 72, 90. Pattern could be: Add 2, then add 4, then add 6, then add 8, ... or: Multiply (term number) (tern number + 1) or: Square the term number and add the term number. b. $12 - 6 = 6$ and $6 - 2 = 4$, so sequence is *not* arithmetic. $12/6 = 2$ and $6/2 = 3$, so sequence is *not* geometric. Neither.

c.

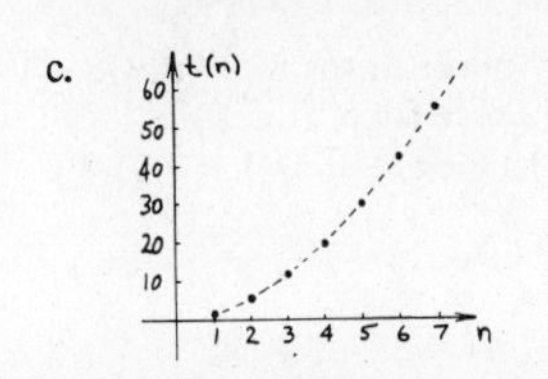

d. $t(n) = (n)(n + 1) = n^2 + n$
c. 490,700

R2. a. i. −148 ii. 73 b. i. $t(31) \approx 4035.397$ ii. 8 c. i. $-120h^3p^7$ ii. $210h^4p^6$ R3. a. 140,500 b. i. 1394.99 ii. 2500 c. 153 R4. a. 17, <u>24-2/3</u>, <u>32-1/3</u>, <u>40</u>, <u>47-2/3</u>, <u>55-1/3</u>, 63 b. 20 or −20
R5. – Swinging a. $t(10) = 23$ cm, $t(20) = 43$ cm b. $t(10) \approx 18.58$ cm, $t(20) \approx 115.01$ cm c. Sitting: $n = 11$ pumps. Standing: $n \approx 11.6$, or 12 pumps d. Sitting: $n = 36$ pumps. Standing: $n \approx 17.65$, or 18 pumps

e.

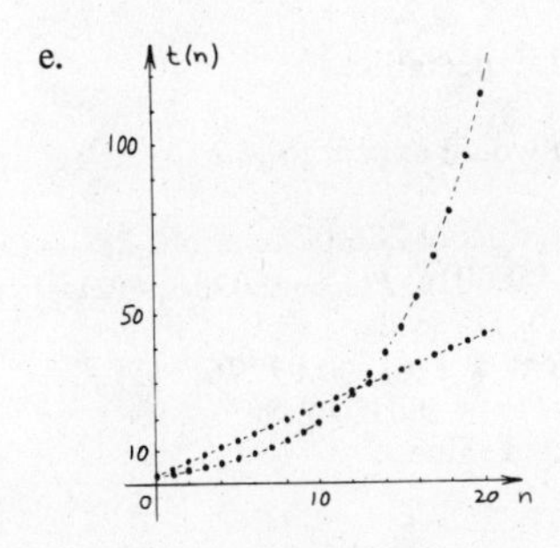

f. If $t(0) = 0$, then $t(n)$ would be: $t(n) = 0 \times 1.2^n = 0$. So you could never start swinging! g. Sitting increases the amplitude faster when you are going *low*, and standing increases it faster when you are going *high*. (Note: The optimum point to change from sitting to standing is at the n for which the *slopes* of the two graphs are equal.)

CHAPTER 12
PROBABILITY, AND FUNCTIONS OF A RANDOM VARIABLE

Exercise 12-1, page 339; Introduction to Probability

1. 1/12 3. 5/6 5. 1/6 7. 5/9 9. 1 11. 1/18 13. 11/36

Exercise 12-2, page 340; Words Associated with Probability

1. a. Random experiment b. 52 outcomes in sample space c. 12 outcomes are faces d. $P(\text{face}) = 12/52 = 3/13$ e. $P(\text{black}) = 26/52 = 1/2$ f. $P(\text{Ace}) = 4/52 = 1/13$ g. $P(3 \le x \le 7) = 20/52 = 5/13$ h. $P(8 \text{ clubs}) = 1/52$ i. $P(\text{in deck}) = 52/52 = 1$ j. $P(\text{Joker}) = 0/52 = 0$, since Joker is *not* in 52-card deck

Exercise 12-3, page 342; Two Counting Principles

1. 345 3. a. 91 b. 20 5. a. 17 b. 60 7. a. 16 b. 55 c. 20 9. 693 11. a. 9 b. 18 c. 504 13. a. 10 ways b. 9 ways c. 90 ways d. 8 ways e. 720 ways f. 3,628,800 ways 17. a. 20 b. 45 c. 1018 d. 630 e. 496

Exercise 12-4, page 346; Probabilities of Various Permutations

1. a. 3024 b. 1320 c. 6720 d. 5040 3. a. 132 b. 11880 c. 479001600 5. 358800 7. a. 210 b. 1716 c. 5040 9. 210 11. a. 720 b. 120 c. 1/6 d. 17% e. 1/720 13. a. 1/30 b. 4/15 c. 1/15 d. 2/15 e. 1/10 15. a. 362880 b. 40320 c. 1/9 d. 11% 17. a. 1/5 b. 1/12 c. 1/84 19. a. 3628800 b. 725760 c. 1/5 21. Permutations with Repeated Elements a. i. 360 iii. 20 v. 5040 b. 210 c. 126

Exercise 12-5, page 353; Probabilities of Various Combinations

1. 10 3. 56 5. 6 7. 330 9. 1 11. 1 13. 360 15. 55440 17. 142506 committees 19. 35 games. Too many games for one evening! 21. 166,167,000 selections 23. a. 20 b. 15 c. 35 d. 1 25. a. 2,598,960 b. 6.35×10^{11} 27. a. 45 b. 252 c. 45. Selecting 8 to *take* is equivalent to selecting 2 *not* to take. 29. a. 10/21 b. 5/21 c. 5/7 d. 1/2 31. a. 56/143 b. 4/715 c. 0 d. 1/5 33. a. 4/7 b. 5/7 c. 2/7 d. $P(\text{none}) = 1 - P(\ge 1)$ 35. a. 225862560 b. 7376656 c. 3.27% d. No. Probability of missing bad bulbs is too high.

Exercise 12-6, page 357; Properties of Probability

1. Calculator Components Problem a. 56% b. 30% c. 20% d. 6% e. 94%
3. Car Breakdown Problem a. 0.9 b. 0.95 c. 0.855 d. 0.005 e. 0.145
5. Traffic Light Problem a. 0.28 b. 0.18 c. 0.12 d. 0.42 e. 0.54
7. Back-Up System Problem 99.96%
9. Basketball Problem a. 0.336 b. 0.024 c. 0.976 d. 0.056
11. Spaceship Problem a. 36.8% b. 99.9895%
13. Football Plays Problem a. 0.12 b. 0.28 c. 0.48 d. 0.12
15. Measles and Chickenpox Problem a. 0.024 b. i. 0.05 ii. 0.9

c. Since $P(C \text{ after } M) < P(C)$, measles seems to give *immunity* to chicken pox. Since $P(M \text{ after } C) > P(M)$, chicken pox seems to *increase* susceptibility to measles

17. Combinations & Binomial Series Problem a. $n(0) = {}_5C_0 = 1$; $n(1) = {}_5C_1 = 5$; $n(2) = {}_5C_2 = 10$; $n(3) = {}_5C_3 = 10$; $n(4) = {}_5C_4 = 5$; $n(5) = {}_5C_5 = 1$ b. $1 + 5 + 10 + 10 + 5 + 1 = 32$ c. $(1 + 1)^5 = 1 + 5 + 10 + 10 + 5 + 1$, which is the *same* as the sum in part b. d. $(1 + 1)^5 = 2^5 = 32$ e. ${}_{10}C_{10} + {}_{10}C_1 + ... + {}_{10}C_{10} = 2^{10} = 1024$

Exercise 12-7, page 362; Functions of a Random Variable

1. Heredity Problem a. 3/4 b. $P(0) = 0.015625$; $P(1) = 0.140625$; $P(2) = 0.421875$; $P(3) = 0.421875$. c. Sum = 1.000000 d.

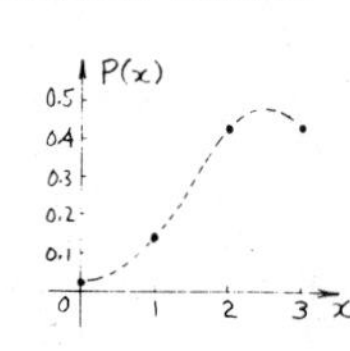

3. Thumbtack Problem a. $P(0) = 0.2401$; $P(1) = 0.4116$; $P(2) = 0.2646$; $P(3) = 0.0756$; $P(4) = 0.0081$; Sum = 1.0000 b.

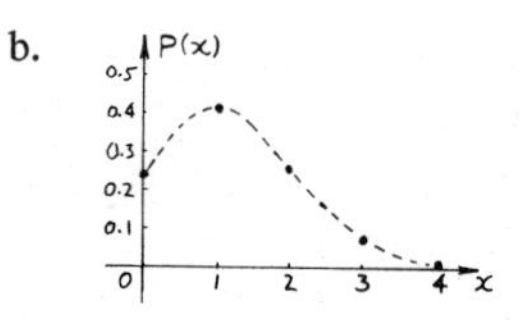

c. $P(> 2 \text{ up}) = 0.6517$ $P(\leq 2 \text{ up}) = 0.3483$ ∴ more than 2 up is more probable.

5. Bull's-Eye Problem a. $P(0) = 0.16807$; $P(1) = 0.36015$; $P(2) = 0.3087$; $P(3) = 0.1323$; $P(4) = 0.02835$; $P(5) = 0.00243$

b.

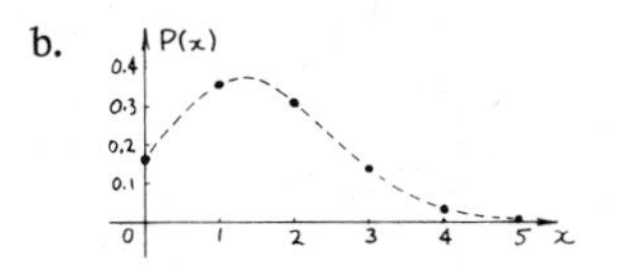

c. 0.47088

7. Color-Blindness Problem a. $P(0) = 0.35848592$; $P(1) = 0.37735360$; $P(2) = 0.18867680$; $P(3) = 0.05958215$.

b.

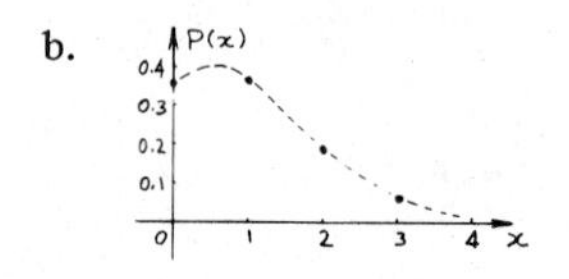

c. 0.00257394

9. Eighteen-Wheeler Problem a. 0.97 i. $P(0) \approx 58\%$ ii. $P(1) \approx 32\%$ ii. $P(2) \approx 8\%$ iv. $P(> 2) \approx 0.0157$ c. 0.9977

11. Airplane Engine Problem a. 0.9 b. $P(0) = 0.6561$; $P(1) = 0.2916$; $P(2) = 0.0486$; $P(3) = 0.0036$; $P(4) = 0.0001$. c. $P(0) + P(1) + P(2) + P(3) + P(4) = 1.0000$ d. 0.9477 e. $P(0) = 0.729$; $P(1) = 0.243$; $P(2) = 0.027$; $P(3) = 0.001$. ∴ $P(\leq 1) = P(0) + P(1) = 0.972$ f. 3-engine plane is *safer* based on this model. (*Actual* p¬obabilities are much *higher!*)

13. Another Dice Problem

a. i.

x	$n(x)$	$P(x)$
2	1	1/36
3	2	1/18
4	3	1/12
5	4	1/9
6	5	5/36
7	6	1/6
8	5	5/36
9	4	1/9
10	3	1/12
11	2	1/18
12	1	1/36

ii.

x	$n(x)$	$P(x)$
−5	1	1/36
−4	2	1/18
−3	3	1/12
−2	4	1/9
−1	5	5/36
0	6	1/6
1	5	5/36
2	4	1/9
3	3	1/12
4	2	1/18
5	1	1/36

iii.

x	$n(x)$	$P(x)$
0	6	1/6
1	10	5/18
2	8	2/9
3	6	1/6
4	4	1/9
5	2	1/18

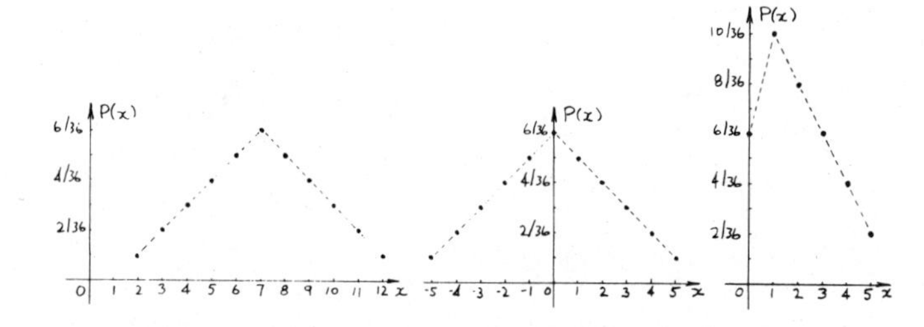

b. i. Most probable is $x = 7$. ii. Most probable is $x = 0$. iii. Most probable is $x = 1$.

15. First Girl Problem a. $P(1) = 0.5$; $P(2) = 0.25$; $P(3) = 0.125$; $P(4) = 0.0625$

b.

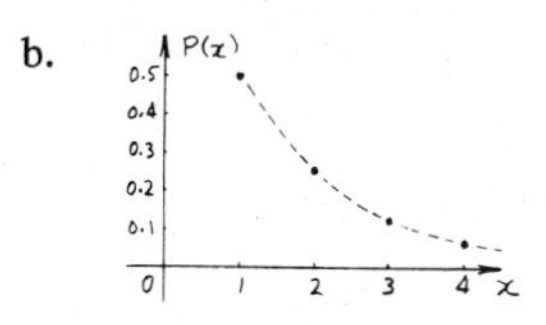

c. No. $P(x)$ never drops to 0. d. Geometric sequence (exp. funct.) e. Sum = $t(1)/(1 - r) = 0.5/(1 - 0.5) = 1$.

17. Lucky Card Problem a. $P(1, 1) = 0.1$ b. $P(2, 1) = 0.09$; $P(3, 1) = 0.081$ c. $P(1, 2) = 0.0729$. ∴ $P(1, n) = 0.1, 0.1(0.9)^3, 0.1(0.9)^6$ which is geometric, with $t(1) = 0.1$ and $r = (0.9)^3 = 0.729$. d. $P(1) \approx 0.369$ e. Second player's probability is a geometric series with $t(1) = 0.09$ and $r = 0.9^3$. Third player's probability is a geometric series with $t(1) = 0.081$ and $r = 0.9^3$. ∴ $P(2) \approx 0.332$; $P(3) \approx 0.299$ f. Answers are reasonable since: (i) First player has greatest probability, etc. (ii) $P(1) + P(2) + P(3) \approx 0.369 + 0.332 + 0.299 = 1.000$

Exercise 12-8, page 368; Mathematical Expectation

1. Card Draw Problem a. 0.01 b. You would expect to *gain* about 1 cent per game.

3. Archery Problem a. 5.96 b. Expected score 286.08 or about 286

5. Seed Germination Problem a. $P(0) = 0.0016$; $P(1) = 0.0256$; $P(2) = 0.1536$; $P(3) = 0.4096$; $P(4) = 0.4096$ b. 3.2 seeds

7. Expectation of a Binomial Experiment a. $P(0) = 0.07776$; $P(1) = 0.2592$; $P(3) = 0.2304$; $P(4) = 0.0768$; $P(5) = 0.01024$. b. E = 2 c. $(0.4)(5) = 2$, which equals the expected value.

9. Dice Game Problem. You *lose* about 72 cents/time.
11. Bad Egg Problem. About \$1.34 profit
13. Nuclear Reactor Problem a. $E(1) = 230$ b. $E(2) = 529$; $E(3) = 1216.7$; $E(4) = 2798.41$ c. 5.3×10^{36} d. 0.4352 e. 100.096 f. 261.05. No. Neutron level rises relatively *slowly*. g. $E(x) = 100 \times 1.00096^x$, which is an exponential function equation.

Exercise 12-9, page 372; Chapter Review

R1. a. i. 220 ii. 31 b. i. 3,921,225 ii. 94,109,400 c. i. 336 ii. 56
R2. a. 1/14 b. i. 28.8% ii. 36.7% c. i. 0.06 ii. 0.56 iii. 0.62 iv. 0.38 d. This is a *binomial* distribution. i. $P(0) = 0.0256$; $P(1) = 0.1536$; $P(2) = 0.3456$; $P(3) = 0.3456$; $P(4) = 0.1296$.

ii.

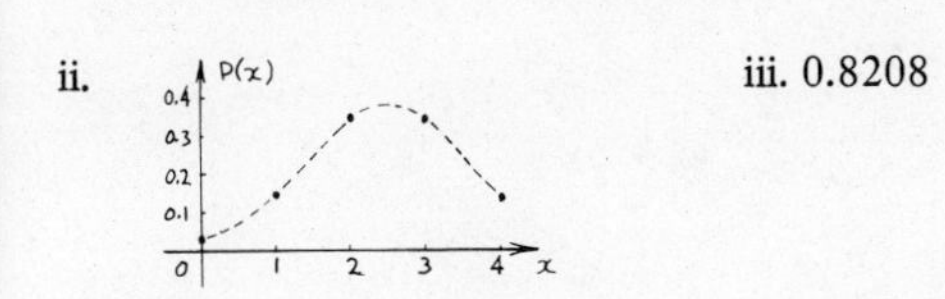

iii. 0.8208

R3. a. Avge. = 87.8 b. Each "weight" is a *fraction* of the total, 1. Each grade is a "payoff." So the weighted average is the *sum* of *fraction* × *payoff.*

Exercise 12-10, page 374; Cumulative Review, Chapters 9 through 12

1. a. $25x^2 + 4y^2 + 150x - 8y + 129 = 0$ b. $x^2 + y^2 + 4x + 10y - 7 = 0$ c. $4x^2 - 9y^2 - 24x - 36y + 63 = 0$ d. $y^2 + x - 4y - 1 = 0$ 2. S = {(−5, 2), (−5, −2), (1, 4), (1, −4)} 3. a. $f(-2) = -54$; $f(-1) = -16$; $f(0) = 3$; $f(1) = 8$; $f(2) = 5$; $f(3) = 0$; $f(4) = -1$; $f(5) = 9$; $f(6) = 33$. b. $f(x) = (x - 3)(x^2 - 4x - 1)$; $x^2 - 4x - 1 = 0 \Rightarrow x = 2 \pm \sqrt{5}$ ∴ zeros are 3, $2 + \sqrt{5}$, and $2 - \sqrt{5}$.

c.

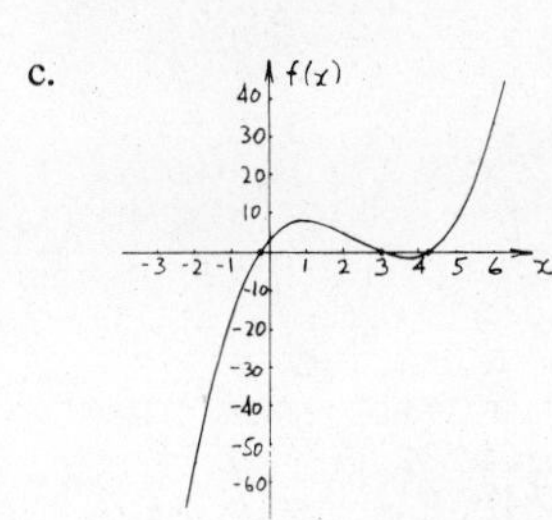

4. a. Let L = no. of lb. wings will lift. Let W = no. of lb. plane weighs. Let f = no. of feet long. $L = (25/8)f^2$, $W = (3/64)f^3$ b. Let $P(f)$ = no. lb. payload. $P(f) = (25/8)f^2 - (3/64)f^3$ c. $P(10) = 265.625$; $P(20) = 875$; $P(30) = 1546.875$; $P(50) = 1953.125$; $P(60) = 1125$; d. $f = 0$, $f = 66\text{-}2/3$. 0 length plane won't lift anything. Above 66-2/3 ft., plane won't fly. f. From graph, max. lift at = 44 ft.

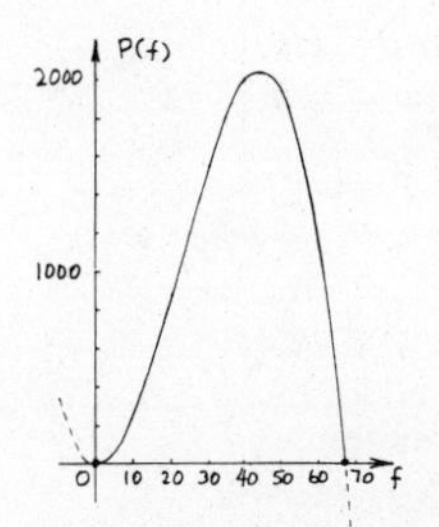

5. a. $-i$ b. $-i\sqrt{6}/2$ c. $2 + 4i$ 6. $x^2 - 6x + 25 = 0$ 7. a. 1/2 b. 3, <u>3-1/2</u>, <u>4</u>, <u>4-1/2</u>, <u>5</u>, <u>5-1/2</u>, <u>6</u>, <u>6-1/2</u>, 7 c. 53 d. 2828 8. a. 5 b. 16/3 9. $-35a^4b^3$ 10. \$2330.48 11. a. 120 b. 24 c. 12 d. 48 e. 30 f. 48 12. Kay: P(3 points) = 0.3456; Hezzy: P(4 points) = 0.2592. So Kay is more likely to win. 13. a. $P(0) = 0.0$ b. 2.9 c. The 2.9 is an *average* number of neutrons per fission, over *many* fissions.

CHAPTER 13
TRIGONOMETRIC AND CIRCULAR FUNCTIONS

Exercise 13-1, page 379; Introduction to Periodic Functions

Note: These graphs are *reasonable.* There are others that may also be reasonable.

1. a.

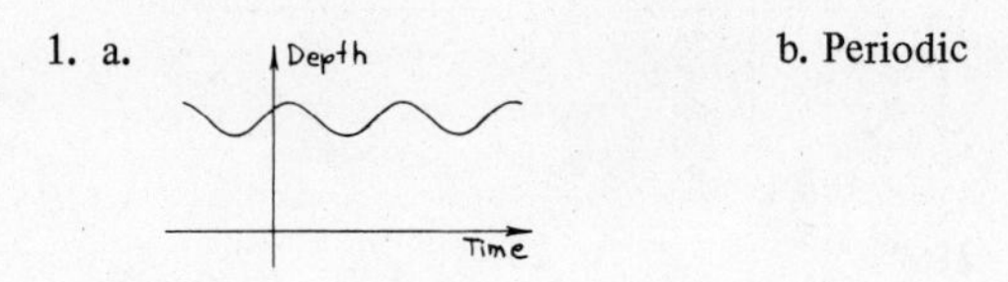

b. Periodic

3. a.

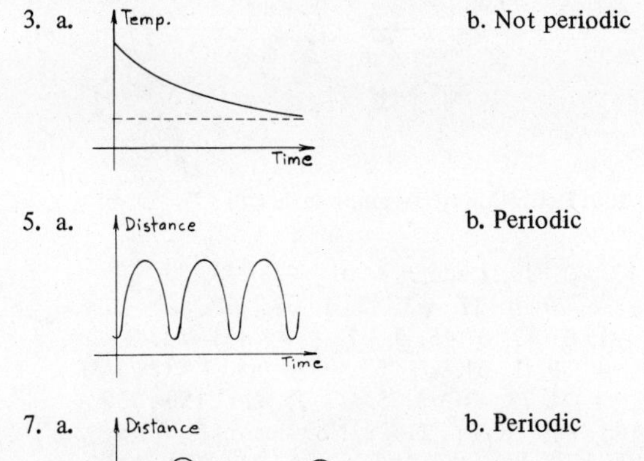

b. Not periodic

5. a. b. Periodic

7. a. 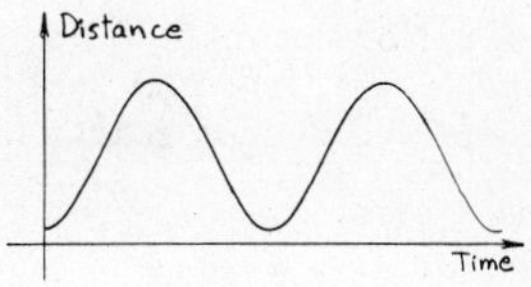 b. Periodic

9. a.

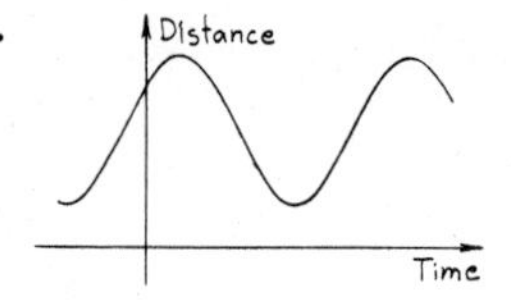

b. Periodic

Exercise 13-2, page 383; Measurement of Arcs and Rotation

1.

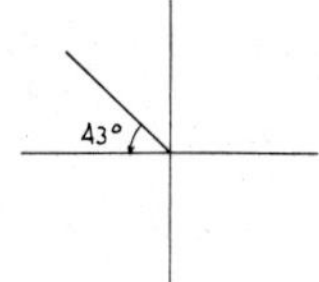

3.

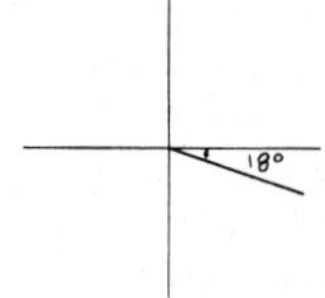

5.

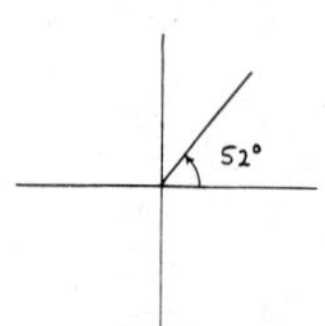

7.

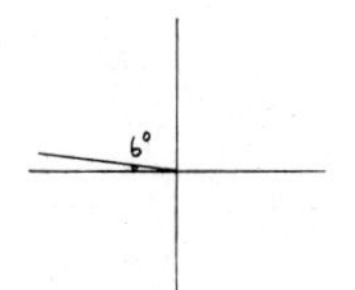

9.

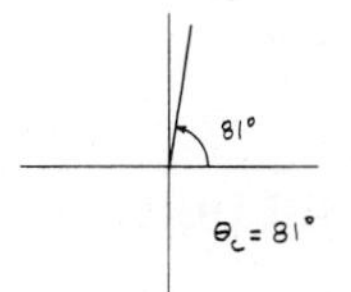

11.

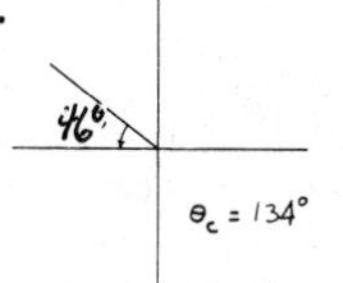

13.

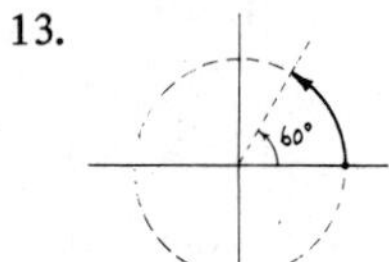

15.

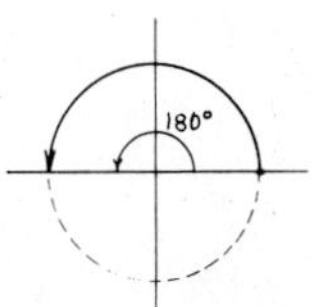

17.

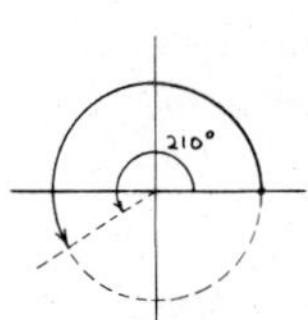

19.

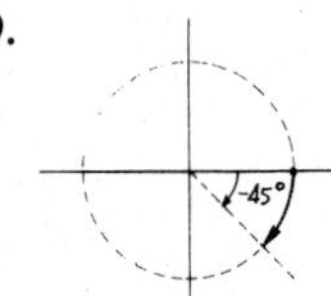

21.

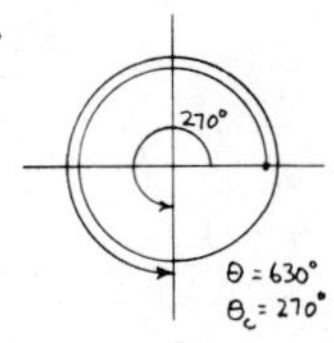

23.

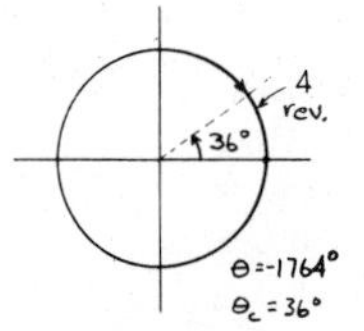

Exercise 13-3, page 388; Definition of Trigonometric and Circular Functions

13. 0 15. −1/2 17. −1 19. Undefined 21. $-2\sqrt{3}/3$ 23. $\sqrt{2}$ 25. $\sqrt{3}/2$ 27. $-\sqrt{2}/2$ 29. 0 31. $\sqrt{3}$ 33. 1 35. $-2\sqrt{3}/3$ 37. 1 39. $-2\sqrt{3}$ 41. 6 43. 0 45. 2 47. 1 49. −1 51. 0 53. 4/3 55. 0 57. 4 59. 1 61. $\sqrt{3}$ 63. −1/2 65. $1 + (2\sqrt{3}/3)$ 67. 1 69. −11/3 71. −1/4 73. $-\sqrt{3}/3$ 75. 4/3 77. a. 0°, 180°, 360° b. 90°, 270° c. 0°, 180°, 360° d. 90°, 270° e. No values of θ f. No values of θ 79. a. $\pi/2$ b. 0, 2π c. $\pi/4, 5\pi/4$ d. $\pi/4, 5\pi/4$ e. 0, 2π f. $\pi/2$ 81. Radians a. $\pi/6$ b. $\pi/4$ c. $\pi/3$ d. $\pi/4$ e. $5\pi/6$ f. π g. $5\pi/3$ h. $-3\pi/2$ i. $50\pi/3$ j. -6π

Exercise 13-4, page 392; Approximate Values of Trig. and Circ. Functions

1. through 8, have answers in the text. 9. 0.5640. 11. 1.736 13. 0.2773 15. 0.9676 17. 1.647 19. 12.17 21. −0.8787 23. 6.537 25. −0.1388 27. a. 0.3467 b. 0.346652545 29. a. 0.4562 b. 0.456202572 31. a. 1.300 b. 1.299827361 33. a. 0.3115 b. 0.311523912 35. a. 0.1397 b. 0.139694526 37. a. 1.115 b. 1.115893459 For 39–50, answers by calculator are in parentheses. 39. 0.9094 (0.909297427) 41. 0.3741 (0.374585640) 43. 3.536 (3.525320097) 45. 0.9599 (0.960170286) 47. 0.2161 (0.215641233) 49. −1.322 (−1.321348710)

Exercise 13-5, page 397; Graphs of Trigonometic and Circular Functions

3.

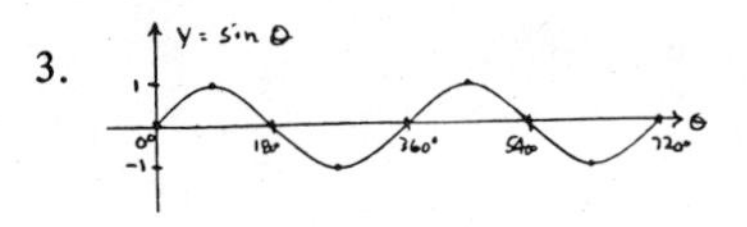

5.

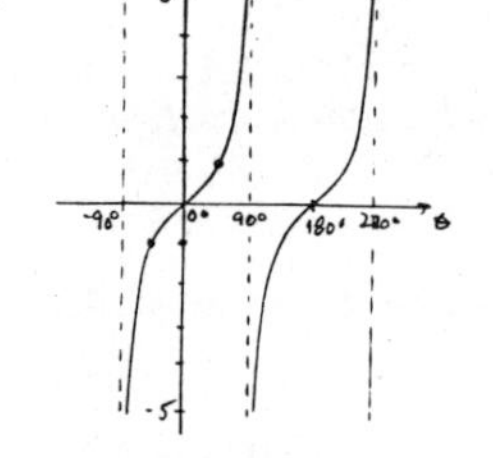

7.

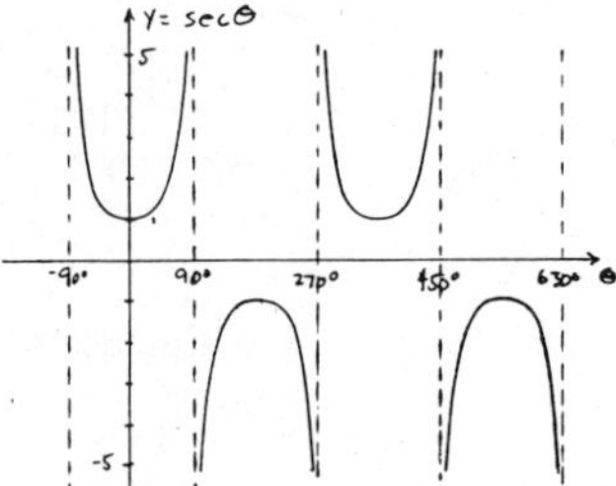

9.

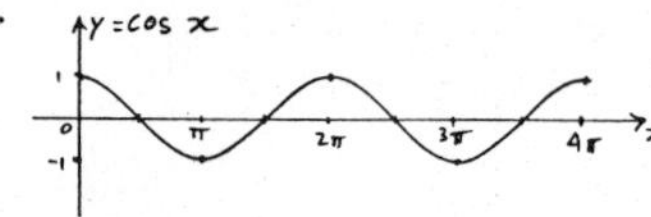

11.

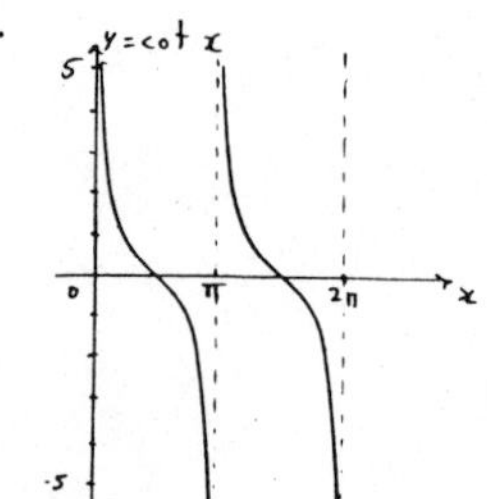

13.

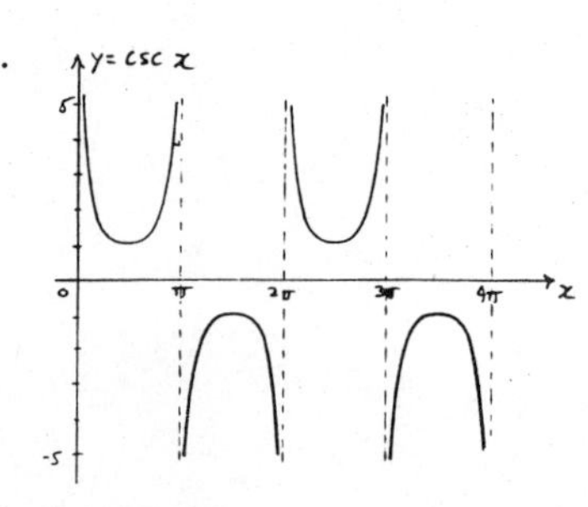

Exercise 13-6, page 402; General Sinusoidal Graphs

1.

$y = 5 + 2 \cos 3(\theta - 20°$
per. = 360°/3 = 120°

3.

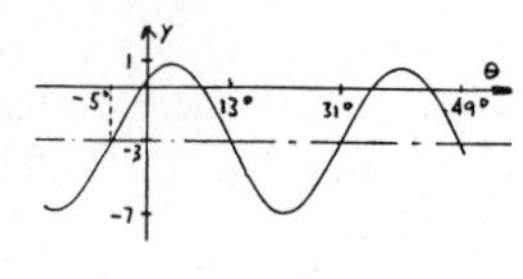

$y = 3 + 4 \sin 10(\theta + 5°)$
per. = 360°/10 = 36°

5. 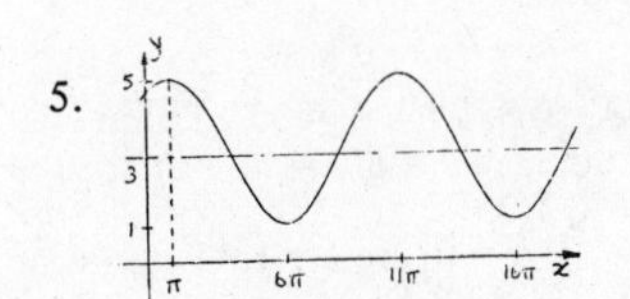

$y = 3 + 2\cos\frac{1}{5}(x - \pi)$

Per. $= \frac{2\pi}{1/5} = 10\pi$

7. 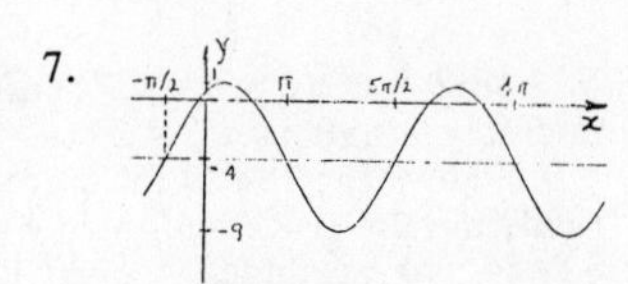

$y = -4 + 5\sin\frac{2}{3}(x + \frac{\pi}{2})$

Per. $= \frac{2\pi}{2/3} = 3\pi$

9. 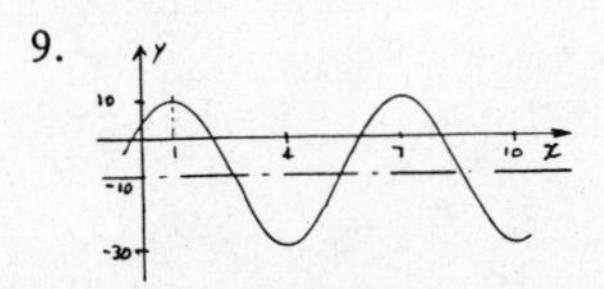

$y = 10 + 20\cos\frac{\pi}{3}(x - 1)$

per. $= 2\pi(\pi/3) = 6$

11. 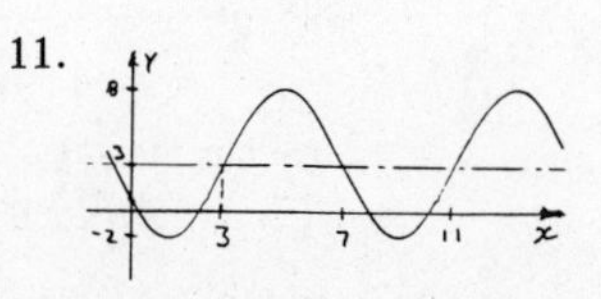

$y = 3 + 5\sin\frac{\pi}{4}(x - 3)$

per. $= 2\pi/(\pi/4) = 8$

13. 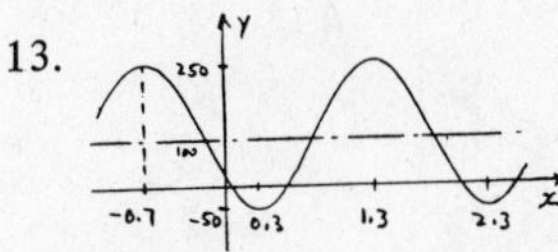

$y = 100 + 150\cos\pi(x + 0.7)$
per. $= 2\pi/\pi = 2$

15. $y = 1 + 4\cos(\pi/10)(x - 2)$

Exercise 13-7, page 403; Equations of Sinusoids from their Graphs

1. $y = 9 + 6\cos 2(\theta - 20°)$ 3. $y = -3 + 5\cos 3(\theta - 10°)$
5. $y = 1.45 + 1.11\cos 10(\theta + 7°)$ 7. $y = \sqrt{3}\cos(\theta - 30°)$
9. $y = 5 + 2\cos 2(x - \pi/6)$ 11. $y = 2 + 5\cos(\pi/15)(x + 5)$
13. $z = -8 + 2\cos 5\pi(t + 0.13)$

15. 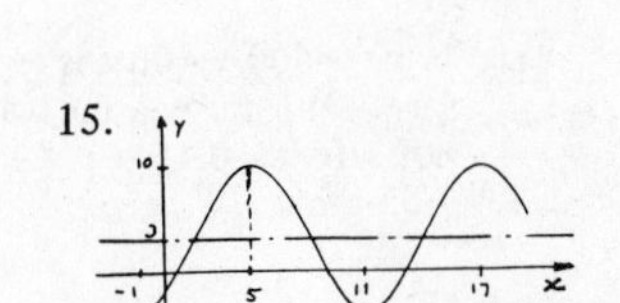

$y = 3 + 7\cos\frac{\pi}{6}(x - 5)$

17. 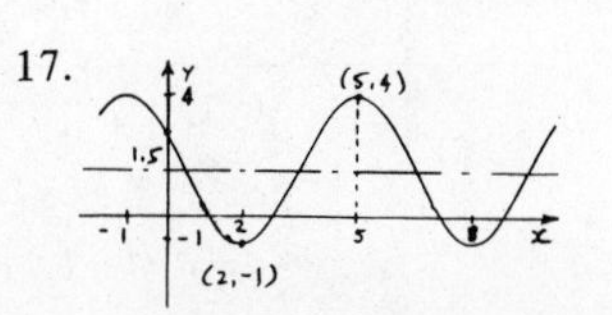

$y = 1.5 + 2.5\cos\frac{\pi}{3}(x - 5)$

Exercise 13-8, page 407; Sinusoidal Functions as Mathematical Models

1. Ferris Wheel Problem

a. 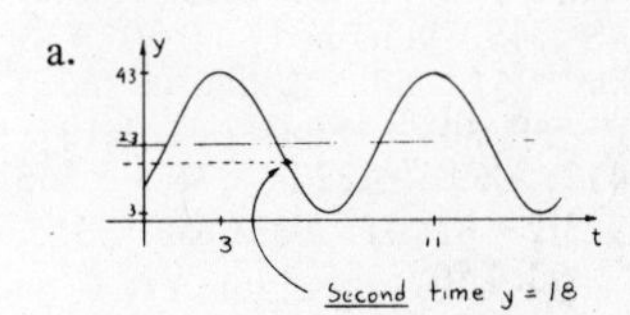

b. The lowest you go is 3 feet above the ground, because seats in a Ferris wheel do not scrape the ground. c. Let y = no. of feet above ground. $y = 23 + 20\cos(\pi/4)(t - 3)$ d. $t = 6 \Rightarrow y = 8.858$ ft. $t = 4\text{-}1/3 \Rightarrow y = 33$ ft. $t = 9 \Rightarrow y = 23$ ft. $t = 0 \Rightarrow y = 8.858$ ft.

3. Tidal Wave Problem. Let y = no. of meters deep. Let t = no. of min. since tsunami first reached pier. $y = 9 - 10\sin(2\pi/15)t$ a. i. 1.569 m ii. The water has receded beyond the point at which the depth is being measured, meaning that the depth at that point is *zero*. iii. 18.5 m b. According to the model, the minimum depth is –1 meter. So there is a time interval, during which the water is all gone. c. 300 km. d. Since the wave length is so long compared to the amplitude, a person on a ship at sea would not even notice that a tsunami had passed. Note that the motion of the water is up-and-down. It is only the location of the crest of the disturbance which moves horizontally at 1200 km per hour.
5. Pebble-in-the-Tire Problem

a. 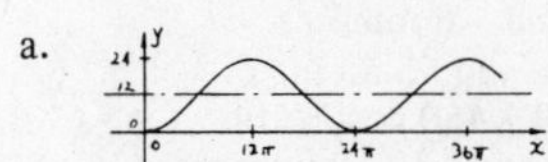

b. Let y = no. of in. from pavement. Let x = no. of in. car travels. $y = 12 - 12\cos(1/12)x$ c. 8.21 in.

7. Electric Current Problem a. $i = 5\cos 120\pi t$ b. –4.045 amperes
9. Roller Coaster Problem $y = 12 + 15\cos(\pi/50)x$ b. i. 27 m ii. 26.53 m iii. 5.61 m c. Total = 323.637 m
11. Sunrise Problem

a. 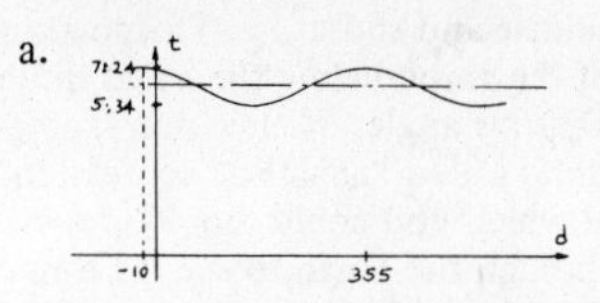

b. $t = 6{:}29 + 55\cos(2\pi/365)(d + 10)$ c. Answers will vary. d. Answers will vary. e. Answers will vary. f. Answers will depend on computer language used.

g. The Earth's orbit is slightly elliptical (see sketch). At the beginning of January, the Earth is *closest* to the Sun. At this time, it has both a *shorter* path and a *higher* speed. So it travels through a quadrant *faster* than predicted by this model.

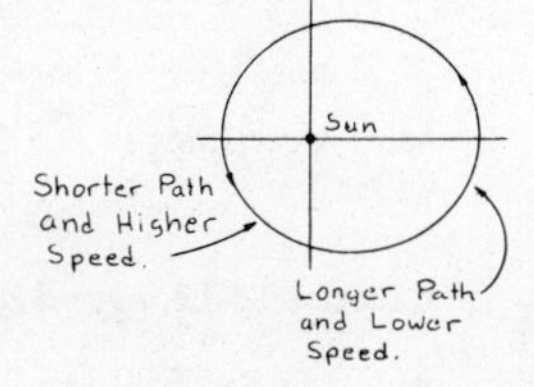

Exercise 13-9, page 417; Inverse Circular Functions

13. $\pi/4 + \pi n$ 15. $\pi/4$ 17. $\pi/6 + 2\pi n$ or $-5\pi/6 + 2\pi n$ 19. $\pi/3$ 21. $\pi/6 + 2\pi n$, $5\pi/6 + 2\pi n$ 23. Undefined 25. $\pi/6 + \pi n$ 27. $\pi/4$ 29. $\pi + 2\pi n$ 31. $\pi/2$ 33. $\pm 5\pi/6 + 2\pi n$ 35. $\pi/4$ 37. $\pm 45° + 360n°$ 39. $-90°$ 41. $-45° + 180n°$ 43. $45°$ 45. Undefined 47. $30° + 360n°$, $150° + 360n°$ 49. $0° + 360n°$ 51. $-60°$ 53. $60° + 360n°$, $120° + 360n°$ 55. $-60°$ 57. $120°$ 59. $30° + 180n°$ 61. 1.101 63. 0.625 65. 0.405 67. -0.412 69. -1.322 71. -0.563 73. 3/4 75. 5/13 77. 15/17 79. 3/2 81. -1 83. $\sqrt{10}$ 85. 5 87. Undefined 89. $\pi/6$ 91. $-17°$ 93. $150°$ 95. $11°$ 99. d. $y = \text{Arctan } x/\sqrt{1-x^2}$ 101. $\text{Arccsc } x = \text{Arctan } 1/\sqrt{x^2-1}$

Exercise 13-10, page 421; Evaluation of Inverse Relations

1. a. $\theta = -10° + (1/2) \arccos (1/4)(y-3)$ b. $20°, 140°, 200°$. 3. a. $x = 3\pi + 4 \arccos (1/2)(y-6)$ b. $(1/3)\pi, (5\text{-}2/3)\pi, (8\text{-}1/3)\pi$ 5. a. $\theta = 40° + (1/4) \arccos (1/10)(y+3)$ b. $30° 47', 49° 13', 120° 47'$ 7. a. $x = -0.7 + (2/\pi) \arcsin (1/5)(y-1)$ b. 0.710, 3.890, 4.710 9. a. $x = -0.2 + (1/\pi) \arccos (1/3)(y-1)$ b. No values of x. 11. a. $x = 0.6 + (4/\pi) \arctan 2(y-2)$ b. i. 1.6 ii. 2.912

Exercise 13-11, page 423; Inverse Circular Relations as Mathematical Models

1. Ferris Wheel Problem. 5.32 sec.
3. Tidal Wave Problem. Between 2.7 and 4.8 minutes
5. Pebble-in-the-Tire Problem 17.85 in. and 55.55 in.
7. Another Spaceship Problem $d = 550 + 450 \cos(\pi/50)t$ b. $t = (50/\pi) \arccos (d - 550)/450$ c. 39.18, 60.82, and 139.18 minutes d. 60.82 minutes
9. Tunnel Problem a. $y = 100 + 150 \cos (\pi/700)(x - 512.6)$ b. $x = 512.6 + (700/\pi) \arccos (y - 100)/150$ c. 1025.2 meters long d. 374.8 meters long e. Tunnel is 950.6 m. Bridge is 449.4 m.
11. Sun Elevation Problem

a.

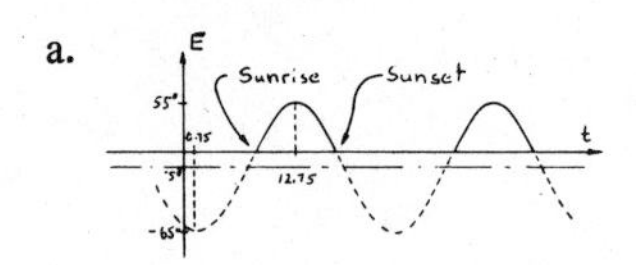

b. The t-intercepts are the times of sunrise and sunset. c. The portion of the graph below the t-axis indicates *negative* angles of elevation (sometimes called "angles of depression") at which you could "look" *down* through the Earth to see the Sun on the other side of the World when it is nighttime where you are. d. $E = -5 + 60 \cos(\pi/12)(t - 12.75)$ i. $34°$ ii. $49°$ e. 7:05 a.m. f. Instead of C being a *constant* in $E = C + A \cos B(t - D)$, make C *vary* sinusoidally with time.

Exercise 13-12, page 426; Chapter Review

Review Problems R1. a. i. $\sin 60° = \sqrt{3}/2$; $\cos(\pi/4) = \sqrt{2}/2$; $\tan(\pi/6) = \sqrt{3}/3$; $\sec 180° = -1$; $\csc 225° = -\sqrt{2}$; $\cot 0$ is undefined. ii. $\cos 57° \approx 0.5446$; $\sin 0.31 \approx 0.3051$; $\cot 1.43 \approx 0.1417$; $\csc 42° 37' \approx 1.477$; $\sec 143° 8' \approx 1.250$; $\tan 2 \approx -2.185$ iii. 6.6252 iv. -1

b. i. $\sin\theta = \sqrt{2}/2 \Rightarrow \theta = 45°$; $\cos\theta = 0.5724 \Rightarrow \theta \approx 55° 5'$; $\tan\theta$ is undef. $\Rightarrow \theta = 90°$; $\sec\theta = 2 \Rightarrow \theta = 60°$; $\csc\theta = 1/2 \Rightarrow$ *no value*; $\cot\theta = 0.2056 \Rightarrow \theta \approx 78° 23'$; ii. $\sin x = \sqrt{3}/2 \Rightarrow x = \pi/3$; $\cos x = 2 \Rightarrow$ *no value;* $\tan x = 1 \Rightarrow x = \pi/4$; $\sec x = 1 \Rightarrow x = 0$; $\csc x = 1.34 \Rightarrow x \approx 0.8424$; $\cot x$ is undef. $\Rightarrow x = 0$ iii. $31°$ iv. $3\frac{2}{3}$

c. i.

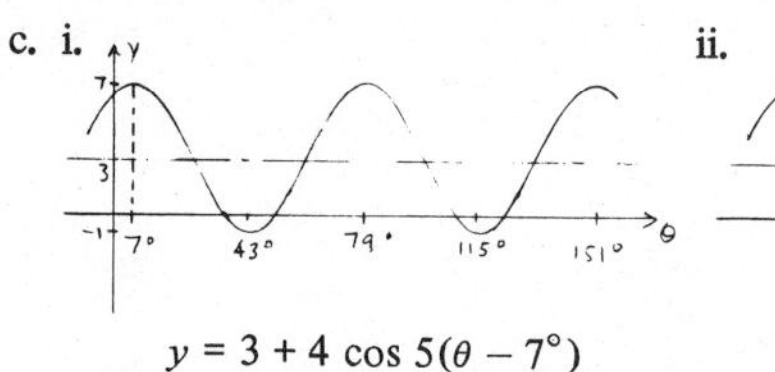

$y = 3 + 4 \cos 5(\theta - 7°)$

ii.

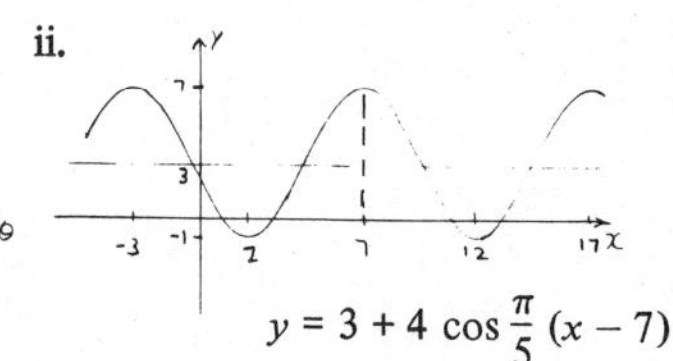

$y = 3 + 4 \cos \frac{\pi}{5}(x - 7)$

iii.

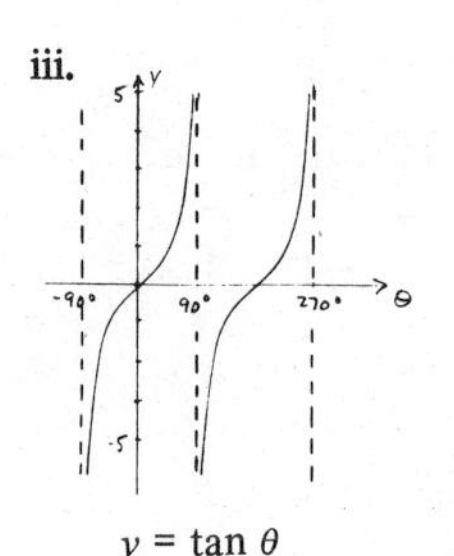

$y = \tan\theta$

iv.

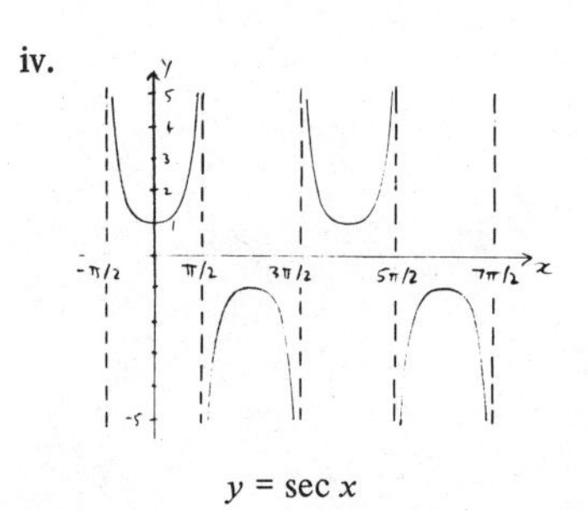

$y = \sec x$

R2. a. $y = -10 + 35 \cos 9(\theta - 3°)$ b. $y = 17 + 2 \cos (\pi/50)(x + 15)$
R3. a. $\text{Arcsin } 0.5 = 30°$ or $\pi/6$; $\arccos(-1) = 180° + 360n°$ or $\pi + 2\pi n$; $\tan^{-1} 0.3241 \approx 17° 57'$ or 0.3134; $\sec^{-1} \sqrt{2} = \pm 45° + 360n°$ or $\pm \pi/4 + 2\pi n$; $\csc^{-1} 0.5$ is *undefined;* $\text{arccot}(-1) = -45° + 180n°$ or $-\pi/4$ πn b. i. $\theta = 77° + (1/3) \arccos (1/4)(y - 5)$. $17°, 137°, 257°$
ii. $x = 3 + (5/\pi) \arctan (1/2)(y - 6)$. 1.1056, 6.1056, 11.1056

c. i. See Section 13-9. ii. See Section 13-9.

R4. Porpoising Problem

a.

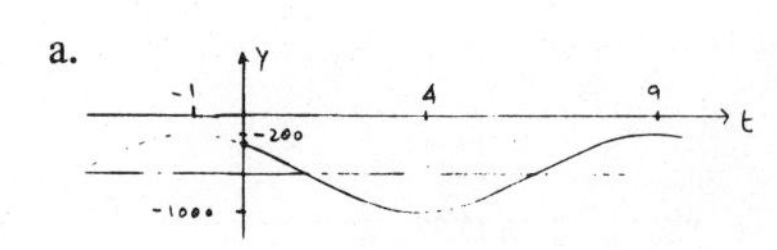

b. $y = -600 + 400 \cos (\pi/5)(t - 9)$ c. Submarine was *not* safe. d. $0.15 \le t \le 7.85$ min.

CHAPTER 14
PROPERTIES OF TRIGONOMETRIC AND CIRCULAR FUNCTIONS

Exercise 14-1, page 433; Three Properties of Trigonometric Functions

Answers for Problems 1 through 26 are contained in the problems. 27. $\sin x \csc x$, $\cos x \sec x$, $\tan x \cot x$, $\cos^2 x + \sin^2 x$, $\csc^2 x - \cot^2 x$, and $\sec^2 x - \tan^2 x$ 29. $\sin x = \sin x$, $\cos x = \pm\sqrt{1-\sin^2 x}$, $\tan x = \pm \sin x/\sqrt{1-\sin^2 x}$, $\cot x = \pm\sqrt{1-\sin^2 x}/\sin x$, $\sec x = \pm 1/\sqrt{1-\sin^2 x}$, $\csc x = 1/\sin x$

Exercise 14-2, page 436; Trigonometric Identities

Answers for Problems 1 through 34 are contained in the problems.

Exercise 14-3, page 442; Properties Involving Functions of More Than One Argument

1. $\cos(60° + 90°) = -\sqrt{3}/2$. $\cos 60° + \cos 90° = 1/2$ 3. $\tan(60° - 90°) = -\sqrt{3}/3$. $\tan 60° - \tan 90°$ is undefined. 5. $\sec(60° + 90°) = -2/\sqrt{3}$. $\sec 60° + \sec 90°$ is undefined. 7. Both expressions equal $\sqrt{3}/2$. b. Both expressions equal 0. 9. a. Both expressions equal 1/2. b. Both expressions equal 1. 11. a. Both expressions equal $\sqrt{3}/3$. b. Both are undefined 21. $(-4\sqrt{3} - 3)/10$ 23. $(-4 - 3\sqrt{3})/10$ 25. $(25\sqrt{3} - 48)/39$ 27. $(\sqrt{6} + \sqrt{2})/4$ 29. $2 - \sqrt{3}$ 31. $\sqrt{6} - \sqrt{2}$ 33. $(\sqrt{6} + \sqrt{2})/4$ 35. $2 - \sqrt{3}$. 37. $\sqrt{6} - \sqrt{2}$. 39. 0.9659258263 41. 0.2679491924 43. 1.035276180 51. $\cos A \cos B \cos C - \sin A \sin B \cos C - \sin A \cos B \sin C - \cos A \sin B \sin C$

Exercise 14-4, page 445; Multiple Argument Properties

1. c. In $\cos 2x$, the "2" affects the *period.* In $2 \cos x$, the "2" is the *amplitude.* 3. $\sin 2(30°) = 2 \sin 30° \cos 30° = 2\cdot(1/2)\cdot(\sqrt{3}/2) = \sqrt{3}/2 = \sin 60°$ 9. $\sin 2A = 24/25$. $\cos 2A = 7/25$. $\tan 2A = 24/7$. 11. $\sin 2A = -24/25$. $\cos 2A = 7/25$. $\tan 2A = -24/7$. 23. $\sin 3x = 3 \sin x - 4 \sin^3 x$ 25. $\cos 4x = 8 \cos^4 x - 8 \cos^2 x + 1$. 27. $\tan 14x = 2 \tan 7x/(1 - \tan^2 7x)$ 29. $\cos 6x = \cos^2 3x - \sin^2 3x$ 31. $\cos 10x = 2 \cos^2 5x - 1$ 33. $\cos x = 2\cos^2 (1/2x - 1)$ 41. $-7/25$. 43. $4\sqrt{2}/9$. 45. $-5/12$ 47. $\sqrt{2}/2$ 49. 11/13. 51. 1/1000 53. 1/4

Exercise 14-5, page 448; Half Argument Properties

1. c. In $\cos(1/2)x$, the "1/2" affects the *period.* In $(1/2)\cos x$, the "1/2" is the *amplitude.*

3. a.

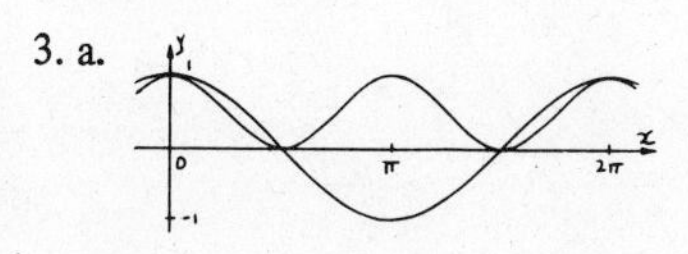

c. $y = (1/2) + (1/2)\cos 2x$ d. $\cos^2 x = (1/2)(1 + \cos 2x)$

15. $\sin(1/2)x = \sqrt{5}/5$. $\cos(1/2)x = 2\sqrt{5}/5$. $\tan(1/2)x = 1/2$. 17. $\sin(1/2)x = 2\sqrt{5}/5$. $\cos(1/2)x = -\sqrt{5}/5$. $\tan(1/2)x = -2$. 19. $\sin(1/2)x = -\sqrt{5}/5$. $\cos(1/2)x = 2\sqrt{5}/5$. $\tan(1/2)x = -1/2$. 33. $\sin 18° = (\sqrt{5} - 1)/4$ f. The author and his students have for several years sought a continuation of the pattern to $(\sqrt{7} - \sqrt{3})/4$, etc., but have been unable to relate the radical expression to the argument of the sine. The author would welcome hearing from anyone who *does* discover such a relationship.

Exercise 14-6, page 451; Sum and Product Properties

1. $\sin 65° + \sin 17°$ 3. $\cos 102° + \cos 4°$ 5. $\sin 7.9 + \sin 0.3$ 7. $-\cos 11.8 + \cos 2.6$ 9. $\sin 8x - \sin 2x$ 11. $\cos 11x + \cos 3x$ 13. $2 \cos 29° \cos 17°$ 15. $2 \sin 38° \cos 16°$ 17. $2 \sin 3.4 \sin 1.0$ 19. $-2 \cos 4 \sin 2$ 21. $2 \sin 6x \cos 3x$ 23. $2 \sin 9x \sin x$

Exercise 14-7, page 453; Linear Combination of Sine and Cosine

1. a. $2\cos(x - 60°)$ b. $2\cos(x - \pi/3)$ 3. a. $5\sqrt{2}\cos(x - 315°) = 5\sqrt{2}\cos(x + 45°)$ b. $5\sqrt{2}\cos(x - 7\pi/4) = 5\sqrt{2}\cos(x + \pi/4)$ 5. a. $13 \cos(x - 157° 23')$ b. $13\cos(x - 2.747)$ 7. a. $17\cos(3x - 208° 04')$ b. $17\cos(3x - 3.632)$ 9. a. $4\cos(x - 15°)$ b. $4\cos(x - 0.2618)$

11.

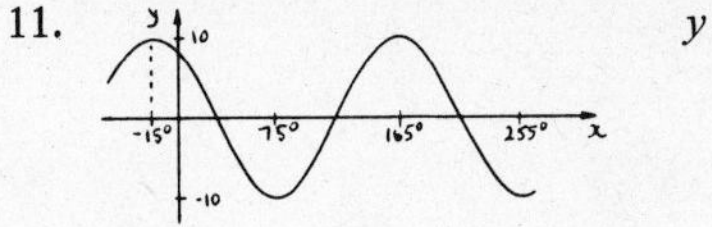

$y = 10\cos 2(x + 15°)$

13.

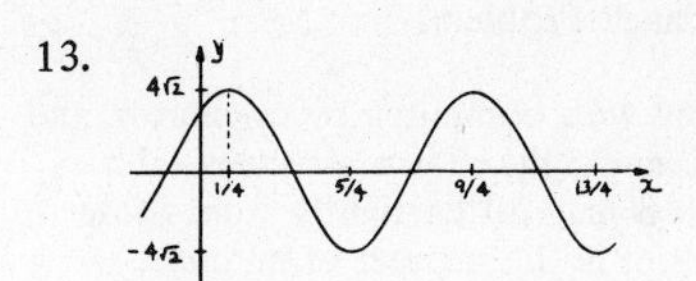

$y = 4\sqrt{2}\cos \pi(x - 1/4)$

15. $5\cos(x - \text{Arctan } 4/3)$ 17. $\sqrt{61}\cos(x - \text{Arctan } 6/5)$ 19. $\cos(x + \text{Arctan } 4/3)$ 21. $-2\sqrt{5}\cos(x - \text{Arctan } 1/2)$ 23. $-\sqrt{53}\cos(x + \text{Arctan } 7/2)$

Exercise 14-8, page 455; Simplification of Trigonometric Expressions

1. $2 \sin x \cos x$ 2. $(1/2)(1 - \cos 2x)$ 3. $1 - \cos^2 x$ 4. $1 - \sin^2 x$ 5. $(1/2)(1 + \cos 2x)$ 6. $2\cos^2 x - 1$ 7. $1 - 2\sin^2 x$ 8. $\cos^2 x - \sin^2 x$ 9. $2\tan x/(1 - \tan^2 x)$ 10. $\pm\sqrt{(1/2)(1 - \cos x)}$ 11. $\pm\sqrt{(1/2)(1 + \cos x)}$ 12. $\pm\sqrt{(1 - \cos x)/(1 + \cos x)}$ 13. $(1 - \cos x)/\sin x = \sin x/(1 + \cos x)$ 14. $3\sin x - 4\sin^3 x$ 15. $2\cos^2 2x - 1$ 16. $2\cos^2 3x - 1$ 17. $(1/2)\sin 2x$ 18. $(1/2)\sin(x + y) + (1/2)\sin(x - y)$ 19. $(1/2)\cos(x + y) + (1/2)\cos(x - y)$ 20. $2\cos(1/2)(x + y)\cos(1/2)(x - y)$ 21. $2\sin(1/2)(x + y)\cos(1/2)(x - y)$ 22. $\sqrt{2}\cos(x - \pi/4) = \sqrt{2}\cos(x - 45°)$ 23. $(-1/2)\cos 10x + (1/2)\cos 4x$ 24. $2\sin 5x\cos 2x$ 25. $2\cos(x + \pi/6) = 2\cos(x + 30°)$ 26. $4\sqrt{2}\cos(x - 5\pi/4) = 4\sqrt{2}\cos(x - 225°)$ 27. a. $\cos x \cos 37° + \sin x \sin 37°$ b. $\cos x \cos y \cos z - \sin x \sin y \cos z - \sin x \cos y \sin z - \cos x \sin y \sin z$ c. $4\cos^3 x - 3\cos x$ 29. a. $(1/2)\sin 2x$ b. $(1/2)(1 + \cos 2x)$ c. $(1/4)\sin 3x + (1/4)\sin x$ d. $(3/8) - (1/2)\cos 2x + (1/8)\cos 4x$. 31. e. $S = \{45°, 135°, 225°, 315°, 0°, 180°\}$

Exercise 14-9, page 459; Trigonometric Equations

1. {120°, 300°} 3. {103°, 343°} 5. {60°, −60°, 120°, −120°} 7. {90° + 180n°, 45° + 360n°, 135° + 360n°} 9. {2π/3 + πn} 11. {0°, −30°, −150°} 13. {π/3, 5π/3} 15. ϕ 17. {π/3, −π/3, −π} 19. {−90°, 0°} 21. {π/6, 7π/6, π/3, 4π/3} 23. {150°, −30°, −210°} 25. {178°, 358°, 28°, 208°} 27. {15°, 75°, −45°} 29. {22-1/2°, −67-1/2°, −22-1/2°, 67-1/2°} 31. {2π, 4π/3} 33. {16°, 34°, 52°, 70°, 88°} 35. {0, 2π, 4π, 3π/2, 7π/2} 37. {30°, −150°, −30°, 150°} 39. {real numbers} 41. {30° 58′, 210° 58′, 135°, 315°} 43. {25° 20′, 228° 24′} 45. {0, 3.1416, 0.3398, 2.8018} 47. {0.6662, 2.475} 49. {5.435, 3.990} 51. {40°, 220°, 160°, 340°} 53. {2.150}

Exercise 14-10, page 460; Chapter Review

R1. a. $\tan x = \sin x/\cos x$, $\cot x = \cos x/\sin x$ b. $\tan x = \sec x/\csc x$, $\cot x = \csc x/\sec x$ c. $\sin x \csc x = 1$, $\cos x \sec x = 1$, $\tan x \cot x = 1$. d. $\sin^2 x = 1 - \cos^2 x$, $\tan^2 x = \sec^2 x - 1$, $\csc^2 x = 1 + \cot^2 x$ R2. See Section 14-3 R3. See Section 14-3. R4. a., b., c.: See Section 14-4. d., e., f.: See Section 14-5. R5. See Section 14-6. R6. $\sqrt{74}\cos(x + 54°\ 28')$

CHAPTER 15
TRIANGLE PROBLEMS

Exercise 15-1, page 467; Right Triangle Problems

Note: The answers to these problems were computed by calculator, and are correct to the number of significant digits shown. Answers calculated using Table III or four-place logs may differ slightly from those shown in the fourth significant digit or in the number of minutes.
1. $m\angle C = 56°$. $a = 17.7$. $b = 9.92$. 3. $m\angle L = 42°\ 28'$. $l = 2.339$. $n = 2.556$. 5. $m\angle Y = 50°\ 29'$. $m\angle Z = 39°\ 31'$. $z = 22.27$. 7. $m\angle R = 77°\ 59'$. $m\angle S = 12°\ 01'$. $r = 46.3$
9. Ladder Problem 6.0 meters
11. Cat Problem $\theta = 75°\ 57'$
13. Moon Crater Problem 142 meters
15. Airplane Landing Problem a. 190.8 km. b. 1° 54′
17. Triangular Block Problem 35° 19′ b. 72 paces
19. Surveying Problem a. 33.5 meters b. 17.5 meters
21. Submarine Problem a. $v \approx 108$ meters. $u \approx 280$ meters b. 2790 meters
23. The Grapevine Problem a. 4° 31′ b. Slope is assumed to be *constant.*

Exercise 15-2, page 474; Oblique Triangles – Law of Cosines

Note: Answers by calculator. See note for Exercise 5-1. 1. 3.978 3. 4.682 5. 49.20 7. 28° 57′ 9. 134° 37′ 11. $\cos Y = 1.0604$. No such triangle. 13. 90°

15.

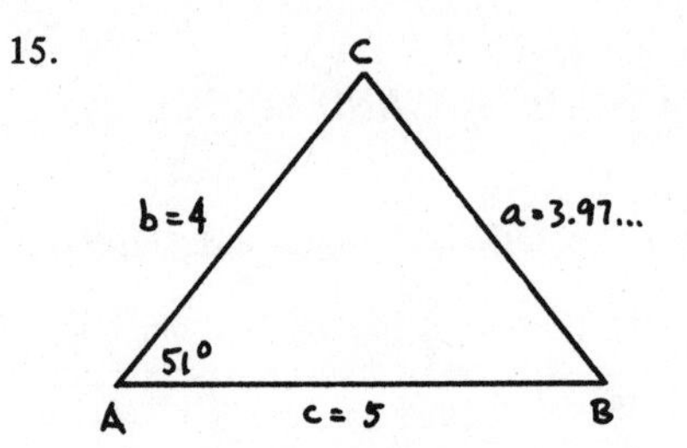

Exercise 15-3, page 475; Area of a Triangle

1. 5.443 3. 6.1 5. 23.66

Exercise 15-4, page 478; Oblique Triangles–Law of Sines

1. a. 5.229 b. 10.08 3. a. 249.9 b. 183.6 5. a. 9.321 b. 4.911 7. a. 214.7 b. 215.3 9. a. A = 32.12...° b. C = 51.31...° (apparently!) c. C = 128.68...° d. C = 180° − 51.31...° = 128.68...° e. Knowing sin C = 0.7806... does not tell whether C is acute or obtuse. Knowing cos C = −0.625 tells that C is *obtuse.*

Exercise 15-5, page 481; The Ambiguous Case

1. 5.315 or 1.317 3. 7.79 5. No values. 7. 5.52 9. Two triangles. 23.00° or 157.00° 11. One triangle. 38.15°

13. a.

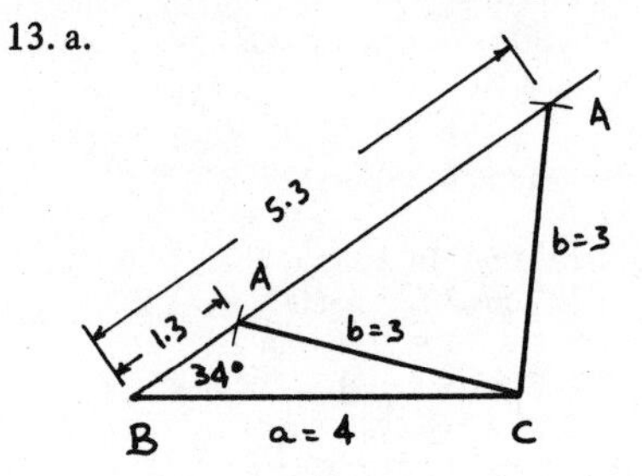

13. b.

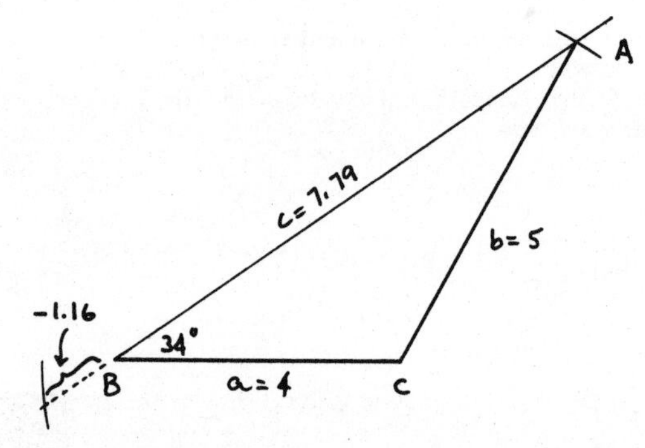

13. c.

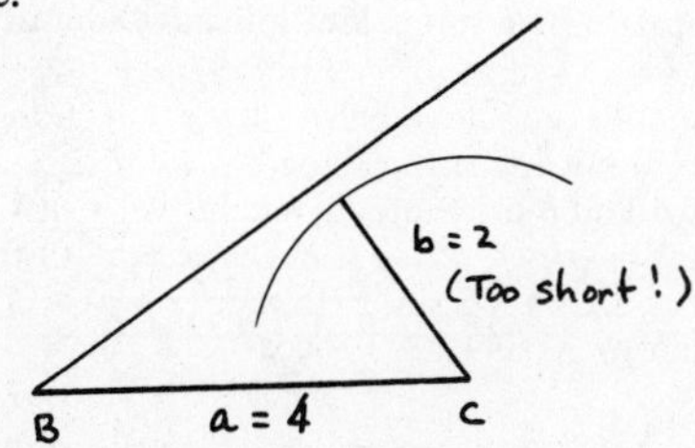

Exercise 15-6, page 482; General Solution of Triangles

1. $c = 4.177$, $m\angle A = 42°\ 59'$ $m\angle B = 65°\ 21'$, Area = 5.695 3. $c = 34.74$, $m\angle A = 20°\ 25'$, $m\angle B = 135°\ 45'$, Area = 363.7 5. $c = 266.0$, $m\angle A = 20°\ 12'$, $m\angle B = 46°\ 28'$, Area = 9641. 7. $m\angle A = 58°\ 25'$, $m\angle B = 73°\ 24'$, $m\angle C = 48°\ 11'$, Area = 26.83 9. $m\angle A = 26°\ 23'$, $m\angle B = 117°\ 17'$, $m\angle C = 36°\ 20'$, Area = 5.333 11. No such triangle 13. $a = 502.5$, $b = 121.5$, $m\angle C = 28°\ 30'$, Area = 14567. 15. $a = 15.78$, $b = 36.76$, $m\angle C = 140°\ 50'$, Area = 183.2 17. $b = 6.507$, $c = 6.190$, $m\angle C = 59°, 10'$, Area = 16.76 19. $c = 11.15$, $m\angle B = 18°\ 08'$, $m\angle C = 136°\ 02'$, Area = 12.15 21. $c = 10.26$, $m\angle B = 37°\ 36'$, $m\angle C = 116°\ 34'$, Area = 15.65; or $c = 2.339$, $m\angle B = 142°\ 24'$, $m\angle C = 11°\ 46'$, Area = 3.567 23. No such Triangle 25. $c = 2.751$, $m\angle B = 34°\ 57'$, $m\angle C = 18°\ 23'$, Area = 5.517 27. $c = 4{,}000$, $m\angle B = 90°$, $m\angle C = 53°\ 08'$, Area = 6.000

Exercise 15-7, page 489; Vectors

1. a. $|\vec{a} + \vec{b}| = 14.66$; $\theta = 45°\ 50'$ b. $|\vec{a} - \vec{b}| = 11.18$; $\theta = 70°\ 13'$ 3. a. $|\vec{a} + \vec{b}| = 11.69$; $\theta = 150°\ 00'$ b. $|\vec{a} - \vec{b}| = 28.73$; $\theta = 11°\ 45'$

5. a. $\vec{r} = 13.55$ at 20.3°
b. $-\vec{r} = 13.55$ at 200.3°
c. Graph

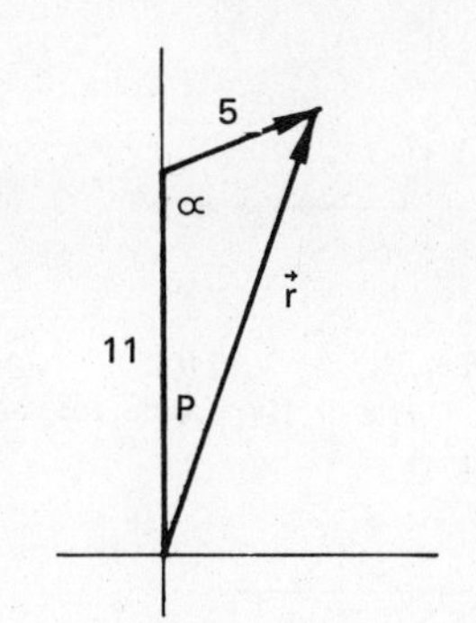

7. a. $\vec{r} = 8.61$ at 123.8°
b. $-\vec{r} = 8.61$ at 303.8°
c. Graph

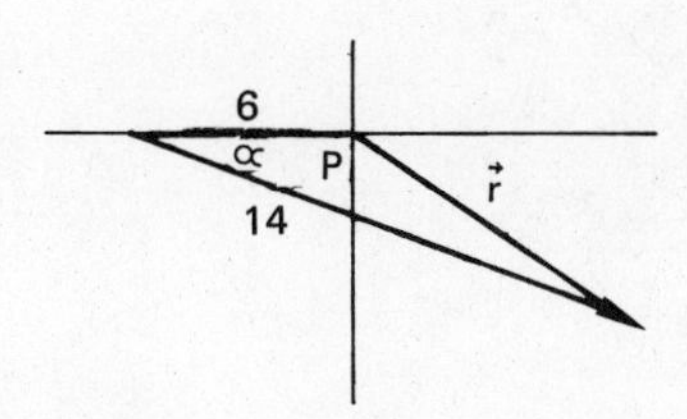

Exercise 15-8, page 493; Vectors–Resolution into Components

1. $6.038\vec{i} - 5.248\vec{j}$ 3. $-6.134\vec{i} + 14.45\vec{j}$ 5. a. $-12.8\vec{i} + 54.4\vec{j}$ b. 55.86 at 103.26° 7. 70 miles at 41.8° 9. 167.8 mph at 304.1°

11. a., b

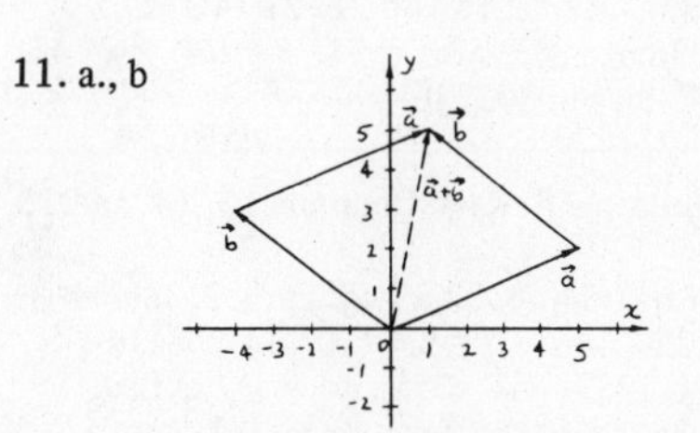

c. From diagram, $\vec{a} + \vec{b} = \vec{b} + \vec{a}$.

13.

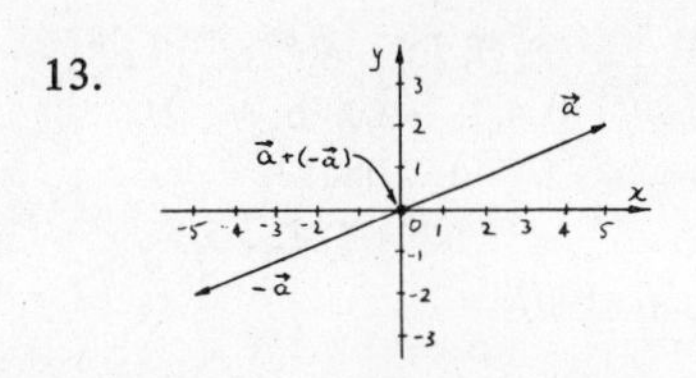

$|\vec{a} + (-\vec{a})| = 0$, so the vector is called the *zero* vector. It has *no* direction.

15. For each real number c, the product $c\vec{a}$ is a *unique vector,* which means that {vectors} is closed under multiplication by a scalar. The zero vector is necessary to insure closure since c could equal zero.

Exercise 15-9, page 494; Real-World Triangle Problems

Note: Answers were computed by calculator, *without* rounding off at intermediate steps, and are correct to the number of significant digits shown. Answers computed using 4-place logarithms or functions from Table III may differ slightly in the fourth significant digit or in the number of minutes.
1. Swimming Problem 1 Velocity ≈ 5.8 kph at 59° to heading
3. Mountain Height Problem 445.7 m
5. Missile Problem a. 0.6327 km b. 0.1265 km/sec = 455.5 kph c. 53° 07′
7. Visibility Problem 358 km (Surprisingly far!)
9. Airplane Lift Problem a. (i) $L = 507{,}713$ lb, $F_h = 88{,}163$ lb (ii) $L = 532{,}089$ lb, $F_h = 181{,}985$ lb (iii) $L = 577{,}350$ lb, $F_h = 288{,}675$ lb (iv) $L = 500{,}000$ lb, $F_h = 0$ lb b. As the angle θ gets larger, the horizontal turning force gets larger, too. Thus, the plane has a larger force pushing it into a tighter circle. c. When the plane is not banking, there is *no* horizontal force to turn the plane. So it flies straight. d. $\theta = 33°\ 33'$ e. (i) F_v may drop below 500,000 lb, causing the plane to start losing altitude. (ii) L might increase above 600,000 lb, causing the plane's wings to tear off!
11. Detour Problem a. 8.69 km further. b. 607.5 sq. km
13. Surveying Problem 2 a. 4476 sq. m b. 137.5 m c. 42° 57′, 58° 03′
15. Ivan Problem 1 $d = 16.52$ km
17. Ivan Problem 3 a. 111° 48′ b. 16.25 sq. km
19. Torpedo Problem a. Torpedoes can be fired from 279° 53′ through 36° 07′ b. Short-range torpedoes will never reach the target's path.

Exercise 15-10, page 499; Chapter Review

R1. a. $m\angle Z = 34° \ 23'$. $m\angle X = 55° \ 37'$. $x = 12.1$ cm b. $m\angle J = 50° \ 06'$. $b = 12.19$ km. $j = 14.58$ km R2. a. 16° 40′ b. 27° 48′ c. i. 475.0 cm ii. 23.30 cm or 10.99 cm R3. Area = 51.76 R4. $|\vec{a} - \vec{b}| = 15.98$. $\theta = 3° \ 45'$. R5. a. 7° 36′ b. 3027 lb

Exercise 15-11, page 501; Comulative Review, Chapters 13, 14, and 15

1. a. 18.73 km on a bearing of 65° 54′ b. 38.97 sq. km 2. a. $d = 80 + 50 \cos (\pi/10)(t - 2)$ b. 50.61 km c. $t = 2 + (10/\pi) \arccos (d - 80)/50$ d. 0.295, 3.705, and 20.295 hours 3. a. $y_1 = y_2$, Q.E.D. b. $y_3 = \cos^2 x = (1/2)(1 + \cos 2x)$. $y_4 = (1/2) \cos 8x + (1/2) \cos 2x$

4.

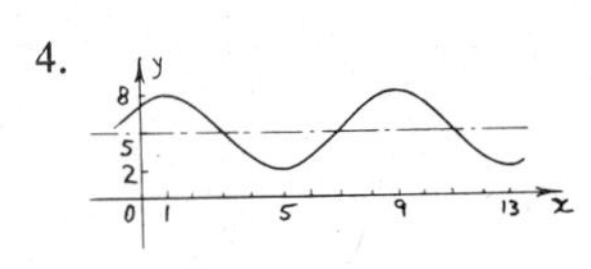

5. $\{-180°, -120°, -60°, 60°, 120°, 180°\}$ 6. a. $-\sqrt{3}/2$ b. $-\sqrt{2}/2$ c. $-6\sqrt{2}/7$ d. 7.08617 e. 30° f. π g. $\pi/3$ h. $5\pi/6$ i. $\sqrt{5}/3$ j. 0.61007

7. 0.059,964,006,479

8. a. $-\pi/3$ b. $-60°$ c. $-\pi/3 + \pi n$ or $-60° + 180n°$ d. $2/\sqrt{5}$ (See sketch)

e.

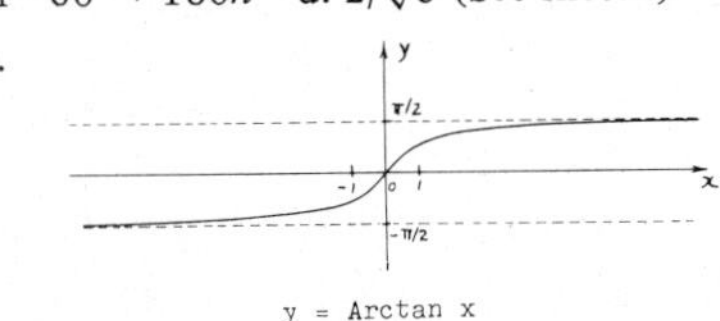

y = Arctan x

9. c. $\tan (1/2)(A - B)/\tan (1/2)(A + B) = (a - b)/(a + b)$, Q.E.D.

Final Examination, page 503

1. a. $y = mx + b$ b. $y = mx$ c. $y = kx^{1.6}$ d. $y = kx/z^2$ e. $y = ax^2 + bx + c$ f. $y = ax^4 + bx^3 + cx^2 + dx + e$ g. $y = a \cdot 10^{kx}$ or $y = a \cdot b^x$ h. $y = P(x)/Q(x)$, where $P(x)$ and $Q(x)$ are polynomials. i. $-(x - h)^2/a^2 + (y - k)^2/b^2 = 1$ j. $P(E) = n(E)/n(\text{Sample Space})$ k. $y = C + A \cos B(x - D)$ l. $y = \sec^{-1} x$ if and only if $x = \sec y$ and $x \in [0, \pi/2) \cup (\pi/2, \pi]$ 2. a. In a geometric sequence, the next term is formed by *multiplying* the preceding term by a constant. b. In an arithmetic sequence, the next term is formed by *adding* a constant to the preceding term. c. i. $t(n) = t(1) \cdot r^{n-1}$ ii. $t(n) = t(1) + (n - 1)d$ d. $t(n) = (n - 1)(n + 1) = n^2 - 1$ 3. a. Exponential function b. Exponential function c. Inverse variation function d. Quadratic function e. Linear function f. Linear relation (inequality) g. Arithmetic sequence h. Geometric sequence i. Rational algebraic function j. Quintic function k. Inverse sine function l. Tangent function 4. a. Associativity for multiplication b. Reflexive axiom c. Distributivity, $\cdot$ over $+$ d. Transitivity for order e. Trichotomy (Comparison Axiom) f. Multiplicative Inverses. g. Multiplicative Identity 5. a. $x^{3/4} = (\sqrt[4]{x})^3$ b. $(x^3)^4 = x^{12}$ c. $(xy)^4 = x^4y^4$ d. $(x + y)^3 = x^3 + 3x^2y + 3xy^2 + y^3$ e. $x^3 + y^3 = (x + y)(x^2 - xy + y^2)$ f. $\log (xy) = \log x + \log y$ g. $\log(x^y) = y \cdot \log x$ h. $r = \log_s t \Leftrightarrow t = s^r$ i. $n(A \text{ or } B) = n(A) + n(B)$ (if A and B are mutually exclusive) $= n(A) + n(B) - n(A \cap B)$ (otherwise) j. $\sqrt{x^2} = |x|$ k. If $rx^2 + sx + t = 0$, then $x = (-s \pm \sqrt{s^2 - 4rt})/2r$ l. $\cos 2x = 1 - 2 \sin^2 x$ m. $\tan (x + y) = (\tan x + \tan y)/(1 - \tan x \tan y)$ n. $\sin^2 x = 1 - \cos^2 x$

6. a.

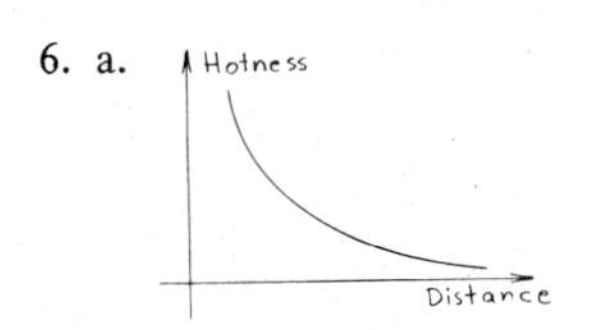

b. Inverse variation (Possibly exponential)

7. a.

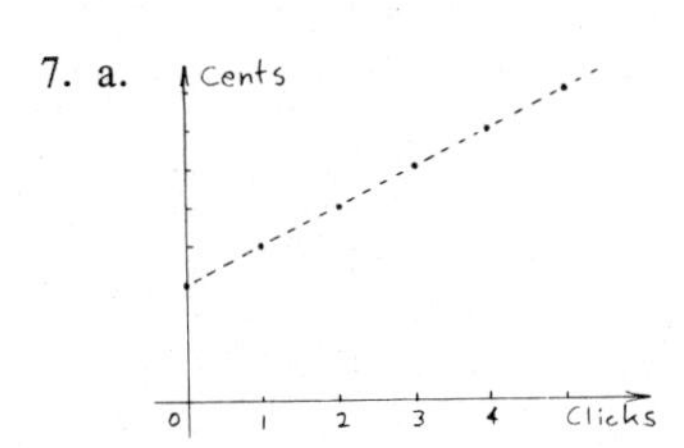

b. This could be an *arithmetic sequence* because the points lie in a straight line, and the domain contains only *integer* numbers of clicks.

8. a. $y = 1350/x^3$ b. $y = 73.10^{0.1003x}$ c. $y = -x^2 + 4x + 5$ d. $y = 6.5 + 4.5 \cos (\pi/8)(x - 5)$
9. a. (−2, 2-1/3)

b.

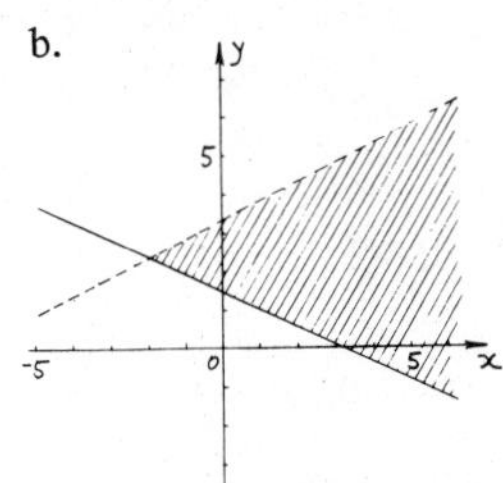

10. a. $f(x) = \sqrt{x} - 3$ b. $f(45) = 3\sqrt{5} - 3$ c. $S = \theta$
11. a. i. Circle ii. Hyperbola iii. Parabola iv. Ellipse b. $(x - 5)^2/2^2 + (y + 3)^2/4^2 = 1$

b.

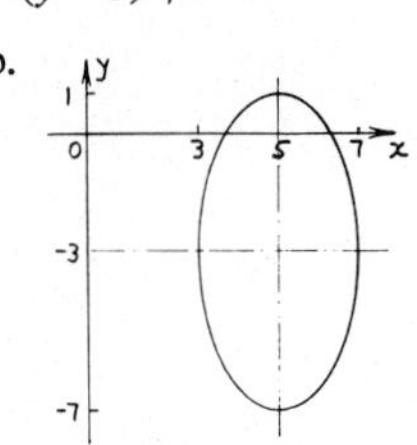

12. a. $3 - 2i$ b. 37 c. i. $3x^3 + x^2 + 9x + 14$ ii. 37 d. Remainder Theorem

13. a.

N	10
R	3
U	8
P	~~1~~ ~~8~~ ~~72~~ 720
F	~~8~~ ~~9~~ 10

Output: 10 P 3 = 120

b. Calculating $_nP_r$ 14. a. $t(1000) = 3004$; $S(1000) = 1505500$ b. $t(100) \approx 5597.6597$; $S(100) = 182{,}186.32$ 15. a. $x^5 - 10ix^4 - 40x^3 + 80ix^2 + 80x - 32i$ b. $-(37!/15!\ 22!)r^{22}s^{15}$ c. 5, $\underline{10.772}$, $\underline{23.2079}$, 50 16. a. 21.79 meters b. 64.95 sq. m

17. Skin Diving Problem a. Let d = no. of feet deep. Let N = no. of mg of nitrogen per ℓ. $N = 3.5d + 120$ b. 470 mg/ℓ c. 160 feet d. $N = 120$ mg/ℓ, the N-intercept

18. Probability Distribution Problem a. 126 ways b., c. 0 boys: 5 ways, probability 5/126; 1 boy: 40 ways, probability 40/126; 2 boys: 60 ways, probability 60/126; 3 boys: 20 ways, probability 20/126; 4 boys: 1 way, probability 1/126 d. 1.0000 = Sum

e.

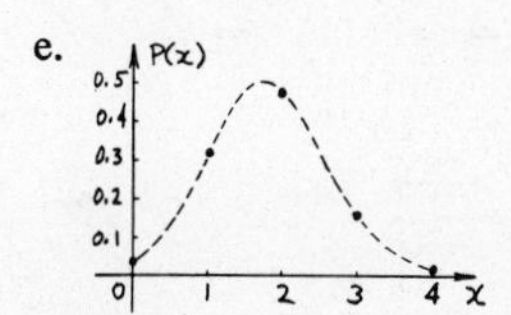

f. 1440 ways g. 2880 ways

APPENDIX A
COMPUTER PROGRAMMING

Exercise A-1, page 512; Reading Flow Charts

1.

S	~~0~~ ~~5~~ 12
X	~~5~~ ~~7~~ 9

Output: 12

3. a.

N	4
P	~~1~~ ~~2~~ ~~6~~ 24
C	~~2~~ ~~3~~ 4

Output: 2 2
3 6
4 24

b.

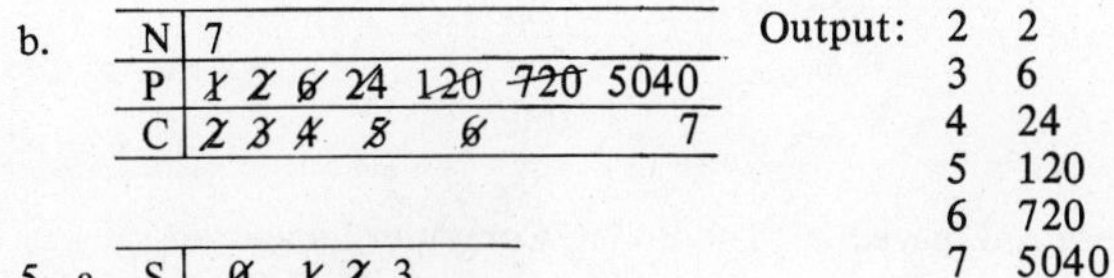

N	7
P	~~1~~ ~~2~~ ~~6~~ ~~24~~ ~~120~~ ~~720~~ 5040
C	~~2~~ ~~3~~ ~~4~~ ~~5~~ ~~6~~ 7

Output: 2 2
3 6
4 24
5 120
6 720
7 5040

5. a.

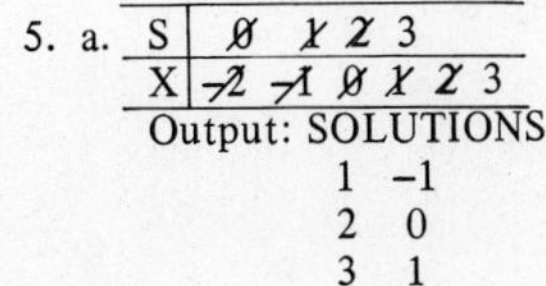

S	~~0~~ ~~1~~ ~~2~~ 3
X	~~-2~~ ~~-1~~ ~~0~~ ~~1~~ ~~2~~ 3

Output: SOLUTIONS
1 -1
2 0
3 1

b.

S	0
X	~~-2~~ ~~-1~~ ~~0~~ ~~1~~ ~~2~~

Output: SOLUTIONS
NONE

Exercise A-2, page 516; Writing Flow Charts

1. Evaluationg Expressions, Prob. 1

a.

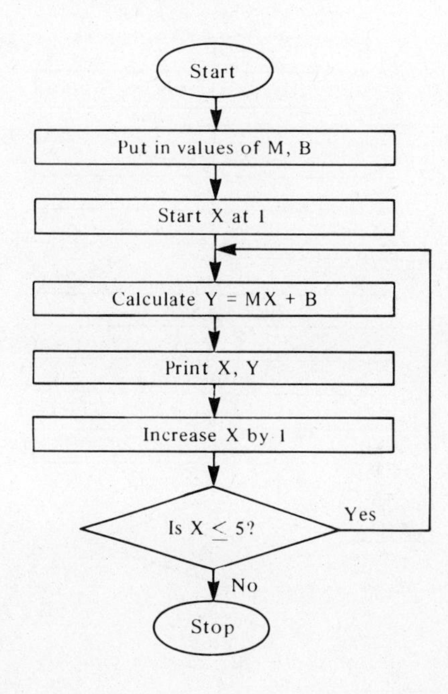

b.

M	7
B	11
X	~~1~~ ~~2~~ ~~3~~ ~~4~~ ~~5~~ 6
Y	~~18~~ ~~25~~ ~~32~~ ~~39~~ 46

Output: 1 18
2 25
3 32
4 39
5 46

7. Sum of the Even Numbers Problem

a.

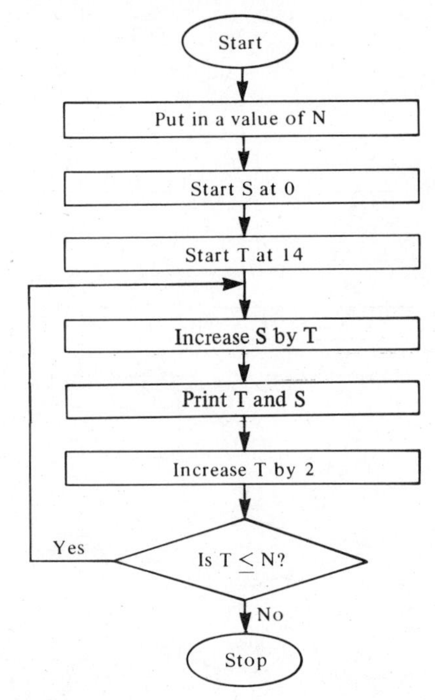

b.

N	20				
S	~~0~~	~~14~~	~~30~~	~~48~~	68
T	~~14~~	~~16~~	~~18~~	~~20~~	22

Output:

14	14
16	30
18	48
20	68

11. Data Search Problem

a.

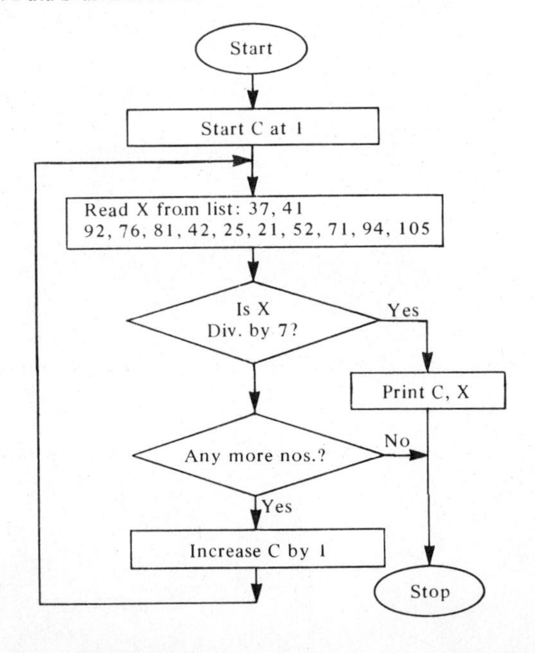

b.

C	~~1~~	~~2~~	~~3~~	~~4~~	~~5~~	6
X	~~37~~	~~41~~	~~92~~	~~76~~	~~81~~	42

Output: 6 42

13. Product of the Integers Problem

a.

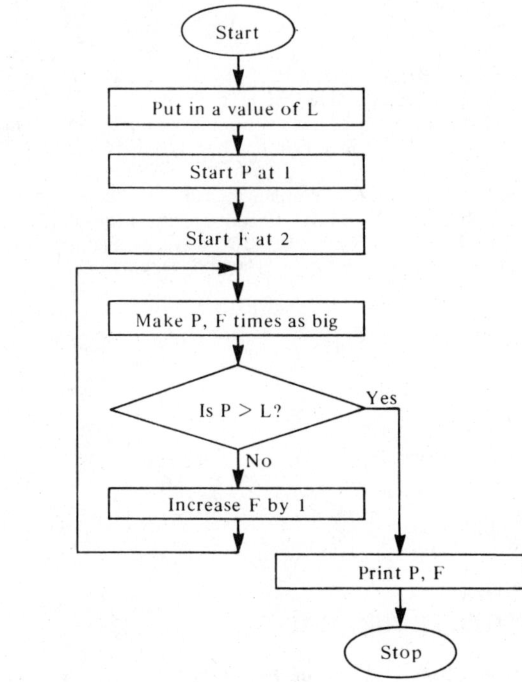

b.

L	100				
P	~~1~~	~~2~~	~~6~~	~~24~~	120
F	~~2~~	~~3~~	~~4~~	5	

Output: 120 5

c.

L	6			
P	~~1~~	~~2~~	~~6~~	24
F	~~2~~	~~3~~	4	

Output: 24 4

Exercise A-3, page 523; The BASIC Computer Language

1.

S	~~0~~	~~13~~	~~21~~	~~24~~	~~22~~	15
X	~~-2~~	~~-1~~	~~0~~	~~1~~	~~2~~	3
Y	~~13~~	~~8~~	~~3~~	~~-2~~	-7	

Output: -7 15

3. a.

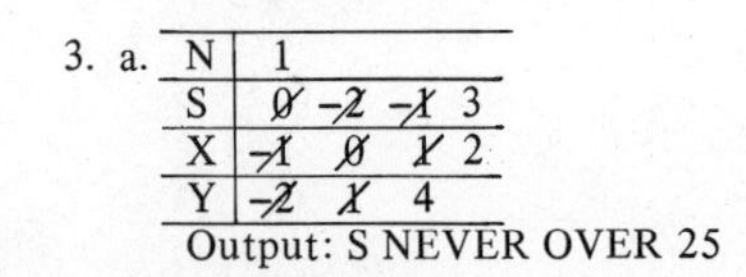

N	1			
S	~~0~~	~~-2~~	~~-1~~	3
X	~~-1~~	~~0~~	~~1~~	2
Y	~~-2~~	~~1~~	4	

Output: S NEVER OVER 25

b.

N	5
S	~~0~~ ~~-2~~ ~~-1~~ ~~3~~ ~~10~~ ~~20~~ 33
X	~~-1~~ ~~0~~ ~~1~~ ~~2~~ ~~3~~ 4
Y	~~-2~~ ~~1~~ ~~4~~ ~~7~~ ~~10~~ 13

Output: S > 25 WHEN X = 4

5.

C	1
X	2
Y	5
E	27
D	~~2~~ 3

Output: 27 DIVISIBLE BY 3

11. Errors:
    1. M is undefined at Line 30.
    2. At Line 30, it should be M*X, not MX.
    3. At Line 40, it should be LET X = X+1.
    4. GO TO 30 creates an infinite loop.
    5. There is no output for the program.
    6. Last instruction must be END, not STOP. (STOP is permitted on some versions of BASIC.)

13. Problem 13

```
10 LET S=0
20 LET X=5
30 LET S=S+X
40 LET X=X+2
50 IF X<8 THEN 30
60 PRINT S
70 END

RUN
 12
```

15. Problem 15

```
10 INPUT N
20 LET P=1
30 LET C=2
40 LET P=P*C
50 IF C>=N THEN 80
60 LET C=C+1
70 GO TO 40
80 PRINT N;P
90 END

RUN
 ? 4
  4   24

RUN
 ? 7
  7   5040
```

17. Problem 17

```
10 LET S=0
20 LET X=-2
30 PRINT"SOLUTIONS:"
40 IF X↑3<>X THEN 70
50 LET S=S+1
60 PRINT S;X
70 LET X=X+1
80 IF X<=2 THEN 40
90 IF S>0 THEN 110
100PRINT"NONE"
110END

RUN
SOLUTIONS:
1  -1
2   0
3   1
```

21. Problem 21

```
10 INPUT M,B
20 FOR X=1 TO 5
30 LET Y=M*X+B
40 PRINT X,Y
50 NEXT X
60 END

RUN
? 7,  11
1              18
2              25
3              32
4              39
5              46
```

27. Problem 27

```
10 INPUT N
20 LET S=0
30 FOR T=14 TO N STEP 2
40 LET S=S+T
50 NEXT T
60 PRINT S
70 END

RUN
?  20
 68
```

31. Problem 31

```
10 LET C=1
20 READ N
30 DATA 37, 41, 92, 76, 81, 42
31 DATA 25, 21, 52, 71, 94, 105
40 IF INT(N/7)=N/7 THEN 80
50 LET C=C+1
60 IF C>12 THEN 90
70 GO TO 20
80 PRINT C, N
90 END

RUN
6                    42
```

33. Problem 33

```
10 INPUT L
20 LET P=1
30 LET F=2
40 LET P=P*F
50 IF P>L THEN 80
60 LET F=F+1
70 GO TO 40
80 PRINT P, F
90 END

RUN
?  100
 120                5

RUN
?  6
 24                 4
```

Appendix B-1, page 529; Matrix Operations

1. $\begin{bmatrix} -2 & 13 \\ 0 & 10 \\ 0 & 11 \end{bmatrix}$ 3. $\begin{bmatrix} 3 & 1 & 0 & 2 \end{bmatrix}$ 5. $\begin{bmatrix} -1 & 59 \\ -22 & -11 \end{bmatrix}$

7. $\begin{bmatrix} 24 & -18 \\ 25 & -17 \end{bmatrix}$ 9. $\begin{bmatrix} 45 & -10 \\ 39 & 22 \\ 9 & 22 \end{bmatrix}$ 11. $\begin{bmatrix} 3 & 22 & 8 \\ -1 & 19 & 12 \\ 11 & 9 & 16 \end{bmatrix}$

13. $\begin{bmatrix} 1 & 4 & 7 \\ 2 & 5 & 8 \\ 3 & 6 & 9 \end{bmatrix}$ 15. $\begin{bmatrix} 1 & 0 \\ 0 & 1 \end{bmatrix}$ 17. $\begin{bmatrix} -2/11 & 7/11 \\ 3/11 & -5/11 \end{bmatrix}$

19. $\begin{bmatrix} 1 & 2 \\ 3 & 4 \end{bmatrix}$ 21. No inverse 23. $\begin{bmatrix} 5 & 3 \\ -1 & 4 \end{bmatrix}$

25. $\begin{bmatrix} 5 & 2 \\ 1 & 1 \end{bmatrix} \begin{bmatrix} x \\ y \end{bmatrix} = \begin{bmatrix} 11 \\ 4 \end{bmatrix}$ S = {(1, 3)}

27. $\begin{bmatrix} 6 & -7 \\ 2 & 5 \end{bmatrix} \begin{bmatrix} x \\ y \end{bmatrix} = \begin{bmatrix} 47 \\ -21 \end{bmatrix}$ S = {(2, –5)}

29. $\begin{bmatrix} 4 & -3 \\ 5 & -6 \end{bmatrix} \begin{bmatrix} x \\ y \end{bmatrix} = \begin{bmatrix} 11 \\ 9 \end{bmatrix}$ S = {(13/3, 19/9)}

31. $\frac{1}{76}\begin{bmatrix} 16 & -10 & -34 \\ 6 & 1 & 11 \\ 14 & -23 & -25 \end{bmatrix}$ 33. $\frac{1}{75}\begin{bmatrix} 17 & 19 & -10 \\ -2 & 11 & 10 \\ -3 & -21 & 15 \end{bmatrix}$

Appendix B-2, page 532; Descartes' Rule of Signs and the Upper Bound Theorem

Prob. #	Pos.	Neg.	Comp.
1.	2	1	0
	0	1	2
3.	3	0	0
	1	0	2
5.	2	2	0
	2	0	2
	0	2	2
	0	0	4
7.	3	1	0
	1	1	2
9.	0	0	8

11. Between –2 and 4.

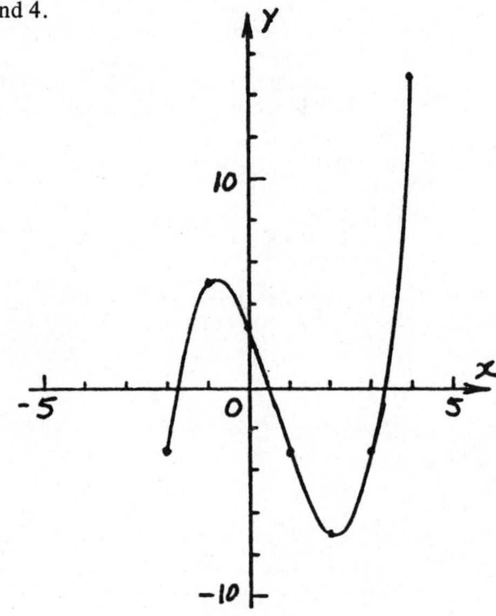

13. Between –1 and 2.

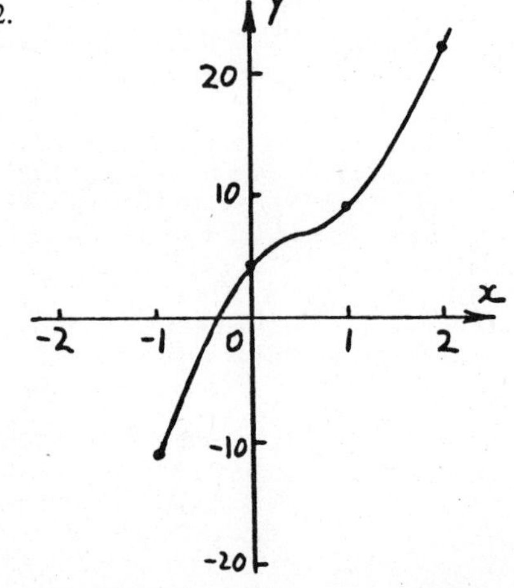

15.a. $c = 1$
b. $c = 3$
c. Since the trinomial is being multiplied by $(x - c)$, and c is positive, the sign of the last term of the product will be *negative*. Since the first term of the product is positive, there must be an *odd* number of sign reversals.

17. a. Even number of positive zeros; even number of negative zeros.
c. Odd number of positive zeros; even number of negative zeros.

19.a. $x = 2$. One sign reversal in $P(x) \Rightarrow$ only *one* positive zero.
No sign reversal in $P(-x) \Rightarrow$ *no* negative zeros.
c. $x = 2$ or -2. One sign reversal in $P(x) \Rightarrow$ only *one* positive zero.
One sign reversal in $P(-x) \Rightarrow$ only *one* negative zero.

Appendix B, page 533; Mathematical Induction

1. Extended Assoc. for Mult. Prove that $a_1a_2a_3 \dots a_{n-1}a_n = (a_1a_2a_3 \dots a_{n-1})a_n$ $\forall n \geq 3$. *Proof:* For $n = 3$, $a_1a_2a_3 = (a_1a_2)a_3$ by the Associative Axiom. Assume that $a_1a_2a_3 \dots a_{k-1}\, a_k = (a_1a_2a_3 \dots a_{k-1})a_k$ for some $k > 3$. Then $a_1a_2a_3 \dots a_{k-1}a_ka_{k+1} = ((\dots((a_1a_2)a_3) \dots a_{k-1})a_k)a_{k+1}$ (Definition) $= ((a_1a_2a_3 \dots a_{k-1})a_k)a_{k+1}$ (Definition) $= (a_1a_2a_3 \dots a_{k-1}a_k)a_{k+1}$ (Ind. Hypoth.) $\therefore a_1a_2a_3 \dots a_{n-1}\, a_n = (a_1a_2a_3 \dots a_{n-1})a_n$ $\forall n \geq 3$, Q.E.D.

7. Exponentiation Distributes over Mult. Prove that $(ab)^n = a^nb^n$ $\forall n \geq 1$. *Proof:* For n=1, $(ab)^1 = ab = a^1b^1$ by def. of x^1. Assume that for $n = k$, $(ab)^k = a^kb^k$. For $n = k$+1, $(ab)^{k+1} = (ab)^k \cdot ab$ (Def. of Exp.) $= a^kb^k \cdot ab$ (Ind. Hypoth.) $= (a^ka)\cdot(b^kb)$ (Comm. & Assoc.) $= a^{k+1} \cdot b^{k+1}$ (Def. of Exp.) $\therefore (ab)^n = a^nb^n$ $\forall n \geq 1$, Q.E.D.

11. n^{th} power of a Number > 1. Prove that if $x > 1$, then $x^n > x^{n-1}$ for all integers $n \geq 1$. *Proof:* If $n = 1$, $x^1 > x^0$ because $x^1 = x$ and $x^0 = 1$, and $x > 1$. Assume that for n=k, $x^k > x^{k-1}$. For $n = k$+1, $x^{k+1} = x^k \cdot x > x^k \cdot 1 = x^k \therefore x^{k+1} > x^k \therefore x^n > x^{n-1}$ for *all* $n \geq 1$.

15. Absolute Value of a Product. Prove that $|a_1a_2a_3 \dots a_n| = |a_1|\,|a_2|\,|a_3| \dots |a_n|$ for all $n > 2$. *Proof:* For n=2, $|a_1a_2| = \sqrt{(a_1a_2)^2} = \sqrt{a_1{}^2a_2{}^2} = \sqrt{a_1{}^2}\,\sqrt{a_2{}^2} = |a_1|\,|a_2|$. Assume that for n=k, $|a_1a_2a_3 \dots a_k| = |a_1|\,|a_2|\,|a_3| \dots |a_k|$. For $n = k + 1$, $|a_1a_2a_3 \dots a_ka_{k+1}| = |(a_1a_2a_3 \dots a_k)a_{k+1}|$ (Extended Assoc.) $= |a_1a_2a_3 \dots a_k|\,|a_{k+1}|$ (by the Anchor) $= |a_1|\,|a_2|\,|a_3| \dots |a_k|\,|a_{k+1}|$ (by the Ind. Hypoth.) $\therefore |a_1a_2a_3 \dots a_n| = |a_1|\,|a_2|\,|a_3| \dots |a_n|$ for *all* $n \geq 2$, Q.E.D.

18. Formula for Triangular Numbers – Prove that $T_n = (n/2)(n + 1)$ $\forall n \geq 1$ *Proof:* For $n = 1$, $T_1 = 1$ and $(1/2)(1 + 1) = 1$, so $T_1 = (1/2)(1 + 1)$. Assume that for $n = k$, $T_k = (k/2)(k + 1)$. For $n = k + 1$, $T_{k+1} = Tk + (k + 1) = (k/2)(k + 1) + (k + 1) = (k + 1)[(k/2) + 1] = \frac{k+1}{2}(k + 2) = \frac{k+1}{2}((k + 1) + 1) \therefore T_n = (n/2)(n + 1)$ for *all* $n \geq 1$, Q.E.D.

22. Sum of the Squares of the Integers. Prove that $1^2 + 2^2 + 3^2 + \dots + n^2 = (n/6)(n + 1)(2n + 1)$ for all $n \geq 1$. *Proof:* For $n = 1$, $1^2 = 1$ and $(1/6)(1 + 1)(2\cdot1+1) = 1$, so $1^2 = (1/6)(1 + 1)(2\cdot1 + 1)$. Assume that for $n = k$, $1^2 + 2^2 + 3^2 + \dots + k^2 = (k/6)(k + 1)(2k + 1)$ For $n = k + 1$, $1^2 + 2^2 + 3^2 + \dots + k^2 + (k + 1)^2 = (1^2 + 2^2 + 3^2 + \dots + k^2) + (k + 1)^2$ (Assoc.) $= (k/6)(k + 1)(2k + 1) + (k + 1)^2$ (Ind. Hypoth.) $= (1/6)(k + 1)(2k^2 + k + 6k + 6)$ (Factoring) $= [(k+1)/6](2k^2 + 7k + 6)$ (Like terms) $= [(k+1)/6]\,(k + 2)(2k + 3)$ (Factoring) $= [(k+1)/6]((k+1) +1)(2(k+1) + 1)$ (Assoc.) $\therefore 1^2 + 2^2 + 3^2 + \dots + n^2 = (n/6)(n + 1)(2n + 1)$ for *all* $n > 1$, Q.E.D.

27. Factorials > Exponentials Prove that $n! > 3^n$ for all $n \geq 7$. *Proof:* For $n = 7$, $7! > 3^7$ ($7! = 5040$, $3^7 = 2187$) Assume that for $n = k$, $k! > 3^k$. For $n = k + 1$, $(k + 1)! = (k + 1)k!$ (Def. of factorials) $> (k + 1)3^k$ (Ind. Hypoth.) $> 3\cdot3^k$ (Since $k > 7$) $= 3^{k+1}$ (Def. of exponents) $\therefore (k + 1)! > 3^{k+1}$ (Transitivity) $\therefore n! > 3^n$ for *all* $n \geq 7$, Q.E.D.

34. Every Positive Integer is Both Odd and Even. Fallacy: Calvin's "proof" has no *anchor.*

Photograph Acknowledgements

Australian Information Service: 165

Joe Baraban: 176

The Bettmann Archive: 219

Daniel S. Brody/Stock, Boston: 378

John Colwell/Grant Heilman Photography: 64, 123, 338

Jose Cuervo: 42

E. R. Degginger/Bruce Coleman Inc.: 233

Diamond Information Center, N.W. Ayer ABH International: 216

Richard H. Dohrmann*: 130; 300 left; 300 right, courtesy Fidelity Savings & Loan Association

Howard Hall/Tom Stack & Associates: 218

Grant Heilman: 168, 213, 235

Ira Kirschenbaum/Stock, Boston: 167

Keith Murakami/Tom Stack & Associates: 359

(c) Bill Pierce/Rainbow: 358

Bil Plummer*: 430

F. Roe/Camera 5: 278

Joe Scherschel, *Life* Magazine, (c) 1957 Time Inc.: 224

Peter Southwick/Stock, Boston: 28

U.S. Department of the Interior, Bureau of Reclamation: 98

Yerkes Observatory: 250

*Photographs provided expressly for the publisher.
All other photographs by Addison-Wesley staff.